Plasma Physics and Engineering

Plasma Physics and Engineering

Alexander Fridman
and Lawrence A. Kennedy

CRC Press
Taylor & Francis Group
Boca Raton London New York

CRC Press is an imprint of the
Taylor & Francis Group, an **informa** business

Third edition published 2021
by CRC Press
6000 Broken Sound Parkway NW, Suite 300, Boca Raton, FL 33487-2742

and by CRC Press
2 Park Square, Milton Park, Abingdon, Oxon, OX14 4RN

First edition published by Taylor and Francis Group 2004
Second edition published by CRC Press 2011
CRC Press is an imprint of Taylor & Francis Group, LLC

Library of Congress Cataloging-in-Publication Data
Names: Fridman, Alexander A., 1953- author. | Kennedy, Lawrence A., author.
Title: Plasma physics and engineering / Alexander Fridman & Lawrence A. Kennedy.
Description: Third edition. | Boca Raton : CRC Press, 2021. | Includes bibliographical references and index. | Summary: "Plasma Physics and Engineering presents basic and applied knowledge on modern plasma physics, plasma chemistry and plasma engineering for senior undergraduate and graduate students as well as for scientists and engineers, working in academia, research labs and industry with plasmas, laser and combustion systems"-- Provided by publisher.
Identifiers: LCCN 2020039494 (print) | LCCN 2020039495 (ebook) | ISBN 9781498772211 (hardback) | ISBN 9781315120812 (ebook)
Subjects: LCSH: Plasma (Ionized gases) | Plasma engineering.
Classification: LCC QC718 .F77 2021 (print) | LCC QC718 (ebook) | DDC 530.4/4--dc23
LC record available at https://lccn.loc.gov/2020039494
LC ebook record available at https://lccn.loc.gov/2020039495

ISBN: 978-1-4987-7221-1 (hbk)
ISBN: 978-0-367-69752-5 (pbk)
ISBN: 978-1-315-12081-2 (ebk)

Typeset in Times
SPi Global, India.

Visit the eResources: https://www.routledge.com/9781498772211

Contents

PART 2 PHYSICS AND ENGINEERING OF ELECTRIC DISCHARGES 355

Preface

Plasma enjoys an important role in a wide variety of industrial processes including material processing, environmental control, electronic chip manufacturing, light sources, green energy, fuel conversion and hydrogen production, bio-medicine, flow control, catalysis, and space propulsion. It is also central to understanding most of the universe outside of earth. As such, the focus of this monograph is to provide a thorough introduction to the subject and to be a valued reference that serves engineers and scientists as well as students. Plasma is not an elementary subject and the reader is expected to have the normal engineering background in thermodynamics, chemistry, physics and fluid mechanics upon which to build an understanding of this subject. The text is organized in two parts. Part I addresses the basic physics of plasma, part II addresses the physics and engineering of electric discharges. The text is adaptable to a wide range of needs. The material has been taught to graduate and senior level undergraduate students from most engineering disciplines, chemistry, and physics. For the latter it can be packaged to focus on the basic physics of plasma with only selections from discharge applications. For graduate courses a faster pace can be set covering Part I and Part II.

This third edition of *Plasma Physics and Engineering* follows the general outline of the first two editions but includes many enhancements and some totally new subjects. A totally new Chapter 13 is introduced, focused on different approaches to generation of plasma inside of different fluids. It includes discharges in bubbles and voids, formation of nano-plasma in water from nano-tip electrodes, and finally nanosecond-pulsed plasma generation in liquids without bubbles. Such liquid plasmas are discussed in the Chapter 13 not only in water and water solutions, but also in dielectric oils, and even in liquid nitrogen. Substantial changes appear in Chapter 11 focused on discharges in aerosols and dusty plasmas. Significant totally new Section 9.7 is aided to Chapter 9 to describe different aspects of physics and engineering of nanosecond and microsecond pulsed dielectric barrier discharges. Significant new material is aided in the third edition to cover new fundamental subjects regarding plasma discharges in water in relation to its cleaning from chemical and biological impurities, as well as in relation with mechanisms and kinetics of plasma-medical processes. In addition, dozens of smaller additions, including new problems and concept questions improve the depth and breadth of the coverage of many subjects.

Alexander Fridman
Lawrence A. Kennedy

Acknowledgements

For preparation of the 3rd edition, we acknowledge the family support; support of John and Chris Nyheim and Kaplan family; support of Drexel University and Nyheim Plasma Institute. For stimulating discussions, we acknowledge Professors A. Rabinovich, D. Dobrynin, D. Vainchtein, C. Sales, A. Starikovsky, A. Gutsol, V. Vasilets, G. Friedman, B. Farouk, Y. Cho, N. Cernansky, D. Miller, G. Fridman, M. Kushner, J. Foster, M. Keidar, A. Saveliev, M. Laroussi, E. Robert, J.-M. Pouvesle, and many others. Special thanks are addressed to Kirill Gutsol for assistance with numerous illustrations.

Authors

Prof. Alexander Fridman is Nyheim Chair Professor of Drexel University and Director of C. & J. Nyheim Plasma Institute. His research focuses on plasma approaches to biology and medicine, to material treatment, fuel conversion, and environmental control. Prof. Fridman has almost 50 years of plasma research in national laboratories and universities of Russia, France, and the United States. He has published 8 books, and received numerous honors for his work, including Stanley Kaplan Distinguished Professorship in Chemical Kinetics and Energy Systems, George Soros Distinguished Professorship in Physics, the State Prize of the USSR, Plasma Medicine Award, Kurchatov Prize, Reactive Plasma Award, and Plasma Chemistry Award. Prof. Fridman is the author of more than 30 patents and fellow of US National Academy, NAI.

Prof. Lawrence A. Kennedy is Dean of Engineering Emeritus and Professor of Mechanical Engineering Emeritus at the University of Illinois at Chicago and Professor of Mechanical Engineering Emeritus at the Ohio State University. His research focuses on chemically reacting flows and plasma processes. He is the author of more than 300 archival publications and 2 books, the editor of three monographs and served as Editor–in-Chief of the *International Journal of Experimental Methods in Thermal and Fluid Science*. Professor Kennedy was the Ralph W. Kurtz Distinguished Professor of Mechanical Engineering at OSU and the Stanley Kaplan University Scholar in Plasma Physics at UIC. Prof. Kennedy is also the recipient of numerous awards such as the American Society of Mechanical Engineers Heat Transfer Memorial Award (2008), and the Ralph Coats Roe Award from ASEE (1993). He is a Fellow of the American Society of Mechanical Engineers, the American Physical Society, the American Institute of Aeronautics and Astronautics and the American Association for the Advancement of Science.

Authors

Prof. Alexander Fridman is Nyheim Chair Professor of Drexel University and Director of C. & J. Nyheim Plasma Institute. His research focuses on plasma approaches to biology and medicine, to material treatment, fuel conversion, and environmental control. Prof. Fridman has almost 50 years of plasma research in national laboratories and universities of Russia, France, and the United States. He has published 8 books and received numerous honors for his work, including Stanley Kaplan Distinguished Professorship in Chemical Kinetics and Energy Systems, George Soros Distinguished Professorship in Physics, the State Prize of the USSR, Plasma Medicine Award, Kurchatov Prize, Plasma Chemistry Award, and Drexel Chemistry Award. Prof. Fridman is the author of more than 30 patents, and fellow of US National Academy, NAI.

Prof. Lawrence A. Kennedy is Dean of Engineering Emeritus and Professor of Mechanical Engineering Emeritus at the University of Illinois at Chicago, and Professor of Mechanical Engineering Emeritus at the Ohio State University. His research focuses on chemically reacting flows and plasma processes. He is the author of more than 300 archival publications and 2 books, the editor of three monographs and served as Editor-in-Chief of the *International Journal of Experimental Methods in Thermal and Fluid Science*. Professor Kennedy was the Ralph W. Kurtz Distinguished Professor of Mechanical Engineering at OSU and the Stanley Kaplan University Scholar in Plasma Physics at UIC. Prof. Kennedy is also the recipient of numerous awards such as the American Society of Mechanical Engineers Heat Transfer Memorial Award (2008), and the Ralph Coats Roe Award from ASEE (1992). He is a Fellow of the American Society of Mechanical Engineers, the American Physical Society, the American Institute of Aeronautics and Astronautics, and the American Association for the Advancement of Science.

Part 1

Fundamentals of Plasma Physics and Engineering

1 Plasma in Nature, in the Laboratory, and in Industry

The term plasma is often referred to as the fourth state of matter. As temperature increases, molecules become more energetic and transform in the sequence: solid, liquid, gas, and plasma. In the latter stages, molecules in the gas dissociate to form a gas of atoms and then a gas of freely moving charged particles, electrons, and positive ions. This state is called the plasma state, a term attributed to Langmuir to describe the region of a discharge that was not influenced by the walls and/or electrodes. It is characterized by a mixture of electrons, ions, and neutral particles moving in random directions that on the average is electrically neutral ($n_e \cong n_i$). In addition, plasmas are electrically conducting due to the presence of these free charge carriers and can attain electrical conductivities larger than metal such as gold and copper. Plasma is not only the most energetic state of matter but also the most challenging for scientists. The temperature of the charged particles in a plasma can be so high that their collisions can result in thermonuclear reactions! Plasmas occur naturally, but also can be man-made. Plasma generation and stabilization in a laboratory and in industrial devices are not easy, but very promising for many modern applications, including thermonuclear synthesis, electronics, lasers, and many others. Most of your computer hardware are made based on plasma technologies and do not forget about the very large and thin TV plasma screens so popular today.

Plasmas offer two main characteristics for practical applications. First, they can have temperatures and energy densities greater than those produced by ordinary chemical means. Second, plasmas are able to produce, even at low temperatures, energetic species that can initiate chemicals reactions that are difficult or impossible to obtain using ordinary chemical mechanisms. The energetic species generated cover a wide spectrum of such species, for example, charged particles including electrons, ions, and radicals, highly reactive neutral species such as reactive atoms (e.g., O, F, etc.), excited atomic states, reactive molecular fragments, and different wavelength photons. Plasmas can also initiate physical changes in material surfaces. Applications of plasma can provide major benefits over existing methods. Often processes can be performed that are not possible in any other manner. Plasma can provide an efficiency increase in processing methods and often can reduce the environmental impact when compared to more conventional processes.

1.1 OCCURRENCE OF PLASMA, NATURAL AND MAN-MADE PLASMAS

Although somewhat rare on Earth, plasmas occur naturally and comprise the majority of the universe encompassing among other phenomena, the solar corona, solar wind, nebula, and the earth's ionosphere. In the earth's atmosphere, plasma is often observed as a transient event in the phenomenon of lightning strokes. Since air is normally nonconducting, large potential differences can be generated between clouds and earth during storms. Lightning discharges occur to neutralize the accumulated charge in the clouds. Such dischargers occur in two phases. First, an initial leader stroke progresses in steps across the potential gap between clouds or the earth. This leader stroke creates a low degree of ionization in the path and provides the conditions for the second phase, the return stroke to occur. The return stroke creates a highly conducting plasma path for the large currents to flow and neutralize the charge accumulation in the clouds. At altitudes of approximately 100 km, the atmosphere no longer remains nonconducting due to its interaction with solar radiation. From the energy absorbed through these solar radiation processes, a significant number of molecules and atoms become ionized and make this region of the atmosphere a plasma. As one progresses further into near-space altitudes, the earth's magnetic field interacts with charged particles streaming from the sun. These particles are diverted and often become trapped by the earth's magnetic field. The trapped particles are most dense near the poles and account for the Aurora Borealis. Lightning and the Aurora Borealis are the most common natural plasma observed on earth and one may think that these are exceptions. However, the earth's atmospheric environment in which we live is an anomaly; we live in a bubble of unionized gas surrounded by plasma. Lightning occurs at relatively high pressures and the Aurora Borealis at low pressures. Pressure has a strong influence on plasma, influencing its respective luminosity, energy or temperature, and the plasma state of its components. Lightning consists of a narrow, highly luminous channel with numerous branches. The Aurora Borealis is a low luminosity, diffuse event. Since plasmas occur over a wide range of pressures, it is customary to classify them in terms of electron temperatures and electron densities. Figure 1.1 provides a representation of the electron temperature (in electron volts, eV) and electron densities (in 1/cm^3) typical of natural and man-made plasmas. The electron temperature is expressed in electron volts (eV); one electron-volt is equal to approximately 11,600 K. Man-made plasma ranges from slightly above room temperature to temperatures comparable to the interior of stars. Electron densities span over 15 orders of magnitudes. However, most plasmas of practical significance have electron temperatures of 1–20 eV with electron densities in the range 10^6–10^{18} 1/cm^3. Not all particles need to be ionized in a plasma; a common condition is for the gases to be partially

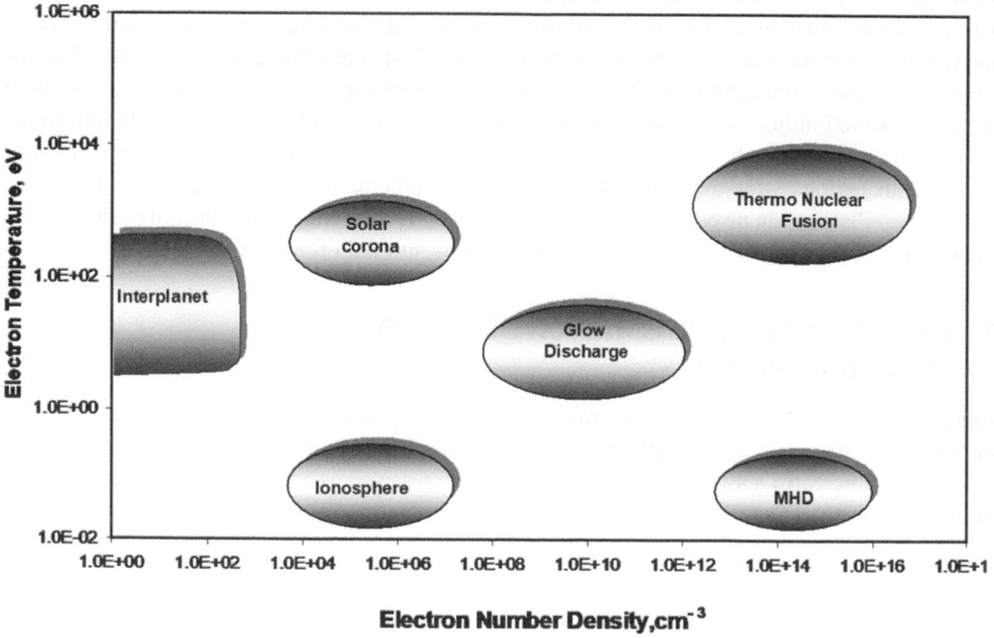

FIGURE 1.1 Operating regions of natural and man made plasma.

ionized. Under the latter conditions, one must examine the particle densities to determine if it is plasma. If there is a very low density of charged particles, the influence of the neutral particles can swamp the effects of interaction between charged ones. Under these conditions, the neutral particles dominate the collision process. The force laws appropriate to the specific particles govern collisions. For neutral particles, collisions are governed by forces that are significant only when the neutrals are in close proximity to each other. Conversely, the longer-range Coulomb Law governs collisions between charged particles. Hence, the behavior of the collision will change as the degree of ionization increases with the Coulomb interactions becoming more important. Coulomb's Law will govern all particles in the fully ionized plasma.

Langmuir was one of the early pioneers who studied gas discharges and defined plasma to be a region not influenced by its boundaries. The transition zone between the plasma and its boundaries was termed the plasma sheath. The properties of the sheath differ from those of the plasma, and these boundaries influence the motion of the charged particles in this sheath. They form an electrical screen for the plasma from the influences of the boundary. As in any gas, the temperature is defined by the average kinetic energy of the particle, molecule, atom, neutral, or charge. Thus, plasmas will typically exhibit multiple temperatures unless there are sufficient collisions between particles to equilibrate. However, since the mass of an electron is much less than the mass of a heavy particle, many collisions are required for this to occur.

The use of electric discharge is one of the most common ways to create and maintain a plasma. Here, the energy from the electric field is accumulated by the electron between collisions that subsequently transfers a portion of this energy to the heavy neutral particles through collisions. Even with a high collision frequency, the electron temperature and heavy particle temperature normally will be different. Since the collision frequency is pressure-dependent, high pressures will increase the collision frequency and the electron's mean free path between collisions will decrease. One can show that the temperature difference between electrons and heavy neutral particles is proportional to the square of the ratio of the energy an electron receives from the electric field (**E**) to the pressure (p). Only in the case of small values of $\mathbf{E}/p$, does the temperatures of electrons and heavy particles approach each other. Thus, this is a basic requirement for *Local Thermodynamic Equilibrium* (LTE) in a plasma. Additionally, LTE conditions require chemical equilibrium as well as restrictions on the gradients. When these conditions are met, the plasma is termed a *thermal plasma*. Conversely, when there are large departures from these conditions, $T_e > T_n$, the plasma is termed a *nonequilibrium plasma* or *nonthermal plasma*. As an example, for plasma generated by glow discharges, the operating pressures are normally less than 1 kPa and have electron temperatures of the order of 10^4 K with ions and neutral temperatures approaching room temperatures. Physics, engineering aspects, and application areas are quite different for thermal and nonthermal plasmas. Thermal plasmas are usually more powerful, while nonthermal plasmas are more selective. However, these two so different types of ionized gases have many more features in common and both are plasmas.

1.2 GAS DISCHARGES

There are many well-established uses of plasma generated by an electric discharge. Historically, it was the study of such discharges that laid the initial foundation of much of our understanding of plasma. In principle, discharges can be simply viewed as two electrodes inserted into a glass tube and connected to a power supply. The tube can be filled with various gases or evacuated. As the voltage applied across the two electrodes increases, the current suddenly increases sharply at a certain voltage characteristic of an electron avalanche. If the pressure is low, of the order of a few Torr, and the external circuit has a large resistance to prohibit a large current flow, a glow discharge develops. This is a low-current, high-voltage device in which the gas is weakly ionized (plasma). Glow discharges are widely used and are a very important type of discharge. They form the basis of fluorescent lighting. Such a discharge is present in every fluorescent lamp and the voltage–current characteristics of such discharges are widely employed in constant voltage gas tubes used in electronic circuits. Other common gas vapor discharges used for lighting are mercury vapor and neon. In power electronics applications, the need for switching and rectification has been addressed with the development of the thyratron and ignitron, which are important constant voltage gas tubes. If the pressure is high, of the order of an atmosphere, and the external circuit resistance is low, a thermal arc discharge can form the following breakdown. Thermal arcs usually carry large currents, greater than 1 A at voltages of the order of tens of volts. Further, they release large amounts of thermal energy. These types of arcs are often coupled with a gas flow from high-temperature plasma jets. A corona discharge occurs only in regions of sharply nonuniform electric fields. The field near one or both electrodes must be stronger than in the rest of the gas. This occurs near sharp points, edges, or small diameter wires. These tend to be low power devices limited by the onset of the electrical breakdown of the gas. However, it is possible to circumvent this restriction through the use of pulsating power supplies. Many other types of discharges are variations or combinations of these. The gliding arc is one such example. Beginning as a thermal arc located at the shortest distance between the electrodes, it moves with the gas flow at a velocity about 10 m/s and the length of the arc column increases together with the voltage. When the length of the gliding arc exceeds its critical value, heat losses from the plasma column begins to exceed the energy supplied by the source, and it is not possible to sustain the plasma in a state of thermodynamic equilibrium. As a result, a fast transition into a nonequilibrium phase occurs. The discharge plasma cools rapidly to a gas temperature of about $T_n = 1{,}000$ K and the plasma conductivity is maintained by a high value of the electron temperature $T_e = 1$ eV (about 11,000 K). After this fast transition, the gliding arc continues its evolution, but under nonequilibrium conditions ($T_e >> T_n$). The specific heat losses in this regime are much smaller than in the equilibrium regime (numerically about three times less). The discharge length increases up to a new critical value approximately three times its original critical length. The main part of the gliding arc power (up to 75%–80%) can be dissipated in the nonequilibrium zone. After the decay of the nonequilibrium discharge, the evolution repeats from the initial break down. This permits the stimulation of chemical reactions in regimes quite different from conventional combustion and environmental situations. It provides an alternate approach to addressing energy conservation and environmental control. Figure 1.2 illustrates the gliding arc. Many of the features in these discharges are also typical for discharges in rapidly oscillating fields where electrodes are not required. As such, the classification of discharges has been developed, which does not involve electrode attributes. Such a classification is based upon two properties: (1) the state of the ionized gas and (2) the

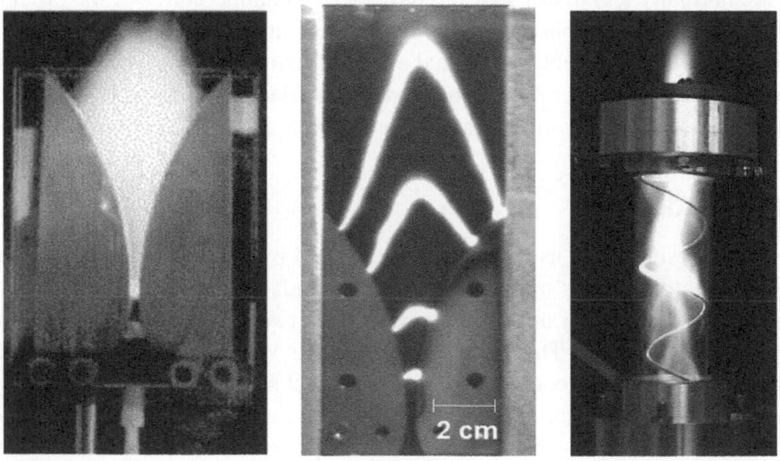

(a) (b) (c)

FIGURE 1.2 (a) Gliding arc viewed with normal photography, (b) Gliding arc viewed with high-speed photography, (c) Vortex gliding arc.

TABLE 1.1
Classification of Discharges

	Breakdown	Nonequilibrium Plasma	Equilibrium Plasma
Constant E Field	Initiation of glow discharge	Glow discharge	High-pressure arc
Radio frequencies	Initiation of RF discharge in rarefied gas	Capacitively coupled RF discharge in rarefied gas	Inductively coupled plasma
Microwave	Breakdown in waveguides	Microwave discharge in rarefied gas	Microwave plasmatron
Optical range	Gas breakdown by laser radiation	Final stage optical breakdown	Continuous optical discharge

frequency range of the electric field. The state of the ionized gas distinguishes between (1) breakdown of the gas, (2) sustaining a nonequilibrium plasma by the electric field, and (3) sustaining an equilibrium plasma. The frequency criteria classify the field into (1) DC, low frequency, and pulsed E fields; (2) radio frequency E fields, (3) microwave fields, and (4) optical fields. These are summarized in Table 1.1.

1.3 PLASMA APPLICATIONS, PLASMAS IN INDUSTRY

Although this text is mainly focused on the fundamentals of plasma physics and engineering, ultimately we are interested in the application of plasmas; see the following book Fridman "Plasma chemistry" (2008) focused on the applications. The number of industrial applications of plasma technologies is extensive and involves many industries, especially electronics, lighting, coatings, treatment and processing of materials, metallurgy, and energy systems. High-energy efficiency, specific productivity, and selectivity may be achieved in plasmas for a wide range of chemical processes. As an example, for CO_2 dissociation in nonequilibrium plasmas under supersonic flow conditions, it is possible to selectively introduce up to 90% of total discharge power in CO production when the vibrational temperature is about 4,000 K and the translational temperature is only 0 (300 K). The specific productivity of such a supersonic reactor achieves 1,000,000 L/h, with power levels up to 1 MW. The key point for practical use of any chemical process in a particular plasma system is to find the proper regime and optimal plasma parameters among the numerous possibilities intrinsic to systems far from equilibrium. In particular, it is desired to provide high operating power for the plasma chemical reactor together with high selectivity of the energy input while simultaneously maintaining nonequilibrium plasma conditions. Generally, two very different kinds of plasmas are used for applications. Thermal plasma generators have been designed for many diverse industrial applications covering a wide range of operating power levels from less than 1 kW to over 50 MW. However, in spite of providing sufficient power levels, these are not well adapted to the purposes of plasma chemistry, where selective treatment of reactants (through the excitation of molecular vibrations or electron excitation) and high efficiency are required. The main drawback of using thermal plasmas is overheating the reaction medium when energy is uniformly consumed by the reagents into all degrees of freedom and hence, high energy consumption required to provide special quenching of the reagents, and so on. Because of that, the energy efficiency and selectivity of such systems are rather small. Promising plasma parameters are achievable in microwave discharges. The skin effect here permits simultaneous achievement of a high level of electron density and a high electric field (and hence a high electronic temperature as well) in the relatively cold gas. Existing super high frequency discharge technology can be used to generate dense ($n_e = 10^{13}$ electrons/cm^3) nonequilibrium plasmas ($T_e = 1$–2 eV, $T_V = 3{,}000$–$5{,}000$ K, $T_n = 800$–$1{,}500$ K, for supersonic flow $T_n \leq 150$ K and less) at pressures up to 200–300 Torr and at power levels reaching 1 MW.

An alternative approach for plasma-chemical gas processing is the nonthermal one. Silent discharges such as glow, corona, short pulse, microwave or radio-frequency (RF) electrical discharges are directly produced in the processed gas, mostly under low pressure. The glow discharge in a low-pressure gas is a simple and inexpensive way to achieve a nonthermal plasma. Here, the ionization processes induced by the electric field dominate the thermal ones and give relatively high energy electrons as well as excited ions, atoms, and molecules which promote selective chemical transitions· However, the power of glow discharges is limited by the glow to arc transition. Gas, initially below 1,000 K, becomes hot (>6,000 K), and the electron temperature, initially sufficiently high (>12,000 K), cools close to the bulk gas temperature. The discharge voltage decreases during such a transition making it necessary to increase the current in order to keep the power on the same level which in turn leads to thermalization of the gas. Thus, cold nonequilibrium plasmas created by conventional glow discharges offer good selectivity and efficiency, but at limited pressure and power levels. In these two general types of plasma discharges, it is impossible to simultaneously keep a high level of nonequilibrium, high electron temperature, and high electron density, whereas most perspective plasma chemical applications simultaneously require high power for high reactor productivity and a high degree of nonequilibrium to support selective chemical process. Recently, a simpler technique offering similar advantages has been proposed, the GlidingArc. Such a gliding arc occurs when the plasma is generated between two or more diverging electrodes placed in a fast gas flow. It operates at atmospheric pressure or higher and the dissipated power at nonequilibrium conditions reaches up to 40 kW per electrode pair. The incontestable advantage of the GlidingArc compared with microwave systems is its cost; it is much less expensive compared to microwave plasma.

1.4 PLASMA APPLICATIONS FOR ENVIRONMENTAL CONTROL

Low-temperature, nonequilibrium plasmas are an emerging technology for abating low VOC emissions. These nonequilibrium plasma processes have been shown to be effective in treating a wide range of emissions including aliphatic hydrocarbons, chlorofluorocarbons, methyl cyanide, phosgene, formaldehyde, as well as sulfur and organophosphorus compounds. Such plasmas may be produced by a variety of electrical discharges or electron beams. The basic feature of nonequilibrium plasma technologies is that they produce plasma in which the majority of the electric energy (more than 99%) goes into the production of energetic electrons instead of heating the entire gas stream. These energetic electrons produce excited species, free radicals, and ions as well as additional electrons through the electron impact dissociation, excitation and ionization of the background molecules. These excited species, in turn, oxidize, reduce, or decompose the pollutant molecules. This is in contrast to the mechanism involved in thermal processes that require heating the entire gas stream in order to destroy pollutants. In addition, the low-temperature plasma technology is highly selective and has relatively low maintenance requirements. Its high selectivity results in relatively low energy costs for emissions control, while low maintenance keeps annual operating expenses low. Furthermore, these plasma discharges normally are very uniform and homogeneous, which results in high process productivity. Although the products of these plasma processes are virtually indistinguishable from incineration products (CO_2, H_2O, SO_2, etc.), chemical reactions occurring in these technologies are substantially different than in incineration. Large electric fields in discharge reactors create conditions for electric breakdowns during which many electron–ion pairs are formed. Primary electrons are accelerated by the electric field and produce secondary ionization, and so forth. Created excited species, atoms, radicals, molecular and atomic ions, electrons, and radicals are capable of interacting to a certain degree with VOCs. Table 1.2 illustrates different plasma approaches to VOC destruction. Electron beams produce a plasma of the best quality with an average electron energy of 5–6 eV. Pulsed corona discharges stand very close (3–5 eV). But corona discharges have the advantage of lower capital cost. Generally, both technologies have already found applications for gas treatment form NO_x and SO_2 emissions in coal power plants. The capability of treating large gas streams makes e-beams and pulsed corona discharges the primary candidates

TABLE 1.2
Comparison of Nonthermal Plasma Technologies

Plasma Technology	Average Electron Energy	Capital Cost, $/W	Unit Capacity, kW	Energy Consumption, kWh/m³	Known Applications	Technical Limitations
Electron beam technology	5–6 eV	2.0	1,000	10–70	Removal of SO_2, NOx in thermal power plants	Initial energy of electrons in the beam is limited by 1.5 MeV due to X-ray production
Gas-Phase Corona Reactor (GPCR)	2–3 eV	2.1	20	6–70	Electrostatic Precipitators	The voltage cannot be as high as in pulsed corona due to spark formation; packed bed limits flow rates.
Pulsed Corona Discharge	3–5 eV	1.0	10	20–150	Removal of polyatomic hydrocarbons in aluminum plant	Voltage is limited by the electrical network coupling and capacitance and inductance of the plasma reactor
Dielectric-Barrier Discharge (DBD)	2–3 eV	0.2	0.1	20–200	Removal of SO_2, NOx, industrial ozone generators	The voltage cannot be as high as in pulsed corona due to spark formation; low unit capacity
Gliding Arc	1–1.5 eV	0.1	120	20–250	Laboratory study on removal of methanol, o-xylene	Discharge is nonuniform, high temperatures of the gas, possible NOx generation.

considered for VOC abatement. Here we just discussed one specific plasma application for environmental control in order to introduce some general parameters characterizing such processes.

1.5 PLASMA APPLICATIONS IN ENERGY CONVERSION

Plasma applications in energy conversion are many. The most publicized are those related to control thermonuclear fusion reactions, magnetohydrodynamic (MHD) power generation, and thermionic energy conversions. Fusion reactions are the source of the sun's energy. It is also the basis of our thermonuclear weapons and the more peaceful application goal of power generation. To undergo fusion reactions, the nuclei must interact very closely by overcoming the electrostatic repulsive force between particles. Thus, they must approach each other at high velocities. If one simply employs opposing beams of particles to do this, too many particles would be scattered to make this a practical approach. One must reflect the scattered particles back into the reactor as often as necessary to generate the fusion reaction. For deuterium, the required temperature exceeds 100 million degrees Kelvin to produce a significant amount of fusion. At these temperatures, any gas is fully ionized and therefore a plasma. The engineering problem for controlled fusion is then to contain this high-temperature gas for a sufficiently long time interval for the fusion reactions to occur and then extract the energy. The only way to contain such a plasma is to exploit its interactions with magnetic fields. This long-term effort is being pursued at national laboratories in many countries.

The MHD generator exploits the flow of plasma interacting with a magnetic field to generate electrical power. From Maxwell's equations, the interaction of a flowing plasma with a magnetic field ($\mathbf{B}$) will give rise to an induced electric field ($\mathbf{E} = \mathbf{v} \times \mathbf{B}$). Energy can be extracted if electrodes are placed across this flow permitting a current flow through the plasma. This in turn gives rise to a force ($\mathbf{J} \times \mathbf{B}$) opposite to the flow which decelerates it. A variation of this concept is the plasma accelerator wherein one configures the applied fields to create an accelerating force. Thermionic generators are another example of a plasma energy converter. The simplest configuration is a basic diode where the anode is cooled and a heated cathode causes electrons to boil off. The region between the electrodes is filled with cesium vapor which can be easily ionized either through contact with the hot cathode or through multiple steps in the gas. The space charge is neutralized by the positive cesium ions and allows high current to flow. Many different configurations are possible and they provide power generation for space vehicles and mobile power generators. They also have been proposed for miniature mobile power units to replace batteries. Serious attention is now directed to plasma applications for hydrogen production from different hydrocarbons, including coal, biomass, and organic wastes. Plasma discharges can be effectively combined in this case with catalytic and different hybrid systems.

1.6 PLASMA APPLICATION FOR MATERIAL PROCESSING

Plasma applications abound in material processing. Initially, thermal plasmas were employed due to the high temperatures they achieved over those obtainable from gaseous flames. While combustion flame temperatures are limited to approximately 3,300 K, a small plasma jet generated by the gas flow through a thermal electric arc provides a working temperature of approximately 6,500 K. When one considers the melting and/or vaporization points of many metals and ceramics, it is obvious that these arcs are extremely useful processing tools. More powerful arcs can generate temperatures to 20–30,000 K which provides environments for activated species surface modifications. The use of electric arcs for welding, melting of metals, high purity metal processing is well established. Plasma is widely employed in the coating industry where its large enthalpy content, high temperatures, and large deposition rates are advantageous for increased throughputs. It provides the ability to mix and blend materials that are otherwise incompatible. Complex alloys, elemental materials, composites, and ceramics can all be deposited. Among others, it is the preferred process for the manufacture of electronic components, wear and heat resistant coatings. With the advent of nanostructured metals, nanopowder production and in-situ coating of these nanoparticles are accomplished with plasma processing.

Surface modification of classes of material is performed using plasma environments. For example, while the choice of a polymer for a specific application is generally driven by the bulk properties of the polymer, many common polymer surfaces are chemically inert and, therefore, pose challenges for use as substrates for applied layers. Typical applications where wetting and adhesion are critical issues are printing, painting, and metallization. Surface modification by plasma treatment of polymer surfaces can improve wetting, bonding, and adhesion characteristics without sacrificing desirable bulk properties of the polymer base material. Plasma treatments have proven useful in promoting adhesion of silver to poly(ethylene terephthalate). In addition, plasma treatments of polyester have been demonstrated to promote adhesion of gelatin containing layers such as used in the production of photographic film. In both of these examples, low-frequency capacitively coupled nitrogen plasmas have proven particularly effective.

1.7 BREAKTHROUGH PLASMA APPLICATIONS IN MODERN TECHNOLOGY

The most exciting applications of plasma are those that have no analogies and no (or almost no) competing processes. A good relevant example is plasma applications in microelectronics, such as etching of deep trenches in single-crystal silicon, which is so important in the fabrication of integrated circuits. Capabilities of plasma processing in microelectronics are extraordinary and unique. We probably would not have achieved such powerful and compact computers and cell phones without plasma processing. When all alternatives fail – plasma remains as a viable and valuable tool. There are other applications where plasma processes are not only highly efficient, but actually unique. For example, there are no other technologies that have competed with plasma for the production of ozone (for more than 100 years); thermonuclear plasma is a unique major future source of energy; low-temperature fuel conversion with the production of hydrogen without CO_2 exhaust is another example of a unique breakthrough energy technology. Another novel and unique plasma application is in water cleaning from PFOS/PFOA, the fluorinated organic compounds, deadly and extremely resistive to existing water cleaning procedures. Again in this case, it looks like plasma and is a unique effective way to do the job.

1.8 PLASMA BIOLOGY AND PLASMA MEDICINE

The application that shifts the paradigm in therapeutics, wound healing, disease control, cosmetics, as well as food protection from pathogens, sterilization, abatement of dangerous microorganisms, including coronavirus is in *the use of plasmas for biological and biomedical fields*. These applications, which are growing at a rapid rate, have opened new possibilities to treat diseases and wounds that are resistant to traditional methods, which includes plasma treatment of complicated ulcers and healing of cancer.

2 Elementary Processes of Charged Species in Plasma

2.1 ELEMENTARY CHARGED PARTICLES IN PLASMA AND THEIR ELASTIC AND INELASTIC COLLISIONS

Because plasma is an ionized medium, the key process in plasma is ionization. This means the conversion of a neutral atom or molecule into a positive ion and also an electron, as illustrated in Figure 2.1a. The main two participants of the ionization process, **electrons and positive ions**, are at the same time, the most important charged particles in plasma. Normally, concentrations or number densities of the electrons and the positive ions are equal or near equal in quasi-neutral plasmas, but in "electronegative" gases, it can be different. Electronegative gases, such as O_2, Cl_2, SF_6, UF_6, and $TiCl_4$, consist of atoms or molecules with a high electron affinity (EA), which means they strongly attract electrons. Electrons stick to such molecules or collide with them with the formation of **negative ions**, the third

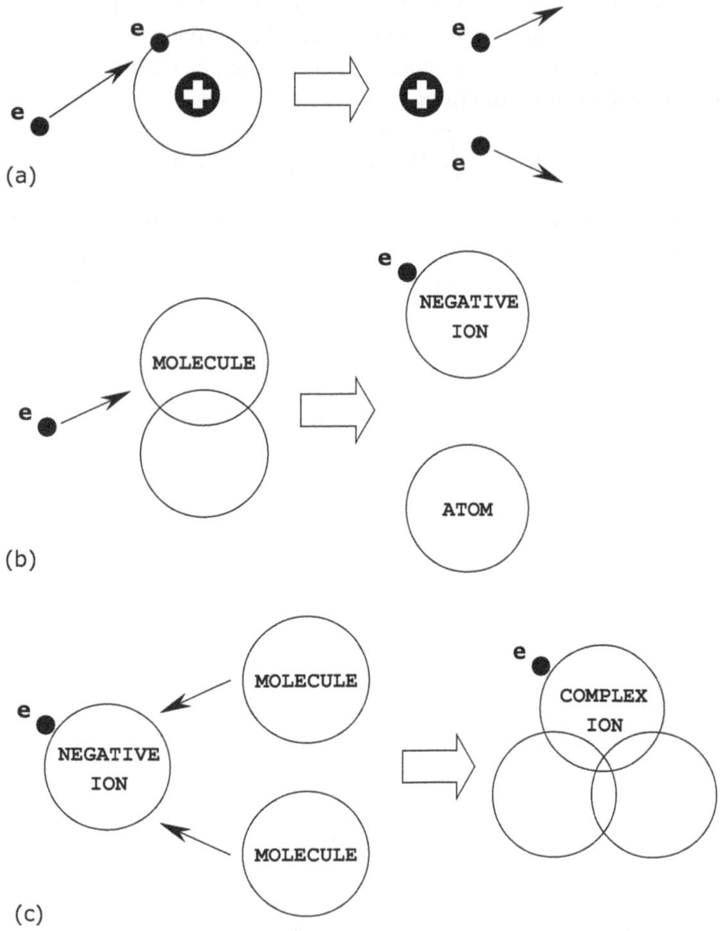

FIGURE 2.1 Illustrations of elementary processes of: (a) ionizations, (b) dissociative attachment and formation of a negative ion, (c) complex ion formation.

important group of charged particles in plasmas. The concentration of negative ions can exceed those of electrons in electronegative gases. Illustrations of electron attachment and negative ion formation are shown in Figure 2.1b. Charged particles can also appear in plasma in more complicated forms. In high-pressure and low-temperature plasmas, the positive and negative ions attach to neutral atoms or molecules to form quite large **complex ions** or ion clusters, for example, $N_2^+N_2$ (N_4^+), O^-CO_2 (CO_3^-), H^+H_2O (H_3O^+), Figure 2.1c. Ion-molecular bonds in complex ions are usually less strong than regular chemical bonds, but stronger than intermolecular bonds in neutral clusters.

2.1.1 Electrons

Electrons are elementary, negatively charged particles with a mass of about 3–4 orders of magnitude less than the mass of ions and neutral particles (electron charge is an elementary one: $e = 1.6\ 10^{-19}$ C, mass $m_e = 9.11\ 10^{-31}$ kg). Because of their lightness and high mobility, electrons are the first particles in receiving energy from electric fields. Afterward, electrons transmit the energy to all other plasma components, providing ionization, excitation, dissociation, etc.

Electrons are energy providers for many plasma chemical processes. The rate of such processes depends on how many electrons have sufficient energy to do the job. This can be described by means of the **electron energy distribution function** $f(\varepsilon)$, which is a probability density for an electron to have the energy ε (probability for an electron to have energy between ε and ε + Δ, divided by Δ). Quite often, this distribution function strongly depends on the electric field and gas composition and can be very far from equilibrium. Sometimes, however, (even in non-equilibrium plasmas of nonthermal discharges), the $f(\varepsilon)$ depends mostly on electron temperature T_e and can then be defined by the quasi-equilibrium **Maxwell–Boltzmann distribution function:**

$$f(\varepsilon) = 2\sqrt{\varepsilon / \left(\pi^3 kT_e\right)^3}\ \exp\left(-\varepsilon / kT_e\right), \tag{2.1}$$

where k is a Boltzmann constant (if temperature is given in energy units, then $k = 1$ and can be omitted, we will follow this rule throughout the book). For this case, the **mean electron energy**, which is the first moment of the distribution function, is proportional to temperature in a traditional way:

$$\langle \varepsilon \rangle = \int_0^\infty \varepsilon f(\varepsilon) dE = \frac{3}{2} T_e. \tag{2.2}$$

Numerically, in most of the plasmas under consideration, the mean electron energy is from 1 to 5 eV.

2.1.2 Positive Ions

Atoms or molecules lose their electrons in ionization processes and form positive ions (Figure 2.1a). In extremely hot thermonuclear plasmas, the ions are multicharged, but in quasi-cold technological plasmas of interest – their charge is usually equal to +1e ($1.6\ 10^{-19}$ C). Ions are heavy particles, so commonly they cannot receive high energy directly from an electric field because of intensive collisional energy exchange with other plasma components. The collisional nature of the energy transfer results in the ion energy distribution function usually being not far from the quasi-equilibrium Maxwellian one (2.1), with the ion temperature T_i close to the neutral gas temperature T_0. However, in some low-pressure discharges and sheaths, ion energy can be quite high.

One of the most important parameters of plasma generation is an ionization energy I, which is the energy needed to form a positive ion. Ionization requires quite large energy and as a rule defines the upper limit of microscopic energy transfer in plasma. Detailed information on ionization energies is available, for example, in V.N. Kondratiev, 1974. For some gases important in applications – the ionization energies are given in Table 2.1. Noble gases like He and Ne have the highest ionization energies I. Alkali metal vapors like Li_2 have the lowest value of I, so even a very small addition of such metals can dramatically stimulate ionization.

TABLE 2.1
Ionization Energies for Different Atoms, Radicals and Molecules

$e + N_2 = N_2^+ + e + e$	I = 15.6 eV	$e + O_2 = O_2^+ + e + e$	I = 12.2 eV
$e + CO_2 = CO_2^+ + e + e$	I = 13.8 eV	$e + CO = CO^+ + e + e$	I = 14.0 eV
$e + H_2 = H_2^+ + e + e$	I = 15.4 eV	$e + OH = OH^+ + e + e$	I = 13.2 eV
$e + H_2O = H_2O^+ + e + e$	I = 12.6 eV	$e + F_2 = F_2^+ + e + e$	I = 15.7 eV
$e + H_2S = H_2S^+ + e + e$	I = 10.5 eV	$e + HS = HS^+ + e + e$	I = 10.4 eV
$e + SF_6 = SF_6^+ + e + e$	I = 16.2 eV	$e + SiH_4 = SiH_4^+ + e + e$	I = 11.4 eV
$e + UF_6 = UF_6^+ + e + e$	I = 14.1 eV	$e + Cs_2 = Cs_2^+ + e + e$	I = 3.5 eV
$e + Li_2 = Li_2^+ + e + e$	I = 4.9 eV	$e + K_2 = K_2^+ + e + e$	I = 3.6 eV
$e + CH_4 = CH_4^+ + e + e$	I = 12.7 eV	$e + C_2H_2 = C_2H_2^+ + e + e$	I = 11.4 eV
$e + CF_4 = CF_4^+ + e + e$	I = 15.6 eV	$e + CCl_4 = CCl_4^+ + e + e$	I = 11.5 eV
$e + H = H^+ + e + e$	I = 13.6 eV	$e + O = O^+ + e + e$	I = 13.6 eV
$e + He = He^+ + e + e$	I = 24.6 eV	$e + Ne = Ne^+ + e + e$	I = 21.6 eV
$e + Ar = Ar^+ + e + e$	I = 15.8 eV	$e + Kr = Kr^+ + e + e$	I = 14.0 eV
$e + Xe = Xe^+ + e + e$	I = 12.1 eV	$e + N = N^+ + e + e$	I = 14.5 eV
$e + C_2H_5OH = e + C_2H_5OH^+ + 2e$	I = 10.5 eV	$e + CH_3COOH = e + CH_3COOH^+ + 2e$	I = 10.4 eV
$e + C_6H_5OH = e + C_6H_5OH^+ + 2e$	I = 8.5 eV	Naphthaldehyde	I = 7.7 eV
$e + (C_2H_5)_2Cr = e + (C_2H_5)_2Cr^+ + 2e$	I = 5.5 eV	4-Methyl-Phenylene-Diamine	I = 6.2 eV

2.1.3 Negative Ions

Electron attachment to atoms or molecules results in the formation of negative ions (Figure 2.1b). Their charge is equal to −1e (1.6 10^{-19} C), attachment of another electron and formation of multicharged negative ions is actually impossible in the gas phase because of electric repulsion. Negative ions are heavy particles, so usually their energy balance is not due to the electric field, but rather due to collisional processes. The energy distribution functions for negative ions (the same as for positive ones) are not far from Maxwellian Eq. (2.1), with temperatures also close to those of the neutral gas.

EA can be defined as the energy release during attachment processes or as the bond energy between the attaching electron and the atom or molecule. EA is usually much lower than ionization energy I, even for very electronegative gases. Numerical values of EA for many different atoms, radicals, and molecules can be found in V.N. Kondratiev, 1974. Information for some gases of interest is presented in Table 2.2. Halogens and their compounds have the highest values of EA. Oxygen, ozone, and some oxides also are strongly electronegative substances with high EA. For this reason, nonthermal plasma generated in air usually contains a large number of negative ions.

2.1.4 Elementary Processes of the Charged Particles

Ionization of atoms and molecules by electron impact, electron attachment to atoms or molecules, and ion-molecular reactions (see Figure 2.1a–c) are examples of elementary plasma-chemical processes, reactive collisions accompanied by the transformation of elementary plasma particles. These elementary reactive collisions, as well as others, for example, electron–ion and ion-ion recombination, excitation and dissociation of neutral species by electron impact, relaxation of excited species, electron detachment and destruction of negative ions, photochemical processes – altogether determine plasma behavior. Elementary processes can be divided into two classes – elastic and nonelastic processes. The elastic collisions are those in which the internal energies of colliding particles do not change, therefore the total kinetic energy is conserved as well. Hence, these processes result only in scattering. Alternately collisions are inelastic. All elementary processes listed in previous

TABLE 2.2
Electron Affinity for Different Atoms, Radicals and Molecules

$H^- = H + e$	EA = 0.75 eV	$O^- = O + e$	EA = 1.5 eV
$S^- = S + e$	EA = 2.1 eV	$F^- = F + e$	EA = 3.4 eV
$Cl^- = Cl + e$	EA = 3.6 eV	$O_2^- = O_2 + e$	EA = 0.44 eV
$O_3^- = O_3 + e$	EA = 2.0 eV	$HO_2^- = HO_2 + e$	EA = 3.0 eV
$OH^- = OH + e$	EA = 1.8 eV	$Cl_2^- = Cl_2 + e$	EA = 2.4 eV
$F_2^- = F_2 + e$	EA = 3.1 eV	$SF_5^- = SF_5 + e$	EA = 3.2 eV
$SF_6^- = SF_6 + e$	EA = 1.5 eV	$SO_2^- = SO_2 + e$	EA = 1.2 eV
$NO^- = NO + e$	EA = 0.024 eV	$NO_2^- = NO_2 + e$	EA = 3.1 eV
$NO_3^- = NO_3 + e$	EA = 3.9 eV	$N_2O^- = N_2O + e$	EA = 0.7 eV
$HNO_3^- = HNO_3 + e$	EA = 2.0 eV	$NF_2^- = NF_2 + e$	EA = 3.0 eV
$CH_3^- = CH_3 + e$	EA = 1.1 eV	$CH_2^- = CH_2 + e$	EA = 1.5 eV
$CF_3^- = CF_3 + e$	EA = 2.1 eV	$CF_2^- = CF_2 + e$	EA = 2.7 eV
$CCl_4^- = CCl_4 + e$	EA = 2.1 eV	$SiH_3^- = SiH_3 + e$	EA = 2.7 eV
$UF_6^- = UF_6 + e$	EA = 2.9 eV	$PtF_6^- = PtF_6 + e$	EA = 6.8 eV
$Fe(CO)_4^- = Fe(CO)_4 + e$	EA = 1.2 eV	$TiCl_4^- = TiCl_4 + e$	EA = 1.6 eV

paragraphs are inelastic ones. Inelastic collisions result in the transfer of energy from the kinetic energy of colliding partners into internal energy. For example, processes of excitation, dissociation, and ionization of molecules by electron impact are inelastic collisions, including the transfer of high kinetic energy of plasma electrons into the internal degrees of freedom of the molecules.

In some instances, the internal energy of excited atoms or molecules can be transferred back into kinetic energy (in particular, into kinetic energy of plasma electrons). Such elementary processes are usually referred as **super-elastic collisions**. According to kinetic theory, the elementary processes can be described in terms of six main collision parameters: (1) cross section, (2) probability, (3) mean free path, (4) interaction frequency reaction rate, and (5) reaction rate coefficient (also termed reaction rate constant).

2.1.5 Fundamental Parameters of Elementary Processes

The most fundamental characteristics of all elementary processes are their cross sections. The **cross section of an elementary process** between two particles can be interpreted as an imaginary circle with area σ, moving together with one of the collision partners (see Figure 2.2a). If the center of the other collision partner crosses the circle, an elementary reaction occurs. The cross sections of elementary processes depend strongly on the energy of colliding species. If two colliding particles can be considered as hard elastic spheres of radii r_1 and r_2, their collisional cross section is equal to $\pi (r_1 + r_2)^2$. Obviously, the interaction radius and cross section can exceed the corresponding geometrical sizes because of long-distance nature of forces acting between electric charges and dipoles. Alternately, if only a few out of many collisions result in a chemical reaction, the cross section is considered to be less than the geometrical one. Typical sizes of atoms and molecules are of the order of 1–3 Å; therefore cross section of simple elastic collisions between

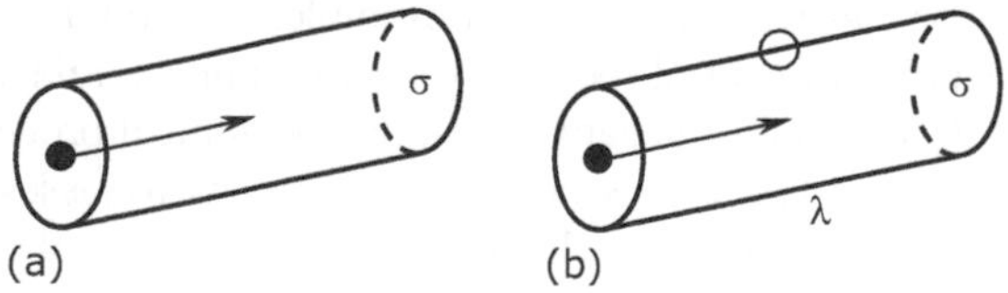

FIGURE 2.2 (a) Cross section of an elementary process and (b) concept of the mean free path.

plasma electrons (energy 1–3 eV) and neutral particles is usually about 10^{-16}–10^{-15} cm^2. Cross sections of inelastic, endothermic, electron–neutral collisions are normally lower. When the electron energy is very large, the cross section can decrease because of a reduction of the interaction time. The ratio of the inelastic collision cross section to the corresponding cross section of an elastic collision under the same condition is often called the dimensionless **probability of the elementary process.**

The mean free path λ of one collision partner A with respect to the elementary process A + B with another collision partner *B* can be calculated as

$$\lambda = 1/\left(n_B \sigma\right), \tag{2.3}$$

where n_B is number density (concentration) of the particles B. During the mean free path λ, the particle A traverses the cylindrical volume $\lambda\sigma$. A reaction occurs if the cylindrical volume transversed by a particle A, contains at least one B-particle, which means $\lambda\sigma n_B = 1$ (see Figure 2.2b), the simple interpretation of Eq. (2.3) is the distance A travels before colliding. In the example of electrons colliding with neutrals, the electron mean free path with respect to elastic collisions with neutrals at atmospheric pressure and $n_B = 3*10^{19}$ cm^{-3} is approximately $\lambda = 1\mu$.

The interaction frequency ν of one collision partner A (e.g., electrons moving with velocity v) with the other collision partner B (e.g., a heavy neutral particle) can be defined as v/λ or taking into account Eq. (2.3) as

$$\nu_A = n_B <\sigma v> \tag{2.4}$$

This relation should be averaged taking into account the velocity distribution function f(v) and the dependence of the cross section σ on the collision partners' velocity. When the collision partners' velocity can be attributed mostly to one light particle (e.g., electron) Eq. (2.4) can be rewritten as:

$$\nu_A = n_B \int f(v)\sigma(v)\, v dv = <\sigma v>\, n_B. \tag{2.5}$$

Numerically, for the abovementioned example of atmospheric pressure, electron neutral, elastic collisions, the interaction frequency is approximately $\nu = 10^{12}$ 1/s.

2.1.6 The Reaction Rate Coefficients

The number of elementary processes, which occur in unit volume per unit time is called the elementary reaction rate. This concept can be used for any kind of reaction: mono-molecular, bimolecular, and three-body reactions. For bimolecular processes A + B, the reaction rate can be calculated by multiplying the interaction frequency of partner A with partner B - "ν_A" and number of particles A in the unit volume (which is their number density- n_A). Thus, $w = \nu_A n_A$. Taking into account Eq. (2.5), this can be rewritten as:

$$w_{A+B} = \langle \sigma v \rangle n_A n_B. \tag{2.6}$$

The factor <σv> is termed the **reaction rate coefficient** (or reaction rate constant), one of the most useful concepts of plasma chemical kinetics. For bimolecular reactions, it can be calculated as

$$k_{A+B} = \int f(v)\sigma(v)\, v dv = <\sigma v> \tag{2.7}$$

In contrast to the reaction cross section σ, which is a function of the partners' energy, the reaction rate coefficient k is an integral factor, which includes information on energy distribution functions and depends on temperatures or mean energies of the collision partners. Equation (2.7) establishes the relation between micro-kinetics (which is concerned with elementary processes) and macro-kinetics (which takes into account real energy distribution functions). Numerically, for the above-considered example of electron neutral, elastic binary collisions, the reaction rate coefficient is approximately $k = 3 * 10^{-8}$ cm^3/s.

The concept of reaction rate coefficients can be applied not only for bimolecular processes Eq. (2.7) but also for mono-molecular (A → products) and three-body processes (A + B + C→ products). Then, the reaction rate is proportional to a product of concentrations of participating particles, and the reaction rate coefficient is just a coefficient of the proportionality:

$$w_A = k_A n_A \ (a); \quad w_{A+B+C} = k_{A+B+C} n_A n_B n_C \ (b). \tag{2.8}$$

Obviously, in these cases Eq. (2.8), the reaction rate coefficients cannot be calculated using Eq. (2.7), they even have different dimensions. Also, for mono-molecular processes Eq. (2.8a), the reaction rate coefficient can be interpreted as the reaction frequency.

2.1.7 Elementary Elastic Collisions of Charged Particles

Elastic collisions (or, elastic scattering) do not result in a change of chemical composition or excitation level of the colliding partners and for this reason, they are not able to influence directly plasma-chemical processes. On the other hand, due to their high values of cross sections, elastic collisions are responsible for kinetic energy and momentum transfer between colliding partners. This leads to their important role in physical kinetics – in plasma conductivity, drift and diffusion, absorption of electromagnetic energy, and in the evolution of energy distribution functions. The typical value for cross sections of the elastic, electron–neutral collisions for electron energy of 1–3 eV is approximately $\sigma = 10^{-16}$–10^{-15} cm^2; the reaction rate coefficient approximately $k = 3*10^{-8}$ cm^3/s. For the elastic, ion-neutral collisions at room temperature, the typical value of cross section is about $\sigma = 10^{-14}$ cm^2; and reaction rate coefficient of about $k = 10^{-9}$ cm^3/s.

Electron–electron, electron–ion, and ion-ion scattering processes are the so-called **Coulomb collisions**. While their cross sections are quite large with respect to those of collisions with neutral partners, these are relatively infrequent processes in discharges with a low degree of ionization. An important feature of the Coulomb collisions is a strong dependence of their cross sections on the kinetic energy of colliding particles. This effect can be demonstrated by a simple analysis, illustrated in Figure 2.3. Here, two particles have the same charge and for simplicity, one collision partner is considered at rest. Scattering takes place if the Coulomb interaction energy (order of $U \sim q^2/b$, where b is the impact parameter) is approximately equal to the kinetic energy ε of a collision partner. Then, the impact parameter $b \sim q^2/\varepsilon$, and the reaction cross section σ can be estimated as πb^2 and

$$\sigma(\varepsilon) \sim \pi q^4/\varepsilon^2 \tag{2.9}$$

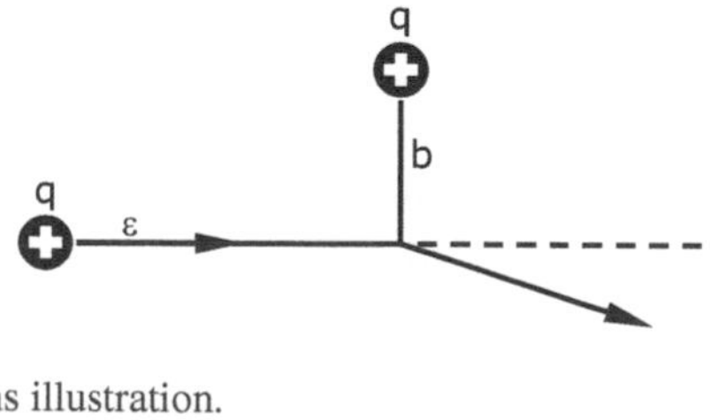

FIGURE 2.3 Coulomb collisions illustration.

The electron–electron scattering cross sections at room temperature is about 1000 times greater than those at electron temperature of 1 eV. Similar consideration of the charged particle scattering on neutral molecules with a permanent dipole momentum (interaction energy $U \sim 1/r^2$) and an induced dipole momentum (interaction energy $U \sim 1/r^4$) gives, respectively, $\sigma(\varepsilon) \sim 1/\varepsilon$ and $\sigma(\varepsilon) \sim 1/\varepsilon^{1/2}$.

Energy transfer during elastic collisions is possible only as a transfer of kinetic energy. The average fraction γ of kinetic energy transferred from one particle of mass m to another one of mass M is:

$$\gamma = 2mM/(m+M)^2 \tag{2.10}$$

In an elastic collision of electrons with heavy neutrals or ions $m \ll M$, and hence $\gamma = 2\, m/M$, which means the fraction of transferred energy is negligible ($\gamma \sim 10^{-4}$). One can find detailed fundamental consideration of the elementary elastic collisions of charged species as well as corresponding experimental data in the books of B.M. Smirnov (1981), E.W. McDaniel (1964, 1989), and Massey et al (1978)

2.2 IONIZATION PROCESSES

The key process in plasma is ionization because it is responsible for plasma generation – the birth of new electrons and positive ions. The simplest ionization process – the ionization by electron impact – is illustrated in Figure 2.1a. In general, all ionization processes can be subdivided into five groups:

The first group, **direct ionization by electron impact**, includes the ionization of neural neutrals and preliminary not-excited atoms, radicals, or molecules by an electron, whose energy is sufficiently high enough to provide the ionization act in one collision. These processes are the most important in cold or nonthermal discharges, where electric fields and hence electron energies are quite high, but the level of excitation of neutral species is relatively moderate. The second group, **stepwise ionization by electron impact**, includes the ionization of preliminary excited neutral species. These processes are important mainly in thermal or energy-intense discharges when the ionization degree (ratio of number densities of electrons and neutrals) and the concentration of highly excited neutral species is quite high. The third group is the **ionization by a collision of heavy particles**. Such processes can take place during ion-molecular or ion-atomic collisions, as well as in collisions of electronically or vibrationally excited species when the total energy of the collision partners exceeds the ionization potential. The chemical energy of the colliding neutral species can be also contributed to ionization via the so-called associative ionization. The fourth group is the **photoionization,** where neutral collisions with photons result in the formation of an electron–ion pair. Photoionization is mainly important in thermal plasmas and some mechanisms of propagation of nonthermal discharges.

2.2.1 THE DIRECT IONIZATION BY ELECTRON IMPACT

This process takes place as a result of the interaction of an incident electron, having a high energy ε, with a valence electron of a preliminary neutral atom or molecule. The act of ionization occurs when energy $\Delta\varepsilon$ transferred between them exceeds the ionization potential I (see Section 2.1.2). Physical picture and good quantitative formulas can be derived from the following classical model, first introduced by Thomson, 1912. The Thomson model supposes that the valence electron is at rest and also neglects the interaction of the two colliding electrons with the rest part of the initially neutral particle. The differential cross section of the incident electron scattering with energy transfer $\Delta\varepsilon$ to the valence electron can be defined by the Rutherford formula:

$$d\sigma_i = \frac{1}{(4\pi\varepsilon_0)^2}\frac{\pi e^4}{\varepsilon(\Delta\varepsilon)^2}d(\Delta\varepsilon) \tag{2.11}$$

Integration of the differential expression Eq. (2.11) over $\Delta\varepsilon$, taking into account that for ionization acts, the transferred energy should exceed the ionization potential $\Delta\varepsilon \geq I$, gives

$$\sigma_i = \frac{1}{(4\pi\varepsilon_0)^2}\frac{\pi e^4}{\varepsilon}\left(\frac{1}{I} - \frac{1}{\varepsilon}\right). \tag{2.12}$$

This relation describes the direct ionization cross section σ_i as a function of an incident electron energy ε and it is known as the **Thomson formula**. Obviously, Eq. (2.12) should be multiplied, in general, by number of valence electrons Z_v. According to the Thomson formula, the direct ionization cross section is growing linearly near the threshold of the elementary process $\varepsilon = I$. In the case of high electron energies $\varepsilon >> I$, the Thomson cross section (2.12) decreases as $\sigma_i \sim 1/\varepsilon$. Quantum mechanics gives a more accurate asymptotic approximation for the high energy electrons $\sigma_i \sim \ln\varepsilon/\varepsilon$. When $\varepsilon = 2I$, the Thomson cross section reaches the maximum value:

$$\sigma_i^{\max} = \frac{1}{(4\pi\varepsilon_0)^2}\frac{\pi e^4}{4I^2} \tag{2.13}$$

The Thomson formula can be rewritten more precisely, taking into account kinetic energy ε_v of the valence electron (B.M. Smirnov, 1981):

$$\sigma_i = \frac{1}{(4\pi\varepsilon_0)^2}\frac{\pi e^4}{\varepsilon}\left(\frac{1}{\varepsilon} - \frac{1}{I} + \frac{2\varepsilon_v}{3}\left(\frac{1}{I^2} - \frac{1}{\varepsilon^2}\right)\right) \tag{2.14}$$

The Thomson formula (2.12) can be derived from Eq. (2.14) by assuming that the valence electron is at rest and $\varepsilon_v = 0$. A useful variation of the Thomson formula can be obtained, assuming the valence electron interaction with the rest of the atom as a Coulomb interaction. Then, according to classical mechanics $\varepsilon_v = I$ and Eq. (2.14) gives another relation for the direct ionization cross section, which takes into account motion of valence electrons:

$$\sigma_i = \frac{1}{(4\pi\varepsilon_0)^2}\frac{\pi e^4}{\varepsilon}\left(\frac{5}{3I} - \frac{1}{\varepsilon} - \frac{2I}{3\varepsilon^2}\right) \tag{2.15}$$

Equations (2.12) and (2.15) describe the direct ionization in a similar way, but the latter predicts a twice larger cross section $\sigma_i(\varepsilon)$ maximum ($\sigma_i^{\max} \sim \pi e^4/2I^2(4\pi\varepsilon_0)^2$) at a slightly lower electron energy $\varepsilon_{\max} = 1.85\ I$. All the discussed relations for the direct ionization cross section can be generalized as:

$$\sigma_i = \frac{1}{(4\pi\varepsilon_0)^2}\frac{\pi e^2}{I^2}Z_v f\left(\frac{\varepsilon}{I}\right), \tag{2.16}$$

where Z_v is a number of valence electrons, and $f(\varepsilon/I) = f(x)$ is a general function, which is common for all atoms. Thus, for the Thomson formula (2.12):

$$f(x) = \frac{1}{x} - \frac{1}{x^2} \tag{2.17}$$

Equation (2.16) is in a good agreement with experimental data for different atoms and molecules (B.M. Smirnov, 1981) if:

$$\frac{10(x-1)}{\pi(x+0.5)(x+8)} < f(x) < \frac{10(x-1)}{\pi x(x+8)} \tag{2.18}$$

Formulas for practical estimations are discussed in Barnett (1989). The semi-empirical formula (2.18) is quite useful for numerical calculations, if the relation $\sigma_i(\varepsilon)$ is unknown experimentally.

2.2.2 The Direct Ionization Rate Coefficient

The ionization rate coefficient k_I (2.7) can be calculated by the integration of the cross section $\sigma_i(\varepsilon)$ over the electron energy distribution function. Electron energy distribution function is determined by electron temperature T_e (or mean energy (2.2)), therefore the ionization rate coefficient $k_I(T_e)$ is also a function of T_e. When the ionization rate coefficient is known, the rate of direct ionization by electron impact w_{ion} can be found as:

$$w_{ion} = k_I(T_e)\, n_e\, n_0. \tag{2.19}$$

In this formula, n_e is the concentration (number density) of electrons, and n_0 is the concentration of neutral gas atoms or molecules. The ionization potential I is usually much greater than the mean electron energy. For this reason, the ionization rate coefficient is very sensitive to the electron energy distribution function. If $T_e << I$, only a small group of electrons can have energy exceeding the ionization potential. Then integrating $\sigma_i(\varepsilon)$ into Eq. (2.7), the linear part of $\sigma_i(\varepsilon)$ near the threshold of ionization $\varepsilon = I$ is sufficient:

$$\sigma_i^{threshold}(\varepsilon) = \frac{1}{(4\pi\varepsilon_0)^2} Z_v \frac{\pi e^4}{I^2}\left(\varepsilon/I - 1\right) = \sigma_0\left(\varepsilon/I - 1\right) \tag{2.20}$$

Here, $\sigma_0 = Z_v \pi e^4 / I^2 (4\pi\varepsilon_0)^2$ is the geometrical atomic cross section. After integrating Eq. (2.7), the direct ionization rate coefficient can be presented as:

$$k_i(T_e) = \sqrt{8T_e / \pi m}\,\sigma_0 \exp\left(-I/T_e\right) \tag{2.21}$$

For numerical estimations, one can take the cross section σ_0 for molecular nitrogen, 10^{-16} cm^2, and for argon $3*10^{-16}$ cm^2. Electron temperature is presented here in energy units, so the Boltzmann coefficient k = 1. Some numerical data on the electron impact direct ionization for different molecular gases – CO_2, H_2, N_2 e.a. – are presented in Figure 2.4 as a function of reduced electric field E/n_0, which is the ratio of the electric field over neutral gas concentration.

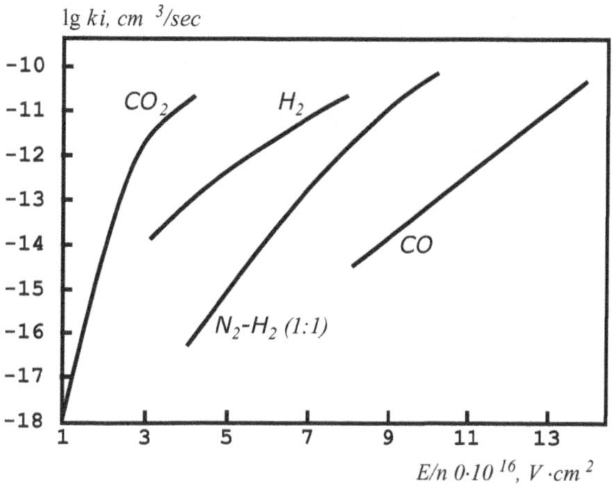

FIGURE 2.4 Ionization rate coefficient in molecular gases as a function of reduced electric field.

2.2.3 Peculiarities of Dissociation of Molecules by Electron Impact. The Frank-Condon Principle and the Process of Dissociative Ionization

First, let us consider a process of non-dissociative ionization of molecules by direct electron impact, taking as an example – the case of ionization of diatomic molecules AB:

$$e + AB \rightarrow AB^{+} + e + e \tag{2.22}$$

This process takes place predominantly when the electron energy exceeds the ionization potential not very essentially. It can be described in the first approximation by the Thomson approach (Section 2.2.1). One can see some peculiarities of ionization of molecules by electron impact, using the illustrative potential energy curves for AB and AB^+, shown in Figure 2.5. The fastest internal motion of atoms inside of molecules is molecular vibration. However, even molecular vibrations have typical times of 10^{-14}–10^{-13} s, which is much longer, than the interaction time between plasma electrons and the molecules $a_0/v_e \sim 10^{-16}$–10^{-15} s (here a_0 is the atomic unit of length, v_e is the mean electron velocity). This means, that all kinds of electronic excitation processes under consideration, induced by electron impact (including the molecular ionization, see Figure 2.5), are much faster, than all kinds of atomic motions inside of the molecules. As a result, all of the atoms inside molecules can be considered as frozen during the process of electronic transition, stimulated by electron impact. This fact is known as the **Frank-Condon Principle.**

According to the Frank-Condon Principle, the processes of collisional excitation and ionization of molecules are presented in Figure 2.5 by vertical lines (inter-nuclear distances are constant). This means that the non-dissociative ionization process Eq. (2.22) usually results in the formation of a vibrationally excited ion $(AB^+)^*$ and requires a little more energy, than corresponding atomic ionization. The phenomena can be described, using the **Frank-Condon factors**. When the electron energy is relatively high and essentially exceeds the ionization potential, the dissociative ionization process can take place:

$$e + AB \rightarrow A + B^{+} + e + e \tag{2.23}$$

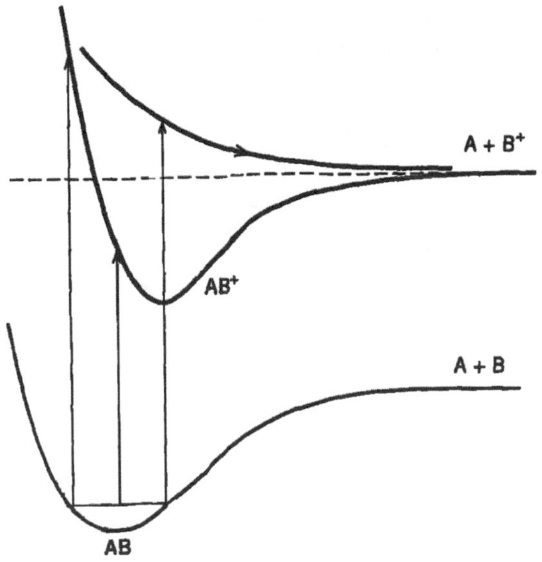

FIGURE 2.5 Molecular and ionic terms illustrating dissociative ionization.

This ionization process corresponds to electronic excitation into a repulsive state of ion $(AB^+)^*$, followed by the decay of this molecular ion. This is also illustrated by the vertical line in Figure 2.5. The energy threshold for the dissociative ionization is essentially greater than for the non-dissociative one.

2.2.4 Stepwise Ionization by Electron Impact

If the number density of electrons and hence the concentration of excited neutral species are sufficiently high, the energy I, necessary for ionization, can be provided in two different ways: (1) by the energy of plasma electrons and (2) by the preliminary electronic excitation of neutrals, which is called the stepwise ionization. The total energy needed for ionization in these two processes is sure the same, but which process is more preferable and which contributes the most in the total rate of ionization? The answer is unambiguous: if the level of electronic excitation is sufficiently high enough, the stepwise ionization is much faster, than the direct one, because the statistical weight of electronically excited neutrals is greater than that of free plasma electrons. In other words, when $T_e << I$, the probability of obtaining the ionization energy I is much lower for free plasma electrons (direct ionization) than for excited atoms and molecules. In contrast with the direct ionization, the stepwise process includes several steps, several electron impacts to provide the ionization act. The first electron – neutral collisions result in the preparation of the highly excited species, and then a final collision with relatively low energy electron provides the actual ionization act.

In the thermodynamic equilibrium, the ionization processes $e + A \rightarrow A^+ + e + e$ are reversed with respect to three-body recombination $A^+ + e + e \rightarrow A^* + e \rightarrow A + e$, which is proceeding through a set of excited states. According to the principle of detailed equilibrium, it means, that the ionization $e + A \rightarrow A^+ + e + e$ should go through the set of electronically excited states as well, which means the ionization should be a stepwise process. The stepwise ionization rate coefficient k_i^s can be found by summation of the partial rate coefficients $k_i^{s,n}$, corresponding to the n-th electronically excited state, over all states of excitation, taking into account their concentrations:

$$k_i^s = \sum_n k_i^{s,n} N_n(\varepsilon_n) / N_0 \tag{2.24}$$

To calculate the maximum stepwise ionization rate (B.M. Smirnov, 1981), let us suppose, that the electronically excited atoms and molecules are in quazi quasi-equilibrium with plasma electrons, and that, the excited species have an energy distribution function, corresponding to the Boltzmann law with the electron temperature T_e:

$$N_n = \left(\frac{g_n}{g_0}\right) N_0 \exp\left(-\frac{\varepsilon_n}{T_e}\right). \tag{2.25}$$

Here N_n, g_n, and ε_n are number density, statistical weight, and energy (with respect to ground state) of the electronically excited atoms, radicals, or molecules; the index "n" is the principal quantum number and actually it shows the particle excitation level. From statistical thermodynamics, the statistical weight of an excited particle $g_n = 2g_i n^2$, where g_i is the statistical weight of ion. N_0 and g_0 – are the concentration and statistical weight of ground state particles. Once more it is noted that the Boltzmann constant can be taken as k = 1 and for this reason omitted if the temperature is given in energy units. We'll follow this rule, in most cases later on in this book.

Typical energy transfer from a plasma electron to an electron, sitting on an excited atomic level is about T_e. This means that excited particles with energy $\varepsilon_n = I - T_e$ make the most important contribution to the sum Eq. (2.24). Taking into account that $I_n \sim 1/n^2$, the number of states with energy about $\varepsilon_n = I - T_e$ and ionization potential about $I_n = T_e$ has an order of n. Thus, from Eqs. (2.24) and Eqs. (2.25):

$$k_i^s \approx \frac{g_i}{g_0} n^3 \langle \sigma v \rangle \exp\left(-\frac{I}{T_e}\right) \tag{2.26}$$

The cross section σ in Eq. (2.26), corresponding to energy transfer (about T_e) between electrons, can be estimated as $e^4/T_e^2(4\pi\varepsilon_0)^2$, velocity v ~ $\sqrt{T_e/m}$ and the quantum number can be taken from:

$$I_n \approx \frac{1}{(4\pi\varepsilon_0)^2} me^4/\hbar n^2 \approx T_e \tag{2.27}$$

As a result, the stepwise ionization rate can be presented based on Eqs. (2.26) and (2.27) as:

$$k_i^s \approx \frac{g_i}{g_0} \frac{1}{(4\pi\varepsilon_0)^5} \left(me^{10}/T_e^3\right) \exp\left(-\frac{I}{T_e}\right) \hbar^3 \tag{2.28}$$

Here $\hbar$ is the Planck's constant. In general, Eq. (2.28) is quite convenient for numerical estimations of the stepwise ionization rate in thermal plasmas, when the electronically excited species are in quasi-equilibrium with the gas of plasma electrons. Also, the expression Eq. (2.28) for stepwise ionization is in a good agreement (in the framework of the principle of detailed equilibrium) with the rate of the reverse reaction – the three-body recombination $A^+ + e + e \leftrightarrow A^* + e \rightarrow A + e$.

Comparing direct ionization Eq. (2.21) with the stepwise ionization Eq. (2.28), one can see, that the second one can be much faster, because of the high large statistical weight of excited species, involved in the stepwise ionization. The ratio of rate coefficients for these two competing mechanisms of ionization can be derived from Eqs. (2.21) and (2.18):

$$\frac{k_i^s(T_e)}{k_i(T_e)} \approx \frac{g_i a_0^2}{g_0 \sigma_0} \left(\frac{1}{(4\pi\varepsilon_0)^2} me^4/\hbar T_e \right)^{7/2} \approx \left(\frac{I}{T_e} \right)^{7/2} \tag{2.29}$$

Here a_0 is the atomic unit of length. We have taken into account in this relation, - the estimations for geometric collisional cross section $\sigma_0 \sim a_0^2$, and for ionization potential: $I \approx \frac{1}{(4\pi\varepsilon_0)^2} me^4/\hbar^2$. For typical discharges with $I/T_e \sim 10$, the stepwise ionization can be 10^3–10^4 times faster, than the direct one. This takes place only in the case of high electron concentration n_e, and high electronic excitation frequency $k_{en}n_e$, which results in the quasi-equilibrium Eq. (2.25) between electronically excited species and plasma electrons. Usually, it can be applied only for quasi-equilibrium plasmas of thermal discharges.

To estimate the contribution of stepwise ionization in non-equilibrium discharges let us make some corrections in Eq. (2.29). If deactivation of the excited species because of radiation, collisional relaxation and losses on the walls has a characteristic time τ_n^* shorter than excitation time $1/k_{en}n_e$, the concentration of excited species will be lower, than the equilibrium one Eq. (2.25). Equation (2.25) can be corrected by taking into account electron concentration and the non-equilibrium losses of the excited species:

$$N_n \approx \frac{g_n}{g_0} N_0 \frac{k_{en} n_e \tau_n^*}{k_{en} n_e \tau_n^* + 1} \exp\left(-\frac{\varepsilon_n}{T_e}\right). \tag{2.30}$$

Then the stepwise ionization rate in non-equilibrium discharges can be rewritten as a function of electron temperature as well as electron concentration:

$$k_i^s\left(T_e,n_e\right) \approx k_i\left(T_e\right)\frac{k_{en}n_e\tau_n^*}{k_{en}n_e\tau_n^*+1}\left(\frac{I}{T_e}\right)^{\frac{7}{2}}. \tag{2.31}$$

One can see, that when the electron concentration is sufficiently high, this expression corresponds to Eq. (2.29) for the stepwise ionization rate in quasi-equilibrium plasmas.

2.2.5 Ionization by High Energy Electron Beams

For many important modern applications, plasma is generated not by electric discharges, but by means of the high energy electron beams. The ionization in this case can be more homogeneous and provide a generation of uniform nonthermal plasmas even in atmospheric pressure systems. The electron beams can be also effectively combined with the electric field in non-self-sustained discharges, where electron beams providing ionization, and the energy consumption is mostly due to the electric field. Peculiarities of ionization, in this case, are due to high energy of electrons in the beams, which usually varies from 50 keV to 1–2 MeV. Typical energy losses of the beams in atmospheric pressure air are of about 1 MeV/m per 1m. The beams with electron energies more than 500 keV are referred to as relativistic electron beams because these energies exceed the relativistic electron energy at rest ($E = mc^2$). In the frameworks of the Born approximation (Landau, 1997), electron energy losses per unit length dE/dx can be evaluated by the non-relativistic Bethe-Bloch formula:

$$-\frac{dE}{dx} = \frac{2\pi Ze^4}{\left(4\pi\varepsilon_0\right)^2 mv^2} n_0 \ln\frac{2mEv^2}{I^2}. \tag{2.32}$$

Here Z is an atomic number of neutral particles, providing the beam stopping; n_0 is their number density; v is the stopping electron velocity. In the case of relativistic electron beams with the electron energies 0.5–1 MeV, the energy losses going to neutral gas ionization can be numerically calculated by the relation:

$$-\frac{dE}{dx} = 2*10^{-22} n_0 Z \ln\frac{183}{Z^{1/3}}, \tag{2.33}$$

where energy losses dE/dx are expressed in MeV/cm, and concentration of neutral particles n_0 is expressed– in cm^{-3}. Equation (2.33) can be rewritten in terms of effective ionization rate coefficient k_i^{eff} for relativistic electrons:

$$k_i^{eff} \approx 3*10^{-10}\left(\mathrm{cm}^3/\mathrm{sec}\right) Z \ln\frac{183}{Z^{1/3}} \tag{2.34}$$

Numerically, this ionization rate coefficient is about 10^{-8}–10^{-7} cm^3/s. The rate of ionization by relativistic electron beams can be expressed in this case as a function of the electron beam concentration n_b or the electron beam current density j_b:

$$q_e = k_i^{eff} n_b n_0 \approx k_i^{eff}\frac{1}{ec} n_0 j_b, \tag{2.35}$$

where c is the sped of light.

2.2.6 Photoionization Processes

Though ionization processes are mostly induced by electron impact, ionization also can be also provided by interaction with high energy photons. The process of photoionization of a neutral particle A with ionization potential I (in eV) by a photon ħω with wavelength λ can be illustrated as:

$$\hbar\omega + A \rightarrow A^{+} + e, \lambda < \frac{12,400}{I(eV)} \text{A} \tag{2.36}$$

To provide ionization, the UV-photon wavelength should be quite low, usually below 100 nm. However, to provide effective ionization of preliminary excited atoms and molecules, photon energy can be lower. The photoionization cross section increases sharply from zero at the threshold energy Eq. (2.36) to quite high values, up to the geometrical cross section. Specific numerical values of the photoionization cross sections (near the threshold of the process) are presented in Table 2.3.

Though the cross sections shown in Table 2.3 are quite high, the contribution of the photoionization process is usually not very significant, because of the low concentration of high energy photons in most situations. However, in the following examples the photoionization plays a very essential role mostly by rapidly supplying seed electrons for the following ionization by electron impact:

1. streamer propagation in nonthermal discharges, where photoionization supplies seed electrons to start electron avalanches;
2. propagation of both nonthermal and thermal discharges in fast flow including supersonic one, where other mechanisms of discharge propagation are too slow;
3. preliminary gas ionization by ultraviolet radiation in non-self-sustained discharges, where UV-radiation is a kind of replacement of the relativistic electron beams.

2.2.7 The Ionization by Collisions of Heavy Particles, Adiabatic Principle, and Parameter Massey

An electron with kinetic energy slightly exceeding the ionization potential is quite effective to perform the ionization act. However, that is not true for ionization by collisions of heavy particles-ions and neutrals. Even when heavy particles have enough kinetic energy, the ionization probability is very low because their velocities are much less than those of electrons in atoms. This effect is a reflection of a

TABLE 2.3
The Photoionization Cross Sections

Atoms or Molecules	Wavelength λ, in A	Cross sections, in cm²
Ar	787	$3.5*10^{-17}$
N_2	798	$2.6*10^{-17}$
N	482	$0.9*10^{-17}$
He	504	$0.7*10^{-17}$
H_2	805	$0.7*10^{-17}$
H	912	$0.6*10^{-17}$
Ne	575	$0.4*10^{-17}$
O	910	$0.3*10^{-17}$
O_2	1020	$0.1*10^{-17}$
Cs	3185	$2.2*10^{-19}$
Na	2412	$1.2*10^{-19}$
K	2860	$1.2*10^{-20}$

general principle of particle interaction. A slow motion is "adiabatic", reluctant to transfer energy to a fast motion. The **adiabatic principle** can be explained in terms of relations between low interaction frequency $\omega_{int} = \alpha v$ (reverse time of interaction between particles) and high frequency of electron transfer in atom $\omega_{tr} = \Delta E \hbar$ Here $1/\alpha$-is a characteristic size of the interacting neutral particles, v is their velocity, ΔE is a change of electron energy in atom during the interaction, $\hbar$ is the Planck's constant. Only fast Fourier components of the slow interaction potential between particles with frequencies about $\omega_{tr} = \Delta E/\hbar$, provide the energy transfer between the interacting particles. The relative contribution of these fast Fourier components is very low if $\omega_{tr} >> \omega_{int}$, numerically it is approximately: $\exp(-\omega_{tr}/\omega_{int})$. As a result, the probability P_{EnTr} and cross sections of energy transfer processes (including the ionization process under consideration) are usually proportional to the so-called Massey parameter:

$$P_{EnTr} \propto \exp\left(-\frac{\omega_{tr}}{\omega_{int}}\right) \propto \exp\left(-\frac{\Delta E}{\hbar \alpha v}\right) = \exp\left(-P_{Ma}\right) \tag{2.37}$$

Here $P_{Ma} = \Delta E \hbar \alpha v$ is the **adiabatic Massey parameter**. If $P_{Ma} >> 1$ the process of energy transfer is adiabatic, and its probability is exponentially low, which occurs in the collisions of energetic heavy neutrals and ions. To get the Massey parameter close to one and eliminate the adiabatic prohibition for ionization, the kinetic energy of the colliding heavy particle has to be about 10–100 KeV, which is about three orders of magnitude greater than ionization potential. Though the kinetic energy of heavy particles in the ground state is ineffective for ionization, the situation can be different if they are electronically excited. If the total electron excitation energy of the colliding heavy particles is close to the ionization potential of one of them, the resonant energy transfer and effective ionization act can take place. Such non-adiabatic ionization processes occurring in the collision of heavy particles will be illustrated below by two specific examples: the Penning ionization effect and associative ionization.

2.2.8 The Penning Ionization Effect and Process of Associative Ionization

If electron excitation energy of a metastable atom A^* exceeds the ionization potential of another atom B, their collision can lead to an act of ionization, the so-called **Penning ionization**. The Penning ionization usually takes place through the intermediate formation of an excited molecule (in the state of auto-ionization), and cross sections of the process can be very large. The cross sections for the Penning ionization of N_2, CO_2, Xe, and Ar by metastable helium atoms $He(2^3S)$ with excitation energy 19.8 eV reach gas-kinetic values of 10^{-15} cm^2. Similar cross sections can be attained in collisions of metastable neon atoms (excitation energy 16.6 eV) with argon atoms (ionization potential 15.8 eV). Exceptionally high cross section $1.4*10^{-14}$ cm^2 (B.M. Smirnov, 1974) can be attained in the Penning ionization of mercury atoms (ionization potential 10.4 eV) by collisions with the metastable helium atoms $He(2^3S$, 19.8 eV). If the total electron excitation energy of colliding particles is not sufficient, an ionization process is possible nevertheless when heavy species stick to each other, forming a molecular ion, and hence their bonding energy can be also contributed to the ionization act. Such a process is called **the associative ionization.** The process differs from the Penning ionization only by the stability of the molecular ion product. A good example of the associative ionization is the process, involving the collision of two metastable mercury atoms:

$$\mathrm{Hg}\left(6^3\mathrm{P}_1, \mathrm{E} = 4.9\,\mathrm{eV}\right) + \mathrm{Hg}\left(6^3\mathrm{P}_0, \mathrm{E} = 4.7\,\mathrm{eV}\right) \rightarrow \mathrm{Hg}_2^+ + \mathrm{e} \tag{2.38}$$

The total electron excitation energy here, 9.6 eV, is less than the ionization potential of mercury atom (10.4 eV), but higher than that for Hg_2 molecule. This is the main "trick" of the associative ionization. Cross sections of the associative ionization (similarly to the Penning ionization) can be high and close to the gas-kinetic one (10^{-15} cm^2). The associative ionization is a reverse process with respect to dissociative recombination e + $AB^+ \rightarrow A^* + B$, the main recombination mechanism for molecular ions. The associative ionization is effective only for such excited species, which can be produced during the dissociative recombination. The relation between the cross sections of associative ionization σ_{ai} and dissociative recombination σ_r^{ei} can be derived (B.M. Smirnov, 1981) from the principle of detailed equilibrium as:

$$\sigma_{ai}\left(v_{rel}\right) = \sigma_r^{ei}\left(v_e\right)\frac{m^2 v_e^2}{\mu^2 v_{rel}^2}\frac{g_e g_{AB^+}}{g_{A^*} g_B} \tag{2.39}$$

In this relation: v_{rel} and v_e, are respectively, the relative velocity of heavy particles and the electron velocity, μ is the reduced mass of heavy particles, g are statistical weights. Ionization (and in particular the associative ionization) can also occur in the collision of vibrationally excited molecules. Cross sections and reaction rate coefficients for such processes are very low because of low Frank-Condon factors in this case. Nevertheless, these processes can make a notable contribution to ionization in the absence of an electric field (V. Rusanov, A. Fridman, 1984), like in the following process:

$$N_2^*\left({}^1\Sigma_g^+, v_1 \approx 32\right) + N_2^*\left({}^1\Sigma_g^+, v_2 \approx 32\right) \rightarrow N_4^+ + e \tag{2.40}$$

While this associative ionization has enough energy accumulated in the nitrogen molecules with 32 vibrational quanta each, but anyway the reaction rate of the process is relatively low: 10^{-15} exp (–2000 K/ T), cm^3/s (I.A. Adamovich et al., 1993, E. Plouges et al., 2001).

2.3 MECHANISMS OF ELECTRON LOSSES: THE ELECTRON–ION RECOMBINATION

The variety of channels of charged particle losses can be subdivided into three major groups. The first group includes different types of **electron–ion recombination processes**, in which collisions of the charged particles in a discharge volume lead to their mutual neutralization. These exothermic processes require consuming in some manner the large release of recombination energy. Dissociation of molecules, radiation of excited particles, or three-body collisions can provide the consumption of the recombination energy. All these different mechanisms of the electron–ion recombination will be discussed in this section.

Electron losses because of their sticking to neutrals and formation of negative ions (see Figure 2.1b) form the second group of volumetric losses, **electron attachment processes**. These processes are often responsible for the balance of charged particles in such electro-negative gases as oxygen (and for this reason air), CO_2 (because of formation of O^-), different halogens, and their compounds. Reverse processes of an electron release from a negative ion are called the **electron detachment**. These processes of electron losses, related to negative ions will be considered in Section 2.4. Note that although electron losses in this second group are due to the electron attachment processes, the actual losses of charged particles take place as a consequence following the fast processes of **ion–ion recombination.** The ion–ion recombination process means neutralization during the collision of negative and positive ions. These processes usually have very high rate coefficients and will be considered in Section 2.5.

Finally, the third group of charged particle losses is not a volumetric one like all those mentioned above, but is due to **surface recombination**. These processes of electron losses are the most important in low-pressure plasma systems, like glow discharges. The surface recombination processes are usually kinetically limited not by the elementary act of the electron–ion recombination on the surface, but by transfer (diffusion) of the charged particles to the walls of the discharge chamber. For this reason, the surface losses of charged particles will be discussed later in Chapter 4, concerning plasma kinetics and transfer phenomena.

2.3.1 Different Mechanisms of Electron–Ion Recombination

The electron–ion recombination is a highly exothermic process like all other recombination processes. The released energy corresponds to the ionization potential and so it is quite high (see Table 2.1). To be effectively realized, such a process should obviously have a channel of accumulation of the energy released during the neutralization of a positive ion and an electron. Taking into account three main channels of consumption of the recombination energy listed below (dissociation, three-body collisions, and radiation), the electron–ion recombination processes can be subdivided into three principal groups of mechanisms. In molecular gases, or just in the presence of molecular ions, the fastest electron neutralization mechanism is **dissociative electron–ion recombination**:

$$e + AB^{+} \rightarrow \left(AB\right)^{*} \rightarrow A + B^{*}. \tag{2.41}$$

In these processes, the recombination energy is usually going via resonance to dissociation of the molecular ion and to excitation of the dissociation products. While these processes principally take place in molecular gases, they can occur even in atomic gases by means of preliminary formation of molecular ions in the **ion conversion processes**: $A^{+} + A + A \rightarrow A_2^{+} + A$.

In atomic gases, in the absence of molecular ions, the neutralization can be due to **three-body electron–ion recombination:**

$$e + e + A^{+} \rightarrow A^{*} + e. \tag{2.42}$$

The excessive energy in this case is going to the kinetic energy of a free electron, which participates in the recombination act as "a third body partner". Heavy particles (ions and neutrals) are unable to accumulate electron recombination energy fast enough in their kinetic energy and are ineffective as the third body partner. Finally, the recombination energy can be converted into radiation in the process of **radiative electron–ion recombination:**

$$e + A^{+} \rightarrow A^{*} \rightarrow A + \hbar\omega \tag{2.43}$$

The cross section of this process is relatively low, and it can be competitive with the three-body recombination only when the plasma density is low.

2.3.2 The Dissociative Electron–Ion Recombination

This resonant recombination process Eq. (2.41) starts with the trapping of an electron on a repulsive auto-ionization level of molecular ion. Then atoms travel apart on the repulsive term, and if the auto-ionization state is maintained, stable products of dissociation are formed. This recombination mechanism is quite fast and plays a major role in the neutralization and decay of plasma in molecular gases. Reaction rate coefficients for most of the diatomic and three-atomic ions are on the level of 10^{-7} cm^3/s, for some important molecular ions the kinetic information can be found in Table 2.4.

In a group of similar ions, like molecular ions of noble gases, the recombination reaction rate is growing with several internal electrons: recombination of Kr_2^{+} and Xe_2^{+} is about 100 times faster than in case of helium. The rate coefficient of dissociative electron–ion recombination (the highly exothermic process) obviously decreases with temperature, as one can see that from Table 2.4. This process has no activation energy, so its dependencies on both electron T_e and gas T_0 temperatures are not very strong. This dependence can be estimated (M. Biondi, 1976, V. Rusanov, A. Fridman, 1984) as:

$$k_r^{ei}\left(T_e, T_0\right) \propto \frac{1}{T_0\sqrt{T_e}}. \tag{2.44}$$

TABLE 2.4

The Dissociative Electron–Ion Recombination Reaction Rate Coefficients at Room Gas Temperature $T_0 = 300$ K (Electron Temperature $T_e = 300$ K and $T_e = 1$ eV)

Electron–Ion Dissociative Recombination Process	Rate Coefficient k_r^{ei}, cm³/s $T_e = T_0 = 300$ K	Rate Coeff. k_r^{ei}, cm³/s $T_e = 1$ eV, $T_0 = 300$ K	Electron–Ion Dissociative Recombination Process	Rate Coefficient k_r^{ei}, cm³/s $T_e = T_0 = 300$ K	Rate Coeff. k_r^{ei}, cm³/s $T_e = 1$ eV, $T_0 = 300$ K
$e + N_2^+ \rightarrow N + N$	$2*10^{-7}$	$3*10^{-8}$	$e + NO^+ \rightarrow N + O$	$4*10^{-7}$	$6*10^{-8}$
$e + O_2^+ \rightarrow O + O$	$2*10^{-7}$	$3*10^{-8}$	$e + H_2^+ \rightarrow H + H$	$3*10^{-8}$	$5*10^{-9}$
$e + H_3^+ \rightarrow H_2 + H$	$2*10^{-7}$	$3*10^{-8}$	$e + CO^+ \rightarrow C + O$	$5*10^{-7}$	$8*10^{-8}$
$e + CO_2^+ \rightarrow CO + O$	$4*10^{-7}$	$6*10^{-8}$	$e + He_2^+ \rightarrow He + He$	10^{-8}	$2*10^{-9}$
$e + Ne_2^+ \rightarrow Ne + Ne$	$2*10^{-7}$	$3*10^{-8}$	$e + Ar_2^+ \rightarrow Ar + Ar$	$7*10^{-7}$	10^{-7}
$e + Kr_2^+ \rightarrow Kr + Kr$	10^{-6}	$2*10^{-7}$	$e + Xe_2^+ \rightarrow Xe + Xe$	10^{-6}	$2*10^{-7}$
$e + N_4^+ \rightarrow N_2 + N_2$	$2*10^{-6}$	$3*10^{-7}$	$e + O_4^+ \rightarrow O_2 + O_2$	$2*10^{-6}$	$3*10^{-7}$
$e + H_3O^+ \rightarrow H_2 + OH$	10^{-6}	$2*10^{-7}$	$e + (NO)_2^+ \rightarrow 2NO$	$2*10^{-6}$	$3*10^{-7}$
$e + HCO^+ \rightarrow CO + H$	$2*10^{-7}$	$3*10^{-8}$			

2.3.3 Ion Conversion Reactions as a Preliminary Stage of the Dissociative Electron–Ion Recombination

It is interesting, that if pressure is sufficiently high, the recombination of atomic ions like Xe^+ is increasing not by means of three-body Eq. (2.42) or radiative Eq. (2.43) mechanisms, but through the preliminary formation of molecular ions, through the so-called **ion conversion reactions** like $Xe^+ + Xe + Xe \rightarrow Xe_2^+ + Xe$. Then the molecular ion can be fast neutralized in the rapid process of dissociative recombination (see Table 2.4). The ion conversion reaction rate coefficients are quite high (B.M. Smirnov, 1974, L. Virin, V. Talrose, et al., 1978), see Table 2.5. When pressure exceeds 10 Torr, the ion conversion is usually faster than the following process of dissociative recombination, which becomes a limiting stage in the overall kinetics. An analytical expression for the ion-conversion three-body reaction rate coefficient can y derived from the dimension analysis (B.M. Smirnov, 1982):

$$k_{ic} \propto \left(\frac{\beta e^2}{4\pi\varepsilon_0}\right)^{5/4} \frac{1}{M^{1/2} T_0^{3/4}} \tag{2.45}$$

In this relation M and β are mass and polarization coefficient of colliding atoms, T_0^- is the gas temperature. The ion conversion effect takes place as a preliminary stage of recombination not only for simple atomic ions, as it was discussed above, but also for some important molecular ions. As is clear from Table 2.4, the polyatomic ions have very high recombination rates, often exceeding 10^{-6} cm^3/s at room temperature. This results in an interesting fact; the recombination of molecular ions like N_2^+, and O_2^+ at elevated pressures sometimes goes through the intermediate formation of dimers like- N_4^+, and O_4^+.

2.3.4 Three-Body Electron–Ion Recombination

This three-body recombination process $e + e + A^+ \rightarrow A^* + e$ is the most important one for high-density equilibrium plasma with temperature about 1 eV when the concentration of molecular ions is very low (because of thermal dissociation). The recombination process starts with a three-body capture of an electron by a positive ion and the formation of a very highly excited atom with a binding energy of about T_e. The initially formed highly excited atom then loses its energy in step-by-step deactivation through electron impacts. A final relaxation step from the lower excited state to the ground state has a relatively high value of energy transfer and it is usually due to radiative transition. The three-body electron–ion recombination process is a reverse one with respect to the stepwise ionization, considered in Section 2.2.4. For this reason, the reaction rate coefficient of this recombination process can be derived from Eq. (2.26) for the stepwise ionization rate coefficient k_i^s and from the Saha equation for ionization/recombination balance and equilibrium electron density:

$$k_r^{eei} = k_i^s \frac{n_0}{n_e n_i} = k_i^s \frac{g_0}{g_e g_i}\left(\frac{2\pi\hbar}{mT_e}\right)^{3/2} \exp\left(\frac{I}{T_e}\right) \approx \frac{e^{10}}{\left(4\pi\varepsilon_0\right)^5 \sqrt{mT_e^9}}. \tag{2.46}$$

TABLE 2.5
Ion Conversion Reaction Rate Coefficients at Room Temperature

Ion conversion process	Reaction rate coefficient	Ion conversion process	Reaction rate coefficient
$N_2^+ + N_2 + N_2 \rightarrow N_4^+ + N_2^+$	$8*10^{-29}$ cm^6/s	$O_2^+ + O_2 + O_2 \rightarrow O_4^+ + O_2^+$	$3*10^{-30}$ cm^6/s
$H^+ + H_2 + H_2 \rightarrow H_3^+ + H_2$	$4*10^{-29}$ cm^6/s	$Cs^+ + Cs + Cs \rightarrow Cs_2^+ + Cs$	$1.5*10^{-29}$ cm^6/s
$He^+ + He + He \rightarrow He_2^+ + H$	$9*10^{-32}$ cm^6/s	$Ne^+ + Ne + Ne \rightarrow Ne_2^+ + Ne$	$6*10^{-32}$ cm^6/s
$Ar^+ + Ar + Ar \rightarrow Ar_2^+ + Ar$	$3*10^{-31}$ cm^6/s	$Kr^+ + Kr + Kr \rightarrow Kr_2^+ + Kr$	$2*10^{-31}$ cm^6/s
$Xe^+ + Xe + Xe \rightarrow Xe_2^+ + Xe$	$4*10^{-31}$ cm^6/s		

Here n_e, n_i, n_0 are number densities of electrons, ions, and neutrals, g_e, g_i, g_0 are their statistical weights, e, m are electron charge and mass, electron temperature T_e is taken as usually in energy units, I is an ionization potential. It is convenient for practical calculations to rewrite Eq. (2.46) as:

$$k_r^{eei}, \frac{cm^6}{\sec} = \frac{\sigma_0}{I} 10^{-14} \left(I \big/ T_e \right)^{4.5} \tag{2.47}$$

In this numerical formula: σ_0, cm^2, is the gas-kinetic cross section, the same as those involved in Eqs. (2.20) and (2.21), I and T_e are the ionization potential and electron temperature, to be taken in eV.

The rate coefficient of the recombination process depends strongly on electron temperature but not exponentially. The typical value of k_r^{eei} at room temperature is about 10^{-20} cm^6/s, at $T_e = 1$ eV this rate coefficient is about 10^{-27} cm^6/s. At room temperature, the three-body recombination is able to compete with dissociative recombination when electron concentration is quite high and exceeds 10^{13} cm^{-3}. If the electron temperature is about 1 eV, three-body recombination can compete with the dissociative recombination only in the case of exotically high electron density exceeding 10^{20} cm^{-3}. As was mentioned earlier, the excessive energy in the process under consideration is going to the kinetic energy of a free electron – the "third body". In this case, heavy particles ion and neutrals are too slow and ineffective as "third body partners. The correspondent rate coefficient of the third-order reaction is about 10^8 times lower.

2.3.5 Radiative Electron–Ion Recombination

This electron–ion recombination process $e + A^+ \rightarrow A + \hbar\omega$ is a relatively slow one because it requires a photon emission during a short interval of the electron–ion interaction. This type of recombination can play a major role in the balance of charged particles only in the absence of molecular ions and in addition if the plasma density is quite low and the three-body mechanisms are suppressed. Typical values of cross sections of the radiative recombination process are about 10^{-21} cm^2, which is low. The reaction rate coefficients are therefore not high and can be simply estimated (Ya. B. Zeldovich, Yu. P. Raizer, 1966) as a function of electron temperature:

$$k_{rad.rec.}^{ei} \approx 3 \cdot 10^{-13} \left(T_e, eV \right)^{-3/4}, \mathrm{cm}^3/\mathrm{s} \tag{2.48}$$

Comparing numerical relations (2.47) and (48), one can see, that the reaction rate of radiative recombination exceeds that of the three-body process when electron concentration is low:

$$n_e < 3 \cdot 10^{13} \left(T_e, eV \right)^{3.75}, \mathrm{cm}^3/\mathrm{s}. \tag{2.49}$$

2.4 ELECTRON LOSSES DUE TO FORMATION OF NEGATIVE IONS: ELECTRON ATTACHMENT AND DETACHMENT PROCESSES

A simple illustration of the electron attachment process and negative ion formation was presented in Figure 2.1b. The electron balance in electro-negative gases, which have high EA (oxygen, different halogens, and their compounds, see Table 2.2) can be strongly affected by these processes. In some other gases, like CO_2 and H_2O, the electron attachment can be equally important, because of high EA of decomposition products and formation of negative ions like O^- and H^- during dissociation. The electron attachment processes are essential in the electron balance in weakly ionized plasma with low electron concentration and a low degree of ionization. This fact is due to the first kinetic order of this process with respect to the concentration of electrons, which means the reaction rate of the dissociative attachment is proportional to electron density. In contrast, the recombination processes are of second or third kinetic order. This means the electron–ion recombination reaction rates are proportional to the square or cube of electron concentration, and therefore they can be neglected with respect to electron attachment if the electron density and ionization degree are low.

The negative ions formed as a result of electron attachment can be neutralized quite easy and extremely fast in the following processes of ion–ion recombination. For this reason, the attachment processes in general can provide significant losses of electrons and charged particles, which can result in the prevention of ignition and propagation of electric discharges. Fortunately for plasma generation, the electron attachment processes can be effectively suppressed by the reverse reactions of electron detachment. These processes of decay of negative ions and restoration of free electrons can be provided by collisions with excited or chemically active species, energetic electrons, etc.

2.4.1 Dissociative Electron Attachment to Molecules

It is important in molecular gases like CO_2, H_2O, SF_6, CF_4, when dissociation products have positive electron affinities, see (Table 2.2):

$$e + AB \rightarrow \left(AB^{-}\right)^{*} \rightarrow A + B^{-} \tag{2.50}$$

The mechanism of this process is somewhat similar to the dissociative recombination and proceeds by intermediate formation of an auto-ionization state $(AB^-)^*$. This excited state is unstable and its decay leads either to the reverse process of auto-detachment (AB + e) or to dissociation ($A + B^-$). During the attachment Eq. (2.50), an electron is captured and not able to provide an energy balance of the elementary process. For this reason, the dissociative attachment is a resonant reaction, which means it requires quite definite values of the electron energy. The most typical potential energy curves, illustrating the dissociative attachment Eq. (2.50), are presented in Figure 2.6a. The electron attachment process starts in this case with a vertical transition from AB molecular ground state electronic term to a repulsive state of AB^- (obviously following the Frank-Condon Principle, Section 2.2.3). During the repulsion, before the $(AB^-)^*$ reaches an intersection point of AB and AB^- electronic terms, the reverse auto-detachment reaction (AB + e) is very possible. But after passing the intersection, the AB potential energy exceeds that of AB^- and further repulsion results in dissociation ($A + B^-$). To estimate the cross section of the dissociative attachment we should take into account, that the repulsion time with possible auto-detachment is proportional to the square root of reduced mass of AB molecule $\sqrt{M_A M_B / \left(M_A + M_B\right)}$ (H. Massey, 1976). The characteristic electron

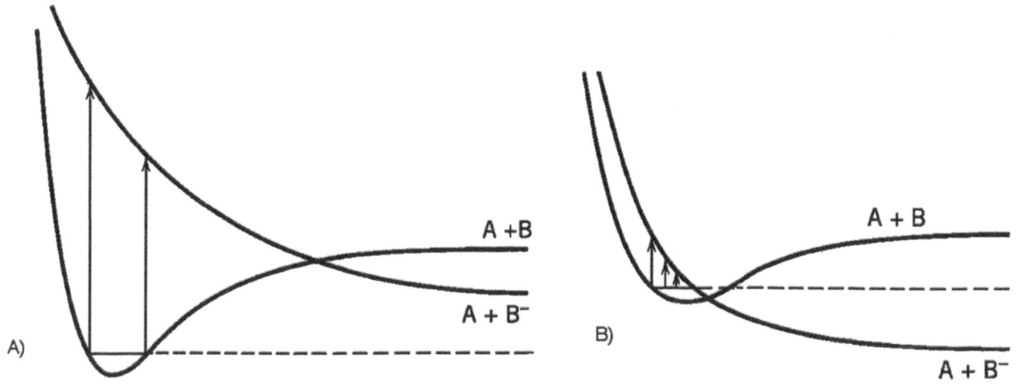

FIGURE 2.6 Dissociative attachment: (a) low electron affinity, (b) high electron affinity.

transition time is much shorter and proportional to the square root of its mass m. For this reason (M.A. Lieberman, A.J. Lichtenberg, 1994), the maximum cross section of dissociative attachment with the described configuration of electronic terms can be estimated as:

$$\sigma_{d.a.}^{\max} \approx \sigma_0 \sqrt{\frac{m(M_A + M_B)}{M_A M_B}} \tag{2.51}$$

The maximum cross section is two orders of value less than the gas-kinetic cross section σ_0, and numerically is of about 10^{-18} cm^2. Some change in the reaction cross section Eq. (2.51) occurs, if the AB^- electronic term corresponds not to the repulsive, but rather to the attractive state (H. Massey, 1976),; this not a common situation. A more interesting process occurs when the EA of a product exceeds the dissociation energy. In particular, it takes place for some halogens and their compounds. Corresponding potential energy curves, illustrating the dissociative attachment, are presented in Figure 2.6b. In this case in contrast to the process depicted by Figure 2.6a, even very low energy electrons can effectively provide the dissociative attachment. Also, if the EA of a product exceeds the dissociation energy, the intersection point of AB and AB^- electronic terms (Figure 2.6b) is actually located inside of the so-called geometrical sizes of the dissociating molecules. As a result, during the repulsion of $(AB^-)^*$, the probability of the reverse auto-detachment reaction (AB + e) is very low and cross section of the dissociative attachment process can in this case reach the gas-kinetic cross section σ_0 of about 10^{-16} cm^2.

Cross sections of the dissociative attachment for several molecular gases are presented in Figure 2.7 as a function of electron energy. Usually, only electrons, having enough energy (more than the difference between dissociation energy and EA), can provide the dissociative attachment. Also, the dissociative attachment cross section as a function of electron energy $\sigma_a(\varepsilon)$ has a resonant structure. It reflects the resonance nature of the process, which can be effective only within a narrow range of electron energy. The resonant structure of $\sigma_a(\varepsilon)$ permits estimating the dissociative attachment rate coefficient k_a as a function of electron temperature T_e as:

$$k_a(T_e) \approx \sigma_{d.a.}^{\max}(\varepsilon_{\max}) \sqrt{\frac{2\varepsilon_{\max}}{m}} \frac{\Delta\varepsilon}{T_e} \exp\left(-\frac{\varepsilon_{\max}}{T_e}\right) \tag{2.52}$$

In this relation the only resonance leading to the dissociative attachment and hence only the resonance of the $\sigma_a(\varepsilon)$ was taken into account, $\varepsilon_{\max}$ and $\sigma^{\max}_{d.a.}$ are the electron energy and maximum cross section (see Eq. (2.51)), corresponding to the resonance, $\Delta\varepsilon$ is its energy width. Equation (2.52) is quite convenient for numerical calculations; the necessary parameters: $\varepsilon_{\max}$, $\sigma^{\max}_{d.a.}$ and $\Delta\varepsilon$

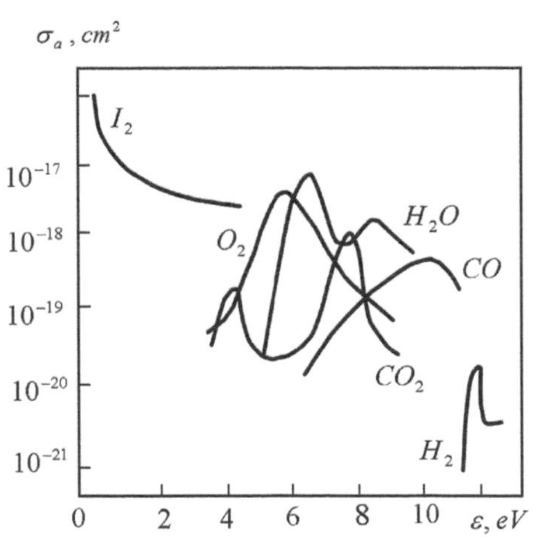

FIGURE 2.7 Cross-sections of dissociative electron attachment to different molecules.

TABLE 2.6
The Resonance Parameters for the Dissociative Attachment of Electrons to Different Molecules

Dissociative Attachment Process	ε_{max}, eV	$\sigma^{max}_{d.a}$, cm^2	$\Delta\varepsilon$, eV
$e + O_2 \rightarrow O^- + O$	6.7	10^{-18}	1
$e + H_2 \rightarrow H^- + H$	3.8	10^{-21}	3.6
$e + NO \rightarrow O^- + N$	8.6	10^{-18}	2.3
$e + CO \rightarrow O^- + C$	10.3	$2*10^{-19}$	1.4
$e + HCl \rightarrow Cl^- + H$	0.8	$7*10^{-18}$	0.3
$e + H_2O \rightarrow H^- + OH$	6.5	$7*10^{-18}$	1
$e + H_2O \rightarrow O^- + H_2$	8.6	10^{-18}	2.1
$e + H_2O \rightarrow H + OH^-$	5	10^{-19}	2
$e + D_2O \rightarrow D^- + OD$	6.5	$5*10^{-18}$	0.8
$e + CO_2 \rightarrow O^- + CO$	4.35	$2*10^{-19}$	0.8

are provided in Table 2.6. As was mentioned above, the information is given in each case only for one resonance, which is supposed to have the maximum contribution to kinetics of the dissociative attachment.

2.4.2 Three-Body Electron Attachment to Molecules

These attachment processes of formation of negative ions in a collision of an electron with two heavy particles (at least one of them is supposed to have positive EA) can be presented as:

$$e + A + B \rightarrow A^- + B \tag{2.53}$$

The three-body electron attachment can be a principal channel of electron losses through formation of negative ions when electron energies are not high enough for the dissociative attachment, and when pressure is elevated (usually more than 0.1 atm) and the third kinetic order processes are preferable. In contrast to the dissociative attachment, the three-body process doesn't require the consumption of electron energy. For this reason, its rate coefficient does not depend strongly on electron temperature (at least, within the temperature range about 1 eV). In this case, heavy particles are responsible for the consumption of energy released during the attachment. Electrons are usually kinetically not effective as a third body B, because of low degree of ionization and low energy release during the attachment (in contrast to the three-body electron–ion recombination Eq. (2.42)).

Atmospheric pressure nonthermal discharges in air are systems, where the three-body attachment plays a key role in the balance of charged particles. That is why the three-body electron attachment to molecular oxygen ($e + O_2 + M \rightarrow O_2^- + M$) can be taken as a good example. The three-body electron attachment to molecular oxygen proceeds by the two-stage **Bloch-Bradbury mechanism** (F. Bloch, N. Bradbury, 1935; N. Alexandrov, 1981). The first stage of the process includes an electron attachment to the molecule with the formation of a negative oxygen ion in an unstable auto-ionization state:

$$e + O_2 \overset{k_{att},\tau}{\leftrightarrow} \left(O_2^-\right)^*. \tag{2.54}$$

Here k_{att} is a rate coefficient of the intermediate electron trapping, τ is lifetime of the excited unstable ion with respect to collisionless decay into the initial state. The second stage of the Bloch-Bradbury mechanism includes collision with the third-body particle M with either relaxation and stabilization of O_2^- (rate coefficient k_{st}) or collisional decay of the unstable ion into initial state (rate coefficient k_{dec}):

$$\left(O_2^-\right)^* + M \xrightarrow{k_{st}} O_2^- + M. \tag{2.55}$$

$$\left(O_2^-\right)^* + M \xrightarrow{k_{dec}} O_2 + e + M. \tag{2.56}$$

Taking into account the steady-state conditions for number density of the intermediate excited ions $(O_2^-)^*$, the rate coefficient for the total process of three-body electron attachment to molecular oxygen $e + O_2 + M \rightarrow O_2^- + M$ can be expressed in the following way:

$$k_{3M} = \frac{k_{att}k_{st}}{\frac{1}{\tau} + \left(k_{st} + k_{dec}\right)n_0}. \tag{2.57}$$

In this relation, n_0 is the concentration of the third-body heavy particles M When the pressure is not too high, $(k_{st} + k_{dec})n_0 << \tau^{-1}$, and Eq. (2.57) for the reaction rate can be significantly simplified:

$$k_{3M} \approx k_{att}k_{st}\tau \tag{2.58}$$

The three-body attachment process has a third kinetic order. It equally depends on the rate coefficient of formation and stabilization of negative ions on a third particle. The latter one strongly depends on the type of the third particle. In general, the more complicated molecule plays a role of the third body (M), the easier it stabilizes the $(O_2^-)^*$ and the higher is the total reaction rate coefficient $k_{3M.}$ Numerical values of the total rate coefficients k_{3M} are presented in Table 2.7 for room temperature:

The three-body attachment rate coefficients are shown on the Figure 2.8 as a function of electron temperature (the gas, in general, is taken to be room temperature). For simple estimations,

TABLE 2.7
Reaction Rate Coefficients of the Electron Attachment to Oxygen Molecules at Room Temperature with Different Third Body Partners

Three-Body Attachment	Rate coefficient	Three-Body Attachment	Rate coefficient
$e + O_2 + Ar \rightarrow O_2^- + Ar$	$3*10^{-32}$ cm^6/s	$e + O_2 + Ne \rightarrow O_2^- + Ne$	$3*10^{-32}$ cm^6/s
$e + O_2 + N_2 \rightarrow O_2^- + N_2$	$1.6*10^{-31}$ cm^6/s	$e + O_2 + H_2 \rightarrow O_2^- + H_2$	$2*10^{-31}$ cm^6/s
$e + O_2 + O_2 \rightarrow O_2^- + O_2$	$2.5*10^{-30}$ cm^6/s	$e + O_2 + CO_2 \rightarrow O_2^- + CO_2$	$3*10^{-30}$ cm^6/s
$e + O_2 + H_2O \rightarrow O_2^- + H_2O$	$1.4*10^{-29}$ cm^6/s	$e + O_2 + H_2S \rightarrow O_2^- + H_2S$	10^{-29} cm^6/s
$e + O_2 + NH_3 \rightarrow O_2^- + NH_3$	10^{-29} cm^6/s	$e + O_2 + CH_4 \rightarrow O_2^- + CH_4$	$>10^{-29}$ cm^6/s

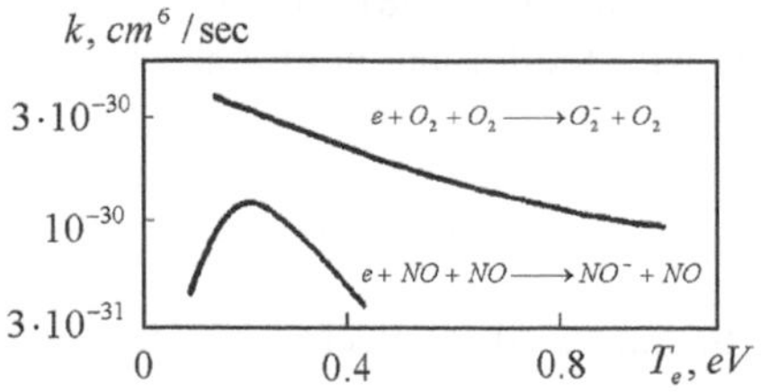

FIGURE 2.8 Rate coefficients of the three body electron attachment to molecules.

when $T_e = 1$ eV, $T_0 = 300$ K, one can take $k_{3M} \approx 10^{-30}$ cm^6/s. The rate of the three-body process is greater than dissociative attachment (k_a) when the gas number density exceeds a critical value $n_0 > k_a(T_e) / k_{3M}$. Numerically in oxygen with $T_e = 1$ eV, $T_0 = 300$ K, this means $n_0 > 10^{18}$ cm^{-3}, or in pressure units $p > 30$ Torr.

2.4.3 Other Mechanisms of Formation of Negative Ions

Let us just mention three other mechanisms of formation of negative ions, which can be usually neglected, but in some specific situations should be taken into account. The first process is **the polar dissociation**:

$$e + AB \rightarrow A^+ + B^- + e. \tag{2.59}$$

This process actually includes both ionization and dissociation, therefore the threshold energy is quite high in this case. On the other hand, the electron is not captured, the process is not a resonant one and can be effective in a wide range of high electron energies. For example in molecular oxygen, the maximum value of cross section of the polar dissociation is about $3*10^{-19}$ cm^2 and corresponds to electron energy 35 eV. To stabilize the formation of a negative ion during electron attachment, the excessive energy of an intermediate excited ion particle can be emitted. As a result, a negative ion can be formed in the process of radiative attachment:

$$e + M \rightarrow \left(M^-\right)^* \rightarrow M^- + \hbar\omega. \tag{2.60}$$

Such an electron capture can take place at low electron energies, but the probability of the radiative process is very low, about 10^{-5}–10^{-7}. Corresponding values of attachment cross sections are about 10^{-21}–10^{-23} cm^2. Finally, some electronegative polyatomic molecules like SF_6 have a negative ion state very close to a ground state (only 0.1 eV in the case of SF_6^-). As a result, the lifetime of such metastable negative ions can be long. Such an attachment process is resonant and for very low electron energies has maximum cross sections of about 10^{-15} cm^2.

2.4.4 Mechanisms of Negative Ion Destruction, Associative Detachment Processes

The negative ions can be then neutralized in the fast ion–ion recombination or can release an electron by its detachment and ion destruction. Competition between these two processes defines the balance of charged particles and often determines regime of the discharges. Different mechanisms can provide the destruction of negative ions with an electron release. These mechanisms are discussed in detail in special books and reviews: H. Massey, 1976; E.W. McDaniel, 1964; B.M. Smirnov, 1978; N. Alexandrov, 1981. Let us consider three detachment mechanisms, which are most important in plasma-chemical systems. In nonthermal discharges, probably the most important one is the **associative detachment**:

$$A^- + B \rightarrow \left(AB^-\right)^* \rightarrow AB + e. \tag{2.61}$$

This process is a reverse one with respect to the dissociative attachment Eq. (2.50) and so it can also be illustrated by the same Figure 2.6. The associative detachment is a non-adiabatic process, which occurs by the intersection of electronic terms of a complex negative ion $A^- - B$ and corresponding molecule AB. For this reason (non-adiabatic reaction), the rate coefficients of such processes are usually quite high. Typical values of the coefficients are about $k_d = 10^{-10}$–10^{-9} cm^3/s, and are not far from those of gas-kinetic collisions. Because the reaction (2.61) corresponds to the intersection of electronic terms of A^-– B and AB, it can have an energy barrier even in exothermic

TABLE 2.8
The Associative Attachment Rate Coefficients at Room Temperature

Associative Attachment Process	Reaction Enthalpy	Rate Coefficient
$H^- + H \rightarrow H_2 + e$	−3.8 eV (exothermic)	$1.3*10^{-9}$ cm³/s
$H^- + O_2 \rightarrow HO_2 + e$	−1.25 eV (exothermic)	$1.2*10^{-9}$ cm³/s
$O^- + O \rightarrow O_2 + e$	−3.8 eV (exothermic)	$1.3*10^{-9}$ cm³/s
$O^- + N \rightarrow NO + e$	−5.1 eV (exothermic)	$2*10^{-10}$ cm³/s
$O^- + O_2 \rightarrow O_3 + e$	0.4 eV (endothermic)	10^{-12} cm³/s
$O^- + O_2(^1\Delta_g) \rightarrow O_3 + e$	−0.6 eV (exothermic)	$3*10^{-10}$ cm³/s
$O^- + N_2 \rightarrow N_2O + e$	−0.15 eV (exothermic)	10^{-11} cm³/s
$O^- + NO \rightarrow NO_2 + e$	−1.6 eV (exothermic)	$5*10^{-10}$ cm³/s
$O^- + CO \rightarrow CO_2 + e$	−4 eV (exothermic)	$5*10^{-10}$ cm³/s
$O^- + H_2 \rightarrow H_2O + e$	−3.5 eV (exothermic)	10^{-9} cm³/s
$O^- + CO_2 \rightarrow CO_3 + e$	(endothermic)	10^{-13} cm³/s
$C^- + CO_2 \rightarrow CO + CO + e$	−4.3 (exothermic)	$5*10^{-11}$ cm³/s
$C^- + CO \rightarrow C_2O + e$	−1.1 (exothermic)	$4*10^{-10}$ cm³/s
$Cl^- + O \rightarrow ClO + e$	0.9 (endothermic)	10^{-11} cm³/s
$O_2^- + O \rightarrow O_3 + e$	−0.6 eV (exothermic)	$3*10^{-10}$ cm³/s
$O_2^- + N \rightarrow NO_2 + e$	−4.1 eV (exothermic)	$5*10^{-10}$ cm³/s
$OH^- + O \rightarrow HO_2 + e$	−1 eV (exothermic)	$2*10^{-10}$ cm³/s
$OH^- + N \rightarrow HNO + e$	−2.4 eV (exothermic)	10^{-11} cm³/s
$OH^- + H \rightarrow H_2O + e$	−3.2 eV (exothermic)	10^{-9} cm³/s

conditions. Rate coefficients of some associative attachment processes are presented in Table 2.8 together with an enthalpy of the reactions.

The associative attachment is quite fast. For example, the reaction $O^- + CO \rightarrow CO_2 + e$ can effectively suppress the dissociative attachment in CO_2 and corespondent electron losses, if CO concentration is high enough. This effect is of importance in nonthermal discharges in CO_2 and in CO_2^- laser mixture. We should note though, that fast three-body cluster formation processes are able to stabilize O^- with respect to associative attachment. The following clusterization reactions:

$$O^- + CO_2 + M \rightarrow CO_3^- + M, k = 10^{-27}\text{cm}^6 / \text{s} \tag{2.62}$$

$$O^- + O_2 + M \rightarrow O_3^- + M, k = 10^{-30}\text{cm}^6 / \text{sec} \tag{2.63}$$

converts O^- ions into CO_3^-, which are more stable with respect to detachment and hence promote ion–ion recombination and loss of charged particles.

2.4.5 Electron Impact Detachment

This detachment process can be described as follows: $e + A^- \rightarrow A + 2e$ and is an essential one in the balance of negative ions when the ionization degree is high. The electron impact detachment is somewhat similar to the direct ionization of neutrals by an electron impact (Thomson mechanism, see section 2.2.1). The main difference in this case is due to repulsive Coulomb force acting between the incident electron and the negative ion. For incident electron energies about 10 eV, the cross section of the detachment process can be high, about 10^{-14} cm². Typical detachment cross section dependence on energy or on the incident electron velocity v_e (M. Inokuti, Y. Kim, R.L. Platzman, 1967) can be illustrated by that for the detachment from negative hydrogen ion ($e + H^- \rightarrow H + 2e$):

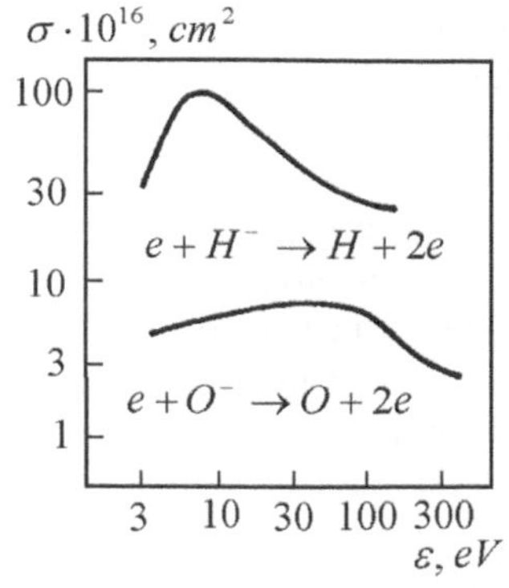

FIGURE 2.9 Negative ion destruction by an electron impact.

$$\sigma(v_e) \approx \frac{\sigma_0 e^4}{(4\pi\varepsilon_0)^2 \hbar^2 v_e^2}\left(-7.5\ln\frac{e^2}{4\pi\varepsilon_0 \hbar v_e} + 25\right) \tag{2.64}$$

In this relation, σ_0 is the geometrical atomic cross section, more exactly defined in Eq. (2.20). The detachment cross section as a function of electron energy is presented in Figure 2.9 for hydrogen and oxygen atomic ions. The maximum cross sections of about 10^{-15}–10^{-14} cm^2 corresponds to electron energies about 10–50 eV. These energies exceed electron affinities more than 10 times in contrast to the Thomson mechanism of the electron impact ionization (which can be explained by the Coulomb repulsion).

2.4.6 Detachment in Collisions with Excited Particles

This detachment process ($A^- + B^* \rightarrow A + B + e$), for the case of using **electronic excitation** energy of particle B, is similar to the Penning ionization (see section 2.2.8). If the electronic excitation energy of a collision partner B exceeds the EA of another particle - A, the detachment process can proceed effectively as an electronically non-adiabatic reaction (without significant energy exchange with translational degrees of freedom of the heavy particles, see section 2.2.7). As an example, consider the exothermic detachment of an electron from an oxygen ion in collision with a metastable electronically excited oxygen (excitation energy 0.98 eV):

$$O_2^- + O_2(^1\Delta_g) \rightarrow O_2 + O_2 + e, \Delta H = -0.6 \text{ eV}. \tag{2.65}$$

The rate coefficient of the detachment reaction is very high, even at room temperature it is about $2{*}10^{-10}$ cm^3/s. Electron detachment also can be effective in collisions with **vibrationally excited molecules**. Consider again as an example the destruction of an oxygen molecular ion: $O_2^- + O_2^*(v > 3) - O_2 + O_2 + e$. Rate coefficients of the detachment process in the quasi-equilibrium systems (characterized by a single temperature) are presented in the following Table 2.9. The first process

TABLE 2.9
Thermal Destruction of an Oxygen Molecular Ion

Detachment Process	Reaction enthalpy	Rate Coefficient, 300K	Rate Coefficient, 600K
$O_2^- + O_2 \rightarrow O_2 + O_2 + e$	0.44 eV	$2.2{*}10^{-18}$ cm^3/s	$3{*}10^{-14}$ cm^3/s
$O_2^- + N_2 \rightarrow O_2 + N_2 + e$	0.44 eV	$<10^{-20}$ cm^3/s	$1.8{*}10^{-16}$ cm^3/s

presented in the Table has the activation energy close to the reaction enthalpy. These quasi-equilibrium detachment processes are using mostly vibrational energy of colliding partners, because in other cases the reaction is strongly adiabatic (see Section 2.2.7). The detachment process stimulated by vibrational excitation of molecules proceeds according to the modified Bloch-Bradbury mechanism Eqs. (2.53–2.55). In this case, the process starts with collisional excitation of O_2^- to the vibrationally excited states with $v > 3$. The excited ions $(O_2^-)^*(v > 3)$ are then in the state of auto-ionization, which results in an electron detachment during a period of time shorter than the interval between two collisions. Excitation of $(O_2^-)^*(v > 3)$ is easier and faster in oxygen, than in nitrogen, where the process is less resonant. It explains the significant difference in detachment rate coefficients in collisions with oxygen and nitrogen, which one can see from Table 2.9. Kinetics of the detachment process can be described in this case in a conventional way for all reactions stimulated by vibrational excitation of molecules (see Chapter 5). The traditional Arrhenius formula: $k_d \propto \exp(-E_a/T_v)$ is applicable here, if molecular vibrations are in internal quasi-equilibrium with the only vibrational temperature T_v. The activation energy of the detachment process can be taken in this case equal to the EA to oxygen molecules ($E_a \approx 0.44$eV) in the same way one can describe detachment kinetics of NO^-, F^- and Br^- (H. Massey, 1976).

2.5 THE ION–ION RECOMBINATION PROCESSES

In electro-negative gases, when the attachment processes are involved in the balance of electrons and ions, the actual losses of charged particles are mostly due to ion–ion recombination. This process means mutual neutralization of positive and negative ions in binary or three-body collisions. The ion–ion recombination can proceed by a variety of different mechanisms, which dominate in different pressure ranges. However, all of them are characterized by very high rate coefficients. At high pressures (usually more than 30 Torr), three-body mechanisms dominate the recombination. The recombination rate coefficient in this case reaches a maximum of about $1–3*10^{-6}$ cm^3/s near atmospheric pressures for room temperature. Traditionally the rate coefficient of ion–ion recombination is recalculated with respect to concentrations of positive and negative ions, which is to the second kinetic order. Due to the three-molecular nature of the recombination mechanism in this pressure range, the recalculated (second kinetic order) recombination rate coefficient depends on pressure. Near atmospheric pressure corresponds to the fastest neutralization. Both an increase and a decrease of pressure results in proportional reduction of the three-body ion–ion recombination rate coefficient. At low pressures, the three-body mechanism becomes relatively slow. In this pressure range, the binary collisions with transfer of energy into electronic excitation make a major contribution to the ion–ion recombination.

2.5.1 Ion–Ion Recombination in Binary Collisions

Such neutralization proceeds in a collision of a negative and positive ion with the released energy going to electronic excitation of a neutral product:

$$A^- + B^+ \rightarrow A + B^* \tag{2.66}$$

This reaction has a second kinetic order and dominates the ion–ion recombination in low-pressure discharges ($p < 10–30$ Torr). Electronic terms, illustrating the recombination, are presented in Figure 2.10. The ions A^- and B^+ are approaching each other following the attractive $A^- – B^+$ Coulomb potential curve. When the distance between the heavy particles is large, this potential curve lies below the term of $A^- – B^+$ and above the final $A – B^*$ electronic term on the energy interval:

$$\Delta E \approx {I_B}/{\eta} - EA_A \quad \Delta E \approx I_B/\eta \tag{2.67}$$

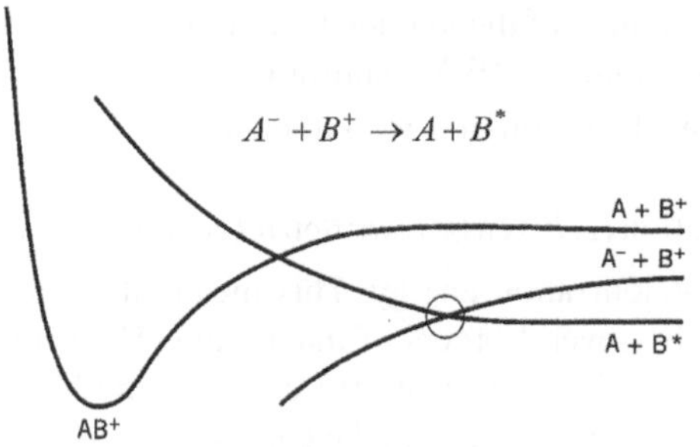

FIGURE 2.10 Terms illustrating the ion-ion recombination.

Here I_B is the ionization potential of particle B. If the principal quantum number of particle B after recombination is high n = 3–4, then ΔE is low. This means that the electronic terms of $A^- - B^+$ and $A - B^*$ are relatively close when the ions A^- and B^+ are approaching each other. When the principal quantum number n is not specifically defined, then the affinity EA_A can be taken as a reasonable estimation for the energy interval Eq. (2.67), which is $\Delta E \approx EA_A$. As can be seen from Figure 2.11, the low value of Δ Eresults in the possibility of effective transition between the electronic terms (from $A^- - B^+$ to $A - B^*$), when distance R_{ii} between ions is still large. Even for this long-distance R_{ii} between the ions the Coulomb attraction energy is already sufficient to compensate the initial energy gap ΔE between the terms:

$$R_{ii} \approx \frac{e^2}{4\pi\varepsilon_0} \frac{1}{\left(I_B/\eta^2 - \mathrm{EA}_A \right)} \approx \frac{e^2}{4\pi\varepsilon_0 \mathrm{EA}_A} \tag{2.68}$$

The high value of R_{ii} results in large cross sections of the ion–ion recombination Eq. (2.66). This can be estimated from the conservation of angular momentum during the attractive Coulomb collision and assuming maximum kinetic energy equal to the potential energy EA_A Eq. (4.5.2). The impact parameter b (see Section 2.1.7) can be found as a function of the ion kinetic energy ε in the center of mass system:

$$b \approx R_{ii} \frac{\sqrt{\mathrm{EA_A}}}{\sqrt{\varepsilon}} \tag{2.69}$$

Based on the expressions for the impact parameter b and for the reactive distance R_{ii}, the formula to calculate the cross section of the ion–ion recombination process can be given as:

$$\sigma^{ii}_{rec} = \pi b^2 \approx \pi \frac{e^4}{\left(4\pi\varepsilon_0\right)^2} \frac{1}{\mathrm{EA}_{\dot{A}}} \frac{1}{\varepsilon} \tag{2.70}$$

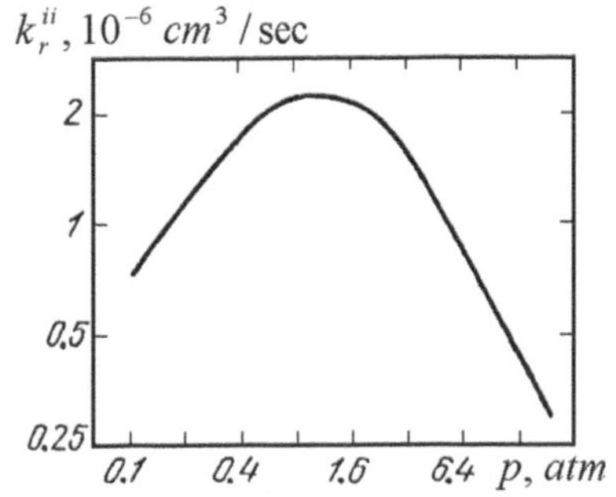

FIGURE 2.11 Ion-ion recombination rate coefficient in air.

Quantum mechanical calculation of the ion–ion recombination across the section should take into account the effect of electron tunneling (B.M. Smirnov, 1978). As a result, it gives the more precise dependence of cross section of the ion–ion recombination in binary collisions of the ion kinetic energy as: $\sigma_{rec}^{ii} \propto 1/\sqrt{\varepsilon}$. Equation (2.70) ca be rewritten referring to the atomic units, $\sigma_{reci}^{ii} \approx \sigma_0 \frac{1}{EA_A} \frac{1}{\varepsilon}$, where I is a typical value of ionization energy. This means that the recombination cross section exceeds the gas-kinetic one by several orders of magnitude. The same is true for the reaction rate coefficients, which usually are of the order of 10^{-7} cm^3/s. Some binary recombination rate coefficients are presented in Table 2.10 together with the energy released in the process.

2.5.2 The Three-Body Ion–Ion Recombination, Thomson's Theory

The neutralization process can effectively proceed in a triple collision of a heavy neutral with a negative and a positive ion:

$$A^- + B^+ + M \rightarrow A + B + M \tag{2.71}$$

This process dominates the ion–ion recombination at moderate and high pressures (p > 10–30 Torr). The three-body reaction has the third kinetic order only in the moderate pressure range, usually less than 1 atm (for high pressures the process is limited by ion mobility, see below). It means that only at moderate pressures, the reaction rate proportional to the product of concentrations of all three collision partners, positive and negative ions, and neutrals. At pressures about (0.01 – 1 atm), the three-body ion–ion recombination can be described in the framework of the Thomson theory (Thomson, 1924). According to this theory, the act recombination takes place if a negative and positive ions approach each other closer than the critical distance $b \approx e^2/4\pi\varepsilon_0 T_0$. In this case, their Coulomb interaction energy reaches the level of thermal energy, which is approximately T_0 the gas temperature or temperature of heavy particles. Then during the collision, the third body (a heavy neutral particle) can absorb the energy of approximately equal to the thermal energy T_0 from an ion, and provide the act of recombination. The collision frequency of an ion with a neutral particle leading to energy transfer of about T_0 between them and subsequently resulting in recombination, can be found as $n_0 \sigma v_t$, where n_0 is the neutrals number density, σ is a typical cross section of ion-neutral elastic scattering, v_t is an average thermal velocity of the heavy particles. After the collision, the

TABLE 2.10
Reaction Rate Coefficients of the Ion–Ion Recombination in Binary Collisions at Room Temperature

Recombination Process	Released Energy, I – EA	Rate Coefficient
$H^- + H^+ \rightarrow H + H$	12.8 eV	3.9*10^{-7} cm^3/s
$O^- + O^+ \rightarrow O + O$	12.1 eV	2.7*10^{-7} cm^3/s
$O^- + N^+ \rightarrow O + N$	13.1 eV	2.6*10^{-7} cm^3/s
$O^- + O_2^+ \rightarrow O + O_2$	11.6 eV	10^{-7} cm^3/s
$O^- + NO^+ \rightarrow O + NO$	7.8 eV	4.9* 10^{-7} cm^3/s
$SF_6 + SF_5^+ \rightarrow SF_6 + SF_5$	15.2 eV	4*10^{-8} cm^3/s
$NO_2^- + NO^+ \rightarrow NO_2 + NO$	5.7 eV	3*10^{-7} cm^3/s
$O_2^- + O^+ \rightarrow O_2 + O$	13.2 eV	2*10^{-7} cm^3/s
$O_2^- + O_2^+ \rightarrow O_2 + O_2$	11.6 eV	4.2*10^{-7} cm^3/s
$O_2^- + N_2^+ \rightarrow O_2 + N_2$	15.1 eV	1.6*10^{-7} cm^3/s

probability P_+ for a positive ion to be closer than b $\approx e^2/4\pi\varepsilon_0 T_0$ to a negative ion, and as a result to be neutralized in the following recombination can be estimated as:

$$P_+ \approx n_- b^3 \approx n_- \frac{e^6}{\left(4\pi\varepsilon_0\right)^3 T_0^3} \tag{2.72}$$

The frequency of the collisions of a positive ion with neutral particles, resulting in ion–ion recombination, can be then be found as $\nu_{ii} = (\sigma v_t n_0)P_+$. Consequently, the total three-body ion–ion recombination rate can be presented according to the Thomson's theory as:

$$w_{ii} \approx \left(\sigma v_t\right) \frac{e^6}{\left(4\pi\varepsilon_0\right)^3 T_0^3} n_0 n_- n_+ \tag{2.73}$$

The process has a third kinetic order with the rate coefficient k_{r3}^{ii} decreasing with temperature:

$$k_{r3}^{ii} \approx \left(\sigma\sqrt{\frac{1}{m}}\right) \frac{e^6}{\left(4\pi\varepsilon_0\right)^3} \frac{1}{T_0^{5/2}} \tag{2.74}$$

At room temperature, the recombination coefficients are about 10^{-25} cm^6/s. For some reactions, these third-order kinetic reaction rate coefficients are presented in Table 2.11. The coefficients in the table are also recalculated to the second kinetic order k_{r2}^{ii}, which means $k_{r2}^{ii} = k_{r3}^{ii} n_0$. Comparing Tables 2.10 and 2.11, the binary and triple collisions are seen to contribute equally to the ion–ion recombination at pressures about 10–30 Torr.

2.5.3 High-Pressure Limit of the Three-Body Ion–Ion Recombination, Langevin Model

According to Eqs. (2.73) and (2.74), the ion–ion recombination rate coefficient recalculated to the second kinetic order $k_{r2}^{ii} = k_{r3}^{ii} n_0$ grows linearly with pressure at fixed temperatures. This growth is limited in the range of moderate pressures (usually not more than 1 atm) by the frameworks of Thomson's theory. Thomson's approach requires the capture distance b to be less than an ion mean free path ($1/n_0\sigma$). This requirement actually prevents using Thomson's theory for a high concentration of neutrals and high pressures (more than 1 atm):

$$n_0 \sigma b \approx n_0 \frac{e^2}{\left(4\pi\varepsilon_0\right) T_0} \sigma < 1 \tag{2.75}$$

TABLE 2.11
Rate Coefficients of the Three-Body Ion–Ion Recombination Processes at Room Temperature and Moderate Pressures (In the Last Column the Coefficients Are Recalculated to the Second Kinetic Order and Pressure of 1 Atm, that Is Multiplied by the Concentration of Neutrals 2.7*10^{19} cm^{-3})

Ion–Ion Recombination	Rate Coefficients, 3rd order	Rate Coefficients, 2-rd order, 1atm
$O_2^- + O_4^+ + O \rightarrow O_2 + O_2 + O_2 + O_2 + O_2$	1.55*10^{-25} cm^6/s	4.2*10^{-6} cm^3/s
$O^- + O_2^+ + O_2 \rightarrow O_3 + O_2$	3.7*10^{-25} cm^6/s	10^{-5} cm^6/s
$NO_2^- + NO^+ + O_2 \rightarrow NO_2 + NO + O_2$	3.4*10^{-26} cm^6/s	0.9*10^{-6} cm^6/s
$NO_2^- + NO^+ + N_2 \rightarrow NO_2 + NO + N_2$	10^{-25} cm^6/s	2.7*10^{-6} cm^6/s

In the opposite case of high pressures, $n_0\sigma b > 1$, the recombination is limited by the phase, in which a positive and negative ion approach each other, moving in electric field overcoming multiple collisions with neutrals. The Langevin model developed in 1903, describes the ion–ion recombination in this pressure range. This motion of a positive and negative ion to collide in recombination can be considered as their drift in the Coulomb field $e/(4\pi\varepsilon_0)r^2$, where r is the distance between them. Use positive and negative ion mobility μ_+ and μ_- (see Chapter 4) as a coefficient of proportionality between ion drift velocity and the strength of the electric field. Then the ion drift velocity to meet each other is:

$$v_d = \frac{e}{(4\pi\varepsilon_0)^2}(\mu_+ + \mu_-). \tag{2.76}$$

Consider sphere with radius r, surrounding positive ion. A flux of negative ions with concentration n_- approaching the positive one and the recombination frequency v_{r+} for the positive ion are as follows:

$$v_{r+} = 4\pi r^2 v_d n_- \tag{2.77}$$

Based on Eq. (2.77) for neutralization frequency with respect to one positive ion, the total ion–ion recombination reaction rate can be found as: $w = n_+ v_{r+} = 4\pi r^2 v_d n_- n_+$. In this case, it shows the second kinetic order of the process with respect to ion concentrations. The final Langevin expression for the ion–ion recombination rate coefficient $k_r^{ii} = w / n_+ n_-$ in the limit of high pressures (usually more than 1 atm) can then be given taking into account Eqs. (2.76) and (2.77), as:

$$k_r^{ii} = 4\pi e(\mu_+ + \mu_-) \tag{2.78}$$

This recombination rate coefficient is proportional to the ion mobility, which decreases proportionally with pressure (see Chapter 4). Comparing the Thomson and Langevin models, it is seen that at first, recombination rate coefficients grow with pressure $k_{r2}^{ii} = k_{r3}^{ii} n_0$ (see Eq. (2.74)), and then at high pressures they begin to decrease as 1/p together with ion mobility Eq. (2.78). The highest recombination coefficient according to Eq. (2.75) corresponds to the concentration of neutrals:

$$n_0 \approx \frac{4\pi\varepsilon_0 T_0}{\sigma e^2} \tag{2.79}$$

At room temperature, this gas concentration corresponds to atmospheric pressure. The maximum value of the ion–ion recombination rate coefficient can be then found by substituting the number density n_0 into $k_{r2}^{ii} = k_{r3}^{ii} n_0$:

$$k_{r,\max}^{ii} \approx \frac{e^4}{(4\pi\varepsilon_0)^2} \frac{v}{T_0^2} \tag{2.80}$$

Numerically, the maximum recombination rate coefficient is about $1\text{–}3*10^{-6}$ cm^3/s. This coefficient decreases for both an increase or decrease of pressure from atmospheric pressure. The total experimental dependence of the k_r^{ii} on pressure is presented in Figure 2.11 The generalized ion–ion recombination model, combining the Thomson's and Langevin approaches for moderate and high pressures, was developed by Natanson (1959).

2.6 THE ION–MOLECULAR REACTIONS

2.6.1 Ion–Molecular Polarization Collisions, the Langevin Rate Coefficient

The ion–molecular processes can start with scattering in the polarization potential, leading to the so-called Langevin capture of a charged particle and formation of an intermediate ion-molecular complex. If a neutral particle itself has no permanent dipole moment, the ion-neutral charge–dipole

TABLE 2.12
Polarizabilities of Atoms and Molecules

Atom or Molecule	A, 10^{-24} cm³	Atom or Molecule	A, 10^{-24} cm³
Ar	1.64	H	0.67
C	1.78	N	1.11
O	0.8	CO	1.95
Cl_2	4.59	O_2	1.57
CCl_4	10.2	CF_4	1.33
H_2O	1.45	CO_2	2.59
SF_6	4.44	NH_3	2.19

interaction and scattering are due to the dipole moment pM, induced in the neutral particle by the electric field E of an ion:

$$p_m = \alpha\varepsilon_0 E = \alpha\frac{e}{4\pi r^2} \tag{2.81}$$

Here r is the distance between the interacting particles, α is the polarizability of a neutral atom or molecule, (see Table 2.12). The typical orbits of relative ion and neutral motion during polarization scattering are shown in Figure 2.12. When the impact parameter is high, the orbit has a hyperbolic character. But when the impact parameter is sufficiently low, the scattering leads to the Langevin polarization capture. The Langevin capture means that the spiral trajectory results in "closer interaction" and formation of the ion-molecular complex, which then can either spiral out or provide inelastic changes of state and formation of different secondary products. This ion-molecular capture process based on polarization occurs when the interaction energy between charge and induced dipole $P_m E = \alpha\frac{e}{4\pi r^2}\frac{e}{4\pi\varepsilon_0 r^2}$ becomes the order of the kinetic energy of the colliding partners $\frac{1}{2}Mv^2$, where M is their reduced mass, and v their relative velocity. From the equality of kinetic and interaction energies, the **Langevin cross section** of the polarization capture can be found as $\sigma_L \approx \pi r^2$:

$$\sigma_L = \sqrt{\frac{\pi\alpha e^2}{\varepsilon_0 M v^2}}. \tag{2.82}$$

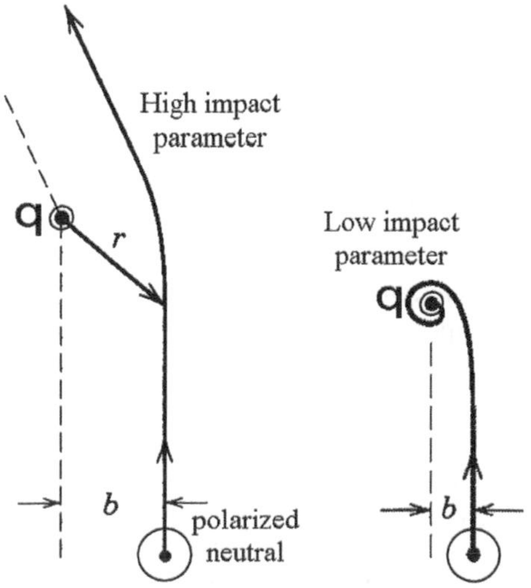

FIGURE 2.12 The Langevin scattering in the polarization potential.

Detailed derivation of the exact formula for Langevin corss section can be found in the book of M.A. Lieberman and A.J. Lichtenberg (1994). The Langevin capture rate coefficient $k_L \approx \sigma_{Lv}$ according to Eq. (2.82) does not depend on velocity, and therefore does not depend on temperature:

$$k_L = \sqrt{\frac{\pi \alpha e^2}{\varepsilon_0 M}} \tag{2.83}$$

It is convenient for calculations to express the polarizability α in 10^{-24} cm^3 (cubic Angstroms, see Table 2.12), reduced mass in atomic mass units (amu), and then to rewrite and use Eq. (2.83) in the following numerical form:

$$k_L^{ion/neutral} = 2.3 \times 10^{-9}\, \mathrm{cm}^3 \Big/ \mathrm{sec}^* \sqrt{\frac{\alpha, 10^{-24}\mathrm{cm}^3}{\mathrm{M, amu}}} \tag{2.84}$$

The typical value of the Langevin rate coefficient for ion-molecular reactions is 10^{-9} cm^3/s, which is 10 times higher than the typical value of gas-kinetic rate coefficient for binary collisions of neutral particles: $k_0 \approx 10^{-10}$ cm^3/s. Also, α/M in Eq. (2.84) reflects the "specific volume" of an atom or molecule, because the polarizability α actually corresponds to the particle volume. This means that the Langevin capture rate coefficient grows with a decrease of "density" of an atom or molecule. The above relations described interactions of a charge with an induced dipole. If an ion interacts with a molecule having not only an induced dipole but also a permanent dipole moment $\mu_{D,}$ then the Langivin capture cross section becomes larger. The corresponding correction of Eq. (2.83) was done by Su and Bowers (1973):

$$k_L = \sqrt{\frac{\pi e^2}{\varepsilon_0 M}} \left(\sqrt{\alpha} + c\mu_D \sqrt{\frac{2}{\pi T_0}} \right) \tag{2.85}$$

Here T_0 is a gas temperature (in energy units) and parameter $0 < c < 1$ describes the effectiveness of a dipole orientation in the electric field of an ion. For molecules like H_20 or HF, having large a permanent dipole moment, the ratio of the second to the first terms in Eq. (2.85) is about $\sqrt{I/T_0}$, where I is ionization potential. It means that the Langevin cross sections and rate coefficients for dipole molecules and radicals can exceed by a factor of 10 the numerical values obtained for pure polarization collisions Eq. (2.84). Derivation of the Langevin formula Eq. (2.82) did not specify the mass of a charged particle. This means that the Langevin cross section and rate coefficients Eqs. (2.82) and (2.83) also can be applied for an electron–neutral interaction. In this case, the reduced mass M is close to an electron mass and numerical formula for the Langevin rate coefficient for the electron–neutral collision can be presented as:

$$\kappa_L^{electron/neutral} = 10^{-7}\, \mathrm{cm}^3 \Big/ \mathrm{s}^* \sqrt{\alpha, 10^{-24}}\, \mathrm{cm}^3 \tag{2.86}$$

Numerical value of the Langevin rate coefficient for the electron–neutral collision is about 10^{-7} cm^3/s and does not depend on temperature.

2.6.2 The Ion-Atom Charge Transfer Processes

An electron can be transferred during a collision from a neutral particle to a positive ion or from a negative ion to a neutral particle. These processes are referred to as charge transfer or charge exchange. The charge exchange reaction without significant defect of the electronic state energy ΔE during collision is called the **resonant charge transfer**. Otherwise, charge transfer is called **non-resonant**. The resonant charge transfer is a non-adiabatic process and usually has a very large cross

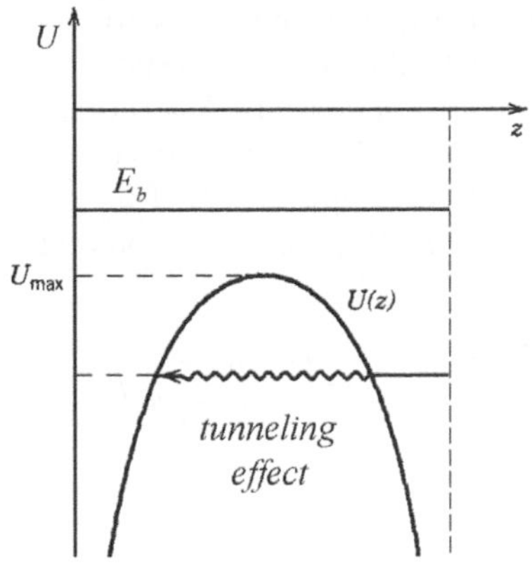

FIGURE 2.13 Energy terms for the resonant charge exchange.

section. Consider a charge exchange between a neutral particle B and a positive ion A^+ assuming the particle B/B^+ being at rest:

$$A^+ + B \rightarrow A + B^+ \tag{2.87}$$

The energy scheme of this reaction is illustrated in Figure 2.13, where one can see the Coulomb potential energy of an electron in the Coulomb field of A^+ and B^+:

$$U(z) = -\frac{e^2}{4\pi\varepsilon_0 z} - \frac{e^2}{4\pi\varepsilon_0 |r_{AB} - z|} \tag{2.88}$$

Here r_{AB} is a distance between the centers of A and B. The maximum value of the potential energy corresponds to $z = r_{AB}/2$ and is equal to:

$$U_{max} = -\frac{e^2}{\pi\varepsilon_0 r_{AB}} \tag{2.89}$$

The charge transfer, Eq. (2.87), is possible in the framework of classical mechanics (see Figure 2.13) if the maximum of potential energy U_{max} is lower than the initial energy E_B of an electron which is going to be transferred from level n of particle B:

$$E_B = -\frac{I_B}{n^2} - \frac{e^2}{4\pi\varepsilon_0 r_{AB}} \geq U_{max} \tag{2.90}$$

Here I_{AB} is the ionization potential of atom B. From Eqs. (2.89) and (2.90), the maximum distance between the interacting heavy particles is found when the charge transfer is still permitted by classical mechanics:

$$r_{AB}^{max} = \frac{3e^2 n^2}{4\pi\varepsilon_0 I_B} \tag{2.91}$$

If the charge exchange is resonant and therefore not limited by the defect of energy, the classical reaction cross section can be found from Eq. (2.91) as πr_{AB}^2:

$$\sigma_{chtr}^{class} = \frac{9e^4 n^4}{16\pi\varepsilon_0^2 I_B^2}. \tag{2.92}$$

The classical cross section of charge exchange does not depend on the kinetic energy of the interacting species and its numerical value for ground state transfer (n = 1) is approximately the gas-kinetic cross section for collisions of neutrals. The actual cross section of a resonant charge transfer can be much higher than Eq. (2.92), taking into account the quantum mechanical effect of electron tunneling from B to A^+ (see Figure 2.13). This effect can be estimated by calculating the electron tunneling probability P_{tunn} across a potential barrier of height about I_B and width d:

$$P_{tunn} \approx \exp\left(-\frac{2d}{\hbar}\sqrt{2meI_B}\right). \tag{2.93}$$

The frequency of electron oscillation in the ground state atom B is about $I_B/\hbar$, so the frequency of the tunneling is $I_B P_{tunn}/\hbar$. Taking into account Eq. (2.93), the maximum barrier width d_{max}, when the tunneling frequency still exceeds the reverse ion-neutral collision time: $I_B P_{tunn}/\hbar > v/d$ and as a result, the tunneling can take place. Here v is the relative velocity of the colliding partners. Then the cross section πd_{max}^2 of the electron tunneling from B to A^+, leading to the resonance charge exchange is:

$$\sigma_{ch.tr}^{tunn} \approx \frac{1}{I_B}\left(\frac{\pi\hbar^2}{8me}\right)\left(\ln\frac{I_B d}{\hbar} - \ln v\right)^2. \tag{2.94}$$

More detailed derivation of the electron tunneling cross section (Rapp, Francis, 1962) results in more complicated expressions, which can be presented in a numerical form very similar to Eq. (2.94):

$$\sqrt{\sigma_{ch.tr}^{tunn}\mathrm{cm}^2} \approx \frac{1}{\sqrt{I_B \mathrm{eV}}}\left(6.5*10^{-7} - 3*10^{-8}\ln v, \frac{\mathrm{cm}}{\mathrm{s}}\right). \tag{2.95}$$

This numerical formula can be applied to calculate the resonant charge transfer cross section in the velocity range v = 10^5–10^8 cm/s. At higher energies, when the velocity v exceeds 10^8 cm/s, the tunneling cross section decreases to the gas-kinetic cross sections (about $3*10^{16}$ cm^2) and then it is stabilized. The stabilized cross section does not depend on energy and is given by the classical expression Eq. (2.92).

At the lower limit of application of Eq. (2.95), v = 10^5 cm/s and the tunneling cross section reaches values as high as 10^{-14} cm^2. Equations (2.94) and (2.95) were derived assuming straight line collision trajectories. Therefore applying Eqs. (2.95) and (2.95) at lower velocities is impossible because when $v \le 10^5 \frac{\mathrm{cm}}{\mathrm{s}} / \sqrt{\mathrm{M}}$ (M is a reduced mass), the collision trajectory is strongly perturbed during the charge exchange. In low energy collision, when $v \le 10^5 \frac{\mathrm{cm}}{\mathrm{s}} / \sqrt{\mathrm{M}}$, the ion and neutral are captured in a complex. In this case, the exchanging electron always has enough time for tunneling and therefore the equal probability ½ are found on either particle. Then the resonant charge exchange cross section can be found as half of the Langevin capture cross section Eq. (2.82):

$$\sigma_L = \frac{1}{2}\sqrt{\frac{\pi\alpha e^2}{\varepsilon_0 \mathrm{M}}}\frac{1}{v}. \tag{2.96}$$

The typical numerical value of this cross section at room temperature is about 10^{-14} cm^2 and higher. It grows as the velocity (and temperature) decreases even faster than in the case of intermediate velocities Eq. (2.95).

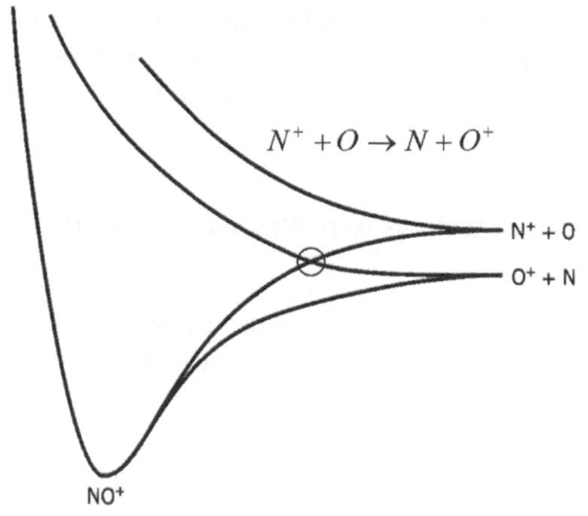

FIGURE 2.14 Terms illustrating the non-resonant charge exchange.

2.6.3 The Non-Resonant Charge Transfer Processes

Consider an electron transfer between oxygen and nitrogen atoms as an example of the non-resonant charge exchange:

$$N^+ + O \rightarrow N + O^+, \quad \Delta E = -0.9 \text{ eV}. \tag{2.97}$$

The principal potential curves, illustrating the nonresonant charge transfer Eq. (2.97) are shown in Figure 2.14. From Table 2.1, the ionization potential of oxygen (I = 13.6 eV) is lower than that of nitrogen (I = 14.5 eV). This is the reason that the electron transfer from oxygen to nitrogen is an exothermic process and the separated $N + O^+$ energy level is located 0.9 eV lower than the separated $O + N^+$ energy level. The reaction (2.97) begins with N^+ approaching O by following the attractive NO^+ term. Then this term is crossing with the repulsive NO^+ term and the system experiences a non-adiabatic transfer, which results in the formation of $O^+ + N$. The excess energy of approximately 1 eV goes into the kinetic energy of the products. The cross section of such exothermic charge exchange reactions at low thermal energies is of the order of the resonant cross sections of tunneling Eq. (2.95) or Langevin capture Eq. (2.96). The endothermic reactions of charge exchange, like the reverse process $N + O^+ \rightarrow N^+ + O$, are usually very slow at low gas temperatures with low energies of the colliding ions and heavy neutral particles.

The important role of the non-resonant charge exchange processes can be illustrated by the so-called effect of **acidic behavior of nonthermal air plasma**. Ionization of air in nonthermal discharges primarily leads to a large amount of N_2^+ ions (with respect to other positive ions) because of the high molar fraction of nitrogen in air. Later, low ionization potential and high dipole moment of water molecules can result in fast charge exchange:

$$N_2^+ + H_2O \rightarrow N_2 + H_2O^+, \quad k(300K) = 2.2 * 10^{-9} \text{cm}^3/\text{s}. \tag{2.98}$$

The whole ionization process can be significantly focused on the formation of water ions H_2O^+ even though the molar fraction of water in air is low. The generated water ions can then react with neutral water molecules in the quite fast ion-molecular reaction:

$$H_2O^+ + H_2O \rightarrow H_3O + OH, \Delta H = -12 \text{ kcal/mole}, k(350\text{ K}) = 0.5 * 10^{-9} \text{ cm}^3/\text{s}.a \tag{2.99}$$

As a result, the production of H_3O^+-ions and OH-radicals can be observed which provides the acidic behavior of the air plasma. The selective generation of OH-radicals in nonthermal air discharges is the fundamental basis for employing discharges for purifying air from different pollutants.

2.6.4 The Ion-Molecular Reactions with Rearrangement of Chemical Bonds

Reaction Eq. (2.99) demonstrates the acidic behavior of nonthermal air plasma. This class of ion-molecular processes consists of very different reactions, which can be subdivided into many groups, including:

- $(A)B^+ + C \rightarrow \mathrm{A} + (C)B^+$, reactions with a positive ion transfer,
- $A(B^+) + C \rightarrow (B^+) + AC$, reactions with a neutral transfer,
- $A(B^+) + (C)D \rightarrow (A)D + (C)B^+$, double exchange reactions,
- $A(B^+) + (C)D \rightarrow AC^+ + BD$, reconstruction processes.

These groups of processes are shown above with positive ions, but similar reactions take place with negative ions. The special feature of the ion-molecular reactions making them very distinctive from regular atom-molecular processes between neutral species can be stated as the following statement: ***Most Exothermic Ion-Molecular Reactions Have No Activation Energy*** (V.L. Talrose, 1952). This is because quantum mechanical repulsion between molecules, which provides the activation barrier even in the exothermic reactions of neutrals, can be suppressed by the charge-dipole attraction in ion-molecular reactions. Thus, the rate coefficients are very large and can be found based on the Langevin relations (2.83) and (2.85). Not every Langevin capture leads to the ion-molecular reactions with complicated rearrangement of chemical bonds or to the orbital symmetry and spin-forbidden reactions. For this reason, the pre-exponential Arrhenius factors of the complicated exchange reaction, orbital symmetry, and spin forbidden reactions can be lower than the Langevin rate coefficients (T. Su, M.T. Bowers, 1995). In general, however, the exothermic ion-molecular reactions with rearrangement of chemical bonds are much faster than the corresponding processes between neutrals. This effect is especially important at low temperatures when even small activation barriers can dramatically slow down exothermic reactions. An extensive listing of the ion-molecular reaction rate coefficients is presented in the monograph of L. Virin, R. Dgagaspanian, G. Karachevtsev, V. Potapov, and V. Talrose (1978).

2.6.5 Ion-Molecular Chain Reactions and Plasma Catalysis

The absence of activation energies in exothermic ion-molecular reactions facilitates the organization of chain reactions in ionized media. Thus, for example, the important industrial process of SO_2 oxidation into SO_3 and sulfuric acid is highly exothermic; however, it can not be arranged as an effective chain process without a catalyst, because of the energy barriers of intermediate elementary reactions between neutral species. However, the long length SO_2 chain oxidation becomes effective in ionized media through the negative ion mechanism in water or in droplets in heterogeneous plasma (Daniel, Jacob, 1986; B. Potapkin, M. Deminsky, A. Fridman, V. Rusanov, 1995).

This chain oxidation process begins with the SO_2 dissolving and forming a negative ion SO_3^{2-} in the case of low acidity (pH > 6.5). Then, a charge transfer process with a OH-radical results in forming an active SO_3^- ion-radical and subsequently initiating a chain reaction:

$$OH + SO_3^{2-} \rightarrow OH^- + SO_3^- \tag{2.100}$$

The active SO_3^- ion-radical becomes the main chain-carrying particle, initiating the chain mechanism. The first reaction of chain propagation is the attachment of an oxygen molecule:

$$SO_3^- + O_2 \rightarrow SO_5^-, \quad k = 2.5 * 10^{-12} cm^3/s. \tag{2.101}$$

Then the chain propagation can go either through the intermediate formation of SO_4^- ion –radical:

$$SO_5^- + SO_3^{2-} \rightarrow SO_4^{2-} + SO_4^-, \quad k = 5 * 10^{-14} cm^3/s. \tag{2.102}$$

$$SO_4^- + SO_3^{2-} \rightarrow SO_4^{2-} + SO_3^-, \quad k = 3.3 * 10^{-12} cm^3/s \tag{2.103}$$

or similarly through the intermediate formation of SO_5^{2-} ion- radical:

$$SO_5^- + SO_3^{2-} \rightarrow SO_5^{2-} + SO_3^-, \quad k = 1.7 * 10^{-14} cm^3/s \tag{2.104}$$

$$SO_5^{2-} + SO_3^{2-} \rightarrow SO_4^{2-} + SO_4^{2-}, \quad k = 2 * 10^{-14} cm^3/s \tag{2.105}$$

Considering all these reactions, one should take into account that the ion SO_4^{2-} is a product of the SO_2 oxidation. This stable ion corresponds to the SO_3^- molecule in the gas phase and to sulfuric acid H_2S in a water solution. Here, the chain termination is due to losses of SO_3^- and SO_5^-, and the chain length easily exceeds 10^3. Thus it can be concluded that plasma as an ionized medium permits avoiding activation energy barriers of elementary exothermic reactions and as a result can operate similarly to traditional catalysts. This phenomenon is called **plasma catalysis** (B. Potapkin, V. Rusanov, S.A. Fridman, 1989).

2.6.6 Ion-Molecular Processes of Cluster Growth, the Winchester Mechanism

From Section 2.6.1, it is clear that ion-molecular reactions are very favorable to clusterization due to the effective long-distance charge-dipole and polarization interaction leading to Langevin capturing. Indeed both positive and negative ion-molecular reactions can make a fundamental contribution in the nucleation and the cluster growth phases of dusty plasma generation in electric discharges (A. Bouchoule, 1993; A. Bouchoule, 1999) and even soot formation in combustion. Specific details of cluster growth in nonthermal discharges will be discussed during consideration of the physics and chemistry of the dusty plasmas. Here the fundamental effects of ion-molecular processes of cluster growth will be pointed out. Besides the effect of Langevin capture (Section 2.6.1), the less known, but very important **Winchester Mechanism** of ion-molecular cluster growth needs to be considered. (L. Boougaenko, L. Polak, 1989). A key point of the Winchester Mechanism is the thermodynamic advantage of the ion-cluster growth processes. If a sequence of negative ions defining the cluster growth is considered:

$$A_1^- \rightarrow A_2^- \rightarrow A_3^- \rightarrow \ldots \rightarrow A_n^- \rightarrow \ldots \tag{2.106}$$

the corresponding electron affinities EA_1, EA_2, EA_3, …, EA_n, … are usually increasing, ultimately reaching the value of the work function (electron extraction energy), which is generally larger than the EA for small molecules. As a result, each elementary step reaction of the cluster growth $A_n^- \rightarrow A_{n+1}^-$ has an apriori tendency to be exothermic. Exothermic ion-molecular reactions have no activation barrier and are usually very fast, thus the Winchester Mechanism explains effective cluster growth based on ion-molecular processes. The phenomenon is important until the cluster becomes too large and the difference in electron affinities $EA^-_{n+1} - EA_n$ becomes negligible. Each elementary step $A^-_n \rightarrow A^-_{n+1}$ includes the cluster rearrangements with an electron usually going to the furthest end of the complex. This explains the origin of the term "Winchester", which sticks to the described mechanism of cluster growth. The Winchester Mechanism is valid not only for negative but also for positive ion clusters. The thermodynamic advantage of ion-cluster growth for positive ion also holds. If a sequence of positive ions defining the cluster growth is considered:

$$A_1^+ \rightarrow A_2^+ \rightarrow A_3^+ \rightarrow \ldots \rightarrow A_n^+ \rightarrow \ldots \tag{2.107}$$

the corresponding ionization energies $I(A_1)$, $I(A_2)$, $I(A_3)$, …, $I(A_n)$, … are usually decreasing to ultimately reach the value of the work function (electron extraction energy), which is generally lower than ionization energy for small molecules. As a result, each elementary step reaction of the cluster growth $A_n^+ \rightarrow A_{n+1}^+$ has an a priori tendency to be exothermic for positive ions as well. Thus, the Winchester Mechanism explains effective cluster growth for both positive and negative ions. As a specific example, one can mention a sequence of ion-molecular reactions during the nucleation stage and cluster growth (dusty plasma formation) in low-pressure silane SiH_4 and silane-argon SiH_4 – Ar discharges (A. Garscadden, 1994; A. A. Howling e.a., 1993a; A. Fridman, L. Boufendi, A. Bouchoule, e.a., 1996a):

$$Si_nH^-_{2n+1} + SiH_4 \rightarrow Si_{n+1}H^-_{2n+3} + H_2 \tag{2.108}$$

The nucleation process can be initiated by a dissociative attachment: $e + SiH_4 + SiH_3^- + H$ (k = 10^{-12} cm^3/s at electron temperature $T_e = 2$ eV) and then continues by the sequence (2.108) of exothermic, thermo-neutral, and slightly endothermic reactions as it is illustrated in Figure 2.15. Thermal effects of the first 4 reactions of Eq. (2.108) demonstrate the Winchester mechanism in silane plasma:

$$SiH_3^- + SiH_4 \rightarrow Si_2H_5^- + H_2 - 0.07eV \tag{2.109}$$

$$Si_2H_5^- + SiH_4 \rightarrow Si_3H_7^- + H_2 + 0.07eV \tag{2.110}$$

$$Si_3H_7^- + SiH_4 \rightarrow Si_4H_9^- + H_2 + 0.07eV \tag{2.111}$$

$$Si_4H_9^- + SiH_4 \rightarrow Si_5H_{11}^- + H_2 + 0.00eV \tag{2.112}$$

The Winchester mechanism shows the tendency of energy effects on ion-cluster growth. Not all reactions of the sequence are exothermic, thermo-neutral and even endothermic stages can be found. This results in "bottle neck" phenomena in nucleation kinetics in low pressure silane discharges.

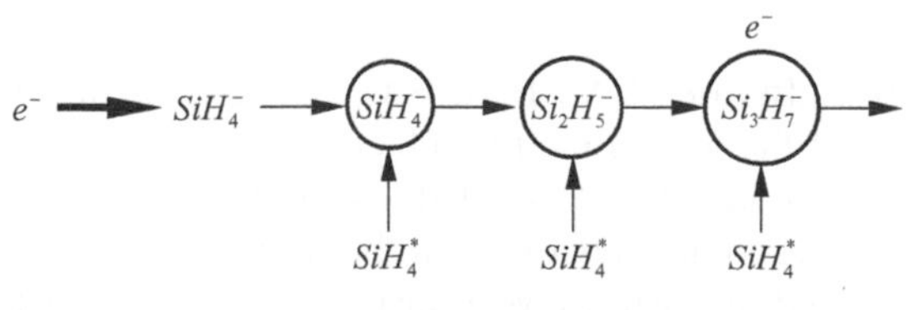

FIGURE 2.15 Illustration of ion-molecular reactions of cluster growth.

2.7 PROBLEMS AND CONCEPT QUESTIONS

2.7.1 Electron Energy Distribution Functions

If the electron energy distribution function is Maxwellian and the electron temperature is $T_e = 1$ eV, (1) what is the mean velocity $<v>$ of the electrons in this case? (2) what is the mean electron energy?, (3) which electron energy $m<v>^2/2$ corresponds to the mean velocity? (4) how can you explain the difference between the two "average electron energies" in terms of standard deviation?

2.7.2 Ionization Potentials and Electron Affinities

Why are most ionization potentials (Table 2.1) greater than electron affinities (Table 2.2)?

2.7.3 Positive and Negative Ions

Why is it very doable to produce the double- or multi-charged positive ions such as A^{++} or A^{+++} in plasma, and impossible to generate in a practical gas discharge plasmas the double or multi-charged negative ions such as A^{-} or A^{---}?

2.7.4 Mean Free Path of Electrons

In which gases are the mean free path of electrons with electron temperature T_e = 1eV longer and why? (helium, nitrogen or water vapor at similar pressure conditions)?

2.7.5 Reaction Rate Coefficients

Recalculate the Maxwellian electron energy distribution function into the electron velocity distribution function f(v) and then find an expression for reaction rate coefficient Eq. (2.7) in the case when $\sigma = \sigma_0 = const.$

2.7.6 Elastic Scattering

Charged particle scattering on neutral molecules with permanent dipole momentum can be characterized by the following cross section dependence on energy: $\sigma(\varepsilon) = Const/\varepsilon$. How does the rate coefficient of the scattering depend on temperature in this case?

2.7.7 Direct Ionization by Electron Impact

Using Eq. (2.18) for the general ionization function f(x), find the electron energy corresponding to the maximum value of direct ionization cross section. Compare this energy with the electron energy optimal for direct ionization according to the Thomson formula.

2.7.8 Comparison of Direct and Stepwise Ionization

Why does the direct ionization make a dominant contribution in nonthermal electric discharges while in thermal plasmas stepwise ionization is more important?

2.7.9 Stepwise Ionization

Estimate a stepwise ionization reaction rate in Ar at electron temperature 1 eV, assuming quasi-equilibrium between plasma electrons and electronic excitation of atoms, and using Eq. (2.29).

2.7.10 Electron Beam Propagation in Gases

How does the propagation length of a high energy electron beam depend on pressure at fixed temperature? How does it depend on the temperature at a fixed pressure?

2.7.11 Photoionization

Why can the photoionization effect play the dominant role in the propagation of both nonthermal and thermal discharges in fast flows including supersonic ones? What is the contribution of photoionization in the propagation of slow discharges?

2.7.12 Massey Parameter

Estimate the adiabatic Massey parameter for ionization of Ar-atom at rest in collision with Ar^+-ion, having kinetic energy twice exceeding the ionization potential. Discuss the probability of ionization in this case.

2.7.13 Ionization in an Ion-Neutral Collision

How large is an H^+-atom energy supposed to be for this ion to reach a velocity the same as an electron velocity in a hydrogen atom? Compare this energy with the H-atom ionization potential. What is your conclusion about the ionization process initiated by ion impact?

2.7.14 Ionization in Collision of Excited Heavy Particles

Why is ionization in the collision of vibrationally excited molecules usually much less effective than in a similar collision of electronically excited molecules.

2.7.15 Ionization in Collision of Vibrationally Excited Molecules

In principle, ionization can take place in collision of nitrogen molecules having 32 vibrational quanta each. Is it possible to increase the probability of such ionization by increasing the vibrational energy of the collision partners or by increasing the electronic energy of the collision partners?

2.7.16 Dissociative Electron–Ion Recombination

According to Eq. (2.44), a rate coefficient of dissociative recombination is reversibly proportional to gas temperature. However in reality, when the electric field is fixed, this rate coefficient is decreasing faster than the reverse proportionality. How can you explain that?

2.7.17 Ion-Conversion, Preceding Electron–Ion Recombination

How large is the relative concentration of complex ions like $N_2^+(N_2)$ supposed to be to compete with conventional electron recombination with N_2^+-ions in the same discharge conditions?

2.7.18 Three-Body Electron–Ion Recombination

Derive the reaction rate coefficient of the three-body recombination, using Eq. (2.26) for the stepwise ionization rate coefficient and the Saha thermodynamic equation for ionization/recombination balance.

2.7.19 Elementary Processes of Charged Species in Plasma

Give an example of discharge parameters (type of gas, range of pressures, plasma densities, etc.), when the radiative electron–ion recombination dominates between other mechanisms of losses of charged particles.

2.7.20 Dissociative Attachment

Calculate the dissociative attachment rate coefficients for molecular hydrogen and molecular oxygen at electron temperatures 1 eV and 5 eV. Compare the results and give your comments.

2.7.21 Negative Ions in Oxygen

What are the discharge and gas parameters supposed to be if the negative oxygen ions are mostly generated by three-body attachment processes, and not by dissociative attachment?

2.7.22 Associative Detachment

Calculate the relative concentration of CO in CO_2-discharge plasma if the main associative detachment process in this discharge dominates over dissociative attachment. Suppose that the electron temperature is 1 eV, and the concentration of negative ions is 10 times less than positive ions.

2.7.23 Detachment by Electron Impact

Detachment of an electron from a negative ion by electron impact can be considered as "ionization" of the negative ion by electron impact. Why can't the Thomson formula be used in this case instead of Eq. (2.64)?

2.7.24 Negative Ions in Thermal Plasma

Why is the concentration of negative ions in thermal plasmas usually very low even in electro-negative gases? Which specific features of thermal plasmas with respect to nonthermal plasmas are responsible for that?

2.7.25 Ion–Ion Recombination in Binary Collisions

Determine how the ion–ion recombination rate coefficient depends on temperature. Use Eq. (2.70) for the cross section of the binary collision recombination rate and assume the Maxwellian distribution function with temperature T_0 for both positive and negative ions.

2.7.26 Three-Body Ion–Ion Recombination

Comparing Thomson's and Langevin's approaches for the three-body ion–ion recombination and correspondent rate coefficients Eqs. (2.74) and (2.62), find a typical value of pressure when the recombination reaches the maximum rate. Consider oxygen plasma with heavy particle temperature 1 eV as an example.

2.7.27 Langevin Cross Section

Calculate the Langevin cross section for Ar^+-ion collision with an argon atom and with a water molecule. Comment on your result. Compare the contribution of the polarization term and the charge-dipole interaction term in the case of Ar^+-ion collision with a water molecule.

2.7.28 Resonant Charge Transfer Process

Is it possible to observe the resonant charge transfer process during an ion-neutral interaction with only a slight perturbation of trajectories of the collision partners?

2.7.29 Tunneling Effect in Charge Transfer

Estimate how strong the tunneling effect can be in resonant charge transfer in ion-neutral collision. For two different velocities compare the cross section of charge exchange between an argon atom and argon positive ion calculated with and without taking into account the tunneling effect.

2.7.30 Non-Resonant Charge Exchange

Most negative ions in nonthermal atmospheric pressure air discharges are in form of O_2^-. What will happen to the negative ion distribution if even a very small amount of fluorine compounds is added to air?

2.7.31 Plasma Catalytic Effect

Compare the catalytic effect of plasma provided by negative ions, positive ions, active atoms, and radicals. All these species are usually quite active, but which one is the most generally relevant to the plasma catalysis.

2.7.32 Winchester Mechanism

Why is the ion mechanism of cluster growth usually less effective in combustion plasma during soot formation than in similar situations in electrical discharge plasma?

3 Elementary Processes of Excited Molecules and Atoms in Plasma

3.1 ELECTRONICALLY EXCITED ATOMS AND MOLECULES IN PLASMA

The extremely high chemical activity of plasma is based on high and quite often a super-equilibrium concentration of active species. The active species generated in plasma include chemically aggressive atoms and radicals, charged particles – electron and ions, and excited atoms and molecules. Elementary processes of atoms and radicals are traditionally considered in a framework of chemical kinetics (H. Eyring, S.H. Lin, S.M. Lin, 1980; V. Kondratiev, E. Nikitin, 1981), elementary processes of the charged particles were considered in Chapter 2, elementary processes of the excited species are considered in this chapter. Excited species, in particular vibrationally excited molecules, are of special importance in plasma chemical kinetics because most of the discharge energy in molecular gases (often more than 95%) can be focused on vibrational excitations by electron impact. Excited species can be subdivided into three groups: electronically excited atoms and molecules, vibrationally excited molecules, and rotationally excited molecules.

3.1.1 Electronically Excited Particles, Resonance, and Metastable States

High electron energies in electric discharges provide a high excitation rate of different electronically excited states of atoms and molecules by electron impact. The energy of the electronically excited particles usually is relatively high (about 5–10 eV), but their lifetime is generally very short (usually about 10^{-8}–10^{-6} s). If a radiative transition to the ground state is not forbidden by quantum mechanical selection rules, such a state is called the **resonance excited state.** The resonance states have the shortest lifetime (about 10^{-8} s) with respect to radiation, and therefore their direct contribution in the kinetics of chemical reactions in plasma is usually small. If the radiative transition is forbidden by selection rules, the lifetime of the excited particles can be much longer because of the absence of spontaneous transition. Such states are called the **metastable excited states**. The energy of the first resonance excited states as well as the energy of the low energy metastable states and their radiative lifetime are given in Table 3.1. Because of their long life with respect to radiation, the metastable electronically excited atoms and molecules are able to accumulate the necessary discharge energy and significantly contribute to the kinetics of different chemical reactions in plasma. The metastable excited particles can lose their energy not only by radiation but also by means of different collisional relaxation processes. The energy of excitation of the metastable states can be quite low (sometimes less than 1 eV) and therefore their concentration in electric discharges can be high.

3.1.2 Electronically Excited Atoms

The collision of a high-energy plasma electron with a neutral atom in a ground state can result in energy transfer from the free plasma electron to a bound electron in the atom. The most important

TABLE 3.1
The Lowest Electronically Excited States and the Lowest Metastable States for Excited Atoms with Their Radiative Lifetimes

Atom and it's Ground State	First Resonance Excited States	Resonance Energy	Low Energy Metastable States	Metastable's Energy	Metastable's Lifetime
He ($1s^2\,{}^1S_0$)	$2p\,{}^1P_1{}^0$	21.2 eV	$2s\,{}^3S_1$	19.8 eV	$2*10^{-2}$ s
He($1s^2\,{}^1S_0$)			$2s\,{}^1S_0$	20.6 eV	$9*10^3$ s
Ne ($2s^2p^6\,{}^1S_0$)	$3s\,{}^1P_1{}^0$	16.8 eV	$4s\,{}^3P_2$	16.6 eV	$4*10^2$ s
Ne ($2s^2p^6\,{}^1S_0$)			$4s\,{}^3P_0$	16.7 eV	20 s
Ar ($3s^2p^6\,{}^1S_0$)	$4s\,{}^2P_1{}^0$	11.6 eV	$4s\,{}^2P_{0,2}{}^0$	11.6 eV	40 s
Kr ($4s^2p^6\,{}^1S_0$)	$5s\,{}^3P_1{}^0$	10.0 eV	$5s\,{}^3P_2$	9.9 eV	2 s
Kr ($4s^2p^6\,{}^1S_0$)			$5s\,{}^3P_0$	10.6 eV	1 s
H ($1s\,{}^2S_{1/2}$)	$2p\,{}^2P_{1/2,3/2}{}^0$	10.2 eV	$2s\,{}^2S_{1/2}$	10.2 eV	0.1 s
N ($2s^2p^3\,{}^4S_{3/2}$)	$3s\,{}^4P_{1/2,3/2,5/2}{}^0$	10.3 eV	$2p^3\,{}^2D_{3/2}$	2.4 eV	$6*10^4$ s
N ($2s^2p^3\,{}^4S_{3/2}$)			$2p^3\,{}^2D_{5/2}$	2.4 eV	$1.4*10^5$ s
N ($2s^2p^3\,{}^4S_{3/2}$)			$2p^3\,{}^2P_{1/2}{}^0$	3.6 eV	40 s
N ($2s^2p^3\,{}^4S_{3/2}$)			$2p^3\,{}^2P_{3/2}{}^0$	3.6 eV	$1.7*10^2$ s
O ($2s^2p^4\,{}^3P_2$)	$3s\,{}^3S_1{}^0$	9.5 eV	$2p^4\,{}^3P_1$	0.02 eV	–
O ($2s^2p^4\,{}^3P_2$)			$2p^4\,{}^3P_0$	0.03 eV	–
O ($2s^2p^4\,{}^3P_2$)			$2p^4\,{}^1D_2$	2.0 eV	10^2 s
O ($2s^2p^4\,{}^3P_2$)			$2p^4\,{}^1S_0$	4.2 eV	1 s
Cl ($3s^2p^5\,{}^2P_{3/2}$)	$4s\,{}^2P_{1/2,3/2}{}^0$	9.2 eV	$3p^5\,{}^3P_{1/2}{}^0$	0.1 eV	10^2 s

growth of energy of a bound electron during the excitation is due to an increase in the **principal quantum number "n"**, but it also usually grows with the value of **angular momentum quantum number "l"**. This effect can be illustrated quantitatively in a case of strong excitation of any atom to a relatively high principal quantum number "n". In this case, one electron is moving quite far from the nucleus and therefore it is almost moving in a Coulomb potential. Such an excited atom is somewhat similar to a hydrogen atom. Its energy can be described by the **Bohr formula with the Rydberg correction** term Δ_l, which takes into account the short-term deviation from the Coulomb potential when the electron approaches the nucleus (see L. Landau, 1997):

$$E = -\frac{me^4}{2\hbar^2\left(4\pi\varepsilon_0\right)^2}\frac{1}{\left(n+\Delta_l\right)^2}. \tag{3.1}$$

The Rydberg correction term $\Delta_l = 0$ in hydrogen atoms, and Eq. (3.1) becomes identical to the conventional Bohr formula. In general, Δ_l does not depend on the principal quantum number "n", but depends on the angular momentum quantum number "l" of the excited electron as well as on the L and S momentum of the whole atom. If the L and S quantum numbers are fixed, the Rydberg correction term Δ_l decreases fast with increasing angular momentum quantum number "l". When the quantum number "l" is growing, the excited electron spends less time in the vicinity of the nucleus and the energy levels are closer to those of hydrogen. This effect can be demonstrated by numerical values of Δ_l for excited helium both in case of S = 0 and S = 1 (see Table 3.2). In general, energy levels of excited atoms depend not only on the principal quantum number "n" of the excited electrons and the **total angular orbital momentum "L"** but also on the **total spin number "S"** and **total momentum quantum number "J"**. Thus, typically, the triplet terms (S = 1) lie below the correspondent singlet terms (S = 0) and hence the energy levels in corresponding excited, atomic states are lower for the higher total spin numbers.

TABLE 3.2
The Rydberg Correction Terms Δ_l for Excited Helium Atoms

	He(S = 0)	He(S = 1)
He (l = 0)	$\Delta_0 = -0.140$	$\Delta_0 = -0.296$
He (l = 1)	$\Delta_1 = +0.012$	$\Delta_1 = -0.068$
He (l = 2)	$\Delta_2 = -0.0022$	$\Delta_2 = -0.0029$

The total orbital angular momentum (corresponding quantum number L) and the total spin momentum (quantum number S) are coupled by weak magnetic forces. This results in splitting of a level with fixed values of L and S into a group of levels with different total momentum quantum numbers J from a maximum J = L + S to a minimum J = $|L - S|$ (altogether there are 2S + 1 energy levels in the multiplet, if $S \leq L$. Also, the energy difference inside of the multiplet between two levels J + 1 and J (with the same L and S) is proportional to J + 1. It should be mentioned that the preceding discussion of the energy hierarchy of electronically excited levels implies the usual case of the so-called **L-S coupling.** The orbital angular momenta of individual electrons, in this case, are strongly coupled among themselves. Then they can be described by only the quantum number L. Similarly, individual electron spin momenta are presented altogether by total quantum number S. The L-S coupling cannot be applied to noble gases and heavy and therefore larger atoms, when coupling between electrons decreases and electron spin–orbit interaction becomes predominant, resulting in the so-called **j–j coupling.**

Before discussing the selection rules for dipole radiation, it would be helpful to review the standard designation of atomic levels. As an example, take the ground state of atomic nitrogen from Table 3.1: N ($2s^2p^3\ ^4S_{3/2}$). This designation includes information about individual outer electrons: $2s^2p^3$ means, that on the outer shell with the principal quantum number n = 2 there are 2 s-electrons (l = 0) and 3 p-electrons (l = 1). The following part of the designation $^4S_{3/2}$ describes the outer electrons of the ground state atomic nitrogen not individually, but rather collectively. The left-hand superscript "4" in $^4S_{3/2}$ denotes the **multiplicity 2S+1**, which corresponds, as it was mentioned before, to the number of energy levels in a multiplet, if $S \leq L$. From the multiplicity, one can obviously determine the total spin number of the ground state atomic nitrogen S = 3/2. The capital letter S in $^4S_{3/2}$ denotes the total angular orbital momentum L = 0 (other values of L = 1, 2, 3, and 4 corresponds to capital letters P, D, F, and G, respectively). The right-hand subscript "3/2" in $^4S_{3/2}$ denotes the total momentum quantum number J = 3/2. One can see from Table 3.1, that some term-symbols, such as the case of the first excited state of helium 2p $^1P_1^0$, also contains a right-hand superscript "o". This superscript is the designation of **parity**, which can be either odd (superscript "o") or even (no right-hand superscript), depending on the odd or even value of the sum of angular momentum quantum numbers for individual electrons in the atom. For completed shells, the parity is obviously even.

To determine which excited states are metastable and which can be easily deactivated by dipole radiation, it is convenient to use the **selection rules**. Actually, these rules indicate whether an electric dipole transition (and hence radiation emission or absorption) is allowed or forbidden. The selection rules allowing the dipole radiation in the case of L-S coupling are:

1. *The parity must change.* Thus, as seen from Table 3.1, ground states and first resonance excited states have different parity.
2. *The multiplicity must remain unchanged.* This rule can also be observed from Table 3.1, where spin-orbit interaction is predominant, resulting not in L-S but j-j coupling. Note that this rule cannot be applied to noble gases.
3. *Quantum numbers J and L must change by +1, −1 or 0 (however transitions 0→0 are forbidden).* It can also be seen in the correspondence between the ground and the first resonance excited states presented in Table 3.1.

3.1.3 Electronic States of Molecules and Their Classification

Classification of electronically, excited states of diatomic and linear polyatomic molecules is somewhat similar to those of atoms. Molecular terms can be specified by the quantum number $\Lambda = 0, 1, 2, 3$ (corresponding Greek symbols Σ, Π, Δ, Φ), which describes the absolute value of the component of the total orbital angular momentum along the internuclear axis. If $\Lambda \neq 0$, the states are doubly degenerate because of two possible directions of the angular momentum component. Molecular terms are then specified by a quantum number S, which designates the total electron spin angular momentum and defines the multiplicity 2S+1. Similar to atomic terms, the multiplicity is written here as a prefixed superscript. Thus, the designation $^2\Pi$ means $\Lambda = 1$, $S = 1/2$. Note that the description and classification of nonlinear polyatomic molecules are more complicated, in particular because of the absence of a single internuclear axis. In the case of Σ-states (e.g. , $\Lambda = 0$), to designate whether the wave function is symmetric or antisymmetric with respect to reflection at any plane including the internuclear axis, the right-hand superscripts "+" and "-" are used. Further, to designate whether the wave function is symmetric or antisymmetric with respect to the interchange of nuclei in homonuclear molecules such as N_2, H_2, O_2, F_2, etc., the right-hand subscripts "g" or "u" are written. Remember, this type of symmetry can be applied only for diatomic molecules. Thus, the molecular term designation $^1\Sigma_g^+$ means $\Lambda = 0$, $S = 0$ and the wave function is symmetric with respect to both reflection at any plane including the internuclear axis and to an interchange of nuclei. To denote the ground state electronic term, the capital X is usually written before the symbol of the term symbol. Capital letters A, B, C, etc. before the main symbol denote the consequence of excited states having the same multiplicity as a ground state. Small letters a, b, c, etc. before the main symbol denote conversely, the consequence of excited states having multiplicity different from that of a ground state. Electronic terms of ground states of some diatomic molecules and radicals are given in Table 3.3 together with the specification of electronic states of atoms, corresponding to dissociation of the diatomic molecules. The majority of chemically stable (saturated) diatomic molecules are seen to have a completely symmetrical normal electronic state with $S = 0$. In other words, a ground electronic state for the majority of diatomic molecules is $X^1\Sigma^+$ or $X\,^1\Sigma_g^+$ for the case of homonuclear molecules. Exceptions of this rule include O_2 (normal term $X\,^3\Sigma_g^-$) and NO (normal term $X^2\Pi$). Obviously, this rule cannot be applied to radicals.

3.1.4 Electronically Excited Molecules, Metastable Molecules

The collisional relaxation processes strongly depend on the degree of the resonance during the collision and the corresponding Massey parameter, see Section 2.2.7. However, the energy transfer for such deactivation and the corresponding Massey parameter are usually very large and the probability of relaxation is low. Nevertheless, the deactivation of molecules can be effective in collision with chemically active atoms, radicals as well as in surface collisions. The most important factor defining the stability of excited molecules is radiation. Electronically excited molecules can easily decay to a lower energy state by a photon emission if the corresponding transition is not forbidden by the selection rules. The selection rules for electric dipole radiation of excited molecules require:

$$\Delta\Lambda = 0, \pm 1; \qquad \Delta S = 0. \tag{3.2}$$

For transitions between Σ-states, and for transitions in the case of homonuclear molecules, additional selection rules require:

$$\Sigma^+ \rightarrow \Sigma^+ \quad \text{or} \quad \Sigma^- \rightarrow \Sigma^-, \quad \text{and} \quad g \rightarrow u \quad \text{or} \quad u \rightarrow g. \tag{3.3}$$

If radiation is allowed, its frequency can be as high as $10^9\ s^{-1}$. Lifetimes for diatomic molecules and radicals are given in Table 3.4 together with the excitation energy of the corresponding states (this energy is taken between the lowest vibrational levels). In contrast to the resonance states, the

TABLE 3.3
Electronic Terms of Non-Excited Diatomic Molecules, Radicals and Products of Their Dissociation

Molecules and Radicals	Ground State Term	Electronic States of Dissociation Products
C_2	$X^1\Sigma_g^+$	$C(^3P) + C(^3P)$
CF	$X^2\Pi$	$C(^3P) + F(^2P)$
CH	$X^2\Pi$	$C(^3P) + H(^2S)$
CN	$X^2\Sigma^+$	$C(^3P) + N(^4S)$
CO	$X^1\Sigma^+$	$C(^3P) + O(^3P)$
BF	$X^1\Sigma^+$	$B(^2P) + F(^2P)$
F_2	$X^1\Sigma_g^+$	$F(^2P) + F(^2P)$
H_2	$X^1\Sigma_g^+$	$F(^2S) + F(^2S)$
HCl	$X^1\Sigma^+$	$H(^2S) + Cl(^2P)$
F	$X^1\Sigma^+$	$H(^2S) + F(^2P)$
Li_2	$X^1\Sigma_g^+$	$Li(^2S) + Li(^2S)$
N_2	$X^1\Sigma_g^+$	$N(^4S^0) + N(^4S^0)$
NF	$X^3\Sigma^-$	$N(^4S^0) + F(^2P)$
NH	$X^3\Sigma^-$	$N(^4S^0) + H(^2S)$
NO	$X^2\Pi$	$N(^4S^0) + O(^3P)$
O_2	$X^3\Sigma_g^-$	$O(^3P) + O(^3P)$
OH	$X^2\Pi$	$O(^3P) + H(^2S)$
SO	$X^3\Sigma^-$	$S(^3P) + O(^3P)$

TABLE 3.4
Life Times and Energies of Electronically Excited Diatomic Molecules and Radicals on the Lowest Vibrational Levels

Molecule or Radical	Electronic State	Energy of the State	Radiative Lifetime
CO	$A^1\Pi$	7.9 eV	$9.5*10^{-9}$ s
C_2	$d^3\Pi_g$	2.5 eV	$1.2*10^{-7}$ s
CN	$A^2\Pi$	1 eV	$8*10^{-6}$ s
CH	$A^2\Delta$	2.9 eV	$5*10^{-7}$ s
N_2	$b^3\Pi_g$	7.2 eV	$6*10^{-6}$ s
NH	$b^1\Sigma^+$	2.7 eV	$2*10^{-2}$ s
NO	$b^2\Pi$	5.6 eV	$3*10^{-6}$ s
O_2	$A^3\Sigma_u^+$	4.3 eV	$2*10^{-5}$ s
OH	$A^2\Sigma^+$	4.0 eV	$8*10^{-7}$ s
H_2	$B^1\Sigma_u^+$	11 eV	$8*10^{-10}$ s
Cl_2	$A^3\Pi$	2.1 eV	$2*10^{-5}$ s
S_2	$B^3\Sigma_u^-$	3.9 eV	$2*10^{-8}$ s
HS	$A^2\Sigma^+$	-	$3*10^{-7}$ s
SO	$A^3\Pi$	4.7 eV	$1.3*10^{-8}$ s

electronically excited, metastable molecules have very long lifetimes: seconds, minutes, and sometimes even hours resulting in the very important role of the metastable molecules in plasma chemical kinetics. These chemically reactive species can be generated in a discharge and then transported to a quite distant reaction zone. Information on excitation energies of some metastable molecules (with respect to the ground electronic state) and their radiative lifetimes are presented in Table 3.5. The

TABLE 3.5
Life Times and Energies of the Metastable Diatomic Molecules (on the Lowest Vibrational Level)

Metastable Molecule	Electronic State	Energy of the State	Radiative Lifetime
N_2	$A^3\Sigma_u^+$	6.2 eV	13 s
N_2	$a'^1\Sigma_u^-$	8.4 eV	0.7 s
N_2	$a^1\Pi_g$	8.55 eV	$2*10^{-4}$ s
N_2	$E^3\Sigma_g^+$	11.9 eV	300 s
O_2	$a^1\Delta_g$	0.98 eV	$3*10^3$ s
O_2	$b^1\Sigma_g^+$	1.6 eV	7 s
NO	$a^4\Pi$	4.7 eV	0.2 s

oxygen molecules have two very low laying (energy about 1 eV) metastable singlet-states with long lifetimes. Such electronic excitation energy is already close to the typical energy of vibrational excitation and therefore the relative concentration of such metastable particles in electric discharges can be very high. Their contribution to plasma chemical kinetics is especially important, taking into account the very fast relaxation rate of vibrationally excited oxygen molecules.

The energy of each electronic state depends on the instantaneous configuration of nuclei in a molecule. In the simplest case of diatomic molecules, the energy depends only on the distance between two atoms. This dependence can be clearly presented by **potential energy curves**. These are very convenient to illustrate different elementary molecular processes like ionization, excitation, dissociation, relaxation, chemical reactions, etc. The potential energy curves were already used in Chapter 2 to describe electron-molecular and other fundamental collisional processes, see for example Figures 2.5, 2.6, 2.10, 2.13, and 2.14. These figures are simplified to illustrate the curves of potential energy and to explain a process, real curves are much more sophisticated. Examples of real potential energy curves are given in Figure 3.1 for some diatomic molecules important in plasma chemical applications, for example, H_2(3.1.1.a), N_2(3.1.1.b), CO(3.1.1.c), NO(3.1.1.d), and O_2(3.1.1.e). Their approximation by the Morse curve is presented in Figure 3.2 which will be discussed in Section 3.2.1.

3.2 VIBRATIONALLY AND ROTATIONALLY EXCITED MOLECULES

When the electronic state of a molecule is fixed, the potential energy curve is fixed as well and it determines the interaction between atoms of the molecules and their possible vibrations. The molecules also participate as a multi-body system in rotational and translational motion. Quantum mechanics of the molecules shows that only discrete energy levels are permitted in molecular vibration and rotation. The structure of these levels both for diatomic and polyatomic molecules will be considered below in connection with the potential energy curves. The vibrational excitation of molecules plays an essential role in plasma physics and chemistry of molecular gases. First of all, this is due to the fact that the largest portion of the discharge energy transfers usually from the plasma electrons primarily to excitation of molecular vibrations, and only after that subsequently to different channels of relaxation and chemical reactions. Several molecules, such as N_2, CO, H_2, CO_2, can maintain vibrational energy without relaxation for a relatively long time, which then can be selectively used in chemical reactions.

3.2.1 Potential Energy Curves for Diatomic Molecules, Morse Potential

The potential curve U(r) for diatomic molecule can be represented by the so-called **Morse potential**:

$$U(r) = D_0\left[1-\exp\left(-\alpha\left(r-r_0\right)\right)\right]^2. \quad (3.4)$$

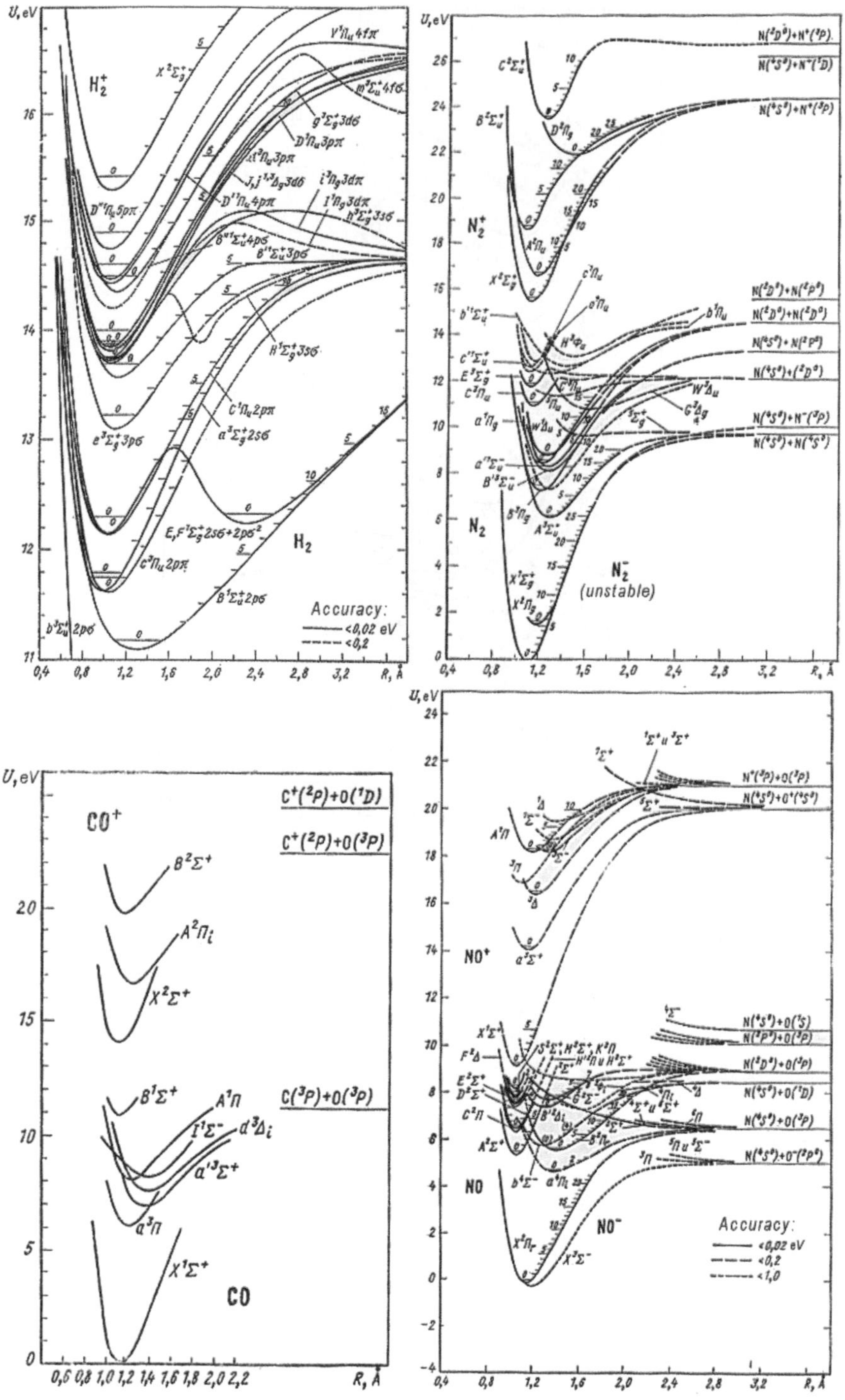

FIGURE 3.1 Electronic terms of different molecules.

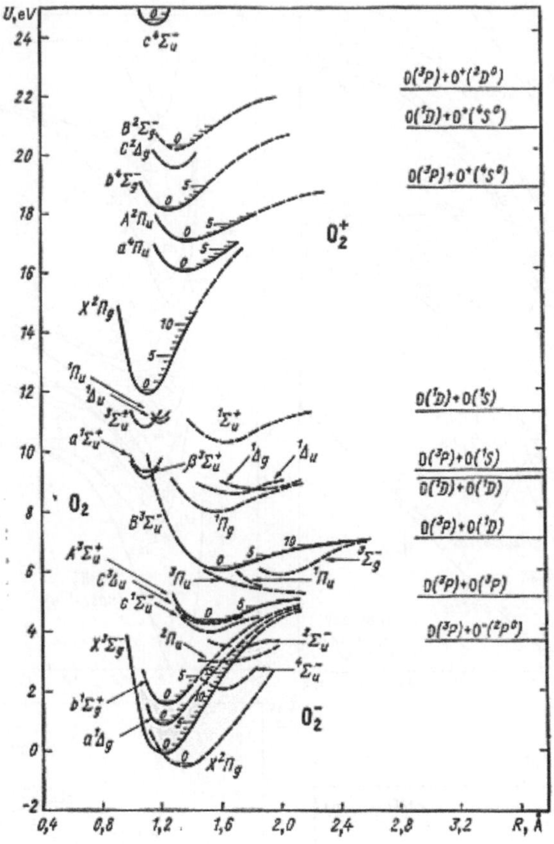

FIGURE 3.1 (Continued)

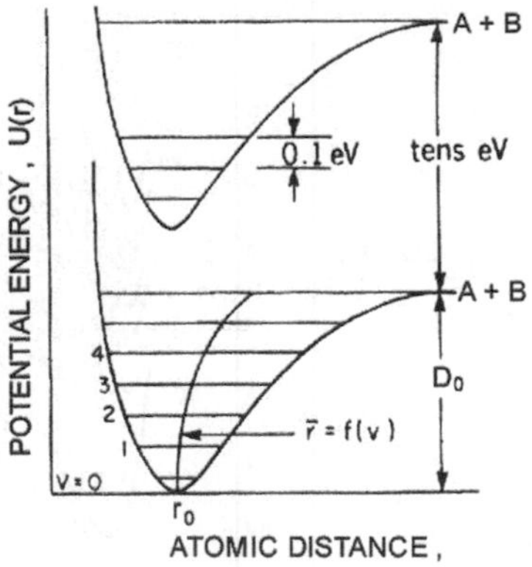

FIGURE 3.2 Potential energy diagram of a diatomic molecule AB in the ground and electronically excited state.

In this expression r is the distance between atoms in a diatomic molecule. Parameters r_0, α, and D_0 are usually referred to as the Morse potential parameters. The distance r_0 is the equilibrium distance between the nuclei in the molecule, the parameter α is a force coefficient of interaction between the nuclei, the energy parameter D_0 is the difference between the energy of the equilibrium molecular configuration ($r = r_0$) and that of corresponding to the free atoms ($r \to \infty$). The Morse parameter D_0 is called the dissociation energy of a diatomic molecule with respect to the minimum energy (see

Figure 3.2). As one can be seen from this figure, the real dissociation energy D is less than the Morse parameter on the so-called value of "zero-vibrations", which is a half of a vibrational quantum ½ $\hbar\omega$:

$$D = D_0 - \frac{1}{2}\hbar\omega. \tag{3.5}$$

This difference between D and D_0 is not large and often can be neglected. Nevertheless, it can be very important sometimes, in particular for isotopic effects in plasma chemical and general chemical kinetics, which will be discussed later. The Morse parameter D_0 is determined by the electronic structure of the molecule, and as a result, it is the same for different isotopes. However, the vibrational quantum ħω depends on the mass of the oscillator, and hence it is different for different isotopes. The difference in vibrational quantum results in differences in the real dissociation energy D and hence differences in the chemical kinetics of dissociation and reaction for different isotopes. The part of the energy curve when $r > r_0$ corresponds to an attractive potential and that of $r < r_0$ corresponds to repulsion between nuclei. Near a U(r) minimum point $r = r_0$, the potential curve is close to a parabolic one $U = D_0\alpha^2(r - r_0)^2$, which corresponds to harmonic oscillations of the molecule. With energy growth, as it is shown in Figure 3.2, the potential energy curve becomes quite asymmetric; the central line demonstrates the increase of an average distance between atoms and the molecular vibration becomes anharmonic.

3.2.2 Vibration of Diatomic Molecules, Model of Harmonic Oscillator

Consider the potential curve of interaction between atoms in a diatomic molecule as: $U = D_0\alpha^2(r - r_0)^2$. Then quantum mechanics predicts for diatomic molecules the sequence of discrete vibrational energy levels:

$$E_v = \hbar\omega\left(v + \frac{1}{2}\right), \quad \omega = \alpha r_0\sqrt{\frac{2D_0}{I}}. \tag{3.6}$$

Here the vibrational energy E_v is taken with respect to the bottom of the potential curve, ħ is the Planck's constant, $I = \frac{M_1M_2}{M_1 + M_2}r_0^2$ is the momentum of inertia of the diatomic molecule with a mass of nuclei M_1 and M_2, ω is the vibration frequency and finally, v is the vibrational quantum number of the molecule (v = 0, 1, 2, 3,…). The sequence of vibrational levels is shown in Figure 3.2. These levels are equidistant in the framework of the harmonic approximation, that is – the energy distance between them is constant and equals to the **vibrational quantum ħω**. The vibrational quantum is actually the smallest portion of vibrational energy, which can be used during an energy transfer or exchange. The value of the quantum is fixed in the framework of harmonic approximation, but it changes (see Figure 3.2) from level to level in the case of an anharmonic oscillator. Even the lowest vibrational level v = 0 is located above the bottom of the potential curve. "No oscillation" maintains some energy there anyway. The level of the "zero-vibrations" corresponds to ½ ħω and defines, as it was mentioned above, the quantum mechanical effect on dissociation energy, Eq. (3.5).

Compare typical frequencies and energies (which are obviously proportional to the frequencies) for electronic excitation and vibrational excitation of molecules. Electronic excitation does not depend on the mass of heavy particles M (but only on electron mass m, see Eq. (3.1)). However, the second one – vibrational excitation (3.6) – is proportional to $\frac{1}{\sqrt{M}}\left(\omega \propto \frac{1}{\sqrt{I}} \propto \frac{1}{\sqrt{M}}\right)$ Eq. (3.6). This means that a vibrational quantum is typically $\sqrt{m/M} \approx 100$ less than characteristic electronic energy (ionization potential I ~ 10–20 eV). As a result, one can easily calculate the typical value of a vibrational quantum, which is usually about 0.1–0.2 eV. On one hand, this energy is relatively low with respect to typical electron energies in electric discharges (1–3 eV) and for this reason, vibrational excitation by electron impact is very effective. On the other hand, the vibrational quantum energy is large enough to provide relatively low gas temperatures, the high values of the Massey parameter (2.2.27) $P_{Ma} = \Delta E/\hbar\alpha v = \omega/\alpha v >> 1$ and to make vibrational deactivation(relaxation) in a collision of heavy particles a slow, adiabatic process (see Section 2.2.7). As a result at least in non-thermal discharges, the molecular vibrations are easy to activate and difficult to deactivate at least in non-thermal discharges, which makes vibrationally excited molecules very special in different applications of plasma chemistry. The actual values of vibrational quantum for some diatomic molecules are presented in Table 3.6. From Table 3.6, and in accordance with Eq. (3.6), the lightest molecule H_2 is seen to have the highest oscillation frequency and hence the highest value of vibrational quantum $\hbar\omega = 0.55$ eV. Detailed information about all parameters characterizing the oscillations of diatomic molecules, including obviously the vibration frequency and anharmonicity can be found in the books of K.P. Huber, G. Herzberg (1979) and A.A. Radzig, B.M. Smirnov (1980).

3.2.3 Vibration of Diatomic Molecules, Model of Anharmonic Oscillator

The parabolic potential: $U = D_0\alpha^2(r - r_0)^2$ and harmonic approximation Eq. (3.6) for vibrational levels are possible to use only for low vibrational quantum numbers, sufficiently far from the level of dissociation. Equation (3.6) is even unable to explain the molecular dissociation itself, vibrational quantum number and vibrational energy can grow according to this formula without any limits. To solve this problem, the quantum mechanical consideration of oscillations of diatomic molecules should be done based on the Morse potential Eq. (3.4). This is an approximation of a diatomic molecule as an anharmonic oscillator. Treating it as an anharmonic oscillator, the discrete vibrational levels of the diatomic molecules have according to exact quantum mechanical considerations, the following energies:

$$E_v = \hbar\omega\left(v + \frac{1}{2}\right) - \hbar\omega x_e\left(v + \frac{1}{2}\right)^2, \quad x_e = \frac{\hbar\omega}{4D_0}, \tag{3.7}$$

These energies of anharmonic vibrational levels are taken in the same manner as previously with respect to the bottom of the potential curve. The Equation (3.7) introduces a new important parameter of a diatomic molecule - x_e, which is a dimensionless coefficient of anharmonicity. As can be seen from Eq. (3.7), the typical value of anharmonicity is $x_e \sim 0.01$. The actual values of anharmonicity for some diatomic molecules are also presented in Table 3.6. Equation (3.7) for harmonic

TABLE 3.6
Vibrational Quantum and Coefficient of Anharmonicity for Diatomic Molecules in Their Ground Electronic States

Molecule	Vibrational Quantum	Coefficient of Anharmonicity	Molecule	Vibrational Quantum	Coefficient of Anharmonicity
CO	0.27 eV	$6*10^{-3}$	Cl_2	0.07 eV	$5*10^{-3}$
F_2	0.11 eV	$1.2*10^{-2}$	H_2	0.55 eV	$2.7*10^{-2}$
HCl	0.37 eV	$1.8*10^{-2}$	HF	0.51 eV	$2.2*10^{-2}$
N_2	0.29 eV	$6*10^{-3}$	NO	0.24 eV	$7*10^{-3}$
O_2	0.20 eV	$7.6*10^{-3}$	S_2	0.09 eV	$4*10^{-3}$
I_2	0.03 eV	$3*10^{-3}$	B_2	0.13 eV	$9*10^{-3}$
SO	0.14 eV	$5*10^{-3}$	Li_2	0.04 eV	$5*10^{-3}$

oscillators obviously coincides with the Eq. (3.6) for anharmonic oscillators if the coefficient of anharmonicity is equal to zero: $x_e = 0$. Also the energy distance between zero -level and the first level of an anharmonic oscillator ($\Delta E_0 = \hbar\omega(1-2x_e)$) is close to that of a harmonic oscillator, which has a vibrational quantum $\hbar\omega$. Vibrational levels are equidistant for harmonic oscillators $\Delta E_v = \hbar\omega$, but it is not the case if anharmonicity is taken into account. For anharmonic oscillators, according to Eq. (3.7), the energy distance $\Delta E_v(v, v + 1)$ between vibrational levels v and v + 1 becomes less and less with an increase of vibrational quantum number v:

$$\Delta E_v = E_{v+1} - E_v = \hbar\omega - 2x_e\hbar\omega(v+1) \quad (3.8)$$

Only a finite number of vibrational levels exist in an anharmonic diatomic molecule. Equation (3.8) provides the possibility to determine the maximum value of the vibrational quantum number v = v_{max}, which corresponds to $\Delta E_v(v, v + 1) = 0$ and hence to dissociation of the diatomic molecule:

$$v_{\max} = \frac{1}{2x_e} - 1. \quad (3.9)$$

Calculating the maximum possible vibrational energy $E_v(v = v_{max})$ based on Eq. (3.7), also leads to a relation similar to Eq. (3.7) between the vibrational quantum, coefficient of anharmonicity, and dissociation energy of a diatomic molecule. The distance between vibrational levels Eq. (3.8) can be also referred to as a "vibrational quantum", which obviously is not a constant, but a function of vibrational quantum number. In this case, the smallest "vibrational quantum" corresponds to the energy difference between the last two vibrational levels $v = v_{max} - 1$ and $v = v_{max}$:

$$\hbar\omega_{\min} = \Delta E_v\left(v_{\max} - 1, v_{\max}\right) = 2x_e * \hbar\omega \quad (3.10)$$

This last vibrational quantum before the level of dissociation of a molecule is the smallest one, with a typical numerical value of about 0.003 eV. Corresponding Massey parameter Eq. (2.2.27) $P_{Ma} = \Delta E/\hbar\alpha v = \omega/\alpha v$ is also relatively low. This means that transition between high vibrational levels during the collision of heavy particles is a fast non-adiabatic process in contrast to adiabatic transitions between low vibrational levels (see Section 2.2.7). Thus, relaxation (deactivation) of highly vibrationally excited molecules is much faster, than the relaxation of molecules with the only quantum. The vibrationally excited molecules can be quite stable with respect to collisional deactivation. On the other hand, their lifetime with respect to spontaneous radiation is also relatively long. The electric dipole radiation, corresponding to a transition between vibrational levels of the same electronic state, is permitted for molecules having permanent dipole moments p_m. In the framework of the model of a harmonic oscillator,- the selection rule requires in this case $\Delta v = \pm 1$. However, the other transitions $\Delta v = \pm 2, \pm 3, \pm 4\ldots$ are also possible in the case of the anharmonic oscillator, though with a much lower probability. The transitions allowed by the selection rule $\Delta v = \pm 1$ provide spontaneous infrared (IR) radiation. The radiative lifetime of a vibrationally excited molecule (as well though as electronically excited ones) can then be found according to the classical formula for an electric dipole p_m, oscillating with frequency ω (here c is the speed of light):

$$\tau_R = 12\pi\varepsilon_0 \frac{\hbar c^3}{p_m^2} \frac{1}{\omega^3} \quad (3.11)$$

The radiative lifetime strongly depends on the oscillation frequency. As it was shown above in Section 3.2.2, the ratio of frequencies corresponding to vibrational excitation and electronic excitation is about $\sqrt{m/M} \approx 100$. Then, taking into account Eq. (3.11), the radiative lifetime of vibrationally excited molecules should be approximately $\left(M/m\right)^{\frac{3}{2}} \approx 10^6$ times longer than that of electronically excited particles. Numerically this means, that even for transitions allowed by the selection rule, the radiative lifetime of vibrationally excited molecules is about 10^{-3}–10^{-2} s, which is quite long with respect to typical time of resonant vibrational energy exchange and some chemical reactions of the molecules.

3.2.4 Vibrationally Excited Polyatomic Molecules, the Case of Discrete Vibrational Levels

Vibration of the polyatomic molecules is more complicated than that of diatomic molecules - because of possible strong interactions between vibrational modes of the multi-body system. The strong interaction between different vibrational modes takes place even at low excitation levels because of degenerate vibrations and intra-molecular resonances. At the high excitation levels, most important for plasma chemistry, this interaction is due to the anharmonisity and leads to a quasi-continuum of vibrational states. Details about this sophisticated subject can be found in G. Herzberg (1945) and A. Fridman, V. Rusanov, G. Sholin (1986). Here, the discussion will just address the most important aspects of the energy characteristics of vibrationally excited, triatomic molecules. The nonlinear triatomic molecules have three vibrational modes (normal vibrations) with three frequencies ω_1, ω_2, ω_3 (which in general are different) and without any degenerate vibrations. When the energy of vibrational excitation is relatively low, the interaction between the vibrational modes is not strong and the structure of vibrational levels is discrete. The relation for vibrational energy of such triatomic molecules at the relatively low excitation levels is a generalization of similar Eq. (3.7) for a diatomic, anharmonic oscillator:

$$\begin{aligned} E_v(v_1,v_2,v_3) &= \hbar\omega_1\left(v_1+\frac{1}{2}\right)+\hbar\omega_2\left(v_2+\frac{1}{2}\right)+\hbar\omega_3\left(v_3+\frac{1}{2}\right)+x_{11}\left(v_1+\frac{1}{2}\right)^2 \\ &+x_{22}\left(v_2+\frac{1}{2}\right)^2+x_{33}\left(v_3+\frac{1}{2}\right)^2+x_{12}\left(v_1+\frac{1}{2}\right)\left(v_2+\frac{1}{2}\right) \\ &+x_{13}\left(v_1+\frac{1}{2}\right)\left(v_3+\frac{1}{2}\right)+x_{23}\left(v_2+\frac{1}{2}\right)\left(v_3+\frac{1}{2}\right) \end{aligned} \quad (3.12)$$

The six coefficients of anharmonicity have energy units in this relation in contrast with those coefficients for diatomic molecules Eq. (3.7). In Table 3.7, information is given about vibrations of some triatomic molecules, including their vibrational quanta, coefficients of anharmonicity as well as the type of symmetry, which describes the types of normal vibrations and, in particular, indicates the presence of degenerate modes. The types of molecular symmetry clarify the peculiarities of vibrational modes of the triatomic molecules. Symbols of molecular symmetry groups, given in the second column of Table 3.7, indicate transformations of coordinates, rotations, and reflections, which keeps the Schroedinger equation unchanged for a triatomic molecule.

1. **The group C_{nv}** includes an nth-order axis of symmetry (which means symmetry with respect to rotation on angle $2\pi/n$ around the axis) and "n" planes of symmetry passing through the axis with an angle π/n between them. The subscript "v" in the symbol C_{nv} of the group reflects that these planes of symmetry are "vertical". In particular, a group $C_{\infty v}$ means complete axial symmetry of a molecule and symmetry with respect to reflection in any plane passing through the axis. This group of symmetry ($C_{\infty v}$) describes linear triatomic molecules like COS or N_2O with different atoms from both sides of a central one. The group of symmetry C_{2v} describes different nonlinear molecules such as H_2O or NO_2 with identical atoms on both sides of a central one.
2. **The group C_{nh}** includes an nth-order axis of symmetry (which means symmetry with respect to rotation on angle $2\pi/n$ around the axis) and planes of symmetry perpendicular to the axis. Similar to above, a subscript "h" in the symbol C_{nh} of the group reminds us that the plane of symmetry is horizontal. In the simplest case of n = 1, the group C_{nh} is usually denoted as C_s.

TABLE 3.7
Parameters of Oscillations of Triatomic Molecules

Molecules & Symmetry		Normal Vibrations & their Quanta, eV			Coefficients of Anharmonicity, 10^{-3} eV					
Molec.	Sym.	v_1	v_2	v_3	x_{11}	x_{22}	x_{33}	x_{12}	x_{13}	x_{23}
NO_2	C_{2v}	0.17	0.09	0.21	−1.1	−0.06	−2.0	−1.2	−3.6	−0.33
H_2S	C_{2v}	0.34	0.15	0.34	−3.1	−0.71	−3.0	−2.4	−11.7	−2.6
SO_2	C_{2v}	0.14	0.07	0.17	−0.49	−0.37	−0.64	−0.25	−1.7	−0.48
H_2O	C_{2v}	0.48	0.20	0.49	−5.6	−2.1	−5.5	−1.9	−20.5	−2.5
D_2O	C_{2v}	0.34	0.15	0.36	−2.7	−1.2	−3.1	−1.1	−10.6	−1.3
T_2O	C_{2v}	0.285	0.13	0.30	−1.9	−0.83	−2.2	−0.76	−7.5	−0.90
HDO	$C_{1h}=C_s$	0.35	0.18	0.48	−5.1	−1.5	−10.2	−2.1	−1.6	−2.5
HTO	$C_{1h}=C_s$	0.29	0.17	0.48	−3.6	−1.3	−10.2	−1.7	−1.3	−2.4
DTO	$C_{1h}=C_s$	0.29	0.14	0.35	−3.6	−0.88	−5.4	1.4	−0.97	−1.4
CO_2	$D_{\infty h}$	0.17	0.085	0.30	−0.47	−0.08	−1.6	0.45	−2.4	−1.6
CS_2	$D_{\infty h}$	0.08	0.05	0.19	−0.13	0.02	−0.64	0.1	−0.61	−0.83
COS	$C_{\infty v}$	0.11	0.06	0.26	−0.50	0.02	−1.4	−0.1	−0.28	−0.91
HCN	$C_{\infty v}$	0.26	0.09	0.43	−0.88	−0.33	−6.5	−0.31	−1.3	−2.4
N_2O	$C_{\infty v}$	0.28	0.07	0.16	−0.65	−0.02	−1.9	−0.06	−3.4	−1.8

The symmetry group C_s includes only symmetry with respect to the plane of a molecule. This group describes nonlinear molecules like HDO, which are actually not symmetrical.

3. **The group D_{nh}** includes an nth-order principal axis of symmetry (which means symmetry with respect to rotation on angle $2\pi/n$ around the axis) and "n" second-order axes crossing at angle π/n and all perpendicular to the principal one. It also includes "n" vertical symmetry planes passing through the principal axis and a second-order one, and a horizontal symmetry plane containing all the "n" second-order axes (this horizontal plane is the reason of "h" in subscript). In particular, a group $D_{\infty h}$ means complete axial symmetry of a molecule, symmetry with respect to reflection in any plane passing through the axis as well as symmetry with respect to reflection in a plane perpendicular to the axis. This group ($D_{\infty h}$) describes "the most symmetrical" – linear triatomic molecules like CO_2 or CS_2 with identical atoms from both sides of a central one.

Triatomic molecules have altogether 9 degrees of freedom, related to different motions of their nuclei. In the general case of nonlinear molecules, these include three translational, three rotational, and three vibrational degrees of freedom. Linear triatomic molecules, however, have only two rotational degrees of freedom because of their axial symmetry (groups $C_{\infty v}$ and $D_{\infty h}$). For this reason, they have four vibrational degrees of freedom (four vibrational modes), but two of them are degenerated. These degenerated vibrational modes have the same frequency, but two different polarization planes, both passing through the molecular axis and perpendicular to each other. In other words, we can say that the linear triatomic molecules have three vibrational modes, but one mode is doubly degenerate. As an example, a linear CO_2 molecule (as well as all other molecules of symmetry group $D_{\infty h}$) has three normal vibrational modes, illustrated in Figure 3.3: asymmetric valence vibration v_3 (see Table 3.7), symmetric valence vibration v_1 and a doubly degenerate symmetric deformation vibration v_2. It should be noted, that there occurs a resonance in CO_2 molecules $v_1 \approx 2v_2$ between the two different types of symmetric vibrations (see Table 3.7). For this reason, symmetric modes are sometimes taken in plasma chemical calculations for simplicity as one triple degenerate vibration. The degenerated symmetric deformation vibrations v_2 can be polarized in two perpendicular planes (see Figure 3.3), which can result after summation in quasi-rotation of the linear molecule around its

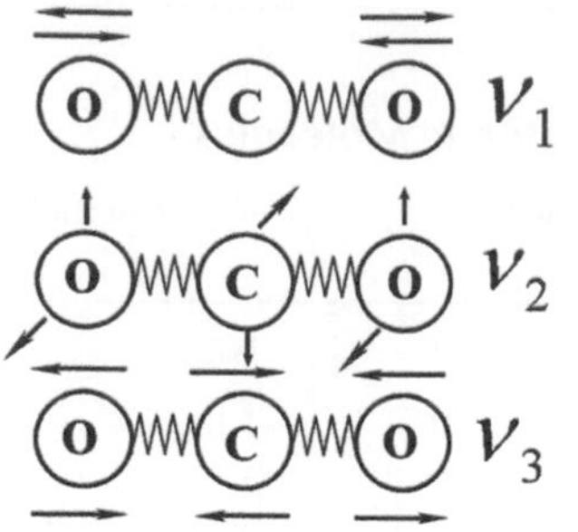

FIGURE 3.3 Illustration of CO_2 vibrational modes.

principal axis. Angular momentum of the quasi-rotation is characterized by the special quantum number "l_2", which assumes the values:

$$l_2 = v_2, v_2 - 2, v_2 - 4, \ldots, 1 \quad \text{or} \quad 0, \tag{3.13}$$

where v_2 is the number of quanta on the degenerate mode. A level of vibrational excitation of the linear triatomic molecule can then be denoted as $CO_2(v_1, v_2^{l_2}, v_3)$. For example, in the molecule $CO_2(0,2^2,0)$, only symmetric deformation vibrations are excited with effective "circular" polarization.

The relation for vibrational energy of the linear triatomic molecules at the relatively low excitation levels is somewhat similar to formula Eq. (3.12) for non-linear molecules. However, it obviously includes also the quantum number l_2 and correspondent coefficient g_{22}, which describe the polarization of the doubly degenerate symmetric deformation vibrations:

$$\begin{aligned} E_v(v_1, v_2, v_3) = {} & \hbar\omega_1\left(v_1 + \frac{1}{2}\right) + \hbar\omega_2(v_2 + 1) + \hbar\omega_3\left(v_3 + \frac{1}{2}\right) + x_{11}\left(v_1 + \frac{1}{2}\right)^2 \\ & + x_{22}(v_2 + 1)^2 + g_{22}l_2^2 + x_{33}\left(v_3 + \frac{1}{2}\right)^2 + x_{12}\left(v_1 + \frac{1}{2}\right)(v_2 + 1) \\ & + x_{13}\left(v_1 + \frac{1}{2}\right)\left(v_3 + \frac{1}{2}\right) + x_{23}(v_2 + 1)\left(v_3 + \frac{1}{2}\right) \end{aligned} \tag{3.14}$$

3.2.5 Highly Vibrationally Excited Polyatomic Molecules, Vibrational Quasi-Continuum

An example of such a case is the vibrational excitation of CO_2 molecules to the level of their dissociation in non-thermal discharges (V. Rusanov, A. Fridman, G. Sholin, 1981). Plasma electrons at an electron temperature 1–2 eV excite mostly the lower vibrational levels of CO_2, and predominantly the asymmetric valence mode of the vibrations. The population of highly excited states, usually involved in plasma chemical reactions, occur in the course of VV-relaxation (resonance or close to resonance intermolecular exchange of vibrational energy). At low levels of vibrational excitation, the VV-exchange occurs independently along the different modes. The asymmetric modes of CO_2 are however better populated in the plasma because they have higher rates of excitation by electron impact, higher rates of the VV-exchange and lower rates of VT-relaxation (quantum losses to the translational degrees of freedom). But as the level of excitation increases, the vibrations of different types are collisionlessly mixed due to the intermode anharmonicity Eq. (3.14) and the Coriolis interaction. The intramolecular VV' quantum exchange results in the **vibrational quasi-continuum** of the highly excited states of CO_2 molecules. E. Shchuryak (1976) proposed the classical description of the transition in terms of beating. The energy on of the asymmetric mode changes in the course of beating corresponding to its interaction with symmetric modes ($v_3 \rightarrow v_1 + v_2$), and as a result the effective frequency of this oscillation mode changes as well:

$$\Delta\omega_3 \approx \left(x_{33}\omega_3\right)^{\frac{1}{3}} \left(A_0\omega_3 n_{sym}\right)^{\frac{2}{3}} \tag{3.15}$$

Here $A_0 \approx 0.03$ is a dimensionless characteristic of the interaction between modes ($v_3 \rightarrow v_1 + v_2$), and n_{sym} is the total number of quanta on the symmetric modes (v_1 and v_2), assuming quasi-equilibrium between modes inside of a CO_2 molecule. As the level of excitation n_{sym} increases, the value of $\Delta\omega_3$ grows and at a certain critical number of quanta n_{sym}^{cr}, it covers the defect of resonance $\Delta\omega$ of the asymmetric-to-symmetric quantum transition:

$$n_{sym}^{cr} \approx \frac{1}{\sqrt{x_{33}A_0^2}} \left(\frac{\Delta\omega}{\omega_3}\right)^{3/2}. \tag{3.16}$$

When the number of quanta exceeds the critical value Eq. (3.16) (fulfillment of the so-called Chirikov stochasticity criterion, G. Zaslavskii, B. Chirikov, 1971), the motion becomes quasi-random and the modes become mixed in the vibrational quasi-continuum. For the general case of polyatomic molecules, the critical excitation level n^{cr} decreases with the number N of vibrational modes: $n^{cr} \approx 100/N^3$. This means, that for molecules with four and more atoms, the transition to the vibrational quasi-continuum takes place at a quite low level of excitation. The polyatomic molecules in the state of vibrational quasi-continuum can be considered as an infinitely high number of small oscillators. These oscillators can be characterized by the distribution function of the squares of their amplitudes $I(\omega)$, which is actually the vibrational Fourier spectrum of the system (A. Fridman, V. Rusanov, G. Sholin, 1986). In this case:

$$I(\omega)\hbar d\omega = \frac{dE}{\omega} \tag{3.17}$$

is an adiabatic invariant (shortened action) for oscillators with energy dE in the frequency range from ω to $\omega + \Delta\omega$ and, according to the correspondence principle, is an analogue of the number of quanta for these oscillators. If a polyatomic molecule is simulated by a set of harmonic oscillators with vibrational frequencies ω_{0i} and quantum numbers n_i, then the vibrational Fourier spectrum $I(\omega)$ of the system can be presented as a following sum of δ-functions:

$$I(\omega) = \sum_i n_i \delta\left(\omega - \omega_{0i}\right). \tag{3.18}$$

Anharmonicity and interaction between modes cause the vibrational spectrum $I(\omega)$ to differ from Eq. (3.18) in two basic ways. First, the anharmonicity can be considered as a shift of the fundamental vibration frequencies:

$$\omega_i\left(n_i\right) = \omega_{0i} - 2\omega_{0i}x_{0i}n_i \tag{3.19}$$

$$x_{0i} = \frac{1}{2}\sum_{j=1}^{j=N} x_{ij}\left(1+\delta_{ij}\right)\frac{n_j}{n_i}. \tag{3.20}$$

In these relations x_{ij} are the anharmonicity coefficients, N is the number of vibrational modes, δ_{ij} is the Kronecker δ-symbol. Second, the energy exchange between modes with characteristic frequency δ_i leads to a broadening of the vibrational spectrum lines (V. Platonenko, N. Sukhareva, 1980, A. Makarov, V. Tyakht, 1982), which can be described by the Lorentz profile:

$$I(\omega) = \frac{1}{\pi}\sum_i \frac{n_i\delta_i}{[\omega - \omega_i(n_i)]^2 + \delta_i^2} \tag{3.21}$$

When the third-order resonances prevail in the intermode exchange of vibrational energy (similar to the earlier case of CO_2-molecules), the frequency δ_i is growing with the growth of vibrational energy as $\delta_i \sim E^{3/2}$. In the case of CO_2, the frequency δ_i can be simply estimated by the intermode anharmonicity $\delta \approx x_{23}n_3n_{sym}/\hbar$. It should be mentioned, that with the growth of the number of atoms in a molecule, – the vibrational energy of transition to the quasi-continuum decreases (see discussion after Eq. (3.16)) and the line width at a fixed energy increases. It is also important to point out, that the energy exchange frequency δ (at least for molecules like CO_2 and not the highest excitation levels) is less than the difference in fundamental frequencies. This means, that in spite of rapid mixing, the modes in vibrational quasi-continuum still keep their individuality.

3.2.6 Rotationally Excited Molecules

The rotational energy of a diatomic molecule with a fixed distance r_0 between nuclei can be found from the Schrodinger equation as a function of the rotational quantum number:

$$E_r = \frac{\hbar^2}{2I}J(J+1) = BJ(J+1) \tag{3.22}$$

Here B is the rotational constant, $I = \frac{M_1M_2}{M_1+M_2}r_0^2$ is the momentum of inertia of the diatomic molecule with a mass of nuclei M_1 and M_2. This relation can be generalized to the case of polyatomic molecules taking into consideration rotation around different principal axes according to the corresponding type of symmetry (see Table 3.7). For example, the linear triatomic molecules with a group of symmetry $D_{\infty h}$ or $C_{\infty v}$ (like CO_2 or N_2O) have only one rotational constant B (because they have two rotational degrees of freedom with the same momentum of inertia). Alternately, the nonlinear triatomic molecules, like H_2O or H_2S, have three rotational degrees of freedom with different values of the momentum of inertia and the rotational constants (see Table 3.8).

TABLE 3.8
Rotational Constants B_e and α_e for a Ground State of Different Molecules (Symbols A, B And C Denote Different Rotational Modes in Nonlinear Triatomic Molecules)

Molecule	B_e, 10^{-4} eV	α_e, 10^{-7} eV	Molecule	B_e, 10^{-4} eV	α_e, 10^{-7} eV
CO	2.39	21.7	HF	26.0	987
H_2	75.5	3796	HCl	12.9	381
F_2	1.09	–	N_2	2.48	21.5
NO	2.11	22.1	O_2	1.79	19.7
OH	23.4	885	NH	20.7	804
CO_2	0.484	–	CS_2	0.135	–
HCN	1.84	–	N_2O	0.520	–
COS	0.252	–	H_2O	A:34.6; B:18.0; C:11.5	–
H_2S	A:12.9; B:11.1; C:5.87	–	NO_2	A:9.92; B:0.53; C:0.51	–
O_3	A:4.40; B:0.55; C:0.48	–	SO_2	A:2.52; B:0.42; C:0.36	–

The momentum of inertia and hence the correct rotational constant B are sensitive to a change of the distance r_0 between nuclei during vibration of a molecule. As a result, the rotational constant B, which has to be used in Eq. (3.22) for diatomic molecules, decreases with a growth of the vibrational quantum number:

$$B = B_e - \alpha_e\left(v + \frac{1}{2}\right). \tag{3.23}$$

In this relation, B_e is the rotational constant corresponding to the zero-vibration level and usually is presented in tables and handbooks (the subscript "e" stands here for "equilibrium"). Another molecular constant α_e is very small compared to B_e and describes the influence of molecular vibration on the momentum of inertia and rotational constant. Let us estimate a typical numerical value of rotational constant and hence a characteristic value of the rotational energy. As was discussed in Section 3.2.2, the value of a vibrational quantum is typically $\sqrt{m/M} \approx 100$ less than characteristic electronic energy, because a vibrational quantum is proportional to $\frac{1}{\sqrt{M}}\left(\omega \propto \frac{1}{\sqrt{I}} \propto \frac{1}{\sqrt{M}}\right)$, where M is a typical mass of heavy particles, and m is an electron mass. In a similar way, the values of rotational constant B and rotational energy are proportional to $\frac{1}{M}\left(B \propto \frac{1}{I} \propto \frac{1}{M}\right)$. This means that the value of the rotational constant is typically $\sqrt{m/M} \approx 100$ less than the value of a vibrational quantum (which is about 0.1 eV) and numerically is about 10^{-3} eV (or even 10^{-4} eV). These rotational energies in temperature units (1eV = 11,600 K) correspond to about 1–10 K, that is the reason why molecular rotation levels are already well populated even at room temperature (in contrast to molecular vibrations). The actual values of the rotational constants B_e and α_e for some diatomic and triatomic molecules are presented in Table 3.8.

The selection rules for rotational transition in molecular spectra require $\Delta J = -1, 0$ or $+1$, taking into account, that transitions from $J = 0$ to $J = 0$ are not allowed. (The different possibilities of coupling, which is usually under detailed consideration in molecular spectroscopy, are not discussed here). Since collisional processes also prefer a transfer of the only rotational quantum (if any), it will be examined in some more details. As seen from relation Eq. (3.22), the energy levels in the rotational spectrum of a molecule are not equidistant. For this reason, the rotational quantum, which is an energy distance between the consequent rotational levels, is not a constant. In contrast to the case of molecular vibrations, the rotational quantum is growing with the increase of quantum number J and hence with the growth of the rotational energy of a molecule. The value of the rotational quantum (in the simplest case of a fixed distance between nuclei) can be easily found from Eq. (3.22):

$$E_r(J+1) - E_r(J) = \frac{\hbar^2}{2I}2(J+1) = 2B(J+1). \tag{3.24}$$

The typical value of a rotational constant is 10^{-3}–10^{-4} eV, which means, that at room temperatures, the quantum number J is about 10. As a result, in this case, even the largest rotational quantum is relatively small, about $5*10^{-3}$ eV. In contrast to the vibrational quantum, the rotational one corresponds to low values of the Massey parameter Eq. (2.37) $P_{Ma} = \Delta E/\hbar\alpha v = \omega/\alpha v$ even at low gas temperatures. This means, that the energy exchange between rotational and translational degrees of freedom is a fast non-adiabatic process (see Section 2.2.7). As a result, the rotational temperature of molecular gas in plasma is usually very close to the translational temperature even in non-equilibrium discharges, while vibrational temperatures can be significantly higher.

3.3 ELEMENTARY PROCESSES OF VIBRATIONAL, ROTATIONAL AND ELECTRONIC EXCITATION OF MOLECULES IN PLASMA

3.3.1 Vibrational Excitation of Molecules by Electron Impact

Vibrational excitation is one of the most important elementary processes in non-thermal molecular plasma. It is responsible for the major portion of energy exchange between electrons and molecules and contributes significantly in the kinetics of non-equilibrium, chemical processes. The elastic collisions of electrons and molecules are not effective in the process of vibrational excitation because of the significant difference in their masses ($m/M \ll 1$). Typical energy transfer from an electron with kinetic energy ε to a molecule in the elastic collision is about $\varepsilon\left(m/M\right)$; vibrational quantum like it was discussed in the previous section: $\hbar\omega \approx I\sqrt{\frac{m}{M}}$. The classical cross section of the vibrational excitation in an elastic electron- molecular collision can be estimated as:

$$\sigma_{vib}^{elastic} \approx \sigma_0 \frac{\varepsilon}{I}\sqrt{\frac{m}{M}}. \tag{3.25}$$

In this relation $\sigma_0 \sim 10^{-16}$ cm^2 is the gas-kinetic cross section, I is an ionization potential. This gives numerical values of the cross section of vibrational excitation ~ 10^{-19} cm^2 for electron energies about 1 eV. Quantum mechanical calculations of the vibrational excitation in elastic scattering result in a similar expression with the selection rule $\Delta v = 1$. However, experiments show that the vibrational excitation cross sections can be much larger, for example, about the same as atomic ones (10^{-16} cm^2). Also, these cross sections are non-monotonic functions of energy and the probability of multi-quantum excitation is not very low. G. J. Schultz (1976) presented a detailed review of these cross sections. This information indicates, that vibrational excitation of a molecule AB from vibrational level v_1 to v_2 is usually not a direct elastic process, but rather a resonance one, proceeding through the formation of an intermediate non-stable negative ion:

$$AB(v_1) + e \overset{\Gamma_{1i},\Gamma_{i1}}{\leftrightarrow} AB^-(v_i) \overset{\Gamma_{i2}}{\rightarrow} AB(v_2) + e. \tag{3.26}$$

In this relation, v_i is the vibrational quantum number of the non-stable negative ion, $\Gamma_{\alpha\beta}$ (measured in s^{-1}) are the probabilities of corresponding transitions between different vibrational states. Cross sections of the resonance vibrational excitation process Eq. (3.26) can be found in the quasi-steady-state approximation using the Breit-Wigner formula:

$$\sigma_{12}(v_i,\varepsilon) = \frac{\pi\hbar^2}{2m\varepsilon}\frac{g_{AB^-}}{g_{AB}g_e}\frac{\Gamma_{1i}\Gamma_{i2}}{\frac{1}{\hbar^2}\left(\varepsilon - \Delta E_{1i}\right)^2 + \Gamma_i^2}. \tag{3.27}$$

Here ε is the electron energy, ΔE_{1i} is energy of transition to the intermediate state $AB(v_1) \rightarrow AB^-(v_i)$; g_{AB^-}, g_{AB}, and g_e are corresponding statistical weights; Γ_i is the total probability of $AB^-(v_i)$ decay through all channels. Equation (3.27) illustrates the resonance structure of vibrational excitation cross section dependence upon electron energy. The energy width of the resonance spikes in this function is $\hbar\Gamma_i$, which is related to the lifetime of the non-stable, intermediate, negative ion $AB^-(v_i)$. The maximum value of the cross section Eq. (3.27) is approximately the atomic value (10^{-16} cm^2). Consider separately the three cases corresponding to different lifetimes of the intermediate ionic states (the so-called "resonances").

3.3.2 Lifetime of Intermediate Ionic States During the Vibrational Excitation

First, consider the so-called **short-lifetime resonances**. In this case, the lifetime of the auto-ionization states $AB^-(v_i)$ is much shorter than the period of oscillation of a molecule's nuclei ($\tau << 10^{-14}$ s). The energy width of the auto-ionization level $\sim\hbar\Gamma_i$ is very large for the short-lifetime resonances in accordance with the Uncertainty Principle. This results in relatively wide maximum spikes (usually several electron-volts) in the dependence of the vibrational excitation cross section on electron energy. There is no fine energy structure of $\sigma_{12}(\varepsilon)$ in this case, which can be seen in Figure 3.4. Because of the short lifetime of the auto-ionization state $AB^-(v_i)$, the displacement of nuclei during the lifetime period is small. As a result, the decay of the non-stable, negative ion leads to excitation

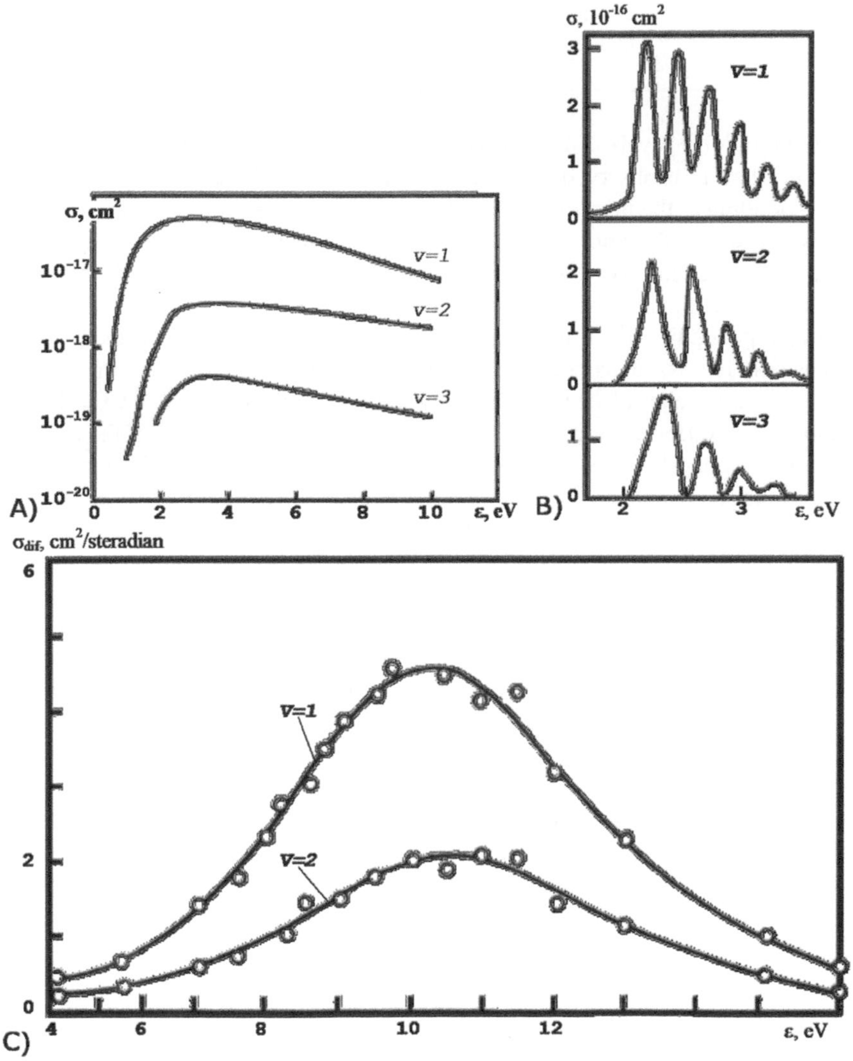

FIGURE 3.4 Cross-sections of vibrational excitation as a functions of electron energy in the case of: (a) short-lifetime resonances (H_2, v = 1,2,3); (b) the boomerang resonances (N_2, v = 1,2,3); (c) long-lifetime resonances (O_2, v = 1,2,3).

TABLE 3.9
Cross Sections of Vibrational Excitation of Molecules by Electron Impact

Molecule	Most effective electron energy	Maximum Cross section	Molecule	Most effective electron energy	Maximum Cross section
N_2	1.7–3.5 eV	$3*10^{-16}$ cm^2	NO	0–1 eV	10^{17} cm^2
CO	1.2–3.0 eV	$3.5*10^{-16}$ cm^2	NO_2	0–1 eV	–
CO_2	3–5 eV	$2*10^{-16}$ cm^2	SO_2	3–4 eV	–
C_2H_4	1.5–2.3 eV	$2*10^{-16}$ cm^2	C_6H_6	1.0–1.6 eV	–
H_2	~3 eV	$4*10^{-17}$ cm^2	CH_4	Thresh. 0.1 eV	10^{-16} cm^2
N_2O	2–3 eV	10^{-17} cm^2	C_2H_6	Thresh. 0.1 eV	$2*10^{-16}$ cm^2
H_2O	5–10 eV	$6*10^{-17}$ cm^2	C_3H_8	Thresh. 0.1 eV	$3*10^{-16}$ cm^2
H_2S	2–3 eV	–	Cyclo-Propane	Thresh. 0.1 eV	$2*10^{-16}$ cm^2
O_2	0.1–1.5 eV	10^{-17} cm^2	HCl	2–4 eV	10^{-15} cm^2

mostly to low vibrational levels. Vibrational excitation through the intermediate formation of the short-lifetime resonances is observed in particular in such molecules as H_2, N_2O, and H_2O. Electron energies corresponding to the most effective vibrational excitation of these molecules as well as the maximum cross sections of the process are presented in Table 3.9. The cross sections of the lowest resonances are presented in this table; after integration over electron energy distribution function, these usually contribute most to the vibrational excitation rate coefficient.

If an intermediate ion lifetime is approximately a molecular oscillation period (~10^{-14} s), such a resonance is usually referred to as **"the boomerang resonance".** In particular, this boomerang type of vibrational excitation takes place for low energy resonances in N_2, CO, and CO_2. The boomerang model, developed by A. Herzenberg (1968), considers formation and decay of the negative ion during one oscillation as interference only of the coming and reflected waves. The interference of the nuclear wave packages results in an oscillating dependence of excitation cross section on the electron energy with typical spikes period about 0.3 eV (see Figure 3.4 and Table 3.9). The boomerang resonances require usually larger electron energies for excitation of higher vibrational levels. For example, the excitation threshold of $N_2(v = 1)$ is 1.9 eV, and that of $N_2(v = 10)$ from ground state is about 3 eV. Excitation of $CO(v = 1)$ requires minimal electron energy of 1.6 eV, while the threshold for $CO(v = 10)$ excitation from the ground state is about 2.5 eV. The maximum value of the vibrational excitation cross section for boomerang resonances usually decreases with growth in the vibrational quantum number v.

The **long-lifetime resonances** correspond to auto-ionization states $AB^-(v_i)$ with much longer lifetime ($\tau = 10^{-14}$–10^{-10} s) than a period of oscillation of the nuclei in a molecule. In particular, this type of vibrational excitation takes place for low energy resonances in such molecules as O_2, NO, and C_6H_6. The long-lifetime resonances usually result in quite narrow isolated spikes (about 0.1 eV) in cross section dependence on electron energy (see Figure 3.4 and Table 3.9). In contrast to the boomerang resonances, the maximum value of the vibrational excitation cross section remains the same for different vibrational quantum numbers.

3.3.3 Rate Coefficients of Vibrational Excitation by Electron Impact, Semi-Empirical Fridman's Approximation

As seen from Table 3.9, the electron energies most effective in vibrational excitation are 1–3 eV, which usually corresponds to the maximum in the electron energy distribution function. The vibrational excitation rate coefficients, which are the results of the integration of the cross sections over the electron energy distribution function, are obviously very large and reach 10^{-7} cm^3/s in this case. Such rate coefficients are given in Table 3.10 for different diatomic or polyatomic molecules and

TABLE 3.10
Rate Coefficients of Vibrational Excitation of Molecules by Electron Impact

Molecule	$T_e = 0.5$ eV	$T_e = 1$ eV	$T_e = 2$ eV
H_2	$2.2*10^{-10}$ cm³/s	$2.5*10^{-10}$ cm³/s	$0.7*10^{-9}$ cm³/s
D_2	–	–	10^{-9} cm³/s
N_2	$2*10^{-11}$ cm³/s	$4*10^{-9}$ cm³/s	$3*10^{-8}$ cm³/s
O_2	–	–	10^{-10} - 10^{-9} cm³/s
CO	–	–	10^{-7} cm³/s
NO	–	$3*10^{-10}$ cm³/s	–
CO_2	$3*10^{-9}$ cm³/s	10^{-8} cm³/s	$3*10^{-8}$ cm³/s
NO_2	–	–	10^{-10} - 10^{-9} cm³/s
H_2O	–	–	10^{-10} cm³/s
C_2H_4	–	10^{-8} cm³/s	–

different electron temperatures T_e. The excitation rate coefficients are quite large. For such molecules as N_2, CO, CO_2, almost every electron-molecular collision leads to vibrational excitation at $T_e = 2$ eV. This explains why such a large fraction of electron energy from non-thermal discharges in several gases goes mostly into the vibrational excitation at electron temperatures $T_e = 1 - 3$ eV.

Vibrational excitation by electron impact is preferably a one-quantum process. Nevertheless, excitation rates of multi-quantum vibrational excitation can also be important. Detailed kinetic information about electron impact excitation rate coefficients $k_{eV}(v_1,v_2)$ for excitation of molecules from an initial vibrational level v_1 to a final level v_2 is very important in numerous plasma chemical and laser problems. The semi-empirical **Fridman's approximation for multi-quantum vibrational excitation** can be very useful. It was first applied for different diatomic and polyatomic gases by V. Rusanov, A. Fridman, G. Sholin, 1981; V. Rusanov, A. Fridman, 1984. The approach permits us to find the excitation rate $k_{eV}(v_1,v_2)$ based on the much better known vibrational excitation rate coefficient $k_{eV}(0,1)$ corresponding to the excitation from ground state to the first vibrational level. (see Table 3.10):

$$k_{eV}\left(v_1,v_2\right) = k_{eV}\left(0,1\right)\frac{\exp\left[-\alpha\left(v_2 - v_1\right)\right]}{1+\beta v_1} \tag{3.28}$$

Parameters α and β for different gases, which are necessary for numerical calculations based on Eq. (3.28), are summarized in Table 3.11. The approximative nature of the approach must be stressed. Thus, sometimes in practical plasma chemical modeling, it is helpful to consider the parameters α and β as functions of electron temperature. Also in non-thermal discharges, the non-Maxwellian behavior of the electron energy distribution function should be taken into account to modify Eq. (3.28).

TABLE 3.11
Parameters of the Multi-quantum Vibrational Excitation by Electron Impact

Molecule	α	β	Molecule	α	β
N2	0.7	0.05	H_2	3	–
CO	0.6	–	O_2	0.7	–
$CO_2(v3)$	0.5	–	NO	0.7	–

3.3.4 Rotational Excitation of Molecules by Electron Impact

If electron energies exceed the values about 1 eV (see Table 3.9), the rotational excitation can proceed resonantly through the auto-ionization state of a negative ion as in the case of vibrational excitation. However, the relative contribution of this multi-stage rotational excitation is small, taking into account the low value of rotational quantum with respect to the vibrational one. The rotational excitation can be relatively important in electron energy balances at lower electron energies when the resonant vibrational excitation is already ineffective, but rotational transitions are still possible by long-distance electron-molecular interaction. The non-resonant rotational excitation of molecules by electron impact can be illustrated using the classical approach. Typical energy transfer from an electron with kinetic energy ε to a molecule in an elastic collision, inducing rotational excitation, is about $\varepsilon\left(m/_{M}\right)$. Typical distance between rotational levels, rotational quantum, is: $I\left(m/_{M}\right)$ where I is the ionization potential. Thus, the classical cross section of the non-resonant rotational excitation can be related to the gas-kinetic collisional cross section with σ_0 ~ 10^{-16} cm^2 - in the following manner:

$$\sigma_{rotational}^{elastic} \approx \sigma_0 \frac{\varepsilon}{I} \tag{3.29}$$

Comparing this relation with Eq. (3.25), it is seen, that the cross section of the non-resonant rotational excitation can exceed that of non-resonant vibrational excitation by a factor of 100. A quantum mechanical approach leads to similar results and conclusions. An electron collision with a dipole molecule induces the rotational transitions with a change of the rotational quantum number ΔJ = 1. Quantum mechanical cross sections of rotational excitation of a linear dipole molecule by a low energy electron can be then calculated as (O.H. Crowford, 1967):

$$\sigma\left(J \rightarrow J+1,\varepsilon\right) = \frac{d^2}{3\varepsilon_0 a_0 \varepsilon} \frac{J+1}{2J+1} \frac{\sqrt{\varepsilon}+\sqrt{\varepsilon'}}{\sqrt{\varepsilon}-\sqrt{\varepsilon'}}. \tag{3.30}$$

In this relation, d is a dipole moment, a_0 is the Bohr radius, $\varepsilon' = \varepsilon - 2B(J+1)$ is the electron energy after collision, B is the rotational constant (see Table 3.8). Numerically, this cross section is approximately $1\text{–}3*10^{-16}$ cm^2 when the electron energy is about 0.1 eV. Homonuclear molecules, such as N_2 or H_2, have no dipole moment, and any rotational excitation of such molecules is due to electron interaction with their quadrupole moment Q. In this case, the rotational transition takes place with the change of rotational quantum number ΔJ = 2. Cross section of the rotational excitation by a low energy electron is obviously lower in this case and can be calculated as (E. Gerjoy, S. Stein, 1955):

$$\sigma\left(J \rightarrow J+2,\varepsilon\right) = \frac{8\pi Q^2}{15e^2 a_0^2} \frac{(J+1)(J+2)}{(2J+1)(2J+3)} \ln\sqrt{\frac{\varepsilon}{\varepsilon'}}. \tag{3.31}$$

Numerically, this cross section of rotational excitation of the homonuclear molecules by electron impact is usually the order of $1\text{–}3*10^{-17}$ cm^2 when the electron energy is about 0.1 eV. To evaluate the elastic or "quasi-elastic" energy transfer from the electron gas to neutral molecules, the rotational excitation can be taken into account combined together with the elastic collisions. The process is then characterized by the gas-kinetic rate coefficient $k_{e0} \approx \sigma_0 \langle v_e \rangle \approx 3\cdot 10^{-8} \frac{\text{cm}^3}{\text{sec}}$ ($\langle v_e \rangle$ is the average thermal velocity of electrons) and each collision is considered as a loss of about $\varepsilon\left(m/M\right)$ of electron energy.

3.3.5 Electronic Excitation of Atoms and Molecules by Electron Impact

In contrast to the vibrational and rotational excitation processes discussed above, the electronic excitation by electron impact needs usually large electron energies ($\varepsilon > 10$ eV). The **Born Approximation** can be applied to calculate the processes' cross sections when electron energies are large enough. For excitation of optically permitted transitions from an atomic state "i" to another state "k", the Born Approximation gives the following process cross section:

$$\sigma_{ik}\left(\varepsilon\right)=4\pi a_0^2 f_{ik}\left(\frac{Ry}{\Delta E_{ik}}\right)^2\frac{\Delta E_{ik}}{\varepsilon}\ln\frac{\varepsilon}{\Delta E_{ik}}. \tag{3.32}$$

In this relation Ry is the Rydberg constant, a_0 is the Bohr radius, f_{ik}- is the force of oscillator for the transition $i \rightarrow k$, ΔE_{ik} is energy of the transition. The relation is valid in the framework of the Born Approximation, for example, for high electron energies $\varepsilon >> \Delta E_{ik}$. The cross sections of electronic excitation of molecules by electron impact should be known for practical calculations and modeling in a wide range of electron energies, starting from the threshold of the process. In this case, semi-empirical formulas can be very useful; two important ones were introduced by H.W. Drawin (1968, 1969):

$$\sigma_{ik}\left(\varepsilon\right)=4\pi a_0^2 f_{ik}\left(\frac{Ry}{\Delta E_{ik}}\right)^2\frac{x-1}{x^2}\ln\left(2.5x\right) \tag{3.33}$$

and by B.M. Smirnov (1968):

$$\sigma_{ik}\left(\varepsilon\right)=4\pi a_0^2 f_{ik}\left(\frac{Ry}{\Delta E_{ik}}\right)^2\frac{\ln\left(0.1x+0.9\right)}{x-0.7}. \tag{3.34}$$

In both relations, $x=\varepsilon/\Delta E_{ik}$. Obviously, the semi-empirical formulas correspond to the Born Approximation at high electron energies ($x >> 1$). Electronic excitation of molecules to the optically permitted states follows relations similar to Eqs. (3.32–3.34), but additionally include the Frank-Condon factors (see Section 2.2.3) and internuclear distances. Numerically, the maximum values of cross sections for excitation of all optically permitted transitions are about the size of atomic gas-kinetic cross sections, $\sigma_0 \sim 10^{-16}$ cm^2. To reach the maximum cross section of the electronic excitation, the electron energy should be greater than the transition energy ΔE_{ik} by a factor of two. The dependence $\sigma_{ik}(\varepsilon)$ is quite different for the excitation of electronic terms from which optical transitions (radiation) are forbidden by selection rules (see Section 3.1). In this case, the exchange-interaction and details of electron shell structure become important. As a result,-the maximum cross section, which is also about the size of the atomic one, $\sigma_0 \sim 10^{-16}$ cm^2, can be reached at much lower electron energies $\varepsilon/\Delta E_{ik} \approx 1.2-1.6$. This leads to an effect of predominant excitation of the optically forbidden and metastable states by electron impact in non-thermal discharges, where electron temperature T_e is usually much less than the transition energy ΔE_{ik}. Obviously, this effect requires the availability of the optically forbidden and metastable states at relatively low energies. Some cross sections of electronic excitation by electron impact are presented in Figure 3.5; detailed information on the subject can be found in the monograph of D.I. Slovetsky (1980).

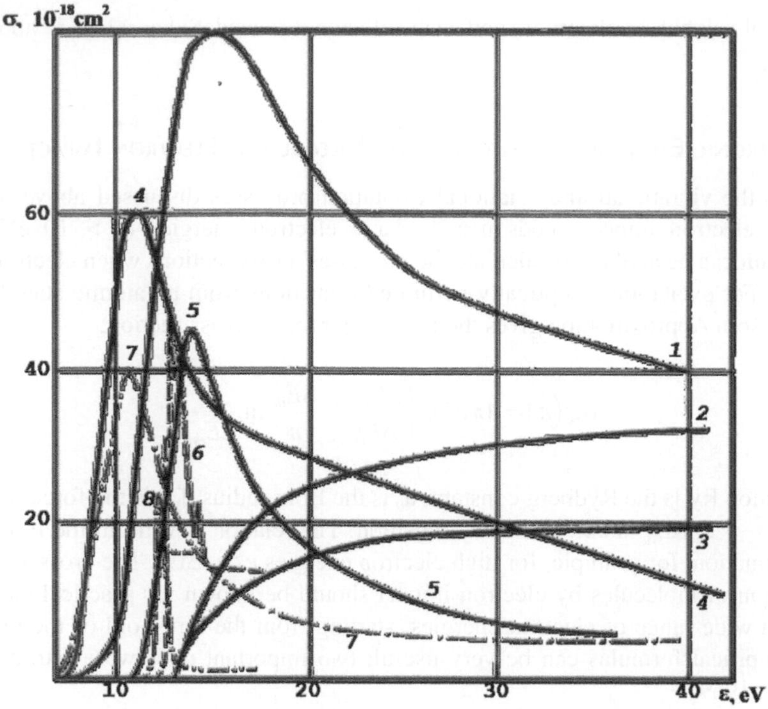

FIGURE 3.5 Cross-section of excitation of different electronic states of $N_2(X^1\Sigma_g^+, v = 0)$ by electron impact: $1 - a^1\Pi_g$; $2 - b^1\Pi_u$ ($v_k = 0–4$); 3 – transitions 12,96 eV; $4 - B^3\Pi_g$; $5 - C^3\Pi_u$; $6 - a''^1\Sigma_g^+$; $7 - A^3\Sigma_u^+$; $8 - E^3\Sigma_g^+$.

3.3.6 Rate Coefficients of Electronic Excitation in Plasma by Electron Impact

The rate coefficients of the electronic excitation can be calculated by integration of the cross sections $\sigma_{ik}(\varepsilon)$ over the electron energy distribution functions, which in the simplest case are Maxwellian. Because electron temperature is usually less than the electronic transition energy ($T_e << \Delta E_{ik}$), the rate coefficient of the process is exponential in the same manner as was the case of ionization by direct electron impact:

$$k_{el.excit.} \propto \exp\left(-\frac{\Delta E_{ik}}{T_e}\right). \tag{3.35}$$

For numerical calculations and modeling it is convenient to use the semi-empirical relation between the electronic excitation rate coefficients and the reduced electric field E/n_0, I.V. Kochetov and V.G. Pevgov (1979). The same formula can be applied to calculate the ionization rate coefficient as well:

$$\lg k_{el.excit.} = -C_1 - \frac{C_2}{E/n_0}. \tag{3.36}$$

Here, the rate coefficient $k_{el.excit.}$ is expressed in cm^3/sec; E is an electric field in V/cm; n_0 is number density in $1/cm^3$. The numerical values of the parameters C_1 and C_2 for different electronically excited states (and also ionization) of CO_2 and N_2 are presented in Table 3.12. Equation (3.36) implies that the electron energy distribution function is not disturbed by the vibrational excitation of molecules in plasma. If the level of vibrational excitation of molecules is quite high, superelastic collisions provide higher electron energies and higher electronic excitation rate coefficients for the

TABLE 3.12
Parameters for Semi-Empirical Approximation of Rate Coefficients of Electronic Excitation and Ionization of CO_2 and N_2 by Electron Impact

Molecule	Excitation level or Ionization	C_1	C_2*10^{16}, V*cm²	Molecule	Excitation level or Ionization	C_1	C_2*10^{16}, V*cm²
N_2	$A^3\Sigma_u^+$	8.04	16.87	N_2	$c^1\Pi_u$	8.85	34.0
N_2	$B^3\Pi_g$	8.00	17.35	N_2	$a^1\Pi_u$	9.65	35.2
N_2	$W^3\Delta_u$	8.21	19.2	N_2	$b'^1\Sigma_u^+$	8.44	33.4
N_2	$B'^3\Sigma_u^-$	8.69	20.1	N_2	$c^3\Pi_u$	8.60	35.4
N_2	$a'^1\Sigma_u^-$	8.65	20.87	N_2	$F^3\Pi_u$	9.30	32.9
N_2	$a^1\Pi_g$	8.29	21.2	N_2	Ionization	8.12	40.6
N_2	$W^1\Delta_u$	8.67	20.85	CO_2	$^3\Sigma_u^+$	8.50	10.7
N_2	$C^3\Pi_u$	8.09	25.5	CO_2	$^1\Delta_u$	8.68	13.2
N_2	$E^3\Sigma_g^+$	9.65	23.53	CO_2	$^1\Pi_g$	8.84	14.8
N_2	$a''^1\Sigma_g^+$	8.88	26.5	CO_2	$^1\Sigma_g^+$	8.23	18.9
N_2	$b^1\Pi_u$	8.50	31.88	CO_2	Other levels	8.34	20.9
N_2	$c'^1\Sigma_u^+$	8.56	35.6	CO_2	Ionization	8.38	25.5

same value of the reduced electric field E/n_0. It can be taken into account for calculating the electronic excitation rate coefficients by including in Eq. (3.36) two special numerical terms related to vibrational temperature T_v:

$$\lg k_{el.excit.} = -C_1 - \frac{C_2}{E/n_0} + \frac{40z + 13z^2}{\left[(E/n_0)*10^{16}\right]^2} - 0.02\left(\frac{T_v}{1000}\right)^{2/3} \quad (3.37)$$

Here, the vibrational temperature T_v is in degrees Kelvin. The Boltzmann factor "z" is:

$$z = \exp\left(-\frac{\hbar\omega}{T_v}\right). \quad (3.38)$$

Equations (3.37 and 3.38) give a good approximation (accuracy about 20%) for the electronic excitation rate coefficients with reduced electric fields $5*10^{-16}$ V*cm² < E/n_0 < $30*10^{-16}$ V*cm² and vibrational temperatures less than 9000 K. At low vibrational temperatures $z \ll 1$ and Eq. (3.37) obviously coincides with the simplified numerical formula Eq. (3.36).

3.3.7 Dissociation of Molecules by Direct Electron Impact

Electron impacts are able to stimulate dissociation of molecules by both vibrational and electronic excitation. Vibrational excitation by electron impact, however, usually results in the initial formation of molecules with only one or few quanta. Dissociation takes place in this case as a non-direct multi-step process, including energy exchange (VV-relaxation) between molecules to collect an amount of vibrational energy sufficient for dissociation. Such processes are effective only for a limited (but very important) group of gases like N_2, CO_2, H_2, CO and will be discussed later. In contrast, the dissociation through electronic excitation of molecules can proceed in just one collision; therefore it is referred to as stimulated by the direct by electron impact. The dissociation through electronic excitation can be observed with any kind of diatomic and polyatomic molecules making this process specifically important in many plasma chemical applications. The dissociation through electronic excitation can proceed as an elementary process by different mechanisms or through different intermediate steps of intramolecular transitions. All these mechanisms are illustrated in Figure 3.6:

- *Mechanism A* begins with the direct electronic excitation of a molecule from the ground state to a repulsive state followed by dissociation. In this case, the required electron energy can significantly (few electron-volts) exceed the dissociation energy. This mechanism generates hot (high-energy) neutral fragments, which for example, could affect surface chemistry in low-pressure non-thermal discharges.
- *Mechanism B* includes the direct electronic excitation of a molecule from the ground state to an attractive state with energy exceeding the dissociation threshold. The excitation results in the following dissociation. As seen from Figure 3.6, the energy of the dissociation fragments is lower in this case.
- *Mechanism C* consists of the direct electronic excitation of a molecule from the ground state to an attractive state corresponding to electronically excited dissociation products. The excitation of the state can lead to a radiative transition to a low-energy repulsive state (see Figure 3.6) followed by dissociation. The energy of the dissociation fragments, in this case, is similar to those of mechanism A.
- *Mechanism D* (similarly to mechanism C) starts with the direct electronic excitation of a molecule from the ground state to an attractive state corresponding to electronically excited dissociation products. In contrast to the previous case, excitation of the state leads to a non-radiative transfer to a highly excited repulsive state (see Figure 3.6 followed by dissociation into electronically excited fragments. This mechanism is usually referred to as **predissociation**.
- *Mechanism E* is similar to the mechanism A and consists of direct electronic excitation of a molecule from a ground state to a repulsive state, but with dissociation into electronically excited fragments. This mechanism requires the largest values of electron energies, and therefore the corresponding rate coefficients are relatively low.

Cross sections of the dissociation of different molecules by direct electron impact are presented as a function of electron energy in Figure 3.7. Some rate coefficients of the process are given in Figure 3.8 as a function of electron temperature.

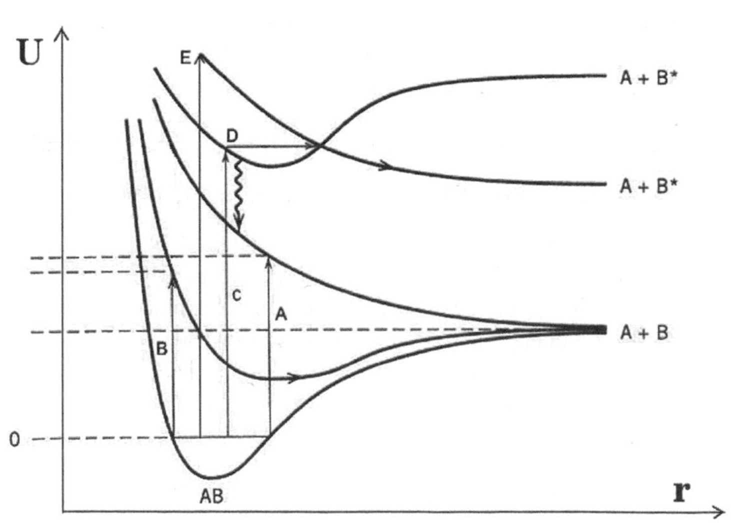

FIGURE 3.6 Mechanisms of dissociation of molecules through electronic excitation.

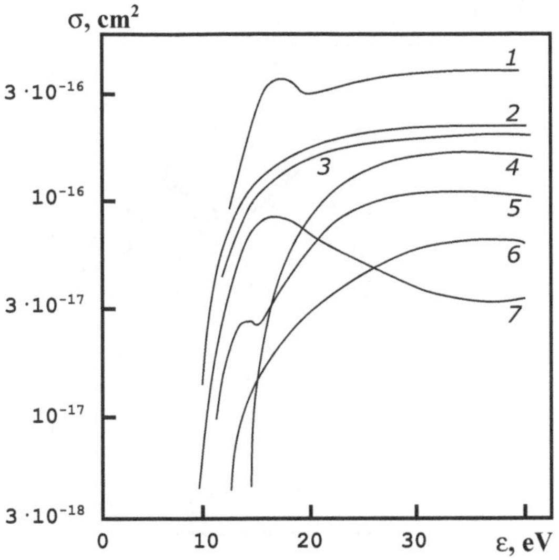

FIGURE 3.7 Cross – sections of dissociation of molecules through electronic excitation as a function of electron energy: 1-CH_4, 2-O_2, 3-NO, 4-N_2, 5-CO_2, 6-CO, 7-H_2.

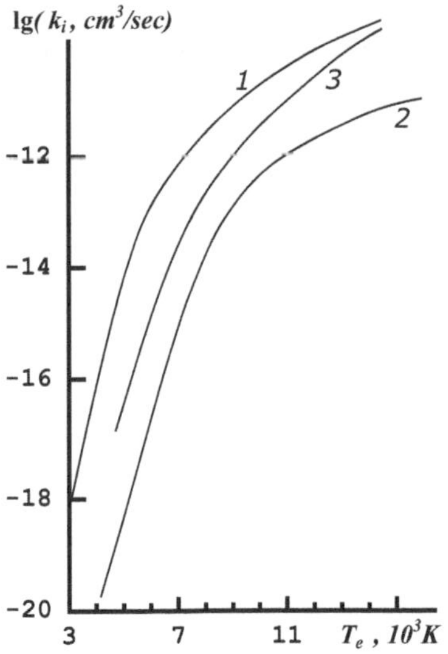

FIGURE 3.8 Rate coefficients of stepwise N_2 electronic state excitation (1), direct dissociation of initially non-excited molecules N_2 (2), N_2 dissociation in stepwise electronic excitation sequence (3).

3.3.8 Distribution of Electrons Energy in Non-Thermal Discharges Between Different Channels of Excitation and Ionization

Most of the electron energy received from an electric field is distributed between elastic energy losses and different channels of excitation and ionization. Such distributions of electron energy in non-thermal discharges as a function of reduced electric field E/n_0 in different atomic and molecular gases are presented in Figures 3.9 to 3.16. The energy distributions between different excitation channels presented on these figures were calculated numerically. Such self-consistent calculations

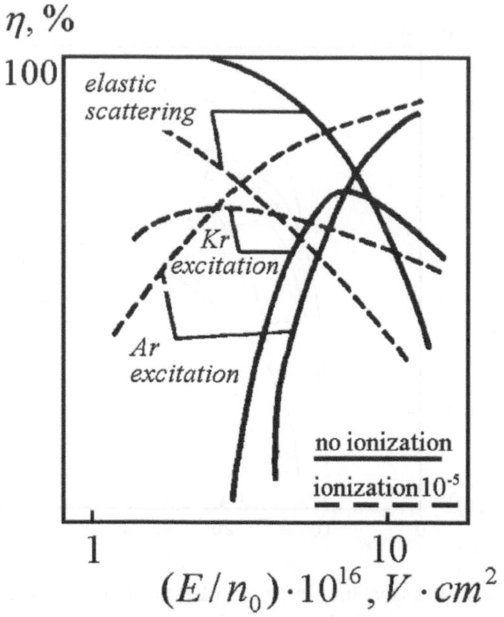

FIGURE 3.9 Electron energy distribution between excitation channels in Kr(5%) - Ar(95%) mixture.

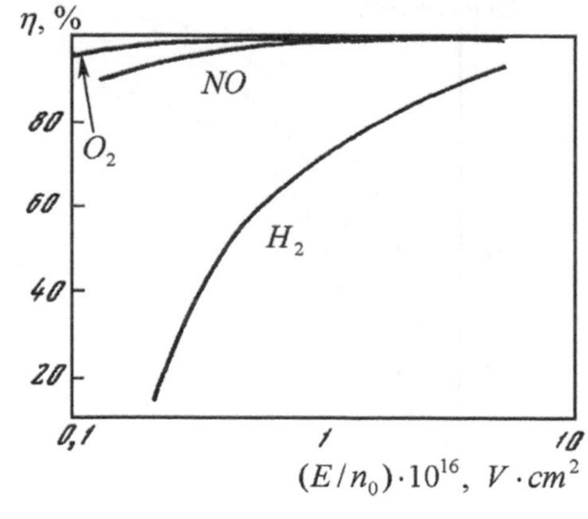

FIGURE 3.10 Fraction of electron energy spent on vibrational excitation of O_2, NO, H_2.

take into account from one hand the influence of different elastic, inelastic, and super-elastic collisions on electron energy distribution function f(ε). On the other hand, this electron energy distribution function f(ε) should be then applied to calculate the rate coefficients of all the mentioned elastic, inelastic, and super-elastic processes. For example, electron excitation rate coefficients on Figures 3.15 and 3.16 were calculated in this quite sophisticated self-consistent way. The energy distributions between different excitation channels have similar general features. For example, the contribution of rotational excitation of molecules and elastic energy losses is significant only at low values of the reduced electric field E/n_0, and hence at low electron temperatures. This is expected since these processes are non-resonant and take place at low electron energies (<<1 eV). At electron temperatures about 1 eV, which are very typical for non-thermal discharges, almost all electron energy and hence most of the discharge power can be localized on vibrational excitation of molecules. This makes the process of vibrational excitation exceptionally important and special in the non-equilibrium plasma chemistry of molecular gases. Obviously, the contribution of electron attachment processes, including the dissociative attachment, can effectively compete with vibrational excitation at

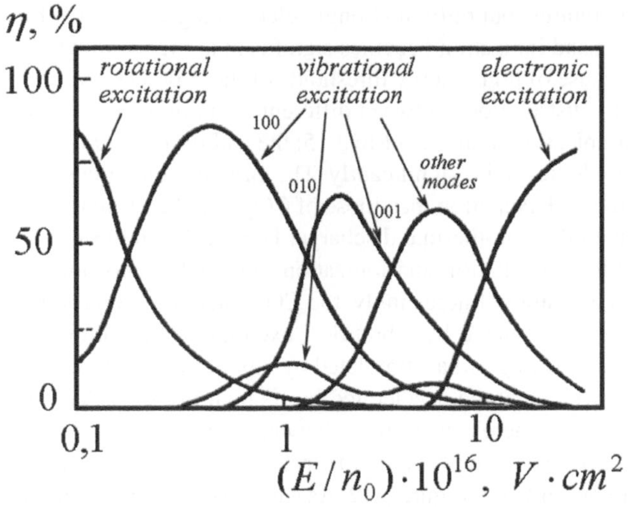

FIGURE 3.11 Electron energy distribution between excitation channels in CO_2.

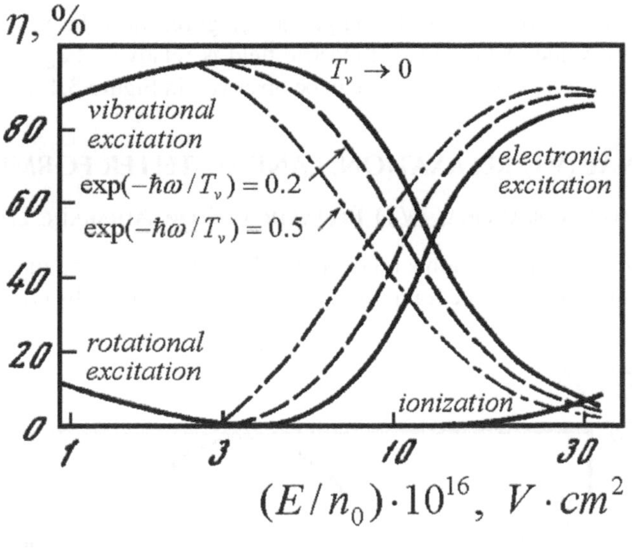

FIGURE 3.12 Electron energy distribution between excitation channels in nitrogen.

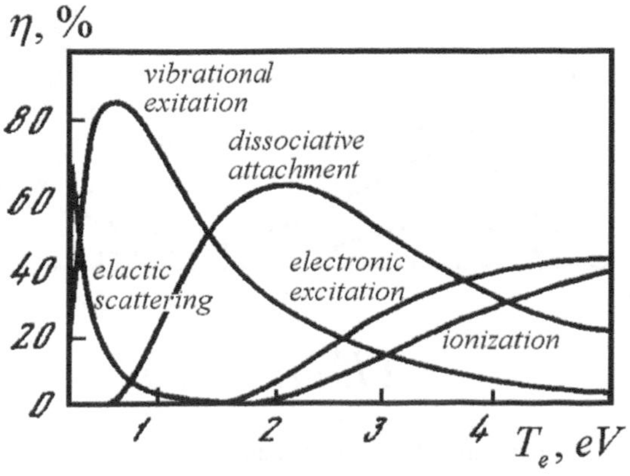

FIGURE 3.13 Electron energy distribution between excitation channels in water vapor.

similar electron temperatures, but only in strongly electro-negative gases. Finally, the contribution of electronic excitation and ionization becomes significant at higher values of E/n_0 and higher electron temperatures, because of high-energy thresholds of these processes.

The electron energy distributions between different excitation channels strongly depend on gas composition. For example, see Figures 3.11 to 3.15; the addition of CO to CO_2 changes the distribution of electron energy losses quite significantly. The carbon monoxide CO molecules have higher cross sections of vibrational excitation than those of CO_2, and therefore they decrease the fraction of high-energy electrons in the non-thermal discharge. Hence, the addition of CO to CO_2 results in a reduction of the electronic excitation and ionization rate coefficients and in the growth of energy going into vibrational excitation. Interestingly, the CO molecules are products of CO_2 dissociation, which can be effectively stimulated by vibrational excitation of the molecules in plasma. For this reason, the CO_2 dissociation in plasma, stimulated by vibrational excitation, can be considered as an autocatalytic process, which can be accelerated by its own products. The influence of vibrational temperature on the electron energy distribution between different excitation channels can be illustrated using a distribution calculated for a non-thermal discharge in molecular nitrogen. The results of the calculations are presented in Figure 3.12. As can be seen from this figure, for higher values of vibrational temperatures, the concentration of the excited molecules is also high and the efficiency of their further vibrational excitation is relatively lowered. This results in an increase in the fraction of high-energy electrons and hence in the intensification of electronic excitation and ionization processes. In other words, the fraction of discharge energy going into the vibrational excitation of molecules at the same value of the reduced electric field is relatively larger when the vibrational temperatures are relatively lower. This effect can be observed in Figure 3.12.

3.4 VIBRATIONAL (VT) RELAXATION, LANDAU-TELLER FORMULA

3.4.1 Vibrational-Translational (VT) Relaxation, Slow Adiabatic Elementary Process

This process is usually called vibrational relaxation or VT-relaxation. The main features of the VT-relaxation can be demonstrated in the framework of classical mechanics by considering the collision

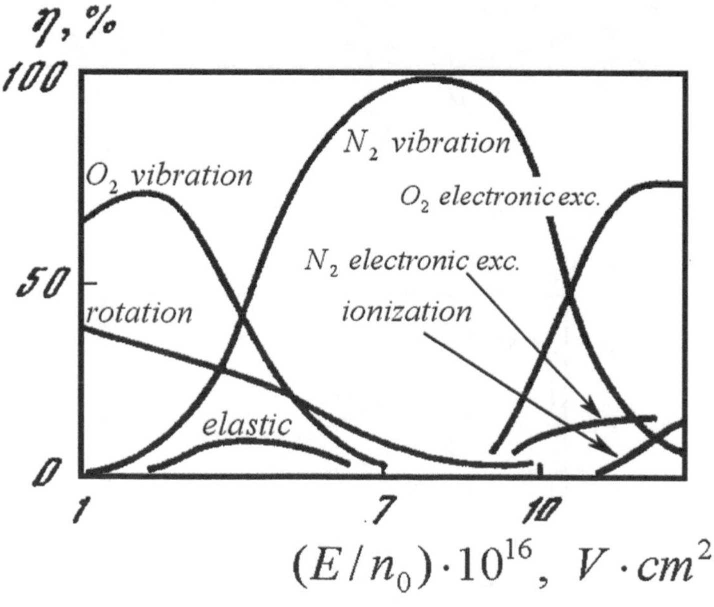

FIGURE 3.14 Electron energy distribution between excitation channels in air.

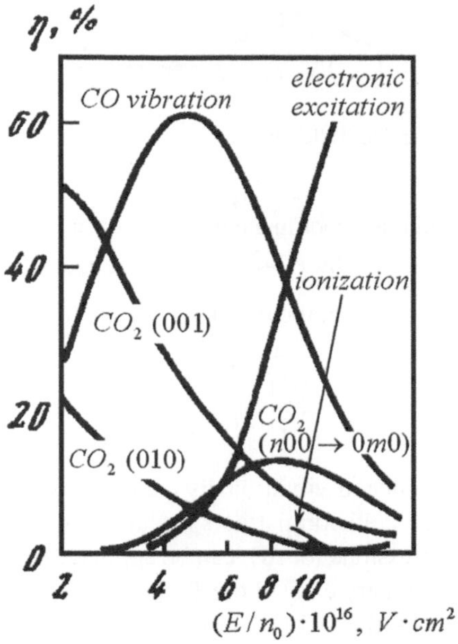

FIGURE 3.15 Electron energy distribution in mixture 50% CO_2 - 50% CO.

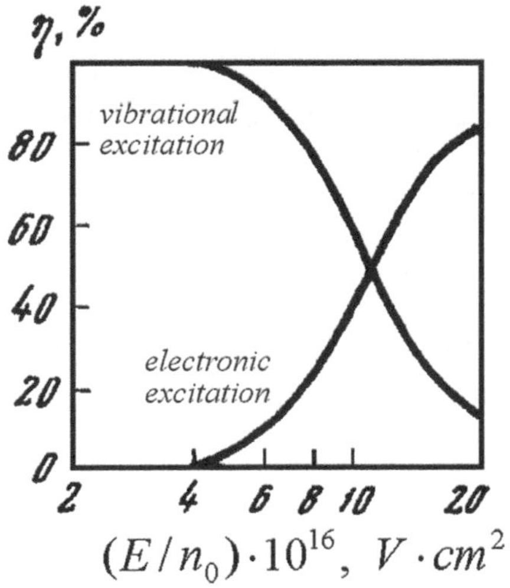

FIGURE 3.16 Electron energy distribution between excitation channels in CO.

of a classical harmonic oscillator with an atom or molecule. The oscillator is considered under the influence of an external force F(t), which represents the intermolecular collision (see Figure 3.17). The one-dimensional motion of the harmonic oscillator can then be described in the center of mass system by the Newton equation:

$$\frac{d^2y}{dt^2} + \omega^2 y = \frac{1}{\mu_0} F(t). \tag{3.39}$$

FIGURE 3.17 Illustration of the Landau-Teller effect.

Here, y is the deviation of vibrational coordinate from equilibrium, ω is the oscillator frequency, μ_0 is its reduced mass. The oscillator is initially ($t \to -\infty$) not excited: $y(t \to -\infty) = 0, \frac{dy}{dt}(t \to -\infty) = 0$. Then the vibrational energy transferred to the oscillator during the collision can be expressed as:

$$\Delta E_v = \frac{\mu_0}{2}\left[\left(\frac{dy}{dt}\right)^2 + \omega^2 y^2\right]_{t=\infty}. \tag{3.40}$$

In the model, the reverse influence of vibrationally excited molecule on "the external force F(t)" is neglected. This means, that the collision partners are taken into account independently, and the same approach and the same expressions for ΔE_v can be applied to also describe the reverse process of energy transfer from the preliminary excited oscillator to "the external force", which is actually another atom or molecule (E.E. Nikitin, 1970; B. Gordietz, A. Osipov, L. Shelepin, 1980). Calculate the energy transfer ΔE_v by introducing instead of the vibrational variable y,a new complex variable $\xi(t) = \frac{dy}{dt} + i\omega y$. Then the oscillator energy transfer Eq. (3.40) can be found as a square of module of this variable:

$$\Delta E_v = \frac{\mu_0}{2}\left|\xi(t)\right|^2_{t=\infty}. \tag{3.41}$$

To find the complex function $\xi(t)$ - the dynamic equation Eq. (3.39) can be rewritten in this case as a first-order equation:

$$\frac{d}{dt}\left(\frac{dy}{dt} + i\omega y\right) - i\omega\left(\frac{dy}{dt} + i\omega y\right) = \frac{1}{\mu_0}F(t); \frac{d\xi}{dt} - i\omega\xi = \frac{1}{\mu_0}F(t) \tag{3.42}$$

The solution of the linear non-uniform differential equation with initial conditions given above $y(t \to -\infty) = 0, \frac{dy}{dt}(t \to -\infty) = 0$ can easily be found as:

$$\xi(t) = \exp(i\omega t)\int_{-\infty}^{t} \frac{1}{\mu_0} F(t')\exp(-i\omega t')\,dt'. \tag{3.43}$$

Substituting the expression for $\xi(t)$ into Eq. (3.41), the energy transfer during the VT-relaxation is:

$$\Delta E_v = \frac{1}{2\mu_0}\left|\int_{-\infty}^{+\infty} F(t)\exp(-i\omega t)\,dt\right|^2. \tag{3.44}$$

This expression shows that the energy transferred from an oscillator (or to an oscillator) is determined by the square of the module of the Fourier-component of the force F(t) on frequency ω corresponding to that of the oscillator. In other words, only the usually small Fourier component of the perturbation force on the oscillator frequency is effective in collisional excitation (or deactivation) of vibrational degrees of freedom of molecules. A simple estimation of the integral can be found by considering time "t" as a complex variable (a trick successfully applied by Landau, see for example L. Landau, 1997), and supposing that $F(t) \to \infty$ in a singularity point $t = \tau + i\tau_{col}$, where $\tau_{col} = 1/\alpha v$ – is the time of the collision, α is the reverse radius of interaction between molecules, v is a relative velocity of the colliding particles. The singularity $t = \tau + i\tau_{col}$ can be considered as the closest one to the real axis. Then the integration of Eq. (3.44) can be accomplished by going around the singularity while shifting the integration line to the upper semi-plane. As a result of the integration, the vibrational energy transfer from (or to) a molecule during an elementary VT-relaxation process can be estimated as:

$$\Delta E_v \propto \exp(-2\omega\tau_{col}). \tag{3.45}$$

This interesting and important relation (L. Landau, E. Teller, 1936) demonstrates the adiabatic behavior of vibrational relaxation. Usually, the Massey parameter (see section 2.2.7) at moderate gas temperatures is fairly high for molecular vibration $\omega\tau_{col} >> 1$, which explains the adiabatic behavior and results in the exponentially low vibrational energy transfer during the VT-relaxation. During the adiabatic collision, a molecule has sufficient time for many vibrations and the oscillator can actually be considered as "structureless", which explains the low level of the energy transfer. The exponentially low probability of the adiabatic VT-relaxation at low gas temperatures as well as the intensive vibrational excitation by electron impact is responsible for the unique role of vibrational excitation in plasma chemistry. Molecular vibrations for some gases such as N_2 CO_2 are able "to trap" energy of non-thermal discharges; it is easy to activate them and difficult to deactivate.

3.4.2 Quantitative Relations for Probability of the Elementary Process of Adiabatic VT Relaxation

Equation (3.45) illustrates the adiabatic behavior of VT-relaxation. To make this formula quantitative, one can specify an interaction potential between a molecule BC with an atom A in the following exponential form:

$$V(r_{AB}) = C\exp(-\alpha r_{AB}) = C\exp[-\alpha(r - \lambda y)]. \tag{3.46}$$

Here it is supposed, that interaction potential is mainly due to interactions between atoms A and B; r is the coordinate of relative motion of mass centers of A and BC; $\lambda = m_C/(m_B + m_C)$; m_B and m_C are masses of the corresponding atoms. Taking into account this interaction potential in the form of Eq. (3.46) and integrating Eq. (3.44) gives:

$$\Delta E_v = \frac{8\pi\omega^2\mu^2\lambda^2}{\alpha^2\mu_0^2}\exp\left(-\frac{2\pi\omega}{\alpha v}\right). \tag{3.47}$$

In this relation μ is a reduced mass of A and BC, μ_0 is a reduced mass of the molecule BC. As can be seen, the exact relation Eq. (3.47) for vibrational energy transfer is in a good agreement with the qualitative expression Eq. (3.45). Equation (3.47) is actually a classical one. However, if the quantum $\hbar\omega$ is the minimum value of vibrational energy transfer, then the probability of transfer of the one quantum can be found based on Eq. (3.47) as:

$$P_{01}^{VT}(v) = \frac{\Delta E_v}{\hbar\omega} = \frac{8\pi^2\omega\mu^2\lambda^2}{\hbar\alpha^2\mu_0^2}\exp\left(-\frac{2\pi\omega}{\alpha v}\right). \tag{3.48}$$

The reverse effect of the vibrationally excited molecule on the "external" force F(t) was neglected. Therefore, the probability Eq. (3.48) can describe both vibrational activation and deactivation. Quantum mechanics generalizes Eq. (3.44) to describe probability $P_{mn}^{VT}(v)$ of an oscillator transition from an initial state with vibrational quantum number "m" to a final one "n". The first order of perturbation theory gives:

$$P_{mn}^{VT}(v) = \frac{\langle m|y|n\rangle^2}{\hbar^2}\left|\int_{-\infty}^{+\infty} F(t)\exp(i\omega_{mn}t)\,dt\right|^2. \tag{3.49}$$

In this relation $\hbar\omega_{mn} = E_m - E_n$ is the transition energy, $\langle m|y|n\rangle$ is a matrix element corresponding to the eigenfunctions (m and n) of non-perturbated Hamiltonian of the oscillator. The integral Eq. (3.49) can be estimated in the same manner as Eq. (3.44), which results in the following qualitative formula for the transition probability:

$$P_{mn}^{VT}(v) \propto \langle m|y|n\rangle^2 \exp(-2|\omega_{mn}|\tau_{col}) \tag{3.50}$$

This formula shows as well as Eq. (3.45) the adiabatic behavior of the vibrational relaxation. Using the specific potential Eq. (3.46), the integration Eq. (3.49) gives the following quantum mechanical expression for the probability of the transition:

$$P_{mn}^{VT}(v) = \frac{16\pi^2\mu^2\omega_{mn}^2\lambda^2}{\alpha^2\hbar^2}\langle m|y|n\rangle^2 \exp\left(-\frac{2\pi|\omega_{mn}|}{\alpha v}\right). \tag{3.51}$$

The quantum mechanical and classical expressions for the probability of vibrational relaxation Eqs. (3.48) and (3.51) are similar, but the quantum mechanical expression additionally includes the square of the matrix elements of relaxation transition. The matrix elements for harmonic oscillators $\langle m|y|n\rangle$ are non-zero only for transitions with $n = m \pm 1$. Then as it is known from quantum mechanics (for example, Landau, 1997):

$$\langle m|y|n\rangle = \sqrt{\frac{\hbar}{2\mu_0\omega}}\left(\sqrt{m}\,\delta_{n,m-1} + \sqrt{m+1}\,\delta_{n,m+1}\right). \tag{3.52}$$

Here the symbol $\delta_{ij} = 1$ if $i = j$, and $\delta_{ij} = 0$ if $i \neq j$. This means, that according to Eqs. (3.51) and (3.52) only the one-quantum VT relaxation processes are possible, However, taking into account the higher powers of "y" in the expansion of $V(r - \lambda y)$ (see Eq. (3.46)), the multi-quantum relaxation processes become possible, but obviously with lower probability. The multi-quantum adiabatic VT-relaxation is also slow, because of much larger values of the Massey parameters in the exponents Eqs. (3.50) and (3.51) in this case. Combining the quantum mechanical relations Eqs. (3.51) and (3.52) gives an expression for the average collisional energy transfer of a quantum oscillator $\Delta E_v = P_{01}^{VT}(v)\cdot\hbar\omega$, which exactly coincides with the corresponding relation for a classical oscillator Eq. (3.47). This demonstrates that vibrational relaxation can be quite often accurately described in the framework of classical mechanics.

3.4.3 VT-Relaxation Rate Coefficients for Harmonic Oscillators, Landau-Teller Formula

To find the rate coefficient of the vibrational VT relaxation of a harmonic oscillator, rewrite the relaxation probability Eqs. (3.51), (3.52) as a function of the relative velocity v of the colliding particles:

$$P_{n+1,n}^{VT}(v) \propto (n+1)\exp\left(-\frac{2\pi\omega}{\alpha v}\right) \tag{3.53}$$

and then integrate it over the Maxwellian distribution function. After such integration, the expression is derived for the averaged probability of VT relaxation as a function of translational temperature T_0 and vibrational quantum number n, which is actually a number of quanta on a molecule:

$$P_{n+1,n}^{VT}(T_0) = (n+1)P_{1,0}^{VT}(T_0). \tag{3.54}$$

The relaxation rate coefficient can be found based on the expression for the relaxation probability $k_{VT}^{n+1,n}(T_0) = P_{n+1,n}^{VT}(T_0) \cdot k_0$, where $k_0 \approx 10^{-10}\frac{\text{cm}^3}{\text{s}}$ is the rate coefficient of gas-kinetic collisions. Equation (3.54) shows the relaxation rate dependence on the number of vibrational quanta. To find the vibrational relaxation temperature dependence Eq. (3.54), take into account that the function under integral of probability Eq. (3.53) over the Maxwellian distribution has a sharp maximum at the velocity:

$$v* = \sqrt[3]{\frac{2\pi T_0}{\mu\alpha}}. \tag{3.55}$$

It helps to find the integral of probability Eq. (3.53) over the Maxwellian distribution and the temperature dependence of vibrational relaxation probability:

$$P_{10}^{VT}(T_0) \propto \exp\left[-\frac{2\pi\omega}{\alpha v^*} - \frac{\mu v^{*2}}{2T_0}\right] \equiv \exp\left(-\frac{3\pi\omega}{\alpha v^*}\right) \equiv \exp\left[-3\left(\frac{\hbar^2\mu\omega^2}{2\alpha^2 T_0}\right)^{1/3}\right]. \tag{3.56}$$

The probability of VT-relaxation exponentially depends on the adiabatic Massey factor calculated for the velocity v*: $\frac{3\pi\omega}{\alpha v^*} \propto \frac{1}{T_0^{1/3}}$. Numerically this adiabatic factor is about 5–15 over a wide range of temperatures. Equation (3.56) obviously can be rewritten for the rate coefficient of the VT-relaxation of the vibrational quantum as a function of translational gas temperature T_0 (Landau, Teller, 1936):

$$k_{VT}^{10} \propto \exp\left(-B/T_0^{1/3}\right), \quad B = \sqrt[3]{\frac{27\hbar^2\mu\omega^2}{2\alpha^2 T_0}}. \tag{3.57}$$

Equation (3.57) shows the exponential growth of the VT-relaxation rate with translational temperature T_0 and plays an important role in plasma chemistry, gas laser, and shock wave kinetics. It is well known as the **Landau-Teller formula**. Generalization of the Landau-Teller formula for better quantitative description of the adiabatic VT-relaxation was developed in the frameworks of SSH-theory (R.N. Schwartz, Z.I. Slawsky, K.F. Herzfeld, 1952) and, in particular, by G. D. Billing (1986). For numerical calculations of the vibrational VT-relaxation rate coefficient, it is convenient to use the following semi-empiric relation, based on the Landau-Teller formula, derived by A. Lifshitz (1974):

$$k_{VT}^{10} = 3.03 * 10^6 (\hbar\omega)^{2.66} \mu^{2.06} \exp\left[-0.492(\hbar\omega)^{0.681} \mu^{0.302} T_0^{-1/3}\right]. \tag{3.58}$$

Here the vibrational relaxation rate coefficient is in cm^3/mole*sec, T_0 and $\hbar\omega$- in degrees Kelvin, and μ the reduced mass of colliding particles - in atomic mass units. Similarly, another semi-empiric formula (R.S. Millican, D.R. White, 1963) describes the VT-relaxation time as a function of pressure:

$$\ln(p\tau_{VT}) = 1.16 * 10^{-3} \mu^{1/2} (\hbar\omega)^{4/3} \left(T_0^{-1/3} - 0.015\mu^{1/4}\right) - 18.42. \tag{3.59}$$

Pressure p is in atm, τ_{VT} in sec, T_0 and $\hbar\omega$ in degrees Kelvin, reduced mass μ - in the atomic mass units. Numerical values of the vibrational relaxation rate coefficients at room temperature can be found in Table 3.13. The temperature dependence of the rate coefficients of the vibrational VT-relaxation for some molecules is given in Table 3.14. Relaxation of most of the molecules considered in this table follows the Landau-Teller tendency in temperature dependence.

3.4.4 The Vibrational VT-Relaxation of Anharmonic Oscillators

The most important peculiarity of relaxation of anharmonic oscillators is due to reduction of the transition energy with an increase of vibrational quantum number:

$$\omega_{n,n-1} = \omega\left(1 - 2x_e n\right). \tag{3.60}$$

Here x_e is the coefficient of anharmonicity of a vibrationally excited diatomic molecule, n is the vibrational quantum number. As seen from Eq. (3.54), the probability and rate coefficient of vibrational VT- relaxation increases with the vibrational quantum number "n" even in the case of harmonic oscillators. Furthermore in the case of anharmonic ones, the increase of vibrational quantum number "n" leads also to a reduction of the Massey parameter, making relaxation less adiabatic and exponentially faster:

$$P_{n+1,n}^{VT}\left(T_0\right) = \left(n+1\right)P_{1,0}^{VT}\left(T_0\right)\exp\left(\delta_{VT} n\right) \tag{3.61}$$

The temperature dependence of the probability of VT-relaxation of anharmonic oscillators Eq. (3.61) is similar to that of harmonic oscillators Eq. (3.54) and corresponds to the Landau-Teller formula. The exponential parameter δ_{VT} can be determined in the case of anharmonic oscillators as:

TABLE 3.13
Vibrational VT-relaxation Rate Coefficients $k_{VT}^{10}\left(T_0 = 300K\right)$ for One-component Gases at Room Temperature

Molecule	$k_{VT}^{10}\left(T_0 = 300\text{ K}\right), \frac{\text{cm}^3}{\text{s}}$	Molecule	$k_{VT}^{10}\left(T_0 = 300\text{ K}\right), \frac{\text{cm}^3}{\text{s}}$
O_2	$5*10^{-18}$	F_2	$2*10^{-15}$
Cl_2	$3*10^{-15}$	D_2	$3*10^{-17}$
Br_2	10^{-14}	$CO_2(01^10)$	$5*10^{-15}$
J_2	$3*10^{-14}$	$H_2O(010)$	$3*10^{-12}$
N_2	$3*10^{-19}$	N_2O	10^{-14}
CO	10^{-18}	COS	$3*10^{-14}$
H_2	10^{-16}	CS_2	$5*10^{-14}$
HF	$2*10^{-12}$	SO_2	$5*10^{-14}$
DF	$5*10^{-13}$	C_2H_2	10^{-12}
HCl	10^{-14}	CH_2Cl_2	10^{-12}
DCl	$5*10^{-15}$	CH_4	10^{-14}
HBr	$2*10^{-14}$	CH_3Cl	10^{-13}
DBr	$5*10^{-15}$	$CHCl_3$	$5*10^{-13}$
HJ	10^{-13}	CCl_4	$5*10^{-13}$
HD	10^{-16}	NO	10^{-13}

TABLE 3.14
Temperature Dependence of the Vibrational VT-relaxation Rate Coefficients $k_{VT}^{10}(T_0)$ for One-component Gases

Molecule	Temperature dependence $k_{VT}^{10}(T_0), \frac{cm^3}{s}$; temperature T_0 in Kelvin degrees
O_2	$10^{-10}\exp\left(-129*T_0^{-1/3}\right)$
Cl_2	$2*10^{-11}\exp\left(-58*T_0^{-1/3}\right)$
Br_2	$2*10^{-11}\exp\left(-48*T_0^{-1/3}\right)$
J_2	$5*10^{-12}\exp\left(-29*T_0^{-1/3}\right)$
CO	$10^{-12}T_0\exp\left(-190*T_0^{-1/3}+1410*T_0^{-1}\right)$
NO	$10^{-12}\exp\left(-14*T_0^{-1/3}\right)$
HF	$5*10^{-10}T_0^{-1}+6*10^{-20}T_0^{2.26}$
DF	$1.6*10^{-5}T_0^{-3}+3.3*10^{-16}T_0$
HCl	$2.6*10^{-7}T_0^{-3}+1.4*10^{-19}T_0^{2}$
F_2	$2*10^{-11}\exp\left(-65*T_0^{-1/3}\right)$
D_2	$10^{-12}\exp\left(-67*T_0^{-1/3}\right)$
$CO_2(01^10)$	$10^{-11}\exp\left(-72*T_0^{-1/3}\right)$

$$\delta_{VT} = 4\gamma_n^{2/3}x_e, \quad if \ \gamma_n \geq 27 \tag{3.62}$$

$$\delta_{VT} = \frac{4}{3}\gamma_n x_e, \quad if \ \gamma_n < 27. \tag{3.63}$$

The adiabatic factor γ_n is actually the Massey parameter for the vibrational relaxation transition $n + 1 \rightarrow n$ with the transition energy $E_{n+1} - E_n$. This adiabatic factor can be calculated as:

$$\gamma_n(n+1 \rightarrow n) = \frac{\pi(E_{n+1} - E_n)}{\hbar\alpha}\sqrt{\frac{\mu}{2T_0}}. \tag{3.64}$$

For numerical calculations of this Massey parameter γ_n it is convenient to use the relation:

$$\gamma_n = \frac{0.32}{\alpha}\sqrt{\frac{\mu}{T_0}}\hbar\omega\left(1 - 2x_e(n-1)\right). \tag{3.65}$$

In this relation the reduced mass μ is in atomic units, reverse radius of interaction between colliding particles α is in A^{-1}, vibrational quantum $\hbar\omega$ and gas temperature T_0 are in degrees Kelvin.

3.4.5 Fast Non-Adiabatic Mechanisms of VT-Relaxation

The vibrational relaxation is quite slow in adiabatic collisions when there is no chemical interaction between colliding partners. The probability of a vibrationally excited N_2 deactivation in collision with another N_2 molecule at room temperature can be as low as 10^{-9}. However, the vibrational relaxation process can be arranged much faster in a non-adiabatic way, with a quantum transfer at almost each collision, if the colliding partners interact chemically. This can happen in collisions of vibrationally excited molecules with active atoms and radicals, in surface collisions, etc. Consider separately the main non-adiabatic mechanisms of VT-relaxation.

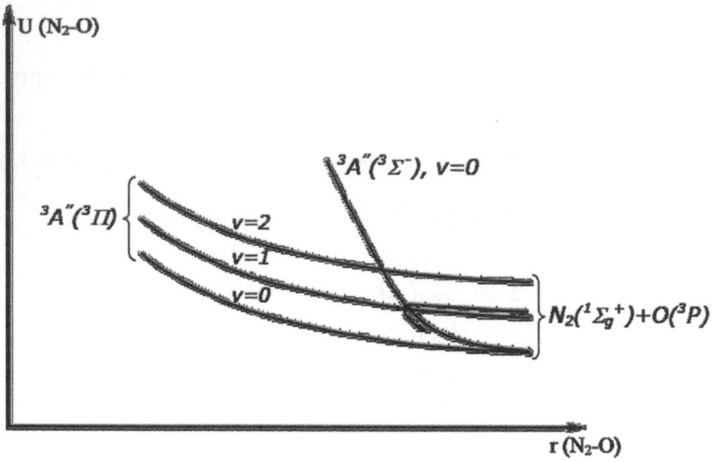

FIGURE 3.18 Illustration of the non-adiabatic VT relaxation of N_2 molecules in collisions with O-atoms.

a. ***VT-relaxation in molecular collisions with atoms and radicals.*** These processes can be illustrated by the relaxation of vibrationally excited N_2-molecules on atomic oxygen, analyzed by E.A. Andreev and E.E. Nikitin (1976), see Figure 3.18. The energy distance initially between degenerated electronic terms grows as a molecule and an atom approach each other. Finally, when this energy of electronic transitions becomes equal to a vibrational quantum, the non-adiabatic relaxation (the so-called vibronic transition, Figure 3.18) can take place. The temperature dependence of the process is due to the low activation energy and is not actually significant. Typical values of the non-adiabatic VT-relaxation rate coefficients are very high, usually about 10^{-13}–10^{-12} cm³/s. Sometimes, as in the case of relaxation of alkaline atoms, the non-adiabatic VT-relaxation rate coefficients reach those for gas-kinetic collisions, for example, about 10^{-10} cm³/sec. (see Eq. 3.97 and Figure 3.20).
b. ***VT-relaxation through the intermediate formation of long-life complexes.*** Fast non-adiabatic relaxation is also possible if a collision results in formatting long-life chemically bonded complexes. This takes place, in particular, in collisions of H_2O^*–H_2O, CO_2^*–H_2O, CH_4^*–CO_2, CH_4^*–H_2O, $C_2H_6^*$–O_2, NO^*–Cl_2, related to important plasma chemical applications. Interaction between the collision partners is based quite often in this case on hydrogen bonds. In this case, relaxation rate coefficients also can reach gas-kinetic values. Temperature dependence is not strong and usually is negative so the relaxation rate coefficients decrease with temperature growth. Another important peculiarity of this relaxation mechanism is its multi-quantum nature; this means that the probabilities of one-quantum and multi-quantum transfer of vibrational energy are relatively close.
c. VT-relaxation in symmetrical exchange reactions. Very fast non-adiabatic relaxation takes place in chemical reactions like:

$$A' + (BA'')*(n = 1) \rightarrow A'' + BA'(n = 0). \tag{3.66}$$

Such processes are effective when activation energies of corresponding chemical processes are low. A very important example is the oxygen exchange reaction, proceeding without any activation energy through the intermediate formation of an excited ozone molecule: $O_2^* + O \rightarrow O_3^* \rightarrow O + O_2$. Here the relaxation rate coefficient is approximately 10^{-11} cm³/s and practically does not depend on temperature. A similar situation takes place in the fast non-adiabatic VT-relaxation processes of H_2-molecules on H-atoms, different halogen molecules on corresponding halogen atoms and hydrogen halides on the hydrogen atoms.

TABLE 3.15
Accommodation Coefficients for the Heterogeneous VT-relaxation

Molecule	Vibration Mode	Quantum Energy, cm^{-1}	Temperature K	Surface	Accommodation Coefficient
CO_2	$v_2 = 1$	667	277–373	Platinum	0.3–0.4
CO_2	$v_2 = 1$	667	297	NaCl	0.22
CO_2	$v_3 = 1$	2349	300–350	Pyrex	0.2–0.4
CO_2	$v_3 = 1$	2349	300	Brass, Teflon, Mylar	0.2
CO_2	$v_3 = 1$	2349	300–1000	Quartz	0.05–0.45
CO_2	$v_3 = 1$	2349	300–560	Molybdenum glass	0.3–0.4
CH_4	$v_4 = 1$	1306	273–373	Platinum	0.5–0.9
H_2	$v = 1$	4160	300	Quartz	$5*10^{-4}$
H_2	$v = 1$	4160	300	Molybdenum glass	$1.3*10^{-4}$
H_2	$v = 1$	4160	300	Pyrex	10^{-4}
D_2	$v = 1$	2990	77–275	Molybdenum glass	$(0.2–8)*10^{-4}$
D_2	$v = 1$	2990	300	Quartz	10^{-4}
N_2	$v = 1$	2331	350	Pyrex	$5*10^{-4}$
N_2	$v = 1$	2331	282–603	Molybdenum glass	$(1–3)*10^{-3}$
N_2	$v = 1$	2331	350	Pyrex	$5*10^{-4}$
N_2	$v = 1$	2331	300	Steel, aluminum, copper	$3*10^{-3}$
N_2	$v = 1$	2331	300	Teflon, alumina	10^{-3}
N_2	$v = 1$	2331	295	Silver	$1.4*10^{-2}$
CO	$v = 1$	2143	300	Pyrex	$1.9*10^{-2}$
HF	$v = 1$	3962	300	Molybdenum glass	10^{-2}
HCl	$v = 1$	2886	300	Pyrex	0.45
OH	$v = 9$	–	300	Boron acid	1
N_2O	$v_2 = 1$	589	273–373	Platinum, NaCl	0.3
N_2O	$v_3 = 1$	2224	300–350	Pyrex	0.2
N_2O	$v_3 = 1$	2224	300–1000	Quartz	0.05–0.33
N_2O	$v_3 = 1$	2224	300–560	Molybdenum glass	0.01–0.03
CF_3Cl	$v_3 = 1$	732	273–373	Platinum	0.5–0.6

d. ***Heterogeneous VT-relaxation, losses of vibrational energy in surface collisions, accommodation coefficients.*** Heterogeneous relaxation is non-adiabatic and a fast process if it proceeds through an adsorption stage and forms intermediate complexes with a surface, which is usually the case (Yu.Gershenson, V. Rosenstein, S. Umansky, 1977). The losses of vibrational energy can be determined in this situation from the probability of the vibrational relaxation calculated with respect to one surface collision. This probability is usually referred to as the **accommodation coefficient**, which can be found in Table 3.15 for different molecules and different surfaces.

3.4.6 VT-Relaxation of Polyatomic Molecules

The peculiar features of the VT relaxation of vibrationally excited polyatomic molecules are related to the fact, that in this case not one but a set of oscillators interacts with an incident atom or molecule. At a low level of excitation, vibrational modes of a polyatomic molecule can be considered separately (see Eqs. (3.12) and (3.14)) and their relaxation can be calculated for each mode following the Landau-Teller approach or a non-adiabatic model described above (see Tables 3.13 and 3.14). For higher levels of excitation, corresponding to the vibrational quasi-continuum of polyatomic molecules (see Section 3.2.4 and Eq. (3.16)), the mean square of the vibrational energy transferred to translational degrees of freedom $\left\langle \Delta E_{VT}^2 \right\rangle$ is obtained by averaging the exponential Landau-Teller factors over the vibrational Fourier spectrum of the system I(ω), which was described in Section 3.2.4: (V.D. Rusanov, A. Fridman, G. Sholin, 1986)

$$\left\langle \Delta E_{VT}^2 \right\rangle = \int_0^\infty \left(\Omega \tau_{col}\right)_{VT}^2 \left(\hbar\omega\right)^2 I(\omega) \exp\left(-\frac{\omega}{\alpha v}\right) d\omega. \tag{3.67}$$

The factor $\left(\Omega \tau_{col}\right)_{VT}^2$ characterizes the smallness of the probability of transition, being due to the smallness of the amplitude of vibrations relative to the interaction radius. Numerically this factor is usually of the order of $\left(\Omega \tau_{col}\right)_{VT}^2 \approx 0.01$. If a polyatomic molecule can be considered as a group of harmonic oscillators with frequencies ω_{0i} and vibrational quantum numbers n_i, then the vibrational Fourier spectrum I(ω) can be presented as a sum of δ-functions Eq. (3.18). Obviously in this case, Eq. (3.67) for the vibrational energy transfer during a collision reduces to the known expression corresponding to the Landau-Teller model:

$$\left\langle \Delta E_{VT}^2 \right\rangle = \sum_i \left(\Omega \tau_{col}\right)_{VTi}^2 n_i \left(\hbar\omega_{0i}\right)^2 \exp\left(-\frac{\omega_{0i}}{\alpha v}\right). \tag{3.68}$$

For the case of vibrationally excited molecules in quasi-continuum, the vibrational Fourier spectrum I(ω) can be described by the Lorentz profile Eq. (3.21) as was discussed in Section 3.2.4. Then the energy transfer from the most rapidly relaxing vibrational mode (with the smallest quantum $\hbar\omega_n$) can be obtained from the integration of Eq. (3.67):

$$\left\langle E_{VT}^2 \right\rangle = \left(\Omega \tau_{col}\right)_{VT}^2 n \left[\frac{\alpha v}{\alpha v + \delta} \exp\left(-\frac{\omega_n - \delta}{\alpha v}\right) + \frac{\delta}{3\pi\omega_n}\left(\frac{\alpha v}{\omega_n}\right)^3\right] \left(\hbar\omega_n\right)^2. \tag{3.69}$$

Here δ is the intermode vibrational energy exchange frequency, considered while discussing Eq. (3.21). Factor "n" is the number of quanta on the mode under consideration. If the low-frequency mode is degenerated or is a part of Fermi-resonance modes, then "n" implies the total number of quanta taking into account the degeneracy. As is seen from Eq. (3.69), the VT-relaxation of polyatomic molecules in quasi-continuum is determined by two effects: an adiabatic effect and a quasi-resonant effect (V.D. Rusanov, A. Fridman, G. Sholin, 1986). The first term is the adiabatic effect and is somewhat similar to the case of diatomic molecules. It, however, is growing faster with "n" because of the effective reduction of the vibrational frequency and the Massey parameter $\frac{\omega_n - \delta(n)}{\alpha v}$ due to the broadening of the given mode line in the vibrational spectrum I(ω). The second term in Eq. (3.69) corresponds to quasi-resonance (non-adiabatic) relaxation of polyatomic molecules at low frequencies $\omega \approx \alpha v$. Excitation at these low frequencies in the vibrational spectrum of polyatomic molecules becomes possible due to the interaction of their fundamental modes. A comparison of these two relaxation effects shows, that the quasi-resonant VT-relaxation has no exponentially small factor and can exceed the adiabatic relaxation. This means that VT-relaxation of polyatomic molecules actually becomes non-adiabatic and very fast when high levels of their vibrational excitation result in a transition to a quasi-continuum of the vibrational spectrum.

3.4.7 Effect of Rotation on the Vibrational Relaxation of Molecules

Interaction between molecular vibrations and rotations is able to accelerate VT-relaxation. This effect is most important for molecules with low momentum of inertia having at the same time a large reduced mass of colliding particles. Vibrational relaxation of methane and hydrogen halides can be taken here as an example. In general, molecules containing a hydrogen atom can lose their vibrational energy through rotation faster (V.N. Kondratiev, E.E. Nikitin 1981). This effect can be explained since the Massey parameter $\omega/\alpha v$ in this case depends not on the relative translational velocity of colliding partners, but rather on the velocity of molecular rotation at the point of the minimal distance between them. Molecules, which consist of both heavy and light atoms, have rotational velocity faster than translational, which results in reducing the Massey parameter and accelerating the vibrational relaxation. Polyatomic molecules have also a specific effect of rotations on vibrational relaxation, the so-called VRT-relaxation process. Degenerated vibrational modes can have "circular" polarization (see section 3.2.3) and angular momentum of the quasi-rotations Eq. (3.13). This opens a fast, non-adiabatic channel for energy transfer from vibrational to translational degrees of freedom through intermediate rotations of polyatomic molecules.

3.5 VIBRATIONAL ENERGY TRANSFER BETWEEN MOLECULES, VV-RELAXATION PROCESSES

Generation of highly vibrationally excited and therefore reactive molecules is usually not due to direct electron impact. The highly excited particles can be effectively formed during collisional energy exchange processes between molecules. These fundamental processes of great importance in plasma chemistry are usually resonant or close to resonance and called VV-relaxation or VV-exchange processes.

3.5.1 Resonant VV-relaxation

These processes usually imply vibrational energy exchange between molecules of the same kind, for example: $N_2^*(v = 1) + N_2(v = 0) \rightarrow N_2(v = 0) + N_2^*(v = 1)$. The resonant VV-exchange between diatomic molecules can be characterized by the probability $q_{mn}^{sl}(v)$ of a collisional transition when one oscillator changes its vibrational quantum number from "s" to "l", and other from "m" to "n". Quantum mechanics gives an expression for the probability $q_{mn}^{sl}(v)$ similar to that obtained for VT-relaxation Eq. (3.49). Instead of ω_{mn}, used in the case of VV-exchange, the frequency describing the change of vibrational energy during the collision should be employed:

$$\omega_{ms,nl} = \frac{1}{\hbar}\left(E_m + E_s - E_l - E_n\right) = \frac{\Delta E}{\hbar}. \tag{3.70}$$

Also, instead of the matrix elements of transition m → n for VT-relaxation Eq. (3.49), the probability of the VV-exchange is determined by a product of matrix elements of transitions m → n and s → l in both two interacting oscillators:

$$q_{mn}^{sl}(v) = \frac{\langle m|y_1|n\rangle^2 \langle s|y_2|l\rangle^2}{\hbar^2}\left|\int_{-\infty}^{+\infty} F(t)\exp\left(i\omega_{ms,nl}t\right)dt\right|^2. \tag{3.71}$$

The squared module of a Fourier-component of interaction force F(t) on a transition frequency Eq. (3.70) characterizes here the level of resonance (of adiabatic behavior) of the process in the same manner as in the case of VT-relaxation. As can be seen from Eq. (3.52), the matrix elements for harmonic oscillators $\langle m|y|n\rangle$ are non- zero only for transitions with $n = m \pm 1$, which means only the one-quantum VV-relaxation processes are possible in this approximation. Expression for the VV-relaxation probability of the one-quantum exchange as a function of translational gas temperature T_0 can be obtained by averaging of the probability $q_{mn}^{sl}(v)$ over the Maxwellian distribution:

$$Q_{n+1,n}^{m,m+1}(T_0) = (m+1)(n+1)Q_{10}^{01}(T_0). \tag{3.72}$$

TABLE 3.16
Resonant VV-relaxation Rate Coefficients at Room Temperature and Their Ratio to Those of Gas-kinetic Collisions (k_{VV}/k_0)

Molecule	k_{VV}, cm³/s	k_{VV}/k_0	Molecule	k_{VV}, cm³/s	k_{VV}/k_0
$CO_2(001)$	$5*10^{-10}$	4	HF	$3*10^{-11}$	0.5
CO	$3*10^{-11}$	0.5	HCl	$2*10^{-11}$	0.2
N_2	10^{-13}–10^{-12}	10^{-3}–10^{-2}	HBr	10^{-11}	0.1
H_2	10^{-13}	10^{-3}	DF	$3*10^{-11}$	–
$N_2O(001)$	$3*10^{-10}$	3	HJ	$2*10^{-12}$	–

Taking into account the higher powers in an expansion of intermolecular interaction potential, the multi-quantum VV-exchange processes become possible even for harmonic oscillators, but obviously with lower probability (E.E. Nikitin, 1970):

$$Q_{0,k}^{m,m-k} \approx \frac{m!}{(m-k)!k!2^{k-1}}\left(\Omega\tau_{col}\right)^{2k}. \tag{3.73}$$

In this relation τ_{col} is time of a collision and Ω is the VV-transition frequency during the collision. The factor $\Omega\tau_{col}$ is usually small due to the smallness of the vibrational amplitude with respect to intermolecular interaction radius in the same manner as in the case of VT-relaxation.VV-relaxation for most molecules is due to the exchange interaction (short distance forces), which results in a short time of the collision τ_{col} and $\Omega\tau_{col} \approx 0.1 - 0.01$. In this case, the probability of the only one quantum transfer is about $Q_{10}^{01} \approx \left(\Omega\tau_{col}\right)^2 \approx 10^{-2} - 10^{-4}$ (see Table 3.16). The probability of multi-quantum VV-exchange in this case according to Eq. (3.73) is very low (e.g. , it is about 10^{-9} even for resonant three-quantum exchanges). However, for some molecules such as CO_2 and N_2O, VV-relaxation is due to dipole or multipole interaction (long-distance forces). Because of the longer interaction distance, the collision time is much longer and both the factor $\Omega\tau_{col}$ and the VV-relaxation probability Q_{10}^{01} become much larger and approaches unity: $Q_{10}^{01} \approx \left(\Omega\tau_{col}\right)^2 \approx 1$ (see Table 3.16). Taking into account Eq. (3.72), it is interesting to note that the cross sections of the resonance VV-exchange between highly vibrationally excited molecules can exceed in this case the gas-kinetic cross section. Also, the multi-quantum resonant VV-exchange Eq. (3.73) is decreasing with the number of transferred quanta, but not as fast as in the case of short-distance exchange interaction between molecules. Numerical values of the resonant VV-exchange rate coefficients are given in Table 3.16 at room temperature and a quantum exchange between the first and zero vibrational levels. Ratios of the rate coefficient to the correspondent gas-kinetic rate can be also found in the table. In general, the resonant VV-exchange is usually much faster at room temperature than VT-relaxation (see Table 3.13). This leads to the possibility of efficient generation of highly vibrationally excited and hence very reactive molecules in non-thermal discharges. The temperature dependence of the VV-relaxation probability $Q_{10}^{01} \approx \left(\Omega\tau_{col}\right)^2$ is different for that provided by dipole and exchange interactions. The transition frequency Ω is proportional to the average interaction energy, hence in the case of the short distance exchange interaction $\Omega \propto T_0$. Taking into account that $\tau_{col} \propto 1/v \propto T_0^{1/2}$, $Q_{10}^{01} \approx \left(\Omega\tau_{col}\right)^2 \propto T_0$. The probability of VV-relaxation provided by short distance forces is proportional to temperature. Conversely, in the case of the long-distance dipole and multipole interactions, the VV-relaxation probability decreases with an increase of the translational gas temperature T_0.

The transition frequency Ω does not depend on temperature in this case and, hence, $Q_{10}^{01} \approx (\Omega\tau_{col})^2 \propto \tau_{col}^2 \propto 1/T_0$.

3.5.2 VV-Relaxation of Anharmonic Oscillators

The influence of the anharmonicity is related to a change of the transition matrix elements, but mostly to the increase of the transition energy. VV-exchange becomes non-resonant, slightly adiabatic and as a result slower than the resonant one. The expression for the transition probability becomes more complicated than Eq. (3.72) and includes the exponential adiabatic factors (E.E. Nikitin, 1970):

$$Q_{n+1,n}^{m,m+1} = (m+1)(n+1)Q_{10}^{01}\exp(-\delta_{VV}|n-m|)\left[2/3 - 1/2\exp(-\delta_{VV}|n-m|)\right]. \tag{3.74}$$

In the same manner as was for the case of VT-relaxation of anharmonic oscillators (see Eqs. (3.4.24) and (3.64)), in this relation:

$$\delta_{VV} = \frac{4}{3}x_e\gamma_0 = \frac{4}{3}\frac{\pi\omega x_e}{\alpha}\sqrt{\frac{\mu}{2T_0}}. \tag{3.75}$$

Note that the adiabatic factor Eq. (3.64) is much lower for a VV-exchange than for VT-relaxation. That is why Eq. (3.62) can be neglected in the case of VV-relaxation, and only Eq. (3.63) is required to be taken into account. In general, in one-component gases $\delta_{VV} = \delta_{VT}$. For numerical calculations of δ_{VV} it is convenient to use the following formula:

$$\delta_{VV} = \frac{0.427}{\alpha}\sqrt{\frac{\mu}{T_0}}x_e\hbar\omega. \tag{3.76}$$

In this relation, the reduced mass of colliding molecules μ should be given in a.m.u. (atomic units), reverse intermolecular interaction radius α - in A^{-1}, translational gas temperature $T_{0,}$ and vibrational quantum $\hbar\omega$ in degrees Kelvin. The above formulas were related to the VV-exchange of anharmonic oscillators due to the short distance forces (the exchange interaction between colliding molecules), which is correct in most cases. Taking into account the long-distance forces leads to modification of Eq. (3.74) (S.D. Rockwood, J.E. Brau, W.A. Proctor, G. H. Canavan, 1973):

$$\begin{aligned} Q_{n+1,n}^{m,m+1} \approx (m+1)(n+1)\Big\{&\left(Q_{10}^{01}\right)_S \exp(-\delta_{VV}|n-m|)\left[3/2 - 1/2\exp(-\delta_{VV}|n-m|)\right] \\ &+\left(Q_{10}^{01}\right)_L \exp\left(-\Delta_{VV}(n-m)^2\right)\Big\}. \end{aligned} \tag{3.77}$$

In this relation, $\left(Q_{10}^{01}\right)_S$ and $\left(Q_{10}^{01}\right)_L$ represent probabilities of the only quantum transfer due respectively to short distance and long distance forces, parameter Δ_{VV} is related to the Massey parameter and determines the adiabatic degree for the VV-exchange of anharmonic oscillators provided by the long-distance forces. When the probability $\left(Q_{10}^{01}\right)_L \ll 1$ is negligible, Eq. (3.77) coincides with the previous one Eq. (3.74). The long collisional time in the case of the VV-exchange provided by long-distance forces results in relatively large Massey parameters $\Delta\omega * \tau_{col}$. As a result, $\Delta_{VV} > \delta_{VV}$, and according to Eq. (3.77), the long-distance forces are able to affect the rate of VV-exchange only close to resonance, which means only for collisions of the anharmonic oscillators with close values of vibrational quantum numbers. Figure 3.19 (W.Q. Geffers, J.D. Kelley, 1971) illustrates this effect by showing the

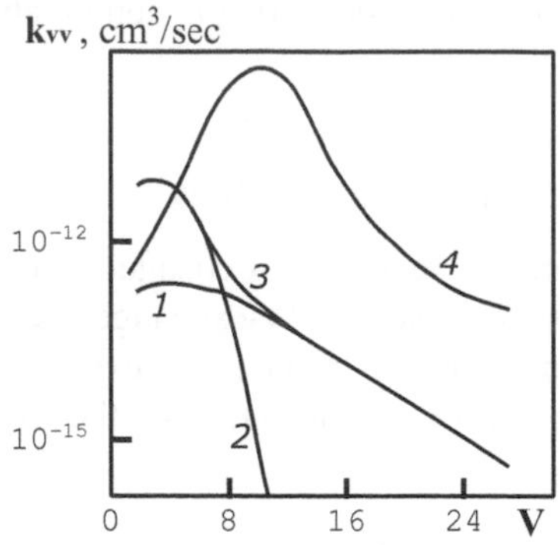

FIGURE 3.19 VV- exchange rate coefficient dependence on vibration quantum number (CO, $T_0 = 300$ K): 1) contribution of short-distance-forces; 2) contribution of long-distance-forces; 3) total curve for transitions ($v \to v-1,\ 0 \to 1$); 4) total curve for transitions ($v \to v-1,\ 8 \to 9$).

dependence of the VV-exchange rate coefficient in pure CO at room temperature on the vibrational quantum number.

Now compare the rate coefficients of VV-exchange and VT relaxation taking into account the anharmonicity. The effect of anharmonicity on the VV-exchange is negligible in resonant collisions, when $|m-n|\ \delta_{VV} << 1$ and Eq. (3.74) actually coincides with Eq. (3.72) for relaxation of harmonic oscillators. Anharmonicity becomes quite necessary in considering VV-exchange processes occurring in collisions of relatively highly excited molecules and molecules on low vibrational levels. Taking into account the effect of anharmonicity, the rate coefficient of the VV-exchange process $k_{VV}(n) = k_0 Q^{0,1}_{n+1,n}$ decreases with the growth of the vibrational quantum number "n"; this is in contrast to the VT-relaxation rate coefficient $k_{VT}(n) = k_0 P^{VT}_{n+1,n}$, which increases with "n" ($k_0 \approx 10^{-10}\ \mathrm{cm}^3/\mathrm{s}$ is the rate coefficient of gas-kinetic collisions). It is interesting to compare the rate coefficients of VV-exchange and VT-relaxation processes for vibrationally excited molecules based on Eqs. (3.74), (3.61), and (Eq. (3.63)):

$$\xi(n) = \frac{k_{VT}(n)}{k_{VV}(n)} \approx \frac{P^{VT}_{10}}{Q^{01}_{10}} \exp\left[\left(\delta_{VV} + \delta_{VT}\right)n\right]. \tag{3.78}$$

As seen from the relation $\xi << 1$ at low excitation levels, the VT-relaxation is much slower than the VV-exchange because $P^{VT}_{10}/Q^{01}_{10} << 1$. This means, that the population of highly vibrationally excited states can increase in non-thermal plasma much faster than losses of vibrational energy. This also explains the high efficiency of this type of excitation in plasma chemistry. The ratio $\xi(n)$, however, is growing exponentially with "n" and the VT-relaxation catches up to the VV-exchange ($\xi = 1$) at some critical value of the vibrational quantum number (and corresponding vibrational energy $E^*(T_0)$). This critical value of vibrational energy shows a maximum level of effective vibrational excitation $E^*(T_0)$ in plasma chemical systems and can be calculated as a function of gas temperature (V. Rusanov, A. Fridman, G. Sholin, 1979):

$$E*(T_0) = \hbar\omega\left(\frac{1}{4x_e} - b\sqrt{T_0}\right). \tag{3.79}$$

The first term in this relation obviously corresponds to the dissociation energy D of a diatomic molecule. Parameter "b" depends only on the molecular gas under consideration. Thus, for

CO-molecules b = 0.90 $K^{-0.5}$, for N_2 – b = 0.90 $K^{-0.5}$, for HCl – b = 0.90 $K^{-0.5}$. The VV-relaxation process between molecules of the same kind can include, taking into account anharmonicity, a multi-quantum exchange. The simplest example here is the two-quantum resonance. This occurs when a molecule with high vibrational energy (e.g. , ≈ $\frac{3}{4}$ D) can resonantly exchange two quanta for the only vibrational quantum of another molecule at a low vibrational level. In general, such processes can be considered as intermolecular VV′-exchange (see below), and their small probability can be estimated by Eq. (3.73).

3.5.3 Intermolecular VV′-Exchange

Vibrational energy exchange between molecules of a different kind is usually referred to as VV′-exchange. Let consider first the VV′-exchange in a mixture of diatomic molecules A and B with only slightly different vibrational quanta $\hbar\omega_A > \hbar\omega_B$. The adiabatic factors determine the small probability of the process. Then Eq. (3.74) can be generalized to describe the VV′-exchange, when a molecule A transfers a quantum ($v_A + 1 \rightarrow v_A$) to a molecule B ($v_B + 1 \rightarrow v_B$):

$$Q_{v_A+1,v_A}^{v_B,v_B+1} = (v_A+1)(v_B+1)Q_{10}^{01}(AB)\exp\left(-\left|\delta_B v_B - \delta_A v_A + \delta_A p\right|\right)\exp(\delta_A p). \quad (3.80)$$

Here $Q_{10}^{01}(AB)$ is the probability of a quantum transfer from A to B for the lowest levels; parameters δ_A and δ_B are related to the process adiabaticity and can be found based on Eqs. (3.75) and (3.76), taking for each molecule a separate coefficient of anharmonicity: x_{eA} and x_{eB}. Parameter $p > 0$ is the vibrational level of the oscillator A, corresponding to the exact resonant transition A(p + 1 → p) – B(0 → 1) of a quantum from molecule A to B:

$$p = \frac{\hbar(\omega_A - \omega_B)}{2x_{eA}\hbar\omega_A}. \quad (3.81)$$

The product $Q_{10}^{01}(AB)\exp(\delta_A p)$ in Eq. (3.80) corresponds to the resonant exchange and does not include the adiabatic smallness, that is $Q_{10}^{01}(AB)\exp(\delta_A p) \approx (\Omega\tau_{col})^2$, where the factor $(\Omega\tau_{col})^2$ is about 10^{-2}–10^{-4} and characterizes the resonant transition of a quantum (see Section 3.5.1). Then the Eq. (3.80) for the one-quantum VV′-exchange can be rewritten for simplicity as:

$$Q_{v_A+1,v_A}^{v_B,v_B+1} = (v_A+1)(v_B+1)(\Omega\tau_{col})^2\exp\left(-\left|\delta_B v_B - \delta_A v_A + \delta_A p\right|\right). \quad (3.82)$$

One can derive from Eq. (3.82) a formula for calculating the non-resonant transition of the only quantum from a molecule A to B:

$$Q_{10}^{01}(AB) = (\Omega\tau_{col})^2\exp(-\delta p). \quad (3.83)$$

For numerical calculations of the non-resonant, one-quantum VV′-exchange between a nitrogen molecule with a similar one, D.J. Rapp (1965) proposed the following semi-empirical formula, corresponding to Eq. (3.83):

$$Q_{10}^{01}(AB) = 3.7 * 10^{-6} T_0 \; ch^{-2}\left(0.174\Delta\omega\hbar / \sqrt{T_0}\right) \quad (3.84)$$

In this relation, the defect of resonance $\hbar\Delta\omega$ should be expressed in cm^{-1}, and temperature T_0 in Kelvin degrees. Equations (3.83) and (3.84) obviously demonstrate, that the probability of VV′-exchange decreases with the growth of the defect of resonance $\hbar\Delta\omega$. Here it should be noted, that when the defect of resonance is quite large $\hbar\Delta\omega \approx \hbar\omega$, different multi-quantum resonant

VV′-exchange processes are able to become of importance. The probability of such multi-quantum resonant VV′-exchange processes can be calculated by the following formula (E.E. Nikitin, 1970):

$$Q_{n,n-s}^{m,m+r} = \frac{1}{r!s!}\frac{n!(m+r)!}{(n-s)!m!} Q_{s0}^{0r}. \tag{3.85}$$

This formula correlates to Eq. (3.72) in the case of transfer of the only quantum r = s = 1. To estimate the probability Q_{s0}^{0r} one should take into account that this transition in the harmonic approximation is due to a term of expansion of the intermolecular interaction potential including coordinates of two oscillators in powers "r" and "s". Each power of these coordinates in the interaction potential corresponds to a small factor $\Omega\tau_{col}$ in the expression for probability. As a result:

$$Q_{s0}^{0r} \propto (\Omega\tau_{col})^{r+s} \quad \text{and} \quad Q_{n,n-s}^{m,m+r} = \frac{1}{r!s!}\frac{n!(m+r)!}{(n-s)!m!} (\Omega\tau_{col})^{r+s}. \tag{3.86}$$

TABLE 3.17
Rate Coefficients of Non-resonant VV′-relaxation, T_0 = 300 K

VV′-Exchange Process	k_{VV}, cm³/sec	VV′-Exchange Process	k_{VV}, cm³/s
H_2(v = 0) + HF(v = 1) → H_2(v = 1) + HF(v = 0)	$3*10^{-12}$	O_2(v = 0) + CO(v = 1) → O_2(v = 1) + CO(v = 0)	$4*10^{-13}$
DCl(v = 0) + N_2(v = 1) → DCl(v = 1) + N_2(v = 0)	$5*10^{-14}$	N_2(v = 0) + CO(v = 1) → N_2(v = 1) + CO(v = 0)	10^{-15}
N_2(v = 0) + HCl(v = 1) → N_2(v = 1) + HCl(v = 0)	$3*10^{-14}$	N_2(v = 0) + HF(v = 1) → N_2(v = 1) + HF(v = 0)	$5*10^{-15}$
NO (v = 0) + CO(v = 1) → NO(v = 1) + CO(v = 0)	$3*10^{-14}$	O_2(v = 0) + COl(v = 1) → O_2(v = 1) + CO(v = 0)	10^{-16}
DCl(v = 0) + CO(v = 1)→ DCl(v = 1) + CO(v = 0)	10^{-13}	CO(v = 0) + D_2(v = 1) → CO(v = 1) + D_2(v = 0)	$3*10^{-14}$
CO(v = 0) + H_2(v = 1) → CO(v = 1) + H_2(v = 0)	10^{-16}	NO(v = 0) + CO(v = 1) → NO(v = 1) + CO(v = 0)	10^{-14}
HCl(v = 0) + CO(v = 1)→ HCl(v = 1) + CO(v = 0)	10^{-12}	HBr(v = 0) + CO(v = 1) → HBr(v = 1) + CO(v = 0)	10^{-13}
HJ(v = 0) + CO(v = 1)→ HJ(v = 1) + CO(v = 0)	10^{-14}	HBr(v = 0) + HCl(v = 1) → HBr(v = 1) + HCl(v = 0)	10^{-12}
HJ(v = 0) + HCl(v = 1)→ HJ(v = 1) + HCl(v = 0)	$2*10^{-13}$	DCl(v = 0) + HCl(v = 1) → DCl(v = 1) + HCl(v = 0)	10^{-13}
O_2(v = 0) + CO_2(01^10)→ O_2(v = 1) + CO_2(00^00)	$3*10^{-15}$	H_2O(000) + N_2(v = 1) → H_2O(010) + N_2(v = 0)	10^{-15}
CO_2(00^00) + N_2(v = 1)→ CO_2(00^01) + N_2(v = 0)	10^{-12}	CO(v = 1) + CH_4 → CO(v = 1) + CH_4*	10^{-14}
CO(v = 1) + CF_4 → CO(v = 1) + CF_4*	$2*10^{-16}$	CO(v = 1) + SF_6 → CO(v = 1) + SF_6*	10^{-15}
CO(v = 1) + SO_2 → CO(v = 1) + SO_2*	10^{-15}	CO(v = 1) + CO_2 → CO(v = 1) + CO_2*	$3*10^{-13}$
CO_2(00^00) + HCl(v = 1)→ CO_2(00^01) + HCl(v = 0)	$3*10^{-13}$	CO_2(00^00) + HF(v = 1)→ CO_2(00^01) + HF(v = 0)	10^{-12}
CO_2(00^00) + DF(v = 1)→ CO_2(00^01) + DF(v = 0)	$3*10^{-13}$	CO(v = 0) + CO_2(00^01)→ CO(v = 1) + CO_2(00^00)	$3*10^{-15}$
H_2O* + CO_2→ H_2O + CO_2*	10^{-12}	O_2(v = 0) + CO_2(10^00)→O_2(v = 1) + CO_2(00^00)	10^{-13}
CS_2(00^00) + CO(v = 1)→ CS_2(00^01) + CO(v = 0)	$3*10^{-13}$	N_2O(00^00) + CO(v = 1)→ N_2O(00^01) + CO(v = 0)	$3*10^{-12}$

This formula for multi-quantum resonant VV′-exchange obviously correlates with that for multi-quantum resonant VV-exchange Eq. (3.73), if we take r = s and m = 0. Rate coefficients of some non-resonant VV′-exchange processes at room temperature are given in Table 3.17.

3.5.4 VV-Exchange of Polyatomic Molecules

Vibrational modes of a polyatomic molecule can be considered separately at a low level of excitation, (see Eqs. (3.12) and (3.14)), and their VV-exchange can be calculated using the same formulas as those described above (see Table 3.17). For higher levels of excitation, corresponding to the vibrational quasi-continuum of polyatomic molecules (see Section 3.2.4 and Eq. (3.16)), the mean square of vibrational energy transferred in VV-exchange $\left\langle \Delta E_{VV}^2 \right\rangle$ can be found similarly to Eq. (3.67) for VT relaxation (V.D. Rusanov, A. Fridman, G. Sholin, 1986):

$$< \Delta E_{VV}^2 >_{12} = \iint_{0,\infty} \left(\Omega\tau_{col}\right)_{VV}^2 I_1(\omega_1) I_2(\omega_2) \exp\left(-\frac{|\omega_1-\omega_2|}{\alpha v}\right)(\hbar\omega_1)^2 \, d\omega_1 d\omega_2. \tag{3.87}$$

In this relation, I(ω) is the vibrational Fourier spectrum of a polyatomic molecule, which can be described by the Lorentz profile Eq. (3.21), indices 1 and 2 are respectively the quantum transferring and quantum accepting molecules. Consider the VV-exchange within one type of vibrations between a molecule from the discrete spectrum region (in the first excited state, frequency ω_0) and a molecule in the quasi-continuum. Taking the vibrational spectra of the two molecules as Eq. (3.18) and Eq. (3.21), and an average factor $(\Omega\tau_{col})^2$ for the chosen mode of vibrations, we can rewrite Eq. (3.87) as:

$$\left\langle \Delta E_{VV}^2 \right\rangle = \left(\Omega\tau_{col}\right)_{VV}^2 (\hbar\omega_0)^2 \int_0^\infty \frac{1}{\pi}\frac{n\delta}{(\omega-\omega_n)^2+\delta^2} \exp\left(-\frac{|\omega_0-\omega|}{\alpha v}\right) d\omega. \tag{3.88}$$

Here "n" and ω_n are the number of quanta and corresponding frequency value Eqs. (3.19), (3.20). If both colliding partners are in the discrete vibrational spectrum region, which corresponds to $\delta \to 0$, then the VV-exchange is obviously non-resonant. Exponential smallness of the transferred energy Eq. (3.88) is then due to the adiabatic factor $\exp\left(-\frac{|\omega_0-\omega|}{\alpha v}\right)$. If one of the molecules is in quasi-continuum, the line width δ exceeds usually the anharmonic shift $|\omega_0 - \omega|$ (A. Makarov, A. Puretzky, V. Tyakht, 1980), and as seen from Eq. (3.88), the VV-exchange is resonant:

$$\left\langle \Delta E_{VV}^2 \right\rangle = \left(\Omega\tau_{col}\right)_{VV}^2 n \frac{\alpha v}{\delta+\alpha v} (\hbar\omega_0)^2. \tag{3.89}$$

In contrast to diatomic molecules (see Eqs. (3.74) and (3.78)), the VV-exchange with a polyatomic molecule in the quasi-continuum is resonant and the corresponding rate coefficient does not decrease with a growth in the excitation level. Now we can compare the rate coefficients of the VV-exchange (with a molecule in the first excited state) and VT relaxation in the same manner as was done for diatomic molecules Eq. (3.78). If the non-resonant term prevails in VT-relaxation Eq. (3.69), the ratio of the rate coefficients for polyatomic molecules can be expressed as:

$$\xi(n) = \frac{\left\langle \Delta E_{VT}^2 \right\rangle(n)}{\left\langle \Delta E_{VV}^2 \right\rangle(n)} \approx \frac{\left(\Omega\tau_{col}\right)_{VT}^2}{\left(\Omega\tau_{col}\right)_{VV}^2} \exp\left(-\frac{\omega_n - \delta(n)}{\alpha v}\right). \tag{3.90}$$

The VV-exchange proceeds faster than the VT-relaxation $\xi(n) << 1$ at low levels of excitation (like the case of diatomic molecules). Intermode exchange frequency δ (n) is growing with "n" faster than $\Delta\omega = \omega_0 - \omega_n$. For this reason $\xi(n)$ reaches unity and the VT-relaxation catches up with the VV-exchange at lower levels of vibrational excitation than in the case of diatomic molecules Eq. (3.79). It is much more difficult to sustain the high population of upper vibrational levels of polyatomic molecules in quasi-continuum, than those of diatomic molecules. VV-quantum transfer from a weakly excited diatomic molecule to a highly excited one results in the transfer of the anharmonic defect of energy $\hbar\omega_0 - \hbar\omega_n$ into translational degrees of freedom. This effect, which is so obvious in the case of diatomic molecules, also takes place in the case of polyatomic molecules in the quasi-continuum. The average value of vibrational energy $<\Delta E_T>$ transferred into translational degrees of freedom in the process of VV-quantum transfer from a molecule in the first (discrete) excited state (frequency ω_0) to a molecule in quasi-continuum within the framework of our approach can be found as:

$$\langle \Delta E_T \rangle = \frac{\left(\Omega\tau_{col}\right)_{VV}^2}{\pi} \iint\limits_{0,\infty} \frac{n\delta(n)}{\left(\omega' - \omega_n\right)^2 + \delta^2(n)} \exp\left(-\frac{|\omega' - \omega''|}{\alpha v}\right) \times \delta\left(\omega'' - \omega_0\right)\hbar\left(\omega'' - \omega'\right) d\omega' d\omega'', \tag{3.91}$$

In the quasi-continuum under consideration, the principal frequency is ω_n, with the effective number of quanta "n". After integrating Eq. (3.91) and taking into account that $\delta > |\omega_n - \omega_0|$, the expression for vibrational energy $<\Delta E_T>$ transferred into translational degrees of freedom during the VV-exchange can be written as:

$$\langle \Delta E_T \rangle = n\left(\Omega\tau_{col}\right)_{VV}^2 \frac{\alpha v}{\alpha v + \delta}\left(\hbar\omega_0 - \hbar\omega_n\right). \tag{3.92}$$

Comparison of Eqs. (3.92) and (3.89) shows that each act of the VV-exchange between the weakly excited diatomic molecule and a highly excited one in quasi-continuum is actually non-resonant and leads to the transfer of the anharmonic defect of energy $\hbar\omega_0 - \hbar\omega_n$ into translational motion. This effect can be interpreted in the quasi-classical approximation as conservation in the collision process of the adiabatic invariant- shortened action, which is the ratio of vibrational energy of each oscillator to its frequency (L.D. Landau, E.M. Lifshitz, 1973):

$$\frac{E_{v1}}{\omega_1} + \frac{E_{v2}}{\omega_2} = const. \tag{3.93}$$

The conservation of the adiabatic invariant corresponds in quantum mechanics to the conservation of the number of quanta during a collision. As it is clear from Eq. (3.93), the transfer of vibrational energy E_v from a higher frequency oscillator to a lower frequency one results in a decrease of the total vibrational energy and hence losses of energy into translational motion Eq. (3.92). The ratio of rate coefficients of the direct (considered above) and reverse processes of the VV-relaxation energy transfer to highly excited molecules can be found based on the relation (3.92) and the principle of detailed balance:

$$\frac{k_{VV}^{+}\left(T_v, T_0\right)}{k_{VV}^{-}\left(T_v, T_0\right)} = \left(1 + \frac{\hbar\omega_0}{E_v}\right)^{s-1} \exp\left(-\frac{\hbar\omega_0}{T_v} + \frac{\hbar\left(\omega_0 - \omega_n\right)}{T_0}\right). \tag{3.94}$$

In this relation, "s" is the effective number of vibrational degrees of freedom of the molecule; T_0 and T_v are the translational and vibrational temperatures. The ratio Eq. (3.94) is indicative of the possibility of super-equilibrium populations of the highly vibrationally excited states at $T_0 < T_v$, which is typical for non-equilibrium discharges in molecular gases.

3.6 PROCESSES OF ROTATIONAL AND ELECTRONIC RELAXATION OF EXCITED MOLECULES

3.6.1 Rotational Relaxation

Both rotational-rotational (RR) and rotational-translational (RT) energy transfer (relaxation) are usually non-adiabatic, because, for not highly excited molecules, rotational quanta are generally small and hence the Massey parameter is also small. As a result, the collision of a rotator with an atom or another rotator can be considered as a free, classical collision accompanied by a neccessary energy transfer. This means that the probability of RT and RR relaxation is actually very high. A formula for calculating the number of collisions necessary for the "total" energy exchange between rotational and translational degrees of freedom (RT-relaxation) was proposed by J.G. Parker (1959) as a function of translational temperature T_0. The Parker formula was later corrected by C.A. Bray, R.M. Jonkman (1970):

$$Z_{rot} = \frac{Z_{rot}^{\infty}}{1+\left(\pi/2\right)^{3/2}\sqrt{\frac{T*}{T_0}}+\left(2+\pi^2/4\right)\frac{T*}{T_0}}. \tag{3.95}$$

In this relation, T* is the depth (energy) of the potential well corresponding to inter-molecular attraction; $Z_{rot}^{\infty} = Z_{rot}\left(T_0 \to \infty\right)$ is the number of collisions necessary for RT-relaxation at very high temperatures. The increase of the collision number Z_{rot} with temperature in the Parker formula becomes clear noting that the inter-molecular attraction accelerates the RT energy exchange. The attraction effect obviously becomes less effective when the translational temperature increases, which explains Eq. (3.6. 1). Numerical values of the collision numbers, which are necessary for the RT relaxation, are given in Table 3.18 for room temperature and for high temperatures $Z_{rot}^{\infty} = Z_{rot}\left(T_0 \to \infty\right)$. These two parameters, presented in Table 3.18, allow determining the energy depth T* of a potential well corresponding to the inter-molecular attraction in the Parker formula. It actually helps to calculate the number of collisions (and hence the rotational relaxation rate coefficient) at any temperature using Eq. (3.95). The factor Z_{rot} is usually small, at room temperature, it is approximately 3–10 for most of the molecules. Exception takes place in the case of hydrogen and deuterium, where Z_{rot} is approximately 200–500. This is due to high values of rotational quanta and relatively high Massey parameters, which make this process slightly adiabatic. The above discussion

TABLE 3.18
Number of Collisions Z_{rot} Necessary for RT-relaxation in One-component Gases

Molecule	$Z_{rot}^{\infty}\left(T_0 \to \infty\right)$	$Z_{rot}(T_0 = 300\ K)$	Molecule	$Z_{rot}^{\infty}\left(T_0 \to \infty\right)$	$Z_{rot}(T_0 = 300\ K)$
Cl_2	47.1	4.9	N_2	15.7	4.0
O_2	14.4	3.45	H_2	–	~500
D_2	–	~200	CH_4	–	15
CD_4	–	12	$C(CH_3)_4$	–	7
SF_6	–	7	CCl_4	–	6
CF_4	–	6	$SiBr_4$	–	5
SiH_4	–	28			

was focused on the RT-relaxation at relatively low levels of rotational excitation when the rotational quantum is very small. We should take into account, however, that the rotational quantum 2B(J + 1) grows with the level of rotational excitation and rotational quantum number (see Section 3.2.5). As a result, the RT-relaxation of the highly rotationally excited molecules can become much more adiabatic and hence much slower than was discussed above and shown in Table 3.18.

In general, RT relaxation (and furthermore RR relaxation) is a fast process usually requiring only several collisions. In most systems under consideration, the rates of the rotational relaxation processes are comparable with the rate of thermalization. Thermalization or TT-relaxation is a process sustaining equilibrium inside of translational degrees of freedom. Hence under most of the conditions taking place even in non-equilibrium discharges, the rotational degrees of freedom can be considered in quasi-equilibrium with the translational degrees of freedom and can be characterized by the same temperature T_0.

3.6.2 Relaxation of Electronically Excited Atoms and Molecules

Electronically excited atoms and molecules have many different channels of relaxation. For example, super-elastic collisions and energy transfer from electronically excited particles back to plasma electrons can be here of great importance. Also, the spontaneous radiation on the permitted transitions makes here much more important contribution, than in the case of vibrational excitation (because of much higher transition frequencies). These processes will be considered later in the book in the chapters related to plasma chemical kinetics. In this section, we are going only to discuss the fundamental basics of the relaxation of electronically excited atoms and molecules in collisions with other heavy neutral particles. The relaxation of electronic excitation, which is usually related to the transfer of several-electron volts into translational degrees of freedom, is a strongly adiabatic process with very high Massey parameters ($\omega\tau_{col} \sim 100$–$1000$). In this case, the relaxation is very slow, with low probability and low values of cross section. Thus, for example, the relaxation process:

$$\mathrm{Na}\left(3^2\mathrm{P}\right) + \mathrm{Ar} \rightarrow \mathrm{Na} + \mathrm{Ar} \tag{3.96}$$

has the cross section not exceeding 10^{-19} cm^2, which corresponds to probability 10^{-9} or less. Sometimes, however, the relaxation processes similar to (3.96) with very high values of the Massey parameters can have pretty high probabilities. Thus relaxation of electronically excited oxygen atoms $O(^1D)$ on heavy atoms of noble gases requires only several collisions. Such an effect can be explained taking into account, that the Massey parameter includes transition frequency in a quasi-molecule formed during a collision. This frequency can be lower than that one directly calculated from energy of the electronically excited atom.

Electronically excited atoms and molecules obviously can transfer their energy not only into translational but also into vibrational and rotational degrees of freedom. These processes can proceed with a higher probability because of the lower energy losses from the internal degrees of freedom and hence smaller Massey parameters. Even stronger effects can be achieved by formation of intermediate ionic complexes. For example, the relaxation probability is close to unity in the case of electronic energy transfer from excited metal atoms Me* to vibrational excitation of nitrogen molecules:

$$\mathrm{Me}* + \mathrm{N}_2\left(\mathrm{v}=0\right) \rightarrow \mathrm{Me}^+\mathrm{N}_2^- \rightarrow \mathrm{Me} + \mathrm{N}_2\left(\mathrm{v}>0\right). \tag{3.97}$$

In this case initial and final energy terms of the interaction Me - N_2 cross the ionic one (see illustration of the process on Figure 3.20), which actually provides the transition between them (A. Bjerre, E.E. Nikitin, 1967). Massey parameters of the transitions near crossings of the terms are low, which makes the relaxation processes non-adiabatic and very fast. The maximum cross sections of such processes of relaxation of excited sodium atoms are given in Table 3.19.

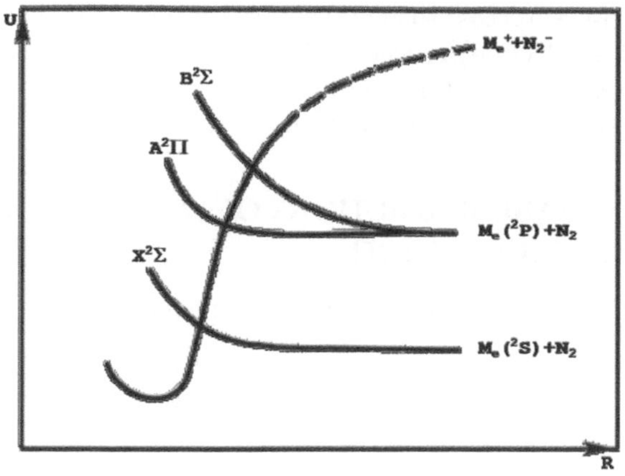

FIGURE 3.20 Illustration of electronic energy relaxation through formation of intermediate ionic complexes.

TABLE 3.19
Cross Sections of Non-adiabatic Relaxation of Electronically Excited Sodium Atoms

Relaxation Processes	Cross Sections, cm^2
$Na^* + N_2(v=0) \rightarrow Na + N_2^*(v>0)$	2 ∗ 10^{-14} cm^2
$Na^* + CO_2(v=0) \rightarrow Na + CO_2^*(v>0)$	10^{-14} cm^2
$Na^* + Br_2(v=0) \rightarrow Na + Br_2^*(v>0)$	10^{-13} cm^2

3.6.3 The Electronic Excitation Energy Transfer Processes

These processes can be effective only very close to resonance, within about 0.1 eV or less. The He-Ne gas laser provides practically important examples of such kind of highly resonant processes:

$$He(^1S) + Ne \rightarrow He + Ne(5s), \tag{3.98}$$

$$He(^3S) + Ne \rightarrow He + Ne(4s). \tag{3.99}$$

These processes lead to a population inversion for the 4 s and 5 s levels of neon atoms. Subsequently, this effect results in coherent radiation from the He-Ne laser. Interaction radius for collisions of electronically excited atoms and molecules is usually large (up to 1 nm). For this reason, the values of cross sections of the electronic excitation transfer obviously can be very large, if these processes are quite close to the resonance. For example, such resonance electronic excitation transfer process from mercury atoms to atoms of sodium has a large value of cross section reaching 10^{-14} cm^2. The electronic excitation transfer in collision of heavy particles could also take place as a fast non-adiabatic transfer inside of the intermediate quasi-molecule formed during the collision. However, this fast non-adiabatic transfer requires, a crossing of the corresponding potential energy surfaces. Such a crossing of terms becomes possible in inter-atoms collisions only at a very close distance between nuclei, which in traditional plasma systems corresponds to non-realistic interaction energies about 10 keV. When the degree of ionization in plasma exceeds values of about 10^{-6}, the electronic excitation transfer can occur faster by interaction with the electron gas than in collisions on heavy particles. Such excitation transfer proceeds through the super-elastic collisions of the excited heavy particles with electrons (deactivation) followed by their excitation through electron impact.

3.7 ELEMENTARY CHEMICAL REACTIONS OF EXCITED MOLECULES, FRIDMAN-MACHERET α-MODEL

3.7.1 Rate Coefficient of the Reactions of Excited Molecules

Formula for calculation of the reaction rate coefficient of an excited molecule with vibrational energy E_v at gas temperature T_0 can be derived from the theoretical-informational approach (R.D. Levine, R. Bernstein, 1978):

$$k_R(E_v, T_0) = k_{R0} \exp\left(-\frac{E_a - \alpha E_v}{T_0}\right)\theta(E_a - \alpha E_v). \tag{3.100}$$

In this relation E_a is the Arrhenius activation energy; coefficient "α"- is the efficiency of vibrational energy in overcoming of the activation energy barrier; k_{R0} is the pre-exponential factor of the reaction rate coefficient; $\theta(x - x_0)$ is the Heaviside function ($\theta(x - x_0) = 1$, when $x > 0$; and $\theta(x - x_0) = 0$ when $x < 0$). According to Eq. (3.100), the reaction rates of vibrationally excited molecules follow the traditional Arrhenius law. The activation energy in the Arrhenius law is the value of vibrational energy taken with efficiency "α". If the vibrational temperature exceeds the translational one $T_v >> T_0$ and the chemical reaction is mostly determined by the vibrationally excited molecules, then Eq. (3.100) can be simplified:

$$k_R(E_v) = k_{R0}\theta(\alpha v - E_a). \tag{3.101}$$

The effective energy barrier of a chemical reaction (or effective activation energy) in this case is equal to E_a/α. The pre-exponential factor can be calculated in the framework of the transition state theory (E.E. Nikitin, 1970), but quite often can be just taken as the gas-kinetic rate coefficient $k_{R0} \approx k_0$. Equation (3.100) describes the rate coefficients of reactions of vibrationally excited molecules after averaging over the Maxwellian distribution function for translational degrees of freedom. The probability of these reactions without the averaging can be found based on the Le Roy formula (R.L. Le Roy, 1969). The Le Roy formula gives the probability $P_v(E_v, E_t)$ of the elementary reaction as a function of vibrational E_v and translational E_t energies of the colliding partners:

$$P_v(E_v, E_t) = 0, \quad if \quad E_t < E_a - \alpha E_v,\ E_a \geq \alpha E_v, \tag{3.102}$$

$$P_v(E_v, E_t) = 1 - \frac{E_a - \alpha E_v}{E_t}, \quad if \quad E_t > E_a - \alpha E_v,\ E_a \geq \alpha E_v, \tag{3.103}$$

$$P_v(E_v, E_t) = 1, \quad if \quad E_a < \alpha E_v, \quad any\ E_t. \tag{3.104}$$

Averaging of $P_v(E_v, E_t)$ over the Maxwellian distribution of the translational energies E_t obviously gives Eq. (3.100), which is actually the most important relation for calculation of elementary reaction of excited particles. Kinetics of chemical reactions of excited molecules mostly depends on the activation energy E_a and the efficiency of vibrational energy α, which will be discussed next.

3.7.2 Potential Barriers to Elementary Chemical Reactions, Activation Energy

Extensive information on chemical kinetic data, including values of the activation energies, was collected in particular by D.L. Baulch (1992–1994) and V.N. Kondratiev (1971). Today, probably, the easiest access to this information is on the web through the NIST Chemical Kinetics Database. However, all these databases are usually not sufficient for describing the detailed mechanism of complicated plasma chemical reactions. In these cases, the activation energies can be found theoretically from the potential energy surfaces, calculated using such sophisticated quantum mechanical methods as LEPS (London-Eyring-Polanyi-Sato method, presented by S.J. Sato, 1955) or DCM (method of diatomic complexes in molecules, H.Eyring, S.H. Lin, S.M. Lin, 1980). Though an example of such a detailed quantum mechanical calculations will be presented later in this section, several simple semi-empirical methods can be successfully applied to find the value of activation energy. The most convenient of them are presented below:

- **The Polanyi-Semenov Rule.** According to this rule (N.N. Semenov, 1958), the activation energies for a group of similar exothermic chemical reactions can be found as:

$$E_a = \beta + \alpha \Delta H. \tag{3.105}$$

In this formula ΔH is a reaction enthalpy (negative for the exothermal reactions), α and β are constants of the model. According to N.N. Semenov, these parameters are equal to α = 0.25–0.27 and β = 11.5 kcal/mole for the following large group of exchange reactions:

$$\begin{aligned}
&H + RH \rightarrow H_2 + R,\\
&D + RH \rightarrow DH + R,\\
&H + RCHO \rightarrow H_2 + RCO,\\
&H + RCl \rightarrow HCl + R,\\
&H + RBr \rightarrow HBr + R;\\
&Na + RCl \rightarrow NaCl + R,\\
&Na + RBr \rightarrow NaBr + R;\\
&CH_3 + RH \rightarrow CH_4 + R,\\
&CH_3 + RCl \rightarrow CH_3Cl + R,\\
&CH_3 + RBr \rightarrow CH_3Br + R;\\
&OH + RH \rightarrow H_2O + R.
\end{aligned}$$

Endothermic chemical reactions where the reaction enthalpy ΔH is positive, can be considered as a reverse process with respect to the exothermic reactions. Hence Eq. (3.105) can be rewritten for them with the same coefficients α and β (N.Tihomirova, V.V. Voevodsky, 1949):

$$E_a = \Delta H + \left(\beta + \alpha\left(-\Delta H\right)\right) = \beta + \left(1 - \alpha\right)\Delta H. \tag{3.106}$$

- **Kagija Method.** This method was developed to find the activation energy of exchange reactions A + BC → AB + C, based on the dissociation energy D of the molecule BC, the reaction enthalpy ΔH and the one semi-empirical parameter "γ" (Kagija, 1969). The relation for activation energy, in this case, is similar to that of the Polanyi-Semenov rule:

$$E_a = \frac{D}{\left(2\gamma D - 1\right)^2} + \frac{2\gamma D\left(2\gamma^2 D^2 - 3\gamma D + 2\right)}{\left(2\gamma D - 1\right)^3}\Delta H. \tag{3.107}$$

The Kagija parameter γ is fixed for the groups of similar reactions:

- for the reactions of the detachment of hydrogen atoms by alkyl radicals γ = 0.019;
- for the reactions of the detachment of hydrogen atoms by halogen atoms γ = 0.025;
- for the reactions of the detachment of halogen atoms from alkyl-halides the parameter γ grows from γ = 0.019 for chlorides to γ = 0.03 for iodides.
- **Sabo Method.** According to this approach, the activation energy can be found based on the sum of bond energies of the initial molecules $\sum D_i$ and the final molecules $\sum D_f$ with one semi-

empirical parameter "a" (Z.G. Sabo, 1966):

$$E_a = \sum D_i - a \sum D_f. \tag{3.108}$$

For exothermic substitution exchange reactions, the Sabo parameter a = 0.83; for endothermic substitution exchange reactions a = 0.96; for reactions of disproportioning a = 0.60 and for exchange reactions of inversion a = 0.84.

- **Alfassi-Benson Method.** This approach is related to the exchange reactions R + AR′→ AR + R′ and claims that the activation of the process depends mainly upon the sum A_e of electron affinities to the particles R and R′ and obviously on the reaction enthalpy ΔH (Z.B. Alfassi, S.W. Benson, 1973). Such analysis of the different reaction of atoms H, Na, O and radicals OH, CH_3, OCH_3, CF_3 results in the following semi-empirical formula:

$$E_a = \frac{14.8 - 3.64 * A_e}{1 - 0.025 * \Delta H}. \tag{3.109}$$

A_e, ΔH, and E_a are in kcal/mole. Alfassi and Benson also proposed several other semi-empirical formulas for calculation of activation energies, based on electron affinities of the radicals R and R′.

3.7.3 The Efficiency "α" of Vibrational Energy in Overcoming the Activation Energy Barrier

The coefficient α is the key parameter in Eq. (3.100), and hence the key parameter describing the influence of excited molecules in plasma on their chemical reaction rates. Database of these important parameters of non-equilibrium kinetics is however relatively limited because methods of their determination are quite sophisticated (some of them are reviewed in particular by A.A. Levitsky and S.O. Macheret, 1983). Numerical values of the coefficient α for different chemical reactions obtained experimentally as well as by detailed modeling applying the method of classical trajectories, are presented in Table 3.20. More detailed database concerning the α-coefficients can be found in the book of V.D. Rusanov and A. Fridman (1984). Information presented in Table 3.20 permits dividing chemical reactions into several classes (endothermic processes, exothermic processes, simple exchange, double exchange, etc.) with the most probable value of the coefficients α in each class. Such a classification (A. Levitsky, S. Macheret, A. Fridman, 1983b) can be expressed in form of a Table 3.22.

3.7.4 The Fridman-Macheret α-Model

This model permits calculating the α-coefficient of the efficiency of vibrational energy in elementary chemical processes based mostly on information about the activation energies of the corresponding direct and reverse reactions (S. Macheret, V.D. Rusanov, A.Fridman, 1984). This model describes the exchange reaction A + BC → AB + C, with the reaction path profile shown in Figure 3.21a,b. Vibration of the molecule BC can be taken into account using the approximation of the vibronic term (E.A. Andreev, E.E. Nikitin, 1976; D.Secrest, 1973); this is also shown in the figure by a dash-dotted line. The energy profile, corresponding to the reaction of a vibrationally excited molecule with energy E_v (the vibronic term), can be obtained in this approach by a parallel shift of the initial profile A + BC up on the value of E_v. The part of the reaction path profile corresponding to products AB + C remains the same, if the products are not excited. As seen from Figure 3.21b, the effective decrease of the activation energy related to the vibrational excitation E_v is equal to:

$$\Delta E_a = E_v \frac{F_{A+BC}}{F_{A+BC} + F_{AB+C}}. \tag{3.110}$$

TABLE 3.20
The Efficiency α of Vibrational Energy in Overcoming the Activation Energy Barrier (α_{exp} – the Coefficient Obtained Experimentally or in Detailed Modeling, α_{MF} – Coefficient Found from the Fridman-Macheret α-model)

Reaction	α_{exp}	α_{MF}	Reaction	α_{exp}	α_{MF}
$F + HF^* \rightarrow F_2 + H$	0.98	0.98	$F + DF^* \rightarrow F_2 + D$	0.99	0.98
$Cl + HCl^* \rightarrow H + Cl_2$	0.95	0.96	$Cl + DCl^* \rightarrow D + Cl_2$	0.99	0.96
$Br + HBr^* \rightarrow H + Br_2$	1.0	0.98	$F + HCl^* \rightarrow H + ClF$	0.99	0.96
$Cl + HF^* \rightarrow ClF + H$	1.0	0.98	$Br + HF^* \rightarrow BrF + H$	1.0	0.98
$J + HF^* \rightarrow JF + H$	1.0	0.98	$Br + HCl^* \rightarrow BrCl + H$	0.98	0.98
$Cl + HBr^* \rightarrow BrCl + H$	1.0	0.97	$J + HCl^* \rightarrow JCl + H$	1.0	0.98
$SCl + HCl^* \rightarrow SCl_2 + H$	0.96	0.98	$S_2Cl + HCl^* \rightarrow S_2Cl_2 + H$	0.98	0.97
$SOCl + HCl^* \rightarrow SOCl_2 + H$	0.95	0.96	$SO_2Cl + HCl^* \rightarrow SO_2Cl_2 + H$	1.0	0.96
$NO + HCl^* \rightarrow NOCl + H$	1.0	0.98	$FO + HF^* \rightarrow F_2O + H$	1.0	0.98
$O_2 + OH^* \rightarrow O_3 + H$	1.0	1.0	$NO + OH^* \rightarrow NO_2 + H$	1.0	1.0
$ClO + OH^* \rightarrow ClO_2 + H$	1.0	1.0	$CrO_2Cl + HCl^* \rightarrow CrO_2Cl_2 + H$	0.94	0.9
$PBr_2 + HBr^* \rightarrow PBr_3 + H$	1.0	0.97	$SF_5 + HBr^* \rightarrow SF_5Br + H$	1.0	0.98
$SF_3 + HF^* \rightarrow SF_4 + H$	0.89	0.98	$SF_4 + HF^* \rightarrow SF_5 + H$	0.97	0.99
$H + HF^* \rightarrow H_2 + F$	1.0	0.95	$D + HF^* \rightarrow HD + F$	1.0	0.95
$Cl + HF^* \rightarrow HCl + F$	0.96	0.97	$Br + HF^* \rightarrow HBr + F$	1.0	0.98
$J + HF^* \rightarrow HJ + F$	0.99	0.99	$Br + HCl^* \rightarrow HBr + Cl$	1.0	0.95
$J + HCl^* \rightarrow HJ + Cl$	1.0	0.98	$J + HBr^* \rightarrow HJ + Br$	1.0	0.96
$OH + HF^* \rightarrow H_2O + F$	1.0	0.95	$HS + HF^* \rightarrow H_2S + F$	1.0	0.98
$HS + HCl^* \rightarrow H_2S + Cl$	1.0	1.0	$HO_2 + HF^* \rightarrow H_2O_2 + F$	1.0	0.98
$HO_2 + HF^* \rightarrow H_2O_2 + F$	1.0	0.98	$NH_2 + HF^* \rightarrow NH_3 + F$	1.0	0.97
$SiH_3 + HF^* \rightarrow SiH_4 + F$	1.0	1.0	$GeH_3 + HF^* \rightarrow GeH_4 + F$	1.0	1.0
$N_2H_3 + HF^* \rightarrow N_2H_4 + F$	0.97	0.98	$CH_3 + HF^* \rightarrow CH_4 + F$	0.98	0.97
$CH_2F + HF^* \rightarrow CH_3F + F$	1.0	0.97	$CH_2Cl + HF^* \rightarrow CH_3Cl + F$	1.0	0.97
$CCl_3 + HF^* \rightarrow CHCl_3 + F$	0.98	0.98	$CH_2Br + HF \rightarrow CH_3Br + F$	1.0	0.97
$C_2H_5 + HF^* \rightarrow C_2H_6 + F$	1.0	0.99	$CH_2CF_3 + HF^* \rightarrow CH_3CF_3 + F$	1.0	0.98
$C_2H_5O + HF^* \rightarrow (CH_3)_2O + F$	1.0	1.0	$C_2H_5Hg + HF^* \rightarrow (CH_3)_2Hg + F$	0.99	0.97
$HCO + HF^* \rightarrow H2CO + F$	1.0	0.99	$FCO + HF^* \rightarrow HFCO + F$	1.0	0.99
$C_2H_3 + HF^* \rightarrow C_2H_4 + F$	0.99	0.97	$C_3H_5 + HF^* \rightarrow C_3H_6 + F$	1.0	0.98
$C_6H_5 + HF^* \rightarrow C_6H_6 + F$	1.0	0.97	$CH_3 + JF^* \rightarrow CH_3J + F$	0.81	0.9
$S + CO^* \rightarrow CS + O$	1.0	0.99	$F + CO^* \rightarrow CF + O$	1.0	1.0
$CS + CO^* \rightarrow CS_2 + O$	0.83	0.9	$CS + SO^* \rightarrow CS_2 + O$	0.90	0.95
$SO + CO^* \rightarrow COS + O$	0.96	0.92	$F_2 + CO^* \rightarrow CF_2 + O$	1.0	1.0
$C_2H_4 + CO^* \rightarrow (CH_2)_2\,C + O$	0.94	0.98	$CH_2 + CO^* \rightarrow C_2H_2 + O$	0.90	0.94
$H_2 + CO^* \rightarrow CH_2 + O$	0.96	–	$OH + OH^* \rightarrow H_2O + O(^1D_2)$	1.0	0.97
$NO + NO^* \rightarrow N_2O + O(^1D_2)$	1.0	–	$CH_3 + OH^* \rightarrow CH_4 + O(^1D_2)$	1.0	1.0
$O + NO^* \rightarrow O2 + N$	0.94	0.86	$H + BaF^* \rightarrow HF + Ba$	0.99	1.0
$O + AlO^* \rightarrow Al + O_2$	0.67	0.7	$O + BaO^* \rightarrow Ba + O_2$	1.0	0.99
$CO + HF^* \rightarrow CHF + O$	1.0	1.0	$CO_2 + HF^* \rightarrow O_2 + CHF$	1.0	–
$CO^* + HF \rightarrow CHF + O$	1.0	1.0	$CF_2O + HF^* \rightarrow CHF_3 + O(^1D_2)$	1.0	1.0
$H_2CO + HF^* \rightarrow CH_3F + O(^1D_2)$	1.0	1.0	$J + HJ^* \rightarrow J_2 + H$	1.0	1.0
$O + N_2 \rightarrow NO + N$ (non-adiabatic)	1.0	1.0	$O + N_2 \rightarrow NO + N$ (adiabatic)	0.6	1.0
$F + ClF^* \rightarrow F_2 + Cl$	1.0	0.93	$O + (CO^+)^* \rightarrow O_2 + C^+$	0.9	1.0
$H + HCl^* \rightarrow H_2 + Cl$	0.3	0.4	$NO + O_2^* \rightarrow NO_2 + O$	0.9	1.0
$O + H_2^* \rightarrow OH + H$	0.3	0.5	$O + HCl^* \rightarrow OH + Cl$	0.60	0.54
$H + HCl^* \rightarrow HCl + H$	0.3	0.5	$H + H_2^* \rightarrow H_2 + H$	0.4	0.5
$NO + O_3^* \rightarrow NO_2(^3B_2) + O_2$	0.5	0.3	$SO + O_3^* \rightarrow SO_2(^1B_1) + O_2$	0.25	0.15
$O^+ + N_2^* \rightarrow NO^+ + N$	0.1	-	$N + O_2^* \rightarrow NO + O$	0.24	0.19
$OH + H_2^* \rightarrow H_2O + H$	0.24	0.22	$F + HCl^* \rightarrow HF + Cl$	0.4	0.1
$H + N_2O^* \rightarrow OH + N_2$	0.4	0.2	$O_3 + OH^* \rightarrow O_2 + O + OH$	0.02	0
$H_2 + OH^* \rightarrow H_2O + H$	0.03	0	$Cl_2 + NO^* \rightarrow ClNO + Cl$	0	0
$O_3 + NO^* \rightarrow NO_2(^2A_1) + O_2$	0	0	$O_3 + OH^* \rightarrow HO_2 + O_2$	0.02	0

In this relation, F_{A+BC} and F_{AB+C} denote the characteristic slopes of the terms A + BC and AB + C presented on the Figure 3.20b. If the energy terms exponentially depend on the reaction coordinates with the decreasing parameters γ_1 and γ_2 (reverse radii of the corresponding exchange forces), then:

$$\frac{F_{A+BC}}{F_{AB+C}} = \frac{\gamma_1 E_a^{(1)}}{\gamma_2 E_a^{(2)}} \tag{3.111}$$

The subscripts "1" and "2" stand here for direct (A + BC) and reverse (AB + C) reactions. The decrease of activation energy Eq. (3.110) together with Eq. (3.111) not only explains the main kinetic relation Eq. (3.100) for reactions of vibrationally excited molecules, but also determines the value of the coefficient α (which is actually equal to $\alpha = \Delta E_a/E_v$):

$$\alpha = \frac{\gamma_1 E_a^{(1)}}{\gamma_1 E_a^{(1)} + \gamma_2 E_a^{(2)}} = \frac{E_a^{(1)}}{E_a^{(1)} + \gamma_1/\gamma_2 \, E_a^{(2)}}. \tag{3.112}$$

Usually the exchange force parameters for direct and reverse reactions γ_1 and γ_2 are close to each other and their ratio is close to unity $\gamma_1/\gamma_2 \approx 1$, which leads to the approximate but very convenient main formula of the Fridman-Macheret α-model:

$$\alpha \approx \frac{E_a^{(1)}}{E_a^{(1)} + E_a^{(2)}}. \tag{3.113}$$

The geometric derivation of this formula is very similar to the corresponding derivation of the Polanyi-Semenov rule Eq. (3.105). Equation (3.113) is in good numerical agreement with the experimental data (see Table 3.20) and reflects the three most important tendencies of the α-coefficient:

- the efficiency α of the vibrational energy is the highest, that is close to unity (100%), for strongly endothermic reactions with activation energies close to the reaction enthalpy;
- the efficiency α of the vibrational energy is the lowest, that is close to zero, for exothermic reactions without activation energies;
- sum of the α-coefficients for direct and reverse reactions is equal to unity $\alpha^{(1)} + \alpha^{(2)} = 1$.

Eq. (3.113) does not include any detailed information on the dynamics of the elementary reaction and type of excitation. For this reason, it can be applied to a wide range of chemical reactions of excited species.

3.7.5 Efficiency of Vibrational Energy in Elementary Reactions Proceeding Through Intermediate Complexes

As an example, such reactions consider the synthesis of lithium hydride:

$$Li\left({}^2S\right) + H_2^*\left({}^1\Sigma_g^+, v\right) \rightarrow LiH\left({}^1\Sigma^+\right) + H\left({}^2S\right). \tag{3.114}$$

The minimum of the potential energy of the system Li-H_2 corresponds to the group of symmetry C_{2v}. The potential energy surface for the reaction with the C_{2v} configuration was calculated using the method of diatomic complexes in molecules DCM (A.Polishchuk, V.D. Rusanov, A.Fridman, 1980) and it is presented in Figure 3.22. The corresponding reaction path profile is shown in Figure 3.23, the potential well energy depth is about 0.4 eV. The potential well makes the lithium hydride synthesis proceed through an intermediate complex, which can be described using the statistical theory of

TABLE 3.21
Classification of Chemical Reactions for Determination of the Vibrational Energy Efficiency Coefficients α

Reaction Type	Simple Exchange Bond break in excited molecule	Simple Exchange Bond break in not-excited molecule Through complex	Simple Exchange Bond break in not-excited molecule Direct	Double Exchange
Endothermic	0.9–1.0	0.8	<0.04	0.5–0.9
Exothermic	0.2–0.4	0.2	0	0.1–0.3
Thermo-neutral	0.3–0.6	0.3	0	0.3–0.5

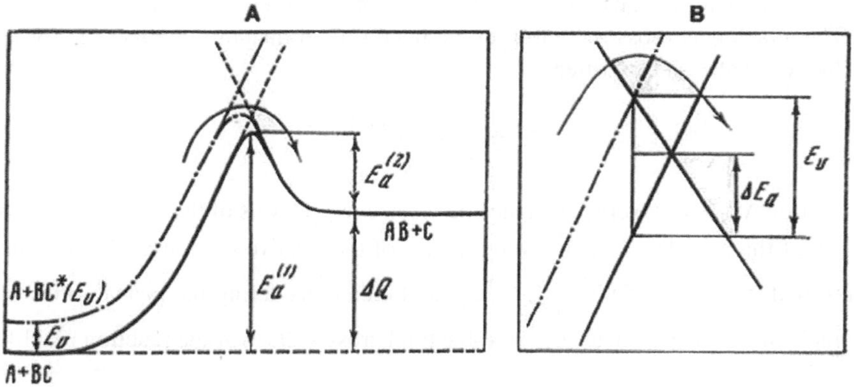

FIGURE 3.21 Efficiency of vibrational energy in a simple exchange reaction $A + BC \rightarrow AB + C$: Solid curve – reaction profile; dashed line – a vibronic term, corresponding to A interaction with a vibration excited molecule $BC^*(E_v)$; Part of the reaction profile near the barrier summit.

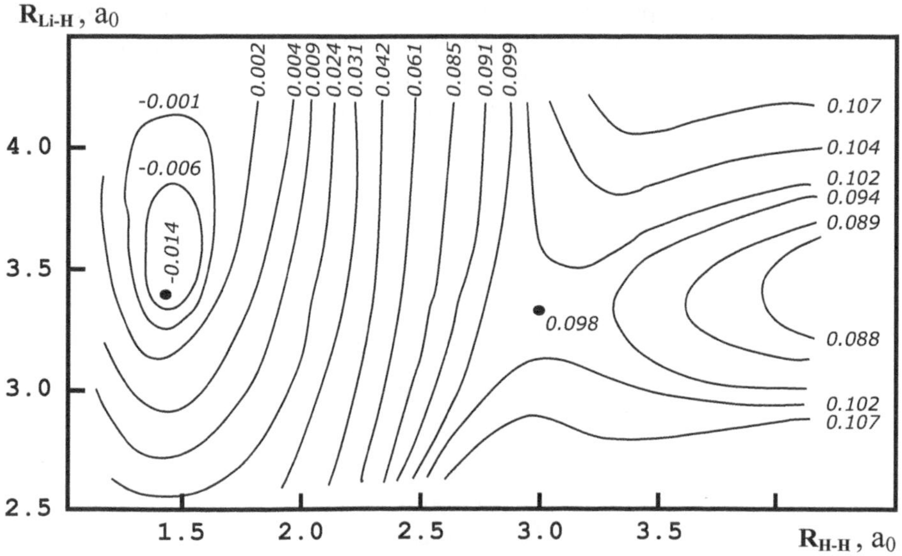

FIGURE 3.22 Li-H_2 Potential energy surface, energy and distances in atomic units.

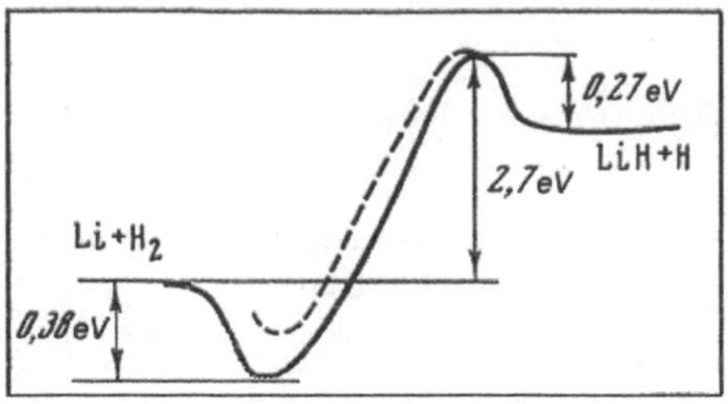

FIGURE 3.23 Reaction profile $Li + H_2 \rightarrow LiH + H$, dashed line illustrates the effect of angular momentum conservation.

chemical processes (A.G. Engelgardt, A.V. Felps, G. G. Risk, 1964). The total cross section of the reaction (3.114) is expressed in the framework of this theory as a product of the cross section of the intermediate complex formation σ_a and the probability of its subsequent decay with the formation of LiH + H. The attachment cross section σ_a at the fixed orbital quantum number "l" for the relative motion of the reactants can be written as:

$$\sigma_a^l = \frac{2\pi}{k^2}(2l+1)\frac{\hbar\Gamma}{\Delta E_k} \tag{3.115}$$

In this relation ΔE_k is the average distance between energy levels of the complex; $\hbar\Gamma$- is the level energy width of the complex; k- is the wave vector of the relative motion of the reactants. Taking into account that in Eq. (3.115) $\hbar\Gamma/\Delta E_k = 1/2\pi$, and then considering the probabilities of different decay channels of the intermediate complex, the total cross section of the reaction (3.114) is:

$$\sigma_R(E) = \sum_l \sigma_a^l \frac{k_1(E,l)}{k_1(E,l)+k_{-1}(E,l)} = \sum_l \frac{\pi}{k^2}(2l+1)A_\omega\left(\frac{E-E_1^a}{E}\right)^{s-1}\theta\left(E-E_1^a\right). \tag{3.116}$$

Here $E_1^a = 2.7$eV is the activation energy of Eq. (3.114); k_1 and k_{-1} are the rate coefficients of the direct and reverse channels of decay of the intermediate complex; s- is the number of vibrational degrees of freedom of the complex; $A_\omega \approx 3$ is the frequency factor; $\theta(E - E_1^a)$ is the Heaviside function ($\theta(x - x_0) = 1$, when $x > 0$; and $\theta(x - x_0) = 0$, when $x < 0$). If E_v, E_r, and E_t are respectively - the vibrational energy of the hydrogen molecule, rotational and translational energies of the reacting particles, then the total effective energy of the intermediate complex E, calculated with respect to the initial level of potential energy, can be found as:

$$E = E_v + \alpha_r E_r + \alpha_t E_t. \tag{3.117}$$

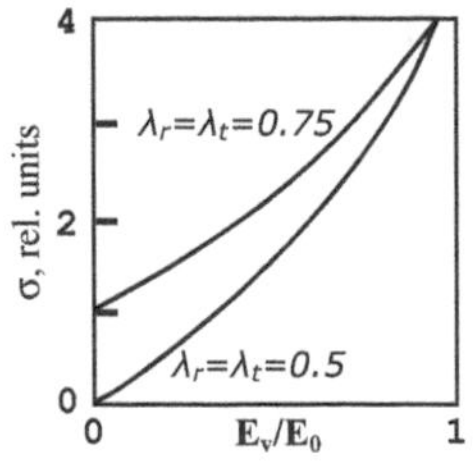

FIGURE 3.24 Contribution of different degrees of freedom in cross-section of the reaction.

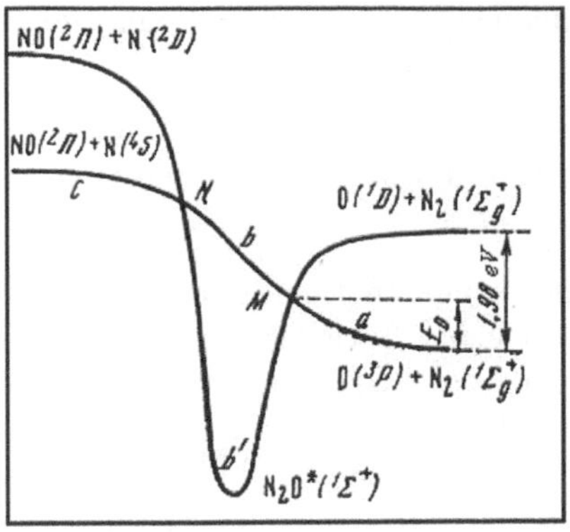

FIGURE 3.25 Reaction profile: $O + N_2 \rightarrow NO + N$.

The efficiencies α_r and α_t of the rotational and translational energies in overcoming the activation energy barrier are less than 100% and can be estimated as:

$$\alpha_r \approx \alpha_t \approx 1 - \frac{r_e^2}{R_a^2}. \tag{3.118}$$

The lower efficiencies of the rotational and translational energies are due to conservation of angular momentum. It results in the fact, that part of the rotational and translational energies remains in rotation at the last moment of decay of the complex (see Figure 3.23) when the characteristic distance between products is equal to R_a. In Eq. (3.118) - r_e is the equilibrium distance between hydrogen atoms in a hydrogen molecule H_2. cross section of the reaction (3.114) is shown on Figure 3.24 as a function of fraction E_v/E_0 of vibrational energy E_v in the total energy E_0, assuming the fixed value of the total energy $E_0 = 2E_1^a = \text{const}$. As one can see from the figure, the vibrational energy of hydrogen is more efficient, than rotational and translational energies in stimulating the typical reaction (3.114), proceeding through the formation of the intermediate complex. According to Eqs. (3.116), (3.117), the efficiency of the vibrational energy in such reactions is close to 100%. Using this fact it is possible to use the general formula Eq. (3.100) with the efficiency coefficient $\alpha \approx 1$ for stimulation the endothermic reactions going through complexes by vibrational excitation of reagents. This was also confirmed in detailed consideration of the non-adiabatic Zeldovich reaction of NO synthesis:

$$O\left({}^3P\right) + N_2^*\left({}^1\Sigma_g^+\right) \Leftrightarrow N_2O*\left({}^1\Sigma^+\right) \rightarrow NO\left({}^2P\right) + N\left({}^4S\right) \tag{3.119}$$

which is proceeding through the formation of the intermediate complex (see Figure 3.25) and can be effectively stimulated ($\alpha = 1$) in plasma by vibrational excitation of nitrogen molecules (V.D. Rusanov, A.Fridman, G. V. Sholin, 1978, see also Table 3.20). In this case, one should take into account that the pre-exponential factor k_{R0} in the relation Eq. (3.100) for such reactions is not constant, but is proportional to the statistical theory factor (E.E. Nikitin, 1970):

$$k_{R0} \approx k_{R00} P_{L-Z} \left(\frac{E_v - E_a}{E_v} \right)^{s-1}. \tag{3.120}$$

The statistical factor in parenthesis (see also Eq. (3.116)) reflects the probability of localization of vibrational energy on the chemical bond to be broken; "s" is a number of degrees of freedom, the Landau-Zener factor P_{LZ} shows the probabilities of transitions between electronic terms during formation and decay of the intermediate complex-molecule (for the non-adiabatic reactions, see Figure 3.25).

3.7.6 Dissociation of Molecules Stimulated by Vibrational Excitation in Non-Equilibrium Plasma

Dissociation of diatomic and polyatomic molecules in non-equilibrium plasma can be effectively stimulated by their vibrational excitation. Such dissociation of diatomic molecules has a trivial elementary mechanism (M. Capitelli, 1986) and it is mostly controlled by vibrational kinetics. A good example of the polyatomic molecule dissociation stimulated by vibrational excitation is the process of CO_2-decomposition in non-thermal discharges (V.K. Givotov, V.D. Rusanov, A. Fridman, 1982, 1984). hereto consider this elementary process it is interesting to analyze the scheme of electronic terms of CO_2, presented in Figure 3.26. The adiabatic dissociation of CO_2 (with spin conservation) in step-by-step vibrational excitation:

$$CO_2 * \left({}^1\Sigma^+ \right) \rightarrow CO\left({}^1\Sigma^+ \right) + O\left({}^1D \right) \tag{3.121}$$

leads to the formation of an oxygen atom in the electronically excited state and requires more than 7 eV of energy. The non-adiabatic transition ${}^1\Sigma^+ \rightarrow {}^3B_2$ in the crossing point of the terms (change of spin during the transition) provides the more effective dissociation stimulated by the step-by-step vibrational excitation:

$$CO_2 * \left({}^1\Sigma^+ \right) \rightarrow CO_2 * \left({}^3B_2 \right) \rightarrow CO\left({}^1\Sigma^+ \right) + O\left({}^3D \right). \tag{3.122}$$

This process results in the formation of an oxygen atom in the electronically ground state 3^P and requires spending exactly OC=O bond energy, which is equal to 5.5 eV. The dissociation processes have high coefficients of efficiency of vibrational energy utilization in overcoming the activation energy barriers (the coefficient α = 1 in relation Eq. (3.100)). The pre-exponential factors for the dissociation processes stimulated by vibrational excitation can be also described in frameworks of

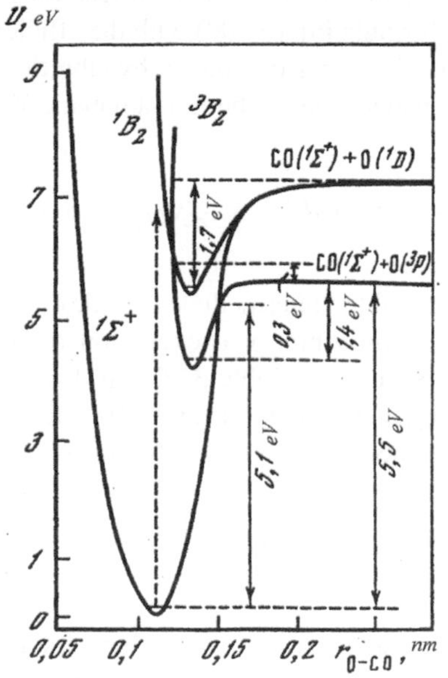

FIGURE. 3.26 Low electronics terms of CO_2.

the statistical theory of the monomolecular reactions (E.E. Nikitin, 1970) or simply by using the relation Eq. (3.120).

3.7.7 Dissociation of Molecules in Non-Equilibrium Conditions with Essential Contribution of Translational Energy

Vibrational temperature in most non-thermal discharges exceeds translational one $T_v > T_0$. Some phenomena such as dissociation of molecules after a shock wavefront, however, are characterized by $T_v < T_0$. In this case, both vibrational and translational temperatures make contributions to stimulation of dissociation. For such cases it is convenient to describe the stimulation using the non-equilibrium factor Z (S.A. Losev, A.L. Sergievska, Rusanov e.a, 1996):

$$Z(T_0,T_V) = \frac{k_R(T_0,T_v)}{k^0(T_0,T_0 = T_v)} = Z_h(T_0,T_v) + Z_l(T_0,T_v). \tag{3.123}$$

In this formula $k_R(T_0, T_v)$ is the dissociation rate coefficient; $k_R^0(T_0, T_v = T_o)$ is the corresponding equilibrium rate coefficient for temperature T_0; the non-equilibrium factors Z_h and Z_l are related to dissociation from high "h" and low "l" vibrational levels. The non-equilibrium factor Z can be found using the **Macheret-Fridman model** (S.O. Macheret, A.Fridman, I.V. Adamovich, J.W. Rich, C.E. Treanor). The model is based on the assumption of classical impulsive collisions. Dissociation is considered to occur mainly through an optimum configuration, which is a set of collisional parameters minimizing the energy barrier. This optimum configuration defines the threshold kinetic energy for dissociation and through it the exponential factor of the rate coefficient. The probability of finding the colliding system near the optimum configuration determines the pre-exponential factor of the rate. In the frameworks of the Macheret-Fridman model, the non-equilibrium factor for dissociation from the high vibrational levels is:

$$Z_h(T_0,T_v) = \left\{ \frac{1-\exp\left(-\frac{\hbar\omega}{T_v}\right)}{1-\exp\left(-\frac{\hbar\omega}{T_0}\right)} (1-L) \right\} \exp\left[-D\left(\frac{1}{T_v} - \frac{1}{T_0}\right)\right], \tag{3.124}$$

where D is the dissociation energy and L is the pre-exponential factor related to the configuration of collisions. For dissociation from the low vibrational levels, which is applicable in this case, the model gives:

$$Z_l(T_0,T_v) = L\exp\left[-D\left(\frac{1}{T_{eff}} - \frac{1}{T_0}\right)\right], \tag{3.125}$$

where the effective temperature is introduced by the following relation (m- is mass of an atom in dissociating molecule, M- is mass of an atom in other colliding partners):

$$T_{eff} = \alpha T_v + (1-\alpha)T_0, \quad \alpha = \left(\frac{m}{M+m}\right)^2. \tag{3.126}$$

In contrast to the general formula Eq. (3.100), showing the efficiency of vibrational energy, the relation Eq. (3.126) shows both efficiency of vibrational energy and efficiency of translational energy in dissociating molecules under non-equilibrium conditions. Experimental values of the non-equilibrium factor Z are presented in Figure 3.27 together with the theoretical predictions. It is interesting to point out, that the non-equilibrium factor $Z(T_0)$ at $T_0 > T_v$ taken as a function of

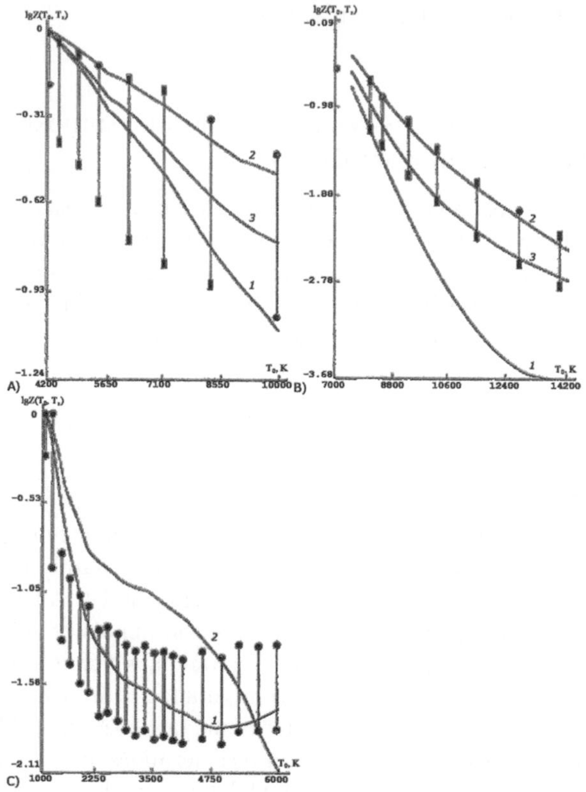

FIGURE. 3.27 Experimental and theoretical values of the non-equilibrium factor Z in oxygen (A), nitrogen (B) and iodine (C): 1) the Macheret-Fridman model; 2) the Kuznetsov model; 3) the Losev model.

temperature T_0 corresponds to a curve with a minimum. When at first T_0 is still close to T_v, the dissociation is mostly due to contributions from high vibrational levels. While the translational temperature is growing in this case, the relative population of the high vibrational levels is decreasing; this results in a reduction of the non-equilibrium factor Z. When the translational temperature becomes very high, the dissociation becomes "direct", for example, mostly due to the strong collisions with vibrationally not-excited molecules. In this case, the non-equilibrium factor Z starts growing. Thus the minimum point of the function $Z(T_0)$ corresponds actually to the transition to direct dissociation mainly from vibrationally non-excited states (see Figure 3.26). Several semi-empirical models were also proposed to make simple estimations of the non-equilibrium dissociation at $T_0 > T_v$. Let us examine some of the most used and compare them with the Macheret-Fridman model.

The Park Model (C.Park, 1987). This model because of its simplicity is widely used in practical calculations. According to this model, the dissociation rate coefficient can be found using the conventional Arrhenius formula for equilibrium reactions, by just replacing the temperature by a new effective value:

$$T_{eff} = T_0^s T_v^{1-s}. \tag{3.127}$$

Here the power "s" is a fitting parameter of the Park model, which was recommended by the author to be taken as s = 0.7 (in first publications, however, the effective temperature was taken approximately as $T_{eff} = \sqrt{T_0 T_v}$). The Macheret-Fridman model permits to find theoretically the Park parameter as:

$$s = 1 - \alpha = 1 - \left(\frac{m}{m+M}\right)^2. \tag{3.128}$$

If two atomic masses are not very different m ~ M, then the parameter s ≈ 0.75 according to (3.128), which is pretty close to the value s = 0.7, recommended by Park.

Losev Model (S.A. Losev, N.A. Generalov, 1961). According to this model, the dissociation rate coefficient can be found using the Arrhenius formula for equilibrium reactions with vibrational temperature T_v and effective value D_{eff} of dissociation energy. To find D_{eff} the actual dissociation energy D should be decreased by the value of translational temperature taken with efficiency β:

$$D_{eff} = D - \beta T_0. \tag{3.129}$$

This relation is similar to Eq. (3.100), with the vibrational and translational energies replacing each other. The coefficient β is the fitting parameter of the Losev model, and for estimations, it should be taken as β ≈ 1.0–1.5. The Macheret-Fridman model permits theoretically finding the Losev parameter as:

$$\beta \approx \frac{D(T_0 - T_v)}{T_0^2}, \quad T_0 > T_v \tag{3.130}$$

As is clear from Eq. (3.130), the Losev parameter β can be considered as a constant only for a narrow range of temperatures.

The Marrone-Treanor Model (P.V. Marrone, C.E. Treanor, 1963). This model supposes the exponential distribution (with the parameter -U) of the probabilities of dissociation from different vibrational levels. If $U \to \infty$, the probabilities of dissociation from all vibrational levels are equal. The non-equilibrium factor Z can be found in the framework of this semi-empirical model as:

$$Z(T_0, T_v) = \frac{Q(T_0)Q(T_f)}{Q(T_v)Q(-U)}, \quad T_0 > T_v. \tag{3.131}$$

The statistical factors $Q(T_j)$ can be defined for the four "temperatures": translational temperature T_0, vibrational temperature T_v, and two special temperatures -U and $T_f = \dfrac{1}{\frac{1}{T_v} - \frac{1}{T_0} - \frac{1}{-U}}$, all denoted below in general as T_j, using the following relation:

$$Q(T_j) = \frac{1 - \exp\left(-\frac{D}{T_j}\right)}{1 - \exp\left(-\frac{\hbar\omega}{T_j}\right)}. \tag{3.132}$$

The negative temperature –U is a fitting parameter of the model. The recommended values of this semi-empirical coefficient U are $0.6T_0 - 3T_0$. The Macheret-Fridman model permits theoretically finding the Marrone-Treanor semi-empirical parameter as:

$$U = \frac{T_0\left[\alpha T_0 + (1-\alpha)T_v\right]}{(T_0 - T_v)(1-\alpha)}. \tag{3.133}$$

More detailed information about the two-temperature kinetics of dissociation for $T_0 > T_v$ can be found in the book of A.L. Sergievska, E.A. Kovach and S.A. Losev, 1995).

3.7.8 Chemical Reactions of Two Vibrationally Excited Molecules in Plasma

The efficiency of vibrational energy in overcoming the activation energy barriers of the strongly endothermic reactions is usually close to 100% (α = 1). It was assumed in this statement, that only one of the reacting molecules keeps the vibrational energy, see Table 3.20. In some plasma chemical processes, two molecules involved in the reaction can be strongly excited vibrationally. An example of such a reaction of two excited molecules is the carbon oxide disproportioning, leading to carbon formation:

$$CO*(v_1)+CO*(v_2)\rightarrow CO_2+C \tag{3.134}$$

The disproportioning is a strongly endothermic process (enthalpy about 5.5 eV) with activation energy 6 eV, which can be stimulated by vibrational energy of both CO-molecules. This elementary reaction was studied by S. Nester, A.V. Demura and A. Fridman (1983) using the vibronic terms approach. According to this approach, the general Eqs. (3.100) and (3.101) also can be applied to the reaction of two excited molecules. In this case, both molecules contribute their energy to decrease the activation energy:

$$\Delta E_a = \alpha_1 E_{v1} + \alpha_2 E_{v2}. \tag{3.135}$$

Here, E_{v1} and E_{v2} are the vibrational energies of the two excited molecules (subscript 1 corresponds to the molecule losing an atom, subscript 2 is related to another one accepting the atom); α_1 and α_2 are coefficients of efficiency of using vibrational energy from these two sources. For the reaction (3.134) it was shown, that $\alpha_1 = 1$and $\alpha_2 \approx 0.2$. Thus vibrational energy of the donating molecule is more efficient in the stimulation of endothermic exchange reactions, than that of the accepting molecule.

3.8 PROBLEMS AND CONCEPT QUESTIONS

3.8.1 Lifetime of Metastable Atoms

Compare the relatively long radiative lifetime of a metastable helium atom He (2s $^3S_1$) with the time interval between collisions of the atom with other neutrals and with a wall of a discharge chamber (radius 3 cm) in a room temperature discharge at a low pressure of 10 Torr. Comment on the result.

3.8.2 The Rydberg Correction Terms in the Bohr Model

Estimate the relative accuracy (in %) of applying the Bohr formula Eq. (3.1) without the Rydberg correction for excited helium atoms with the following quantum numbers: (a) n = 2, l = 0, S = 1, and (b) n = 3, l = 2, S = 0. Explain the large difference in accuracy.

3.8.3 Metastable Atoms and Selection Rules

Using Table 3.1, check the change in parity and quantum numbers S, J and L between excited metastable states and corresponding ground states for different atoms. Compare these changes in parity and quantum numbers with the selection rules for dipole radiation.

3.8.4 Molecular Vibration Frequency

Consider a diatomic homonuclear molecule as a classical oscillator with mass M of each atom. Take interaction between the atoms in accordance with the Morse potential with known parameters. Based on that, find a vibration frequency of the diatomic homonuclear molecule. Compare the frequency with an exact quantum mechanical frequency given by Eq. (3.6).

3.8.5 Anharmonicity of Molecular Vibrations

Equation (3.7) gives a relation between anharmonicity coefficient, vibrational quantum, and the dissociation energy for diatomic molecules. Using this relation and numerical data from Table 3.6, find dissociation energies of molecular nitrogen, oxygen, and hydrogen. Compare the calculated values with actual dissociation energies of these molecules (for $N_2 - D = 9.8$ eV; for $O_2 - D = 5.1$ eV; for $H_2 - D = 4.48$ eV).

3.8.6 Maximum Energy of an Anharmonic Oscillator

Calculate based on Eq. (3.7) the maximum vibrational energy $E_v(v = v_{max})$ of an anharmonic oscillator, corresponding to the maximum vibrational quantum number Eq. (3.9). Compare this energy with the dissociation energy of the diatomic molecule, calculated based on Eq. (3.7) between vibrational quantum and coefficient of anharmonicity. Comment the role of Eq. (3.5) in this case.

3.8.7 Isotopic Effect in Dissociation Energy of Diatomic Molecules

Dissociation of which diatomic molecule-isotope – light one or heavy one – is a kinetically faster process in the thermal quasi-equilibrium discharges. Make your decision based on Eqs. (3.5) and (3.6) for dissociation energy vibration frequency.

3.8.8 Parameter Massey for Transition Between Vibrational Levels

Estimate the Parameter Massey Eq. (2.37) for a process of loss of one vibrational quantum from an excited nitrogen molecule in a collision with another heavy particle at room temperature. Consider the loss of the highest Eq. (3.10) and the lowest $\hbar\omega$ vibrational quantum. Compare and comment on the results.

3.8.9 Anharmonicity of Vibrations of Polyatomic Molecules

Why can the coefficients of anharmonicity be both positive and negative for polyatomic molecules, while for diatomic molecules they are always effectively negative?

3.8.10 Symmetric Modes of CO_2 Vibrations

In Eqs. (3.15) and (3.16), n_{sym} – is the total number of quanta on the symmetric modes (ν_1 and ν_2) of CO_2. It can be presented as, $n_{sym} = 2n_1 + n_2$ taking into account the relation between frequencies of the symmetric valence and deformation vibrations ($\nu_1 \approx 2\nu_2$). Estimate the values of vibrational quantum and anharmonicity for such a combined symmetric vibrational mode.

3.8.11 Transition to the Vibrational Quasi-Continuum

Based on the Eq. (3.16), find out numerically – at which level of vibrational energy CO_2 modes can not be considered any more individually?

3.8.12 Effect of Vibrational Excitation on Rotational Energy

Calculate, based on the Eq. (3.23), the maximum possible relative decrease of the rotational constant B of N_2 molecule due to the effect of vibration.

3.8.13 Cross Section of Vibrational Excitation by Electron Impact

Using the Breit-Wigner formula, estimate the half-width of a resonance pike in the vibrational excitation cross section dependence on electron energy. Discuss the values of the energy half-width for the resonances with different range of lifetime.

3.8.14 Multi-Quantum Vibrational Excitation by Electron Impact

Based on the Fridman's approximation and taking into account known values of the basic excitation rate coefficient $k_{eV}(0,1)$ as well as parameters α and β of the approximation, calculate the total rate coefficient of the multi-quantum vibrational excitation from an initial state v_1 to all other vibrational levels $v_2 > v_1$. Compare the total vibrational excitation by electron impact rate coefficient with the basic one $k_{eV}(0,1)$ for different gases, using Table 3.11.

3.8.15 Rotational Excitation of Molecules by Electron Impact

Compare qualitatively the dependences of rotational excitation cross section on electron energy in the cases of excitation of dipole and quadrupole molecules (Eqs. (3.30) and (3.31)). Discuss the qualitative difference of the cross sections behavior near the process threshold.

3.8.16 Cross Sections of Electronic Excitation by Electron Impact

Find electron energies (in units of ionization energy, $x = \varepsilon/I$), corresponding to the maximum value of cross sections of the electronic excitation of permitted transitions. Use semi-empirical relations of H.W. Drawin Eq. (3.33) and B.M. Smirnov Eq. (3.34), compare the results.

3.8.17 Electronic Excitation Rate Coefficients

Using Table 3.12 for nitrogen molecules, calculate and draw the dependence of rate coefficients of electronic excitation of $A^3\Sigma_u^+$ and ionization in a wide range of reduced electric fields E/n_0 typical for non-thermal discharges. Discuss the difference in behavior of these two curves.

3.8.18 Influence of Vibrational Temperature on Electronic Excitation Rate Coefficients

Calculate the relative increase of electronic excitation rate coefficient of molecular nitrogen in a non-thermal discharge with reduced electric field $E/n_0 = 3*10^{-16}$ Vcm^2, when the vibrational temperature increases from room temperature to $T_v = 3000$ K.

3.8.19 Dissociation of Molecules Through Electronic Excitation by Direct Electron Impact

Explain why the electron energy threshold of the dissociation through electronic excitation almost always exceeds the actual dissociation energy in contrast to dissociation, stimulated by vibrational excitation, where the threshold is equal quite often to the dissociation energy. Use the dissociation of diatomic molecules as an example.

3.8.20 Distribution of Electron Energy Between Different Channels of Excitation and Ionization

Analyzing Figures 3.9 to 3.16, find the molecular gases with the highest possibility to localize selectively discharge energy on vibrational excitation. What is the highest percentage of electron energy, which can be selectively focused on vibrational excitation? Which values of reduced electric field or electron temperatures are optimal to reach this selective regime?

3.8.21 Probability of VT-relaxation in Adiabatic Collisions

Probability of the adiabatic VT-relaxation (for example, in expression Eq. (3.48)) exponentially depends on the relative velocity "v" of colliding partners, which is actually different before and after collision (because of the energy transfer during the collision). Which one from the two translational velocities should be taken to calculate the Massey parameter and relaxation probability?

3.8.22 VT-relaxation Rate Coefficient as a Function of Vibrational Quantum Number

The VT-relaxation rate grows with the number of quantum for two reasons (see for example Eq. (3.53)). First- the matrix element increasing as (n + 1), and second-vibrational frequency decreases because of anharmonicity. Show numerically – which effect is dominating the acceleration of VT-relaxation with the level of vibrational excitation.

3.8.23 Temperature Dependence of VT-relaxation Rate

The vibrational relaxation rate coefficient for adiabatic collisions usually decreases with the reduction of translational gas temperature in good agreement with the Landau-Teller formula. In practice, however, this temperature dependence function has a minimum, usually at about room temperature, and then the relaxation accelerates with further cooling. Explain the phenomena.

3.8.24 Semi-Empirical Relations for VT-relaxation Rate Coefficients

Calculate vibrational relaxation rate coefficients for a couple of molecular gases at room temperature, using the semi-empirical formula Eq. (3.58). Compare your results with data obtained from Table 3.13, estimate the relative accuracy of using the semi-empirical relation.

3.8.25 VT-relaxation in Collision with Atoms and Radicals

How small a concentration of oxygen atoms should be in non-thermal air-plasma at room temperature to neglect the effect of VT-relaxation of vibrationally excited nitrogen molecules in collision with the atoms. Make the same estimation for non-equilibrium discharges in pure oxygen, explain difficulties in the accumulation of essential vibrational energy in such discharges.

3.8.26 Surface Relaxation of Molecular Vibrations

Estimate pressure, at which heterogeneous vibrational relaxation in a room temperature nitrogen discharge becomes dominant with respect to volume relaxation. Take the discharge tube radius about 3 cm, suppose molybdenum glass as the discharge wall material, and estimate the diffusion coefficient of nitrogen molecules in the system by the simplified numerical formula $D \approx \frac{0.3}{p(\text{atm})} \text{cm}^3/\text{sec}$.

3.8.27 Vibrational Relaxation of Polyatomic Molecules

How does the relative contribution of resonant and non-resonant (adiabatic) relaxation of polyatomic molecules depend on the level of their vibrational excitation? Consider cases of excitation of independent vibrational modes and also transition to the quasi-continuum. Use the relation Eq. (3.69) and discussions concerning the transition to the quasi-continuum in section 3.2.4.

3.8.28 The Resonant Multi-Quantum VV-exchange

Estimate the smallness of the double-quantum resonance VV-exchange $A_2^*(v=2)+A_2(v=0)\rightarrow A_2(v=0)+A_2^*(v=2)$ and triple-quantum resonance VV-exchange $A_2^*(v=3)+A_2(v=0)\rightarrow A_2(v=0)+A_2^*(v=3)$ in diatomic molecules, taking the matrix element factor $(\Omega\tau_{col})^2 = 10^{-3}$.

3.8.29 The Resonant VV-relaxation of Anharmonic Oscillator

How does the probability and hence rate coefficient of the resonant VV-exchange in the case of relaxation of identical diatomic molecules $A_2^*(v=n+1)+A_2^*(v=n)\rightarrow A_2^*(v=n)+A_2^*(v=n+1)$ depends on their vibrational quantum number.

3.8.30 Comparison of Adiabatic Factors for Anharmonic VV- and VT-relaxation Processes

As it was shown in the discussion after Eqs. (3.74) and (3.75), the adiabatic factors for anharmonic VV- and VT-relaxation processes are equal in one-component gases $\delta_{VV} = \delta_{VT}$. What is the reason for difference of the adiabatic factors in the case of collisions of non-identical anharmonic diatomic molecules?

3.8.31 VV-exchange of Anharmonic Oscillators, Provided by Dipole Interaction

Why should the long-distance forces usually be taken into account in the VV-relaxation of anharmonic oscillators only for exchange close to resonance?

3.8.32 The Non-Resonant One-Quantum VV′-exchange

Estimate the rate coefficient of the non-resonant VV′-exchange between a nitrogen molecule and another one (presented in Table 3.17) using the semi-empirical Rapp- approximation. Compare the obtained result with experimental data, which can be found in Table 3.17.

3.8.33 VV-relaxation of Polyatomic Molecules

Explain based on Eq. (3.90), why excitation of high vibrational levels of polyatomic molecules in non-thermal discharges is much less effective than similar excitation of diatomic molecules.

3.8.34 Rotational RT-relaxation

Using the Parker formula and data from Table 3.18, calculate the energy depth of the intermolecular attraction potential T* and probability of the RT-relaxation in pure molecular nitrogen at translational gas temperature 400K.

3.8.35 LeRoy Formula and α-Model

Derive the general Eq. (3.100), describing stimulation of chemical reactions by vibrational excitation of molecules, averaging the LeRoy formula over the different values of translational energy E_t. Use the Maxwellian function for distribution of molecules over the translational energies.

3.8.36 Semi-Empirical Methods of Determination of Activation Energies

Estimate activation energy of at least one reaction from the group CH3 + RH → CH4 + R using a) the Polanyi-Semenov rule, b) the Kagija method, and c) the Sabo method. Compare the results and comment on them; which method looks more reliable in this case?

3.8.37 Efficiency of Vibrationally Excited Molecules in Stimulation of Endothermic and Exothermic Reactions

Using the potential energy surfaces of exchange reactions A + BC → AB + C, illustrate the conclusion of the Fridman-Macheret α-model, stating that vibrational excitation of molecules is more efficient in the stimulation of endothermic (than exothermic) reactions.

3.8.38 Accuracy of the Fridman-Macheret α-Model

Using Table 3.20 of the α-efficiency coefficients, determine the typical values of the relative accuracy of application the Fridman-Macheret α-model for a) endothermic, b) exothermic and c) thermo-neutral reactions. Which of these three groups of chemical reactions of vibrationally excited molecules can be described by this approach more accurately and why?

3.8.39 Contribution of Translational Energy in Dissociation of Molecules under Non-Equilibrium Conditions

Using the expression for the non-equilibrium factor Z of the Fridman-Macheret model, derive the formula for dissociation of molecules at a translational temperature much exceeding the vibrational one.

3.8.40 Efficiency of Translational Energy in Elementary Endothermic Reactions

According to the Eq. (3.118) and to common sense (conservation of momentum and angular momentum), efficiency of translational energy in endothermic reactions should be less than 100%. It means, that the effective dissociation energy in this case should exceed the real one. On the other hand, the Fridman-Macheret model permits the dissociation when total translational energy is equal to D. Explain the "contradiction".

3.8.41 The Park Model

Is it possible to apply the Park model for the semi-empirical description of dissociation of diatomic molecules in non-equilibrium plasma, when $T_v > T_0$. Analyze the effect of vibrational and translational temperatures in this case.

4 Plasma Statistics and Kinetics of Charged Particles

4.1 STATISTICS AND THERMODYNAMICS OF EQUILIBRIUM AND NON-EQUILIBRIUM PLASMAS, THE BOLTZMANN, SAHA, AND TREANOR DISTRIBUTIONS

4.1.1 Statistical Distribution of Particles over Different States. The Boltzmann Distribution

Consider an isolated system with total energy E, consisting of a big number N of particles in different states (i). Number "n_i" of particles is in the state "i", defined by a set of quantum numbers and energies E_i:

$$N = \sum_i n_i, \quad E = \sum_i E_i n_i. \tag{4.1}$$

Collisions of the particles lead to the continuous change of populations n_i, but the total group follows the conservation laws Eq. (4.1) and the system is in thermodynamic equilibrium. The objective of the statistical approach in this case is to find the distribution function of particles over the different states "i" without details of probabilities of transitions between the states. Such statistical function can be derived in general, because the probability to find n_i particles in the state "i" is proportional to the number of ways in which this distribution can be arranged. The so-called **thermodynamic probability** $W(n_1, n_2,\ldots, n_i,\ldots)$ is the probability to have n_1 particles in the state "1", n_2 particles in the state "2" etc. It can be found taking into account, that N particles can be arranged in N! different ways, but - because n_i particles have the same energy - this number should be divided by the relevant factor to exclude repetitions:

$$W\left(n_1,n_2,\ldots,n_i,\ldots\right) = A\frac{N!}{N_1!N_2!\ldots N_i!\ldots} = A\frac{N!}{\prod_i n_i!}, \tag{4.2}$$

where A – is a factor of normalization. Let us find the most probable numbers of particles $\overline{n_i}$, when the probability W Eq. (4.2) as well as its logarithm:

$$\ln W\left(n_1,n_2,\ldots,n_i,\ldots\right) = \ln\left(AN!\right) - \sum_i \ln n_i! \approx \ln\left(AN!\right) - \sum_i \int_0^{n_i} \ln x dx \tag{4.3}$$

have a maximum. Achievement of the maximum of the function lnW of many variables at a point, where n_i is equal to the most probable numbers of particles $\overline{n_i}$ requires:

$$0 = \sum_i \left(\frac{\partial \ln W}{\partial n_i}\right)_{n_i=\overline{n_i}} dn_i = \sum_i \ln \overline{n_i}\, dn_i. \tag{4.4}$$

Differentiation of the conservation laws Eq. (4.1) gives:

$$\sum_i dn_i = 0, \quad \sum_i E_i dn_i = 0. \tag{4.5}$$

Multiplying the two equations (4.5) respectively by parameters –lnC and 1/T and then adding the Eq. (4.4), yields:

$$\sum_i \left(\ln \overline{n_i} - \ln C + \frac{E_i}{T} \right) dn_i = 0. \tag{4.6}$$

This sum Eq. (4.6) is supposed to be equal to zero at any independent values of dn_i. This is possible only if the expression is equal to zero, which brings us to the **Boltzmann distribution function:**

$$\overline{n_i} = C \exp\left(-\frac{E_i}{T} \right), \tag{4.7}$$

where C is the normalizing parameter related to the total number of particles, and T- is the statistical temperature of the system, related to the average energy per one particle. As always in this book, the same units are used for energy and temperature. The expression Eq. (4.7) was derived for the case when the subscript "i" was related to the only state of a particle. If the state is degenerated, we should add into Eq. (4.7) the statistical weight "g", showing the number of states with the given quantum number:

$$\overline{n_j} = C\, g_j \exp\left(-\frac{E_j}{T} \right). \tag{4.8}$$

In this case, the subscript "j" corresponds to the group of states with the statistical weight g_j and energy E_j. The Boltzmann distribution can be expressed then in terms of the number densities N_j and N_0 of particles in j-states and the ground state (0):

$$N_j = N_0 \frac{g_j}{g_0} \exp\left(-\frac{E_j}{T} \right). \tag{4.9}$$

In this relation, g_j and g_0 are the statistical weights of the "j" and ground states. The general Boltzmann distribution Eqs. (4.7), (4.9) can be applied to derive many specific distribution functions, such as the Maxwell-Boltzmann (or just the Maxwellian) distribution Eq. (2.1).

4.1.2 The Equilibrium Statistical Distribution of Diatomic Molecules Over Vibrational-Rotational States

Vibrational levels of diatomic molecules are not degenerated and their energy with respect to vibrationally ground state is $E_v = \hbar\omega v$, when the vibrational quantum numbers are low (see Eq. (3.6)). Then according to Eq. (4.9) the number density of molecules with "v" vibrational quanta is:

$$N_v = N_0 \exp\left(-\frac{\hbar\omega v}{T} \right). \tag{4.10}$$

The total number density of molecules N is a sum of densities in different vibrational states:

$$N = \sum_{v=0}^{\infty} N_v = \frac{N_0}{1 - \exp\left(-\frac{\hbar\omega}{T} \right)}. \tag{4.11}$$

Based on the relation Eq. (4.11), the distribution of molecules over the different vibrational levels can be renormalized in a traditional way with respect to the total number density N of molecules:

$$N_v = N\left[1-\exp\left(-\frac{\hbar\omega}{T}\right)\right]\exp\left(-\frac{\hbar\omega v}{T}\right). \quad (4.12)$$

Now find the Boltzmann vibrational–rotational distribution $N_{vJ}\left(\sum_J N_{vJ} = N_v\right)$. We should take into account Eq. (3.22) for rotational energy (B << T), and the statistical weight for a rotational state with quantum number J, which is equal to 2J + 1. Thus based on the general relation Eq. (4.9):

$$N_{vJ} = N\frac{B}{T}(2J+1)\left[\left(1-\exp\left(-\frac{\hbar\omega}{T}\right)\right)\right]\exp\left(-\frac{\hbar\omega v + BJ(J+1)}{T}\right). \quad (4.13)$$

4.1.3 The Saha Equation for Ionization Equilibrium in Thermal Plasma

The Boltzmann distribution Eq. (4.9) determines the ratio of number of electrons and ions $\overline{n_e} = \overline{n_i}$ to number of atoms $\overline{n_a}$ in a ground state, or in other words to describe the ionization equilibrium $A^+ + e \Leftrightarrow A$ in plasma:

$$\frac{\overline{n_i}}{\overline{n_a}} = \frac{g_e g_i}{g_a}\int\left[\frac{d\vec{p}d\vec{r}}{(2\pi\hbar)^3}\exp\left(-\frac{I+\frac{p^2}{2m}}{T}\right)\right]. \quad (4.14)$$

In this relation, I is the ionization potential; p is momentum of a free electron, then $I+\frac{p^2}{2m}$ is the energy necessary to produce the electron; g_a, g_i, and g_e are the statistical weights of atoms, ions, and electrons; $\frac{d\vec{p}d\vec{r}}{(2\pi\hbar)^3}$ is the statistical weight corresponding to continuous spectrum, which is the number of states in a given element of the phase volume $d\vec{p}d\vec{r} = dp_x dp_y dp_z dxdydz$. Integration of Eq. (4.14) over the electron momentum gives:

$$\frac{\overline{n_i}}{\overline{n_a}} = \frac{g_e g_i}{g_a}\left(\frac{mT}{2\pi\hbar^2}\right)^{3/2}\exp\left(-\frac{I}{T}\right)\int d\vec{r}. \quad (4.15)$$

Here m is an electron mass, and $\int d\vec{r} = V/\overline{n_e}$ is volume corresponding to one electron, V-total system volume. Introducing in Eq. (4.15) the number densities of electrons $N_e = \overline{n_e}/V$, ions $N_i = \overline{n_i}/V$, and atoms $N_a = \overline{n_a}/V$, yields the **Saha equation**, describing the ionization equilibrium:

$$\frac{N_e N_i}{N_a} = \frac{g_e g_i}{g_a}\left(\frac{mT}{2\pi\hbar^2}\right)^{3/2}\exp\left(-\frac{I}{T}\right). \quad (4.16)$$

As one can estimate from Eq. (4.16), the effective statistical weight of the continuum spectrum is very high. As a result, the Saha equation predicts the very high values of degree of ionization $N_e/$

N_a, which can be close to unity even when the temperature is still much less, than ionization potential $T << I$.

4.1.4 Dissociation Equilibrium in Molecular Gases

The Saha equation (4.16) derived above for the ionization equilibrium $A^+ + e \Leftrightarrow A$ can be generalized to describe the dissociation equilibrium $X + Y \Leftrightarrow XY$. Relation between densities N_X of atoms X, N_Y of atoms Y and N_{XY} of the molecules XY in the ground vibrational-rotational state can be written based on Eq. (4.16) as:

$$\frac{N_X N_Y}{N_{XY}(v=0, J=0)} = \frac{g_X g_Y}{g_{XY}} \left(\frac{\mu T}{2\pi\hbar^2}\right)^{3/2} \exp\left(-\frac{D}{T}\right). \tag{4.17}$$

In this relation g_X, g_Y, and g_{XY} are the corresponding statistical weights, related to their electronic states; μ is the reduced mass of atoms X and Y; D- is dissociation energy of the molecule XY. Most of the molecules are not in a ground state but in excited one at the high temperatures typical for thermal plasmas.

Substituting the ground state concentration $N_{XY}(v = 0, J = 0)$ in the relation Eq. (4.17) by the total N_{XY} concentration, using Eq. (4.13):

$$N_{XY}(v=0, J=0) = \left[1 - \exp\left(-\frac{\hbar\omega}{T}\right)\right] \frac{B}{T} N_{XY} \tag{4.18}$$

the final statistical relation for the equilibrium dissociation of molecules is obtained.

$$\frac{N_X N_Y}{N_{XY}} = \frac{g_X g_Y}{g_{XY}} \left(\frac{\mu T}{2\pi\hbar^2}\right)^{3/2} \frac{B}{T} \left[1 - \exp\left(-\frac{\hbar\omega}{T}\right)\right] \exp\left(-\frac{D}{T}\right). \tag{4.19}$$

4.1.5 Equilibrium Statistical Relations for Radiation, the Planck Formula, and the Stefan-Boltzmann Law

According to the general expression of the Boltzmann distribution function Eq. (4.7), the relative probability to have "n" photons in a state with frequency ω is equal in equilibrium to $\exp(-\hbar\omega n)$. The average number of photons in the state with frequency ω can be found then in form of the so-called **Plank distribution**:

$$\overline{n_\omega} = \frac{\sum_n n \exp\left(-\frac{\hbar\omega n}{T}\right)}{\sum_n \exp\left(-\frac{\hbar\omega n}{T}\right)} = \frac{1}{\exp\frac{\hbar\omega}{T} - 1}. \tag{4.20}$$

The electromagnetic energy in the frequency interval $(\omega, \omega + d\omega)$ and in volume V can be then expressed as the energy of the $\overline{n_\omega}$ photons in one state - $\hbar\omega\overline{n_\omega}$, multiplied by the number of states in the given volume of the phase space $2\frac{V d\vec{k}}{(2\pi)^3}$ (factor 2 is related here to the number of polarizations of electromagnetic wave):

$$dE(V, \omega \div (\omega + d\omega)) = \hbar\omega\overline{n_\omega} 2 \frac{V d\vec{k}}{(2\pi)^3}. \tag{4.21}$$

Using the relation Eq. (4.21) and the dispersion relation $\omega = kc$ (c is the speed of light), the Plank distribution Eq. (4.20) can be rewritten in terms of the spectral density of radiation $U_\infty = \frac{1}{V}\frac{dE(V,\omega \div (\omega + d\omega))}{d\omega}$, which is the electromagnetic radiation energy taken per unit volume and unit interval of frequencies:

$$U_\omega = \frac{\hbar\omega^3}{\pi^2 c^3}\overline{n}_\omega = \frac{\hbar\omega^3}{\pi^2 c^3\left(\exp\frac{\hbar\omega}{T} - 1\right)}. \tag{4.22}$$

This is the **Planck formula** for the spectral density of radiation. The Planck formula at a high-temperature limit $\hbar\omega/T << 1$ gives the classical **Rayleigh-Jeans formula,** which does not contain the Planck constant $\hbar$:

$$U_\omega = \frac{\omega^2 T}{\pi^2 c^3}, \tag{4.23}$$

and at a low-temperature limit $\hbar\omega/T << 1$ it gives the **Wien formula**, describing the exponential decrease of radiation energy:

$$U_\omega = \frac{\hbar\omega^3}{\pi^2 c^3}\exp\left(-\frac{\hbar\omega}{T}\right). \tag{4.24}$$

The Planck formula Eq. (4.22) permits calculating the total equilibrium radiation flux falling on the surface. In the case of a black body this is equal to the radiation flux from the surface:

$$J = \frac{c}{4}\int_0^\infty U_\omega d\omega = \frac{\pi^2 T^4}{60c^2\hbar^3} = \sigma T^4 \tag{4.25}$$

This is the **Stefan–Boltzmann law**, with the Stefan-Boltzmann coefficient σ:

$$\sigma = \frac{\pi^2}{60c^2\hbar^3} = 5.67 * 10^{-12}\frac{\mathrm{W}}{\mathrm{cm^2K^4}}. \tag{4.26}$$

4.1.6 Concepts of the Complete Thermodynamic Equilibrium (CTE) and the Local Thermodynamic Equilibrium (LTE) for Plasma Systems

In plasma systems, **complete thermodynamic equilibrium** (CTE) is related to uniform homogeneous plasma, in which kinetic and chemical equilibrium as well as all the plasma properties are unambiguous functions of temperature. This temperature is supposed to be homogeneous and the same for all degrees of freedom, all the plasma system components, and all their possible reactions. In particular, the following five equilibrium statistical distributions described above obviously should take place with the same temperature T:

1. The Maxwell-Boltzmann velocity or translational energy distributions Eq. (2.1) for all neutral and charged species, that exists in the plasma.
2. The Boltzmann distribution Eqs. (4.9), (4.13) for the population of excited states for all neutral and charged species, that exist in the plasma.
3. The Saha equation (4.16) for ionization equilibrium to relate the number densities of electrons, ions, and neutral species.

4. The Boltzman distribution Eq. (4.19) for dissociation or more generally the thermodynamic relations for chemical equilibrium.
5. The Plank distribution Eq. (4.20) and Plank formula Eq. (4.22) for spectral density of electromagnetic radiation.

Plasma in the CTE-conditions cannot be practically realized in the laboratory. Nevertheless, thermal plasmas sometimes are modeled in this way for simplicity. To imagine a plasma in the CTE-conditions - one should consider a very large plasma volume such that the central part is homogeneous and does not sense the boundaries. In this case, the electromagnetic radiation of the plasma can be considered as black body radiation with plasma temperature. Actual (even thermal) plasmas are quite far from these ideal conditions. Most plasmas are optically thin over a wide range of wavelengths and as a result, the plasma radiation is much less than that of a black body. Plasma non-uniformity leads to irreversible losses related to conduction, convection, and diffusion, which also disturb the complete thermodynamic equilibrium (CTE). A more realistic approximation is the so-called **local thermodynamic equilibrium (LTE).** According to the LTE-approach, a thermal plasma is considered optically thin and hence radiation is not required to be in equilibrium. However, the collisional (not radiative) processes are required to be locally in equilibrium similar to that described above for CTE but with a temperature T, which can differ from point to point in space and time, see the book of M.I. Boulos, P. Fauchais, and E. Pfender (1994).

4.1.7 Partition Functions

Determination of the thermodynamic plasma properties, free energy, enthalpy, entropy, etc., at first requires the calculation of the partition functions, which are the links between microscopic statistics and macroscopic thermodynamics. Their calculation is an important statistical task (J.E. Mayer and G.M. Mayer, 1966), which should be accomplished to simulate different plasma properties. The partition function Q of an equilibrium particle system at the equilibrium temperature T in general can be expressed as a statistical sum over different states "s" of the particle with energies E_s and statistical weights g_s:

$$Q = \sum_s g_s \exp\left(-\frac{E_s}{T}\right). \tag{4.27}$$

The translational and internal degrees of freedom of the particles can be considered as independent. Then their energy can be expressed as the sum $E_s = E^{tr} + E^{int}$ and the partition function as a product of translational and internal partition functions. In this case for particles of a chemical species "i" and plasma volume V:

$$Q_i = Q_i^{tr} Q_i^{\text{int}} = \left(\frac{m_i T}{\hbar^2}\right)^{3/2} V Q^{\text{int}}. \tag{4.28}$$

If the translational partition function in Eq. (4.28) implies calculation of the continuous spectrum statistical weight, the partition functions of internal degrees of freedom (including molecular rotation and vibration) depend on the system characteristics (L. Landau, E. Lifshitz, 1967). A particular numerical example of the partition functions of nitrogen atoms and ions is presented in Figure 4.1.

4.1.8 Thermodynamic Functions of Thermal Plasma Systems

If the total partition function Q_{tot} of all the particles is known, the different thermodynamic functions can be calculated. For example, the Helmholtz free energy F, related to a reference-free energy F_0 is (L. Landau, E. Lifshitz, 1967):

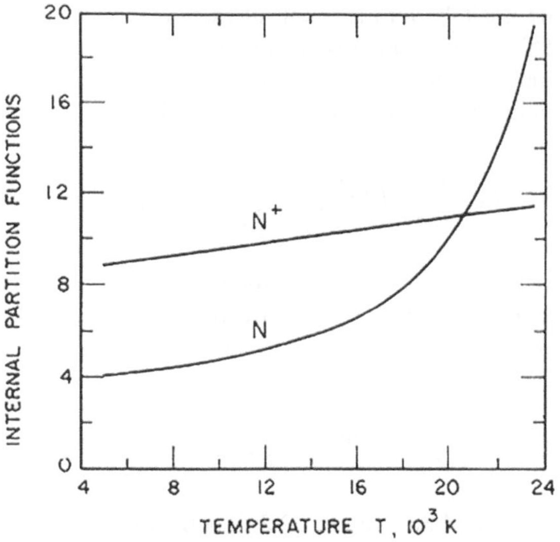

FIGURE 4.1 Partition function of nitrogen atoms and ions. (From *Thermal Plasma*, M. Boules, P. Fauchais and E. Phender.)

$$F = F_0 - T\ln Q_{tot}. \tag{4.29}$$

If the system particles are non-interacting (the ideal gas approximation), then the total partition function can be expressed as a product of partition functions Q_i of a single particle of a component "i":

$$Q_{tot} = \frac{\prod_i Q_i^{N_i}}{\prod_i N_i!}, \tag{4.30}$$

where N_i is the total number of particles of the species "i" in the system. Using Stirling's formula for factorial ($\ln N! = N\ln N - N$), and substituting the expression for the partition function Eq. (4.30) into Eq. (4.29), the final formula for calculation of the Helmholtz free energy of the system of non-interacting particles ("e" is the base of natural logarithms is obtained):

$$F = F_0 - \sum_i N_i T\ln\frac{Q_i e}{N_i}. \tag{4.31}$$

Other thermodynamic functions can be easily calculated based on the relation Eq. (4.31). For example, the internal energy U can be found as:

$$U = U_0 + T^2\left(\frac{\partial(F/T)}{\partial T}\right)_{V,N_i} = \sum_i N_i\left[\frac{3}{2}T + T^2\left(\frac{\partial\ln Q_i^{\text{int}}}{\partial V}\right)_{T,N_i}\right]. \tag{4.32}$$

Also, pressure "p" can be found from Eq. (4.31) and the differential thermodynamic relations:

$$p = -\left(\frac{\partial(F/T)}{\partial V}\right)_{T,N} = \sum_i N_i T\left(\frac{\partial\ln Q_i}{\partial V}\right)_{T,N_i}. \tag{4.33}$$

The above relations do not take into account interactions between particles. Ion and electron number densities grow at higher temperatures of thermal plasmas, the Coulomb interaction becomes important and should be added to the thermodynamic functions. The Debye Model (H.W. Drawin, 1972) provides the Coulomb interaction factor term to Eq. (4.31):

$$F = F_0 - \sum_i N_i T \ln \frac{Q_i e}{N_i} - \frac{TV}{12\pi}\left(\frac{e^2}{\varepsilon_0 TV}\sum_i Z_i^2 N_i\right)^{3/2}. \tag{4.34}$$

Here Z_i is the number of charges of species "i" (for electrons $Z_i = -1$). This **Debye correction** in the Helmholtz free energy F obviously leads to corrections of the other thermodynamic functions. Thus, for example, the Gibbs energy G with the Debye correction becomes:

$$G = G_0 - \sum_i N_i T \ln \frac{Q_i}{N_i} - \frac{TV}{8\pi}\left(\frac{e^2}{\varepsilon_0 TV}\sum_i Z_i^2 N_i\right)^{3/2}. \tag{4.35}$$

The relation for pressure Eq. (4.33) with the Debye correction should also be modified from the case of ideal gas as:

$$p = \frac{NT}{V} - \frac{T}{24\pi}\left(\frac{e^2}{\varepsilon_0 TV}\sum_i Z_i^2 N_i\right)^{3/2}. \tag{4.36}$$

Numerically the Debye correction becomes relatively important (typically around 3%) at temperatures of thermal plasma about 14,000–15,000 K. The thermodynamic functions help to calculate the chemical and ionization composition of thermal plasmas. Examples of such calculations can be found in a book of M.I. Boulos, P. Fauchais, and E. Pfender (1994). Numerous thermodynamic results related to the equilibrium chemical and ionization composition can be found through the use of handbooks (A. L. Souris (1985), or S. Nester, B. Potapkin, A. Levitsky, V. Rusanov, B. Trusov, A. Fridman, 1988).

4.1.9 Non-Equilibrium Statistics of the Thermal and Non-Thermal Plasmas

The correct description of the non-equilibrium plasma systems and processes requires the application of detailed kinetic models. The application of statistical models, in this case, can lead to huge deviations from reality, as was demonstrated by D.I. Slovetsky (1980). However, statistical approaches can not only be simple but also quite successful in the description of the non-equilibrium plasma systems. Several examples of the subject will be discussed below. The first example is related to thermal discharges with a small deviation of the electron temperature from the temperature of heavy particles. Such a situation can take place in boundary layers separating quasi-equilibrium plasma from electrodes and walls. In this case, **two-temperature statistics and thermodynamics** can be developed (M.I. Boulos, P. Fauchais, E. Pfender, 1994). These models suppose that partition functions depend on both electron temperature and temperature of heavy particles. Electron temperature in this case determines the partition functions related to ionization processes; chemical processes are in turn determined by the temperature of heavy particles. The partition functions found in such a way can be then used to calculate the free energy, Gibbs potential and, finally, composition and thermodynamic properties of the two-temperature plasma. Example of such a calculation of composition in two-temperature Ar- plasma is given in Figure 4.2. This approach is valid only for a low level of non-equilibrium. The second example of the application of non-equilibrium statistics is related to non-thermal plasma gasification of solid surfaces

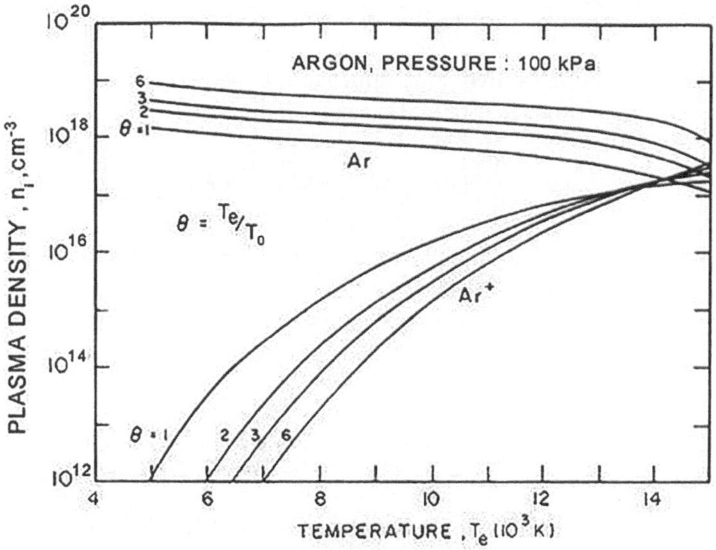

FIGURE 4.2 Dependence of the number densities of Ar and Ar+ on electron temperature at atmospheric pressure for two-temperature plasma with $\boldsymbol{\theta = T_e/T_0} = 1, 2, 3, 6$.

$$A(solid) + bB*(gas) \rightarrow cC(gas), \tag{4.37}$$

stimulated by particles B excited in non-thermal discharges, predominantly in one specific state (S. Veprek, 1972; V.A. Legasov, V.D. Rusanov, A. Fridman, 1978). In this case, deviation from the conventional equation for the equilibrium constant

$$K = \frac{(Q_C)^c}{Q_A (Q_B)^b} \tag{4.38}$$

is related only in a more general form of definition the partition functions. Differences in temperatures: translational T_t, rotational T_r, and vibrational T_v as well the non-equilibrium distribution function (population) $f_e^X(\varepsilon_e^X)$ of electronically excited states should be taken into account for each reactant and products X = A, B, C:

$$Q_X = \sum_e g_e^X f_e^X(\varepsilon_e^X) \sum_{(t,r,v)} \prod_{k=t,r,v} g_{ke}^X \exp\left(-\frac{\varepsilon_{ke}^X}{T_k^X}\right). \tag{4.39}$$

Here g and ε are the statistical weights and energies of corresponding states. The introduction of three temperatures can lead to internal inconsistencies. The statistical approach can be applied in a consistent way for the reactions Eq. (4.37), if the non-thermal plasma stimulation of the process is limited to the electronic excitation of a single state of a particle B with energy E_b. The population of the excited state can then be expressed by δ-function and the Boltzmann factor with an effective electronic temperature T*:

$$f_e^B(\varepsilon_e^B) = \delta(\varepsilon_e^B - E_B)\exp\left(-\frac{E_B}{T*}\right). \tag{4.40}$$

Other degrees of freedom of all the reaction participants are considered actually in quasi-equilibrium with the gas temperature T_0:

$$f_e^{A,C}\left(\varepsilon_e^{A,C}\right)=1, \quad T_{k=t,r,v}^{X}=T_0. \tag{4.41}$$

In the framework of **a single excited state approach**, in particular, the "quasi-equilibrium constant of the non-equilibrium process" Eq. (4.37) can be estimated as:

$$K \approx \exp\left\{-\frac{1}{T_0}\left[\left(\Delta H-\Delta F^A\right)-bE_B\right]\right\}=K_0\exp\frac{bE_B}{T_0}. \tag{4.42}$$

In this relation, ΔH- is the reaction enthalpy, ΔF_A- is the free energy change corresponding to heating to T_0, K_0 is the equilibrium constant at temperature T_0. As one can see from Eq. (4.42), the equilibrium constant can be significantly increased due to the electronic excitation, and equilibrium significantly shifted towards the reaction products.

4.1.10 Non-Equilibrium Statistics of Vibrationally Excited Molecules, the Treanor Distribution

The equilibrium distribution of diatomic molecules over the different vibrationally excited states follows the Boltzmann formula Eqs. (4.10) and (4.12) with the temperature T. Vibrational excitation of diatomic molecules in non-thermal plasma can be much faster, than the vibrational VT-relaxation rate. As a result, the vibrational temperature T_v of the molecules in plasma can essentially exceed the translational one T_0. The vibrational temperature, in this case, is usually defined in accordance with Eq. (4.10) by the ratio of populations of ground state and the first excited level:

$$T_v=\frac{\hbar\omega}{\ln\left(N_0/N_1\right)}. \tag{4.43}$$

The diatomic molecules can be considered as harmonic oscillators Eq. (3.2.3) for not very large levels of vibrational excitation. In this case, the VT relaxation can be neglected with respect to the VV-exchange of vibrational quanta and the VV-exchange is absolutely resonant and not accompanied by energy transfer to translational degrees of freedom. As a result, the vibrational degrees of freedom are independent of the translational one, and the vibrational distribution function follows the same Boltzmann formula Eqs. (4.10), (4.12) but with the vibrational temperature T_v Eq. (4.43) exceeding the translational one T_0. An interesting non-equilibrium phenomenon takes place if the anharmonicity of the diatomic molecules is taken into account. In this case, the one-quantum VV-exchange is not completely resonant, the translational degrees of freedom become involved in the vibrational distribution, which results in a strong deviation from the Boltzmann distribution. Considering vibrational quanta as quasi-particles, and using the Gibbs distribution with a variable number of quasi-particles "v", the relative population of the vibrational levels can be expressed (N.M. Kuznetzov, 1971) as:

$$N_v=N_0\exp\left(-\frac{\mu v-E_v}{T_0}\right). \tag{4.44}$$

In this relation, the parameter μ is the chemical potential; E_v is the energy of the vibrational level "v" taken with respect to the zero-level. Comparing Eqs. (4.44) and (4.45), the effective chemical potential of a quasi-particle μ is found:

$$\mu=\hbar\omega\left(1-\frac{T_0}{T_v}\right). \tag{4.45}$$

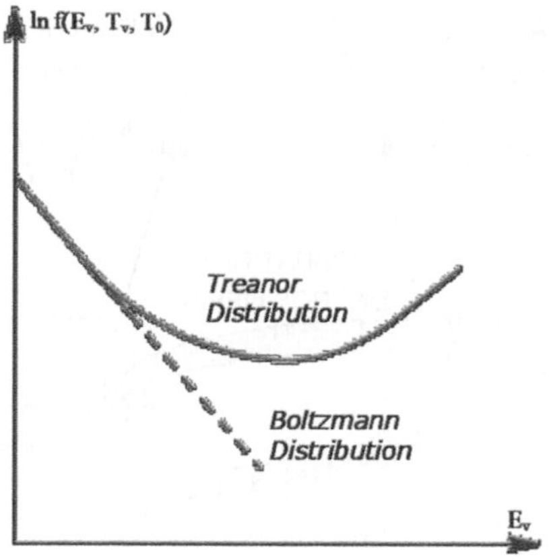

FIGURE 4.3 Comparison of the Treanor and Boltzmann distribution functions.

From the Gibbs distribution Eq. (4.44), chemical potential Eq. (4.45) and definition Eq. (4.43) of the vibrational temperature, the distribution of vibrationally excited molecules over different vibrational levels can be expressed in form of the famous **Treanor distribution function**:

$$f\left(v,T_v,T_0\right) = B\exp\left(-\frac{\hbar\omega v}{T_v} + \frac{x_e\hbar\omega v^2}{T_0}\right). \tag{4.46}$$

Here x_e is the coefficient of anharmonicity and B- is the normalizing factor of the distribution. Comparison of the parabolic-exponential Treanor distribution with the linear-exponential Boltzmann distribution is illustrated in Figure 4.3. The Treanor distribution function was first derived statistically by C.E.Treanor, I.W.Rich and R.G.Rehm (1968). When $T_v > T_0$ it shows that the population of the highly vibrationally excited levels can be many orders of value higher than predicted by the Boltzmann distribution even at vibrational temperature (Figure 4.3). The Treanor distribution (or the Treanor effect) explains in particular the very high rates and high energy efficiency of chemical reactions stimulated by vibrational excitation in plasma. The Treanor distribution function Eq. (4.46) can be transformed back to the traditional form of the Boltzmann distribution in the following two situations. First, if molecules can be considered as harmonic oscillators and the anharmonicity coefficient is zero $x_e = 0$, then Eq. (4.46) gives the Boltzmann distribution $f\left(v\right) = B\exp\left(-\frac{\hbar\omega v}{T_v}\right) = B\exp\left(-\frac{E_v}{T_v}\right)$ even if the vibrational and translational temperatures are different. Second, if the translational and vibrational temperatures are equal $T_v = T_0 = T$, the Treanor distribution Eq. (4.46) converts back to the Boltzmann function:

$$f\left(v,T\right) = B\exp\left(-\frac{\hbar\omega v}{T} + \frac{x_e\hbar\omega v^2}{T}\right) = B\exp\left(-\frac{\hbar\omega v - x_e\hbar\omega v^2}{T}\right) \propto \exp\left(-\frac{E_v}{T}\right). \tag{4.47}$$

Physical interpretation of the Treanor effect can be illustrated by the collision of two anharmonic oscillators, presented in Figure 4.4. First, recall the VV-exchange of harmonic molecules. Here, the resonant inverse processes of a quantum transfer from a lower excited harmonic oscillator to a higher excited one and back are not limited by translational energy and as a result have the same probability. That is the reason why the population of vibrationally excited, harmonic molecules keeps following the Boltzmann distribution even when the vibrational temperature in non-thermal plasma greatly exceeds the translational one. In contrast, the VV-exchange of anharmonic oscillators (see Figure 4.4) under conditions of a deficiency of translational energy ($T_v > T_0$) gives priority to a

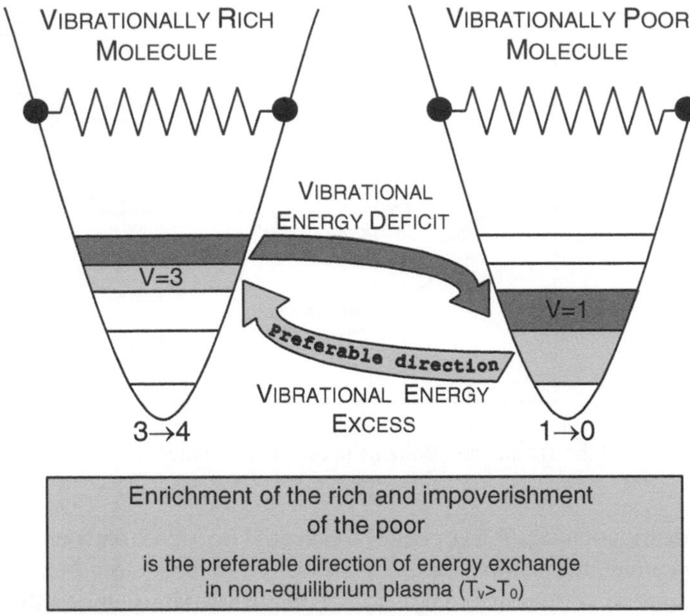

FIGURE 4.4 The overpopulation of highly vibrationally excited states (Treanor effect) in non-equilibrium plasma chemistry.

quantum transfer from a lower excited anharmonic oscillator to a higher excited one. The reverse reaction (a quantum transfer from a higher excited oscillator to a lower excited one) is less favorable because to cover the anharmonic defect of vibrational energy requires translational energy which cannot be provided effectively if $T_v > T_0$. Thus, the molecules having a larger number of vibrational quanta keep receiving more than those with a small number of quanta. In other words, the rich molecules become richer, and the poor molecules become poorer. That's why the Treanor effect of overpopulation of highly excited vibrational states in non-equilibrium conditions is sometimes referred to as "*the capitalism in molecular life*". The exponentially-parabolic Treanor distribution function has a minimum at a specific number of vibrational quanta:

$$v_{\min}^{Tr} = \frac{1}{2x_e}\frac{T_0}{T_v}. \tag{4.48}$$

The "Treanor minimum" corresponds to low levels of vibrational excitation only if the vibrational temperature greatly exceeds the translational one. In equilibrium $T_v = T_0 = T$, and point of the Treanor minimum goes beyond the physically possible range of the vibrational quantum numbers Eq. (3.9) for diatomic molecules. The population of the vibrational level, corresponding to the Treanor minimum, can be found from Eqs. (4.46) and (4.48) as:

$$f_{\min}^{Tr}\left(T_v,T_0\right) = B\exp\left(-\frac{\hbar\omega T_0}{4x_eT_v^2}\right). \tag{4.49}$$

At vibrational levels higher than the Treanor minimum Eq. (4.48), the population of vibrational states according to the Treanor distribution Eq. (4.46) becomes inverse, which means, that the vibrational levels with higher energy are more populated than those with lower energy. This interesting effect of the inverse population takes place in non-thermal discharges of molecular gases and, in particular, plays an important role in CO-lasers. According to the Treanor distribution, the population of states with high vibrational quantum numbers is growing without limitation (see Figure 4.3). This unlimited growth is not physically realistic. Deriving the Treanor distribution, the VT-relaxation with respect to the VV-exchange was neglected, which is correct only at relatively low levels of vibrational excitation (see Eqs. (3.78) and (3.79)). The VT-relaxation prevails at large vibrational energies and causes the vibrational distribution function to fall exponentially in accordance with the Boltzmann distribution function corresponding to the translational temperature T_0. The description of this effect is beyond the statistical approach and requires analysis of the kinetic equations for vibrationally excited molecules.

4.2 THE BOLTZMANN AND FOKKER-PLANCK KINETIC EQUATIONS, ELECTRON ENERGY DISTRIBUTION FUNCTIONS

4.2.1 The Boltzmann Kinetic Equation

Consider the evolution of a distribution function $f\left(\vec{r},\ \vec{v},\ t\right)$ in the six-dimensional **phase space** $\left(\vec{r},\ \vec{v}\right)$ of particle positions and velocities. Quantum numbers related to internal degrees of freedom are not included because of primary interest in the electron distributions. In the absence of collisions between particles:

$$\frac{df}{dt} = \frac{f\left(\vec{r}+\overrightarrow{dr},\ \vec{v}+\overrightarrow{dv},t+dt\right)-f\left(\vec{r},\vec{v},t\right)}{dt}. \tag{4.50}$$

Taking into account, that in the collisionless case – number of particles in a given state is fixed: $\frac{df}{dt}=0,\ \frac{\overrightarrow{dv}}{dt}=\frac{\vec{F}}{m}$ and $\frac{\overrightarrow{dr}}{dt}=\vec{v}$, where m is the particle mass and $\vec{F}$ is an external force, we can rewrite the Eq. (4.50) as:

$$\frac{df}{dt} = \frac{\partial f}{\partial t} + \vec{v}\frac{\partial f}{\partial \vec{r}} + \frac{\vec{F}}{m}\frac{\partial f}{\partial \vec{v}} = 0. \tag{4.51}$$

The external electric E and magnetic B forces for electrons can be introduced in the relation as: $\vec{F} = -e\left(\vec{E}+\vec{v}\times\vec{B}\right)$. It brings us from (4.51) to the **collisionless Vlasov equation**:

$$\frac{\partial f}{\partial t} + \vec{v}*\nabla_r f - \frac{e}{m}\left(\vec{E}+\vec{v}\times\vec{B}\right)*\nabla_v f = 0. \tag{4.52}$$

Here the operators ∇_r and ∇_v denote the electron distribution function gradients related to space and velocity coordinates. Binary collisions between particles are very important in kinetics. Since they occur over a very short time, velocities can be changed practically "instantaneously", and $\frac{df}{dt}$ no longer is zero. The corresponding evolution of the distribution function can be described by adding the special term $I_{col}(f)$, the so-called collisional integral, to the kinetic equation (4.51):

$$\frac{df}{dt} = \frac{\partial f}{\partial t} + \vec{v}*\nabla_r f + \frac{\vec{F}}{m}*\nabla_v f = I_{col}\left(f\right). \tag{4.53}$$

This is the well-known **Boltzmann kinetic equation**, which is used the most to describe the evolution of distribution functions and to calculate different macroscopic kinetic quantities. The collisional integral of the Boltzmann kinetic equation (E.M. Lifshitz, L.P. Pitaevsky, 1979) is a non-linear function of $f(\vec{v})$ and can be presented as:

$$I_{col}(f) = \int \left[f(v')f(v_1') - f(v)f(v_1) \right] |v - v_1| \frac{1}{\varepsilon} d\omega dv_1, \tag{4.54}$$

where v_1 and v are velocities before collision, v_1' and v'-velocities after collision (all the velocities in this expression are vectors), ε- is a parameter proportional to the mean free path ratio to the typical system size, $d\omega$ -is area element in the plane perpendicular to the $(v_1 - v)$ vector. In equilibrium, the collisional integral Eq. (4.54) should be equal to zero, which requires the relation between velocities:

$$f(v')f(v_1') = f(v)f(v_1) \quad \text{or} \quad \ln f(v') + \ln f(v_1') = \ln f(v) + \ln f(v_1). \tag{4.55}$$

This relation corresponds to the conservation of kinetic energy during the elastic collision:

$$\frac{m(v')^2}{2} + \frac{m(v_1')^2}{2} = \frac{mv^2}{2} + \frac{mv_1^2}{2}, \tag{4.56}$$

and explains the proportionality $\ln f \propto \frac{mv^2}{2}$, which leads to the equilibrium Maxwell distribution function of particles at equilibrium:

$$f_0(\vec{v}) = B \exp\left(-\frac{mv^2}{2T} \right). \tag{4.57}$$

After calculation of the normalizing factor B, the final form of the equilibrium Maxwellian (or Maxwell-Boltzmann) distribution of particles over translational energies is:

$$f_0(\vec{v}) = \left(\frac{m}{2\pi T} \right)^{3/2} \exp\left(-\frac{mv^2}{2T} \right). \tag{4.58}$$

This relation corresponds to the earlier considered Maxwell-Boltzmann distribution of electrons over translational energies Eq. (2.1). The non-equilibrium ideal gas entropy (including with some restrictions the case of gas of the plasma electrons) can be expressed by the following integral:

$$S = \int f \ln \frac{e}{f} \overrightarrow{dv}. \tag{4.59}$$

Differentiating the entropy Eq. (4.59) and taking into account the Boltzmann kinetic equation Eq. (4.53) the time evolution of the entropy is as:

$$\frac{dS}{dt} = -\int \ln f * I_{col}(f) \overrightarrow{dv}. \tag{4.60}$$

In equilibrium, when the collisional integral I_{col} (f) Eq. (4.54) is equal to zero, the entropy reaches its maximum value, which is known as the **Boltzmann H-Theorem.**

4.2.2 The τ-Approximation of the Boltzmann Kinetic Equation

As seen from Eqs. (4.53), (4.54), the most informative part of the Boltzmann equation is the collisional integral. Unfortunately, the expression Eq. (4.54) of the collisional integral is quite complicated. Several approximations have been developed to simplify the calculations. The simplest one is so-called **τ-approximation**:

$$I_{col}(f) = -\frac{f - f_0}{\tau}. \tag{4.61}$$

The τ-approximation is based on the fact, that the collisional integral should be equal to zero when the distribution function is Maxwellian Eq. (4.58) or Eq. (2.1). The characteristic time of evolution of the distribution function is the collisional time τ (the reverse collisional frequency Eq. (2.4)). The Boltzmann equation in the frameworks of the τ-approximation becomes simple:

$$\frac{\partial f}{\partial t} = -\frac{f - f_0}{\tau}, \tag{4.62}$$

with the solution exponentially approaching the Maxwellian distribution function f_0:

$$f(\vec{v},t) = f_0 + \left[f(\vec{v},0) - f_0\right]\exp\left(-\frac{t}{\tau}\right). \tag{4.63}$$

Here $f(\vec{v},0)$ is the initial distribution function. Obviously, the τ-approximation is just a very simplified but useful approach. Details about the Boltzmann equation and its solutions can be found in particular in a book by E.M. Lifshitz and L.P. Pitaevsky (1979).

4.2.3 Macroscopic Equations Related to the Kinetic Boltzmann Equation

Integrate the kinetic Boltzmann equation Eq. (4.53) over the particle velocity. Then the right-hand side of the resulting equation (the integral of I_{col} over velocities) represents the total change of the particles number in a unit volume per unit time "G-L" (generation rate minus rate of losses):

$$\int \overrightarrow{dv}\frac{\partial f}{\partial t} + \int \overrightarrow{dv} * \left(\vec{v}\frac{\partial f}{\partial \vec{r}}\right) + \frac{\vec{F}}{m}\int \frac{\partial f}{\partial \vec{v}}\overrightarrow{dv} = G - L. \tag{4.64}$$

If a given volume does not include particle formation or loss, the right side of the equation is equal to zero. The distribution function is normalized here on the number of particles, then $\int f\overrightarrow{dv} = n$ (where n is the number density of particles) and the first term in Eq. (4.64) corresponds to $\frac{\partial n}{\partial t}$. The particle flux is defined as $\int \vec{v} f\overrightarrow{dv} = n\vec{u}$, where $\vec{u}$ is the particle mean velocity, which is normally referred to as the **drift velocity.** In this case, the second term in Eq. (4.64) is equal to $\frac{\partial}{\partial \vec{r}}(n\vec{u})$. The third term in Eq. (4.64) is equal to zero, because at infinitely high velocities – the distribution function is equal to zero. As a result, the relation Eq. (4.64) can be expressed as:

$$\frac{\partial n}{\partial t} + \nabla_r(n\vec{u}) = G - L \tag{4.65}$$

This relation reflects the particle conservation and it is known as **the continuity equation**. Again the right-hand side of the equation is equal to zero if there are no processes of particle generation and loss. **The momentum conservation equation** can be derived in a similar manner. In this case, the Boltzmann equation should be multiplied by $\vec{v}$ and then integrated over velocity, which for the gas of charged particles gives as (see derivation, for example, in the book of N.A. Krall and A.W. Trivelpiece, 1973):

$$mn\left[\frac{\partial \vec{u}}{\partial t} + (\vec{u} * \nabla)\vec{u}\right] = qn\left(\vec{E} + \vec{u} \times \vec{B}\right) - \nabla * \Pi + F_{col}. \tag{4.66}$$

The second term from the right-hand side is the divergence of the pressure tensor Π_{ij}, which can be replaced in plasma by pressure gradient ∇p; q is the particle charge, E and B –are the electric and magnetic fields. The last term F_{col} represents the collisional force, which is the time rate of momentum transfer per unit volume due to collisions with other species (i). The collisional force for plasma electrons and ions are usually due to collisions with neutrals. The **Krook collision operator** can approximate the collisional force by summation over the other species "i":

$$\overrightarrow{F_{col}} = -\sum_i mn\nu_{mi}\left(\vec{u} - \vec{u_i}\right) - m\vec{u}\left(G - L\right). \tag{4.67}$$

In this relation ν_{mi} is the momentum transfer frequency for collisions with the species "i", $\vec{u}$ and $\vec{u_i}$ are the mean or drift velocities. The first term in the Krook collision operator can be interpreted as "friction". The second term corresponds to the momentum transfer due to the creation or destruction of particles and is generally small. Equation (4.66) for plasma in the electric field can be simplified considering in the Krook operator only neutral species, assuming they are at rest ($u_i = 0$); and also neglecting the inertial force $\left(\vec{u} * \nabla\right)\vec{u} = 0$. This results in the most common form of the **momentum conservation equation**:

$$m\frac{\partial\vec{u}}{\partial t} = q\vec{E} - \frac{1}{n}\nabla p - m\nu_m\vec{u}. \tag{4.68}$$

Finally, **the energy conservation equation** for the electron and ion fluids can be derived by multiplying the Boltzmann equation by $\frac{1}{2}mv^2$ and integrating over velocity:

$$\frac{\partial}{\partial t}\left(\frac{3}{2}p\right) + \nabla * \frac{3}{2}\left(p\vec{u}\right) + p\nabla * \vec{u} + \nabla * \vec{q} = \frac{\partial}{\partial t}\left(\frac{3}{2}p\right)_{col} \tag{4.69}$$

In this relation $\vec{q}$ is the heat flow vector, $\frac{3}{2}p$- corresponds to the energy density, and the term $\frac{\partial}{\partial t}\left(\frac{3}{2}p\right)_{col}$ represents changes in the energy density due to collisional processes, including ohmic heating, excitation, and ionization. The energy balance of the steady-state discharges can be usually described by the simplified form of the equation Eq. (4.69):

$$\nabla * \frac{3}{2}\left(p\vec{u}\right) = \frac{\partial}{\partial t}\left(\frac{3}{2}p\right)_{col}, \tag{4.70}$$

which balances the macroscopic energy flux with the rate of energy change due to collisional processes.

4.2.4 The Fokker-Planck Kinetic Equation for Determination of the Electron Energy Distribution Functions

The Boltzmann kinetic equation obviously can be used for determinating the electron energy distribution functions in different non-equilibrium conditions (see, for example, appendix to the book of M.A. Lieberman, A.J. Lichtenberg (1994)). A better physical interpretation of the evolution of the electron energy distribution function evolution can be presented using the Fokker-Planck approach. The evolution of the electron energy distribution function is considered in the framework of this kinetic approach as an electron diffusion and drift in the space of electron energy (Yu. P. Raizer, 1991; V.D. Rusanov, A. Fridman, 1984). To derive the kinetic equation for the electron energy distribution function *f*(ε) in the Fokker-Planck approach, consider the

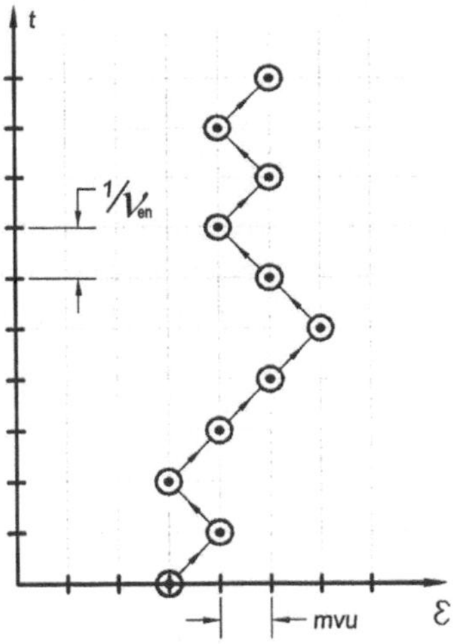

FIGURE 4.5 Electron diffusion and drift in energy space.

dynamics of energy transfer from the electric field to electrons. Suppose that most of the electron energy obtained from the electric field is then transferred to neutrals through elastic and inelastic collisions. Let an electron has a velocity $\vec{v}$ after a collision with a neutral particle. Then the electron receives an additional velocity during the free motion between collisions (see illustration on the Figure 4.5):

$$\vec{u} = -\frac{e\vec{E}}{m\nu_{en}}, \tag{4.71}$$

which corresponds to its drift in the electric field $\vec{E}$. Here ν_{en} is the frequency of the electron-neutral collisions, -e and m are an electron charge and mass. The corresponding change of the electron kinetic energy between two collisions is equal to:

$$\Delta\varepsilon = \frac{1}{2}m\left(\vec{v}+\vec{u}\right)^2 - m\vec{v}^2 = m\vec{v}\vec{u} + \frac{1}{2}m\vec{u}^2. \tag{4.72}$$

Usually the drift velocity is much lower than the thermal velocity u << v, and the absolute value of the first term $m\vec{v}\vec{u}$ in Eq. (4.72) is much greater than that of the second one. However, the thermal velocity vector $\vec{v}$ is isotropic, and the value of $m\vec{v}\vec{u}$ can be positive and negative with the same probability. The average contribution of the term into Δε is equal to zero, and the average electron energy increase between two collisions is related only to the square of drift velocity:

$$\langle\Delta\varepsilon\rangle = \frac{1}{2}m\vec{u}^2, \tag{4.73}$$

Comparing the relations Eqs. (4.72) and (4.73) brings us to the conclusion, that *the electron motion along the energy spectrum (or in energy space) can be considered as a diffusion process*. An electron with energy $\varepsilon = \frac{1}{2}mv^2$ receives or loses per one collision an energy portion about "mvu", depending on the direction of its motion - along or opposite to the electric field (see Figure 4.5). The energy portion "mvu" can be considered in this case as the electron "mean free path along the energy spectrum" (or in the energy space). Taking into account also the possibility of an electron motion across the electric field, the corresponding coefficient of electron diffusion along the energy spectrum can be introduced as:

$$D_\varepsilon = \frac{1}{3}\left(mvu\right)^2 \nu_{en} = \frac{2}{3} mu^2 \varepsilon \nu_n \tag{4.74}$$

Besides the diffusion along the energy spectrum, there is also drift in the energy space, related to the permanent average energy gain and losses. Such energy consumption from the electric field is described by Eq. (4.73). The average energy losses per one collision are mostly due to the elastic scattering $\frac{2m}{M}\varepsilon$ and the vibrational excitation $P_{eV}(\varepsilon)\hbar\omega$ of molecular gases. Here M and m are the neutral particles and electron mass, $P_{eV}(\varepsilon)$ is the probability of vibrational excitation by electron impact. The mentioned three effects define the electron drift velocity in the energy space neglecting the super-elastic collisions:

$$u_\varepsilon = \left[\frac{mu^2}{2} - \frac{2m}{M}\varepsilon - P_{eV}(\varepsilon)\hbar\omega\right]\nu_{en}. \tag{4.75}$$

The electron energy distribution function $f(\varepsilon)$ can be considered as a number density of electrons in the energy space and can be found from the continuity equation Eq. (4.65) sure also in the energy space. Based upon the expressions for diffusion coefficient Eq. (4.74) and drift velocity Eq. (4.75), such a *continuity equation for an electron motion along the energy spectrum* can be presented as the **Fokker-Planck kinetic equation**:

$$\frac{\partial f(\varepsilon)}{\partial t} = \frac{\partial}{\partial \varepsilon}\left[D_\varepsilon \frac{\partial f(\varepsilon)}{\partial \varepsilon} - f(\varepsilon)u_\varepsilon\right]. \tag{4.76}$$

The Fokker-Plank kinetic equation for electrons is obviously much easier to interpret and to solve for different non-equilibrium plasma conditions than the general Boltzmann kinetic equation Eq. (4.53).

4.2.5 Different Specific Electron Energy Distribution Functions, Druyvesteyn Distribution

The steady-state solution of the Fokker-Planck equation Eq. (4.76) for the electron energy distribution function in non-equilibrium plasma, corresponding to $f(\varepsilon \to \infty) = 0$, $\frac{df}{d\varepsilon}(\varepsilon \to \infty) = 0$ is:

$$f(\varepsilon) = B\exp\left\{\int_0^\varepsilon \frac{u_\varepsilon}{D_\varepsilon} d\varepsilon'\right\}, \tag{4.77}$$

where B is the pre-exponential normalizing factor. Using in the relation Eq. (4.77) the expressions Eqs. (4.74), (4.75) for the diffusion coefficient and drift velocity in the energy space, we can rewrite the electron energy distribution function in the following integral form:

$$f(\varepsilon) = B\exp\left[-\int_0^\varepsilon \frac{3m^2}{Me^2E^2}\nu_{en}^2\left(1 + \frac{M}{2m}\frac{\hbar\omega}{\varepsilon'}P_{eV}(\varepsilon)\right)d\varepsilon'\right]. \tag{4.78}$$

This distribution function for the constant electric field can be generalized to the alternating one. In this case, the electric field strength E should be replaced in Eq. (4.78) by the effective strength:

$$E_{eff}^2 = E^2 \frac{\nu_{en}^2}{\omega^2 + \nu_{en}^2}. \tag{4.79}$$

In this relation E is the average value of the electric field strength, and ω is the field frequency. From Eq. (4.79), the electric field can be considered as quasi-constant at room temperature and pressure 1 Torr if frequencies ω << 2,000 MHz. Consider four specific distribution functions, which can be derived after integration of Eq. (4.78), and corresponding to specific energy dependences $\nu_{en}(\varepsilon)$ and $P_{eV}(\varepsilon)$:

1. **Maxwellian distribution.** For atomic gases, the elastic collisions dominate electron energy losses, $(P_{eV} << \dfrac{2m}{M}\dfrac{\varepsilon}{\hbar\omega})$; and the electron-neutral collision frequency can be approximated as constant $\nu_{en}(\varepsilon) = const$, then integration of Eq. (4.78) gives the Maxwellian distribution Eq. (2.1) with:

$$T_e = \frac{e^2 E^2 M}{3m^2 \nu_{en}^2}. \tag{4.80}$$

2. **Druyvesteyn distribution.** If the elastic collisions dominate the electron energy losses; but rather than the collision frequency, the electrons mean free path λ is taken constant $\nu_{en} = \dfrac{v}{\lambda}$, then integration of Eq. (4.78) gives the exponential-parabolic Druyvesteyn distribution function, derived in 1930:

$$f(\varepsilon) = B\exp\left[-\frac{3m}{M}\frac{\varepsilon^2}{(eE\lambda)^2}\right]. \tag{4.81}$$

The Druyvesteyn distribution is decreasing with energy much faster, than the Maxwellian one for the same mean energy. It is well illustrated in Figure 4.6. Equation (4.81) also leads to an interesting conclusion. The value eEλ in the Druyvesteyn exponent corresponds to the energy, which an electron receives from the electric field during one mean free path λ along the field. As can be seen from Eq. (4.81), this requires an electron about $\sqrt{\dfrac{M}{m}} \approx 100$ of the mean free paths λ along the electric field to get the average electron energy.

3. **Margenau distribution.** This distribution takes place in conditions similar to those of the Druyvesteyn distribution, but in an alternating electric field. Based on Eqs. (4.78) and (4.79),

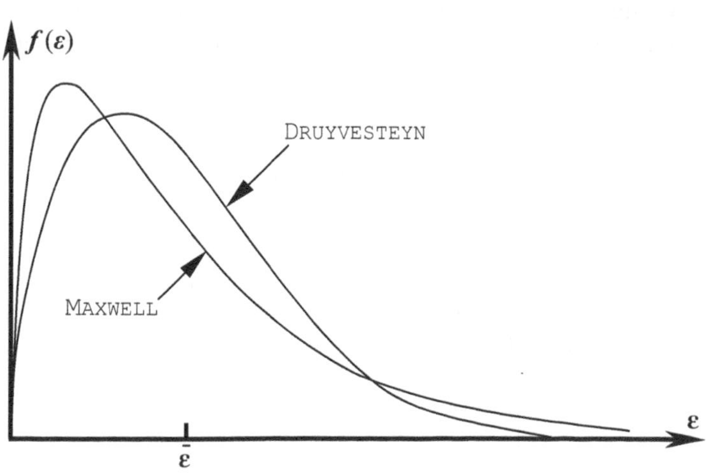

FIGURE 4.6 Maxwell and Druyvesteyn distribution functions at the same value of mean energy (statistical weight effect related to the pre-exponential factor B and resulting in f(0) = 0 are also taken here into account).

integration yields the Margenau distribution function for electrons in high-frequency electric fields:

$$f(\varepsilon) = B\exp\left[-\frac{3m}{M}\frac{1}{(eE\lambda)^2}\left(\varepsilon^2 + \varepsilon m\omega^2\lambda^2\right)\right]. \tag{4.82}$$

This was first derived in 1946. The Margenau distribution function Eq. (4.82) obviously corresponds to the Druyvesteyn distribution Eq. (4.81), if the electric field frequency is equal to zero $\omega = 0$.

4. ***Distributions, controlled by vibrational excitation.*** The influence of vibrational excitation of molecules by electron impact on the electron energy distribution function is very strong in molecular gases ($P_{eV} >> \frac{2m}{M}\frac{\varepsilon}{\hbar\omega}$). If it is assumed that $P_{eV}(\varepsilon) = const$ over some energy interval $\varepsilon_1 < \varepsilon < \varepsilon_2$, and similarly to the Druyvesteyn conditions: λ = const ($\nu_{en} = v/\lambda$), then – the electron energy distribution function in the energy interval $\varepsilon_1 < \varepsilon < \varepsilon_2$ can be presented as:

$$f(\varepsilon) \propto \exp\left[-\frac{6P_{eV}\hbar\omega}{(eE\lambda)^2}\varepsilon\right]. \tag{4.83}$$

Outside of this energy interval – the Druyvesteyn distribution Eq. (4.123) takes place. Comparing the relations Eqs. (4.81) and (4.83), it should be taken into account that the vibrational excitation of molecules provides higher electron energy losses than elastic collisions ($P_{eV}\hbar\omega >> \frac{m}{M}\varepsilon$). This means, that there is a reduction of the electron energy distribution function in the energy interval $\varepsilon_1 < \varepsilon < \varepsilon_2$, related to vibrational excitation. This sharp decrease of the distribution function can be described by the following small factor:

$$\alpha_v = \frac{f(\varepsilon_2)}{f(\varepsilon_1)} \approx \exp\left(-\frac{2M}{m}P_{eV}\hbar\omega\frac{\varepsilon_2 - \varepsilon_1}{<\varepsilon>^2}\right), \tag{4.84}$$

where $<\varepsilon>$ is the average electron energy of the Druyvesteyn distribution. The parameter α_v of the distribution function reduction actually describes the probability for an electron to come through the energy interval of intensive vibrational excitation. It is proportional to the probability for the electron to participate in the high-energy processes of electronic excitation and ionization. In other words, the ionization and electronic excitation rate coefficients are proportional to the parameter α_v. For this reason, the parameter α_v can be used to analyze the influence of the vibrational temperature T_v on the ionization and electronic excitation rate coefficients. The probability P_{eV} is proportional to the number density of vibrationally non-excited molecules $N(v=0) \propto 1 - \exp\left(-\frac{\hbar\omega}{T_v}\right)$ (see Eq. (4.12)), taking into account the possibility of super-elastic collisions. As a result, the influence of the vibrational temperature T_v on the ionization and electronic excitation rate coefficients can be characterized as the double exponential function:

$$\alpha_v \approx \exp\left[-\frac{2M}{m}P_{eV}^0\hbar\omega\frac{\varepsilon_2 - \varepsilon_1}{<\varepsilon>^2}\left(1 - \exp\left(-\frac{\hbar\omega}{T_v}\right)\right)\right]. \tag{4.85}$$

Here P_{eV}^0 is the vibrational excitation probability of non-excited molecules ($T_v = 0$). Equation (4.85) illustrates the strong increase of electron energy distribution function as well as ionization and electronic excitation rate coefficients with vibrational temperature.

4.2.6 The Electron Energy Distribution Functions in Different Non-Equilibrium Discharge Conditions

The quantitative distributions found either experimentally or from detailed kinetic modeling are more sophisticated; see Figure 4.7. Discuss shortly general features of such electron energy distribution functions. The electron energy distribution function in non-equilibrium discharges in noble gases are usually close to the Druyvesteyn distribution Eq. (4.81), if the ionization degree is not high enough for the essential contribution of electron–electron collisions. For the same mean energies, $f(\varepsilon)$ in molecular gases is much closer to the Maxwellian function. Quite strong deviations of $f(\varepsilon)$ in molecular gases from the Maxwellian distribution take place only for high energy electrons at relatively low mean energies about 1.5 eV. Even a small admixture of molecular gas (about 1%) into a noble gas dramatically changes the electron energy distribution function, strongly decreasing the fraction of high-energy electrons. The influence of such a small molecular gas admixture is the strongest at relatively low values of the reduced electric field E/n_0 (n_0 is the number density of gas) when the Ramsauer effect is important and essential. The addition of only 10% of air into argon makes the electron energy distribution function appears like that for molecular gas (D.I Slovetsky, 1980). The electron–electron collisions were not taken into account above in the kinetic equation (4.76). They can influence the distribution function and make it Maxwellian when the ionization degree n_e/n_0 is high. This effect is convenient to characterize using the following numerical factor (I.V. Kochetov, V.G. Pevgov, L.S. Polak, D.I. Slovetsky, 1979):

$$a = \frac{\nu_{ee}}{\delta \nu_{en}} \approx 10^8 \frac{n_e}{n_0} \times \frac{1}{<\varepsilon, \mathrm{eV}>^2} \times \frac{10^{-16}\mathrm{cm}^2}{\sigma_{en}, \mathrm{cm}^2} \times \frac{10^{-4}}{\delta}. \tag{4.86}$$

In this expression ν_{ee} and ν_{en} are frequencies of electron–electron and electron–neutral collisions, corresponding to the average electron energy $<\varepsilon>$; σ_{en} is the cross section of the electron–neutral

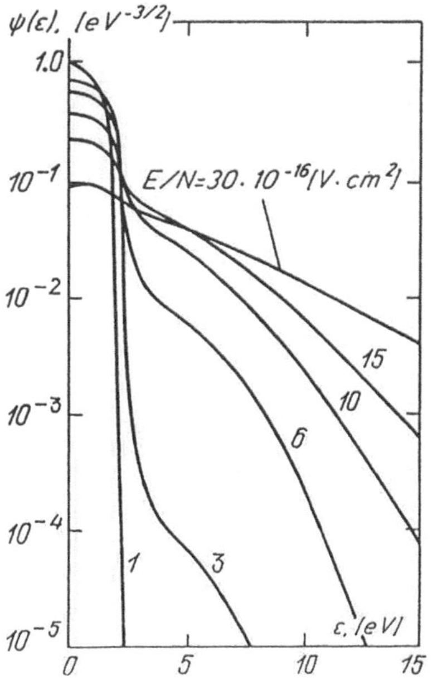

FIGURE 4.7 Electron distribution function $\psi(\varepsilon) = n(\varepsilon)/(n_e\sqrt{\varepsilon})$ in nitrogen. (From "Gas Discharge Physics" Yu. P. Raizer)

collisions at the same electron energy; δ is the average fraction of electron energy transferred to a neutral particle during their collision. The electron–electron collisions make the electron energy distribution function Maxwellian at large high values of the factor a >> 1. Note that the Maxwellization of high-energy electrons ($\varepsilon >> <\varepsilon>$) becomes possible at higher levels of the degree of ionization than Maxwellization of electrons with average energy. The Maxwellization of the high-energy electrons with energy ε by means of the electron–electron collisions requires $a >> \varepsilon^2 / T_e^2$. As seen from Eq. (4.86), the Maxwellization of electrons in a plasma of noble gases (smaller δ) requires lower degrees of ionization than in case of molecular gases (higher δ). For example, the effective Maxwellization of the electron energy distribution function in argon plasma at $E/n_0 = 10^{-16} V * cm^2$ begins at the ionization degree $n_e/n_0 = 10^{-7} \div 10^{-6}$. For molecular nitrogen at the same reduced electric field and ionization degrees up to 10^{-4}, the deviations of the electron distribution provided by the electron–electron collisions are weak.

4.2.7 Relations between Electron Temperature and the Reduced Electric Field

The electron energy distribution functions permit finding the relation between the reduced electric field and average electron energy, which then can be recalculated to an effective electron temperature: $\langle \varepsilon \rangle = \frac{3}{2} T_e$ even for non-Maxwellian distributions. Such relation can be derived by averaging the electron drift velocity in energy space Eq. (4.75) over the entire energy spectrum, taking into account Eq. (4.71):

$$\frac{e^2 E^2}{m v_{en}^2} = \delta \frac{3}{2} T_e. \tag{4.87}$$

Factor δ characterizes the fraction of electron energy lost in a collision with a neutral particle:

$$\delta \approx \frac{2m}{M} + \langle P_{eV} \rangle \frac{\hbar \omega}{\langle \varepsilon \rangle}. \tag{4.88}$$

In atomic gases δ = 2 m/M, and the relation Eq. (4.88) permits finding the exact value of the factor δ for the formula Eq. (4.87). In molecular gases, however, this factor is usually considered as a semi-empirical parameter. Thus, in typical conditions of non-equilibrium discharges in nitrogen, one can take for numerical calculations: $\delta \approx 3 * 10^{-3}$. The further analysis of Eq. (4.87) can be done, taking into account the following relations between parameters:

$$v_{en} = n_0 \langle \sigma_{en} v \rangle, \quad \lambda = \frac{1}{n_0 \sigma_{en}}, \quad \langle v \rangle = \sqrt{\frac{8 T_e}{\pi m}}. \tag{4.89}$$

A combination of these formulas with (4.87) gives the relation between electron temperature, electric field and the mean free path of electrons:

$$T_e = \frac{eE\lambda}{\sqrt{\delta}} \sqrt{\pi / 12}. \tag{4.90}$$

This relation is in agreement with Maxwell (4.80) and Druyvesteyn (4.81) distributions. It is convenient to rewrite (4.90) as a relation between electron temperature and the reduced electric field E/n_0:

$$T_e = \left(\frac{E}{n_0} \right) \frac{e}{\langle \sigma_{en} \rangle} \sqrt{\frac{\pi}{12 \delta}}. \tag{4.91}$$

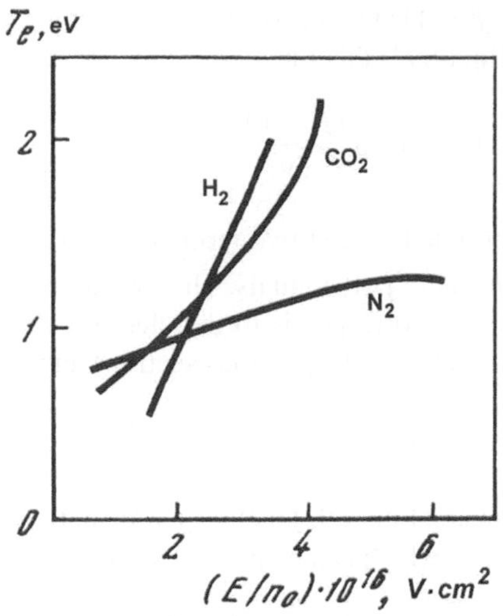

FIGURE 4.8 Electron temperature as a function of reduced electric field.

This linear relation between electron temperature and the reduced electric field is obviously only a qualitative one. As one can see from Figure 4.8, the relation $T_e\left(\frac{E}{n_0}\right)$ can be more complicated in reality.

4.3 ELECTRIC AND THERMAL CONDUCTIVITY IN PLASMA, DIFFUSION OF CHARGED PARTICLES

4.3.1 Isotropic and Anisotropic Parts of the Electron Distribution Functions

In the previous section, we focused on the electron energy distribution functions, which are related to the isotropic part of the general electron velocity distribution $f(\vec{v})$. The electron velocity distribution $f(\vec{v})$ is, obviously, anisotropic in electric field because of the peculiarity of motion in the direction of the field. This anisotropy determines the electric current, plasma conductivity, and all related effects. To describe the anisotropy of the electron velocity distribution, we should recall that an electron receives an additional velocity $\vec{u}$ during the free motion between collisions (see Eq. (4.71) and illustration in Figure 4.5). If the anisotropy is not very strong (which is the normal case) and u<<v, it can be assumed that the fraction of electrons in the point $\vec{v}$ of the real anisotropic distribution f is directly related to the fraction of electrons in the point $\vec{v}-\vec{u}$ of the correspondent isotropic distribution $f^{(0)}$. In this case, taking into account, that directions of $\vec{E}$ and $\vec{u}$ are opposite for electrons, the electron velocity distribution function is:

$$f(\vec{v}) = f^{(0)}(\vec{v}-\vec{u}) \approx f^{(0)}(\vec{v}) - \vec{u}\frac{\partial}{\partial \vec{v}} f^{(0)}(\vec{v}) = f^{(0)}(\vec{v}) + u\cos\theta \frac{\partial f^{(0)}(v)}{\partial v}. \tag{4.92}$$

Here θ is the angle between directions of the velocity $\vec{v}$ and the electric field $\vec{E}$. Taking into account the azimuthal symmetry of $f(\vec{v})$, and omitting the dependence on θ for the isotropic function $f^{(0)}$, Eq. (4.92) can be rewritten in the following scalar form:

$$f(v,\theta) = f^{(0)}(v) + \cos\theta * f^{(1)}(v). \tag{4.93}$$

In this relation, the function $f^{(1)}(v)$ is the one responsible for the anisotropy of the electron velocity distribution. According to Eq. (4.92) this function can be expressed as:

$$f^{(1)}(v) = u\frac{\partial f^{(0)}(v)}{\partial v} = \frac{eE}{m\nu_{en}}\frac{\partial f^{(0)}(v)}{\partial v}. \tag{4.94}$$

Equation (4.93) can be interpreted as first two terms of the series expansion of $f(\vec{v})$ using the ortho-normalized system of Legendre polynomials. The first term of the expansion is the isotropic part of the distribution function. It corresponds to the electron energy distribution functions (see (2.1.1) and other functions considered in the previous section) and related to them as:

$$f(\varepsilon)d\varepsilon = f^{(0)}(v)4\pi v^2 dv. \tag{4.95}$$

The second term of the expansion Eq. (4.93) includes information on the electric field and, hence, it is related to the electron current, which can be presented as:

$$j = -e\int \vec{v}f(\vec{v})d\vec{v} = -\frac{4\pi}{3}e\int_0^{\infty} v^3 f^{(1)}(v)dv. \tag{4.96}$$

The results on anisotropy of the electron distribution function could be generalized for the high-frequency fields. It is sufficient to replace the electric field by the effective one Eq. (4.79). The Eq. (4.94) permits to compare the anisotropic and isotropic parts of the electron velocity distribution function:

$$f^{(1)} = u\frac{\partial f^{(0)}}{\partial v} \sim \frac{u}{\langle v\rangle} f^{(0)}. \tag{4.97}$$

This shows that the smallness of the anisotropy of the distribution function $f^{(1)} << f^{(0)}$, which determines actually the framework of presented consideration, is directly related to smallness of the electron drift velocity with respect to the thermal velocity.

4.3.2 Electron Mobility and Plasma Conductivity

Taking into account Eq. (4.94) for the anisotropic part of distribution function $f^{(1)}(v)$ and the relation between current density and the strength of electric field $j = \sigma E$, the formula for the **electron conductivity** in plasma follows:

$$\sigma = \frac{4\pi e^2}{3m}\int_0^{\infty} \frac{v^3}{\nu_{en}(v)}\left[-\frac{\partial f^{(0)}(v)}{\partial v}\right]dv. \tag{4.98}$$

In the simple case, when $\nu_{en}(v) = \nu_{en} = const$, the integration Eq. (4.98) by parts gives the well known relation between the electric conductivity of the plasma and the electron concentration:

$$\sigma = \frac{n_e e^2}{m\nu_{en}}. \tag{4.99}$$

The following numerical formula based on Eq. (4.99) can be used for practical calculations of the conductivity of the weakly ionized plasma under consideration:

$$\sigma = 2.82 * 10^{-4}\frac{n_e\left(\text{cm}^{-3}\right)}{\nu_{en}\left(\text{s}^{-1}\right)},\ \text{Ohm}^{-1}\text{cm}^{-1} \tag{4.100}$$

Equations (4.99) and (4.100) imply, that electrons mostly provide the plasma conductivity, which is correct in most of cases. Only when the ions concentration much exceeds that of electrons, the ions contribution becomes important as well. For this case, a corresponding expression for ions similar to Eq. (4.100) should be added to calculate the total conductivity. Plasma conductivity presented in form Eq. (4.100) depends on the electron concentration n_e and frequency of electron–neutral collisions ν_{en} (calculated with respect to the momentum transfer). The electron concentration for an equilibrium thermal plasma can be taken in this case from the Saha equation (4.16); for non-equilibrium, non-thermal plasma it can be found from a balance of charged particles (see, for example, Section 4.5). The frequency of electron–neutral collisions ν_{en} is proportional to the gas number density (or to pressure) and can be found numerically for some specific gases in Table 4.1. In general, as seen from Eq. (4.100), the conductivity is proportional to the degree of ionization $\frac{n_e}{n_0}$ in a plasma. Equations (4.99) and (4.100) can be, obviously, applied in particular to calculate the power transferred from an electric field to plasma electrons. This leads to other well-known formula of the so-called **Joule heating**:

$$P = \sigma E^2 = \frac{n_e e^2 E^2}{m\nu_{en}}. \tag{4.101}$$

Also the Eqs. (4.99) and (4.100) can be used to determine **the electron mobility** μ_e, which is the coefficient of proportionality between the electron drift velocity v_d and electric field:

$$v_d = \mu_e E, \quad \mu_e = \frac{\sigma_e}{en_e} = \frac{e}{m\nu_{en}}. \tag{4.102}$$

To calculate the electron mobility the numerical relation similar to Eq. (4.100) can be used:

$$\mu_e = \frac{1.76 * 10^{15}}{\nu_{en}\left(\text{sec}^{-1}\right)}, \frac{\text{cm}^2}{\text{Vs}}. \tag{4.103}$$

Equation (4.102) demonstrates the proportionality of the drift velocity and reduced electric field E/n_0, which is not far from reality (see Figure 4.9) if the electron–neutral collision cross section is not changing significantly. It is not always true. This cross section, for example, decreases significantly in noble gases at low electron energies, which is known as the **Ramsauer effect**. As a result,

TABLE 4.1

Estimated Similarity Parameters, Describing the Electron Neutral Collision Frequency, Electron Mean-Free-Path, Electron Mobility and Conductivity at $\frac{E}{p} = 1 \div 30 \frac{\text{V}}{\text{cm.Torr}}$

Gas	$\lambda * p$, 10^{-2}cmTorr	$\frac{\nu_{en}}{p}$, $10^9 s^{-1} Torr^{-1}$	$\mu_e * p$, $10^6 \frac{cm^2 Torr}{Vs}$	$\frac{\sigma * p}{n_e}$, $10^{-13} \frac{Torr cm^2}{Ohm}$
Air	3	4	0.45	0.7
N_2	3	4	0.4	0.7
H_2	2	5	0.4	0.6
CO_2	3	2	1	2
CO	2	6	0.3	0.5
Ar	3	5	0.3	0.5
Ne	12	1	1.5	2.4
He	6	2	0.9	1.4

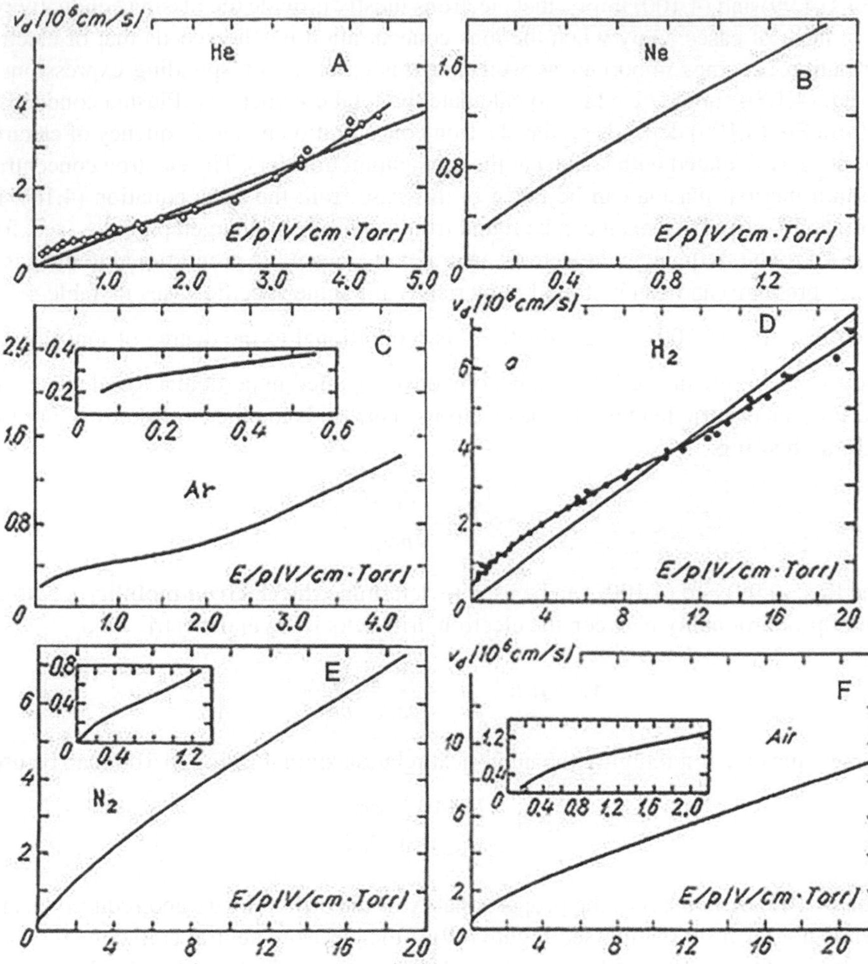

FIGURE 4.9 Electron drift velocities in different inert and molecular gases. (From "Gas Discharge Physics" Yu. P. Raizer)

the effective electron mobility in noble gases can be relatively high at low values of the reduced electric field E/n_0 or E/p (in argon it takes place at E/p about $10^{-3}\,\frac{\text{V}}{\text{cm Torr}}$). The reduced electric field is presented in Figure 4.9 as E/p not as E/n_0. In the same way in Table 4.1, the gas concentration n_0 is also replaced by pressure. Such a way of presentation is historically typical for the description of the room temperature non-thermal discharges, where $n_0(\text{cm}^{-3}) = 3.295 * 10^{16} * p(\text{Torr})$. If the gas temperature T_0 of a non-thermal discharge differs from the room temperature one T_{00}, the pressure should be recalculated to the gas concentration n_0 or at least replaced by the effective pressure $p_{eff} = p\,{}^{T_{00}}\!/_{T_0}$.

4.3.3 The Similarity Parameters, Describing Electron Motion in Non-Thermal Discharges

Taking into account Eqs. (4.99), (4.102) and the above remark about effective pressure, it is convenient to construct the so-called **similarity parameters** $\frac{\nu_{en}}{p}, \lambda p, \mu_e p, \frac{\sigma p}{n_e}$. These similarity

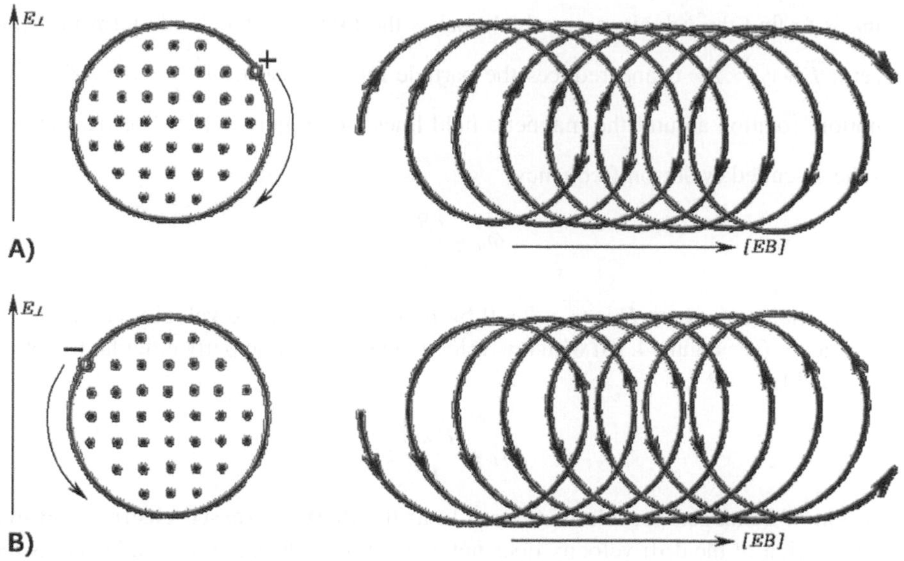

FIGURE 4.10 Illustration of positive (A) and negative (B) particles drift in perpendicular electric and magnetic fields.

parameters are approximately constant for each gas, at low-temperature conditions and $\frac{E}{p} = 1 \div 30 \frac{V}{\text{cm.Torr}}$. Hence, they can be applied for simple determination of the electron neutral collision frequency ν_{en}, electron mean-free-path λ, electron mobility μ_e and conductivity σ. Numerical values of these similarity parameters for several gases are collected in Table 4.1.

4.3.4 Plasma Conductivity in the Perpendicular Static Uniform Electric and Magnetic Fields

The direction of current in the presence of a magnetic field becomes not collinear to the electric field and the conductivity should be considered as the tensor σ_{ij}:

$$j_i = \sigma_{ij} E_j. \tag{4.104}$$

This motion in general is quite sophisticated and discussed in detail, for example, in the book of M.A. Uman (1963). To illustrate the physics of the phenomena let us consider at first the motion of a particle with electric charge q in the static uniform perpendicular electric E and magnetic B fields (E << cB, here c is the speed of light) in the absence of collisions. This motion of a charged particle is shown in Figure 4.10 and can be described by the equation:

$$m\frac{\overrightarrow{dv}}{dt} = q\left(\vec{E} + \vec{v} \times \vec{B}\right). \tag{4.105}$$

Analyze this motion in a reference frame, moving with respect to the initial laboratory reference frame with some velocity $\overrightarrow{v_{EB}}$. The particle velocities in the new and old reference frames are obviously related as: $\vec{v}' = \vec{v} + \overrightarrow{v_{EB}}$. Then Eq. (4.105) in the new moving reference frame can be rewritten as:

$$m\frac{\overrightarrow{dv'}}{dt} = q\left(\vec{E} + \overrightarrow{v_{EB}} \times \vec{B}\right) + q\vec{v}' \times \vec{B}. \tag{4.106}$$

The trick is to find the velocity $\overrightarrow{v_{EB}}$, which makes the first right-hand-side term in Eq. (4.106) equal to zero: $\vec{E}+\overrightarrow{v_{EB}}\times\vec{B}=0$ and reduces the particle motion just to $m\frac{\overrightarrow{dv'}}{dt}=q\overrightarrow{v'}\times\vec{B}$, that is to a circular motion, rotation around the magnetic field lines (see Figure 4.10). The frequency of this rotation is the so-called cyclotron frequency:

$$\omega_B=\frac{eB}{m}. \tag{4.107}$$

Multiplying the requirement $\vec{E}+\overrightarrow{v_{EB}}\times\vec{B}=0$ by $\vec{B}$, a formula for the velocity $\overrightarrow{v_{EB}}$can be derived. The "spiral" moves (see Figure 4.10) with this velocity in the direction perpendicular to both electric and magnetic fields:

$$\overrightarrow{v_{EB}}=\frac{\vec{E}\times\vec{B}}{B^2}. \tag{4.108}$$

This motion of a charged particle is referred to as the **drift in crossed electric and magnetic fields**. The direction of the drift velocity does not depend on a charge sign and, hence, is the same for electrons and ions. Equations (4.107) and (4.108) describe the collisionless drift of charged particles in crossed electric and magnetic fields. Collisions with neutral particles make the charged particles to move additionally along the electric field. This combined motion can be described by the conductivity tensor Eq. (4.104), which includes two conductivity components: the first one represents conductivity along the electric field $\sigma_{\|}$, and the second one corresponds to the current perpendicular to both electric and magnetic fields $\sigma_{\perp}$:

$$\sigma_{||}=\frac{\sigma_0}{1+\left(\frac{\omega_B}{\nu_{en}}\right)^2},\quad \sigma_{\perp}=\sigma_0\frac{\left(\frac{\omega_B}{\nu_{en}}\right)}{1+\left(\frac{\omega_B}{\nu_{en}}\right)^2}. \tag{4.109}$$

In these relations σ_0- is the conventional conductivity Eq. (4.99) in absence of a magnetic field. When the magnetic field is low or pressure high, - then $\omega_B << \nu_{en}$. In this case, according to Eq. (4.109) the transverse conductivity $\sigma_{\perp}$ can be neglected, and the longitudinal conductivity $\sigma_{\|}$ actually coincides with the conventional one σ_0. Conversely, in the case of high magnetic fields and low pressures ($\omega_B >> \nu_{en}$), first of all electrons become "trapped" by the magnetic field and start drifting across the electric and magnetic fields. The longitudinal conductivity then can be neglected, and the transverse conductivity becomes independent of n pressure and mass of a charged particle:

$$\sigma_{\perp}\approx\frac{n_e e^2}{m\omega_B}=\frac{n_e e}{B}. \tag{4.110}$$

The transverse conductivity Eq. (4.109) corresponds to the collisionless drift velocity Eq. (4.110) in the crossed electric and magnetic fields. Note, that this relation for plasma conductivity is valid when the cyclotron velocity slightly exceeds the electron neutral collision frequency ($\omega_B >> \nu_{en}$) but not too much. If the magnetic field is much larger, than it is enough for $\omega_B >> \nu_{en}$, then the ion motion can also become quasi-collisionless, and according to Eq. (4.108) - ions and electrons move together without any current.

4.3.5 Conductivity of the Strongly Ionized Plasma

The electric conductivity in weakly ionized plasma (Eqs. (4.99) and (4.100)) was always related to the resistance provided by electron–neutral collisions. In strongly ionized plasma with higher degrees of ionization, the electron–ion scattering also has to be taken into account. In this case, the electron–neutral collision frequency ν_{en} in Eqs. (4.99), (4.100) has to be replaced by the total frequency, including the electron–neutral and electron–ion collisions:

$$\nu_{\Sigma} = \nu_{en} + n_e \langle v \rangle \sigma_{Coul}. \tag{4.111}$$

It is assumed, that the ion and electron densities are equal (n_e); $\langle v \rangle$ is the average electron velocity; and σ_{Coul} is the averaged **Coulomb cross section** of the electron–ion collisions:

$$\sigma_{Coul} = \frac{4\pi}{9} \frac{e^4 \ln \Lambda}{\left(4\pi\varepsilon_0 T_e\right)^2} = \frac{2.87 * 10^{-14} \ln \Lambda}{\left(T_e, \mathrm{eV}\right)}, \mathrm{cm}^2. \tag{4.112}$$

Though trajectory deflection in the electron–ion interaction is low when they are relatively far apart one from another, such large distance collisions make an important contribution to the momentum transfer cross section because of the long-range nature of Coulomb forces. This effect is taken into account in relation Eq. (4.112) by multiplication of the natural cross section for the interaction of charged particles $\frac{e^4}{\left(4\pi\varepsilon_0 T_e\right)^2}$ by the so-called **Coulomb logarithm:**

$$\ln \Lambda = \ln\left[\frac{3}{2\sqrt{\pi}} \frac{\left(4\pi\varepsilon_0 T_e\right)^{3/2}}{e^3 n_e^{1/3}}\right] = 13.57 + 1.5 \log\left(T_e, \mathrm{eV}\right) - 0.5 \log n_e. \tag{4.113}$$

Based upon Eqs. (4.112) and (4.113) one can calculate, that the electron–ion collisions become significant in the total electron–collision frequency Eq. (4.111) and, hence, in the electric conductivity when the degree of ionization exceeds some critical value about: $\frac{n_e}{n_0} \geq 10^{-3}$. When the degree of ionization much exceeds the critical value, the electron–ion collisions become dominant in Eq. (4.111). The electric conductivity in this case (see Eq. (4.111)) becomes almost independent on the electron concentration n_e (only through the Coulomb logarithm lnΛ) and reaches its maximum value:

$$\sigma = \frac{9\varepsilon_0 T_e^2}{m \langle v \rangle e^2 \ln \Lambda} = 1.9 * 10^2 \frac{\left(T_e, \mathrm{eV}\right)^{3/2}}{\ln \Lambda}, \mathrm{Ohm}^{-1}\mathrm{cm}^{-1}. \tag{4.114}$$

4.3.6 Ion Energy and Ion Drift in Electric Field

Relation between the ions average energy $\langle \varepsilon_i \rangle$, gas temperature T_0 and electric field E is (Yu. P. Raizer (1991):

$$\langle \varepsilon_i \rangle = \frac{3}{2} T_0 + \frac{M}{2M_i}\left(1 + \frac{M_i}{M}\right)^3 \frac{e^2 E^2}{M_i \nu_{in}^2}. \tag{4.115}$$

Here M_i and M are masses of an ion and a neutral particle; ν_{in} frequency of ion-neutral collisions corresponding to momentum transfer. If the reduced electric field in plasma is not too strong (usually E/p < 10 V/cmTorr), the ion energy Eq. (4.115) only slightly exceeds that of neutrals. For electrons, because of their low mass, such a situation takes place only at very low reduced electric fields E/p < 0.01 V/cmTorr. When the reduced electric fields are much stronger than mentioned above (E/p >> 10 V/cmTorr) and the second term in Eq. (4.115) exceeds the first one, then the ion velocity, collision frequency ν_{in} and the ion energy are growing with the electric field:

$$\langle \varepsilon_i \rangle \approx \frac{1}{2}\sqrt{\frac{M}{M_i}}\left(1+\frac{M_i}{M}\right)^{3/2} eE\lambda, \tag{4.116}$$

here λ is the ion mean-free-path. In the case of weak and moderate electric fields (E/p < 10 V/cm Torr), the ion drift velocity is proportional to the electric field and the ion mobility is constant:

$$\overrightarrow{v_{id}} = \frac{e\vec{E}}{v_{in} * MM_1/(M+M_1)}, \quad \mu_i = \frac{e}{v_{in} * MM_1/(M+M_1)}. \tag{4.117}$$

The ion mobility is actually constant as long as the ion-neutral collision frequency ν_{in} is constant, which takes place in the case of polarization-induced collisions (see the Langevin model, relations Eqs. (2.81)–(2.84)). The following convenient numerical relation can be used for calculations of the ion mobility in frameworks of the Langevin model:

$$\mu_i = \frac{2.7 * 10^4 \sqrt{1+M/M_i}}{p(Torr)\sqrt{A*(\alpha/a_0^3)}}. \tag{4.118}$$

In this relation α is the polarizability of a neutral particle (see Section 2.6); "A" is its molecular mass; a_0 is the Bohr radius. At relatively strong electric fields (E/p >> 10 V/cm Torr), the ion energies are higher and the ion-neutral collision cross section is limited by the gas-kinetic one. Similar to Eq. (4.116) - not ν_{in} but λ is better to be considered as constant. Taking into account Eq. (4.116), we can find the drift velocity in the case of strong electric fields as:

$$v_{id} \approx \sqrt[4]{\frac{M_i}{M}\left(1+\frac{M_i}{M}\right)}\sqrt{\frac{eE\lambda}{M_i}}. \tag{4.119}$$

The linear dependence of the ion drift velocity on the electric field for weak fields is changing to the square root proportionality at strong electric fields, which is illustrated in Figure 4.11.

4.3.7 Free Diffusion of Electrons and Ions

If electron and ion concentrations are quite low- their diffusion in plasma can be considered as independent and described, for example, by the continuity equation Eq. (4.65). The total flux of charged particles ($\vec{\Phi} = n\vec{u}$, which in Eq. (4.65) was referred to as "drift") includes in this case the actual drift in the electric field described above in Section (4.93), and diffusion, following the Fick's law:

$$\overrightarrow{\Phi_{e,i}} = \pm n_{e,i}\mu_{e,i}\vec{E} - D_{e,i}\frac{\partial n_{e,i}}{\partial \vec{r}}. \tag{4.120}$$

The sings "+" and "–" here are corresponding to the charge of particles. The continuity equation (4.65) then can be rewritten as:

$$\frac{\partial n_{e,i}}{\partial t} + \frac{\partial}{\partial \vec{r}}\left(\pm n_{e,i}\mu_{e,i}\vec{E} - D_{e,i}\frac{\partial n_{e,i}}{\partial \vec{r}}\right) = G - L. \tag{4.121}$$

The characteristic diffusion time corresponding to the diffusion distance R can be estimated based on the relation (4.121) as $\tau_D \approx R^2/D_{e,i}$. This simple relation is widely used to estimate all kinds of diffusion processes. The diffusion coefficients for both electrons and ions can be estimated by the following formula including the corresponding values of thermal velocity, mean free path and collisional frequency with neutrals ($<v_{e,i}>$, $\lambda_{e,i}$, $\nu_{en,in}$):

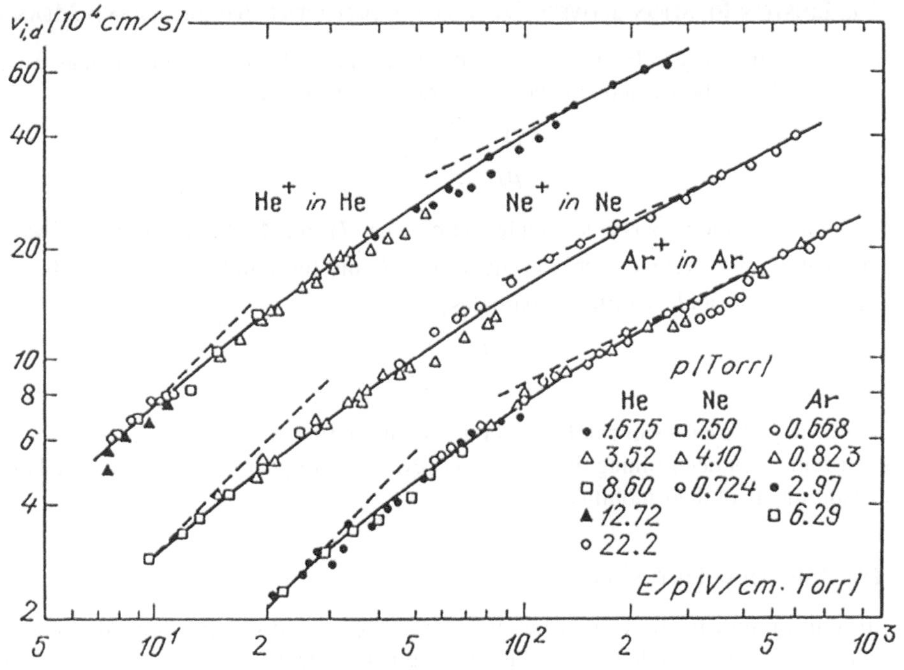

FIGURE 4.11 Drift velocities of ions in inert gases T = 300K: $v_{id} \propto E$-dashed line on the left; $v_{id} \propto \sqrt{E}$-dashed line on the right. (From "Gas Discharge Physics" Yu. P. Raizer)

TABLE 4.2
Coefficients of the Free Diffusion D of Electrons and Ions at Room Temperature in Different Gases

Diffusion	$D * p, \frac{cm^2}{s}$ Torr	Diffusion	$D * p, \frac{cm^2}{s}$ Torr	Diffusion	$D * p, \frac{cm^2}{s}$ Torr
"e" in He	$2.1*10^5$	"e" in Ne	$2.1*10^6$	"e" in Ar	$6.3*10^5$
"e" in Kr	$4.4*10^4$	"e" in Xe	$1.2*10^4$	"e" in H_2	$1.3*10^5$
"e" in N_2	$2.9*10^5$	"e" in O_2	$1.2*10^6$	N_2^+ in N_2	40

$$D_{e,i} = \frac{\langle v_{e,i}^2 \rangle}{3 v_{en,in}} = \frac{\lambda_{e,i} \langle v_{e,i} \rangle}{3}. \tag{4.122}$$

As it is clear from the relation Eq. (4.122), the diffusion coefficients are inversely proportional to the gas number density or pressure. This means that the similarity parameter $D_{e,i} * p$ (see Section 4.3.3) can be used to easily calculate easily the coefficients of free diffusion at room temperature. Some numerical values of the similarity parameters for both electrons and ions are presented in Table 4.2. In accordance with Eq. (4.132), the coefficient of free electron diffusion grows with temperature in non-thermal constant pressure discharges as $D_e \propto T_e^{1/2} T_0$. Assuming that ions and neutrals have the same temperature, the coefficient of free ions diffusion grows with temperature in non-thermal discharges as $D_i \propto T_0^{3/2}$.

4.3.8 The Einstein Relation between Diffusion Coefficient, Mobility, and Mean Energy

Taking the electron mobility as Eq. (4.102) and the electron free diffusion coefficient as Eq. (4.122), one can find the relation between them and the average electron energy:

$$\frac{D_e}{\mu_e} = \frac{\frac{2}{3}\langle \varepsilon_e \rangle}{e}. \tag{4.123}$$

A similar relation for ions can be derived from Eqs. (4.117) and (4.122). In the case of Maxwellian distribution function, the average energy is equal to $\frac{3}{2}T$ and formula Eq. (4.123) can be expressed in a more general form valid for different particles:

$$\frac{D}{\mu} = \frac{T}{e}, \tag{4.124}$$

This is known as **the Einstein relation**. The Einstein relations Eqs. (4.123) and (4.124) are very useful, in particular, for determining the diffusion coefficients of charged particles, based on experimental measurements of their mobility.

4.3.9 The Ambipolar Diffusion

The electron and ion diffusion cannot be considered "free" and independent like it was earlier assumed when the ionization degree is relatively high**.** The electrons are moving faster than ions and form the charge separation zone with a strong polarization field, that adjusts the electron and ion fluxes. This electric polarization field accelerates ions and slow down electrons, which makes all of them diffuse together "as a team". This phenomenon is known as the **ambipolar diffusion** and is illustrated in Figure 4.12. If the separation of charges is small, the electron and positive ion concentrations are approximately equal $n_e \approx n_i$ as well as their fluxes. The equality of the fluxes, including diffusion and drift in the polarization field, can be expressed based on Eq. (4.120) as:

$$\overrightarrow{\Phi_e} = -\mu_e \vec{E} n_e - D_e \frac{\partial n_e}{\partial \vec{r}}, \quad \overrightarrow{\Phi_i} = \mu_i \vec{E} n_i - D_i \frac{\partial n_i}{\partial \vec{r}}, \quad \overrightarrow{\Phi_e} = \overrightarrow{\Phi_i}. \tag{4.125}$$

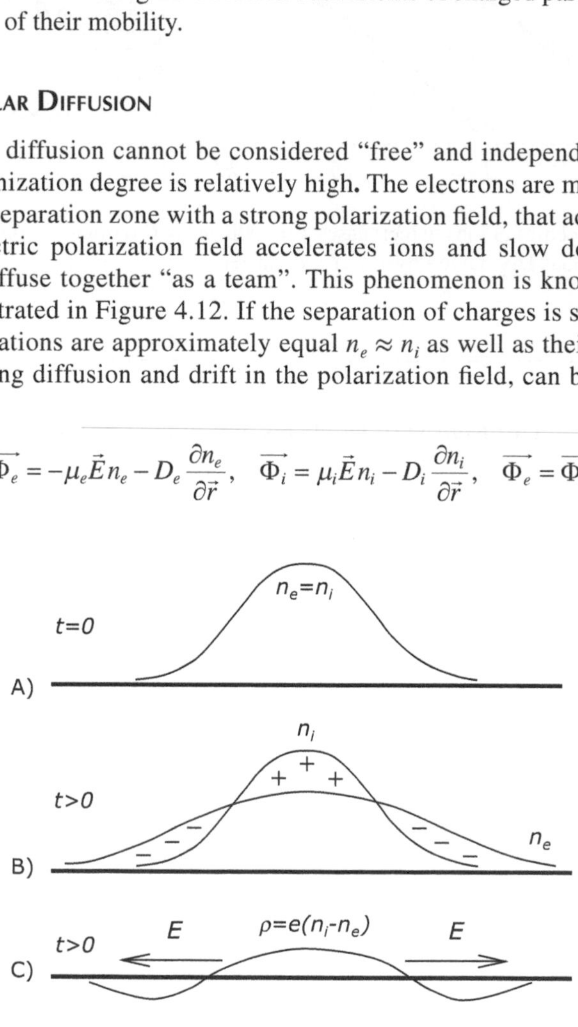

FIGURE 4.12 Illustration of the ambipolar diffusion effect and plasma polarization in the presence of electron and ion density gradients.

Dividing the first relation by μ_e, second one by μ_i and adding up the results gives the general relation for both electron and ion fluxes:

$$\overrightarrow{\Phi_{e,i}} = -\frac{D_i\mu_e + D_e\mu_i}{\mu_e + \mu_i}\frac{\partial n_{e,i}}{\partial \vec{r}}. \tag{4.126}$$

This permits the introduction of the coefficient of ambipolar diffusion $\overrightarrow{\Phi_{e,i}} = -D_a \frac{\partial n_{e,i}}{\partial \vec{r}}$, which describes the collective diffusive motion of electrons and ions:

$$D_a = \frac{D_i\mu_e + D_e\mu_i}{\mu_e + \mu_i}. \tag{4.127}$$

Noting that $\mu_e >> \mu_i$ and $D_e >> D_i$, Eq. (4.127) can be rewritten as $D_a = D_i + D_e {\mu_i}/{\mu_e}$ and lead to the conclusion that ambipolar diffusion is greater than that of ions and less than that of electrons. Hence the ambipolar diffusion "speeds up" ions and "slow down" electrons. Taking into account also the Einstein relation Eq. (4.124) yields:

$$D_a \approx D_i + \frac{\mu_i}{e}T_e. \tag{4.128}$$

It follows that in equilibrium plasmas with equal electron and ion temperatures that $D_a = 2D_i$. For non-equilibrium plasmas with $T_e >> T_i$, the ambipolar diffusion $D_a = \frac{\mu_i}{e}T_e$ corresponds to the temperature of the fast electrons and mobility of the slow ions.

4.3.10 Conditions of Ambipolar Diffusion, the Debye Radius

To determine the conditions of the ambipolar diffusion with respect to the free diffusion of electrons and ions, at first estimate the absolute value of the polarization field from the equations Eq. (4.125):

$$E \approx \frac{D_e}{\mu_e}\frac{1}{n_e}\frac{\partial n_e}{\partial r} = \frac{T_e}{e}\frac{\partial \ln n_e}{\partial r} \propto \frac{T_e}{eR}. \tag{4.129}$$

Here R is the characteristic length for change of the electron concentration. From another view, the difference between the ion and electron concentrations $\Delta n = n_i - n_e$, which characterizing the space charge, is related to the electric field by the Maxwell equation: $\frac{\partial}{\partial \vec{r}}\vec{E} = \frac{e\Delta n}{\varepsilon_0}$. Simplifying the Maxwell equation for estimations as: $\frac{E}{R} \propto \frac{e\Delta n}{\varepsilon_0}$. and combining this relation with Eq. (4.129), yields the relative deviation from quasi-neutrality:

$$\frac{\Delta n}{n_e} \approx \frac{T_e}{e^2 n_e}\frac{1}{R^2} = \left(\frac{r_D}{R}\right)^2, \quad r_D = \sqrt{\frac{T_e\varepsilon_0}{e^2 n_e}}. \tag{4.130}$$

Here r_D is the so-called **Debye radius**. This the important plasma parameter characterizing the quasi-neutrality. The Debye radius is the characteristic of strong charge separation and plasma polarization. If the electron concentration is high and the Debye radius is small $r_D << R$, then - according to Eq. (4.130) – the deviation from quasi-neutrality is small, the electrons and ions move "as a group" and the diffusion should be considered as ambipolar. Conversely, if the electron concentration is relatively low and the Debye radius is large $r_D \geq R$, then the plasma is not quasi-neutral, the electrons and ions move separately and their diffusion should be considered as free. For calculations of the Debye radius it is convenient to use the following numerical formula:

$$r_D = 742\sqrt{\frac{T_e, eV}{n_e, \text{cm}^{-3}}}, \text{cm}. \tag{4.131}$$

For example, if the electron temperature is 1eV and the electron concentration exceeds 10^8cm^{-3}, then according to Eq. (4.131), the Debye radius is less than 0.7mm and diffusion should be considered as ambipolar for gradients scale more greater than 3 mm–1 cm. A quite large collection of numerical data concerning diffusion and drift of electron and ions can be found in books of Earl W. MacDaniel, E.A. Mason (1973) and L.G.H. Huxley, R.W. Crompton (1974).

4.3.11 Thermal Conductivity in Plasma

Heat transfer and, in particular, the behavior of the thermal conductivity in a plasma is quite a complex subject of plasma engineering, books of M.I. Boulos, P. Fauchais and E. Pfender (1994) or A.V. Eletsky, L.A. Palkina, B.M. Smirnov (1975). The equation of heat transfer due to the thermal conductivity in moving one-component gas can be expressed similarly to the continuity equation (see Section 4.2.3 and specifically relation Eq. (4.65)):

$$\frac{\partial}{\partial t}\left(n_0\langle\varepsilon\rangle\right) + \nabla * \vec{q} = 0. \tag{4.132}$$

Here n_0 and $<\varepsilon>$-are the gas number density and average energy of a molecule; $\vec{q}_c$is the heat flux. In Eq. (4.132) replace the average energy of a molecule by the equilibrium temperature T_0 and specific heat c_v with respect to one molecule. Also let us take into account, that the heat flux includes two terms: one related to the gas motion with velocity $\vec{u}$ and another one related to the thermal conductivity with a coefficient κ: $\vec{q} = \langle\varepsilon\rangle n_0\vec{u} - \kappa\nabla * T_0$. This gives a convenient equation, describing the thermal conductivity in moving quasi-equilibrium gas:

$$\frac{\partial}{\partial t}T_0 + \vec{u}\nabla * T_0 = \frac{\kappa}{c_v n}\nabla^2 T_0. \tag{4.133}$$

The thermal conductivity coefficient in a one-component gas without dissociation, ionization, and chemical reactions can be estimated similarly to Eq. (4.122) as:

$$\kappa \approx \frac{1}{3}\lambda\langle v\rangle n_0 c_v \propto \frac{c_v}{\sigma}\sqrt{\frac{T_0}{M}}. \tag{4.134}$$

Here σ is a typical cross section of the molecular collisions, and M- is the molecular mass. According to Eq. (4.134), - the thermal conductivity coefficient does not depend on gas density and grows slowly with temperature. The growth of thermal conductivity with temperature in plasma at high temperatures, however, can be much faster than given by Eq. (4.134), because of the influence of dissociation, ionization, and chemical reactions. For example, consider the effect of dissociation and recombination $2A \Leftrightarrow A_2$ on accelerating the temperature dependence of the thermal conductivity. Molecules are mostly dissociated into atoms in a zone with the higher temperature (see Figure 4.13) and much less dissociated in lower temperature zones. Then the quasi-equilibrium diffusion of the molecules (D_m) to the higher temperature zone leads to their intensive dissociation, consumption of dissociation energy E_D and to the related big heat flux:

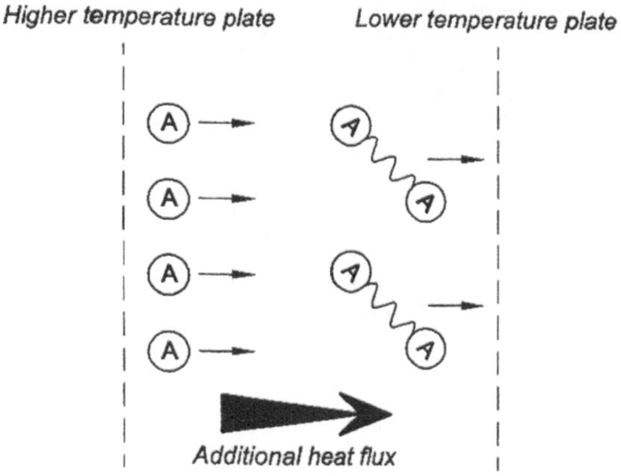

FIGURE 4.13 Illustration of recombination contribution into the high temperature thermal conductivity.

$$\overrightarrow{q_D} = -E_D D_m \nabla n_m = -\left(E_D D_m \frac{\partial n_m}{\partial T_0} \right) \nabla T_0 \tag{4.135}$$

which can be interpreted as accelerating the thermal conductivity. When the concentration of molecules is less than that of atoms concentration $n_m << n_a$ and $T_0 << E_D$, the equilibrium relation Eq. (4.19) permits evaluation of the molecules density derivative $\frac{\partial n_m}{\partial T_0} = \frac{E_D}{T_0^2} n_m$. Substituting this derivative into Eq. (4.135), the coefficient of thermal conductivity related to the dissociation of molecules is given by:

$$\kappa_D = D_m \left(\frac{E_D}{T_0} \right)^2 n_m. \tag{4.136}$$

Compare the temperature dependence of the coefficient of thermal conductivity related to the dissociation of molecules with the one (4.134) not taking into account any reactions Eq. (4.134). Based on Eqs. (4.136), (4.122) and (4.134), the ratio of the two coefficients can be expressed as:

$$\frac{\kappa_D}{\kappa} \approx \frac{1}{c_v} \left(\frac{E_D}{T_0} \right)^2 \frac{n_m}{n_a}. \tag{4.137}$$

The contribution of dissociation to the temperature dependence of the thermal conductivity related to dissociation can be very significant since $T_0 << E_D$. Obviously, the strongest effect takes place within the relatively narrow range of typical dissociation temperatures, when the concentrations of atoms and molecules are of the same order and very sensitive to temperature. Relations similar to Eq. (4.137) can be derived for showing the contributions of ionization and chemical reactions to the temperature dependence of the thermal conductivity. These effects also are able to increase the thermal conductivity, because they are related to the transfer of relatively large quantities of energy. Both ionization energy and typical chemical reaction enthalpy usually much exceeds temperature. However, the relative influence of these effects depends on the relative concentration of corresponding species, which could be pretty low even in the high-temperature thermal plasmas. Specific examples of the temperature dependence of the thermal conductivity for different inert gases are presented in Figure 4.14, and for different molecular gases presented on Figure 4.15. The influence of different effects on the total thermal conductivity can be very well illustrated by the "roller coaster like" $\kappa(T_0)$ dependence in air (see Figure 4.15).

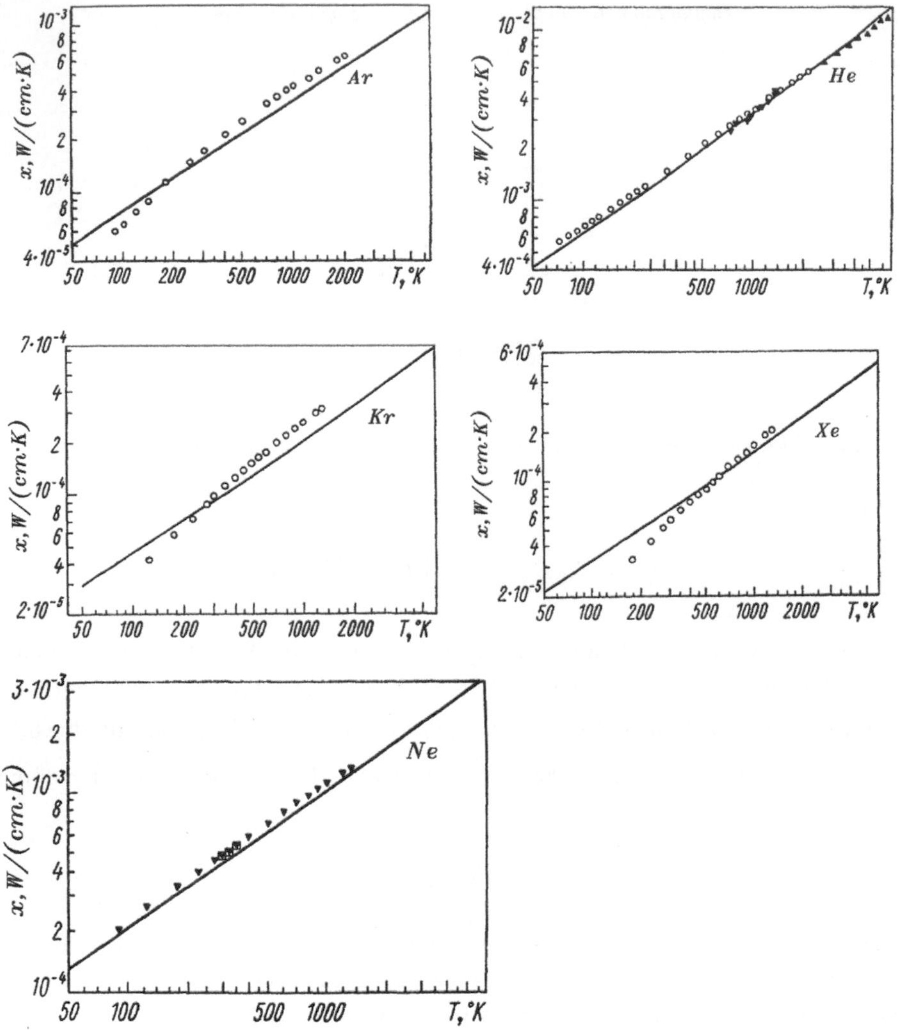

FIGURE 4.14 Thermal conductivity of inert gases.

4.4 BREAKDOWN PHENOMENA: THE TOWNSEND AND SPARK MECHANISMS, AVALANCHES, STREAMERS, AND LEADERS

4.4.1 Electric Breakdown of Gases, the Townsend Mechanism

The electric breakdown is a complicated multi-stage threshold process, which occurs when the electric field exceeds some critical value. During the short breakdown period, usually $0.01 - 100\,\mu s$, the non-conducting gas becomes conductive, as a result generating different kind of plasmas. The breakdown mechanisms can be very sophisticated, but all of them usually start with the electron avalanche, for example, multiplication of some primary electrons in cascade ionization. Consider first the simplest breakdown in a plane gap with electrode separation d, connected to a DC power supply with voltage V, which provides the homogeneous electric field $E = V/d$ (Figure 4.16). It is apparent that occasional formation of primary electrons near cathode occurs which provides the very low initial current i_0. Each primary electron drifts to the anode, concurrently ionizing the gas and generating an avalanche. The avalanche evolves both in time and in space because the

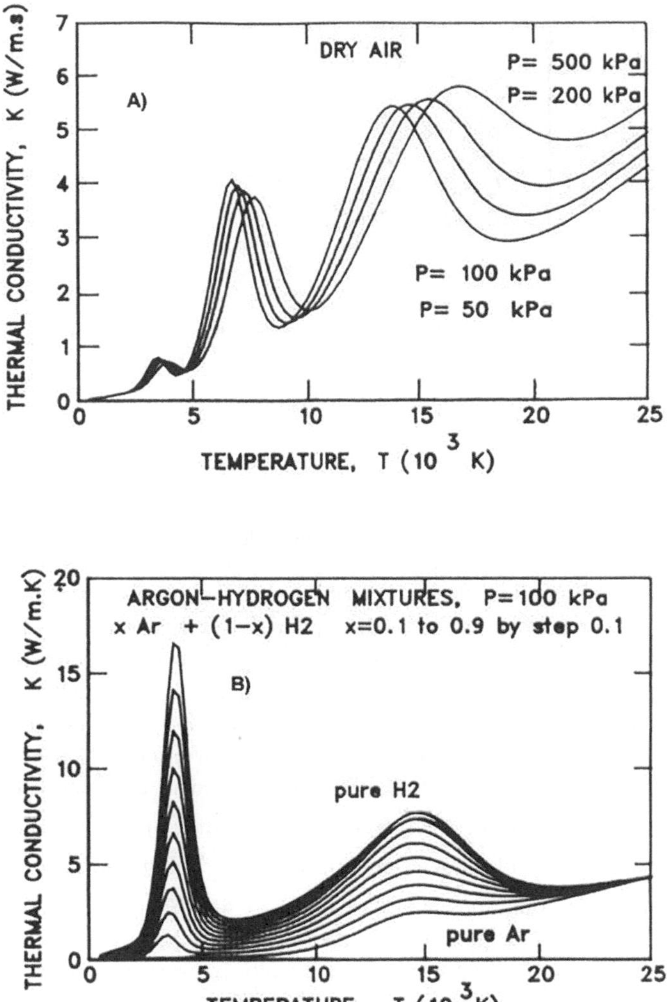

FIGURE 4.15 Temperature dependence of the thermal conductivity of plasmas in air and Ar-H_2. (From Thermal Plasma, M. Boules, P. Fauchais and E. Phender)

multiplication of electrons proceeds along with their drift from the cathode to anode (Figure 4.16). It is convenient to describe the ionization in an avalanche not by the ionization rate coefficient, but rather by the **Townsend ionization coefficient α**, that shows electron production per unit length or the multiplication of electrons (initial density n_{e0}) per unit length along the electric field: $dn_e/dx = \alpha n_e$ or the same $n_e(x) = n_{e0} \exp(\alpha x)$. The Townsend ionization coefficient is related to the ionization rate coefficient $k_i\left(E/n_0\right)$ (Section 2.2.1) and electron drift velocity v_d as:

$$\alpha = \frac{v_i}{v_d} = \frac{1}{v_d} k_i\left(E/n_0\right) n_0 = \frac{1}{\mu_e} \frac{k_i(E/n_0)}{E/n_0}, \tag{4.138}$$

where ν_i is the ionization frequency with respect to one electron, μ_e is the electron mobility. Noting that breakdown begins at room temperature and that the electron mobility is inversely proportional to pressure, it is convenient to present the Townsend coefficient α as the similarity parameter α/p

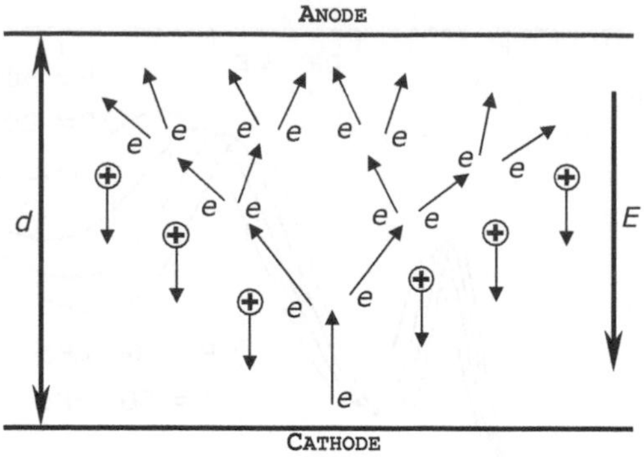

FIGURE 4.16 Illustration of the Townsend breakdown gap.

depending on the reduced electric field E/p. Such dependence $\alpha/p = f\left(E/p\right)$ is presented in Figure 4.17 for different inert and molecular gases. According to the definition of the Townsend coefficient α, each one primary electron generated near the cathode produces $\exp(\alpha d) - 1$ positive ions in the gap (Figure 4.16). Here the electron losses due to recombination and attachment to electronegative molecules were neglected. All the $\exp(\alpha d) - 1$ positive ions produced in the gap per one electron are moving toward the cathode, and altogether eliminate $\gamma * [\exp(\alpha d) - 1]$ electrons from the cathode in the process of secondary electron emission. Here another Townsend coefficient γ is the secondary emission coefficient, characterizing the probability of a secondary electron generation on the cathode by an ion impact. The **secondary electron emission coefficient γ** depends on cathode material, state of surface, type of gas, and reduced electric field E/p (defining the ions energy). Relevant numerical data for different conditions are given in Figure 4.18. Typical values of γ in electric discharges is 0.01–0.1; the effect of photons and meta-stable atoms and molecules (produced in an avalanche) on the secondary electron emission is usually incorporated in the same "effective" γ coefficient. Taking into account the current of primary electrons i_0 and electron current due to the secondary electron emission from the cathode, the total electronic part of the cathode current i_{cath} is:

$$i_{cath} = i_0 + \gamma i_{cath}\left[\exp(\alpha d) - 1\right]. \tag{4.139}$$

The total current in the external circuit is equal to the electronic current at the anode because of the absence of ion current. The total current can be found as $i = i_{cath} \exp(\alpha d)$, which taking Eq. (4.139) into account, to **the Townsend formula**, first derived in 1902 to describe the ignition of electric discharges:

$$i = \frac{i_0 \exp(\alpha d)}{1 - \gamma\left[\exp(\alpha d) - 1\right]}. \tag{4.140}$$

The current in the gap is non-self-sustained as long as the denominator in Eq. (4.140) is positive. As soon as the electric field, and hence, the Townsend α coefficient becomes sufficiently high the denominator in Eq. (4.140) goes to zero and transition to self-sustained current (breakdown!) takes place. Thus the simplest breakdown condition in the gap can be expressed as:

$$\gamma\left[\exp(\alpha d) - 1\right] = 1, \quad \alpha d = \ln\left(\frac{1}{\gamma} + 1\right). \tag{4.141}$$

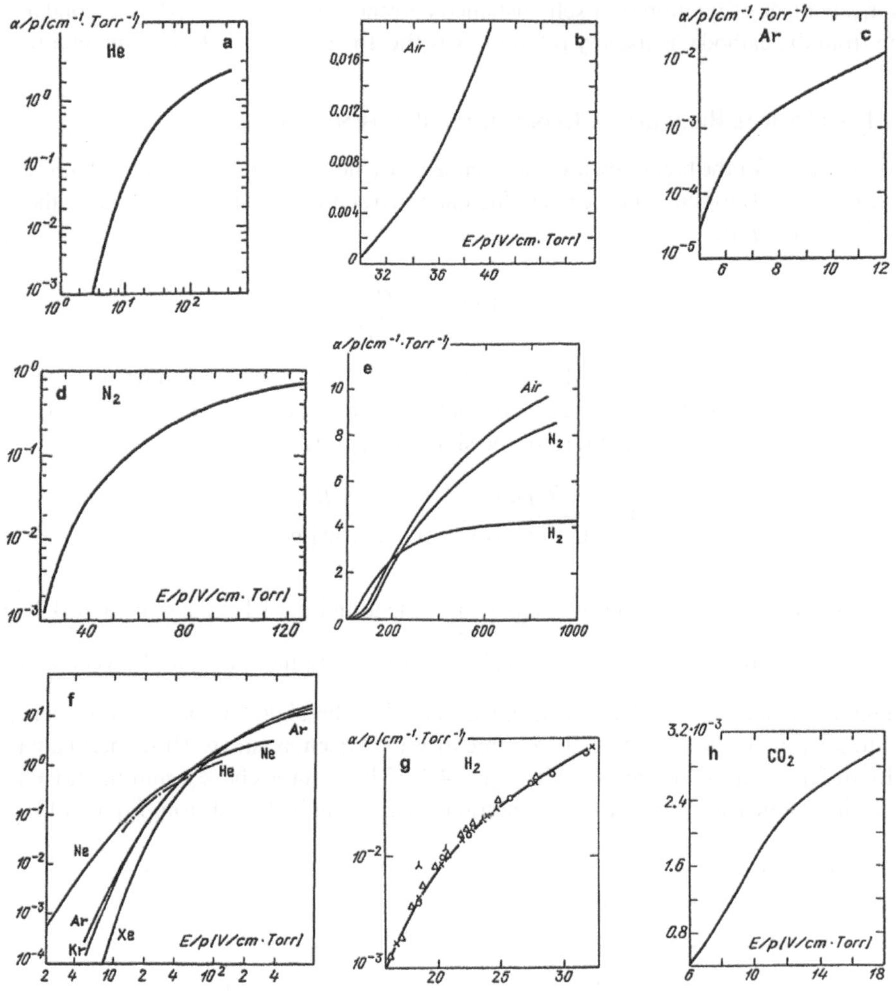

FIGURE 4.17 Ionization coefficients in different molecular gases.

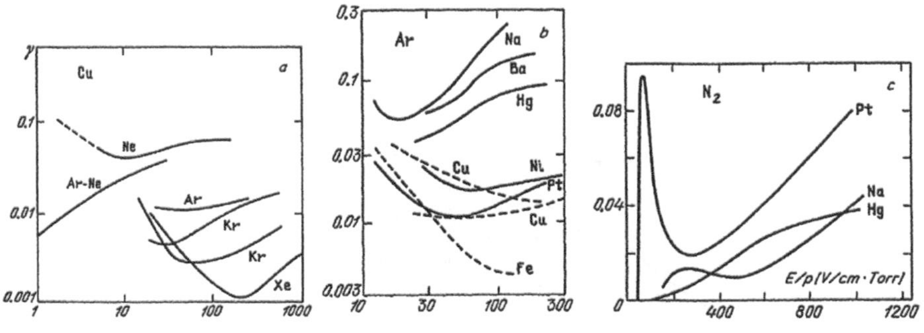

FIGURE 4.18 Effective secondary emission coefficient: (a) copper cathode in inert gases, (b) various metals in Ar, (c) various metals in N_2 (From "Gas Discharge Physics" Yu. P. Raizer)

This mechanism of ignition of a self-sustained current in gap, controlled by secondary electron emission from the cathode, is usually referred to as **the Townsend breakdown mechanism**.

4.4.2 The Critical Breakdown Conditions, Paschen Curves

To derive relations for the breakdown voltage and electric field based on Eq. (4.140), it is convenient to rewrite Eq. (4.138) for the Townsend coefficient α in a semi-empirical way, relating the similarity parameters α/p and E/p:

$$\frac{\alpha}{p} = A\exp\left(-\frac{B}{E/p}\right). \tag{4.142}$$

The parameters A and B of Eq. (4.142) at $E/p = 30 \div 500$ V/cm Torr, are given in Table 4.3. Combining Eqs. (4.141) and (4.142) gives formulas for calculating breakdown voltage and breakdown reduced electric field as functions of the similarity parameter pd:

$$V = \frac{B(pd)}{C + \ln(pd)}, \quad \frac{E}{p} = \frac{B}{C + \ln(pd)}. \tag{4.143}$$

In these relations, parameter B is obviously the same as in Eq. (4.142) and in Table 4.3. However, the parameter A, is replaced by $C = \ln A - \ln\ln\left(\frac{1}{\gamma} + 1\right)$, taking into account the very weak, double logarithmic influence of secondary electron emission. The breakdown voltage dependence (4.143) on the similarity parameter pd is usually referred to as the **Paschen curve**. The experimental Paschen curves for different gases are presented in Figure 4.19. These curves have a minimum voltage point, corresponding to the easiest breakdown conditions, which can be found from Eq. (4.143):

TABLE 4.3
Numerical Parameters A and B for Calculation of the Townsend Coefficient α

Gas	$A, \frac{1}{\text{cmTorr}}$	$B, \frac{\text{V}}{\text{cmTorr}}$	Gas	$A, \frac{1}{\text{cmTorr}}$	$B, \frac{\text{V}}{\text{cmTorr}}$
Air	15	365	N_2	10	310
CO_2	20	466	H_2O	13	290
H_2	5	130	He	3	34
Ne	4	100	Ar	12	180
Kr	17	240	Xe	26	350

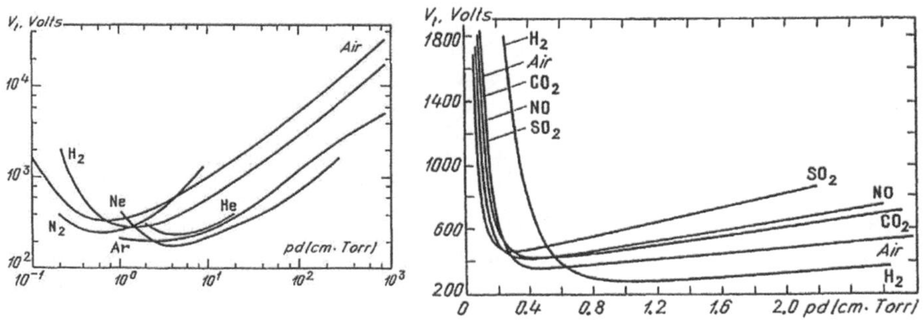

FIGURE 4.19 Paschen curves for different gases. (From "Gas Discharge Physics" Yu. P. Raizer)

$$V_{\min} = \frac{eB}{A}\ln\left(1+\frac{1}{\gamma}\right), \quad \left(\frac{E}{p}\right)_{\min} = B, \quad (pd)_{\min} = \frac{e}{A}\ln\left(1+\frac{1}{\gamma}\right), \tag{4.144}$$

where e = 2.72 is the base of natural logarithms. The typical value of minimum voltage is about 300V, corresponding to reduce an electric field about 300 V/cm Torr and the parameter pd about 0.7 cm Torr. The right-hand branch of the Paschen curve (pressure greater than 1 Torr for a gap about 1 cm) is related to a case when the electron avalanche has enough distance and gas pressure to provide intensive ionization even at moderate electric fields. In this case, the reduced electric field is almost fixed; it is just slowly reducing logarithmically with pd-growth. Conversely, the left-hand branch of the Paschen curve is related to the case when ionization is limited by the avalanche size and gas pressure. Obviously in this latter case, the ionization sufficient for breakdown can be provided in such a situation only by very high electric fields. The reduced electric field at the Paschen minimum corresponds to the so-called **Stoletov constant** $\left(\frac{E}{p}\right)_{\min} = B$, which is the minimum discharge energy necessary to produce one electron–ion pair (minimum price of ionization). Here, the price of ionization can be expressed as $W = \frac{eE}{\alpha}$, and its minimum which is the Stoletov constant, is equal to $W_{\min} = \frac{2.72 * eB}{A}$. The Stoletov constant usually exceeds the ionization potentials several times, because electrons dissipate energy to vibrational and electronic excitation per each act of ionization. Numerical estimation for the minimum ionization price in electric discharges with high electron temperatures is about 30eV. Analyzing Eq. (4.144), it is seen that the reduced electric field at the Paschen minimum $\left(\frac{E}{p}\right)_{\min} = B$ does not depend on γ and hence, on a cathode material in contrast to the minimum voltage $V_{\min}$ and the corresponding similarity parameter $(pd)_{\min}$.

4.4.3 The Townsend Breakdown of Larger Gaps, Specific Behavior of Electronegative Gases

The Townsend mechanism of breakdown is relatively homogeneous and includes the development of independent avalanches; it usually occurs at $pd < 4000\,\text{Torr} \times \text{cm}$ (at atmospheric pressure d < 5 cm). For larger gaps (more than 6 cm at atmospheric pressure) the avalanches disturb the electric field and are no longer independent. This leads to the spark mechanism of breakdown. Consider relatively large gaps, but still not sufficiently large ($pd < 4000\,\text{Torr} \times \text{cm}$) for sparks. The reduced electric field E/p necessary for breakdown Eq. (4.143) is slightly logarithmically decreasing with pd in the framework of the Townsend breakdown mechanism ($pd < 4000\,\text{Torr} \times \text{cm}$). This is illustrated by the E(d) dependence in atmospheric air, presented in Figure 4.20. For larger gaps and the larger avalanches, E/p is less sensitive to the secondary electron emission and cathode material. This explains the E/p decrease with pd. The reduction in electronegative gases is limited by electron attachment processes (usually dissociative attachment, Section 2.4.1, Eq. (2.4.3)). The influence of attachment processes can be taken into account in a similar manner as with ionization Eq. (4.138) by introducing an additional Townsend coefficient β:

$$\beta = \frac{\nu_a}{v_d} = \frac{1}{v_d} k_a\left(E/n_0\right) n_0 = \frac{1}{\mu_e}\frac{k_a(E/n_0)}{E/n_0}. \tag{4.145}$$

In this relation $k_a\left(E/n_0\right)$ and ν_a are the attachment rate coefficient and attachment frequency with respect to an electron. Thus altogether we have three Townsend coefficients α, β and γ, describe the

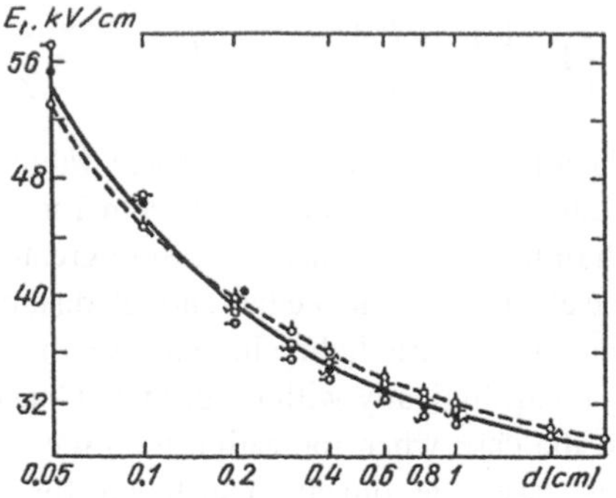

FIGURE 4.20 Breakdown electric field in atmospheric air.

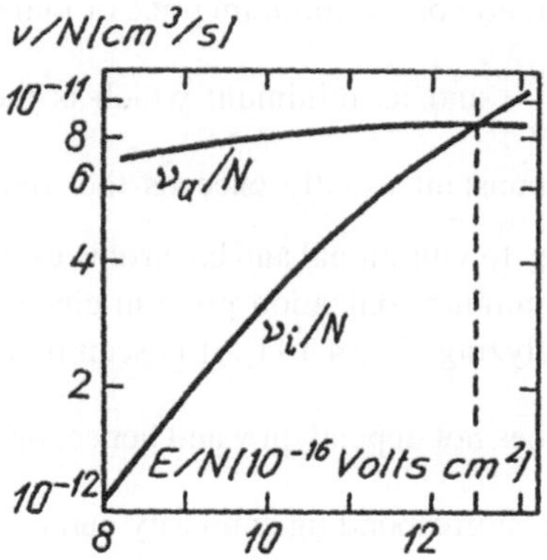

FIGURE 4.21 Frequencies of ionization and electron attachment in air.

Townsend mechanism of electric breakdown. The Townsend coefficient β shows the electron losses due to attachment per unit length. Combining α (see Eq. (4.138)) and β gives:

$$\frac{dn_e}{dx} = (\alpha - \beta)n_e, \quad n_e(x) = n_{e0}\exp\left[(\alpha - \beta)x\right]. \tag{4.146}$$

The Townsend coefficient β α is an exponential function of the reduced electric field in the same way as a. However, this function is not as sharp (see Figure 4.21). For this reason, ionization much exceeds attachment at relatively high values of reduced electric fields, and in this case the β coefficient can be neglected with respect to α. When the gaps are relatively large (centimeters range at atmospheric pressure), the Townsend breakdown electric field in electronegative gases actually becomes constant and limited by attachment processes. Obviously in this case, the breakdown of electronegative gases requires much higher values of the reduced electric fields. The breakdown electric fields at high pressures for both electronegative and non-electronegative gases are presented in Table 4.4.

4.4.4 Sparks Versus Townsend Breakdown Mechanism

The Townsend quasi-homogeneous breakdown occurs only in low pressures and short discharge gaps ($pd < 4000$ Torr × cm, at atmospheric pressure d < 5cm). Another breakdown mechanism, the so-called spark, takes place in larger gaps at high pressures. **Sparks** in contrast to the

TABLE 4.4
Electric Fields Sufficient for the Townsend Breakdown of Centimeters-Size Gaps at Atmospheric Pressure

Gas	E/p, kV/cm	Gas	E/p, kV/cm	Gas	E/p, kV/cm
Air	32	O_2	30	N_2	35
H_2	20	Cl_2	76	CCl_2F_2	76
CSF_8	150	CCl_4	180	SF_6	89
He	10	Ne	1.4	Ar	2.7

Townsend mechanism provide a breakdown in a local narrow channel, without direct relation to electrode phenomena and with very high currents (up to 10^4–10^5 A) and current densities. Sparks as well as Townsend breakdown are primarily related to avalanches. In large gaps, they cannot be considered as independent and stimulated by electron emission from the cathode. The spark breakdown at high *pd* and overvoltage develops much faster than the time necessary for ions to cross the gap and provide the secondary emission. The high conductivity spark channel can be formed faster than electron drift time from cathode to anode. In this case, the breakdown voltage is independent of the cathode material, which is also a qualitative difference between Townsend and spark breakdowns. The mechanism of spark breakdown is based on the concept of **a streamer.** This thin ionized channel grows fast along the positively charged trail left by an intensive primary avalanche between electrodes. This avalanche also generates photons, which in turn initiate numerous secondary avalanches in the vicinity of the primary one. Electrons of the secondary avalanches are pulled by the strong electric field into the positively charged trail of the primary avalanche, creating a streamer propagating fast between electrodes. The streamer theory was originally developed by H. Raether (1964), L.B. Loeb (1960) and J.M. Meek (1978). If the distance between electrodes is multi-meter or even kilometers long, as in the case of lightning, the individual streamers are not sufficient to provide large-scale spark breakdown. In this case the so-called **leader** is moving from one electrode to another. The leader as well as streamer is also a thin channel but much more conductive than an individual streamer. Leaders include the streamers as its elements.

4.4.5 Physics of the Electron Avalanches

According to Eq. (4.146) and taking into account possible electron attachment, the increase of the total number of electrons N_e, positive N_+ and negative N_- ions in an avalanche moving along the axis x follow the equations:

$$\frac{dN_e}{dx} = (\alpha - \beta)N_e, \quad \frac{dN_+}{dx} = \alpha N_e, \quad \frac{dN_-}{dx} = \beta N_e, \tag{4.147}$$

where α and β are the ionization and attachment Townsend coefficients. If the avalanche starts from the only primary electron, the numbers of charged particles, electrons, positive and negative ions, can be found from Eq. (4.147) as:

$$N_e = \exp\left[(\alpha - \beta)x\right], \quad N_+ = \frac{\alpha}{\alpha - \beta}(N_e - 1), \quad N_- = \frac{\beta}{\alpha - \beta}(N_e - 1). \tag{4.148}$$

The electrons in an avalanche move altogether along the direction of the non-disturbed electric field E_0 (axis x) with the drift velocity $v_d = \mu_e E_0$. Concurrently, free diffusion (D_e) spread the group of electrons around the axis x in radial direction r. Taking into account both the drift and the diffusion, the electron density in an avalanche $n_e(x, r, t)$ can be found from Eq. (4.121) in the following form (E.D. Lozansky, O.B. Firsov, 1975):

$$n_e(x,r,t) = \frac{1}{(4\pi D_e t)^{3/2}} \exp\left[-\frac{(x-\mu_e E_0 t)^2 + r^2}{4D_e t} + (\alpha - \beta)\mu_e E_0 t\right]. \tag{4.149}$$

From Eq. (4.149), the electron density decreases with distance from the axis x following the Gaussian law. Then, the avalanche radius r_A (where the electron density is e times less than on the axis x) grows with time and the distance x_0 of the avalanche propagation:

$$r_A = \sqrt{4D_e t} = \sqrt{4D_e \frac{x_0}{\mu_e E_0}} = \sqrt{\frac{4T_e}{eE_0} x_0}. \tag{4.150}$$

In Eq. (4.150) the Einstein relation between electron mobility and free diffusion coefficient Eq. (4.124) is taken into account. The space distribution of positive and negative ion densities during the short interval of avalanche propagation, when the ions remain at rest, can be calculated using Eq. (4.149):

$$n_+(x,r,t) = \int_0^t \alpha\mu_0 E_0 n_e(x,r,t')dt', \quad n_-(x,r,t) = \int_0^t \beta\mu_0 E_0 n_e(x,r,t')dt'. \tag{4.151}$$

A simplified expression for the space distribution of positive ion density not too far from the axis x can be derived based on Eqs. (4.151) and (4.149) in absence of attachment and in the limit $t \to \infty$ (E.D. Lozansky, O.B. Firsov, 1975) as:

$$n_+(x,r) = \frac{\alpha}{\pi r_A^2(x)} \exp\left[\alpha x - \frac{r^2}{r_A^2(x)}\right], \tag{4.152}$$

where $r_A(x)$ is the avalanche radius. This distribution reflects the fact that the ion concentration in the trail of the avalanche grows along the axis in accordance with the multiplication and exponential increase of the number of electrons Eq. (4.148). Concurrently, the radial distribution is Gaussian with an effective avalanche radius $r_A(x)$ growing with x Eq. (4.150). An avalanche propagation is illustrated in Figure 4.22. Though the avalanche radius Eq. (4.150) is growing parabolically (proportionally to $\sqrt{x}$), the visible avalanche outline is wedge-shaped. This means that the visible avalanche radius is growing linearly (proportionally to x, see picture on Figure 4.21). This happens because the visible avalanche radiation is determined by the absolute density of excited species that is approximately proportional to the exponential factor Φ in the Eq. (4.152) and obviously grows with x. The visible avalanche radius r(x) can then be expressed from Eq. (4.152), taking into account the smallness of r at small x, as:

$$\frac{r^2(x)}{r_A^2(x)} = \alpha x - \ln\Phi, \quad r(x) \approx r_A(x)\sqrt{\alpha x} = \sqrt{\frac{4T_e x}{eE_0} * \alpha x} = x\sqrt{\frac{4T_e \alpha}{eE_0}}. \tag{4.153}$$

This explains the linearity of r(t) and wedge-shape form of an avalanche (Figure 4.21). The qualitative change in avalanche behavior takes place when the charge amplification exp(αx) is large. In this case, the considerable space charge is created with its own significant electric field $\vec{E_a}$, which should be added to the external field $\vec{E_0}$. The electrons are in the head of avalanche while the positive ions remains behind, creating a dipole with the characteristic length $\frac{1}{\alpha}$ and charge $N_e \approx \exp(\alpha x)$. For breakdown fields about 30 kV/cm in atmospheric pressure air, the α-coefficient is approximately 10 cm^{-1} and the characteristic ionization length can be estimated as $1/\alpha \approx 0.1$cm. The external electric

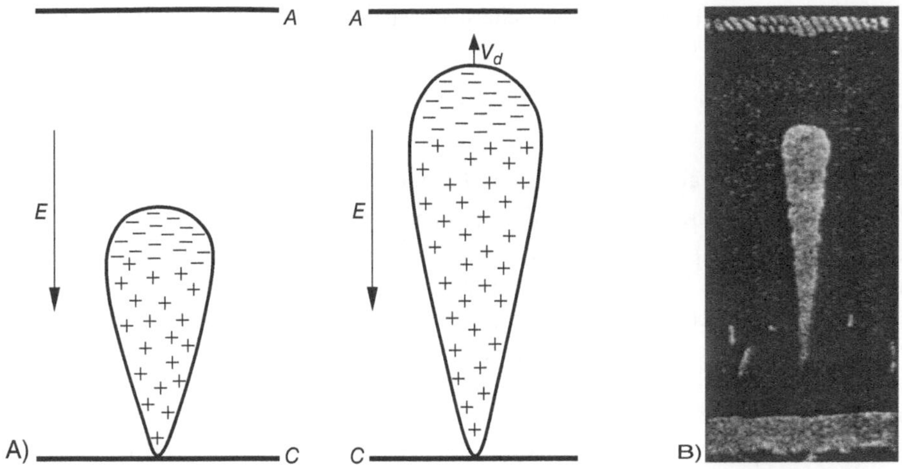

FIGURE 4.22 Avalanche evolution: (A)schematic, and (B) photograph. (From "Gas Discharge Physics" Yu. P. Raizer)

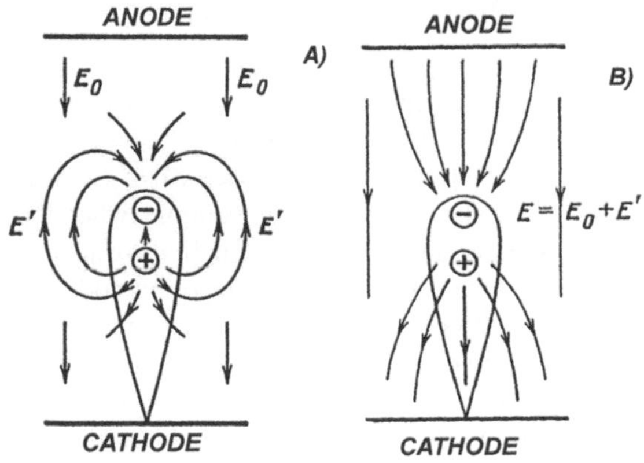

FIGURE 4.23 Electric field distribution in an avalanche: external and space charge fields are shown (a) separately, and (b) combined

field distortion due to the space charge of the dipole is shown in Figure 4.23. In front of the avalanche head (and behind the avalanche) the external $\vec{E_0}$ and internal $\vec{E_a}$ electric fields add to make a total field stronger, which in turn accelerates ionization. Conversely, in between the separated charges or "inside of avalanche" the total electric field is lower than the external one, which slows down the ionization. The space charge creates a radial electric field (see Figure 4.23). At a distance of approximately the avalanche radius Eq. (4.150) the electric field of the charge $N_e \approx \exp\,(\alpha x)$ reaches the value of the external field $\vec{E_0}$ at some critical value of αx. For example, during a 1 cm-gap breakdown in air, the avalanche radius is about $r_A = 0.02\,\text{cm}$ and the critical value of $\alpha x$when the avalanche electric field becomes comparable with E_0 is $\alpha x = 18$ (note this corresponds to the Meek criterion of streamer formation, see below). When $\alpha x \geq 14$ the radial growth of an avalanche due to repulsion drift of electrons exceeds the diffusion effect and should be taken into account. In this case the avalanche radius grows with x as:

$$r = \sqrt[3]{\frac{3e}{4\pi\varepsilon_0 \alpha E_0}} \exp\frac{\alpha x}{3} = \frac{3}{\alpha}\frac{E_a}{E_0}. \tag{4.154}$$

This fast growth of the transverse avalanche size restricts the electron density in the avalanche to the maximum value:

$$n_e = \frac{\varepsilon_0 \alpha E_0}{e}. \tag{4.155}$$

When the transverse avalanche size reaches the characteristic ionization length $1/\alpha \approx 0.1\text{cm}$ (the avalanche "dipole" size), the broadening of the avalanche head dramatically slows down. Obviously, the avalanche electric field is about the external one in this case (see Eq. 4.154). Typical values of maximum electron density in an avalanche are about $10^{12} \div 10^{13}\,\text{cm}^{-3}$. As soon as the avalanche head reaches the anode, the electrons sink into the electrode and it is mostly the ionic trail that remains in the discharge gap. The electric field distortion due to the space charge in this case is shown in Figure 4.24. Because electrons are no longer present in the gap, the total electric field is due to the external field, the ionic trail and also the ionic charge "image" in the anode (see Figure 4.24). The resulting electric field in the ionic trail near the anode is less than the external electric field, but further away from the electrode, it exceeds E_0. The total electric field reaches the maximum value at the characteristic ionization distance $\frac{1}{\alpha} \approx 0.1\text{cm}$ from the anode.

4.4.6 Cathode Directed and Anode Directed Streamers

A strong primary avalanche is able to amplify the external electric field and form a thin, weakly ionized plasma channel, the so-called **streamer**, growing very fast between electrodes. When the streamer channel connects the electrodes, the current may be significantly increased to form the spark. The avalanche-to-streamer transformation takes place when the internal field of an avalanche becomes comparable with the external electric field, for example, when the amplification parameter αd is sufficiently large. If the discharge gap is relatively small, the transformation occurs only when the avalanche reaches the anode. Such a streamer growth from anode to cathode is known as the **cathode-directed or positive streamer**. If the discharge gap and overvoltage are large, the avalanche-to-streamer transformation can take place quite far from the anode. In this case, the so-called **anode-directed or negative streamer** is able to grow actually toward both electrodes.

The mechanism of formation of a cathode-directed streamer is illustrated in Figure 4.25. High-energy photons emitted from the primary avalanche provide photo-ionization in the vicinity, which initiates secondary avalanches. Electrons of secondary avalanches are pulled into the ionic trail of the primary one (see the electric field distribution in Figure 4.24) and create a quasi-neutral plasma

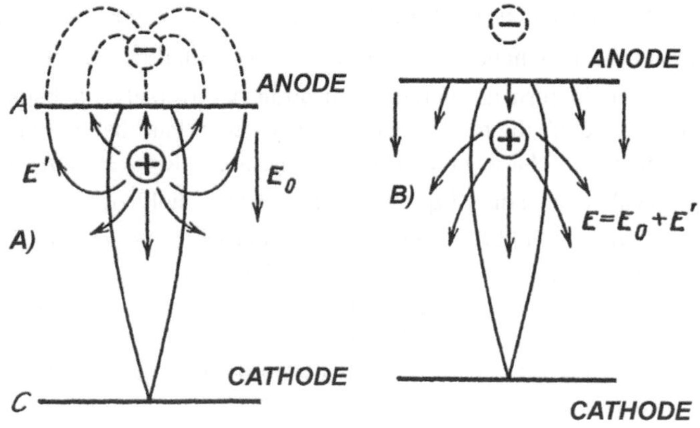

FIGURE 4.24 Electric field distribution when the avalanche reaches the anode: external and space charge fields are shown (a) separately, and (b) combined

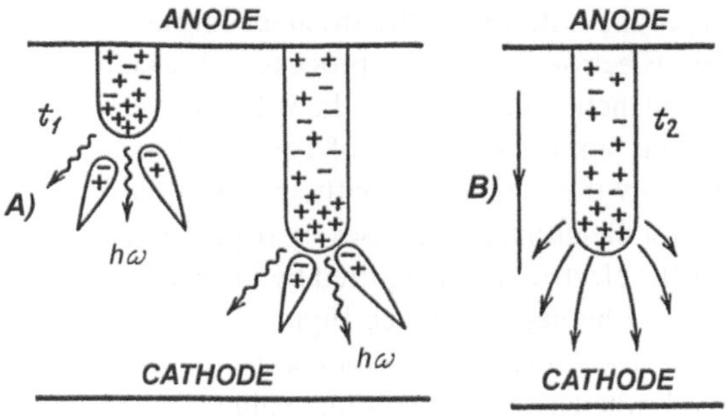

FIGURE 4.25 Illustration of a cathode-directed streamer: (a)propagation, (b)electric field near the streamer head.

channel. Subsequently, the process repeats, providing growth of the streamer. The cathode-directed streamer begins near the anode, where the positive charge and electric field of the primary avalanche is the highest. The streamer appears and operates as a thin conductive needle growing from the anode. Obviously, the electric field at the tip of the "anode needle" is very large which stimulates the fast streamer propagation in the direction of the cathode. Usually, streamer propagation is limited by neutralization of the ionic trail near the tip of the needle. Here the electric field is so high that provides electron drift with velocities about 10^8 cm/s. This explains the high speed of streamer growth, which is also about 10^8 cm/s and exceeds by a factor of 10 the typical electron drift velocity in the external breakdown field. The latter is usually about 10^7 cm/s. All streamer parameters are in some way related to those of the primary avalanche. Thus the diameter of the streamer channel is about 0.01–0.1 cm and corresponds to the maximum size of a primary avalanche head, which can be estimated as the ionization length $1/\alpha$ (see Section 4.4.5). Plasma density in the streamer channel also corresponds to the maximum electron concentration in the head of a primary avalanche, which is about $10^{12} \div 10^{13}$ cm^{-3}. It is important to note that the specific energy input (electron energy transferred to one molecule) in a streamer channel is small during the short period (~30 ns) of the streamer growth between electrodes. In molecular gases, it is about 10^{-3} eV/mol, which in temperature units corresponds to ~ 10K. The anode directed streamer occurs between electrodes if the primary avalanche becomes sufficiently strong even before reaching the anode. Such streamer, which grows in two directions, is illustrated in Figure 4.26. The mechanism of the streamer propagation in direction of the cathode is obviously the same as that of cathode-directed streamers. The mechanism of the streamer growth in direction of the anode is also similar, but in this case electrons from primary avalanche head neutralize the ionic trail of secondary avalanches. The secondary avalanches could be initiated here not only by photons, but also by some electrons moving in front of the primary avalanche.

4.4.7 Criterion of Streamer Formation, the Meek Breakdown Condition

Formation of a streamer, as was shown in Sections 4.4.5 and 4.4.6, requires the electric field of space charge in the avalanche E_a to be of the order of the external field E_0:

$$E_a = \frac{e}{4\pi\varepsilon_0 r_A^2} \exp\left[\alpha\left(\frac{E_0}{p}\right)x\right] \approx E_0. \tag{4.156}$$

Assuming the avalanche head radius to be the ionization length: $r_a \approx 1/\alpha$, the criterion of streamer formation in the gap with the distance d between electrodes can be presented as the requirement for the avalanche amplification parameter αd to exceeds the critical value:

$$\alpha\left(\frac{E_0}{p}\right)d = \ln\frac{4\pi\varepsilon_0 E_0}{e\alpha^2} \approx 20, \quad N_e = \exp(\alpha d) \approx 3\cdot 10^8. \tag{4.157}$$

This convenient and important criterion of the streamer formation is known as the **Meek breakdown condition** ($\alpha d \geq 20$). Electron attachment processes in electronegative gases slow down the electron multiplication in avalanches and increase the value of electric field required for a streamer formation. The situation here is similar to the case of Townsend breakdown mechanism (see Section 4.4.3 and Eq. (4.146)). Actually the ionization coefficient α in the Meek breakdown condition Eq. (4.157) should be replaced for electronegative gases by α-β. However, when the discharge gaps are not large (in air $d \leq 15\,\text{cm}$), the electric fields required by the Meek criterion are relatively high; then $\alpha >> \beta$ and the attachment can be neglected (see Figure 4.21). Increasing the distance d between electrodes in electronegative gases does lead to gradual decreases of the electric field necessary for streamer formation, but it is limited by some minimum value. The minimal electric field required for streamer formation can be found from the ionization-attachment balance $\alpha\,(E_0/p) = \beta\,(E_0/p)$ (see Figure 4.21). In atmospheric pressure air, this limit is about 26 kV/cm. At atmospheric pressure in such a strongly electronegative gas as SF_6, which is commonly used for electric insulation, the balance $\alpha\,(E_0/p) = \beta\,(E_0/p)$ requires a very high electric field 117.5 kV/cm. Electric field non-uniformity has a strong influence on breakdown conditions and an avalanche transformation into a streamer. Quite obviously, the non-uniformity as in the case of corona discharge (see next chapters), non-uniformities decreases the breakdown voltage for a given distance between electrodes. This is clear because the breakdown condition Eq. (4.157) requires not αd but the integral $N_e = \int_{x_1}^{x_2} \alpha(E)dx$ to exceed the certain threshold. The function $\alpha(E)$ is strongly exponential (see Eq. (4.142)), which results in significantly higher values of the integral in a non-uniform field with respect to a uniform one. In other words, the voltage applied to the rod-electrode due to the non-uniform electric field should provide the intensive electron multiplication only near the electrode to initiate a streamer. Once the plasma channel is initiated, its growth is then controlled mostly by the high electric field of its own streamer tip. Such a situation is considered below in the next section as propagation of the quasi-self-sustained streamer. In this case for very long (≥1 m) and non-uniform systems, the average electric field can be as low as 2–5 kV/cm (see Figure 4.27). The breakdown threshold in the non-uniform constant electric field also depends on the polarity of the principal electrode, where the electric field is higher. Then the threshold voltage in a long gap between a negatively charged rod and a plane is about twice as large as in the case of a negatively charged rod (see Figure 4.27). This polarity effect is due to the non-uniformity of the electric field near the electrodes. In the case of rod anode the avalanches approaches the anode where the electric field becomes stronger and stronger, which facilitates the avalanche-streamer transition. Also the avalanche electrons easily sink into the anode in this case, leaving near the electrode the ionic trail enhancing the electric field. In the case

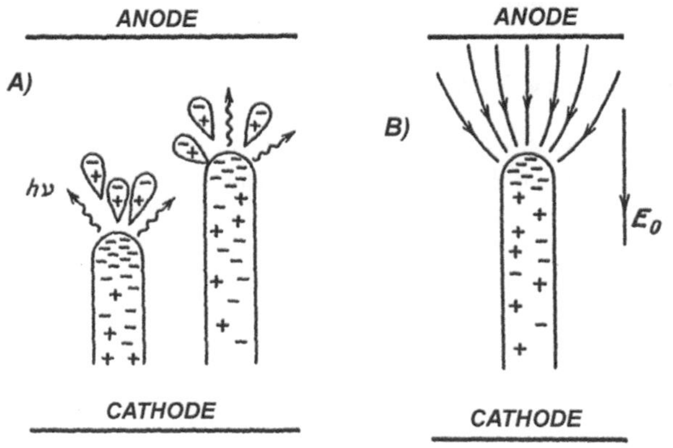

FIGURE 4.26 Illustration of an anode-directed streamer: (a)propagation, (b)electric field near the streamer head.

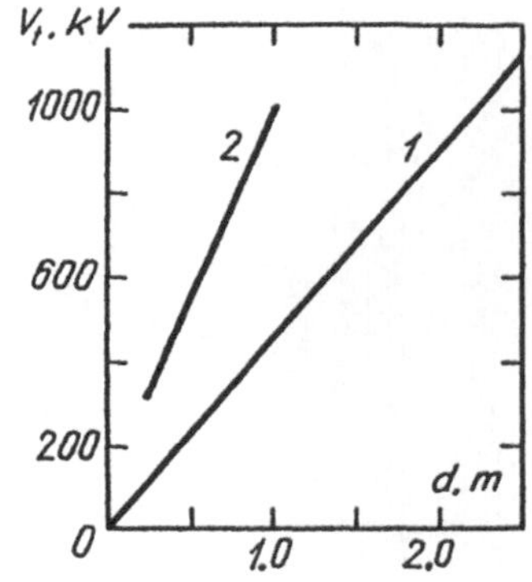

FIGURE 4.27 Breakdown voltage in air at 50Hz: (1) rod-rod gap, (2) rod-plane gap.

of rod cathode the avalanches move from the electrode into the low electric field zone, which requires higher voltage for the avalanche-streamer transition.

4.4.8 Streamer Propagation Models

The model of propagation of the **quasi-self-sustained streamers** was proposed by G.A. Dawson and W.P. Winn (1965) and then developed by I. Gallimberti (1972). This model, illustrated in Figure 4.28, assumes very low conductivity of a streamer channel, and makes the streamer propagation autonomous and independent from the anode. Photons initiate avalanche at a distance x_1 from the center of the positive charge zone of radius r_0. According to this model the avalanche then develops in the autonomous electric field of the positive space charge $E(x) = \frac{eN_+}{4\pi\varepsilon_0 x^2}$, the number of electrons is increasing by ionization as: $N_e = \int_{x_1}^{x_2} \alpha(E)dx$, and the avalanche radius grows by the mechanism of free diffusion as:

$$\frac{dr^2}{dt} \approx 4D_e, \quad r(x_2) = \left[\int_{x_1}^{x_2} \frac{4D_e}{\mu_e E(x)} dx\right]^{1/2}. \tag{4.158}$$

To provide continuous and steady propagation of the self-sustained streamer, its positive space charge N_+ should be compensated by the negative charge of avalanche head $N_e = N_+$ at the meeting point of the avalanche and streamer: $x_2 = r_0 + r$. Also the radii of the avalanche and streamer should be correlated at this point $r = r_0$. These equations permit describing the streamer

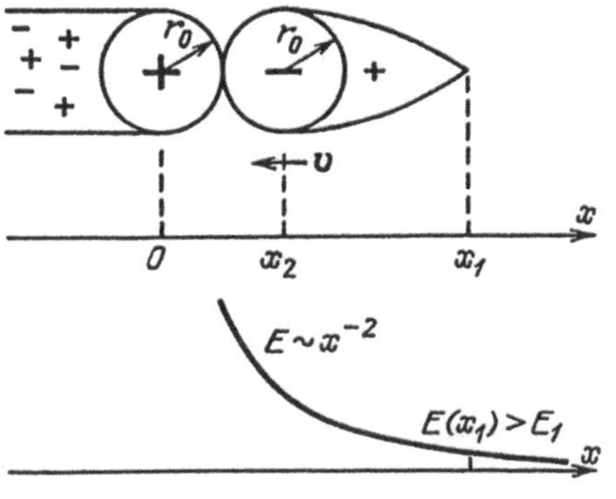

FIGURE 4.28 Illustration of a self-sustaining streamer.

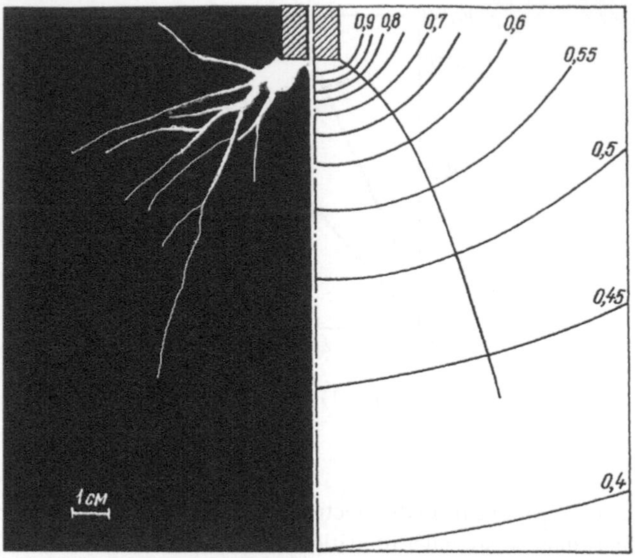

FIGURE 4.29 Streamer propagation from a positive 2 cm rod to a plane at distance of 150 cm, constant voltage 125 kV; equipotential surfaces are shown at right.

parameters including the propagation velocity, which can be found as x_2 divided by time of the avalanche displacement from x_1 to x_2. Comparison of the I.Gallimberti model of streamers in the strongly non-uniform electric field with an experimental photograph of a streamer corona is presented in Figure 4.29. The model of quasi-self-sustained streamer is helpful in describing the breakdown of long gaps with high voltage and low average electric fields (see the previous section and Figure 4.27).

A qualitatively different model of streamer propagation was developed by R.D. Klingbeil, A. Tidman, R.F. Fernsler (1972), and E.D. Lozansky and O.B. Firsov (1975). In contrast to the above approach of the self-sustained propagation of streamers, this model considers the streamer channel as an ideal conductor connected to the anode (see Section 4.4.6). In this model, the **ideally conducting streamer channel** is considered as an anode elongation in the direction of external electric field E_0 with the shape of an ellipsoid of revolution. For ideally conducting streamers, the streamer propagation at each point of the ellipsoid is normal to its surface. The propagation velocity is equal to the electron drift velocity in the appropriate electric field. To calculate the streamer growth velocity, a convenient formula for the maximum electric field E_m on the tip of the streamer with length l and radius r was proposed by E.M. Baselyan, A. Yu. Goryunov (1986):

$$\frac{E_m}{E_0} = 3 + \left(\frac{l}{r}\right)^{0.92}, \quad 10 < \frac{l}{r} < 2000. \tag{4.159}$$

The model of the ideally conducting streamer is also in good agreement with experimental results. Other more detailed numerical streamer propagation models taking into account the finite conductivity of the streamer channel will be discussed during consideration of corona and dielectric barrier discharges.

4.4.9 Concept of a Leader, Breakdown of Multi-Meter and Kilometer Long Gaps

The spark or streamer breakdown mechanism discussed above can be summarized as a sequence of three processes: (1) development of an avalanche and the avalanche-streamer transformation; (2) the

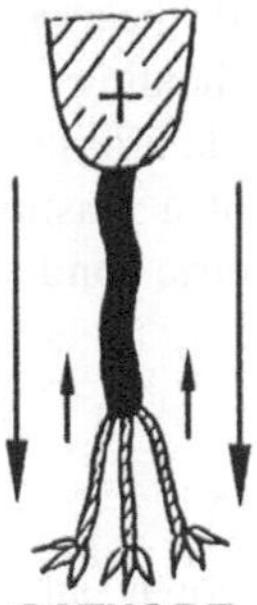

FIGURE 4.30 Illustration of a leader propagation.

streamer growth from anode to cathode; and (3) triggering of a return wave of intense ionization, which results in a spark formation. However, this breakdown mechanism cannot be applied directly to very long gaps particularly in electronegative gases (including air). The spark cannot be formed by the described mechanism in long gaps, because the streamer channel conductivity is not sufficiently high to transfer the anode potential close to the cathode and there stimulate the return wave of intense ionization and spark. This effect is especially strong in electronegative gases where the streamer channel conductivity is low. Also in non-uniform electric fields, the streamer head grow from the strong to the weak field region, which delays its propagation. The streamers just stop in the long non-uniform air gaps without reaching the opposite electrode (see Figure 4.29). Usually, the streamer length is about 0.1–1 m. Breakdown of longer gaps, including those with the multi-meter and kilometer long inter-electrode distances, are related to the formation and propagation of the leaders. The leader is highly ionized and highly conductive (with respect to streamer) plasma channel and grows from the active electrode along the path prepared by the preceding streamers. Because of the high conductivity the leaders are much more effective, than streamers in transferring the anode potential close to the cathode and there stimulating the return wave of intense ionization and spark. The leaders were first investigated in connection with the natural phenomenon of lightning, their propagation is illustrated in Figure 4.30. The leaders have high conductivity and are able to make "ideally conductive channels" much longer than streamers. This results in a stronger electric field Eq. (4.159) on the leader head, which is able to create new streamers growing from the head and preparing the path for further propagation of the leader. The difference between leaders and streamers is more quantitative than qualitative: longer length, higher degree of ionization, higher conductivity, higher electric field; a streamer absorb avalanches, a leader absorb streamers (Figure 4.30). Heating effects of the relatively short centimeters long streamers was estimated in Section 4.4.6 as about 10K. For meters long channels at least near the active electrode the heating effect reaches 3000K. This heating together with the corresponding high level of non-equilibrium excitation of atoms and molecules probably explain the transformation of a streamer channel into the leader. A temperature 3000K is not sufficient for thermal ionization of air, but this temperature together with the elevated non-equilibrium excitation level is sufficiently high for other mechanisms of increase of electric conductivity in the plasma channel.

I. Gallimberti (1977) assumed a mechanism for streamer-to-leader transition in air related to thermal detachment of electrons from the negative ions O_2^-, which are the main products of electron attachment in the electronegative gas. The effective destruction of the ions O_2^- and as a result compensation of electron attachment becomes possible if the temperature exceeds 1500K in dry air and 2000K in humid air (where the electron affinity of complexes $O_2^-(H_2O)_n$ is higher). Such temperatures are available in the plasma channel and can provide the formation of a high conductivity leader in the electronegative gas. During the evolution of a streamer in air, at first, the Joule heat is stored in vibrational excitation of N_2-molecules. While the temperature of air is increasing, the VT-relaxation grows exponentially providing the explosive heating of the plasma channel. As an example, typical parameters of a leader growing from an anode are: leader current is about 100 A (for comparison streamer current is 0.1–10 mA), diameter of a plasma channel about 0.1 cm, quasi-equilibrium plasma temperature is 20,000–40,000 K, electric conductivity is about 100 $Ohm^{-1}cm^{-1}$, propagation velocity is $2 * 10^6$ cm/s.

4.4.10 Streamers and Microdischarges

Streamers are elements of the spark breakdown, and their visual observation can be related to dielectric barrier (DBD) and corona discharges. DBD gap (from 0.1 mm to 3 cm) usually includes one or more dielectric layers located in the current path between metal electrodes; typical frequency is 0.05–100 kHz; voltage is about 10 kV at atmospheric pressure. In most cases, DBDs are not uniform and consist of numerous microdischarges built from streamers and distributed in the discharge gap. Electrons in the conducting plasma channel established by the streamers dissipate from the DBD gap in about 40 ns, while slowly drifting ions remain there for several microseconds. Deposition of electrons from the conducting channel onto the anode dielectric barrier results in charge accumulation and prevents new avalanches and streamers nearby until the cathode and anode are reversed. Usual DBD operation frequency is around 20 kHz, therefore the voltage polarity reversal occurs within 25 μs. After the voltage polarity reverses, the deposited negative charge facilitates the formation of new avalanches and streamers in the same spot. As a result, a multi-generation family of streamers is formed that is macroscopically observed as a bright spatially localized **filament**. An initial electron starting from some point in the discharge gap (or from cathode or dielectric that covers the cathode) produces secondary electrons by direct ionization and develops an electron avalanche. If the avalanche is big enough (see Meek condition), the cathode-directed streamer is initiated. The streamer bridges the gap in few nanoseconds and forms a conducting channel of weakly ionized plasma. Intensive electron current flows through this plasma channel until the local electric field collapses. Collapse of the local electric field is caused by the charges accumulated on the dielectric surface and ionic space charge (ions are too slow to leave the gap for the duration of this current peak). Group of local processes in the discharge gap initiated by an avalanche and developed until electron current termination usually called **microdischarge**. After electron current termination, there is no longer an electron–ion plasma in the main part of microdischarge channel, but high levels of vibrational and electronic excitation in channel volume along with charges deposited on the surface and ionic charges in the volume allow us to separate this region from the rest of the volume and call it **microdischarge remnant**. Positive ions (or positive and negative ions in the case of electronegative gas) of the remnant slowly move to electrodes resulting in low and very long (~10 μs for 1 mm gap) falling ion current. Microdischarge remnant will facilitate the formation of new microdischarge in the same spot as the polarity of the applied voltage changes. That is why it is possible to see single filaments in DBD. If microdischarges would form at a new spot each time the polarity changes, the discharge would appear uniform. Thus filament in DBD is a group of microdischarges that form on the same spot each time polarity is changed. The fact that microdischarge remnant is not fully dissipated before the formation of next microdischarge is called the **memory effect**.

4.4.11 Interaction of Streamers and Microdischarges

The mutual influence of microdischarge in DBD is related to their electric interaction with residual charges left on the dielectric barrier, as well as with the influence of excited species generated in one microdischarge on the formation of another microdischarge. The interaction of streamers and microdischarges is responsible for formation of microdischarge patterns reminiscent of 2D-crystals. The interaction of microdischarges is due to the following effect. Positive charge (or dipole field in the case of deposited negative charge) intensifies electric field in the cathode area of the neighboring microdischarge and decreases electric field in the anode area. Since the avalanche-to-streamer transition depends mostly on near-anode electric field (from which new streamers originate), formation of neighboring microdischarges is actually prevented, and microdischarges effectively repel each other. The **quasi-repulsion between microdischarges** leads to formation of short-range order that is related to a characteristic repulsion distance between microdischarges. Observation of this cooperative phenomenon depends on several factors, including number of the microdischarges and the operating frequency. For example, when the number of microdischarges is not large enough (when the average distance between microdischarges is larger than characteristic interaction radius), no significant microdischarge interaction is observed. When the AC frequency is too low to keep the microdischarge remnants from dissipation (low frequency means that period is longer than typical lifetime of microdischarge remnant or "memory effect" lifetime) microdischarge repulsion effects are not observed. DBD cells operating at very high (MHz) frequencies don't exhibit microdischarge repulsion because the very high-frequency switching of the voltage interferes with ions still moving to electrodes. The interaction of streamers in DBD can lead to the formation of an organized structure of microdischarges similar to the Coulomb crystals.

4.5 STEADY-STATE REGIMES OF NON-EQUILIBRIUM ELECTRIC DISCHARGES

4.5.1 Steady-State Discharges Controlled by Volume and Surface Recombination Processes

The steady-state regimes of non-equilibrium discharges are provided by a balance of generation and losses of charged particles. The generation of electrons and positive ions are mostly due to volume ionization processes. To sustain the steady-state plasma, the ionization should be quite intensive; this usually requires the electron temperature to be on the level of $1/10$ of the ionization potential (~ 1 eV). The losses of charged particles can be related to volume processes of recombination (Section 2.3) or attachment (Section 2.4), but they can be also provided by diffusion of charged particles to the walls with subsequent surface recombination. These two mechanisms of charge losses define two different regimes of sustaining the steady-state discharge: the first controlled by volume processes and the second controlled by diffusion to the walls. If the degree of ionization is relatively high and diffusion considered as ambipolar, the frequency of charge losses due to the diffusion to the walls can be described as:

$$\nu_D = \frac{D_a}{\Lambda_D^2}. \tag{4.160}$$

In this relation D_a is the coefficient of ambipolar diffusion (see Eqs. (4.127), (4.128)); and Λ_D is the characteristic diffusion length, which can be calculated for different shapes of the discharge chambers:

- for a cylindrical discharge chamber of radius R and length L:

$$\frac{1}{\Lambda_D^2} = \left(\frac{2.4}{R}\right)^2 + \left(\frac{\pi}{L}\right)^2, \tag{4.161}$$

- for a parallelepiped with side lengths L_1, L_2, L_3:

$$\frac{1}{\Lambda_D^2} = \left(\frac{\pi}{L_1}\right)^2 + \left(\frac{\pi}{L_2}\right)^2 + \left(\frac{\pi}{L_3}\right)^2, \tag{4.162}$$

- for a spherical discharge chamber with radius R:

$$\frac{1}{\Lambda_D^2} = \left(\frac{\pi}{R}\right)^2. \tag{4.163}$$

Based on Eqs. (4.160)–(4.163) for the diffusion charge losses, a criterion of predominantly volume-process-related charge losses and hence, a criterion for the volume-process-related steady-state regime of sustaining the non- equilibrium discharges is obtained:

$$k_i(T_e)n_0 >> \frac{D_a}{\Lambda_D^2}. \tag{4.164}$$

In this relation $k_i(T_0)$ is the ionization rate coefficient (see Eqs. (2.21) or (2.28)), n_0 is the neutral gas density. The criterion Eq. (4.164) actually restricts pressure, because $D_a \propto 1/p$ and $n_0 \propto p$. When pressure in a discharge chamber exceeds 10–30 Torr (range of moderate and high pressures), the diffusion is relatively slow and the balance of charge particles is due to volume processes. In this case, the kinetics of electrons as well as positive and negative ions can be characterized by the set of equations:

$$\frac{dn_e}{dt} = k_i n_e n_0 - k_a n_e n_0 + k_d n_0 n_- - k_r^{ei} n_e n_+, \tag{4.165}$$

$$\frac{dn_+}{dt} = k_i n_e n_0 - k_r^{ei} n_e n_+ - k_r^{ii} n_+ n_-, \tag{4.166}$$

$$\frac{dn_-}{dt} = k_a n_e n_0 - k_d n_0 n_- - k_r^{ii} n_+ n_-. \tag{4.167}$$

In this set of equations: n_+, n_- are concentrations of positive and negative ions; n_e, n_0 are concentrations of electrons and neutral species; rate coefficients $k_i, k_a, k_d, k_r^{ei}, k_r^{ii}$ are related to the processes of ionization by electron impact, dissociative or other electron attachment, electron detachment from negative ions, electron–ion and ion–ion recombination. The rate coefficients of processes involving neutral particles (k_i, k_a, k_d) are expressed in the system Eqs. (4.165)–(4.167) with respect to the total gas density. If a moderate or high-pressure gas is not electronegative, then the volume balance of electrons and positive ions in the non-equilibrium discharge obviously can be reduced to the simple ionization-recombination balance. However, for electronegative gases, two qualitatively different self-sustained regimes can be achieved (at different effectiveness of electron detachment): one controlled by recombination, and the other controlled by electron attachment.

4.5.2 Discharge Regime Controlled by Electron–Ion Recombination

In some discharges, destruction of negative ions by, for example, the electron detachment is faster than ion-ion recombination:

$$k_d n_0 >> k_r^{ii} n_+. \tag{4.168}$$

In this case, the losses of charged particles are due to the electron–ion recombination similarly to non-electronegative gases. Such situations can take place in plasma-chemical processes of CO_2 and H_2O dissociation, and NO-synthesis in air. In these systems the associative electron detachment processes:

$$O^- + CO \rightarrow CO_2 + e, \quad O^- + NO \rightarrow NO_2 + e, \quad O^- + H_2 \rightarrow H_2O + e \tag{4.169}$$

proceed very fast; these require about 0.1 μs at concentrations of the CO, NO, and H_2 molecules $\sim 10^{17}cm^{-3}$. This electron liberation time (at relatively low degrees of ionization) is shorter than time of ion-ion recombination, and according to Eq. (4.168), the discharge is controlled by electron–ion recombination. The electron attachment and detachment are in the dynamic quasi-equilibrium in the recombination regime Eq. (4.168) during the time sufficient for electron detachment ($t >> 1/k_d n_0$). Then the concentration of negative ions is in dynamic quasi-equilibrium with electron density:

$$n_- = \frac{k_a}{k_d} n_e = n_e \varsigma. \tag{4.170}$$

Using the quasi-constant parameter $\varsigma = {k_a}/{k_d}$, reduce the set of Eqs. (4.165)–(4.167) to the kinetic equation for electron concentration. This equation can be derived by taking into account the plasma quasi-neutrality $n_+ = n_e + n_- = n_e(1 + \varsigma)$, and substituting Eq. (4.170) into (4.166) for positive ions:

$$\frac{dn_e}{dt} = \frac{k_i}{1+\varsigma} n_e n_0 - \left(k_r^{ei} + \varsigma k_r^{ii}\right) n_e^2. \tag{4.171}$$

The parameter $\varsigma = {k_a}/{k_d}$ shows the detachment ability to compensate the electron losses due to attachment. If $\varsigma << 1$, the attachment influence on the electron balance is negligible and the kinetic Eq. (4.171) becomes equivalent to that for non-electronegative gases including only ionization and electron–ion recombination. The kinetic equation includes the effective rate coefficient of ionization $k_i^{eff} = {k_i}/{1+\varsigma}$, which interprets the fraction of electrons lost in the attachment process as not generated by ionization at all. In the same manner, the coefficient $k_r^{eff} = k_r^{ei} + \varsigma k_r^{ii}$ in Eq. (4.171) can be interpreted as the effective coefficient of recombination. This effective coefficient of recombination describes both the direct electron losses through the electron–ion recombination, and the indirect electron losses through the attachment and following ion-ion recombination. Equation (4.171) describes the electron concentration evolution to the steady-state n_e magnitude of the recombination-controlled regime:

$$\frac{n_e}{n_0} = \frac{k_i^{eff}(T_e)}{k_r^{eff}} = \frac{k_i}{\left(k_r^{ei} + \varsigma k_r^{ii}\right)(1+\varsigma)}. \tag{4.172}$$

The important peculiarity of the recombination-controlled regime of a non-equilibrium discharge is the fact, that there is the steady-state degree of ionization (n_e/n_0) for each value of electron temperature T_e. The criterion of recombination-controlled regime Eq. (4.168) can be rewritten using only rate coefficients, taking into account the quasi-neutrality $n_+ = n_e + n_- = n_e(1 + \varsigma)$ and Eq. (4.172):

$$k_d >> \frac{\left(k_i - k_a\right) k_r^{ii}}{k_r^{ei}}. \tag{4.173}$$

4.5.3 Discharge Regime Controlled by Electron Attachment

This regime takes place if the balance of charged particles is due to volume processes and the discharge parameters correspond to inequalities opposite to Eqs. (4.168) and (4.173). In this case, the negative ions produced by electron attachment go almost instantaneously into ion–ion recombination,

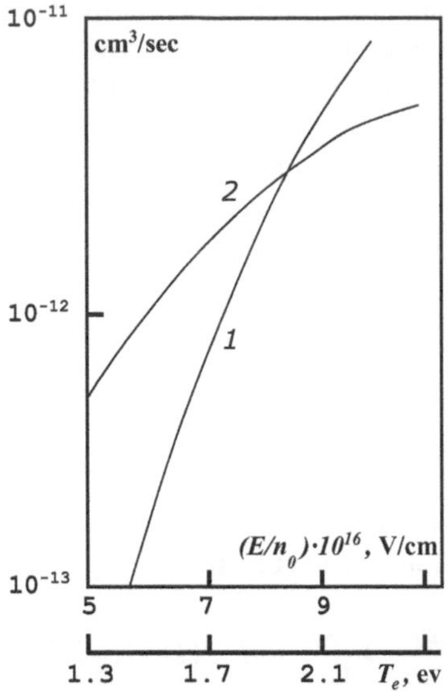

FIGURE 4.31 Rate coefficients of ionization (1) and dissociative attachment (2) for CO_2.

and electron losses are mostly due to the attachment. The steady-state solution of Eq. (4.166) for the attachment-control regime can be presented as:

$$k_i\left(T_e\right) = k_a\left(T_e\right) + k_r^{ei}\frac{n_+}{n_0}. \tag{4.174}$$

In the attachment-controlled regime, the electron attachment is usually faster than recombination and Eq. (4.174) actually requires $k_i(T_e) \approx k_a(T_e)$. The exponential functions $k_i(T_e)$ and $k_a(T_e)$ usually appear as shown in Figure 4.31 (see also Figure 4.21), and have the only crossing point T_{st}. This crossing point determines the steady-state electron temperature in the non-equilibrium discharge, self-sustained in the attachment-controlled regime. In contrast to the recombination-controlled regime, the steady-state non-equilibrium discharge can be controlled by electron attachment only at high electron temperatures $T_e \geq T_{st}$. From Eq. (4.174), the generation of electrons cannot compensate for their losses at lower values of electron temperature without detachment.

4.5.4 Discharge Regime Controlled by Charged Particles Diffusion to the Walls, the Engel-Steenbeck Relation

When gas pressure is relatively low and the inequality opposite to Eq. (4.164) is valid, then the balance of charged particles governed by the competition between ionization in volume and diffusion of charged particles to the walls Eqs. (4.160)–(4.163). The balance of direct ionization by electron impact Eq. (2.21) and ambipolar diffusion to the walls of a long discharge chamber of radius R (see Eqs. (4.160), (4.161) and (4.127)) gives the relation between electron temperature and pressure:

$$\left(\frac{T_e}{I}\right)^{1/2}\exp\left(\frac{I}{T_e}\right) = \frac{\sigma_0}{\mu_i p}\left(\frac{8I}{\pi m}\right)^{1/2}\left(\frac{n_0}{p}\right)(2.4)^2\left(pR\right)^2. \tag{4.175}$$

This is the **Engel-Steenbeck relation** for the diffusion-controlled regime of non-equilibrium discharges. Here I is the ionization potential, μ_i is the ion mobility, σ_0 is the electron–neutral

TABLE 4.5
The Numerical Parameters of the Engel-Steenbeck Relation

Gas	C, Torr^{-2}cm^{-2}	c, Torr^{-1}cm^{-1}	Gas	C, Torr^{-2}cm^{-2}	c, Torr^{-1}cm^{-1}
N_2	2*10^4	4*10^{-2}	Ar	2*10^4	4*10^{-2}
He	2*10^2	4*10^{-3}	Ne	4.5*10^2	6*10^{-3}
H_2	1.25*10^3	10^{-2}			

gas-kinetic cross section, m is an electron mass. If the gas temperature is fixed, the parameters $\mu_i p$ and n_0/p in Eq. (4.175) are constant and the Engel-Steenbeck relation can be written as:

$$\sqrt{\frac{T_e}{I}}\exp\left(\frac{I}{T_e}\right)=C\left(pR\right)^2. \tag{4.176}$$

The constant C only depends on the type of gas. Table 4.5 provides values of C for some gases. The universal relation between T_e/I and the similarity parameter cpR for the diffusion-controlled regime is usually also presented as a graph (see Figure 4.32). The gas type parameters c for this graph are also given in Table 4.5. From the Engel-Steenbeck curve, the electron temperature in the diffusion-controlled regimes decreases with increases of pressure and radius of the discharge tube. In contrast to the regimes sustained by volume processes, the diffusion-controlled regimes are sensitive to the radial density distribution of charged particles. Such radial distribution for a long cylindrical discharge follows the Bessel function:

$$n_e\left(r\right)\propto J_0\left(\frac{r}{R}\right). \tag{4.177}$$

4.5.5 Propagation of Electric Discharges

The electric fields required to initiate a discharge usually exceed those necessary to sustain the non-equilibrium steady-state plasma. This can be explained by the formation of excited and chemically active species in plasma, which facilitates the non-thermal ionization mechanisms, facilitates destruction of negative ions and finally reduces the requirements for the electric field. Thermal plasma propagation is related to the heat transfer processes providing high temperatures sufficient for thermal ionization of the incoming gas (Yu. P. Raizer, 1974). The non-thermal plasma propagation at high

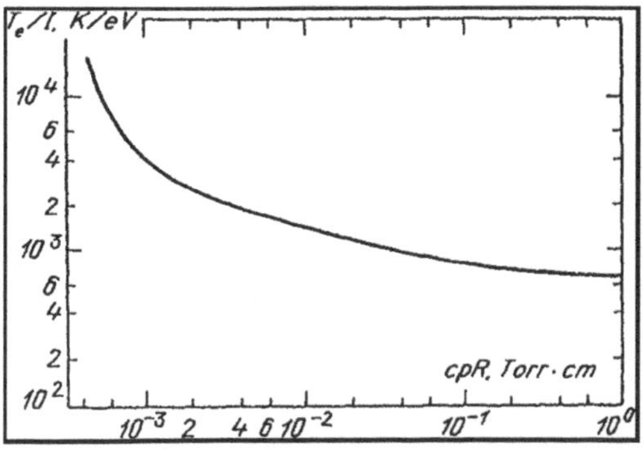

FIGURE 4.32 Universal relation between electron temperature, pressure and discharge tube radius.

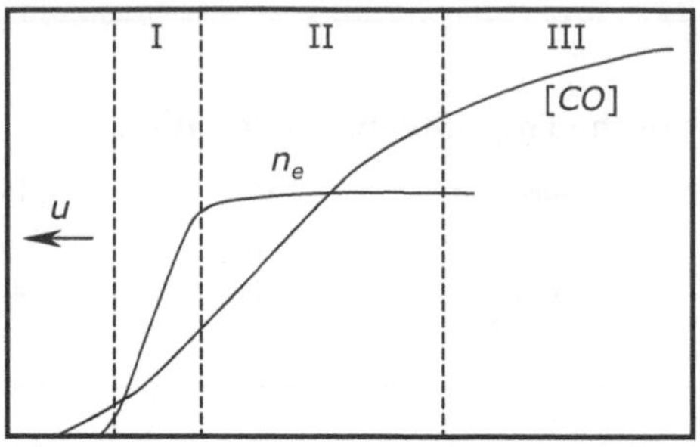

FIGURE 4.33 Electron and CO density distributions in the front of propagating discharge. I-low electron concentration zone; II-discharge zone where CO-diffusion provides effective detachment and sufficient electron density; III-effective CO_2 dissociation zone.

electric fields can be provided by electron diffusion in front of the discharge (V. Yu. Baranov, A.A. Vedenov, V.G. Niziev, 1972; R. Munt, R.S.B. Ong, D.L. Turcotte, 1969). Specific mechanism of discharge propagation in plasma-chemical systems is related to preliminary propagation of such discharge facilitating species as excited atoms and molecules, or products of chemical reactions providing the effective electron detachment (V. Liventsov, V. Rusanov, A. Fridman, G. Sholin, 1981). Analyze this mechanism, taking as an example the non-thermal discharge one-dimensional propagation in CO_2 in a uniform electric field, Te about 1 eV (see Figure 4.33). The breakdown of CO_2 is controlled by the dissociative attachment $e + CO_2 \rightarrow CO + O^-$and requires high electric fields and electron temperatures exceeding 2 eV (see Figure 4.31). However, CO_2 dissociation in plasma produces sufficient CO to provide effective electron detachment Eq. (4.169) and the recombination-controlled regime corresponding to the lower electric fields under consideration. Parameters of the CO_2 discharge under consideration, propagating in fast gas flow are: $n_e = 10^{13}\ \text{cm}^{-3}$, $n_0 = 3 \cdot 10^{18}\ \text{cm}^{-3}$, $T_e = 1\ eV$, $T_0 \approx 700\ \text{K}$ (Yu.P. Boutylkin, V.K. Givotov, E.G. Krasheninnikov e.a., 1981). Based on Eqs. (4.168) and (4.172), the critical value of CO-concentration, separating the attachment and recombination-controlled regimes is:

$$\left[\text{CO}\right]_{cr} = \frac{k_a k_r^{ii}}{k_d k_r^{ei}} n_0 \approx \left(10^{-3} \div 10^{-4}\right) \cdot n_0. \tag{4.178}$$

If the CO-concentration exceeds the critical value$[\text{CO}] > [\text{CO}]_{cr}$, the recombination-controlled balance gives the relatively high electron density:

$$n_e\left(T_e\right) = \frac{k_i\left(T_e\right)}{k_r^{ei}} n_0. \tag{4.179}$$

Conversely, if the carbon monoxide concentration is below the critical limit, $[\text{CO}] < [\text{CO}]_{cr}$, then the electron concentration is very low, controlled by the dissociative attachment and is also proportional to the CO concentration:

$$n_e = \frac{k_i k_d}{k_a k_r^{ii}} \left[\text{CO}\right]. \tag{4.180}$$

Thus propagation of the electron concentration and of the discharge in general is related to the propagation of the CO-concentration. Most of CO production in this system is due to the dissociation of vibrationally excited CO_2 molecules and takes place in the main plasma zone III (Figure 4.33). CO diffusion from zone III into zone II provides the high CO concentration for sustaining the high electron concentration Eq. (4.179), which subsequently provides the vibrational excitation and CO_2 dissociation in zone III. Further decrease of the CO concentration below the critical value in the zone I corresponds to a dramatic fall of the electron concentration, Eq. (4.157).

4.5.6 Propagation of the Non-Thermal Ionization Wave, Self-Sustained by Diffusion of Plasma Chemical Products

To determine the evolution of the profile of electron concentration and the velocity of the ionization wave, the linear one-dimensional propagation of CO concentration along the axis x in the plasma-dissociating CO_2 was described by a differential equation with only the variable $\xi = x + ut$ (V. Liventsov, V. Rusanov, A. Fridman, G. Sholin, 1981):

$$u\frac{\partial[\mathrm{CO}]}{\partial\xi} = D\frac{\partial^2[\mathrm{CO}]}{\partial\xi^2} + g(\xi, T_v, T_0, n_e, n_0). \tag{4.181}$$

In this equation u is the velocity of ionization wave (discharge propagation); T_v, T_0 are vibrational and translational gas temperatures; n_e, n_0 are electron and neutral gas concentrations; D is the diffusion coefficient of CO molecules, and g is a model source of CO as a result of CO_2 dissociation:

$$g(\xi) = \frac{n_0}{\tau_{eV}}\exp\left\{-\frac{[\mathrm{CO}]_{cr}}{[\mathrm{CO}](\xi - u\tau_{eV})} - \alpha\frac{[\mathrm{CO}](\xi)}{[\mathrm{CO}]_0}\right\}. \tag{4.182}$$

Here $\tau_{eV} = 1/k_{eV}n_e^{II}$ is the vibrational excitation time in the zone II; $[\mathrm{CO}]_0 = [\mathrm{CO}](\xi_0)$ is the maximum concentration of CO at the end of zone III $\xi_0 = u\tau_{chem}$, τ_{chem} is the total chemical reaction time in zone III; parameter $\alpha \approx 3$ shows the exponential smallness of the dissociation rate at the end of zone III $\xi \to \xi_0$, $t \to \tau_{chem}$, when the process is actually completed. The boundary conditions for the equation Eq. (4.181) should be taken as: $[\mathrm{CO}](-\infty) = 0$, $[\mathrm{CO}](\xi_0) = [\mathrm{CO}]_0$. The source g(ξ) is not powerful at the negative values of ξ, and perturbation theory can be applied to solve the nonlinear Eqs. (4.181), (4.182). The non-perturbed equation (g = 0) gives the solution $[\mathrm{CO}] = [\mathrm{CO}]_0\exp\left(\frac{\xi u}{D}\right)$. Contribution of the source g(ξ) in the first order of the perturbation theory leads to the linear equation:

$$\frac{\partial}{\partial\xi}\left\{u[CO](\xi) - D\frac{\partial}{\partial\xi}[\mathrm{CO}](\xi)\right\} = \frac{n_0}{\tau_{eV}}\exp\left[-\aleph\exp\left(-\frac{\xi u}{D}\right)\right], \tag{4.183}$$

with the numerical parameter ℵ, which is equal to:

$$\aleph = \frac{[\mathrm{CO}]_{cr}}{[\mathrm{CO}]_0}\exp\frac{u^2\tau_{eV}}{D}. \tag{4.184}$$

Taking into account the asymptotical decrease of the right-hand part of Eq. (4.183) as $\xi \to -\infty$, the solution of this equation can be presented as:

$$[\mathrm{CO}](\xi) = [\mathrm{CO}]_0\exp\left(\frac{\xi u}{D}\right)\left\{1 - \frac{D}{u^2\tau_{eV}}\frac{n_0}{[\mathrm{CO}]_0}\exp\left(-\frac{\xi u}{D} - \aleph\exp\left(-\frac{\xi u}{D}\right)\right)\right\}. \tag{4.185}$$

First-order perturbation theory permits solving Eqs. (4.181), (4.182) in the interval $0 \le \xi \le \xi_0$, because as $\xi \to \xi_0$ the source g(ξ) also become very small. First-order perturbation theory gives:

$$[\mathrm{CO}](\xi) = [\mathrm{CO}]_0\left\{1 + \frac{n_0}{[\mathrm{CO}]_0}\frac{\xi - \xi_0}{u\tau_{eV}}\exp(-\alpha)\right\}. \tag{4.186}$$

Equations (4.185) and (4.186) give solutions for the concentration profiles for both positive and negative magnitudes of the auto-model variable ξ. To find the united solution, Eqs. (4.185) and

(4.186) are matched at the wavefront $\xi = 0$. Such cross-linking of the solutions is possible only at the similar value of the parameter u^2, for example, only at the magnitude of the velocity of the ionization wave:

$$u^2 = \frac{D}{\tau_{eV}} \ln\left[\gamma \ln\left(\frac{De^{\alpha}}{u^2 \tau_{chem}}\right)\right], \tag{4.187}$$

where $\gamma = [CO]_0/[CO]_{cr} \approx 10^3 - 10^4$ (see Eq. (4.178)). If the parameter $\alpha \sim 3$, the solution of the transcendent equation (4.187) for the velocity of the ionization wave can be expressed as:

$$u^2 \approx \frac{D}{\tau_{eV}} \ln \gamma. \tag{4.188}$$

This velocity of the ionization wave and non-thermal discharge propagation can be physically interpreted as the velocity of diffusion transfer of the detachment active heavy particles (CO) ahead of the discharge front on a distance necessary for effective vibrational excitation of CO_2 molecules with their further dissociation. For numerical calculations, it is convenient to rewrite Eq. (4.188) in terms of speed of sound c_s, Mach number M and the ionization degree in plasma n_e/n_0:

$$u \approx 30 c_s \sqrt{\frac{n_e}{n_0}}, \quad M = 30\sqrt{\frac{n_e}{n_0}}. \tag{4.189}$$

From Eqs. (4.188) and (4.189), the velocity of the non-equilibrium ionization wave propagation actually does not strongly depend on the details of the propagation mechanism. It depends mostly on the degree of ionization in the main plasma zone and also on the critical amount ($1/\gamma$) of the ionization active species (in this case CO), which should be transported in front of the discharge to facilitate ionization. The degree of ionization in discharges under consideration is moderate and the discharge velocity Eq. (4.189) is always lower than the speed of sound. Discharge propagation in very fast flows, in particular in supersonic flows, requires special consideration, (Yu. P. Raizer, 1974).

4.5.7 Non-Equilibrium Behavior of Electron Gas, Difference Between Electron and Neutral Gas Temperatures

There are two principal aspects of the non-equilibrium behavior of an electron gas: the first is related to temperature differences between electrons and heavy particles, the second is concerned with a significant deviation of the degree of ionization from that predicted by the Saha equilibrium. Ionization is mostly provided by electron impact and the ionization process should be quite intensive to sustain the steady-state plasma. This usually requires the electron temperature T_e to be at least on the level of $1/10$ of the ionization potential (~1 eV). Indeed, the electron temperature in plasma is usually about 1–3 eV, for both for thermal equilibrium and cold non-equilibrium discharges. It is the gas temperature T_0, which determines the equilibrium or non-equilibrium plasma behavior; in thermal discharges T_0 is high and the system can be close to equilibrium, in non-thermal discharges T_0 is low and the degree of non-equilibrium T_e/T_0 can be high, sometimes up to 100. The effect of non-equilibrium $T_e/T_0 >> 1$ is due to the fact that for some reason, the neutral gas is not heated in the discharge. Thus in low pressure discharges it is related to intensive heat losses to the discharge chamber walls. The difference between the gas temperature in plasma T_0 and room temperature T_{00} in such discharges can be estimated from the simple relation:

$$\frac{T_0 - T_{00}}{T_{00}} = \frac{P}{P_0}, \tag{4.190}$$

In this relation: P is the discharge power per unit volume (specific power); P_0 is the critical value of specific power, corresponding to the gas temperature increase over $T_{00} = 300$ K. Taking into account that both the thermal conductivity and adiabatic discharge heating per one molecule are reversibly proportional to pressure, practically the critical specific power does not depend on pressure and numerically is $P_0 = 0.1 \div 0.3\,\text{W/cm}^3$. In moderate and high-pressure non-equilibrium discharges (usually more than 20–30 Torr) where heat losses to the wall are low, the neutral gas overheating can be prevented either by high velocities and low residence times in the discharge or by a short time of discharge pulses in non-steady-state systems. In both cases this restricts the specific energy input E_v, that is the energy transferred from the electric discharge into the neutral gas calculated per one molecule (it can be expressed in eV/mol or in J/cm^3). The relation similar to Eq. (4.190), estimating over-heating for moderate and high-pressure non-equilibrium discharges, can be presented as:

$$\frac{T_0 - T_{00}}{T_{00}} = \frac{E_v(1-\eta)}{c_p T_{00}}. \tag{4.191}$$

Here c_p is the specific heat; η is the energy efficiency of plasma chemical process, which is the fraction of discharge energy spent to perform a chemical reaction. The energy efficiency η in moderate and high-pressure plasma chemical systems stimulated by vibrational excitation, can be large, up to 70–90% in the particular case of CO_2 dissociation, which facilitates the conditions necessary for sustaining the non-equilibrium. In this case, Eq. (4.191) restricts the specific energy input by pretty high critical value $E_v \approx 1.5$ eV/mol. However, the non-equilibrium discharge can be overheated and converted into a thermal one even at specific energy inputs due to possible plasma instabilities. Another important element of the non-equilibrium behavior of electron gas in electric discharges is the strong deviation of the electron energy distribution function from the Maxwellian distribution. This effect is also primarily due to the low neutral gas temperature. Hence, this leads to a low concentration of electronically excited atoms and molecules, which causes electrons to establish their energy balance directly with the electric field.

4.5.8 Non-Equilibrium Behavior of Electron Gas, Deviations from the Saha – Degree of Ionization

The quasi-equilibrium electron concentration and degree of ionization can be found as the function of one temperature, based on the Saha formula (4.16). Although the ionization processes (both in thermal and non-thermal discharges) are provided by the electron gas, for non-equilibrium discharges the Saha formula with electron temperature T_e gives the ionization degree n_e/n_0 several orders of value higher than the real one. Obviously, the Saha formula assuming the neutral gas temperature gives even much fewer electron concentrations and much worse agreement with reality. This non-equilibrium effect is due to the presence of additional channels of charged particles losses in a cold gas. These are much faster than those reverse processes of recombination leading to the Saha-equilibrium. The Saha equilibrium actually implies the balance of ionization by electron impact and the double-electron three-body recombination:

$$e + A \Leftrightarrow A^+ + e + e. \tag{4.192}$$

In reality, however, the charged particles losses in moderate and high-pressure systems are mostly due to the dissociative recombination:

$$e + AB^+ \rightarrow A + B^*$$

This recombination process is much faster than the three-body recombination process while at the same time it is not compensated by the reverse process of associative ionization because of the relatively low density of electronically excited atoms. The low (under-equilibrium) density of the excited species in non-thermal plasma is due to intensive energy transfer in other degrees of freedom. Further, in non-equilibrium low-pressure discharges, the ionization is not compensated by three-body recombinations but rather by the diffusion to the walls, which is also much faster and obviously not balanced by electron emission.

4.6 PROBLEMS AND CONCEPT QUESTIONS

4.6.1 Average Vibrational Energy and Vibrational Specific Heat

Based on the equilibrium Boltzmann distribution function Eq. (4.12), find the average value of vibrational energy and related specific heat of a diatomic molecule. Analyze the result for the case of high ($T >> \hbar\omega$) and low ($T >> \hbar\omega$) temperatures.

4.6.2 Ionization Equilibrium, the Saha Equation

Using the Saha equation, estimate the degree of ionization in the thermal Ar-plasma at atmospheric pressure and temperature T = 20,000 K. Explain, why the high ionization degree can be reached at temperatures much less than the ionization potential.

4.6.3 Statistics of Plasma Radiation at the Complete Thermodynamic Equilibrium (CTE)

Calculate the total radiation of the Ar-arc discharge plasma with a surface area of10 cm^2 at an equilibrium temperature T = 20,000 K assuming CTE conditions. Does such thermal radiation power look reasonable for the arc discharge?

4.6.4 The Debye Corrections of Thermodynamic Functions in Plasma

Estimate the Debye correction for the Gibbs potential per unit volume of a thermal Ar-plasma at atmospheric pressure and equilibrium temperature T = 20,000 K. The ionization degree can be calculated using the Saha equation.

4.6.5 The Treanor Distribution Function

Estimate the normalizing factor B for the Treanor distribution function, taking into account only relatively low levels of vibrational excitation. Find a criterion for application of such a normalizing factor (most of the molecules should be located on the vibrational levels lower, than the Treanor minimum).

4.6.6 Average Vibrational Energy of Molecules, Following the Treanor Distribution

Based on the non-equilibrium Treanor distribution function, find the average value of vibrational energy taking into account only relatively low vibrational levels. Find an application criterion for the result (most of the molecules should be located on the vibrational levels lower than the Treanor minimum).

4.6.7 The Boltzmann Kinetic Equation

Prove that the Maxwell distribution function is an equilibrium solution of the Boltzmann equation. Compare the Maxwell distribution functions for 3D-velocities Eq. (4.58), the absolute value of velocities and for translational energies Eq. (2.1).

4.6.8 Entropy of Electrons in Non-Equilibrium Plasma

Using Eq. (4.59), estimate the entropy of plasma electrons, following the Druyvesteyn electron energy distribution function, as a function of the reduced electric field in non-thermal discharges.

4.6.9 The Krook Collisional Operator for Momentum Conservation Equation

Compare the frictional and particle generation terms in the Krook collisional operator, and explain why the frictional term usually dominates in non-thermal discharges.

4.6.10 The Fokker-Planck Kinetic Equation

Using the Fokker-Planck kinetic equation in the form Eq. (4.76), estimate the time which is necessary to reestablish the steady-state electron energy distribution function after fluctuation of the electric field.

4.6.11 The Druyvesteyn Electron Energy Distribution Function

Calculate the average electron energy for the Druyvesteyn distribution. Define the effective electron temperature of the distribution and compare it with that of the Maxwellian distribution function.

4.6.12 The Margenau Electron Energy Distribution Function

Compare the behavior of Margenau and Druyvesteyn distribution functions. Estimate the average electron energy and effective electron temperature for the Margenau electron energy distribution function. Consider the case of very high frequencies, when the Margenau distribution becomes exponentially linear with respect to electron energy.

4.6.13 Influence of Vibrational Temperature on Electron Energy Distribution Function

Simplify the general relation Eq. (4.85), describing the influence of vibrational temperature on the electron energy distribution function, for the case of high vibrational temperatures $T_v >> \hbar\omega$. Using this relation, estimate the typical acceleration of the ionization rate coefficient corresponding to a 10% increase of vibrational temperature.

4.6.14 Influence of Molecular Gas Admixture to a Noble Gas on Electron Energy Distribution Function

Using Eq. (4.78), estimate the percentage of nitrogen to be added to a non-thermal discharge initially sustained in Ar to change the behavior of the electron energy distribution function.

4.6.15 Electron–Electron Collisions and Maxwellization of Electron Energy Distribution Function

Calculate the minimum ionization degree $\frac{n_e}{n_0}$, when the Maxwellization provided by electron–electron collisions becomes essential in establishing the electron energy distribution function in the case of (1) argon-plasma, $T_e = 1$ eV and (2) nitrogen-plasma $T_e = 1$ eV.

4.6.16 Relation Between Electron Temperature and Reduced Electric Field

Calculate the reduced electric field necessary to provide an electron temperature $T_e = 1$ eV in the case of (1) argon-plasma and (2) nitrogen-plasma. Compare and discuss the result.

4.6.17 Electron Conductivity

Calculate the electron conductivity in plasma using the general Eq. (4.98) assuming a constant value of the electron mean-free-path λ. Compare the result with Eq. (4.99) corresponding to a constant value of the electron–neutral collision frequency.

4.6.18 Joule Heating

Using Eq. (4.101) for the Joule heating, prove the stability of a thermal plasma sustained at a constant current density $\vec{j} = \sigma \vec{E}$ with respect to the fluctuation of plasma temperature.

4.6.19 Similarity Parameters

Based on the similarity parameters, presented in Table 4.1, find the electron mean-free-pass, electron–neutral collision frequency, and electron mobility in atmospheric pressure air at 1) room temperature, 2) at 600 K.

4.6.20 Electron Drift in the Crossed Electric and Magnetic Fields

Explain why the electron drift in crossed electric and magnetic fields is impossible at fixed values of the magnetic field and relatively high values of the electric field. Estimate the criteria for the critical electric field. Illustrate the electron motion if the electric field exceeds the critical one.

4.6.21 Electric Conductivity in the Crossed Electric and Magnetic Fields

Calculate the lower and upper application limits of the ratio $\frac{\omega_B}{\nu_{en}}$ for the transverse plasma conductivity Eq. (4.110) in a strong magnetic field. Discuss the difference in magnetic fields and gas pressures necessary for trapping plasma electrons and ions in the magnetic field.

4.6.22 Plasma Rotation in the Crossed Electric and Magnetic Fields, Plasma Centrifuge

Describe plasma motion in the electric field $\vec{E}(r)$ created by a long charged cylinder and uniform magnetic field $\vec{B}$ parallel to the cylinder. Derive a relation for the maximum operation pressure of such a plasma centrifuge. Is it necessary or not to trap ions in the magnetic field to provide the plasma rotation as a whole in this system?, Is it necessary or not to trap electrons in the magnetic field?

4.6.23 Electric Conductivity of Strongly Ionized Plasma

In calculating the electric conductivity of a strongly ionized plasma Eq. (4.111), why is the electron–neutral and electron–ion but not electron–electron collisions are taken into account?

4.6.24 Free Diffusion of Electrons

Using the similarity parameters presented in Table 4.2, estimate the free diffusion time of electrons with temperature about 1 eV to the wall of the 1cm diameter cylindrical chamber filled with atmospheric pressure nitrogen.

4.6.25 Ambipolar Diffusion

Estimate the ambipolar diffusion coefficient in room temperature atmospheric pressure nitrogen at an electron temperature of about 1 eV. Compare this diffusion coefficient with that for free diffusion of electrons and free diffusion of nitrogen molecular ions (take the ion temperature as room temperature).

4.6.26 Debye Radius and Ambipolar Diffusion

Estimate the Debye radius and, hence, the size scale at which diffusion of charged particles should be considered not as ambipolar but free in the non-thermal discharge. Take as an example the electron temperature of 1 eV and electron concentration of 10^{12} cm^{-3}. Which specific plasma systems and plasma chemical processes can involve such size scales?

4.6.27 Thermal Conductivity in Plasma, Related to Dissociation and Recombination of Molecules

Based on Eq. (4.137) for the effect of dissociation on thermal conductivity and Eq. (4.19) for equilibrium degree of dissociation, determine a general expression for the temperature range, corresponding to the strong influence of the dissociation on thermal conductivity. Compare the result with numerical data presented in Figure 4.15.

4.6.28 Thermal Conductivity in Plasma, Related to Ionization and Charged Particle Recombination

Discuss the effect of ionization-recombination and different chemical reactions on the total coefficient of thermal conductivity in high-temperature quasi-equilibrium plasma. In the same way, as was done for the effect of dissociation Eqs. (4.135) and (4.136), derive a formula for calculating the coefficient of thermal conductivity related to the ionization process. Compare the influence of these two effects on the total thermal conductivity.

4.6.29 The Townsend Breakdown Mechanism

Based on Eqs. (4.141), (4.143) and Table 4.3, compare the Townsend breakdown conditions in molecular gases (take air as an example) and monatomic gas (take argon as an example). How can you explain the difference between these two cases?

4.6.30 The Stoletov Constant and Energy Price of Ionization

Using the electron energy distribution between ionization and different channels of excitation, presented in Figures 3.8–3.15, estimate the energy price of ionization in electric discharges as a function of the reduced electric field. Compare this ionization cost with the Stoletov constant, and explain the similarity and difference between them.

4.6.31 Effect of Electron Attachment on Breakdown Conditions, Formation of Streamers and Leaders

Which attachment mechanism, the dissociative attachment or three body attachment, is more important in preventing different breakdown related phenomena? Is there a difference from this perspective in preventing a Townsend breakdown and the formation of streamers and leaders?

4.6.32 Radial Growth of an Avalanche due to the Repulsion of Electrons

Based on consideration of the radial electron drift in the avalanche head, derive Eq. (4.154) for the transverse growth of the avalanche due to the repulsion of electrons. Estimate the critical value of αx when this repulsion effect exceeds that of the free electron diffusion.

4.6.33 Limitation of Electron Density in Avalanche due to Electron Repulsion

Analyzing the electric charge of an avalanche head and its size (when it is controlled by the electron repulsion), show that the electron density in the head reaches a maximum value which does not depend on x. Estimate and discuss the numerical value of the maximum electron concentration.

4.6.34 Energy Input and Temperature in a Streamer Channel

As it was shown, the specific energy input (electron energy transferred to one molecule) in a streamer channel is small during the short period (~30 ns) of the streamer growth between electrodes. In molecular gases, it is about 10^{-3} eV/mol, which in temperature units corresponds to ~ 10K. In this case, estimate the vibrational temperature in the streamer channel taking a vibrational quantum as $\hbar\omega = 0.3$ eV.

4.6.35 Streamer Propagation Velocity

In frameworks of the model of ideally conducting streamer channel Eq. (4.159) estimate the difference between streamer velocity and electron drift velocity. Explain the streamer propagation velocity dependence on its length at other parameters fixed.

4.6.36 Leader

Estimate the electric field on the head of a 2-m long leader and calculate the correspondent electron drift velocity. Compare the electron drift velocity with a leader propagation velocity and explain the difference (for streamer channel growth - these velocities are usually supposed to be equal).

4.6.37 Steady-State Non-Thermal Discharge Regime in Non-Electronegative Gas

Taking into account the absence of negative ions in non-electronegative gases, simplify the set of Eqs. (4.165) to (4.167) and find the steady-state concentrations of charged particles. Compare the results with density of charged particles in the recombination-controlled regime of a discharge in electronegative gas.

4.6.38 Recombination-Controlled Regime of Steady-State Non-Thermal Discharge

Analyzing Eq. (4.172), determine how the ionization degree in the recombination-controlled regime depends on the concentration of species, providing electron detachment from negative ions. Take as an example the steady-state non-thermal discharge in CO_2, with small additions of CO, providing the effective detachment (4.169).

4.6.39 Attachment-Controlled Regime of Steady-State Non-Thermal Discharge

In the attachment controlled regime Eq. (4.174), the electron temperature and reduced electric field are actually fixed by the ionization-attachment balance $k_i(T_e) \approx k_a(T_e)$. What is the restriction on electron concentration and degree of ionization in this case? How to calculate these? From this point of view, analyze the difference between the non-thermal glow and microwave discharges.

4.6.40 The Engel-Steenbeck Model, Diffusion-Controlled Discharges

Calculate the electron temperature in non-thermal sustained discharge in nitrogen with a pressure 1-Torr and radius 1 cm, based on the Engel-Steenbeck model. Estimate the reduced electric field corresponding to the electron temperature. Compare this reduced electric field with that required for nitrogen breakdown in similar conditions.

4.6.41 Propagation of Non-Thermal Discharges

Imagine a steady-state non-thermal discharge in gas flow with the ionization wave propagation velocity exceeding the gas flow velocity. Discuss.

4.6.42 Ionization Wave Propagation

According to Eq. (4.189), what level of electron concentration and degree of ionization is necessary to provide the velocity of the ionization wave close to the speed of sound. Explain why such discharge conditions are physically non-realistic for the type of ionization wave under consideration.

5 Kinetics of Excited Particles in Plasma

5.1 VIBRATIONAL DISTRIBUTION FUNCTIONS IN NON-EQUILIBRIUM PLASMA, THE FOKKER–PLANCK KINETIC EQUATION

5.1.1 NON-EQUILIBRIUM VIBRATIONAL DISTRIBUTION FUNCTIONS, GENERAL CONCEPT OF THE FOKKER-PLANK EQUATION

The statistical approaches provide only a qualitative analysis of the vibrational kinetics. Only the kinetic equations permit describing the evolution of the vibrational distributions, taking into account all the variety of relaxation processes and chemical reactions. The electrons in non-thermal discharges mostly provide excitation of low vibrational levels, which determines vibrational temperature. Formation of the highly excited and chemically active molecules depends on many different processes in plasma, but mostly on the competition of multi-steps VV-exchange processes and vibrational quanta losses in VT-relaxation. This competition should be described to determine the population of the highly excited states. Different kinetic approaches were developed to describe the evolution of the vibrational distributions in non-equilibrium systems (see e.g., E.E. Nikitin, A.I. Osipov, 1977; A.A. Likalter, G. V. Naidis, 1981; M. Capitelli, 1986; B.F. Gordiets, S. Zhdanok, 1986). The vibrational kinetics in plasma chemical systems includes not only the relaxation processes in the collision of heavy particles and chemical reactions but also the direct energy exchange between the electron gas and excited molecules (the so-called eV-processes). The necessity of simultaneous consideration of so many kinetic processes with different natures suggests using the Fokker-Plank approach. This was developed for vibrational kinetics by V.D. Rusanov, A. Fridman, G. V. Sholin (1979). In the framework of this approach, *the evolution of vibrational distribution function is considered as diffusion and drift of molecules in the space of vibrational energies (or in other words as the diffusion and drift of the molecules along the vibrational spectrum).* The distribution $f(E)$ of molecules over vibrational energies E (the vibrational distribution) can be considered as the density in energy space and determined from the continuity equation:

$$\frac{\partial f(E)}{\partial t} + \frac{\partial}{\partial E} J(E) = 0. \tag{5.1}$$

In this continuity equation J(E) is the flux of molecules in the energy space, which additively includes all relaxation and energy exchange processes, while taking into account chemical reactions from different vibrationally excited states.

5.1.2 THE ENERGY-SPACE-DIFFUSION RELATED VT-FLUX OF EXCITED MOLECULES

Consider the vibrational kinetics for a diatomic molecular gas plasma. The VT relaxation of molecules with vibrational energy E and vibrational quantum $\hbar\omega$ is conceptually viewed as diffusion

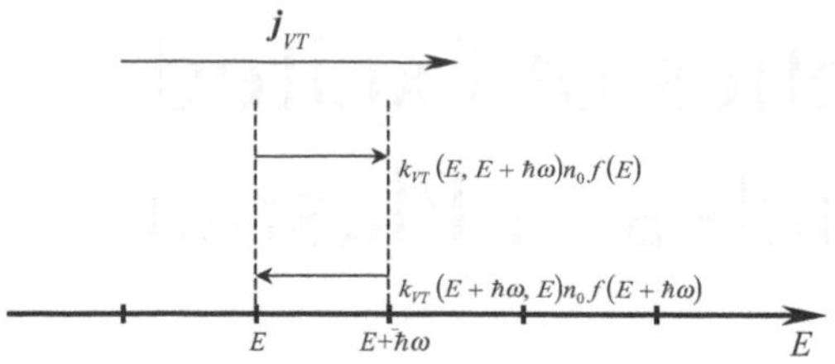

FIGURE 5.1 Illustration of VT-flux in energy space.

along the vibrational energy spectrum and is illustrated in Figure 5.1. The VT-flux then can be expressed as:

$$j_{VT} = f(E)k_{VT}(E, E+\hbar\omega)n_0\hbar\omega - f(E+\hbar\omega)k_{VT}(E+\hbar\omega, E)n_0\hbar\omega. \tag{5.2}$$

Here $k_{VT}(E+\hbar\omega, E)n_0$ and $k_{VT}(E, E+\hbar\omega)n_0$ are the frequencies of the direct and reverse processes of vibrational relaxation, whose ratio is $\exp\frac{\hbar\omega}{T_0}$; n_0 is the neutral gas density. Expanding the vibrational distribution function in series $f(E+\hbar\omega) = f(E) + \hbar\omega\frac{\partial f(E)}{\partial E}$, and denoting the relaxation rate coefficient $k_{VT}(E+\hbar\omega, E) \equiv k_{VT}(E)$, rewrite Eq. (5.2) in the final form of the energy-space diffusion VT-flux:

$$j_{VT} = -D_{VT}(E)\left[\frac{\partial f(E)}{\partial E} + \tilde{\beta}_0 f(E)\right]. \tag{5.3}$$

In this relation, the diffusion coefficient D_{VT} of excited molecules in the vibrational energy space, related to VT-relaxation, was introduced:

$$D_{VT}(E) = k_{VT}(E)n_0 \cdot (\hbar\omega)^2. \tag{5.4}$$

This definition of the diffusion coefficient along the energy spectrum is quite natural and corresponds to that for the conventional diffusion coefficient in coordinate space (see Eq. (4.122)). Obviously, the mean free path λ in the conventional coordinate space corresponds to the vibrational quantum $\hbar\omega$ in energy space; and the frequency of collisions ν in the coordinate space corresponds to the quantum transfer frequency $k_{VT}(E)n_0$ in energy space. The energy-space diffusion coefficient $D_{VT}(E)$ grows exponentially with increasing vibrational energy according to the Landau-Teller relation Eq. (3.61). The translational temperature parameter $\tilde{\beta}_0$ in Eq. (5.3) is defined by the following relation:

$$\tilde{\beta}_0 = \frac{1-\exp\left(-\frac{\hbar\omega}{T_0}\right)}{\hbar\omega}. \tag{5.5}$$

If the translational gas temperature is relatively high $T_0 >> \hbar\omega$, then $\tilde{\beta}_0 = \beta_0 = 1/T_0$. The first term in the flux-relation Eq. (5.3) can be interpreted as diffusion and the second term as the drift in energy space; then $\tilde{\beta}_0 = \beta_0 = 1/T_0$ is the ratio of diffusion coefficient over mobility completely in accordance with the Einstein relation. However, for lower translational temperatures $T_0 \le \hbar\omega$, $\tilde{\beta}_0 < \beta_0 = 1/T_0$. This corresponds to the quantum-mechanical effect of a VT-flux decrease with respect to the case of the quasi-continuum vibrational spectrum. The Eq. (5.3) can only be applied for not very abrupt changes of the vibrational distribution function:

$$\left| \hbar\omega \frac{\partial \ln f(E)}{\partial E} \right| << 1, \tag{5.6}$$

which is necessary to make the f(E) series expansion. When VT relaxation is the dominating process, the kinetic equation for vibrational distribution based on Eqs. (5.1), (5.3) and (5.6) can be presented as:

$$\frac{\partial f(E)}{\partial t} = \frac{\partial}{\partial E} \left\{ D_{VT}(E) \left[\frac{\partial f(E)}{\partial E} + \frac{1}{T_0} f(E) \right] \right\}. \tag{5.7}$$

At steady-state conditions ($\partial/\partial t = 0$), after integration Eq. (5.7) gives $\frac{\partial f(E)}{\partial E} + \frac{1}{T_0} f(E) = const(E)$. Taking into account the boundary conditions at $E \to \infty$: $\frac{\partial f(E)}{\partial E} = 0, f(E) = 0$, yields $const(E) = 0$. As a result, the solution of Eq. (5.17), corresponds to the steady-state domination of the VT-relaxation, which leads to the quasi-equilibrium exponential Boltzmann distribution with temperature T_0: $f(E) \propto \exp(-E/T_0)$.

5.1.3 The Energy-Space-Diffusion Related VV-flux of Excited Molecules

The VV exchange in contrast to VT-relaxation involves two vibrationally excited molecules. Therefore, VV-flux is non-linear with respect to the vibrational distribution function (A.V. Demura, S.O. Mahceret, A. Fridman, 1984):

$$j_{VV} = k_0 n_0 \hbar\omega \int_0^{\infty} \left[Q_{E+\hbar\omega,E}^{E',E'+\hbar\omega} f(E+\hbar\omega) f(E') - Q_{E,E+\hbar\omega}^{E'+\hbar\omega,E'} f(E) f(E'+\hbar\omega) \right] dE'. \tag{5.8}$$

The VV-exchange probabilities Q in this integral correspond to those presented in Eq. (3.74), k_0 is the rate coefficient of neutral-neutral gas-kinetic collisions. Taking into account the defect of vibrational energy in the VV-exchange process $2x_e(E' - E)$ (x_e is the coefficient of anharmonicity), the probabilities Q are related to each other by the detailed equilibrium relation:

$$Q_{E,E+\hbar\omega}^{E'+\hbar\omega,E'} = Q_{E+\hbar\omega,E}^{E',E'+\hbar\omega} \exp\left[-\frac{2x_e(E'-E)}{T_0} \right]. \tag{5.9}$$

The Treanor distribution function Eq. (4.46), derived in Section 4.1.10 as the non-equilibrium statistical distribution makes the VV-flux Eq. (5.8) equal to zero. The Treanor distribution function Eq. (4.46) is a steady-state solution of the Fokker-Planck kinetic equation Eq. (5.1) if VV-exchange is the dominating process and the vibrational temperature T_v exceeds the translational temperature T_0. Replacing vibrational quantum numbers by vibrational energy, rewrite the Treanor distribution function as:

$$f(E) = B \exp\left(-\frac{E}{T_v} + \frac{x_e E^2}{T_0 \hbar\omega} \right), \quad \frac{1}{T_v} = \frac{\partial \ln f(E)}{\partial E} \Big|_{E \to 0}. \tag{5.10}$$

The exponentially parabolic Treanor distribution describes significant overpopulation of the highly vibrationally excited states and is illustrated in Figure 4.3 in comparison with the Boltzmann distribution. The physical nature of the Treanor distribution, related to the predominant quantum transfer from vibrationally poor to vibrationally rich molecules (effect of the "capitalism in molecular life"), was discussed in Section 4.1.10 and illustrated in Figure 4.4. To analyze the complicated VV-flux along the vibrational energy spectrum, it is convenient to divide the total VV-flux and, hence, the total integral Eq. (5.18) into two parts, corresponding to linear $j^{(0)}$ and non-linear $j^{(1)}$ flux-components:

$$j_{VV}(E) = j_{VV}^{(0)}(E) + j_{VV}^{(1)}(E). \tag{5.11}$$

The linear flux-component $j_{VV}^{(0)}(E)$ corresponds to the non-resonant VV-exchange collisions of a highly vibrationally excited molecule of energy E with the bulk of low vibrational energy molecules, which is related in Eq. (5.8) to the integration domain $0 < E' < T_v$. The non-linear flux-component $j_{VV}^{(1)}(E)$ corresponds to the resonant VV-exchange collisions of two highly vibrationally excited molecules. Vibrational energies of the two molecules are mostly in the interval confined by the adiabatic parameter δ_{VV} (Eq. 3.75): $|E - E'| \le \delta_{VV}^{-1}$), and the integral Eq. (5.18) domain $T_v < E' < \infty$.

5.1.4 Linear VV-Flux along the Vibrational Energy Spectrum

The linear flux-component $j_{VV}^{(0)}(E)$ corresponds to the non-resonant VV-exchange with the bulk of low vibrational energy molecules with $0 < E' < T_v$. The distribution function for low vibrational energy is $f(E') \approx \frac{1}{T_v}\exp\left(-\frac{E'}{T_v}\right)$, and: $f(E+\hbar\omega) = f(E) + \hbar\omega\frac{\partial f(E)}{\partial E}$, $\exp\left(-\frac{\hbar\omega}{T_v} + \frac{2x_eE}{T_0}\right) \approx 1 - \frac{\hbar\omega}{T_v} + \frac{2x_eE}{T_0}$. Then Eq. (5.8) yields the linear component of the VV-flux along the vibrational energy spectrum:

$$j_{VV}^{(0)} = -D_{VV}(E)\left[\frac{\partial f(E)}{\partial E} + \left(\frac{1}{T_v} - \frac{2x_eE}{T_0\hbar\omega}\right)f(E)\right]. \tag{5.12}$$

Similar to Eq. (5.3), the diffusion coefficient D_{VV} in the vibrational energy space, related to the non-resonant VV-exchange of a molecule of energy E with the bulk of low energy molecules is:

$$D_{vv}(E) = k_{VV}(E)n_0(\hbar\omega)^2. \tag{5.13}$$

The relevant VV-exchange rate coefficient can be expressed based on Eq. (3.74) as: $k_{vv}(E) \approx \frac{ET_v}{(\hbar\omega)^2}Q_{01}^{10}k_0\exp(-\delta_{vv}E)$, where k_0 is the rate coefficient of neutral-neutral gas-kinetic collisions, and the adiabatic parameter δ_{VV} Eq. (3.75) is recalculated with respect to the vibrational energy rather than the vibrational quantum number. The definition of the linear VV-diffusion coefficient along the energy spectrum Eq. (5.13) corresponds to the diffusion coefficient in the coordinate space (4.122). The mean free path λ in this case is again the vibrational quantum $\hbar\omega$, and collisional frequency ν is the non-resonant quantum exchange frequency $k_{VV}(E)n_0$. The solution of the linear kinetic equation $j_{VV}^{(0)}(E) = 0$ (with flux Eq. (5.12)) leads to the Treanor distribution function Eq. (5.10).

5.1.5 Non-Linear VV-Flux along the Vibrational Energy Spectrum

The non-linear flux-component $j_{VV}^{(1)}(E)$ is related to the resonant VV-exchange $|E - E'| \le \delta_{VV}^{-1}$. It is convenient for the domain $T_v < E' < \infty$ to rewrite the integral Eq. (5.18) in the integral-differential form:

$$\begin{aligned} j_{VV}^{(1)} = k_0n_0Q_{10}^{01}\frac{E+\hbar\omega}{\hbar\omega}\int_{T_v}^{\infty}\frac{E'+\hbar\omega}{\hbar\omega}\exp\left(-\delta_{vv}|E'-E|\right) \\ \times\left\{\hbar\omega\left[f(E')\frac{\partial f(E)}{\partial E} - f(E)\frac{\partial f(E')}{\partial E'}\right] + \frac{2x_e(E'-E)}{T_0}f(E)f(E')\right\}\hbar\omega dE'. \end{aligned} \tag{5.14}$$

When the contribution of the quasi-resonance VV-exchange processes is dominant, the under-integral function has a sharp maximum at $E = E'$. In this case, assume $f(E) \approx f(E')$ and $\frac{\partial \ln f(E')}{\partial E'} - \frac{\partial \ln f(E)}{\partial E} \approx \frac{\partial^2 \ln f(E)}{\partial E^2}(E' - E)$. Integration of (5.4) over $(E - \delta_{VV}^{-1}, E + \delta_{VV}^{-1})$ gives the final differential expression of the non-linear VV-flux along the vibrational energy spectrum:

$$j_{VV}^{(1)} = -D_{VV}^{(1)} \frac{\partial}{\partial E}\left[f^2(E) E^2 \left(\frac{2x_e}{T_0} - \hbar\omega \frac{\partial^2 \ln f(E)}{\partial E^2} \right) \right]. \tag{5.15}$$

The energy-space diffusion coefficient $D_{VV}^{(1)}$ describes the resonance VV-exchange:

$$D_{VV}^{(1)} = 3k_0 n_0 Q_{10}^{01} \left(\delta_{VV} \hbar\omega \right)^{-3}. \tag{5.16}$$

In the same manner, as in the case of the linear kinetic equation $j_{VV}^{(0)}(E) = 0$ with the flux Eq. (5.12), one solution of the non-linear kinetic equation $j_{VV}^{(1)}(E) = 0$ with flux Eq. (5.15) is again the Treanor distribution function Eq. (5.10). Indeed, the Treanor distribution satisfies the equality: $\frac{2x_e}{T_0} - \hbar\omega \frac{\partial^2 \ln f(E)}{\partial E^2} = 0$. However, the Treanor distribution function is not the only solution to the non-linear kinetic equation $j_{VV}^{(1)}(E) = 0$; the other plateau-like distribution will be discussed next.

5.1.6 Equation for Steady-State Vibrational Distribution Function, Controlled by VV- and VT Relaxation Processes

The vibrational distribution functions in non-equilibrium plasma are usually controlled by VV-exchange and VT relaxation processes. The vibrational excitation of molecules by electron impact, chemical reactions, radiation etc. is mainly related to the averaged energy balance and temperatures. Eqs. (5.3), (5.12) and (5.15) for VT- and VV-fluxes permit determining the vibrational distribution functions provided by these relaxation processes. At steady-state conditions, the general Fokker-Planck kinetic equation (5.1) gives J(E) = const. Taking into account that at $E \to \infty$: $\frac{\partial f(E)}{\partial E} = 0, f(E) = 0$, yields *const*(*E*) = 0. As a result, the steady-state vibrational distribution function, provided by VT- and VV-relaxation processes can be found from the equation:

$$j_{VV}^{(0)}(E) + j_{VV}^{(1)}(E) + j_{VT}(E) = 0. \tag{5.17}$$

Taking into account the specific relations Eqs. (5.3), (5.12) and (5.15) for the fluxes, the Fokker-Planck kinetic equation can be written as:

$$\begin{gathered} D_{VV}(E)\left(\frac{\partial f(E)}{\partial E} + \frac{1}{T_v} f(E) - \frac{2x_e E}{T_0 \hbar\omega} f(E) \right) + D_{VV}^{(1)} \frac{\partial}{\partial E} \\ \times \left[f(E)^2 E^2 \left(\frac{2x_e}{T_0} - \hbar\omega \frac{\partial^2 \ln f(E)}{\partial E^2} \right) \right] + D_{VT}(E)\left(\frac{\partial f(E)}{\partial E} + \tilde{\beta}_0 f(E) \right) = 0. \end{gathered} \tag{5.18}$$

The first two terms in the kinetic Eqs. (5.17) and (5.18) are related to VV relaxation and prevail at low vibrational energies, the third term is related to VT-relaxation and dominates at higher energies (see Eqs. 3.78 and 3.79). The linear part of equation Eq. (5.18), including the first and third terms, is easy to solve, but the second non-linear term makes a solution much more complicated. It is helpful to point out three cases of strong, intermediate, and weak excitation, corresponding to different contributions of the non-linear term (which is the most important at higher excitation levels). The distinction between the three regimes is determined by two dimensionless parameters: $\delta_{VV}T_v$ and $\frac{x_e T_v^2}{\hbar\omega T_v}$.

5.1.7 Vibrational Distribution Functions, the Strong Excitation Regime

The regime of strong excitation takes place in plasma at high vibrational temperature:

$$\delta_{VV}T_v \geq 1. \tag{5.19}$$

The adiabatic parameter δ_{VV} was discussed in Section 3.5.2. Its typical value at the room temperature is $\delta_{VV} \approx (0.2 \div 0.5)/\hbar\omega$ (in this chapter, δ is expressed with respect to vibrational energy, not a quantum number). This means that the strong excitation regime requires high vibrational temperatures exceeding 5,000–10,000 K. When vibrational temperatures are so high, the resonant VV-exchange between highly excited molecules (the non-linear VV-flux) dominates over the non-resonant VV-exchange of the excited molecules with the bulk of low excited molecules (the linear VV-flux). The kinetic Eq. (5.18) at moderate vibrational energies (e.g., when the VT-flux can still be neglected) can be simplified to:

$$E^2 f^2(E)\left(\frac{2x_e}{T_0} - \hbar\omega\frac{\partial^2 \ln f(E)}{\partial E^2}\right) = F. \tag{5.20}$$

Here F is the constant proportional to the quantum flux from low levels (where they appear due to excitation by electron impact) to high levels (where they disappear in VT relaxation and chemical reactions). More detailed solution of Eq. (5.20) can be found in the publications of A.A. Likalter, G. V. Naidis (1981), and V.D. Rusanov, A. Fridman (1984). However, the following simple analysis of Eq. (5.20) permits the solution. At low vibrational energies, when the product $Ef(E)$ is large,: $\frac{2x_e}{T_0} - \hbar\omega\frac{\partial^2 \ln f(E)}{\partial E^2} \approx 0$ and the vibrational distribution function is close to the Treanor distribution. For larger energies, $Ef(E)$ becomes smaller and $\frac{2x_e}{T_0} - \hbar\omega\frac{\partial^2 \ln f(E)}{\partial E^2} = \frac{F}{E^2 f^2(E)}$ grows and cannot be taken as zero as was previously assumed. As a result, the vibrational distribution deviates from the Treanor distribution becoming flatter, the derivative $\frac{\partial^2 \ln f(E)}{\partial E^2}$ decreases and finally leads to the so-called hyperbolic plateau-distribution:

$$\frac{2x_e}{T_0} = \frac{F}{E^2 f^2(E)}, \quad f(E) = \frac{C}{E}. \tag{5.21}$$

The "plateau level" C can be found from the power P_{eV} of vibrational excitation in plasma, calculated per molecule:

$$C(P_{eV}) = \frac{1}{\hbar\omega}\sqrt{\frac{P_{eV}T_0(\delta_{VV}\hbar\omega)^3}{4x_e k_0 n_0 Q_{10}^{01}}}. \tag{5.22}$$

The vibrational distribution in the strong excitation regime first follows the Treanor function at relatively low energies:

$$E < E_{Tr} - \hbar\omega\sqrt{\frac{T_0}{2x_e\hbar\omega}}. \tag{5.23}$$

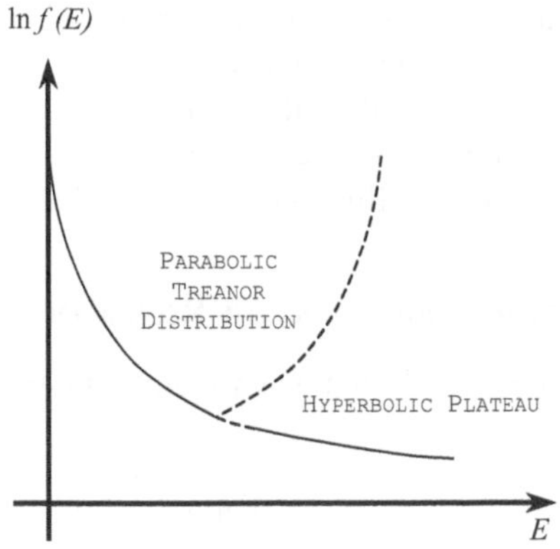

FIGURE 5.2 Vibrational energy distribution function due to VV-exchange, strong and intermediate excitation regimes.

Here $E_{Tr} = \hbar\omega \cdot v_{\min}^{Tr}$ is the Treanor minimum energy, corresponding to the minimum of the Treanor distribution function (see Eq. 4.48). As the vibrational energies exceed the Treanor minimum point ($E > E_{Tr}$), the distribution function becomes the hyperbolic plateau:

$$f(E) = B\frac{E_{Tr}}{E}\exp\left(-\frac{T_0\hbar\omega}{4x_eT_v^2} - \frac{1}{2}\right), \tag{5.24}$$

where B is the Treanor normalization factor. The transition from the plateau (5.24) to the rapidly decreasing Boltzmann distribution with temperature T_0 takes place at higher energies exceeding the critical one:

$$E(plateau - VT) = \frac{1}{\delta_{VT}}\left[\ln\frac{k_0Q_{10}^{01}}{k_{VT}(E=0)}\frac{E_{Tr}}{\hbar\omega} - \frac{T_0\hbar\omega}{4x_eT_v^2} - \frac{1}{2}\right]. \tag{5.25}$$

This plateau-Boltzmann transitional energy corresponds to the equality of probabilities of the vibrational VT-relaxation and the resonance VV-exchange. The typical behavior of the vibrational distribution in the regime of strong excitation in non-thermal molecular plasma is illustrated in Figure 5.2.

5.1.8 Vibrational Distribution Functions, the Intermediate Excitation Regime

The vibrational distribution in this regime is similar to that presented in Figure 5.2. It also includes the Treanor function Eq. (5.10) at $E \leq E_{Tr}$, the hyperbolic plateau Eqs. (5.21), (5.24) at the energies exceeding the Treanor minimum ($E > E_{Tr}$), and the sharp plateau-Boltzmann falls at higher energies Eq. (5.25). The intermediate excitation regime takes place when the vibrational temperature is sufficiently high to satisfy the inequality Eq. (5.19) (it is less than the so-called VV-exchange radius: $T_v < \delta_{VV}^{-1}$, which is about 5,000–10,000 K), but sufficiently high to provide the conditions for the Treanor effect:

$$\frac{x_eT_v^2}{T_0\hbar\omega} \geq 1. \tag{5.26}$$

Under such conditions, the population of the vibrationally excited states at the Treanor minimum $E = E_{Tr}$ is quite high Eq. (4.49), the non-linear resonance VV-exchange dominates and provides the plateau at $E > E_{Tr}$ even though $T_v < \delta_{VV}^{-1}$. At low levels of vibrational energy $E < E_{Tr}$, the linear non-resonant VV-exchange dominates over the non-linear one. However, the point is that the vibrational distribution function does not change because both components of VV-exchange the non-resonant Eq. (5.12) and resonant (5.15) result in the same Treanor distribution at $E < E_{Tr}$. The strong and intermediate excitation regimes for two different reasons and at two different conditions lead to the same vibrational distribution function illustrated in Figure 5.2.

5.1.9 Vibrational Distribution Functions, the Regime of Weak Excitation

This regime takes place in non-thermal plasma when vibrational temperature is not very high:

$$\delta_{VV} T_v < 1, \quad \frac{x_e T_v^2}{T_0 \hbar\omega} < 1. \tag{5.27}$$

The non-linear term in Eq. (5.18) can be neglected and the Fokker-Planck equation simplified:

$$\frac{\partial f(E)}{\partial E}\left(1+\xi(E)\right)+f(E)\cdot\left(\frac{1}{T_v}-\frac{2x_e E}{T_0\hbar\omega}+\xi(E)\tilde{\beta}_0\right)=0. \tag{5.28}$$

Here $\xi(E)$ is the ratio of the rate coefficients of VT-relaxation and non-resonant VV-exchange, determined by the exponential relation Eq. (3.78). The solution of the Eq. (5.28) gives after integration:

$$f(E)=B\exp\left[-\frac{E}{T_v}+\frac{x_e E^2}{T_0\hbar\omega}-\frac{\tilde{\beta}_0-\dfrac{1}{T_v}}{2\delta_{VV}}\ln\left(1+\xi(E)\right)\right]. \tag{5.29}$$

This continuous vibrational distribution function $f(E)$ for the weak excitation regime was derived by V.D. Rusanov, A. Fridman, G. V. Sholin (1979) and corresponds to the discrete distribution $f(v)$ over the vibrational quantum numbers:

$$f(v)=f_{Tr}(v)\prod_{i=0}^{v}\frac{1+\xi_i\exp\dfrac{\hbar\omega}{T_0}}{1+\xi_i}. \tag{5.30}$$

This is usually referred to as the Gordiets vibrational distribution (B.F. Gordiets, A.I. Osipov, E.B. Stupochenko, L.A. Shelepin, 1972). In this relation, $f_{Tr}(v)$ is the discrete form of the Treanor distribution. From Eqs. (5.29) and (5.30), at low vibrational energies ($E < E*(T_0)$, (Eq. 3.79)): $\xi(E) << 1$ and the vibrational distribution is close to the Treanor distribution. Conversely at high energies $E < E*(T_0)$: $\xi(E) >> 1$ and the vibrational distribution is exponentially decreasing according to the Boltzmann law with temperature T_0. The vibrational distribution in nitrogen at $T_v = 3000$ K and different translational temperatures is shown in Figure 5.3. Comparison of different types of vibrational distribution functions at the same conditions is presented in Figure 5.4 (in both the figures the dashed lines stand for $\xi(E) = 1$) Obviously, this figure demonstrates the significant difference between the Treanor and Boltzmann vibrational distribution functions. Also Figure 5.4 illustrates two less trivial effects:

i. the Bray distribution (1968), only taking into account VT-relaxation when $\xi(E) \geq 1$,- overestimates the actual $f(E)$; and
ii. the completely classical ($\tilde{\beta}_0 = \beta_0$) Bray distribution (1972) conversely underestimates the actual distribution function $f(E)$.

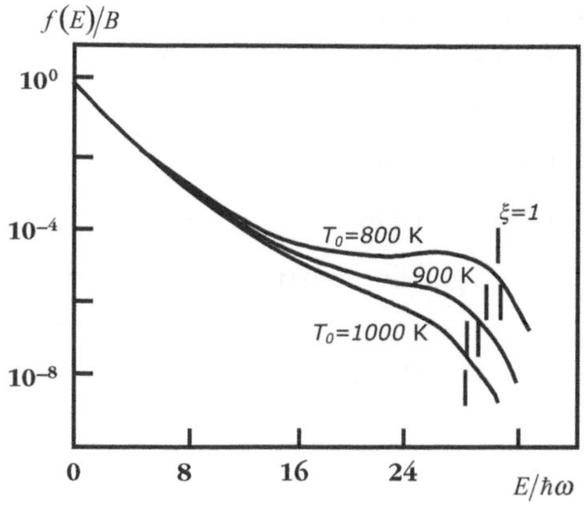

FIGURE 5.3 Non-equilibrium vibrational distributions in nitrogen, $T_v = 3000$ K.

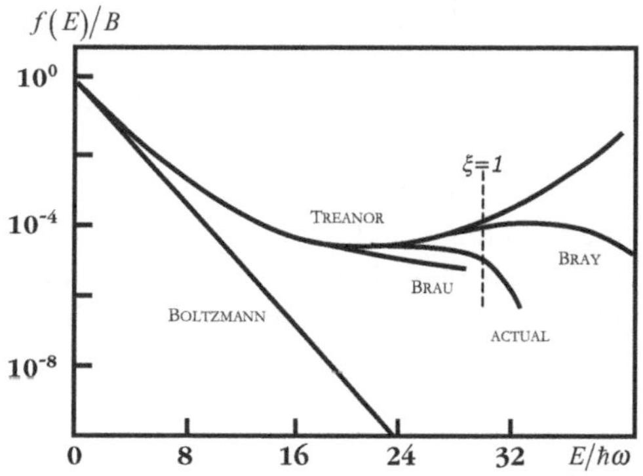

FIGURE 5.4 Different models of vibrational distribution in nitrogen, $T_v = 3000$ K, $T_0 = 800$ K.

Equation (5.29) permits stressing the following qualitative features of the vibrational distribution function $f(E)$, controlled by the VV and VT-relaxation processes in the regime of weak excitation:

1. The logarithm of the vibrational distribution function $\ln f(E)$ always has an inflection point $\left(\frac{\partial^2 \ln f(E)}{\partial E^2} = 0\right)$, corresponding to the vibrational energy:

$$E_{\mathrm{infl}} = E*(T_0) - \frac{1}{2\delta_{VV}} \ln \frac{\delta_{VV} T_0}{x_e} < E*(T_0). \tag{5.31}$$

Here $E*(T_0)$ is the critical vibrational energy when the resonant VV- and VT-relaxation rate coefficients are equal. Obviously, because $E_{\mathrm{infl}} < E*(T_0)$, the VV-exchange is still faster than VT-relaxation at the inflection point:

$$\xi(E_{\mathrm{infl}}) = \frac{x_e}{\delta_{vv} T_0} < 1. \tag{5.32}$$

In particular, for nitrogen at $T_0 = 1000K$: $\xi(E_{infl}) \approx 0.1$. It should be noted, that the inflection point on the function $\ln f(E)$exists even if the function is continuously decreasing with vibrational energy (see Figure 5.3).

2. Sometimes the vibrational distribution function includes the domain of inverse population ($\frac{\partial \ln f(E)}{\partial E} > 0$), see Figure 5.3. The existence criterion of the inverse population can be presented as the requirement of a positive logarithmic derivative at the inflection point:

$$2\delta_{VV}\left(E*(T_0)-E_{Tr}\right) > \ln\frac{x_e}{\delta_{VV}T_0}. \tag{5.33}$$

3. If the inequality (5.33) is valid and the function $f(E)$ has the domain of inverse population, then the vibrational distribution function has a maximum point E_{max}, which can be found from the following equation:

$$\frac{2x_e}{T_0}\left(E_{max}-E_{Tr}\right) = \xi\left(E_{max}\right). \tag{5.34}$$

As can be seen from the Eq. (5.34), the parameter $\xi(E_{max})$ is usually small numerically ($\xi(E_{max}) < 1$). This means that $E_{max} < E * (T_0)$. The qualitative characteristics of the vibrational distribution $f(E)$, the inflection and maximum energies E_{infl} and E_{max} are both related to the influence of the VT-relaxation and both are less than $E * (T_0)$. This leads to the conclusion that VT relaxation already has a significant effect on the vibrational distribution function $f(E)$, when the VT relaxation rate is slower with respect to VV-exchange ($\xi < 1$). This effect is well illustrated in Figures 5.3 and 5.4.

5.2 NON-EQUILIBRIUM VIBRATIONAL KINETICS EV-PROCESSES, POLYATOMIC MOLECULES, NON-STEADY-STATE REGIMES

5.2.1 The eV-Flux along the Vibrational Energy Spectrum

Vibrational excitation of molecules by electron impact (eV-relaxation process) was indirectly discussed in Section 5.1. There the assumed effective excitation of only the low vibrational levels that determined vibrational temperature T_v. of the plasma was assumed. The evolution of the vibrational distribution function $f(E)$ in this approach was controlled by VV- and VT-relaxations and described by Fokker-Planck kinetic equation with vibrational temperature T_v as a parameter. Such an approach of the indirect influence of the electron gas on $f(E)$ is relevant when VV-exchange is much faster than the vibrational excitation (eV-processes), which then can be considered only as boundary conditions. However, at high degrees of ionization, the frequency of eV-processes becomes comparable with that of VV-exchange and the eV-flux along the vibrational energy spectrum must be taken into account. The eV-flux in energy space describes the direct influence of the energy exchange of excited molecules with the electron gas on the vibrational distribution $f(E)$. In contrast to VV-relaxation and VT-exchange, the eV-relaxation can be effective not only as a one-quantum but also as multiquantum process. The probability of the multi-quantum eV-processes, described by Fridman's approximation (Eq. 3.28) can be large. The eV-flux along the vibrational energy spectrum can be subdivided into two terms, corresponding to two mentioned types of eV-relaxation processes (S.O. Macheret, V.D. Rusanov, A. Fridman, G. V. Sholin, 1979):

$$j_{eV}(E) = j_{eV}^{(1)}(E) + j_{eV}^{(0)}(E). \tag{5.35}$$

The linear eV-flux $j_{eV}^{(1)}$ describes the eV-processes with the transfer of one or a few quanta. This flux in the energy space can be expressed similarly to the linear VV- and VT-fluxes in the Fokker-Planck form:

$$j_{eV}^{(1)}(E) = -D_{eV}\left(\frac{\partial f(E)}{\partial E} + \frac{1}{T_e} f(E)\right). \tag{5.36}$$

Here D_{eV} is the one-quantum, vibrational excitation diffusion coefficient in energy space:

$$D_{eV} = \lambda k_{eV}^0 n_e (\hbar\omega)^2, \quad \lambda \approx \frac{2}{\alpha^3}. \tag{5.37}$$

The one-quantum excitation rate coefficient $k_{eV}^0 = k_{eV}(0,1)$ and parameter α correspond to the Fridman's approximation Eq. (3.28), numerically $\alpha \approx 0.5 \div 0.7$. The factor λ accounts for the transfer of a few quanta and numerically in nitrogen $\lambda \approx 10$. The eV-flux-component $j_{eV}^{(0)}(E)$ in Eq. (5.35) is related to the multi-quantum excitation of molecules from low levels to energy E and can be expressed as:

$$j_{eV}^{(0)}(E) = -\int_E^{\infty} k_{eV}^0 n_e f(0) \exp\left(-\alpha\frac{E'}{\hbar\omega}\right)\left[1 - \frac{f(E')}{f(0)}\exp\frac{E'}{T_e}\right] dE'. \tag{5.38}$$

This integral flux is actually a source (either positive or negative) of vibrationally excited molecules. The multi-quantum excitation flux Eq. (5.38) as well as the one-quantum excitation flux becomes equal to zero when the vibrational distribution $f(E)$ is the Boltzmann function with temperature T_e: $f(E) \propto \exp(-E/T_e)$. When the eV de-excitation processes (super-elastic collisions) can be neglected, the integral expression for the multi-quantum excitation eV-flux can be simplified. Such simplification can be done, if $f(E) << f(0)\exp\left(-\frac{E}{T_e}\right)$. Then after integration, the simplified expression for the eV-flux-component can be written as:

$$j_{eV}^{(0)}(E) = -D_{eV}\frac{\alpha^2}{2\hbar\omega} f(0)\exp\left(-\frac{E\alpha}{\hbar\omega}\right). \tag{5.39}$$

Comparing the eV-flux-components Eqs. (5.36), (5.38) and (5.39) illustrates that the one-quantum eV-processes Eq. (5.36) dominate ($j_{eV}^{(1)} >> j_{eV}^{(0)}$), if the vibrational distribution function $f(E)$ decreases with energy slower than $\exp\left(-\frac{\alpha E}{\hbar\omega}\right)$. Conversely, in the typical case of $T_v < \frac{\hbar\omega}{\alpha}$, the multi-quantum processes dominate the eV-relaxation at least at low vibrational energies.

5.2.2 Influence of eV-Relaxation on Vibrational Distribution at High Degrees of Ionization

In this case, the criteria of high electron density and high degree of ionization means the domination of the vibrational excitation frequency over the frequency of VV-exchange even at low vibrational levels:

$$\frac{n_e}{n_0} >> \frac{k_{eV}^0}{k_{VV}^0}, \quad k_{VV}^0 = Q_{01}^{10} k_0. \tag{5.40}$$

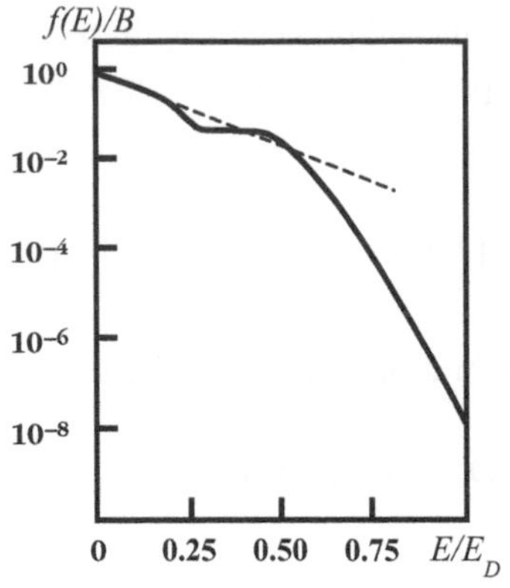

FIGURE 5.5 Vibrational distribution in N_2:Ar = 1:100 mixture, T_e = 1 eV, T_0 = 750 K, ionization degree 10^{-6}.

Numerically, this requires a fairly high degree of ionization in non-thermal plasma, typically exceeding $10^{-4} - 10^{-3}$. Here most of the vibrationally excited molecules are in quasi-equilibrium with the electron gas, which can be characterized by the electron temperature T_e. The vibrational distribution function $f(E)$ is close to the Boltzmann function with temperature T_e for a wide range of vibrational energies from the lowest to a high critical one:

$$E_{VT-eV} \approx \frac{1}{\delta_{VT}} \ln \frac{k_{eV} n_e}{k_{VT}(0) n_0}. \tag{5.41}$$

At this critical energy, the VT-relaxation rate matches the rate of the eV-processes and the vibrational distribution function falls exponentially according to Boltzmann distribution with temperature T_0. The vibrational distribution controlled by eV- and VT-relaxation processes at high degrees of ionization is shown in Figure 5.5. This figure presents the results of the numerical calculations of P.A. Sergeev, D.I. Slovetsky, 1979). The $f(E)$ behavior around the transition energy Eq. (5.41) is not trivial and includes the "micro-plateau", which can be seen in Figure 5.5. One can find the relevant analysis and explanation of the phenomenon as well as other details of eV-processes in vibrational kinetics in the publication of A.V. Demura, S.O. Mahceret and A. Fridman (1984).

5.2.3 Influence of eV-Relaxation on Vibrational Distribution at Intermediate Degrees of Ionization

If in contrast to Eq. (5.40) $\frac{n_e}{n_0} << \frac{k_{eV}^0}{k_{vv}^0}$ and the vibrational excitation by electron impact is much slower than VV-exchange, obviously, the direct influence of eV-processes on $f(E)$ can be neglected. The case of intermediate ionization degree implies $\frac{n_e}{n_0} \sim \frac{k_{eV}^0}{k_{vv}^0}$. This means, that at low levels of vibrational excitation, VV-relaxation is still sufficiently strong to build the Treanor distribution, but at higher energies, eV-processes dominate creating the Boltzmann distribution with temperature T_e. Obviously at higher values of vibrational energy, VT-relaxation prevails (see Eq. (5.41)) leading to the Boltzmann distribution with temperature T_0. The direct influence of eV-processes on distribution function $f(E)$ is not possible when VV processes are dominated by the resonance exchange (strong excitation regime). This can easily be illustrated in the case of multi-quantum eV-relaxation when the kinetic equation: $j_{VV}^{(1)} + j_{eV}^{(0)} = 0$, after integration can be written as:

$$D_{VV}^{(1)}E^2 f^2\left(\frac{2x_e}{T_0}-\hbar\omega\frac{\partial^2 \ln f}{\partial E^2}\right)=-D_{eV}\frac{\alpha f(0)}{2}\left[1-\exp\left(\frac{-\alpha E}{\hbar\omega}\right)\right]\approx -D_{eV}\frac{\alpha f(0)}{2}. \quad (5.42)$$

Eq. (5.42) is identical to (5.20), because $\exp\left(\frac{-\alpha E}{\hbar\omega}\right) << 1$. As the result, the solution of Eq. (5.42) is the hyperbolic plateau Eq. (5.21), which includes only indirect information on eV-processes. Thus, the direct influence of eV-processes on $f(E)$ at intermediate degrees of ionization takes place only in the regime of weak excitation. The vibrational distribution in this case is controlled by eV- and non-resonant VV-relaxation at relatively low energies and can be found from the equation:

$$D_{VV}(E)\left(\frac{\partial f}{\partial E}+\frac{f}{T_v}-\frac{2x_e E}{T_0\hbar\omega}f\right)+D_{eV}\left(\frac{\partial f}{\partial E}+\frac{f}{T_e}\right)+D_{eV}\frac{\alpha^2}{2\hbar\omega}f(0)\exp\left(-\frac{\alpha E}{\hbar\omega}\right)=0. \quad (5.43)$$

The solution of this equation can be presented in the following integral form:

$$f(E)=\phi(E)\left[1-\int_0^E\frac{\alpha^2 f(0)\eta_{eV}(E')\exp\left(-\frac{\alpha E'}{\hbar\omega}\right)}{2\hbar\omega\phi(E')\left(1+\eta_{eV}(E')\right)}dE'\right]. \quad (5.44)$$

In this expression, the factor $\eta_{eV}(E) = D_{eV}(E)/D_{VV}(E)$ is proportional to the degree of ionization in the plasma and shows the relation between eV- and non-resonant VV-processes. The special function $\phi(E)$ is given by the following integral, which is similar to that corresponding to the weak excitation regime controlled by VV- and VT-relaxation (see Eqs. (5.28), (5.29)):

$$\phi(E)=\frac{1}{T_v}\exp\left[-\int_0^E\frac{\frac{1}{T_v}-\frac{2x_e E'}{T_0\hbar\omega}+\frac{1}{T_e}\eta_{eV}(E')}{1+\eta_{eV}(E')}dE'\right]. \quad (5.45)$$

If the vibrational temperature is sufficiently high $T_v > \hbar\omega/\alpha$ (this assumption does not change qualitative conclusions), then the integral in the expression Eq. (5.44) is small and the vibrational distribution function $f(E) \approx \phi(E)$ can be found from Eq. (5.45). Equation (5.45) permits analyzing the vibrational distribution function $f(E) \approx \phi(E)$ in the case of intermediate degrees of ionization. At low vibrational energies, when $\eta_{eV}(E) \leq 1$, the vibrational distribution is close to the Treanor function. The factor $\eta_{eV}(E)$ grows with energy, and at some point, the function $f(E) \approx \phi(E)$ becomes the Boltzmann distribution with temperature T_e. A detailed discussion of this f(E) behavior can be found in the book of V.D. Rusanov and A. Fridman (1984). The relevant graphic illustration is presented in Figure 5.6.

Transition from the Treanor to Boltzmann distribution with high electron temperature T_e can be interpreted also as a transition to the plateau (see Figure 5.6). Remember, however, that this plateau has nothing in common with the hyperbolic plateau related to the resonance VV-exchange.

5.2.4 Diffusion in Energy Space and Relaxation Fluxes of Polyatomic Molecules in Quasi-Continuum

If polyatomic molecules are not strongly excited, their vibrational levels can be also considered discrete rather than continuous (see Section 3.2.3). Vibrational kinetics of polyatomic molecules in this

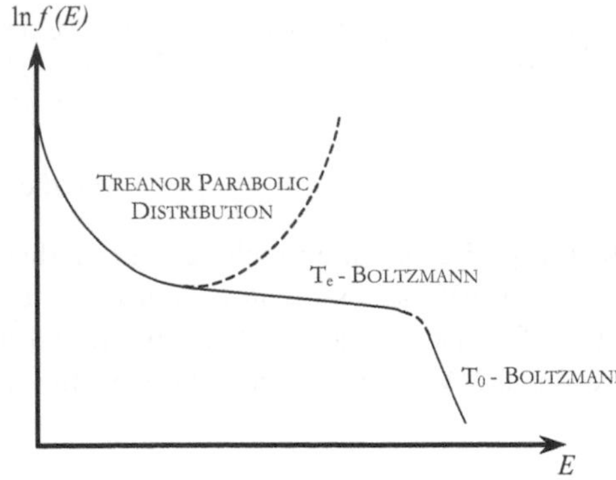

FIGURE 5.6 eV-Processes influence on vibrational distribution at intermediate ionization degrees.

case is quite similar to that of diatomic molecules (A.A. Likalter 1975a, 1976). An interesting example of such discrete vibrational distribution functions is presented in Figure 5.7, where even the Treanor effect can be seen. The specific features of polyatomic molecules manifest themselves at higher excitation levels when the interaction between vibrational modes is strong and the molecules are in the state of vibrational quasi-continuum (Section 3.2.4). VT- and VV-relaxation processes for polyatomic molecules in quasi-continuum were discussed in Sections 3.4.6 and 3.5.4. Fluxes related to the intermode energy exchange do not appear directly in the Fokker–Planck equation because they are very fast and compensate each other. The distribution function $f(E)$ of the polyatomic molecules in quasi-continuum over the total vibrational energy E can be then found from the kinetic equation:

$$\frac{\partial f(E)}{\partial E}+\frac{\partial}{\partial E}\left(j_{VV}^{poly}+j_{VT}^{poly}\right)=0. \tag{5.46}$$

The VT-relaxation flux in energy space can be expressed for polyatomic molecules similarly to Eqs. (5.3), (5.4) as (V.D. Rusanov, A. Fridman, G. V. Sholin, B.V. Potapkin, 1985):

$$j_{VT}^{poly}=-\sum_{i=1}^{N}D_{VT}^{i}(E)\rho(E)\left[\frac{\partial}{\partial E}\left(\frac{f(E)}{\rho(E)}\right)+\frac{1}{T_0}\frac{f(E)}{\rho(E)}\right]. \tag{5.47}$$

The main peculiarity of this relation with respect to the similar one for diatomic molecules Eq. (5.3) is the presence of the statistic factor $\rho(E) \propto E^{s-1}$, showing the density of vibrational states and taking into account the effective number s of vibrational degrees of freedom. $\tilde{\beta}_0 \approx 1/T_0$, which reflects the relative smallness of vibrational quantum in quasi-continuum with respect to temperature. The summation is taken over all N vibrational modes and the diffusion coefficient in energy space, D_{VT}^{i} is related to each of the modes "i":

$$D_{VT}^{i}(E)=\left\langle\left(E_{VT}^{i}\right)^{2}\right\rangle k_0 n_0. \tag{5.48}$$

Here k_0 is the rate coefficient of gas-kinetic collisions; n_0 gas density; and the averaged square of VT- energy transfer from a mode in quasi-continuum, $\left\langle\left(E_{VT}^{i}\right)^{2}\right\rangle$ is determined by Eq. (4.121). It must be emphasized, that the flux Eq. (5.47) is equal to zero for the Boltzmann distribution function with the statistical weight factor: $f(E) \propto \rho(E)\exp(-E/T_0)$. To introduce the VV-flux j_{VV}^{poly} in Eq. (5.12) for polyatomic molecules in quasi-continuum, note at first, that this flux in contrast to diatomic molecules Eq. (5.11) has only the linear component $j_{VV}^{(0)}$. This flux corresponds to the VV-exchange

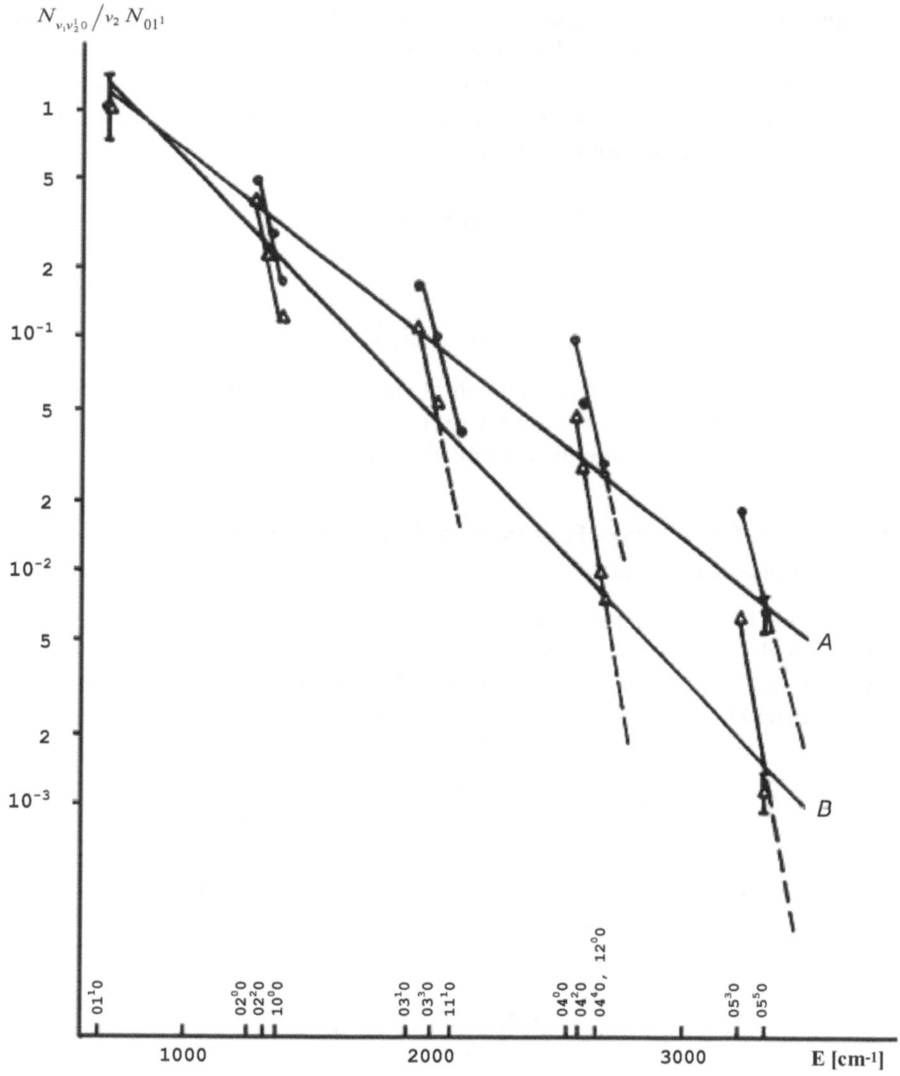

FIGURE 5.7 Vibrational distribution of CO_2 in symmetric valence and deformation modes: (a) $T_{v1} = 780 \pm 40$ K, $T_{r1} = 150 \pm 15$ K; (b) $T_{v2} = 550 \pm 40$ K, $T_{r2} = 110 \pm 10$ K.

between excited molecules of high vibrational energy E with low excited molecules of the "thermal reservoir". The non-linear VV-flux component $j_{VV}^{(1)}$ can be neglected because of the resonant nature of the VV-exchange of polyatomic molecules in quasi-continuum Eq. (3.89), even if they have different vibrational energies. Then the VV-flux for the polyatomic molecules in quasi-continuum can be presented in the following linear differential form (V.D. Rusanov, A. Fridman, G. V. Sholin, 1986):

$$j_{VV}^{poly}(E) = -\sum_{i=1}^{N} D_{VV}^{i}(E)\rho(E)\left[\frac{\partial}{\partial E}\left(\frac{f(E)}{\rho(E)}\right) + \frac{1}{T_v^i}\left(\frac{f(E)}{\rho(E)}\right) - \left(\frac{f(E)}{\rho(E)}\right)\frac{E}{sT_0\hbar\omega_{0i}}\sum_{i=1}^{N} x_{ij}\left(1+\delta_{ij}\right)g_i\frac{\omega_{0i}}{\omega_{0j}}\right]. \tag{5.49}$$

Here, T_{vi} and $\hbar\omega_{oi}$ are vibrational temperatures and the first quantum for different modes; s is the number of effective vibrational degrees of freedom; δ_{ij} is the Kronecker delta-symbol; x_{ij} are the anharmonicity coefficients Eq. (3.20); g_i is the degree of degeneracy of a vibrational mode. Here the diffusion coefficient in energy space D_{VV}^{i} is related to the VV-exchange between excited molecules with molecules of the thermal reservoir with temperature T_v:

$$D_{VT}^{i}\left(E\right)=\left\langle \left(E_{VV}^{i}\right)^{2}\right\rangle k_{0}n_{0}, \tag{5.50}$$

where the averaged square of VV-energy transfer $\left\langle \left(E_{VV}^{i}\right)^{2}\right\rangle$ is determined by Eq. (3.89). The most intriguing part of the VV-flux Eq. (5.49) is the third term leading to the Treanor effect. This term arose from the balance of direct and reverse VV-exchange processes Eq. (3.94). The Treanor effect is still valid for polyatomic molecules though vibrations are in quasi-continuum.

5.2.5 Vibrational Distribution Functions of Polyatomic Molecules in Non-Equilibrium Plasma

The kinetic equation (5.46), or in particular the steady-state equation $j_{VV}^{poly}+j_{VT}^{poly}=0$, can be solved with Eqs. (5.47) and (5.49) for VV- and VT-fluxes to find the vibrational distribution function *f*(*E*). First, analyze the steady-state case controlled only by the VV-exchange ($j_{VV}^{poly}=0$). Assuming single mode primarily determining *f*(*E*) in quasi-continuum, integration of Eq. (5.49) gives:

$$\frac{f\left(E\right)}{\rho\left(E\right)}=B\exp\left[-\frac{E}{T_{Va}}+\frac{E^{2}}{2sT_{0}}\sum_{j=1}^{N}\frac{x_{aj}}{\hbar\omega_{0j}}\left(1+\delta_{aj}\right)g_{j}\right]. \tag{5.51}$$

This is a generalization of the Treanor distribution Eq. (5.10) for polyatomic molecules in quasi-continuum. Parameter B is the normalization factor. Statistical weight factor $\rho(E)\propto E^{s-1}$ characterizes the density of the vibrational states; and also the effective coefficient of anharmonicity:

$$x_{m}=\frac{1}{2s}\sum_{j=1}^{N}\frac{x_{aj}}{\hbar\omega_{0j}}\left(1+\delta_{aj}\right)g_{j}. \tag{5.52}$$

These effective anharmonicity coefficients for polyatomic molecules are usually less than those for diatomic molecules. This makes the Treanor effect weaker for polyatomic molecules: the more vibrational degrees of freedom the weaker is the Treanor effect. Taking the VT-flux given by Eq. (5.47) into consideration leads to the following integral form of the vibrational distribution function:

$$f\left(E\right)=B\rho\left(E\right)\exp\left(-\int_{E_{c}}^{E}\frac{\frac{1}{T_{va}}-\frac{2x_{m}E'}{T_{0}\hbar\omega_{0a}}+\frac{\xi\left(E'\right)}{T_{0}}}{1+\xi\left(E'\right)}dE'\right). \tag{5.53}$$

Here E_c designates the energy at which a polyatomic molecule enters the quasi-continuum. Somewhat similarly to Eq. (5.28): the factor $\xi\left(E\right)=\sum_{i}D_{VT}^{i}/D_{VV}^{a}$ characterizes the ratio of VT and VV-relaxation rates for polyatomic molecules in quasi-continuum. The factor $\xi(E)$ was analyzed in

Section 3.5.4. (see relation Eq. (3.90)), where it was shown, that $\xi(E)$ is much larger for polyatomic molecules, than for diatomic ones. This results in important peculiarity of the vibrational distribution of polyatomic molecules. Although the Treanor function can be observed in quasi-continuum, transition to Boltzmann distribution with translational temperature T_0 takes place at a lower level of vibrational excitation.

5.2.6 Non-Steady-State Vibrational Distribution Functions

Analytical solutions of the non-steady-state kinetic equation are known only for a few specific problems, see S.A. Zhdanok, A.P. Napartovich, A.N. Starostin (1979), and S.O. Macheret, V.D. Rusanov, A. Fridman, G. V. Sholin (1979). As an example consider the non-steady VV-exchange with a variable diffusion coefficient in energy space: $D_{VV}(E) = D^{(0)} \exp(-\delta_{VV}E)$. Neglecting the Treanor term in the Fokker-Planck equation gives:

$$\frac{\partial f}{\partial t} = D^{(0)} \exp(-\delta_{VV}E)\left(\frac{\partial f}{\partial E} + \frac{1}{T_v} f\right). \tag{5.54}$$

This partial differential equation can be reduced to an ordinary differential equation by the introduction of the following new variable:

$$Z(E,t) = D^{(0)} t \delta_{VV}^2 \exp(-\delta_{VV}E). \tag{5.55}$$

$Z(E, t) = 1$ describes the propagation of the front of the Boltzmann distribution function. In terms of this new variable $Z(E, t)$ the non-steady-state kinetic equation (5.54) can be rewritten as a second-order ordinary differential equation for the distribution function $f(Z)$:

$$Z^2 f''_{ZZ} + f'_Z\left[1 - \left(2 - \frac{1}{\delta_{VV}T_v}\right)Z\right] - \frac{1}{\delta_{VV}T_v} f(Z) = 0. \tag{5.56}$$

For $Z(E, t) << 1$, corresponding to low time intervals from the starting point of VV-exchange and to high vibrational energies, the asymptotic solution for the non-steady-state vibrational distribution is:

$$f(E,Z) \propto \exp\left(-\frac{1}{Z} - \frac{1}{T_v}E\right). \tag{5.57}$$

Equation (5.57) can be applied for estimations of the evolution of the Boltzmann function during the VV-exchange also for not very low values of $Z(E, t) << 1$. The non-steady-state distribution Eq. (5.57) is illustrated in Figure 5.8.

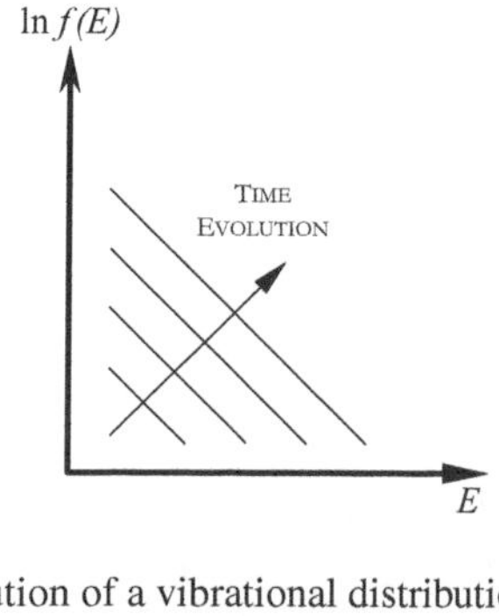

FIGURE 5.8 Illustration of time evolution of a vibrational distribution function.

5.3 MACROKINETICS OF CHEMICAL REACTIONS AND RELAXATION OF VIBRATIONALLY EXCITED MOLECULES

5.3.1 Chemical Reaction Influence on the Vibrational Distribution Function, the Weak Excitation Regime

The macrokinetic rates of reactions of vibrationally excited molecules are self-consistent with the influence of the reactions on the vibrational distribution functions $f(E)$. This chemical reaction effect on $f(E)$ can be taken into account by introducing into the Fokker-Planck kinetic equation (5.1) an additional flux related to the reaction:

$$j_R(E) = -\int_E^{\infty} k_R(E')n_0 f(E')dE' = -J_0 + n_0\int_0^E k_R(E')f(E')dE'. \tag{5.58}$$

Here $J_0 = -j_R(E=0)$ is the total flux of the molecules in the chemical reaction (this means, that the total reaction rate $w_R = n_0 J_0$); $k_R(E)$ is the microscopic reaction rate coefficient, given by Eq. (3.100). In the weak excitation regime controlled by non-resonant VV- and VT-relaxation processes which is usually the case in plasma chemistry, this leads to the equation: $j_{VV}^{(0)} + j_{VT} + j_R = 0$. Taking into account the specific fluxes Eqs. (5.3), (5.12) and (5.58) results in the kinetic equation:

$$\frac{\partial f(E)}{\partial E}\left(1+\xi(E)\right) + f(E)\cdot\left(\frac{1}{T_v} - \frac{2x_e E}{T_0\hbar\omega} + \tilde{\beta}_0\xi(E)\right) = \frac{1}{D_{VV}(E)} j_R(E). \tag{5.59}$$

The exact solution of this non-uniform linear equation can be found in form of $f(E) = C(E) * f^{(0)}(E)$, where $f^{(0)}(E)$ is the solution Eq. (5.29) of the corresponding uniform Eq. (5.28). In other words, the function $f^{(0)}(E)$ makes the left-hand-side of Eq. (5.59) equals to zero. The function C(E) can be then found from the equation:

$$D_{VV}(E)\left(1+\xi(E)\right)f^{(0)}(E)\frac{\partial C(E)}{\partial E} = j_R(E). \tag{5.60}$$

After the integration of Eq. (5.60), the vibrational distribution function perturbed by plasma chemical reaction can be expressed in the following integral equation:

$$f(E) = f^{(0)}(E)\left[1 - \int_0^E \frac{-j_R(E')dE'}{D_{VV}(E')f(E')\left(1+\xi(E')\right)}\right]. \tag{5.61}$$

The function $-j_R(E)$ determines the flux of molecules along the energy spectrum, which are going to participate in a chemical reaction at $E \geq E_a$, E_a – is activation energy. At relatively low energies $E < E_a$, reaction can be neglected: $-j_R(E) = \int_{E_a}^{\infty} k_R(E')n_0 f(E')dE' = J_0 = const$. At these energies ($E < E_a$), the perturbation of the vibrational distribution $f^{(0)}(E)$ by reaction Eq. (5.61) is:

$$f(E) = f^{(0)}(E)\left[1 - J_0\int_0^E \frac{dE'}{D_{VV}(E')f^{(0)}(E')\left(1+\xi(E')\right)}\right]. \tag{5.62}$$

The total flux J_0, which is taken here as a constant parameter, is related to the total reaction rate and will be calculated later on Eq. (5.67).

At high energies ($E \geq E_a$), according to Eq. (5.58) $j_R(E) \approx -k_R(E)n_0 f(E)\hbar\omega$. Then the integral equation (5.61) can be converted at $E \geq E_a$ into the following one:

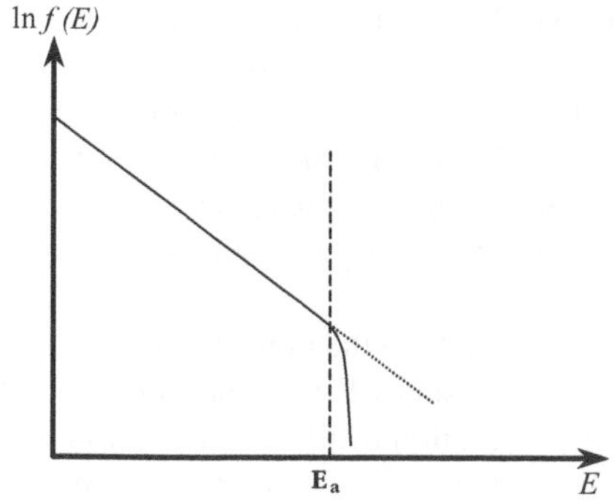

FIGURE 5.9 Reaction influence on vibrational distribution in the weak excitation regime.

$$\left\{\frac{f(E)}{f^{(0)}(E)}\right\} = \int_{E_a}^{E} \left\{\frac{f(E)}{f^{(0)}(E)}\right\} \frac{k_R(E')\hbar\omega dE'}{D_{VV}(E')(1+\xi(E'))}. \tag{5.63}$$

Equation (3.75) can be easily solved:

$$f(E) \propto f^{(0)}(E)\exp\left[-\int_{E_a}^{E} \frac{k_R(E')\hbar\omega dE'}{D_{VV}(E')(1+\xi(E'))}\right] \tag{5.64}$$

and determines the decrease of the vibrational distribution function at $E \geq E_a$, for example, in the domain of fast chemical reactions. The total vibrational distribution function taking into account chemical reactions is the combination of Eqs. (5.62) and (5.64). Such a function is illustrated in Figure 5.9. A significant influence of chemical reaction on vibrational population takes place at energies, where the reaction is already effective.

5.3.2 Macrokinetics of Reactions of Vibrationally Excited Molecules, the Weak Excitation Regime

Equations (5.61) and (5.64) can be applied to calculate rates of the reactions of vibrationally excited molecules in the specific limits of slow and fast reactions:

1. The fast reaction limit implies, that the chemical reaction is fast for $E \geq E_a$:

$$D_{VV}(E = E_a) << n_0 \cdot k_R(E + \hbar\omega)\cdot(\hbar\omega)^2, \tag{5.65}$$

and the chemical process in general is limited by the VV-diffusion along the vibrational spectrum to the threshold $E = E_a$. In this case according to Eq. (5.64),- the distribution function $f(E)$ decays very fast at $E > E_a$, and one can take $f(E = E_a) = 0$ in Eq. (5.61):

$$1 = \int_0^{E_a} \frac{-j_R(E')dE'}{D_{VV}(E')f^{(0)}(E')(1+\xi(E'))}. \tag{5.66}$$

This equation allows finding the total chemical process rate for the fast reaction limit, taking into account that $-j_R(E) = J_0 = const$ at $E < E_a$:

$$w_R = n_0 J_0 = n_0 \left\{\int_0^{E_a} \frac{dE'}{D_{VV}(E')f^{(0)}(E')(1+\xi(E'))}\right\}^{-1}. \tag{5.67}$$

As can be seen from Eq. (5.67), the chemical reaction rate in this case is determined by the frequency of the VV-relaxation and by the non-perturbed vibrational distribution function $f^{(0)}(E)$. The rate given by Eq. (5.67) in the fast reaction limit actually is not sensitive to the detailed characteristics of the elementary chemical reaction once it is sufficiently large. Such a situation in practical plasma chemistry takes place, for example, in CO_2 and H_2O monomolecular dissociation processes, proceeding as second kinetic order reactions (A. Fridman, V.D. Rusanov, 1994).

2. The slow reaction limit corresponds to the opposite inequality in Eq. (5.65). In this case, the population of the highly reactive states $E > E_a$, provided by VV-exchange takes place faster than the elementary chemical reaction itself. According to Eqs. (5.61), (5.64), the vibrational distribution function is almost non-perturbed by the chemical reaction $f(E) \approx f^{(0)}(E)$, and the total macroscopic reaction rate coefficient can be found as:

$$k_R^{macro} = \int_0^{\infty} k_R(E')f(E')dE'. \tag{5.68}$$

The above consideration assumed a high efficiency of vibrational energy in the chemical reaction $\alpha = 1$. This can be generalized by using the microscopic rate coefficient $k_R(E)$ in form Eq. (3.100) with an arbitrary value of α. Then integration of Eq. (5.68) over the distribution function which is mostly controlled by VV-exchange leads to the following approximation of the macroscopic rate coefficient (V.K. Givotov, V.D. Rusanov, A. Fridman, 1985):

$$k_R(T_v,T_0) = k_R^{(0)} \exp\left(-\frac{E_a}{T_0}\right) + k_R^{(v)} \exp\left(-\frac{E_a}{\alpha T_v} + \frac{x_e E_a^2}{T_0 \hbar\omega \alpha^2}\right). \tag{5.69}$$

In this relation $k_R^{(0)}$ and $k_R^{(v)}$ are the pre-exponential factors of the reaction rate coefficient. According to Eq. (5.69) if $\alpha T_v < T_0$ the chemical reaction proceeds by the quasi-equilibrium mechanism related to the translational temperature $k_R \propto \exp\left(-\frac{E_a}{T_0}\right)$. Conversely high vibrational temperatures ($\alpha T_v > T_0$) correspond to effective stimulation of chemical processes by vibrational excitation, and the macroscopic reaction rate is related to the population of vibrational levels with energy exceeding E_a/α. Actually, the energy threshold E_a/α can be interpreted as the effective activation energy for reactions stimulated by vibrational excitation. In contrast to the fast reaction limit Eq. (5.67), in the slow reaction limit Eq. (5.69) the elementary chemical process is explicitly presented in the rate coefficient. Thus, the pre-exponential factors $k_R^{(0)}$ and $k_R^{(v)}$ in Eq. (5.69) for chemical reactions proceeding through long-life-time complexes include the factor $(T_v/E_a)^{s-1}$. This factor corresponds before averaging to the statistical theory factor $\left(\frac{E - E_a}{E}\right)^{s-1}$ in the microscopic reaction rate (A.A. Levitsky, S.O. Macheret, A. Fridman et al., 1983). Similarly, the pre-exponential factors $k_R^{(0)}$ and $k_R^{(v)}$ in eq. (5.69) for electronically, non-adiabatic chemical reactions include the relevant Landau-Zener transition factors (Landau, 1997).

5.3.3 Macrokinetics of Reactions of Vibrationally Excited Molecules in Regimes of Strong and Intermediate Excitation

For this case, it is logical to assume that the distribution function $f(E)$ is controlled at the activation energy ($E = E_a$) by the resonance VV-exchange processes. The chemical reaction influence on the plateau-distribution ($E_{Tr} < E < E_a < E(plateau - VT)$) can be described by the Fokker-Planck kinetic equation derived from Eqs. (5.1), (5.15) and (5.58):

$$\frac{\partial}{\partial E}\left(D_{VV}^{(1)}E^2 f^2 \frac{2x_e}{T_0}\right) = J_0. \tag{5.70}$$

In the slow reaction limit, this means:

$$D_{VV}^{(1)}E_{Tr}^2 f^2\left(E_{Tr}\right)\frac{2x_e}{T_0} >> k_R\left(E+\hbar\omega\right)n_0\hbar\omega. \tag{5.71}$$

The distribution function is actually not perturbed by the chemical process, and the macroscopic reaction rate can be found by integration Eq. (5.68) as:

$$k_R^{macro} = k_R\left(E \geq E_a\right)E_{Tr} f\left(E_{Tr}\right)\ln\frac{E\left(plateau - VT\right)}{E_a}. \tag{5.72}$$

In this relation $E(plateau - VT)$ is the transition energy Eq. (5.25) from the hyperbolic plateau to the Boltzmann distribution with temperature T_0 Eq. (5.25); $k_R(E \geq E_a)$ is the microscopic reaction rate coefficient at vibrational energies exceeding the activation energy; $E_{Tr} = \frac{\hbar\omega}{2x_e}\frac{T_0}{T_v}$ – is the Treanor minimum point. In the fast reaction limit, when the inequality opposite to Eq. (5.71) is valid,-the vibrational population at high energies $E \geq E_a$ is negligible. Then one can take $f(E = E_a) = 0$ as a boundary condition for the kinetic equation (5.71) to derive the vibrational distribution function strongly affected by the chemical reaction:

$$f(E) = f\left(E_{Tr}\right)\frac{E_{Tr}}{E}\sqrt{\frac{E_a - E}{E_a - E_{Tr}}}. \tag{5.73}$$

The perturbation of vibrational distribution functions in the case of strong excitation is illustrated in Figure 5.10. The macroscopic rate coefficient can be then expressed as:

$$k_R^{macro} = \frac{2x_e\left(\hbar\omega\right)^2}{T_0\left(E_a - E_{Tr}\right)\left(\delta_{VV}\hbar\omega\right)^3}Q_{01}^{10}k_0 E_{Tr}^2 f^2\left(E_{Tr}\right). \tag{5.74}$$

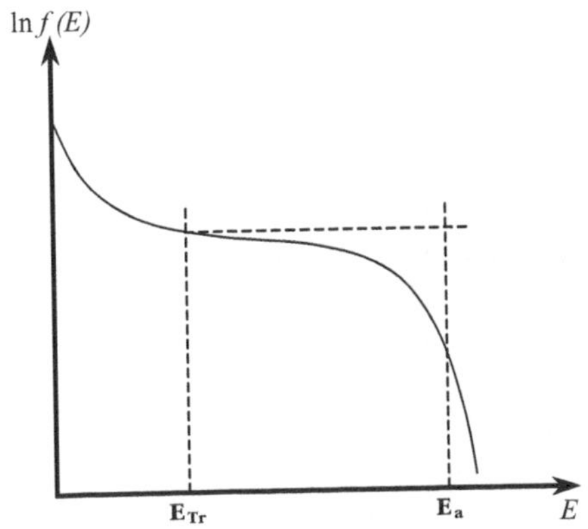

FIGURE 5.10 Reaction influence on vibrational distribution in the strong excitation regime.

5.3.4 Macrokinetics of Reactions of Vibrationally Excited Polyatomic Molecules

The kinetic equation describing the reaction influence on the vibrational distribution function of polyatomic molecules can be derived based on Eqs. (5.46), (5.47), (5.49) and (5.58) neglecting the anharmonicity as (B. Potapkin, V.D. Rusanov, A.E. Samarin, A. Fridman, 1980):

$$D_{VV}\rho(E)\left[\frac{\partial}{\partial E}\left(\frac{f(E)}{\rho(E)}\right)(1+\xi)+\frac{f(E)}{\rho(E)}\left(\frac{1}{T_v}+\frac{\xi(E)}{T_0}\right)\right]=j_R(E). \tag{5.75}$$

The solution $f^{(0)}(E)$ of the uniform equation, related to the linear non-uniform equation (5.75), is the vibrational distribution Eq. (5.53) without anharmonicity. Equation (5.61) remains true in this case and permits calculating both the distribution function perturbed by the reaction and macroscopic reaction rate coefficient. For example, the plasma chemical dissociation of polyatomic molecules such as CO_2 and H_2O, effectively stimulated by vibrationally excitation of the molecules, corresponds to the fast reaction limit. The macroscopic dissociation rate coefficients of these processes can be derived from Eq. (5.75), (5.53) and (5.61) as (B. Potapkin, V.D. Rusanov, A.E. Samarin, A. Fridman, 1980):

$$k_R^{macro}=\frac{k_{VV}^0}{\Gamma(s)}\frac{\hbar\omega}{T_v}\left(\frac{E_a}{T_v}\right)^s\exp\left(-\frac{E_a}{T_v}\right)\sum_{r=0}^{\infty}\frac{(s+r-1)!}{(s-1)!r!}\frac{\gamma(r+1,E_a/T_v)}{(E_a/T_v)^r}. \tag{5.76}$$

In this relation, $\Gamma(s)$ is the gamma-function; $\gamma(r+1, E_a/T_v)$ is the incomplete gamma-function; s is the number of vibrational degrees of freedom; k_{VV}^0 and $\hbar\omega$ are the lowest vibrational quantum and the corresponding low energy VV-exchange rate coefficient. The sum in Eq. (5.76) is not a strong function of E_a/T_v, and for numerical calculations can be taken approximately as 1.1–1.3 if $E_a = 3 \div 5\,eV$ and $T_v = 1000 \div 4000\,K$. Detailed analysis of chemical reactions of vibrationally excited polyatomic molecules can be found in the review of V.D. Rusanov, A. Fridman, G. Sholin (1986).

5.3.5 Macrokinetics of Reactions of Two Vibrationally Excited Molecules

Microkinetics of elementary reactions of two vibrationally excited molecules, for example $CO* + CO* \rightarrow CO_2 + C$ was considered in Section 3.7.8. Assume that the effective reaction takes place, when the sum of vibrational energies of two partners exceeds the activation energy: $E' + E'' \geq E_a$ (compare with (Eq. 3.135)). Then we can find the total reaction rate ($w_R = n_0 J_0$) for the vibrational distribution function $f^{(0)}(E)$ (5.29) not perturbed by the chemical process:

$$J_0=\iint_{E'+E''\geq E_a} f^{(0)}(E')f(E'')k_R(E',E'')n_0\,dE'dE''. \tag{5.77}$$

If $E_a > 2E^*(T_0)$, the process stimulation by vibrational excitation is ineffective because population of excited states is low due to VT-relaxation (here $E^*(T_0)$- is the critical vibrational energy, when rates of VT and VV-relaxation become equal, relation (Eq. 3.79)). If $E_a < E^*(T_0)$, the Treanor effect is stronger if all the vibrational energy is located on just one molecule, and the reaction of two excited molecules actually uses effectively only one of them. Indeed, according to (V.D.Rusanov, A.Fridman, G. V.Sholin, 1977), when total vibrational energy is fixed $E' + E'' = const$, than $(E')^2 + (E'')^2 = (E' + E'')^2 - 2E'E''$and the Treanor effect is the strongest if $E'' = 0$. As the result, the non-trivial kinetic effect can be obtained from (5.77), if only $E*(T_0) < E_a < 2E^*(T_0)$. The integration (5.77) determines the optimal vibrational energies of reaction partners E_{opt} and $E_a - E_{opt}$, which make the biggest contribution to the total reaction rate:

$$E_{opt}=E*(T_0)-\frac{1}{2\delta_{VV}}\ln\frac{T_0}{2x_e\left[2E*(T_0)-E_a\right]}. \tag{5.78}$$

The rate coefficient for the reaction of two vibrationally excited molecules can be then presented as:

$$k_R^{macro}\left(T_v,T_0\right)\propto\exp\left[-\frac{E_a}{T_v}+\frac{x_e}{T_0\hbar\omega}\left(E_{opt}^2+\left(E_a-E_{opt}\right)^2\right)-\frac{\tilde{\beta}_0x_e}{\delta_{VV}T_0}\left(2E*\left(T_0\right)-E_a\right)\right] \quad (5.79)$$

Comparison of relations (5.78) and (5.31) shows, that the optimal energy exceeds the inflection energy-point of $f(E)$: $E_{infl} < E_{opt} < E^*\ (T_0)$. Thus although $E_a > E^*\ (T_0)$, the VT-relaxation does not significantly slow down the chemical reaction of two excited molecules because $E_{opt} < E*\ (T_0)$. The above considered slow reaction limit is most typical for the reactions of two vibrationally excited molecules. The distribution function is almost not disturbed by the reaction in this case. Consideration of less typical fast reaction limit and influence of reactions of two excited molecules on distribution function can be found in a book of V.D. Rusanov, A. Fridman (1984).

5.3.6 Vibrational Energy Losses Due To VT-Relaxation

Calculate the vibrational relaxation averaged over the vibrational distribution $f(E)$. The average vibrational energy is defined as:

$$\varepsilon_v=\int_0^\infty Ef\left(E\right)dE. \quad (5.80)$$

Multiplying the equation (5.1) by E and then integrating, results in the balance relation:

$$\frac{d\varepsilon_v}{dt}=\int_0^\infty j_{VV}\left(E\right)dE+\int_0^\infty j_{VT}\left(E\right)dE. \quad (5.81)$$

We have taken into account here the vibrational energy losses related to only VT and VV- relaxation processes. Taking the VV-flux from (5.3) and the VT-diffusion coefficient in energy space as:

$$D_{VT}\left(E\right)=k_{VT}\left(E\right)n_0\left(\hbar\omega\right)^2=k_{VT}^0\left(\frac{E}{\hbar\omega}+1\right)\exp\left(\delta_{VT}E\right)n_0\left(\hbar\omega\right)^2, \quad (5.82)$$

we can derive based on (5.81) the formula for vibrational energy losses due to VT-relaxation:

$$\left(\frac{d\varepsilon_v}{dt}\right)^{VT}=D_{VT}\left(0\right)f\left(0\right)-\int_0^\infty D_{VT}\left(0\right)\left[\frac{\left(\tilde{\beta}_0-\delta_{VT}\right)E-1}{\hbar\omega}-\left(\tilde{\beta}_0-\delta_{VT}\right)\right]\exp\left(\delta_{VT}E\right)fdE. \quad (5.83)$$

These vibrational energy losses can be subdivided into two classes – first one, related to the low vibrational levels, and prevailing in conditions of weak excitation; and second one, related to the high vibrational levels, and dominating in conditions of strong excitation.

5.3.7 VT-Relaxation Losses from Low Vibrational Levels, the Losev Formula and the Landau-Teller Relation

Based on (5.83), - the VT-losses, related to the low vibrational levels, can be expressed by the Losev formula (S.A. Losev, O.P. Shatalov, M.S. Yalovik, 1970):

$$\left(\frac{d\varepsilon_v}{dt}\right)_L=-k_{VT}\left(0\right)n_0\left[1-\exp\left(-\frac{\hbar\omega}{T_0}\right)\right]\cdot\left[\frac{1-\exp\left(-\frac{\hbar\omega}{T_v}\right)}{1-\exp\left(-\frac{\hbar\omega}{T_v}+\delta_{VT}\hbar\omega\right)}\right]^2\left(\varepsilon_v-\varepsilon_{v0}\right). \quad (5.84)$$

Here $\varepsilon_{v0} = \varepsilon_v(T_v = T_0)$. Neglecting the effect of anharmonicity ($\delta_{VT} = 0$), the Losev formula can be rewritten in the well-known Landau-Teller relation:

$$\left(\frac{d\varepsilon_v}{dt}\right)_L = -k_{VT}(0)n_0\left[1-\exp\left(-\frac{\hbar\omega}{T_0}\right)\right](\varepsilon_v - \varepsilon_{v0}). \tag{5.85}$$

Obviously, in equilibrium between vibrational and translational degrees of freedom, when $T_v = T_0$ and, hence: $\varepsilon_v = \varepsilon_0$, the losses of vibrational energy become equal to zero.

5.3.8 VT-Relaxation Losses from High Vibrational Levels

The contribution of the high levels into the VT-losses of vibrational energy is usually related to the highest vibrational levels before the fast fall of the distribution function due to the VT-relaxation or chemical reaction. The contribution of vibrational levels, corresponding to the Boltzmann distribution with temperature T_0, is small because of the exponential decrease of the product $k_{VT}(E) \cdot f(E)$ with energy ($\delta_{VT}T_0 << 1$). The vibrational population fall at $E > E_a$ in the fast reaction limit leads to an even faster decrease of $k_{VT}(E) \cdot f(E)$. The VT-losses from the high vibrational levels in *the fast reaction limit*, when the above-mentioned fast fall of $f(E)$ is related to a chemical reaction, can be calculated from the relation (5.83) as ($T_0 << \hbar\omega$, $\delta_{VT}T << 1$):

$$\left(\frac{d\varepsilon_v}{dt}\right)_H^{VT} \approx -D_{VT}(0)f(0) - \left[\frac{(\tilde{\beta}_0 - \delta_{VT})E_a - 1}{\hbar\omega} - (\tilde{\beta}_0 - \delta_{VT})\right]$$
$$\times D_{VT}(0)\exp(\delta_{VT}E_a)f^{(0)}(E_a)\Delta \approx -k_{VT}(E_a)n_0\hbar\omega f^{(0)}(E_a)\Delta. \tag{5.86}$$

Here $f^{(0)}(E)$- is the vibrational distribution function not perturbed by chemical reaction, and the effective integration domain: $\Delta = \left|\frac{1}{T_v} - \delta_{VT} - \frac{2x_e E_a}{T_0\hbar\omega}\right|^{-1}$. Based on (5.67), (5.86) and assuming $\xi(E_a) << 1$, the VT-losses per one act of fast chemical reaction are:

$$\Delta\varepsilon_{VT} \approx \frac{k_{VT}(E_a)}{k_{VV}(E_a)}\Delta. \tag{5.87}$$

This relation reflects the fact, that fast reaction and VT-relaxation from high levels are related to actually the same excited molecules with energies slightly exceeding E_a. Frequencies of the processes are, however, different – and proportional respectively to k_{VV} and k_{VT}. Also the relation (5.87) shows, that if $\xi(E_a) << 1$ - the VT-relaxation from high levels does not affect much the energy efficiency of plasma chemical processes stimulated by vibrational excitation. The relation (5.87) was derived for the case of predominantly non-resonance VV-exchange, - if VV-relaxation is mostly resonant – (5.87) can be used as the higher limit of the losses. The VT-losses from the high vibrational levels in *the slow reaction limit*, when the above-mentioned fast fall of $f(E)$ is related to VT-relaxation reaction, should be calculated directly from Eq. (5.83). These losses specifically in the strong excitation regime can be higher than those considered above (A.V. Demura, S.O. Mahceret, A. Fridman, 1981).

5.3.9 Vibrational Energy Losses Due to the Non-Resonance Nature of VV-Exchange

The VV-losses per one act of chemical reaction can be calculated in frameworks of the model, illustrated in Figure 5.11. A diatomic molecule is excited by electron impact from the zero-level to the

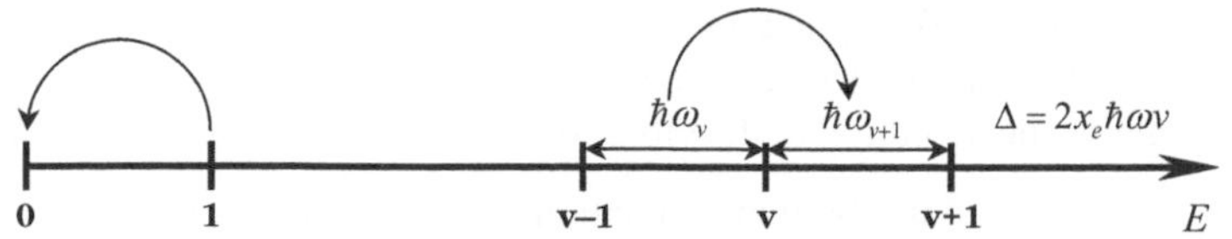

FIGURE 5.11 Vibrational energy loses in non-resonant VV-exchange.

first vibrational level; so a quantum comes to the system as a $\hbar\omega$. Further population of the higher excited vibrational levels is due to the one-quantum VV-exchange. Quanta become smaller and smaller during the VV-exchange due to anharmonicity (Eq. 3.8). So each step up on the "vibrational ladder" requires the resonance-defect energy transfer $2x_e\hbar\omega v$ from vibrational to translational degrees of freedom. Thus the total losses corresponding to excitation of a molecule to the n-th vibrational level can be found as the following sum:

$$\Delta\varepsilon^{VV}(n) = \sum_{v=0}^{v=n-1} 2x_e\hbar\omega v = x_e\hbar\omega(n-1)n \tag{5.88}$$

For dissociation of diatomic molecules, stimulated by vibrational excitation: $n = n_{max} \approx 1/2x_e$. In this case (see Eq. 3.7) the losses of vibrational energy per one act of dissociation, associated with anharmonicity and non-resonance nature of VV-relaxation, are equal to the dissociation energy:

$$\Delta\varepsilon_D^{VV}(n = n_{max}) \approx \frac{\hbar\omega}{4x_e} = D_0. \tag{5.89}$$

Taking into account these losses, the total vibrational energy necessary for dissociation provided by the VV-exchange is equal to not D_0, but $2D_0$. This amazing fact was first mentioned by P.A. Sergeev and D.I. Slovetsky (1979), and then described by A.V. Demura, S.O. Mahceret, A. Fridman (1981). If a reaction stimulated by vibrational excitation and has activation energy $E_a << D_0$, the VV-losses are:

$$\Delta\varepsilon_R^{VV}(E_a) \approx \frac{1}{4}D_0\left(\frac{E_a}{D_0}\right)^2 = x_e\frac{E_a^2}{\hbar\omega}. \tag{5.90}$$

Consider, as an example, plasma chemical NO synthesis in air, stimulated by vibrational excitation of N_2. According to the Zeldovich mechanism, this synthesis is limited by the reaction:

$$O + N_2* \rightarrow NO + N, \quad E_a = 1.3eV, \quad D_0 = 10eV. \tag{5.91}$$

Energy losses (5.90) in this reaction are equal to 0.28 eV per NO molecule, which results in a 14% decrease in the energy efficiency of the total plasma chemical process (S. Macheret, V.D. Rusanov, A. Fridman, G. V. Sholin, 1980a). Mention below some other related phenomena, which could be significant in special conditions (details on the subject can be found in the book of V.D. Rusanov, A. Fridman, 1984):

1. The relations (5.88)-(5.90) determine the losses of vibrational energy in steady-state systems. The non-steady-state initial establishment of the vibrational distribution function also includes the conversion of the first "big quanta" $\hbar\omega$ into the smaller ones. It leads according to (5.88) to the additional VV-energy losses of about $\frac{x_eT_v^2}{\hbar\omega}$.
2. Each act of VT-relaxation from the highly excited states with "smaller quantum" $\hbar\omega_s$ corresponds to VV-losses of vibrational energy $\hbar\omega - \hbar\omega_s$ into translational degrees of freedom.

Obviously, this effect takes place only for the population of the highly excited states of diatomic molecules due to the one-quantum VV-relaxation.

3. We implied above the evolution of the vibrational distribution $f(E)$ mostly due to one-quantum VV-exchange. However at vibrational quantum numbers about $v \approx \frac{1}{2}v_{\max} = \frac{1}{4x_e}$, the value of vibrational quantum becomes exactly twice less than the initial value - $\hbar\omega$, which makes the double-quantum exchange possible (though still with low probability). Obviously such kind of VV-exchange decreases the VV-losses of vibrational energy.
4) At higher ionization degrees, when the evolution of vibrational distribution function is provided mostly by direct interaction with electron gas (eV-processes), the losses of vibrational energy due to VV-exchange become much less significant.

5.4 VIBRATIONAL KINETICS IN GAS MIXTURES, ISOTROPIC EFFECT IN PLASMA CHEMISTRY

The population of vibrationally excited molecular states in gas mixtures is provided not only by VV-, VT- and eV-relaxation but also by the non-resonant vibration-vibration *VV′*-exchange between different components of the molecular gas. Even if the difference in oscillation frequencies of two components of the gas mixture are small (e.g., isotopes), the *VV′*-exchange can result in significant differences in the level of their vibrational excitation. Usually a gas-component with a low value of vibrational quantum becomes excited to the higher vibrational levels (J.W. Rich, R.C. Bergman, 1986; Sh.S. Mamedov, 1979).

5.4.1 Kinetic Equation and Vibrational Distribution in Gas Mixture

Consider the kinetic equation for the vibrational distributions $f_i(E)$ of double-component mixture: subscripts $i = 1, 2$ correspond to the 1-st (lower oscillation frequency) and 2-nd (higher frequency) molecular components:

$$\frac{\partial f_i(E)}{\partial t} + \frac{\partial}{\partial E}\left[j_{VV}^{(i)}(E) + j_{VT}^{(i)}(E) + j_{eV}^{(i)}(E) + j_{VV'}^{(i)}(E)\right]. \tag{5.92}$$

The VV-, VT- and eV-fluxes in energy space $j_{VV}^{(i)}(E), j_{VT}^{(i)}(E), j_{eV}^{(i)}(E)$ are similar to those for a one-component gas and can be taken respectively for example as (5.12), (5.3), and (5.36). The *VV′*-relaxation flux provides a predominant population of a component with lower oscillation frequency (i = 1), can be expressed (S.O. Macheret, A. Fridman, V.D. Rusanov, G.V. Sholin, 1980a) as:

$$j_{VV'}^{(i)} = -D_{VV'}^{(i)}\left(\frac{\partial f_i}{\partial E} + \beta_i f_i - 2x_e^{(i)}\beta_0 \frac{E}{\hbar\omega_i} f_i\right). \tag{5.93}$$

In this relation $x_e^{(i)}, \omega_i$ are the anharmonicity coefficient and the oscillation frequency for the molecular component "i". The correspondent diffusion coefficient in energy space is:

$$D_{VV'}^{(i)} = k_{VV'}^{(i)} n_{l \neq i}\left(\hbar\omega_i\right)^2. \tag{5.94}$$

Here $k_{VV'}^{(i)}$ corresponds to the inter-component *VV′* - exchange and is related to the rate coefficient $k_{VV}^{(0)}$ of the resonant VV-exchange at the low vibrational levels of the component "i" as:

$$k_{VV'}^{(i)} = k_{VV}^{(0)} \exp\left[\frac{\delta_{VV}\hbar\left(\omega_i - \omega_l\right)}{2x_e^{(i)}} - \delta_{VV}E\right]. \tag{5.95}$$

The concentration $n_{l\neq i}$ in Eq. (5.94) represents the number density of a component "l" of the gas mixture, which interacts with the component "i" ($i, l = 1, 2$). The reverse temperature factor β_i in the flux Eq. (5.93) can be found from the relation:

$$\beta_i = \frac{\omega_l}{\omega_i}\beta_{vl} + \frac{\omega_i - \omega_l}{\omega_i}\beta_0. \tag{5.96}$$

The other reverse temperature parameters in Eq. (5.96) are quite traditional: $\beta_0 = T_0^{-1}, \beta_{vi} = \left(T_{vi}\right)^{-1}$. A practically interesting case is that of isotope-mixtures which consists of a large fraction of a light gas component (higher frequency of molecular oscillations, concentration n_2), and only a small fraction of heavy component (lower oscillation frequency, concentration n_1). Usually:

$$\frac{n_2}{n_1} << \exp\left[-\frac{\hbar\delta_{VV}}{2x_e}\left(\omega_2 - \omega_1\right)\right]. \tag{5.97}$$

In this case, the steady-state solution of the Fokker-Planck kinetic equation (5.92) with the relaxation fluxes Eqs. (5.12), (5.3), (5.36), and (5.93) gives the following vibrational distribution functions for two components of a gas-mixture:

$$f_{1,2}\left(E\right) = B_{1,2}\exp\left[-\int_0^E \frac{\beta_{1,v2} - 2x_e^{(1,2)}\beta_0 \dfrac{E'}{\hbar\omega_{1,2}} + \tilde{\beta}_0^{(1,2)}\xi_{1,2} + \beta_e\eta_{1,2}}{1 + \xi_{1,2} + \eta_{1,2}} dE'\right]. \tag{5.98}$$

Here $B_{1,2}$ are the normalization factors; reverse electron temperature parameter $\beta_e = T_e^{-1}$; factors ξ_i and η_i describe the relative contribution of VT- and eV-processes with respect to VV (VV') – exchange (see Sections 5.1 and 5.2):

$$\xi_i\left(E\right) = \xi_i\left(0\right)\exp\left(2\delta_{VV}E\right), \quad \eta_i\left(E\right) = \eta_i\left(0\right)\exp\left(\delta_{VV}E\right). \tag{5.99}$$

5.4.2 The Treanor Isotopic Effect in Vibrational Kinetics

VT- and eV-relaxation can be neglected ($\xi_i << 1, \eta_i << 1$) on the low energy part of the vibrational distributions $f_{1,2}(E)$. In this case, Eq. (5.98) gives for both gas components (i = 1, 2) the Treanor distributions Eq. (5.10) with the same translational temperature T_0, but with different vibrational temperatures T_{v1} and T_{v2}:

$$\frac{\omega_1}{T_{v1}} - \frac{\omega_2}{T_{v2}} = \frac{\omega_1 - \omega_2}{T_0}. \tag{5.100}$$

This relation is known as the Treanor formula for isotopic mixture. The Treanor formula shows, that under non-equilibrium conditions ($T_{v1,\,v2} > T_0$) of isotopic mixture, the component with the lower oscillation frequency (heavier isotope) has a higher vibrational temperature. In this case, the predominant VV-transfer of vibrational energy from the molecules with larger vibrational quanta to those with smaller quanta is similar to the main Treanor effect (Section 4.1.10) providing overpopulation of highly excited states with lower values of vibrational quanta. Experimental illustration of this Treanor effect in the isotopic gas mixture of $^{12}C^{16}O$ and $^{12}C^{18}O$ is presented in Figure 5.12 (R.C. Bergman, G. F. Homicz, J.W. Rich, G. L. Wolk, 1983). The Treanor effect can be applied for isotope separation in plasma chemical reactions, stimulated by vibrational excitation. It was first suggested by E.M. Belenov, E.P. Markin, A.N. Oraevsky, V.I. Romanenko (1973). The ratio of rate coefficients of chemical reactions of two different vibrationally excited isotopes $\kappa = k_R^{(1)} / k_R^{(2)}$ is proportional to

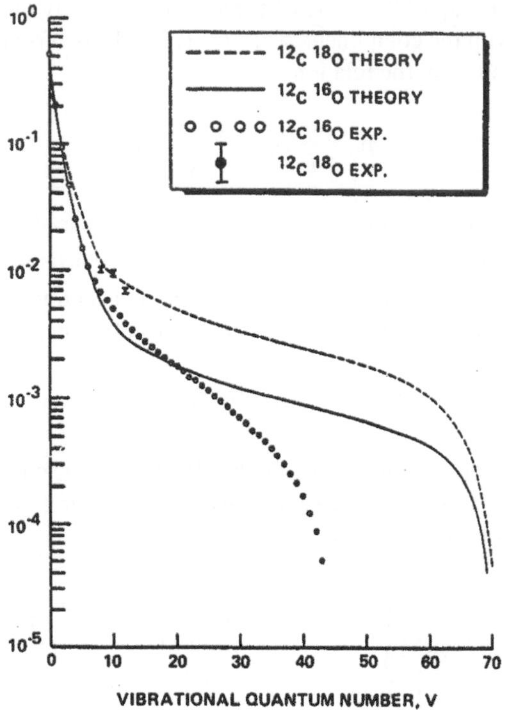

FIGURE 5.12 Isotopic effect in vibrational population for 12C18O/12C16O mixture. (From "Non Equilibrium Vibrational Kinetics" Topics in Current Physics, Ed. M. Capitelli.)

the ratio of the population of vibrational levels $E \geq E_a$ for the two isotopes (see Section 5.3). This so-called coefficient of selectivity in both the cases of slow and fast reactions can be expressed as:

$$\kappa = \frac{f_1(E_a)\Delta E_1}{f_2(E_a)\Delta E_2}. \tag{5.101}$$

Here $\Delta E_1 \approx \Delta E_2$ are the parameters of the $f_{1,2}(E)$exponential decrease at $E = E_a$. Assuming the Treanor functions for distributions $f_{1,2}(E)$ of both gas components, the coefficient of selectivity is:

$$\kappa \approx \exp\left[\frac{\Delta\omega}{\omega} E_a \left(\frac{1}{T_0} - \frac{1}{T_{v2}}\right)\right], \tag{5.102}$$

where the relative defect of resonance $\frac{\Delta\omega}{\omega} = \frac{\omega_2 - \omega_1}{\omega_2}$. It is very interesting that the coefficient of selectivity does not depend on the vibrational temperature in the non-equilibrium conditions ($T_v > T_0$), but depends only on the translational temperature T_0, the relative defect of resonance $\Delta\omega/\omega$, and activation energy E_a. Another remark is related to the "direction" of the isotopic effect in vibrational kinetics, which is opposite to one conventional in the quasi-equilibrium chemical kinetics. The dissociation energy, see (Eq. 3.5), and similarly the activation energy are sensitive to the "zero-vibration level"- $\frac{1}{2}\hbar\omega$. Heavy isotopes with the lower value of $\frac{1}{2}\hbar\omega$ have higher activation (and dissociation) energies and as a result, their quasi-equilibrium reactions are slower. This is the conventional isotopic effect. Conversely, in vibrational kinetics heavy isotopes react faster due to the Treanor effect. Numerical values of the coefficient of selectivity for different plasma chemical

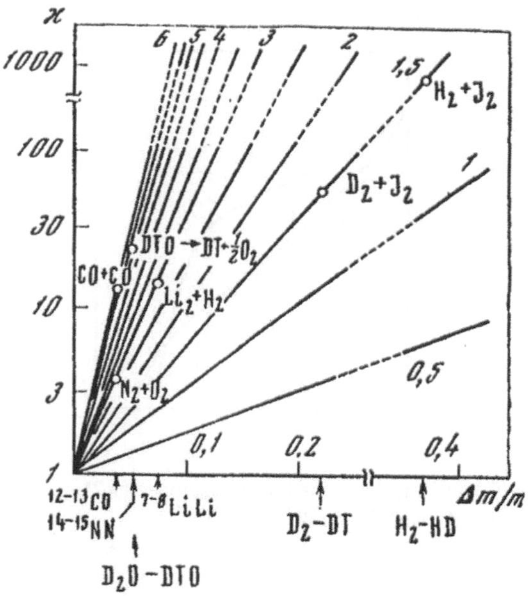

FIGURE 5.13 Isotopic effect for different molecules; numbers on the curves represent activation energies of specific reactions.

processes of isotope separation, stimulated by vibrational excitation of the molecules-isotopes, are presented in Figure 5.13. The coefficients of selectivity are shown in the figure in a convenient form as the function Eq. (5.102) of the defect of mass $\Delta m/m$ and the process activation energy. The related detailed calculations of the isotope separation using the Treanor effect were carried out: for nitrogen- and carbon monoxide- isotopes by V.M. Akulintsev, V.M. Gorshunov, Y.P. Neschimenko (1977a, 1983), and for hydrogen-isotopes by A.V. Eletsky, N.P. Zaretsky (1981) and A.D. Margolin, A.V. Mishchenko, V.M. Shmelev (1980).

5.4.3 Influence of VT-Relaxation on Vibrational Kinetics of Mixtures, the Reverse Isotopic Effect

Next, take into account the contribution of VT-relaxation in the isotopic effect, but still neglect the direct influence of the eV-processes. The direct influence of eV-processes in the general expression Eq. (5.98) for the distribution functions in mixtures at all vibrational energies, can be neglected if the ionization degree in plasma is relatively low:

$$\left(\frac{n_e}{n_0}\right)^2 << \frac{k_{VV'}(0)k_{VT}(0)}{k_{eV}^2}. \tag{5.103}$$

Here $n_0 = n_1 + n_2 \approx n_2$ the total concentration of the gas-isotope, k_{eV}, $k_{VT}(0)$, $k_{VV}'(0)$ are respectively the rate coefficients of eV-relaxation, and VT-, VV'-relaxation processes at low vibrational levels. At the not very high electron concentrations Eq. (5.103), the eV-factor $\eta_{1,2}(E)$ can be neglected and the integral (5.98) can be taken:

$$f_{1,2}(E) = B_{1,2}\exp\left[-\beta_{1,v2}E + \frac{x_e^{(1,2)}\beta_0 E^2}{\hbar\omega_{1,2}} - \frac{\tilde{\beta}_0^{(1,2)}}{2\delta_{VV}}\ln(1+\xi_{1,2})\right]. \tag{5.104}$$

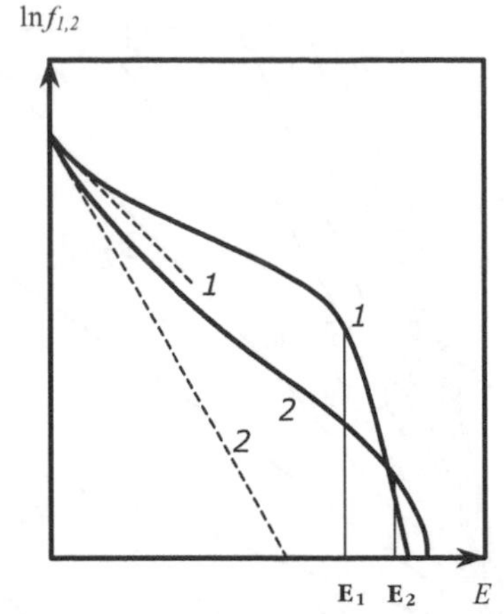

FIGURE 5.14 Vibrational distribution functions for two isotopes (1,2) without significant influence of eV-processes; dashed-lines represent the Boltzmann distribution.

The vibrational distribution functions Eq. (5.104) for two isotopes are illustrated in Figure 5.14. As can be seen from the figure, the population $f_1(E)$ of the same low vibrational levels is higher for the heavier isotope ("1", usually small additive). It obviously corresponds to the Treanor effect and Eq. (5.100). However, the situation becomes completely opposite at higher levels of excitation, where the vibrational population of the relatively light isotope exceeds the population of the heavier one. This phenomenon is known as the reverse isotopic effect in vibrational kinetics (S.O. Macheret, A. Fridman, V.D. Rusanov, G. V. Sholin, 1980a). The physical basis of the reverse isotopic effect is quite clear. The vibrational distribution function $f_1(E)$ for the heavier isotope (small additive) is determined by the VV'-exchange, which is slower than VV-exchange because of the defect of resonance. As a result, the VT-relaxation makes the vibrational distribution $f_1(E)$ start falling at lower energies $E_1(\xi_1 = 1)$ with respect to the distribution function $f_2(E)$ of the main isotope, which is determined by VV-exchange and starts falling at the higher vibrational energy $E_2(\xi_2) > E_1$. From Eq. (5.104):

$$E_1 = \frac{1}{2\delta_{VV}} \ln \frac{k_{VV}(0)}{k_{VT}(0)} - \frac{\hbar\Delta\omega}{4x_e}, \quad E_2 = \frac{1}{2\delta_{VV}} \ln \frac{k_{VV}(0)}{k_{VT}(0)}. \tag{5.105}$$

Thus, the reverse isotopic effect takes place if $E_1 < E_a < E_2$ (see Figure 5.14) and the light isotope is excited stronger and reacts faster than the heavier one. The coefficient of selectivity for the reverse isotopic effect can be calculated from Eqs. (5.104) and (5.105) (S.O. Macheret, A. Fridman, V.D. Rusanov, G. V. Sholin, 1980a) and expressed as:

$$\kappa \approx \exp\left(-\frac{\Delta\omega}{\omega}\frac{D_0}{T_0}\right). \tag{5.106}$$

Here D_0 is the dissociation energy of the diatomic molecules in the manner as that for the direct effect Eq. (5.102). The selectivity coefficient Eq. (5.106) for the reverse isotopic effect, does not depend directly on the vibrational temperature. It's even more interesting, that although the coefficient of selectivity was derived for a chemical reaction with activation energy $E_a < D_0$, the activation energy E_a is not explicitly presented in Eq. (5.106). Taking into account $E_a < D_0$, the reverse isotopic effect is much stronger than the direct one and can be achieved in a narrow range of translational temperatures:

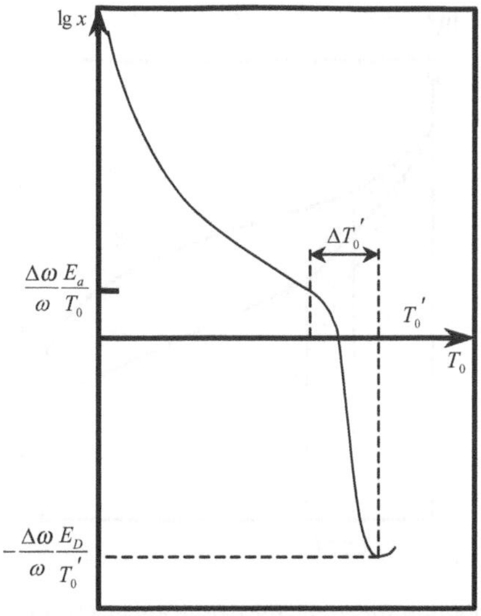

FIGURE 5.15 Isotopic effect dependence on translational temperature T_0; temperature T_0' corresponds to the maximum value of the inverse isotopic effect coefficient.

$$\frac{\Delta T_0}{T_0} = 2\frac{\Delta\omega}{\omega}\left(1 - \frac{E_a}{D_0}\right)^{-1}. \tag{5.107}$$

The selectivity coefficient dependence on translational temperature is illustrated in Figure 5.15. As one can see from the figure, the direct effect takes place at relatively low translational temperatures. By increasing the translational temperature one can find the narrow temperature range Eq. (5.107) of the reverse effect, where the isotopic effect changes "direction" and becomes much stronger.

5.4.4 Influence of eV-Relaxation on Vibrational Kinetics of Mixtures and the Isotopic Effect

Consider the vibrational kinetics of the isotopic mixture at high degrees of ionization, when the inequality opposite to Eq. (5.103) is valid. Vibrational distributions then at not very high energies, are controlled by VV- (*VV'*-) and eV-relaxation. The distributions for two isotopes can be found by integrating Eq. (5.98):

$$f_{1,2}(E) = B_{1,2}\exp\left[-\beta_{1,v2}E + \frac{x_e^{(1,2)}\beta_0 E^2}{\hbar\omega_{1,2}} + \frac{\beta_{1,v2} - \beta_e}{\delta_{VV}}\ln\left(1 + \eta_{1,2}\right)\right]. \tag{5.108}$$

The vibrational distributions for two isotopes are illustrated in Figure 5.16. At relatively high energies, when eV-processes are dominating: $f_{1,2}(E) \propto \exp(-\beta_e E)$, which can be considered as the plateau on the vibrational distribution (see Section 5.2). The *VV'*-exchange processes are slower than VV-exchange because of the defect of resonance. For this reason, the transition from the Treanor distribution to the eV-plateau (see Figure 5.16) takes place for heavy isotope at lower vibrational energies:

$$E_{eV}^{(2)} - E_{eV}^{(1)} = \frac{\hbar\Delta\omega}{2x_e}. \tag{5.109}$$

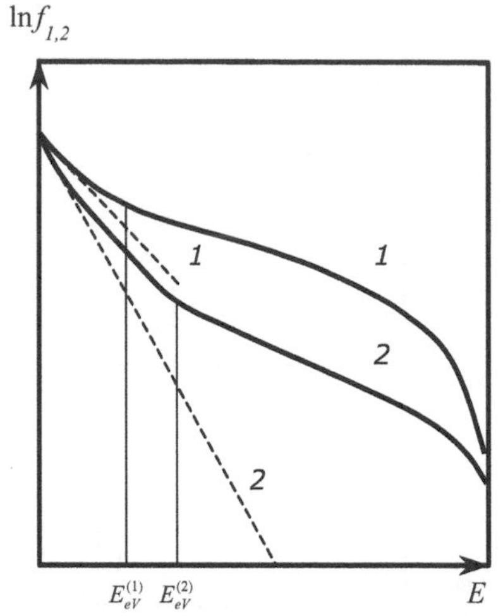

FIGURE 5.16 Vibrational distribution functions for two isotopes (1,2) with significant influence of eV-processes; dashed-lines represent the Boltzmann distribution.

As a result, the direct isotopic effect can be higher than Eq. (5.102), not taking into account eV-processes. The coefficient of selectivity κdependence on the degree of ionization is presented in Figure 5.17. At low degrees of ionization, the coefficient of selectivity corresponds to Eq. (5.102). Then κ grows with the degree of ionization. The selectivity coefficient reaches a high value:

$$\kappa \approx \exp\left(\frac{\Delta\omega}{\omega}\frac{E_{eV}^{(1)}}{T_0}\right), \tag{5.110}$$

when E_a corresponds to the eV-plateau for both isotopes (see Figure 5.16). Here, the vibrational energy $E_{eV}^{(1)}$ corresponds to the vibrational distribution transition from the Treanor function to the eV-plateau. The vibrational energy $E_{eV}^{(1)}$ is reversibly proportional to $\frac{n_e}{n_0}$, and the selectivity

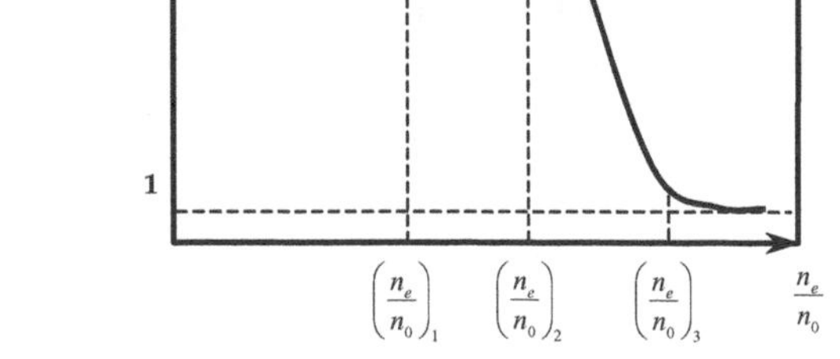

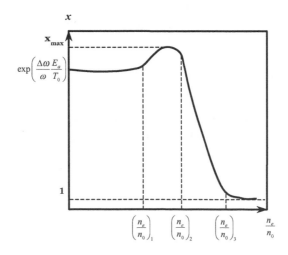

FIGURE 5.17 Isotopic effect coefficient dependence on ionization degree.

coefficient also decreases with the degree of ionization. The maximum coefficient of selectivity occurs at the degree of ionization:

$$\frac{n_e}{n_0} \approx \frac{k_{VV}(0)}{k_{eV}} \exp(-\delta_{VV} E_a). \tag{5.111}$$

At higher electron concentrations, both vibrational distributions follow the Boltzmann functions with temperature Te from the relatively low vibrational levels and the isotopic effect decreases with the ionization and finally disappears (see Figure 5.17).

5.4.5 Integral Effect of Isotope Separation

The foregoing selectivity coefficients described the ratio of rate coefficients of chemical reactions for different molecules-isotopes. Practical calculations require, however, the determination of the separation coefficient:

$$R = \frac{n_1 / n_2}{(n_1 / n_2)_0}. \tag{5.112}$$

This integral coefficient of isotope separation determines the change of molar fractions of different isotopes in mixtures and, hence, describes the effect of isotope enrichment in a system. The calculation of this coefficient is more complicated. Relatively simple expression for the separation coefficient can be found if the main channels of VT-relaxation and chemical reaction are related to the same molecules (S.O. Macheret, A. Fridman, V.D. Rusanov, G.V. Sholin, 1980a). Such a situation takes place in the plasma chemical systems $H_2 - J_2$ and $N_2 - O_2$, considered for the separation of hydrogen and nitrogen isotopes in reactions stimulated by vibrational excitation. The particle balance equation in these mentioned kinetic systems (assuming the small difference of isotope mass $\frac{\Delta m}{m} << \frac{2T_0}{E_a}$) gives the coefficient of isotope separation:

$$R \approx \exp\left[\frac{1}{P_{VT}(T_0)} \exp\left(-\frac{E_a}{T_{v1}}\right) * \frac{E_a}{2T_0} \frac{\Delta m}{m}\right]. \tag{5.113}$$

Here $P_{VT}(T_0)$ is the averaged probability of VT-relaxation. According to Eq. (5.113), the coefficient of isotope separation R depends on the vibrational temperature much stronger than on the mass defect $\Delta m/m$. This provides the possibility to reach high values of the isotope separation coefficient R even at low values of the mass defect $\Delta m/m$, for example, for separation of heavy molecular isotopes of special importance. However, the strongest isotopic effect of that kind can be reached, for the separation of light isotopes. An important example is H_2-HD, where the selectivity in reactions $H_2 + J_2 \rightarrow 2HJ$ and $H_2 + Br_2 \rightarrow 2HBr$ reaches 1000.

5.5 KINETICS OF ELECTRONICALLY AND ROTATIONALLY EXCITED STATES, NON-EQUILIBRIUM TRANSLATIONAL DISTRIBUTIONS, RELAXATION AND REACTIONS OF "HOT ATOMS" IN PLASMA

5.5.1 Kinetics of Population of Electronically Excited States, the Fokker–Planck Approach

Transfer of electronic excitation energy in collisions of heavy particles is effective in contrast to VV-exchange only for a limited number of specific electronically excited states. Even for high levels of electronic excitation, the transitions between electronic states are mostly due to collisions with plasma electrons at degrees of ionization exceeding 10^{-6}. In relatively low-pressure plasma systems, radiation transition also can be significant as well (D.I. Slovetsky, 1980). Models describing the population of electronically excited species in plasma were developed by L.M. Biberman, V.S. Vorobiev, I.T. Yakubov, 1982 and L.I. Gudzenko, S.I. Yakovlenko, 1978. Description of the highly electronically, excited states can be accomplished in the framework of the Fokker-Planck diffusion approach (S.T. Beliaev, G.I. Budker, 1958; L.I. Pitaevsky, 1962), similar to that considered for vibrational excitation. The possibility to apply the diffusion in energy space approach is due to: 1) low energy intervals between the highly electronically excited states; and 2) transitions occurring mostly between close levels in the relaxation collisions with plasma electrons. The population of these highly electronically excited states in plasma $n(E)$, due to energy exchange with electron gas, can be then found in the framework of the diffusion approach from the Fokker-Planck kinetic equation, similar to the kinetic equation (5.1) for vibrationally excited states:

$$\frac{\partial n(E)}{\partial t} = \frac{\partial}{\partial E}\left[D(E)\left(\frac{\partial n(E)}{\partial E} - \frac{\partial \ln n^0}{\partial E} n(E)\right)\right]. \tag{5.114}$$

In this relation $n^0(E)$ is the quasi-equilibrium population of the electronically excited states, corresponding to the electron temperature T_e:

$$n^0(E) \propto E^{-\frac{5}{2}} \exp\left(-\frac{E_1 - E}{T_e}\right). \tag{5.115}$$

Here E is the absolute value of the bonded electron energy; transition to continuum corresponds to the zero electron energy E = 0; E_1 is the ground state energy ($E_1 \geq E$). Note, that for $E >> T_e$ - $\frac{\partial \ln n^0(E)}{\partial E} = \frac{1}{T_e}$ in Eq. (5.1). The diffusion coefficient D(E) in energy space, related to the energy exchange between bonded electrons of highly electronically excited particles with plasma electrons, can be expressed (N.M. Kuznetzov, Yu.P. Raizer, 1965) as:

$$D(E) = \frac{4\sqrt{2\pi}\, e^4 n_e E}{3\sqrt{mT_e}\,(4\pi\varepsilon_0)^2}\Lambda, \tag{5.116}$$

where Λ is the Coulomb logarithm for the electronically excited state with ionization energy E.

5.5.2 Simplest Solutions of Kinetic Equation for the Electronically Excited States

To solve the kinetic equation in quasi-steady-state conditions, it is convenient to introduce a new variable, the relative dimensionless population of electronically excited states:

$$y(E) = \frac{n(E)}{n^0(E)}. \tag{5.117}$$

Boundary conditions for Eq. (5.114) can be taken as $y(E_1) = y_1$, $y(0) = y_e y_i$. Parameters: y_e, y_i are the electron and ion densities in plasma, divided by the corresponding equilibrium values. Then the quasi-steady-state solution of the kinetic equation (5.114) can be expressed in the following way:

$$y(E) = \frac{y_1 \chi\left(\frac{E}{T_e}\right) + y_e y_i\left[\chi\left(\frac{E_1}{T_e}\right) - \chi\left(\frac{E}{T_e}\right)\right]}{\chi\left(\frac{E_1}{T_e}\right)}. \tag{5.118}$$

Here $\chi(x)$ is a function, determined by the integral:

$$\chi(x) = \frac{4}{3\sqrt{\pi}} \int_0^x t^{\frac{3}{2}} \exp(-t)dt, \tag{5.119}$$

and having the following asymptotic approximations:

$$\chi(x) \approx 1 - \frac{4}{3\sqrt{\pi}} e^{-x} x^{3/2}, \quad if\, x >> 1, \tag{5.120}$$

$$\chi(x) \approx \frac{1}{2\sqrt{\pi}} x^{5/2}, \quad if\, x << 1. \tag{5.121}$$

For electronically excited levels with energies $E << T_e << E_1$ close to continuum, this gives the relative population:

$$y(E) \approx y_e y_i \left[1 - \frac{1}{2\sqrt{\pi}} \left(\frac{E}{T_e}\right)^{5/2}\right] + y_1 \frac{1}{2\sqrt{\pi}} \left(\frac{E}{T_e}\right)^{5/2} \to y_e y_i. \tag{5.122}$$

The population of electronically excited states Eqs. (5.117), (5.122) is decreasing exponentially with the effective Boltzmann temperature T_e and the absolute value corresponding to equilibrium with continuum $y(E) \to y_e y_i$. For the opposite case $E >> T_e$, the population of electronically excited states far from the continuum, that can be found as:

$$y(E) = y_1 + y_e y_i \frac{4}{3\sqrt{\pi}} \exp\left(-\frac{E}{T_e}\right) \cdot \left(\frac{E}{T_e}\right)^{3/2} \to y_1. \tag{5.123}$$

In this range of the electronic excitation energy (far from continuum) the population is also exponential with effective temperature T_e, but the absolute value corresponds to equilibrium with the ground state. Note that the Fokker-Planck approach, assuming diffusion of neutral particles in plasma along the energy spectrum, is much less accurate in describing the population of lower electronic levels, which are quite remote one from another. More accurately, the "modified" diffusion approach, including discrete consideration of the lower levels, was developed by L.M. Biberman, V.S.Vorobiev, and I.T. Yakubov (1979).

Practically, the Boltzmann distribution of electronically excited states with the temperature equal to the temperature of plasma electrons requires a very high degree of ionization in plasma $n_e/n_0 \geq 10^{-3}$ (although the domination of energy exchange with electron gas requires only $n_e/n_0 \geq 10^{-6}$). This is mostly due to the influence of some resonance transitions and the non-Maxwell Ian behavior of electron energy distribution function at the lower degree of ionization. The radiative deactivation of electronically excited particles is required at low pressures, usually when $p < 1 \div 10\, Torr$. Contribution of the radiative processes decreases with the growth of excitation energy, for example, approaching continuum. This can be explained by the reduction of the intensity of the radiative processes and, conversely, intensification of collisional energy exchange when the electron bonding energy in atom becomes smaller. It follows that the numerical formula for a critical value of the electron bonding energy applies:

$$E_R, eV = \left(\frac{n_e, cm^{-3}}{4.5 \cdot 10^{13} cm^{-3}}\right)^{1/4} * (T_e, eV)^{-1/8}. \tag{5.124}$$

Collisional energy exchange dominates when the excitation level is higher and electron binding energy is lower ($E < E_R$). When the excitation level is not high and the bonding energy is significant $E > E_R$, then even though the electronic excitation taking place still occur collisional processes, the deactivation is mostly due to radiation.

5.5.3 Kinetics of the Rotationally Excited Molecules, Rotational Distribution Functions

Even in non-thermal discharges, the rotational and translational degrees of freedom are usually in equilibrium between them and can be characterized by the same temperature T_0. Consider the kinetics and evolution of the rotational energy distribution functions in non-equilibrium plasma conditions. Consider the rotational and translational relaxation of small admixture of relatively heavy diatomic molecules (m_{BC}) in a light inert gas (m_A). This was first described by M.N. Safarian, E.V. Stupochenko (1964). Low values of energy transferred per one collision permit the use in this case the Fokker-Planck kinetic equation:

$$\frac{\partial f(E_t,E_r,t)}{\partial t}=\frac{\partial}{\partial E_t}\left[bE_t\left(\frac{\partial f}{\partial E_t}-f\frac{\partial \ln f^{(0)}}{\partial E_t}\right)\right]+\frac{\partial}{\partial E_r}\left[bE_r\left(\frac{\partial f}{\partial E_r}-f\frac{\partial \ln f^{(0)}}{\partial E_r}\right)\right]. \tag{5.125}$$

In this relation $f(E_t, E_r, t)$ is the distribution function related to translational E_t and rotational E_r energies; $f^{(0)}(E_t, E_r)$ is the equilibrium distribution function corresponding to temperature T_0 of the monatomic gas; the diffusion coefficient in energy space is characterized by parameter:

$$b=\frac{32}{3}\frac{m_A}{m_{BC}}N_A T_0 \Omega_{col}, \tag{5.126}$$

where N_A. is the density of the monatomic gas particles (which is actually close to the total gas density), and Ω_{col} is the dimensionless collisional integral. The kinetic equation (5.125) independently represents the rotational and translational relaxation processes. Then the averaged rotational distribution:

$$f(E_r,t)=\int_0^\infty f(E_t,E_r,t)\,dE_t \tag{5.127}$$

can be described by Eq. (5.125), the simplified Fokker-Planck equation including only rotational energy:

$$\frac{\partial f(E_r,t)}{\partial t}=\frac{\partial}{\partial E_r}\left[bE_r\left(\frac{\partial f(E_r,t)}{\partial E_r}-\frac{1}{T_0}f(E_r,t)\right)\right]. \tag{5.128}$$

Multiplication of Eq. (5.128) by E_r and following integrating from 0 to ∞ results in the macroscopic relation for the total rotational energy E_r^{total}:

$$\frac{dE_r^{total}}{dt}=\frac{E_r^{total}-E_{r,0}^{total}(T_0)}{\tau_{RT}},\quad \tau_{RT}=\frac{T_0}{b}. \tag{5.129}$$

Here τ_{RT} is the RT-relaxation time, $E_r^{total}(T_0)$ is the equilibrium value of rotational energy at the inert gas temperature T_0. It is interesting to note that Eq. (5.129) is valid for any kind of initial rotational distribution functions $f(E_r, t = 0)$. The general solution of the kinetic equation (5.128) at the arbitrary initial rotational distribution functions $f(E_r, t = 0)$ can be presented as the series:

$$f(E_r,t)=\Sigma c_n \exp\left(-\frac{nt}{\tau_{RT}}\right)\cdot L_n\left(\frac{E_r}{T_0}\right)\cdot\exp\left(-\frac{E_r}{T_0}\right), \tag{5.130}$$

where $L_n(x)$ are the Laguerre polynomials. Take the initial rotational distribution as Boltzmann:

$$f(E_r,t)=\frac{N_{BC}}{T_R(t)}\exp\left[-\frac{E_r}{T_r(t)}\right], \tag{5.131}$$

with the initial rotational temperature $T_r(t=0)\neq T_0$. Then the solution of kinetic equation (5.128) shows that the rotational distribution function $f(E_r,t)$ maintains the same form of the Boltzmann distribution (5.131) during the relaxation to equilibrium, but obviously with changing the value of the rotational temperature approaching the translational one:

$$T_r(t)=T_0+\left[T_r(t=0)-T_0\right]\exp\left(-\frac{t}{\tau_{RT}}\right). \tag{5.132}$$

This important property of always maintaining the same Boltzmann distribution function during relaxation to equilibrium (with only temperature change) is usually referred to as canonical invariance (H.C. Anderson, I. Oppenheim, K.E. Shuler, 1964).

5.5.4 Non-Equilibrium Translational Energy Distribution Functions, Effect of "Hot Atoms"

Relaxation of the translational energy of neutral particles requires only couple of collisions and usually determines the shortest time-scale in plasma chemical systems. The assumption of a local quasi-equilibrium is usually valid for the translational energy sub-system of neutral particles, even in strongly non-equilibrium plasma-chemical systems. The translational energy distributions in most of discharge conditions are Maxwellian with one local temperature T_0 for all neutral components participating in plasma chemical processes. However, this rule has some exceptions because of the possible formation of high-energy neutral particles in plasma, which strongly perturb the conventional Maxwellian distribution. Generation of the high-energy "hot" atoms can be related either to fast exothermic chemical reaction or to fast vibrational relaxation processes, if their frequencies can somewhat exceed the very high frequency of Maxwellization. Consider separately these two possible sources of "hot atoms" in plasma chemical systems.

5.5.5 Kinetics of "Hot Atoms" in Fast VT-Relaxation Processes, the Energy-Space Diffusion Approximation

Different elementary processes of the fast non-adiabatic VT-relaxation were discussed in Section 3.4.5. Probably the fastest of these are the relaxation processes of molecules, Mo, on alkaline atoms, Me, proceeding through intermediate formation of ionic complexes $[Me^+ Mo^-]$. Thus molecular nitrogen N_2* vibrational relaxation on atoms of Li, K and Na actually takes place at each collision which makes these alkaline atoms "hot" when they are added to the non-equilibrium nitrogen plasma ($T_v >> T_0$). Consider the evolution of the translational energy distribution function $f(E)$ of a small admixture of alkaline atoms into a non-equilibrium molecular gas ($T_v >> T_0$), for simplicity a diatomic gas. Under these conditions the translational energy distribution is determined by the kinetic competition of the fast VT-relaxation energy exchange between the alkaline atoms and the diatomic molecules and the Maxwellization TT-processes in collisions of the same partners. The exponential part of the steady-state distribution function $f(E)$ then can be found from the Fokker-Planck kinetic equation describing the atoms diffusion along the translational energy spectrum (A.K. Vakar, V.K. Givotov, A. Fridman, et al. 1981):

$$D_{VT}\left(\frac{\partial f}{\partial E}+\frac{f}{T_v}\right)+D_{TT}\left(\frac{\partial f}{\partial E}+\frac{f}{T_0}\right)=0. \tag{5.133}$$

This continuous approach can be applied only for high energies: $E >> \hbar\omega$. Consideration of lower energies requires more complicated discrete models. In Eq. (5.133); D_{VT} and D_{TT} are the diffusion coefficients of the alkaline atoms, Me, along the energy spectrum respectively related to VT-relaxation and Maxwellization. The translational energy distribution function of the alkaline atoms $f(E)$ depends upon the ratio of the diffusion coefficients $\mu(E) = D_{TT}(E)/D_{VT}(E)$, which is proportional to the ratio of the corresponding relaxation rate coefficients. Integration of the linear differential equation (5.133) gives:

$$f(E) = B\exp\left[-\int \frac{\frac{1}{T_v} + \frac{\mu(E)}{T_0}}{1+\mu(E)} dE\right], \tag{5.134}$$

where B is the normalization factor. The Maxwellization is much faster than VT-relaxation and $\mu(E) >> 1$. According to Eq. (5.134), this means, that the distribution $f(E)$ is Maxwellian with a temperature equal to $T_{0,}$ the translational temperature of the molecular gas. However, the situation can be different for the admixture of light alkaline atoms (e.g., Li, atomic mass m) into a relatively heavy molecular gas (e.g., N_2 or CO_2, molecular mass $M >> m$), taking into account that the VT and TT-relaxation frequencies are almost equal for this mixture. Because of the large difference in masses, the energy transfer during Maxwellization can be lower than $\hbar\omega$ at low energies $\mu << 1$. In accordance with Eq. (5.134), the translational temperature of the light alkaline atoms can be equal not to the translational T_0 but rather to the vibrational T_v (!) temperature of the main molecular gas (see illustration in Figure 5.18). For further calculation of the distribution function, the factor $\mu(E)$ when $m << M$ can be expressed as:

$$\mu(E) \approx \frac{E+T_v}{\hbar\omega}\left(\frac{m}{M}\frac{E}{T_v}\right)^2. \tag{5.135}$$

Taking into account the relations Eqs. (5.134) and (5.135) it can be seen, that the translation distribution function $f(E)$ can be Maxwellian with vibrational temperature T_v only at relatively low translational energies $E < E*$. At higher energies $E > E*$, the factor is always large $\mu >> 1$ and the exponential decrease of $f(E)$ always corresponds to the temperature T_0. The critical value of the translational energy $E*(T_v)$ can be found from the relations Eqs. (5.134) and (5.135) as:

$$E*(T_v) = \frac{M}{m}\sqrt{T_v\hbar\omega}, \quad if\ T_v > \hbar\omega\left(\frac{M}{m}\right)^2 \tag{5.136a}$$

$$E*(T_v) = (\hbar\omega)^{1/3}\left(\frac{T_v M}{m}\right)^{2/3}, \quad if\ T_v < \hbar\omega\left(\frac{M}{m}\right)^2. \tag{5.136b}$$

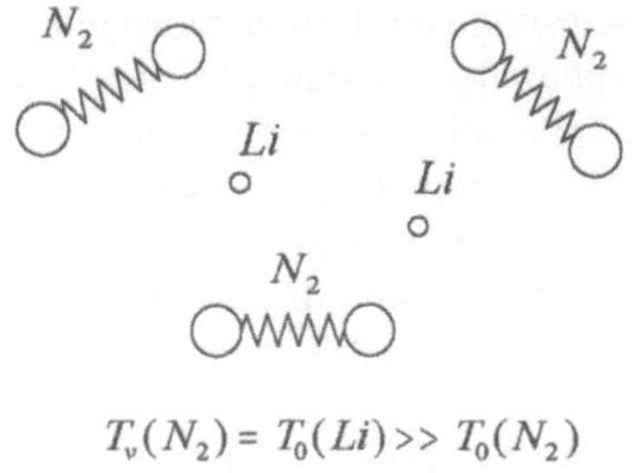

FIGURE 5.18 Equilibrium between N_2-vibrational degrees of freedom and Li-translational degrees of freedom.

Note that the visible $f(E)$ decrease with temperature T_v is possible only at $E > T_v$, which requires $E * > T_v$. Taking into account Eq. (5.136), it can be seen, that alkaline atoms can have "vibrational" temperature only if this temperature is not very high large:

$$T_v < \hbar\omega\left(\frac{M}{m}\right)^2. \tag{5.137}$$

5.5.6 "Hot Atoms" in Fast VT-Relaxation Processes, Discrete Approach, and Applications

The Fokker-Planck energy-space diffusion approximation applied above is valid only for energy intervals exceeding $\hbar\omega$. Determination of $f(E)$ on the smaller scale $E < \hbar\omega$ is more complicated and requires discrete consideration. In the most interesting conditions when $\mu << 1$ and $f(E)$ is dominated mostly by the process of VT-relaxation ($T_0 << \hbar\omega$, $m << M$), the distribution function is actually almost discrete: the only possible translational energies are: $0 \cdot \hbar\omega$, $1 \cdot \hbar\omega$, $2 \cdot \hbar\omega$, ... The Maxwellization process (TT-relaxation) and the partial energy transfer into molecular rotation make the translational distribution function more smooth. Quite sophisticated mathematical consideration of the discrete random motion of the atoms along the translational energy spectrum results in the following analytical distribution function at $E \le \hbar\omega$ (A.K. Vakar, V.K. Givotov, A. Fridman, et al. 1981):

$$f(E) = \frac{2}{\sqrt{\pi \frac{M}{m}\left(1 - \frac{E}{\hbar\omega}\right)}}\left\{1 - \Phi\left[\frac{\hbar\omega}{2T_v}\sqrt{\frac{M}{m}\left(1 - \frac{E}{\hbar\omega}\right)}\right]\right\}\exp\left[\left(\frac{\hbar\omega}{2T_v}\right)^2 \frac{M}{m}\left(\frac{E}{\hbar\omega} - 1\right)\right]. \tag{5.138}$$

Here $\Phi(x)$ is the normal distribution function. In this case, as one can see from Eq. (5.138), the translational energy distribution $f(E)$ at $E \le \hbar\omega$ has nothing in common with any kind of Maxwellian distribution functions. Hence if $T_v < \hbar\omega$, the alkaline atoms distribution cannot be described by the vibrational temperature. From the restriction Eq. (5.137), one can conclude that the temperature of the light alkaline atoms corresponds to the vibrational temperature of molecular gas only if:

$$\hbar\omega \le T_v \le \hbar\omega\left(\frac{M}{m}\right)^2. \tag{5.139}$$

Within this framework, Doppler broadening measurements of the alkaline atoms can be used to determine the vibrational temperature of the molecular gas. According to Eq. (5.139), this is possible only if $M/m >> 1$. The heavier the alkaline atoms, the larger is the difference between their temperature and the vibrational temperature of molecules. This method of diagnostics has been applied, to measure the vibrational temperature of CO_2 in a non-equilibrium microwave discharge by adding a small amount of lithium and sodium atoms (V.K. Givotov, V.D. Rusanov, A. Fridman, 1985). The Doppler broadening was observed by the Fourier-analysis of the Li-spectrum line (transition $2^2s_{1/2} - 2^2p_{1/2}$, $\lambda = 670.776$ nm) and the Na-spectrum line (transition $3^2s_{1/2} - 3^2p_{1/2}$, $\lambda = 588.995$ nm). Some results of these experiments are presented in Figure 5.19 to demonstrate the effect of "hot" atoms, related to fast VT-relaxation. Although the gas temperature in the non-thermal discharge was less than 1000 K, the alkaline temperature (Figure 5.19) was up to 10 times larger. The sodium temperature (M/m = 1.9) was always lower than the temperature of lithium atoms (M/m = 6.3). Similar measurements in non-equilibrium nitrogen plasma using in-resonator-laser-spectroscopy of Li, Na, and Cs additives were accomplished by R.A. Ahmedganov, Yu.V. Bykov, A.V. Kim and A. Fridman (1986). The relaxation-induced effect of hot atoms could be observed not only in the case of alkaline atoms additives. As discussed in Section 3.4.5, the fast non-adiabatic VT-relaxation with frequencies comparable to Maxwellization can be generated by symmetrical exchange reactions (Eq. 3.66). Even the adiabatic Landau-Teller mechanism of vibrational relaxation can provide

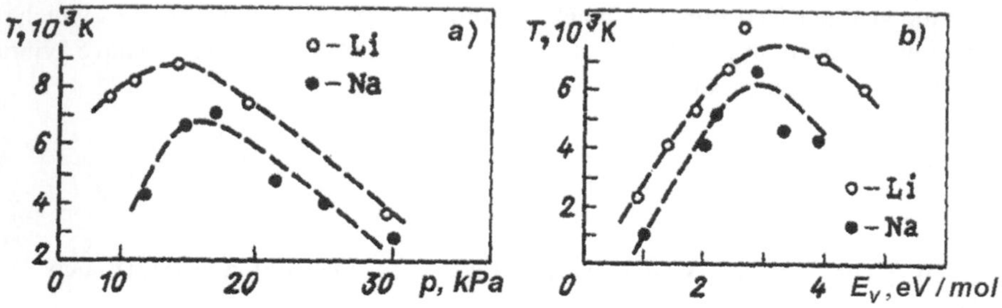

FIGURE 5.19 Li, Na – atoms temperature in CO_2 microwave discharge as function of: (a) – pressure (energy input 3 J/cm3) ; (b) – energy input (pressure 15.6 kPa).

the hot atoms in some special conditions. This effect is due to the exponential acceleration of VT-relaxation with the translational energy of atoms. Being accelerated once, a group of atoms can support their high energy in fast Landau-Teller VT-relaxation collisions because of their high velocity and low Massey-parameters.

5.5.7 "Hot Atoms" Formation in Chemical Reactions

If fast atoms generated in exothermic reactions react again before Maxwellization, they are able to significantly perturb the translational distribution function. The formed group of hot atoms is able to accelerate the exothermic chemical reactions. The fast laser-chemical chain reaction, stimulated in $H_2 - F_2$

$$H + F_2 \to HF * + F, \quad F + H_2 \to HF * + H \tag{5.140}$$

is a good example of such a generation of hot atoms in fast exothermic chemical reactions (M.E. Piley, M.K. Matzen, 1975). In non-equilibrium, plasma-chemical systems the "hot atoms" can be generated in endothermic chemical reactions as well. For example, as shown in Section 5.3.2, vibrationally excited molecules participate in the endothermic reaction with some excess of energy, which depends on the slope of the vibrational distribution function $f(E)$ and in average equals to:

$$\langle \Delta E_v \rangle = \left| \left[\frac{\partial \ln f}{\partial E} (E = E_a) \right]^{-1} \right|. \tag{5.141}$$

This energy can be large and much exceeds the value of a vibrational quantum. This situation occurs mostly in the slow reaction limit. Vibrationally excited molecules in non-thermal discharges are very effective in stimulating endothermic chemical reactions. However, exothermic elementary reactions with activation barriers are not effectively stimulated by the molecular vibrations (see Section 3.7.3) and can retard the whole plasma-chemical process. In this case, hot atoms can make a difference due to their high efficiency in stimulating of exothermic processes. Plasma-chemical NO-synthesis in non-thermal air plasma is a good relevant example. The NO-synthesis in moderate, atmospheric pressure, and non-thermal air plasma can be effectively stimulated by vibrational excitation of N_2-molecules. The synthesis proceeds following the Zeldovich mechanism including the chain reaction (V.D. Rusanov, A. Fridman, 1976)

$$O + N_2* \to NO + N, \quad E_a^{(1)} = 3.2eV, \quad \Delta H = 3.2eV, \tag{5.142}$$

$$N + O_2 \to NO + O, \quad E_a^{(2)} = 0.3eV, \quad \Delta H = -1.2eV. \tag{5.143}$$

The fast reverse reaction of NO-destruction:

$$N + NO \rightarrow N_2 + O \tag{5.144}$$

has no activation energy and as an alternative reaction of atomic nitrogen strongly competes with the direct synthesis reaction Eq. (5.144). Comparing the rates of Eqs. (5.143) and (5.144) it can be seen, that competition between these reactions restricts the concentration ratio $[NO]/[O_2]$ and hence, the degree of conversion of air into NO because the reaction Eq. (5.143) has activation energy and the reaction (5.144) does not. Numerically, this prevents formation of more than approximately 0.5%–1% of NO in non-equilibrium discharges with $T_v \approx 3000 \div 5000$ K and $T_0 \approx 600 \div 800$ K. Vibrational excitation is not effective to stimulate the reaction Eq. (5.143), but the "hot nitrogen atoms", generated in Eq. (5.142) with reasonably high energy Eq. (5.141), are able to accelerate Eq. (5.143) while influencing little the rate of the reverse reaction (T.A. Grigorieva, A.A. Levitsky, S.O. Macheret, A. Fridman, 1984). The kinetic equation for the nitrogen atoms can be solved taking into account their formation in the endothermic reactions Eq. (5.142) stimulated by vibrational excitation, Maxwellization in TT-relaxation, and their losses in chemical processes, under the simplest conditions the following distribution function is obtained:

$$f_N(E) \approx f_N^{(0)}(E) + \frac{\nu_R}{\nu_{\max}} f_{N_2}^{vib}\left(E_a^{(1)} + \frac{E}{\beta}\right). \tag{5.145}$$

Here $f_N^{(0)}(E)$ is the nitrogen atoms distribution function not perturbed by the hot atoms effect; $\nu_R/\nu_{\max}$ is a frequency ratio of Eq. (5.134) to the Maxwellization; $f_{N_2}^{vib}(E)$ is vibrational distribution function of nitrogen molecules; $E_a^{(1)}$ is activation energy of the reaction; coefficient β shows the fraction of the excess vibrational energy released in the reaction (5.134) and going to translational energy of a nitrogen atom. The interpretation of the distribution Eq. (5.145) is clear. The hot atoms are represented by the second term and actually copy the vibrational distribution function of nitrogen molecules after adjusting the activation barrier with the scaling factor β. The hot atoms are able to increase the maximum conversion of air into NO. This effect is shown in Figure 5.20 (T.A. Grigorieva, A.A. Levitsky, S.O. Macheret, A. Fridman, 1984) and becomes stronger at higher oxygen fractions. The effect of hot atoms is of special interest for plasma processes in supersonic gas flows. In these systems, the gas temperature can be so low that several secondary exothermic processes are stopped even though they have very small activation energy.

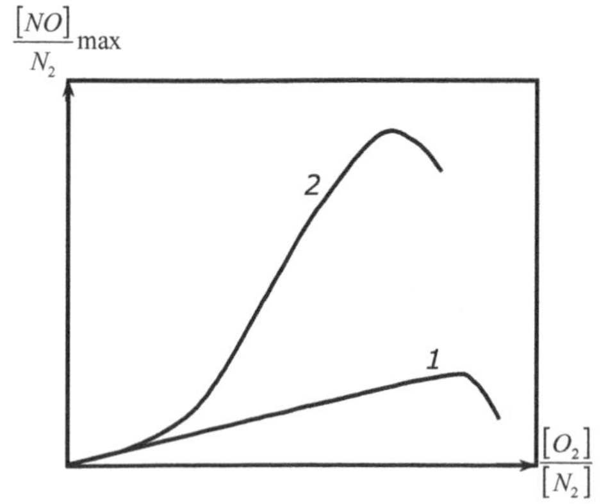

FIGURE 5.20 Maximum NO-yield at low translational temperature ($T_0 < 1500$ K) without (1) and with (2) contribution of the hot atoms effect.

5.6 ENERGY EFFICIENCY, ENERGY BALANCE AND MACROKINETICS OF PLASMA-CHEMICAL PROCESSES

Energy cost and energy efficiency of a plasma-chemical process are related to its mechanism. The same process in different discharge systems or under different conditions (corresponding to different mechanisms) requires very different expenses of energy. For example, plasma chemical purification of air from a small amount of SO_2 using pulse corona discharge requires 50–70 eV/mol. The same process stimulated under the special plasma conditions required by relativistic electron beams requires about 1 eV/mol and hence requires two orders of magnitude less of electrical energy so almost hundred times less electric energy (E.I. Baranchicov, V.P. Denisenko, V.D. Rusanov et al. 1990).

5.6.1 Energy Efficiency of Quasi-Equilibrium and Non-Equilibrium Plasma-Chemical Processes

The energy efficiency η is the ratio of the thermodynamically minimal energy cost of the plasma-chemical process (which is usually the reaction enthalpy ΔH) to the actual energy consumption W_{plasma}:

$$\eta = \Delta H \Big/ W_{plasma}. \tag{5.146}$$

The energy efficiency of the quasi-equilibrium plasma chemical systems organized in thermal discharges is usually relatively low (less than 10–20%). This is due to two major factors:

1. Thermal energy in the quasi-equilibrium plasma is distributed over all components and all degrees of freedom of the system, but most of them are useless in stimulating the plasma chemical reaction under consideration. Obviously, this makes products of chemical processes using thermal plasma relatively more expensive, if heat recuperation is not arranged.
2. If the high-temperature gas generated in a thermal plasma and containing the process products, is slowly cooled afterward, the process products will be converted back into the initial substances by the reverse reactions. Conservation of products of the quasi-equilibrium plasma-chemical processes requires applying quenching, for example, very fast cooling of the product to avoid reverse reactions. The quenching is an important factor limiting energy efficiency.

The energy efficiency of the quasi-equilibrium plasma chemical processes intrinsically depends on the type of quenching employed. The energy efficiency of the non-equilibrium plasma-chemical systems can be much higher. If correctly organized, the non-thermal plasma-chemical process' discharge power can be selectively focused on the chemical reaction of interest without heating the whole gas. Also, low bulk gas temperatures automatically provide products stability with respect to reverse reactions without any special quenching. The energy efficiency in this case strongly depends on the specific mechanism of the process.

5.6.2 Energy Efficiency of Plasma-Chemical Processes Stimulated by Vibrational Excitation of Molecules

This mechanism can provide the highest energy efficiency of endothermic plasma-chemical reactions in non-equilibrium conditions due to the following four factors:

1. The major fraction (70–95%) of the discharge power in most molecular gases (including N_2, H_2, CO, CO_2 ea) at $T_e \approx 1\,eV$ can be transferred from the plasma electrons to the vibrational excitation of molecules (see Figures 3.8–3.15).

2. The rate of VT relaxation is usually low at low gas temperatures. For this reason, the optimal choice of the degree of ionization and the specific energy input permits spending most of the vibrational energy on stimulating chemical reactions.
3. The vibrational energy of molecules is the most effective in stimulating endothermic chemical processes (see Section 3.7.3).
4. The vibrational energy necessary for an endothermic reaction is usually equal to the activation barrier of the reaction and is much less than the energy threshold of the corresponding processes proceeding through electronic excitation. For example, the dissociation of H_2 through vibrational excitation requires 4.4 eV. The same process proceeding through excitation of electronically excited state $^3\Sigma_u^+$ requires more than twice the energy: 8.8 eV (see Figure 3.1a).

Some processes, stimulated by vibrational excitation, are able to consume most of the discharge energy. Probably the best example is the dissociation of CO_2, which can be arranged in non-thermal plasma with energy efficiency up to 90% (V.K. Givotov, M.F. Krotov, V.D. Rusanov, A. Fridman, 1981; R. Asisov, A. Vakar, V.K. Givotov et al. 1985). In this case, almost all of the discharged power is spent selectively on the chemical process. No other mechanism provides such high energy efficiency.

5.6.3 Dissociation and Reactions of Electronically Excited Molecules and Their Energy Efficiency

None of the four kinetic factors listed above to positively influence the energy efficiency can be applied to plasma-chemical processes proceeding through electronic excitation. The energy cost of chemical products generated this way is relatively high. Numerically, the energy efficiency of this type of plasma-chemical processes usually does not exceed 20–30%. Plasma-chemical processes stimulated by electronic excitation can be energy effective if they initiate chain reactions. For example, such a situation takes place in NO-synthesis, where the Zeldovich mechanism (5.142), (5.143) can be effectively initiated by dissociation of molecular oxygen through electronic excitation.

5.6.4 Energy Efficiency of Plasma-Chemical Processes, Proceeding Through Dissociative Attachment

The energy threshold of the dissociative attachment is lower than the threshold of dissociation into neutrals. When the electron affinity of the products is large (e.g. some halogens and their compounds), the dissociative attachment even can be exothermic and has no energy threshold at all. This means not only a low energy price for the reaction but also the transfer of the important part of electron energy into dissociative attachment. These are very positive factors in obtaining low energy cost of products through the dissociative attachment. However, the energy efficiency of the dissociative attachment is strongly limited by the energy cost of producing an electron, which is lost during attachment and following ion-ion recombination. The energy cost of an electron can be easily calculated from Figures 3.8 to 3.15 and numerically is approximately 30–100 eV (at high electron temperatures typical for strongly electro-negative gases it's about 30 eV). Consider as an example the plasma-chemical dissociation of NF_3. The process is promoted by dissociative attachment with an energy price of 30 eV/mol, which corresponds to the energy cost of an electron in the NF_3 discharge (S. Conti, A. Fridman, S. Raoux, 1999). The energy price of dissociative attachment, which is controlled by the cost of an electron (about 30–100 eV), is obviously high and much exceeds the dissociation energy. Therefore, the plasma-chemical process based on dissociative attachment can be energy effective, if the same one electron, generated in plasma, is able to participate in the reaction many times. In such chain reactions, the detachment process and liberation of the electron from negative ion should be faster than the loss of the charged particle in ion-ion recombination. As an example of such chain process, consider the dissociation of water in a non-thermal plasma, which at high degrees of ionization can be arranged as a chain reaction (V.P. Bochin, V.A. Legasov, V.D. Rusanov, A. Fridman, 1977):

$$e + H_2O \rightarrow H^- + OH, \tag{5.147}$$

$$e + H^- \rightarrow H + e + e. \tag{5.148}$$

The first reaction in the chain is the dissociative attachment and the second is the electron impact detachment process. The rate coefficient of the detachment is fairly large and at electron temperatures of about 2 eV can reach 10^{-6} cm^3/s. The energy efficiency of this chain process of water dissociation can be relatively high (40%–50%, see Figure 5.24). However, achieving such energy efficiency requires very high values of the degree of ionization typically exceeding 10^{-4} (V.P. Bochin, V.A. Legasov, V.D. Rusanov, A. Fridman, 1979). The required high value of the degree of ionization is due to the kinetic competition of the detachment reaction (5.148) with the fast ion-molecular reaction:

$$H^- + H_2O \rightarrow H_2 + OH^-. \tag{5.149}$$

The negative hydroxyl ion OH^- produced in the ion-molecular reaction (5.149) is stable with respect to detachment and participates in fast ion-ion recombination, which in turn leads to loss of the charged particle and to chain termination.

5.6.5 Methods of Stimulation of the Vibrational-Translational Non-Equilibrium in Plasma

- *Vibrational-Translational Non-Equilibrium, Provided by High Degree of Ionization.* This is the most typical way of vibrational excitation in non-thermal plasma. It requires the vibrational excitation frequency (which is proportional to the electron concentration n_e) to be faster than VT-relaxation frequency (which is proportional to gas concentration n_0, or in some more specific cases to the concentration of neutral species the most active in the relaxation). For this reason, significant vibrational-translational non-equilibrium can be reached when the degree of ionization exceeds the critical value:

$$\frac{n_e}{n_0} >> \frac{k_{VT}^{(0)}(T_0)}{k_{eV}(T_e)}. \tag{5.150}$$

  Here $k_{eV}(T_e)$ is the rate coefficient of excitation of the lower vibrational levels in plasma by electron impact; $k_{VT}^{(0)}(T_0)$ is the VT relaxation rate coefficient from the low vibrational levels (correlated with the concentration n_0). To achieve $T_v >> T_0$ in room temperature nitrogen plasma (with $T_e \approx 1\,eV$), Eq. (5.150) requires $n_e/n_0 >> 10^{-9}$; at the same conditions in CO_2-plasma it requires $n_e/n_0 >> 10^{-6}$.
- *Vibrational-Translational Non-Equilibrium, Provided by Fast Gas Cooling.* The vibrational-translational non-equilibrium $T_v >> T_0$ can be achieved not only by an increase of vibrational temperature but also by rapidly decreasing the translational temperature. Such stimulation of the vibrational-translational non-equilibrium can be organized by using the expansion in the supersonic nozzle for the rapid cooling of the gas highly preheated in thermal plasma. This effect plays the key role in providing non-equilibrium and subsequent radiation in the gas-dynamic lasers (S.A. Losev, 1977).

The third method of stimulating the vibrational-translational non-equilibrium is related to the *fast transfer of vibrational energy in non-equilibrium systems*. This phenomenon, which is a generalization of the Treanor effect on vibrational energy transfer, is more sophisticated and will be discussed next.

5.6.6 Vibrational-Translational Non-Equilibrium, Provided by Fast Transfer of Vibrational Energy

The Treanor Effect in Vibrational Energy Transfer. Cold gas flowing around the high-temperature plasma zone provides the vibrational-translational non-equilibrium in the area of their contact (Yu.V. Kurochkin, L.S. Polak, A.V. Pustogarov et al., 1978). In particular, this effect is due to the higher rate of vibrational energy transfer from the quasi-equilibrium high-temperature zone with respect to the rate of translational energy transfer (S.V. Dobkin, E.E. Son, 1982). The possibility of fast transfer of vibrational energy in non-equilibrium conditions is illustrated in Figure 5.21. The average value of a vibrational quantum is lower because of the anharmonicity at higher vibrational temperatures. The fast VV-exchange during the transfer of the vibrational quanta makes the vibrational quanta motion from high T_v to lower T_v, more preferable than in the opposite direction. This effect can be interpreted as the domination of the vibrational energy transfer over the transfer of translational energy, and can also be interpreted as an additional VV-flux of vibrational energy (V.V. Liventsov, V.D. Rusanov, A. Fridman, 1983, 1984):

$$J_{VV} \approx -un_0\hbar\omega \iint Q_{v_2,v_2+1}^{v_1,v_1-1} \frac{2x_e\hbar\omega}{T_0}(v_2 - v_1) f_1(v_1) f_2(v_2) dv_1 dv_2. \tag{5.151}$$

In this relation u is the average thermal velocity of molecules; $f(v)$ is the vibrational distribution function expressed with respect to the number v of vibrational quanta; Q are the probabilities of VV-exchange; n_0 is the gas density; subscripts "1" and "2" are related to two planes (Figure 5.21) with the distance between them equal to the mean free path. The described phenomenon of preferable transfer of vibrational energy in non-equilibrium conditions is actually a generalization of the Treanor effect (see Figure 4.4) on the case of space transfer of vibrational energy. Integrating Eq. (5.151) using Eq. (3.74) for the probabilities Q of VV-exchange between anharmonic oscillators and taking into account only the relaxation processes close to resonance results in the VV-vibrational energy transfer flux:

$$J_{VV} = -un_0\hbar\omega \frac{2x_e\hbar\omega}{T_0} Q_{10}^{01} \left(\langle v_1^2 \rangle \langle v_2 \rangle - \langle v_1 \rangle \langle v_2^2 \rangle \right), \tag{5.152}$$

where $<v>$ and $<v^2>$ are the first and second momentum of the vibrational distribution function. The additional VV-vibrational energy flux Eq. (5.152) can be recalculated into the relative increase of the

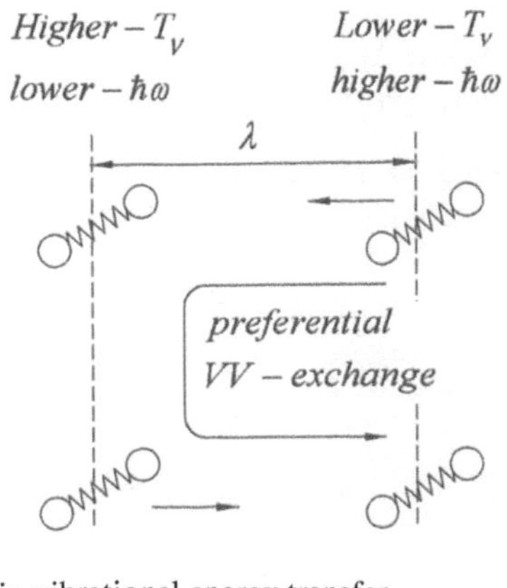

FIGURE 5.21 The Treanor effect in vibrational energy transfer.

coefficient of vibrational temperature-conductivity ΔD_v with respect to the conventional coefficient of temperature-conductivity D_0. It can be done based on Eq. (5.152), and taking the vibrational distribution as the initial part of the Treanor function (V.V. Liventsov, V.D. Rusanov, A. Fridman, 1983, 1984):

$$\frac{\Delta D_v}{D_0} \approx 4Q_{01}^{10} q \frac{1+30q+72q^2}{(1+2q)^2}, \quad q = \frac{x_e T_v^2}{T_0 \hbar\omega}. \tag{5.153}$$

As seen from Eq. (5.153), the effect is strong for molecules such as CO_2, CO, N_2O with VV-exchange provided by long-distance-forces, which results in $Q_{01}^{10} \approx 1$. As always with the Treanor effects, anharmonicity $x_e \neq 0$, $q \neq 0$ is necessary for the phenomenon to occur. In room temperature CO_2-plasma with T_v = 6000 K: the transfer of vibrational energy (D_v) according to Eq. (5.153) can exceed the transfer of translational energy (D_0) almost a hundred times. Such a large difference in coefficients of vibrational and translational energy conductivity is able to support a high level of the vibrational-translational non-equilibrium. The Treanor effect Eq. (5.153) is strong in developing of non-equilibrium only if the factor q is large beforehand. This means that some non-equilibrium occurs preliminary in a different way. Such initial vibrational-translational non-equilibrium can be due to electron impact but at relatively low degrees of ionization this can be also be provided by the following three effects:

1. *Effect of non-resonant VV-exchange*. Vibrational quantum move in a system with a temperature gradient from the higher T_v-zone (where average quantum is lower because of anharmonicity) to the lower T_v-zone (where average quantum is larger). During the vibrational thermal conductivity, the vibrational quanta have a tendency to grow, taking a defect of energy from the translational degrees of freedom in the non-resonant VV-relaxation. Even in quasi-equilibrium systems, the vibrational conductivity leads to some small but notable gas cooling ΔT and, hence, some vibrational-translational non-equilibrium on the level:

$$\frac{\Delta T_o}{T_0} \approx \frac{T_v - T_0}{T_v} \approx x_e \frac{T}{\hbar\omega}. \tag{5.153}$$

2. *Recombination Effect.* Diffusion of atoms and radicals from the high temperature zone of thermal discharges to the low temperature zone results in recombination and hence, an over-equilibrium concentration of vibrationally excited molecules.
3. *Specific heat effect.* Mixing of quasi-equilibrium hot and cold gases averages their energies and translational temperatures, but not their vibrational temperatures (because of the non-linear dependence of vibrational energy on temperature, and growth of specific heat with temperature). The growth of the vibrational specific heat with temperature results in vibrational-translational non-equilibrium $T_v > T_0$ with a temperature difference of about $1/2\,\hbar\omega$.

5.6.7 Energy Balance and Energy Efficiency of Plasma-Chemical Processes, Stimulated by Vibrational Excitation of Molecules

The vibrational energy balance in the plasma-chemical process can be illustrated by the simplified, one-component equation taking into account vibrational excitation by electron impact, VT-relaxation, and chemical reaction:

$$\frac{d\varepsilon_v(T_v)}{dt} = k_{eV}(T_e) n_e \hbar\omega \theta\left(E_v - \int_0^t k_{eV} n_e \hbar\omega dt\right) - k_{VT}^{(0)}(T_0) n_0 \left[\varepsilon_v(T_v) - \varepsilon_{v0}(T_v = T_0)\right] - k_R(T_V, T_0) n_0 \Delta Q_\Sigma. \tag{5.154}$$

The left side of this equation represents the change of the vibrational energy per molecule, which in the one mode approximation and $x_e T_v^2 / T_0 \hbar\omega < 1$ can be expressed by the Planck formula:

$$\varepsilon_v(T_v) = \frac{\hbar\omega}{\exp\left(\frac{\hbar\omega}{T_v}\right) - 1}. \tag{5.155}$$

The first term on the right side of Eq. (5.154) describes the vibrational excitation (rate coefficient k_{eV}, density n_e). The Heaviside function in this term $\theta(x)$: $[\theta = 1, if\, x \geq 0;\ \theta = 0, if\, x < 0]$ restricts the vibrational energy input in the discharge per one molecule by the value of E_v. If most of the discharge power is going into vibrational excitation of molecules, the specific energy input E_v per one molecule is equal to the ratio of the discharge power over the gas flow rate through the discharge. It is related with the energy efficiency (η) and the degree of conversion (χ):

$$\eta = \frac{\chi \times \Delta H}{E_v}. \tag{5.156}$$

where ΔH is the enthalpy of the plasma-chemical reaction. The second term on the right side of Eq. (5.154) describes the VT-relaxation from lower vibrational levels; n_0 is the gas density. The third term on the right-side is related to vibrational energy losses in a chemical reaction (reaction rate $k_R(T_v, T_0)$) and the relaxation channels having rates corresponding to those of the chemical reaction. In this term, the total losses of ΔQ_Σ vibrational energy per one act of chemical reaction are:

$$\Delta Q_\Sigma = E_a/\alpha + \langle \Delta E_v \rangle + \Delta\varepsilon_{VT} + \Delta\varepsilon_R^{VV}. \tag{5.157}$$

Here E_a/α is the vibrational energy necessary to overcome the activation barrier (α is the efficiency of vibrational energy, see Section 3.7.3); $<\Delta E_v>$ characterizes the excess of vibrational energy in chemical reaction Eq. 5.141); $\Delta\varepsilon_{VT}$ and $\Delta\varepsilon_R^{VV}$ are losses of vibrational energy due respectively to VT-relaxation from high levels and to non-resonance of VV-exchange, expressed with respect to one act of chemical reaction (Eqs. (5.87) and (5.90)).

5.6.8 Energy Efficiency as a Function of Specific Energy Input and Degree of Ionization

The reaction rate coefficient depends more strongly on the vibrational temperature T_v than all other terms in the balance equation (5.154). In the case of weak excitation, the rate coefficient can be estimated as:

$$k_R(T_v, T_0) = A(T_0)\exp\left(-\frac{E_a}{T_v}\right). \tag{5.158}$$

The pre-exponential factor $A(T_0)$ differs from the conventional Arrhenius one A_{Arr}. For example, in the case of the Treanor distribution function: $A(T_0) = A_{Arr}\exp\left(\frac{x_e E_a^2}{T_0 \hbar\omega}\right)$. Based on Eq. (5.154), the critical value of vibrational temperature, when the reaction and VT-relaxation rates are equal is given by:

$$T_v^{\min} = E_a \ln^{-1} \frac{A(T_0)\Delta Q_\Sigma}{k_{VT}^{(0)}(T_0)\hbar\omega}. \tag{5.159}$$

An increase of vibrational temperature results in much stronger exponential acceleration of the chemical reaction with respect to vibrational relaxation. This means that at high vibrational temperatures $T_v > T_v^{\min}$ almost all the vibrational energy is going to chemical reaction according to Eq. (5.154). Conversely, at lower vibrational temperatures $T_v < T_v^{\min}$ almost all vibrational energy can be lost in vibrational relaxation. The critical vibrational temperature $T_v^{\min}$ determines the threshold dependence of the energy efficiency of plasma-chemical processes $\eta(E_v)$, stimulated by vibrational excitation, on the specific energy input E_v (see Eq. (5.154)):

$$\left(E_v\right)_{threshold} = \varepsilon_v\left(T_v^{\min}\right) = \frac{\hbar\omega}{\exp\left(\dfrac{\hbar\omega}{T_v^{\min}}\right) - 1}. \tag{5.160}$$

If the specific energy input is lower than the threshold value $E_v < (E_v)_{threshold}$, the vibrational energy cannot reach the critical value $T_v < T_v^{\min}$ and the energy efficiency is very low. The typical threshold values of energy input for plasma-chemical reactions stimulated by vibrational excitation are approximately 0.1 – 0.2 eV/mol. The threshold in the dependence $\eta(E_v)$ is a very important qualitative feature of the plasma-chemical processes stimulated by vibrational excitation. For example, to avoid intensive dissociation of carbon dioxide in a gas-discharge CO_2-laser, the specific energy input is always limited by the threshold value. The dependences $\eta(E_v)$ are shown in Figures 5.22, 5.23 and 5.24 for specific plasma-chemical processes of dissociation of CO_2, H_2O and NO-synthesis, stimulated by vibrational excitation (V.D. Rusanov, A. Fridman, 1984). Another important feature of the plasma-chemical processes stimulated by vibrational excitation, which can be observed in Figures 5.22, 5.23 and 5.24, is

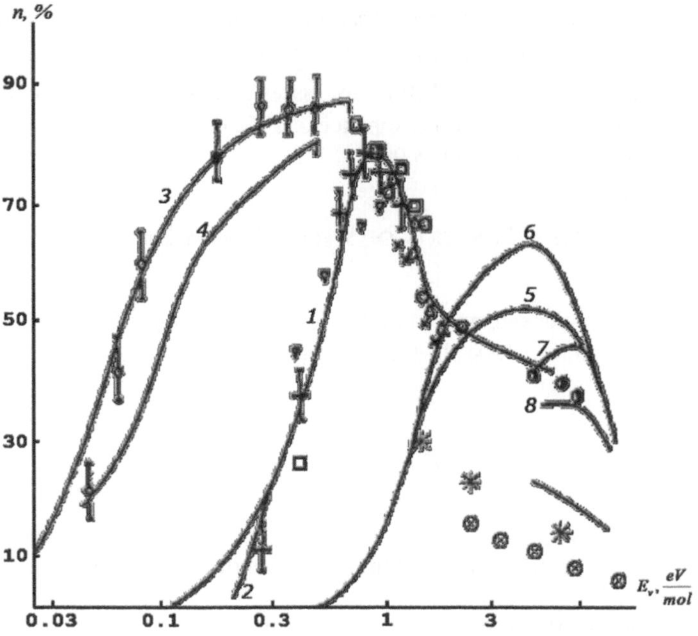

FIGURE 5.22 Energy efficiency of CO_2-dissociation as a function of energy input: 1,2 – non-equilibrium calculations in 1-T_V and 2-T_V approximations; non-equilibrium calculations for supersonic flows: 3 – M = 5, 4 – M = 3.5; calculations of thermal dissociation with ideal (5) and super-ideal (6) quenching; thermal dissociation with quenching rates 10^9 K/sec (7), 10^8 K/sec (8), 10^7 K/sec (9). Different experiments in microwave discharges: ⌖, ⫯, Δ, x. Experiments in supersonic microwave discharges: ⌖. Experiments in different RF-CCP discharges: O, ∇. Experiments in RF-ICP discharges: ◑. Experiments in different arc discharges: ⊗, *.

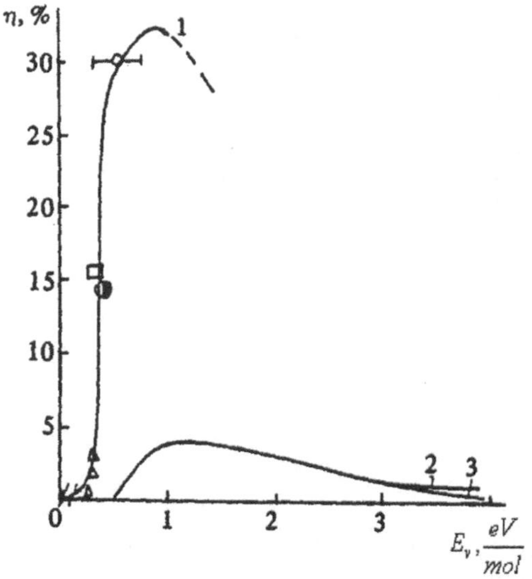

FIGURE 5.23 Energy efficiency of NO-synthesis in air as a function of energy input: 1 – non-equilibrium process stimulated by vibrational excitation; 2,3 – thermal synthesis with ideal (2) and absolute (3) quenching. Experiments with microwave discharges: ⊢○⊣, ⌇; with discharges sustained by electron beams: ◑, Δ, ○.

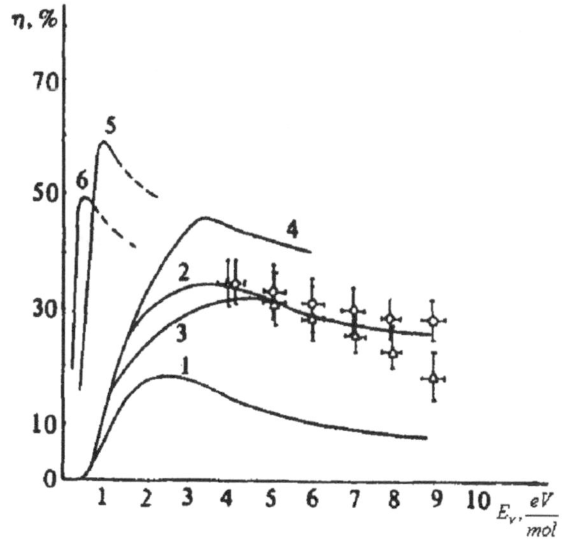

FIGURE 5.24 Energy efficiency of water vapor dissociation as a function of energy input: 1,2 – thermal dissociation with absolute and ideal quenching; 3,4 – super-absolute and super-ideal quenching mechanisms; 5-non-equilibrium decomposition due to dissociative attachment; 6-non-equilibrium dissociation stimulated by vibrational excitation. Different experiments in microwave discharges: ⊢○⊣, ⊢Δ⊣.

the maximum in the dependence $\eta(E_v)$. The optimal value of the specific energy input is usually about 1 eV. At larger values of the specific energy input, the translational temperature increases, which accelerates VT relaxation and decreases the energy efficiency. Also at large energy inputs E_v, the major fraction of the discharge energy can be spent on useless excitation of products. Even if the plasma-chemical process degree of conversion χ approaches 100% at high energy inputs, the energy efficiency (see (5.156) decreases hyperbolically with the energy input $\eta \propto \Delta H/E_v$. If the vibrational temperature exceeds the critical one $T_v > T_v^{min}$, the relaxation losses in Eq. (5.154) can be neglected

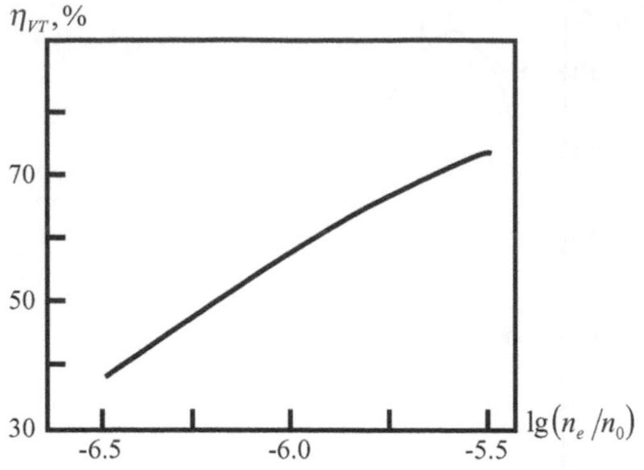

FIGURE 5.25 Dependance of CO_2-dissociation energy efficiency on ionization degree, $E_v = 0.5$ eV/mol.

with almost entire energy going through vibrational excitation into the chemical reaction. The steady-state value of vibrational temperature then can be found from the balance equation (5.154) taking into account Eq. (5.158) as:

$$T_v^{st} = E_a \ln^{-1}\left(\frac{A(T_0)\Delta Q_\Sigma}{k_{eV}\hbar\omega} \cdot \frac{n_0}{n_e}\right). \tag{5.161}$$

This steady-state vibrational temperature exceeds the critical temperature Eq. (5.159) when the electron concentration exceeds the corresponding critical value Eq. (5.150). In general, the energy efficiency of plasma-chemical processes stimulated by vibrational excitation grows with the degree of ionization. Example of the dependence $\eta(n_e/n_0)$ is presented in Figure 5.25 for the plasma-chemical process of CO_2 dissociation stimulated by vibrational excitation.

5.6.9 Components of the Total Energy Efficiency: Excitation, Relaxation, and Chemical Factors

The total energy efficiency of any plasma-chemical process can be subdivided into the three main components: 1) excitation factor (η_{ex}), 2) relaxation factor (η_{rel}) and 3) chemical factor (η_{chem}):

$$\eta = \eta_{ex} \times \eta_{rel} \times \eta_{chem}. \tag{5.162}$$

The excitation factor shows the fraction of discharge energy directed to producing the principal agent of the plasma-chemical reaction. For example, in the plasma-chemical process of NF_3 dissociation for cleaning the CVD chamb er (S. Conti, A. Fridman, S. Raoux, 1999), this factor shows the efficiency of generation of F-atoms responsible for cleaning. In plasma-chemical processes stimulated by vibrational excitation, the excitation factor (η_{ex}) illustrates the fraction of electron energy (or discharge energy, which is often the same), going to vibrational excitation of molecules. This factor depends on the electron temperature and gas composition; at $T_e \approx 1\,eV$ it can reach 90% and above.

The relaxation factor (η_{rel}) is the efficiency, related to conservation of the principal active species (like the F-atoms or vibrationally excited molecules in the above examples) with respect to their losses in relaxation, recombination, etc. In plasma-chemical processes stimulated by vibrational excitation, the factor η_{rel} shows the fraction of vibrational energy which can avoid vibrational relaxation and be spent in the chemical reaction of interest. Taking into account that vibrational energy losses into VT-relaxation are small inside of discharge zone if $T_v > T_v^{\min}$, we can express the factor η_{rel} as a function of specific energy input E_v near the threshold value (5.160):

$$\eta_{rel} = \frac{E_v - \varepsilon_v\left(T_v^{\min}\right)}{E_v}. \tag{5.163}$$

Equation (5.163) includes the threshold of the dependence $\eta(E_v)$ (see Figures 5.22, 5.23, and 5.24), and again shows that when the vibrational temperature finally becomes lower than the critical one, most of the vibrational energy going to VT-relaxation come from lower levels. The relaxation factor (η_{rel}) strongly depends on the temperature in accordance with the Landau-Teller theory (see Section 3.4). For example, at very low translational temperatures $T_0 \approx 100$ K in a supersonic microwave discharge, the relaxation factor can be very high (more than 97%, V.D. Rusanov, A. Fridman, G. V. Sholin, 1982).

The chemical factor (η_{chem}) is a component of Eq. (5.162), which shows the efficiency of the application of the principal active discharge species in a chemical reaction. In the example of NF_3 dissociation and F-atoms generation for cleaning CVD-chambers, the chemical factor (η_{chem}) shows the efficiency of atomic fluorine in the gasification of silicon and its solid compounds. In reactions stimulated by vibrational excitation, this factor based on the balanced equation can be presented as:

$$\eta_{chem} = \frac{\Delta H}{E_a/\alpha + \langle \Delta E_v \rangle + \Delta\varepsilon_{VT} + \Delta\varepsilon_R^{VV}}. \tag{5.164}$$

The chemical factor restricts the energy efficiency mainly because the activation energy quite often exceeds the reaction enthalpy: $E_a > \Delta H$. For example, in the plasma-chemical process of NO-synthesis, proceeding by the Zeldovich mechanism Eqs. (5.142)–(5.143), the second reaction is significantly exothermic. It results in a low value of the chemical factor $\eta_{chem} \approx 50\%$ (R.I. Asisov, V.K. Givotov, V.D. Rusanov, A. Fridman, 1980) and restricts the total energy efficiency, see Figure 5.23.

5.7 ENERGY EFFICIENCY OF QUASI-EQUILIBRIUM PLASMA-CHEMICAL SYSTEMS, ABSOLUTE, IDEAL AND SUPER-IDEAL QUENCHING

5.7.1 Concepts of Absolute, Ideal, and Super-Ideal Quenching

Thermal plasma-chemical processes can be subdivided into two phases. In the first one, reagents are heated to high temperatures necessary to shift the chemical equilibrium in direction of products. In the second phase called quenching, temperature decreases fast to protect the products produced on the first high-temperature phase from reverse reactions. Absolute quenching means that the cooling process is sufficiently fast to save all products formed in the high-temperature zone. In the high-temperature zone, initial reagents are partially converted into the products of the process, but also into some unstable atoms and radicals. In the case of absolute quenching, stable products are saved, but atoms and radicals during the cooling process are converted back into the initial reagents. Ideal quenching means that the cooling process is very effective and able to maintain the total degree of conversion on the same level as was reached in the high-temperature zone. The ideal quenching not only saves all the products formed in the high-temperature zone, but also provides conversion of all the relevant atoms and radicals into the process products. It is interesting that during the quenching phase, the total degree of conversion can be not only saved, but even increased. Such quenching is usually referred to as the super-ideal one. Super-ideal quenching permits increasing the degree of conversion during the cooling stage, using the chemical energy of atoms and radicals as well as the excitation energy accumulated in molecules. In particular, the super-ideal quenching can be organized when the gas cooling is faster than VT relaxation, and the vibrational-translational, non-equilibrium $T_v > T_0$ can be achieved during the quenching. In this case, direct endothermic reactions are stimulated by vibrational excitation, while reverse exothermic reactions related to translational degrees of freedom proceeds slower. Such an unbalance between direct and reverse reactions provides additional conversion of the initial substances, or in other words, provides super-ideal quenching.

5.7.2 Ideal Quenching of CO_2-Dissociation Products in Thermal Plasma

Thermodynamically equilibrium composition of the CO_2-dissociation products in thermal plasma depends only on temperature and pressure. This temperature dependence, which is much stronger than the pressure dependence, is presented in Figure 5.26. The main products of the plasma-chemical process under consideration:

$$CO_2 \rightarrow CO + \frac{1}{2}O_2, \quad \Delta H_{CO_2} = 2.9eV \tag{5.165}$$

are the saturated molecules CO and O_2. Also, the high-temperature heating provide a significant concentration of atomic oxygen and also atomic carbon at very high T_0. If quenching of the products is not sufficiently fast, the slow cooling could be quasi-equilibrium and reverse reactions would return the composition to the initial one, for example, pure CO_2. Fast cooling rates (>10^8 K/s) permits saving CO in the products (A.A. Levitsky, L.S. Polak, I.M. Rytova, D.I. Slovetsy, 1981). In this case, atomic oxygen recombines $O + O + M \rightarrow O_2 + M$ faster than reacting with carbon monoxide $O + CO + M \rightarrow CO_2 + M$, which maintains the CO_2 degree of conversion and hence, the conditions for the ideal quenching. In this case, this is the same as the absolute quenching. Energy efficiency of the quasi-equilibrium process with ideal quenching can be calculated in this case as follows. First, the energy input per one initial CO_2-molecule required to heat the dissociating CO_2 at constant pressure to temperature T can be expressed as:

$$\Delta W_{CO_2} = \frac{\Sigma x_i I_i(T)}{x_{CO_2} + x_{CO}} - I_{CO_2}(T = 300K). \tag{5.166}$$

In this relation $x_i(p, T) = n_i/n$ is the quasi-equilibrium relative concentration of "i" component of the mixture, $I_i(T)$ is the total enthalpy of the component. If β is the number of CO-molecules

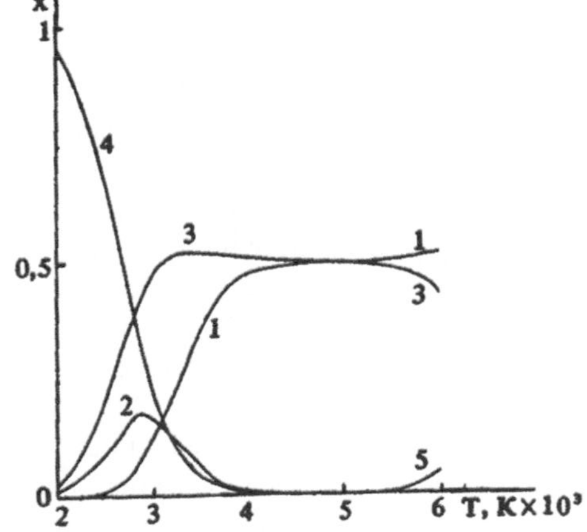

FIGURE 5.26 Equilibrium composition for CO_2 thermal decomposition (p = 0.16 atm): 1 – O, 2 – O_2, 3 – CO, 4 – CO_2, 5 – C.

produced from one initial CO_2-molecule taking into account quenching, then the total energy efficiency of the quasi-equilibrium plasma-chemical process can be given as:

$$\eta = \frac{\Delta H_{CO_2}}{\left(\Delta W_{CO_2}/\beta\right)}. \tag{5.167}$$

The conversion is equal to $\beta^0 = \frac{x_{CO}}{x_{CO_2} + x_{CO}}$ in the case of ideal quenching, which leads to the final expression of energy efficiency:

$$\eta = \frac{\Delta H_{CO_2} x_{CO}}{\Sigma x_i I_i(T) - (x_{CO_2} + x_{CO}) I_{CO_2}(T = 300K)}. \tag{5.168}$$

The energy efficiency of CO_2-dissociation in plasma with ideal quenching is presented in Figure 5.27. Maximum energy efficiency 50% is reached at T = 2900 K. The energy cost of CO-production in quasi-equilibrium p CO_2 plasma is shown in Figure 5.28 as a function of the cooling rate (L.S. Polak, D.I. Slovetsky, Yu.P. Butylkin, 1977). High cooling rate of 10^8 K/s is required for the ideal quenching.

5.7.3 Non-Equilibrium Effects during Product Cooling, Super-Ideal Quenching

Vibrational-translational non-equilibrium effects during cooling can provide the additional conversion. In CO_2 dissociation, this effect of super-ideal quenching is related to shifting equilibrium of the reaction:

$$O + CO_2 \Leftrightarrow CO + O_2, \quad \Delta H = 0.34eV. \tag{5.169}$$

This direct endothermic reaction of CO-formation can be effectively stimulated by vibrational excitation of CO_2-molecules, and not balanced by reverse exothermic reaction at $T_v > T_0$. The energy efficiency of the thermal CO_2-dissociation with super-ideal quenching can be found in this case as:

$$\eta = \frac{\Delta H_{CO_2}}{\Delta W_{CO_2}} \frac{x_{CO} + x_O}{x_{CO_2} + x_{CO}}, \quad if \ x_{CO_2} > x_O, \tag{5.170}$$

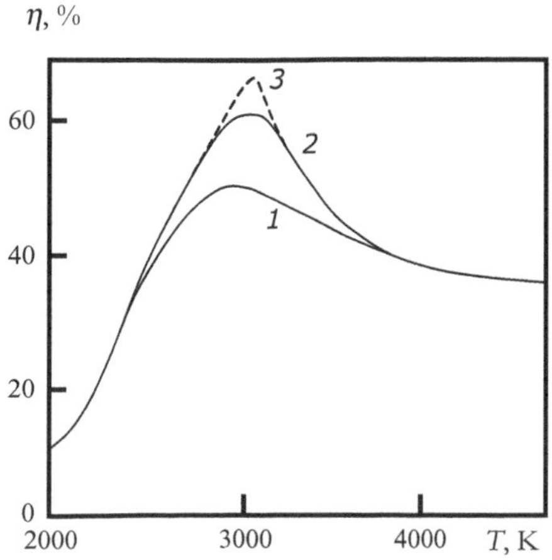

FIGURE 5.27 Energy efficiency of CO_2 thermal dissociation as a function of heating temperature, p = 0.16 atm: 1 – ideal quenching; 2 – super-ideal quenching; 3 – upper limit of the super-ideal quenching.

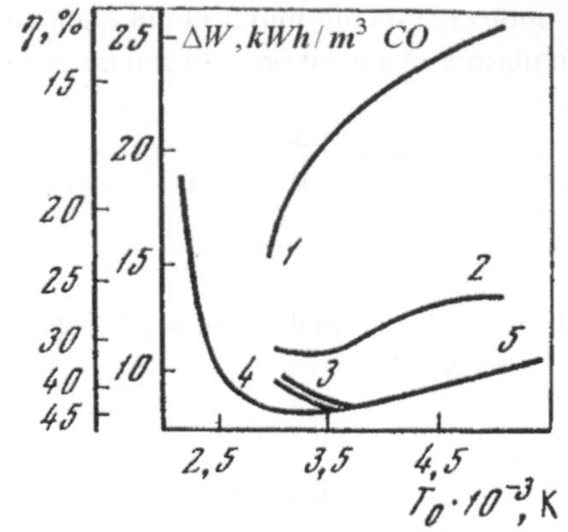

FIGURE 5.28 Dependence of energy efficiency η and energy cost ΔW of CO_2 thermal dissociation on heating temperature at p = 1 atm and different quenching rates: 1 – 10^6 K/sec, 2 – 10^7 K/sec, 3 – 10^8 K/sec, 4 – 10^9 K/sec, 5 – instantaneous cooling.

$$\eta = \frac{\Delta H_{CO_2}}{\Delta W_{CO_2}}, \quad \text{if } x_{CO_2} < x_O. \tag{5.171}$$

The energy efficiency of super-ideal quenching $\eta(T)$as a function of temperature is shown in Figure 5.27. The maximum efficiency $\eta = 64\%$ is reached at a temperature in the hot zone about T = 3000 K. This efficiency is 14% higher than the maximum one for ideal quenching (but still much less than that of the pure non-equilibrium process, see Figure 5.25). Detailed analysis of the super-ideal quenching under different conditions can be found in B.V. Potapkin, V.D. Rusanov, A.A. Fridman (1984).

5.7.4 Mechanisms of Absolute and Ideal Quenching for H_2O-Dissociation in Thermal Plasma

Composition of the H_2O-dissociation products in thermal plasma is presented in Figure 5.29 as a function of temperature. The main products of the plasma-chemical process under consideration:

$$H_2O \rightarrow H_2 + \frac{1}{2}O_2, \quad \Delta H_{H_2O} = 2.6eV \tag{5.172}$$

are the saturated molecules H_2 and O_2. In contrast to the case of CO_2, thermal dissociation of water vapor results in a variety of atoms and radicals. As seen in Figure 5.29., concentrations of O, H, and OH are significant in this process. Even when H_2 and O_2 initially formed in the high-temperature zone are saved from reverse reactions, the active species O, H, and OH can be converted either into products (H_2 and O_2) or back into H_2O. This qualitatively different behavior of radicals determines the key difference between the absolute and ideal mechanisms of quenching. In the case of the absolute quenching, when the active species are converted back to H_2O, the energy efficiency of the process (5.172) of hydrogen production is:

$$\eta = \frac{\Delta H_{H_2O}}{\left(\Delta W_{H_2O} / \beta_{H_2^0}\right)}. \tag{5.173}$$

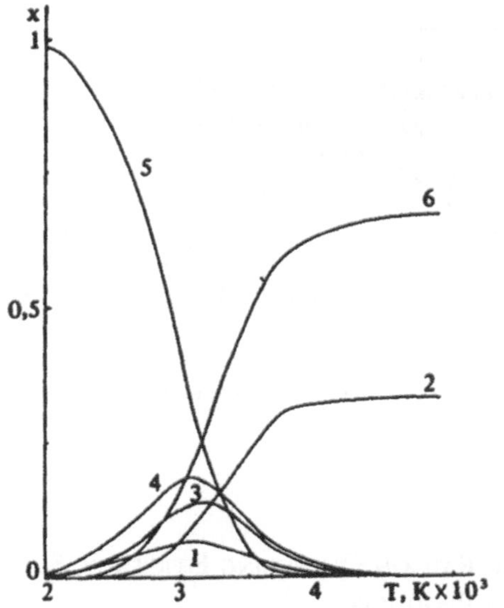

FIGURE 5.29 Equilibrium composition for H_2O thermal decomposition: 1 – O_2, 2 – O, 3 – OH, 4 – H_2, 5 – H_2O, 6 – H.

Here $\beta^0_{H_2}(T)$ is the conversion of water into hydrogen. In other words, $\beta^0_{H_2}(T)$ is the number of hydrogen molecules formed in the quasi-equilibrium phase calculated per one initial H_2O-molecule:

$$\beta^0_{H_2} = \frac{x_{H_2}}{x_{H_2O} + x_O + 2x_{O_2} + x_{OH}}. \tag{5.174}$$

The discharge energy input per one initial H_2O-molecule to heat the dissociating water at constant pressure to temperature T can be expressed similarly to Eq. (5.166) as:

$$\Delta W_{H_2O} = \frac{\Sigma x_i I_i(T)}{x_{H_2O} + x_O + 2x_{O_2} + x_{OH}} - I_{H_2O}(T = 300K). \tag{5.175}$$

Based on Eqs. (5.173, 5.174, 5.175), the formula for the energy efficiency of dissociating water and hydrogen production in thermal plasma with absolute quenching can be presented as:

$$\eta = \frac{\Delta H_{H_2O} x_{H_2}}{\Sigma x_i I_i(T) - (x_{H_2O} + x_O + 2x_{O_2} + x_{OH}) I_{H_2O}(T = 300K)}. \tag{5.176}$$

The energy efficiency of the thermal water dissociation in case of ideal quenching can be calculated in a similar way, taking into account the additional conversion of active species, H, OH, and O into product of the process (H_2 and O_2):

$$\eta = \Delta H_{H_2O}\left[\frac{\Sigma x_i I_i(T)}{x_{H_2} + 1/2(x_H + x_{OH})} - \frac{x_{H_2O} + x_O + 2x_{O_2} + x_{OH}}{x_{H_2} + 1/2(x_H + x_{OH})} I_{H_2O}(T = 300K)\right]^{-1} \tag{5.177}$$

The energy efficiencies of water dissociation in thermal plasma for absolute and ideal quenching are presented in Figure 5.30. From the figure, it is seen that complete usage of atoms and radicals to form products (ideal quenching) permit increasing energy efficiency by almost a factor of two.

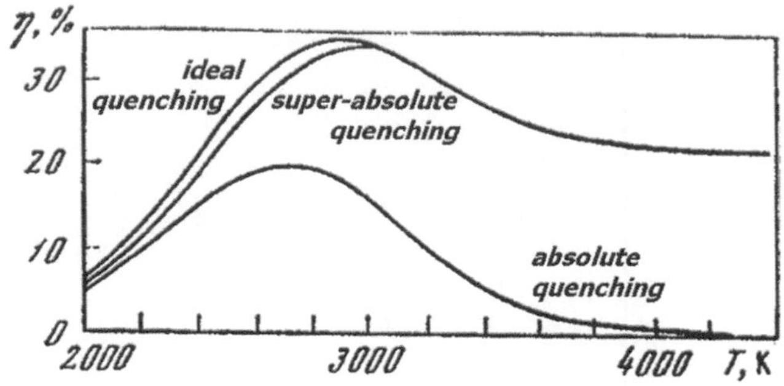

FIGURE 5.30 Energy efficiency of H_2O thermal dissociation as function of heating temperature, p = 0.05 atm.

5.7.5 Effect of Cooling Rate on Quenching Efficiency, Super-Ideal Quenching of H_2O-Dissociation Products

The slow cooling shifts the equilibrium between direct and reverse reactions in the exothermic direction of the destruction of molecular hydrogen:

$$O + H_2 \rightarrow OH + H, \tag{5.178}$$

$$OH + H_2 \rightarrow H_2O + H. \tag{5.179}$$

The following three-body recombination leads to the formation of water molecules:

$$H + OH + M \rightarrow H_2O + M, \tag{5.180}$$

and its reaction rate coefficient is about 30-times larger than the rate coefficient of the alternative three-body recombination with the formation of molecular hydrogen:

$$H + H + M \rightarrow H_2 + M. \tag{5.181}$$

The reactions (5.178, 5.179, 5.180, 5.181) explain the destruction of hydrogen and the reformation of water molecules during the slow cooling of the water dissociation products. If the cooling rate is sufficiently high ($>10^7$ K/s), the reactions of O and OH with saturated molecules very soon become less effective. This provides conditions for absolute quenching. Instead of participating in processes (5.178) and (5.179), the active species OH and O react to form mostly atomic hydrogen:

$$OH + OH \rightarrow O + H_2O, \tag{5.182}$$

$$O + OH \rightarrow O_2 + H. \tag{5.183}$$

The atomic hydrogen (being in excess with respect to OH) recombines in molecular form (5.181), providing at higher cooling rates ($>2 \cdot 10^7$ K/s) the additional conversion of radicals into the stable process products. This means that the super-absolute (almost ideal, see Figure 5.30) quenching can be achieved at these cooling rates. Energy efficiency of thermal water dissociation and hydrogen production as a function of the cooling rate is presented in Figure 5.31 (V.K. Givotov, S.Yu. Malkov, V.D. Rusanov, A. Fridman, 1983). Vibrational-translational non-equilibrium during fast cooling of the water dissociation products can lead to the surer-ideal quenching effect in a similar way as was described in Section 5.7.3 regarding CO_2-dissociation. The super-ideal quenching effect can be related in this case to the shift of quasi-equilibrium (at $T_v > T_0$) of the reactions:

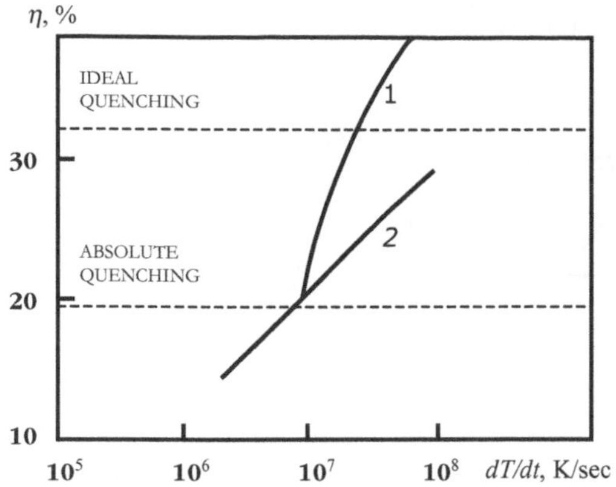

FIGURE 5.31 Energy efficiency of H_2O thermal dissociation as a function of quenching rate, heating temperature 2800 K, p = 0.05 atm: 1 – vibrational-translational non-equilibrium, 2 – vibrational-translational equilibrium.

$$H + H_2O \Leftrightarrow H_2 + OH, \quad \Delta H = 0.6eV. \tag{5.184}$$

The direct endothermic reaction of hydrogen production is stimulated here by vibrational excitation, while the reverse reaction stays relatively slow. The energy efficiency of water dissociation in such non-equilibrium quenching conditions are presented in Figure 5.31 as a function of cooling rate (V.K. Givotov, S.Yu. Malkov, B.V. Potapkin et al., 1984). From this figure, the super-ideal quenching of the water dissociation products requires cooling rates exceeding $5 \cdot 10^7$ K/s and provides energy efficiency up to approximately 45%.

5.7.6 Mass and Energy Transfer Equations in Multi-Component Quasi-Equilibrium Plasma-Chemical Systems

To analyze the influence of transfer phenomena on the energy efficiency of plasma-chemical processes begin with some general aspects of mass and energy transfer. The conservation equations, describing enthalpy (I) and mass transfer in multi-component (number of component N) quasi-equilibrium reacting gas are similar in thermal plasma and combustion systems and can be presented (F.A. Williams, 1985; D.A. Frank-Kamenetsky, 1971) as:

- Conservation equation for total enthalpy:

$$\rho \frac{dI}{dt} = -\nabla \vec{q} + \frac{dp}{dt} + \Pi_{ik} \frac{\partial v_i}{\partial x_k} + \rho \sum_{\alpha=1}^{N} Y_\alpha \overrightarrow{v_\alpha} \overrightarrow{f_\alpha}. \tag{5.185}$$

- Continuity equation for chemical components:

$$\rho \frac{dY_\alpha}{dt} = -\nabla \left(\rho Y_\alpha \overrightarrow{v_\alpha} \right) + \omega_\alpha. \tag{5.186}$$

The thermal flux $\vec{q}$ in Eq. 5.185) neglecting radiation heat transfer can be expressed as:

$$\vec{q} = -\lambda \nabla T + \rho \sum_{\alpha=1}^{N} I_\alpha Y_\alpha \overrightarrow{v_\alpha}. \tag{5.187}$$

The diffusion velocity $\overrightarrow{v_\alpha}$ for α-chemical component can be found from the following relation:

$$\nabla x_\alpha = \sum_{\beta=1}^{N} \frac{x_\alpha x_\beta}{D_{\alpha\beta}} \left(\overrightarrow{v_\beta} - \overrightarrow{v_\alpha}\right) + \left(Y_\alpha - x_\alpha\right)\frac{\nabla p}{p} + \frac{\rho}{p}\sum_{\beta=1}^{N} Y_\alpha Y_\beta \left(\overrightarrow{f_\alpha} - \overrightarrow{f_\beta}\right)$$
$$+ \sum_{\beta=1}^{N}\left[\frac{x_\alpha x_\beta}{D_{\alpha\beta}} \frac{1}{\rho}\left(\frac{D_{T,\beta}}{Y_\beta} - \frac{D_{T,\alpha}}{Y_\alpha}\right)\right]\frac{\nabla T}{T}. \quad (5.188)$$

In the above relations: $I = \sum_{\alpha=1}^{N} Y_\alpha I_\alpha$ is the total enthalpy per unit mass of the mixture including the enthalpy of formation; Π_{ik} is the tensor of viscosity; $\overrightarrow{f_\alpha}$ is external force per unit mass of the component α; ω_α is the rate of mass change of the component α per unit volume as a result of chemical reactions; λ, $D_{\alpha\beta}$, $D_{T,\alpha}$ are the coefficients of thermal conductivity, binary diffusion, and thermodiffusion respectively; $x_\alpha = n_\alpha/n$, $Y_\alpha = \rho_\alpha/\rho$ are the molar and mass fractions of the component α; v_i ith component of hydro-dynamic velocity; p is the pressure of the gas mixture. To simplify Eqs. (5.185, 5.186, 5.187, 5.188) neglect the total enthalpy change due to viscosity; assume the forces $\overrightarrow{f_\alpha}$ the same for all components; and take the binary diffusion coefficients as $D_{\alpha\beta} = D(1 + \delta_{\alpha\beta})$, $\delta_{\alpha\beta} < 1$. After such simplification, Eqs. (5.185, 5.186, 5.187, 5.188), describing the energy and mass transfer, can be rewritten as:

$$\rho\frac{dI}{dt} = -\nabla\vec{q} + \frac{dp}{dt}, \quad (5.189)$$

$$\vec{q} = -\frac{\lambda}{c_p}\nabla I + (1 - Le)\sum_{\alpha=1}^{N} I_\alpha D \rho_\alpha \nabla \ln Y_\alpha + \sum_{\alpha=1}^{N}\rho_\alpha I_\alpha \overrightarrow{v_\alpha^c} + \sum_{\alpha=1}^{N}\rho_\alpha I_\alpha \overrightarrow{v_\alpha^g}, \quad (5.190)$$

$$\overrightarrow{v_\alpha} = -D\nabla \ln Y_\alpha + \overrightarrow{v_\alpha^c} + \overrightarrow{v_\alpha^g}, \quad (5.191)$$

plus the continuity Eq. (5.186). $Le = \frac{\lambda}{\rho c_p D}$ is the Lewis number, and $c_p = \sum_{\alpha=1}^{N} Y_\alpha \frac{\partial I_\alpha}{\partial T}$ is the specific heat of the unit mass. The component of diffusion velocity in equations (5.190) and (5.191):

$$\overrightarrow{v_\alpha^c} = D\sum_{\beta=1}^{N} Y_\beta\left(\frac{\overrightarrow{F_\beta}}{x_\beta} - \frac{\overrightarrow{F_\alpha}}{x_\alpha}\right) \quad (5.192)$$

is related to baro- and thermodiffusion. Another component of the diffusion velocity in (5.190, 5.191):

$$\overrightarrow{v_\alpha^g} = \sum_{\beta=1}^{N} x_\beta \delta_{\alpha\beta}\left(\overrightarrow{v_\beta^{(0)}} - \overrightarrow{v_\alpha^{(0)}}\right) + \sum_{\gamma=1}^{N} Y_\gamma \sum_{\beta=1}^{N}\delta_{\beta\gamma}\left(\overrightarrow{v_\beta^{(0)}} - \overrightarrow{v_\gamma^{(0)}}\right) \quad (5.193)$$

is related to the difference in the binary diffusion coefficients. The α-component force in Eq. (5.192) and the α-component velocities in Eq. (5.193) are given by the expressions:

$$\overrightarrow{F_\alpha} = \left(Y_\alpha - x_\alpha\right)\frac{\nabla p}{p} + \sum_{\gamma=1}^{N}\frac{x_\alpha x_\gamma}{\rho D_{\alpha\beta}}\left(\frac{D_{T,\gamma}}{Y_\gamma} - \frac{D_{T,\alpha}}{Y_\alpha}\right)\frac{\nabla T}{T}, \quad (5.194)$$

$$\overrightarrow{v_\alpha^{(0)}} = -D\nabla \ln Y_\alpha + D\sum_{\beta=1}^{N} Y_\beta\left(\frac{\overrightarrow{F_\beta}}{x_\beta} - \frac{\overrightarrow{F_\alpha}}{x_\alpha}\right). \quad (5.195)$$

5.7.7 Influence of Transfer Phenomena on Energy Efficiency of Plasma-Chemical Processes

Based on these energy and mass transfer equations, first establish the important rule. The transfer phenomena don't change the limits of energy efficiency of thermal plasma-chemical processes (which are absolute and ideal quenching) if the two following requirements are satisfied:

1. The binary diffusion coefficients are fixed, equal to each other and equal to the reduced coefficient of thermal conductivity (the Lewis number $Le = 1,\ \delta_{\alpha\beta} = 0$).
2. The effective pressure in the system is constant implying the absence of external forces and large velocity gradients. Taking into account the above conditions, rewrite the enthalpy balance equation (5.189) and the continuity equation (5.186) for the steady-state one-dimensional case as:

$$\rho v \frac{\partial}{\partial x} I = \frac{\partial}{\partial x}\left(\rho D \frac{\partial}{\partial x} I\right), \tag{5.196}$$

$$\rho v \frac{\partial}{\partial x} Y_\alpha = \frac{\partial}{\partial x}\left(\rho D \frac{\partial}{\partial x} Y_\alpha\right) + \omega_\alpha. \tag{5.197}$$

Introduce new dimensionless functions describing reduced enthalpy and reduced mass fractions:

$$\xi = \frac{I - I^r}{I^l - I^r}, \quad \eta = \frac{Y_\alpha - Y_\alpha^r}{Y_\alpha^l - Y_\alpha^r}, \tag{5.198}$$

where Y_α^l, I^l and Y_α^r, I^r are the mass fractions and enthalpy respectively on the left (l) and right (r) boundaries of the region under consideration. With these new functions, the enthalpy conservation (5.196) and continuity (5.197) equations can be rewritten as:

$$\rho v \frac{\partial \xi}{\partial x} = \frac{\partial}{\partial x} \rho D \frac{\partial \xi}{\partial x}, \tag{5.199}$$

$$\rho v \frac{\partial \eta_\alpha}{\partial x} = \frac{\partial}{\partial x} \rho D \frac{\partial \eta_\alpha}{\partial x} + \omega_\alpha \left(Y_\alpha^l - Y_\alpha^r\right). \tag{5.200}$$

The boundary conditions for these equations on the left and right sides of the region are:

$$\xi(x = x_l) = \eta(x = x_l) = 1, \quad \xi(x = x_r) = \eta(x = x_r) = 0. \tag{5.201}$$

Chemical reactions (ω_α in the continuity equation (5.200)) during the diffusion process can increase or decrease the concentration of the reacting components. However, considering the endothermic processes (which is usually the case in thermal plasma), chemical reactions only decrease the concentration of products. To determine the maximum yield of the products, neglect the reaction rate during diffusion, assuming $\omega_\alpha = 0$ in the continuity equation (5.200). In this case, equations and boundary conditions for η and ξ are completely identical, which means they are equal and gradients of total enthalpy I and mass fraction Y_α are related by the formula (B.V. Potapkin, V.D. Rusanov, A. Fridman, 1985):

$$\left(Y_\alpha^l - Y_\alpha^r\right)^{-1} \frac{\partial}{\partial x} Y_\alpha = \left(I^l - I^r\right)^{-1} \frac{\partial}{\partial x} I. \tag{5.202}$$

From Eq. (5.202) it can be concluded that the minimum ratio of the enthalpy flux to the flux of products, taking into account effect of chemical reactions, is equal to:

$$A = \frac{I^l - I^r}{Y_\alpha^l - Y_\alpha^r}. \tag{5.203}$$

The ratio Eq. (5.203) gives the energy price of the process product and completely correlates with the expressions for energy efficiency of quasi-equilibrium plasma-chemical processes considered in Sections 5.7.1, 5.7.2, and 5.7.4. In absence of external forces and when the diffusion and reduced thermal conductivity coefficients are equal (the Lewis number $Le = 1$) the minimum energy cost of products of plasma-chemical reaction is determined by the product formation in the quasi-equilibrium high-temperature zone. This minimum energy cost (maximum energy efficiency) corresponds to the limits of absolute and ideal quenching and taking into account the non-uniformity of heating can be calculated as:

$$\langle A \rangle = \frac{\int y(T)\left[I(T) - I_0\right]dT}{\int y(T)\chi(T)dT}. \tag{5.204}$$

Here y(T) is the mass fraction of initial substance heated to temperature T; I(T) and I_0 are the total enthalpies of the mixture at temperature T and initial temperature respectively; $\chi(T)$ is the degree of conversion of the initial substance (for ideal quenching) or degree of conversion into the final product (for absolute quenching), achieved in the high temperature zone. The possibility to use the thermodynamic relation (5.204) for calculating the energy cost of chemical reactions in thermal plasma under conditions of intensive heat and mass transfer is a consequence of the similarity principle of concentrations and temperature fields. This principle is valid at the Lewis number $Le = 1$ in absence of external forces, and is widely applied in combustion theory (F.A. Williams, 1985; D.A. Frank-Kamenetsky, 1971). If the Lewis number differs from unity ($Le = \frac{\lambda}{\rho c_p D} \neq 1$), the maximum energy efficiency can be different from that of the ideal quenching. If light and fast hydrogen atoms are products of dissociation, then the correspondent Lewis number is relatively small ($Le \ll 1$). In this case, diffusion of products is faster than the energy transfer and energy cost of the products can be less with respect to ideal quenching Eq. (5.204). The statistical approach (taking into account kinetic and transfer process limitations) was applied in the book of S. Nester, B. Potapkin, A. Levitsky, V. Rusanov, B. Trusov, A. Fridman (1988) to analyze products composition, degree of conversion and energy efficiency of 156 endothermic plasma processes. Some examples of thermal plasma processes, described this way are presented in Figs. 5.32–5.38.

5.8 SURFACE REACTIONS OF PLASMA-EXCITED MOLECULES IN CHEMISTRY, METALLURGY, AND BIO-MEDICINE

5.8.1 Surface Relaxation of Excited Molecules, Non-Equilibrium Surface Heating, and Evaporation in Non-Thermal Discharges

From a variety of surface reactions of excited molecules, the most general is their relaxation (see Table 3.15). If molecular plasma is characterized by vibrational-translational non-equilibrium

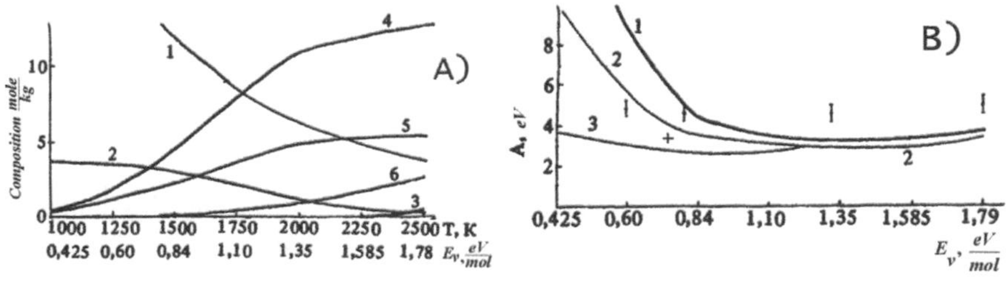

FIGURE 5.32 Thermal reaction in mixture S + CO_2. Composition (A): 1 – CO_2, 2 – S_2, 3 – S, 4 – CO, 5 – SO_2, 6 – SO. Energy price (B) of CO formation: 1 – ideal quenching; 2 – super-ideal quenching; 3 – process stimulated by CO_2 vibrational excitation. Experiments with sulfur vapor (ϕ), with sulfur powder (+).

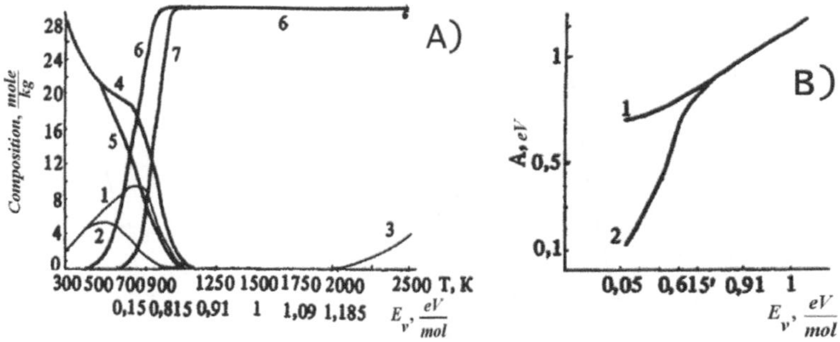

FIGURE 5.33 Thermal reaction in mixture C + H_2O. Composition (A): 1 – CO_2, 2 – CH_4, 3 – H, 4 – C(solid), 5 – H_2O, 6 – H_2, 7 – CO. Energy price (B) of CO/H_2 formation: 1 – absolute quenching; 2 – ideal quenching.

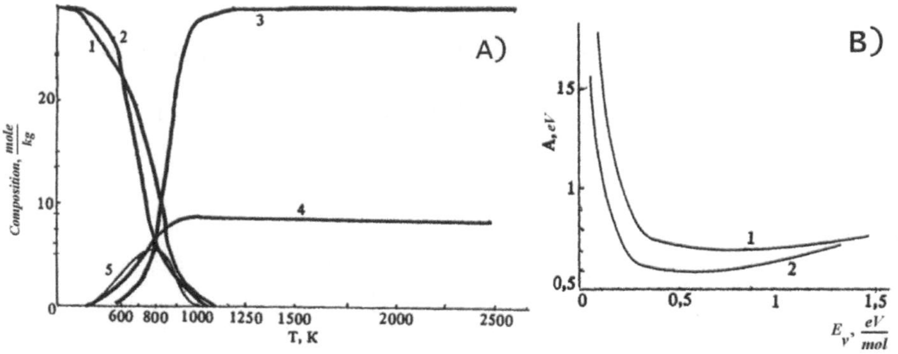

FIGURE 5.34 Thermal reaction in mixture CH_4 + H_2O. Gas phase composition (A): 1 – H_2O, 2 – CH_4, 3 – CO , 4 – H_2/10, 5 – CO_2. Energy price (B) of CO/H_2 formation: 1 – absolute quenching; 2 – ideal quenching.

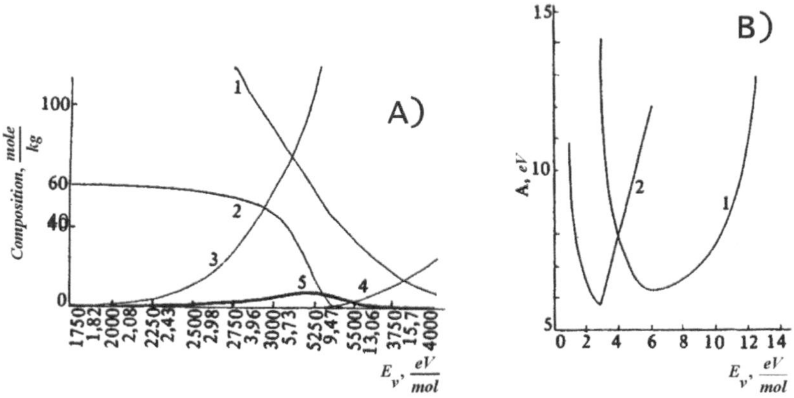

FIGURE 5.35 Thermal dissociation of CH_4. Composition (A): 1 – H_2, 2 – C(solid), 3 – H , 4 – C, 5 – C_2H_2. Energy price (B) of C_2H_2 formation: 1 – absolute quenching A/10; 2 – ideal quenching.

($T_e > T_v >> T_0$), the surface temperature (T_S) can significantly exceed the translational gas temperature (T_0) because surface VT-relaxation is much faster than that in the gas phase. This overheating ($T_e > T_v > T_S >> T_0$) can be accompanied by surface evaporation, desorption of chemisorbed complexes and passivating layers. It is called the non-equilibrium surface heating and evaporation effect. This effect can be significant for plasma interaction with powders, and flat surfaces especially in plasma metallurgy. The preferential macro-particles surface heating without essential gas heating

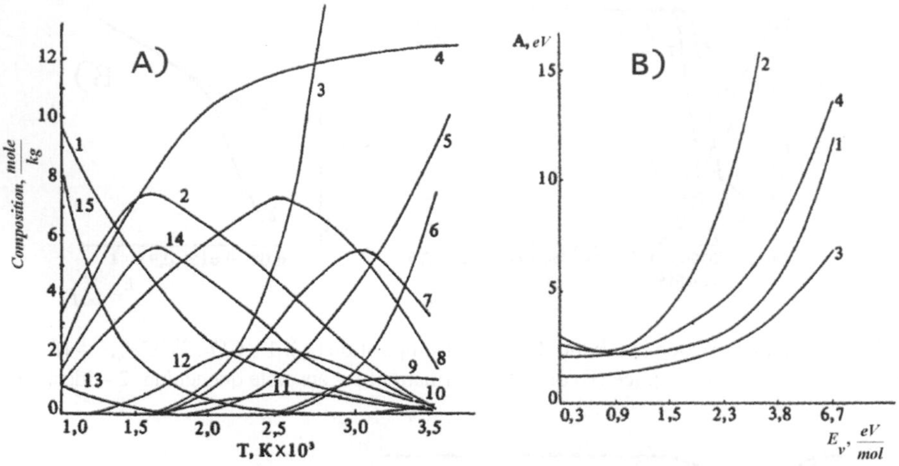

FIGURE 5.36 Thermal reaction in mixture $H_2S + CO_2$. Composition (A): 1 – CO_2, 2 – H_2O, 3 – H , 4 –CO, 5 – S, 6 – O, 7 – SO, 8 – H_2, 9 – OH, 10 – O_2, 11 – HS, 12 – SO_2, 13 – COS, 14 – S_2, 15 – H_2S. Energy price (B) of CO/H_2 formation: 1 – absolute quenching; 2 – absolute quenching with respect to atomic sulfur; 3 – ideal quenching , 4 – ideal quenching with respect to atomic sulfur.

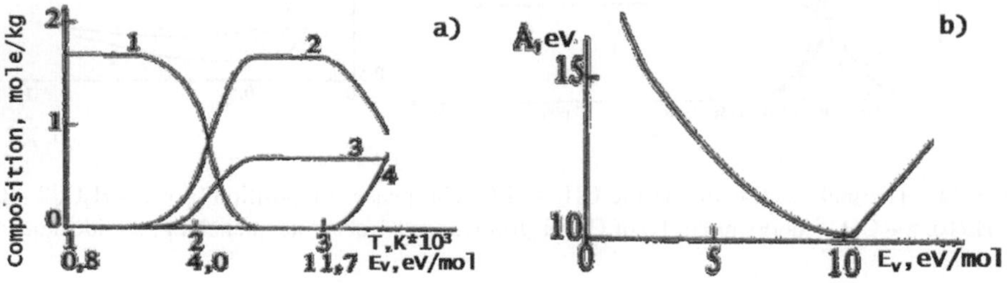

FIGURE 5.37 Thermal decomposition of ZrI_4. Composition (a): 1 – ZrI_4; 2 – Zr.

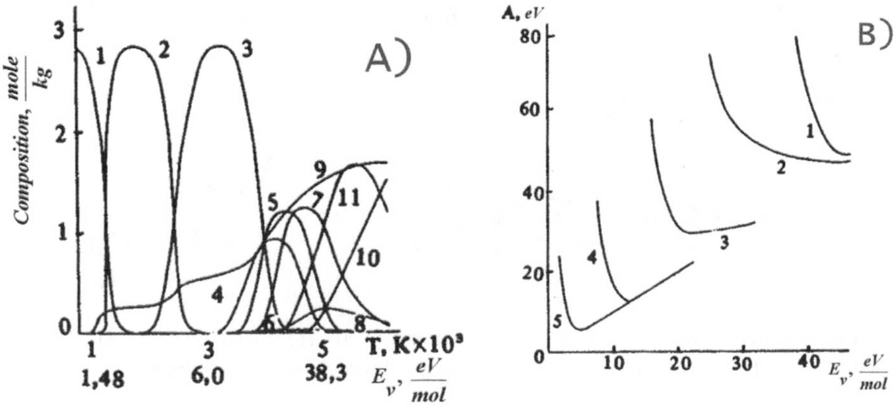

FIGURE 5.38 Thermal dissociation of UF_6. Composition (A): 1 – UF_6, 2 – UF_5, 3 – UF_4, 4 – UF_3, 5 – UF_2, 6 – UF_2^+, 7 – UF, 8 – UF^+, 9 – F/10, 10 – U^+, 11 – U. Energy price (B) of an uranium atom formation: 1 – absolute quenching ; 2 – ideal quenching – formation of UF_3 from UF_4 ; 3 – ideal quenching – formation of UF_4 from UF_5 ; 4 – ideal quenching – formation of UF_5 from UF_6 ; 5 – ideal quenching.

becomes possible in molecular gases, when the total surface area of the macro-particles is so high that vibrational VT-relaxation on the surface dominates over VT-relaxation in plasma volume:

$$4\pi r_a^2 n_a \geq \frac{k_{VT} n_0}{P_{VT}^S v_T}. \tag{5.205}$$

In this relation: n_a and r_a are the density and radius of macro-particles; $k_{VT}(T_0)$ is the rate coefficient of gas-phase vibrational VT relaxation; n_0 is the gas density; v_T is the thermal velocity of molecules; P_{VT}^S is the probability of vibrational VT relaxation on the macro-particle surface, called the accommodation coefficient. We can take for a numerical example the following parameters: $P_{VT}^S = 3 \cdot 10^{-3}$; $k_{VT} = 10^{-17}$ cm^3/s; $n_0 = 3 \cdot 10^{19}$cm^{-3},$v_T = 10^5$ cm/s. In this case, according to (5.205), the non-equilibrium heating requires: $n_a r_a^2 > 0.1$ cm^{-1}. Vibrational energy accumulated in gas is sufficient to evaporate a surface layer of the macro-particles with depth $r_S < r_a$, if the total volume fraction of the macro-particles is small:

$$n_a r_a^3 \leq \frac{n_0 T_v}{n_k \left(T_S + r_S \varepsilon_k / r_a\right)}. \tag{5.206}$$

Here, T_v is the vibrational temperature; n_k is number density of atoms in a macro-particle; ε_k is the energy required for evaporation of an atom. If thickness of the reduced layer is relatively low:

$$r_S < r_a \frac{T_S}{\varepsilon_k}, \tag{5.207}$$

then the major portion of vibrational energy input in (5.206) is related to heating up of the macro-particles to temperature T_S. Numerically, in this case, the limitation (5.206) means $n_a r_a^3 \leq 10^{-3}$, for the conventional conditions: $n_0 = 3 \cdot 10^{19}$ cm^{-3}, $n_k = 3 \cdot 10^{22}$ cm^{-3}, $T_v = 0.3$ eV, $T_S = 0.3$ eV.

5.8.2 Surface Reactions of Excited Hydrogen Molecules in Formation of Hydrides by Gasification of Elements and Thin-Film Processing in Non-Thermal Plasma

Between a variety of surface reactions of plasma-excited molecules in chemistry and electronics, first consider gasification of silicon, germanium, and boron by vibrationally excited hydrogen with the direct production of hydrides:

$$(Si)_{solid} + H_2^*\left(X^1\Sigma_g^+, v\right) \to (SiH_4)_{gas}. \tag{5.208}$$

$$(Ge)_{solid} + H_2^*\left(X^1\Sigma_g^+, v\right) \to (GeH_4)_{gas}. \tag{5.209}$$

$$(B)_{solid} + 1.5 \cdot H_2^*\left(X^1\Sigma_g^+, v\right) \to 0.5 \cdot (B_2H_6)_{gas}. \tag{5.210}$$

Gasification stimulated by non-thermal plasma excitation and subsequent dissociation of H_2 was also observed in experiments with the production of hydrides of sulfur, arsenic, tin, tellurium, and selenium. Mechanism of hydrides production includes in this case surface dissociation of excited hydrogen molecules, chemisorption of atomic hydrogen, surface reactions, and desorption of produced volatile hydrides. Non-thermal plasma-stimulated synthesis of metal hydrides can also take place in thin surface films, which can be exemplified by the synthesis of aluminum hydride AlH_3 from the elements:

$$(Al)_{solid} + 1.5 \cdot H_2^*\left(X^1\Sigma_g^+, v\right) \to (Al)_{solid} + 3H \to (AlH_3)_{solid}. \tag{5.211}$$

This process proceeds through the intermediate formation of atomic hydrogen but requires low temperatures because AlH_3 is stable only at temperatures below 400 K. The formation of atomic hydrogen stimulated by vibrational excitation takes place mostly on the surface of aluminum where it requires only about 2 eV/mol. Gradual hydrogenation of aluminum in a thin film is determined by a competition between solid-state reactions of hydrogen attachment and hydrogen exchange:

$$H + AlH_k \rightarrow AlH_{k+1}, \quad k = 0,\ 1,\ 2. \tag{5.212}$$

$$H + AlH_k \rightarrow AlH_{k-1} + H_2, \quad k = 1,\ 2,\ 3. \tag{5.213}$$

The average quasi-stationary concentration of AlH_3 in the film can be estimated as:

$$\gamma \approx 1 - \exp\left(-\tilde{E}_a / T_0\right). \tag{5.214}$$

$\tilde{E}_a$ is the activation barrier of (5.213); T_0 is solid-state temperature. Formation of the hydrides occurs in a layer whose thickness is determined by losses of atomic hydrogen in solid-state recombination:

$$H + H + AlH_k \rightarrow H_2 + AlH_k, \tag{5.215}$$

as well as in reactions (5.213). The depth δ_1 of these reactions can be found from the 1D diffusion equation for H atoms density [H] in a solid matrix of aluminum during the synthesis:

$$D_H \frac{\partial^2}{\partial x^2}\left[H\right] - \nu_0 \exp\left(-\tilde{E}_a / T_0\right) = 0. \tag{5.216}$$

Here, D_H is the coefficient of diffusion of H atoms in a crystal structure; ν_0 is the "jumping" frequency. The depth δ_1 of synthesis of the aluminum hydride can be then found as:

$$\delta_1 \approx \lambda \sqrt{f} \exp \frac{\tilde{E}_a - E_a^D}{2T_0}. \tag{5.217}$$

5.8.3 Effect of Vibrational Excitation of CO Molecules on Direct Surface Synthesis of Metal Carbonyls in Non-Thermal Plasma

Metal carbonyls can be directly produced by reactions with CO molecules excited in non-thermal plasma. It is especially important in metallurgy for the synthesis of carbonyls of chromium, molybdenum, manganese, and tungsten, which cannot be produced without solvents:

$$(\mathrm{Cr})_{solid} + 6(\mathrm{CO})_{gas} \rightarrow \left[\mathrm{Cr}(\mathrm{CO})_6\right]_{gas}. \tag{5.218}$$

$$(\mathrm{Mo})_{solid} + 6(\mathrm{CO})_{gas} \rightarrow \left[\mathrm{Mo}(\mathrm{CO})_6\right]_{gas}. \tag{5.219}$$

$$(\mathrm{Mn})_{solid} + 5(\mathrm{CO})_{gas} \rightarrow 1/2\left[\mathrm{Mn}_2(\mathrm{CO})_{10}\right]_{gas}. \tag{5.220}$$

$$(W)_{solid} + 6(\mathrm{CO})_{gas} \rightarrow \left[W(\mathrm{CO})_6\right]_{gas}. \tag{5.221}$$

The gaseous metal carbonyls are unstable at temperatures exceeding 600K, which leads to restrictions in performing of non-thermal discharges. The mechanism for plasma-chemical formation of metal carbonyls can be considered taking as an example the synthesis of $Cr(CO)_6$ (5.218) from chromium and CO vibrationally excited in a non-thermal discharge. The process includes the following four stages:

1. Chemisorption of plasma-excited CO molecules on a chromium surface:

$$(\mathrm{Cr})_{solid} + k \cdot (\mathrm{CO})_g \Leftrightarrow \mathrm{Cr}\left[k \cdot \mathrm{CO}\right]_{chemisorbed}. \tag{5.222}$$

The chromium crystal structure is characterized by a cubic lattice with valence 6 and coordination number 8. Each Cr surface atom is able to attach $k \leq 3$ carbonyl groups without destruction of the lattice.

2. Formation of higher carbonyls ($k > 3$) requires taking a chromium atom with attached CO-groups from the crystal lattice to a chemisorbed state:

$$\mathrm{Cr}\left[m \cdot \mathrm{CO}\right]_{chemisorbed} + (\mathrm{Cr})_{solid}\,\mathrm{CO} \Leftrightarrow \mathrm{Cr}\left[\mathrm{Cr}(\mathrm{CO})_{m+1}\right]_{chemisorbed}. \tag{5.223}$$

The characteristic time of taking out the chromium atom from the crystal lattice $\tau_{to}$to form the chemisorbed carbonyl complex ($T_v >> T_0$) can be estimated as:

$$\tau_{to} \approx \frac{1}{a\Phi} \exp\left(\frac{E_{to}}{T_v}\right). \tag{5.224}$$

Here, E_{to} is the activation barrier of (5.223); Φ is the flux of CO molecules to the surface; a is the surface area of a chemisorbed complex; T_v is vibrational temperature of CO. The dissociation kinetics of the chemisorbed complex $\mathrm{Cr}[\mathrm{Cr}(\mathrm{CO})_{m+1}]_{chemisorbed}$ is determined by the surface temperature T_S.

3. Formation of chromium hexa-carbonyl $\mathrm{Cr(CO)_6}$ adsorbed on the surface) takes place in the surface reactions of intermediate chemisorbed chromium carbonyl complexes:

$$\mathrm{Cr}\left[\mathrm{Cr}(\mathrm{CO})_l\right]_{chemisorbed} + (6-l) \cdot \mathrm{Cr}\left[\mathrm{CO}\right]_{chemisorbed} \Leftrightarrow \mathrm{Cr}\left[\mathrm{Cr}(\mathrm{CO})_6\right]_{adsorbed}. \tag{5.225}$$

The characteristic time of the surface reactions with activation barriers $E_{sr,\,l}$ is:

$$\tau_{sr,l} \approx \frac{2\pi\hbar b_0}{T_S \cdot \left[\mathrm{CO}\right]} \exp\left(\frac{E_{sr,l}}{T_S}\right). \tag{5.226}$$

Here, b_0 is the reverse adsorption coefficient at the surface coverage $\xi = 0.5$.

4. The final stage is desorption of volatile chromium hexa-carbonyl $\mathrm{Cr(CO)_6}$ into gas phase:

$$\mathrm{Cr}\left[\mathrm{Cr}(\mathrm{CO})_6\right]_{adsorbed} \to (\mathrm{Cr})_{solid} + \left[\mathrm{Cr}(\mathrm{CO}_6)\right]_{gas}. \tag{5.227}$$

In the case of $T_v >> T_0$, the desorption rate can be calculated as:

$$w_{des} \approx \xi \left[\frac{\omega_L}{a} \exp\left(-\frac{E_a^{des}}{T_S}\right) + \Phi \exp\left(-\frac{E_a^{des}}{T_v}\right)\right] = \xi \cdot k_{des}\left(T_v,\, T_S\right). \tag{5.228}$$

Here, ξ is the surface fraction covered by complexes; ω_L is the characteristic lattice frequency; E_a^{des} is the activation energy of the desorption process; $k_{des}(T_v,\ T_S)$ is the desorption rate coefficient.

5.8.4 Reactions of Plasma-Generated Singlet Oxygen and Other Reactive Oxygen Species (ROS) on Bio-Active Surfaces

Atmospheric pressure cold air plasma effectively generates reactive oxygen species (ROS), including singlet oxygen $O_2(^1\Delta_g)$, superoxide O_2^-, OH, and peroxides, which significantly affect various types of living tissues, stimulating sterilization (deactivation) of bacteria, viruses, cell apoptosis, as well as cell migration, cell proliferation, signaling, release of growth factors, etc. The superoxide, which is an important ROS generated in the direct plasma discharges, is a precursor to additional production of singlet oxygen and peroxynitrite in the tissue. O_2^-, also acts as a signaling molecule for the regulation of cellular processes. In tissues with a sufficient level of acidity, the superoxide is converted into hydrogen peroxide and oxygen in a reaction called the superoxide dismutation:

$$2O_2^- + 2H^+ \rightarrow H_2O_2 + O_2. \tag{5.229}$$

The superoxide dismutation can be spontaneous or can be catalyzed by the enzyme superoxide dismutase (SOD). The superoxide not only generates hydrogen peroxide (H_2O_2) but also stimulate its conversion into OH-radicals, which are actually extremely strong oxidizers very effective in sterilization through a chain oxidation mechanism. The H_2O_2 conversion into OH known as the Fenton reaction proceeds as a redox process provided by oxidation of a metal ions (e.g., Fe^{2+}):

$$H_2O_2 + Fe^{2+} \rightarrow OH + OH^- + Fe^{3+}. \tag{5.230}$$

Restoration of the Fe^{2+} ions takes place in the Fe^{3+} reduction process provided by O_2^-:

$$Fe^{3+} + O_2^- \rightarrow Fe^{2+} + O_2. \tag{5.231}$$

Superoxide can also react in water with nitric oxide NO (usually generated in plasma at higher temperatures) producing peroxynitrate ($OONO^-$), which is another highly reactive oxidizing ion-radical:

$$O_2^- + NO \rightarrow OONO^-. \tag{5.232}$$

The discussed ROS reactions occurring on the surface of living tissues in contact with non-thermal discharges play a significant role in plasma medicine (A. Fridman, 2008).

5.9 PROBLEMS AND CONCEPT QUESTIONS

5.9.1 Diffusion of Molecules Along the Vibrational Energy Spectrum

Considering VV-exchange with frequency $k_{vv}n_0$ as diffusion of molecules along the vibrational energy spectrum, estimate time for a molecule to get the vibrational energy, corresponding to the Treanor minimum point E_{Tr}.

5.9.2 VT-Relaxation Flux in the Energy Space

The kinetic equation (5.7) is often used to describe the vibrational distribution at low temperatures ($T_0 < \hbar\omega$) when the criterion Eq. (5.6) is not valid. Analyze the steady-state solution of the equation (5.7) in this case, and explain the contradiction.

5.9.3 The Treanor Distribution Function and Criterion of the Fokker-Plank Approach

Using the criterion Eq. (5.6), determine the conditions for the application of the Fokker-Planck approach to the case of the Treanor vibrational distribution function.

5.9.4 Flux of Molecules and Flux of Quanta Along the Vibrational Energy Spectrum

Prove, that the flux of molecules along the vibrational energy spectrum is equal to the derivative $\partial/\partial v$ of the flux of quanta in the energy space. Based on this explain why the constant F in the kinetic equation (5.20) is proportional to the flux of quanta in the energy space. Find the coefficient of proportionality between them.

5.9.5 The Hyperbolic Plateau Distribution

Derive the relation between the plateau coefficient C (5.22) and the degree of ionization in non-thermal plasma n_e/n_0. Compare the result with the plateau distribution Eq. (5.24) to find a relation between vibrational temperature and the degree of ionization.

5.9.6 Vibrational Distribution Functions in the Strong and Intermediate Excitation Regimes

Compare numerically the criteria of the strong excitation ($\delta_{VV}T_v \geq 1$) and intermediate excitation ($\frac{x_e T_v^2}{T_0 \hbar\omega} \geq 1$). Which one requires the larger value of vibrational temperature at different T_0?

5.9.7 The Gordiets Vibrational Distribution Function

Compare the discrete Gordiets vibrational distribution with the continuous distribution function Eq. (5.29). Analyze the decrease of the vibrational distribution at high vibrational energies in the case of low translational temperatures $T_0 < \hbar\omega$.

5.9.8 The eV-Flux Along the Vibrational Energy Spectrum

Prove, that one-quantum Eq. (5.36) and integral multi-quantum Eq. (5.38) eV-fluxes in the energy space becomes equal to zero if the vibrational distribution is the Boltzmann function with temperature T_e: $f(E) \propto \exp(-E/T_e)$.

5.9.9 eV-Processes at Intermediate Ionization Degrees

The vibrational distribution for this case is $f(E) = \phi(E)$ Eq. (5.45) at the weak excitation regime and $T_v > \hbar\omega/\alpha$. Using more general Eq. (5.44),- estimate the relevant vibrational distribution at higher vibrational temperatures $T_v < \hbar\omega/\alpha$.

5.9.10 Treanor Effect for Polyatomic Molecules

Explain how the Treanor effect, which is especially related to the discrete anharmonic structure of vibrational levels, can be achieved in polyatomic molecules in the quasi-continuum of vibrational states, where there is no discrete structure of the levels.

5.9.11 The Treanor–Boltzmann Transition in Vibrational Distributions of Polyatomic Molecules

Explain why the transition from the Treanor or Boltzmann distribution with temperature T_v (at low energies) to the Boltzmann distribution with T_0 (at high energies) takes place in polyatomic molecules at lower vibrational levels than the same transition in the case of diatomic molecules.

5.9.12 Non-Steady-State Vibrational Distribution Function

Using the approximation Eq. (5.57), estimate the evolution of the average vibrational energy during the early stages of the VV-exchange process, stimulated by the electron impact excitation of the lowest vibrational levels.

5.9.13 Reactions of Vibrationally Excited Molecules, Fast Reaction Limit

Calculate the integral (5.67), neglecting VT-relaxation ($\xi(E) << 1$), determine the reaction rate of the chemical process, stimulated by vibrational excitation and limited by VV-relaxation to reach the activation energy.

5.9.14 Reactions of Vibrationally Excited Molecules, Slow Reactions Limit

Equation (5.69) describes the reaction rates of excited molecules non-symmetric with respect to the contribution of vibrational and translational degrees of freedom: vibrationally excited molecules have activation energy E_a/α, while other degrees of freedom have the activation energy E_a. Does this mean that the vibrational energy is less effective than translational energy? Surely, not. Explain the "contradiction".

5.9.15 Reactions of Vibrationally Excited Polyatomic Molecules

The rate coefficient of the polyatomic molecule reactions Eq. (5.76) can be presented over a narrow range of vibrational temperatures in the simple Arrhenius manner $k_R^{macro} \propto \exp\left(-\frac{E_a}{T_v}\right)$. Determine the effective activation energy in this case as a function of vibrational temperature T_v and number of degrees of freedom "s".

5.9.16 Macrokinetics of Reactions of Two Vibrationally Excited Molecules

Show for the case of the Treanor distribution, that the main contribution to the rate of reaction of two vibrationally excited molecules $E' + E'' = E_a$ is made by collisions when entire energy is located on one molecule.

5.9.17 Vibrational Energy Losses Due To VT-Relaxation from High Levels

Estimate the VT-losses from high vibrational levels per one act of chemical reaction in the regime of strong excitation, when the reaction is controlled by the resonant VV-exchange. Compare the result with Eq. (5.87).

5.9.18 Contribution of High and Low Levels in Total Rate of VT-Relaxation

Contribution of high and low vibrational levels in VT-relaxation depends on the degree of ionization. Estimate the maximum degree of ionization when the total VT-relaxation rate is controlled by the low excitation levels.

5.9.19 Vibrational Energy Losses Due to Non-Resonant VV-Exchange

Derivation of Eq. (5.88) for calculating the VV-exchange vibrational energy losses assumed predominantly non-resonant VV-relaxation. However, this relation is correct for any kind of one-quantum

VV-exchange. Derive Eq. (5.88) in a general way comparing the starting "big quantum" after excitation by electron impact and the final "smaller one" after series of one-quantum exchange processes.

5.9.20 VV- and VT-Losses of Vibrational Energy of Highly Excited Molecules

Calculate the ratio of the vibrational energy losses related to VT-relaxation of highly excited diatomic molecules to the correspondent losses during the series of the one-quantum VV-relaxation process, which provided the formation of the highly excited molecule. What is the theoretically minimum value of the ratio? At which vibrational quantum number these losses are equal?

5.9.21 Vibrational Energy Losses Related to Double-Quantum VV-Exchange

Estimate the vibrational energy losses related to VV-exchange per one act of plasma chemical dissociation of a diatomic molecule stimulated by vibrational excitation. Assume domination of the one-quantum VV-relaxation at lower vibrational levels until the double quantum resonance becomes possible (assume no VV-losses at higher vibrational energies).

5.9.22 Treanor Formula for Isotopic Mixtures

Based on the Treanor formula (5.100) and assuming for an isotopic mixture $\frac{\Delta\omega}{\omega} = \frac{1}{2}\frac{\Delta m}{m}$, estimate the difference in vibrational temperatures of nitrogen molecules-isotopes at room temperature and $T_v \approx 3000\ K$.

5.9.23 Isotopic Effects in Vibrational Kinetics and in Conventional Quasi-Equilibrium Kinetics

Demonstrate that the isotopic effect in vibrational kinetics is usually stronger than the conventional one. Compare the selectivity coefficients for these two isotopic effects. Is it possible for these two isotopic effects to take place together and to compensate each other? Give a numerical example.

5.9.24 Coefficient of Selectivity for Separation of Heavy Isotopes

Based on Eq. (5.102) and Figure 5.13, estimate the level of the selectivity coefficient for separation of uranium isotopes, assuming a chemical process with UF_6 at room temperature. Assume the activation energy of about 5 eV.

5.9.25 Reverse Isotopic Effect in Vibrational Kinetics

For the hydrogen-isotope separation (H_2-HD) plasma chemical process stimulated by vibrational excitation at room temperature find: (a) the typical value of the selectivity coefficient for the conventional direct isotopic effect in vibrational kinetics ($E_a = 2$ eV); (b) the typical value of the selectivity coefficient for the reverse isotopic effect; (c) the relative temperature interval necessary for the achievement of the reverse effect.

5.9.26 Effect of eV-Processes on Isotope Separation in Plasma

Based on Eq. (5.110) estimate the maximum selectivity coefficient determined by the influence of eV-processes in the hydrogen-isotope mixture (H_2-HD) at room temperature. Compare the result with the typical values of the separation coefficient for conventional direct and reverse isotopic effects in non-equilibrium vibrational kinetics.

5.9.27 Canonical Invariance of Rotational Relaxation Kinetics

Based on Eq. (5.128) prove the canonical invariance of relaxation to the equilibrium of the initial Boltzmann distribution with rotational temperature exceeding the translational one. Show that the rotational distribution $f(E_r, t)$ always maintains the same form of the Boltzmann distribution Eq. (5.131) during the relaxation to equilibrium, but obviously with changing values of the rotational temperature as it approaches the translational temperature.

5.9.28 "Hot Atoms" Generated by Fast VT-Relaxation

In non-thermal plasma of molecular gases ($T_v >> T_0$) admixture of the light alkaline atoms are able to establish the quasi-equilibrium with vibrational (not translational) degrees of freedom. This can be explained by the high rate of VT-relaxation and the large difference in masses, which slows down the Maxwellization (TT-exchange) processes. Why the same effect cannot be achieved with addition of relatively heavy alkaline atoms when the Maxwellization process is also somewhat slower for the same reason of large differences in masses?

5.9.29 Relation Between Translation Temperature of Alkaline Atoms and Vibrational Temperature of Molecular Gas

Using Eq. (5.134) for the translational energy distribution function of the small addition of alkaline atoms in molecular gas, derive the relation between the corresponding temperatures: T_v, T_0, T_{Me}, which could be measured experimentally. Estimate the Treanor effect of the molecular gas anharmonicity on the relation between the temperatures.

5.9.30 "Hot Atoms", Generated in Fast Endothermic Plasma-Chemical Reactions, Stimulated by Vibrational Excitation

Compare energy of hot atoms, generated in the fast endothermic plasma-chemical reactions, in the cases of strong, intermediate, and weak excitation. Use Eq. (5.141), and assume the slow reaction limit in analyzing the type of the vibrational energy distribution function.

5.9.31 Energy Efficiency of Quasi-Equilibrium and Non-Equilibrium Plasma-Chemical Processes

Explain why the energy efficiency of the non-equilibrium plasma-chemical process can be higher than that of the quasi-equilibrium process. Is it a contradiction with the thermodynamic principles?

5.9.32 Plasma-Chemical Reactions Controlled by Dissociative Attachment

For water molecules dissociation in plasma Eqs. (5.147), (5.148), determine the chain length as a function of the degree of ionization. Estimate the energy cost for this mechanism as a function of the ionization degree.

5.9.33 The Treanor Effect in Vibrational Energy Transfer

Estimate, using the relation (5.153), how big should be the parameter $q = \frac{x_e T_v^2}{T_0 \hbar\omega}$ in nitrogen plasma ($Q_{01}^{10} \approx 3 \cdot 10^{-3}$) to observe the Treanor effect in vibrational energy transfer ($\Delta D_v \approx D_0$).

5.9.34 Stimulation of Vibrational-Translational Non-Equilibrium by the Specific Heat Effect

Assuming no energy exchange between vibrational and other degrees of freedom, calculate temperatures T_v and T_0 after mixing 1 mole of air at room temperature with an equal amount of air at 1000 K. Explain and discuss the difference in the temperatures.

5.9.35 Plasma-Chemical Processes, Stimulated by Vibrational Excitation of Molecules

Explain why the plasma-chemical reactions can be effectively stimulated by vibrational excitation of molecules only if the specific energy input exceeds the critical value, while reactions related to electronic excitation or dissociative attachment can proceeds effectively at any levels of the specific energy input.

5.9.36 Absolute and Ideal Quenching of Products of Chemical Reactions in Thermal Plasma

Explain why energy efficiencies of the absolute and ideal quenching of the CO_2-dissociation products are identical, while in the case of dissociation of water molecules they are significantly different.

5.9.37 Super-Ideal Quenching Effect Related to Vibrational-Translational Non-Equilibrium

How can the vibrational-translational non-equilibrium be achieved during the quenching, if the degree of ionization is not sufficient to provide vibrational excitation faster than vibrational relaxation?

5.9.38 Super-Ideal Quenching Effects Related to Selectivity of Transfer Processes

Compare the conditions necessary for effective super-ideal quenching related to the following three different causes: (1) vibrational-translational non-equilibrium during the cooling process; (2) the Lewis number significantly differs from unity during separation and collection of products; (3) cluster-products move relatively fast from the rotating high-temperature discharge zone.

6 Electrostatics, Electrodynamics and Fluid Mechanics of Plasma

6.1 ELECTROSTATIC PLASMA PHENOMENA: DEBYE-RADIUS AND SHEATHS, PLASMA OSCILLATIONS, AND PLASMA FREQUENCY

6.1.1 IDEAL AND NON-IDEAL PLASMAS

In most plasmas, electrons and ions move straightforward between collisions, which make them similar to ideal gas. This means that the inter-particle potential energy $U \propto e^2/4\pi\varepsilon_0 R$, corresponding to the distance between electrons and ions $R \approx n_e^{-1/3}$, is much less than their relative kinetic energy (which is approximately temperature T_e, expressed in energy units):

$$\frac{n_e e^6}{\left(4\pi\varepsilon_0\right)^3 T_e^3} << 1. \tag{6.1}$$

Plasma satisfying this condition is called the **ideal plasma.** The non-ideal plasma (corresponding to the inverse of inequality (6.1) and very high density of charged particles) is not found in nature. Even creation of such plasma in a laboratory is problematic.

6.1.2 PLASMA POLARIZATION, "SCREENING" OF ELECTRIC CHARGES AND EXTERNAL ELECTRIC FIELDS

The Debye Radius in Two-Temperature Plasma. External electric fields induce plasma polarization, which prevents penetration of the field inside of plasma. Plasma "screens" the external electric field, which means this field exponentially decreases (protecting in this way the quasi-neutrality of plasma). Such an effect of the electrostatic "screening" of the electric field around a specified charged particle is illustrated in Figure 6.1. To describe the space evolution of the potential φ Poisson's equation can be used:

$$div\vec{E} = -\Delta\phi = \frac{e}{\varepsilon_0}\left(n_i - n_e\right). \tag{6.2}$$

Here n_e and n_i are the number densities of electrons and positive ions; E is the electric field strength. One can assume the Boltzmann distribution for electrons (temperature T_e) and ions (temperature T_i), and the same quasi-neutral plasma concentration n_{e0} for electrons and ions:

$$n_e = n_{e0}\exp\left(+\frac{e\phi}{T_e}\right), \quad n_i = n_{e0}\exp\left(-\frac{e\phi}{T_i}\right). \tag{6.3}$$

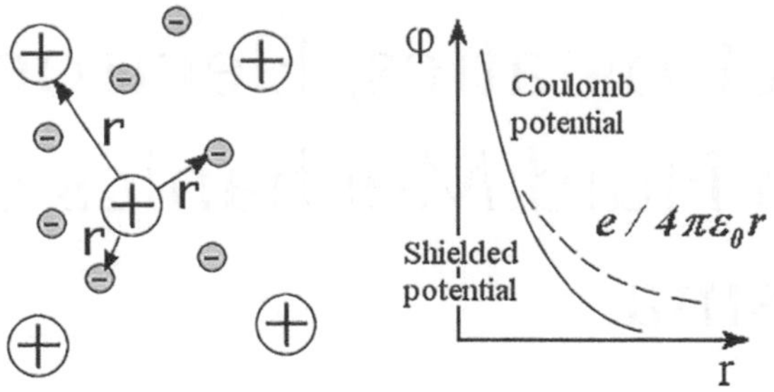

FIGURE 6.1 Plasma polarization around a charged particle.

Combining the Poisson's equation with the Boltzmann distribution Eq. (6.3), results in:

$$\Delta\phi = \frac{\phi}{r_D^2}, \quad r_D = \sqrt{\frac{\varepsilon_0}{n_{e0}e^2\left(1/T_e + 1/T_i\right)}}. \tag{6.4}$$

In this equation r_D is **the Debye radius**, an important electrostatic plasma parameter (see Section 4.3.10). In 1D case, Eq. (6.4) becomes: $d^2\phi/d^2x = \phi/r_D^2$, and describes penetration of electric field (E_0 on the plasma boundary at x = 0) along the axis "x" ($x > 0$) perpendicular to the plasma boundary:

$$\vec{E} = -\nabla\phi = \vec{E_0}\exp\left(-\frac{x}{r_D}\right). \tag{6.5}$$

In a similar manner, reduction of the electric field of a specified charge "q" located in plasma (see Figure 6.1) can be also described by Eq. (6.4), but in spherical symmetry:

$$\Delta\phi \equiv \frac{1}{r}\frac{d^2}{dr^2}(r\phi) = \frac{1}{r_D^2}\phi. \tag{6.6}$$

Taking into account the boundary condition: $\phi = q/4\pi\varepsilon_0 r$ at $r \to 0$, the solution of Eq. (6.6) again results in the exponential decrease of electric potential (and electric field):

$$\phi = \frac{q}{4\pi\varepsilon_0 r}\exp\left(-\frac{r}{r_D}\right). \tag{6.7}$$

Thus, the Debye radius gives the characteristic of the plasma size necessary for screening or suppressing the external electric field. Obviously, the same distance is necessary to compensate the electric field of one specified charged particle in plasma (Figure 6.1). The same Debye radius indicates the scale of plasma quasi-neutrality Eq. (4.130). There is some physical discrepancy between one-temperature (T_e) and two temperature (T_e, T_i) relations for the Debye radius: Eqs. (4.130) and (6.4). According to Eq. (6.4), if $T_e >> T_i$, the Debye radius depends mostly on the lower (ion) temperature, while according to Eq. (4.130) and to common sense it should depend on electron temperature. In reality the heavy ions at low T_i are unable to establish the quasi-equilibrium Boltzmann distribution. It is more correct to assume at $T_e >> T_i$, that the ions are at rest and $n_i = n_{e0} = const$. In this case, the term $1/T_i$ negligible and the Debye radius can be found from Eq. (4.130) or, numerically, from Eq. (4.130).

6.1.3 Plasmas and Sheaths

Not all ionized gases are plasma. Plasma is supposed to be quasi-neutral ($n_e \approx n_i$) and provide the screening of an external electric field and the field around a specified charged particle. To obtain

TABLE 6.1
Debye Radius and the Typical Size of Different Plasma Systems

Type of Plasma	Typical n_e, cm^{-3}	Typical T_e, eV	Debye Radius, cm	Typical Size, cm
Earth Ionosphere	10^5	0.03	0.3	10^6
Flames	10^8	0.2	0.03	10
He-Ne Laser	10^{11}	3	0.003	3
Hg-lamp	10^{14}	4	$3 \cdot 10^{-5}$	0.3
Solar Chromosphere	10^9	10	0.03	10^9
Lightning	10^{17}	3	$3 \cdot 10^{-6}$	100

qualities, the typical size of plasma should much larger than the Debye radius. In Table 6.1 the characteristic parameters of some plasma systems, including comparison of their size and Debye radius are given. Plasma ($n_e \approx n_i$) interacts with a wall across positively charged thin layers called **sheath**. The example of a sheath between plasma and zero-potential surfaces is illustrated in Figure 6.2. Formation of the positively charged sheaths is due to the fact that electrons are much faster than ions, and the electron thermal velocity $\sqrt{T_e/m}$ is about 1000 times faster than the ions thermal velocity $\sqrt{T_i/M}$. The fast electrons are able to stick to the wall surface leaving the area near the walls for positively charged ions alone. The positively charged sheath results in a typical potential profile for the discharge zone which is, also presented in Figure 6.2. The bulk of the plasma is quasi-neutral and hence, it is iso-potential ($\phi = const$) according to the Poisson's Eq. (6.2). Near the discharge walls, the positive potential falls sharply, providing a high electric field, accelerating the ions and deceleration of electrons, which again sustains the plasma quasi-neutrality. Because of the ion acceleration in the sheath, the energy of ions bombarding the walls corresponds not to the ion temperature, but to the temperature of electrons.

6.1.4 Physics of the DC Sheaths

If a discharge operates at low gas temperatures ($T_e >> T_0$, T_i) and low pressures, the sheath can be considered as collisionless. Then basic 1D equation governing the DC sheath potential φ in the direction perpendicular to the wall can be obtained from Poisson's equation, energy conservation for the ions and Boltzmann distribution for electrons:

$$\frac{d^2\phi}{dx^2} = \frac{en_s}{\varepsilon_0}\left[\exp\frac{\phi}{T_e} - \left(1 - \frac{e\phi}{E_i}\right)^{-1/2}\right]. \tag{6.8}$$

In this equation: n_s is plasma density at the sheath edge; $E_{is} = \frac{1}{2}Mu_{is}^2$ is the initial energy of an ion entering the sheath (u_{is} is the corresponding velocity); and the potential is assumed to be zero ($\phi = 0$) at the sheath edge (x = 0). Multiplying Eq. (6.8) by $d\phi/dx$ and then integrating (assuming boundary conditions: $\phi = 0$, $d\phi/dx = 0$ at x = 0, see Figure 6.3) permits finding the electric field in the sheath (the potential gradient) as a function of potential:

$$\left(\frac{d\phi}{dx}\right)^2 = \frac{2en_s}{\varepsilon_0}\left[T_e \exp\left(\frac{e\phi}{T_e}\right) - T_e + 2E_{is}\left(1 - \frac{e\phi}{E_{is}}\right)^{1/2} - 2E_{is}\right]. \tag{6.9}$$

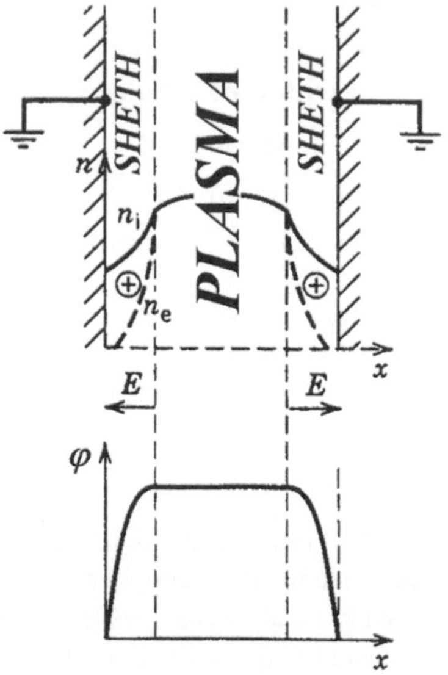

FIGURE 6.2 Illustration of plasma and sheathes.

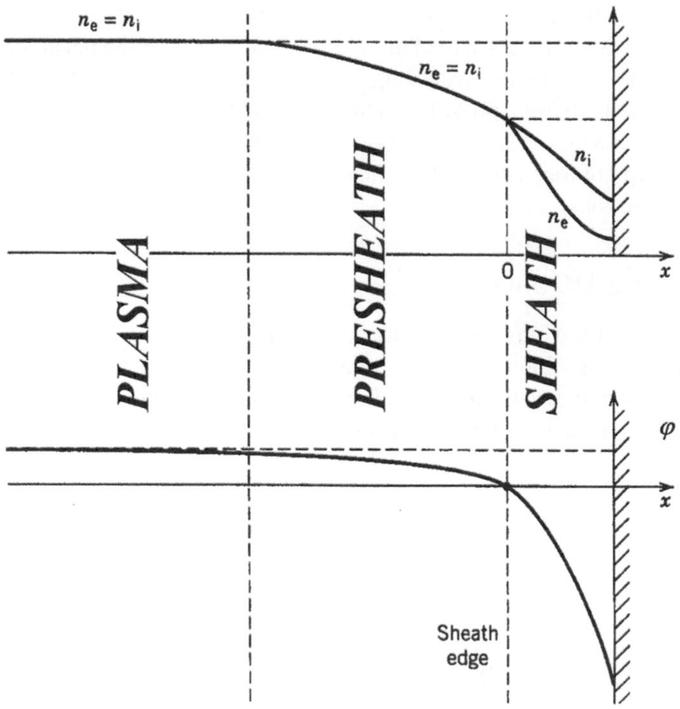

FIGURE 6.3 Illustration of a sheath and a presheath in contact with a wall.

The solution of Eq. (6.9) can exist only if its right-hand-side is positive. Expanding Eq. (6.9) to the second order in a Taylor series, leads to the conclusion, that the sheath can exist only if the initial ion velocity exceeds the critical one; this is known as the **Bohm velocity u_B:**

$$u_{is} \geq u_B = \sqrt{T_e/M}. \tag{6.10}$$

The Bohm velocity is equal to the velocity of ions having energy corresponding to the electron temperature. The condition Eq. (6.10) of a sheath existence is usually referred to as the **Bohm sheath criterion**. To provide ions with the energy and directed velocity necessary to satisfy the Bohm criterion, there must be a quasi-neutral region (wider, than the sheath, e.g several r_D) with some electric field. This region, illustrated in Figure 6.3, is called the **pre-sheath.** Minimum presheath potential is:

$$\phi_{presheath} \approx \frac{1}{2e} M u_B^2 = \frac{T_e}{2e}. \tag{6.11}$$

Balancing ion and electron fluxes to the wall gives change of potential across the sheath:

$$\Delta\phi = \frac{1}{e} T_e \ln \sqrt{M/2\pi\, m}. \tag{6.12}$$

Because the ion-to-electron mass ratio M/m is large, potential across the sheath (even at a floating wall) exceeds 5–8 times the potential across the pre-sheath. Typical sheath is $s \approx 3r_D$.

6.1.5 High Voltage Sheaths, Matrix and Child Law Sheath Models

Equation (6.12) was related to the floating potential which exceeds the electron temperatures 5–8 times. However in general, the change of potential across a sheath (sheath voltage V_0) is often driven to be very large compared to electron temperature T_e/e. In this case the electron concentration in the sheath can be neglected and only ions need be taken into account. As an interesting consequence of the electron absence, the sheath region appears dark when visually observed. The simplest model of such a high voltage sheath assumes uniformity of the ion density in the sheath. This sheath is usually referred to as the **matrix sheath**. In the framework of the simple matrix sheath model, the sheath thickness can be expressed in terms of the Debye radius r_D corresponding to plasma concentration at the sheath edge:

$$s = r_D \sqrt{\frac{2V_0}{T_e}}. \tag{6.13}$$

When voltage is high enough numerically the matrix sheath thickness can be large and exceed the Debye radius 10–50 times. The more accurate approach to describing the sheath should take into account the decrease of the ion density as the ions accelerate across the sheath. This is done in the model of the so-called **Child law sheath**. In the framework of this model, the ion current density $j_0 = n_s e u_B$ is taken equal to that of the well-known **Child law of space-charge-limited current** in a plane diode:

$$j_0 = n_s e u_B = \frac{4\varepsilon_0}{9}\sqrt{\frac{2e}{M}}\frac{1}{s^2} V_0^{3/2}. \tag{6.14}$$

Considering the Child law Eq. (6.14) as an equation with respect to "s", to the relation for the thickness of the Child law sheath is:

$$s = \frac{\sqrt{2}}{3} r_D \left(\frac{2V_0}{T_e}\right)^{3/4}. \tag{6.15}$$

Numerically, the Child law sheath can be of the order of 100 Debye lengths in typical low pressure discharges applied for surface treatment. More details regarding sheaths, including collisional sheaths, sheaths in electronegative gases, radio-frequency plasma sheaths and pulsed potential sheathes can be found for example, in the text by M.A. Lieberman, A.J. Lichtenberg (1994).

6.1.6 Electrostatic Plasma Oscillations; Langmuir or Plasma Frequency

Consider the electrostatic plasma oscillations, illustrated in Figure 6.4. Assume in the 1D approach that all electrons at $x > 0$ are initially shifted to the right on the distance x_0 while heavy ions are not perturbed and located in the same position. This will result in the electric field pushing the electrons back. If $E = 0$ at $x < 0$, this electric field restoring the plasma quasi-neutrality can be found at $x > x_0$ from the 1D Poisson's equation as:

$$\frac{dE}{dx} = \frac{e}{\varepsilon_0}\left(n_i - n_e\right), \quad E = -\frac{e}{\varepsilon_0} n_{e0} x_0 \;\left(at \;\; x > x_0\right). \tag{6.16}$$

This electric field pushes all the initially shifted electrons to move back to the left (see Figure 6.4) together with their boundary (at $x = x_0$). The resulting electron motion is described by the equation of electrostatic plasma oscillations:

$$\frac{d^2 x_0}{dt^2} = -\omega_p^2 x_0, \quad \omega_p = \sqrt{\frac{e^2 n_e}{\varepsilon_0 m}}. \tag{6.17}$$

Here the parameter ω_p is the **Langmuir frequency or plasma frequency**. Comparing Eqs. (6.7) and (4.130) for plasma frequency and Debye radius, it is interesting to note, that:

$$\omega_p \times r_D = \sqrt{2T_e/m}. \tag{6.18}$$

The product of the Debye radius and the plasma frequency gives the thermal velocity of electrons. The time of a plasma reaction to an external perturbation ($1/\omega_p$) corresponds to the time required by a thermal electron (with velocity $\sqrt{2T_e/m}$) to travel the distance r_D necessary to provide screening of the external perturbation. Since the plasma (or Langmuir frequency) depends only on the plasma density it can conveniently be calculated by the following numerical formula:

$$\omega_p\left(\sec^{-1}\right) = 5.65 \cdot 10^4 \sqrt{n_e\left(cm^{-3}\right)}. \tag{6.19}$$

While different types of electrical discharges have different plasma frequencies, these usually reside in the microwave frequency region of 1–100 GHz.

6.1.7 Penetration of Slow Changing Fields into a Plasma, Skin Effect

Consider the penetration of a changing field into a plasma with a frequency less than the plasma frequency ($\omega < \omega_p$). In this case, Ohm's law can be applied in the simple form:

$$\vec{j} = \sigma \vec{E}, \tag{6.20}$$

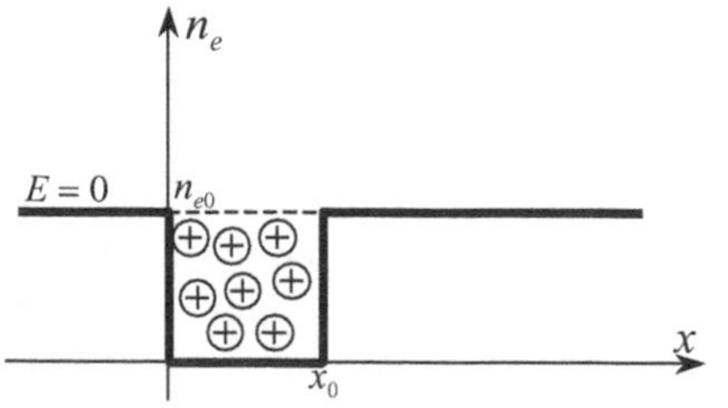

FIGURE 6.4 Electron density distribution in plasma oscillations

where $\vec{j}$ is the current density in plasma, $\vec{E}$ is the electric field strength, σ is plasma conductivity corresponding to a constant electric field (see Section 4.3.2). To describe the evolution of the field (), penetrating into plasma, the Maxwell equation for the curl of the magnetic field must be taken into account

$$curl\vec{H} = \vec{j} + \varepsilon_0 \frac{\partial \vec{E}}{\partial t}. \tag{6.21}$$

Assume the frequency is also low with respect to the plasma conductivity and the displacement current (second current term in Eq. (6.21)) can be neglected. Combining Eqs. (6.20) and (6.21) gives:

$$curl\vec{H} = \sigma\vec{E}. \tag{6.22}$$

Taking the electric field from Eq. (6.22) and substituting into the Maxwell equation:

$$curl\vec{E} = -\mu_0 \frac{\partial \vec{H}}{\partial t}, \tag{6.23}$$

which leads to the amplitude of the electromagnetic field decrease during penetration into a plasma:

$$\frac{\partial \vec{H}}{\partial t} = -\frac{1}{\mu_0\sigma} curl\, curl\, \vec{H} = -\frac{1}{\mu_0\sigma}\nabla\left(div\vec{H}\right) + \frac{1}{\mu_0\sigma}\Delta\vec{H} = \frac{1}{\mu_0\sigma}\Delta\vec{H}. \tag{6.24}$$

A similar equation can be derived for an electric field penetrating into a plasma. Equation (6.24) describes the decrease of the amplitude of low frequency electric and magnetic fields during their penetration into a plasma. The characteristic space scale of this decrease can be easily found from Eq. (6.24) as:

$$\delta = \sqrt{\frac{2}{\omega\mu_0\sigma}}. \tag{6.25}$$

If this space-scale δ is small with respect to the plasma size, then the external fields and currents are located only on the plasma surface layer (within the depth δ). This effect is known as the **skin effect**, and the boundary layer, where the external fields penetrate and where plasma currents are located, is usually referred to as the **skin-layer**. As one can see from Eq. (6.25), the depth of the skin layer depends on the frequency of the electromagnetic field ($f = \omega/2\pi$) and the plasma conductivity. For calculating the skin layer depth, it is convenient to use the following numeric formula:

$$\delta(\text{cm}) = \frac{5.03}{\sigma^{1/2}(1/\text{Ohm}\times\text{cm})\cdot f^{1/2}(\text{MHz})}. \tag{6.26}$$

6.2 MAGNETO-HYDRODYNAMICS OF PLASMA

6.2.1 Equations of Magneto-Hydrodynamics

H. Alfven (1950) first pointed out the behavior of plasma in magnetic fields. Motion of high-density plasma (the degree of ionization is assumed high in this section) induces electric currents, which together with the magnetic field influence the motion of the plasma. Such phenomena can be described by the system of magneto-hydrodynamics equations including:

a. Navier-Stokes equation neglecting viscosity, but taking into account the magnetic force on the plasma current with density $\vec{j}$, B is magnetic induction, M is the mass of ions, $n_e = n_i$, $Mn_e = \rho$:

$$Mn_e\left[\frac{\partial \vec{v}}{\partial t} + (\vec{v}\nabla)\vec{v}\right] + \nabla p = \left[\vec{j}\vec{B}\right], \tag{6.27}$$

b. Continuity equations for electrons and ions, moving together with macroscopic velocity $\vec{v}$:

$$\frac{\partial n_e}{\partial t} + div\left(n_e\vec{v}\right) = 0, \tag{6.28}$$

c. Maxwell equations for the magnetic field, neglecting the displacement current:

$$curl\vec{H} = \vec{j}, \quad div\vec{B} = 0, \tag{6.29}$$

d. Maxwell equation for electric field $curlE = -\frac{\partial \vec{B}}{\partial t}$ together with (6.29) and Ohm's law $(\vec{j} = \sigma\left(\vec{E} + \left[\vec{v}\vec{B}\right]\right)$, for plasma with conductivity σ) give the relation for the magnetic field:

$$\frac{\partial \vec{B}}{\partial t} = curl\left[\vec{v}\vec{B}\right] + \frac{1}{\sigma\mu_0}\Delta\vec{B}. \tag{6.30}$$

Magneto-hydrodynamics (MHD) is able to describe plasma equilibrium and confinement in magnetic fields, which is of great importance in problems related to the thermonuclear plasma systems (M.D. Kruscal, R.M. Kulsrud, 1958; B.B. Kadomtsev, 1958).

6.2.2 Magnetic Field "Diffusion" in a Plasma, Effect of Magnetic Field Frozen in a Plasma

If plasma is at rest ($\vec{v} = 0$), Eq. (6.30) can be reduced to the diffusion equation:

$$\frac{\partial \vec{B}}{\partial t} = D_m\Delta\vec{B}, \quad D_m = \frac{1}{\sigma\mu_0}. \tag{6.31}$$

The factor D_m can be interpreted as a coefficient of "diffusion" of the magnetic field into the plasma. Sometimes it is also called the **magnetic viscosity**. If the characteristic time of the magnetic field change is $\tau = 1/\omega$, then the characteristic length of magnetic field diffusion according to Eq. (6.31) is $\delta \approx \sqrt{2D_m\tau} = \sqrt{2/\sigma\omega\mu_0}$, which again is the skin-layer depth (6.25). Equation (6.31) can be applied to describe the damping time τ_m of currents and magnetic fields in a conductor with characteristic size L:

$$\tau_m = \frac{L^2}{D_m} = \mu_0\sigma L^2. \tag{6.32}$$

According to Eq. (6.32), the damping time is infinitely high for superconductors. The magnetic field damping time is also very long for large special objects; for solar spots this time exceeds 300 years. Equations (6.31) and (6.32) can be interpreted in another way. If the plasma conductivity is high ($\sigma \to \infty$), the diffusion coefficient of the magnetic field is small ($D_m \to 0$) and the magnetic field is unable to "move" with respect to the plasma. Thus, the magnetic field sticks to plasma or in other words the: **magnetic field is frozen in the plasma**. Consider displacement (correspondent to time interval dt) of the surface element ΔS, moving with velocity $\vec{v}$ together with plasma (see Figure 6.5). The effect takes place when the plasma conductivity is high ($\sigma \to \infty$) and the second left-hand side term of Eq. (6.30) can be neglected:

$$\frac{\partial \vec{B}}{\partial t} = curl\left[\vec{v}\vec{B}\right]. \tag{6.33}$$

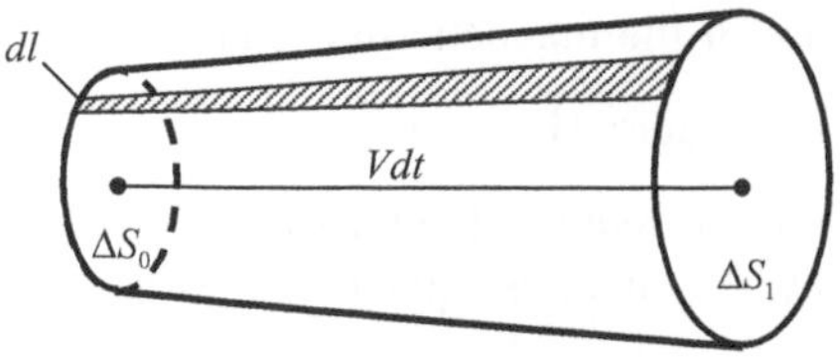

FIGURE 6.5 Illustration of magnetic field frozen in plasma.

Because of $div\vec{B} = 0$, difference in fluxes $\vec{B}\frac{d\vec{S}}{dt}$ through surface elements ΔS_0 and ΔS_1 is equal to the flux $\oint \vec{B}\left[\vec{v}\,d\vec{l}\,\right] = \oint \left[\vec{B}\vec{v}\,\right]d\vec{l}$ through the side surface of the fluid element in Figure 6.5. Based on this fact and Eq. (6.33), the time derivative of the magnetic field flux $\int \vec{B}\,d\vec{S}$ through the moving surface element ΔS can be expressed as:

$$\frac{d}{dt}\int \vec{B}d\vec{S} = \int \frac{\partial \vec{B}}{\partial t}d\vec{S} + \int \vec{B}\frac{d\vec{S}}{dt} = \int \frac{\partial \vec{B}}{\partial t}d\vec{S} + \oint \left[\vec{B}\vec{v}\,\right]d\vec{l} = \int curl\left[\vec{v}\vec{B}\,\right]d\vec{S} + \oint \left[\vec{B}\vec{v}\,\right]d\vec{l}. \tag{6.34}$$

According to Stokes theorem, the last sum of equality (6.34) is equal to zero:

$$\frac{d\Phi}{dt} = \frac{d}{dt}\int \vec{B}\,d\vec{S} = 0, \quad \Phi = const. \tag{6.35}$$

Equation (6.35) shows that magnetic flux Φ through any surface element moving with the plasma is constant. It again shows that the magnetic field is "frozen" in moving plasma with high conductivity.

6.2.3 Magnetic Pressure, Plasma Equilibrium in Magnetic Field

Under steady-state plasma conditions ($d\vec{v}/dt = 0$), the Navier-Stokes equation (6.27) can be simplified to:

$$grad\,p = \left[\vec{j}\vec{B}\,\right]. \tag{6.36}$$

This equation can be interpreted as a balance of hydrostatic pressure p and ampere force. Taking into account the first of Eq. (6.29), eliminate the current from the balance of forces Eq. (6.36):

$$\nabla p = \left[\vec{j}\vec{B}\,\right] = \mu_0\left[curl\vec{H} \times \vec{H}\,\right] = -\frac{\mu_0}{2}\nabla H^2 + \mu_0\left(\vec{H}\nabla\right)\vec{H}. \tag{6.37}$$

Combination of the gradients in Eq. (6.37) leads to the following equation describing plasma equilibrium in the magnetic field:

$$\nabla\left(p + \frac{\mu_0 H^2}{2}\right) = \mu_0\left(\vec{H}\nabla\right)\vec{H} = \frac{\mu_0 H^2}{R}\vec{n}. \tag{6.38}$$

Here R is the radius of curvature of the magnetic field line; $\vec{n}$ is the unit normal vector to the line (directed inside to the curvature center). Hence, the force $\frac{\mu_0 H^2}{R}\vec{n}$ is related to the bending of the magnetic field lines and can be interpreted as the **tension of magnetic lines**. This tension tends to make the magnetic field lines straight, and it is equal to zero when the lines are straight. The pressure term $\frac{\mu_0 H^2}{2}$ is usually called **magnetic pressure.** The sum of the hydrostatic and magnetic pressures $p+\frac{\mu_0 H^2}{2}$ is referred to as the total pressure. Equation (6.38) for plasma equilibrium in the magnetic field can be considered as the dynamic balance of the gradient of total pressure and the tension of magnetic lines. If the magnetic field lines are straight and parallel, then $R \to \infty$ and the "tension" of magnetic lines is equal to zero. In this case, Eq. (6.38) gives the equilibrium criterion:

$$p+\frac{\mu_0 H^2}{2} = const. \tag{6.39}$$

Plasma equilibrium in a magnetic field means a balance of pressure in plasma with the outside magnetic pressure. Plasma equilibrium is also possible sometimes for special configurations of magnetic fields, when $p \ll \frac{\mu_0 H^2}{2}$. In this case, the outside magnetic pressure is compensated by the "tension" of magnetic lines. Such equilibrium configurations of the magnetic field are called forceless configurations.

6.2.4 The Pinch-Effect

This effect means self-compression of plasma in its own magnetic field. Consider the Pinch effect in a long cylindrical discharge plasma with the electric current along the axis of the cylinder (see Figure 6.6). This is the so-called Z-pinch (W.P. Allis, 1960). Equilibrium of the completely ionized

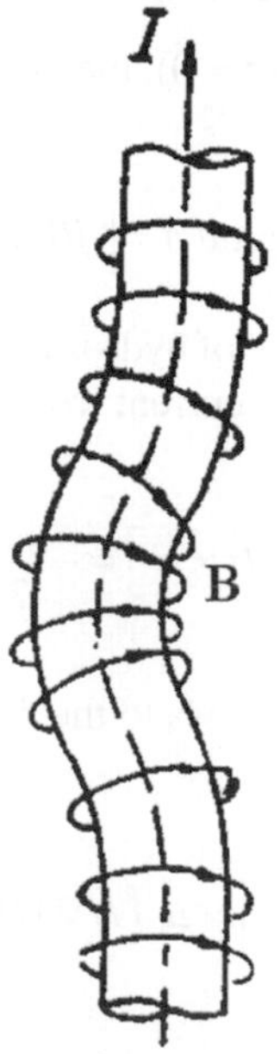

FIGURE 6.6 Illustration of the pinch effect, when a discharge channel is bent.

Z-pinch plasma is determined by Eqs. (6.36) or (6.38), combined with the first Maxwell equation (6.29) for the self magnetic field of the plasma column:

$$N_L T = \frac{\mu_0}{8\pi} I^2. \tag{6.40}$$

This equilibrium criterion is called the **Bennet relation** (W.H. Bennett, 1934). In this relation N_L is the plasma density per unit length of the cylinder; T is the plasma temperature. The Bennett relation shows that plasma temperature should grow proportionally to the square of the current to provide a balance of plasma pressure and magnetic pressure of the current. Thus, to reach thermonuclear temperatures of about 100 keV in a plasma with density 10^{15} cm^{-3} and cross section 1 cm^2, the necessary current according to Eq. (6.40) should be about 100 kA. A current of O(100 KA) can be achieved in Z-pinch discharges, which in the early fifties stimulated enthusiasm in the controlled thermonuclear fusion research. However, hopes for easy controlled fusion in Z-pinch discharges were shattered, .because very fast instabilities arose to destroy the plasma. These very fast instabilities of Z-pinch, related to plasma bent and non-uniform plasma compression, are illustrated in Figure 6.6. If the discharge column is bent, the magnetic field and magnetic pressure become larger on the concave side of the plasma, which leads to a break of the channel. This instability of Z-pinch is usually referred to as the "wriggle" instability. Similar to the "wriggle" instability, if the discharge channel becomes locally thinner, the magnetic field ($B \propto 1/r$) and magnetic pressure at this point grow leading to further compression and to a subsequent break of the channel.

6.2.5 Two-Fluid Magneto-Hydrodynamics, the Generalized Ohm's Law

In the magneto-hydrodynamic approach, the electron $\vec{v_e}$ and ion $\vec{v_i}$. velocities were considered equal. Actually, this is in contradiction with the presence of electric current $\vec{j} = en_e\left(\vec{v_i} - \vec{v_e}\right)$ in the quasi-ncutral plasma with density n_e. Thus, it is clear that effects related to the separate motion of electrons and ions require consideration under a two-fluid model of magneto-hydrodynamics. In the two-fluid MHD model, the Navier-Stokes equation for electrons is somewhat similar to Eq. (6.1), but includes electron's mass m, pressure p_e, velocity, and additionally takes into account the friction between electrons and ions (which corresponds to the last term in the equation, where ν_e is the frequency of electron collisions):

$$mn_e \frac{d\vec{v}_e}{dt} + \nabla p_e = -en_e\vec{E} - en_e\left[\vec{v}_e\vec{B}\right] - mn_e\nu_e\left(\vec{v}_e - \vec{v}_i\right). \tag{6.41}$$

A similar Navier-Stokes equation for ions includes the same friction term, but with opposite sign. First term in Eq. (6.41) is related to the electron inertia and can be neglected because of the very low electron mass. Also denoting the ion's velocity as $\vec{v}$ and the plasma conductivity as $\sigma = n_e e^2/m\nu_e$, we can rewrite the equation (6.41) in the form known as the **generalized Ohm's law**:

$$\vec{j} = \sigma\left(\vec{E} + \left[\vec{v}\vec{B}\right]\right) + \frac{\sigma}{en_e}\nabla p_e - \frac{\sigma}{en_e}\left[\vec{j}\vec{B}\right]. \tag{6.42}$$

The generalized Ohm's law in contrast to the conventional one takes into account the electron pressure gradient and the $[\vec{j}x\vec{B}]$ term related to the **Hall effect**. Both these two additional terms show, that the current direction is not always straight correlated with the direction of the electric field (even corrected by $[\vec{v}\vec{B}]$). The Hall effect is related to the electron conductivity in the presence of a magnetic field, which provides electric current in the direction perpendicular to the electric field. This effect was discussed earlier in Section 4.3.4, (see Eq. (4.109)) in terms of plasma drift in crossed electric and magnetic fields. The generalized Ohm's law can be applied as well as the equations of motion for individual species to analyze different types of charged particle drifts in magnetic fields (see W.B. Thompson, 1962). The solution of Eq. (6.42) with respect to electric current is complicated because of the presence of current in two terms of the generalized Ohm's law. This equation can be solved if the plasma conductivity is sufficiently large. In this case ($\sigma \to \infty$), the generalized Ohm's law can be expressed in the form (B.B. Kadomtsev, 1976):

$$\vec{E} + \left[\vec{v}\vec{B}\right] + \frac{1}{en_e}\nabla p_e = \frac{1}{en_e}\left[\vec{j}\vec{B}\right]. \tag{6.43}$$

If the electron temperature is uniform, the electron hydrodynamic velocity $\vec{v}$ can be used and the generalized Ohm's law Eq. (6.43) rewritten as:

$$\vec{E} = -\left[\vec{v}_e\vec{B}\right] - \frac{1}{e}\nabla\left(T_e \ln n_e\right). \tag{6.44}$$

Using this expression for the electric field in the Maxwell equation $curl E = -\frac{\partial \vec{B}}{\partial t}$, and taking into account that curl of the gradient is equal to zero,:

$$\frac{\partial \vec{B}}{\partial t} = curl \geq \left[\vec{v}_e\vec{B}\right]. \tag{6.45}$$

This form of two-fluid magneto-hydrodynamics is similar to Eq. (6.33).

6.2.6 Plasma Diffusion across Magnetic Field

The generalized Ohm's law Eq. (6.42) of the two-fluid magneto-hydrodynamics model is able to describe the electrons and ions diffusion in a direction perpendicular to the magnetic field. Calculations of the diffusion coefficients for electrons and ions in this direction in the framework of this model gives (S.I. Braginsky, 1963):

$$D_{\perp,e} = \frac{D_e}{1+\left(\frac{\omega_{B,e}}{\nu_e}\right)^2}, \quad D_{\perp,i} = \frac{D_i}{1+\left(\frac{\omega_{B,i}}{\nu_i}\right)^2}. \tag{6.46}$$

In this relation: D_e and D_i are the coefficients of free diffusion of electrons and ions in plasma without magnetic field (see Section 4.3.7); ν_e and ν_i are the collisional frequencies of electrons and ions; $\omega_{B,e}$ and $\omega_{B,i}$ are the electron and ion cyclotron frequencies (see (4.107)):

$$\omega_{B,e} = \frac{eB}{m}, \quad \omega_{B,i} = \frac{eB}{M}, \tag{6.47}$$

which show the frequencies of electron and ion collisionless rotation in magnetic field. Since ions are much heavier than electrons (M >> m), the electron-cyclotron frequency much greater than the ion-cyclotron frequency in the same magnetic field B. If diffusion is ambipolar, which is actually the case in the highly ionized plasma under consideration, then Eq. (4.127) can still be applied to find D_a in a magnetic field. Obviously, the free diffusion coefficients for electrons and ions in Eq. (4.127) should be replaced by those in the magnetic field Eq. (6.46). The regular electron and ion mobilities μ_e, μ_i in Eq. (4.127) also should be replaced by those corresponding to the drift perpendicular to the magnetic field Eq. (4.109). This leads to the following coefficient of ambipolar diffusion perpendicular to the magnetic field:

$$D_{\perp} = \frac{D_a}{1+\frac{\omega_{B,i}^2}{\nu_i^2}+\frac{\mu_i}{\mu_e}\left(1+\frac{\omega_{B,e}^2}{\nu_e^2}\right)}. \tag{6.48}$$

Here D_a is the regular coefficient of ambipolar diffusion Eq. (4.127) in plasma without a magnetic field. Equation (6.48) for ambipolar perpendicular to the magnetic field can be simplified when electrons are magnetized $\omega_{B,e}/\nu_e >> 1$, trapped in the magnetic field, and rotating around the magnetic lines. If at the same time heavy ions are not magnetized ($\omega_{B,i}/\nu_i << 1$), then Eq. (6.48) can be rewritten as:

$$D_\perp = \frac{D_a}{1 + \frac{\mu_i}{\mu_e}\frac{\omega_{B,e}^2}{\nu_e^2}}. \tag{6.49}$$

Under strong magnetic fields when $\frac{\mu_i}{\mu_e}\frac{\omega_{B,e}^2}{\nu_e^2} >> 1$ the relation for ambipolar diffusion is:

$$D_\perp \approx D_a \frac{\mu_e}{\mu_i}\frac{\nu_e^2}{\omega_{B,e}^2} \approx D_e \frac{\nu_e^2}{(e/M)^2}\frac{1}{B^2}. \tag{6.50}$$

In strong magnetic fields, the coefficient $D_\perp$ significantly decreases with the strength of the magnetic field: $D_\perp \propto 1/B^2$. The magnetic field is not transparent for the plasma and can be used to prevent the plasma from decay. This as well as the MHD plasma equilibrium is very important for plasma confinement in a magnetic field. This was experimentally demonstrated by N.D'Angelo and N. Rynn (1961). The slow ambipolar diffusion across the strong magnetic field can be interpreted using the illustration in Figure 6.7. The magnetized electron is trapped by the magnetic field and rotates along the Larmor circles until a collision pushes the electron to another Larmor circle. The **electron Larmor radius**:

$$\rho_L = \frac{v_\perp}{\omega_{B,e}} = \frac{1}{eB}\sqrt{2T_e m}. \tag{6.51}$$

is the radius of the circular motion of a magnetized electron (see Figure 6.7); here $v_\perp$ is the component of the electron thermal velocity perpendicular to the magnetic field. In the case of diffusion across the strong magnetic field, the Larmor radius plays the same role as the mean free path plays in diffusion without a magnetic field. This can be illustrated by rewriting Eq. (6.50) in terms of the electron Larmor radius:

$$D_\perp \approx D_e \frac{\nu_e^2}{\omega_{B,e}^2} \approx \rho_L^2 \nu_e. \tag{6.52}$$

Eq. (6.52) for the diffusion coefficient across the strong magnetic field is similar to Eq. (4.122) for free diffusion without a magnetic field, but with the mean free path replaced by the Larmor radius. When the magnetic field is high and electrons are magnetized ($\omega_{B,i}/\nu_i << 1$), then the electron Larmor radius is shorter than the electron mean free path ($\rho_L << \lambda_e$). Plasma diffusion across

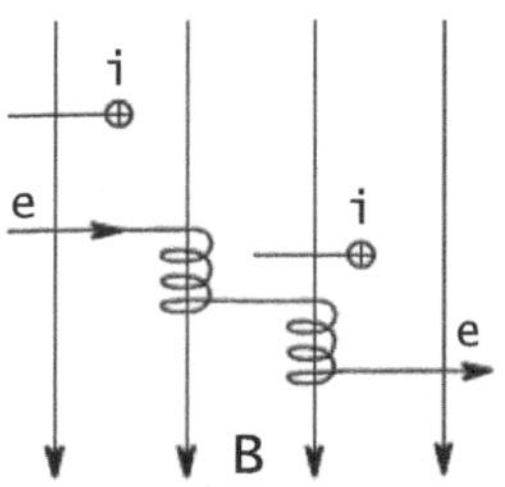

FIGURE 6.7 Ambipolar diffusion across magnetic field.

the strong magnetic field then is slower than without the magnetic field and decreases with the strength of the field.

6.2.7 Conditions for Magneto-Hydrodynamic Behavior of Plasma: the Alfven Velocity and the Magnetic Reynolds Number

Determine the plasma conditions, when the magneto-hydrodynamic effects take place, for example, when the fluid dynamics is strongly coupled with the magnetic field. Magneto-hydrodynamics demands the "diffusion" of the magnetic field to be less than the "convection", that is the curl in the right-hand side of Eq. (6.30) should exceed the Laplacian (the space-scale of plasma is L):

$$\frac{vB}{L} >> \frac{1}{\sigma\mu_0}\frac{B}{L^2}, \quad \text{or} \quad v >> \frac{1}{\sigma\mu_0 L}. \tag{6.53}$$

Using the concept of magnetic viscosity D_m Eq. (6.31), we can rewrite the condition Eq. (6.53) of magneto-hydrodynamic behavior as:

$$\mathrm{Re}_m = \frac{vL}{D_m} >> 1. \tag{6.54}$$

Here $\mathbf{Re_m}$, is the **magnetic Reynolds number** that is somewhat similar to the conventional Reynolds number, but with kinematic viscosity replaced by the magnetic viscosity. The physical interpretation of the magnetic can be further clarified that the plasma velocity v in magneto-hydrodynamic systems usually satisfies the approximate balance of dynamic and magnetic pressures:

$$\frac{\rho v^2}{2} \propto \frac{\mu_0 H^2}{2}, \quad \text{or} \quad v \propto v_A = \frac{B}{\sqrt{\rho\mu_0}}. \tag{6.55}$$

$\rho = Mn_e$ is the plasma density. The characteristic plasma velocity v_A corresponding to the equality of the dynamic and magnetic pressures is called the **Alfven velocity**. The criterion of magneto-hydrodynamic behavior can then be written:

$$\mathrm{Re}_m = \frac{v_A L}{D_m} = BL\sigma\sqrt{\frac{\mu_0}{\rho}} >> 1. \tag{6.56}$$

This form of the magnetic Reynolds number and criterion for magneto-hydrodynamic behavior was derived by S. Lundquist (1952). The magnetic Reynolds numbers for some laboratory and nature plasmas are presented in Table 6.2, see F. Chen (1984), P.H. Rutherford, R.J. Goldston (1995).

TABLE 6.2
The Magnetic Reynolds Numbers in Different Laboratory and Nature Plasmas

Type of plasma	B, Tesla	Space-scale, m	ρ, kg/m³	σ, 1/Ohm × cm	R_m
Ionosphere	10^{-5}	10^5	10^{-5}	0.1	10
Solar atmosphere	10^{-2}	10^7	10^{-6}	10	10^8
Solar corona	10^{-9}	10^9	10^{-17}	10^4	10^{11}
Hot interstellar gas	10^{-10}	10 light yr	10^{-21}	10	10^{15}
Arc discharge plasma	0.1	0.1	10^{-5}	10^3	10^3
Hot confined plasma, $n = 10^{15}\text{cm}^{-3}$, $T = 10^6$ K	0.1	0.1	10^{-6}	10^3	10^4

6.3 INSTABILITIES OF LOW-TEMPERATURE PLASMA

6.3.1 Types of Instabilities of Low-Temperature Plasmas, Peculiarities of Plasma-Chemical Systems

The instabilities of hot confined plasmas, are not subject of this book, see D.B. Melrose (1989) and A.B. Mikhailovskii (1998a). The most serious instabilities of non-thermal, non-equilibrium plasmas are related to the system's tendency to restore thermodynamic quasi-equilibrium between different degrees of freedom. Typically in non-equilibrium plasmas: electron, vibrational and translational temperatures remain in the following hierarchy: $T_e >> T_v >> T_0$. The homogeneous state of non-equilibrium discharges is possible only in a limited range of parameters. Usually this means low pressures, low specific energy inputs and low specific powers (which is power per unit volume). Small fluctuation of the plasma parameters beyond this stability range can increase exponentially, resulting in the discharge transition into a non-homogeneous form. The optimal specific energy input for quite a few non-equilibrium plasma-chemical processes is about $E_v \approx 1\,eV/mol$. This specific energy input exceeds about 10 times the optimal one ($E_v \approx 0.1\,eV/mol$) for highly effective non-equilibrium discharges applied for stimulation of gas lasers. The requirements of $E_v \approx 0.1\,eV/mol$ is related to the final gas heating, while the optimal energy input $E_v \approx 1\,eV/mol$ is related to the initial energy going mostly into vibrational excitation. If the energy efficiency of the plasma-chemical process reaches 90%, the requirement can be met. Plasma-chemical systems in contrast to gas lasers can be effective in non-homogeneous systems as well if the required level of the plasma non-equilibrium remains (V.K. Givotov, I.A. Kalachev, E.G. Krasheninnikov e.a., 1983a).

Striation is an instability related to the formation of plasma structure, that looks like a series of alternating light and dark layers along the discharge current. The general appearance of a striated discharge is shown in Figure 6.8. Normally, the striations are able to move fast with velocities up to 100 m/sec, but also can be at rest usually the conditions of their appearance do not depend on pressure and specific energy input. It is very important, that the striations do not significantly affect plasma parameters. The characteristics of plasma-chemical processes in non-equilibrium discharges with and without striations are close. Often the striations are even not detectable by the naked eye. The physical phenomenon of striation is related to ionization instability (see below) and can be interpreted as ionization oscillations and waves (A.V. Nedospasov, 1968; P.S. Landa, N.A. Miskinova, Yu. V. Ponomarev, 1978). **Contraction** is an instability related to plasma "self-compression" into one or several bright current filaments. The contraction takes place when the pressure or specific energy input exceeds some critical values. In contrast to striation, contraction significantly changes plasma parameters. The plasma filament formed as a result of contraction is close to

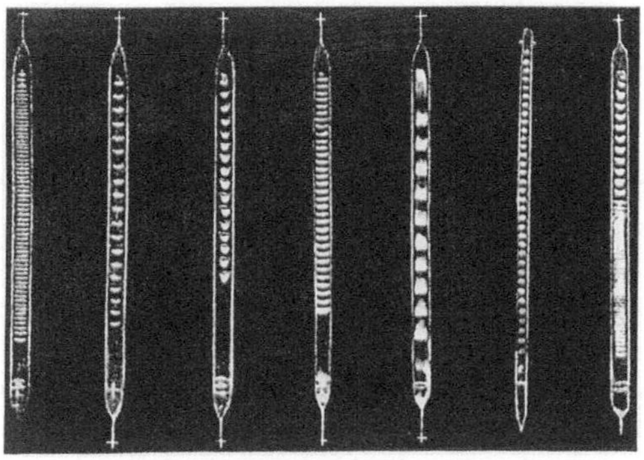

FIGURE 6.8 General view of striated discharges.

quasi-equilibrium. For this reason, contraction is a serious factor limiting power and efficiency of gas lasers and non-equilibrium plasma-chemical systems.

6.3.2 The Thermal (Ionization-Overheating) Instability in Monatomic Gases

Consider the ionization-overheating instability (often called "thermal" instability) for plasma of monatomic gases. The instability is due to the strong exponential dependence of the ionization rate coefficient and hence the electron concentration on the reduced electric field E_0/n_0. The thermal instability can be illustrated by the following closed chain of causal links, which can start from the fluctuation of the electron concentration:

$$\delta n_e \uparrow \to \delta T_0 \uparrow \to \delta n_0 \downarrow \to \delta\left(\frac{E}{n_0}\right) \uparrow \to \delta n_e \uparrow. \tag{6.57}$$

The local increase of electron concentration δn_e leads to the intensification of gas heating by electron impact and, hence, to an increase of temperature δT_0. Taking into account that pressure $p = n_0 T_0$ is constant, the local increase of temperature δT_0 leads to a decrease in gas density δn_0 and to an increase of the reduced electric field $\delta\left(\frac{E}{n_0}\right)$ (electric field E = const). Finally, the increase of the reduced electric field $\delta\left(\frac{E}{n_0}\right)$ results in a further increase in electron concentration δn_e, which makes the chain Eq. (6.57) closed and determines the positive feedback. The sequence Eq. (6.57) gives the physical meaning of thermal instability. Because of the strong dependence of the ionization rate on E/n_0, a small local initial overheating δT_{00} according to the mechanism (6.57) grows up exponentially:

$$\delta T_0(t) = \delta T_{00} \cdot \exp \Omega t. \tag{6.58}$$

The exponentially increased in overheating leads to the formation of the hot filament, or in other words to contraction. Here δT_{00} is the initial perturbation of temperature. The parameter of the exponential growth of the initial perturbation is called **the instability increment**. If the system is stable, the instability increment is negative ($\Omega > 0$), the positive increment ($\Omega > 0$) means the actual instability. The increment of thermal instability Eq. (6.57) can be found by the linearization of the differential equations of heat and ionization balance. In the case of high or moderate pressure discharges, when the influence of walls can be neglected, the thermal instability increment is equal to (see, for example, - W.L. Nighan, 1976):

$$\Omega = \hat{k}_i \frac{\sigma E^2}{n_o c_p T_0} = \hat{k}_i \frac{\gamma - 1}{\gamma} \frac{\sigma E^2}{p}. \tag{6.59}$$

Here the dimensionless factor $\tilde{k}_i = \frac{\partial \ln k_i}{\partial \ln T_e}$ is sensitivity of the ionization rate to electron temperature (directly related to E/n_0); this factor is usually about 10; σ is plasma conductivity, p is pressure, c_p and γ are specific heat and specific heat ratio. $\nu_{Tp} = \frac{\gamma - 1}{\gamma} \frac{\sigma E^2}{p}$ is frequency of gas heating by electric current at constant pressure, therefore the instability increment Eq. (6.59) is determined by the heating frequency and exceeds ν_{Tp} because of the strong sensitivity of ionization to electron temperature. Equationss (6.59) show that steady-state discharges at high and moderate pressures are always unstable with respect to the considered thermal instability. The thermal instability Eq. (6.57) can be suppressed by intense cooling as in the case of heat losses to the walls at low discharge pressures.

Also one can avoid thermal instability if the specific energy input is low and does not provide sufficient energy for contraction. In other words the thermal instability can be neglected, if the gas residence time in discharge (or discharge duration) is small with respect to the thermal instability time ($1/\Omega$).

6.3.3 The Thermal (Ionization-Overheating) Instability in Molecular Gases with Effective Vibrational Excitation

The key difference of the thermal instability in molecular gases is related to the fact that fluctuation of electron density δn_e does not lead directly to heating δT_0, but rather through intermediate vibrational excitation. Vibrational VT relaxation is relatively slow ($\tau_{VT}\nu_{Tp} >> 1$, τ_{VT}-is the VT-relaxation time) in the highly effective laser and plasma-chemical systems. In this case, the thermal instability becomes more sensitive to VT-relaxation than to excitation (ν_{VT}) itself as was the case in monatomic gases Eq. (6.59). In the most interesting case of fast vibrational excitation and slow relaxation ($\tau_{VT}\nu_{Tp} >> 1$), the increment of thermal instability in the molecular gas plasma can be expressed as (R. Haas, 1973; W.L. Nighan, W.G. Wiegand, 1974):

$$\Omega_T = \frac{b}{2} \pm \sqrt{\frac{b^2}{4} + c}, \tag{6.60}$$

where the parameters "b" and "c" are:

$$b = \frac{1}{\tau_{VT}}\left(1 - \frac{\partial \ln \tau_{VT}}{\partial \ln \varepsilon_v}\right) + \nu_{Tp}\left(2 + \hat{\tau}_{VT}\right), \tag{6.61}$$

$$c = \frac{\nu_{Tp}}{\tau_{VT}}\left(-\frac{\partial \ln n_e}{\partial \ln n_0}\right)\left(1 - \frac{\partial \ln \tau_{VT}}{\partial \ln \varepsilon_v}\right). \tag{6.62}$$

In these relations the dimensionless factor $\hat{\tau}_{VT} = \frac{\partial \ln \tau_{VT}}{\partial \ln T_0}$ is the logarithmic sensitivity of the vibrational relaxation time to translational temperature; numerically this factor is usually about $-3 \div -5$; ε_v is vibrational energy of a molecule. Analysis of Eqs. (6.60)–(6.62) shows that thermal instability in molecular gases can proceed in two different ways, thermal and vibrational modes:

1. **The thermal mode of instability** corresponds to the condition: $b < 0$. Taking into account the relatively high rate of vibrational excitation ($\nu_{Tp}\tau_{VT} >> 1$), one can see that $b^2 >> c$. Then Eq. (6.60) gives the instability increment for thermal mode:

$$\Omega_T = |b| \approx -\nu_{Tp}\left(2 + \hat{\tau}_{VT}\right) = k_{VT} n_0 \frac{\hbar\omega}{c_p T_0}\left(\hat{k}_{VT} - 2\right). \tag{6.63}$$

   In this relation, the dimensionless factor $\hat{k}_{VT} = \frac{\partial \ln k_{VT}}{\partial \ln T_0}$ is the logarithmic sensitivity of the vibrational relaxation rate coefficient to translational temperature T_0. Numerically this sensitivity factor is usually about $3 \div 5$. As seen from Eq. (6.63), the increment of instability in the thermal mode corresponds to the frequency of heating due to vibrational VT-relaxation, multiplied by $\left(\hat{k}_{VT} - 2\right) >> 1$ because of the strong exponential dependence of the Landau-Teller relaxation rate coefficient on gas temperature (see Section 3.4.3). The pure thermal mode does not include ionization (only overheating, no factors directly related to n_e). Hence, it is the "overheating" part of the whole "ionization-overheating" instability.
2. **The vibrational mode** corresponds to the condition: $c > 0$ (at any sign of the parameter b).

The instability increment at a high rate of excitation ($\nu_{Tp}\tau_{VT} >> 1$) can be expressed for this mode as:

$$\Omega_T = \frac{c}{|b|} \approx k_{VT} n_0 \frac{\hat{k}_i}{\hat{k}_{VT} - 2 - \left(\nu_{Tp}\tau_{Tp}\right)^{-1}}. \tag{6.64}$$

The instability increment for the vibrational mode is mostly characterized by the frequency of vibrational relaxation. In contrast to thermal mode, the sensitivity term in Eq. (6.64) does include factors related to the electron concentration and the overheating effect on the ionization rate. This makes the instability similar to that in monatomic gases Eq. (6.59). Thus, it is the "ionization" part of the whole "ionization-overheating" instability. In the case of slow excitation ($\nu_{Tp}\tau_{VT} << 1$, the instability increment is:

$$\Omega_T = \sqrt{\frac{\nu_{Tp}}{\tau_{VT}}\left(-\frac{\partial \ln n_e}{\partial \ln n_0}\right)\left(1 - \frac{\partial \ln \tau_{VT}}{\partial \ln \varepsilon_v}\right)}. \tag{6.65}$$

6.3.4 Physical Interpretation of Thermal and Vibrational Instability Modes

The different modes of thermal instability discussed above are illustrated in Figure 6.9. In the same manner, as in the case of monatomic gases Eq. (6.57), local decreases of gas density in molecular gases leads to an increase of the reduced electric field and electron temperature. An increase of electron temperature results in an acceleration of ionization, a growth of electron concentration, and vibrational temperature. Growth of the vibrational temperature makes VT-relaxation and heating more intensive, which leads to an increase of the translational temperature and to a decrease of the gas density (because $p = n_0 T$ = const). This is the mechanism of the vibrational mode of instability, which is somewhat similar to Eq. (6.57). Thermal mode, which is also shown in Figure 6.9, is not directly related to electron density, temperature and to ionization. This is due to the strong exponential dependence of the VT-relaxation rate coefficient on translational temperature. Even small increases of the translational temperature lead to a significant acceleration of the VT-relaxation rate (Landau-Teller mechanism, see Section 3.4.3), intensification of heating and then to further growth of the translational temperature. This phenomenon is called sometimes

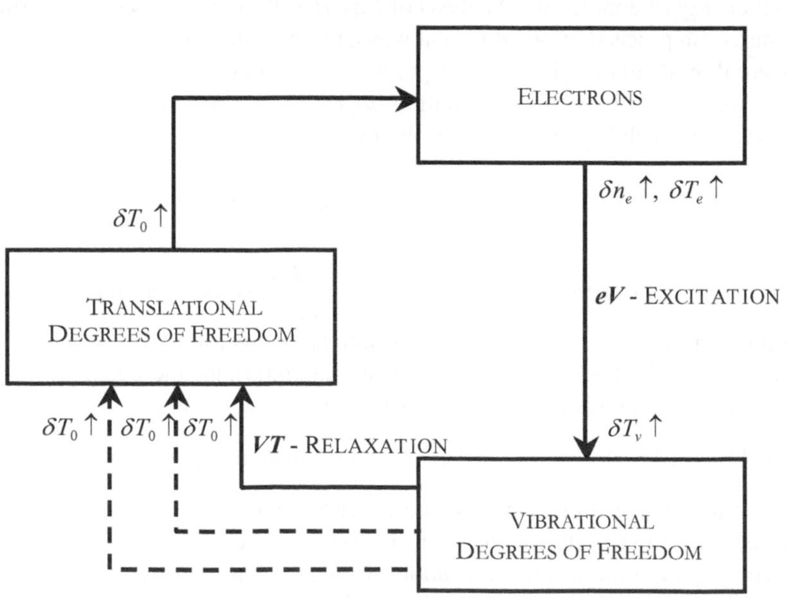

FIGURE 6.9 Thermal instability in molecular gases: solid lines – ibrational mode, dashed line – thermal mode.

– the *thermal explosion of vibrational reservoir*. An increase of temperature in the thermal mode, obviously, also stimulates the growth of the electron density, but it is not included inside of the principal chain of events in this mode of instability. The instability increment in both thermal Eq. (6.63) and vibrational Eq. (6.64) modes usually has close numerical values. Sometimes, however, even the qualitative behavior of these two instability modes is different. For example, vibrational energy losses into translational degrees of freedom at strong vibrational non-equilibrium are mostly provided by VV and VT relaxation from high vibrational levels (see Sections 5.3.8 and 5.3.9). In this case, an increase of translational temperature T_0 leads to a decrease in the effective VT relaxation and gas heating. The thermal instability mode is impossible under such conditions. At the same time, the increment of the vibrational mode even increases. The opposite situation takes place in non-equilibrium discharges sustained in supersonic flows (see next section). In this case, gas heating leads not to a decrease but to an increase of gas density, and to a reduction of electron temperature and reduction of further heating. This means the plasma is stable with respect to the vibrational mode, but the thermal mode of instability is still in place. Plasma-chemical processes have both stabilizing and destabilizing effects on instabilities. A significant part of the vibrational energy in plasma-chemical systems can be consumed in endothermic reactions instead of heating. This provides plasma stabilization with respect to the ionization-overheating instability. Destabilization in turn is due to fast heat release in exothermic reactions.

6.3.5 Non-Equilibrium Plasma Stabilization by Chemical Reactions of Vibrationally Excited Molecules

Both modes Eqs. (6.63) and (6.64) of the strong thermal instability, which leads to discharge contraction, have typical time comparable with the time of VT relaxation ($\tau_{VT} = 1/k_{VT}n_0$). This makes the thermal instability less dangerous for effective plasma-chemical processes, where the reaction time is approximately the time of vibrational excitation ($\tau_{eV} = 1/k_{eV}n_e$) and is shorter than the time of VTrelaxation (see Eq. (5.150)). An effective plasma-chemical process can be completed before the development of the strong but slow thermal instability. One of such **fast ionization instabilities** with a frequency approximately $\tau_{eV} = 1/k_{eV}n_e$ is due to a direct increase of T_e and acceleration of ionization by an increase of T_v (A.A. Vedenov, 1973). The effect is provided by super-elastic collisions of electrons with vibrationally excited molecules (Section 4.2.5 Eq. (4.85)). The scheme of fast ionization instability is shown in Figure 6.10. The initial small increase of electron

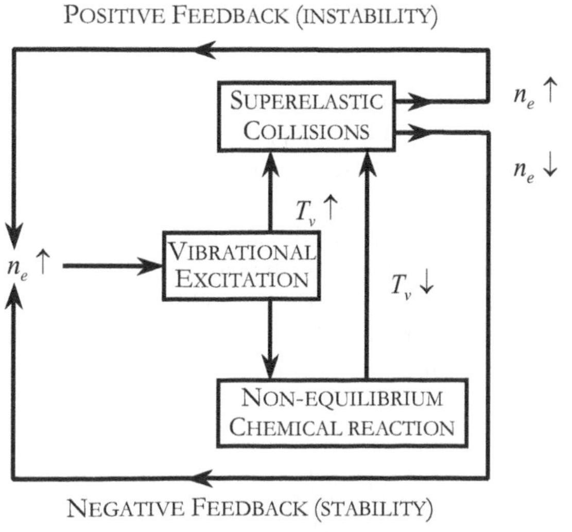

FIGURE 6.10 Ionization instability of molecular gas in chemically active plasma.

concentration leads to an intensification of vibrational excitation and growth of the vibrational temperature. Higher vibrational temperatures because of super-elastic collisions result in acceleration of ionization and a further increase in the electron concentration. This instability includes neither VT relaxation nor any heating, and it is fast (controlled by vibrational excitation $\tau_{eV} = 1/k_{eV}n_e$). This instability can be stabilized by endothermic chemical reactions consuming vibrational energy and decreasing T_v, which is also illustrated in Figure 6.10. Increment of this fast ionization instability can be calculated by linearization of the differential equations for the vibrational energy balance and balance of electrons, taking into account the endothermic reactions stimulated by vibrational excitation (I.A. Kirillov, V.D. Rusanov, A. Fridman, 1984e):

$$\Omega_v = k_{eV} n_e \frac{\hbar\omega}{c_v^v T_v}\left(\tilde{k}_i - \tilde{k}_r\right). \tag{6.66}$$

Here $c_v^v = \dfrac{\partial \varepsilon_v}{\partial T_v}$ is the vibrational heat capacity; $\tilde{k}_i = \dfrac{\partial \ln k_i}{\partial \ln T_v}$ is the dimensionless sensitivity of the ionization rate coefficient to the vibrational temperature; $\tilde{k}_r = \dfrac{\partial \ln k_r}{\partial \ln T_v}$ is the dimensionless sensitivity of the chemical reaction rate to vibrational temperature (in the case of weak excitation $\tilde{k}_r = \dfrac{E_a}{T_v}$, at strong excitation $\tilde{k}_r = \dfrac{T_o \hbar\omega}{x_e T_v^2}$, see Section 5.3). The factor $\tilde{k}_i = \dfrac{\partial \ln k_i}{\partial \ln T_v}$ can be found taking into account the influence of the vibrational temperature and hence, super-elastic processes on the ionization rate (see Section 4.2.5 and Eq. (4.85)) as:

$$\tilde{k}_i = \left(\frac{\hbar\omega}{T_e}\right)^2 \frac{\Delta\varepsilon}{T_v}. \tag{6.67}$$

In the above: $\Delta\varepsilon \approx 1 \div 3\, eV$ is the energy range of effective vibrational excitation (see Table 3.9) and E_a is the activation energy. Based on Eqs. (6.66) and (6.67), the fast ionization instability increment for the case of not very strong excitation can be expressed as:

$$\Omega_v = k_{eV} n_e \frac{\hbar\omega E_a}{c_v^v T_v^2}\left(\frac{\hbar^2\omega^2}{T_e^2}\frac{\Delta\varepsilon}{E_a} - 1\right). \tag{6.68}$$

6.3.6 Destabilizing Effect of Exothermic Reactions and Fast Mechanisms of Chemical Heat Release

Chemical reactions of vibrationally excited molecules stabilize the perturbations of ionization Eq. (6.68), but unfortunately, they are also able to amplify instabilities related to the plasma direct overheating. As was discussed in Section 5.6.7 (see Eqs. (5.154) and (5.157)), each act of chemical reaction stimulated by vibrational excitation is accompanied by transfer of energy $\xi E_a = \Delta Q_\Sigma - \Delta H$ into heat. This so-called *chemical heat release* includes the effect of exothermic reactions, non-resonant VV-exchange, and VT-relaxation from high vibrational levels. The scheme of the ionization-overheating instability is illustrated in Figure 6.11. The instability of chemical heat release is the thermal (ionization-overheating) instability. It is much faster because of the fast heating controlled by the fast chemical reactions themselves. The increment of the instability (I.A. Kirillov, L.S. Polak, A. Fridman, 1984a) is:

$$\Omega_T = k_{eV} n_e \frac{\hbar\omega}{c_p T_o}\left\{\xi\left(\hat{k}_r - 2\right) - \tilde{k}_r \frac{c_p T_0}{c_v^v T_v} + \sqrt{\left[\xi\left(\hat{k}_r - 2\right) + \tilde{k}_r \frac{c_p T_0}{c_v^v T_v}\right]^2 + 4\xi \tilde{k}_r \frac{c_p T_0}{c_v^v T_v}\left(\hat{k}_i - \hat{k}_r\right)}\right\}. \tag{6.69}$$

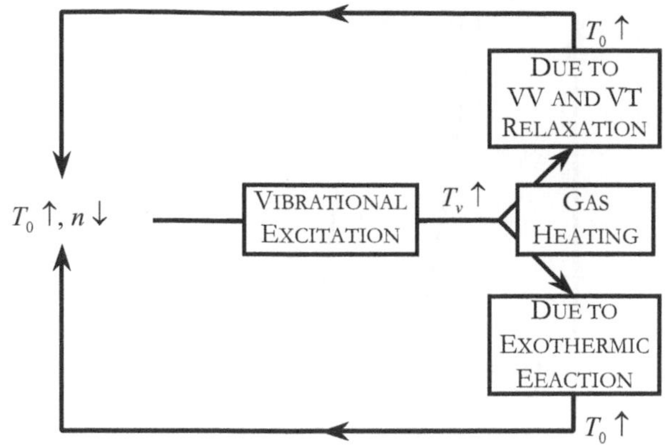

FIGURE 6.11 Instability related to the "chemical" heat release.

Here $\hat{k}_r = \dfrac{\partial \ln k_r}{\partial \ln T_0}$ is the sensitivity of the chemical reaction rate to translational temperature. Equation (6.69) demonstrates that instability time is comparable with the time of vibrational excitation. This establishes a limit on the maximum specific energy input into non-equilibrium discharge:

$$E_v^{\max} = k_{eV}\frac{\hbar\omega}{\Omega_T} = c_pT_0\left\{\xi\left(\hat{k}_r - 2\right) - \tilde{k}_r\frac{c_pT_0}{c_v^vT_v}\right.$$
$$\left.+\sqrt{\left[\xi\left(\hat{k}_r - 2\right) + \tilde{k}_r\frac{c_pT_0}{c_v^vT_v}\right]^2 + 4\xi\tilde{k}_r\frac{c_pT_0}{c_v^vT_v}\left(\hat{k}_i - \hat{k}_r\right)}\right\}^{-1}. \quad (6.70)$$

This strong restriction of the plasma parameters required for stability of a moderate or high-pressure discharge (with chemical reactions stimulated by vibrational excitation) is shown numerically in Figure 6.12 in coordinates of vibrational and translational temperatures. In most conditions, the homogeneous, steady-state, non-equilibrium plasma-chemical discharges are unstable with respect to overheating, if pressure is moderate or high, and there is no discharge stabilization by walls. This explains why most eprocesses, stimulated by vibrational excitation of molecules at high or moderate pressures, are experimentally observed only in space-non-uniform, non-homogeneous or non-steady-state discharges (V.D. Rusanov, A. Fridman, 1984. This ionization-overheating instability does not affect plasma-chemical discharges in supersonic flows. In that case, a small increase in temperature leads not to reduction but to the growth of the gas density (see Figure 6.11), which provides plasma stability.

6.3.7 Electron Attachment Instability

The attachment instability takes place if electron detachment essentially compensates electron attachment. This can be observed in glow discharges when the perturbation of n_e does not affect current. The sequence of perturbations can illustrate the instability:

$$\delta n_e \uparrow \to \delta T_e \downarrow \to \nu_a \downarrow \to \delta n_e \uparrow. \quad (6.71)$$

Analyze the chain of causal links Eq. (6.71). The small increase of electron concentration leads to a decrease of local electric field (current is not perturbed) and, hence a decrease of electron temperature. Obviously, it intensifies ionization, but the effect of the weakening of attachment may have a

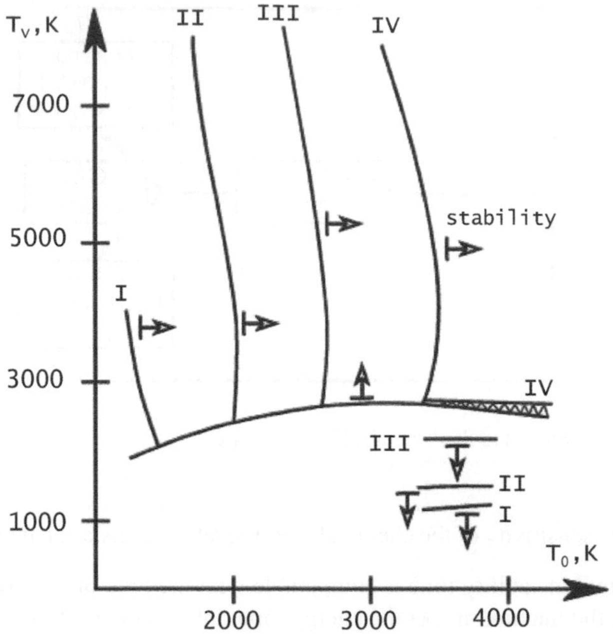

FIGURE 6.12 Thermal instability limits. Specific energy input: I – 0.1 eV/mol, II – 0.3 eV/mol; III – 0.5 eV/mol; IV – 0.7 eV/mol. Range of stability parameters is crossed out.

stronger effect (because the ionization rate is much less than the rate of attachment in the presence of intensive detachment process, see below). Before the perturbation, attachment and detachment processes were in balance. Therefore weakening of the attachment rate at a constant level of detachment results in an increase of electron concentration and, finally, in the instability. Increment of the attachment instability is:

$$\Omega_a \approx k_a n_0 \left[\hat{k}_a \left(1 - \frac{k_i}{k_a} \frac{\hat{k}_i}{\hat{k}_a} \right) - \frac{n_+}{n_-} \right]. \tag{6.72}$$

k_a is the rate coefficient of electron attachment; dimensionless factor $\hat{k}_a = \frac{\partial \ln k_a}{\partial \ln T_e}$ is the sensitivity of this rate coefficient to electron temperature; n_+ and n_- are concentrations of positive and negative ions respectively. From Eq. (6.72), the characteristic time of attachment instability is, obviously the time of electron attachment. The attachment instability does not take place in the discharge regime controlled by electron attachment where $k_a \approx k_i$ (see Section 4.5.3). The instability increment is negative in this case because ionization is more sensitive to electron temperature than attachment ($\hat{k}_i > \hat{k}_a$). The attachment instability can be observed only in the presence of intensive detachment (recombination regime, see Section 4.5.2), where $k_a >> k_i$ and parenthesis in the Eq. (6.69) becomes positive. The attachment instability leads to the formation of **electric field domains**, which are a form of striations. For example, an initial local fluctuation $\delta n_e > 0,\ \delta T_e < 0$ in the presence of a high concentration of negative ions results in their decay, growth of electron concentration and further decrease of a local electric field. This is called a *weak field domain*. An opposite local fluctuation $\delta n_e < 0,\ \delta T_e > 0$ leads to formation of a *strong field domain*. The domains usually move towards an anode much slower than the electron drift velocity.

6.3.8 Other Instability Mechanisms in Low-Temperature Plasma

a. ***Ionization Instability Controlled by Dissociation of Molecules***. As shown in Section 4.2.6, the effective electron temperature is significantly higher in monatomic gases than in corresponding molecular gases at the same value of reduced electric field. This is due to a significant reduction of electron energy distribution function in an energy interval corresponding to intensive vibrational excitation (see Eq. (4.85)). This effect explains the non-equilibrium discharge instability, related to the dissociation of molecular gases, which can be illustrated by the scheme (J. Meyer, 1969; M.M. Kekez, M.R. Barrault, J.D. Craggs, 1970; J.D. Chalmers, 1972):

$$\delta n_e \uparrow \rightarrow \delta\left(dissociation\right)\uparrow \rightarrow \delta T_e \uparrow \rightarrow \delta n_e \uparrow . \tag{6.73}$$

The increase in electron concentration (or temperature) leads to the intensification of dissociation, conversion of molecules into atoms, and then results in further growth of the electron concentration and temperature. Increment of the instability Eq. (6.73) is determined by the dissociation time.

b. ***The Stepwise Ionization Instability***. This instability is similar to the fast ionization instability related to vibrational excitation, see 6.3.5. Here also the increase of electron concentration leads to additional population of excited species (but in this case electronically excited particles), which provides faster ionization and a further increase of electron concentration. The increment of this instability is approximately the frequency of electronic excitation. In contrast to the instability related to vibrational excitation (Section 6.3.5), this one cannot be as effectively stabilized by chemical reactions.

c. ***Electron Maxwellization Instability***. Electron energy distribution functions are restricted at high energies by a variety of channels of inelastic electron collisions. Maxwellization of electrons at higher electron densities provides larger amounts of high-energy electrons Eq. (4.128) and a related stimulate ionization in this way. Instability can be illustrated by the following sequence of events:

$$\delta n_e \uparrow \rightarrow \delta\left(Maxwellization\right)\uparrow \rightarrow \delta f\left(E\right)\uparrow \rightarrow \delta n_e \uparrow . \tag{6.74}$$

An increase of electron concentration leads to Maxwellization, growth of the electron energy distribution function at high energies, intensification of dissociation and finally to further growth of electron concentration. This instability mechanism (as well as stepwise ionization) may lead to striation or contraction at sufficiently high electron densities.

d. ***Instability in fast oscillating fields***. An example of instability in oscillating fields is the ionization instability of microwave plasma in low-pressure monatomic gases $\nu_{en} << \omega$ (V.B. Gildenburg, 1981). The mechanism of this instability is similar to the modulation instability of hot plasma. An increase of electron density in a layer (perpendicular to the electric field) provides growth of the plasma frequency, which approaches the microwave frequency ω and leads t an increase of the field. The growth of the electric field results in intensification of ionization and further increase in electron density, which determines the instability. At higher pressures ($\nu_{en} >> \omega$), temperature perturbation also plays a role in the development of the ionization instability. The instability results in the formation of overheated filaments parallel to the electric field E. The maximum increment of the instability corresponds to perturbations with the wave vector:

$$\kappa_m = \left(\frac{\partial \ln k_i}{\partial \ln E}\frac{n_e e^2}{\varepsilon_0 m\left(\omega^2+v_{en}^2\right)}\frac{k_i n_e}{D_a}\frac{\omega^2}{c^2}\right)^{1/4}. \tag{6.75}$$

Here D_a is the coefficient of ambipolar diffusion; ν_{en} is the frequency of electron-neutral collisions; c is the speed of light. The distance between the filaments formed as a result of the instability can be estimated $l \approx 2\pi/\kappa_m$. For electron temperatures about 1 eV, electron density close to the critical one (when microwave frequency is close to plasma frequency), pressure of air 200 Torr and wave length $\lambda = 1$ cm, the distance between filaments is about 0.2–0.3 cm.

6.4 NON-THERMAL PLASMA FLUID MECHANICS IN FAST SUBSONIC AND SUPERSONIC FLOWS

6.4.1 Non-Equilibrium Supersonic and Fast Subsonic Plasma-Chemical Systems

The optimal energy input in energy-effective plasma-chemical systems should be about 1 eV/mol at moderate and high pressures. This determines the proportionality between discharge power and gas flow rate in the optimal regimes of plasma-chemical processes: high discharge power requires high flow rates through the discharge. The gas flow rates through the discharge plasma zone are limited. Instabilities restrict space-sizes and pressures of the steady-state discharge systems; velocities are usually limited by the speed of sound. As a result, the flow rate and, hence, power are restricted by some critical value, see N.P. Ageeva, G.I. Novikov, V.K. Raddatis ea., 1986. Higher powers of steady-state, non-equilibrium discharges can be reached in supersonic gas flows. Under such conditions, the maximum power of a steady-state, non-equilibrium microwave discharge was increased up to 1 MW (V.A. Legasov, R.I. Asisov, Yu.P. Butylkin ea., 1983). A simplified scheme and flow parameters of such a discharge system (power 500 kW, Mach number M = 3) is presented in Figure 6.13. Ignition of the non-equilibrium discharge takes place in this case after a supersonic nozzle in a relatively low-pressure zone. In the after-discharge zone, the pressure is restored in a diffuser, so initial and final pressures are above the atmospheric. Besides the two mentioned technological advantages of the non-thermal, supersonic discharges: high values of power and pressure (before and after the discharge), two other more fundamental points of interest should be mentioned. Gas temperature in the discharge is much less than room temperature (down to 100 K, see Figure 6.13, which provides low VT relaxation rates and much higher levels of non-equilibrium. Also, the high flow velocities make this discharge more stable with respect to contraction at higher pressures and energy inputs (A.S. Provorov, V.P. Chebotaev, 1977). For example, homogeneous glow discharges can be organized in atmospheric pressure, fast flows at energy input 0.5 kJ/g (A.E. Hill, 1971; W.E. Gibbs, R. McLeary, 1971).

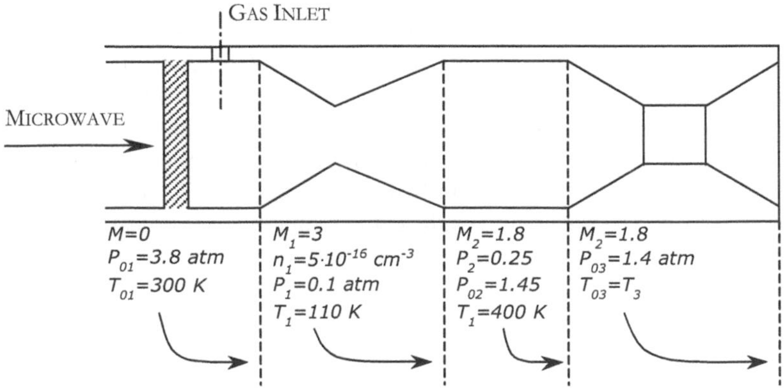

FIGURE 6.13 Typical parameters of a nozzle system. Subscripts 1,2 are related to inlet and exit of a discharge zone; subscript 3 is related to exit from the nozzle system; subscript 0 is related to stagnation pressure and temperature.

6.4.2 Gas Dynamic Parameters of Supersonic Discharges, the Critical Heat Release

Plasma and therefore, the heat release zone in supersonic discharge (Figure 6.13) is located between the nozzle and diffuser. The gas flow beyond the heat release zone (and shock in the diffuser) can be considered as isentropic at high Reynolds numbers and smooth duct profile. Gas pressure p, density $n_{0,}$ and temperature T are related in the isentropic zones to the Mach number M as:

$$p\left(1+\frac{\gamma-1}{2}M^2\right)^{\frac{\gamma}{\gamma-1}}=const, \tag{6.76}$$

$$n_0\left(1+\frac{\gamma-1}{2}M^2\right)^{\frac{1}{\gamma-1}}=const, \tag{6.77}$$

$$T\left(1+\frac{\gamma-1}{2}M^2\right)=const. \tag{6.78}$$

Here γ is the specific heat ratio. The Mach number M, is determined by the variation of the cross section S of plasma-chemical system (duct):

$$SM\left(1+\frac{\gamma-1}{2}M^2\right)^{\frac{\gamma+1}{2(1-\gamma)}}=const. \tag{6.79}$$

The above equations permit calculating the supersonic gas flow parameters at the beginning of the discharge as a function of gas parameters in the initial tank and Mach number after the nozzle. Data for CO_2 are presented in Figure 6.14. For example, if initial tank pressure is about 5 atm at room temperature and M = 0.05, then in the supersonic flow after nozzle and before the discharge the gas pressure is 0.1 atm, Mach number M ~ 3 and gas temperature T ~ 100 K. The supersonic gas motion in the discharge zone and afterward strongly depends on the plasma heat release q. This plasma heat release leads to an increase of stagnation temperature $\Delta T_0 = q/c_p$, and therefore decreases the velocity according to the relations for the supersonic flow in the duct with the constant cross section:

$$\frac{\sqrt{T_0}}{f(M)}=const, \quad f(M)=\frac{M\sqrt{1+\frac{\gamma-1}{2}M^2}}{1+\gamma M^2}. \tag{6.80}$$

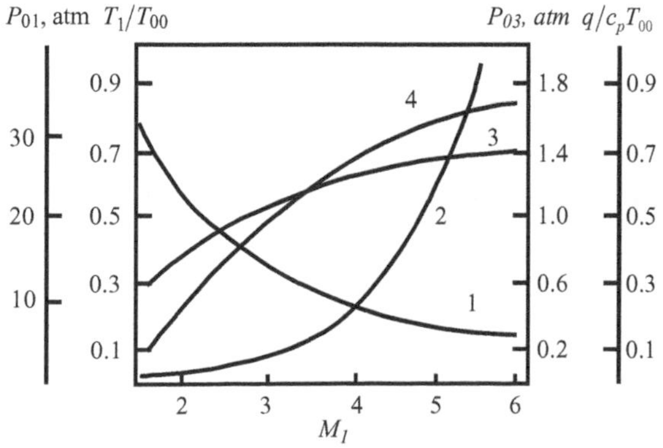

FIGURE 6.14 Gas dynamic characteristics of a discharge in supersonic flow. Discharge inlet temperature T_I(1); initial tank pressure p_{0I} (2); exit pressure p_{03} at the critical heat release (3); critical heat release q (4) as functions of Mach number in front of discharge. Initial gas tank temperature T_{00} = 300 K; static pressure in front of discharge p_I = 0.1 atm.

The equation (6.80) permits determining **the critical heat release** for the supersonic reactor with constant cross section, which corresponds to the decrease of the initial Mach number from M > 1 before the discharge to M = 1 afterward:

$$q_{cr} = c_p T_{00}\left[\frac{\left(1+\gamma M^2\right)^2}{2(\gamma+1)M^2\left(1+\frac{\gamma-1}{2}M^2\right)}-1\right]. \tag{6.81}$$

Here T_{00} is the initial gas temperature in tank, the reactor cross section in the discharge zone is considered constant. As seen from Eq. (6.81), if the initial Mach number is not very close to unity, the critical heat release can be estimated as $q_{cr} \approx c_p T_{00}$. The critical heat release at different initial Mach numbers for the supersonic CO_2 flow can be also found in Figure 6.14. Further increase of the heat release over the critical value leads to the formation of non-steady flow perturbations like shock waves which are detrimental to non-equilibrium plasma systems. Such a generation of shock waves will be discussed in the last section of the chapter. Even taking into account the high-energy efficiency of chemical reactions in supersonic flows, the critical heat release restricts the specific energy input and subsequently the degree of conversion of the plasma-chemical process. The maximum degree of conversion of CO_2 dissociation in a supersonic microwave discharge (T_{00} = 300 K and M = 3) does not exceed 15–20%, even at the extremely high energy efficiency of about 90% (N.A. Zyrichev, S.M. Kulish, V.D. Rusanov ea., 1984).

6.4.3 Supersonic Nozzle and Discharge Zone Profiling

The effect of critical heat release on stability and efficiency of supersonic plasma-chemical systems can be mitigated by an increase of initial temperature T_{00} or by reagents dilution in noble gases. However, the most effective approach for suppressing the critical heat is by special profiling of the supersonic nozzle and discharge zone (B.V. Potapkin, V.D. Rusanov, A. Fridman, 1983). The supersonic reactor can be designed to maintain a constant Mach number during the heat release in the discharge zone. The duct's cross section of the duct should gradually increase to accelerate the supersonic flow and compensate for the deceleration effect related to the heat release q. The heat release in this case has no limit. To provide the constant Mach number and conditions of unrestricted heat release, the reactor cross section S should increase with q as:

$$S = S_0\left[1+\frac{q}{T\left(c_p+\gamma M_0^2/2\right)}\right]^{\frac{\gamma M_0^2+1}{2}}. \tag{6.82}$$

Here S_0, M_0, T are the reactor cross section, Mach number and temperature in the beginning of the discharge zone; specific heat is dimensionless because energy, heat and temperature are considered in the same units. With sufficiently large Mach numbers Eq. (6.82) can be simplified and used for estimations as:

$$S \approx S_0\left(1+\frac{q}{c_p T_{00}}\right)^{\frac{\gamma M_0^2+1}{2}}. \tag{6.83}$$

Although the idea to keep Mach number high without restrictions of heat release looks very attractive, the required increase of the discharge cross section is too large and not realistic for significant values of $q/c_p T_{00}$. Reasonable increases in the cross section S/S_0 requires non-constant Mach number with constant pressure during the heat release in the supersonic reactor. The heat release in this case is not unlimited, but the critical limit here is not as serious as that for the reactor with constant cross section:

$$q_{cr}\left(p = const\right) = c_p T_{00} \frac{M_0^2 - 1}{1 + \frac{\gamma - 1}{2} M_0^2}. \tag{6.84}$$

At relatively high Mach numbers, this critical heat release at constant pressure is approximately $q_{cr}(p = const) \approx 5c_pT_{00}$, for example, five times less great than for the case of constant cross section. This is enough to permit relatively high specific energy input in the plasma and degree conversion (see discussion after Eq. (6.81)). To provide the conditions for the constant pressure plasma-chemical process, the required increase of the reactor cross section is much less than Eq. (6.83) and can be expressed as:

$$S = S_0\left(1 + \frac{q}{c_p T}\right) = S_0\left[1 + \frac{q}{c_p T_{00}}\left(1 + \frac{\gamma - 1}{2} M_0^2\right)\right]. \tag{6.85}$$

For this constant pressure system, in order to obtain heat release values twice that of the critical value at S = const; it is necessary to increase the reactor cross section in the discharge zone by a factor of 6.

6.4.4 Pressure Restoration in Supersonic Discharge Systems

The technological advantage of the supersonic plasma systems is the possibility to operate a discharge at moderate pressures while keeping the system inlet and exit at high pressures. This requires an effective supersonic diffuser to restore the pressure after the discharge zone. Some relevant information on the pressure restoration after the supersonic discharge with different Mach numbers is presented in Figure 6.15. For this case, the initial pressure before the supersonic nozzle was 3.8 atm at room temperature and the heat release was about half of the critical value. As can be seen from the figure, the pressure recovery can be quite decent (up to 1.5–2.5 atm), though heating and stopping of the supersonic flow in this way are obviously not optimal for pressure recovery. Energy efficiency of CO_2 dissociation in a supersonic discharge is presented in Figure 6.16, taking into account the energy spent on gas compression for non-converted gas recirculation (N.A. Zyrichev, S.M. Kulish, V.D. Rusanov ea., 1984). Comparing Figure 6.16 with Figure 5.6.2 (pure plasma-chemical efficiency of the process) shows that energy spent on compression in this supersonic system is about 10% of the total process energy cost. As seen from Figure 6.16, the energy cost of compression makes the process less effective at high Mach numbers (M > 3). This occurs even

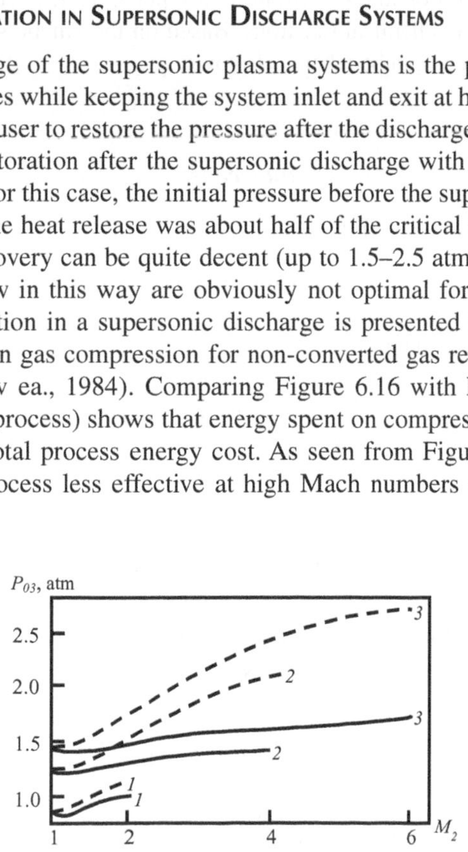

FIGURE 6.15 Pressure restoration in a diffuser (p_{03}). M_1, M_2 are Mach numbers in the discharge inlet and exit. M_1 = 2 (1), 4 (2), 6 (3). Dashed lines – ideal diffuser, solid lines take into account non-ideal shock waves; static pressure at the discharge inlet 0.1 atm.

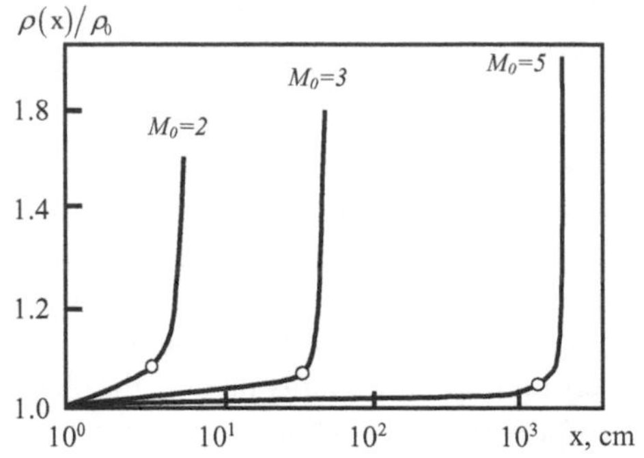

FIGURE 6.16 Gas density profiles in plasma-chemical reaction zone at different Mach numbers.

though the plasma-chemical efficiency increases with Mach number because of the reduction of vibrational relaxation losses.

6.4.5 Fluid Mechanic Equations of Vibrational Relaxation in Fast Subsonic and Supersonic Flows of Non-Thermal Reactive Plasma

Consider the vibrational relaxation in fast subsonic and supersonic flows, taking into account gas compressibility and heat release in chemical reactions. Based on the kinetic scheme discussed in Section 5.6.7, the relevant 1D equations can be taken as follows:

1. Continuity equation:

$$\rho u = const. \tag{6.86}$$

2. Momentum conservation equation:

$$p + \rho u^2 = const. \tag{6.87}$$

3. Translational energy balance:

$$\rho u \frac{\partial}{\partial x}\left(\frac{R}{\mu}\frac{\gamma}{\gamma - 1}T_0 + \frac{u^2}{2}\right) = \xi P_R(T_v) + P_{VT}. \tag{6.88}$$

4. Vibrational energy balance:

$$\rho u \frac{\partial}{\partial x}\left[\varepsilon_v(T_v)\right] = P_{ex} - P_R(T_v) - P_{VT}. \tag{6.89}$$

These equations include specific powers of a chemical reaction (P_R), VT-relaxation (P_{VT}), and vibrational excitation (P_{ex}), which can be expressed as:

$$P_R = k_R(T_v)\rho^2\left(\frac{N_A}{\mu}\right)^2 \Delta Q, \tag{6.90}$$

$$P_{VT} = k_{VT}(T_0)\rho^2\left(\frac{N_A}{\mu}\right)^2\left[\varepsilon_v(T_v) - \varepsilon_v(T_0)\right], \tag{6.91}$$

$$P_{ex} = k_{eV} n_e \rho \frac{N_A}{\mu}\hbar\omega. \tag{6.92}$$

In these relations: $k_R(T_v)$, $k_{VT}(T_0)$, $k_{eV}(T_e)$ are the rate coefficients of chemical reaction, VTrelaxation and vibrational excitation by electron impact; n_e is electron density; $\gamma = c_p/c_v$ is the specific heat ratio, taking into account, that the heat capacities include in this case only translational and rotational degrees of freedom; p, ρ, u, μ are pressure, density, velocity and molecular mass of gas; N_A is the Avogadro number; ΔQ is vibrational energy spent per one act of chemical reaction; ξ is the fraction of this vibrational energy, which goes into translational degrees of freedom; T_v, T_0 are vibrational and translational temperatures. Introduce a new dimensionless density variable $y = \rho_0/\rho(x)$, assume $T_v \approx \hbar\omega > T_0$ and also use the ideal gas equation of state:

$$p = \frac{\rho}{\mu} R T_0. \tag{6.93}$$

Then rewrite the system of relaxation dynamic Eqs. (6.86)–(6.89) in the form:

$$u = u_0 y, \tag{6.94}$$

$$T_0 = T_{00}\left[\left(1+\gamma M_0^2\right)y - \gamma M_0^2 y^2\right], \tag{6.95}$$

$$y^2 \frac{\partial}{\partial x}\left[y\left(1+\gamma M_0^2\right) - \frac{\gamma+1}{2} M_0^2 y^2\right] = \xi Q_R + Q_{VT}, \tag{6.96}$$

$$y^2 \frac{\partial}{\partial x}\varepsilon_v\left(T_v\right) = \frac{k_{eV} n_e}{u_0}\hbar\omega y - c_p T_{00} Q_R - c_p T_{00} Q_{VT}. \tag{6.97}$$

In this system of equations:

$$Q_R = \frac{k_R\left(T_v\right) n_0}{u_0}\frac{\Delta Q}{c_p T_{00}}, \quad Q_{VT} = \frac{k_{VT}\left(T_{00}\right) n_0}{u_0}\frac{\hbar\omega}{c_p T_{00}}, \tag{6.98}$$

M is the Mach number; n_0 is gas concentration; the subscript "0" means that the corresponding parameter is related to the inlet of the plasma-chemical reaction zone. Analyze the solution of the system of Eqs. (6.94)–(6.95) for the cases of the strong and weak contribution of the chemical heat release (ξ).

6.4.6 Dynamics of Vibrational Relaxation in Fast Subsonic and Supersonic Flows

If the contribution of the chemical heat release can be neglected ($\xi \Delta Q << Q_{VT}$), then Eq. (6.96) can be solved with respect to the reduced gas density $y = \rho(x)/\rho_0$ and analyzed in the following integral form (I.A. Kirillov, B.V. Potapkin, A. Fridman ea., 1984b):

$$\int_1^y \frac{\left(y'\right)^2\left[1+\gamma M_0^2 - (\gamma+1) M_0^2 y'\right] dy'}{\exp\left[\frac{B}{T_{00}^{1/3}}\left(y'\left(1+\gamma M_0^2\right) - \left(y'\right)^2 \gamma M_0^2\right)^{1/3}\right]} = \frac{k_0 n_0}{u_0}\frac{\hbar\omega}{c_p T_{00}}. \tag{6.99}$$

In the integration of the Eq. (6.96) the vibrational relaxation rate coefficient was taken as $k_{VT}\left(T_0\right) = k_0 \exp\left(-\frac{B}{T_0^{1/3}}\right)$ based on the Landau-Teller approach (see Section 3.4.3). The dependence of $\rho(x)$, calculated from Eq. (6.99), is presented in Figure 6.16 for supersonic flows with different initial Mach numbers. As can be seen from the figure, the growth of density during vibrational relaxation in supersonic flow is "explosive", which reflects the thermal mode of the overheating instability of vibrational relaxation discussed in Section 6.3.4 and illustrated in Figure 6.9. The density during the explosive overheating is a specific feature of supersonic flow. Explosions" similar to those shown in Figure 6.16 can be observed in the growth of temperature and pressure and in the decrease of velocity during the vibrational relaxation in supersonic flow. Linearization of the system of Eqs (6.94)–(6.96) allows finding the increment Ω (see Section 6.3.) describing the common time scale for the explosive temperature growth and corresponding explosive changes of density and velocity:

$$\Omega_{VT} = k_{VT}\left(T_{00}\right)n_0 \frac{\hbar\omega}{c_p T_{00}} \frac{2+\hat{k}_{VT}\left(\gamma M_0^2 - 1\right)}{M_0^2 - 1}. \tag{6.100}$$

Here T_{00}, n_0, and M_0 are the initial translational temperature, gas density, and Mach number at the beginning of the relaxation process; $\hat{k}_{VT} = \frac{\partial \ln k_{VT}}{\partial \ln T_0}$ the logarithmic sensitivity of vibrational relaxation rate coefficient to translational temperature (normally $\hat{k}_{VT} >> 2$, see below). Equation (6.100) shows that the dynamics of vibrational relaxation is qualitatively different at different Mach numbers:

1. ***For subsonic flows with low Mach numbers***:

$$M_0 < \sqrt{\frac{1}{\gamma}\left(1 - \frac{2}{\hat{k}_{VT}}\right)}. \tag{6.101}$$

the increment Eq. (6.100) is positive ($\Omega_{VT} > 0$) and the explosive heating takes place. At very low Mach numbers ($M << 1$), Eq. (6.100) for the increment can be simplified:

$$\Omega_{VT} = k_{VT}\left(T_{00}\right)n_0 \frac{\hbar\omega}{c_p T_{00}}\left(\hat{k}_{VT} - 2\right). \tag{6.102}$$

This expression coincides with Eq. (6.63) describing the increment of ionization-overheating instability in thermal mode. The dependence of the overheating increment on Mach number $\Omega_{VT}(M_0)$ at fixed values of stagnation temperature and initial gas density is shown in Figure 6.17. The same dependence $\Omega_{VT}(M_0)$, but at fixed values, of thermodynamic temperature and initial gas density, is presented in Figure 6.18. From these figures, the frequency (increment) of explosive vibrational relaxation decreases for subsonic flows as the Mach number increases. This phenomenon is due to the fact that the relaxation heat release at higher Mach numbers contributes more to flow acceleration than temperature growth.

2. ***For transonic flows with Mach numbers***:

$$\sqrt{\frac{1}{\gamma}\left(1 - \frac{2}{\hat{k}_{VT}}\right)} < M_0 < 1. \tag{6.103}$$

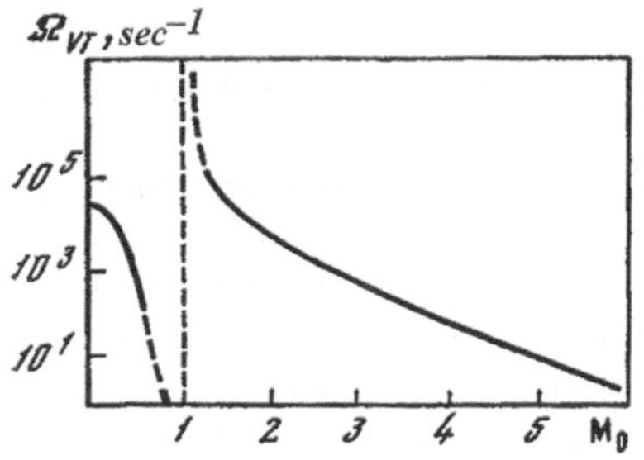

FIGURE 6.17 Relaxation frequency Ω_{VT} as a function of Mach number. Stagnation temperature T_{00} = 300 K , density $n_0 = 3*10^{18}$ cm $^{-3}$.

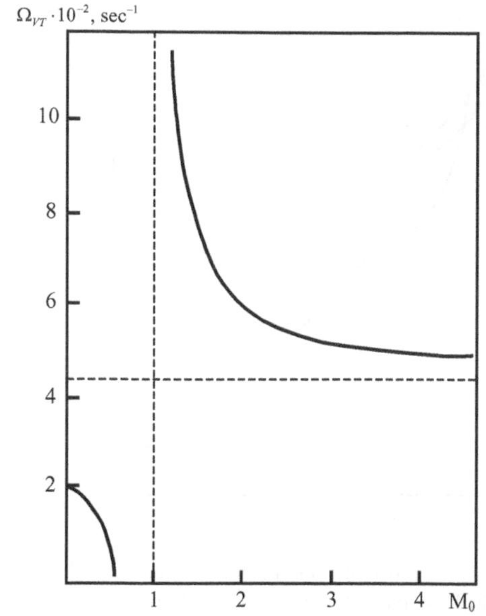

FIGURE 6.18 Relaxation frequency Ω_{VT} as a function of Mach number at the fixed thermodynamic temperature $T_{00} = 100$ K, density $n_0 = 3 * 10^{18}$ cm^{-3}.

the increment Eq. (6.100) is negative ($\Omega_{VT} < 0$) and the effect of explosive heating does not take place. The system is stable with respect to VT-relaxation overheating.

3. ***For supersonic flows*** ($M > 1$) the increment Eq. (6.100) again becomes positive ($\Omega_{VT} > 0$) and the effect of explosive overheating. takes place. At high Mach numbers:

$$\Omega_{VT} \approx k_{VT}\left(T_{00}\right)n_0 \frac{\hbar\omega}{c_p T_{00}} \gamma \hat{k}_{VT}, \tag{6.104}$$

which exceeds the increment Eq. (6.102) for subsonic flows. Hence the overheating instability is faster in the supersonic flows. This can be explained by noting that heating of the supersonic flow not only increases its temperature directly but also decelerates the flow, which leads to additional temperature growth. This effect can be seen in Figure 6.18, where the initial thermodynamic temperature is fixed. If the stagnation temperature is fixed, then the increase in Mach number results first in significant cooling. According to the Landau-Teller approach, this provides low values of the relaxation rate coefficient k_{VT} and significant reduction of the overheating instability increment Eq. (6.104), which can be seen in Figure 6.17.

6.4.7 Effect of Chemical Heat Release on Dynamics of Vibrational Relaxation in Supersonic Flows

Linearization of the system of equations describing vibrational relaxation in fast flows, taking into account the chemical heat release, leads to the overheating instability increment:

$$\Omega_{VT} = k_{VT}\left(T_{00}\right)n_0 \frac{\hbar\omega}{c_p T_{00}} \frac{2+\hat{k}_{VT}\left(\gamma M_0^2 - 1\right)}{M_0^2 - 1} + k_R\left(T_v\right)n_0 \frac{2\xi \Delta Q}{\left(M_0^2 - 1\right)c_p T_{00}}. \tag{6.105}$$

This expression for increment coincides with the corresponding Eq. (6.100) in absence of the chemical heat release ($\xi = 0$). It is interesting that the chemical heat release stabilizes the overheating in

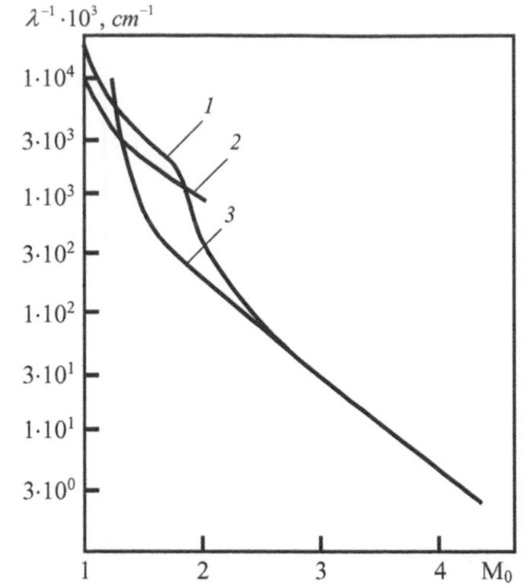

FIGURE 6.19 Inverse length of vibrational relaxation as a function of Mach number at the discharge inlet, stagnation temperature $T_{00} = 300$ K is fixed. 1 – numerical calculation; 2,3 – analytical model.

subsonic flows where the second term in Eq. (6.105) is negative. This effect has the same explanation as that for stabilizing overheating in transonic flows (see the above Section 6.4.6). In supersonic flows, the chemical heat release obviously accelerates the overheating. Note that the plasma-chemical reaction rate coefficient is a function of vibrational temperature, which was considered in this case as unperturbed. Equation (6.105) implies that the plasma-chemical reaction and hence, the chemical heat release takes place throughout the relaxation process. This is not the case in processes, stimulated by vibrational excitation, where the reaction time is shorter than relaxation (see Section 5.6.7). Taking this fact into account, time averaging of Eq. (6.105) gives the following corrected expression for the increment of overheating:

$$\langle \Omega_{VT} \rangle = k_{VT}\left(T_{00}\right) n_0 \frac{\hbar\omega}{c_p T_{00}} \frac{2 + \hat{k}_{VT}\left(\gamma M_0^2 - 1\right)}{M_0^2 - 1 - 2\dfrac{q_R}{c_p T_{00}}}. \tag{6.106}$$

Here q_R is the total integral chemical heat release per one molecule, which is total energy transfer into translational degrees of freedom related to VVexchange, VTrelaxation from high vibrational levels and heating due to exothermic chemical reactions. When the chemical heat release can be neglected, Eq. (6.106) also coincides with Eq. (6.100). The values of the overheating increment Ω_{VT} are recalculated often into the length $1/\Lambda$ or reverse length Λ of vibrational relaxation. Typical dependence of the reverse length Λ of vibrational relaxation taking into account the chemical heat release on the initial Mach number M_0 is presented in Figure 6.19. This figure also permits comparing results of detailed modeling with analytical formulas of the above-considered linear approximation.

6.4.8 Spatial Non-Uniformity of Vibrational Relaxation in Chemically Active Plasma

Assume, 1D distribution of vibrational $T_v(x)$ and translational $T_0(x)$ temperatures with fluctuations:

$$T_0\left(t = 0,\ x\right) = T_{00} + g\left(x\right), \quad T_v\left(t = 0,\ x\right) = T_{v0} + h\left(x\right). \tag{6.107}$$

Because of the strong exponential temperature dependence of the VTrelaxation rate, heat transfer is unable to restore spatial uniformity of the temperatures when the density of vibrationally excited molecule is relatively high and heterogeneous vibrational relaxation can be neglected. The effect of convective heat transfer on the spatial non-uniformity of vibrational relaxation in chemically active plasma is quite sophisticated (I.A. Kirillov, B.V. Potapkin, M.I. Strelkova, 1982). However, at low Peclet numbers, the problem can be simplified taking into consideration only vibrational and translational energy conduction, VT relaxation, and chemical reactions:

$$n_0 c_v \frac{\partial T_0}{\partial t} = \lambda_T \frac{\partial T_0}{\partial x^2} + k_{VT} n_0^2 \left[\varepsilon_v(T_v) - \varepsilon_v(T_0)\right] + \xi k_R n_0^2 \Delta Q, \tag{6.108}$$

$$n_0 c_v^v \frac{\partial T_v}{\partial t} = \lambda_V \frac{\partial T_v}{\partial x^2} - k_{VT} n_0^2 \left[\varepsilon_v(T_v) - \varepsilon_v(T_0)\right] - k_R n_0^2 \Delta Q. \tag{6.109}$$

Here, $\lambda_V, \lambda_T, c_v^v, c_v$ are vibrational and translational coefficients of thermal conductivities and specific heats. To analyze the time and spatial evolution of small perturbations $T_1(x,t)$, $T_{v1}(x,t)$ of the translational and vibrational temperatures, the system of Eqs. (6.108), (6.109) can be linearized:

$$\frac{\partial T_1}{\partial t} = \omega_{TT} T_1 + \omega_{TV} T_{v1} + D_T \frac{\partial^2 T_1}{\partial x^2}, \tag{6.110}$$

$$\frac{\partial T_{v1}}{\partial t} = \omega_{VT} T_1 + \omega_{VV} T_{v1} + D_V \frac{\partial^2 T_{v1}}{\partial x^2}. \tag{6.111}$$

Here the following frequencies have been introduced:

$$\omega_{TT} = k_{VT} n_0 \left[\frac{\varepsilon_v(T_v) - \varepsilon_v(T_0)}{c_v T_0} \hat{k}_{VT} - \frac{c_v^v(T_0)}{c_v}\right], \tag{6.112}$$

$$\omega_{VV} = -k_{VT} n_0 - k_R n_0 \frac{\Delta Q}{c_v^v T_v} \hat{k}_R, \tag{6.113}$$

which characterize the changes of translational temperature due to perturbations of T_0 and changes of vibrational temperature to perturbations of T_v. In a similar way, two other frequencies:

$$\omega_{TV} = k_{VT} n_0 \frac{c_v^v(T_v)}{c_v} + \xi k_R n_0 \frac{\Delta Q}{c_v T_v} \hat{k}_R, \tag{6.114}$$

$$\omega_{TT} = -k_{VT} n_0 \frac{c_v}{c_v^v(T_v)} \left[\frac{\varepsilon_v(T_v) - \varepsilon_v(T_0)}{c_v T_0} \hat{k}_{VT} - \frac{c_v^v(T_0)}{c_v}\right]. \tag{6.115}$$

describe changes in translational temperature due to perturbations of T_v and changes of vibrational temperature due to perturbations of T_0. Here, logarithmic sensitivity factors are $\hat{k}_R = \frac{\partial \ln k_R}{\partial \ln T_v}$ and $\hat{k}_{VT} = \frac{\partial \ln k_{VT}}{\partial \ln T_0}$; $D_T = \lambda_T / n_0 c_v$ and $D_V = \lambda_V / n_0 c_v^v$ are the reduced coefficients of translational and vibrational thermal conductivity. To obtain the dispersion equation for the evolution of fluctuations of vibrational and translational temperatures, consider the temperature perturbations in the linearized system of Eqs. (6.110) and (6.111) in the exponential form with amplitudes A, B as:

$$T_1(x,t) = A \cos kx \cdot \exp(\Lambda t), \quad T_{v1}(x,t) = B \cos kx \cdot \exp(\Lambda t). \tag{6.116}$$

The system of Eqs. (6.110) and (6.111) then leads to the following dispersion equation relating increment Λ of amplification of the perturbations with their wave number k:

$$\left(\Lambda - \Omega_T^k\right)\left(\Lambda - \Omega_V^k\right) = \omega_{VT} \omega_{TV}. \tag{6.117}$$

In this dispersion equation: Ω_T^k and Ω_V^k are the so-called thermal and vibrational modes, which are determined by the wave number k as:

$$\Omega_T^k = \omega_{TT} - D_T k^2, \tag{6.118}$$

$$\Omega_V^k = \omega_{TT} - D_V k^2. \tag{6.119}$$

One can note, that $\omega_{VT}\omega_{TV}$ is the nonlinear coupling parameter between the vibrational and thermal modes. If the product is equal to zero, the modes Eqs. (6.118) and (6.119) are independent. The solution of Eq. (6.117) with respect to the increment $\Lambda(k)$ can be presented as:

$$\Lambda_\pm(k) = \frac{\Omega_T^k + \Omega_V^k}{2} \pm \sqrt{\frac{\left(\Omega_T^k - \Omega_V^k\right)^2}{4} + \omega_{VT}\omega_{TV}}. \tag{6.120}$$

This increment describes the initial linear phase of the spatial non-uniform vibrational relaxation in chemically active plasma, which results in different effects on structure formation in the discharge.

6.4.9 Space Structure of Unstable Vibrational Relaxation

Analysis of different spatial-non-uniform VT-relaxation instability spectra, expressed by the dispersion Eq. (6.120) (I.A. Kirillov, B.V. Potapkin, M.I. Strelkova, ea., 1984) shows a variety of possible phenomena in the system, including the interaction between the instability modes, wave propagation, etc. The following relation between the frequencies can usually characterize plasma-chemical processes stimulated by vibrational excitation of molecules: $\omega_{TT} > 0$, $\omega_{VV} < 0$, $\omega_{VT}\omega_{TV} < 0$, $\max\{\omega_{VV}^2, \omega_{TT}^2\} > |\omega_{VT}\omega_{TV}|$, which greatly simplified the analysis of the dispersion equation. In this case, the growth of the initial temperature perturbations Eq. (6.116) is controlled by the thermal mode and takes place under the following condition:

$$\Omega_T^k = \omega_{TT} - D_T k^2 = k_{VT} n_0 \left[\frac{\varepsilon_v(T_v) - \varepsilon_v(T_0)}{c_v T_0}\hat{k}_{VT} - \frac{c_v^v(T_0)}{c_v}\right] - D_T k^2 > 0. \tag{6.121}$$

Stability diagrams or dispersion curves $\mathrm{Re}\,\Lambda(k^2)$, which illustrates the evolution of temperature perturbations at $D_T > D_V$ and frequency relations mentioned above, are shown in Figure 6.20. Obviously, $\mathrm{Re}\,\Lambda(k^2) > 0$ means amplification of initial perturbations, and $\mathrm{Re}\,\Lambda(k^2) < 0$, their stabilization. As can be seen from Figure 6.20, the unstable harmonics Eq. (6.116) correspond to the

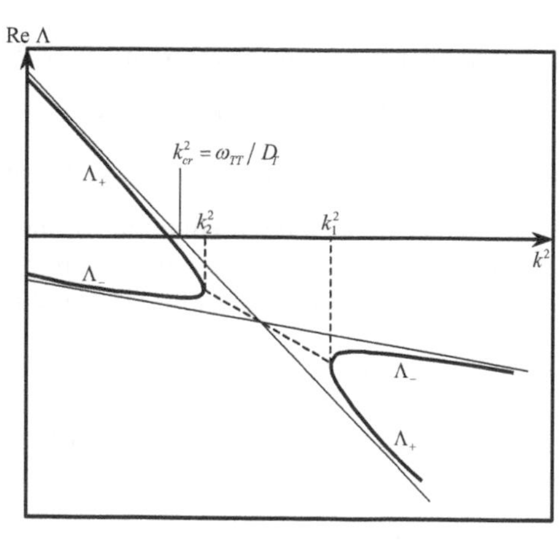

FIGURE 6.20 Re Λ as a function of perturbation wave number.

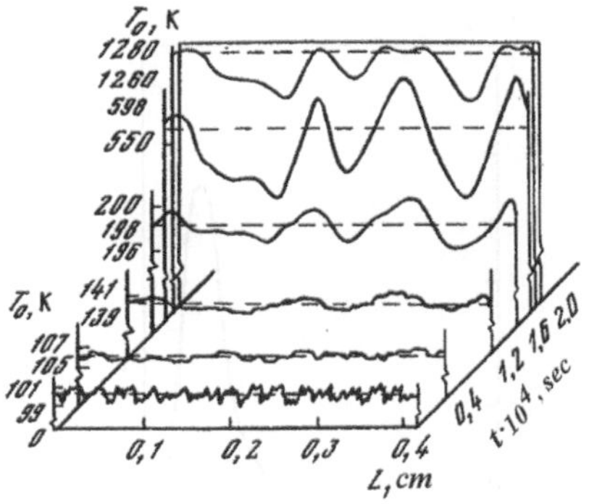

FIGURE 6.21 Dynamics of the space non-uniform relaxation; dashed line – uniform solution.

thermal mode Eq. (6.121), which is related to Λ_+ line on the figure. Thus, during the linear phase of the relaxation process: short-scale perturbations $k > k_{cr} = \sqrt{\omega_{TT}/D_T}$ decrease and disappear because of thermal conductivity, while perturbations with longer wavelength $\lambda > \lambda_{cr} = 2\pi/k_{cr}$ grow exponentially. When amplitudes of the temperature perturbations become sufficiently large, the regime of the vibrational relaxation is nonlinear. This results in the formation of a high-gradient spatial structures, which consists of periodical temperature zones with quite different relaxation times. The minimal distance between such zones can be determined by the above-calculated critical wavelength:

$$\lambda_{cr} = 2\pi\sqrt{\frac{D_T c_v T_0}{k_{VT} n_0 \hat{k}_{VT}\left(\varepsilon_v(T_v) - \varepsilon_v(T_0)\right)}}. \tag{6.122}$$

Formation of the spatial structure of temperature provided by vibrational relaxation on the nonlinear phase of evolution is shown in Figure 6.21 (I.A. Kirillov, B.V. Potapkin, M.I. Strelkova, ea., 1984) For CO_2 ($n_0 = 3 \cdot 10^{18}$ cm^{-3}) with the following initial and boundary ($x = 0, L$) conditions:

$$T_0(x,t=0) = 100K,\ T_v(x,t=0) = 2500K,\ \frac{\partial T_0}{\partial x}(x=0;L;t) = \frac{\partial T_v}{\partial x}(x=0,L;t) = 0.$$

Initial perturbations of vibrational temperature were taken as "white noise" with the mean-square deviation $<\delta T_0^2> = 0.8\cdot 10^{-4} K^2$. As seen from Figure 6.21, the evolution of perturbations can be subdivided into two phases: linear ($t < 5 \cdot 10^{-4}$ sec) and nonlinear ($t > 5 \cdot 10^{-4}$ sec). The linear phase can be characterized by the damping of the short-scale perturbations and, hence, by an approximate 100 times decrease of the mean-square level of fluctuations $\langle \delta T_0^2 \rangle$ (see Figure 6.22). In the nonlinear phase, perturbations grow significantly (see Figure 6.22), forming the structures with characteristic sizes about 0.1–0.2 cm. in accordance with Eq. (6.122).

6.4.10 Plasma Interaction with High-Speed Flows and Shocks

Weak ionization of gases results in large changes in the standoff distance ahead of a blunt body in ballistic tunnels, in reduced drag, and in modifications of traveling shocks. These plasma effects are of practical importance for high-speed aerodynamics and flight control. It is clear that energy addition to the flow results in an increase in the local sound speed that leads to expected modifications of the flow and changes to the pressure distribution around a vehicle due to the decrease in local Mach number. Intensive research, however, has been focused recently on finding out specific strong plasma effects influencing high-speed flows and shocks, which are the most attractive for applications. It has been demonstrated that although the heating in many cases is global; experiments with positive columns,

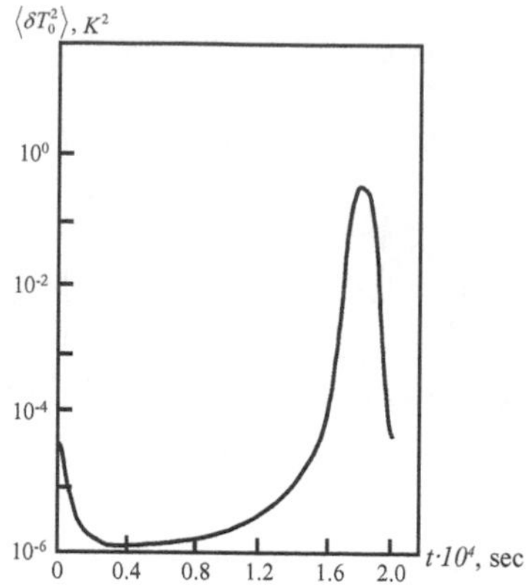

FIGURE 6.22 Time evolution of translational temperature mean-square-deviation.

dielectric barrier discharges, and focused microwave plasmas can produce localized special energy deposition effects more attractive for energy efficiency in flow control. Numerous schemes have been proposed recently for modifying and controlling the flow around a hypersonic vehicle. These schemes include approaches for plasma generation, magneto-hydrodynamic (MHD) flow control and power generation, and other purely thermal approaches. Two major physical effects of heat addition to the flowfield lead to drag reduction at supersonic speeds: (1) reduction of density in front of the body due to the temperature increase (that assumes either constant heating or a pulsed heating source having time to equilibrate in pressure before impinging on the surface); (2) coupling of the low density wake from the heated zone with the flowfield around the body, which can lead to a dramatically different flowfield (the effect is very large for blunt bodies, changing their flowfield into something more akin to the conical flow). The efficiency of the effect grows with Mach number.

6.4.11 Aerodynamic Effects of Surface and Dielectric Barrier Discharges (DBD), Aerodynamic Plasma Actuators

Surface discharges permit rapid and selective heating of a boundary layer at or slightly off the surface, as well as creating plasma zones for effective magneto-hydrodynamic (MHD) interactions. The surface discharges for the supersonic and hypersonic aerodynamic applications have been organized using surface microwave discharges, DC and pulsed discharges between surface-mounted electrodes and internal volumes, and sliding discharges. Surface discharges are effective for adding thermal energy to a boundary layer. Asymmetric dielectric barrier discharges (DBD), with one electrode located inside (beneath) the barrier and another one mounted on top of the dielectric barrier and shifted aside, stimulate intensive "ion wind" along the barrier surface, influence boundary layers, and can be used as aerodynamic actuators. Aerodynamic plasma actuators attract significant interest lately in relation to flow control above different surfaces (aircraft wings, turbine blades, etc.) The ion wind pushes neutral gas with not very high velocities (typically, 5–10 m/sec at atmospheric pressure). However, even those values of the DBD stimulated gas velocities near the surface are able to make significant changes in the flow near the surface, especially when the flow speed is not high. The major contribution to the "ion wind" drag in the DBD plasma actuators is mostly due not to the DBD streamers themselves, but to the ion motion between the streamers when ions move along the surface from one electrode to another similar to the case of corona discharges. During the phase when the electrode mounted on the barrier is positive, the positive ions create the wind and drag the flow. It is interesting that during the following phase when the electrode mounted on the barrier is negative, the negative ions formed by electron attachment to oxygen (in air) make a significant contribution to total ion wind and drag the flow in the same direction. Applicational of the asymmetric nano-second pulsed DBD permitted to A. Starikovsky and his group achieving strong plasma effect on the boundary layer at gas velocities close to the speed of sound. The effect in this case is due to not the ion wind, but to ultra-fast boundary layer heating and shock wave generation.

6.5 ELECTROSTATIC, MAGNETO-HYDRODYNAMIC, AND ACOUSTIC WAVES IN PLASMA

6.5.1 Electrostatic Plasma Waves

Electrostatic plasma oscillations discussed in Section 6.1.6 are able to propagate in plasma as longitudinal waves. Electric fields in these longitudinal waves are in the direction of the wave propagation, which is the direction of wave vector $\vec{k}$. To analyze these electrostatic plasma waves, the amplitude A^1 of oscillations of any macroscopic parameter A(x,t) can be considered small ($A^1 << A_0$). Then the oscillations $A(x,t)$ for the linearized relevant equations can be expressed as:

$$A = A_0 + A^1 \exp\left[i(kx - \omega t)\right], \tag{6.123}$$

where A_0 is the value of the macroscopic parameter in absence of oscillations, k is the wave number, ω is the wave frequency. To obtain the dispersion equation of the electrostatic plasma waves, consider the linearized system of Eq. (6.123), including continuity Eq. (4.65) for electrons, momentum conservation Eq. (4.68) for the electron gas without dissipation, the adiabatic relation for the gas and Poisson equation:

$$-i\omega n_e^1 + ikn_e u^1 = 0. \tag{6.124}$$

$$-i\omega u^1 + ik\frac{p^1}{mn_e} + \frac{eE^1}{m} = 0, \tag{6.125}$$

$$\frac{p^1}{p_0} = \gamma\frac{n_e^1}{n_e}, \tag{6.126}$$

$$ikE^1 = -\frac{1}{\varepsilon_0}en_e^1. \tag{6.127}$$

Here: n_e^1, u^1, p^1, E^1 are the amplitude of oscillations of electron concentration, velocity, pressure and electric field, n_e is the unperturbed plasma density; e, m, γ are charge, mass and specific heat ratio for an electron gas; $p_0 = n_e m\langle v_x^2\rangle = n_e T_e$ is the electron gas pressure in the absence of oscillations; $\langle v_x^2\rangle$ averaged square of electron velocity in direction of oscillations; T_e is the electron temperature. The dispersion equation for the electrostatic plasma waves can be derived as:

$$\omega^2 = \omega_p^2 + \frac{\gamma T_e}{m}k^2, \quad \omega_p = \sqrt{\frac{n_e e^2}{\varepsilon_0 m}}. \tag{6.128}$$

Here ω_p is the plasma frequency. From Eq. (6.128), the electrostatic wave frequency is close to the plasma frequency, if the wavelength $2\pi/k$ exceeds the Debye radius. In the opposite limit of short wavelengths, the phase velocity of the electrostatic waves corresponds to the thermal speed of electrons.

6.5.2 Collisional Damping of the Electrostatic Plasma Waves in Weakly Ionized Plasma

In the dispersion Eq. (6.128) the electron-neutral collisions (frequency ν_{en}), has been neglected. In weakly ionized plasma, electron and neutral collisions can influence the plasma oscillations quite significantly. Plasma oscillations actually do not exist when $\omega < \nu_{en}$. At higher frequencies $\omega > \nu_{en}$, the electron-neutral collisions provide damping of the electrostatic plasma oscillations by corrected Eq. (6.128):

$$\omega = \sqrt{\omega_p^2 + \frac{\gamma T_e}{m}k^2} - i\nu_{en}. \tag{6.129}$$

The amplitude of plasma oscillations Eqs. (6.123), (6.129) decays exponentially as a function of time: $\propto \exp(-\nu_{en}t)$. This effect is called the collisional damping of the electrostatic plasma waves. The wave frequencies are usually near the Langmuir frequency, the numerical criterion of the existence of electrostatic plasma waves with respect to the collisional damping can be expressed based on Eq. (6.129) as:

$$\frac{\sqrt{n_e, \text{cm}^{-3}}}{n_0, \text{cm}^{-3}} >> 10^{-12}\text{cm}^{3/2}. \tag{6.130}$$

Here n_0 is the neutral gas density. In atmospheric pressure room temperature discharges $n_0 = 3 \cdot 10^{19}$ cm^{-3} and the criterion Eq. (6.130) requires high electron concentrations $n_e >> 10^{15}$ cm^{-3}.

6.5.3 Ionic Sound

Electrostatic plasma oscillation related to the motion of ions is called the ionic sound. The ionic sound waves are longitudinal as the electrostatic plasma waves, the direction of the electric field in the wave coincides with the direction of the wave vector $\vec{k}$. To derive the dispersion equation of the ionic sound, consider the Poisson equation for the potential φ of the plasma oscillations:

$$\frac{\partial^2 \phi}{\partial x^2} = \frac{e}{\varepsilon_0}(n_e - n_i). \tag{6.131}$$

Electrons quickly correlate their local instantaneous concentration in the wave $n_e(x,t)$ with the potential φ (x,t) of the plasma oscillations in accordance with the Boltzmann distribution:

$$n_e = n_p \exp\left(+\frac{e\phi}{T_e}\right) \approx n_p\left(1 + \frac{e\phi}{T_e}\right). \tag{6.132}$$

Here the unperturbed density of the homogeneous plasma is n_p. Linearizing (6.123) permits obtaining from Eqs. (6.131) and (6.132), the expression for the amplitude of oscillations of ion density:

$$n_i^1 = n_p \frac{e\phi}{T_e}\left(1 + k^2 \frac{\varepsilon_0 T_e}{n_p e^2}\right). \tag{6.133}$$

Combining Eq. (6.133) with the linearized motion equation for ions in the electric field of the wave:

$$M\frac{d\vec{u}_i}{dt} = e\vec{E} = -e\nabla\phi, \quad M\omega u_i^1 = ek\phi, \tag{6.134}$$

and with the linearized continuity equation for ions:

$$\frac{\partial n_i}{\partial t} + \nabla(n_i\vec{u}_i) = 0, \quad \omega n_i^1 = kn_p u_i^1, \tag{6.135}$$

yields the dispersion equation for the ionic sound:

$$\left(\frac{\omega}{k}\right)^2 = c_{si}^2 \frac{1}{1 + k^2 r_D^2}, \quad c_{si} = \sqrt{\frac{T_e}{M}}. \tag{6.136}$$

Here c_{si} is the speed of ionic sound; M is the mass of an ion; $\vec{u}_i$ and u_i^1 are the ionic velocity and amplitude of its oscillation; r_D is the Debye radius Eq. (4.130). From Eq. (6.136), the ionic sound waves $\omega/k = c_{si}$ propagate in plasma with wavelengths exceeding the Debye radius ($k r_D \ll 1$). For shorter wavelengths ($k r_D \gg 1$), the Eq. (6.136) describes plasma oscillations with the frequency:

$$\omega_{pi} = \frac{c_{si}}{r_D} = \sqrt{\frac{n_p e^2}{M \varepsilon_0}}, \tag{6.137}$$

known as the plasma-ion frequency.

6.5.4 Magneto-Hydrodynamic Waves

Special types of waves occur in the plasma magnetic field. Dispersion equations for such waves can be derived by the conventional routine of linearization of magneto-hydrodynamic equations. It is interesting, to describe these magneto-hydrodynamic waves just by physical analysis of the oscillations of the magnetic field frozen in plasma. Consider plasma with high magnetic Reynolds number ($\mathrm{Re}_m \gg 1$), where the magneto-hydrodynamic approach can be applied and the magnetic field is frozen in plasma. Any displacement of the magnetic field $\vec{H}$ leads then to plasma displacement, plasma oscillations, and propagation of specific magneto-hydrodynamic waves. The propagation velocity of the elastic oscillations can be determined using the conventional relation for speed of sound: $v_A = \sqrt{\frac{\partial p}{\partial \rho}}$, where p is pressure and $\rho = Mn_p = Mn_e$ is the plasma's density. The total pressure of the relatively cold plasma is equal to its magnetic pressure $p = \frac{\mu_0 H^2}{2}$, therefore:

$$v_A = \sqrt{\frac{\partial\left(\mu_0 H^2/2\right)}{\partial\left(Mn_p\right)}} = \sqrt{\frac{\mu_0 H}{M}\frac{\partial H}{\partial n_p}}. \tag{6.138}$$

Taking into account that the magnetic field is frozen in the plasma, the derivative in the Eq. (6.138) is: $\frac{\partial H}{\partial n_p} = \frac{H}{n_p}$ and the propagation velocity of the magneto-hydrodynamic waves in plasma is:

$$v_A = \sqrt{\frac{\mu_0 H^2}{\rho}} = \frac{B}{\sqrt{\mu_0 \rho}}. \tag{6.139}$$

This wave propagation velocity is known as the **Alfven velocity.** There are two types of plasma oscillations in the magnetic field, which have the same Alfven velocity, but different directions of propagation. The plasma oscillations propagating along the magnetic field like a wave propagating along an elastic string are usually referred to as the **Alfven wave** or magneto-hydrodynamic wave. Also, the oscillation of a magnetic line induces oscillation of other nearby magnetic lines. This results in propagation perpendicularly to the magnetic field of another wave with the same Alfven velocity. This wave is called the **magnetic sound**. These are the main features of the phenomena; more details regarding magneto-hydrodynamic, electrostatic, and other plasma waves can be found, for example, in T.H. Stix (1992).

6.5.5 Collisionless Interaction of Electrostatic Plasma Waves with Electrons

Consider the effect of energy exchange between electrons and plasma oscillations without collisions. It is convenient to illustrate the collisionless energy exchange in a reference frame moving with a wave (where the wave is at rest, see Figure 6.23). An electron can be trapped in a potential

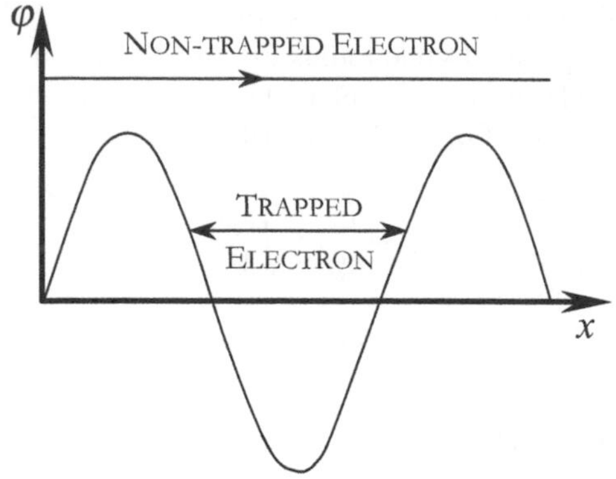

FIGURE 6.23 Electron interaction with plasma oscillations.

well created by the wave, which leads to their effective interaction. If an electron moves in the reference frame of an electrostatic wave with velocity "u" along the wave, and after reflection changes to the opposite direction and its velocity to "-u", then change of electron energy due to the interaction with the electrostatic wave is:

$$\Delta\varepsilon = \frac{m\left(v_{ph}+u\right)^2}{2} - \frac{m\left(v_{ph}-u\right)^2}{2} = 2mv_{ph}u. \tag{6.140}$$

Here $v_{ph} = \omega/k$ is the phase velocity of the wave. If ϕ is the amplitude of wave, then the typical velocity of a trapped electron in the reference frame of plasma wave is $u \approx \sqrt{e\phi/m}$ and typical energy exchange between electrostatic wave and a trapped electron can be estimated as:

$$\Delta\varepsilon \approx v_{ph}\sqrt{me\phi} = e\phi\sqrt{\frac{mv_{ph}^2}{e\phi}}. \tag{6.141}$$

In contrast to Eq. (6.140), the typical energy exchange between an electrostatic wave and a non-trapped electron is approximately $e\phi$. This means that at low amplitudes of oscillations, the collisionless energy exchange is mostly due to the trapped electrons. The described collisionless energy exchange trends to equalize the electron distribution function $f(v)$ at the level of electron velocities near to their resonance with the phase velocity of the wave. Actually, the equalization (trend to form plateau on $f(v)$) takes place in the range of electron velocities between $v_{ph} - u$ and $v_{ph} + u$. It has a typical frequency, corresponding to the frequency of a trapped electron oscillation in the potential well of the wave:

$$v_{ew} \approx uk \approx k\sqrt{\frac{e\phi}{m}} \approx \sqrt{\frac{eE^1k}{m}} \approx \sqrt{\frac{e^2n_e^1}{\varepsilon_0 m}} = \omega_p\sqrt{\frac{n_e^1}{n_e}}. \tag{6.142}$$

Here k is the wave number; m is the electron mass; E^1 is the amplitude of electric field oscillation; ω_p is the plasma frequency; n_e, n_e^1 are electron density and amplitude of its oscillation. In Eq. (6.142). the linearized Maxwell relation Eq. (6.127) between amplitudes of oscillations of electric field and electron density was taken into account. While the electrons interaction with the Langmuir oscillations trends to form a plateau on $f(v)$ at electron velocities close to $v_p = \omega/k$, the electron–electron Maxwellization collisions trends to restore the electron distribution function $f(v)$. The frequency of the electron–electron Maxwellization process can be found based on estimations of the thermal electron temperature v_{Te} and the electron–electron Coulomb cross section σ_{ee} as:

$$\nu_{ee} = n_e v_{Te} \sigma_{ee} \approx n_e \sqrt{\frac{T_e}{m}} \cdot \frac{e^4}{(4\pi\varepsilon_0)^2 T_e^2}. \tag{6.143}$$

If the Maxwellization frequency exceeds the electron–wave interaction frequency $\nu_{ee} >> \nu_{ew}$, the electron distribution function is almost non-perturbed by the collisionless interaction with the electrostatic wave. Comparing the frequencies Eqs. (6.142) and (6.143) shows, that changes in the electron distribution functions $f(E)$ during the collisionless interaction can be neglected, if relative perturbations of electron density are small:

$$\frac{n_e^1}{n_e} << \frac{n_e e^6}{(4\pi\varepsilon_0)^3 T_e^3}. \tag{6.144}$$

It is clear from Figure 6.23 and Eq. (6.140) that electrons with velocities $v_{ph} + u$ transfer their energy to electrostatic plasma wave, while electrons with velocities $v_{ph} - u$ receive energy from the plasma wave. Thus the collisionless damping of electrostatic plasma oscillations takes place when:

$$\frac{\partial f}{\partial v_x}(v_x = v_{ph}) < 0. \tag{6.145}$$

Here, v_x is the electron velocity component in the direction of wave propagation; the derivative of the electron distribution function is taken at the electron velocity equal to the phase velocity $v_{ph} = \omega/k$ of the wave. The electron distribution functions $f(v)$decreases at high velocities corresponding to v_{ph}, the inequality Eq. (6.145) is satisfied and the collisionless damping of the electrostatic plasma waves takes place. The opposite situation of amplification of the electrostatic plasma waves due to the interaction with electrons is also possible. Injection of an electron beam in plasma creates a distribution function, where the derivative Eq. (6.145) is positive (see Figure 6.24); this corresponds to energy transfer from the electron beam and amplification of electrostatic plasma oscillations. The electron beam keeps transferring energy to plasma waves until the total electron distribution function becomes always decreasing (see Figure 6.24). This so-called **beam instability** is used for stimulation of non-equilibrium plasma-beam discharge.

6.5.6 The Landau Damping

If the inequality (6.145) is valid, the electrostatic plasma oscillations transfer energy to electrons as was described above. Estimate the rate of this energy transfer from the plasma oscillations per unit time and per unit volume as:

$$\frac{\partial W}{\partial t} \approx \left[f(v+u) - f(v-u)\right] u n_e \cdot v_{ew} \cdot \Delta\varepsilon \approx \frac{\partial f}{\partial v_x}\frac{n_e e^2}{m}\omega\phi^2. \tag{6.146}$$

Taking into account, that $\phi^2 \approx (E^1)^2/k^2 \approx W/\varepsilon_0 k^2$, rewrite Eq. (6.146) in a conventional form for the decreasing specific energy W of damping oscillations:

$$\frac{\partial W}{\partial t} = -2\gamma W, \quad \gamma \approx -\frac{\omega_p^2 \omega}{k^2}\frac{\partial f}{\partial v_x}\left(v_x = \frac{\omega}{k}\right) > 0. \tag{6.147}$$

This damping of the electrostatic plasma waves $W = W_0 \exp(-2\gamma t)$ due to their collisionless interaction with electrons is known as the **Landau damping** (see, T.M. O'Neil, 1965; B.B. Kadomtsev, 1968). The existence criterion for electrostatic plasma oscillations can be expressed as $\gamma << \omega$. This can be rewritten based on Eq. (6.147) for plasma oscillations ($\omega \sim \omega_p$) and a Maxwellian distribution function as: $k r_D << 1$, where r_D is the Debye radius. When this criterion is valid, the phase velocity of the plasma wave exceeds the thermal velocity of electrons. Then the plasma wave is able to trap only a small fraction of electrons from the tail of the electron distribution function $f(v)$ and the damping of the wave is not strong.

6.5.7 The Beam Instability

Consider the dissipation of an electron beam injected into plasma. Assume that the velocity of the beam electrons exceeds the thermal velocity of plasma electrons and that the electron beam density is much less than plasma density ($n_b << n_p$). Dissipation and stopping of the electron beam obviously can be provided by collisions with electrons, ions, or neutral particles. However, the dissipation and stopping of an electron beam sometimes can be faster than that, which is known as **the beam instability or the Langmuir paradox**. The effect of beam instability is actually opposite with respect to Landau damping. This was mentioned above and is illustrated in Figure 6.24. Electron beam (where the inequality (6.145) has an opposite direction) transfers energy to excitation and amplification of plasma oscillations. This collisionless dissipation of electron beam energy explains the Langmuir paradox. The dispersion equation for plasma waves in the presence of the electron beam can be derived by linearization of the continuity equation, momentum conservation, and Poisson equation:

$$1 = \frac{\omega_p^2}{\omega^2} + \frac{\omega_p^2}{(\omega - ku)^2}\frac{n_b}{n_p}. \tag{6.148}$$

Here u is the electron beam velocity, ω_p is the plasma frequency. The strongest collisionless interaction of the electron beam is related to plasma waves with phase velocity close to the electron beam velocity $|\omega/k - u| << \omega/k$. Also, the oscillation frequency ω is close to the plasma frequency ω_p, because the electron beam density is much less than the plasma density ($n_b << n_p$). Taking into account these two facts and assuming $\omega = \omega_p + \delta$, $|\delta/\omega_p| << 1$, the dispersion equation:

$$\delta = \omega_p \left(\frac{n_b}{2n_p}\right)^{1/3} e^{2\pi i m/3}, \tag{6.149}$$

where m is an integer. The amplification of the specific energy $W \propto (E^1)^2$ of plasma oscillations corresponds in accordance with Eq. (6.123), to the imaginary part of ω (and hence δ). This amplification of the Langmuir oscillation energy W again can be described by Eq. (6.147) but with a negative coefficient γ.

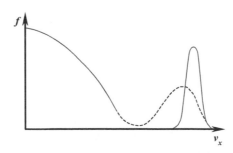

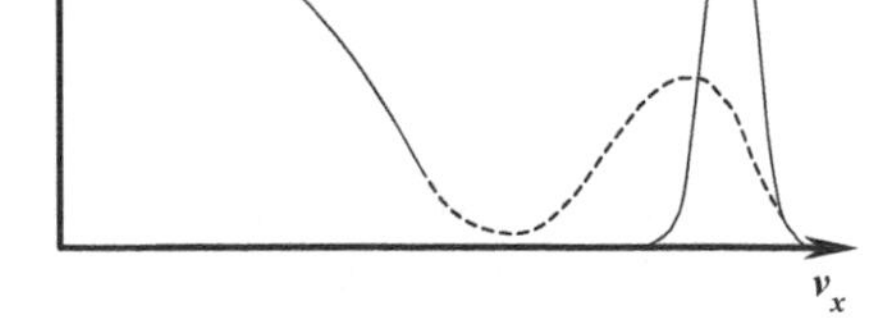

FIGURE 6.24 Electron energy distribution function for electron beam interaction with plasma: solid curve – initial distribution, dashed curve – resulting distribution.

$$\frac{\partial W}{\partial t}=-2\gamma W,\quad \gamma\approx-\frac{\sqrt{3}}{2}\left(\frac{n_b}{2n_p}\right)^{1/3}\omega_p=-0.69\left(\frac{n_b}{n_p}\right)^{1/3}\omega_p<0. \tag{6.150}$$

Thus an electron beam is able to effectively transfer its energy into electrostatic plasma oscillations, which thereafter can dissipate to other plasma degrees of freedom, see A.A. Ivanov (1977).

6.5.8 The Buneman Instability

The beam instability considered above is an example of the so-called kinetic instabilities, where amplification of plasma oscillations is due to the difference in the motion of different groups of charged particles (beam electrons and plasma electrons in this case). Another example of such kinetic instabilities is the Buneman instability, which occurs if the average electron velocity differs from that of ions. To describe the Buneman instability assume that all plasma ions are at rest while all plasma electrons are moving with velocity u. As in the beam instability, electron energy dissipates to generate and amplify electrostatic plasma oscillations. The dispersion equation for the Buneman instability is similar to Eq. (6.148), however in this case, concentrations of two groups of charged particles (electrons and ions, no beam) are equal but their mass is very different ($m/M \ll 1$):

$$1=\frac{m}{M}\frac{\omega_p^2}{\omega^2}+\frac{\omega_p^2}{(\omega-ku)^2}. \tag{6.151}$$

Taking into account $m/M \ll 1$, Eq. (6.151) predicts that $\omega - ku$ is close to the plasma frequency. Then assuming: $\omega = \omega_p + ku + \delta$, $|\delta/\omega_p| \ll 1$, rewrite the dispersion Eq. (6.151) as:

$$\frac{2\delta}{\omega_p}=\frac{m}{M}\frac{\omega_p^2}{(\omega_p+ku+\delta)^2}. \tag{6.152}$$

The electrons interact most efficiently with a wave having the wave number: $k = -\omega_p/u$. For this wave based on Eq. (6.152):

$$\delta=\left(\frac{m}{2M}\right)^{1/3}\omega_p e^{2\pi i m/3}, \tag{6.153}$$

where m is an integer (the strongest plasma wave amplification corresponds to $m = 1$). Similar to Eqs. (6.149), (6.150), the coefficient of amplification γ of the electrostatic plasma oscillation amplitude is:

$$-\gamma = Im\delta=\frac{\sqrt{3}}{2}\left(\frac{m}{2M}\right)^{1/3}\omega_p=0.69\left(\frac{m}{M}\right)^{1/3}\omega_p. \tag{6.154}$$

Because $k = -\omega_p/u$, the frequency of the amplified plasma oscillations for the Buneman instability is close to the coefficient of wave amplification: $\omega \approx |\gamma|$, see A.B. Mikhailovski (1998).

6.5.9 Dispersion and Amplification of Acoustic Waves in Non-Equilibrium Weakly Ionized Plasma, General Dispersion Equation

Non-equilibrium plasma of molecular gases is able to amplify acoustic waves due to different relaxation mechanisms. In principle, amplification of sound, related to phased heating has been known since 1878 (J.W.S. Rayleigh, 1878). The Rayleigh mechanism of acoustic instability in the weakly ionized plasma due to Joule heating was analyzed by different authors, see J.M. Jacob and S.A.

Mani, 1975. Consider the dispersion $k(\omega)$ and amplification of acoustic waves in non-equilibrium chemically active plasma, taking into account a heat release provided by vibrational relaxation and chemical reactions. Chemical reactions are assumed stimulated by the vibrational excitation of molecules. Linearizing the continuity equation, momentum conservation and balance of translational and vibrational energies, the dispersion equation for acoustic waves Eq. (6.123) in the vibrationally non-equilibrium chemically active plasma can be given (I.A. Kirillov, B.V. Potapkin, V.D. Rusanov, A. Fridman, 1983) as:

$$\frac{k^2 c_s^2}{\left(\omega - k\vec{v}\right)^2} = 1 - \frac{i\left(\omega - k\vec{v}\right)\left[e_T(\gamma-1) + e_n\right] - \left(e_v\bar{e}_T - e_T\bar{e}_v\right)\gamma(\gamma-1) + \left(e_n\bar{e}_v - e_v\bar{e}_n\right)\gamma}{\left(\omega - k\vec{v}\right)^2 + i\left(\omega - k\vec{v}\right)\left(e_n - e_T - \gamma\bar{e}_v\right) + \gamma\left(e_v\bar{e}_T - e_T\bar{e}_v\right) + \gamma\left(e_n\bar{e}_v - e_v\bar{e}_n\right)} \tag{6.155}$$

In this dispersion equation: $c_s = \sqrt{\gamma T_0/M}$ is the "frozen" speed of sound (which means, that vibrational degrees of freedom are "frozen" and don't follow variations of T_0); M is mass of heavy particles; γ is the specific heat ratio; $\vec{v}$ is the gas velocity; $e_{n,T,v}$ and $\bar{e}_{n,T,v}$ are the of translational (e) and vibrational ($\bar{e}$) temperatures changes related to perturbations of respectively gas concentration no, translational T_0 and vibrational T_v temperatures:

$$e_n = 2\left(\nu_{VT} + \xi\nu_R\right), \tag{6.156}$$

$$e_T = \hat{k}_{VT}\nu_{VT} - k_{VT}n_0\frac{c_v^v(T_0)}{c_v}, \tag{6.157}$$

$$e_v = k_{VT}n_0\frac{c_v^v(T_v)T_v}{c_v T_0} + \hat{k}_R\xi\nu_R, \tag{6.158}$$

$$\bar{e}_n = \nu_{eV}\frac{c_v T_0}{c_v^v(T_v)T_v}\left(1 + \frac{\partial \ln n_e}{\partial \ln n_0}\right) - 2\nu_{VT}\frac{c_v T_0}{c_v^v(T_v)T_v} - 2\xi\nu_R\frac{c_v T_0}{c_v^v(T_v)T_v}, \tag{6.159}$$

$$\bar{e}_T = -\hat{k}_{VT}\nu_{VT}\frac{c_v T_0}{c_v^v(T_v)T_v} + k_{VT}n_0\frac{T_0}{T_v}, \tag{6.160}$$

$$\bar{e}_v = -k_{VT}n_0 - \hat{k}_R\nu_R\frac{c_v T_0}{c_v^v(T_v)T_v}. \tag{6.161}$$

The characteristic frequencies of vibrational (VT) relaxation, chemical reaction and vibrational excitation of molecules by electron impact Eqs. (6.156)–(6.161) can be expressed respectively by:

$$\nu_{VT} = \frac{\gamma - 1}{\gamma}\cdot\frac{k_{VT}n_0^2\left[\varepsilon_v(T_v) - \varepsilon_v(T_0)\right]}{p}, \tag{6.162}$$

$$\nu_R = \frac{\gamma - 1}{\gamma}\cdot\frac{k_R n_0^2\Delta Q}{p}, \tag{6.163}$$

$$\nu_{eV} = \frac{\gamma - 1}{\gamma}\cdot\frac{k_{eV}n_e n_0\hbar\omega}{p}. \tag{6.164}$$

Here $\varepsilon_v(T_v) = \dfrac{\hbar\omega}{\exp(\hbar\omega/T_v) - 1}$ is the average vibrational energy of molecules; c_v, c_v^v are translational and vibrational heat capacities; $p = n_0T_0$ is the gas pressure; k_{eV}, k_{VT} and k_R are rate coefficients of vibrational excitation, vibrational relaxation and chemical reaction; n_e and n_0 are electron and gas concentrations; ΔQ is the vibrational energy consumption per one act of chemical reaction; ξ is that part of the energy which is going into translational degrees of freedom (the chemical heat release); the dimensionless factors $\hat{k}_R = \frac{\partial \ln k_R(T_v)}{\partial T_v}$, $\hat{k}_{VT} = \frac{\partial \ln k_{VT}}{\partial \ln T_0}$ represents before the logarithmic sensitivities of rate coefficients k_R and k_{VT} to vibrational and translational temperatures respectively. Eq. (6.155) describes not only propagation of sound but evolution of aperiodic perturbations as well. If $\omega = 0$, Eq. (6.155) leads to Eq. (6.100) for frequency and length of relaxation in a fast flow. Without heat release, the dispersion Eq,. (6.155) gives: $v_{ph} = v \pm c_s$ and describes sound waves propagating along gas flow.

6.5.10 Analysis of Dispersion Equation for Sound Propagation in Non-Equilibrium Chemically Active Plasma

Analyze the dispersion Eq. (6.155) in gas at rest ($\vec{kv} = 0$, though some results can be generalized). The frequency ω will be considered a real number, while the wave $k = k_0 - i\delta$ will be considered as a complex number. Then k_0 characterizes the acoustic wavelength, and δ is related to the space amplification of sound.

1. ***Acoustic wave in molecular gas at equilibrium***. There is no especial vibrational excitation in this case and without sound $T_v = T_0$. If the VT-relaxation time is constant ($\tau_{VT} = const$), then Eq. (6.155) gives the well-known **dispersion equation of relaxation gas-dynamics**:

$$\frac{k^2 c_s^2}{\omega^2} = 1 + \frac{\frac{c_S^2}{c_e^2} - 1}{1 - i\omega\tau_{VT}}. \tag{6.165}$$

At the high-frequency limit, the acoustic wave propagates with the "frozen" sound velocity c_s (vibrational degrees of freedom are frozen, that is not participating in the sound propagation). The acoustic wave velocity at low frequencies ($\omega\tau_{VT} << 1$) is the conventional equilibrium speed of sound c_e. The sound waves are always damping Eq. (6.165), but the maximum damping takes place when $\omega\tau_{VT} \approx 1$. The phase velocity v_{ph} and damping δ/k_0 as functions of $\omega\tau_{VT}$ are shown in Figure 6.25.

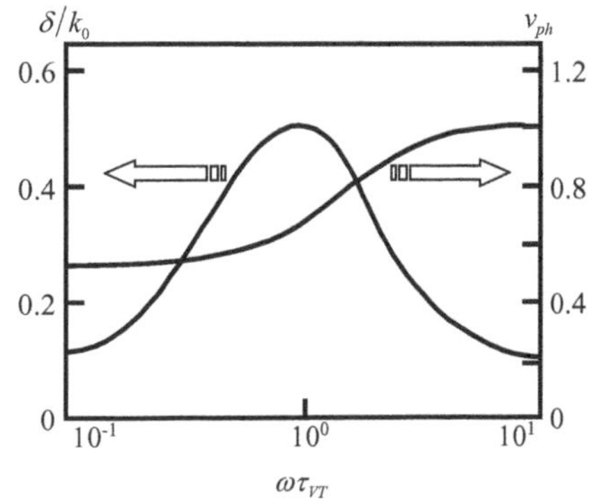

FIGURE 6.25 Phase velocity (in speed of sound units) and attenuation coefficient as functions of relaxation parameter.

2. ***Acoustic wave in high-frequency limit***. Gas is initially non-equilibrium $T_v > T_0$, and the sound frequency exceeds the relaxation frequencies Eqs. (6.156)–(6.161): $\omega >> \max\{\overline{e}_n, \overline{e}_T, \ldots, e_v\}$. The dispersion Eq. (6.155) can be simplified at these conditions and expressed as:

$$\frac{k^2 c_s^2}{\omega^2} = 1 - i\frac{(\gamma - 1)\hat{k}_{VT}\nu_{VT} + 2(\nu_{VT} + \xi\nu_R)}{\omega}. \quad (6.166)$$

The wave propagates in this limit at the "frozen" sound velocity with slight dispersion:

$$v_{ph} = \frac{\omega}{k_0} = c_s\left[1 - \frac{\left((\gamma - 1)\hat{k}_{VT}\nu_{VT} + 2(\nu_{VT} + \xi\nu_R)\right)^2}{8\omega^2}\right]. \quad (6.167)$$

The initial vibrational-translational non-equilibrium results in the amplification of acoustic waves (in contrast to the relaxation gas-dynamics Eq. (6.165)). The amplification coefficient (increment) of the acoustic waves in the non-equilibrium gas at the high-frequency limit can be obtained from Eq. (6.166) as:

$$\frac{\delta}{k_0} = \frac{(\gamma - 1)\hat{k}_{VT}\nu_{VT} + 2(\nu_{VT} + \xi\nu_R)}{2\omega}. \quad (6.168)$$

3. ***Acoustic wave dispersion in the presence of intensive plasma-chemical reaction***. This case is of the most practical interest in plasma-chemical applications. It implies that the reaction frequency greatly exceeds the frequency of vibrational relaxation, heating is mostly related to the chemical heat release ($\xi\nu_R >> \nu_{VT}$), and the steady-state T_v is balanced by vibrational excitation and chemical reaction (see Section 5.6.7). The general dispersion Eq. (6.155) can be simplified in this case to:

$$\frac{k^2 c_s^2}{\omega^2} = 1 - \frac{i\omega a\nu_R + b\nu_R^2}{\omega^2 + i\omega c\nu_R + b\nu_R^2}. \quad (6.169)$$

The dimensionless parameters a, b, and c can be found from the relations:

$$a = 2\xi, \quad b = -\xi\gamma\frac{c_v T_0}{c_v^v T_v}\frac{\partial \ln k_R}{\partial \ln T_v}\left(1 + \frac{\partial \ln n_e}{\partial \ln n_0}\right), \quad c = 2\xi + \gamma\frac{c_v T_0}{c_v^v T_v}\frac{\partial \ln k_R}{\partial \ln T_v}. \quad (6.170)$$

TABLE 6.3
Characteristic Frequencies of Chemical Reaction ν_R, sec^{-1} and VT-relaxation ν_{VT}, sec^{-1} at Different T_0 and T_v; $n_0 = 3 \cdot 10^{18}$ cm^{-3}, $E_a = 5.5$ eV, $B = 72\ K^{1/3}$ (B is the Landau-Teller Coefficient Eq. (3.57) for Vibrational Relaxation)

$T_0\downarrow$	T_v = 2500 K	T_v = 3000 K	T_v = 3500 K	T_v = 4000 K
100 K	$\nu_R = 7\cdot10^3$	$\nu_R = 3\cdot10^5$	$\nu_R = 4\cdot10^6$	$\nu_R = 3\cdot10^8$
	$\nu_{VT} = 3\cdot10^2$	$\nu_{VT} = 4\cdot10^2$	$\nu_{VT} = 5\cdot10^2$	$\nu_{VT} = 7\cdot10^2$
300 K	$\nu_R = 2\cdot10^3$	$\nu_R = 10^5$	$\nu_R = 10^6$	$\nu_R = 10^8$
	$\nu_{VT} = 1.4\cdot10^4$	$\nu_{VT} = 1.6\cdot10^4$	$\nu_{VT} = 1.9\cdot10^4$	$\nu_{VT} = 2.7\cdot10^4$
700 K	$\nu_R = 10^3$	$\nu_R = 4\cdot10^4$	$\nu_R = 5\cdot10^5$	$\nu_R = 5\cdot10^7$
	$\nu_{VT} = 10^5$	$\nu_{VT} = 1.2\cdot10^5$	$\nu_{VT} = 1.3\cdot10^5$	$\nu_{VT} = 1.9\cdot10^5$

Take parameters for the non-equilibrium supersonic discharge in CO_2: $T_v = 3500\ K$, $T_0 = 100\ K$, $E_a = 5.5eV, \dfrac{\partial \ln k_R}{\partial \ln T_v} \approx \dfrac{E_a}{T_v} = 18, \xi = 10^{-2}, \dfrac{\partial \ln n_e}{\partial \ln n_0} \approx (-3) \div (-5)$. In this case according to Eq. (6.170): $a \approx b \approx 0.2$, $c \approx 0.4$. Numerical values for characteristic frequencies of chemical reaction ν_R Eq.

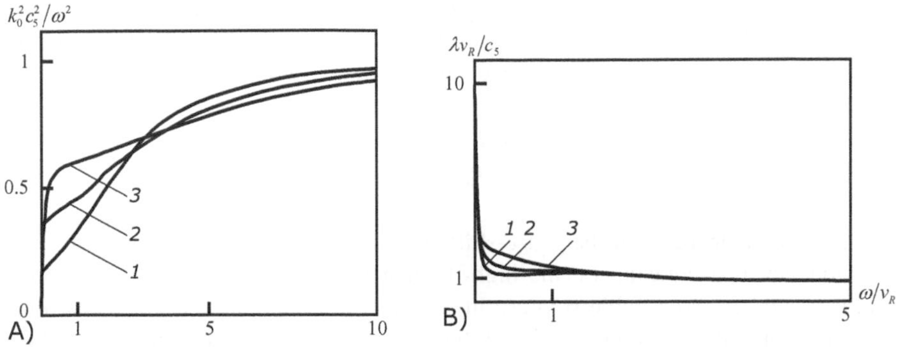

FIGURE 6.26 Refraction index (A) and wavelength (B) as functions of frequency. Translational temperature: 1 – 100 K, 2 – 300 K, 3 – 700 K.

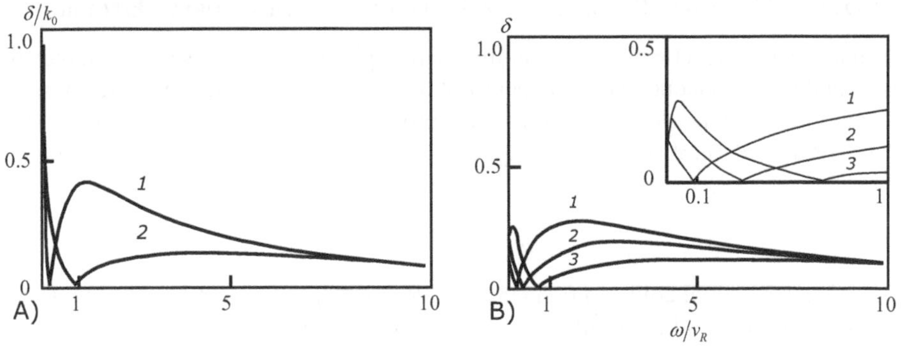

FIGURE 6.27 Wave amplification coefficients (A, B) as functions of frequency. Translational temperature: 1A, 1B – 100 K; 2A, 3B – 700 K; 2B – 300 K.

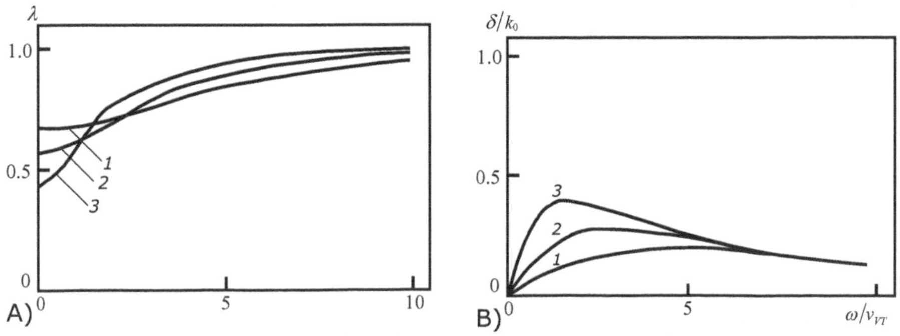

FIGURE 6.28 Wavelength λ (A, in units ω/c_s) and amplification coefficient δ/k_0 (B) as functions of frequency. Translational temperature: 1 – 100 K, 2 – 300 K, 3 – 700K.

(6.163) and VT-relaxation ν_{VT} Eq. (6.162), corresponding to the example of CO_2 dissociation stimulated in plasma by vibrational excitation, are given in Table 6.3. Different dispersion curves for acoustic waves in non-equilibrium plasma, illustrating relation (6.155), are presented in Figures 6.26–6.28. Phase velocities and wavelengths as functions of frequency are given in Figure 6.26 for the case of intensive chemical reaction Eq. (6.169). The acoustic wave amplification coefficients as a function of frequency are presented for the same case in Figure 6.27. Wavelength and amplification coefficient as functions of frequency are illustrated in Figure 6.28 for case when the heat release is mostly due to vibrational relaxation ($\nu_{VT} >> \xi\nu_R$). An interesting physical effect can be seen in Figure 6.27, illustrating the dependence of $\delta/k_0(\omega)$. When frequency ω is decreasing, the amplification coefficient at first grows then passes maximum, anddecreases reaching zero ($\delta = 0$) at frequency:

$$\omega = \nu_R \gamma \frac{c_v T_0}{c_v^v T_v} \sqrt{\xi \left| 1 + \frac{\partial \ln n_e}{\partial \ln n_0} \right|}. \tag{6.171}$$

This effect is due to an increase of the phase-shift between oscillations of heating rate and pressure when the frequency is decreasing (at the high-frequency limit this phase-shift is about $2\xi\nu_R/\omega$). Typical values of ν_R, ν_{VT} (see Table 6.3) are in the ultra-sonic range of frequencies ($\omega > 10^5 Hz$, $\lambda < 0.1\,\text{cm}$). This provides possibilities to use the acoustic dispersion (Figure 6.26–6.28) in ultrasonic diagnostics of plasma-chemical systems (V.K. Givotov, V.D. Rusanov, A. Fridman, 1985).

6.6 PROPAGATION OF ELECTRO-MAGNETIC WAVES IN PLASMA

6.6.1 Complex Dielectric Permittivity of Plasma in High-Frequency Electric Fields

Consider first, electromagnetic waves in non-magnetized plasma. In this case, the complex dielectric permittivity and its components: high-frequency plasma conductivity and dielectric constant are the key concepts in describing electromagnetic wave propagation. 1D electron motion in the electric field $E = E_0 \cos \omega t = \text{Re}(E_0 e^{i\omega t})$ of electromagnetic wave with frequency ω can be described as:

$$m\frac{du}{dt} = -eE - mu\nu_{en}, \tag{6.172}$$

where ν_{en} is electron-neutral collision frequency, $u = \text{Re}\,(u_0 e^{i\omega t})$ is the electron velocity, E_0 and u_0 are amplitudes of the corresponding oscillations. The relation between the amplitudes of electron velocity and electric field is complex and based on Eq. (6.172) can be expressed as:

$$u_0 = -\frac{e}{m}\frac{1}{\nu_{en} + i\omega}E_0. \tag{6.173}$$

The imaginary part of the coefficient between u_0 and E_0 (complex electron mobility) reflects a phase shift between them. Alternately, the Maxwell equation:

$$curl\vec{H} = \varepsilon_0 \frac{\partial \vec{E}}{\partial t} + \vec{j} \tag{6.174}$$

allows the total current density to be presented as the sum:

$$\vec{j}_t = \varepsilon_0 \frac{\partial \vec{E}}{\partial t} + \vec{j}. \tag{6.175}$$

The first current density component in Eq. (6.175) is related to displacement and its amplitude in complex form is $\varepsilon_0 i\omega E_0$. The second current density component in Eq. (6.175) corresponds to the conduction current and has amplitude $-en_e u_0$. Then the amplitude of the total current is:

$$j_{t0} = i\omega\varepsilon_0 E_0 - en_e u_0. \tag{6.176}$$

Taking into account the complex electron mobility Eqs. (6.173), (6.176) for the total current density and, hence, the Maxwell Eq. (6.174) can be rewritten as:

$$j_{t0} = i\omega\varepsilon_0\left[1 - \frac{\omega_p^2}{\omega(\omega - i\nu_{en})}\right]E_0, \quad curl\vec{H}_0 = i\omega\varepsilon_0\left[1 - \frac{\omega_p^2}{\omega(\omega - i\nu_{en})}\right]\vec{E}_0. \tag{6.177}$$

Here ω_p is the electron plasma frequency. When considering electromagnetic waves propagation, it is always convenient to keep the Maxwell Eq. (6.174) for the amplitude of magnetic field H_0 in the form $curl\vec{H}_0 = i\omega\varepsilon_0\varepsilon E_0$. In this case, the conduction current is included in displacement current through the introduction of the complex dielectric constant. According to Eq. (6.177), this complex dielectric constant (or dielectric permittivity) in plasma can be presented as:

$$\varepsilon = 1 - \frac{\omega_p^2}{\omega(\omega - i\nu_{en})}. \tag{6.178}$$

This complex relation can be used for high frequency, dielectric permittivity, and conductivity.

6.6.2 High-Frequency Plasma Conductivity and Dielectric Permittivity

The complex dielectric constant Eq. (6.178) can be rewritten with specified real and imaginary parts:

$$\varepsilon = \varepsilon_\omega + i\frac{\sigma_\omega}{\varepsilon_0\omega}. \tag{6.179}$$

Here, ε_ω the real component of Eq. (6.178), is the high-frequency dielectric constant of plasma:

$$\varepsilon_\omega = 1 - \frac{\omega_p^2}{\omega^2 + \nu_{en}^2}. \tag{6.180}$$

Plasma can be considered a dielectric material. However, in contrast to conventional dielectrics with $\varepsilon > 1$, a plasma is characterized by $\varepsilon < 1$. This can be explained by noting that according to Eqs. (6.172) and (6.173), a free electron (without collisions) oscillates in phase with the electric field and in counter-phase with the electric force. As a result, polarization in plasma is negative. Negative polarization and $\varepsilon < 1$ is typical not only for plasma but also for metals and in the case of extremely high frequencies when electrons can be considered free. The imaginary component of Eq. (6.178) corresponds to the high-frequency conductivity, which is based on Eq. (6.179) can be expressed as:

$$\sigma_\omega = \frac{n_e e^2 \nu_{en}}{m(\omega^2 + \nu_{en}^2)}. \tag{6.181}$$

Equations (6.180) and (6.181) for the high frequency, dielectric permittivity and conductivity can be simplified in two cases, in the so-called collisionless plasma and in the static limit. The **collisionless plasma** limit means $\omega >> \nu_{en}$. For example, a microwave plasma can be considered collisionless at low pressures (about 3 Torr and less). In this case:

$$\sigma_\omega = \frac{n_e e^2 \nu_{en}}{m\omega^2}, \quad \varepsilon_\omega = 1 - \frac{\omega_p^2}{\omega^2}. \tag{6.182}$$

The conductivity in a collisionless plasma is proportional to the electron-neutral collision frequency, and dielectric constant does not depend on the frequency ν_{en}. The ratio of conduction current to polarization current (which actually corresponds to displacement current) can be estimated as:

$$\frac{j_{conduction}}{j_{polarization}} = \frac{\sigma_\omega}{\varepsilon_0 \omega |\varepsilon_\omega - 1|} = \frac{\nu_{en}}{\omega}. \quad (6.183)$$

The polarization current in collisionless plasma ($\nu_{en} << \omega$) exceeds the conductivity current. In the opposite case of the **static limit** $\nu_{en} >> \omega$, the conductivity and dielectric permittivity are:

$$\sigma_\omega = \frac{n_e e^2}{m\nu_{en}}, \quad \varepsilon = 1 - \frac{\omega_p^2}{\nu_{en}^2}. \quad (6.184)$$

In the static limit, the conductivity coincides with the conventional conductivity in DC-conditions (see Eq. (4.99)). In this case, the dielectric permittivity also does not depend on frequency.

6.6.3 Propagation of Electromagnetic Waves in Plasma

Electromagnetic wave propagation in plasmas can be described by the conventional wave equation derived from the Maxwell equations:

$$\Delta\vec{E} - \frac{\varepsilon}{c^2}\frac{\partial^2 \vec{E}}{\partial t^2} = 0, \quad \Delta\vec{H} - \frac{\varepsilon}{c^2}\frac{\partial^2 \vec{H}}{\partial t^2} = 0. \quad (6.185)$$

Plasma peculiarities with respect to electromagnetic wave propagation are related to the complex dielectric permittivity ε Eqs. (6.178), (6.179). The dispersion equation for electromagnetic wave propagation

$$\frac{kc}{\omega} = \sqrt{\varepsilon} \quad (6.186)$$

is valid in a plasma, again with the complex dielectric permittivity ε in form Eqs. (6.178), (6.179). In the wave, electric and magnetic fields can be considered as $\vec{E}, \vec{H} \propto \exp\left(-i\omega t + i\vec{k}\vec{r}\right)$. Assuming the electromagnetic wave frequency ω as real, the wave number k is complex, because ε is complex in Eq. (6.186). Separate the real and imaginary components of the wave number k, based on Eq. (6.186):

$$k = \frac{\omega}{c}\sqrt{\varepsilon} = \frac{\omega}{c}(n + i\kappa). \quad (6.187)$$

From this relation, the physical meaning of the parameter n is **the refractive index** of an electromagnetic wave in plasmas. The phase velocity of the wave is $v = \frac{\omega}{k} = \frac{c}{n}$, the wavelength in the plasma is $\lambda = \lambda_0/n$ (where λ_0 is the corresponding wavelength in vacuum). The wave number κ characterizes the **attenuation of electromagnetic wave in plasma**; the wave amplitude decreases e^κ times on the length $\lambda_0/2\pi$. Taking into account Eq. (6.179) for ε, from Eq. (6.187) the relation between the refractive index and attenuation of electromagnetic wave n, κ with high-frequency dielectric permittivity ε_ω and conductivity σ_ω is obtained

$$n^2 - \kappa^2 = \varepsilon_\omega, \quad 2n\kappa = \frac{\sigma_\omega}{\varepsilon_0 \omega}. \quad (6.188)$$

Solving these equations, an explicit expression for the electromagnetic wave attenuation coefficient is:

$$\kappa = \sqrt{\frac{1}{2}\left(-\varepsilon_\omega + \sqrt{\varepsilon_\omega^2 + \frac{\sigma_\omega^2}{\varepsilon_0^2 \omega^2}}\right)}. \quad (6.189)$$

From Eq. (6.189), the attenuation of an electromagnetic wave in plasma is due to the conductivity if $\sigma_\omega << \varepsilon_\omega \varepsilon_0 \omega$; the electromagnetic field damping can be neglected. The explicit expression for the electromagnetic wave refractive index also can be obtained by solving the system of Eq. (6.188):

$$n = \sqrt{\frac{1}{2}\left(\varepsilon_\omega + \sqrt{\varepsilon_\omega^2 + \frac{\sigma_\omega^2}{\varepsilon_0^2 \omega^2}}\right)}. \tag{6.190}$$

If the conductivity is negligible, the refractive index $n \approx \sqrt{\varepsilon_\omega}$. The plasma polarization is negative and $\varepsilon_\omega < 1$, which means at the low conductivity limit $n < 1$. This results in an interesting conclusion: the velocity of an electromagnetic wave exceeds the speed of light $v_{ph} = c/n > c$. The group velocity is obviously less than the speed of light. The above expression for the refractive index $n \approx \sqrt{\varepsilon_\omega}$ in the case of negligible conductivity leads together with Eq. (6.182) to the well-known form of dispersion equation for electromagnetic waves in collisionless plasma:

$$\frac{k^2 c^2}{\omega^2} = 1 - \frac{\omega_p^2}{\omega^2}, \quad \omega^2 = \omega_p^2 + k^2 c^2. \tag{6.191}$$

The corresponding dispersion curve for electromagnetic waves in plasma is compared in Figure 6.29 with the dispersion curve for electrostatic plasma waves. From Eq. (6.191) and Figure 6.29, it is seen that the phase velocity in this case exceeds the speed of light as was mentioned above. Differentiation of the dispersion Eq. (6.191) gives the relation between the phase and group velocities of electromagnetic waves:

$$\frac{\omega}{k} \times \frac{d\omega}{dk} = v_{ph} v_{gr} = c^2. \tag{6.192}$$

6.6.4 Absorption of Electromagnetic Waves in Plasmas, the Bouguer Law

The energy flux of electromagnetic waves can be described by the Pointing-vector:

$$\vec{S} = \varepsilon_0 c^2 \left[\vec{E} \times \vec{B}\right], \tag{6.193}$$

where the electric and magnetic fields obviously should be averaged over the oscillation period. Electric and magnetic fields in the waves are related according to the Maxwell equations (when $\mu = 1$) as:

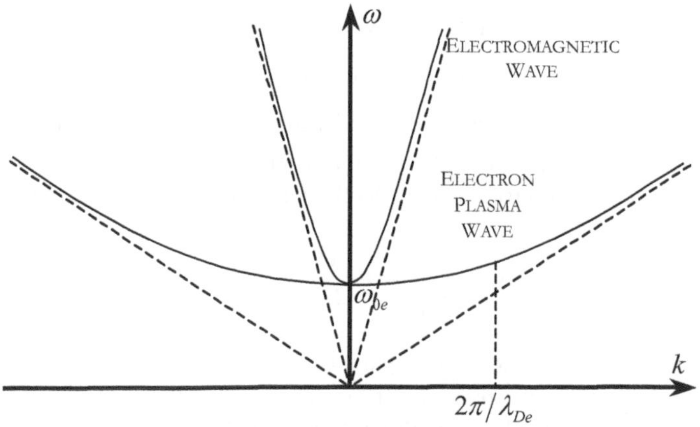

FIGURE 6.29 Electromagnetic and electrostatic plasma wave dispersion.

$$\varepsilon\varepsilon_0 E^2 = \mu_0 H^2. \tag{6.194}$$

Thus, the damping of the electric and magnetic field oscillations are proportional to each other and can be described by the attenuation coefficient Eq. (6.187). As a result, attenuation of the electromagnetic energy flux Eq. (6.193) in plasma follows the **Bouguer law**:

$$\frac{dS}{dx} = -\mu_\omega S, \quad \mu_\omega = \frac{2\kappa\omega}{c} = \frac{\sigma_\omega}{\varepsilon_0 nc}. \tag{6.195}$$

In the Bouguer law, $\mu_\omega$is called the absorption coefficient. The energy flux S (the Pointing-vector) decreases by a factor "e" over the length $1/\mu_\omega$. The product $\mu_\omega S$ in the Bouguer law is actually the electromagnetic energy dissipated per unit volume of plasma, which obviously corresponds to energy dissipation according to the Joule heating law:

$$\mu_\omega S = \varepsilon_0 c^2 \langle EB \rangle = \sigma \langle E^2 \rangle. \tag{6.196}$$

In general, the absorption coefficient μ_ω should be calculated from Eq. (6.195), taking the conductivity σ_ω, refractive index n, and coefficient κ from Eqs. (6.181), (6.190) and (6.189). If the plasma degree of ionization and absorption is relatively low: $n \approx \sqrt{\varepsilon} \approx 1$, then the expression for the absorption coefficient can be simplified and based on Eqs. (6.195) and (6.181) expressed as:

$$\mu_\omega = \frac{n_e e^2 \nu_{en}}{\varepsilon_0 mc\left(\omega^2 + \nu_{en}^2\right)}. \tag{6.197}$$

For practical calculations of electromagnetic wave absorption, it is convenient to use the following numerical formula, corresponding to Eq. (6.197):

$$\mu_\omega, \mathrm{cm}^{-1} = 0.106 n_e\left(\mathrm{cm}^{-3}\right) \frac{\nu_{en}\left(\sec^{-1}\right)}{\omega^2\left(\sec^{-1}\right) + \nu_{en}^2}. \tag{6.198}$$

As seen from Eqs. (6.197) and (6.198), at high frequencies $\omega >> \nu_{en}$ the absorption coefficient is proportional to the square of the wavelength $\mu_\omega \propto \omega^{-2} \propto \lambda^2$. This means that short electromagnetic wave propagates through plasma better than longer ones.

6.6.5 Total Reflection of Electromagnetic Waves from Plasma, Critical Electron Density

If the plasma conductivity is not high $\sigma_\omega << \omega\varepsilon_0|\varepsilon|$, the electromagnetic wave propagates in plasma quite easily if the frequency is sufficiently high. When frequency decreases, the dielectric permittivity $\varepsilon_\omega = 1 - \frac{\omega_p^2}{\omega^2}$ becomes negative and the electromagnetic wave is unable to propagate. According to Eqs. (6.189) and (6.190), the negative dielectric permittivity makes the refractive index equal to zero ($n = 0$) and attenuation coefficient $\kappa \approx \sqrt{|\varepsilon|}$. The phase velocity tends to infinity and the group velocity of the electromagnetic field is equal to zero. Then the depth of electromagnetic wave penetration in plasma:

$$l = \frac{\lambda_0}{2\pi\sqrt{|\varepsilon_\omega|}} = \frac{\lambda_0}{2\pi}\left|1 - \frac{\omega_p^2}{\omega^2}\right|^{-1/2} \tag{6.199}$$

does not depend on conductivity and is not related to energy dissipation. Such not-dissipative stopping corresponds to the total wave reflection from plasma. To illustrate the reflection, consider a wave propagating in non-uniform plasma without significant dissipative absorption (see Figure 6.30).

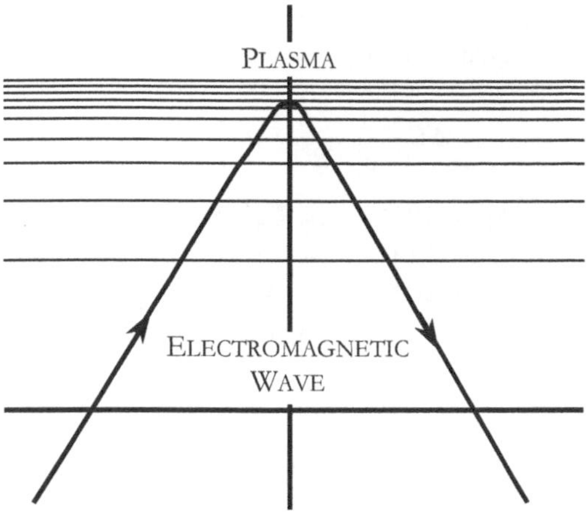

FIGURE 6.30 Electromagnetic wave reflection from plasma with density growing in vertical direction.

The electromagnetic wave propagates from areas with low electron density to areas where the density increases. The wave frequency is obviously fixed, but the plasma frequency increases together with the electron density leading to a decrease of ε_ω. At the point when dielectric permittivity $\varepsilon_\omega = 1 - \frac{\omega_p^2}{\omega^2}$ becomes equal to zero, the total reflection takes place. The total reflection of electromagnetic waves takes place when the electron density reaches the critical value, which can be found from $\omega = \omega_p$ as:

$$n_e^{crit} = \frac{\varepsilon_0 m\omega^2}{e^2}, \quad n_e\left(\text{cm}^{-3}\right) = 1.24 \cdot 10^4 \cdot \left[f\left(\text{MHz}\right)\right]^2. \tag{6.200}$$

An electromagnetic wave with frequency 3 GHz reflects from plasma with a density exceeding the critical value, which in this case equals about 10^{11} cm^{-3}. Atmospheric air is slightly ionized by solar radiation at heights exceeding 100 km; electron density in the ionosphere is about $10^4 \div 10^5$ cm^{-3}. This plasma reflects radio waves with frequencies of about 1 MHz, providing their long-distance transmission.

6.6.6 Electromagnetic Wave Propagation in Magnetized Plasma

If the electric field direction of a plane-polarized wave coincides with the direction of the external uniform magnetic field, the magnetic field does not influence the propagation of the electromagnetic wave. Interesting dispersion can be observed in electromagnetic wave propagation along the uniform magnetic field. Neglecting collisions ($\nu_{en} << \omega$), an electron motion equation in the transverse wave propagating along the magnetic field B_0 is given as:

$$\frac{d\vec{v}}{dt} = -\frac{e}{m}\left(\vec{E} + \left[\vec{v} \times \vec{B}_0\right]\right). \tag{6.201}$$

If ion motion is neglected, the motion Eq. (6.201) can be rewritten using $\vec{j} = -n_e e\vec{E}$:

$$\frac{d\vec{j}}{dt} = \frac{n_e e^2}{m}\vec{E} - \frac{e}{m}\left[\vec{j} \times \vec{B}_0\right]. \tag{6.202}$$

To describe the electromagnetic wave dispersion in magnetized plasma, consider Eq. (6.202) together with the electromagnetic wave equation in the conventional form:

$$\Delta E - \frac{1}{c^2}\frac{\partial^2 E}{\partial t^2} - \mu_0 \frac{\partial \vec{j}}{\partial t} = 0. \tag{6.203}$$

In contrast to the similar wave Eq. (6.185), in the Eq. (6.203) the complex dielectric permittivity is replaced by an additional term related to conduction current. Next look for a solution of the Eqs. (6.202), (6.203) for a circularly polarized electromagnetic wave, when the electric vector rotates in a plane (x,y) perpendicular to direction z of the wave propagation:

$$E_x = E_0 \cos(\omega t - kz), \quad E_y = \pm E_0 \sin(\omega t - kz). \tag{6.204}$$

Here E_0 is the amplitude of electric field oscillations in the wave; signs (+) and (−) corresponds to the rotation of the electric field vector in opposite directions. Current density components can be also expressed in a similar way:

$$j_x = j_0 \cos(\omega t - kz), \quad j_y = \pm j_0 \sin(\omega t - kz), \tag{6.205}$$

where j_0 is amplitude of current density oscillations. Projection of Eqs. (6.202) and (6.203) on the axis "x" can be presented as:

$$\frac{dj_x}{dt} = \varepsilon_0 \omega_p^2 E_x - \omega_B j_y, \tag{6.206}$$

$$\frac{\partial^2 E_x}{\partial z^2} - \frac{1}{c^2}\frac{\partial^2 E_x}{\partial t^2} - \mu_0 \frac{\partial j_x}{\partial t} = 0. \tag{6.207}$$

Here ω_p is plasma frequency; $\omega_B = eB_0/m$ is the electron-cyclotron frequency, which is the frequency of an electron rotation around magnetic lines. Applying the Eqs. (6.204), (6.205) for components of the electric field and current density, rewrite the Eqs. (6.206) and (6.207) as:

$$-\omega j_0 \sin(\omega t - kz) = \varepsilon_0 \omega_p^2 E_0 \cos(\omega t - kz) \pm \omega_B j_0 \sin(\omega t - kz), \tag{6.208}$$

$$-k^2 E_0 \cos(\omega t - kz) + \frac{\omega^2}{c^2} E_0 \cos(\omega t - kz) + \mu_0 \omega j_0 \sin(\omega t - kz) = 0. \tag{6.209}$$

This system of equations has a non-trivial solution only if the following relation between wavelength and frequency is valid:

$$\frac{k^2 c^2}{\omega^2} = 1 - \frac{\omega_p^2}{\omega^2} \frac{1}{\left(1 \pm \frac{\omega_B}{\omega}\right)}. \tag{6.210}$$

This is the dispersion equation for the electromagnetic wave propagation in collisionless plasma along a magnetic field. In the absence of a magnetic field, when the electron-cyclotron frequency is equal to zero $\omega_B = eB_0/m = 0$, the dispersion Eq. (6.210) coincides with Eq. (6.191) for a collisionless plasma.

6.6.7 Propagation of Ordinary and Extra-Ordinary Polarized Electromagnetic Waves in Magnetized Plasma

Two signs (+) and (−) corresponds to two different directions of rotation of vector $\vec{E}$. The (−) sign is related to the **right-hand-side circular polarization** of electromagnetic waves when the direction of $\vec{E}$ rotation coincides with the direction of an electron gyration in the magnetic field. In optics, such

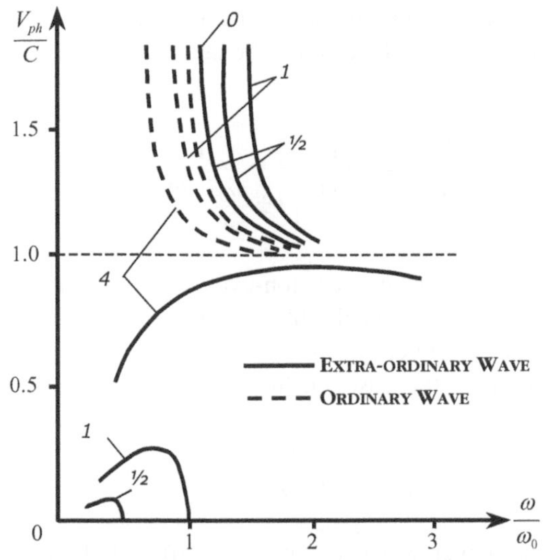

FIGURE 6.31 Dispersion of a transverse electromagnetic wave, propagating in plasma along magnetic field; numbers correspond to values of ω_H/ω_p.

a wave is usually referred to as the **extra-ordinary wave**. The extra-ordinary wave has $\omega = \omega_B$ as the resonant frequency, when the denominator in the dispersion equation tends to zero. This **electron-cyclotron resonance (ECR)** provides effective absorption of electromagnetic waves and is used as a principal physical effect in the ECR-discharges. The (+) sign in the dispersion Eq. (6.211) is related to the **left-hand-side circular polarization** of electromagnetic wave. In optics, such a wave is usually referred to as the **ordinary wave** and it is not a resonant one. Based on the dispersion Eq. (6.210), the phase velocity of the electromagnetic waves, propagating along the magnetic field, can be presented as:

$$v_{ph} = c\left[1 - \frac{\omega_p^2}{\omega^2}\frac{1}{\left(1 \pm \frac{\omega_B}{\omega}\right)}\right]^{-1/2}. \tag{6.211}$$

Propagation of electromagnetic waves in the absence of a magnetic field is possible according to Eqs. (6.211) and (6.191) only with frequencies exceeding the plasma frequency $\omega > \omega_p$. This situation becomes very different in the magnetic field. The dispersion curves corresponding to Eq. (6.211) are shown in Figure 6.31. As is seen from the curves, propagation of both ordinary and extra-ordinary waves is possible at low frequencies ($\omega < \omega_p$). Also, in contrast to $B_0 = 0$, where phase velocities always exceed the speed of light, the electromagnetic waves can be "slower" than the speed of light in the presence of magnetic field.

6.6.8 Influence of Ion Motion on Electromagnetic Wave Propagation in Magnetized Plasma

The extra-ordinary waves are able to propagate in plasma even at low frequencies (see Figure 6.10). In this case, the ion motion can also be important the ion-cyclotron frequency should be taken into account:

$$\omega_{Bi} = \frac{e}{M}B, \tag{6.212}$$

which corresponds to ions (mass M) gyration in magnetic field B_0. The dispersion equation for wave propagating along a uniform magnetic field B_0 can be derived taking into account the ion motion:

$$\frac{k^2c^2}{\omega^2}=1-\frac{\omega_p^2}{\omega^2}\frac{1}{\left(1\pm\frac{\omega_B}{\omega}-\frac{\omega_B\omega_{Bi}}{\omega^2}\right)}. \tag{6.213}$$

If the electromagnetic wave frequency exceeds ion-cyclotron frequency ($\omega >> \omega_B$), the influence of the term $\omega_B\omega_{Bi}/\omega^2$ in the denominator is negligible and the dispersion Eq. (6.213) coincides with Eq. (6.210). Conversely, if the wave frequency is low ($\omega >> \omega_B$), then the term $\omega_B\omega_{Bi}/\omega^2$ becomes dominant in the dispersion equation. In this case, the dispersion Eq. (6.213) can be rewritten as:

$$\frac{c^2}{(\omega/k)^2}=1+\frac{\omega_p^2}{\omega_B\omega_{Bi}}=1+\frac{\mu_0\rho}{B^2}\approx\frac{c^2}{v_A^2}. \tag{6.214}$$

This introduces again the Alfven wave velocityv_A, Eq. (6.139); here $\rho = n_eM$ is the mass density in completely ionized plasma, see V.L. Ginsburg (1960), and V.L. Ginsburg, A.A. Rukhadze (1970).

6.7 EMISSION AND ABSORPTION OF RADIATION IN PLASMA, CONTINUOUS SPECTRUM

6.7.1 Classification of Radiation Transitions

Radiation occurs due to transitions between different energy levels of a quantum system: transition up corresponds to the absorption of a quantum $E_f - E_i = \hbar\omega$, transition down the spectrum corresponds to emission $E_i - E_f = \hbar\omega$ (see Section 6.7.2). From the point of classical electrodynamics, radiation is related to the nonlinear change of dipole momentum, actually with the second derivative of dipole momentum. Neither emission nor absorption of radiation is impossible for free electrons. Electron collisions are necessary in this case. It will be shown in Section 6.7.4, that electron

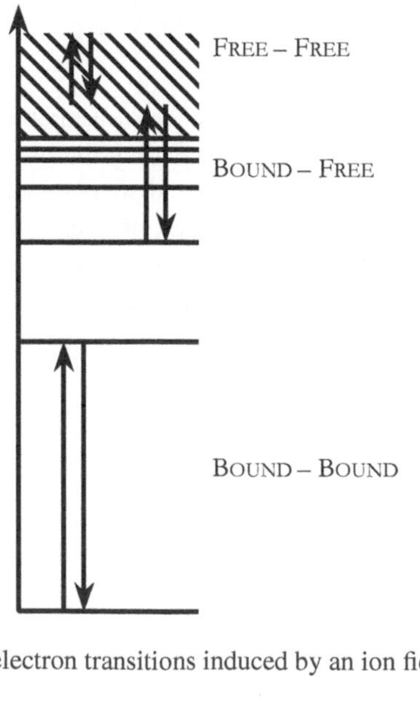

FIGURE 6.32 Energy levels and electron transitions induced by an ion field.

interaction with a heavy particle, ion or neutral, is able to provide emission or absorption, but electron–electron interaction cannot. It is convenient to classify different types of radiation according to the different types of an electron transition from one state to another. Electron energy levels in the field of an ion as well as transitions between the energy levels are illustrated in Figure 6.32. The case when both initial and final electron states are in the continuum is called the **free-free transition**. A free electron in this transition loses part of its kinetic energy in the Coulomb field of a positive ion or in interaction with neutrals. The emitted energy in this case is a continuum usually infrared and called **bremsstrahlung** (direct translation - stopping radiation). The reverse process is the bremsstrahlung absorption. Electron transition between a free state in continuum and a bound state in atom (see Figure 6.32) is usually referred to as the **free-bound transition**. The free-bound transitions correspond to processes of the radiative electron-ion recombination (see Section 2.3.5) and the reverse one of photoionization (see Section 2.2.6). Such kind of transitions also could take place in electron-neutral collisions. In this case these are related to photo-attachment and photo-detachment processes of formation and destruction of negative ions (see Section 2.4.). The free-bound transitions correspond to continuum radiation. Finally **the bound-bound transitions** mean transition between discrete atomic levels (see Figure 6.32) and result in emission and absorption of spectral lines. Molecular spectra are obviously much more complex than those of single atoms because of possible transitions between different vibrational and rotational levels.

6.7.2 Spontaneous and Stimulated Emission, the Einstein Coefficients

Consider transitions between two states (upper "u" and ground "0") of an atom or molecule with emission and absorption of a photon $\hbar\omega$, which is illustrated in Figure 6.33. The probability of photon absorption by an atom per unit time (and, hence, atom transition "0" → "u") can be expressed as:

$$P\left(\text{“0”}, n_\omega \rightarrow \text{“}u\text{”}, n_\omega - 1\right) = A n_\omega. \tag{6.215}$$

Here n_ω is the number of photons; A is **the Einstein coefficient**, which depends on atomic parameters and does not depend on electromagnetic wave characteristics. Similarly, the probability of atomic transition with a photon emission is:

$$P\left(\text{“}u\text{”}, n_\omega \rightarrow \text{“0”}, n_\omega + 1\right) = \frac{1}{\tau} + B n_\omega. \tag{6.216}$$

Here $1/\tau$ is the frequency of **spontaneous emission**, which takes place without direct relation to external fields; **Einstein coefficient B** characterizes emission induced by an external electromagnetic field. The factors B and τ as well as A depend only on atomic parameters. If the first right-side term in Eq. (6.216) corresponds to spontaneous emission, the second term is related to the **stimulated emission**. To find relations between the Einstein coefficients A, B and the spontaneous emission frequency $1/\tau$ analyze the thermodynamic equilibrium of radiation with the atomic system. In this case, the densities of atoms in lower and upper states (Figure 6.33) are related in accordance with the Boltzmann law Eq. (4.9) as:

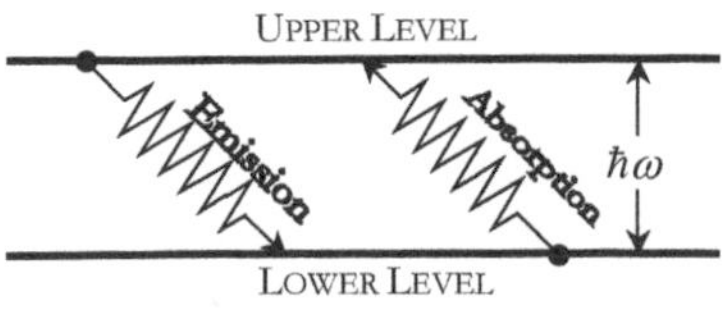

FIGURE 6.33 Radiative transition between two energy levels.

$$n_u = \frac{g_u}{g_0} n_0 \exp\left(-\frac{\hbar\omega}{T}\right). \tag{6.217}$$

where $\hbar\omega$ is the energy difference between the two states, g_u, g_0 are their statistical weights. According to the Planck distribution, the averaged number of photons $\overline{n}_\omega$ in one state can be determined as:

$$\overline{n}_\omega = 1/\left(\exp\frac{\hbar\omega}{T} - 1\right). \tag{6.218}$$

Taking into account the balance of photon emission and absorption, see Figure 6.33:

$$n_0 P(\text{“}0\text{”}, n_\omega \rightarrow \text{“}u\text{”}, n_\omega - 1) = n_u P(\text{“}u\text{”}, n_\omega \rightarrow \text{“}0\text{”}, n_\omega + 1), \tag{6.219}$$

which can be rewritten based on Eqs. (6.215) and (6.216) as:

$$n_0 A \overline{n}_\omega = n_u \left(\frac{1}{\tau} + B\overline{n}_\omega\right). \tag{6.220}$$

The relations between the Einstein coefficients A, B, and the spontaneous emission frequency $1/\tau$ can be expressed based on Eqs. (6.217), (6.218) and (6.220) as:

$$A = \frac{g_u}{g_0}\frac{1}{\tau}, \quad B = \frac{1}{\tau}. \tag{6.221}$$

The emission probability can be rewritten from Eqs. (6.216) and (6.221) as:

$$P\left(\text{“}u\text{”}, n_\omega \rightarrow \text{“}0\text{”}, n_\omega + 1\right) = \frac{1}{\tau} + n_\omega \frac{1}{\tau}. \tag{6.222}$$

The contribution of the stimulated emission $n_\omega$times exceeds the spontaneous times. Although the expressions (6.221) for the Einstein coefficients were derived from equilibrium thermodynamics, the final Eq. (6.222) can be applied for the non-equilibrium number of photons n_ω.

6.7.3 General Approach to the Bremsstrahlung Spontaneous Emission, Coefficients of Radiation Absorption and Stimulated Emission During Electron Collisions with Heavy Particles

The bremsstrahlung emission is related to the free-free electron transitions and does not depend on external radiation like the spontaneous emission. An electron slows down in a collision with a heavy particle, ion or neutral, and loses kinetic energy, which then partially goes to radiation. A free electron cannot absorb or emit a photon because of the requirements of momentum conservation. Describe the bremsstrahlung emission at a specific frequency ω in terms of the **differential spontaneous emission cross section**: $d\sigma_\omega(v_e) = \frac{d\sigma_\omega}{d\omega} d\omega$, which is a function of electron velocity v_e. The emission probability per one electron of quanta with energies $\hbar\omega \div \hbar\omega + d(\hbar\omega)$can be given by the conventional definition:

$$dP_\omega = \frac{1}{\hbar\omega} dQ_\omega = v_e n_0 d\sigma_\omega(v_e) = v_e n_0 \frac{d\sigma_\omega}{d\omega} d\omega. \tag{6.223}$$

Here dQ_ω is the emission power per one electron in the frequency interval $d\omega$; n_0 is the density of heavy particles. The differential cross section of bremsstrahlung emission $d\sigma_\omega/d\omega(v_e)$ characterizes the spontaneous emission of plasma electrons during their collisions with heavy particles in the same way as the frequency $1/\tau$ characterizes spontaneous emission of an atom Eq. (6.216). Also, similar to the Einstein coefficients A and B for atomic systems (see relations (6.215) and (6.216)), introduce coefficients $a_\omega(v_e)$ and $b_\omega(v_e)$ for radiation absorption and stimulated emission during the electron collision with heavy particles. To introduce the coefficients $a_\omega(v_e)$ and $b_\omega(v_e)$, first define the radiation intensity $I(\omega)$ as the emission power in spectral interval $\omega \div \omega + d\omega$ per unit area and within the unit solid angle (Ω, measured in Steradians). The product $I_\omega d\omega\, d\Omega$ actually describes the density of the radiation energy flux. Then the radiation energy from the interval $d\omega\, d\Omega$ absorbed by electrons with velocities $v_e \div v_e + dv_e$ in unit volume per unit time can be presented as:

$$\left(I_\omega d\omega d\Omega\right) n_0 \left[f\left(v_e\right) dv_e\right] \times a_\omega\left(v_e\right). \tag{6.224}$$

This is the definition of the **absorption coefficient** $a_\omega(v_e)$, which is actually calculated with respect to one electron and one atom. In its definition Eq. (6.224): n_0 . is the density of heavy particles, and $f(v_e)$is the electron distribution function. Similar to Eq. (6.224), the stimulated emission of quanta related to electron collisions with heavy particles can be expressed as:

$$\left(I_\omega d\omega d\Omega\right) n_0 \left[f\left(v_e'\right) dv_e'\right] \times b_\omega\left(v_e'\right). \tag{6.225}$$

Here the **coefficient of stimulated emission** $b_\omega\left(v_e'\right)$ is introduced with respect to one electron and one atom. In accordance with the energy conservation for absorption Eq. (6.224) and stimulated emission Eq. (6.225) – the electron velocities v_e and v_e' are related to each other as:

$$\frac{m\left(v_e'\right)^2}{2} = \frac{mv_e^2}{2} + \hbar\omega. \tag{6.226}$$

Similar to Eq. (6.221), the coefficients of absorption $a_\omega(v_e)$ and stimulated emission $b_\omega\left(v_e'\right)$ of an electron during its collision with a heavy particle are related through the **Einstein formula**:

$$b_\omega\left(v_e'\right) = \frac{v_e}{v_e'} a_\omega\left(v_e\right) = \left(\frac{\varepsilon}{\varepsilon + \hbar\omega}\right)^{1/2} a_\omega\left(v_e\right). \tag{6.227}$$

Here ε is the electron energy (as well as $\varepsilon + \hbar\omega$). The stimulated emission coefficient is related to the spontaneous bremsstrahlung emission cross section as:

$$b_\omega\left(v_e'\right) = \frac{\pi^2 c^2 v_e'}{\omega^2} \frac{d\sigma_\omega\left(v_e'\right)}{d\omega}. \tag{6.228}$$

6.7.4 Bremsstrahlung Emission due to Electron Collisions with Plasma Ions and Neutrals

According to classical electrodynamics, emission of a system of electric charges is determined by the second derivative of its dipole momentum $\ddot{\vec{d}}$. In the case of an electron (characterized by a radius-vector $\vec{r}$) scattering by a heavy particle at rest: $\ddot{\vec{d}} = -e\ddot{\vec{r}}$. In the case of electron–electron collisions: $\ddot{\vec{d}} = -e\ddot{\vec{r}}_1 - e\ddot{\vec{r}}_2 = -\frac{e}{m} \cdot \frac{d}{dt}\left(m\vec{v}_1 + m\vec{v}_2\right) = 0$ because of momentum conservation. This relation explains the absence of bremsstrahlung emission in the electron–electron collisions. Total energy, emitted by an electron during interaction with a heavy particle, can be expressed by (L.D. Landau, 1982):

$$E = \frac{e^2}{6\pi\varepsilon_0 c^3}\int_{-\infty}^{+\infty}\left|\ddot{\vec{r}}(t)\right|^2 dt. \tag{6.229}$$

The Fourier expansion of electron acceleration $\ddot{\vec{r}}(t)$ in Eq. (6.229) gives the emission spectrum of the bremsstrahlung, the electron energy emitted per one collision in frequency interval $\omega \div \omega + d\omega$:

$$dE_\omega = \frac{e^2}{3\pi\varepsilon_0 c^3}\left|\int_{-\infty}^{+\infty}\ddot{\vec{r}}(t)\exp(-i\omega t)\,dt\right|^2 d\omega. \tag{6.230}$$

Taking into account that the time of effective electron interaction with heavy a particle is shorter than the electromagnetic field oscillation time, the integral in Eq. (6.230) can be estimated as the electron velocity change $\Delta\vec{v}_e$ during scattering. This allows simplification of (6.230):

$$dE_\omega = \frac{e^2}{6\pi^2\varepsilon_0 c^3}\left(\Delta v_e^2\right)d\omega. \tag{6.231}$$

Averaging Eq. (6.232) over all electron collisions with heavy particles (frequency ν_m, electron velocity v_e), yields the formula for dQ_ω (see (6.223)), which is the bremsstrahlung emission power per one electron in the frequency interval $d\omega$:

$$dQ_\omega = \frac{e^2 v_e^2 \nu_m}{3\pi^2\varepsilon_0 c^3}d\omega. \tag{6.232}$$

Obviously, this formula of classical electrodynamics can only be used at relatively low frequencies $\omega < mv_e^2/2\hbar$, when electron energy exceeds the emitted quantum of radiation. At higher frequencies $\omega > mv_e^2/2\hbar$, we should assume $dQ_\omega = 0$ must be assumed to avoid the "ultra-violet catastrophe". Based on Eqs. (6.232) and (6.223), the cross section of the spontaneous bremsstrahlung emission of quantum $\hbar\omega$ by an electron with velocity v_e can be expressed as:

$$d\sigma_\omega = \frac{e^2 v_e^2 \sigma_m}{3\pi^2\varepsilon_0 c^3\hbar\omega}d\omega, \quad \hbar\omega \le \frac{mv_e^2}{2}. \tag{6.233}$$

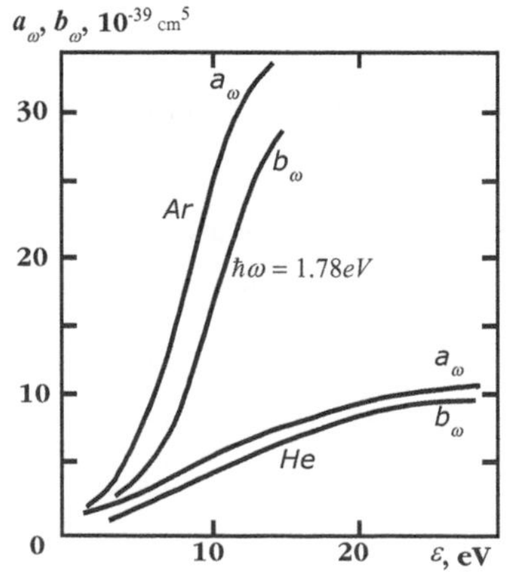

FIGURE 6.34 Absorption coefficients per one electron and per one Ar, He atom.

Bremsstrahlung emission is related to an electron interaction with heavy particles, therefore its cross section Eq. (6.233) is proportional to the cross section $\sigma_m$of electron-heavy particle collisions. Equations (6.227) and (6.228) then permit the derivation of formulas for the quantum coefficients of stimulated emission b_ω and absorption a_ω. For example, the absorption coefficient $a_\omega(v_e)$ can be presented in this way at $\omega >> \nu_m$as:

$$a_\omega\left(v_e\right)=\frac{2}{3\varepsilon_o}\frac{e^2 v_e}{mc\omega^2}\frac{\left(\varepsilon+\hbar\omega\right)^2}{\varepsilon\hbar\omega}\sigma_m\left(\varepsilon+\hbar\omega\right), \tag{6.234}$$

where ε is electron energy before the absorption. Absorption and stimulated emission coefficients per one electron and one atom of Ar and He as a function of electron energy at a frequency of a ruby laser are presented in Figure 6.34. Plasma radiation is more significant in thermal plasma, where the major contribution in ν_m and σ_m is provided by electron-ion collisions. Taking into account the Coulomb nature of the collisions, the cross section of the bremsstrahlung emission Eq. (6.233) is:

$$d\sigma_\omega=\frac{16\pi}{3\sqrt{3}\left(4\pi\varepsilon_0\right)^3}\frac{Z^2e^6}{m^2c^3v^2\hbar\omega}. \tag{6.235}$$

Here Z is the charge of an ion; in most of our systems under consideration $Z = 1$, but relation (6.235) in general can be applied to multi-charged ions as well. Based on Eq. (6.235), the total energy emitted in unit volume per unit time by the bremsstrahlung mechanism in the spectral interval $d\omega$ is:

$$J_\omega^{brems}d\omega=\int_{v_{\min}}^{\infty}\hbar\omega n_i n_e f\left(v_e\right)v_e dv_e d\sigma_\omega\left(v_e\right). \tag{6.236}$$

Here n_i and n_e are concentrations of ions and electrons; $f(v_e)$ is the electron velocity distribution function, which can be taken here as Maxwellian; $v_{\min}=\sqrt{2\hbar\omega/m}$ is the minimum electron velocity sufficient to emit a quantum $\hbar\omega$. After integration the spectral density of bremsstrahlung emission ($Z = 1$) per unit volume, Eq. (6.237) finally gives:

$$\begin{aligned}J_\omega^{brems}d\omega&=\frac{16}{3}\left(\frac{2\pi}{3}\right)^{1/2}\frac{e^6 n_e n_i}{m^{3/2}c^3\left(4\pi\varepsilon_0\right)^3T^{1/2}}\exp\left(-\frac{\hbar\omega}{T}\right)d\omega=C\frac{n_i n_e}{T^{1/2}}\exp\left(-\frac{\hbar\omega}{T}\right)d\omega\\&=1.08\cdot10^{-45}W\cdot\mathrm{cm}^3\cdot K^{1/2}\times\frac{n_i\left(1/\mathrm{cm}^3\right)\cdot n_e\left(1/\mathrm{cm}^3\right)}{\left(T,K\right)^{1/2}}\exp\left(-\frac{\hbar\omega}{T}\right)d\omega.\end{aligned} \tag{6.237}$$

6.7.5 Recombination Emission

This emission occurs during the radiative electron-ion recombination. According to the classification of emission in plasma, the recombination emission is related to free-bound transitions because as a result of this process, a free plasma electron becomes trapped in a bound atomic state with negative discrete energy E_n. This recombination leads to emission of a quantum:

$$\hbar\omega=\left|E_n\right|+\frac{mv_e^2}{2}. \tag{6.238}$$

Correct quantum mechanical derivation of the recombination emission cross section is complicated, but an approximate formula can be derived using the quasi-classical approach. Equation (6.235) was derived for bremsstrahlung emission when the final electron is still free and has positive energy. To calculate the cross section σ_{RE} of the recombination emission, generalize the quasi-classic relation (6.235) for electron transitions into discrete bound states. It can be done because the energy distance between high electronic levels is quite small and hence quasi-classic. The electronic levels of bound atomic states are discrete, so here the spectral density of the recombination cross section should be redefined taking into account number of levels Δn per small energy interval $\Delta E = \hbar\Delta\omega$ (around E_n):

$$\frac{d\sigma_\omega}{d\omega}\Delta\omega = \sigma_{RE}\Delta n, \quad \sigma_{RE} = \frac{d\sigma_\omega}{d\omega}\frac{1}{\hbar}\frac{\Delta E}{\Delta n}. \tag{6.239}$$

Consider the hydrogen-like atoms (where an electron moves in the field of a charge Ze), then $E_n = -\frac{I_H Z^2}{n^2}$. Here $I_H = \frac{me^4}{2\hbar^2(4\pi\varepsilon_0)^2} \approx 13.6eV$ is the hydrogen atom ionization potential, n is the principal quantum number. The ratio $\Delta E/\Delta n$ characterizing the density of energy levels is:

$$\frac{\Delta E}{\Delta n} \approx \left|\frac{dE_n}{dn}\right| = \frac{2I_H Z^2}{n^3}. \tag{6.240}$$

Based on Eq. (6.236) and taking into account Eqs. (6.239) and (6.240), the formula for the cross section of the recombination emission in an electron-ion collision with the formation of an atom in an excited state with the principal quantum number n is obtained:

$$\sigma_{RE} = \frac{4}{3\sqrt{3}}\frac{e^{10}Z^4}{(4\pi\varepsilon_0)^5 c^3\hbar^4 m v_e^2\omega}\frac{1}{n^3} = 2.1\cdot 10^{-22}\text{cm}^2\cdot\frac{\left(I_H Z^2\right)^2}{\varepsilon\hbar\omega n^3}. \tag{6.241}$$

Here ε is the initial electron energy. The total energy $J_{\omega n}d\omega$, emitted as a result of photo-recombination of electrons with velocities in interval $v_e \div v_e + dv_e$ and ions with $Z = 1$, leading to the formation of an excited atom with the principal quantum number n, per unit time and per unit volume is:

$$J_{\omega n}d\omega = \hbar\omega n_i n_e \sigma_{RE} f(v_e) v_e dv_e = C\frac{n_i n_e}{T^{1/2}}\frac{2I_H}{Tn^3}\exp\left(\frac{I_H}{Tn^2} - \frac{\hbar\omega}{T}\right)d\omega. \tag{6.242}$$

The parameter $C = 1.08 \cdot 10^{-45} W \cdot \text{cm}^3 \cdot K^{1/2}$ is the same factor as in Eq. (6.237); $f(v_e)$ is the electron velocity distribution, which was taken as Maxwellian; the relation between electron velocity and radiation frequency from Eq. (6.238) was written as: $mv_e dv_e = \hbar d\omega$. Emission of quanta with the same energy $\hbar\omega$ can be provided by trapping electrons on different excitation levels with different principal quantum numbers n (obviously, the electron velocities should be relevant and determined by Eq. (6.238)). The total spectrum of recombination emission is the sum of similar but shifted terms:

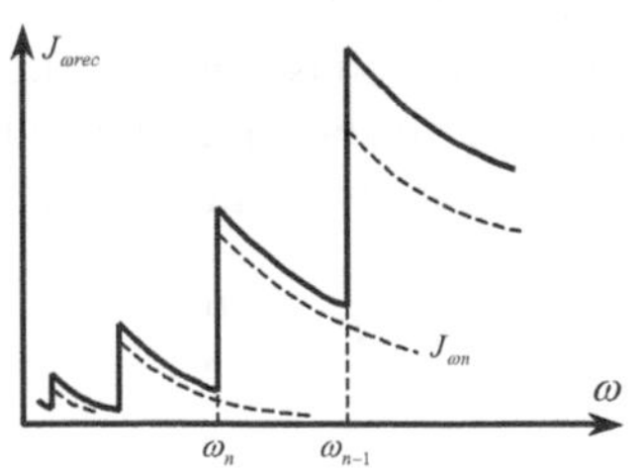

FIGURE 6.35 Recombination emission spectrum.

$$J_{\omega}^{recomb} d\omega = \sum_{n*(\omega)}^{\infty} J_{\omega n}. \tag{6.243}$$

Each of the individual terms in Eq. (6.243) is related to the different principal quantum number n but giving the same frequency ω, and each is proportional to $\exp\left(-\frac{\hbar\omega}{T}\right)$. The lowest possible principal quantum number $n*(\omega)$ can be defined from the condition $|E_{n*}| < \hbar\omega < |E_{n*-1}|$. The spectrum of recombination emission corresponding to the sum Eq. (6.243) is illustrated in Figure 6.35.

6.7.6 Total Emission in Continuous Spectrum

The total spectral density of plasma emission in continuous spectrum consists of the bremsstrahlung and recombination components, Eqs. (6.237) and (6.243). These two emission components can be combined in one approximate but reasonably accurate general formula for the total continuous emission:

$$J_{\omega} d\omega = C \frac{n_i n_e}{T^{1/2}} \Psi\left(\frac{\hbar\omega}{T}\right) d\omega. \tag{6.244}$$

The parameter $C = 1.08 \cdot 10^{-45} W \cdot cm^3 \cdot K^{1/2}$ is the same factor as in Eqs. (6.237) and (6.243); $\Psi(x)$ is the dimensionless function, which can be approximated as:

$$\Psi(x) = 1, \quad if\ x = \frac{\hbar\omega}{T} < x_g = \frac{|E_g|}{T}, \tag{6.245}$$

$$\Psi(x) = \exp[-(x - x_g)], \quad if\ x_g < x < x_1, \tag{6.246}$$

$$\Psi(x) = \exp\left[-(x - x_g)\right] + 2x_1 \exp\left[-(x - x_1)\right], \quad if\ x = \frac{\hbar\omega}{T} > x_1 = \frac{I}{T}. \tag{6.247}$$

Here $|E_g|$ is the energy of the first (lowest) excited state of atom calculated with respect to transition to continuum; I is the ionization potential. When the radiation quanta and hence frequencies are not very large ($\hbar\omega < |E_g|$), the contributions of free-free transitions (bremsstrahlung) and free-bound transitions (recombination) in the total continuous emission are related to each other as:

$$J_{\omega}^{recomb} / J_{\omega}^{brems} = \exp\frac{\hbar\omega}{T} - 1. \tag{6.248}$$

According to Eq. (6.248) emission $\hbar\omega < 0.7T$ is mostly due to the bremsstrahlung mechanism, while the emission of larger quanta $\hbar\omega > 0.7T$ is mostly due to recombination mechanism. At plasma temperatures of 10,000 K, this means that only infrared radiation ($\lambda > 2\mu m$) is provided by bremsstrahlung, all other emission spectrum is due to the electron-ion recombination. To obtain the total radiation losses, integrate the spectral density Eq. (6.244) over all the emission spectrum (contribution of quanta $\hbar\omega > I$ can be neglected because of their intensive reabsorption). These total plasma energy losses per unit time and unit volume can be expressed after integration by the following numerical formula:

$$J, \frac{kW}{cm^3} = 1.42 \cdot 10^{-37} \sqrt{T, K}\, n_e n_i \left(cm^{-3}\right) \cdot \left(1 + \frac{|E_g|}{T}\right). \tag{6.249}$$

6.7.7 Plasma Absorption of Radiation in Continuous Spectrum, the Kramers, and Unsold-Kramers Formulas

The differential cross section of bremsstrahlung emission Eq. (6.235) with Eqs. (6.227), (6.228) peSrmits determining the coefficient of **bremsstrahlung absorption** of a quantum $\hbar\omega$ calculated per one electron having velocity v_e and one ion:

$$a_\omega(v_e) = \frac{16\pi^3}{3\sqrt{3}} \frac{Z^2 e^6}{m^2 c (4\pi\varepsilon_0 \hbar\omega)^3 v_e}. \tag{6.250}$$

This is the so-called **Kramers formula** of the quasi-classical theory of plasma continuous radiation. Multiplying the Kramers formula by $n_e n_i$ and integrating over the Maxwellian distribution $f(v_e)$, yields the coefficient of bremsstrahlung absorption in plasma (the reverse length of absorption):

$$\kappa_\omega^{brems} = C_1 \frac{n_e n_i}{T^{1/2} \nu^3}, \quad C_1 = \frac{2}{3}\left(\frac{2}{3\pi}\right)^{1/2} \frac{e^6}{(4\pi\varepsilon_0)^3 m^{3/2} c\hbar}. \tag{6.251}$$

To use Eq. (6.251) for numerical calculations of the absorption coefficient $\kappa_\omega^{brems}\left(\text{cm}^{-1}\right)$, the coefficient C_1 can be taken as: $C_1 = 3.69 \cdot 10^8 \dfrac{\text{cm}^5}{\text{sec}^3 \cdot K^{1/2}}$; here the frequency $\nu = \omega/2\pi$, and $Z = 1$. Another mechanism of plasma absorption of radiation in continuous spectrum is **photoionization**. The photoionization cross section can be easily calculated, taking into account that it is a reverse process with respect to recombination emission. Based on a detailed balance of the photoionization and recombination emission, Saha Eq. (4.16) and Eq. (6.241), the expression for the photoionization cross section for a photon $\hbar\omega$ and an atom with principal quantum number n is:

$$\sigma_{\omega n} = \frac{8\pi}{3\sqrt{3}} \frac{e^{10} m Z^4}{(4\pi\varepsilon_0)^5 c\hbar^6 \omega^3 n^5} = 7.9 \cdot 10^{-18} \text{cm}^2 \frac{n}{Z^2}\left(\frac{\omega_n}{\omega}\right)^3. \tag{6.252}$$

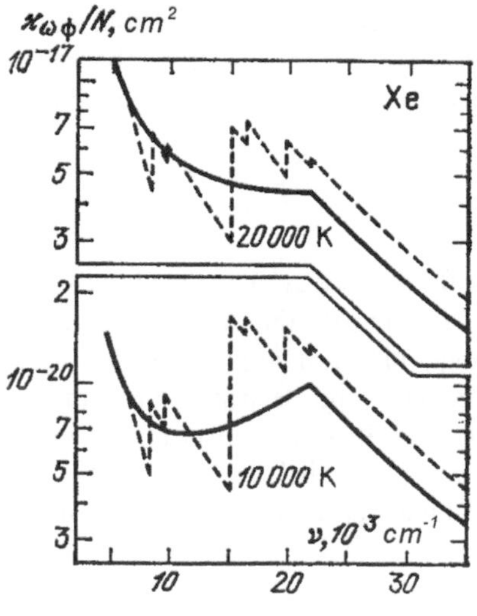

FIGURE 6.36 Absorption coefficient per one Xe atom, related to photo-ionization; dashed line – summation over actual energy levels, solid line – replacement of the summation by integration.

Here $\omega_n = |E_n|/\hbar$ is the minimum frequency sufficient for photoionization from the electronic energy level with energy E_n and principal quantum number n. As seen from Eq. (6.252), the photoionization cross section decreases with frequency as $1/\omega^3$ at $\omega > \omega_n$. To calculate the **total plasma absorption coefficient in a continuum** it is needed to add to the bremsstrahlung absorption Eq. (6.250), the sum $\sum n_{0n}\sigma_{\omega n}$ related to the photoionization of atoms in different states of excitation (n) with concentrations n_{0n}. Replacing the summation by integration, the total plasma absorption coefficient in continuum can be calculated for $Z = 1$as:

$$\kappa_\omega = C_1 \frac{n_e n_i}{T^{1/2}\nu^3} e^x \Psi(x) = 4.05 \cdot 10^{-23} \text{cm}^{-1} \frac{n_e n_i \left(\text{cm}^{-3}\right)}{(T,K)^{7/2}} \frac{e^x \Psi(x)}{x^3}. \tag{6.253}$$

Here factor C_1 is the same as in Eq. (6.251); parameter $x = \hbar\omega/T$; and the function $\Psi(x)$ is defined by Eqs. (6.245)–(247). Obviously, the replacement of summation over discrete levels by integration makes the resulting approximate dependence $\kappa_\omega(\nu)$ smoother than in reality; this is illustrated in Figure 6.36 (L.M. Biberman, G.E. Norman, 1967). The relative contribution of photoionization and bremsstrahlung mechanisms in total plasma absorption of radiation in continuum at the relatively low frequencies $\hbar\omega < |E_g|$ ($|E_g|$ is the energy of the lowest excited state with respect to continuum) can be characterized by the same ratio as in the case of emission Eq. (6.248). The product $n_e n_i$ in Eq. (6.253) can be replaced at not very high temperatures by the gas density n_0 using the Saha Eq. (4.16). At the relatively low frequencies $\hbar\omega < |E_g|$ the total plasma absorption coefficient in the continuum is:

$$\kappa_\omega = \frac{16\pi}{3\sqrt{3}} \frac{e^6 T n_0}{\hbar^4 c\omega^3 (4\pi\varepsilon_0)^3} \frac{g_i}{g_a} \exp\left(\frac{\hbar\omega - I}{T}\right) = 1.95 \cdot 10^{-7} \text{cm}^{-1} \frac{n_0 \left(\text{cm}^{-3}\right)}{(T,K)^2} \frac{g_i}{g_a} \frac{e^{-(x_1 - x)}}{x^3}. \tag{6.254}$$

This relation for the absorption coefficient is usually referred to as the **Unsold–Kramers formula**, where: g_a, g_i are statistical weights of an atom and an ion, $x = \hbar\omega/T$, $x_1 = I/T$ and I is the ionization potential. As an example, the absorption length κ_ω^{-1} of red light $\lambda = 0.65\mu m$ in an atmospheric pressure hydrogen plasma at 10,000 K can be calculated from Eq. (6.254) as $\kappa_\omega^{-1} \approx 180m$. Thus this plasma is fairly transparent in continuum at such conditions. It is interesting to note, that the Unsold-Kramers formula can be used only for quasi-equilibrium conditions, while Eq. (6.253) can be applied for non-equilibrium plasma as well.

6.7.8 Radiation Transfer in Plasma

The intensity of radiation I_ω decreases along its path "s" due to absorption (scattering in plasma neglected) and increases because of spontaneous and stimulated emission. The **radiation transfer equation** in a quasi-equilibrium plasma can be then given as:

$$\frac{dI_\omega}{ds} = \kappa'_\omega \left(I_{\omega e} - I_\omega\right), \quad \kappa'_\omega = \kappa_\omega \left[1 - \exp\left(-\frac{\hbar\omega}{T}\right)\right]. \tag{6.255}$$

Here the factor $I_{\omega e}$ is the quasi-equilibrium radiation intensity:

$$I_{\omega e} = \frac{\hbar\omega^3}{4\pi^3 c^2} \frac{1}{\exp(\hbar\omega/T) - 1}, \tag{6.256}$$

which can be obtained from the Planck formula for the spectral density Eq. (4.22). Introduce the **optical coordinate** ξ, calculated from the plasma surface $x = 0$ (with positive x directed into the plasma):

$$\xi = \int_0^x \kappa'_\omega(x)dx, \quad d\xi = \kappa'_\omega(x)dx. \tag{6.257}$$

Then the radiation transfer equation can be rewritten in terms of the optical coordinate as:

$$\frac{dI_\omega(\xi)}{d\xi} - I_\omega(\xi) = -I_{\omega e}. \tag{6.258}$$

Assuming, that in direction of a fixed ray, the plasma thickness is "d"; $x = 0$ corresponds to the plasma surface; and there is no source of radiation at $x > d$, then the radiation intensity on the plasma surface $I_{\omega 0}$ can be expressed by the solution of the radiation transfer equation as:

$$I_{\omega 0} = \int_0^{\tau_\omega} I_{\omega e}\left[T(\xi)\right]\exp(-\xi)d\xi, \quad \tau_\omega = \int_0^d \kappa'_\omega dx \tag{6.259}$$

The quasi-equilibrium radiation intensity $I_{\omega e}[T(\xi)]$ is shown here as a function of temperature and therefore an indirect function of the optical coordinate. In Eq. (6.259) another radiation transfer parameter is introduced: τ_ω is the **optical thickness of plasma.**

6.7.9 Optically Thin Plasmas and Optically Thick Systems, the Blackbody Radiation

First assume that optical thickness is small $\tau_\omega << 1$; this is usually referred to as **transparent or optically thin plasma**. In this case the radiation intensity on the plasma surface $I_{\omega 0}$ Eq. (6.259) is:

$$I_{\omega 0} = \int_0^{\tau_\omega} I_{\omega e}\left[T(\xi)\right]d\xi = \int_0^{\tau_\omega} I_{\omega e}\kappa'_\omega dx = \int_0^d j_\omega dx. \tag{6.260}$$

Here the emissivity term $j_\omega = I_{\omega e}\kappa'_\omega$ corresponds to spontaneous emission. Equation (6.260) shows that radiation of optically thin plasma is the result of the summation of independent emission from different intervals dx along the ray. In this case, all radiation generated in plasma volume is able to leave it. If plasma parameters are uniform, then the radiation intensity on the plasma surface can be expressed as:

$$I_{\omega 0} = j_\omega d = I_{\omega e}\kappa'_\omega d = I_{\omega e}\tau_\omega << I_{\omega e}. \tag{6.261}$$

From Eq. (6.261), the radiation intensity of the optically thin plasma is seen to be much less ($\tau_\omega << 1$) than the equilibrium value Eq. (6.256) corresponding to the Planck formula. The opposite case takes place when the optical thickness is high $\tau_\omega >> 1$. This is usually referred to as the **non-transparent or optically thick systems.** If the quasi-equilibrium temperature can be considered constant in the emitting body, then for the optically thick systems Eq. (6.259) gives: $I_{\omega 0} = I_{\omega e}(T)$. The emission density of the entire surface in all directions is the same and equal to the equilibrium Planck value Eq. (6.256). This is the case of the quasi-equilibrium **blackbody emission**, discussed in Section 4.1.5. Unfortunately, it cannot be directly applied to continuous plasma radiation (where the plasma is usually optically thin). The total black body emission per unit surface and unit time can be found by

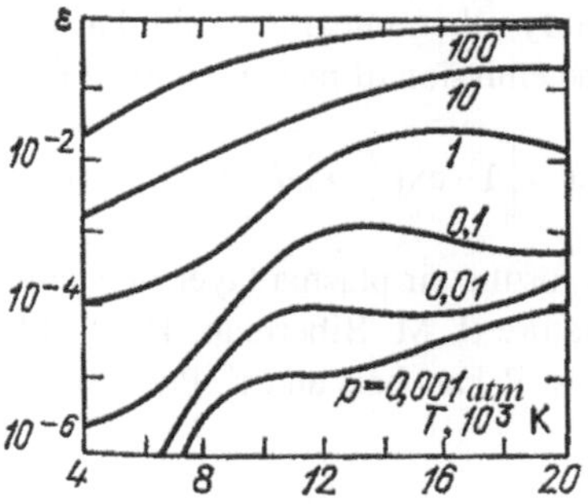

FIGURE 6.37 Emissivity of 1 cm air-layer at different pressures.

integration of the quasi-equilibrium radiation intensity Eq. (6.256) over all frequencies and all directions of hemisphere solid angle (2π):

$$J_e = \int d\omega \int_{2\pi} I_{\omega e} \cos\vartheta d\Omega = \int \pi I_{\omega e}(\omega,T) d\omega = \sigma T^4. \tag{6.262}$$

This is the **Stefan–Boltzmann law** of blackbody emission. σ is the Stephan-Boltzmann coefficient (4.26); ϑ is the angle between the ray (Ω) and normal vector to the emitting surface.

6.7.10 Reabsorption of Radiation, Emission of Plasma as the Gray Body, the Total Emissivity Coefficient

Plasmas are usually optically thin for radiation in the continuous spectrum, and the Stephan-Boltzmann law cannot be applied without special corrections. However, plasma is not absolutely transparent, the total emission can be affected by **reabsorption of radiation**. This can be illustrated by Eq. (6.259) for the radiation intensity $I_{\omega 0}$ assuming fixed values of the quasi-equilibrium temperature and absorption coefficient Eq. (6.253):

$$I_{\omega 0} = I_{\omega e}\left[1 - \exp(-\tau_\omega)\right] = I_{\omega e}\left[1 - \exp(-\kappa'_\omega d)\right] = I_{\omega e}\varepsilon. \tag{6.263}$$

Here ε, the total emissivity coefficient, which characterizes the plasma as a gray body. It is interesting to note, that in optically thin plasmas, the total emissivity coefficient coincides with the optical thickness: $\varepsilon = \tau_\omega$. If the total emissivity coefficient ε dependence on frequency is neglected

TABLE 6.4
Parameters of Some Strong Atomic Lines in the Visible Spectrum

Atomic Line	Wavelength, λ, nm	Oscillator Power, f	Coefficient A_{nk}, sec^{-1}
H_α	656.3	0.641	$4.4 \cdot 10^7$
H_β	486.1	0.119	$8.4 \cdot 10^6$
He	587.6	0.62	$7.1 \cdot 10^7$
Ar	696.5	-	$6.8 \cdot 10^6$
Na	589.0	0.98	$6.2 \cdot 10^6$
Hg	579.1	0.7	$9.0 \cdot 10^7$

(assumption of the **ideal gray body**), then integration similar to Eq. (6.262) leads to the corrected Stefan-Boltzmann formula for the emission of plasma in the continuous spectrum:

$$J = \left[1 - \exp\left(-\kappa'_\omega d\right)\right]\sigma T^4 = \varepsilon\sigma T^4. \tag{6.264}$$

The total emissivity coefficients for the air plasma layer of 1 cm are presented in Figure 6.37 as a function of pressure and temperature (L.M. Biberman, 1970). More details on the subject can be found in the books of M.I. Boulos, P. Fauchais and E. Pfender (1994), B.H. Armstrong (1990) and M.F. Modest (1993).

6.8 SPECTRAL LINE RADIATION IN PLASMA

6.8.1 Probabilities of Radiative Transitions and Intensity of Spectral Lines

Spontaneous discrete transition from an upper state E_n into a state E_k results in emission of a quantum (see Figure 6.32):

$$\hbar\omega_{nk} = \frac{2\pi\hbar c}{\lambda_{mn}} = E_n - E_k. \tag{6.265}$$

This radiation energy quantum determines a spectral line with frequency ω_{nk} and wavelength λ_{mn}. The intensity of the spectral line S can be characterized by the energy emitted by an excited atom or molecule per unit time in the line; this can be found in frameworks of quantum mechanics as:

$$S = \hbar\omega_{nk}A_{nk} = \frac{1}{3\varepsilon_0}\frac{\omega_{nk}^4}{c^3}\left|\vec{d}_{nk}\right|^2. \tag{6.266}$$

Here $\overrightarrow{d_{nk}}$ is the matrix element of the dipole momentum of the atomic system, A_{nk} is the probability of the bound-bound transition in frequency units (A_{nk}^{-1} can be interpreted as the atomic lifetime with respect to the radiative transition $n \to k$). For the simplest atomic systems, the dipole momentum matrix elements and hence factors A_{nk} can be quantum mechanically calculated; however, in most of the cases, the transition probabilities A_{nk} should be found experimentally. Typically the transition probabilities for the intensive spectral lines are approximately 10^8 sec^{-1}, numerical values of the coefficients A_{nk} for some strong atomic lines are given in Table 6.4. The intensity of the spectral line can also be calculated in the framework of classical electrodynamics, assuming that an electron is oscillating in an atom around the equilibrium position ($\vec{r}$ is the electron radius-vector). This leads to the following classical expression for the intensity of the spectral line, which is actually equivalent to the quantum-mechanical one Eq. (6.266):

$$S = \frac{1}{6\pi\varepsilon_0 c^3}\left\langle \ddot{\vec{d}}^2 \right\rangle = \frac{1}{6\pi\varepsilon_0}\frac{\omega_0^4}{c^3}\left\langle \vec{d}^2 \right\rangle = \frac{e^2\omega_0^4}{6\pi\varepsilon_0 c^3}\left\langle \vec{r}^2 \right\rangle. \tag{6.267}$$

Here: c is the speed of light, and ω_0 is the frequency of the electron "oscillation" in an atom.

6.8.2 Natural Width and Profile of a Spectral Line

Excited states of an atom with energy E_n has a finite lifetime with respect to radiation $\tau \approx A_{nk}^{-1}$. Hence according to the quantum-mechanical uncertainty principle, the energy level E_n actually is not absolutely thin but has a characteristic width of $\Delta E \propto \hbar/\tau \approx \hbar A_{nk}$. As a result, a spectral line related to the

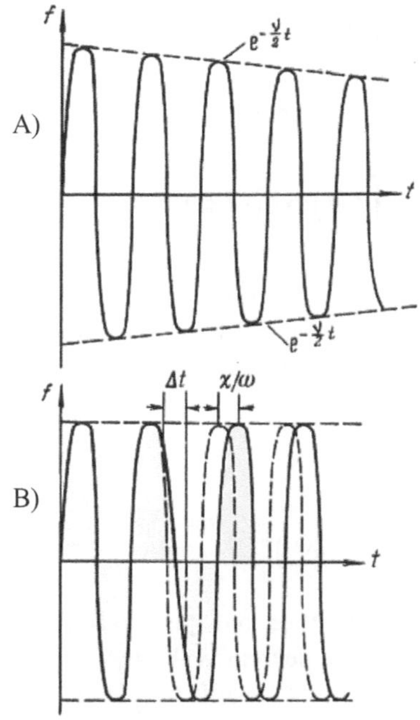

FIGURE 6.38 Time evolution of electromagnetic field amplitude: (a) finite state's life-time; (b) collision of an emitting and perturbing atoms; Δt – collision time; χ – collisional phase shift.

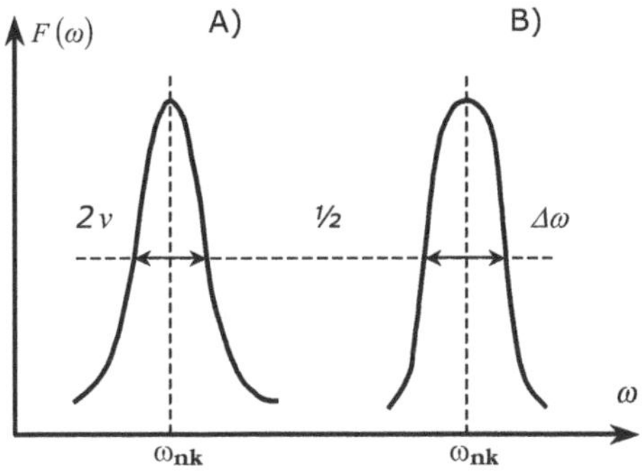

FIGURE 6.39 "Wide wings" Lorentz (A) and "narrow wings" Gaussian (B) profiles of spectral lines.

transition from this atomic energy level has a specific width $\Delta\omega \approx A_{nk}$, which is independent of external conditions and called the **natural spectral line width.** Typically the natural width is very small with respect to the characteristic radiation frequency for electronic transitions $\Delta\omega \approx 10^8 \text{sec}^{-1} << \omega_{nk} \approx 10^{15} \text{sec}^{-1}$. Profile of spectral lines can be determined as the photon distribution functions $F(\omega)$ over the radiation frequencies, which is usually normalized as: $\int_{-\infty}^{+\infty} F(\omega) d\omega = 1$. To describe the natural profile of the spectral line $F(\omega)$, express the corresponding oscillations of the electric and magnetic fields in an electromagnetic wave $f(t)$, taking into account the finite lifetime of both initial τ_n and final τ_k atomic states (B.M. Smirnov, 1977):

$$f(t) \propto \exp(i\omega_{nk}t - \nu t), \quad 2\nu = \frac{1}{\tau_n} + \frac{1}{\tau_k} = \frac{1}{\tau} \approx A_{nk}. \tag{6.268}$$

The frequency $\nu = \frac{1}{2} A_{nk}$ characterizes the photon emission probability Eq. (6.266) and describes the attenuation of the electromagnetic wave because of the finite lifetime of atomic states. Frequency distribution for the electromagnetic wave can be found as a Fourier component f_ω of the function $f(t)$:

$$f_\omega = \frac{1}{2\pi} \int_{-\infty}^{+\infty} f(t) \exp(-i\omega t)\, dt \propto \frac{1}{\nu + i(\omega - \omega_{nk})}. \tag{6.269}$$

The radiation intensity of a spectral line at a fixed frequency ω and the photon distribution function $F(\omega)$ are both proportional to the square of the Fourier component $F(\omega) \propto |f_\omega|^2$. Then taking into account the normalization of the function $F(\omega)$, the following expression is obtained for the photon distribution function and the natural profile of a spectral line related to the finite lifetime of excited states:

$$F(\omega) = \frac{\nu}{\pi} \frac{1}{\nu^2 + (\omega - \omega_{nk})^2}, \quad \nu = \frac{1}{2}\left(\frac{1}{\tau_n} + \frac{1}{\tau_k}\right) = \frac{1}{2\tau} \approx \frac{1}{2} A_{nk}. \tag{6.270}$$

This shape of the photon distribution function and spectral line is usually referred to as the **Lorentz spectral line profile**. The Lorentz profile decreases slowly (hyperbolically) with deviation from the principal frequency one $\omega = \omega_{nk}$. The Lorentz profile has "wide wings" (hyperbolic wings), which is illustrated in Figure 6.39. The parameter ν in Eq. (6.270) is the so-called **spectral line half-width.** In this case, the width of a line at half the maximum intensity is presented as 2ν, which is a double half-width or the entire width of a spectral line (see Figure 6.39). It is interesting to remark that the natural width of a spectral line in the wavelength-scale is independent of frequency and can be estimated as:

$$\Delta\lambda = \frac{e^2}{3\varepsilon_0 mc^2} = 1.2 \cdot 10^{-5}\, nm. \tag{6.271}$$

6.8.3 The Doppler Broadening of Spectral Lines

The broadening of spectral lines is due to the thermal motion of the emitting particles, to collisions perturbing the emitting particles, and to the Stark-effect. Electromagnetic wave emitted by a moving atom or molecule with frequency ω_{nk}, is observed by a detector (at rest) at the frequency shifted according to the Doppler effect formula:

$$\omega = \omega_{nk}\left(1 + \frac{v_x}{c}\right), \tag{6.272}$$

where v_x is the emitter velocity in the direction of wave propagation; c is the speed of light. The Doppler profile of a spectral line then can be found based on Eq. (6.271), and assuming a Maxwellian distribution for atomic velocities $f(v_x)$. Taking into account that $F(\omega)\, d\omega = f(v_x)\, dv_x$, the Doppler profile is:

$$F(\omega) = \frac{1}{\omega_{mn}} \sqrt{\frac{Mc^2}{\pi T_0}} \exp\left[-\frac{Mc^2}{T_0} \frac{(\omega - \omega_{nk})^2}{\omega_{nk}^2}\right]. \tag{6.273}$$

Here M is the mass of a heavy particle; T_0 is gas temperature; c is the speed of light. In contrast to the Lorentzian hyperbolic natural profile of spectral lines,- the Doppler profile decreases much faster (exponentially) with frequency deviation from the center of the line. Such shape of the curve $F(\omega)$ with relatively "narrow wings" is called the **spectral line Gaussian profile**, and it is also illustrated in Figure 6.39 to compare with the "wide wings" Lorentz profile. The entire width (double half-width) of the Doppler profile of a spectral line can be expressed from Eq. (6.273) as:

$$\frac{\Delta\omega}{\omega_{nk}} = \frac{\Delta\lambda}{\lambda_{nk}} = 7.16 \cdot 10^{-7}\sqrt{T_0(K)/A}, \tag{6.274}$$

where A is the atomic mass of the emitting particle. The Doppler broadening of spectral lines is most significant at high temperatures and for light atoms like hydrogen (for H_β line, $\lambda = 486.1$ nm, the Doppler width at a high gas temperature of 10,000 K is $\Delta\lambda = 0.035$ nm).

6.8.4 Pressure Broadening of Spectral Lines

The influence of collisions on electromagnetic wave emission by excited atoms is illustrated in Figure 6.38. The phase of electromagnetic oscillations changes randomly during the collision and therefore the oscillations can be considered as harmonic only between collisions. The effect of collisions is somewhat similar to the effect of spontaneous radiation (natural broadening) because it also restricts the interval of harmonic oscillations. For this reason collisions with all plasma components, including heavy neutrals, leads to this broadening effect. This broadening can be described by the same Lorentz profile Eq. (6.270) as in natural broadening, though in this case with a frequency ν related to the frequency of the emitting atom collisions with other species (B.M. Smirnov, 1977):

$$F(\omega) = \frac{n_0\langle\sigma v\rangle}{\pi}\frac{1}{\left(n_0\langle\sigma v\rangle\right)^2 + (\omega - \omega_{nk})^2}. \tag{6.275}$$

This effect is usually referred to as the **pressure broadening (or impact broadening)** of spectral lines. In Eq. (6.275): n_0 is the neutral gas density; σ is the cross section of the perturbing collision (which is usually close to the gas-kinetic cross section); and v is the velocity of the related collision partners (which is usually the thermal velocity of heavy particles). At high pressures, the collisional frequency obviously can exceed the frequency of spontaneous radiation; this can make the pressure broadening much more significant than the natural width of spectral lines.

6.8.5 Stark Broadening of Spectral Lines

When the density of charged particles in a plasma is sufficiently large, degree of ionization not less than about 10^{-2}), their electric micro-fields perturb the atomic energy levels providing the Stark broadening of spectral lines. The Stark effect is the most significant for hydrogen atoms (H_β line, 486.1 nm, is widely applied in related plasma diagnostics) and some helium levels, where the effect is strong and proportional to the first power of electric field. In most other cases, the Stark effect is proportional to square of electric field and is not strong. The Stark broadening induced by electric micro-fields of ions is qualitatively different from that of electrons. For slowly moving ions, Stark broadening can be described based on the statistical distribution of the electric micro-fields. This approach is called the **quasi-static approximation.** In this case, the electric micro-fields are provided by the closest charged particles and can be estimated as: $E \propto e^2/r^2 \propto n_e^{2/3}$, where $r \propto n_e^{-1/3}$ is the average distance between the charged particles. In the case of the strong linear Stark effect, the relevant width of a spectral line is also proportional to the factor $n_e^{2/3}$. The Stark broadening induced by the electric micro-fields of ions in terms of wavelength $\Delta\lambda_S$ and in terms of frequencies $\Delta\omega_S$ can be estimated in the specific case of the H_β line (486.1 nm, transition from $n = 4$ to $n = 2$) as:

$$\Delta\lambda_S = \Delta\omega_S\frac{\lambda_{nk}^2}{2\pi c} \approx \frac{3e^2 a_0 n(n-1)\lambda_{nk}^2}{8\pi^2\varepsilon_0\hbar c}n_e^{2/3} = 5 \cdot 10^{-12}\,\text{nm} \cdot \left(n_e, \text{cm}^{-3}\right)^{2/3}, \tag{6.276}$$

where a_0 is the Bohr radius, λ_{nk} is the spectral line principal wavelength. The Stark broadening induced by electric micro-fields of the fast-moving electrons can be described in the framework of the **impact approximation**. In this case, the emitting atom is unperturbed most of the time and broadening is due to impacts well separated in time. The impact Stark broadening is proportional to the electron concentration and an additional factor decreasing with n_e. The entire Stark broadening, including the effects of ions and electrons, can be still approximated as proportional to the factor $n_e^{2/3}$. In the electron temperature range 5,000 $K \div$ 40,000 K and electron concentrations in the interval $10^{14} \div 10^{17}\text{cm}^{-3}$, the total Stark broadening again of the H_β line, 486.1 nm, can be found using the following numeric formula:

$$\Delta\lambda_S \approx (1.8 \div 2.3)\cdot 10^{-11}\text{nm}\cdot\left(n_e,\text{cm}^{-3}\right)^{2/3}. \tag{6.277}$$

The Stark broadening results in the Lorentzian profile of spectral in the same way as the pressure and natural broadenings. Obviously, different Lorentzian half-widths ($\Delta\omega_L/2$ or $\Delta\lambda_L/2$) should be used for describing different types of the Lorentzian broadenings: pressure, Stark or natural. The total Lorentzian profile is determined then by the strongest of the three broadening effects.

6.8.6 Convolution of Lorentzian and Gaussian Profiles, the Voigt Profile of Spectral Lines

The three mechanisms of spectral line broadening in a plasma (Stark, pressure, and natural) lead to the same "wide wing" Lorentzian profile, and only the Doppler broadening results in the "narrow wing" Gaussian profile (see Figure 6.39). If the characteristic widths of the Lorentzian $\Delta\lambda_L$ and Gaussian $\Delta\lambda_G$ profiles are of the same order of value, the resulting profile can be not a trivial one. The central part of a spectral line can be related to Doppler broadening, while the wide wings can be provided by the Stark effect. The final shape of a spectral line is the result of the convolution of the Gaussian and Lorentzian profiles. Assuming that the Doppler and Stark broadenings are independent, the result of the convolution is the Voigt profile of spectral lines. The Voigt profile can be presented in terms of the distribution over wavelength λ as:

$$F(\lambda) = \frac{1}{\Delta\lambda_G\sqrt{\pi}}\int_{-\infty}^{+\infty}\frac{a}{\pi}\frac{\exp\left(-y^2\right)}{(b-y)^2+a^2}dy. \tag{6.278}$$

The Voigt profile is defined by two parameters "a" and "b", which are related to the spectral line wavelength λ_{mn}, the line shift Δ, and the widths of the Lorentzian $\Delta\lambda_L$ and Gaussian $\Delta\lambda_G$ broadenings:

$$a = \frac{\Delta\lambda_L}{\Delta\lambda_G}(\ln 2)^{1/2}, \quad b = \frac{2\ln 2}{\Delta_G}(\lambda - \lambda_{nk} - \Delta). \tag{6.279}$$

Details on plasma spectroscopy and its application to plasma diagnostics can be found in the books of H. Griem (1964, 1974), F. Cabannes and J. Chapelle (1971), and G. Traving (1968).

6.8.7 Spectral Emissivity of a Line, Constancy of a Spectral Line Area

Emission in a spectral line per one atom was described above by the spectral line intensity S. Then the total energy emitted in a spectral line, corresponding to the transition $n \to k$, in unit volume per unit time can be expressed as:

$$J_{nk} = \hbar\omega_{nk}A_{nk}n_n, \tag{6.280}$$

where n_n is the concentration of atoms in the state "n". Taking into account the profile shape function $F(\omega)$, the spectral emissivity (spectral density of J_{nk} taken per unit solid angle) can be presented as:

$$j_\omega d\omega = \frac{1}{4\pi}\hbar\omega_{nk}A_{nk}n_n F(\omega)d\omega. \tag{6.281}$$

The profile $F(\omega)$ is normalized: $\int_{-\infty}^{+\infty} F(\omega)d\omega = 1$. Therefore, *the spectral line broadening does not change the area of the spectral line*. Broadening makes a spectral line wider, but lower.

6.8.8 Selective Absorption of Radiation in Spectral Lines, Absorption of One Classical Oscillator

Absorption of radiation by atoms and molecules are selective in spectral lines. The efficiency of the absorption obviously depends on the broadening of the spectral lines. If the broadening is provided by collisions, the cross section of electromagnetic wave absorption by one oscillator (which represent an atom in classical electrodynamics) depends on the frequency near the resonance as:

$$\sigma_\omega = \frac{e^2}{4mc\varepsilon_0}\frac{\tau^{-1} + n_0\langle\sigma v\rangle}{(\omega-\omega_{nk})^2 + \frac{1}{4}\left(\tau^{-1} + n_0\langle\sigma v\rangle\right)^2}. \tag{6.282}$$

This absorption – frequency curve also has the Lorentz profile (compare with Eq. (6.275)) in accordance with Kirchoff's law, which requires a correlation between emission and absorption processes; $n_0 < \sigma v>$ is the collisional frequency, τ^{-1} is the frequency of the spontaneous electronic transition $n \to k$, $\hbar\omega_{nk}$ is energy of this transition. Equation (6.282) shows that the maximum resonance absorption cross section Eq. (6.282) in the absence of collisions ($n_0 < \sigma v > << \tau^{-1}$, $\omega = \omega_{nk}$) is large for visible light:

$$\sigma_\omega^{\max} = \frac{3\lambda_{nk}^2}{2\pi} \approx 10^{-9}\,\text{cm}^2. \tag{6.283}$$

If collisional broadening of a spectral line is sufficiently large, the corresponding absorption cross section decreases with pressure: $\sigma_\omega \propto 1/n_0 < \sigma v > \propto 1/p$. The area of the absorption spectral line can be found by integrating the absorption cross section σ_ω (Eq. (6.282)) over frequencies (traditionally ν):

$$\int \sigma_\omega d\nu = \frac{1}{2\pi}\int \sigma_\omega d\omega = \frac{e^2}{4mc\varepsilon_0} = 2.64\cdot 10^{-2}\,\frac{\text{cm}^2}{\text{sec}}. \tag{6.284}$$

This area describes the absorption in a spectral line of one classical oscillator.

6.8.9 The Oscillator Power

The total radiation energy absorbed in a spectral line per unit time and unit volume is independent of broadening and can be expressed as:

$$\int 4\pi\kappa_\omega I_\omega d\omega = 4\pi I_{\omega nk} n_k \int \sigma_\omega d\omega. \tag{6.285}$$

Here I_ω is the radiation intensity, κ_ω is the absorption coefficient (characterizing reverse length of absorption); n_k is the density of the light-absorbing atoms. According to Eq. (6.285), the total radiation energy absorbed in a spectral line is determined by the area of the absorption line $\int \sigma_\omega d\omega$. It should be noted, that the absorption factor $B_{kn} = \frac{1}{\omega_{nk}} \int \sigma_\omega d\nu$ as well as A_{nk} are also usually called the Einstein coefficients. Based on Eq. (6.281) and Kirchhoff's law, the area of the absorption line can be found as:

$$\sigma_\omega = \frac{g_n}{g_k}\frac{\pi^2 c^2}{\omega^2} A_{nk} F(\omega), \quad \int \sigma_\omega d\nu = \frac{g_n}{2g_k}\frac{\pi c^2}{\omega^2} A_{nk}. \tag{6.286}$$

Here, g_n, g_k are the statistical weights of the upper and lower electronic states; A_{nk} is the Einstein coefficient for spontaneous emission in the spectral line (see (6.266)). The actual area of the absorption line Eq. (6.286) differs from the classical one Eq. (6.283) corresponding to the absorption of one classical oscillator. It is convenient to characterize the absorption in a spectral line by the ratio of the actual area of the absorption line Eq. (6.286) over the area Eq. (6.283) corresponding to one classical oscillator:

$$f = \frac{g_n}{2g_k}\frac{4m\varepsilon_0 \pi c^3}{\omega^2 e^2} A_{nk}. \tag{6.287}$$

This ratio is called the oscillator power and illustrates the number of classical oscillators providing the same absorption as the actual spectral line. Examples of the oscillator powers together with the Einstein coefficients A_{nk} for some spectral lines are presented in Table 6.4. The oscillator powers for the one-electron transitions are less than unity, and they approach the unity for the strongest spectral lines.

6.8.10 Radiation Transfer in Spectral Lines, Inverse Population of the Excited States, and Principle of Laser Generation

The radiation transfer Eq. (6.255) for a spectral line in terms of radiative transitions $n \Leftrightarrow k$ in an atom or molecule can be presented as:

$$\frac{dI_\omega}{ds} = j_\omega + n_n \sigma_{b\omega} I_\omega - n_k \sigma_{a\omega} I_\omega. \tag{6.288}$$

In this equation: n_n, n_k are concentrations of particles in the upper "n" and lower "k" states; I_ω is the radiation intensity; the coordinate "s" is the path along a ray propagation; j_ω is the spontaneous emissivity on the frequency ω; and cross sections of the radiation absorption $\sigma_{a\omega}$ and stimulated emission $\sigma_{b\omega}$, related to each other in accordance with the Kirchhoff and Boltzmann laws as:

$$\sigma_{b\omega} = \left(\frac{n_k}{n_n}\right)_{eq} \exp\left(-\frac{\hbar\omega}{T}\right) \sigma_{a\omega} = \frac{g_k}{g_n}\sigma_{a\omega}. \tag{6.289}$$

Here $(n_k/n_n)_{eq}$ is the equilibrium ratio of populations of lower and higher energy levels; g_k, g_n are the corresponding statistical weights. Taking into account Eq. (6.289), the radiation transfer equation (6.288) can be rewritten as:

$$\frac{dI_\omega}{ds} = j_\omega + (N_2 - N_1) g_k \sigma_{a\omega} I_\omega. \tag{6.290}$$

In this relation $N_2 = n_n/g_n$, $N_1 = n_k/g_k$ are the concentrations of atoms in the specific quantum states related to energy levels "n" and "k". These factors correspond in equilibrium to the exponential factors of the Boltzmann distribution. In equilibrium gases $N_2 = N_1 \exp(-\hbar\omega/T) < N_1$ and the radiation transfer Eq. (6.290) results in absorption corresponding to Eq. (6.255). However, in non-equilibrium conditions, the population of the higher energy states can exceed the population of those with lower energy $N_2 > N_1$. Such non-equilibrium conditions are usually referred to as the **inverse population of excited states**. In the case of the inverse population, according to Eq. (6.290), the non-equilibrium medium does not provide absorption but rather significant amplification of radiation in a spectral line. This is the basis of the laser. In a homogeneous medium with $N_2 - N_1 = const > 0$, this exponential amplification of the radiation intensity along a ray can be expressed as:

$$I_\omega = I_{\omega 0} \exp\left[\left(N_2 - N_1\right) g_k \sigma_{a\omega} s\right]. \tag{6.291}$$

The factor $(N_2 - N_1)g_k\sigma_{a\omega}$ in Eq. (6.291) is called the **laser amplification coefficient**. The inverse population could be quite easily achieved in the non-equilibrium plasma due to intensive excitation of atoms and molecules by electron impact. This provides a physical basis for gas-discharge lasers. The non-equilibrium Treanor distribution, discussed in the Sections 4.1.10 and 5.1.3, can be considered as one of the interesting examples of plasma-induced inverse population of excited states. The Treanor effect provides, in particular, the physical conditions for generation in the CO-laser, see E.B. Bradley (1980).

6.9 NONLINEAR PHENOMENA IN PLASMA

6.9.1 Nonlinear Modulation Instability, the Lighthill Criterion

Consider a perturbation of some plasma parameter as a "wave package", which is a group of waves with different but close wave vectors ($\Delta k << k$). The perturbation at a point "x" can be presented as:

$$a(x,t) = \sum_k a(k) \exp(ikx - i\omega t), \tag{6.292}$$

where $a(k)$ is the amplitude of the wave with wave vector k. To further analyze the evolution of the perturbation we can assume the following general dispersion relation $\omega(k)$ for the wave:

$$\begin{aligned} \omega(k) &= \omega(k_0) + \frac{\partial \omega}{\partial k}(k = k_0)\cdot(k - k_0) + \frac{1}{2}\frac{\partial^2 \omega}{\partial k^2}(k = k_0)\cdot(k - k_0)^2 \\ &= \omega_0 + v_{gr}(k - k_0) + \frac{1}{2}\frac{\partial v_{gr}}{\partial k}(k - k_0)^2. \end{aligned} \tag{6.293}$$

Here k_0 is the average value of the wave vector for the group of waves; ω_0 is frequency corresponding to k_0;v_{gr} is group velocity of the wave. Because of different group velocities for the waves with different values of wave vectors ($\partial v_{gr}/\partial k \neq 0$), the group of waves (the initial perturbation) grows in size. If the waves are not interacting with each other (linear waves), the initial perturbation grows to the size $\Delta x \approx 1/\Delta k$ during the period of time about: $\tau \approx \left(\Delta k^2 \frac{\partial v_{gr}}{\partial k}\right)^{-1}$. Take into account the nonlinearity by introducing frequency dependence on the local value of amplitude $E(x)$:

$$\omega = \omega_0 - \alpha E^2, \tag{6.294}$$

where ω_0 is the wave frequency in the low amplitude limit. Using Eqs. (6.293) and (6.294) for frequency deviations, we can finally rewrite the expression (6.292) for plasma perturbation as:

$$a(x,t) = \sum_k a(k)\exp\left[i(k-k_0)(x-x_0) - ik_0x_0 - i(k-k_0)^2\frac{\partial v_{gr}}{\partial k}t - i\alpha E^2(x)t\right]. \quad (6.295)$$

Eq. (6.295) shows the modulation of the group of waves, which is called the **modulation instability**. At some specific modes of the modulation instability, the whole "wave package" decays into smaller groups of waves or can be "compressed" and converted into the specific single wave called the **solitone**. Competition between two last terms in the exponent Eq. (6.295) determines the type of evolution of the wave package. The term $i(k-k_0)^2\frac{\partial v_{gr}}{\partial k}t$ leads to expansion of the wave package, while the nonlinear term $i\alpha E^2(x)t$ can compensate the previous one. To compensate expansion of a perturbation, and provide an opportunity of the wave package decay into smaller groups of waves, compression or formation of solitones, the mentioned two terms should at least have different sings. This leads to the requirement:

$$\alpha \cdot \frac{\partial v_{gr}}{\partial k} < 0, \quad (6.296)$$

which is called the **Lighthill criterion**. This criterion is necessary for the modulation instability and for the n-linear evolution of perturbations (or "the wave packages") in plasma.

6.9.2 The Korteweg-de Vries Equation

The influence of a weak nonlinearity on the wave package expansion due to dispersion $\omega(k)$ can be described using the Korteweg-de Vries equation. Consider propagation of the longitudinal long-wavelength oscillations (for example, ionic sound). These oscillations at relatively long wavelength ($r_0k << 1$) can be described by the following dispersion relation:

$$\omega = v_{gr}k\left(1 - r_0^2k^2\right). \quad (6.297)$$

In the case of the ionic wave, the size-parameter r_0 corresponds to the Debye radius. Begin with the Euler equation for particle velocities in the longitudinal wave under consideration:

$$\frac{\partial v}{\partial t} + v\frac{\partial v}{\partial x} - \frac{F}{M} = 0. \quad (6.298)$$

Here $v(x, t)$ is particle velocity in the longitudinal wave along the axis "x"; F is force acting on the plasma particle with mass M. For the linear approximation, assume $v = v_{gr} + v'$, where $v' << v_{gr}$ is the plasma particle velocity with respect to the wave. Euler equation can be presented then as a linear one:

$$\frac{\partial v'}{\partial t} + v_{gr}\frac{\partial v'}{\partial x} - \frac{F}{M} = 0, \quad (6.299)$$

with the function F/m considered as a linear operator of v'. In the harmonic approximation for the plasma particle velocities: $v' \propto \exp(-i\omega t + ikx)$, choose the linear operator $\frac{F}{m}(v')$ in such a way to get from the linear equation (6.299) the dispersion relation (6.297). This leads to the linear operator $\frac{F}{m}(v') = -r_0^2 v_{gr}\frac{\partial^3 v'}{\partial x^3}$ and to the linear equations of motion in the form:

$$\frac{\partial v'}{\partial t} + v_{gr}\left(\frac{\partial v'}{\partial x} + r_0^2\frac{\partial^3 v'}{\partial x^3}\right) = 0. \quad (6.300)$$

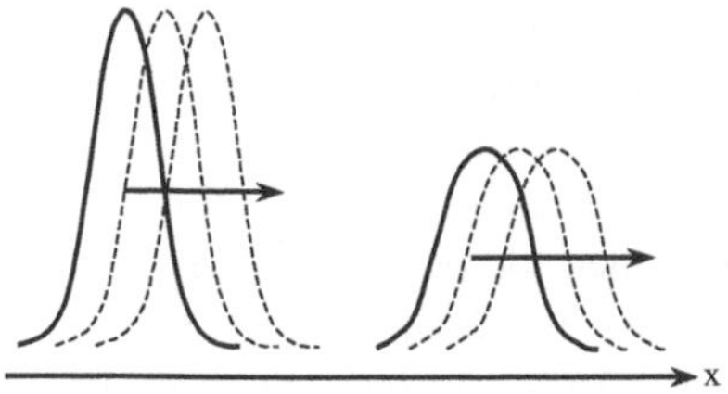

FIGURE 6.40 Illustration of a solitone propagation.

The last term in this linear equation describes the dispersion of the long-wavelength oscillations. To take into account the nonlinearity, replace v' by the plasma particle velocity v (the reverse step with respect to the transition from Eqs. (6.298) to (6.299)). This results in the nonlinear equation for plasma particles:

$$\frac{\partial v}{\partial t} + v\frac{\partial v}{\partial x} + v_{gr} r_0^2 \frac{\partial^3 v}{\partial x^3} = 0. \tag{6.301}$$

This motion equation for plasma particles is known as the **Korteweg-de Vries**. The Korteweg-de Vries equation is especially interesting for the description of nonlinear dissipative processes because it takes into account both nonlinearity and dispersion of waves. It successfully describes the propagation in plasma of different long-wavelength oscillations corresponding to the dispersion Eq. (6.297).

6.9.3 Solitones as Solutions of the Korteweg-de Vries Equation

The Korteweg-de Vries equation has a physically very interesting solution that corresponds to a lone or solitary non-harmonic wave, which is called solitones. Solitones are particular in ability to maintain a fixed shape during propagation in a medium with dispersion. Although shorter harmonics are moving slower the nonlinearity compensates this effect and the wave package does not expand but rather keeps its shape. To describe the solitones, consider a wave propagating with velocity "u". Dependence of the plasma particle velocities on coordinate and time can be presented as $v = f(x - ut)$. In this case $\frac{\partial v}{\partial t} = -u\frac{\partial v}{\partial x}$ and the Korteweg-de Vries equation can be rewritten as the following third order ordinary differential equation:

$$\left(v - u\right)\frac{dv}{dx} + v_{gr} r_0^2 \frac{d^3 v}{dx^3} = 0. \tag{6.302}$$

Integration of Eq. (6.302), taking into account absence of the plasma particles velocity at infinity ($v = 0$, $dv^2/dx^2 = 0$ at $x \to \infty$), leads to the second order equation:

$$v_{gr} r_0^2 \frac{d^2 v}{dx^2} = uv - \frac{v^2}{2}. \tag{6.303}$$

The solitone, the solitary wave, is one of the solutions of the nonlinear Eq. (6.303):

$$v = \frac{3u}{\cosh^2 \dfrac{x}{2r_0}\sqrt{\dfrac{u}{v_{gr}}}}, \tag{6.304}$$

where $\cosh\alpha$ is the hyperbolic cosine. Profiles of solitones with the same fixed parameters u, v_{gr}, and different amplitudes are illustrated in Figure 6.40. From the figure and Eq. (6.304), the solitary wave becomes narrower with increases of its amplitude. The product of the solitone's amplitude and square of solitone's width remains constant during its evolution. If the initial non-solitone perturbation has relatively low amplitude, then the perturbation expands with time because of dispersion until the product of its amplitude and width square corresponds to that one for solitones (6.304). Then during its evolution, the perturbation converts into a solitone. Conversely, if the amplitude of an initial perturbation is relatively high, then during the evolution of the perturbation it decays to form several solitones.

6.9.4 Formation of the Langmuir Solitones in Plasma

Formation of solitones in plasma is due to electric fields induced by a wave, which confines a plasma perturbation in a local domain. The higher the wave amplitude, the stronger the electric field can be induced by this wave and the stronger effect of perturbation compression. It can be illustrated by plasma oscillation leading to the formation of Langmuir solitones. The energy density of Langmuir oscillations at a point "x" is a function of time-averaged electric field:

$$W(x) = \frac{\varepsilon_0 \langle E^2(x,t) \rangle}{2}. \tag{6.305}$$

Assume the electron and ion temperatures equal to the uniform value T, and assume plasma to be completely ionized and quasi-neutral. The total effective pressure is also constant in space for the long-wavelength plasma oscillations with propagation velocities much less than the speed of sound:

$$2n_e(x)T + W(x) = 2n_{e0}T. \tag{6.306}$$

Here n_{e0} is electron density at infinity, where there are no plasma oscillations. The dispersion equation for plasma oscillations Eq. (6.128) can be then rewritten taking into account the relation (6.306) in a nonlinear way, including the effect of amplitude of electric field:

$$\omega^2(x) = \omega_{p0}^2 \left(1 - \frac{\varepsilon_0 \langle E^2 \rangle}{4n_{e0}T}\right) + \frac{\gamma T}{m}k^2. \tag{6.307}$$

Here, ω_{p0} is plasma frequency far from the perturbation, T is plasma temperature, and γ is the specific heat ratio. From Eq. (6.307), taking into account that $E = E_0 \cos\omega t$, $\langle E^2 \rangle = \frac{1}{2}E_0^2$:

$$\omega(x) = \omega_{p0}\left(1 - \frac{\varepsilon_0 E_0^2}{16n_{e0}T} + \gamma r_D^2 k^2\right). \tag{6.308}$$

.The second and third terms are much less than unity; only the low levels of nonlinearity of the oscillations are considered. Eq. (6.308) satisfies the Lighthill criterion (6.296). Differentiating Eq. (6.304):

$$\alpha \frac{\partial v_{gr}}{\partial k} = \alpha \frac{\partial^2 \omega}{\partial k^2} = -\frac{\gamma \varepsilon_0 \omega_{p0}}{8mn_{e0}} < 0. \tag{6.309}$$

This explains formation of the solitary wave in plasma, called in this case the Langmuir solitone.

6.9.5 Evolution of Strongly Nonlinear Oscillations, the Nonlinear Ionic Sound

Analyze the strong nonlinearity using ionic sound as an example. To describe the nonlinear ionic sound, use the the Euler equation of ionic motion, the continuity equation for ions and the Poisson equation:

$$\frac{\partial v_i}{\partial t}+v_i\frac{\partial v_i}{\partial x}+\frac{e}{M}\frac{\partial \phi}{\partial x}=0, \tag{6.310}$$

$$\frac{\partial n_i}{\partial t}+\frac{\partial}{\partial x}(n_i v_i)=0, \tag{6.311}$$

$$\frac{\partial^2\phi}{\partial x^2}=\frac{e}{\varepsilon_0}(n_e-n_i). \tag{6.312}$$

Here: v_i is the ion velocity in wave; $\phi$is the electric field potential; n_e, n_i- are the electron and ion densities; *e*- is an electron charge; *M*- is an ion mass. Consider ionic motion as a wave propagating with velocity "u"; then the plasma parameters (v_i, n_i, ϕ) depend on coordinate x and time t as $f(x-ut)$. Also take into account the high electron mobility, which results in their Boltzmann quasi-equilibrium with electric field: $n_e=n_{e0}\exp\left(\frac{e\phi}{T_e}\right)$, where T_e is the electron temperature and n_{e0} is the average density of charged particles. Then rewrite the system of Eqs. (6.310)–(6.312) as:

$$-u\frac{dv_i}{dx}+v_i\frac{dv_i}{dx}+\frac{e}{M}\frac{d\phi}{dx}=0, \tag{6.313}$$

$$\frac{d}{dx}\left[n_i(v_i-u)\right]=0, \tag{6.314}$$

$$\frac{\partial^2\phi}{\partial x^2}=\frac{e}{\varepsilon_0}\left[n_{e0}\exp\left(\frac{e\phi}{T_e}\right)-n_i\right]. \tag{6.315}$$

Integrate Eqs. (6.313), (6.314) assuming that there are no perturbations far enough from the wave ($n_i=n_{e0}$, $v_i=0$ at $x\to\infty$). This leads to the following two non-differential relations:

$$\frac{v_i^2}{2}-uv_i+\frac{e\phi}{M}=0, \tag{6.316}$$

$$n_i=n_{e0}\frac{u}{u-v_i}. \tag{6.317}$$

Based on these two relations, the Poisson equation (6.313) can be expressed as a second order non-linear differential relation between the electric field potential and wave velocity:

$$\frac{d^2\phi}{dx^2}=\frac{en_{e0}}{\varepsilon_0}\left[\exp\left(\frac{e\phi}{T_e}\right)-\frac{u}{\sqrt{u^2-\frac{2e\phi}{M}}}\right]. \tag{6.318}$$

It can be converted into a first order differential equation by multiplying Eq. (6.318) by $\frac{d\phi}{dx}$, integrating, and assuming electric field and potential far from the wave equal to zero $\phi = 0,\ d\phi/dx = 0$:

$$\frac{1}{2}\left(\frac{d\phi}{dx}\right)^2 + \frac{n_{e0}T_e}{\varepsilon_0}\left[1-\exp\left(\frac{e\phi}{T_e}\right)\right] + \frac{n_{e0}Mu^2}{\varepsilon_0}\left(1-\sqrt{1-\frac{2e\phi}{Mu^2}}\right) = 0. \quad (6.319)$$

This nonlinear differential equation describes the solitary ionic sound waves, the ionic sound solitones, without limitations on their amplitudes. This equation can illustrate the effects of strong nonlinearity. For example, analyze the relation between the maximum potential in the solitary wave $\phi = \phi_{max}$ and its propagation velocity u. To do that assume in Eq. (6.319): $\phi = \phi_{max},\ d\phi/dx = 0$ and introduce the dimensionless variables: the dimensionless maximum potential $\xi = e\phi_{max}/T_e$ and the dimensionless square of the wave velocity $\eta = Mu^2/2T_e$. Then rewrite Eq. (6.319) as the algebraic relation:

$$1-\exp(\xi)+2\eta\left(1-\sqrt{1-\xi/\eta}\right). \quad (6.320)$$

If the amplitude of the ionic-sound wave is small ($\xi \to 0$), Eq. (6.320) gives $\eta = 1/2$ and $u = \sqrt{T_e/M}$. Obviously, this corresponds to the conventional formula for velocity of ionic sound Eq. (6.136). In the opposite limit of the maximum amplitude for a solitone, the ion energy in the solitary wave exactly corresponds to the potential energy in the wave $\xi = \eta$. In this case, the critical value of ξ can be determined based on Eq. (6.320) from the equation:

$$1-\exp(\xi)+2\xi = 0. \quad (6.321)$$

Solving this equation yields: $\xi = 1.26,\ e\phi_{max} = 1.26T_e,\ u = 1.58\sqrt{T_e/M}$. At amplitudes above the critical one, ions "are reflected" by the wave. This results in the decay of the strongly nonlinear wave into smaller separate wave packages. The solitones only exist at some limited levels of the wave amplitude. Large amplitudes and strong nonlinearity leads to decay of the solitary waves.

6.9.6 Evolution of Weak Shock Waves in Plasma

A weak shock wave means gas-dynamic perturbation related to a sharp change of derivatives of gas-dynamic variables, while the change of these variables themselves is continuous. The evolution of shock waves can be stimulated by VT relaxation and by chemical heat release in reactions of vibrationally excited molecules (B.S. Parity, 1967; Sydney, 1970; M. Capitelli, 2000. The phenomena can be described by the transport equation for the amplitude of weak shock wave $\alpha(x)$, derived based on the method of characteristics (I.A. Kirillov, V.D. Rusanov, A. Fridman, 1983). The weak shock wave $\alpha(x)$ amplitude is introduced by the relative change of space derivatives of gas-dynamic functions (gas density n_0, velocity v_0, and pressure p) on the shock wave front:

$$n'^{(+)}_{0x} - n'^{(-)}_{0x} = \alpha(x)n_0, \quad (6.322)$$

$$v'^{(+)}_{0x} - v'^{(-)}_{0x} = \alpha(x)c_s, \quad (6.323)$$

$$p'^{(+)}_{x} - p'^{(-)}_{x} = \alpha(x)Mn_0c_s^2. \quad (6.324)$$

Here c_s is the speed of sound in non-perturbed flow; n_0 is gas density in the flow; and M is the mass of molecules. One should note that not only the average vibrational energy of the molecules but also its derivative is continuous across the wave front in the weak shock waves (J.F. Clarke, 1978).

Solution of the transport equation (B.L. Rogdestvensky, N.N. Yanenko, 1978) describes the evolution of the amplitude of a weak shock wave propagating along the plasma flow (I.A. Kirillov, V.D. Rusanov, A. Fridman, 1983b):

$$\alpha(x) = \alpha_i \xi(x)\left[1 \pm \frac{\gamma+1}{2}\alpha_i \int_{x_0}^{x} \xi(x') \frac{dx'}{M \pm 1}\right]^{-1}. \quad (6.325)$$

In this relation α_i is initial amplitude of a weak shock wave ($x = x_0$); the sign "+" corresponds to a shock wave propagating downstream; the sign "–" corresponds to a shock wave propagating upstream; γ is the specific heat ratio; M is the Mach number; the special function $\xi(x)$ is defined by:

$$\xi(x) = \left(\frac{c_{si}}{c_s}\right)^{5/2} \left(\frac{M_i \pm 1}{M \pm 1}\right)^2 \exp \int_{x_0}^{x} \left\{\delta_{lin}(x') + \frac{1}{c_s}\frac{\partial v}{\partial x'}\left(\frac{\pm \gamma M - 1}{M \pm 1}\right)\right\} \frac{dx'}{M \pm 1}. \quad (6.326)$$

Here c_{si}, M_i are the initial values of the speed of sound and Mach number (at $x = 0$); c_s, M are the current values of speed of sound and Mach number at any arbitrary x; v is the gas velocity; $\delta_{lin}(x)$ is the increment of amplification of gas-dynamic perturbations in linear approximation, determined by Eq. (6.168). The nonlinear relations (6.325) and (6.326) take into account non-uniformity of the medium where the weak shock wave propagates. The non-uniformity is taken into account by means of the pre-exponential factor and by the second term under the integral in Eq. (6.326). If $dv/dx = 0$ and the coordinate x is close to the initial one x_0, then the nonlinear relations (6.325) and (6.326) obviously become identical to the result of linear approach Eq. (6.168).

6.9.7 Transition From a Weak to a Strong Shock Wave

Equations (6.325), (6.326) permit describing the transition from a weak to a strong shock wave, which corresponds to $|\alpha(x)| \to \infty$. Strong shock wave appears at a critical point x_{cr}, where the denominator of Eq. (6.325) becomes equal to zero:

$$\frac{\gamma+1}{2}\alpha_i \int_{x_0}^{x} \xi(x') \frac{dx'}{M \pm 1} = \mp 1. \quad (6.327)$$

The strong shock wave can be generated only from waves of compression (for compression waves moving downstream $\alpha_i < 0$, upstream $\alpha_i > 0$) with initial amplitude exceeding the critical one:

$$|\alpha_i| > \alpha_{cr} = \left|\frac{\gamma+1}{2} \int_{x_0}^{\infty} \xi(x) \frac{dx}{M \pm 1}\right|^{-1}. \quad (6.328)$$

The threshold for generation of a strong shock wave, Eq. (6.328) , depends on the gradient of the background flow velocity dv/dx, where the perturbation propagates (see Eq. (6.326)).

6.10 PROBLEMS AND CONCEPT QUESTIONS

6.10.1 Ideal and Non-Ideal Plasmas

Based on Eq. (6.1), calculate the minimum electron density necessary to reach conditions of the non-ideal plasma: 1). at electron temperature 1 eV, 2) at electron temperature equal to room temperature.

6.10.2 Derivation of Debye Radius

To derive the formula for the Debye radius, it was assumed that $e\phi/T << 1$ to expand the Boltzmann distribution in the Taylor series. Show that this assumption is equivalent to the requirement of plasma ideality.

6.10.3 Number of Charged Particles in Debye Sphere

The number of particles necessary for "screening" of the electric field of a charged particle (Figure 6.1) can be found as the number of charged particles in a Debye sphere. Show that this number is $\sqrt{T_e^3\left(4\pi\varepsilon_0\right)^3/e^6 n_e}$, and is large in ideal plasma.

6.10.4 The Bohm Sheath Criterion

Based on Eq. (6.9) for potential distribution in sheath, show that the ion velocity on the steady-state plasma-sheath boundary exceeds the critical Bohm velocity.

6.10.5 Floating Potential

Micro-particles or aerosols are usually negatively charged in steady-state plasma and have negative floating potential (6.12) with respect to plasma. Estimate the negative charge of such particles as a function of their radius. Assume that micro-particles are spherical and located in non-thermal plasma with electron temperature is about 1 eV.

6.10.6 Matrix and Child Law Sheaths

Calculate the matrix and Child law sheaths for non-thermal plasma with electron temperature 3 eV, electron concentration $10^{12} cm^{-3}$ and sheath voltage 300 V. Compare the results obtained for the models of matrix and Child law sheaths.

6.10.7 Plasma Oscillations and Plasma Frequency

Calculate the plasma density necessary to resonance between plasma oscillations (Langmuir frequency Eq. (6.19)) and vibration of molecules. Is it possible to use such resonance for vibrational excitation of molecules in plasma without electron impacts?

6.10.8 Skin-Layer Depth as a Function of Frequency and Conductivity

Both skin-layer depth and Debye radius reflect plasma tendency to "screen" itself from external electric field. Compare these two effects qualitatively as well as numerically for typical parameters of non-thermal plasma.

6.10.9 Magnetic Field Frozen in Plasma

The magnetic field becomes frozen in plasma when conductivity is high. Find the minimum value of conductivity necessary to provide the effect, calculate corresponding level of plasma density. Use for calculations the magnetic Reynolds number criterion.

6.10.10 Magnetic Pressure and Plasma Equilibrium in Magnetic Field

Estimate the magnetic field necessary to provide a sufficient level of magnetic pressure to balance the hydrostatic pressure of hot confined plasma. For calculations take typical parameters of the hot confined plasma from Table 6.2.

6.10.11 The Pinch-Effect, Bennet Relation

Derive the Bennet relation (6.40) for equilibrium of the completely ionized Z-pinch plasma using Eqs. (6.36) or (6.38), combined with the first Maxwell equation Eq. (6.29) for self magnetic field of the plasma column. Calculate how the radius of a plasma cylinder in the Z-pinch depends on the discharge current.

6.10.12 Two-Fluid Magneto-Hydrodynamics

Give a physical interpretation of Eq. (6.45) between the magnetic field and electron gas velocity, derived in the frameworks of the two-fluid magneto-hydrodynamics. This relation implies that the magnetic field is frozen specifically in plasma electrons.

6.10.13 Plasma Diffusion across Magnetic Field

Comparing Eqs. (6.48) and (6.49), discuss the contribution of ions in plasma diffusion across the magnetic field. Consider cases of magnetized and non-magnetized electrons and ions.

6.10.14 The Larmor Radius and Diffusion of Magnetized Plasma

Derive Eq. (6.52) for the diffusion coefficient of plasma across the uniform magnetic field in terms of the Larmor radius. Find the criterion for the minimum magnetic field necessary to decrease the diffusion across the magnetic field.

6.10.15 The Magnetic Reynolds Number

Compare magnetic and kinematic viscosity in plasma and discuss the difference between magnetic and conventional Reynolds numbers. Compare the consequences of high magnetic Reynolds number in plasma and high Reynolds number in non-magnetic fluid mechanics.

6.10.16 Thermal Instability in Monatomic Gases

The ionization-overheating instability can be explained by the chain of causal links Eq. (6.57). Explain why the sequence of events shown in Eq. (6.57) cannot take place in non-equilibrium supersonic discharges?

6.10.17 Thermal and Vibrational Modes of the Ionization-Overheating Instability

Compare Eqs. (6.63) and (6.64) for the instability increment in thermal and vibrational modes. Explain why one of them is proportional and other inversely proportional to the logarithmic sensitivity of VT relaxation rate.

6.10.18 Electron Attachment Instability

Why does the electron attachment instability affect the plasma-chemical process much less than thermal (ionization-overheating) instability? Compare typical frequencies of the electron attachment instability with those of vibrational excitation by electron impact and VT-relaxation at room temperature.

6.10.19 Critical Heat Release in Supersonic Flows

Estimate the maximum conversion degree for the plasma-chemical process with $\Delta H \approx 1\,eV$ in supersonic flow in a constant cross section reactor with energy efficiency of about 30% and initial stagnation temperature 300 K. Assume the maximum heat release as half of the critical one; assume typical values of specific heats as those for diatomic gases.

6.10.20 Profiling of Non-Thermal Discharges in Supersonic Flow

The effect of critical heat release in supersonic flow can be suppressed by duct profiling and keeping the Mach number constant. Using Eqs. (6.82) and (6.83), calculate the increase of the reactor cross section in the discharge zone to provide the conditions of the constant Mach number for M = 3 in CO_2 and initial temperature $T_{00} = 300$ K.

6.10.21 Dynamics of Vibrational Relaxation in Transonic Flows

Transonic flows can be stable with respect to thermal instability of vibrational relaxation at Mach numbers less but close to unity (see Eq. (6.103)). However in this case, the critical heat release is small because of the nearness of speed of sound. Estimate the maximum possible heat release for the transonic flows with the Mach numbers sufficient to observe the stabilization effect.

6.10.22 Space-Non-Uniform Vibrational Relaxation

Explain the qualitative physical difference between vibrational and translational modes of the non-uniform spatial vibrational relaxation. Why is the vibrational mode stable in most of the practical discharge conditions?

6.10.23 Electrostatic Plasma Waves

Using the dispersion Eq. (6.127) shows that product of phase and group velocities of the electrostatic plasma waves corresponds to the square of thermal electron velocities at any values of wave number. Which of the two characteristic wave velocities are larger?

6.10.24 Ionic Sound

Based on the dispersion Eq. (6.136) derive relation for the group velocity of ionic sound. Compare the group velocity with the phase velocity of the ionic sound.

6.10.25 Criterion of Collisionless Damping of Plasma Oscillations

The criterion Eq. (6.145) of the collisional damping of electrostatic plasma oscillations was derived in the case when the electron distribution function is not perturbed by interaction with waves. Prove that this criterion of the collisionless damping can be applied in general case, even when the relative level of perturbations of plasma density is high and inequality (6.144) is not valid.

6.10.26 Landau Damping

Estimate the coefficient γ for Landau damping for electrostatic plasma oscillations (in ω_p units) in a typical range of microwave frequencies. Assume that the plasma oscillation frequency is close to the Langmuir frequency, $kr_D \approx 0.1$.

6.10.27 Beam Instability

Estimate the increment γ of Langmuir oscillations in electron beam instability **Eq.** (6.150) assuming plasma density 10^{12} cm^{-3}, electron beam density 10^9 cm^{-3}. Estimate the gas pressure at which this increment exceeds the frequency of electron-neutral collisions responsible for collisional damping of plasma oscillations Eq. (6.129).

6.10.28 Amplification of Acoustic Waves in Non-Equilibrium Plasma

Based on the complex dispersion Eq. (6.169), derive a relation for the coefficient of amplification of acoustic waves in non-equilibrium plasma in the presence of intensive plasma-chemical reactions.

6.10.29 High-Frequency Dielectric Permittivity of Plasma

Explain why the high-frequency dielectric permittivity of plasma is less than one, while for conventional dielectric materials the dielectric constant is greater than one. What is the principal physical difference between these two cases?

6.10.30 Attenuation of Electromagnetic Waves in Plasma

Based on Eq. (6.189), estimate the attenuation coefficient κ of electromagnetic waves in plasma in the case of low conductivity $\sigma_\omega << \varepsilon_\omega \varepsilon_0 \omega$. Compare the result with relation (6.188).

6.10.31 Electromagnetic Waves in Magnetized Plasma

Based on Eq. (6.213), find the resonance frequencies for electromagnetic waves propagating along the magnetic field, taking into account the ionic motion. Compare the frequencies with the electron-cyclotron frequency ($\omega = \omega_B$).

6.10.32 Emission and Absorption of Radiation by Free Electrons

Based on the energy and momentum conservation laws for the collision of an electron and photon, prove that neither emission nor absorption of radiation is possible for free electrons. Analyze the role of the third heavy collision partner. Explain why a second free electron cannot be an effective third partner.

6.10.33 Spontaneous and Stimulated Emission, the Einstein Relation

Using the relation between the Einstein coefficients, analyze the relative contribution of spontaneous and stimulated emission in quasi-equilibrium plasma conditions. Assume the quasi-equilibrium temperature 10,000 K.

6.10.34 Cross Section of the Bremsstrahlung Emission

Using Eqs. (6.233) and (6.235), estimate numerically typical values of the bremsstrahlung emission cross sections per unit interval of radiation frequencies for typical thermal plasma parameters. Compare the relative contributions of electron-ion and electron-neutral collisions in the bremsstrahlung emission in thermal plasma.

6.10.35 Total Emission in Continuous Spectrum

Total spectral density of plasma emission in continuous spectrum consists of bremsstrahlung and recombination components. After integration over the spectrum, these total plasma radiative energy losses per unit time and unit volume can be calculated using Eq. (6.249). Calculate radiative power per unit volume for typical conditions of thermal plasma with T = 10,000 K and compare with the typical value of specific thermal plasma power of about 1 kW/cm.

6.10.36 Total Plasma Absorption in Continuum, the Unsold-Kramers Formula

Using the Unsold-Kramers formula, discuss plasma absorption in the continuum as a function of temperature. Estimate the quasi-equilibrium plasma temperature, when maximum absorption of red light can be achieved.

6.10.37 Optical Thickness and Emissivity Coefficient

In optically thin plasmas, the total emissivity coefficient coincides with the optical thickness: $\varepsilon = \tau_\omega$. Derive the relation between these two parameters in the general case. Discuss their dependence on radiation frequency.

6.10.38 Probability of the Bound-Bound Transition, Intensity of Spectral Line

Using Eq. (6.266) for the intensity of spectral lines, estimate the ratio of the spontaneous emission frequencies for the cases of transitions between different electronic states and transitions between vibrational levels for the same electronic state. Estimate absolute spontaneous emission frequencies, using data from Table 6.4.

6.10.39 Natural Profile of a Spectral Line

The photon distribution function and natural profile of spectral line, related to finite lifetime of excited states, can be described by the Lorentz profile Eq. (6.270). Show that the Lorentzian profile satisfies the normalization criterion: $\int_{-\infty}^{+\infty} F(\omega)d\omega = 1$.

6.10.40 Doppler Broadening of Spectral Lines

The Doppler broadening of spectral lines depends on gas temperature and atomic mass, while natural width does not. Estimate minimal temperature when Doppler broadening of argon line $\lambda = 696.5\,\text{nm}$ exceeds the natural width of the spectral line.

6.10.41 Pressure Broadening of Spectral Lines

Using Eq. (6.275) for the pressure broadening profile, compare the half-widths for this case (at different pressures and temperatures) with those typical for the Doppler broadening of spectral lines in thermal plasmas, as well as with the natural half-width.

6.10.42 The Stark Broadening of Spectral Lines

Based on Eqs. (6.277) and (6.275) compare the half-width for the total Stark and pressure broadening effects. For typical conditions of non-thermal plasma, determine the degree of ionization when these two broadening effects are the same order of magnitude. How does the critical ionization degree depend on pressure?

6.10.43 The Voigt Profile of Spectral Lines

The Voigt profile is a result of the convolution of the Gaussian and Lorentzian profiles. Prove that the integral Voigt profile Eq. (6.278) becomes either Gaussian or Lorentzian in the cases of significant prevailing of the Doppler or Stark (or pressure) broadening effects.

6.10.44 Absorption of Radiation in a Spectral Line by One Classical Oscillator

By integrating Eq. (6.284), prove that the spectral line area $\int \sigma_\omega d\nu$ characterizing the absorption of one oscillator is constant, and does not depend on the frequency $n_0 < \sigma v >$, τ^{-1} related to broadening.

6.10.45 The Oscillator Power

Give a physical interpretation to the fact that the oscillator powers for the one-electron transitions are less than unity and they approach the unity for the strongest spectral lines. Analyze Eq. (6.287) and data presented in Table 6.4.

6.10.46 Inverse Population of the Excited States and the Laser Amplification Coefficient

Based on Eq. (6.291) estimate the laser amplification coefficient, assuming that the inverse population of excited states is due to the Treanor effect (see Sections 4.1.10 and 5.1.3).

6.10.47 The Carteveg-de Vris Equation and Dispersion Equation of the Ionic Sound

Compare the dispersion (6.297) used for derivation of the Carteveg-de Vris equation with the dispersion equation for the ionic sound (6.136). Determine the criterion of the approximation. Is this criterion specific for the ionic sound or it can be generalized for consideration of nonlinear behavior of other waves?

6.10.48 Solitones as Solutions of the Korteweg-de Vries Equation

Prove that solitones Eq. (6.304) are solutions of the Korteweg-de Vries equation in the forms Eqs. (6.303) and (6.301). Using the general expression Eq. (6.304) for solitones, analyze a relation between velocity amplitude in the solitary wave and characteristic width of the solitone. Prove that product of the solitone's amplitude and square of the solitone's width has a constant value during its evolution.

6.10.49 The Langmuir Solitones

Analyze the Lighthill criterion for plasma oscillations, using the nonlinear dispersion equation Eq. (6.307). Compare this criterion with Eq. (6.309) derived from the dispersion Eq. (6.308), where the low level of nonlinearity of plasma oscillations was directly assumed.

6.10.50 Nonlinear Ionic Sound

Analyzing Eqs. (6.313)–(6.317) describing the nonlinear ionic sound show, that in this case the ionic velocity is always less than velocity of wave propagation ($v_i < u$). Based on the same equations prove, that electric field potential in the ionic sound wave is always positive with respect to the potential at infinity.

6.10.51 Velocity of the Nonlinear Ionic-Sound Waves

Analyzing Eq. (6.320) prove, that the velocity of ionic-sound at low amplitudes is equal to $c_{si} = \sqrt{T_e/M}$. During the derivation pay attention to the fact that the second order of expansion in series is necessary to obtain the result.

6.10.52 The Ionic-Sound Solitones

Based on Eq. (6.320) determine the maximum possible amplitude of the potential in the solitary ionic-sound wave. Analyze the dependence of amplitude of the wave and its velocity near the critical value of the amplitude. How does the nonlinear ionic-sound wave velocity depend on amplitude of the wave?

6.10.53 Evolution of Weak Shock Waves in Plasma

As was mentioned during consideration of Eqs. (6.245–247) not only the average vibrational energy of molecules itself, but also its space derivative is continuous across the wave front in the weak shock waves in plasma.

6.10.54 Comparison of Linear and Nonlinear Approaches to Evolution of Perturbations

Analyzing Eqs. (6.325) and (6.326) prove that the nonlinear Eqs. (6.325) and (6.326) become identical to the result of linear approach Eq. (6.168), if flow can be considered as uniform $dv/dx = 0$ and the coordinate x is close to the initial one x_0 (which means small shock wave amplitude).

6.10.55 Generation of Strong Shock Waves and Detonation Waves in Plasma

Analyze the behavior of plasma with strong vibrational-translational non-equilibrium sustained in supersonic flow when the heat release approaches the critical value. Apply Eq. (6.328) to describe a strong shock wave. Is it possible to reach the conditions necessary for the propagation of a detonation wave in this case?

Part 2

Physics and Engineering of Electric Discharges

7 Glow Discharge

7.1 STRUCTURE AND PHYSICAL PARAMETERS OF GLOW DISCHARGE PLASMA, CURRENT–VOLTAGE CHARACTERISTICS; COMPARISON OF GLOW AND DARK DISCHARGES

7.1.1 General Classification of Discharges, Thermal, and Non-Thermal Discharges

Electric discharges in gases are generators of plasma. The term "gas discharge" initially defined the process of "discharge" of a capacitor into a circuit containing a gas gap between two electrodes. If the voltage between the electrodes is sufficiently large, breakdown occurs in the gap, the gas becomes a conductor, and the capacitor discharges. Now the term gas discharge is applied more generally to any system with ionization in a gap induced by the electric field. Even electrodes are not necessary for these systems since discharges can occur just by the interaction of electromagnetic waves with gas. Variety of electric discharges can be classified according to their physical features and peculiarities in many different ways, for example:

- ***High-pressure discharges (usually atmospheric ones, like arcs or corona) and low-pressure discharges (10 Torr and less, like glow discharges).*** Differences between these are related mostly to how discharge walls are involved in the kinetics of charged particles, energy, and mass balance. Low-pressure discharges are usually cold and not very powerful, while high-pressure discharges can be both, very hot and powerful (arc) as well as cold and weak (corona).
- ***Electrode discharges (like glow and arc) and electrodeless discharges (like inductively coupled radio-frequency RF and microwave).*** Differences between these are related mostly in the manner that electrodes contribute to sustaining electric current by different surface ionization mechanisms and as a result closing the electric circuit. The electrodeless discharges play an important role in "clean" technologies, where material erosion from electrodes is undesirable.
- ***Direct current (DC) discharges (like arc, glow, and pulsed corona) and non-DC discharges (like RF, microwave, and most of dielectric barrier discharges - DBD).*** The DC-discharges can have either constant current (arc, glow) or can be sustained in a pulse-periodic regime (pulsed corona). The pulsed periodic regime permits providing higher power in cold discharges at atmospheric pressure. Non-DC-discharges can be either low or high frequencies (including radio-frequency and microwave). Microwave discharges because of the skin effect can be sustained continuously in a non-equilibrium way at moderate pressures at extremely high power levels up to 1 MW.
- ***Self-sustained and non-self-sustained discharges***. Non-self-sustained discharges can be externally supported by electron beams and ultraviolet radiation. Probably the most important aspect of these discharges is the independence of ionization and energy input to the plasma. This permits achieving large energy inputs at high pressures without serious instabilities.
- ***Thermal (quasi-equilibrium) and non-thermal (non-equilibrium) discharges.*** As is clear from their names, the thermal discharges are hot (10,000–20,000 K), while the non-thermal

discharges operate close to room temperature. The difference between these two qualitatively different types of electric discharges is primarily related to different ionization mechanisms. Ionization in non-thermal discharges is mostly provided by direct electron impact (electron collisions with "cold" non-excited atoms and molecules), in contrast to thermal discharges where ionization is due to electron collisions with preliminary excited hot atoms and molecules. Thermal discharges (the most typical example is an electric arc) are usually powerful, easily sustained at high pressures, but operate close to thermodynamic equilibrium and are not chemically selective. The non-thermal discharges (the most typical example is the glow discharge) are very selective with respect to plasma-chemical reactions. The non-thermal discharges can operate very far from thermodynamic equilibrium with very high energy efficiency, but usually with limited power.

7.1.2 Glow Discharge: General Structure and Configurations

The name "glow discharge" appeared just to point out the fact that plasma of the discharge is luminous (in contrast to the relatively low power dark discharge). According to a more descriptive physical definition: ***a glow discharge is the self-sustained continuous DC discharge having a cold cathode, which emits electrons as a result of secondary emission mostly induced by positive ions.*** A schematic drawing of typical normal glow discharge is shown in Figure 7.1. An important distinctive feature of the general structure of glow discharge is **the cathode layer** (see Figure 7.1) with a large positive space charge, strong electric field with a potential drop of about 100–500 V. The thickness of the cathode layer is inversely proportional to gas density and pressure. If the distance between electrodes is sufficiently large, quasi-neutral plasma with a low electric field, the so-called positive column, is formed between the cathode layer and anode (see Figure 7.1). The **positive column** of glow discharge is the most traditional example of weakly ionized non-equilibrium low-pressure plasma. The positive column is separated from anode by anode layer. The **anode layer** (see Figure 7.1) is characterized by a negative space charge, a slightly elevated electric field, and also special potential drops. The most conventional configuration of glow discharge is considered above the discharge tube with typical parameters given in Table 7.1. This classical discharge tube was widely investigated and industrially used for several decades in fluorescent lamps as a lighting device. Other glow discharge configurations, applied in particular for thin film deposition and electron bombardment are shown in Figure 7.2. The coplanar magnetron glow discharge convenient for plasma-assisted sputtering and deposition includes a magnetic field for plasma confinement (Figure 7.2a). The configuration, optimized as an electron bombardment plasma source (Figure 7.2b), is coaxial and includes the hollow cathode ionizer as well as a diverging magnetic field. The strongly non-equilibrium glow discharge plasma has been successfully applied at different power levels as the active medium for different gas lasers. Attempts to increase specific (per unit volume) and the total power of the lasers also resulted in significant modifications of the glow discharge configuration (see

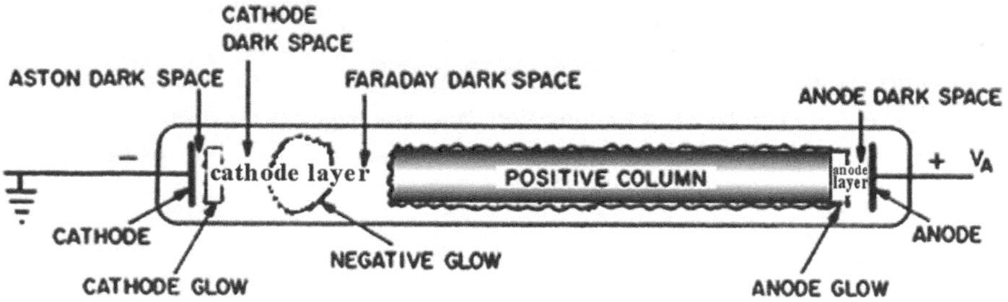

FIGURE 7.1 General Structure of the glow discharge.

TABLE 7.1
Characteristic Parameter Ranges of the Conventional Glow Discharge in a Tube

Glow Discharge Parameter:	Typical Values:
Discharge tube radius	0.3–3 cm
Discharge tube length	10–100 cm
Plasma volume	About 100 cm^3
Gas pressure	0.03–30 Torr
Voltage between electrodes	100–1000 V.
Electrode current	10^{-4}–0.5 A
Power level	Around 100 W
Electron temperature in positive column	1–3 eV
Electron density in positive column	$10^9 - 10^{11}$ cm^{-3}

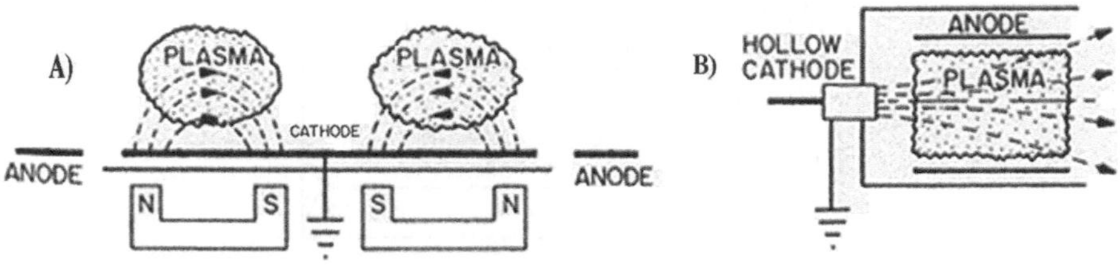

FIGURE 7.2 Magnetron (A) and hollow cathode (B) glow discharge configurations.

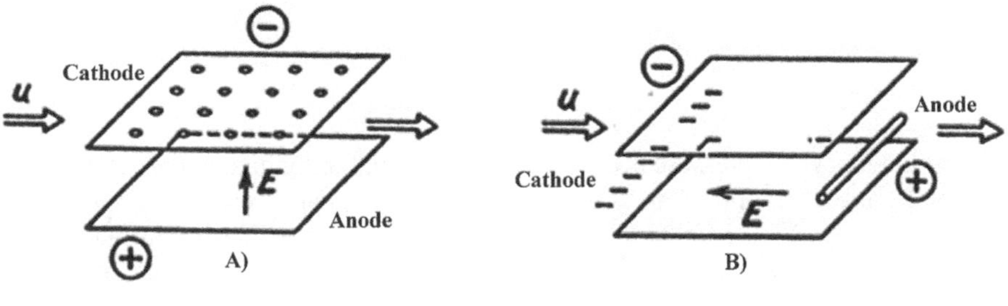

FIGURE 7.3 Transverse and longitudinal configurations of glow discharges in gas flow.

Figure 7.3). At high power levels, it is a parallel plate discharge with a gas flow. The discharge can be transverse with the electric current perpendicular to the gas flow (Figure 7.3a), or longitudinal if these are parallel to each other (Figure 7.3b). These powerful glow discharges operate at higher levels of current and voltage, reaching values of 10 – 20 A and 30 – 50 kV respectively.

7.1.3 Glow Pattern and Distribution of Plasma Parameters along the Glow Discharge

Analyze the pattern of light emission in a classical low-pressure discharge in a tube. Along such a discharge tube, a sequence of dark and bright luminous layers is seen (Figure 7.4a). The typical size-scale of glow discharge structure is usually proportional to electron mean free path $\lambda \propto 1/p$, and hence inversely proportional to pressure. For this reason, it is easier to observe the glow pattern at low pressures, when the distance between layers is sufficiently large. Thus the layered pattern

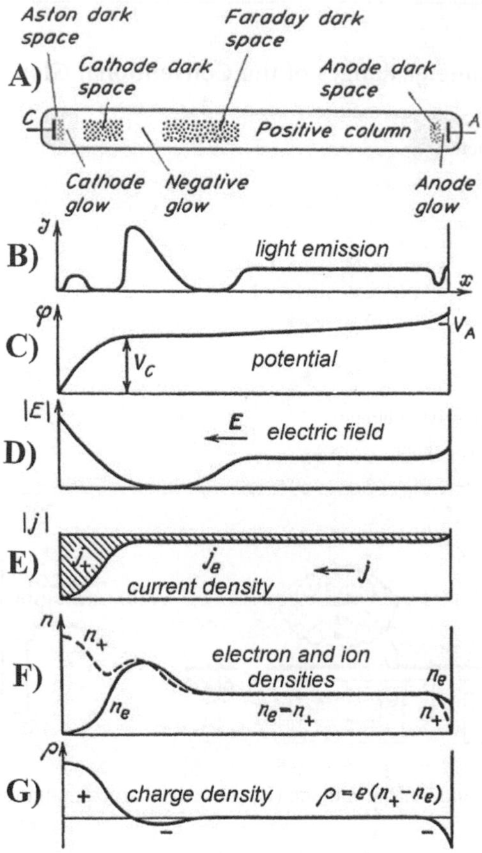

FIGURE 7.4 Physical parameters distribution in a glow discharge.

extends to centimeters when pressure is about 0.1 Torr. Special individual names were given to each layer shown in Figure 7.4a. Immediately adjacent to the cathode is a dark layer known as the ***Aston dark space***. Then there is a relatively thin layer of the ***cathode glow***. This is followed by the ***dark cathode space***. The next zone is the so-called ***negative glow***, which is sharply separated from the dark cathode space. The negative glow gradually decreases in brightness toward the anode, becoming the ***Faraday dark space***. Only after that does the positive column begins. The ***positive column*** is bright (though not as bright as the negative glow), uniform, and relatively long if discharge tubes are length is sufficiently large. In the area of the anode layer, the positive column is transferred first into the ***anode dark space***, and finally into the narrow ***anode glow*** zone. The described layered pattern of glow discharge can be interpreted based on the distribution of the discharge parameters shown in Figure 7.4 b-g. Electrons are ejected from the cathode with relatively low energy (about 1 eV) insufficient to excite atoms; this explains the Aston dark space. Then electrons obtained from the electric field with sufficient energy for electronic excitation provides the cathode glow. Further acceleration of electrons in the cathode dark space leads mostly not to electronic excitation but to ionization (see Section 3.3.8). This explains the low level of radiation and the significant increase of electron density in the cathode dark space. Slowly moving ions have relatively high concentrations in the cathode layer and provide most of the electric current. The high electron density at the end of the cathode dark space results in a decrease of the electric field, and hence a decrease of the electron energy and ionization rate, but leads to a significant intensification of radiation. This is the reason the transition to the brightest layer of the negative glow. Moving further and further from cathode, electron energy decreases, which results in transition from the negative glow into the Faraday dark space. Plasma density decreases in the Faraday dark space and electric field again grows finally establishing the positive column. The average electron energy in positive column is about 1–2 eV, which provides light emission in this major part of the glow discharge. Notice that the cathode layer structure remains the same if electrodes are moved closer at fixed pressure, while the positive column shrinks. The positive column can be extended between connecting electrodes. The anode repels ions and removes electrons from the positive column, which creates the negative space charge and leads to some increase of electric field in the anode layer. Reduction of the electron density in this zone explains the anode dark space, while the electric field increase explains the anode glow.

7.1.4 General Current–Voltage Characteristic of Continuous Self-Sustained DC Discharges Between Electrodes

If the voltage between electrodes exceeds the critical threshold value V_t necessary for breakdown, a self-sustained discharge can be ignited. The current current–voltage characteristic of such a discharge is illustrated in Figure 7.5 for a wide range of currents I. The electric circuit of the discharge gap also includes an external ohmic resistance R. Then the Ohm's law for the circuit can be presented as:

$$EMF = V + RI, \tag{7.1}$$

where EMF is the electromotive force, V is the voltage on the discharge gap. Equation (7.1) is usually referred to as the **load line**. This straight line corresponds to Ohm's law and is also shown in Figure 7.5. The intersection of the current–voltage characteristic and the load line gives the actual value of current and voltage in a discharge. If the external ohmic resistance is sufficiently large and the current in the circuit is very low (about $10^{-10} - 10^{-5}A$), then the electron and ion densities are so negligible that perturbations of the external electric field in plasma can be neglected. Such a discharge is known as the **dark Townsend discharge**. The voltage necessary to sustain this discharge does not depend on current and coincides with the breakdown voltage. The dark Townsend discharge corresponds to the plateau BC in Figure 7.5. An increase of the EMF or decrease of the external ohmic resistance R leads to the growth of the discharge current and plasma density, which results in significant reconstruction of the electric field. This leads to a reduction of voltage with current (interval CD in Figure 7.5) and to transition from a dark to a glow discharge. This low current version of glow discharge is called the **sub-glow discharge**. Further EMF increase or R reduction leads to the lower voltage plateau DE on the current–voltage characteristic, corresponding to the **normal glow discharge**. The normal glow discharge exists over a range of currents $10^{-4} - 0.1\ A$.

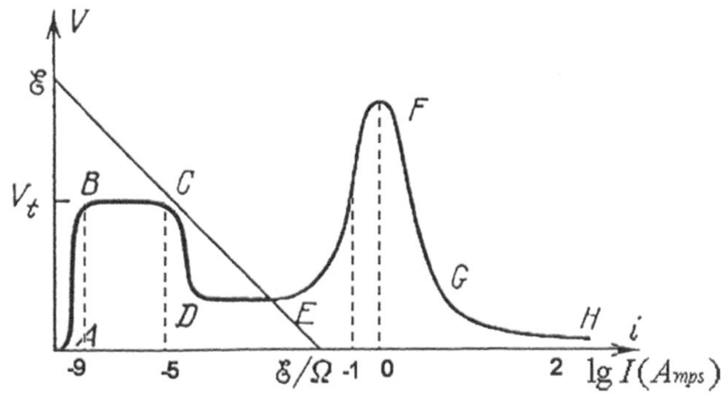

FIGURE 7.5 Current voltage characteristic of DC discharges.

The current density on the cathode is fixed in normal glow discharges. An increase in the total discharge current occurs only by growth of the so-called cathode spot through which the current flows. Only when the current is so large that no additional free surface remains on the cathode, does further current growth requires a voltage increase to provide higher values of current density. Such a regime is called the abnormal glow discharge and corresponds to the interval EF on the current–voltage characteristics in Figure 7.5). Further increase of current accompanied by voltage growth in the abnormal glow regime leads to higher power levels and transition to arc discharge. The glow-to-arc transition usually occurs at currents about 1A.

7.1.5 The Dark Discharge Physics

The distinctive feature of the dark discharge is the smallness of its current and plasma density, which keeps the external electric field almost unperturbed. In this case, the steady-state continuity equation for charged particles can be expressed taking into account their drift in the electric field and ionization as:

$$\frac{dj_e}{dx} = \alpha\, j_e, \quad \frac{dj_+}{dx} = -\alpha\, j_e. \tag{7.2}$$

The direction from cathode to anode is chosen as the positive one; j_e and j_+ are the electron and positive ion current densities; α is the Townsend coefficient the characterizing rate ionization per unit length (see Section 4.4.1). Adding Eqs. (7.2) gives: $j_e + j_+ = j = const$, which reflects the constancy of the total current. The boundary conditions on the cathode ($x = 0$) relate the ion and electron currents on the surface due to the secondary electron emission (with coefficient γ, see Section 8.2.5):

$$j_{eC}(x=0) = \gamma\, j_{+C}(x=0) = \frac{\gamma}{1+\gamma}\, j. \tag{7.3}$$

Boundary conditions on the anode ($x = d_0$, d_0 is the inter-electrode distance) reflect the fact of absence of ion emission from the anode:

$$j_{+A}(x=d_0) = 0, \quad j_{eA}(x=d_0) = j. \tag{7.4}$$

Solution of Eq. (7.2) taking into account the boundary condition on cathode Eq. (7.3) can be expressed assuming the constancy of the Townsend coefficient α as:

$$j_e = \frac{\gamma}{1+\gamma}\, j\exp(\alpha x), \quad j_+ = j\left(1 - \frac{\gamma}{1+\gamma}\exp(\alpha x)\right). \tag{7.5}$$

Equation (7.5) should satisfy the anode boundary conditions. This is possible if the electric field and hence the Townsend coefficient α is sufficiently large:

$$\alpha(E)\, d_0 = \ln\frac{\gamma+1}{\gamma}. \tag{7.6}$$

This formula describing the dark discharge self-sustainment coincides with the breakdown condition in the gap (see Section 4.4.1, Eq. (4.141)). Taking into account the discharge self-sustainment condition Eq. (7.6), the relation between currents Eq. (7.5) can be rewritten as:

$$\frac{j_e}{j} = \exp\left[-\alpha(d_0 - x)\right], \quad \frac{j_+}{j} = 1 - \exp\left[-\alpha(d_0 - x)\right], \quad \frac{j_+}{j_e} = \exp\left[\alpha(d_0 - x)\right] - 1. \tag{7.7}$$

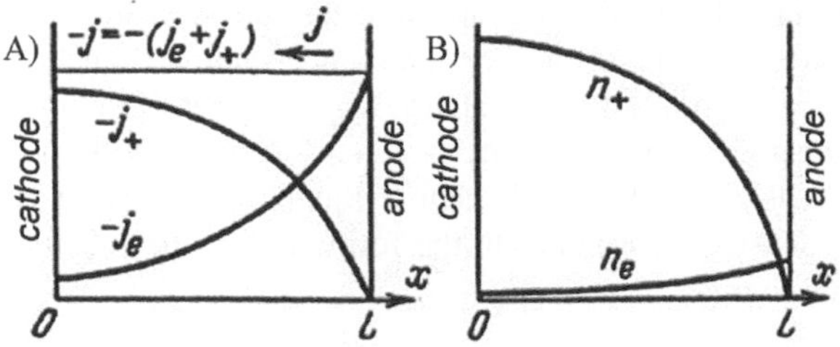

FIGURE 7.6 Current density (A) and electron/ion density (B) distributions in a dark discharge.

According to Eq. (7.6): $\alpha d_0 = \ln\frac{\gamma+1}{\gamma} >> 1$, because $\gamma << 1 (\alpha d_0 = 4.6$, if $\gamma = 0.01)$. From Eq. (7.7), the ion current exceeds electron current over the major part of discharge gap (Figure 7.6). The electron and ion currents become equal only near the anode ($j_e = j_+$ at $x = 0.85 d_0$). The difference in electrons and ions density is even stronger because of large differences in electron and ion mobilities (μ_e, μ_+). Electron and ion densities become equal at a point close to the anode (see Figure 7.6), where:

$$1 = \frac{n_+}{n_e} = \frac{\mu_e}{\mu_+}\frac{j_+}{j_e} = \frac{\mu_e}{\mu_+}\left[\exp\alpha\left(d_0 - x\right) - 1\right]. \tag{7.8}$$

Assuming $\mu_e/\mu_+ \approx 100$, from Eq. (7.8) the electron and ion concentrations become equal at $x = 0.998$. Almost the entire gap is charged positively in a dark discharge. However, the absolute value of the positive charge is not high because of low current and hence, low ion density in this discharge.

7.1.6 Transition of Townsend Dark to Glow Discharge

Transition from the dark to glow discharge is due to the growth of the positive space charge and distortion of the external electric field, which results in the formation of the cathode layer. To describe this transition, use the Maxwell equation:

$$\frac{dE}{dx} = \frac{1}{\varepsilon_0} e\left(n_+ - n_e\right). \tag{7.9}$$

Taking into account that $n_+ \approx j/e\mu_+ E >> n_e$, Eq. (7.9) gives the distribution of electric field:

$$E = E_c\sqrt{1 - \frac{x}{d}}, \quad d = \frac{\varepsilon_0 \mu_+ E_c^2}{2j}. \tag{7.10}$$

Here E_c is the electric field at the cathode. The electric field decreases near the anode with respect to the external field and grows in the vicinity of the cathode; see Figure 7.7. As seen from Eq. (7.10)

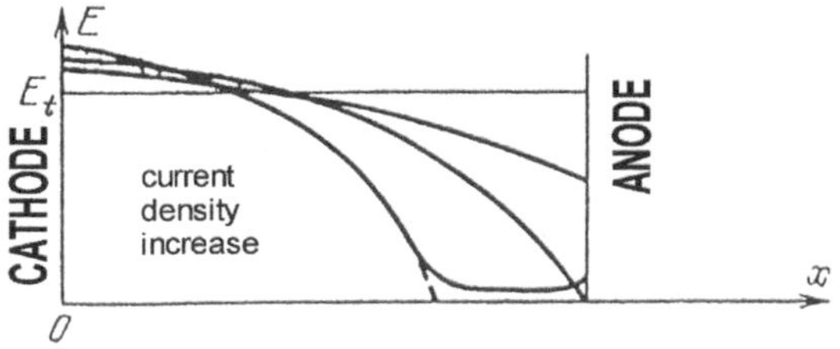

FIGURE 7.7 Electric field evolution in a dark discharge.

and Figure 7.7, higher current densities lead to more distortion of the external electric field. The parameter "d" in Eq. (7.10) corresponds to a virtual point, where the electric field becomes equal to zero. This point is located far beyond the discharge gap ($d >> d_0$) at low currents typical for dark discharges. However, when the current density becomes sufficiently high, this imaginary point of zero electric field can reach the anode ($d = d_0$). This critical current density is maximum for the dark discharge. This current density corresponds to the formation of the cathode layer and to transition from the dark to glow discharge:

$$j_{\max} = \frac{\varepsilon_0 \mu_+ E_c^2}{2d_0}. \tag{7.11}$$

For numerical calculations based on Eq. (7.11), the electric field in the cathode's vicinity is estimated as the breakdown electric field. For example, in nitrogen at a pressure 10 Torr, inter-electrode distance 10 cm, electrode area 100 cm^2, and secondary electron emission coefficient $\gamma = 10^{-2}$, the maximum dark discharge current can be estimated from as $j_{\max} \approx 3 \cdot 10^{-5} A$. When the current density of a dark discharge is relatively high and the electric field non-uniform, Eq. (7.10) should be taken into account and the dark discharge self-sustainment condition Eq. (7.6) must be rewritten in integral form:

$$\int_0^{d_0} \alpha\left[E(x)\right]dx = \ln\frac{\gamma+1}{\gamma}. \tag{7.12}$$

Typical voltage of glow discharge is lower than the value of a dark discharge (see Figure 7.5). Growth of the positive space charge during the dark-to-glow transition results in redistribution of the initially uniform electric field: it becomes stronger near the cathode and lower near the anode. However, the increase of the exponential function $\alpha[E(x)]$ on the cathode side is more significant than its decrease on the anode side. The electric field non-uniformity facilitates the breakdown condition Eq. (7.12), which explains why the voltage of glow discharge is lower than in a dark discharge.

7.2 CATHODE AND ANODE LAYERS OF A GLOW DISCHARGE

7.2.1 The Engel-Steenbeck Model of a Cathode Layer

When voltage is applied to a discharge gap, the uniform distribution of the electric field is not optimal to sustain the discharge. It is easier to satisfy the above criterion Eq. (7.12), if the sufficiently high potential drop occurs in the vicinity of the cathode, for example, in the cathode layer. As was mentioned in Section 7.1, the required electric field non-uniformity can be provided in the discharge gap by positive space charge formed near the cathode due to relatively low ion mobility. The qualitative theory of a cathode layer was developed by A. von Engel and M. Steenbeck (1934). The electric field $E(x = d)$ on the "anode end" of a cathode layer ($x = d$, see Figure 7.7) is much less than near the cathode $E(x = 0) = E_c$. Also the ion current into a cathode layer from a positive column can be neglected due to relatively low ion mobility ($\mu_+/\mu_e \propto 10^{-2}$). For this reason, the Engel-Steenbeck model assumes zero electric field at the end of a cathode layer $E(x = d) = 0$ and consider the anode layer as an independent system of length d (defined by Eq. (7.10)), where the condition Eq. (7.12) of the discharge self-sustainment should be valid. In this case, the cathode potential drop is:

$$V_c = \int_0^d E(x)dx. \tag{7.13}$$

The Engel-Steenbeck model solves the system of equations (7.9), (7.12), and (7.13), taking the Townsend coefficient dependence on the electric field $\alpha(E)$ in the form Eq. (4.142), and assuming a linear decrease of the electric field along the cathode layer (see Figure 7.4):

$$E(x) = E_c(1 - x/d) \text{ if } 0 < x < d, \quad E = 0 \text{ if } x > d. \tag{7.14}$$

Here E_c is the electric field at the cathode. The integral Eq. (7.12) with electric field Eq. (7.14) and exponential relation $\alpha(E)$ cannot be found analytically. Simplified solutions (but with quite sufficient accuracy) can be found assuming of the electric field constant over the cathode layer $E(x) = E_c = const$ in Eq. (7.12). Equation (7.12) can be then simplified to Eq. (7.6) with the inter-electrode distance d_0 replaced by the length d of the cathode layer; the Townsend coefficient $\alpha(E)$ can be expressed by Eq. (4.142), and Eq. (7.2) can be simplified to the product: $V_c = E_c d$. In this case, the Engel-Steenbeck model leads to the relations between the electric field E_c, the cathode potential drop V_c, and the length of cathode layer pd, which are similar to those describing breakdown of a gap Eq. (4.143):

$$V_c = \frac{B(pd)}{C + \ln(pd)}, \quad \frac{E_c}{p} = \frac{B}{C + \ln(pd)}. \tag{7.15}$$

Here $C = \ln A - \ln\ln\left(\frac{1}{\gamma} + 1\right)$; A and B – are the pre-exponential and exponential parameters of the function $\alpha(E)$ (see Eq. (4.142) and Table 4.3). Here the cathode potential drop V_c, electric field E_c, and the similarity parameter pd depend on the discharge current density j (which is close to the ion current density because $j_+ >> j_e$ near the cathode, see Figure 7.4). To determine this dependence according to the Engel-Steenbeck model, first determine the positive ion density $n_+ >> n_e$. The positive ion density can be found based on the Maxwell relation (7.9), taking into account the linear decrease of electric field $E(x)$ along the cathode layer from $E(x = 0) = E_c$ to $E(x = d) = 0$:

$$n_+ \approx \frac{\varepsilon_0}{e}\left|\frac{dE(x)}{dx}\right| \approx \frac{\varepsilon_0 E_c}{ed}. \tag{7.16}$$

Total current density in the cathode's vicinity is close to the current density of positive ions:

$$j = en_+\mu_+ E \approx \frac{\varepsilon_0 \mu_+ E_c^2}{d} \approx \frac{\varepsilon_0 \mu_+ V_c^2}{d^3}. \tag{7.17}$$

7.2.2 Current–Voltage Characteristic of Cathode Layer

The cathode potential drop V_c as a function of the similarity parameter pd corresponds in frameworks of the Engel-Steenbeck approach to the Paschen curve (Figure 4.19) for breakdown. This function $V_c(pd)$ has a minimum V_n, corresponding to the minimum breakdown voltage (see Eq. (4.144)). Taking into account Eq. (7.17), the cathode potential drop V_c as a function of current density j also has the same minimum point V_n. It is convenient to express the relations between V_c, E_c, pd and j using the following dimensionless parameters:

$$\tilde{V} = \frac{V_c}{V_n}, \quad \tilde{E} = \frac{E_c/p}{E_n/p}, \quad \tilde{d} = \frac{pd}{(pd)_n}, \quad \tilde{j} = \frac{j}{j_n}. \tag{7.18}$$

Here electric field E_n/p and cathode layer length $(pd)_n$ correspond to the minimum point of the cathode voltage drop V_n. The subscript "n" stands here to denote the "normal" regime of a glow discharge. These three "normal" parameters: E_n/p, $(pd)_n$ as well as V_n can be found using Eq. (4.144) originally derived in Section 4.4.2 for an electric breakdown as parameters of the Paschen curve. Corresponding values of the normal current density can be expressed based on Eq. (7.17) using the following numeric formula constructed with similarity parameters:

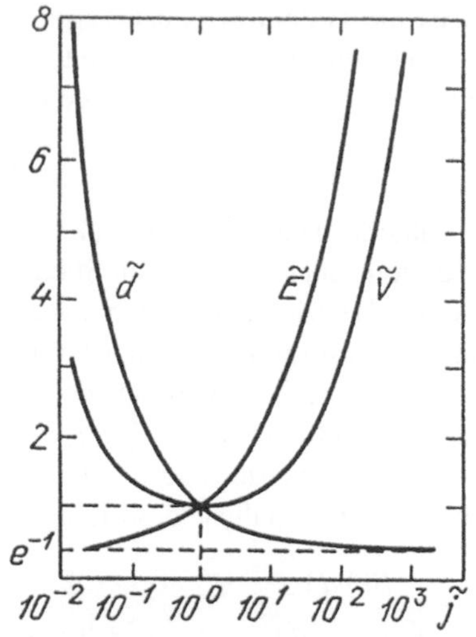

FIGURE 7.8 Dimensionless parameters of a cathode layer.

$$\frac{j_n, \mathrm{A/cm^2}}{(p, \mathrm{Torr})^2} = \frac{1}{9 \cdot 10^{11}} \frac{(\mu_+ p), \mathrm{cm^2 Torr/Vsec} \times (V_n, V)^2}{4\pi \left[(\mathrm{pd})_\mathrm{n}, \mathrm{cmTorr}\right]^3}. \tag{7.19}$$

Relations between V_c, E_c, j and the cathode layer length pd can be expressed as:

$$\tilde{V} = \frac{\tilde{d}}{1 + \ln \tilde{d}}, \quad \tilde{E} = \frac{1}{1 + \ln \tilde{d}}, \quad \tilde{j} = \frac{1}{\tilde{d}\left(1 + \ln \tilde{d}\right)^2}. \tag{7.20}$$

Dimensionless voltage $\tilde{V}$, electric field $\tilde{E}$, cathode layer length $\tilde{d}$ are presented in Figure 7.8 as a function of dimensionless current density. It is the cathode layer's current–voltage characteristic.

7.2.3 Normal Glow Discharge: Normal Cathode Potential Drop, Normal Layer Thickness, and Normal Current Density

According to the current–voltage characteristic Eq. (7.20) any current densities are possible in a glow discharge. In reality, this discharge "prefers" to operate only at one value of current density, the normal one j_n, which corresponds to the minimum of the cathode potential drop and can be calculated using Eq. (7.19). The total glow discharge current I is usually controlled by the external resistance and the load line Eq. (7.1). The current conducting discharge channel occupies on the cathode surface a spot with area $A = I/j_n$, which provides the required normal current density. This is called the cathode spot. Other current densities are unstable when the cathode surface area sufficiently large. If because of some perturbation $j > j_n$, the cathode spot grows until current density becomes normal; if $j < j_n$, the cathode spot decreases to reach the same normal current density. The normal current density can be reached if the cathode surface area is sufficiently large, which means $A > I/j_n$. This permits accommodating on the cathode the necessary current spot. The glow discharge with normal current density on its cathode is usually referred to as the normal glow discharge. The normal glow discharges have fixed current density j_n, and fixed cathode layer thickness ($(pd)_n$) and voltage V_n, which only depend at room temperature on gas composition and cathode material. Typical values of these normal glow discharge parameters are presented in Table 7.2. Typical normal cathode current density is about 100 μA/cm² at pressure about 1 Torr; typical thickness of the normal cathode layer at this pressure is about 0.5 cm, typical normal cathode potential drop is 200 V and does not depend on pressure and temperature.

TABLE 7.2
Normal Current Density j_n/p^2, $\mu A/cm^2\ Torr^2$, Normal Thickness of Cathode Layer $(pd)_n$, cm Torr, and Normal Cathode Potential Drop V_n, V for Different Gases and Cathode Materials at Room Temperature

Gas	Cathode Material	Normal Current Density	Normal Thickness of Cathode Layer	Normal Cathode Potential Drop
Air	Al	330	0.25	229
Air	Cu	240	0.23	370
Air	Fe	–	0.52	269
Air	Au	570	–	285
Ar	Fe	160	0.33	165
Ar	Mg	20	–	119
Ar	Pt	150	–	131
Ar	Al	–	0.29	100
He	Fe	2.2	1.30	150
He	Mg	3	1.45	125
He	Pt	5	–	165
He	Al	–	1.32	140
Ne	Fe	6	0.72	150
Ne	Mg	5	–	94
Ne	Pt	18	–	152
Ne	Al	–	0.64	120
H_2	Al	90	0.72	170
H_2	Cu	64	0.80	214
H_2	Fe	72	0.90	250
H_2	Pt	90	1.00	276
H_2	C	–	0.90	240
H_2	Ni	–	0.90	211
H_2	Pb	–	0.84	223
H_2	Zn	–	0.80	184
Hg	Al	4	0.33	245
Hg	Cu	15	0.60	447
Hg	Fe	8	0.34	298
N_2	Pt	380	–	216
N_2	Fe	400	0.42	215
N_2	Mg	–	0.35	188
N_2	Al	–	0.31	180
O_2	Pt	550	–	364
O_2	Al	–	0.24	311
O_2	Fe	–	0.31	290
O_2	Mg	–	0.25	310

7.2.4 Mechanism Sustaining the Normal Cathode Current Density

When the normal glow discharge total current is changing, the current density remains always the same (see Table 7.2). This impressive effect was investigated by numerous scientists (V.L. Granovsky, 1971, A. von Engel, 1965), and was explained in different ways: analyzing current stability, modeling ionization kinetics and applying the minimum power principle. Initially, Engel and Steenbeck's explanation of this effect was related to the instability of the cathode layer with $j < j_n$, where the current–voltage characteristic is "falling" (see Figure 7.8). The falling current–voltage characteristics are generally unstable in non-thermal discharges. For example, if a fluctuation results in a local increase of current density in some area of a cathode spot, the necessary voltage to sustain ionization in this area decreases. The actual voltage in this area then exceeds the required one, which leads to a further increase of current until it becomes normal $j = j_n$. If a fluctuation results in a local decrease of current density, the discharge extinguishes for the same reason in this local area of a cathode spot. Such a mechanism is able to only explain the instability at $j < j_n$ and the growth of current density to reach the normal one. A more detailed model, able to describe establishing the normal cathode current density $j = j_n$ starting from both lower $j < j_n$ and higher $j > j_n$ current densities, was developed by A.A. Vedenov (1982) and Yu.P. Raizer, S.T. Surzhikov (1987, 1988). This model describes the phenomenon in terms of the charge reproduction coefficient:

$$\mu = \gamma \left\{ \exp \int_0^{d(r)} \alpha \left[E(l) \right] dl - 1 \right\}. \tag{7.21}$$

Integration is along an electric current line "l", which crosses the cathode in some point "r". The charge reproduction coefficient Eq. (7.21) shows the multiplication of charge particles during a single cathode layer ionization cycle. This cycle includes the multiplication of the primary electron formed on a cathode in an avalanche moving along a current line across the cathode layer, the return of positive ions formed in the layer back to the cathode to produce new electrons due to secondary electron emission with the coefficient γ. Sustaining the steady-state cathode layer Eq. (7.12) requires $\mu = 1$. Excessive ionization then corresponds to $\mu > 1$, $\mu < 1$ means extinguishing the discharge. The curve $\mu = 1$on the "voltage-cathode layer thickness" diagram represents the cathode layer potential drop V_c, with the near-minimum point $(pd)_n$ corresponding to the normal voltage V_n (see Figure 7.9, where the current density axis is also shown). All the area on the V-pd diagram above the cathode potential drop curve corresponds to $\mu > 1$; the area under this curve means $\mu < 1$. If the current density is less than normal $j < j_n(d > d_n$, for example point "1" in Figure 7.9), fluctuations can destroy the cathode layer by the instability mechanism described above. At the edges, however, the cathode layer decays in this case even without any fluctuations. Positive space charge is much less at the edges and the same potential corresponds to points located further from the cathode, which can be illustrated in Figure 7.9 as moving to the right from point "1" in the area where $\mu < 1$. As a result, current disappears from the edges, and the discharge voltage increases in accordance with the load line (7.1). The increase of voltage over the line $\mu = 1$ in a central part of the cathode layer leads to

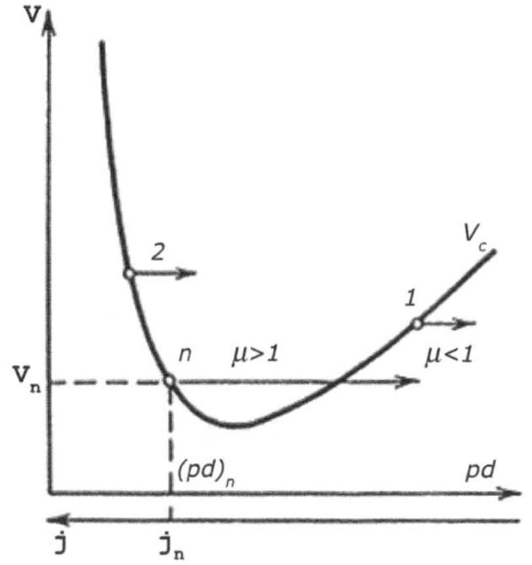

FIGURE 7.9 Explanation of the normal current density effect.

growth of current density until it reaches the normal value j_n. Similarly can be considered the cathode layer with supernormal current density $j > j_n$, which corresponds to point "2" in Figure 7.9. In this case, the central part of the channel is stable with respect to fluctuations (see the Engel-Steenbeck stability analysis for $j < j_n$). At the edges of the channel, space charge is smaller and effective pd is greater for the fixed voltage (see Figure 7.9), which brings us to the area of the V-pd diagram with the charge reproduction coefficient $\mu > 1$. The condition $\mu > 1$leads to breakdown at the edges of the cathode spot, to an increase of total current, to a decrease of total voltage across the electrodes (in accordance with the load line Eq. (7.1)), and finally to a decrease of current density in the major part of a cathode layer until it reaches the normal value j_n. Any deviations of current density from $j = j_n$ stimulate ionization p at the edges of cathode spot, which bring current density to the normal value.

7.2.5 The Steenbeck Minimum Power Principle, Application to Effect of Normal Cathode Current Density

The effect of normal current density can be illustrated using the minimum power principle. The total power released in the cathode layer can be found as:

$$P_c(j) = A\int_0^d jE\,dx = AjV_C(j) = IV_c(j). \tag{7.22}$$

In this relation A is area of a cathode spot, I is the total current. The total current in glow discharge is mostly determined by external resistance (see the load line Eq. (7.1)) and can be considered as fixed. In this case, the cathode spot area can be varied together with the current density keeping fixed their product $j \times A = const$. According to the minimum power principle, the current density in glow discharge should minimize the power $P_c(j)$. As seen from Eq. (7.22), the minimization of the power $P(j)$ at constant current $I = const$ requires minimization of $V_c(j)$, which according to the definition corresponds to the normal current density $j = j_n$. The minimum power principle was proposed in 1932 by Steenbeck and is useful for illustrating discharge phenomena, including striation, thermal arcs, and the normal current in glow discharge. The minimum power principle cannot be derived from fundamental physical laws and hence, should be used mostly for illustrations.

7.2.6 Glow Discharge Regimes Different from a Normal One: the Abnormal Discharge, the Subnormal Discharge, and the Obstructed Discharge

Increase of current in a normal glow discharge is provided by the growth of the cathode spot area at $j = j_n = const$. As soon as the entire cathode is covered by the discharge, further current growth results in an increase of current density over the normal value. This discharge is called **abnormal glow discharge**. The abnormal glow discharge corresponds to the right-hand-side branches ($j > j_n$) of the dependences presented in Figure 7.8. The current–voltage characteristic of the abnormal discharge $\tilde{V}(\tilde{j})$ is growing. It corresponds to the interval EF on the general current–voltage characteristic shown in Figure 7.5. According to Eq. (7.18), when the current density grows further ($j \to \infty$), the cathode layer thickness decreases asymptotically to a finite value $\tilde{d} = 1/e \approx 0.37$; while the cathode potential drop and electric field grow as:

$$\tilde{V} = \frac{1}{e^{3/2}}\sqrt{\tilde{j}}, \quad \tilde{E} \approx \frac{1}{e^{1/2}}\sqrt{\tilde{j}}. \tag{7.23}$$

Actual growth of the current and cathode voltage are limited by cathode overheating. Significant cathode heating at voltages about 10 kV and current densities 10 - 100 A/cm^2 results in transition of the abnormal glow discharge into an arc discharge. Normal glow discharge transition to a dark discharge takes place at low currents (about $10^{-5} A$) and starts with the so-called **subnormal discharge**.

The subnormal discharge corresponds to the interval CD on the general current–voltage characteristic shown in Figure 7.5. The size of the cathode spot at low currents becomes large and comparable with the total cathode layer thickness. This results in electron losses with respect to normal glow discharge and hence, requires higher values of voltage to sustain the discharge (Figure 7.5). Another glow discharge regime, different from the normal one, takes place at low pressures and narrow gaps between electrodes when their product pd_0 is less than normal value $(pd)_n$ for a cathode layer. This discharge mode is called the **obstructed glow discharge**. Conditions in the obstructed discharge correspond to the left-hand branch of the Paschen curve (Figure 4.19), where voltage exceeds the minimum value V_n. Since short inter-electrode distance in the obstructed discharge is not sufficient for effective multiplication of electrons, to sustain the mode, the inter-electrode voltage should be greater than the normal one.

7.2.7 Negative Glow Region of Cathode Layer, the Hollow Cathode Discharge

The negative glow and the Faraday dark space complete the cathode layer and provide a transition to the positive column. As was mentioned in Section 7.1.3, the negative glow region is a zone of intensive ionization and radiation (see Figure 7.4). Most electrons in the negative glow have moderate energies. However, quite a few electrons in this area are very energetic even though the electric field is relatively low. These energetic electrons are formed in the vicinity of the cathode and cross the cathode layer with only a few inelastic collisions. They provide a non-local ionization effect and lead to electron densities in a negative glow even exceeding those in a positive column (Figure 7.4). Details on the non-local ionization effect, formation and propagation of energetic electrons across the cathode into the negative glow and the Faraday dark space can be in particular found in publications of P. Gill, C.E. Webb (1977), J.P. Boeuf, E. Marode (1982) and S.Ya. Bronin, V.M. Kolobov (1983). The effect of intensive "non-local" ionization in a negative glow can be applied to form an effective electron source, the so-called **hollow cathode discharge**. Imaging a glow discharge with a cathode arranged as two parallel plates with the anode on the side. If the distance between the cathodes gradually decreases, at some point the current grows 100–1000 times without a change of voltage. This effect takes place when two negative glow regions overlap, accumulating energetic electrons from both cathodes. Strong photoemission from cathodes in this geometry also contributes to the intensification of ionization in the hollow cathode, see B.I. Moskalev (1969). Effective accumulation of the high negative glow current can be reached if the cathode is arranged as a hollow cylinder and an anode lies further along the axis. Pressure is chosen in such a way to have the cathode layer thickness comparable with the internal diameter of the hollow cylinder. The most traditional configuration of the system is the **Lidsky hollow cathode**, which is shown in Figure 7.10. The Lidsky hollow cathode is a narrow capillary-like nozzle, which operates with axially flowing gas. The hollow cathode is usually operated with the anode located about 1 cm downstream of the capillary nozzle. It can provide high electron currents with densities exceeding those corresponding to Child's law (see Section 6.1.5, Eq. (6.14)). The Lidsky hollow cathode is hard to initiate and maintain in the steady state. For this reason, different modifications of the hollow cathode with external heating were developed to raise the cathode to incandescence and provide long-time steady operation (R.L. Poeschel, J.R. Beattle, P.A. Robinson, J.W. Ward, 1979; A.T. Forrester, 1988).

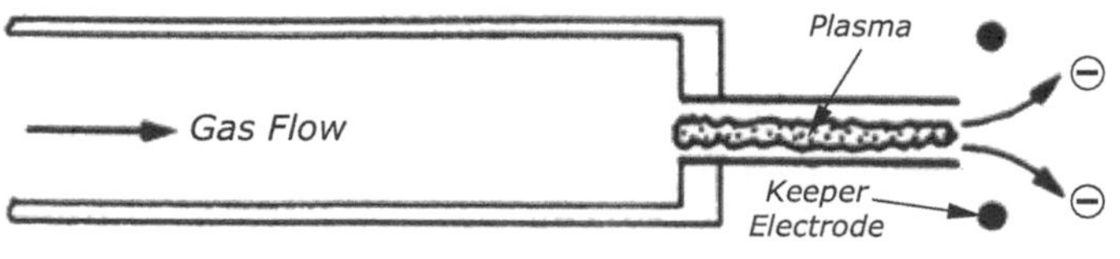

FIGURE 7.10 Schematic of the Lidsky capillary hollow cathode.

7.2.8 Anode Layer

Because positive ions are not emitted (but repelled) by the anode, their concentration at the surface of this electrode is equal to zero. Thus a negatively charged zone, called the anode layer, exists between the anode and positive column (Figure 7.4). The ionic current density in the anode layer grows from zero at the anode to its value $j_{+c} = \frac{\mu_+}{\mu_e} j$ in positive column (here j is the total current density; μ_+, μ_e are mobilities of ions and electrons). In terms of the Townsend coefficient α:

$$\frac{dj_+}{dx} = \alpha\, j_e \approx \alpha\, j, \quad j_{+c} \approx j \times \int \alpha\, dx. \tag{7.24}$$

From Eq. (7.24), it is sufficient for one electron to provide only a very small number of ionization acts to establish the necessary ionic current:

$$\int \alpha\, dx = \frac{\mu_+}{\mu_e} << 1. \tag{7.25}$$

Number of generations of electrons produced in the anode layer Eq. (7.1.13) is about three orders of magnitude smaller than the corresponding number of generations of electrons produced in the cathode layer. For this reason, the anode layer's potential drop is less than the potential drop across the cathode layer (see Figure 7.4). Numerically the value of the anode potential drop is approximately the value of the ionization potential of the gas in the discharge system. However, the anode voltage slightly grows with pressure in the range of moderate pressures around 100 Torr (the dependence is a little bit stronger in electronegative gases, see Figure 7.11). Typical values of the reduced electric field E/p in the anode layer are about $E/p \approx 200 - 600\ V/\text{cm Torr}$. Thickness of the anode layer is of order of the electron mean free path and can be estimated in simple numerical calculations as:

$$d_A(\text{cm}) \approx 0.05\,\text{cm}/p(\text{Torr}). \tag{7.26}$$

The current density j/p^2 at the anode is independent of the value of current in the same manner as for the normal cathode layer. Numerical values of the current density in anode layer are about 100 $\mu A/\text{cm}^2$ at gas pressures about 1 Torr and actually coincide with those of the cathode layer (see Table 7.2). Actually the normal cathode layer imposes the value of current density j on the entire discharge column including the anode. Glow discharge modeling including both cathode and anode layers (Yu.P. Raizer, S.T. Surzhikov, 1987, 1988) shows that even radial distribution of the current density $j(r)$ originating at the cathode can be repeated at the anode. Although many similarities take place between current density in the cathode and anode layers, theoretically these phenomena are quite different (Yu.S. Akishev, A.P. Napartovich, 1980). For example, in contrast to the cathode, the anode quite often is covered by an electric current not continuously, but in a group of spots not directly

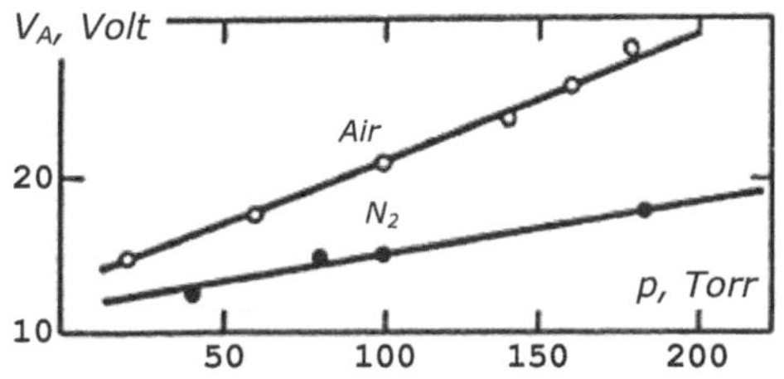

FIGURE 7.11 Anode fall for discharges in nitrogen and air as a function of pressure.

connected with each other. Also in general, processes taking place in the anode layer can play a significant role in developing instabilities in glow discharges, see S.C. Brown (1959, 1966), Yu.P. Raizer (1991), A.M. Howatson (1976).

7.3 POSITIVE COLUMN OF GLOW DISCHARGE

7.3.1 General Features of the Positive Column, Balance of Charged Particles

The positive column of a normal glow discharge is the most conventional source of non-equilibrium non-thermal plasma. Plasma behavior and parameters in the long positive column are independent of the phenomena in cathode and anode layers. The state of plasma in the positive column is determined by local processes of charged particles formation and losses, and by the electric current, which is actually controlled by external resistance and EMF (see the load line Eq. (7.1)). Losses of charged particles in the positive column due to volume and surface recombination should be balanced in a steady-state regime by ionization. The ionization rate sharply depends on electron temperature and reduced electric field. Therefore, the electric field in the positive column is more or less fixed, which determines the longitudinal potential gradient and the voltage difference across a column of a given length. Generation of electrons and ions in a positive column is always due to volumetric ionization processes, so classification of the steady-state regimes is determined by the dominant mechanisms of losses of charged particles, see Section 4.5.

7.3.2 General Current–Voltage Characteristics of a Positive Column and of a Glow Discharge

If the balance of charged particles is controlled by their diffusion to the wall (Section 4.5.4), then the electric field E in the positive column is determined by the Engel-Steenbeck relation Eq. (4.175) and does not depend on electron density and electric current. This is due in this case to the fact that both ionization rate and rate of the diffusion losses are proportional to n_e; thus electron density can be just canceled from the balance of electron generation and losses. In this case, the current–voltage characteristic of the positive column and of a normal glow discharge is almost a horizontal straight line (see Figure 7.5):

$$V(I) = V_n + E d_c \approx const. \tag{7.27}$$

Here, d_c is the length of the positive column; V_n is the normal potential drop in the cathode layer; E is the electric field in column Eq. (4.5.15), which only depends on type of gas, pressure, and radius of discharge tube. When the current and hence electron density grows, the contribution of volumetric electron-ion recombination becomes significant, see Section 4.5.2. In this case electron density $n_e(E/p)$ follows the formula Eq. (4.172), and the relation between current density and the reduced electric field E/p is:

$$j = (\mu_e p)\frac{E}{p} e \frac{k_i(E/p)\, n_0}{\left(k_r^{ei} + \varsigma k_r^{ii}\right)(1+\varsigma)}. \tag{7.28}$$

Here $\mu_e p$ is the electron mobility presented as a similarity parameter (see Table 4.1); $k_i(E/p)$ is the ionization coefficient; n_0 is the gas density; k_r^{ei}, k_r^{ii} are coefficients of electron-ion and ion-ion recombination; k_a, k_d are coefficients of electron attachment and detachment; $\varsigma = k_a/k_d$ is the factor characterizing the balance between electron attachment and detachment. Taking into account the sharp exponential behavior of the function $k_i(p)$, one can conclude from Eq. (7.28) that electric field E grows very slowly with current density in the recombination regime. Thus the current–voltage characteristic of the positive column and of a whole normal glow discharge is almost a horizontal straight

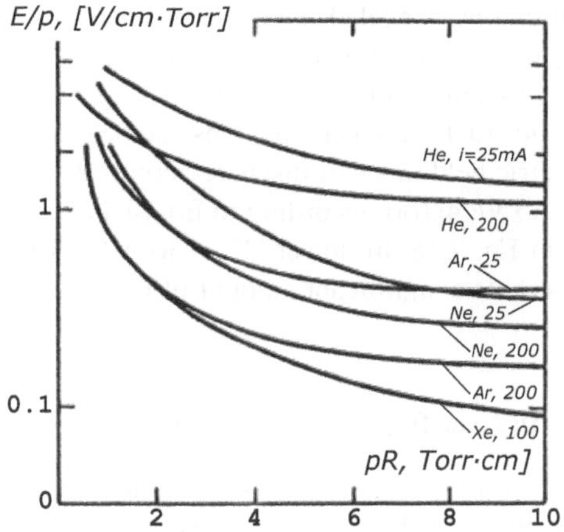

FIGURE 7.12 Measured values of E/p for positive column in inert gases.

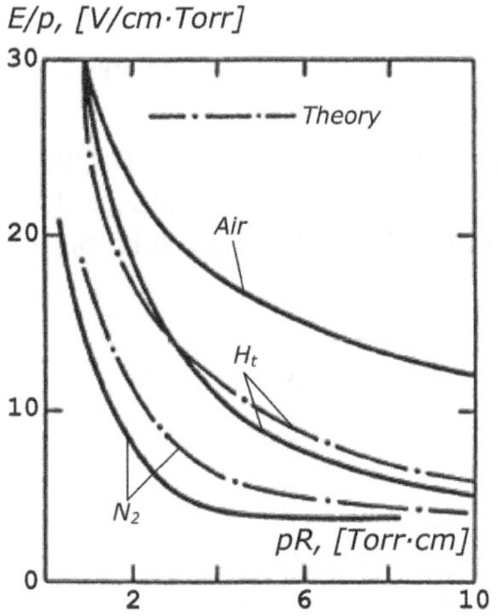

FIGURE 7.13 Measured values of E/p for the positive column in molecular gases.

line in this case as well as in the diffusion-controlled discharge mode (see Eq. (7.27), and Figure 7.5). Examples of the experimental dependences of reduced electric field E/p on the similarity parameter pR (p is pressure, R is the radius of a discharge tube) are presented in Figs. 7.12, 7.13. Some E/p decrease with the current is due to increases in the gas temperature (ionization actually depends on E/n_0; increase in temperature leads to the reduction of n_0 and growth of E/n_0 at constant pressure and electric field). Comparing Figs. 7.12 and 7.13, it is seen that reduced electric fields E/p are about 10 times less in inert gases relative to molecular gases. This is related to the effects of inelastic collisions (mainly vibrational excitation), which significantly reduces the electron energy distribution function at the same values of the reduced electric field (see Eqs. (4.84), (4.85)). The electric fields necessary to sustain glow discharge are essentially less than those necessary for breakdown (see Section 4.4.2). In particular, this is due to the fact that electron losses to the walls during breakdown are provided by free diffusion, while the corresponding discharge is controlled by much slower ambipolar diffusion. If electron detachment from negative ions is negligible, the discharge regime can be controlled by electron attachment. This discharge mode was considered in Section 4.5.3. Both ionization rate and electron attachment rate are proportional to n_e, therefore in this case the electron density actually can be canceled from the balance of electron generation and losses (similar to the discharge regime controlled by diffusion). Obviously the current–voltage characteristic in the attachment regime is again an almost horizontal straight line $V(I) = const$ (see Figure 7.5). The electron detachment and destruction of negative ions can be effective in steady state discharges. For example, the reduced electric field in an air discharge controlled by electron attachment to oxygen should be equal to E/p = 35 V/cmTorr according to Eq. (4.174). However, experimentally measured E/p values, presented in Eq. 7.18 are lower: 12 – 30 V/cmTorr. This shows that the steady state discharge is able to accumulate significant amount of particles efficient for the destruction of negative ions.

7.3.3 Heat Balance and Plasma Parameters of Positive Column

The power, which electrons receive from the electric field and then transfer through collisions to atoms and molecules, can be found as Joule heating (Section 4.3.2, Eq. (4.101)): $jE = \sigma E^2$. If the electron energy consumption into chemical processes is neglected (Section 5.6.7), the entire Joule heating should be balanced by conductive and convective energy transfer. At steady-state conditions this is expressed as:

$$w = jE = n_0 c_p \left(T - T_0\right) \nu_T, \tag{7.29}$$

where w is the discharge power per unit volume; c_p is the specific heat per one molecule; T is the gas temperature in the discharge; T_0 is the room temperature; ν_T is the heat removal frequency. Note, the heat removal frequency from the cylindrical discharge tube of radius R and length d_0 can be determined as:

$$\nu_T = \frac{8}{R^2} \frac{\lambda}{n_0 c_p} + \frac{2u}{d_0}, \tag{7.30}$$

where λ is the coefficient of thermal conductivity, and u is the gas flow velocity. The first term in the Eq. (7.30) is related to heat removal due to thermal conductivity, the second term describes the convective heat removal. Based on Eqs. (7.29) and (7.30), estimate plasma parameters of a positive column. If heat removal is controlled by thermal conduction, combination of Eqs. (7.29) and (7.30) gives the typical discharge power per unit volume, which doubles the gas temperature in the discharge ($T - T_0 = T_0$):

$$w = jE = \frac{8\lambda T_0}{R^2}. \tag{7.31}$$

The thermal conductivity coefficient λ does not depend on pressure and can be estimated as $\lambda \approx 3 \cdot 10^{-4}$ W/cm · K. Hence based on Eq. (7.31) it can be concluded that specific discharge power also does not depend on pressure and for tubes with radius R = 1 cm it can be estimated as 0.7 W/cm^3. Higher values of specific power result in higher gas temperatures and hence, in contraction of a glow discharge. Current density in the positive column with the heat removal controlled by conduction is inversely proportional to pressure and can be estimated based on Eq. (7.31) as:

$$j = \frac{8\lambda T_0}{R^2} \frac{1}{\left(E/p\right)} \frac{1}{p}. \tag{7.32}$$

Assuming the reduced electric field as E/p = 3 – 10 V/cmTorr, Eq. (7.32) gives numerically: j, mA/cm^2 $\approx 100/p$, Torr. Electron concentration in the positive column can be calculated from Eq. (7.32), taking into account Ohm's law $j = \sigma E$ and Eq. (4.99) for the electric conductivity σ. In this case, the electron concentration is easily shown to be inversely proportional to gas pressure:

$$n_e = \frac{w}{E^2}\frac{m\nu_{en}}{e^2} = \frac{w}{(E/p)^2}\frac{mk_{en}}{e^2T_0}\frac{1}{p}. \tag{7.33}$$

Here ν_{en}, k_{en} are the frequency and rate coefficient of electron-neutral collisions. Numerically for the above-assumed values of parameters, the electron concentration in a positive column with conductive heat removal can be estimated as: n_e, cm^{-3} = $3 \cdot 10^{11}/p$, Torr. The electron concentration and current densities both fall linearly with pressure growth. The reduction of the plasma degree of ionization with pressure is even more significant: $\frac{n_e}{n_0} \propto \frac{1}{p^2}$. Hence, low pressures are in general more favorable for sustaining the steady-state homogeneous non-thermal plasma.

7.3.4 Glow Discharge in Fast Gas Flows

The convective heat removal (second term in Eq. (7.30)) promotes pressure and power increases in non-thermal discharges. Gas velocities in such systems are 50 – 100 m/sec. Based on Eqs. (7.29), (7.30), the current and electron densities corresponding to doubling of the gas temperature, do not depend on gas pressure:

$$j = \frac{2uc_p}{d_0(E/p)}, \quad n_e = \frac{2uc_p mk_{en}}{e^2 d_0 T_0 (E/p)^2}. \tag{7.34}$$

Assuming $d_0 = 10\,\text{cm}$, $u = 50\ \text{m/sec}$, $E/p = 10$ V/cmTorr, the typical values of current density and electron concentration in the positive column under consideration are: $j \approx 40$ mA/cm^2, $n_e = 1.5 \cdot 10^{11}$ cm^{-3}. The specific discharge power in the positive column with convective heat removal grows proportionally to pressure:

$$w = jE = c_p T_0 \frac{2u}{d_0} \cdot \frac{p}{T_o}. \tag{7.35}$$

For such parameters, the specific power can be calculated as w, W/cm^3 = $0.4 \cdot p$, Torr. High gas flow velocity also results in voltage and reduced electric field growth (to values of about $E/p = 10 - 20$ V/cmTorr) to intensify ionization and compensate charge losses. Significant increases of charge losses in fast gas flows can be related to turbulence, which accelerates the effective charged particle diffusion to the discharge tube walls. The acceleration of charged particle losses can be estimated by replacing the ambipolar diffusion coefficient D_a by an effective one including a special turbulent term:

$$D_{eff} = D_a + 0.09Ru, \tag{7.36}$$

where u is the gas flow velocity, and R is the discharge tube radius (or half-distance between walls in plane geometry). The elevated values of reduced electric field and electron temperature can be useful to improve the efficiency of several plasma-chemical processes, to improve discharge stability, and to provide higher limits of the stable energy input. To elevate the electric field, small-scale turbulence is often deliberately introduced, especially in high power gas-discharge lasers. The increase of voltage and reduced electric field in fast flow discharges can be directly provided by convective charged particle losses, especially if the discharge length along the gas flow is small (S.V. Pashkin, P.I. Peretyatko, 1978). Also, the effect of E/p increasing in the fast flows can be explained by convective losses of active species responsible for electron detachment from negative ions (E.P. Velikhov, V.S. Golubev, S.V. Pashkin, 1982).

7.3.5 Heat Balance and It's Influence on Current–Voltage Characteristic of Positive Column

The current–voltage characteristic of glow discharges controlled by diffusion is slightly decreasing: a current increase leads to some reduction of voltage. This effect is due to Joule heating. The increase of current leads to some growth of gas temperature T_0, which at constant pressure results in a decrease in gas density n_0. The ionization rate is a function of E/n, which is often only "expressed" as E/p assuming room temperature. For this reason, the decrease of gas density at a fixed ionization rate leads to a decrease in electric field and voltage, which finally explains the decreasing current–voltage characteristics. Analytically this slightly decreasing current–voltage characteristic can be described based on Eq. (7.29) and the similarity condition: $E/n_0 \propto ET_0 \approx const$, which is valid at the approximately fixed ionization rate. In this case, the relation between current density and electric field can be expressed as:

$$\frac{j}{j_0} = \left(\frac{E_0}{E}\right)^{3/2}\left(\frac{E_0}{E} - 1\right). \tag{7.37}$$

Here E_0 is the electric field, which is necessary to sustain low discharge current $j \to 0$, when gas heating is negligible; j_0 is the typical value of current density, which corresponds to Eq. (7.32) and can be also expressed as (w_0 is typical a typical value of specific power, see Eq. (7.31)):

$$j_0 = n_0 c_p T_0 \frac{v_T}{E_0} = \frac{w_0}{E_0}. \tag{7.38}$$

Note that this slightly decreasing current–voltage characteristic can be crossed by the load line Eq. (7.1) not in one, but in two points. This situation is illustrated in Figure 7.14. In this case, only one state, namely the lower one, is stable; the upper one is unstable. If current grows slightly in some perturbation δI with respect to the upper crossing point (Figure 7.14), the plasma voltage, corresponding to the load line, exceeds the voltage which is necessary to sustain the steady-state discharge. As a result of the overvoltage the degree of ionization and total current will grow further until the second (lower) crossing point is reached. If the current fluctuation from the upper crossing point is negative $\delta I < 0$, the plasma voltage becomes less than required to sustain steady-state ionization, and the discharge extinguishes. Similar reasoning proves the stability of the lower crossing point (Figure 7.14). Current fluctuation $\delta I < 0$ results in plasma overvoltage, acceleration of ionization and restoration of the steady-state conditions.

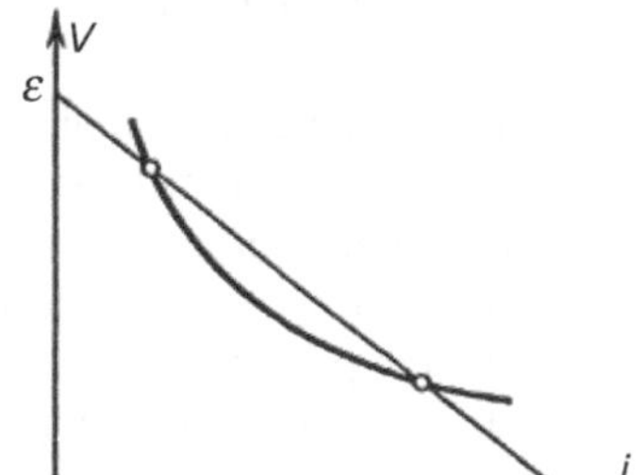

FIGURE 7.14 Illustration of the decreasing current-voltage characteristic.

7.4 GLOW DISCHARGE INSTABILITIES

7.4.1 Contraction of the Positive Column

Contraction is a non-thermal plasma instability related to instantaneous self-compression of a discharge column into one or several bright current filaments. This instability occurs when pressure and current is attempted to be increased while maintaining limits on the specific energy input and specific power of a non-thermal discharge. The positive column contraction can be illustrated by the example of a glow discharge in neon at pressures 75–100 Torr sustained in a 2.8 cm, radius tube with walls at room temperature (Yu. B. Golubovsky, A.K. Zinchenko, Yu. M. Kagan, 1977). The current–voltage characteristic of this discharge in Figure 7.15 demonstrates contraction which takes place when the current exceeds a critical value of about 100 mA. The current–voltage characteristic is decreasing at current values close to the critical one, which demonstrates the strong effect of Joule heating. The transition between the diffusive and contracted modes demonstrates hysteresis. At the critical current, the electric field in the positive column abruptly decreases with a related sharp transition of the discharge regime from the initial strongly non-equilibrium diffusive mode into the contracted mode. A brightly luminous filament appears along the axis of the discharge tube while the rest of the discharge becomes almost dark. Average current density at the transition point is 5.3 mA/cm^2, the corresponding current density on the axis of the discharge tube is 12 mA/cm^2 with an electron density on the axis approximately 10^{11} cm^{-3}. Radial distributions of the relative electron density before and after contraction in the same glow discharge system at a fixed pressure $p = 113$ Torr are shown in Figure 7.16. Corresponding changes of discharge parameters (electron and gas temperatures, electron concentration) during the contraction transition on axis of the discharge tube are presented in Table 7.3. From Figure 7.16, the diameter of a filament formed as a result of contraction

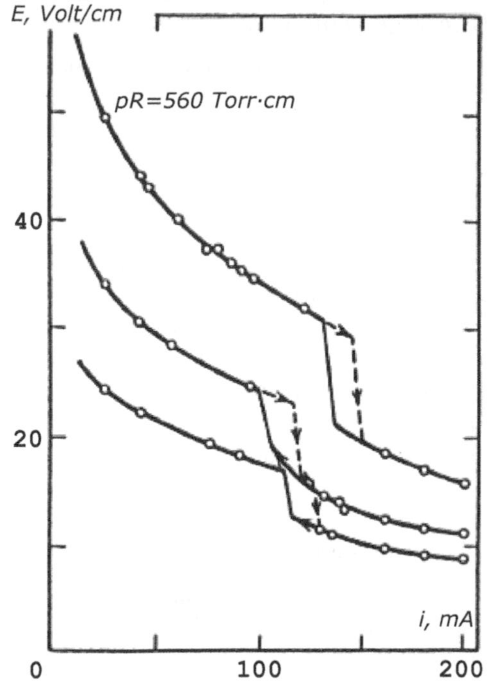

FIGURE 7.15 *V-i* characteristic of discharge in a tube containing neon in the region of transition from diffuse to contracted form. Tube radius R = 2.8 cm; (1) pR = 210 Torr•cm; (2) pR = 316 Torr•cm; (3) pR = 560 Torr•cm. The solid curve in the region of jump was recovered as current was decreased, and dashed curve, as it was raised.

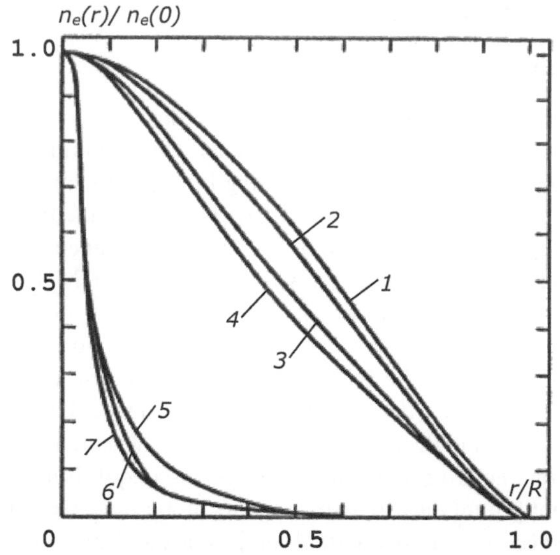

FIGURE 7.16 Profiles of n_e measured under conditions of Figure 7.15. (1) $i/R = 4.8$ mA/cm; (2) 15.4; (3) 26.8; (4) 37.5; (5) 42.9; (6) 57.2; (7) 71.5 mA/cm. The transition state occurred in the region between (4) and (5).

TABLE 7.3
Change of Electron Temperature, Gas Temperatures and Electron Density During the Glow Discharge Contraction

Plasma Parameters	Before contraction (I = 96 mA)	After contraction (I = 120 mA)
Electron Temperature	3.7 eV	3.0 eV
Gas Temperature	930 K	1200 K
Electron density	$1.2 \cdot 10^{11}$ cm^{-3}	$5.4 \cdot 10^{12}$ cm^{-3}

is almost two orders of magnitude smaller than the diameter of a tube initially completely filled with the diffusive glow discharge. Electron concentration increases about 50 times. Gas temperature increases due to localized heat release, and the electron temperature decreases because of reduction of the electric field (Figure 7.15). Thus after contraction, a glow discharge is no longer strongly non-equilibrium (see Table 3.13). This is the reason the contraction phenomenon is sometimes referred to as arcing. However, this term is not completely suitable for contraction because the contracted glow discharge filaments are still not in quasi-equilibrium.

7.4.2 Glow Discharge Conditions Resulting in Contraction

Instability mechanisms leading to glow discharge contraction were considered in Section 6.3. Specifically, the principal causes of the glow discharge contraction were the thermal instability (Subsections 6.3.2, 6.3.3), the stepwise ionization instability (Subsection 6.3.8b) and the electron Maxwellization instability (Subsection 6.3.8c). All these mechanisms provide non-linear growth of ionization with electron density, which is the main physical cause of contraction. These instability mechanisms become significant at electron concentrations exceeding critical values of about 10^{11}cm^{-3}, and specific powers of about 1 W/cm^3. It corresponds to the contraction of glow discharges with heat removal controlled by diffusion. Glow discharges in fast flows have similar mechanisms of contraction. However the heat balance and hence the gas temperature and overheating of a glow discharge in fast flow is controlled by convection, which determines the gas residence time in the discharge zone. Transition to the contracted mode takes place in the fast flow glow discharges not when the specific power (W/cm^3), but rather the specific energy input (discharge energy released per one molecule, eV/mol or J/cm^3) and hence, discharge overheating exceeds the critical value (see Section 6.3.6, Figure 6.12). Maximizing the specific energy input in homogeneous, non-thermal discharges is an especially important engineering problem for powerful gas lasers. Numerically, contraction of glow discharges usually takes place in the case of fast flows, when the specific energy input corresponds to a gas temperature growth of about 100–300 K (N.A. Generalov, V.D. Kosynkin, V.P. Zimakov, Yu.P. Raizer, 1980; E.P. Velikhov, A.S. Kovalev, A.T. Rakhimov, 1987).

7.4.3 Comparison of Transverse and Longitudinal Instabilities, Observation of Striations in Glow Discharges

Contraction of glow discharge is the **transverse instability;** plasma parameters are changed across the electric field. Taking into account that the tangential component of the electric field is always continuous, a sharp decrease of the electric field in the central filament occurs as a consequence of contraction and results in an overall voltage decrease (kind of a short circuit) and loss of non-equilibrium in the discharge as a whole (see Figure 7.17). That is why the transverse instability, contraction, is so harmful to strongly non-equilibrium glow discharges. Striations are related to longitudinal perturbations of plasma parameters, changes of plasma parameters occur along a positive column. In this case, the electric current (and the current density in 1D approximation) remains fixed during a local perturbation of electron density δn_e and temperature δT_e. Local growth of electron density n_e (and electric conductivity σ) induces a local decrease of electric field E and vice versa; while the current density remains the same:

$$j = \sigma E \propto n_e E = const, \quad \frac{\delta n_e}{n_e} = -\frac{\delta E}{E}. \tag{7.39}$$

The mechanism of the electric field reduction in perturbations with elevated electron density and vice versa is illustrated in Figure 7.18. The shift of electron density with respect to ion density is due to the electron drift in the electric field. As is seen, the direction of the polarization field δE is opposite to the direction of the external electric field E if the fluctuation of electron concentration is positive. For some reason, if the electron density is reduced the polarization field is added to the external one. Such instability is unable to destroy the non-equilibrium discharge as a whole. Striations, see Figure 6.8 can move fast (up to 100 m/sec from anode to cathode at pressures 0.1–10 Torr in inert gases), or remain at rest. The stationary striations are usually formed if some strongly fixed perturbation is present in or near the positive column (e.g., an electric probe, or even the cathode layer). The fixed striations build up away from the perturbation towards the anode and gradually vanish. Plasma

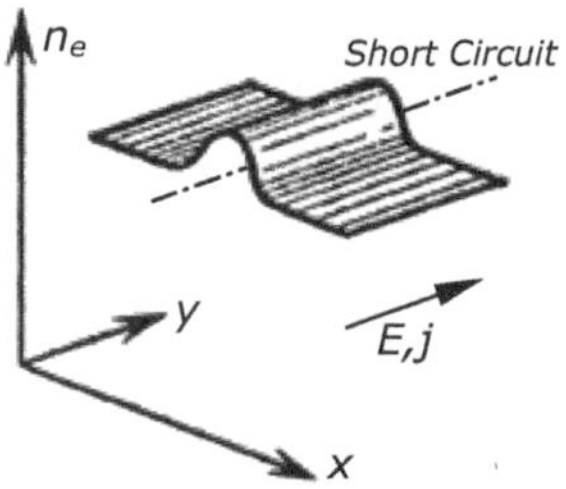

FIGURE 7.17 Electron density perturbation in transverse instabilities.

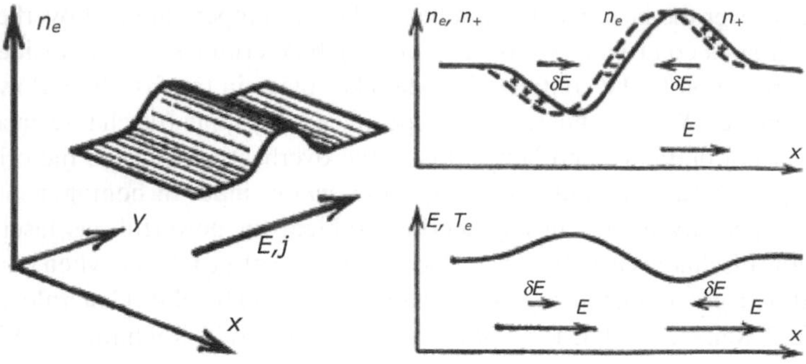

FIGURE 7.18 Illustration of a longitudinal instability.

parameters of a glow discharge with and without striations are nearly the same. It is important to point out that this type of instability, in contrast to the above-discussed contraction, does not significantly affect the non-thermal discharge. Striations exist for a limited range of current, pressure and radius of a discharge tube. The electric current, gas pressure and tube radius also determine the amplitude of luminosity oscillations, the striations wavelength and their propagation velocity. The striations normally behave as linear waves of low amplitude, but large amplitude non-linear striations are possible. Additional details about the observation of this phenomenon can be found in publications of G. Francis (1956), L. Pekarek (1968) and A. Garscadden (1978).

7.4.4 Analysis of Longitudinal Perturbations Resulting in Formation of Striations

From a physical point of view, the striations can be considered as ionization oscillations and waves. The striations can be initiated by the stepwise ionization instability. In this case, an increase of electron density leads to a growth in the concentration of excited species, which accelerates stepwise ionization and results in a further increase of the electron density. When the electron concentration becomes too large, super-elastic collisions deactivate the excited species. Further non-linear growth of the ionization rate is usually due to the Maxwellization instability (see Subsection 6.3.8c). Both of these ionization instability mechanisms, involved in striations, stepwise ionization and Maxwellization, are not directly related to gas overheating. Remember that overheating requires relatively high values of specific power and specific energy input. As a result, striations can be observed at less intensive plasma parameters (electron concentration, electric current, specific power) than those related to the thermal (ionization overheating) instability, which finally is responsible for contraction. Change of the electric field is a strong stabilizing factor for striations, which is an important peculiarity of this kind of instability. Auto-acceleration of ionization and non-linear growth of electron density in striations induces the reduction of the electric field in accordance with Eq. (7.39). The reduction of the electric field leads to a reduction of the effective electron temperature after a short delay $\tau_f = 1/(\nu_{en}\delta)$ related to establishing the corresponding electron energy distribution function (here ν_{en} is the frequency of electron-neutral collisions, δ is the average fraction of electron energy transferred to a neutral particle during the collision, see Eq. (4.88)). The reduction of electron temperature results in an exponential decrease of ionization and in stabilization of the instability. This stabilization effect suppresses striations if the characteristic length of the longitudinal perturbations $2\pi/k_s$ is sufficiently large (or alternately, the wave numbers of ionization wave k_s is small enough). Hence, the striations cannot be observed if there is sufficient time for electrons to establish the new corrected energy distribution function (corresponding to the new electric field) during the electron drift along the perturbation:

$$k_s \cdot v_d \tau_f \approx k_s \lambda / \sqrt{\delta} << 1. \tag{7.40}$$

Here λ is the electron mean free path. In this inequality, it was taken into account that $v_d\tau_f \approx \lambda/\sqrt{\delta}$, which can be derived from the definition of τ_f given above, Eq. (4.102) for drift velocity, and Eq. (4.87) between the electric field and electron temperature. In the opposite case when perturbation wavelengths are short, electron temperature and related ionization effects do not have sufficient time to follow the changing electric field and stabilize striations. Thus striations can be effectively generated at relatively low wavelength. However, the striations wavelengths cannot be very small (shorter than the discharge tube radius R) because of stabilization due to electron losses in longitudinal ambipolar diffusion. The most favorable conditions for generation of striations can be expressed in terms of the striations wave numbers as (A.V. Nedospasov, V.V. Khait, 1979):

$$k_s\lambda/\sqrt{\delta} \approx 5 \div 10, \quad k_s R \propto 1. \tag{7.41}$$

Eliminating the perturbation wavelength k_s from Eq. (7.41), rewrite the conditions favorable for generation of striations as a relation between the electron mean free path and the discharge tube radius as:

$$\frac{\lambda}{R} \approx (5 \div 10) \cdot \sqrt{\delta}. \tag{7.42}$$

The factor δ is small Eq. (4.88) in inert gases for which the conditions Eq. (7.41) were actually proposed. In this case, the condition (7.42) can be satisfied at pressures about 0.1–1 Torr, which actually corresponds to typical pressures necessary for generation of striations. In molecular gases, the factor δ is much larger Eq. (4.88), and this is the reason it is so difficult to observe striations in molecular gases.

7.4.5 Propagation Velocity and Oscillation Frequency of Striations

Striations are usually moving in direction from the anode to cathode. Physical interpretation of this motion is illustrated in Figure 7.19. In the case of relatively short wavelengths typical for actual striations, the gradients of electron density in a perturbation δn_e are quite significant, and charge separation is mostly due to electron diffusion. The electric field of polarization δE, occurring as a result of this electron diffusion, actually determines the oscillation of the total electric field. Maximum of the electric field oscillations δE_{max} corresponds to the points on the wave where the

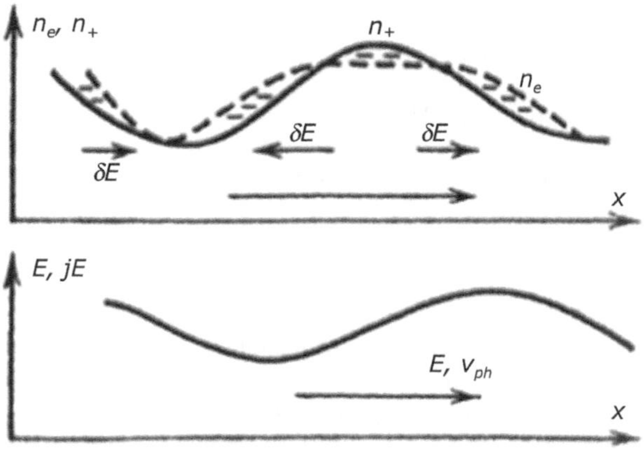

FIGURE 7.19 Propagation of striations.

electron density is not perturbed $\delta n_e = 0$. The maximum of the electric field oscillations δE_{max} is shifted with respect to the maximum of plasma density δn_e oscillations by one-quarter of a wavelength towards the cathode (see Figure 7.19). The ionization rate is fastest at the point of maximum electric field (δE_{max}), resulting in moving the point of maximum plasma density δn_e towards the cathode. The striations propagate from anode to cathode as the ionization waves. To determine the striations velocity as velocity of the ionization wave, assume perturbations of electric field, electron density, and temperature change in a harmonic way: $\delta E,\ \delta n_e,\ \delta T_e \propto \exp[i(\omega t - k_s x]$. Then based on Eq. (4.99), the relation between perturbations of electric field and electron density is:

$$\delta E = ik_s \frac{T_e}{e} \frac{\delta n_e}{n_e}. \tag{7.43}$$

Electrons receive a portion of energy about T_e from the electric field during their drift over the length needed to establish the electron energy distribution function ($v_d \tau_f \approx \lambda / \sqrt{\delta}$, where λ is the electron mean free path, δ is the fraction of electron energy transferred during a collision). Thus taking into account that $eE\left(\lambda/\sqrt{\delta}\right) \approx T_e$, Eq. (7.43) can be rewritten as:

$$\frac{\delta E}{E} \approx ik_s \frac{\lambda}{\sqrt{\delta}} \frac{\delta n_e}{n_e}. \tag{7.44}$$

Relation between perturbations of electric field and electron temperature can be derived from the balance of Joule heating and electron thermal conductivity (with coefficient λ_e) $j \cdot \delta E = k_s^2 \lambda_e\ \delta T_e$:

$$\frac{\delta T_e}{T_e} \approx \frac{1}{\left(k_s \lambda / \sqrt{\delta}\right)^2} \frac{\delta E}{E}. \tag{7.45}$$

Here the Einstein relation, $\lambda_e/\mu_e = T_e/e$, the Ohm's law $j = en_e\mu_e E$, and the relation $eE\left(\lambda/\sqrt{\delta}\right) \approx T_e$ was taken into account. Acceleration of the ionization rate $\partial n_e/\partial t$ in striations related to the electron temperature increase δT_e can be expressed as:

$$\delta\left(\frac{\partial n_e}{\partial t}\right) \approx n_e n_0 \frac{\partial k_i}{\partial T_e} \delta T_e = k_i n_e n_0 \frac{\partial \ln k_i}{\partial \ln T_e} \frac{\delta T_e}{T_e}, \tag{7.46}$$

where n_e, n_0 are concentrations of electrons and neutral species; k_i is the ionization rate coefficient; and $\partial \ln k_i/\partial \ln T_e \approx I/T_e >> 1$ is the logarithmic sensitivity of the ionization rate coefficient to the electron temperature. As was illustrated in Figure 7.19, the electron density grows with the amplitude perturbation δn_e during a quarter of a period (about $1/k_s v_{ph}$, where v_{ph} is the phase velocity of ionization wave, k_s is the wave number). This means: $\delta\left(\frac{\partial n_e}{\partial t}\right) \times \frac{1}{k_s v_{ph}} \approx \delta n_e$. Combining this relation with Eq. (7.46), a formula for the phase velocity of the ionization wave is derived as:

$$v_{ph} \approx \frac{k_i n_0}{k_s} \frac{\partial \ln k_i}{\partial \ln T_e} \left(\frac{\delta T_e}{T_e} / \frac{\delta n_e}{n_e}\right). \tag{7.47}$$

Taking into account Eqs. (7.44) and (7.45) for the relative perturbations of electron temperature and concentration, rewrite Eq. (7.47) into the final expression for the phase velocity of the ionization wave, for example, the phase velocity of striations:

$$v_{ph} = \frac{\omega_s}{k_s} = \frac{1}{k_s^2 \lambda / \sqrt{\delta}} k_i n_0 \frac{\partial \ln k_i}{\partial \ln T_e}. \tag{7.48}$$

Obviously, this is a simple derivation of the striations phase velocity. Detailed derivation (A.V. Nedospasov, Yu.B. Ponomarenko, 1965; L.D. Tsendin, 1970) results in an expression similar to Eq. (7.48), see also J.R. Roth (1967,1969). Thus the striations velocity is proportional to the square of the wavelength; numerically their typical value is about 100 m/sec. The frequency of oscillations of electron density, temperature and other plasma parameters in the striations can then be found from the expression:

$$\omega = \frac{1}{k_s \lambda / \sqrt{\delta}} k_i n_0 \frac{\partial \ln k_i}{\partial \ln T_e}. \tag{7.49}$$

The oscillation frequency in striations is proportional to wavelength ($2\pi/k_s$), and is about the ionization frequency $10^4 \div 10^5$ sec^{-1}. According to the dispersion equation, the absolute value of group velocity of striations (v_{gr}) is equal to that of the phase velocity. Directions of these two velocities are opposite:$v_{ph} = \omega/k = -d\omega/dk = -v_{gr}$. For this reason, some special discharge marks (e.g., bright pulsed perturbations) move to the anode, in opposite direction with respect to striations themselves.

7.4.6 The Steenbeck Minimum Power Principle, Application to Striations

The effect of striations was explained based on physical kinetics and the discharge electrodynamics. In a similar manner to the case of the normal cathode current density (see Section 7.2.5), striations can be illustrated using the Steenbeck minimum power principle. If the discharge current is fixed, the voltage drop related to a wavelength of striations is less than the corresponding voltage of a uniform discharge. This can be explained by the strong exponential dependence of the ionization rate on the electric field value. Because of this strong exponential dependence, an oscillating electric field provides a more intensive ionization rate than an electric field fixed at the average value. Hence, to provide the same ionization level in the discharge with striations requires less voltage and consequently lower power at the same current. The Steenbeck minimum power principle is only an illustration of the phenomenon, which is actually determined and controlled by the earlier discussed ionization instabilities of non-thermal plasma.

7.4.7 Some Approaches to Stabilization of the Glow Discharge Instabilities

Suppression of the non-thermal discharge instabilities is the most important problem in these discharge systems at elevated currents, powers, pressures, and volumes. The most energy-intensive regimes of the glow discharge can be achieved in fast gas flows. Several approaches were developed to suppress contraction. The most applied approach is **segmentation of the cathode**. If a high conductivity plasma filament (contraction) occurs between two points on two large electrodes, current grows and the discharge voltage immediately drops. This can be somewhat suppressed by segmentation of an electrode, usually the cathode. Voltage is applied to each segment independently through an individual external resistance. If a filament occurs at one of the cathode segments, the discharge voltage related to other segments does not drop significantly. Another reason for cathode segmentation is due to the relation between current density in the positive column and the normal cathode current density. As shown in Section 7.3.4, establishing glow discharges in fast flows provides higher specific power at relatively elevated pressures (see Eq. (7.35)). Thus a pressure increase is preferable for different plasma-chemical and laser applications where high levels of power are desirable. Current density in the positive column of a fast flow glow discharges does not depend on pressure (Eq. (7.34)), and at typical discharge parameters can be estimated as 40 mA/cm^2. Alternately, the normal cathode current density is proportional to the square of gas pressure and according to Table 7.2 can be estimated as $0.1 \div 0.3$ mA/cm^2 x p^2(Torr). This yields a normal cathode current density of about $300 \div 500$ mA/cm^2 for typical pressures of 40 Torr, which exceeds the positive column current density by an order of

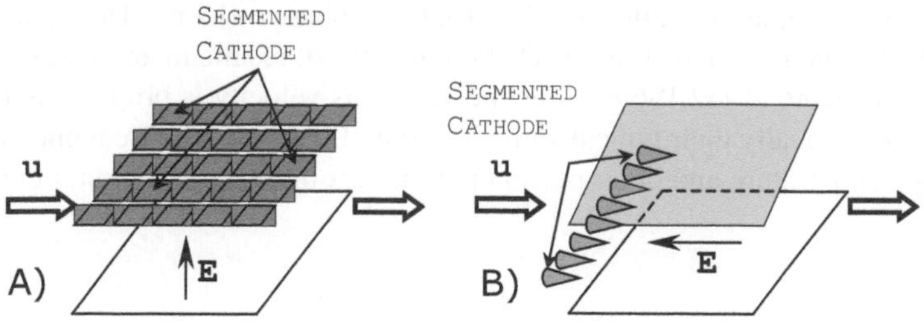

FIGURE 7.20 Cathode segmentation in transverse longitudinal discharge configuration.

magnitude. Glow discharges usually operate in a transition between normal and abnormal regimes when all the cathode area is covered by the electric current Then cathode segmentation is useful to provide the above-mentioned 10-times difference in current density. Some practical ways of the cathode segmentation are illustrated in Figure 7.20. In the case of transverse discharges, the cathode segments are spread over a dielectric plate; in longitudinal discharges, the segments are arranged as a group of cathode rods at the gas inlet to the discharge chamber. Another helpful method of suppressing the glow discharge contraction is related to the **gas flow in the discharge chamber**. Making the velocity field as uniform as possible prevents the inception of instabilities. Usually, high gas velocities also stabilize a discharge because of reduced residence time to values insufficient to contraction. Finally, discharge stabilization can be achieved by utilizing **intensive small-scale turbulence**, which provides damping of incipient perturbations, see G.A. Abilsiitov, E.P. Velikhov, V.S. Golubev e.a, 1984 and A.A. Vedenov, 1982.

7.5 DIFFERENT SPECIFIC GLOW DISCHARGE PLASMA SOURCES

7.5.1 Glow Discharges in Cylindrical Tubes, in Parallel Plates Configuration, in Fast Longitudinal and Transverse Flows, and with Hollow Cathodes

The normal glow discharge in a cylindrical tube is the most widely used. Glow discharges in fast longitudinal and transverse fast gas flows, especially of interest for laser applications were considered in Section 7.3.4. Segmentation of cathodes in these discharge systems, to achieve higher values of specific power without contraction, was discussed in Section 7.4.7. The physical basis and general principles of a hollow cathode discharge were considered in Section 7.2.7. Consider now some other configurations of practical interest.

7.5.2 The Penning Glow Discharges

The special feature of the Penning discharge is its strong magnetic field (up to 0.3 Torr), which permits magnetizing both electrons and ions. This discharge was proposed by F.M. Penning (1936, 1937) and further developed by J.R. Roth (1966). The classical configuration of the Penning discharge is shown in Figure 7.21. To effectively magnetize charge gas particles in this discharge requires low gas pressures, $10^{-6} - 10^{-2}$ Torr. The two cathodes in this scheme are grounded and the cylindrical anode has a voltage about $0.5 - 5\,kV$. Even though the gas pressure in the Penning glow discharge is low, plasma densities in these systems can be relatively high, up to $6 \cdot 10^{12}\,cm^{-3}$. The plasma is so dense because radial electron losses are reduced by the strong magnetic field and axially the electrons are trapped in an electrostatic potential well (see Figure 7.21). Although the configuration of the Penning discharge is markedly different from the traditional glow discharge in cylindrical tube, and it is still a glow discharge because the electrode current is sustained by secondary electron emission from cathode provided by energetic ions. The ions in the Penning discharge are so energetic that they usually cause intensive sputtering from the cathode surface. Ion

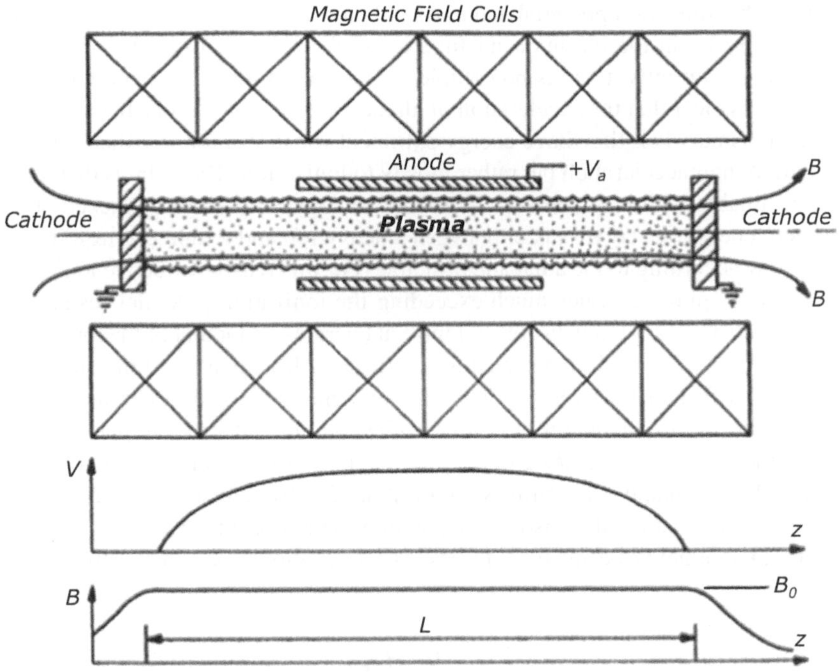

FIGURE 7.21 The classical Penning discharge with uniform magnetic induction and electrostatic trapping of electrons. (From J.R. Roth, Industrial plasma Engineering)

energy in these systems can reach several kilo-electron-volts and greatly exceeds the electron energy. Plasma between the two cathodes is almost equipotential (see Figure 7.21), so it actually plays the role of a second electrode inside of the cylindrical anode. As a result, the electric field inside of the cylindrical anode is close to radial. Both electrons and ions can be magnetized in the Penning discharge, which leads to azimuthal drift of charged particles in the crossed fields: radial electric E_r and axial magnetic B. The tangential velocity v_{EB} of the azimuthal drift can be found from Eq. (4.108), and it is the same for electrons and ions. As a result, the kinetic energy $E_K(e, i)$ of electrons and ions is proportional to their mass $M_{e,i}$:

$$E_K(e,i) = \frac{1}{2} M_{e,i} v_{EB}^2 = \frac{1}{2} M_{e,i} \frac{E_r^2}{B^2}. \tag{7.50}$$

This explains why the ion temperature in the Penning discharge much exceeds the electronic temperature. The electron temperature in a Penning discharge is approximately 3 – 10 eV, while the ion temperature can be an order of magnitude higher (30 – 300 eV), see J.R. Roth (1973, 2000).

7.5.3 Plasma Centrifuge

The effect of azimuthal drift in the crossed electric and magnetic fields in the Penning discharge and the related fast plasma rotation allows the creation of plasma centrifuges. In these systems electrons and ions circulate around the axial magnetic field with very large velocities Eq. (4.108). This "plasma wind" is able to drag neutral particles transferring to them the high energies of the charged particles. It is interesting to note that in the collisional regime, the gas rotation velocity is usually limited by the kinetic energy corresponding to ionization potential:

$$v_{rA} = \sqrt{2eI/M}. \tag{7.51}$$

Here I and M are the ionization potential and mass of heavy neutral particles. The maximum neutral gas rotation velocity v_{rA} in the plasma centrifuge is usually referred to as the **Alfven velocity for plasma centrifuge**. Currently, there is no complete explanation for the phenomenon of the critical Alfven velocity. It is clear that the acceleration of already very energetic ions becomes impossible in weakly ionized plasma when they have energy sufficient for ionization. Further energy transfer to ions does not go to their acceleration but rather mostly to ionization. The point is that ions in contrast to electrons are unable to ionize neutrals when their energy only slightly exceeds the ionization potential. Effectiveness of energy transfer between a heavy ion and a light electron inside of neutral particle is very low according to the adiabatic principle. For this reason, effective ionization by direct ion impact usually requires energies much exceeding the ionization potential (usually it is in keV energy range). Electrons are obviously able to ionize at energies about that of the ionization potential, but their average energies in a plasma centrifuge are usually less than those of ions Eq. (7.50). The explanation of the Alfven critical velocity is related in some way to energy transfer from energetic ions to plasma electrons responsible for ionization. Gas rotation velocities in plasma centrifuges are very high and reach $2 - 3 \cdot 10^6$ cm/sec in the case of light atoms. For comparison, the maximum velocities in similar mechanical centrifuges are of about $5 \cdot 10^4$ cm/sec. The fast gas rotation in a plasma centrifuge can be applied for isotope separation, which occurs as a result of a regular diffusion coefficient, D, in a field of centrifugal forces. The separation time can be estimated as:

$$\tau_S \approx \frac{T_0 R_C^2}{D \Delta M v_\phi^2}, \tag{7.52}$$

where T_0 is the gas temperature, v_ϕ is the maximum gas rotation velocity, R_C is the centrifuge radius, and $\Delta M = |M_1 - M_2|$ is the atomic mass difference of isotopes or components of gas the mixture. Because of the high rotation velocities v_ϕ, the separation time in plasma centrifuges is low. Because of high rotation velocities v_ϕ, the steady-state separation coefficient for binary mixture is significant even for isotopes with relatively small difference in their atomic masses $M_1 - M_2$:

$$R = \frac{(n_1/n_2)_{r=r_1}}{(n_1/n_2)_{r=r_2}} = \exp\left[(M_1 - M_2)\int_{r_1}^{r_2} \frac{v_\phi^2}{T_0}\frac{dr}{r}\right]. \tag{7.53}$$

Here $(n_1/n_2)_r$ is the concentration ratio of the binary mixture components at the radius "r". For plasma centrifuges with $n_e/n_0 \approx 10^{-4} - 10^{-2}$, the parameter $\frac{Mv_\phi^2}{2} / \frac{3}{2}T_0$ is about 3 (V.D. Rusanov, A. Fridman, 1984). The separation coefficient for mixture He-Xe in such centrifuges exceeds 300, for mixture $^{235}U - {}^{238}U$ it is about 1.1. Radial distribution of partial pressures of gas components for deuterium- neon separation in the plasma centrifuge is presented in Figure 7.22. The plasma centrifuges are able to provide chemical process and product separation at once. Water can be dissociated this way $H_2O \rightarrow H_2 + \frac{1}{2}O_2$ with simultaneous separation of hydrogen and oxygen (this separation coefficient is shown in Figure 7.23 in comparison with H_2-Ne separation), see N.P. Poluektov, N.P. Efremov (1998).

7.5.4 Magnetron Discharges

This glow discharge configuration is mostly applied for the sputtering of cathode material and film deposition. A general schematic of the **magnetron discharge with parallel plate electrodes** is shown in Figure 7.24. To provide effective sputtering and film deposition, the mean free path of the sputtered atoms must be large enough and hence, gas pressure should be sufficiently low ($10^{-3} - 3$ Torr). However in this system, because electrons are trapped in the magnetic field by the

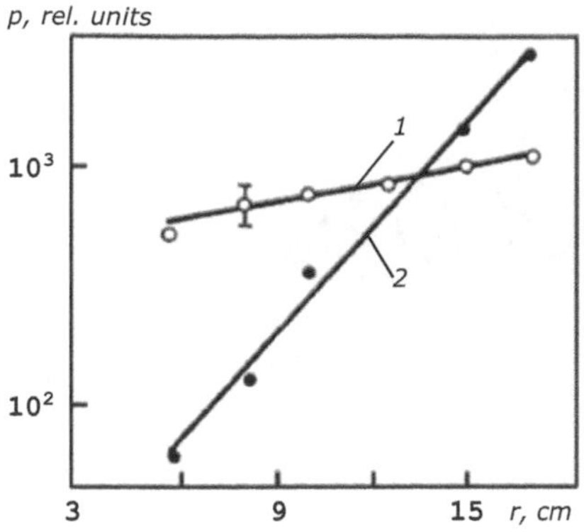

FIGURE 7.22 Radial distribution of partial pressures of D_2(1) and Ne(2) in the mixture.

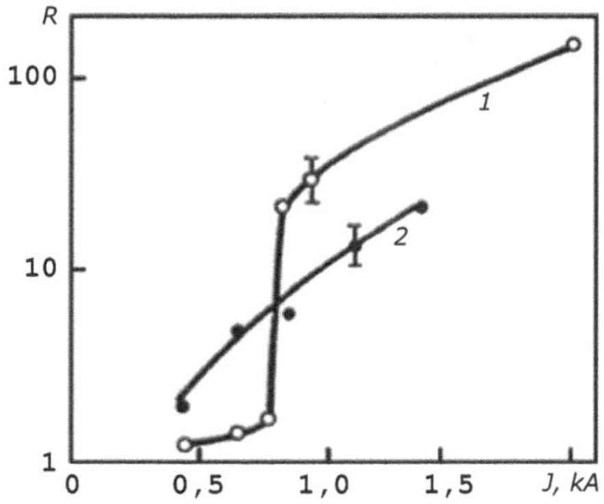

FIGURE 7.23 Separation coefficient for water decomposition products (1) and for H_2-Ne mixture ($p = 0.3$ Torr, $H = 5$ kG, $\tau = 3$ msec).

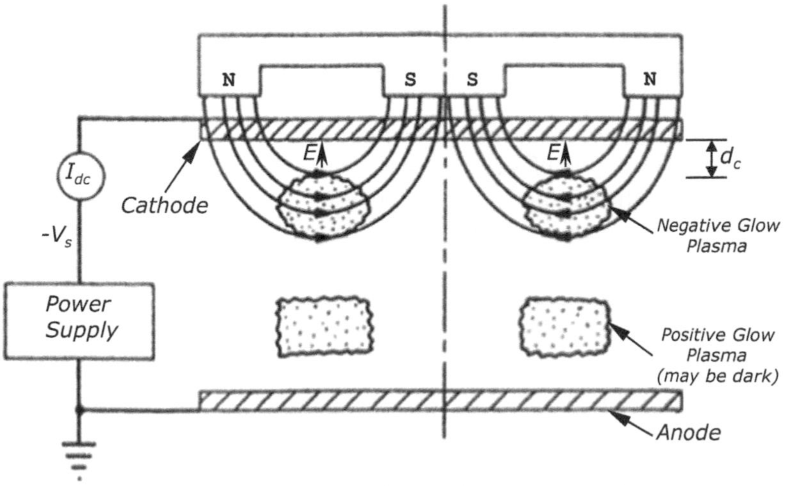

FIGURE 7.24 Glow discharge plasma formation in the parallel plate magnetron. The negative plasma glow is trapped in the magnetic mirror formed by magnetron magnets.

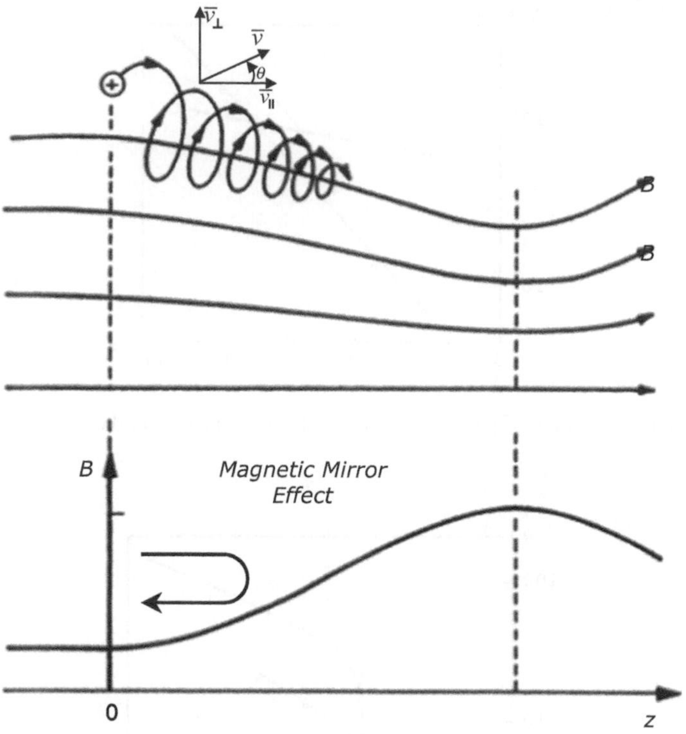

FIGURE 7.25 Magnetic mirror effect.

magnetic mirror, a plasma density on the level of 10^{10} cm^{-3} is achieved. Ions are not supposed to be magnetized in this system to provide sputtering. Typical voltage between electrodes is several hundred volts, magnetic induction approximately 5 – 50 mT. Negative glow electrons are trapped in the magnetron discharge by the **magnetic mirror**. The effect of the magnetic mirror causes "reflection" of electrons from areas with an elevated magnetic field (see Figure 7.25). The magnetic mirror is actually one of the simplest systems for plasma confinement in a magnetic field (see Section 6.2). The magnetic mirror effect is based on the fact, that if spatial gradients of the magnetic field are small, the magnetic moment of a charged particle gyrating around the magnetic lines is an approximate constant of the motion. The particle motion in such magnetic field is said to be the **adiabatic motion.** The electric field between the cathode and the negative glow zone is relatively strong, which provides ions with the energy necessary for effective sputtering of the cathode material. It is important that the drift in crossed electric and magnetic fields cause the plasma electrons in this system to drift around the closed plasma configuration. This makes the plasma of the magnetron discharge quite uniform, which is important for sputtering and film deposition. The magnetron discharges can be arranged in various configurations. For example, if the cathode location prevents effective deposition, it can be relocated. This leads to the so-called **co-planar configuration of the magnetron discharge**, shown in Figure 7.26. Other possible geometries of the magnetron discharge are reviewed in a book of J.R. Roth (2000); from the physical point of view these are almost identical.

7.5.5 Magnetic Mirror Effect in Magnetron Discharges

The magnetic mirror effect plays a key role in magnetron discharges, trapping the negative glow electrons and providing sufficient plasma density for effective sputtering at relatively low pressures of the discharge. Due to its importance, this effect will be discussed in more detail. The magnetic moment μ

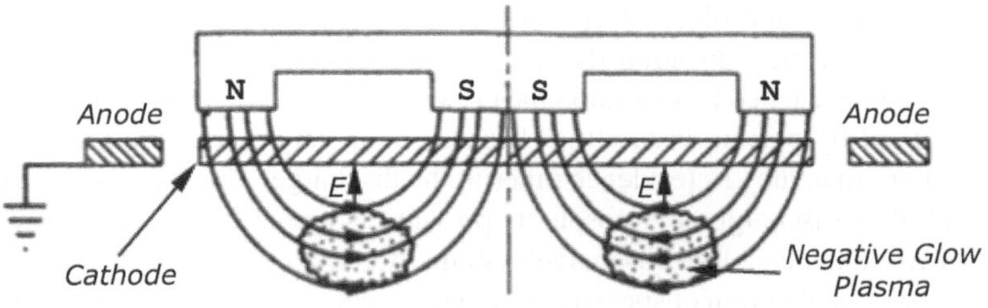

FIGURE 7.26 Co-planar magnetron configuration. Only the negative glow plasma, trapped in the magnetic pole pieces, is normally visible in this configuration. (From J.R. Roth, Industrial Plasma Engineering)

of a charged particle (in this case it is an electron, mass m) gyrating in a magnetic field B is defined as the current I related to this circular motion multiplied by the enclosed area $\pi\rho_L^2$ of the orbit:

$$\mu = I \cdot \pi\rho_L^2 = \frac{mv_\perp^2}{2B} = const. \tag{7.54}$$

In this relation for the magnetic moment μ: ρ_L is the electron Larmor radius (see relations (6.2.25), (6.2.21)), $v_\perp$ is the component of electron thermal velocity perpendicular to the magnetic field (see Figure 7.25 and Eq. (6.51)). The magnetic mirror effect can be understood from Figure 7.25. When a gyrating electron moves adiabatically towards higher electric fields, its normal velocity component $v_\perp$ is growing proportional to the square root of the magnetic field in accordance with Eq. (7.54). The growth of $v_\perp$ is obviously limited by energy conservation. For this reason, the electron drift into the zone with elevated magnetic field also should be limited leading to the reflection of the electron back to the area with lower values of the magnetic field. This generally explains the magnetic mirror effect. If the magnetic lines in the plasma zone are parallel to electrodes (see Figure 7.24), energy transfer from the electric field in this zone can be neglected in the zeroth-approximation. In this model, the kinetic energy of the gyrating electron $mv^2/2$ and total electron velocity v can be considered as constant. Taking into account that $v_\perp = v \sin\theta$ (where θ is the angle between total electron velocity and magnetic field direction, see Figure 7.25), the constant magnetic momentum Eq. (7.54) can be rewritten for these simplified conditions as:

$$\frac{\sin^2\theta}{B(z)} = \frac{2\mu}{mv^2} = const. \tag{7.55}$$

Thus the angle θ between total electron velocity and magnetic field direction grows during the electron penetration into areas with higher values of magnetic field $B(z)$until it reaches the "reflection" point $\theta = \pi/2$. From Eq. (7.55), electrons can be reflected if their initial angle θ_i is sufficiently large:

$$\sin\theta_i > \sqrt{B_{min}/B_{max}}, \tag{7.56}$$

where B_{min}, B_{max} are the maximum and minimum magnetic fields in the mirror (see Figure 7.25). If the initial angle θ_i is not sufficiently large and criterion Eq. (7.56) is not valid, electrons are not reflected by the mirror and are able to escape. Therefore the minimum angle is referred as the **escape cone angle**.

7.5.6 Glow Discharges at Atmospheric Pressure

Glow discharges usually operate at low gas pressures. If the steady-state discharge cooling is controlled by conduction (the diffusive regime), the pressure increase is limited by overheating. The maximum current density and electron concentration proportionally decrease with pressure (see Eqs. (7.32), (7.33)), and the ionization degree decreases as the square of pressure $n_e/n_0 \propto 1/p^2$. Higher pressures can be reached by operating the glow discharges in fast flows. However, these pressures are also limited by plasma instabilities because their very short induction times should be comparable or longer than the gas residence times in the discharge. Thus operating a stable continuous, non-thermal glow discharge at atmospheric pressure is a challenging task. However, this has been accomplished in some special discharge systems, for example, glow discharges in transonic and supersonic flows. Application of special aerodynamic techniques permits sustaining the uniform steady-state glow discharges at atmospheric pressure and specific energy input up to 500 J/g (V.P. Chabotaev, 1972; A.E. Hill, 1971; W.E. Gibbs, R. McLeary, 1971). The glow discharge de-contraction at atmospheric pressure becomes possible due to suppressing the transverse diffusion influence on the temperature and current density distribution.

Another interesting application of the glow discharge at atmospheric pressure is related to using special gas mixtures as a working fluid, and additionally elaborating some special types of electrodes. Such gas mixtures are supposed to be able to provide the necessary level of the ionization rate at relatively low values of reduced electric field E/p and are suitable to sustain glow discharge operation at atmospheric pressure. Gases for such discharges usually include different inert gas mixtures. Comparing Figure 7.12 and 7.13, it is seen that the reduced electric field necessary to sustain a glow discharge in inert gases can be greater than 30 times less than that for molecular gases; this is due to the absence of electron energy losses to vibrational excitation. Mixtures of helium or neon with argon or mercury are very effective for ionization when the Penning effect takes place. In this case, hundreds of volts are sufficient to operate a glow discharge at atmospheric pressure. The use of helium is also helpful to increase heat exchange and cooling of the systems. The normal cathode current density is proportional to the square of pressure and becomes large at elevated pressures. For this reason, special types of electrodes should be applied, for example, fine wires or barrier discharge type electrodes. These are necessary to avoid overheating and correlate current density on the cathode and in the discharge volume. While atmospheric pressure glow discharges are physically similar to such traditional non-thermal atmospheric pressure discharges as corona or dielectric barrier discharge (DBD), their voltage can be much less, see S. Kanazava, M. Kogoma, T. Moriwaki, S. Okazaki (1987), T. Yokayama, M. Kogoma, S. Okazaki e.a. (1990), S. Okazaki, M. Kogoma (1994), Y. Babukutty, R. Prat, K. Endo e.a. (1999). Some special APG configurations will be discussed below considering atmospheric pressure non-thermal discharges.

7.5.7 Some Energy Efficiency Peculiarities of Glow Discharge Application for Plasma-Chemical Processes

Glow discharges are widely used as light sources, as an active medium for gas lasers, for treatment of different surfaces, sputtering, film deposition, etc. Traditional glow discharges controlled by diffusion are of interest only for such chemical applications where electric energy cost-effectiveness is not an issue. Application of glow discharges as the active medium for highly energy effective plasma-chemical processes is limited by the following three major factors:

1. Specific energy input in the glow discharges controlled by diffusion are of about 100 eV/mol and exceeds the optimal value of this discharge parameter $E_v \approx 1\,\text{eV/mol}$. Taking into account that the energy necessary for one act of chemical reaction is usually about 3 eV/mol, it is clear that the maximum possible energy efficiency in these systems is about 3% even at complete 100%-conversion. These high specific energy inputs (about 100 eV/mol) are related to relatively low gas flows passing through the discharge. In the optimal case when a molecule receives energy $E_v \approx 1\,\text{eV/mol} \approx 3\hbar\omega$, it is supposed to leave the discharge zone. From a

viewpoint of energy balance, this means that plasma cooling should be controlled by convection, which takes place only in fast flow glow discharges.
2. The most energy effective processes require high ionization degrees, see Eq. (5.6.5). Taking into account that $n_e/n_0 \propto 1/p^2$ (see Eq. (7.33)), the requirement of a high degree of ionization leads to low gas pressures and hence, to further growth of the specific energy input and decrease of energy efficiency. The reduction of pressure necessary for the increase of the degree of ionization also results in an increase of the reduced electric field and hence, an increase of electron temperature. Such electron temperature increase is also not favorable for effective plasma-chemical processes.
3. Specific power of the conventional glow discharges controlled by diffusion does not depend on pressure and is actually quite low. Numerically, the glow discharge power per unit volume is about $0.3 - 0.7\,W/cm^3$ (see Eq. (7.31)). For this reason, the specific productivity of such plasma-chemical systems is also relatively low. Possible increase of the specific power and specific productivity of related plasma-chemical systems can be achieved by increasing the gas pressure and applying the fast flow glow discharges with convective cooling. Thus energy efficiency of plasma-chemical processes in conventional glow discharges is not very high with respect to other non-thermal discharges.

7.6 PROBLEMS AND CONCEPT QUESTIONS

7.6.1 Space Charges in Cathode and Anode Layers

Space charges are formed near the cathode and the anode. Explain why the space charge of the cathode layer is much greater than in the anode layer.

7.6.2 Radiation of Plasma Layers Immediately Adjacent to Electrodes

Explain why the plasma layer immediately adjacent to the cathode is dark (the Aston dark space), while the plasma layer immediately adjacent to the anode is bright (the anode glow).

7.6.3 The Seeliger's Rule of Spectral Line Emission Sequence in Negative and Cathode Glows

The negative glow first reveals (closer to cathode) spectral lines emitted from higher excited atomic levels and then spectral lines related to lower excited atomic level. This sequence is reversed with respect to the order of spectral line appearance in the cathode glow. Explain this so-called Seeliger's rule.

7.6.4 Glow Discharge in Tubes of Complicated Shapes

Glow discharges can be maintained in tubes of very complicated shapes. This effect is widely used in luminescent lamps. Explain the mechanism of sustaining the glow discharge uniformity in such a case.

7.6.5 Current–Voltage Characteristic of DC Discharges Between Electrodes

Analyze the current–voltage characteristic in Figure 7.5; explain why the dark discharge cannot exist at small currents below the critical one (interval AB in the figure). Estimate this minimal current for a dark discharge.

7.6.6 Space Distribution of Ion and Electron Currents in Dark Discharge

Based on the continuity equations for electron and ions, derive Eq. (7.2) for electron and ion current distributions. Estimate the accuracy of the equation due to neglecting diffusion and recombination of charged particles.

7.6.7 Maximum Current of Dark Discharge

Analyze Eq. (7.11)and show that the maximum current of a dark discharge is proportional to the square of gas pressure. Derive a relation between the relevant similarity parameters: the dark discharge maximum current j_{max}/p^2, ion mobility μ_+p, reduced electric field E/p and inter-electrode distance pd_0.

7.6.8 Comparison of Typical Voltages in Dark and Glow Discharges

Typical voltage of a glow discharge is lower than that in a dark discharge because of the strongly exponential dependence of the Townsend coefficient on the electric field in a gap $\alpha[E(x)]$. However, it cannot be applied, to very large electric fields when the dependence $\alpha(E)$ is close to saturation and not strong. Determine the maximum electric field necessary for the effect, based on Eq. (4.142) for the Townsend coefficient $\alpha(E)$.

7.6.9 The Engel-Steenbeck Model of a Cathode Layer

Analyze the solution of the system of Eqs. (7.9), (7.12) and (7.13), taking the Townsend coefficient dependence on the electric field $\alpha(E)$ in the form Eq. (4.142), and assuming a linear decrease of electric field along the cathode layer Eq. (7.14). Find possible analytical approximations of the solution. Discuss the accuracy of replacing in Eq. (7.12) the linear expression for electric field Eq. (7.14) by a constant electric field.

7.6.10 Normal Cathode Potential Drop, Normal Current Density and Normal Thickness of Cathode Layer

Analyzing Eq. (4.4.7) prove that the normal cathode potential drop does not depend on either pressure or on gas temperature. Determine dependence of normal current density and normal thickness of cathode layer on temperature at constant pressure, and on pressure at constant temperature.

7.6.11 Stability of Normal Current Density

Prove that central quasi-homogeneous region of cathode layer is stable even if current density exceeds the normal one $j > j_n$. Show that the periphery of the cathode spot is unstable in the same time, which finally leads to a decrease of current density until it reaches the normal value $j = j_n$.

7.6.12 The Steenbeck Minimum Power Principle

Analyze the minimum power principle and explain why it is not related to fundamental principles. How you can explain the wide applicability of the minimum power principal to different specific problems in gas discharge physics. Prove this principle for the specific problem of establishing normal current density in the cathode layer of a glow discharge.

7.6.13 Abnormal Glow Discharge

Determine the dependence of the conductivity in the cathode layer on total current for the abnormal regime of glow discharge. Compare this conductivity with that one for the cathode layer of a normal glow discharge. Estimate the total resistance of the cathode layer per unit area of the cathode surface in the abnormal glow discharge; compare it with that for the normal glow discharge.

7.6.14 Glow Discharge with Hollow Cathode

For the Lidsky hollow cathode (Figure 7.10), explain how the electric field becomes able to penetrate into the capillary thin hollow cathode. Take into account that plasma formed inside of the capillary tube is a good conductor. Explain how to start the glow discharge inside of the hollow capillary metal tube, where initially there is almost no electric field.

7.6.15 Anode Layer of a Glow Discharge

Compare the thickness of the anode and cathode layers and their dependences on pressure for a normal glow discharge.

7.6.16 Current–Voltage Characteristics of a Glow Discharge in Recombination Regime

Based on Eq. (7.28), analyze the influence of electron attachment and detachment on the current–voltage characteristic of a glow discharge in recombination regime. Analyze different effects promoting slight growing and slight decreasing of this almost horizontal current–voltage characteristic.

7.6.17 Conductive and Convective Mechanisms of Heat Removal from Positive Column of a Glow Discharge

Based on Eq. (7.30), determine the critical gas velocity for the convective mechanism to dominate the heat removal from the positive column of a glow discharge. Consider gases with high and low values of heat conductivity coefficients, cases of capillary-thin and relatively thick discharge tubes.

7.6.18 Glow Discharge in Fast Gas Flows

The effect of E/p increase in the fast laminar flows can be explained not only by direct convective losses of charged particles but also by convective losses of active species responsible for electron detachment from negative ions. Compare the effectiveness of these two mechanisms in the reduced electric field E/p increase.

7.6.19 Joule Heating Influence on Current–Voltage Characteristic of Glow Discharge

Derive Eq. (7.37), describing the slightly decreasing current–voltage characteristic of a glow discharge. Take into account the heat balance Eq. (7.29) and the similarity condition: $E/n_0 \propto ET_0 \approx const.$

7.6.20 Contraction of Positive Column of Glow Discharge Controlled by Diffusion

Explain why in addition to the non-linear mechanism of ionization growth with electron density, the volumetric character of electron losses is necessary to provide contraction of a glow discharge. Based on relations from Sections 4.5 and 6.3, estimate the critical electron concentrations and specific powers typical for glow discharge transition to the contracted mode.

7.6.21 Contraction of Glow Discharge in Fast Gas Flow

Explain why the transition to the contracted mode takes place in fast flow glow discharges when not specific power (W/cm^3), but rather the specific energy input (eV/mol or J/cm^3) exceeds the critical value (see Section 6.3.6, Figure 6.12). Explain why the fast flow mode of a glow discharge is preferable from the point of view of total power with respect to one controlled by diffusion.

7.6.22 Limitation of the Striation Wavelength

Derive the criterion Eq. (7.40) for suppression of striations by ionization effects that are based on the relation: $v_d \tau_f \approx \lambda / \sqrt{\delta}$. Use the definition of the time interval τ_f, necessary for building up the electron energy distribution function ($\tau_f = 1/(\nu_{en}\delta)$), Eq. (4.102) for drift velocity, and Eq. (4.87) between electric field and electron temperature. Analyze the possible wavelength of striations in inert and molecular gases.

7.6.23 Phase and Group Velocity of Striations

Considering striations as ionization waves following the dispersion Eq. (7.49), prove that their group and phase velocities have the same absolute value, but opposite signs. Give a physical interpretation of different directions of the phase and group velocities for the wave with frequencies proportional to wavelength.

7.6.24 Striations from the View Point of the Steenbeck Principle of Minimum Power

Based on the strong exponential dependence of ionization rate on the value of electric field $k_i(E/p)$, prove that if the discharge current is fixed, the voltage drop related to a wavelength of striations is less than the corresponding voltage of a uniform discharge for the same length.

7.6.25 Cathode Segmentation

Using Eq. (7.34) for current density in the positive column of the fast flow glow discharge, and Table 7.2 for the normal cathode current density, determine the discharge and cathode material conditions when the cathode segmentation is necessary to correlate the current densities in cathode layer and positive column.

7.6.26 The Penning Discharge

Analyze the classical configuration of the Penning discharge and explain why the electric field in this system can be considered as radial. Compare the Penning discharge with the hollow cathode glow discharge, where the electric field configuration is also quite sensitive to the plasma presence. The classical configuration of the hollow cathode glow discharge is difficult to ignite, does the Penning discharge have a similar problem?

7.6.27 The Alfven Velocity in Plasma Centrifuge

Estimate electric and magnetic fields in plasma centrifuge, when the ionic drift velocity in the crossed electric and magnetic fields reaches the critical Alfven velocity. Give your interpretation of the Alfven velocity, taking into account that direct ion impact ionization is an adiabatic process and requires much higher energy than the ionization potential.

7.6.28 Isotope Separation in Plasma Centrifuge

Using Eq. (7.53) and typical parameters of a plasma centrifuge given in Section 7.5.3, estimate the value of the separation coefficient for uranium isotopes $^{235}U - {}^{238}U$. Compare your estimations with the experimental value of the coefficient, see 7.5.3.

7.6.29 Magnetron Discharge

Taking into account typical values of the magnetic field induction in the magnetron discharges given in Section 7.5.4, estimate the interval of pressures when electrons are magnetized but ions are not. Compare your estimation pressures in the magnetron discharges.

7.6.30 Adiabatic Motion of Electrons in Magnetic Mirror

Based on Eqs. (6.51), (6.47) for the Larmor radius and cyclotron frequency, derive Eq. (7.54) for the magnetic momentum of an electron gyrating around a magnetic line in the magnetron discharge. Explain the magnetic moment constancy for the adiabatic motion of electrons in a magnetic mirror, and its relation to magnetic plasma confinement.

7.6.31 The Escape Cone Angle in Magnetic Mirror

If the initial angle θ_i between electron velocity and magnetic field direction is not sufficiently large, electrons are not reflected by the magnetic mirror and are able to escape. Derive Eq. (7.56) for the escape cone angle as a function of the maximum and minimum values of the magnetic field in the magnetic mirror.

7.6.32 Atmospheric Pressure Glow Discharges

Explain the problems to organize glow discharges at atmospheric pressure. What is the difference between traditional non-thermal atmospheric pressure discharges (corona, DB) and the atmospheric pressure glow discharges?

8 Arc Discharges

8.1 PHYSICAL FEATURES, TYPES, PARAMETERS AND CURRENT–VOLTAGE CHARACTERISTICS OF ARC DISCHARGES

8.1.1 General Characteristic Features of Arc Discharges

Arcs are self-sustaining DC discharges, but in contrast to glow, they have a low cathode fall voltage of about 10 eV, which corresponds to the ionization potential. Arc cathodes emit electrons by intensive ***thermionic and field emission***; these are able to provide high cathode current already close to the total discharge current. Because of the high cathode current, there is no need for a high cathode fall voltage to multiply electrons in the cathode layer to provide the necessary discharge current. Arc cathodes receive large amounts of Joule heating from the discharge current and therefore are able to reach very high temperatures in contrast to the glow discharges, which are actually cold. The high temperature leads to evaporation and erosion of electrodes. The main arc discharge zone located between electrode layers is called positive column. The positive column can be either quasi-equilibrium or non-equilibrium depending on gas pressure. Non-equilibrium DC-plasma can be generated not only in glow discharges but also in arcs at low pressures, while quasi-equilibrium DC-plasma can be generated only in electric arcs.

8.1.2 Typical Ranges of Arc Discharge Parameters

The thermal and non-thermal regimes of arc discharges have many peculiarities and quite different parameters. The principal cathode emission mechanism is thermionic in non-thermal regimes and mostly field emission in thermal arcs. Also, the reduced electric field E/p is low in thermal arcs and relatively high in non-thermal arcs. The total voltage in any kind of arc is usually relatively low; in some special forms, it can be only couple of volts. Ranges of plasma parameters typical for the thermal and non-thermal arc discharges are outlined in Table 8.1. Thermal arcs operating at high pressures are much more energy-intensive. These have higher currents and current densities plus higher power per unit length. For this reason, these discharges are sometimes referred to as high-intensity arcs. the division of arc discharges into two groups is simplified. The following classification according to specific peculiarities of the cathode processes, peculiarities of the positive column, and peculiarities of the working fluid can be more informative.

TABLE 8.1
Typical Ranges of the Thermal and Non-Thermal Arc Discharge Plasma Parameters

Discharge Plasma Parameter	Thermal arc Discharge	Non-Thermal Arc Discharge
Gas Pressure	0.1 – 100 atm	10^{-3} – 100 Torr
Arc Current	30*A* – 30 kA	1 – 30 A
Cathode Current Density	10^4 – 10^7 A/cm^2	10^2 – 10^4 A/cm^2
Voltage	10 – 100 V	10 – 100 V
Power per Unit Length	>1 kW/cm	<1 kW/cm
Electron Density	10^{15} – 10^{19} cm^{-3}	10^{14} – 10^{15} cm^{-3}
Gas Temperature	1 – 10 eV	300 – 6,000 K
Electron Temperature	1 – 10 eV	0.2 – 2 eV

8.1.3 Classification of Arc Discharges

Different DC-discharges with low cathode fall voltage are considered as arc discharges. They can be classified by the cathode and positive column mechanisms:

- ***Hot Thermionic Cathode Arcs***. In such arcs, a cathode has temperatures of 3000 K and greater which provides a high current due to thermionic emission. These arcs are stationary to a fixed and quite large cathode spot. Current is distributed over a relatively large cathode area and therefore its density is not high, about $10^2 - 10^4$ A/cm^2. Only special refractory materials like carbon, tungsten, molybdenum, zirconium, tantalum, etc. can withstand such high temperatures and be used in these types of arc discharges. The hot thermionic cathode can be heated to sufficiently high temperatures not only by the arc current but also in a non-self-sustained manner from an external source of heating. Such cathodes are utilized in low-pressure arcs, and in particular in thermionic converters. Cathodes in such arc discharges are usually activated to decrease the temperature of thermionic emission.
- ***Arcs with Hot Cathode Spots.*** If a cathode is made from low-melting-point metals like copper, iron, silver, or mercury, the high temperature necessary for emission cannot be sustained permanently. Electric current flows in this case through hot spots, which appear, move fast and disappear on the cathode surface. Current density in the spots is high about $10^4 - 10^7$ A/cm^2. This leads to localized intensive, short heating and evaporation of the cathode material, while the rest of the cathode actually stays cold. The principal mechanism of electron emission from the spots is thermionic field emission to provide a high current density at temperatures limited by melting point. Note that cathode spots appear not only in the case of the low-melting-point cathode materials but also on refractory metals at low currents and low pressures.
- ***Vacuum Arcs***. This type of low-pressure arc, operating with the cathode spots, is special because the gas-phase working fluid is provided by intensive erosion and evaporation of electrode material. The vacuum arc operates in a dense metal vapor, which is self-sustained in the discharge. This type of arc is of importance in high-current electrical equipment, for example, high current vacuum circuit breakers and switches.
- ***High-Pressure Arc Discharges***. An arc positive column plasma is a quasi-equilibrium one at pressures exceeding 0.1–0.5 atm. Most traditional thermal arcs obviously operate at atmospheric pressure in the open air. The main parameters of such arcs, current, voltage, temperature, and electron density, were shown in Table 8.1. Thermal arcs operating at very high pressures exceeding 10 atm are a special example. In this case, thermal plasma is so dense, that most of the discharge power, 80–90%, is converted into radiation, which is much greater than at atmospheric pressure. Such types of arcs in xenon and in mercury vapors are applied as special sources of radiation.
- ***Low-Pressure Arc Discharges.*** Positive column plasmas of arc discharges at low pressures of about $10^{-3} - 1$ Torr are non-equilibrium and quite similar to that in glow discharges. However it should be pointed out, that ionization degree in non-thermal arcs is higher than in glow discharges because arc currents are much larger. Typical parameters of such non-thermal plasma, current, voltage, temperature, and electron density, were presented in Table 8.1.

8.1.4 Current–Voltage Characteristics of Arc Discharges

General current–voltage characteristic of continuous self-sustained DC discharges for a wide range of currents was discussed in Section 7.1.4 and illustrated in Figure 7.5. Transition from a glow to arc discharge corresponds to the interval FG in the figure. Current density of abnormal glow discharge

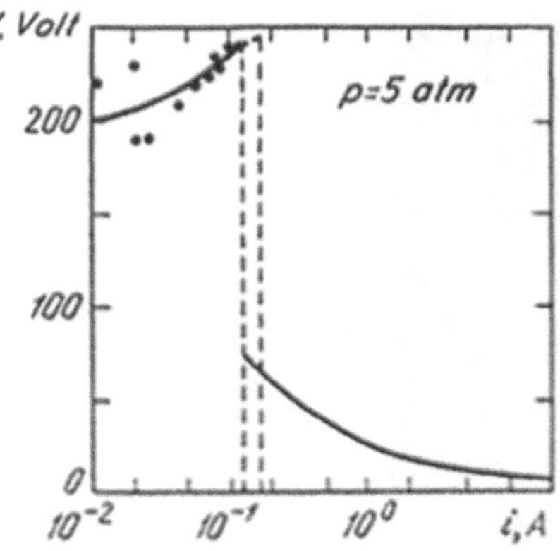

FIGURE 8.1 $V - i$ characteristic of a xenon lamp, $p = 5atm.$, in the region of transition from glow to arc discharge.

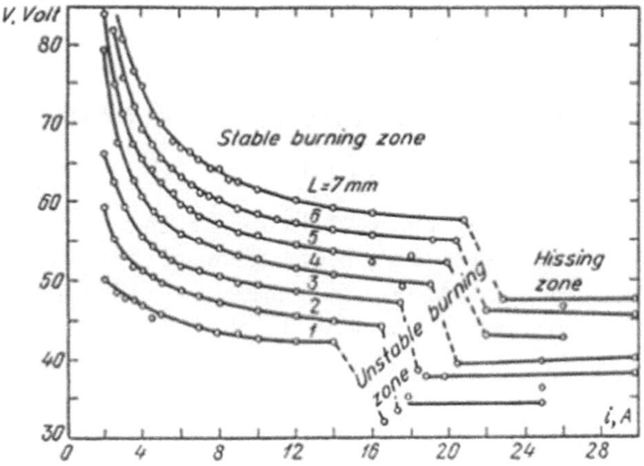

FIGURE 8.2 $V - i$ characteristic of carbon arc in air. Values of L indicate the distance between electrodes.

increases resulting in cathode heating and growth of thermionic emission, which determines the glow-to-arc transition. The glow-to-arc transition is continuous (Figure 7.5) in the case of thermionic cathodes made from refractory metals, and takes place at currents about 10 A. Cathodes made from low-melting-point metals, provide the transition at lower currents 0.1–1 A. This transition is sharp, unstable, and accompanied by the formation of hot cathode spots. An example of the current-density characteristic for such a glow-to-arc transition is presented in Figure 8.1. An example of the current–voltage characteristic (Figure 8.2) for an actual arc discharge, corresponds to the interval GF in Figure 7.5. This example corresponds to the most classical type of arc discharges, **the voltaic arc** (Figure 8.3), which is a carbon arc in atmospheric air (arc discharges were first discovered in this form). Cathode and anode layer voltages in the voltaic arc are both about 10 V, the balance of voltage (see Figure 8.2) corresponds to a positive column. The increase of the discharge length leads to linear growth of voltage, which means that the reduced electric field in the arc is constant at fixed current. When the discharge current grows, the electric field and voltage gradually decrease until a critical point of sharp explosive voltage reduction (see Figure 8.2); this is followed by an almost horizontal current–voltage characteristic. The transition is accompanied by specific hissing noises associated with the formation of hot anode spots with intensive evaporation. Although arc discharges may have

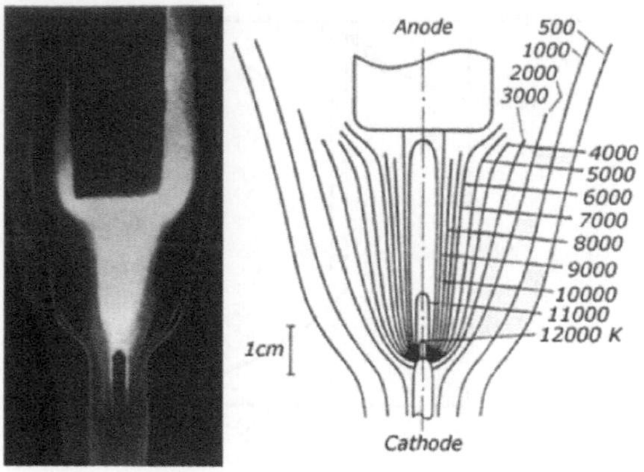

FIGURE 8.3 Carbon arc in air at a current of 200 A: a Toepler photograph and measured temperature field.

very different configurations, in general their structure includes cathode layer, positive column, and anode layer.

8.2 MECHANISMS OF ELECTRON EMISSION FROM CATHODE

8.2.1 Thermionic Emission, the Sommerfeld Formula

Thermionic emission refers to the phenomena of electron emission from a high-temperature metal surface (e.g., hot cathode), which is due to the thermal energy of electrons located in metal. Emitted electrons can remain in the surface vicinity creating there a negative space charge, which prevents further electron emission. However, the electric field in the cathode vicinity is enough to push the negative space charge out of the electrode and reach the saturation current density. This saturation current density is the main characteristic of the cathode thermionic emission. To derive the expression for the saturation current of thermionic emission, consider the electron distribution function in metals, which is the density of electrons in velocity interval $v_x \div v_x + dv_x$, etc. The electron distribution in metals is essentially a quantum-mechanical one, and can be described by the Fermi function (see, for example, Landau, Lefshits, 1999):

$$f\left(v_x, v_y, v_z\right) = \frac{2m^3}{\left(2\pi\hbar\right)^3} \frac{1}{1+\exp\dfrac{\varepsilon-\mu\left(T, n_e\right)}{T}}. \tag{8.1}$$

Here $\mu(T, n_e)$ is the chemical potential, which can be found from the normalization of the Fermi distribution function. This means that the integral of the distribution Eq. (8.1) over all velocities should be equal to the total electron density n_e in metal. Also in Eq. (8.1): m, ε are electron mass and total energy. The most energetic electrons are able to leave the metal if their kinetic energy $mv_x^2/2$ in the direction "x" perpendicular to the metal surface exceeds the absolute value of the potential energy $|\varepsilon_p|$. The electric current density of these energetic electrons leaving the metal surface can be found by integrating the Fermi distribution function Eq. (8.1):

$$j = e\int_{-\infty}^{+\infty} dv_y \int_{-\infty}^{+\infty} dv_z \int_{\sqrt{2|\varepsilon_p|/m}}^{+\infty} v_x f\left(v_x, v_y, v_z\right) dv_x. \tag{8.2}$$

The energy distance from the highest electronic level in metal (the Fermi level) to the continuum is called ***the work function*** W; this actually corresponds to the minimum energy necessary to extract

TABLE 8.2
The Work Functions of Some Cathode Materials

Material	C	Cu	Al	Mo
Work Function	4.7 eV	4.4 eV	4.25 eV	4.3 eV
Material	W	Pt	Ni	W/ThO$_2$
Work Function	4.54 eV	5.32 eV	4.5 eV	2.5 eV

an electron from the metal. In terms of the work function, the integration Eq. (8.2) leads us to **the Sommerfeld formula** describing the saturation current density for thermionic emission:

$$j = \frac{4\pi me}{(2\pi\hbar)^3} T^2 (1-R) \exp\left(-\frac{W}{T}\right). \tag{8.3}$$

Here R is a quantum mechanical coefficient describing the reflection of electrons from the potential barrier related to the metal surface. It is convenient for practical calculations to use the numerical value of the Sommerfeld constant: $\frac{4\pi me}{(2\pi\hbar)^3} = 120 \frac{A}{\text{cm}^2\text{K}^2}$, and to take into account typical values of the reflection coefficient $R = 0 \div 0.8$. Numerical values of the work function W for some cathode materials are given in Table 8.2. A tungsten cathode covered by thorium oxide has a lower work function than pure tungsten cathode, almost half. The thermionic emission current from the oxide cathode exponentially exceeds that for a cathode made from pure refractory metal.

8.2.2 The Schottky Effect of Electric Field on Work Function and Thermionic Emission Current

The thermionic current grows with electric field until the negative space charge near the cathode is eliminated and saturation is achieved. However, this saturation is rather relative. Further increase of the electric field gradually leads to an increase in the saturation current level, which is related to the reduction of work function. The effect of the electric field on the work function is known as the **Schottky effect.** The work function W is the binding energy of an electron to a metal surface. Neglecting the external electric field this is work $W_0 = e^2/(4\pi\varepsilon_0)4a$ against the attractive image force $e^2/(4\pi\varepsilon_0)(2r)^2$ (between the electron and its mirror image inside of metal). Here a is the inter-atomic distance in metal. In the presence of an external electric field E, extracting electrons from the cathode, the total electric field applied to the electron can be expressed as a function of its distance from the metal surface:

$$F(r) = \frac{e^2}{16\pi\varepsilon_0 r^2} - eE. \tag{8.4}$$

The extraction occurs if electron distance from the metal exceeds critical: $r_{cr} = \sqrt{e/16\pi\varepsilon_0 E}$, and attraction to the surface changes to repulsion. The work function can be calculated as the integral:

$$W = \int_a^{r_{cr}} F(r)dr \approx \frac{e^2}{16\pi\varepsilon_0 a} - \frac{1}{\sqrt{4\pi\varepsilon_0}} e^{3/2}\sqrt{E}. \tag{8.5}$$

TABLE 8.3
The Current Densities of Thermionic, Field Electron and Thermionic Field Emissions as a Function of Electric Field *E* (The Following Values of Electrode Temperature, Work Function, Fermi Energy, and Pre-Exponential Factor of The Sommerfeld Relation Are Taken in the Example of Numerical Calculations: $T = 3000$ K, $W = 4$ eV, $\varepsilon_F = 7$ eV, $A_0(1 - R) = 80$ A/cm² K²)

Electric Field, 10^6 V/cm	Schottky Decease of W, V	Thermionic Emission, j, A/cm²	Field Emission j, A/cm²	Thermionic Field Emission, j, A/cm²
0	0	$0.13 \cdot 10^3$	0	0
0.8	1.07	$8.2 \cdot 10^3$	$2 \cdot 10^{-20}$	$1.2 \cdot 10^4$
1.7	1.56	$5.2 \cdot 10^4$	$2.2 \cdot 10^{-4}$	$1.0 \cdot 10^5$
2.3	1.81	$1.4 \cdot 10^5$	1.3	$2.1 \cdot 10^5$
2.8	2.01	$3.0 \cdot 10^5$	130	$8 \cdot 10^5$
3.3	2.18	$6.0 \cdot 10^5$	$4.7 \cdot 10^3$	$2.1 \cdot 10^6$

The Schottky relation Eq. (8.5) returns in the absence of an electric field $E = 0$ to $W = W_0$. The work function decrease in an external electric field, the Schottky relation Eq. (8.5), can be rewritten as:

$$W, eV = W_0 - 3.8 \cdot 10^{-4} \cdot \sqrt{\text{E, V/cm}}. \tag{8.6}$$

From Eq. (8.6), the decrease of work function is small at reasonable values of the electric field. However, the Schottky effect can result, in a major change of the thermionic current, because of its strong exponential dependence on the work function in accordance with the Sommerfeld formula Eq. (8.3). The thermionic emission current density dependence on the electric field is given in Table 8.3 (Yu.P. Raizer, 1997). 4-times change of electric field results in an 800-times increase of the thermionic current density.

8.2.3 The Field Electron Emission in Strong Electric Fields, the Fowler-Nordheim Formula

If the external electric fields are very high (about $1 - 3 \cdot 10^6$ V/cm), they are able not only to decrease the work function but also directly extract electrons from cold metal due to the quantum-mechanical tunneling. A simplified triangular potential barrier for electrons inside metal, taking into account the external electric field E but neglecting the mirror forces is presented in Figure 8.4. Electrons are able to escape from metal across the barrier due to the tunneling, which is called the field emission. The field electron emission current density in the approximation of the triangular barrier (Figure 8.4) was first calculated by Fowler and Nordheim in 1928, and can be calculated by the following Fowler-Nordheim formula:

$$j = \frac{e^2}{4\pi^2\hbar} \frac{1}{(W_0 + \varepsilon_F)} \sqrt{\frac{\varepsilon_F}{W_0}} \exp\left[-\frac{4\sqrt{2m}\,W_0^{3/2}}{3e\hbar E}\right]. \tag{8.7}$$

Here ε_F is the Fermi energy of a metal, W_0 is the work function not perturbed by the external electric field (see (4.2.6)). As seen from the Fowler-Nordheim formula, the field emission current is sensitive to small changes in the electric field and work function, including those related to the Schottky effect Eq. (8.6). Electron tunneling across the potential barrier influenced by the Schottky effect and the corresponding field emission is illustrated in Figure 8.5. The Fowler-Nordheim formula can be then

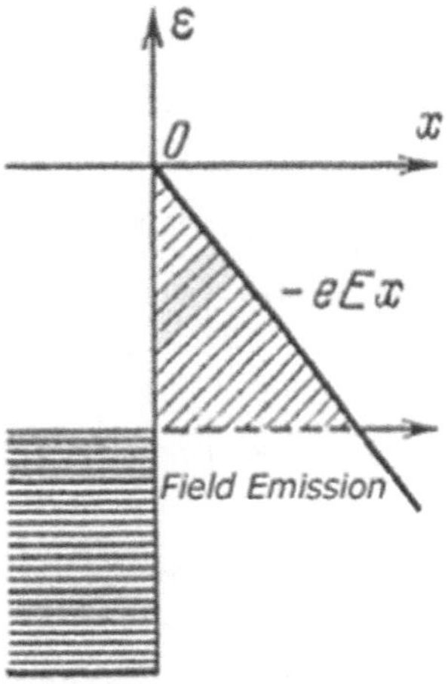

FIGURE 8.4 Electron potential energy on the metal surface when electric field is applied.

corrected by factor $\xi(\Delta W/W_0)$ depending on the relative Schottky decrease of work function Eq. (8.6):

$$j = 6.2 \cdot 10^{-6}\,\mathrm{A/cm^2} \times \frac{1}{\left(W_0,eV + \varepsilon_F,eV\right)} \sqrt{\frac{\varepsilon_F}{W_0}} \exp\left[-\frac{6.85 \cdot 10^7 W_0^{3/2}\left(eV\right) \cdot \xi}{E,V/cm}\right]. \tag{8.8}$$

The correction factor $\xi(\Delta W/W_0)$ is given in Table 8.4 (V.L. Granovsky, 1971). Numerical examples of the field emission calculations based on the corrected Fowler-Nordheim relation Eq. (8.8) are presented in Table 8.3. The field emission current density dependence on the electric field is really strong in this case: a 4-times change of electric field results in an increase of the field emission current density by more than 23 orders of magnitude. According to the corrected Fowler-Nordheim formula Eq. (8.8) and Table 8.3, the field electron emission becomes significant when the electric field exceeds 10^7 V/cm. However, the electron field emission already makes a significant contribution to electric fields about $3 \cdot 10^6$ V/cm because of the field enhancement at the microscopic protrusions on metal surfaces.

8.2.4 Thermionic Field Emission

When cathode temperature and external electric field are high, both thermionic and field emission make significant contribution to the current of electrons escaping the metal. This emission mechanism is usually referred to as the thermionic field emission and plays an important role in cathode spots. To compare the emission mechanisms it is convenient to subdivide electrons escaping the metal surfaces into four groups, illustrated in Figure 8.5. Electrons of the 1-st group have energies below the Fermi-level, so they are able to escape metal only through tunneling, or in other words by the field emission mechanism. Electrons of the fourth group leave metal by the thermionic emission mechanism without any support from the electric field. These two groups of electrons present extremes in the electron emission mechanisms. Electrons of the third group overcome the potential energy barrier because of its reduction in the external electric field. This Schottky effect of the electric field is obviously a pure classical one. The second group of electrons is able to escape the metal only quantum-mechanically by tunneling similar to those from the first group. However in this case the potential barrier of tunneling is not so large in this case because of the relatively high thermal

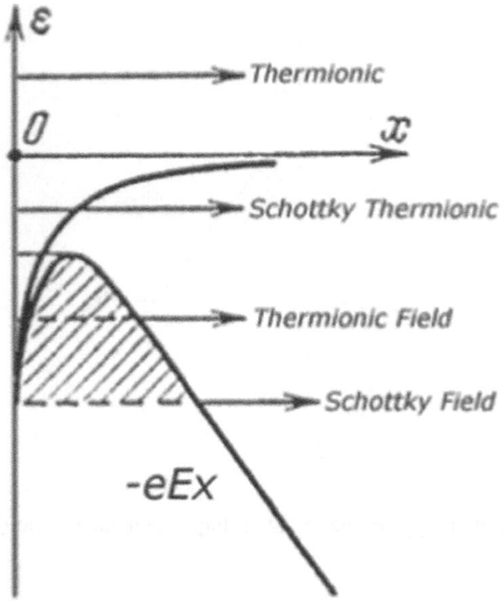

FIGURE 8.5 Illustration of thermionic and field emission; Schottky effect.

energy of the 2nd group electrons. These electrons escape the cathode by the mechanism of the thermionic field emission. Because the thermionic emission is based on the synergetic effects of temperature and electric field, these two key parameters of electron emission need only to be reasonably high to provide the significant emission current. Results of calculations of the thermionic field emission (V.L. Granovsky, 1971) are also presented above in Table 8.3. The thermionic field emission dominates over other mechanisms at T = 3000 K and $E > 8 \cdot 10^6$V/cm. Note that at high temperatures but lower electric fields $E < 5 \cdot 10^6$V/cm, electrons of the 3-rd group usually dominate the emission, which follows, in this case, the Sommerfeld relation Eq. (8.3) with the work function diminished by the Schottky effect Eq. (8.6).

8.2.5 About the Secondary Electron Emission

The thermionic and field emissions play the most important role in the cathode processes of arc discharges. There are other mechanisms of electron emission from solids, related to surface bombardment by different particles. These mechanisms are called secondary electron emissions. The secondary electron emission does not make major contributions in electrode kinetics of arc discharges (with exception of gliding arcs), but are important in other discharges, especially in glow discharges. **The secondary ion-electron emission** is induced by ion impact. The secondary ion-electron emission is the principal distinctive feature of the glow discharges. The direct ionization in collisions of ions with neutral atoms is not effective because of the adiabatic principle. Heavy ions are unable to transfer energy to light electrons to provide ionization. This general statement also can be applied to the direct electron emission from solid surfaces induced by ion impact. As seen from Figure 8.6 (L.N. Dobretsov, M.V. Gomounova, 1966), the secondary electron emission coefficient γ (electron yield per one ion, see Section 4.4.1) effectively starts growing with ion energy only at very high energies exceeding 1 keV, when the Massey parameter becomes large. Although the secondary ion-electron emission coefficient γ is indeed lower (on the level of 0.01–0.1) at lower ion energies, it is not negligible (see Figure 8.6). This value of the coefficient γ stays almost constant at ion energies below the kilovolt-range. This can be explained by **the Penning mechanism of the secondary ion-electron emission**. According to the Penning mechanism, which is also called the potential

TABLE 8.4
Correction Factor $\xi(\Delta W/W_0)$ in the Fowler-Nordheim Formula for Field Emission

Relative Schottky decrease of work function, $\Delta W/W_0$	0	0.2	0.3	0.4	0.5
Fowler-Nordheim correction factor $\xi(\Delta W/W_0)$	1	0.95	0.90	0.85	0.78
Relative Schottky decrease of work function, $\Delta W/W_0$	0.6	0.7	0.8	0.9	1
Fowler-Nordheim correction factor $\xi(\Delta W/W_0)$	0.70	0.60	0.50	0.34	0

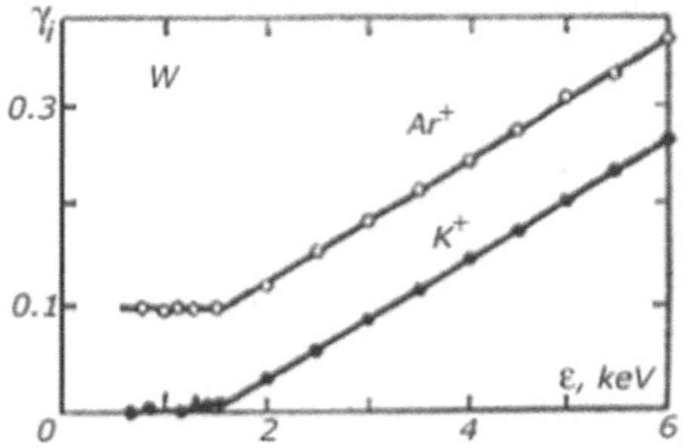

FIGURE 8.6 Secondary electron emission from tungsten as a function of ion energy.

TABLE 8.5
The Secondary Emission Coefficient γ for the Potential Electron Emission Induced by Collisions with Meta-Stable Atoms

Meta-stable Atom	Surface Material	Secondary Emission Coefficient γ
He(2^3S)	Pt	0.24 electron/atom
He(2^1S)	Pt	0.4 electron/atom
Ar*	Cs	0.4 electron/atom

mechanism, ions approaching a surface extract an electron because the ionization potential I exceeds the work function W. The defect of energy $I - W$ is usually enough ($I - W > W$) to provide an escape of one more electron from the surface. Such a process is non-adiabatic (Section 2.2.7), so its probability is not negligible. If surfaces are clean, the secondary ion-electron emission coefficient γ can be estimated as:

$$\gamma \approx 0.016(I - 2W). \tag{8.9}$$

If surfaces are not clean the γ coefficients are lower and grow with ion energy even when they are not high enough for adiabatic emission to take place. Another secondary electron emission mechanism is related to the surface bombardment by excited meta-stable atoms with excitation energy exceeding the surface work function. This so-called **potential electron emission induced by meta-stable atoms** can have quite a high secondary emission coefficient γ. Some of these are presented in Table 8.5. The term "potential" is applied to this emission mechanism to contrast it with the kinetic one, related to not very effective adiabatic ionization or emission induced by the kinetic energy of the heavy particles. Secondary electron emission also can be provided by a photo-effect metallic surface. The **photo-electron emission** is usually characterized by the **quantum yield** $\gamma_{\hbar\omega}$, which is

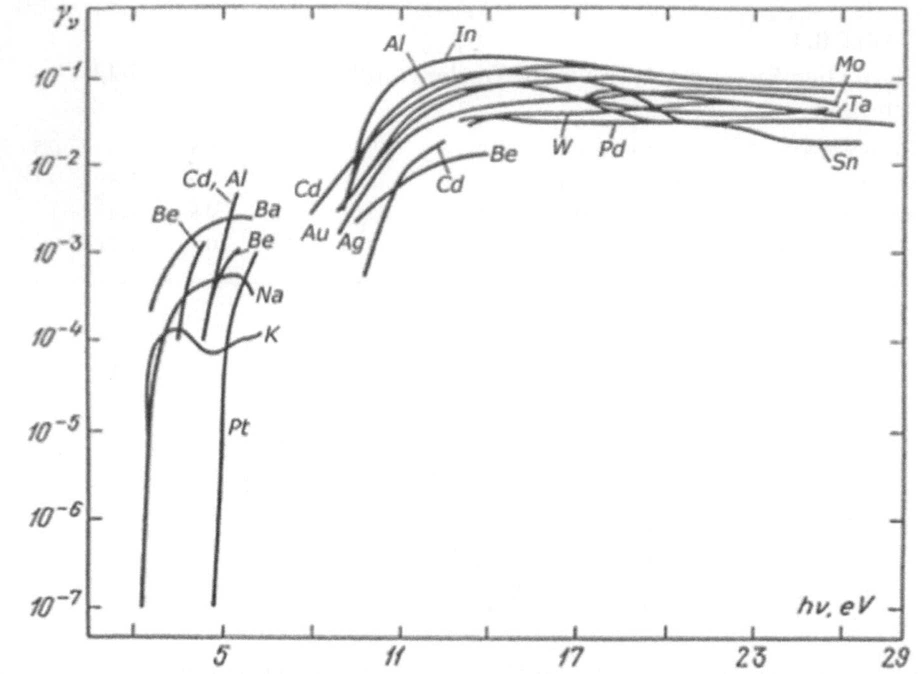

FIGURE 8.7 Photoelectron emission coefficients as a function of photon energy.

similar to the secondary emission coefficient and shows the number of emitted electrons per one quantum $\hbar\omega$ of radiation. The quantum yields as a function of photon energy $\hbar\omega$ for different metal surfaces (E.D. Lozansky, O.B. Firsov, 1975) are shown in Figure 8.7. From this figure, the visual light and low energy UV-radiation give the quantum yield $\gamma_{\hbar\omega} \approx 10^{-3}$, which is sensitive to the quality of the surface. High energy UV-radiation provides emission with the quantum yield in the range 0.01–0.1 and is less sensitive to the surface characteristics. **The secondary electron-electron emission** is electron emission from a solid surface induced by electron impact. This emission mechanism does not play any significant role in DC-discharges. The plasma electrons move to the anode where electron emission is not effective in the presence of an electric field pushing them back to the anode. However, the secondary electron-electron emission can be important, in the case of high-frequency breakdown of discharge gaps at very low pressures, and also in heterogeneous discharges. The secondary electron-electron emission is characterized by the multiplication coefficient γ_e, which shows the number of emitted electrons produced by the initial one. Dependence of the multiplication coefficient γ_e on electron energy for different metals and dielectrics (L.N. Dobretsov, M.V. Gomounova, 1966; G. Fransis, 1960) is shown in Figure 8.8, see A. Modinos, 1984, and S.A. Komolov, 1992.

8.3 CATHODE AND ANODE LAYERS IN ARC DISCHARGES

8.3.1 General Features and Structure of the Cathode Layer

The general function of the cathode layer is to provide the high current necessary for electric arc operation. Electron emission from the cathode in arcs is due to thermionic and field emission mechanisms. These are much more effective with respect to secondary ion-electron emission which dominates in glow discharges. In the case of thermionic emission, ion bombardment provides cathode heating, which then leads to the escape of electrons from the surface. The secondary emission usually gives about $\gamma \approx 0.01$ electrons per one ion, while thermionic emission can generate $\gamma_{eff} = 2 - 9$

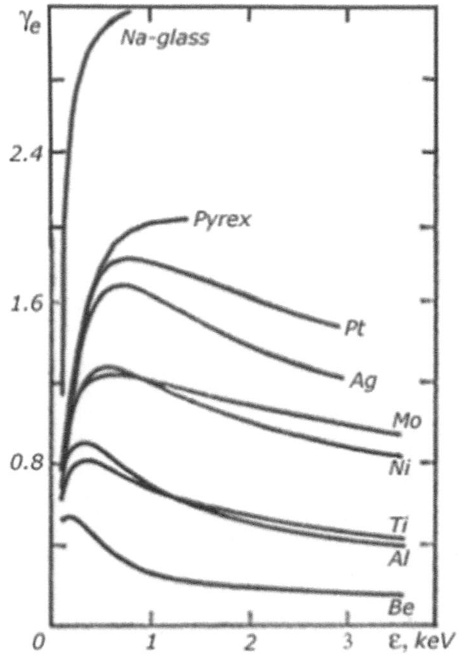

FIGURE 8.8 Secondary electron emission as a function of bombarding electron energy.

electrons per one ion. The fraction of electron current near the cathode in glow discharge is small $\frac{\gamma}{\gamma+1} \approx 0.01$. The fraction of electron current in the cathode layer of an arc is:

$$S = \frac{\gamma_{eff}}{\gamma_{eff}+1} \approx 0.7-0.9. \tag{8.10}$$

Eq. (8.10) shows that thermionic emission from the cathode actually provides most of the electric current in the arc discharge. On the other hand, it is known that the electric current in the positive column of both arc and glow discharges is almost completely provided by electrons because of their very high mobility with respect to ions. In glow discharges most of this current $(1-\frac{\gamma}{\gamma+1} \approx 99\%)$ is generated by electron-impact gas-phase ionization in the cathode layer. For this reason, the cathode layer voltage should be quite high- on the level of hundreds of volts – to provide several generations of electrons (see Section 7.2.3, Table 7.2). In contrast, the electron-impact gas-phase ionization in the cathode layer of arc discharges should provide only a minor fraction of the total discharge current $1 - S \approx 10 - 30\%$. This means that less than one generation of electrons should be born in the arc cathode layer. The necessary cathode drop voltage in this case is relatively low, about or even less than ionization potential. The cathode layer in arc discharges has several self-consistent specific functions. First, a sufficient number density of ions should be generated there to provide the necessary cathode heating in the case of thermionic emission. Gas temperature near the cathode is the same as the cathode surface temperature and couple of times less than the temperature in the positive column (see Figure 8.9). For this reason, thermal ionization mechanism is unable to provide the necessary degree of ionization, which requires the necessity of non-thermal ionization mechanisms (direct electron impact, etc.) and hence, in the necessity of elevated electric fields near the cathode (see Figure 8.9). The elevated electric field in the cathode vicinity stimulates electron emission by a

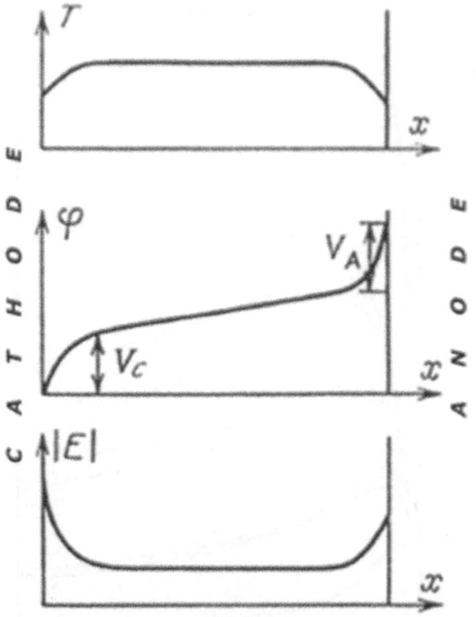

FIGURE 8.9 Distributions of temperature, potential, and electric field from cathode to anode.

decrease of work function, the Schottky effect, as well as by contributions of field emission. On the other hand, intensive ionization in the cathode vicinity leads to a high concentration of ions in the layer and in the formation of a positive space charge, which actually provides the elevated electric field. The general distribution of arc parameters, temperature, voltage, and electric field, along the discharge from cathode to anode is illustrated in Figure 8.9. The structure of the cathode layer by itself is illustrated in Figure 8.10. Large positive space charge, with high electric fields and most of

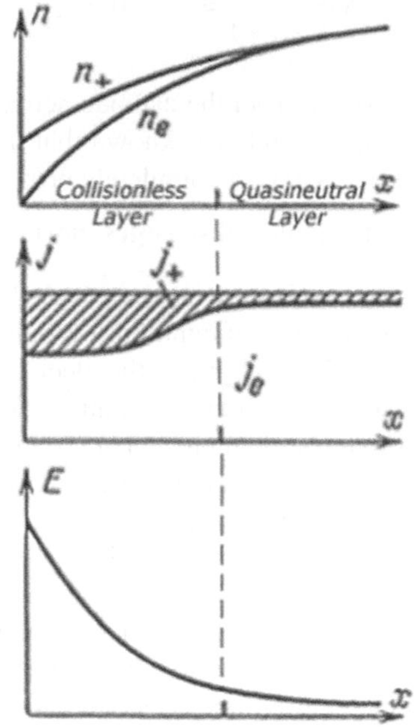

FIGURE 8.10 Distributions of charge density, current, and field in the cathode layer.

the cathode voltage drop, is located in the very narrow layer near the cathode. This layer is actually even shorter than the ions and electrons mean free path, so it is usually referred to as **the collisionless zone of the cathode layer.** Between the narrow collisionless layer and positive column, the longer quasi-**neutral zone of the cathode layer** is located. While the electric field in the quasi-neutral layer is not so high, the ionization is quite intensive there because electrons retain the high energy received in the collisionless layer. Most of ions carrying current and energy to the cathode are generated in this quasi-neutral zone of the cathode layer. From Figure 8.10, the electron and ion components of the total discharge current are constant in the collisionless layer, where there are no sources of charge particles. In the following quasi-neutral zone of cathode layer, fraction of electron current grows from $S \approx 0.7 - 0.9$ (see Eq. (8.10)) to almost unity in the positive column (to ratio of mobilities $\mu_+/(\mu_e + \mu_+)$). Plasma density $n_e \approx n_+$ in the quasi-neutral zone of cathode layer steadily grows in the direction of the positive column because of intensive formation of electrons and ions.

8.3.2 Electric Field in the Cathode Vicinity

Consider the collisionless zone of the cathode layer. Electron j_e and ion j_+ components of the current density j are fixed in this zone:

$$j_e = S \cdot j = n_e e v_e, \quad j_+ = (1-S)\, j = n_+ e v_+. \tag{8.11}$$

Electron and ion velocities v_e, v_+ are functions of voltage V, assuming $V = 0$ at the cathode and $V = V_C$ at the end of the collisionless layer (m and M are masses of electrons and positive ions):

$$v_e = \sqrt{2eV/m}, \quad v_+ = \sqrt{2e(V_C - V)/M}. \tag{8.12}$$

Based on Eqs. (8.11) and (8.12), the Poisson's equation in the collisionless layer is:

$$-\frac{d^2V}{dx^2} = \frac{e}{\varepsilon_0}(n_+ - n_e) = \frac{j}{\varepsilon_0\sqrt{2e}}\left[\frac{(1-S)\sqrt{M}}{\sqrt{V_C - V}} - \frac{S\sqrt{m}}{\sqrt{V}}\right]. \tag{8.13}$$

Taking into account that $\dfrac{d^2V}{dx^2} = \dfrac{1}{2}\dfrac{dE^2}{dV}$, integrate the Poisson's equation (8.13) assuming as a boundary condition that at the positive column side electric field is relatively low: $E \approx 0$ at $V = V_C$. This leads to the relation between the electric field near the cathode, current density, and the cathode voltage drop (which can be represented by V_C):

$$E_C^2 = \frac{4j}{\varepsilon_0\sqrt{2e}}\left[(1-S)\sqrt{M} - S\sqrt{m}\right]\sqrt{V_C}. \tag{8.14}$$

The first term in Eq. (8.14) is related to the ions contribution in the formation of positive space charge and enhancement of the electric field near the cathode. The second term is related to electrons' contribution to compensation of the ionic space charge. From Eq. (8.10), the fraction of electron current $S = 0.7 - 0.9$, and the second term in Eq. (8.14) can be neglected. Eq. (8.14) can be then rewritten as:

$$E_c, \text{V/cm} = 5 \cdot 10^3 \cdot A^{1/4}(1-S)^{1/2}(V_C, \text{V})^{1/4}\left(j, \text{A/cm}^2\right)^{1/2}, \tag{8.15}$$

where A is the atomic mass of ions in a.m.u. For example: for an arc discharge in nitrogen (A = 28) at typical values of current density for hot cathodes $j = 3 \cdot 10^3$ A/cm^2, cathode voltage drop $V_C = 10$ eV and $S = 0.8$- gives according to Eq. (8.15) the electric field near cathode: $E_c = 5.7 \cdot 10^5$ V/cm. This electric field also provides a reduction of the cathode work function Eq. (8.6) of about 0.27 eV, which permits the thermionic emission at 3000 K to triple. Integration of the Poisson's equation (8.13), neglecting the second term related to the effect of electrons on the space charge, permits finding an expression for the length of the collisionless zone of the cathode layer as:

$$\Delta l = 4V_C/3E_C. \tag{8.16}$$

Numerically, for this example, it gives the length of the collisionless layer as $\Delta l \approx 2 \cdot 10^{-5}$ cm.

8.3.3 Cathode Energy Balance and the Electron Current Fraction on Cathode (the S-Factor)

The cathode layer characteristics depend on the S-factor, showing the fraction of the electron current in the total current on the cathode. Simple numerical phenomenological estimation of the S-factor was given by Eq. (8.10). The S-factor depends on a detailed energy balance on the arc discharge cathode. Reasonable qualitative estimations can be done assuming that the energy flux brought to the cathode surface by ions goes completely to provide electron emission from the cathode. Each ion brings to the surface its kinetic energy (which is of the order of cathode voltage drop V_C) and also the energy released during neutralization (which is equal to difference $I - W$ between ionization potential and work function, necessary to provide an electron for the neutralization). This simplified cathode energy flux balance, neglecting conductive and radiation heat transfer components can be expressed as:

$$j_e \cdot W = j_+ \left(V_C + I - W\right). \tag{8.17}$$

The balance (8.17) leads to a relation for the S-factor, the fraction of electron current on cathode:

$$S = \frac{j_e}{j_e + j_+} = \frac{V_C + I - W}{V_C + I}. \tag{8.18}$$

If W = 4 eV, I = 14 eV and V_C = 10eV, according to Eq. (8.17) the fraction of electron current is S = 0.83. More detailed calculations, taking into account additionally conductive and radiative heat transfer do not essentially change the result (M.F. Zhukov, 1982).

8.3.4 About Cathode Erosion

The high energy flux to the cathode results obviously not only in thermionic emission Eq. (8.17), but also in erosion of the electrode material. The erosion is very sensitive to the presence of oxidizers: even 0.1% of oxygen or water vapor makes a significant effect. The erosion effect is usually characterized by the **specific erosion**, which shows the loss of electrode mass per unit charge passed through the arc. The specific erosion of tungsten rod cathodes at moderate and high pressures of inert gases and a current of about hundred amperes is about 10^{-7} *g/C*. The most intensive erosion takes place in the hot cathode spots. At low pressures of about 1 Torr and less, the cathode spots are formed even on refractory materials. For this reason, the rod cathodes of refractory metals are usually used only at high pressures. At low pressures, the hollow cathode configuration can be effectively used from this point of view (see Section 7.2.7). The arc is anchored to the inner surface of the hollow cathode tube, where the gas flow rate is sufficiently high. Specific erosion of such a hollow cathode made from refractory metals can reach a very low level of $10^{-9} - 10^{-10}$ *g/C* (M.F. Zhukov, 1982).

8.3.5 The Cathode Spots

The cathode spots are localized current centers, which can appear on the cathode surface when significant current should be provided in the discharge, but the entire cathode cannot be heated enough to make it. The most typical cause of the cathode spots is the application of metals with relatively low melting point. The cathode spots can be caused as well by relatively low levels of the arc current which are able to provide the necessary electron emission only when concentrated to a small area of the cathode spot. The cathode spots appear also at low gas pressures even in the case of cathodes made of refractory metals. At the low gas pressures (usually less than 1Torr), metal vapor from the cathode provides atoms to generate enough positive ions, which bring their energy to the cathode to sustain electron emission. To provide the required evaporation of cathode material, the electric current should be concentrated in the cathode spots. At low pressures (less than 1 Torr) and currents 1–10 A, the cathode spots appear even on refractory metals (on low-melting-point metals the cathode spots appear at any pressures and currents). Initially, the cathode spots are formed pretty small ($10^{-4} - 10^{-2}$ cm) and move very fast ($10^3 - 10^4$ cm/sec). These primary spots are non-thermal, relevant erosion is not significant and probably is related to micro-explosions due to the localization of current on tiny protrusions on the cathode surface. After a time interval of about 10^{-4} sec, the small primary spots merge into larger spots ($10^{-3} - 10^{-2}$ cm). These matured cathode spots can have temperatures of 3000 K and greater, and provide conditions for intensive thermal erosion mechanisms. They also move much slower 10 – 100 cm/sec. Typical current through an individual spot is 1–300 A. The growth of current leads to splitting of the cathode spots and their multiplication in this way. The minimum current through a single spot is about $I_{min} \approx 0.1 - 1A$. The arc as a whole extinguishes at lower currents. This critical minimum current though an individual cathode spot for non-ferromagnetics can be found as:

$$I_{\min}, A \approx 2.5 \cdot 10^{-4} \cdot T_{boil}(K) \cdot \sqrt{\lambda, \text{W/cmK}}. \tag{8.19}$$

In this relation T_{boil}(*in K*) is the boiling temperature of the cathode material, and λ (*in* W/cm K) is the heat conduction coefficient. The cathode spots are sources of intensive jets of metal vapor. Emission of 10 electrons corresponds approximately to an erosion of one atom. The metal vapor jet velocities can be extremely high: $10^5 - 10^6$ cm/sec, see J.M. Lafferty, 1980 and G. A. Lyubimov, V.I. Rachovsky, 1978. Some of the data are summarized in Table 8.6. The current density in cathode spots can reach extremely high levels of 10^8 A/cm^2. Such large values of electron emission current density can be explained only by thermionic field emission (see Section 8.2.4). Contribution into initial high current densities in a spot can also be due to **the explosive electron emission**, related to localization of strong electric fields and following explosion of micro-protrusions on the cathode surface. This phenomenon, which plays a key role in the pulse-breakdown of vacuum gaps, can influence the early stages of the cathode spot evolution (Yu.D. Korolev, G. A. Mesiatz, 1982). Although extensive experimental and theoretical research has been made (J.M. Lafferty, 1980, G. A. Lyubimov, V.I. Rachovsky, 1978), several problems related to the cathode spot phenomenon are not completely solved. Some of them sound like paradoxes. For example, the current–voltage characteristics of vacuum arcs (where the cathode spots are usually observed) are not decreasing, as is traditional for arcs, but rather increasing. Further, there is no complete explanation even of a mechanism for the cathode spot motion and splitting. The most intriguing cathode spot paradox is related to the

TABLE 8.6
Typical Characteristics of Cathode Spots

Cathode Material	Cu	Hg	Fe	W	Ag	Zn
Minimum Current through a Spot, A	1.6	0.07	1.5	1.6	1.2	0.3
Average Current through a Spot, A	100	1	80	200	80	10
Current Density, A/cm^2	$10^4 \div 10^8$	$10^4 \div 10^6$	10^7	$10^4 \div 10^6$	-	$3 \cdot 10^4$
Cathode Voltage Drop, V	18	9	18	20	14	10
Specific Erosion at 100–200A, g/C	10^{-4}	-	-	10^{-4}	10^{-4}	-
Vapor Jet Velocity, 10^5 cm/sec	1.5	1	0.9	3	0.9	0.4

direction of its motion in external magnetic field. If the external magnetic field is applied along a cathode surface, the cathode spots move in the direction opposite to that corresponding to the magnetic force $\vec{I} \times \vec{H}$. Although quite a few hypothesis and models were proposed on the subject, a consistent explanation of the paradox is still absent.

8.3.6 External Cathode Heating

If a cathode is externally heated, it is not necessary to provide its heating by ion current. In this case, the main function of a cathode layer is the acceleration of thermal electrons to energies sufficient for ionization and sustaining the necessary level of plasma density. Losses of charged particles in such thermal plasma systems cannot be significant, which results in low values of cathode voltage drop. The cathode voltage drop in such systems is often lower than ionization (and even electronic excitation). If the discharge chamber is filled with low pressure (about 1 Torr) inert gas, the voltage to sustain the positive column is also low, about 1 V. Anode voltage drop is also not large. The arc discharges with a low total voltage of about 7–8 V is usually referred to as **the low-voltage arc**. For example, such low values of voltage are sufficient for the non-self-sustained arc discharge in a spherical chamber with a radius of 5 cm in argon at 1–3 Torr pressure and currents 1–2 A. Such kinds of low-pressure gas discharges are used in particular in diodes and thyratrons.

8.3.7 Anode Layer

Similarl to the case of cathode, arc can connect to the anode in two different ways, by the diffuse connection or by the anode spots. The diffuse connection usually occurs on large area anodes; current density in this case is about $100\,A/cm^2$. The anode spots usually appear on relatively small and non-homogeneous anodes, current density in the spots is about $10^4 - 10^5\,A/cm^2$. The number of spots grows with total current and pressure. Sometimes the anode spots are arranged in regular patterns and move along regular trajectories. The anode voltage drop consists of two components. The first is related to a negative space charge near the anode surface, which obviously repels ions. This small voltage drop (of about the ionization potential and less at low currents) stimulates some additional electron generation to compensate for the absence of ion current in the region. The second component of the anode voltage drop is related to the arc discharge geometry. If the anode surface area is smaller than the positive column cross-section or the arc channel is contracted on the anode surface, electric current near the electrode should be provided only by electrons. This requires higher values of electric fields near the electrode and additional anode voltage drop, sometimes exceeding the space charge voltage by a factor of two. Each electron brings to the anode an energy of about 10 eV, which consists of the kinetic energy obtained in the pre-electrode region and the work function. Hence the energy flux in an anode spot at current density $10^4 - 10^5\,A/cm^2$ is about $10^5 - 10^6\,W/cm^2$. Temperatures in anode spots of vacuum metal arcs are about 3000 K and in carbon arcs about 4000 K.

8.4 POSITIVE COLUMN OF ARC DISCHARGES

8.4.1 General Features of Positive Column of High-Pressure Arcs

The Joule heat released per unit length of the positive column in the high-pressure arcs is quite significant, usually $0.2 - 0.5\,kW/cm$. This heat release can be balanced in three different ways, which defines three different manners of arc stabilization. If the Joule heat is balanced by heat transfer to the cooled walls, the arc is referred to as wall stabilized. If the Joule heat is balanced by intensive – often rotating – gas flow, the arc is referred to as flow stabilized. Finally, if heat transfer to electrodes balances Joule heat in the short positive column, the arc is referred to as electrode stabilized. The arc discharge plasma of molecular gases at high pressures ($p \geq 1\,atm$) is always in

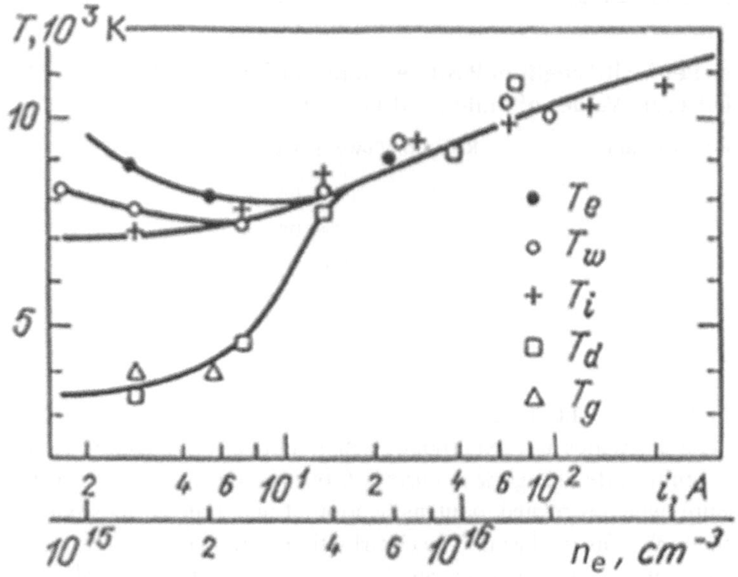

FIGURE 8.11 Temperature separation in the positive arc column in argon, or Ar with an admixture of H_2 at p = 1atm as a function of the current density or electron density. T_e is the electron temperature, T_w corresponds to the population of the upper levels, the ion temperature T_i is related to n_e by the Saha formula, T_g is the gas temperature, and T_d corresponds to the population of the lower levels.

quasi-equilibrium at any values of current. In the case of inert gases, the electron-neutral energy exchange is less effective and requires relatively high values of current and electron density to reach quasi-equilibrium at atmospheric pressure (V.L. Granovsky, 1971, see Figure 8.11). Temperatures and electron concentrations reach their maximum on the axis of the positive column and decrease towards the walls; the electron density obviously decays exponentially faster than temperatures. Also, the temperatures of electrons and neutrals can differ considerably in arc discharges at low pressures ($p \leq 0.1$ Torr) and currents ($I \approx 1A$). Typical current–voltage characteristics of arc discharges are illustrated in Figure 8.12 for different pressures (V.L. Granovsky, 1971).

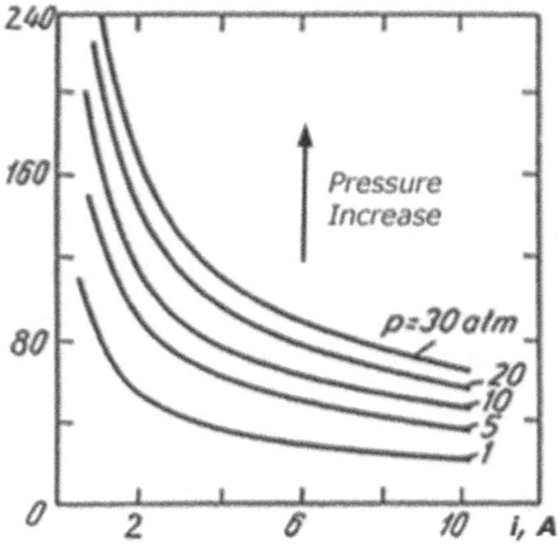

FIGURE 8.12 V – i characteristics of positive arc columns in air at various pressures.

TABLE 8.7
Radiation Power Per Unit Length of Positive Column of Arc Discharges at Different Pressures and Different Values of Joule Heat Per Unit Length $w = EI$, W/cm

Gas	Pressure, atm	Radiation Power per Unit Length, W/cm	$w = EI$, W/cm
Hg	≥1	$0.72 \cdot (w - 10)$	–
Xe	12	$0.88 \cdot (w - 24)$	>35
Kr	12	$0.72 \cdot (w - 42)$	>70
Ar	1	$0.52 \cdot (w - 95)$	>150

The electric field E is constant along the positive column, so it actually describes voltage. The current–voltage characteristics are hyperbolic, which indicates that Joule heat per unit length $w = EI$ does not significantly change with current I. Further the Joule heat per unit length $w = EI$ grows with pressure, which is related to intensification of heat transfer mostly due to the radiation contribution of the high-density plasma. The contribution of radiation increases somewhat proportionally to the square of the plasma density, and hence grows with pressure. The arc radiation losses in atmospheric air are only about 1%, but become quite significant at pressures higher than 10 atm and high arc power. The highest level of radiation can be reached in Hg, Xe and Kr. This effect is practically applied in mercury and xenon lamps. Convenient empirical formulas for calculating plasma radiation in different gases (V.L. Granovsky, 1971) are presented in Table 8.7. From the table, the percentage of arc power conversion into radiation is very high in mercury and xenon even at relatively low values of the Joule heat per unit length.

8.4.2 Thermal Ionization in Arc Discharges, the Elenbaas-Heller Equation

Quasi-equilibrium plasma of arc discharges at high pressures and high currents have a wide range of applications (see, for example, publications of L.S. Polak, A.A. Ovsiannikov, D.I. Slovetsky, F.B. Vursel, 1975, and E. Pfender, 1978). In contrast to non-thermal plasmas, which are sensitive to the details of the discharge kinetics (see chapters 4 and 5), characteristics of the quasi-equilibrium plasma (density of charged particles, electric and thermal conductivity, viscosity etc.) are determined only by its temperature and pressure. This makes the quasi-equilibrium thermal plasma systems much easier to describe. Gas pressure is fixed by experimental conditions, so actually description of the arc positive column requires only description of temperature distribution. Such distribution can be found in particular from the Elenbaas-Heller equation, which is derived considering a long cylindrical steady-state plasma column stabilized by walls in a tube of radius R. In the framework of the Elenbaas–Heller approach, pressure and current are supposed to be not too high and plasma temperature does not exceed 1 eV. Radiation can be neglected in this case and heat transfer across the positive column can be reduced just to heat conduction with the coefficient $\lambda(T)$. According to the Maxwell equation, $curl\, E = 0$. For this reason, the electric field in a long homogeneous arc column is constant across its cross-section. Radial distributions of electric conductivity $\sigma(T)$ (see Section 4.3.2), current density $j = \sigma(T)E$, and Joule heating density $w = jE = \sigma(T)E^2$ are determined only by the radial temperature distribution $T(r)$. Then plasma energy balance can be expressed by the heat conduction equation with the Joule heat source:

$$\frac{1}{r}\frac{d}{dr}\left[r\lambda(T)\frac{dT}{dr}\right] + \sigma(T)E^2 = 0. \tag{8.20}$$

This equation is known as the Elenbaas-Heller equation. Boundary conditions for the Elenbaas-Heller equation are: $dT/dr = 0$ at $r = 0$, and $T = T_w$ at $r = R$ (the wall temperature T_w can be considered as zero taking into account that the plasma temperature is much larger). The electric field E is a parameter of Eq. (8.20), but the experimentally controlled parameter is not electric field but rather current, which is related to electric field and temperature as:

$$I = E\int_0^R \sigma\left[T(r)\right]\cdot 2\pi r dr. \tag{8.21}$$

Taking into account Eq. (8.21), the Elenbaas-Heller equation permits calculating the function $E(I)$, which is the current–voltage characteristic of plasma column. In this case, electric conductivity $\sigma(T)$ and thermal conductivity $\lambda(T)$ remain as two material functions, which determine the current–voltage characteristic. To reduce the number of material functions to the only one, it is convenient to introduce instead of temperature **the heat flux potential** $\Theta(T)$ as an independent parameter:

$$\Theta = \int_0^T \lambda(T)dT, \quad \lambda(T)\frac{dT}{dr} = \frac{d}{dr}\Theta\sqrt{b^2 - 4ac}. \tag{8.22}$$

Using the heat flux potential, the Elenbaas-Heller Eq. (8.20) can be rewritten as:

$$\frac{1}{r}\frac{d}{dr}\left[r\frac{d\Theta}{dr}\right] + \sigma(\Theta)E^2 = 0. \tag{8.23}$$

The Elenbaas-Heller equation includes only the material function $\sigma(\Theta)$. The temperature dependence $\Theta(T)$ determined by Eq. (8.22) is smoother with respect to the temperature dependence of $\lambda(T)$ (see Figure 8.13, V. Penski, 1968), which makes the material function $\mu(\Theta)$ also smooth enough. The Elenbaas-Heller Eq. (8.21) was derived above for electric arcs stabilized by walls. Nevertheless, it has wider applications, including the important case of arcs stabilized by gas flow (see Section 8.4.1), because the temperature distribution in vicinity of the discharge axis is not very sensitive to external conditions.

8.4.3 The Steenbeck "Channel" Model of Positive Column of Arc Discharges

The Elenbaas-Heller Eq. (8.23) cannot be solved analytically because of the complicated non-linearity of the material function $\sigma(\Theta)$, see S.V. Dresvin (1977). A qualitative and quantitative description of the positive column can be achieved, applying the Elenbaas-Heller equation in the analytical, channel model proposed by Steenbeck. The Steenbeck approach, illustrated in Figure 8.14, is based on the very strong exponential dependence of electric conductivity on the plasma temperature related to the Saha equation (see sections 4.1.3, 4.3.2, and Eqs. (4. 16), (4.99)). At relatively low

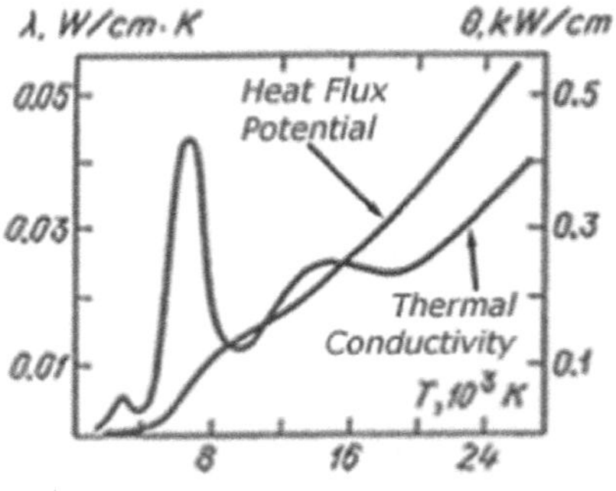

FIGURE 8.13 Thermal conductivity λ and heat flux potential θ in air at 1atm.

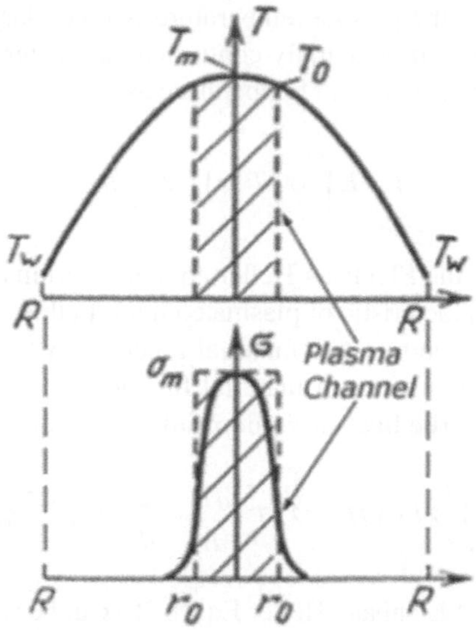

FIGURE 8.14 Illustration of the Steenbeck channel model.

temperatures (less than 3000 K), the quasi-equilibrium plasma conductivity is small. The electric conductivity grows significantly when the temperature exceeds 4000–6000 K. From Figure 8.14, the radical temperature decrease $T(r)$ from the discharge axis to the walls is quite gradual, while the electric conductivity change with radius $\sigma[T(r)]$ is very sharp. This leads to the conclusion that the arc current is located only in a channel of radius r_0, which is the principal physical basis of the Steenbeck channel model. Temperature and electric conductivity can be considered as the constant inside of the arc channel and taken equal to their maximum value on the discharge axis: T_m and $\sigma(T_m)$. In this case, the total electric current of the arc can be expressed as:

$$I = E\sigma(T_m)\cdot \pi r_0^2. \tag{8.24}$$

Outside of the arc channel $r > r_0$ the electric conductivity can be neglected as well as the current and the Joule heat release. The Elenbaas–Heller equation can be integrated outside of the arc channel with boundary conditions: $T = T_m$ at $r = r_0$, and $T = 0$ by the walls at $r = R$. The integration leads us to the relation between the heat flux potential $\Theta_m(T_m)$ in the arc channel, related to the plasma temperature in the positive column, and the discharge power (the Joule heating) per unit arc length $w = EI$:

$$\Theta_m(T_m) = \frac{w}{2\pi}\ln\frac{R}{r_0}. \tag{8.25}$$

The heat flux potential $\Theta_m(T_m)$ in the arc channel and the discharge power per unit arc length $w = EI$ are defined in Eq. (8.25) as:

$$w = \frac{I^2}{\pi r_0^2 \sigma_m(T_m)}, \quad \Theta_m(T_m) = \int_0^{T_m} \lambda(T)\,dT. \tag{8.26}$$

The channel model of an arc includes three discharge parameters, which are assumed determined: plasma temperature T_m, arc channel radius r_0, and electric field E. Electric current I and discharge tube radius R are experimentally controlled parameters. To find the three unknown discharge parameters: T_m, r_0 and E, the Steenbeck channel model has only Eqs. (8.24) and (8.25). Steenbeck suggested the **principle of minimum power** to provide a lacking third equation to complete the system. According to the Steenbeck principle of minimum power, the plasma temperature T_m and the arc channel radius r_0 should minimize the specific discharge power w and electric field $E = w/I$ at fixed values of current I and discharge tube radius R. Application of the minimization requirement $\left(\frac{dw}{dr_0}\right)_{I=const} = 0$ to the functional Eqs. (8.24) and (8.25) gives the necessary third equation of the Steenbeck channel model:

$$\left(\frac{d\sigma}{dT}\right)_{T=T_m} = \frac{4\pi\lambda_m(T_m)\sigma_m(T_m)}{w}. \tag{8.27}$$

Arc discharge modeling, based on the system of Eqs. (8.24), (8.25) and (8.27), gives excellent agreement with experimental data (see, for example, W. Finkelburg, H. Maecker (1956)). However, the validity of the minimum power principle requires a non-equilibrium thermodynamic proof. For the case of the arc discharge, the principle was proved (M.O. Rozovsky, 1972).

8.4.4 The Raizer "Channel" Model of Positive Column

In 1972 Yu.P. Raizer showed the channel model does not necessarily require the minimum power principle to justify and complete the system of Eqs. (8.24), (8.25) and (8.27). The "third" Eq. (8.27) can be derived by analysis of the conduction heat flux J_0 from the arc channel: $w = J_0 \cdot 2\pi r_0$. This flux is provided by the actual temperature difference $\Delta T = T_m - T_0$ across the arc channel (see Figure 8.14) and can be estimated as:

$$J_0 \approx \lambda_m(T_m)\cdot\frac{\Delta T}{r_0} = \lambda_m(T_m)\cdot\frac{T_m - T_0}{r_0}. \tag{8.28}$$

Eq. (8.28) can be replaced by more accurate relation integrating the Elenbaas-Heller Eq. (8.23) inside of arc channel $0 < r < r_0$ and still assuming homogeneity of the Joule heating σE^2:

$$4\pi\,\Delta\Theta = w \approx 4\pi\lambda_m\Delta T, \quad \Delta\Theta = \Theta_m - \Theta_0. \tag{8.29}$$

The key point of Raizer modification of the channel model is the definition of an arc channel as a region where electric conductivity decreases not more than "e" times with respect to the maximum value at the axis of the discharge. This definition permits specifying the arc channel radius r_0 and gives the "third" Eq. (8.27) of the channel model. The electric conductivity of the quasi-equilibrium plasma in the arc channel can be expressed from the Saha Eq. (4.1.15) and Eq. (4.3.8) as the function of temperature:

$$\sigma(T) = C\exp\left(-\frac{I_i}{2T}\right), \tag{8.30}$$

where I_i is the ionization potential of gas in the arc, and C is the conductivity parameter which is approximately constant. To be accurate it should be noted that Eq. (8.30) is valid at moderate currents and temperatures when the degree of ionization is not too high and electron-atomic collisions dominate the collisional frequency in Eq. (4.99). However, Eq. (8.30) can be applied even at high temperatures and currents, by only replacing the ionization potential by the effective potential. For example, electric conductivity in air, nitrogen, and argon at atmospheric pressure and temperatures $T = 8{,}000 \div 14{,}000\,\text{K}$ can be expressed by the same numerical formula:

$$\sigma(T), \text{Ohm}^{-1}\text{cm}^{-1} = 83 \cdot \exp\left(-\frac{36{,}000}{T, K}\right). \tag{8.31}$$

which corresponds to the effective ionization potential $I_{eff} \approx 6.2\,\text{eV}$. Taking the electric conductivity of the arc discharge channel in the form of Eq. (8.32) and assuming $I/2T >> 1$, the "e" times decrease of conductivity corresponds to the following small temperature decrease:

$$\Delta T = T_m - T_0 = \frac{2T_m^2}{I_i}. \tag{8.32}$$

Combination of Eqs. (8.29) and (8.32) gives the "third" equation of the arc channel model:

$$w = 8\pi\lambda_m(T_m)\frac{T_m^2}{I_i}. \tag{8.33}$$

The third equation of the Raizer channel model Eq. (8.33) completely coincides with that of the Steenbeck model Eq. (8.27), which was based on the principle of minimum power. In this case, the electric conductivity in the Steenbeck Eq. (8.27) obviously should be taken in form Eq. (8.30).

8.4.5 Plasma Temperature, Specific Power, and Electric Field in Positive Column According to the Channel Model

Eq. (8.33) determines temperature in the arc channel as a function of power w per unit length (with ionization potential I_i and thermal conductivity coefficient λ_m as parameters):

$$T_m = \sqrt{w \cdot \frac{I_i}{8\pi\lambda_m}}. \tag{8.34}$$

Plasma temperature does not depend directly on the discharge tube radius and mechanisms of the discharge cooling outside of the arc channel. The plasma temperature depends only on the specific power w, which in turn depends on the intensity of the arc cooling. Dependence of plasma temperature T_m on the specific power is not very strong, less than $\sqrt{w}$ because the heat conductivity $\lambda(T)$ grows with T (see Figure 8.13). The principal controlled parameter of an arc discharge is the electric current, which can vary easily by changing external resistance (see Eq. (7.1)). Express the parameters of the arc channel as functions of current. Assuming for simplicity: $\lambda = const$, $\Theta = \lambda T$ in Eqs. (8.25), (8.30) and (8.33), the conductivity in the arc channel is almost proportional to electric current:

$$\sigma_m = I \cdot \sqrt{\frac{I_i C}{8\pi^2 R^2 \lambda_m T_m^2}}. \tag{8.35}$$

Obviously, plasma temperature in the arc grows with current I, but only as logarithm:

$$T_m = \frac{I_i}{\ln\left(8\pi^2\lambda_m C T_m^2 / I_i\right) - 2\ln(I/R)}. \tag{8.36}$$

The fact of near-constant thermal arc temperature reflects the strong exponential dependence of the degree of ionization on the temperature. The weak logarithmic growth of temperature in the channel with electric current leads (according to Eq. (8.33)) to the similar weak logarithmic dependence on the electric current of the arc discharge power w per unit length:

$$w \approx \frac{const}{\left(const - \ln I\right)^2}. \tag{8.37}$$

Experimental data demonstrating the relatively weak dependence of the arc discharge power w per unit length on arc conditions are presented in Figure 8.15. This important effect will be also discussed later in this chapter regarding the evolution of the gliding arcs. The logarithmic constancy of the arc power $w = EI$ per unit length results in the approximate to the hyperbolic decrease of the electric field with current I:

$$E = \frac{8\pi\lambda_m T_m^2}{I_i} \cdot \frac{1}{I} \approx \frac{const}{I \cdot \left(const - \ln I\right)^2}. \tag{8.38}$$

This relation explains the hyperbolic decrease of current–voltage characteristics typical for thermal arc discharges. The radius of the arc discharge channel may be found in the framework of this model as:

$$r_0 = R\sqrt{\frac{\sigma_m}{C}} = R\sqrt{\frac{I}{R}}\sqrt[4]{\frac{I_i}{8\pi^2\lambda_m T_m^2 C}}. \tag{8.39}$$

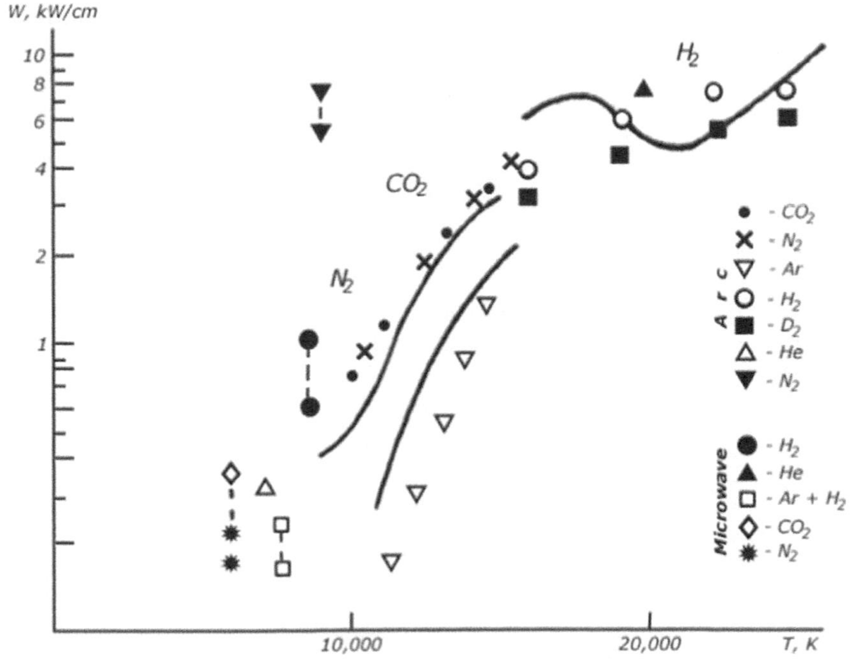

FIGURE 8.15 Dissipated power per unit length versus maximum temperature for different types of discharges: solid lines – numerical calculation.

From Eq. (8.39), the arc channel radius grows as the square root of the discharge current I. This can be interpreted as $I \propto r_0^2$, which means that the increase of current mostly leads to the growth of the arc channel cross-section, while the current density is logarithmically fixed in the same manner as plasma temperature. All the relations include current in combination with the discharge tube radius R (the similarity parameter I/R). This is due to the initial assumption of "wall stabilization arc " in the Elenbaas-Heller equation. To generalize the results to the case of gas flow stabilized arcs the radius R should be replaced by an effective value describing the actual mechanism of the arc cooling. Intensive cooling in fast gas flows corresponds to small values of the effective radius, which according to Eqs. (8.36) and (8.33) leads to an increase in plasma temperature.

8.4.6 Possible Difference between Electron and Gas Temperatures in Thermal Discharges

Plasma in the channel model was taken in quasi-equilibrium, which corresponds to high pressures and high currents (see Section 8.4.1). However, at lower pressures and lower currents (when the electric field E is higher, see Eq. (8.38)), the electron temperature T_e can exceed gas temperature T (see Figure 8.11). This effect can be described by the simplified balance equation for electron temperature (see Section 4.2.7):

$$\frac{3}{2}\frac{dT_e}{dt} = \left[\frac{e^2E^2}{mv_{en}^2} - \delta\left(T_e - T\right)\right]\cdot v_{en}. \tag{8.40}$$

where ν_{en} is the frequency of electron-neutral collisions, and the factor δ characterizes the fraction of electron energy transferred to neutrals during collisions ($\delta = 2m/M$ in monatomic gases, and higher in molecular gases, see (4.2.39)). The balance Eq. (8.40) shows that electrons receive energy from the electric field and transfer it to neutrals if $T_e > T$. If the temperature difference $T_e - T$ is not large, the excited neutrals are able to transfer energy back to electrons in super-elastic collisions. Eq. (8.40) gives quasi-equilibrium $T_e = T$ in the steady state and the absence of electric field $E = 0$. In steady-state conditions with electric field E, the balance Eq. (8.40) gives the difference between temperatures:

$$\frac{T_e - T}{T} = \frac{2e^2E^2}{3\delta T_e mv_{en}^2}. \tag{8.41}$$

Assume monatomic gas ($\delta = 2m/M$), and replace the electron-neutral frequency ν_{en} by the electron mean free path $\lambda = v_e/\nu_{en}$ (v is the average electron velocity). Then rewrite Eq. (8.41) for the difference between electron and gas temperatures in a numerical way:

$$\frac{T_e - T}{T_e} \approx 200A\cdot\left[\frac{E(\text{V/cm})\cdot\lambda(\text{cm})}{T_e(\text{eV})}\right]^2. \tag{8.42}$$

Here A is the atomic mass of the monatomic gas in the discharge. If some amount of molecules is present in the discharge, the coefficient in Eq. (8.42) is less than 200 and the degree of non-equilibrium is lower. Equation (8.42) explains the essential temperature difference at low pressures when the mean free path λ is high, and at low currents when the electric field is elevated (see Figure 8.11). The electric fields necessary to sustain thermal arc are always lower than those to sustain non-equilibrium discharges. Electric fields in thermal discharges are responsible to provide the necessary Joule heating, while electric fields in non-thermal discharges are "directly" responsible for ionization by electron impact.

8.4.7 Dynamic Effects in Electric Arcs

High arc currents can induce relatively high magnetic fields and hence, discharge compression effects in the magnetic field (see pinch effect Section 6.2.4). The body forces acting on axisymmetric arcs are illustrated in Figure 8.16. Assuming that the current density j in the arc channel is constant, the azimuthal magnetic field B_θ inside of arc can be expressed as:

$$B_\theta(r) = \frac{1}{2}\mu_0 j r, \quad r \leq r_0. \tag{8.43}$$

This magnetic field results in a radial body force (per unit volume) directed inward and tending to pinch the arc (see Figure 8.16). The magnetic force can be found as a function of distance "r" from the axis:

$$\vec{F} = \vec{j} \times \vec{B}, \quad F_r(r) = -\frac{1}{2}\mu_0 j^2 r. \tag{8.44}$$

A cylindrical arc channel, where plasma pressure p is balanced by the inward radial magnetic body force Eq. (8.44), is called **the Bennet pinch**. This balance can be expressed as:

$$\nabla p = \vec{j} \times \vec{B}, \quad \frac{dp}{dr} = -jB_\theta. \tag{8.45}$$

Together with formula Eq. (8.43) for the azimuthal magnetic field B_θ, this leads to the relation for pressure distribution inside of the Bennet pinch of radius r_0:

$$p(r) = \frac{1}{4}\mu_0 j^2 \left(r_0^2 - r^2\right). \tag{8.46}$$

The corresponding maximum pressure (on axis of the arc channel) can be rewritten in terms of total current $I = j \cdot \pi r_0^2$ as:

$$p_a = \frac{1}{4}\mu_0 j^2 r_0^2 = \frac{\mu_0 I^2}{4\pi^2 r_0^2}. \tag{8.47}$$

According to Eq. (8.47), it is necessary to have a very high arc current, almost 10,000 A to provide an axial pressure of 1 atm in a channel with radius $r_0 = 0.5$ cm. Currents in most arc discharges are

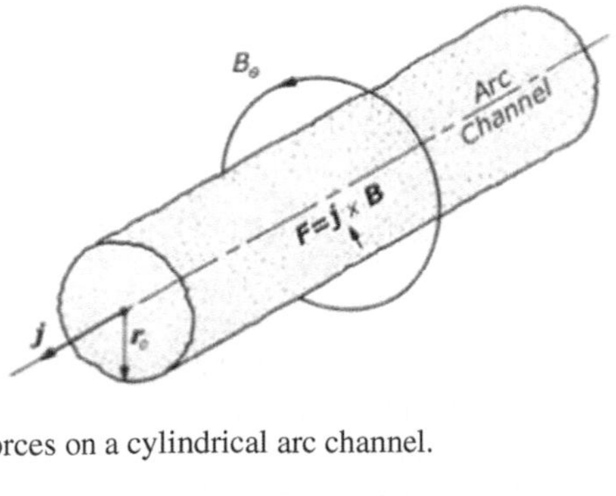

FIGURE 8.16 Radial body forces on a cylindrical arc channel.

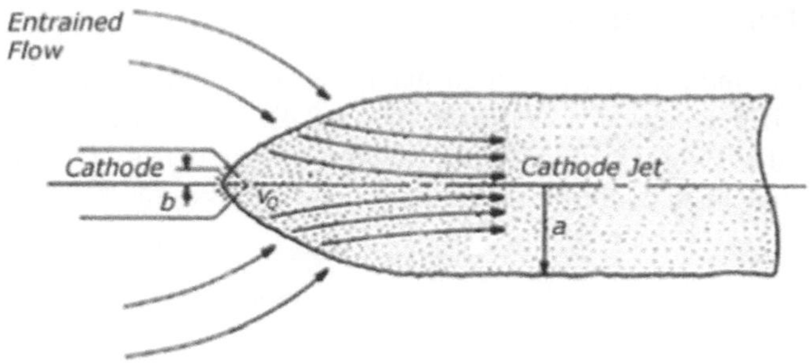

FIGURE 8.17 Illustration of the electrode jet formation.

much less than those of the Bennet pinch (defined by Eq. (8.47)), so the arc is stabilized primarily not by magnetic body forces, but by other factors.

8.4.8 The Bennet Pinch Effect and Electrode Jet Formation

Although the Bennet pinch effect pressure is usually small with respect to total pressure, it can initiate electrode jets, intensive gas streams, which flow away from electrodes. The physical nature of the electrode jet formation is illustrated in Figure 8.17. Additional gas pressure related to the Bennet pinch effect Eq. (8.47) is inversely proportional to the square of the arc radius. Also, the radius of the arc channel attachment to electrode ($r_0 = b$) is less than that one corresponding to the positive column ($r_0 = a$, see Section 8.3 and Figure 8.17). This results in the development of an axial pressure gradient, which drives neutral gas along the arc axis away from electrodes:

$$\Delta p = p_b - p_a = \frac{\mu_0 I^2}{4\pi^2}\left(\frac{1}{b^2} - \frac{1}{a^2}\right). \tag{8.48}$$

Assuming that $b << a$, the jet dynamic pressure and the jet velocity can be found from:

$$\Delta p \approx \frac{\mu_0 I^2}{4\pi^2 b^2} = \frac{1}{2}\rho v_{jet}^2, \tag{8.49}$$

where ρ is the plasma density. The electrode jet velocity then can be expressed as the function of arc current and radius of arc attachment to the electrode:

$$v_{jet} = \frac{I}{\pi b}\sqrt{\frac{\mu_0}{2\rho}}. \tag{8.50}$$

The electrode jet velocities, Eq. (8.50), are $v_{jet} = 3 - 300$ m/sec, see L. Tonks, I. Langmuir (1929), W.H. Bennet (1934), B. Gross, B. Grycz, K. Miklossy (1969) and M.N. Hirsh, H.J. Oscam (1978).

8.5 DIFFERENT CONFIGURATIONS OF ARC DISCHARGES

8.5.1 Free-Burning Linear Arcs

These axisymmetric arcs between two electrodes can be arranged in both horizontal and vertical configurations (see Figure 8.18). Sir Humphrey Davy first observed such a horizontal arc in the beginning of the 19th century. The buoyancy of hot gases in the horizontal free-burning liner arc leads to bowing up or "arcing" of the plasma channel, which explains the origin of the name "arc".

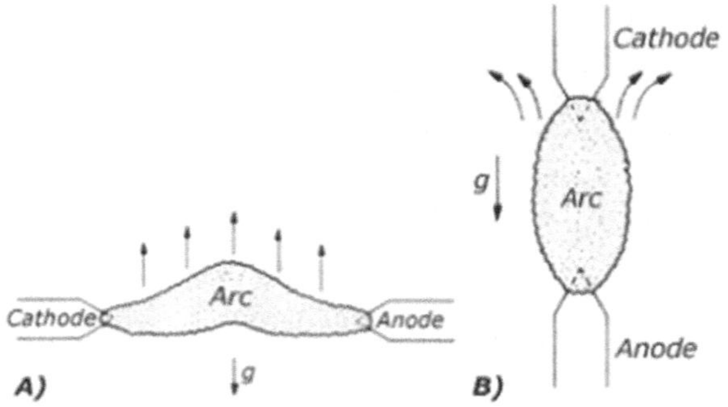

FIGURE 8.18 Illustration of horizontal (a) and vertical (b) free-burning arc.

If the free-burning arc has the vertical configuration (see Figure 8.18b), the cathode is usually placed at the top of the discharge. In this case buoyancy provides more intensive heating of cathode, which sustains more effective thermionic emission from the electrode. The length of free-burning linear arcs typically exceeds their diameter. Sometimes, however, the arcs are shorter than their diameter (see Figure 8.19) and the "arc channel" actually does not exist. Such discharges are referred to as the **obstructed arcs**. The distance between electrodes in such discharges is typically about 1 mm, voltage between electrodes nevertheless exceeds anode and cathode drops. The obstructed arcs are usually electrode-stabilized.

8.5.2 Wall-Stabilized Linear Arcs

This type of arcs is most widely used for gas heating (see sections 8.4.1 and 8.4.2). A simple configuration of the **wall-stabilized arc with a unitary anode** is presented in Figure 8.20. The cathode is axial in this configuration of linear arc discharges and the unitary anode is hollow and coaxial. In this discharge, the arc channel is axisymmetric and stable with respect to asymmetric perturbations.

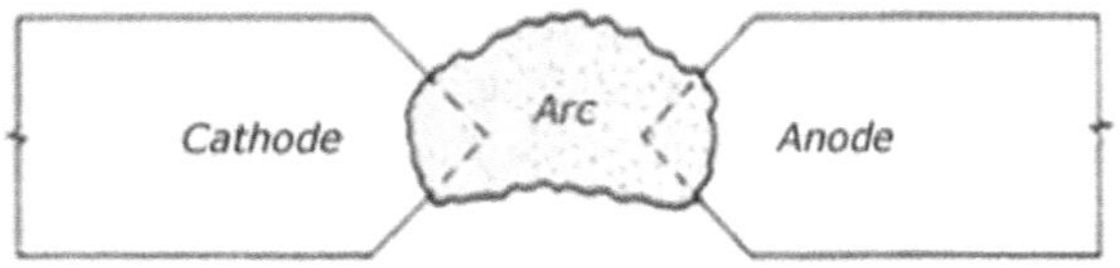

FIGURE 8.19 Illustration of the obstructed electrode-stabilized arc.

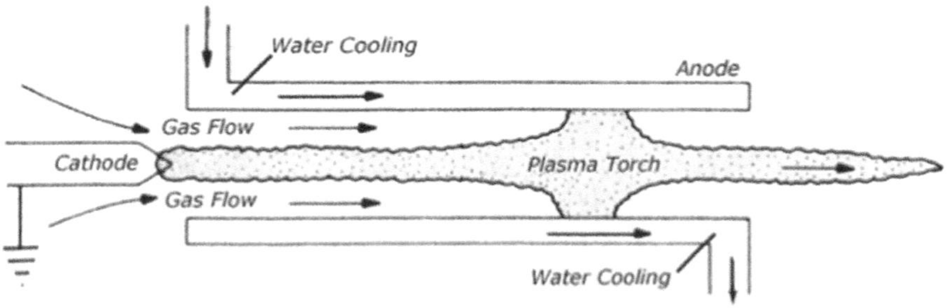

FIGURE 8.20 Illustration of the wall-stabilized arc

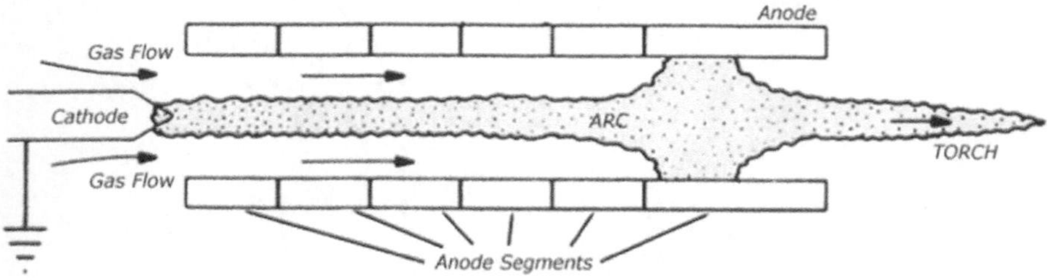

FIGURE 8.21 Wall-stabilized arc with segmented anode.

If the arc channel is asymmetrically perturbed and approaches a coaxial anode, it leads to the intensification of the discharge cooling and to a temperature increase on the axis. The increase of temperature results in the displacement of the arc channel back on the axis of the discharge tube, which stabilizes the linear arc. That the anode of the high power arcs can be water-cooled (see Figure 8.20). This arc discharge can attach to the anode at any point along the axis, which makes the discharge system not regular and less effective for several applications. Although the coaxial gas flow is able to push the arc-to-anode attachment point to the far end of the anode cylinder, the better-defined discharge arrangement can be achieved in the so-called **segmented wall-stabilized arc configuration.** The wall-stabilized linear arc, organized with the segmented anode, is illustrated in Figure 8.21. The anode walls are usually water-cooled and electrically segmented and isolated. Such a configuration provides a linear decrease of axial voltage and forces the arc attachment to the wider anode segment, the furthest one from the cathode (see Figure 8.21). The length of the other segments is taken sufficiently small to avoid breakdown between them.

8.5.3 The Transferred Arcs

The linear transferred arcs with water-cooled non-consumable cathodes are illustrated in Figure 8.22. Generation of electrons on inner walls of the hollow cathodes is provided by field emission, which permits operating the transferred arcs at multi-megawatt power during thousands of hours. The electric circuit in the discharge is completed by transferring the arc to an external anode. In this case, the external anode is actually a conducting material where the high arc discharge energy is supposed to be applied. The arc "transferring" to an external anode gave the name to this arc configuration, which is quite effective in metal melting and refining industry. The arc root in the transferred arcs is also able to move over the cathode surface, which increases the lifetime of the cathode.

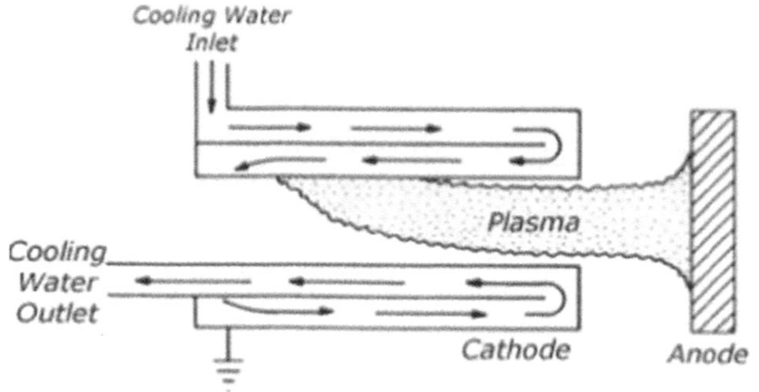

FIGURE 8.22 Illustration of the transferred arc configuration.

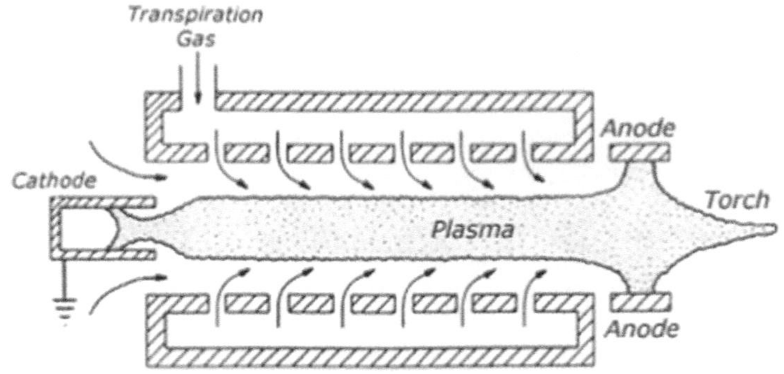

FIGURE 8.23 Illustration of the transpiration-stabilized arc.

8.5.4 Flow-Stabilized Linear Arcs

The arc channel can be stabilized on the axis of discharge chamber by radially inward injection of cooling water or gas. Such a configuration, illustrated in Figure 8.23, is usually referred to as the **transpiration-stabilized arc**. This discharge system is similar to the segmented wall-stabilized arc (see Section 8.5.2), but the transpiration of cooling fluid through annular slots between segments makes lifetime of the interior segments longer. Another linear arc configuration providing high power discharge is the so-called **coaxial flow stabilized arc**, illustrated in Figure 8.24. In this case, the anode is located too far from the main part of the plasma channel and cannot provide the wall stabilization. Instead of a wall, the arc channel is stabilized by a coaxial gas flow moving along the outer surface of the arc. Such stabilization is effective without heating of the discharge chamber walls if the coaxial flow is laminar. Similar arc stabilization can be achieved using a flow rotating fast around the arc column. Different configurations of **the vortex-stabilized arcs** are shown in Figure 8.25(a,b). The arc channel is stabilized in this case by a vortex gas flow, which is introduced from a special tangential injector. The vortex gas flow cools the edges of the arc and maintains the arc column confined to the axis of the discharge chamber. The vortex flow is very effective in promoting the heat flux from the thermal arc column to the walls of the discharge chamber.

8.5.5 Non-Transferred Arcs, Plasma Torches

A non-linear non-transferred wall-stabilized arc is shown in Figure 8.26. This discharge system consists of a cylindrical hollow cathode and coaxial hollow anode located in a water-cooled

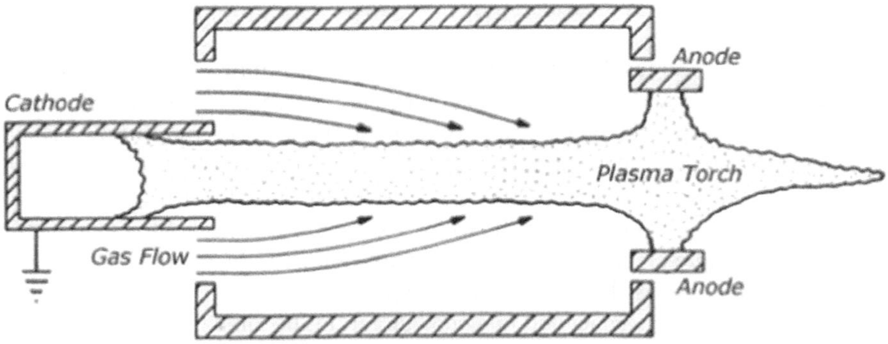

FIGURE 8.24 Illustration of the coaxial flow stabilized arc.

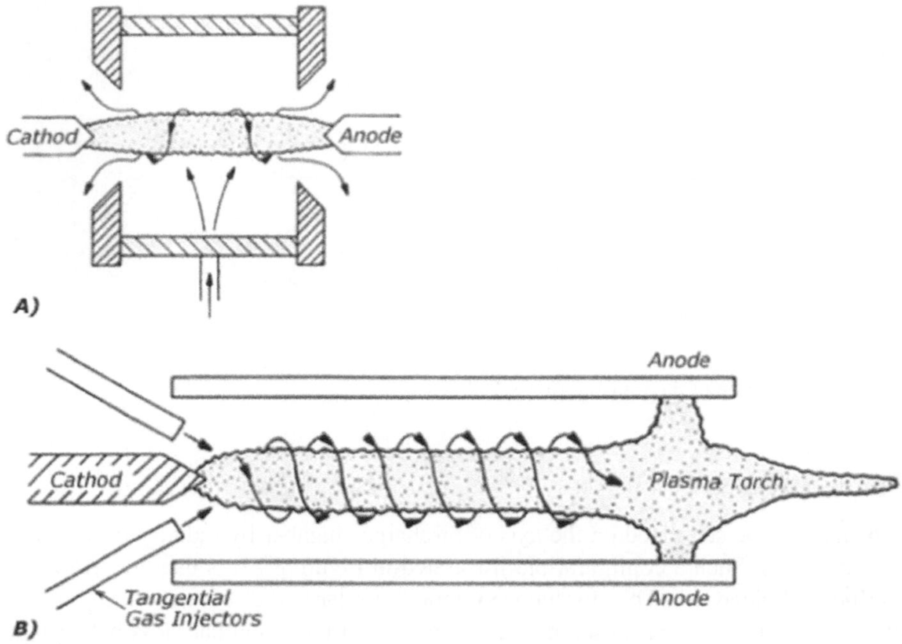

FIGURE 8.25 Illustration of vortex-stabilized arcs.

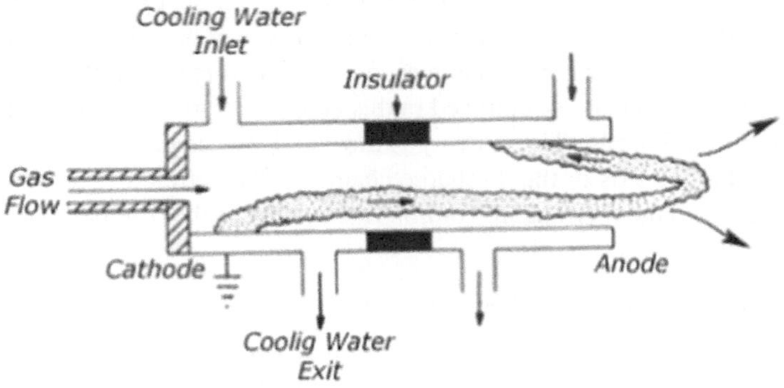

FIGURE 8.26 Illustration of the non-transferred arc.

chamber and separated by an insulator. Gas flow blows the arc column out of the anode opening to heat material which is supposed to be treated. In contrast to the transferred arcs, the treated material is not supposed to operate as an anode. This explains the name "non-transferred" given to these arcs. Magnetic $\vec{I} \times \vec{B}$ forces in these discharge systems cause the arc roots to rotate around electrodes (see Figure 8.26). This magnetic effect provides a longer lifetime of the cathodes and anodes with respect to more traditional incandescent electrodes. In this case generation of electrons on the cathode is provided completely by field emission. An axisymmetric version of the non-transferred arc is illustrated in Figure 8.27. This discharge configuration is usually referred to as the **plasma torch** or the **arc jet**. The arc is generated in a conical gap in the anode, and is pushed out of this opening by the gas flow. The heated gas flow forms a very high-temperature arc jet sometimes at supersonic velocities.

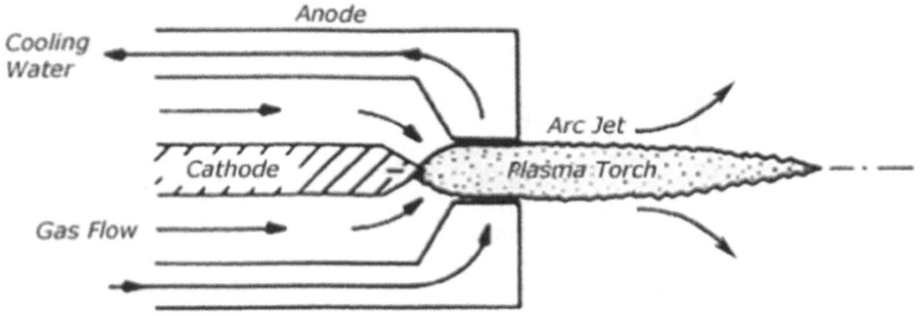

FIGURE 8.27 Illustration of the plasma torch.

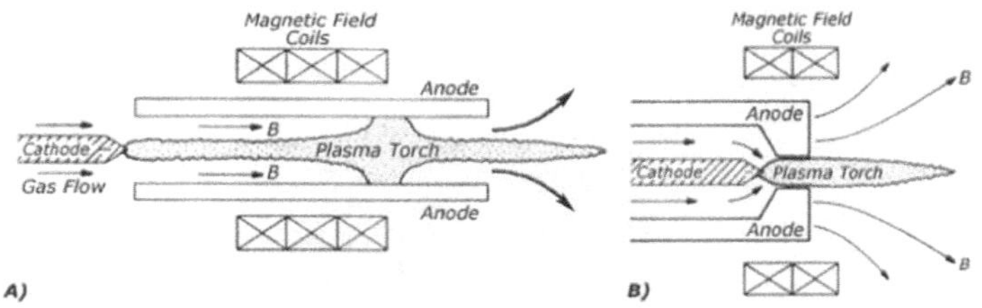

FIGURE 8.28 Illustration of magnetically stabilized arcs.

8.5.6 Magnetically Stabilized Rotating Arcs

Such configurations of arc discharges are illustrated in Figure 8.28(a,b). The external axial magnetic field provides $\vec{I} \times \vec{B}$ forces, which cause very fast rotation of the arc discharge and protect the anode from intensive local overheating. Figure 8.28(a) shows the case of additional magnetic stabilization of a wall-stabilized arc (see Section 8.5.2). Figure 8.28(b) presents a practically very important configuration of the magnetically stabilized plasma torch. The effect of magnetic stabilization is an essential supplement to wall or gas flow stabilization.

8.6 GLIDING ARC DISCHARGE

8.6.1 General Features of the Gliding Arc

The gliding arc discharge is an auto-oscillating periodic phenomenon developing between at least two diverging electrodes submerged in a laminar or turbulent gas flow. Picture and illustration of a gliding arc are shown in Figure 8.29 (a,b). Self-initiated in the upstream narrowest gap, the discharge forms the plasma column connecting the electrodes. This column is further dragged by the gas flow towards the diverging downstream section. The discharge length grows with the increase of inter-electrode distance until it reaches a critical value, usually determined by the power supply limits. After this point, the discharge extinguishes but momentarily reignites itself at the minimum distance between the electrodes, and a new cycle starts. The time-space evolution of the gliding arc is illustrated by a series of snap-shots presented in Figure 8.30. Plasma generated by the gliding arc has thermal or non-thermal properties depending on the system parameters such as power input and flow rate. Along with completely thermal and completely non-thermal modes of the discharge, it is possible to define the transition regimes of the gliding arc. In this most interesting regime, the discharge starts as a thermal, but during space and time evolution becomes a non-thermal one. The powerful and energy-efficient transition discharge combines the benefits of both equilibrium and non-equilibrium discharges in a single structure. They can provide plasma conditions typical for

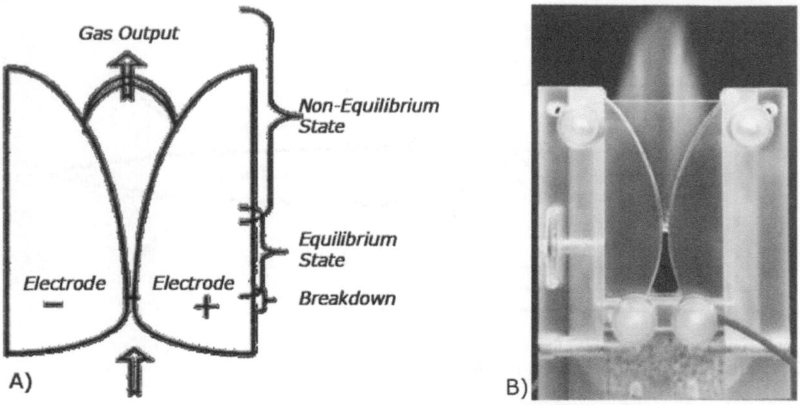

FIGURE 8.29 Illustration (A) and picture (B) of gliding arc discharge.

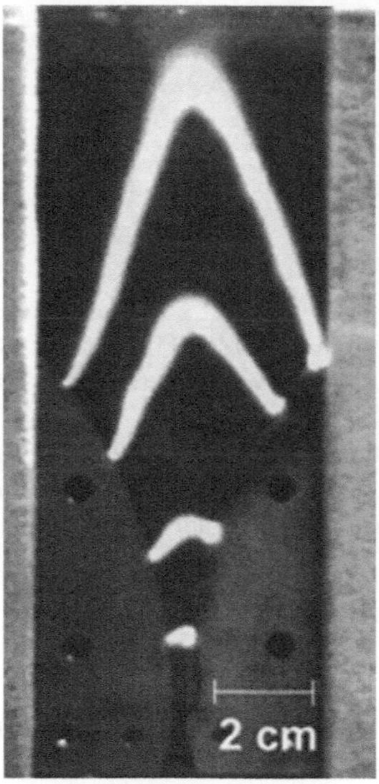

FIGURE 8.30 UIC's Gliding arc evolution shown with 20 ms separation between snapshots at flow rate 110SLM.

non-equilibrium cold plasmas but at elevated power levels. Two very different kinds of plasmas are used for chemical applications. Thermal plasma generators have been designed for many diverse industrial applications covering a wide range of operating power levels from less than 1 kilowatt to over 50 Megawatts. However, in spite of providing sufficient power levels, these appear not to be well adapted to the purposes of plasma chemistry, where selective treatment of reactants (through the excitation of molecular vibrations or electron excitation) and high efficiency are required. An alternative approach for plasma-chemical gas processing is the non-thermal one. The glow discharge in a low-pressure gas seems to be a simple and inexpensive way to achieve the non-thermal plasma conditions. Indeed, the cold non-equilibrium plasmas offer good selectivity and energy efficiency of chemical processes, but they are generally at limited to low pressures and powers. In these two general types of plasma discharges it is impossible to simultaneously maintain a high level of non-equilibrium, high electron temperature, and high electron density, whereas the most perspective plasma chemical applications require high power for high reactor productivity and a high degree of non-equilibrium to support selective chemical process at the same time. However, this can be achieved in the transition regime of the gliding arc which for this reason attracts interest to this type of non-equilibrium plasma generators. Gliding arcs operate at atmospheric pressure or higher; the dissipated power at non-equilibrium conditions can reach up to 40 kW per electrode pair. The gliding arc configuration is actually well known for more than a hundred years. In form of the Jacob's ladder, it is often used in different science exhibits. The gliding arc was first used in chemical applications in the beginning of the 1900s for the production of nitrogen-based fertilizers (A.A. Naville, C.E. Guye, 1904). An important recent contribution to the development of fundamental and applied aspects of the gliding arc discharge was made by A. Czernichowski and his colleagues (H. Lesueur, A. Czernichowski, J. Chapelle, 1990; A. Czernichowski, 1994). The non-equilibrium nature of the transition regime of the gliding arc was first reported by A. Fridman, A. Czernichowski, J. Chapelle, J.M. Cormier, H. Lesuer, J. Stevefelt (1994) and V.D. Rusanov, A.S. Petrusev, B.V. Potapkin, A. Fridman, A. Czernichowski, J. Chapelle (1993).

8.6.2 Physical Phenomenon of the Gliding Arc

Consider a simple case of a direct current gliding arc in air, driven by two generators. A typical electrical scheme of the circuit is shown in Figure 8.31. One generator is a high voltage (up to 5000 V) used to ignite the discharge and the second one is a power generator (with the voltage up to 1 kV, and a total current I up to 60 A). A variable resistor $R = 0–25\ \Omega$ is in series with a self-inductance $L = 25$ mH. More advanced schemes such as an AC Gliding Arc, 3-phase Gliding arc, and the arc with several parallel or serial electrodes, etc. could be also configured.

Initial breakdown of the gas begins the cycle of the gliding arc evolution. The discharge starts at the shortest distance (1–2 mm) between two electrodes (see Figure 8.32). The high voltage generator provides the necessary electric field to break down the air between the electrodes. For atmospheric pressure air and a distance between electrodes of about 1 mm the breakdown voltage V_b is approximately 3 kV. The characteristic time of the arc formation τ_i can be estimated from the kinetic equation:

$$\frac{dn_e}{dt} = k_i n_e n_0 = \frac{n_e}{\tau_i}, \tag{8.51}$$

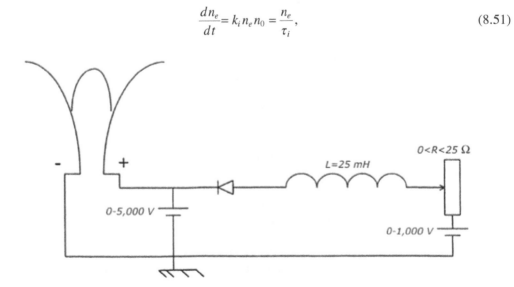

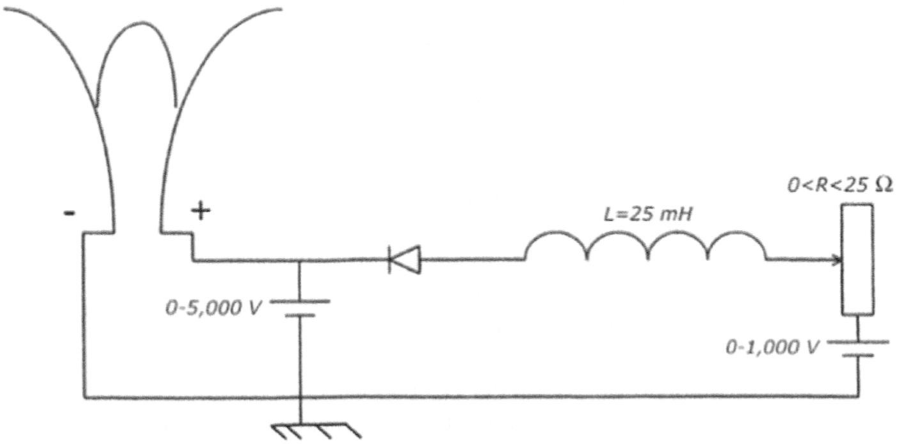

FIGURE 8.31 Typical gliding arc discharge electrical scheme.

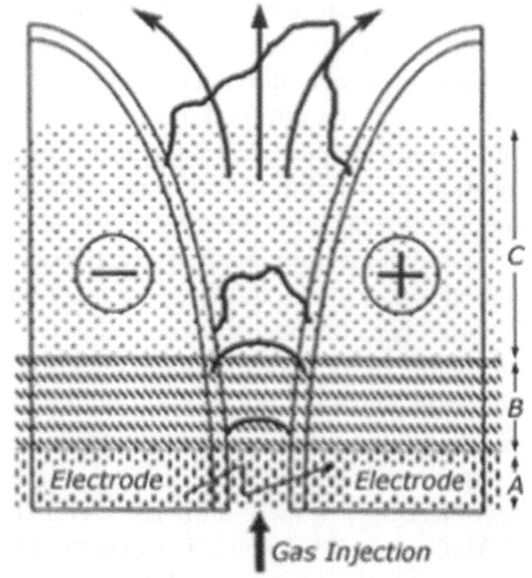

FIGURE 8.32 Phases of gliding arc evolution: (A) reagent gas breakdown; (B) equilibrium heating phase; (C) non-equilibrium reaction phase;

where k_i is the ionization rate coefficient; n_e and n_0 are electron and gas concentrations. The estimation for τ_I in clean air and total arc electrical current $J = 1$ A gives the value $\tau_i \approx 1\ \mu s$. Within a time of about 1 μs a low resistance plasma is formed and the voltage between the electrodes falls.

The Equilibrium stage takes place after the formation of a stable plasma channel. The gas flow pushes the small equilibrium plasma column with the velocity about 10 m/sec, and the length l of the arc channel increases together with the voltage. Initially the electric current increases during the formation of the quasi-equilibrium channel up to its maximum value of $I_m = V_0/R \approx 40A$. The current time dependence during the current growth phase is sensitive to the inductance L:

$$I(t) = (V_0/R)\left(1 - e^{-t/\tau_L}\right), \tag{8.52}$$

where $\tau_L = L/R \approx 1\,m\text{sec}$. In this stage, the equilibrium gas temperature T_0 does not change drastically and lies in the range $7000 \le T_0 \le 10{,}000$ K for our numerical example (A. Fridman, A. Czernichowski, J. Chapelle, J.-M. Cormier, H. Lesueur, J. Stevefelt, 1993). For the experiments with lower electric current and lower power, the temperature could be different. For example at power level 200 W and electric current 0.1 A, the gas temperature is as low as 2500 K (A. Fridman, S. Nester, O. Yardimci, A. Saveliev, L. A. Kennedy, 1997). The quasi-equilibrium arc plasma column moves with the gas flow. At a power level 2 kW, the difference in velocities grows from 1 to 10 m/s with an increase in the inlet flow rate from 830 to 2200 cm^3/sec (F. Richard, M. Cormier, S. Pellerin, J. Chapelle, 1996). In the reported case, the flow velocity was 30 m/sec and the arc velocity was 24 m/sec. In the experiments reported by M. A. Deminsky, B. V. Potapkin, J. -M. Cormier ea. (1977), the difference in velocities was up to 30 m/sec for high flow rates (2200 cm^3/sec) and small arc lengths (5 cm). This difference decreases several meters per second when arc length increases up to 15 cm. In general, the difference in arc and flow velocities increases with the increase in the arc power and absolute values of flow rates and, hence, flow velocities. It decreases with an increase of the arc length. The length of the column l during this movement and hence the power consumed by the discharge from the electrical circuit increases up to the moment when the electrical power reaches the maximum value available from the power supply source P_{max}.

The ***Non-equilibrium stage*** begins when the length of the gliding arc exceeds its critical value l_{crit}. Heat losses from the plasma column begin to exceed the energy supplied by the source, and it is not possible to sustain the plasma in the state of thermodynamic equilibrium. As a result, the discharge plasma cools rapidly to about $T_0 = 2000$ K, and the plasma conductivity is maintained by a high value of the electron temperature $T_e = 1$ eV and step-wise ionization. After the decay of the non-equilibrium discharge, a new breakdown takes place at the shortest distance between electrodes, and the cycle repeats.

8.6.3 Equilibrium Phase of the Gliding Arc

The parameters of the equilibrium phase are similar to those for the conventional atmospheric pressure arc discharges. After the ignition and formation of a steady arc column, the energy balance can be described by the Elenbaas-Heller equation (8.20). Then the application of the Raizer "channel" model of the positive column (see Section 8.4.4) leads to Eq. (8.33) describing the power w dissipated per unit length of the gliding arc. As was discussed in Section 8.4.5 (in particular see Eq. (8.37) and Figure 8.15), the specific power w does not change significantly. In the temperature range of 6000 K – 12,000 K, the characteristic values of the specific power w is about 50–70 kW/m. Assuming a constant specific power w permits describing the evolution of current, voltage, and power during the gliding arc quasi-equilibrium phase. Neglecting the self-inductance L, the Ohm's law can be written for the circuit including the plasma channel, active resistor, and power generator as:

$$\mathrm{V_0 = R\,I + w \cdot l/I.} \tag{8.53}$$

Here V_0, R, and J are respectively the open-circuit voltage of the power supply, the external serial resistance, and current. The arc current can be determined from the Ohm's law Eq. (8.53) and presented as a function of the arc length l, which is growing during the arc evolution:

$$\mathrm{I = \left(V_0 \pm \left(V_0^2 - 4\,w\,l\,R\right)^{1/2}\right)/2R.} \tag{8.54}$$

The solution with the sign "+" describes the steady state of the arc; and the solution with $\mathrm{J < V_0/2R}$ corresponds to negative differential resistance ρ ($\rho = \mathrm{dV/dJ}$) of and to an unstable arc regime:

$$\rho = R - wl/I^2 = 2R - V_0/I < 0. \tag{8.55}$$

From Eq. (8.54), the current slightly decreases during the quasi-equilibrium period. At the same time, the arc voltage is growing as $W\,l/I$, and the total arc power $P = Wl$ increases almost linearly with the length l. According to Eq. (8.55), the absolute value of the differential resistance of the gliding arc grows. Arc discharges are generally stable and have descending volt-ampere characteristics. This means that the differential resistance (du/dj) of an arc as a part of the electric circuit is negative. To provide a stable regime of the circuit operation, the total differential resistance of the whole circuit should be positive. When the power dissipated by arc achieves its maximum, the differential resistance of arc becomes equal to the differential resistance of the external part of circuit (see Eq. (8.55)) and the electric circuit loses its stability. This leads to a change in the electrical parameters of the circuit and affects the arc parameters. The power supply cannot provide the increasing arc

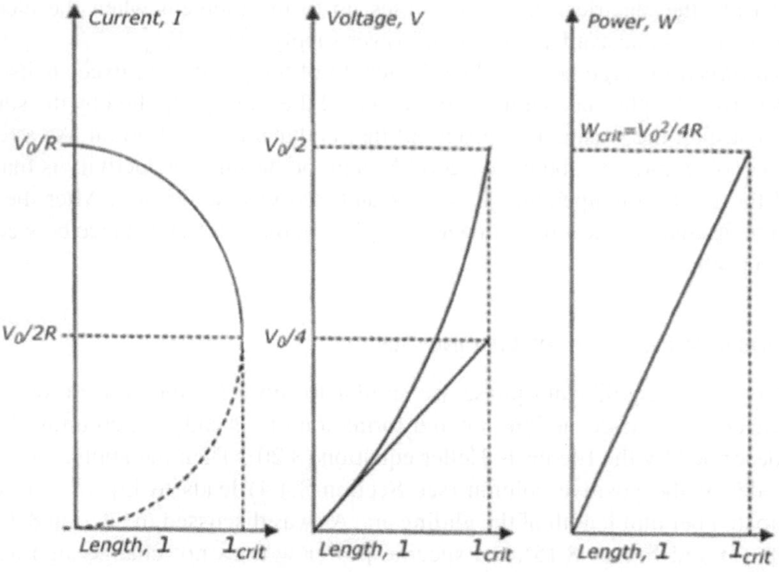

FIGURE 8.33 Evolution of current, voltage and power versus length.

power, and the system transfers to a non-equilibrium state. The current I, voltage V and power P in the equilibrium phase are shown in Figure 8.33.

8.6.4 Critical Parameters of the Gliding Arc

The quasi-equilibrium evolution of gliding arc is terminated when the arc length approaches the critical value:

$$l_{crit} = V_0^2/(4wR) \tag{8.56}$$

and the square root in Eq. (8.54) (that is the differential resistance of the whole circuit) becomes equal to zero. The current falls to its minimum $I_{crit} = V_0/2R$, which is half of the initial value. Plasma voltage, electric field E and total power at the same critical point approach their maximum:

$$V_{crit} = V_0/2, E_{crit} = w/J_{crit}, W_{crit} = V_0^2/4R. \tag{8.57}$$

At the critical point Eqs. (8.56), (8.57), a growing plasma resistance becomes equal to the external one, and, obviously the maximum value of discharge power corresponds to half of the maximum power of generator. The numerical values of the critical parameters for pure air conditions are: $J_{crit} = 20$ A, $V_{crit} = V_0/2 = 400$ V, $W_{crit} = 8$ kW (with a characteristic equilibrium temperature of the gliding arc discharge $7000 \leq T \leq 10{,}000$ K), $l_{crit} = 10$ cm. Eq. (8.56) for the gliding arc critical length can be made more specific by taking into account the channel model relation (8.33) for the specific power w:

$$l_{crit} = V_0^2\, I_i/32\pi R\lambda_m\, T^2. \tag{8.58}$$

This relation gives the similar estimation for the critical length of the gliding arc $l_{crit} \sim 10$ cm.

8.6.5 Fast Equilibrium → Non-Equilibrium Transition (FENETRe - Phenomenon)

When the length of the gliding arc exceeds the critical value $l > l_{crit}$, the heat losses wl continue to grow. But since the electrical power from the power supply cannot further increase, it is no longer possible to sustain the arc in the thermodynamic quasi-equilibrium. The gas temperature falls rapidly during the conduction time (about $r^2/D \approx 10^{-4}$ sec), and the plasma conductivity can be still maintained only by a high value of the electron temperature $T \approx 1$ eV and stepwise mechanism of ionization. The phenomenon of the "Fast Equilibrium → Non-Equilibrium Transition" (the so-called FENETRe) in the gliding arc is kind of "discharge window" from the thermal plasma zone to a relatively cold one (A. Fridman, S. Nester, L. Kennedy, A. Saveliev, O. Mutaf-Yardimci, 1999). The arc instability is due to the slow increase of the electric field $E = w/I$ and corresponding increase of electron temperature T_e during the gliding arc evolution:

$$T_e = T_0\left(1 + E^2/E_i^2\right). \tag{8.59}$$

Here the electric field E_i corresponds to the transition from thermal to direct electron impact ionization. The equilibrium arc discharge is usually stable. This thermal ionization stability is due to a peculiarity of both the heat and the electric current balances in the arc discharge. Indeed, a small temperature increase in the arc discharge leads to a growth of both electron concentration and electric conductivity σ. Taking into account the fixed value of current density $j = \sigma E$, the conductivity increase results in a reduction of the electric field E and the heating power σE^2, and hence results in stabilizing the initial temperature perturbations. In contrast to that, the direct electron impact ionization (typical for non-equilibrium plasma) is usually not stable. A small temperature increase leads, for a fixed pressure (Pascal law) and constant electric field, to a reduction of the gas concentration n_0 and to an increase of both the specific electric field E/n_0 and the ionization rate. This results in an increase of the electron concentration n_e, the conductivity σ, the Joule heating power σE^2 and hence in an additional temperature increase, and finally in total discharge instability. When the conductivity σ depends mainly on the gas translational temperature, the discharge is stable. When the electric field is relatively high and the conductivity σ becomes depending mainly on the specific electric field, the ionization becomes unstable. The gliding arc passes such a critical point during its evolution and the related electric field grows. The electric field growth during the gliding arc evolution could be written based on (8.54) as:

$$E = 2wR/\left(V_0 + \left(V_0^2 - 4wlR\right)^{0,5}\right). \tag{8.60}$$

For a high electric field, the electric conductivity $\sigma(T_0, E)$ begins depending not only on the gas temperature as in quasi-equilibrium discharges but also on the field E. The corresponding logarithmic sensitivity of the electric conductivity to the gas temperature corresponds to the Saha ionization:

$$\sigma_T = \partial \ln\sigma\left(T_0, E\right)/\partial \ln T_0 \approx I_i/2T_0, \tag{8.61}$$

where I_i is the ionization potential. Taking into account Eq. (8.59), the logarithmic sensitivity of electric conductivity to the electric field could be written as:

$$\sigma_E = \partial \ln\sigma\left(T_0, E\right)/\partial \ln E = I_i E^2/E_i^2 T_0. \tag{8.62}$$

To analyze the arc stability consider Ohm's law Eq. (8.53), the logarithmic sensitivities Eqs (8.61), (8.62), and the thermal balance equation:

$$n_0 c_p \, dT_0/dt = \sigma \left(T_0, E\right) E^2 - 8\pi \; \lambda \; T_0^2 / I_i S, \tag{8.63}$$

where S is the arc cross-section, n_0 and c_p - are the gas concentration and specific heat. The linearization procedure permits describing the time evolution of temperature fluctuation ΔT in the exponential form:

$$\Delta T(t) = \Delta T_0 \exp(\Omega\, t), \tag{8.64}$$

where ΔT_0 is an initial temperature fluctuation; and exponential frequency parameter Ω - is an instability decrement, that is a frequency of temperature fluctuation disappearance. In this case the negative value of the decrement $\Omega < 0$ corresponds to discharge stability, while the positive one $\Omega > 0$ corresponds to a case of instability. The linearization gives the following instability decrement for the gliding arc discharge (A. Fridman, S. Nester, L. Kennedy, A. Saveliev, O. Mutaf-Yardimci, 1999):

$$\Omega = -\omega \, \frac{\sigma_T}{1+\sigma_E} \left(1 - \frac{E}{E_{crit}}\right). \tag{8.65}$$

In this relation ω is the thermal instability frequency factor, similar to that one described in Section 6.3.2, and defined by the relation:

$$\omega = \frac{\sigma \; E^2}{n_0 c_p T_0}. \tag{8.66}$$

Development of the ionization instability finally results either in transition to non-equilibrium regime or just in break of the discharge.

8.6.6 Gliding Arc Stability Analysis

The scheme of the gliding arc ionization instability is presented in Figure 8.34. Initially the gliding arc could remain stable ($\Omega < 0$) at relatively small electric field $E < E_{crit}$ ($l < l_{crit}$), but it becomes completely unstable ($\Omega > 0$) when the electric field grows stronger Eq. (8.60). The gliding arc can lose stability even in the "theoretically stable" regime (see Figure 8.34):

$$E_i \left(T_0 / I_i\right)^{0,5} < E < E_{crit}, \tag{8.67}$$

Although such an electric field is less than a critical one (stability in the linear approximation Eq. (8.65)), its influence on electric conductivity becomes dominant, which strongly perturbs stability. In this regime of the "quasi-instability" (8.67), the stability factor Ω is still negative, but decreases 10–30 times (compared to one in the initial regime of the gliding arc) due to the influence of the logarithmic sensitivity σ_E (factor $1/(1+\sigma_E)$). This means that the gliding arc discharge actually becomes unstable in such electric fields. This effect of quasi-instability may be better illustrated using a non-linear analysis, where the instability decrement in addition to Eq. (8.65) also depends on a level of temperature perturbation. The results of this non-linear analysis of the differential equations can be represented by a critical value of temperature fluctuation, corresponding to the gliding arc transition to an unstable form:

$$T_{cr} = \left(4T_0^2 / I_i\right)\left(1 - E/E_{crit}\right) / \left(1 + \sigma_E\right). \tag{8.68}$$

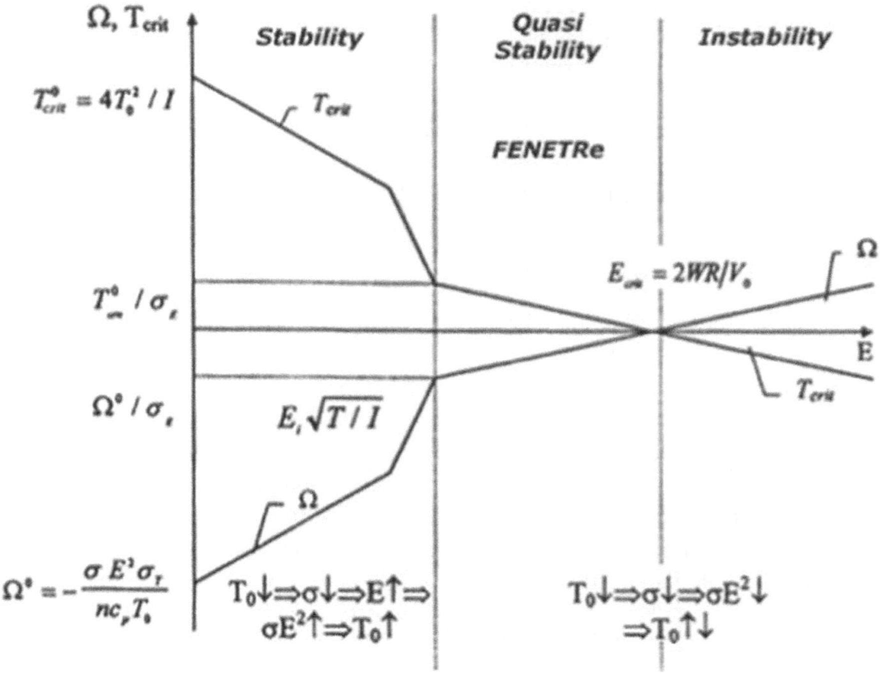

FIGURE 8.34 Characteristics of gliding arc instability: instability decrement and critical thermal fluctuation

In a regular arc and in the initial equilibrium stage of the gliding arc, the electric field and the sensitivity of conductivity to electric field is relatively small $E << E_{crit}$, $\sigma_E << 1$. Then the maximum permitted temperature perturbation to sustain the arc stability of the discharge is :

$$T_{cr}\left(equil.\right) = T_0^{\,2}/I_i. \tag{8.69}$$

This maximum temperature perturbation is about 1000 K, and is high enough to guarantee the arc's stability. In the transitional stage Eq. (8.67), the critical temperature fluctuation T decreases to:

$$T_{cr}\left(non-eq\right) = T_0^{\,3}\,E_i^{\,2}/I_i^2\,E^2, \tag{8.70}$$

which corresponds numerically to about 100 K. In this case, a small temperature perturbation leads to arc instability and to the FENETRe phenomenon. The main qualitative results of the linear and the non-linear analysis of the gliding arc evolution, including the instability decrement and the critical temperature fluctuation as a function of the electric field, are presented in Figure 8.34. During the FENETRe- stage, the electric field and the electron temperature T_e in the gliding arc are slightly increasing, numerically from $T_e \approx 1\,eV$ in the quasi-equilibrium zone to approximately $T_e \approx 1\,eV$ in the non-equilibrium zone. At the same time, the translational gas temperature decreases about three times from its initial equilibrium value of approximately 0.5 eV. The electron concentration falls to about 10^{12} cm^{-3}. The decrease of the electron concentration and conductivity σ leads to the reduction of specific discharge power σE^2. It cannot be compensated by the electric field growth as in the case of the stable quasi-equilibrium arc, because of the strong sensitivity of the conductivity to the electric field (see Eq. (8.62)). This is the main physical reason for the quasi-unstable discharge behavior in FENETRe conditions (see Figure 8.34). During the transition phase, the specific heat losses w Eq. (8.33) also decrease (to a smaller value w_{non-eq}) as well as the specific discharge power σE^2 with the reduction of the translational temperature T_0. However according to Ohm's law ($E < E_{crit}$), the total discharge power increases due to the growth of plasma resistance. This discrepancy between the total power and specific power (per unit of length or volume) results in a possible "explosive" increase in the arc length. This increase of length can be easily estimated from Eq. (8.56) as:

$$l_{max}/l_{crit} = w/w_{non\text{-}eq} \approx 3. \tag{8.71}$$

This was experimentally observed by J.-M. Cormier, F. Richard, J. Chapelle, M. Dudemaine, 1993.

8.6.7 Non-Equilibrium Phase of the Gliding Arc

After the fast FENETRe-transition, the gliding arc is able to continue its evolution under the non-equilibrium conditions $T_e >> T_0$ (V.D. Rusanov, A.S. Petrusev, B.V. Potapkin, A. Fridman, A. Czernichowski, J. Chapelle, 1993; L. Kennedy, A. Fridman, A. Saveliev, S. Nester, 1997). Up to 70–80% of the total gliding arc power can be dissipated after the critical transition resulting in plasma where the gas temperature is approximately 1,500–3,000 K and the electron temperature is about 1 eV. Effective gliding arc evolution in the non-equilibrium phase is possible only if the electric field during the transition is sufficiently high, which is possible when the arc current is not too large. The critical gliding arc parameters before the transition in atmospheric air are shown in Table 8.8 as a function of the initial electric current. Under conditions of the decaying arc, even the relatively small electric field $\left(\frac{E}{n_0}\right)_{crit} \approx (0.5\text{–}1.0) \times 10^{-16}\ \text{V} \times \text{cm}^2$ is sufficient to sustain the non-equilibrium phase of a gliding arc, because of the influence of stepwise and possibly Penning ionization mechanism. Then taking into account data from Table 8.8, one concludes:

i. if $I_0 > 5\text{–}10$ A, then the arc discharge extinguishes after reaching the critical values;
ii. alternatively for $J_0 < 5\text{–}10$ A, the value of reduced electric field $\left(\frac{E}{n_0}\right)_{crit}$ is sufficient for maintaining the discharge in a new non-equilibrium ionization regime.

It is possible to observe three different types of gliding arc discharge. At relatively low currents and high gas flow rates, the gliding arc discharge is non-equilibrium throughout all stages of its development. Alternately at high currents and low gas flow rates, the discharge is thermal (quasi-equilibrium) and just breaks out at the critical point. Only at the intermediate values of currents and flow rates does the FENETRe-transition takes place. Experimental data and comparative analysis regarding all these three types of gliding arcs are discussed in the publication of O. Mutaf-Yardimci, A. Saveliev, A. Fridman, L. Kennedy, 1999. The detailed electron T_e, vibrational T_v, and translational T_0 temperature measurements in the non-equilibrium regimes of gliding arc were reported by A. Czernichowski

TABLE 8.8
The Critical Gliding Arc Parameters before Transition in Dependence on Magnitude of the Initial Current I_0

I_{crit}, A	50	40	30	20	10	5	1	0.5	0.1
I_0, A	100	80	60	40	20	10	2	1	0.2
T_0, K	10,800	10,300	9700	8900	7800	6900	5500	5000	4100
w_{crit}, W/cm	1600	1450	1300	1100	850	650	400	350	250
E_{crit}, V/cm	33	37	43	55	85	130	420	700	2400
$\left(\frac{E}{n_0}\right) 10^{-16}$ Vcm²	0.49	0.51	0.57	0.66	0.9	1.3	3.1	4.8	14

ea.,1996. Typical temperatures were: $T_e \approx 10000K$, $T_v \approx 2000-3000K$, $T_0 \approx 800-2100K$, see also V. Dalaine, J.-M. Cormier, S. Pellerin, P. Lefaucheux (1998); S. Pellerin, J.-M. Cormier, F. Richard, K. Musiol, J. Chapelle (1996), O. Mutaf-Yardimci, A. Saveliev, A. Fridman, L. Kennedy (1998a).

8.6.8 Effect of Self-Inductance on Gliding Arc Evolution

Introducing a self-inductance into the gliding arc circuit (see Figure 8.31) decelerates the rate of current decrease and thereby prolongs the time of evolution of the non-equilibrium arc. Taking into account the self-inductance of the electric circuit the Ohm's law can be written as:

$$V_0 = RI + L\,{}^{dI}\!/_{dt} + P/I, \tag{8.72}$$

where power P in the non-equilibrium regime can be presented in the form:

$$P = P_{crit} + w_{crit}\,2v\alpha\, t; \tag{8.73}$$

P_{crit} and w_{crit} are the total and critical discharge power; v is the velocity of gliding arc; 2α is angle between the diverging electrodes; $t = 0$ at the transition point. Using dimensionless variables:

$$i = I/I_{crit}, \qquad \tau = t/\tau_L, \qquad \lambda = 2v\tau_L\alpha/l_{max}, \tag{8.74}$$

rewrite Ohm's law as follows:

$$(i-1)^2 + idi/d\tau + \lambda t = 0, \tag{8.75}$$

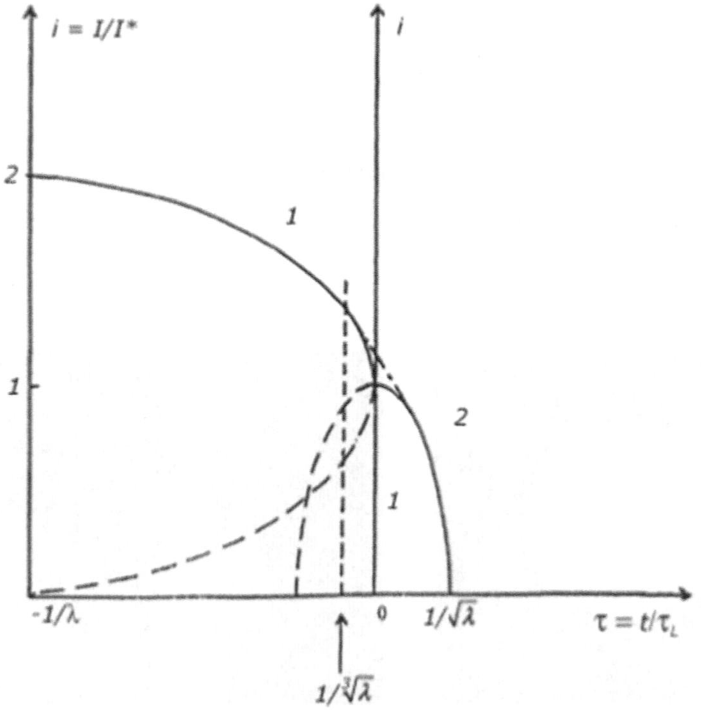

FIGURE 8.35 Effect of the self-inductance on the transient phase of gliding arc.

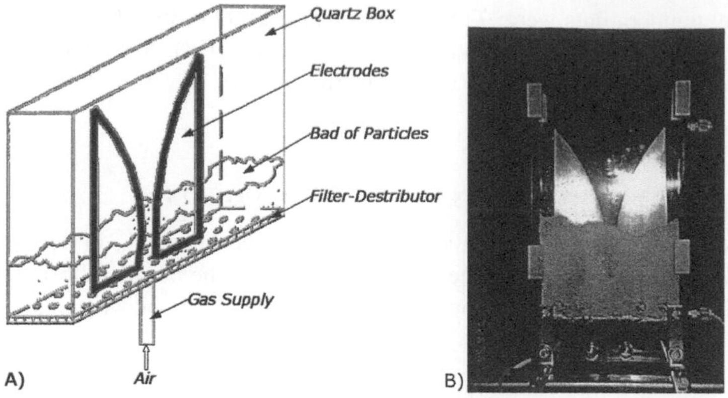

FIGURE 8.36 UIC's Gliding arc-fluidized bed system: (A) schematic and (B) picture.

where $\tau_L = L/R$ is the characteristic time of the electric circuit. For the typical conditions of the gliding arc, the magnitude of λ is rather small ($\lambda = 0.003$). At the beginning of the process when $\tau < 0$ ($|\tau| >> 1$) and neglecting the derivative $di/d\tau$, the approximation for i can be written as

$$i = 1 + \left(-\lambda\tau\right)^{1/2}, \tag{8.76}$$

which corresponds to the current evolution shown in Figure 8.33. When $\tau > -1/\lambda^{1/3}$, Eq. (8.75) becomes:

$$i^2 + \lambda\tau^2 = 1. \tag{8.77}$$

The electric current corresponding to solutions of Eqs. (8.75)–(8.77) is shown in Figure 8.35. Introducing the self-inductance prolongs the transition phase by the time $\tau = 1/\lambda^{1/2}$. As a result, the power dissipated throughout the transition phase can also be increased (numerically on about 20–50%).

8.6.9 Special Configurations of Gliding Arcs

The gliding arc discharges can be organized in various ways for different practical applications. Of special interest are gliding arc discharges organized in a fluidized bed (see Figure 8.36), and also

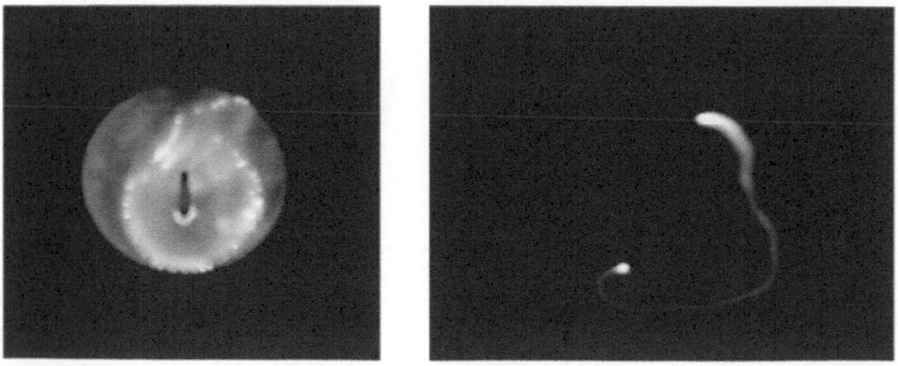

FIGURE 8.37 UIC's Gliding arc rotating in magnetic field.

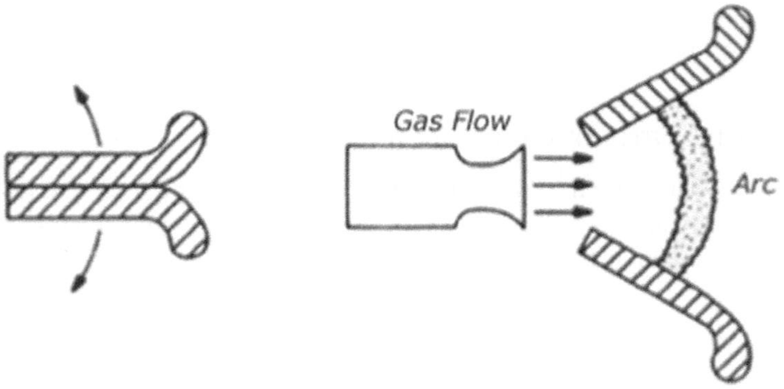

FIGURE 8.38 Switches as gliding arcs.

gliding arc discharges in a fast rotating in the magnetic field (see Figure 8.37). An expanding arc configuration similar to the gliding arc is used in switchgears, as shown in Figure 8.38. The contacts start out closed. When the contactors open, an expanding arc is formed and glides along the opening of increasing length until it extinguishes. Extinction of the gliding arc in this case is often promoted by using strongly electro-negative gases like $SF_{6.}$

8.6.10 Gliding Arc Stabilized in Reverse Vortex (Tornado) Flow

Interesting for applications is gliding arc stabilization in the reverse vortex (tornado) flow. This approach is opposite to the conventional **forward-vortex stabilization**, where the swirl generator is placed upstream with respect to discharge and the rotating gas provides the walls protection from the heat flux. Some reverse axial pressure gradient and central reverse flow appear due to fast flow rotation and strong centrifugal effect near the gas inlet, which becomes a bit slower and weaker downstream. The hot reverse flow mixes with incoming cold gas and increases heat losses to the walls, which makes the discharge walls insulation less effective. More effective wall insulation is achieved by the **reverse-vortex stabilization**. In this case, the outlet of the plasma jet is directed along the axis to the swirl generator side. Cold incoming gas moves at first by the walls providing their cooling and insulation, and only after that it goes to the central plasma zone and becomes hot. Thus in the case of reverse-vortex stabilization, the incoming gas is entering the discharge zone from all directions except the outlet side, which makes it effective for gas heating/conversion efficiency and for discharge walls protection. A picture of the gliding arc discharge trapped in the reverse vortex (tornado) flow is shown on the cover of this book.

8.7 PROBLEMS AND CONCEPT QUESTIONS

8.7.1 The Sommerfeld Formula for Thermionic Emission

Using the Sommerfeld formula (8.3) calculate the saturation current densities of thermionic emission for tungsten cathode at 2500 K. Compare the calculated current densities with those presented in Table 8.1.

8.7.2 The Richardson Relation for Thermionic Emission

Derive the thermionic current density dependence on temperature assuming in the integration of Eq. (8.2) not Fermi, but Maxwell distribution for electron velocities. In 1908 Richardson derived this relation with the pre-exponential factor proportional not to T^2 but to T. Explain the difference between Richardson and Sommerfeld formulas.

8.7.3 The Schottky Effect of Electric Field on the Work Function

Derive the Schottky relation by integrating Eq. (8.5) with the electric force Eq. (8.4) applied to an electron. Estimate the relative accuracy of this relation. Calculate the maximum Schottky decrease of work function, corresponding to very high electric fields sufficient for field emission.

8.7.4 The Field Electron Emission

The field electron emission is characterized by very sharp current dependence on the electric field (see, for example, Table 8.3). As a result of this strong dependence, one can determine the critical value of the electric field when the contribution of the field electron emission becomes significant. Based on the Fowler-Nordheim formula (8.7), determine the critical value of the electric field necessary for the field electron emission and its dependence on the work function.

8.7.5 Secondary Electron Emission

Explain why the coefficient γ of the secondary electron emission induced by the ion impact is almost constant at relatively low energies when it is provided by the Penning mechanism. Estimate the probability of the electron emission following the Penning mechanism. Explain why the coefficient γ of the secondary electron emission induced by the ion impact grows linearly with ion energy when the ion energy is sufficiently high.

8.7.6 The Secondary Ion-Electron Emission Coefficient

γ. Based on Eq. (8.9) estimate the coefficient γ of secondary electron emission induced by the ion impact at relatively low ion energies for ions with different ionization potentials and surfaces with different work functions. Compare the obtained result with the experimental data presented in Figure 8.6.

8.7.7 Electric Field in the Vicinity of Arc Cathode

Integrating the Poison's Eq. (8.13), derive Eq. (8.14) for the electric field near the cathode of arc discharge. Show that the first term in this formula is related to the contribution of ions in forming the positive space charge, and the second term to the effect of electrons in compensating the positive space charge. Compare the numerical contribution of these two terms.

8.7.8 Structure of the Arc Discharge Cathode Layer

Integrating the Poisson's equation (8.13) twice and neglecting the second term related to the effect of electrons on the space charge, derive Eq. (8.16) for the length of the collisionless zone of the cathode layer. Give a physical interpretation of the fact that most of the cathode voltage drop takes place in the extremely short collisionless zone of the cathode layer.

8.7.9 Erosion of Hot Cathodes

Based on the numerical example of specific erosion given in Section 8.3.4, estimate the rate of cathode mass losses per hour for the tungsten rod cathode at atmospheric pressure of inert gases and current about 300 A.

8.7.10 Cathode Spots

The critical minimum current through an individual cathode spot for different non-ferromagnetics can be found from (8.19). Using this relation calculate the minimum current density for copper and silver electrodes and compare the results with data presented in Table 8.6.

8.7.11 Radiation of the Arc Positive Column

Based on the formulas, presented in Table 8.7, calculate the percentage of arc power conversion into radiation for different gases at low and high values of the Joule heat per unit length $w = EI$, W/cm. Make your conclusion about the effectiveness of different gases as the arc radiation sources at the different levels of specific discharge power.

8.7.12 The Elenbaas-Heller Equation

The Elenbaas-Heller Eq. (8.23) includes the material function $\sigma(\Theta)$, expressing the dependence of the quasi-equilibrium plasma conductivity on the heat flux potential Eq. (8.22). Taking into account that thermal conductivity dependence on temperature $\lambda(T)$ is non-monotonic (see Figure 8.13), explain why the material function $\sigma(\Theta)$ nevertheless is smooth enough.

8.7.13 The Steenbeck Channel Model of Arc Discharges

Integrating the Elenbaas-Heller equation in the framework of the Steenbeck channel model, derive the relation between the heat flux potential $\Theta_m(T_m)$, related to the plasma temperature, and the discharge power per unit arc length.

8.7.14 Principle of Minimum Power for Positive Column of Arc Discharges

According to the Steenbeck principle of minimum power, the plasma temperature T_m and the arc channel radius r_0 should minimize the specific discharge power w and electric field $E = w/I$ at fixed values of current I and discharge tube radius R. Apply the minimization requirement $\left(\dfrac{dw}{dr_0}\right)_{I=const} = 0$ to the functional relations (8.24), (8.25) and derive the third equation (8.27) of the Steenbeck channel model.

8.7.15 The Raizer Channel Model of Arc Discharges

Prove the equivalence of the Raizer and Steenbeck approaches to the channel model. Show that the third equation of the Raizer channel model Eq. (8.33) completely coincides with that one of the Steenbeck model Eq. (8.27) taking the electric conductivity in form Eq. (8.30). The Steenbeck approach was based on the principle of minimum power. Is it possible to interpret the agreement between two modifications of the channel model as a proof of validation of the principle of minimum power for arc discharges?

8.7.16 Arc Temperature in Frameworks of the Channel Model

Derive Eq. (8.36) for the plasma temperature in arc discharges based on the principal equations of the arc channel model. Explain why the arc temperature is close to a constant with only weak logarithmic dependence on current and radius R (characterizing cooling of the discharge). Compare this conclusion with a similar one for non-thermal discharges (like a glow), where the range of changes of electron temperature is also not wide.

8.7.17 Modifications of the Arc Channel Model for the Discharge Stabilization by Gas Flow

Generalize the principal results of the channel model Eq. (8.36)–(8.39) for the case of the arc stabilization by fast gas flow, replacing the radius R by an effective one R_{eff}, which takes into account convective arc cooling. Use the derived relations to explain the increase of plasma temperature in the channel by intensive arc cooling in fast gas flows.

8.7.18 Difference between Electron and Gas Temperatures in Arc Discharges

Using Eq. (8.42), estimate the critical electric current when the difference between electron and gas temperatures in atmospheric pressure argon arc becomes essential. Compare the results with Figure 8.11.

8.7.19 The Bennet Pinch Pressure Distribution

Based on the dynamic balance of expansionary kinetic pressure of plasma p and inward radial magnetic body force Eq. (8.45), together with Eq. (8.43) for the azimuthal magnetic field B_θ, derive the relation for pressure distribution inside of the Bennet pinch of radius r_0. Compare arc stabilization in the Bennet pinch and in industrial arcs with lower currents.

8.7.20 Electrode Jet Formation

Based on Eq. (8.50), estimate the electrode jet velocity corresponding to parameters of cathode spots given in Section 8.3 for atmospheric pressure arc discharges. Discuss possible consequences of the high-speed cathode jets on the pre-electrode arc behavior.

8.7.21 Stabilization of Linear Arcs Near Axis of the Discharge Tube

Based on the Elenbaas-Heller equation (see Section 8.4.2), explain why: if the arc is cooled on its edges, the thermal discharge temperature on axis rises. Use this effect to interpret the stability of linear arcs near the axis of the discharge tube.

8.7.22 Critical Length of Gliding Arc Discharge

Analyze the relation (8.54) for electric current in the quasi-equilibrium phase of gliding arc, and asymptotically simplify if in the vicinity of the critical point when the discharge power approaches its maximum value. Why this relation is unable to describe the current evolution for bigger lengths of the arc?

8.7.23 Quasi-Unstable Phase of Gliding Arc Discharge

Based on the stability diagram shown in Figure 8.34, explain the mechanism leading to instability of the formally stable regime of arc with length lower than the critical one. Compare the quasi-unstable phase of the gliding arc with totally stable and totally unstable regimes of the discharge.

8.7.24 Discharge Power Distribution Between Quasi-Equilibrium and Non-Equilibrium Phases of Gliding Arc

Assuming that the power of the gliding arc remains on the maximum level after the FENETRe-transition and using Eq. (8.71) for the maximum equilibrium and non-equilibrium lengths, derive formula describing the fraction of the total gliding arc energy released in the non-equilibrium phase.

9 Non-Equilibrium Cold Atmospheric Pressure Discharges

9.1 THE CONTINUOUS CORONA DISCHARGE

9.1.1 General Features of the Corona Discharge

Corona is a weakly luminous discharge, which appears at atmospheric pressure near sharp points, edges, or thin wires where the electric field is sufficiently large. Coronas are always non-uniform: strong electric field, ionization, and luminosity are located in the vicinity of one electrode. Weak electric fields drag charged particles to another electrode to close the circuit. No radiation appears from the "outer region" of the corona. Corona can be observed in air around high voltage transmission lines, around lightning rods, and even masts of ships, where they are called "Saint Elmo's fire". This is the origin of the corona name, which means "crown". High voltage is required to ignite the corona, which occupies the region around one electrode. If the voltage grows even larger, the remaining part of the discharge gap breaks down and the corona transfers into the spark, see L.B. Loeb (1965), R. Bartnikas, E.J. McMahon (1979) and M. Goldman, N. Goldman (1978).

9.1.2 Electric Field Distribution in Different Corona Configurations

The corona discharges occur only if the electric field is essentially non-uniform. The electric field in the vicinity of one or both electrodes should be much stronger than in the rest of the discharge gap. It occurs if the characteristic size of an electrode r is much smaller than the characteristic distance d between electrodes. For example, corona discharge in air between parallel wires occurs only if the inter-wire distance is sufficiently large $d/r > 5.85$. Otherwise an increase of voltage results not in a corona, but in a spark. Engineering calculations of coronas require information on the electric field distribution in the system. Neglecting effects of the plasma, these distributions can be found in the framework of electrostatics. For simple discharge gap geometries (see Figure 9.1a–g), such electric field distributions can be found analytically:

a. The electric field in the space between ***coaxial cylinders*** of radii r (internal) and R (external) at a distance x from the axis (Figure 9.1a) can be expressed as:

$$E = \frac{V}{x\ln(R/r)}, \quad E_{\max}(x = r) = \frac{V}{x\ln(R/r)}, \tag{9.1}$$

where V is the voltage between two electrodes.

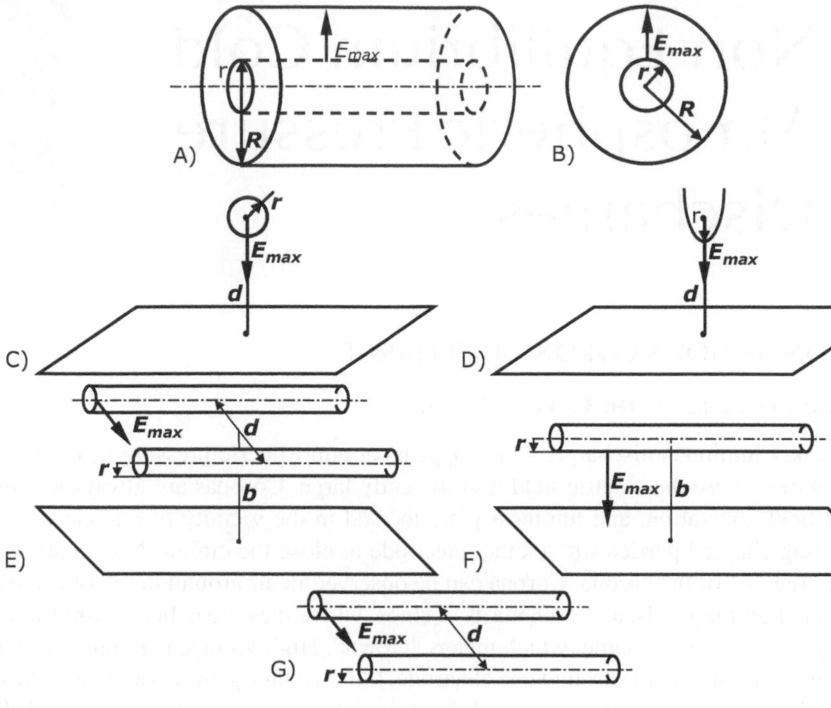

FIGURE 9.1 Illustration of different corona configurations.

b. The electric field in the space between ***concentric spheres*** also of radii r (internal) and R (external) at a distance x from the center (Figure 9.1b) is:

$$E = V\frac{rR}{x^2(R-r)}, \quad E_{\max} \approx \frac{V}{r} \; (\text{if } R >> r). \tag{9.2}$$

c. The electric field in the space between ***a sphere of radius r and a remote plane*** $d/r \to \infty$ (Figure 9.1c) can be written as a function of distance x from the sphere center:

$$E \approx V\frac{r}{x^2}, \quad E_{\max} = \frac{V}{r}. \tag{9.3}$$

d. The electric field in the space between ***a parabolic tip with curvature radius r and a plane perpendicular to it*** and located at a distance d from the tip (Figure 9.1d) can be expressed as the following function of distance x from the tip:

$$E = \frac{2V}{(r+2x)\ln(2d/r+1)}, \quad E_{\max} \approx \frac{2V}{r\ln(2d/r)}. \tag{9.4}$$

e. If the corona is organized between ***two parallel wires of radius r separated on the distance d between each other, and located on distance b from the earth*** (Figure 9.1e), the maximum electric field is obviously achieved near the surface of the wires and equals to:

$$E_{\max} = \frac{V}{r\ln\left[d/\left\{r\cdot\sqrt{1+(d/2b)^2}\right\}\right]}. \tag{9.5}$$

f. Based on Eq. (9.5), the maximum electric field induced by the single wire and a parallel plane located at a distance b ($d \to \infty$, (see Figure 9.1f) can be presented as:

$$E_{max} = \frac{V}{r\ln(2b/r)}. \tag{9.6}$$

g. One can also derive from Eq. (9.5) the maximum electric field between two parallel wires spaced by a distance d between each other ($b \to \infty$, see Figure 9.1g):

$$E_{max} = \frac{V}{r\ln(d/r)}. \tag{9.7}$$

Obviously, the above Eqs. (9.1)–(9.7) describe electric fields in the corona discharge system before the generated plasma perturbs it.

9.1.3 Negative and Positive Corona Discharges

Mechanism for sustaining the continuous ionization level in a corona depends on the polarity of electrode where the high electric field is located. If the high electric field zone is located around the cathode, such corona discharge is referred to as the **negative corona**. Conversely, if the high electric field is concentrated in the region of the anode, such a discharge is called the **positive corona.** Ionization in the negative corona is due to the multiplication of avalanches. Continuity of electric current from the cathode into the plasma is provided by secondary emission from the cathode. Ignition of the negative corona actually has the same mechanism as the Townsend breakdown, generalized taking into account non-uniformity and possible electron attachment processes:

$$\int_0^{x_{max}} \left[\alpha(x) - \beta(x)\right] dx = \ln\left(1 + \frac{1}{\gamma}\right). \tag{9.8}$$

In this equation $\alpha(x)$, $\beta(x)$ and γ are the first, second, and third Townsend coefficients, describing respectively ionization, electron attachment, and secondary electron emission from the cathode; x_{mam} corresponds to the distance from the cathode, where the electric field becomes low enough and $\alpha(x_{max}) = \beta(x_{max})$, which means that no additional electron multiplication takes place. The equality $\alpha(x_{max}) = \beta(x_{max})$ actually corresponds to the breakdown electric field E_{break} in electronegative gases (see Section 4.4). If the gas is not electro-negative ($\beta = 0$), integration of Eq. (9.8) is formally not limited; however, due to the exponential decrease of $\alpha(x)$ it is. The critical distance $x = x_{max}$ determines not only the ionization, but also the electronic excitation zone, and hence the zone of plasma luminosity. This means that the critical distance $x = x_{max}$ can be considered as the visible size of the corona (or an active corona volume). Ionization in the positive corona cannot be provided by the cathode phenomena, because in this case, the electric field at the cathode is low. Here ionization processes are related to the formation of the cathode-directed (or the so-called positive) streamers, see Figure 4.25. Ignition conditions can be described for the positive corona using the criteria of cathode-directed streamer formation. In this case, the generalization of the Meek breakdown criterion Eq. (4.157) is a quite a good approximation, taking into account the non-uniformity of the corona and possible contributions of electron attachment:

$$\int_0^{x_{max}} \left[\alpha(x) - \beta(x)\right] dx \approx 18 - 20. \tag{9.9}$$

Comparing the ignition criteria Eqs. (9.8) and (9.9), the minimal amplification coefficients should be 2–3 times lower to provide the ignition of negative corona (because $\ln(1/\gamma) \approx 6–8$). However, the critical values of the electric field for the ignition of positive and negative coronas are pretty close even though these are related to very different breakdown mechanisms. This can be explained by the strong exponential dependence of the amplification coefficients on the electric field value.

9.1.4 Corona Ignition Criterion in Air, the Peek Formula

According to Eqs. (9.8) and (9.9), ignition for both positive and negative coronas mostly is determined by the maximum electric field in the vicinity of the electrode, where the discharge is to be initiated. The critical value of the igniting electric field (near the electrode) for the case of coaxial electrodes in air can be calculated using the Peek formula:

$$E_{cr}, \frac{\text{kV}}{\text{cm}} = 31\delta\left(1 + \frac{0.308}{\sqrt{\delta r(\text{cm})}}\right). \quad (9.10)$$

In the Peek formula, δ is the ratio of air density to the standard one; r is the radius of the internal electrode. The formula can be applied for pressures 0.1–10 atm, polished internal electrodes with radius $r \approx 0.01-1\,\text{cm}$, with both direct current and AC with frequencies up to 1 kHz. Roughness of the electrodes decreases the critical electric field by 10–20%. The Peek formula Eq. (9.10) was derived for the coaxial cylinders, but it can be used also for other corona configurations with slightly different values of coefficients. The critical corona-initiating electric field for two parallel wires can be calculated as:

$$E_{cr}, \frac{\text{kV}}{\text{cm}} = 30\delta\left(1 + \frac{0.301}{\sqrt{\delta r(\text{cm})}}\right). \quad (9.11)$$

The empirical Eq. (9.11) does not differ much from the Peek formula. Both relations correspond to the formula for the Townsend coefficient α in air at reduced electric fields $E/p < 150\,\text{V/cm} \cdot \text{Torr}$:

$$\alpha, \frac{1}{\text{cm}} = 0.14 \cdot \delta\left[\left(\frac{E, \text{kV/cm}}{31\delta}\right)^2 - 1\right]. \quad (9.12)$$

Equations (9.10) and (9.11) determine the critical corona electric field. Voltage necessary to ignite the corona discharge then can be found based on such relations as Eqs. (9.1)–(9.7). In this case, the critical value of the electric field is supposed to be reached in the close vicinity of an active electrode.

9.1.5 Active Corona Volume

Ionization and effective generation of charged particles take place in corona discharges only in the vicinity of an electrode where the electric field is sufficiently high. This zone is usually referred to as the active corona volume (see Figure 9.2). The active corona volume is the most important part of the discharge from the point of view of plasma-chemical applications because most excitation and reaction processes take place in this zone. External radius (or more general size) of the active corona volume is determined by the value of the electric field, which should correspond to breakdown value E_{break} on the boundary of the active volume, see discussion after Eq. (9.8). Consider the generation of a corona discharge in air between a thin wire electrode of radius $r = 0.1\,\text{cm}$ and coaxial cylinder external electrode with radius $R = 10\,\text{cm}$. According to the Peek formula, the critical igniting electric field in the vicinity of the internal electrode is about 60 kV/cm. In this case, Eq. (9.1) gives the minimum value of voltage required for the ignition of the corona to be about 30 kV. The electric field near

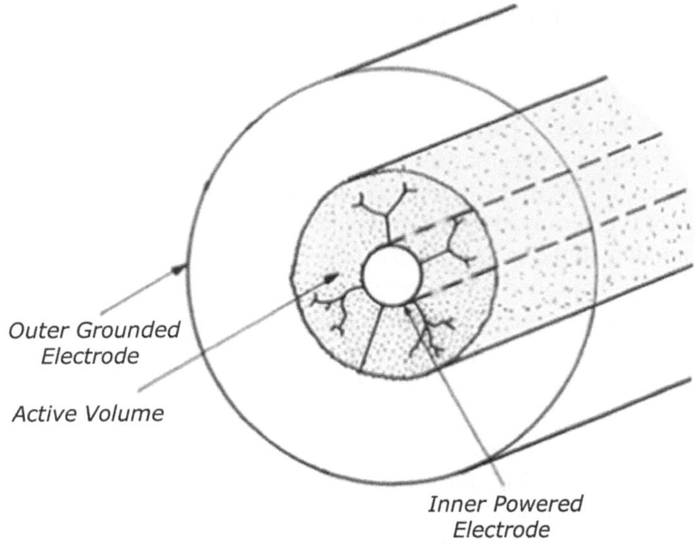

FIGURE 9.2. Illustration of active corona volume.

the external electrode is very low $E(R) \approx 0.6\,\text{kV/cm}$, $E(R)/p \approx 0.8\,\text{V/cm} \cdot \text{Torr}$, and not sufficient for ionization. Effective multiplication of charges requires the breakdown of the electric field, which can be estimated as $E_{break} \approx 25\,\text{kV/cm}$. This determines the external radius of the active corona volume case as $r_{AC} = rE_{cr}/E_{break} \approx 0.25\,\text{cm}$. Hence, the active corona volume occupies the cylindrical layer $0.1\,\text{cm} < x < 0.25\,\text{cm}$ around the thin wire. In general, the external radius of the active corona volume around the thin wire can be determined based on Eq. (9.1) as:

$$r_{AC} = \frac{V}{E_{break} \ln\left(R/r\right)}, \tag{9.13}$$

where V is the voltage applied to sustain the corona discharge. As is seen from Eq. (9.13), the radius of active corona volume is increasing with applied voltage because values of R, r and E_{break} are actually fixed by the discharge geometry and gas composition. Similarly to Eq. (9.13), the external radius of the active corona volume generated around a sharp point can be expressed according to Eq. (9.2) as:

$$r_{AC} \approx \sqrt{\frac{rV}{E_{break}}}. \tag{9.14}$$

Using (9.13) and (9.14), compare the active radii of corona around thin wire and sharp point:

$$\frac{r_{AC}\left(wire\right)}{r_{AC}\left(point\right)} \approx \frac{1}{\ln\left(R/r\right)}\sqrt{\frac{V}{aE_{break}}} = \frac{r_{AC}\left(point\right)}{r\ln\left(R/r\right)}. \tag{9.15}$$

This ratio is about 3, which illustrates the advantage of corona generated around a thin wire, to produce a larger volume of non-thermal atmospheric pressure plasma effective for different applications.

9.1.6 Space Charge Influence on Electric Field Distribution in a Corona Discharge

The generation of charged particles takes place only in the active corona volume in the vicinity of an electrode. Thus, the electric current to the external electrode outside of the active volume is provided by the drift of charged particles (generated in the active volume) in the relatively low electric field. In the positive corona, these drifting particles are positive ions, in the negative corona, these are negative ions (or electrons, if corona is generated in non-electronegative gas mixtures). The discharge current is determined by the difference between the applied voltage V and the critical one V_{cr}, corresponding to the critical electric field E_{cr} and to ignition of the corona (see Section 9.1.4). The current value is limited by the space charge outside of the active corona volume. The active corona volume is able to generate high current, but the current of charged particles is partially reflected back by the space charge formed by these particles. The phenomenon is somewhat similar to the phenomenon of current limitation by space charge in sheathes, or in vacuum diodes (see sections 6.1.4, 6.1.5). However, in the case under consideration, the motion of charged particles is not collisionless but determined by drift in the electric field. To analyze the space charge influence on the electric field distribution, consider the example of a corona generated between coaxial cylinders with radii R and r (e.g. , corona around a thin wire). The electric current per unit length of the wire i is constant outside of active corona volume, where there is no charge multiplication:

$$i = 2\pi x \cdot en \cdot \mu E = const. \tag{9.16}$$

Here x is the distance from corona axis; n is the number density of charged particles providing electric conductivity outside of the active volume; μ is the mobility of the charged particles. The space charge perturbation of the electric field is not very strong, and the number density distribution $n(x)$ can be found based on Eq. (9.16) and non-perturbed electric field distribution Eq. (9.1):

$$n(x) = \frac{i}{2\pi e\mu Ex} = \frac{i\ln(R/r)}{2\pi e\mu V} = const. \tag{9.17}$$

This expression for the charged particles density distribution $n(x)$, then can be applied to find the electric field distribution $E(x)$, using the Maxwell equation for the case of cylindrical symmetry:

$$\frac{1}{x}\frac{d[xE(x)]}{dx} = \frac{1}{\varepsilon_0} en(x), \quad \frac{1}{x}\frac{d[xE(x)]}{dx} = \frac{i\ln(R/r)}{2\pi\varepsilon_0\mu V}. \tag{9.18}$$

The Maxwell equation can be integrated, taking into account that the electric field distribution should follow Eq. (9.1) with the critical value of voltage V_{cr} (corresponding to corona ignition) at the low current limit $i \to 0$. It gives the electric field distribution taking into account current and the space charge:

$$E(x) = \frac{V_{cr}\ln(R/r)}{x} + \frac{i\ln(R/r)}{2\pi\varepsilon_0\mu V} \cdot \frac{x^2 - r^2}{2x}. \tag{9.19}$$

Obviously, in this relation $\int_r^R E dx = V$. Also, the distribution Eq. (9.19) is valid only in the case of small electric field perturbations due to the space charge outside of the active corona volume. Expressions similar to Eq. (9.19), describing the influence of electric current and space charge on the electric field distribution could be derived for other corona configurations (see J.R. Roth, 2000).

9.1.7 Current–Voltage Characteristics of a Corona Discharge

Integration of the expression for the electric field Eq. (9.19) over the radius x taking into account, that in most of the corona discharge gap $x^2 >> r^2$, gives the relation between current (per unit length) and voltage of the discharge, which is the current–voltage characteristic of corona generated around a thin wire:

$$i = \frac{4\pi\varepsilon_0 \mu V\left(V - V_{cr}\right)}{R^2 \ln\left(R/r\right)}. \tag{9.20}$$

The corona current depends on the mobility of the main charge particles providing conductivity outside of the active corona volume. Noting that mobilities of positive and negative ions are nearly equal, the electric currents in positive and negative corona discharges are also close. Negative corona in gases without electron attachment (e.g., noble gases) provides much larger currents because electrons are able to rapidly leave the discharge gap without forming a significant space charge. Even a small admixture of an electro-negative gas decreases the corona current. The parabolic current–voltage characteristic Eq. (9.20) is valid not only for thin wires but for other corona configurations. Obviously, the coefficients before the quadratic form $V(V - V_{cr})$ are different for different geometries of corona discharges:

$$I = CV\left(V - V_{cr}\right). \tag{9.21}$$

In this relation I is the total current in corona. For example, the current–voltage characteristic for the corona generated in atmospheric air between a sharp point cathode with radius $r = 3$–$50\,\mu m$ and a perpendicular flat anode located on the distance of $d = 4$–$16\,mm$ can be expressed as:

$$I, \mu A = \frac{52}{\left(d, \text{mm}\right)^2}\left(V, \text{kV}\right)\left(V - V_{cr}\right). \tag{9.22}$$

In this empirical relation I is the total corona current from the sharp point cathode. The critical corona ignition voltage V_{cr} in this case can be taken as $V_{cr} \approx 2.3\,kV$ and does not depend on the distance d (M. Goldman, N. Goldman, 1978).

9.1.8 Power Released in the Continuous Corona Discharge

Based on Eq. (9.20), the electric power released in the continuous corona discharge can be determined for the case of a long thin wire as:

$$P = \frac{4\pi L\varepsilon_0 \mu V\left(V - V_{cr}\right)}{R^2 \ln\left(R/r\right)}, \tag{9.23}$$

where L is the length of the wire. In more general cases, the corona discharge power can be determined based on the current–voltage characteristic Eq. (9.21) as:

$$P = CV\left(V - V_{cr}\right). \tag{9.24}$$

For the corona generated in atmospheric air between the sharp-pointed cathode with radius $r = 3 - 50\,\mu m$ and a perpendicular flat anode located at a distance of $d = 1\,cm$, the coefficient $C \approx 0.5$ if the voltage is expressed in kV and power is expressed in mW. This yields the corona power as about 0.4 W at the voltage of about 30 kV. This power is low. Similarly, corona discharges generated in atmospheric pressure air around the thin wire ($r = 0.1\,\text{cm}$, $R = 10\,\text{cm}$, $V_{cr} = 30\,\text{kV}$) with voltage 40 kV releases power of about 0.2 W per cm of the discharge. The power of the continuous corona discharges is low and not acceptable for many applications. Recall that further increases of voltage and current leads to corona transition into sparks. However, these can be prevented organizing the corona discharge in a pulse-periodic mode. Although, the corona power is relatively low per unit length of a wire, the total corona power becomes significant when the wires are very long. Such situations takes place in the case of high voltage overland transmission lines, where coronal losses are significant. In humid and snow conditions, these can exceed the resistive losses. In the case of high voltage overland transmission lines, that the two wires generate corona discharges of opposite polarity. Electric currents outside of active volumes of the opposite polarity corona discharges are provided by positive and negative ions moving in opposite directions. These positive and negative ions meet and neutralize each other between wires, which results in a decrease of the space charge and an increase of the corona current which relates to phenomenon power losses. The coronal power losses for a transmission line wire with 2.5 cm diameter and voltage of 300 kV are about 0.8 kW per km in the case of fine and sunny weather. In the case of rainy or snowy conditions, the critical voltage V_{cr} is lower, and coronal losses grow significantly. For this case, the same loss of about 0.8 kW per km corresponds to a voltage about 200 kV.

9.2 THE PULSED CORONA DISCHARGE

9.2.1 Why the Pulsed Corona?

To increase the corona current and power, the voltage and electric field should be increased. However, as the electric field increases, the active corona volume would grow until it occupied the entire discharge gap. When streamers are able to reach the opposite electrode, the formation of a spark channel occurs which subsequently results in local overheating and plasma non-uniformity that are not acceptable for applications. Increasing of corona voltage and power without spark formation becomes possible by using pulse-periodic voltages. Today–the pulsed corona, one of the most promising atmospheric pressure, non-thermal discharges. Streamer velocity is about 10^8 cm/s and exceeds by a factor of 10 the typical electron drift velocity in an avalanche. If the distance between electrodes is about 1–3 cm, the total time necessary for the development of avalanches, avalanche-to-streamer transition, and streamer propagation between electrodes is about 100–300 ns. This means that voltage pulses of this duration range are able to sustain streamers and effective power transfer into non-thermal plasma without streamer transformations into sparks. For the pulsed corona discharges the key point is to make relevant pulse power supplies, which are able to generate sufficiently short voltage pulses with steep front and very short rise times.

9.2.2 Corona Ignition Delay

Numerous experimental data regarding ignition delay of the continuous corona are available. These are quite helpful in understanding pulsed corona discharge. For example, the ignition delay of the continuous negative corona strongly depends on cathode conditions and varies from one experiment to another. Such facility-specific characteristics are one reason why pulsed coronas are more often organized as positive ones. Typical ignition delay in the case of positive corona is about 100 ns (30–300 ns). This interval is much longer than streamer propagation time; hence it is related to the time for initial electrons formation and propagation of initial avalanches. Note that the initial electrons are not formed near the cathode but rather in the discharge gap. Random electrons in the atmosphere usually exist in the form of negative ions O_2^-, their effective detachment is due to ion-neutral collisions and effectively takes place at electric fields about 70 kV/cm. If humidity is high and the negative ions O_2^- are hydrated, the electric field necessary for detachment and formation of a free electron is slightly higher. Experimental data related to the ignition delay of the continuous corona actually indicates the same limits for pulse duration in pulsed corona discharges. This indicates some advantages of the positive corona and the cathode-directed streamers, and also shows the electric fields necessary for effective release of initial free electrons by detachment from negative ions.

9.2.3 Pulse-Periodic Regime of the Positive Corona Discharge Sustained by Continuous Constant Voltage, Flashing Corona

Corona discharges sometimes operate in form of periodic current pulses even at constant voltage conditions. The frequency of these pulses can reach 10^4 Hz in the case of positive corona, and 10^6 Hz for negative corona. This self-organized pulsed corona discharge is obviously unable to overcome the current and power limitations of the continuous corona discharges because continuous high voltage still promotes the corona-to-spark transition. However, it is an important step toward the non-steady-state coronas with higher voltages, higher currents, and higher power. Consider at first a positive corona discharge formed between a sharp point anode of radius 0.17 mm and flat cathode located 3.1 cm apart. Discharge ignition takes place in this system at the critical voltage $V_{cr} \approx 5\,kV$. This corona operates in the pulse-periodic regime starting from the ignition voltage to a voltage $V_1 \approx 9.3\,kV$ (M. Goldman, N. Goldman, 1978). Near the boundary voltage values (V_{cr}, V_1), the frequency of the current pulses is low. The frequency of the pulses reaches the maximum value of about 6.5 kHz in the middle of the interval (V_{cr}, V_1). This pulsing discharge is usually referred to as a **flashing corona**. The mean current value in this regime reaches 1 μA at the voltage $V_1 \approx 9.3\,kV$. Increasing the voltage increase from $V_1 \approx 9.3\,kV$ to $V_2 \approx 16\,kV$ stabilizes the corona; the discharge operates in steady state without pulses. Current grows in this regime from 1 μA to 10 μA. Further increase of voltage from $V_2 \approx 16\,kV$ up to the corona transition into a spark at $V_t \approx 29\,kV$ leads again to the pulse-periodic regime. Frequency grows up in this regime to about 4.5 kHz; mean value of current grows during the same time, to about 100 μA. The flashing corona phenomenon can be explained by the effect of positive space charge, which is created when electrons formed in streamers decrease fast at the anode but slow positive ions remain in the discharge gap. The growing positive space charge decreases electric field near anode and prevents new streamers generation. Positive corona current is suppressed until the positive space charge goes to the cathode and clears up the discharge gap. After that a new corona ignition takes place and the cycle can be repeated again. The flashing corona phenomenon does not occur at intermediate voltages ($V_1 < V < V_2$), when the electric field outside of the active corona volume is sufficiently high to provide effective steady-state clearance of positive ions from the discharge gap but not too high to provide intensive ionization. It is interesting to note that the electric current in the flashing corona regime does not fall to zero between pulses, some constant component of the corona current is continuously present. One can calculate from this experimental data, the maximum power of pulse-periodic corona around a sharp point to be about 3 W. This is more than 10 times higher than the maximum power in a continuous regime for the same corona system which is 0.2 W (see also numerical example after Eq. (9.24)). Thus the pulse-periodic regime leads to a fundamental increase of corona power. However, the power increase in this system is still limited by spark formation because the applied voltage is continuous.

9.2.4 Pulse-Periodic Regime of the Negative Corona Discharge Sustained by Continuous Constant Voltage, Trichel Pulses

Negative corona discharges sustained by continuous voltage also can operate in the pulse-periodic regime at relatively low value of voltages close to the ignition value. The frequency of the pulses is much higher in this case ($10^5 - 10^6$ Hz) relative to the positive corona; the pulse duration is short, approximately 100 ns. If the mean corona current is 20 μA, the peak value of current in each pulse can reach 10 mA. The pulses disappear at higher voltages, and in contrast to the case of positive corona ma, the steady-state discharge exists till transition to spark. The pulse-periodic regime of the negative corona discharge is usually referred to as **Trichel pulses**. General physical causes of the Trichel pulses are similar to those of the flashing corona discussed above, though with some peculiarities. The growth of avalanches from the cathode leads to the formation of two charged layers: (a) an internal one which is positive and consists of positive ions and (b) external one which is negative and consists of either negative ions (in air or other electronegative gases) or electrons in the case of electropositive gases. In electropositive gases, like nitrogen or argon, the Trichel pulses are not generated at all. Because of their high mobility, electrons reach the anode quite fast. As a result, the density of the space charge of electrons in the external layer is very low and as a result, the electric field near the cathode is not suppressed. The positively charged internal layer even increases the electric field in the vicinity of the cathode and provides even better conditions for the active corona volume. Thus the Trichel pulses take place only in electronegative gases. In air and other electronegative gases, negative ions are able to form significant negative space charge around the cathode. This negative space charge cannot be compensated by narrow the layer of positive ions, which are effectively neutralized at the nearby cathode. Thus the space charge of the negative ions suppresses the electric field near cathode and, hence, suppresses the corona current. Subsequently, when the ions leave the discharge gap and are neutralized on the electrodes, the negative corona can be reignited, and the cycle again can be repeated.

9.2.5 Pulsed Corona Discharges Sustained by Nano-Second Pulse Power Supplies

The key element of this corona discharge systems is the nanosecond pulse power supply, which generates pulses with duration in the range 100–300ns, sufficiently short to avoid the corona-to-spark transition. Also, the power supply should provide a pretty high voltage rise rate (0.5–3 kV/ns), which results in higher corona ignition voltage, and higher power. As an illustration of this effect, Figure 9.3 shows the corona inception voltage as a function of the voltage rise rate (F. Mattachini, E. Sani, G. Trebbi, 1996). The high voltage rise rates and, hence higher voltages and mean electron energies in the pulsed corona also result in better efficiency of several plasma-chemical processes requiring higher electron energies. In these plasma-chemical processes, such as plasma cleansing of gas and liquid steams, high values of mean electron energy are necessary to decrease the fraction of

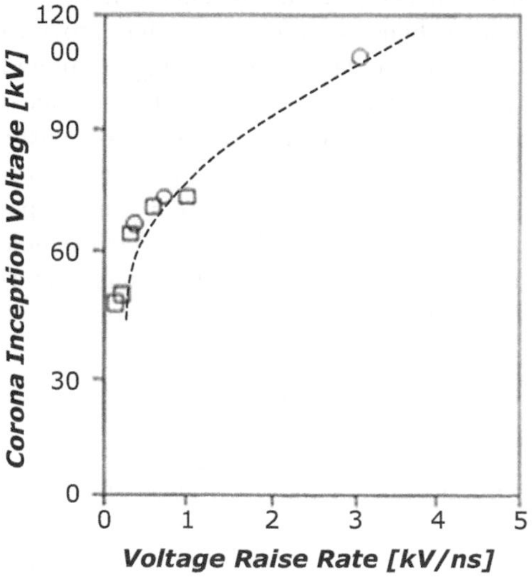

FIGURE 9.3 Corona inception voltage as a function of the voltage raise rate.

the discharge power going to vibrational excitation of molecules and stimulate ionization and electronic excitation and dissociation of molecules. The nanosecond pulse power supply relevant to application in pulse corona discharges can be organized in many different ways. Especially of note is the application of Marx generators, simple and rotating spark gaps, electronic lamps, thyratrons, and thyristors with possible further magnetic compression of pulses (see, for example, Y.K. Pu, P.P. Woskov, 1996). Application of transistors for the high voltage pulse generation is also done. Consider the pulse power supply for a corona discharge based on the thyristor triggered spark gap, which switches the capacitor (E.M. van Veldhuizen, W.R. Rutgers, V.A. Bityurin, 1996; L.M. Zhou, E.M. van Veldhuizen, 1996). The general scheme of this power supply is illustrated in Figure 9.4; the generated voltage and current wave-shapes are shown in Figure 9.5. The high voltage transformer provides a 70 kV voltage pulse with a rise time about 100 ns and half-width 180 ns. The maximum pulse voltage rise rate was not very high in these experiments (0.7 kV/ns) due to the large inductance of the transformer; this also resulted in a long and decayed oscillation on the tail of the voltage pulse. Compare –this with the voltage rise rate claimed by F. Mattachini, E. Sani, G. Trebbi, 1996, which exceeded 3 kV/ns. The half-widths of their corona current pulses were 140–150 ns, which was sufficient to prevent the corona-to-spark transition. To reach higher current pulses and increase the pulse energy, a DC-bias voltage can be effectively added (see Figure 9.4). At a DC-bias voltage of about 20 kV, the single pulse energy of the corona is about 0.3–1 J/m (L.M. Zhou, E.M. van Veldhuizen, 1996; F. Mattachini, E. Sani, G. Trebbi, 1996). The power of a pulsed corona is usually varied by changing the pulse frequency. In the referred systems, frequencies were up to 300 kHz, which yields the power of 3 W per 1 cm of wire. This specific power of the pulsed corona discharge is 15 times larger than that of the continuous corona. The repetition frequency of high voltage pulses could be higher, which provides the possibility of further increase of specific power of pulsed corona discharges. The power supply with a magnetic pulse compression system delivers up to 35 kV, 100 ns pulses at repetition rates up to 1.5 kHz (B.M. Penetrante 1996). In general, the total power of the pulsed coronas can be increased by increasing the length of the wire and reach high values for non-thermal atmospheric pressure discharges. Thus the pulsed corona discharge described by F. Mattachini, E. Sani, G. Trebbi, 1996, should operate at powers exceeding 10 kW, and that one presented by S. Korobtsev, D. Medvedev, V. Rusanov, V. Shiryaevsky, 1997, is able to operate at power level of 30 kW. The pulsed corona can be relatively powerful, luminous and quite nice looking, which is more or less illustrated in Figure 9.6. More information about the physical aspects and applications of the pulsed corona discharge can be found in the publications of B. Penetrante, S. Schulteis (1993), and S. Masuda (1998).

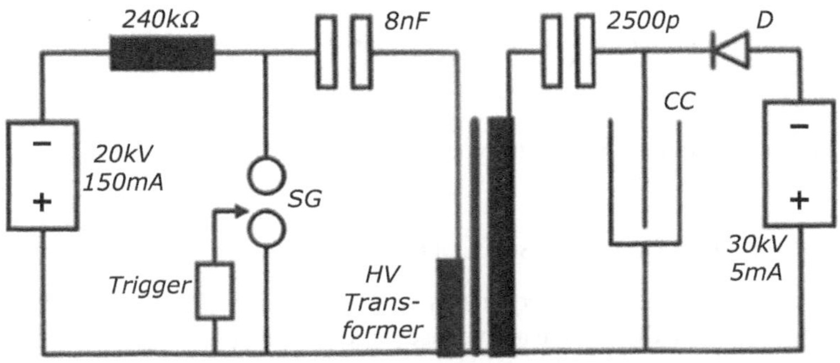

FIGURE 9.4 The electrical circuit for the production of high-voltage pulses with a DC bias.

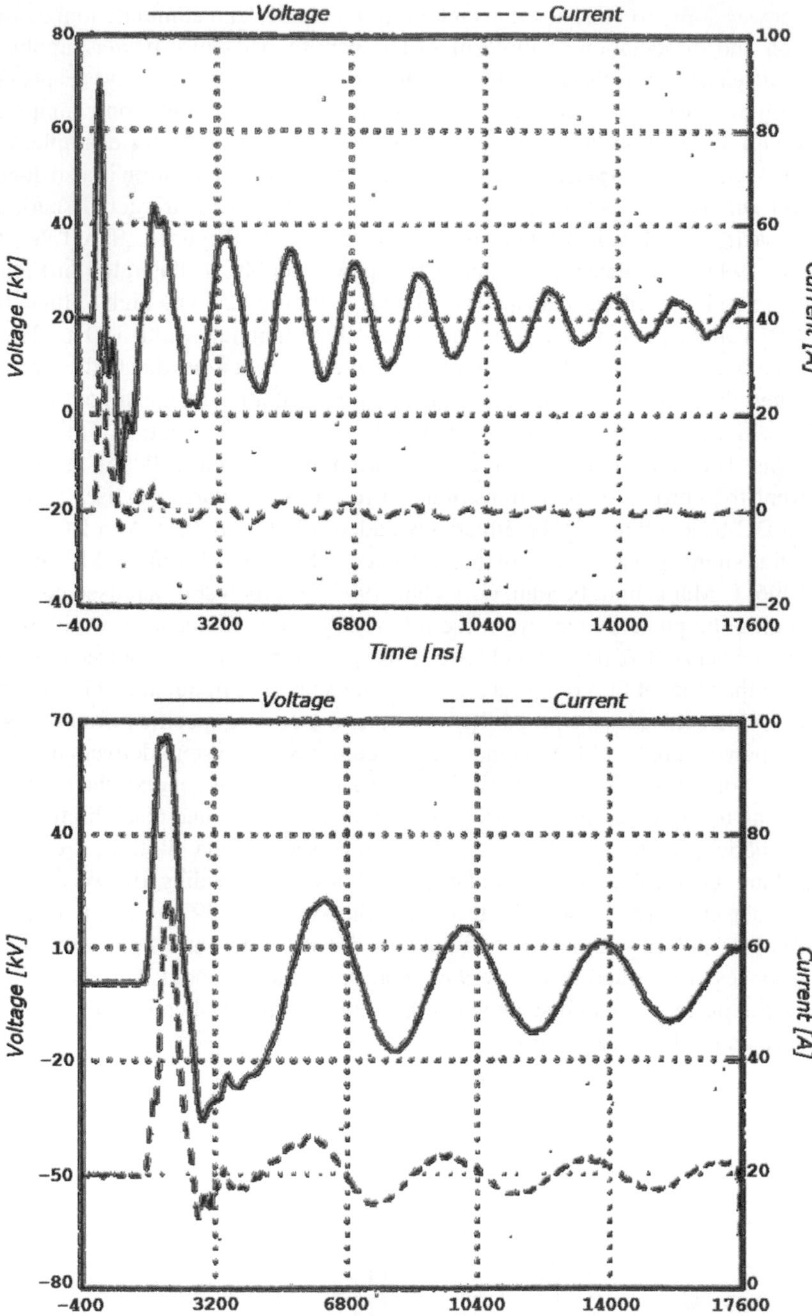

FIGURE 9.5 Current and voltage evolution in pulsed corona.

9.2.6 Specific Configurations of the Pulsed Corona Discharges

The most typical configurations of the pulsed corona as well as continuous corona discharges are based on using thin wires, which maximizes the active discharge volume (see Section 9.1.5). One of these con urations is illustrated in Figure 9.7. Limitations of the wire configuration of the corona are related by the durability of the electrodes and also by non-optimal interaction of discharge volume

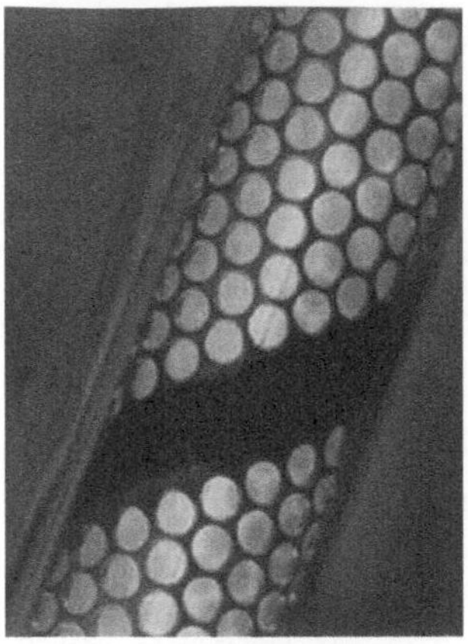

FIGURE 9.6 Pulsed corona discharge.

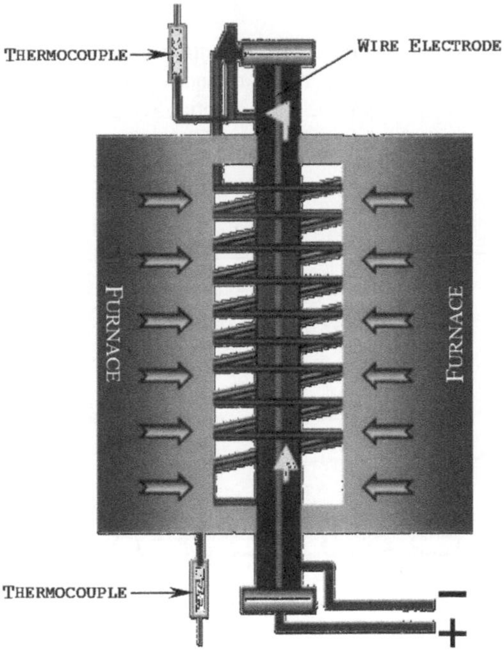

FIGURE 9.7 Pulsed corona discharge in wire cylinder configuration with preheating, developed in the University of Illinois at Chicago.

with incoming gas flow. Another corona configuration, based on multiple stages of pin-to-plate electrodes is useful, see Figure 9.8 (M. Park, D. Chang, M. Woo, G. Nam, S. Lee, 1998). This system is durable, see Figure 9.9, and able to provide good interaction of the incoming gas stream with the active corona volume formed between the electrodes with pins and holes. The pulse corona was successfully combined with catalysis to achieve improved results in the plasma treatment of automotive exhausts (B.M. Penetrante 1998), and for hydrogen production from heavy hydrocarbons (M. Sobacchi 2001). Another technological hybrid is the pulsed corona coupled with water flow. Such a system can be arranged either in form of shower, which is called the **spray corona** or with a thin water film on the walls, which is usually referred to as the **wet corona** (see Figure 9.10).

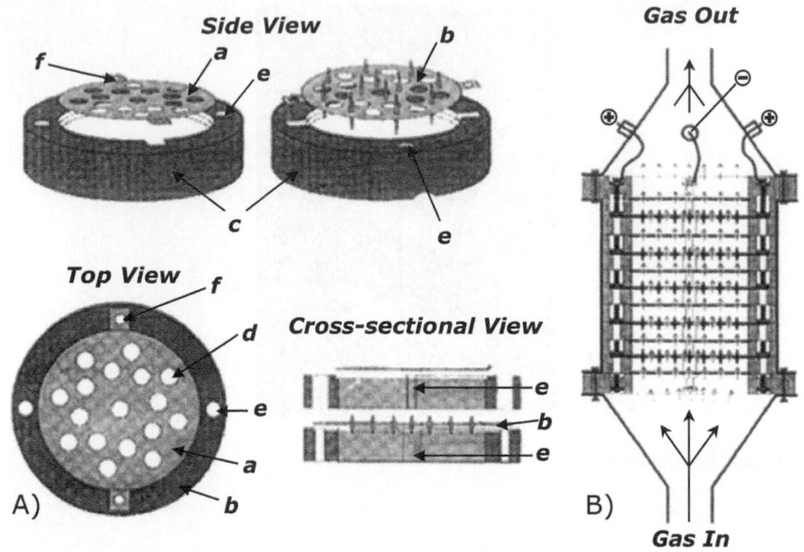

FIGURE 9.8 Schematic diagram of electrodes and mounting blocks (A) and cross-sectional view of the assembled plasma reactor and incoming gas stream (B). a – anode plates; b – cathode plate; c – mounting block; d – holes for gas flow; holes for connecting post; f – connection wings; (From SAE SP-1395 Plasma Exhaust After Treatment).

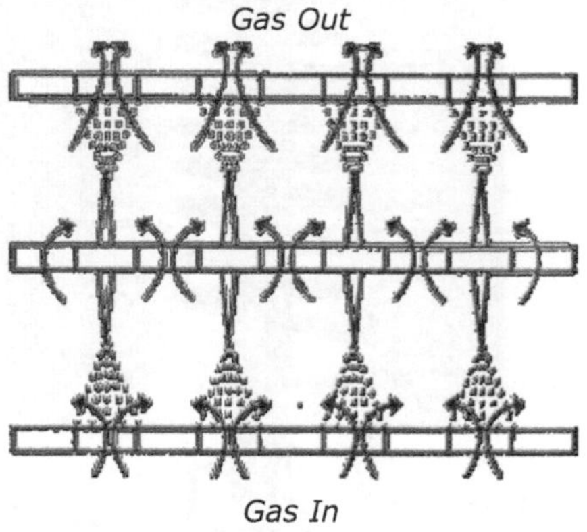

FIGURE 9.9 Illustration of interaction of incoming gas stream with corona discharge formed in between the electrodes with pins and holes.

9.3 DIELECTRIC-BARRIER DISCHARGE

9.3.1 General Features of the Dielectric-Barrier Discharge

The corona-to-spark transition at high voltage is prevented in pulsed corona discharges by employing special nano-second pulse power supplies. An alternate approach to avoid formation of sparks and current growth in the channels formed by streamers is to place a dielectric barrier in the discharge gap. This is the principal idea of the **dielectric barrier discharges** (DBD). The presence of a dielectric barrier in the discharge gap precludes DC operation of DBD and which usually operates at frequencies between 0.5 and 500 kHz. Sometimes dielectric-barrier discharges are also called **silent discharges**. This is due to the absence of sparks, which are accompanied by local overheating,

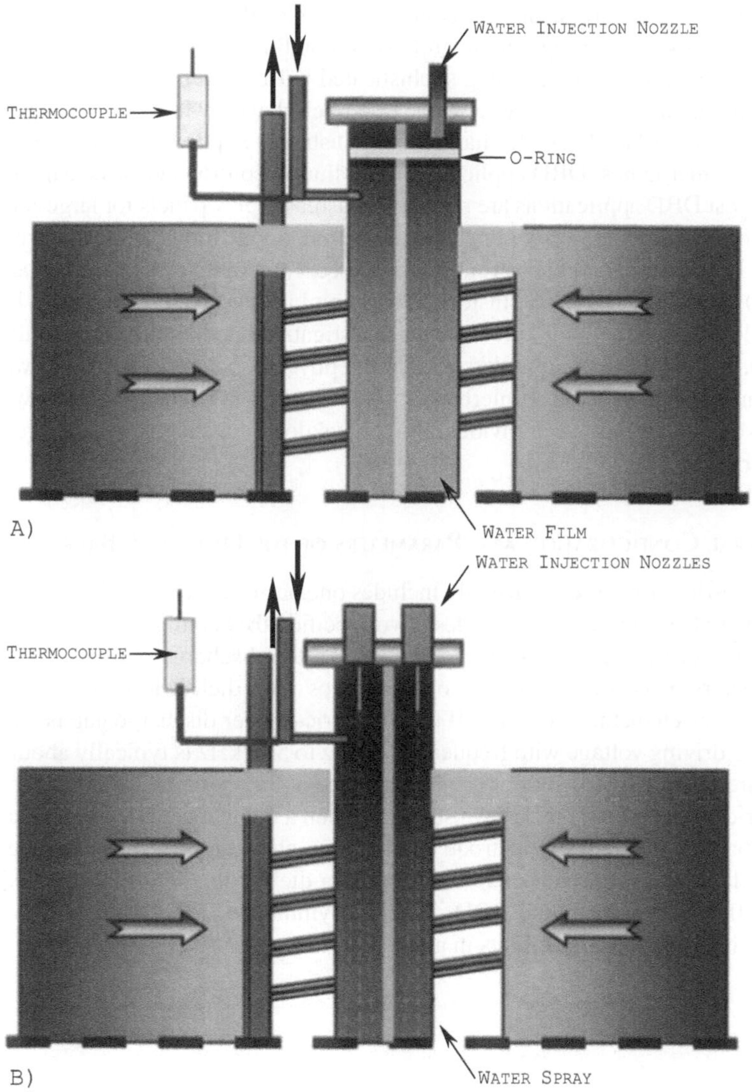

FIGURE 9.10 Wet (a) and spray (b) corona discharge developed in the University of Illinois at Chicago.

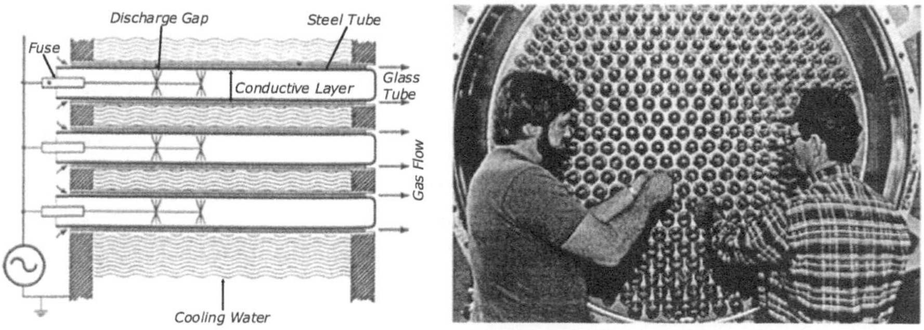

FIGURE 9.11 Schematic diagram of discharge tubes and photograph of large ozone generator at the Los Angeles Aqueduct Filtration Plant (*From K. L. Rakness, J. dePhysique IV, V7 Colloque C4 1977*).

generation of local shock waves, and noise. an advantage of the DBD is the simplicity of its operation. It can be employed in strongly non-equilibrium conditions at atmospheric pressure and at reasonably high power levels, without using sophisticated pulse power supplies. Today the DBD finds large-scale industrial use for ozone generation (U. Kogelschatz, 1988; U. Kogelschatzs, B. Eliasson, 1995; also see Figure 9.11). These discharges are industrially applied as well in CO_2 lasers, and as a UV-source in excimer lamps. DBD application for pollution control and surface treatment is promising, but the largest DBD applications are related to plasma display panels for large-area flat television screens. Contributions to understanding and industrial applications of DBD were made by U. Kogelschatzs, B. Eliasson, and their group at ABB (see, for example, U. Kogelschatzs, B. Eliasson, W. Egli, 1997). The DBD has a long history. It was first introduced by W. Siemens in 1857 to create ozone which determined the main direction for investigations and applications of this discharge for many decades. Important steps in understanding the physical nature of the DBD were made by K. Bussin 1932 and A. Klemenc, H. Hinterberger, H. Hofer in 1937. Their work showed that this discharge occurs in a number of individual tiny breakdown channels, which are referred to as **micro-discharges**.

9.3.2 General Configuration and Parameters of the Dielectric Barrier Discharges

The dielectric barrier discharge gap usually includes one or more dielectric layers, which are located in the current path between metal electrodes. Two specific DBD configurations, planar and cylindrical are illustrated in Figure 9.12. Typical clearance in the discharge gaps varies from 0.1 mm to several centimeters. Breakdown voltages of these gaps with dielectric barriers are practically the same as those between metal electrodes. If the dielectric-barrier discharge gap is a few millimeters, the required AC driving voltage with frequency 500 Hz to 500 kHz is typically about 10 kV at atmospheric pressure. The dielectric barrier can be made from glass, quartz, ceramics, or other materials of low dielectric loss and high breakdown strength. Then a metal electrode coating can be applied to the dielectric barrier. The barrier-electrode combination also can be arranged in the opposite manner, for example, metal electrodes can be coated by a dielectric. Steel tubes coated by an enamel layer can be effectively used in the DBD. Special asymmetric DBD configuration is discussed in Section 6.4.11 regarding plasma effects in aerodynamics.

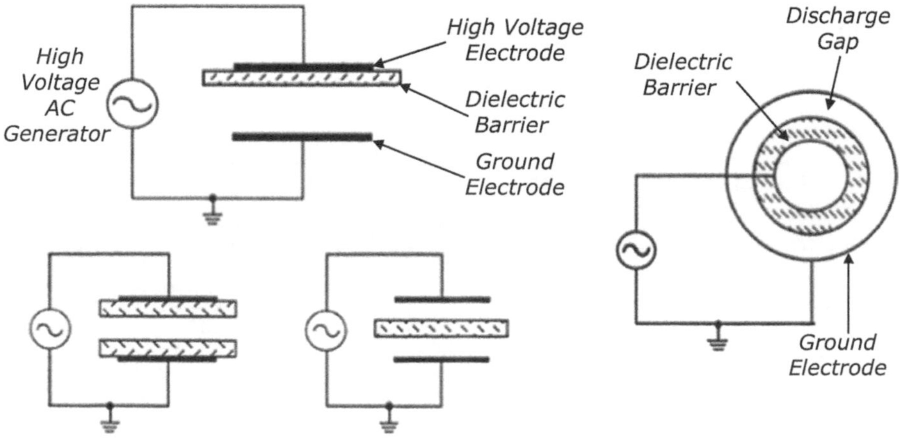

FIGURE 9.12 Common dielectric-barrier discharge configurations (*From K. L. Rakness, J. dePhysique IV, V7 Colloque C4 1977*).

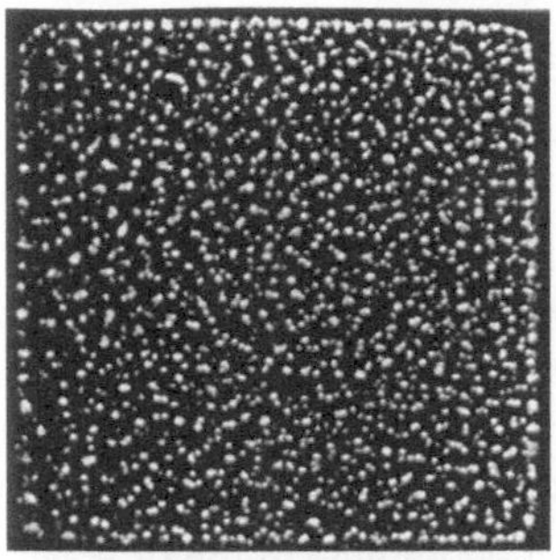

FIGURE 9.13 End-on view of micro discharges; original size: 6 × 6 cm, exposure time: 20 ms. (*From K. L. Rakness, J. dePhysique IV, V7 Colloque C4 1977*).

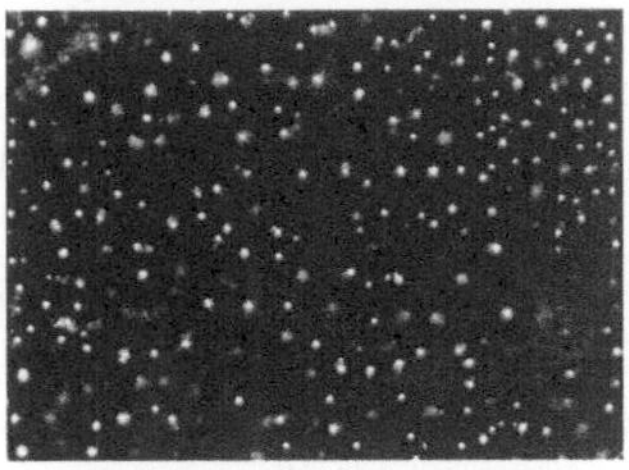

FIGURE 9.14 Lightenberg figure showing footprints of individual micro discharges; original size: 7 × 10 cm. (*From K. L. Rakness, J. dePhysique IV, V7 Colloque C4 1977*).

9.3.3 Micro-Discharge Characteristics

The dielectric-barrier discharge proceeds in most gases through a large number of independent current filaments usually referred to as micro-discharges. From a physical point of view, these micro-discharges are actually streamers that are self-organized taking into account charge accumulation on the dielectric surface. Typical characteristics of the DBD micro-discharges in a 1-mm gap in atmospheric air are summarized in Table 9.1. The snapshot of the micro-discharges in a 1-mm DBD air gap photographed through a transparent electrode is shown in Figure 9.13. As seen, the micro-discharges are spread over the whole DBD-zone quite uniformly. Footprints of the micro-discharges left on a photographic plate with the emulsion facing the discharge gap and the glass plate serving as the dielectric barrier are shown in Figure 9.14. The extinguishing voltage of micro-discharges is not far below the voltage of their ignition. Charge accumulation on surface of the dielectric barrier reduces the electric field at the location of a micro-discharge. This results in current termination within just several nanoseconds after breakdown (see Table 9.1). The short duration of micro-discharges leads to very low overheating of the streamer channel, and the DBD plasma remains

TABLE 9.1
Typical Parameters of a Microdischarge

Lifetime	1–20 ns	Filament radius	50/100 μm
Peak current	0.1 A	Current density	0.1/1 kA cm^{-2}
Electron density	1014/1015 cm^{-3}	Electron energy	1–10 eV
Total transported charge	0.1–1 nC	Reduced electric field	$E/n = (1/2)(E/n)_{Paschen}$
Total dissipated energy	5 μJ	Gas temperature	Close to advantage, about 300 K
Overhearing	5 K		

strongly non-thermal. New micro-discharges then occur at new positions because the presence of residual charges on the dielectric barrier reduced the electric fields at the locations where micro-discharges have already occurred. However, when the voltage is reversed, the next micro-discharges will be formed for the same reason in the old locations. As a result, the high voltage low-frequency DBD-s have a tendency of spreading micro-discharges, while low voltage high-frequency DBD-s tend to reignite the old micro-discharge channels every half-period. The AC-plasma displays uses this memory phenomenon based on charge deposition on the dielectric barrier. The principal micro-discharge properties for most of frequencies do not depend on the characteristics of the external circuit, but only on thetas composition, pressure and the electrode configuration. An increase of power just leads to generation of a larger number of micro-discharges per unit time, which simplifies scaling of the dielectric-barrier discharges. Modeling of the micro-discharges is related to the analysis of the avalanche-to-streamer transition and streamer propagation, see Section 4.4.2 D-modeling of formation and propagation of streamers can be found in E.E. Kunchardt, Y. Tzeng, 1988; E. Marode ea., 1995; S.K. Dali, P.F. Williams, 1987; K. Yoshida, H. Tagashira, 1979; A.A. Kulikovsky, 1994; N. Babaeba, G. Naidis, 1996; R. Morrow, J.J. Lowke, 1997; P.A. Vitello, B.M. Penetrante, J.N. Bardsley, 1994). More detailed DBD analysis of streamers includes charge accumulation on the dielectric barrier (see B. Eliasson, U. Kogelschatz, 1991; G. J. Pietsch, D. Braun, V.I. Gibalov, 1993). It shows that the arrival of a cathode-directed streamer to the dielectric barrier creates within a fraction of ns, a cathode layer with a thickness about 200 μm and an extremely high electric field of several thousands V/cm · Torr. The time and space evolution of such a streamer is presented in Figure 9.15. An interesting phenomena can occur due to the mutual influence of streamers in a DBD. These are related to the electrical interaction of streamers with residual charge left on the dielectric barrier and with the influence of excited species generated in one streamer on the propagation of another streamer (see, X.P. Xu, M.J. Kuchner, 1998; K. Iskenderova 2001).

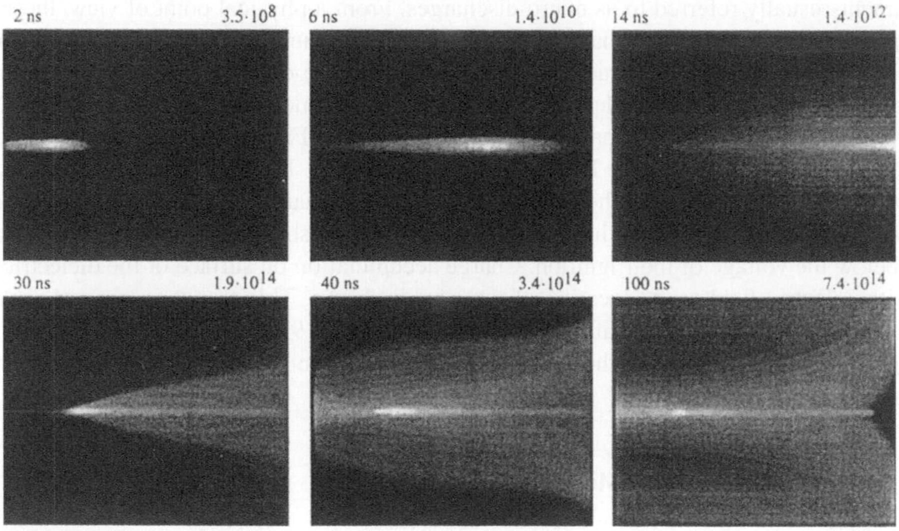

FIGURE 9.15 Development of a micro discharge in an atmospheric pressure H_2/CO_2 mixture (4/1). The 1 mm discharge gap is bounded by a plane metal cathode (left) and a 0.8 mm thick dielectric of $\varepsilon = 3$ (right). A constant voltage is applied, that results in an initially homogeneously reduced field of 125 Td in the gas space which corresponds to an overvoltage of 90% in this mixture. The numbers in the right upper corner indicates the maximum electron density in cm^{-3} reached in that picture. The maximum current of 35 mA is reached at 40 ns. (*From K. L. Rakness, J. dePhysique IV, V7 Colloque C4 1977*).

9.3.4 Surface Discharges

Related to the DBD are surface discharges, generated at dielectric surfaces imbedded by metal electrodes in a different way. The dielectric surface essentially decreases the breakdown voltage in such systems, because of the creation of significant non-uniformities of electric field and hence creates a local overvoltage. The surface discharges, as well as DBD, can be supplied by AC or pulsed voltage. A very effective decrease of the breakdown voltage can be reached in the surface discharge configuration, where one electrode just lays on the dielectric plate, with another one partially wrapped around as it is shown in Figure 9.16a (B.M. Borisov, O.F. Khristoforov, 2001). This discharge is called **sliding discharge**. It can be pretty uniform in some regimes on the dielectric plates of high surface area with linear sizes over 1m at voltages not exceeding 20 kV (see Figure 9.16b). The component of electric field E_y normal to the dielectric surface plays an important role in generation the pulse-periodic sliding discharge that does not depend essentially on the distance l between electrodes along the dielectric (axis x in Figure 9.16a). For this reason, breakdown voltages of the sliding discharge don't follow the Paschen law (see Section 4.4.2). As an example, the breakdown curve for the sliding discharge in air is given in Figure 9.17. One can achieve two qualitatively different modes of the surface discharges by changing the applied voltage amplitude: (1.) complete one (sliding surface spark) and (2.) incomplete one (sliding surface corona). Pictures of both complete and incomplete surface discharges are presented in Figure 9.18. The sliding surface corona discharge takes place at voltages below the critical breakdown value. Current in this discharge regime is low and limited by charging the dielectric capacitance. Active volume and luminosity of this discharge is localized near the igniting electrode and does not cover all the dielectric. The sliding surface spark, (or the complete surface discharge) takes place at voltages exceeding the critical one corresponding to breakdown. In this case, the formed plasma channels actually connect electrodes of the surface discharge gap. At low overvoltages, the breakdown delay is of about 1 μs. In this case the

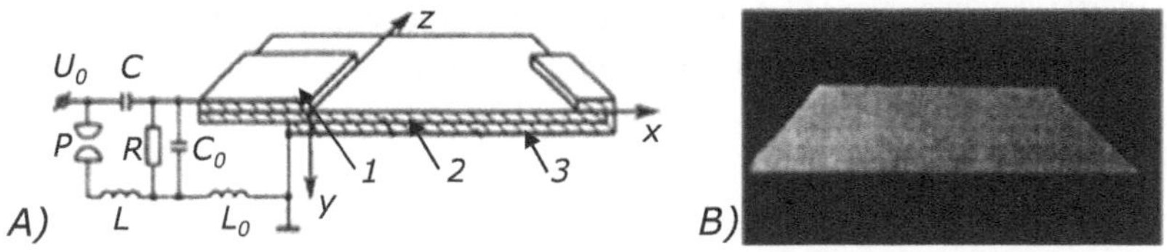

FIGURE 9.16 Schematic (A) and emission picture (B) of a pulse surface discharge: 1 – initiating electrode, 2 – dielectric, 3 – shielding electrode. (*From Encyclopedia of Low Temperature Plasma, Ed. V.E Fortov, Nauka, Moscow, 2002*).

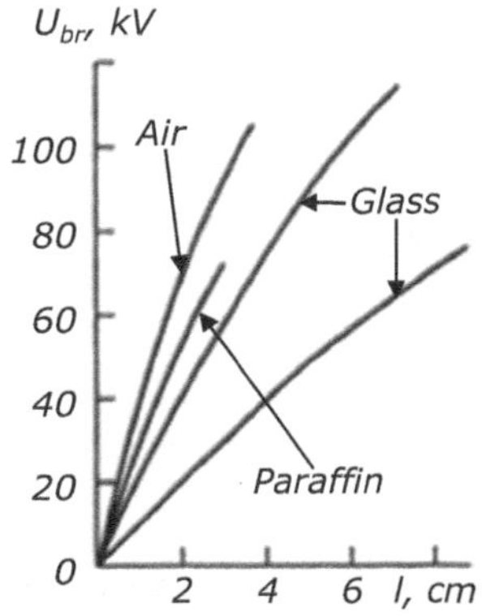

FIGURE 9.17 Breakdown voltage in air along different insulation cylinders (d = 50 mm, f = 50 Hz).

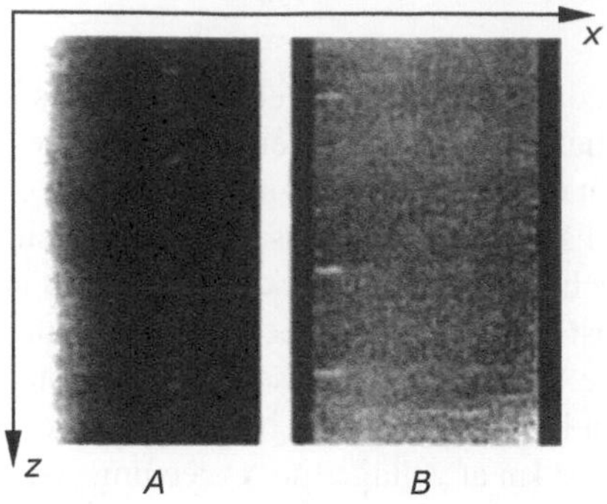

FIGURE 9.18 Complete (B) and incomplete (A) surface discharges (He, $p = 1$ atm, $\varepsilon \approx 5$, $d = 0.5$ mm, pulse frequency 6×10^{13} Hz). (*From Encyclopedia of Low Temperature Plasma, Ed. V.E Fortov, Nauka, Moscow, 2002*).

many step breakdown phenomenon starts with the propagation of a direct ionization wave, which is followed by a possible more intense reverse wave related to the compensation of charges left on the dielectric surface. After about 0.1 μs, the complete surface discharge covers the entire electrodes of the discharge gap. The sliding spark at low overvoltage usually consists of only one or two current channels. At higher overvoltages, the breakdown delay becomes shorter reaching the ns timerange. In this case the complete discharge regime takes place immediately after the direct ionization wave reaches the opposite electrode. The surface discharge consists of many current channels in this regime. In general, the sliding spark surface discharge is able to generate the luminous current channels of very sophisticated shapes, usually referred to as **the Lichtenberg figures**. Some simple, but non-typical examples of the Lichtenberg figures are shown on Figure 9.14. The number of the channels r depends on the capacitance factor ε/d (ratio of dielectric permittivity over thickness of dielectric layer), which determines the level of electric field on the sliding spark discharge surface. This effect is illustrated in Figure 9.19 and is important in the formation of large area surface discharges with homogeneous luminosity, see E.M. Baselyan and Yu.P. Raizer, 1977.

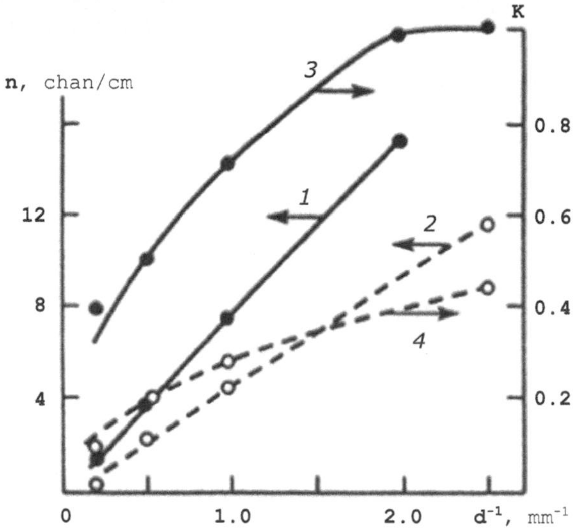

FIGURE 9.19 Linear density of channels n (1,2) and surface coverage by plasma K (3,4) as function of d^{-1} inverse dielectric thickness: He (1,3), air (2,4). (*From Encyclopedia of Low Temperature Plasma, Ed. V.E Fortov, Nauka, Moscow, 2002*).

9.3.5 The Packed-Bed Corona Discharge

The packed-bed corona is an interesting combination of the dielectric barrier discharge DBD and the sliding surface discharge. In this system, high AC voltage (about 15–30 kV) is applied to a packed bed of dielectric pellets and creates a non-equilibrium plasma in the void spaces between the pellets (W.O. Heath, J.G. Birmingham,1995; J.G. Birmingham, R.R. Moore, 1990). The pellets effectively refract the high voltage electric field, making it essentially non-uniform and stronger than the externally applied field by a factor of 10 to 250 times depending on the shape, porosity and dielectric constant of the pellet material.A typical scheme for organizing a packed-bed corona is shown in Figure 9.20a; picture of the discharge is presented in Figure 9.20b. The discharge chamber shown in the figure is shaped as coaxial cylinders with an inner metal electrode and an outer tube made of glass. The dielectric pellets are placed in the annular gap. A metal foil or screen in contact with the outside surface of the tube serves as the ground electrode. The inner electrode is connected to a high voltage AC power supply operated on the level of 15–30 kV at a fixed frequency of 60 Hz or at variable frequencies. the glass tube serves in this discharge system as a dielectric barrier to inhibit direct charge transfer between electrodes and as a plasma-chemical reaction vessel.

9.3.6 Atmospheric Pressure Glow Modification of the Dielectric Barrier Discharge

Principal concepts and physical effects regarding the atmospheric pressure glow discharges were already discussed in Section 7.5.6. Such discharges can be effectively organized in a DBD configuration. In this case the key difference is related to using only special gases, for example, helium. The atmospheric pressure glow DBD modification permits arranging the barrier discharge homogeneously without streamers and other spark-related phenomena. Practically it is important that the glow modification of DBD can be operated at much lower voltages (down to hundreds volts) with respect to those of traditional DBD conditions. A detailed explanation of the special functions of helium in the atmospheric pressure glow discharge is not known. However it is clear that these are related mostly to the following effects. First, it is related to the high electronic excitation levels of helium and the absence of electron energy losses on vibrational excitation. This leads to high values of electron temperatures at lower levels of the reduced electric field. To see this effect compare the reduced electric fields necessary to sustain glow discharges in inert and molecular gases, shown in Figures 7.12 and 7.13. Second, it is related to heat and mass transfer processes that are relatively fast in helium. This prevents contraction and other instability effects in the glow discharge at high pressures. One can state that streamers are overlapping in this case. The same processes can be important in preventing

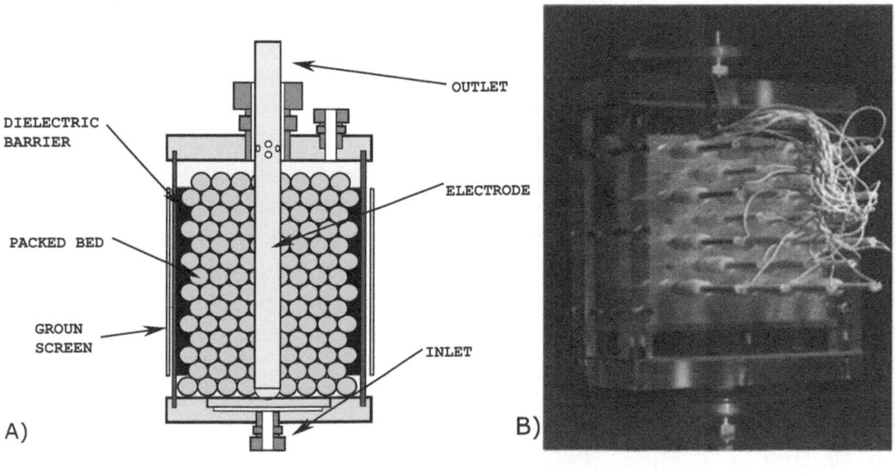

FIGURE 9.20 UIC's Packed-bed corona: scheme (a) and picture (b).

the generation of space-localized streamers and sparks. An important role in avoiding narrow streamers is played by the "memory effect"; this is the influence of particles generated in a previous streamer on a subsequent streamer. The memory effect can be related to metastable atoms and molecules and to electrons deposited on the dielectric barrier, see S. Kanazawa, M. Kogoma, T. Moriwaki, S. Okazaki, 1988; B. Lacour, C. Vannier, 1987; Y. Honda, F. Tochikubo, T. Watanabe, 2001.

9.3.7 Ferroelectric Discharges

Special properties of DBD can be revealed by using ferroelectric ceramic materials of a high dielectric permittivity (ε above 1000) as the dielectric barriers (see, for example, A. Szymanski, 1985; T. Opalinska, A. Szymanski, 1996). Today, ceramics based on $BaTiO_3$ are most employed ferroelectric material for DBD. To illustrate the peculiarities of the ferroelectric discharge, recall that the mean power of the dielectric-barrier discharges can be determined by the equation (see, for example, V.G. Samoylovich, V.I. Gibalov, K.V. Kozlov, 1989):

$$P = 4fC_dV\left(V - V_{cr}\frac{C_d + C_g}{C_d}\right). \tag{9.25}$$

Here f is the frequency of applied AC voltage; V is amplitude value of the voltage; V_{cr} is the critical value of the voltage corresponding to breakdown of the discharge gap; C_d and C_g are the electric capacities of a dielectric barrier and gaseous gap respectively. When the dielectric barrier capacity exceeds that of the gaseous gap $C_d >> C_g$, the relation for discharge power can be simplified:

$$P = 4fC_dV\left(V - V_{cr}\right). \tag{9.26}$$

From Eq. (9.26), high values of dielectric permittivity of the ferroelectric materials are helpful in providing relatively high discharge power at relatively low values of frequency and applied voltage. The ferroelectric discharge based on $BaTiO_3$ which employs a ceramic barrier with a dielectric permittivity ε exceeding 3000, thickness 0.2–0.4 cm and gas discharge gap 0.02 cm operates effectively at an AC frequency of 100 Hz, voltages below 1 kV and electric power of about 1W (T. Opalinska, A. Szymanski, 1996). Such discharge parameters are of particular interest for special practical applications such as medical ones, where low voltage and frequency are very desirable for safety and simplicity reasons, and high discharge power is not required. Physical peculiarities of the ferroelectric discharges are related to the physical nature of the ferroelectric materials, which in a given temperature interval can be spontaneously polarized. Such spontaneous polarization means that the ferroelectric materials can have a non-zero dipole moment even in the absence of external electric field. The electric discharge phenomena accompanying contact of a gas with a ferroelectric sample were first observed in detail by G. D. Robertson and N.A. Baily (1965). The first qualitative description of this sophisticated phenomenon was developed by J. Kusz, 1978. The long-range correlated orientation of dipole moments can be destroyed in ferroelectrics by thermal motion. The temperature at which the spontaneous polarization vanishes is called the temperature of ferroelectric phase transition or the **ferroelectric Curie point**. When the temperature is below the ferroelectric Curie point, the ferroelectric sample is divided into macroscopic uniformly polarized zones called the **ferroelectric domains** and illustrated in Figure 9.21. From Figure 9.21, the directions of the polarization vectors of individual domains in the equilibrium state are set up in the way to minimize internal energy of the crystal and to make polarization of the sample as a whole close to zero. The application of an external AC voltage leads to overpolarization of the ferroelectric material and reveals strong local electric fields on the surface. As it was shown by H. Hinazumi, M. Hosoya, T. Mitsui, 1973, this local surface electric fields can exceed 10^6 V/cm, which stimulate the discharge on ferroelectric surfaces. Thus the active volume of the ferroelectric discharge is located in the vicinity of the dielectric barrier, which is essentially the narrow inter-electrode gaps typical for the discharge. In this case scaling of the ferroelectric discharge can be achieved by using some special configurations. One such special discharge configuration is comprised of a series of parallel thin ceramic plates. High dielectric permittivity of ferroelectric ceramics enables such multi-layer sandwich to be supplied by only two edge electrodes. Another interesting configuration can be arranged by using a packed bed of the ferroelectric pellets. In the same manner as was described in Section 9.3.5, non-equilibrium plasma is created in such system in the void spaces between the pellets.

9.4 SPARK DISCHARGES

9.4.1 Development of a Spark Channel, Back Wave of Strong Electric Field, and Ionization

When streamers provide an electric connection between electrodes and neither a pulse power supply nor a dielectric barrier prevents further growth of current, it opens an opportunity for the development of a spark. The initial streamer channel does not have very high conductivity and usually provides only a very low current of about 10 mA. The potential of the head of the cathode-directed streamer is close to the anode potential. This is the region of strong an electric field around the streamer's head. While the streamer approaches the cathode, this electric field is obviously growing. It stimulates the intensive formation of electrons on the cathode surface and its vicinity and subsequently their fast multiplication in this elevated electric field. New ionization waves more intense than the original streamer start propagating along the streamer channel but in opposite direction from the cathode to the anode. This is referred to as the **back ionization wave** and propagates back to the anode with high velocity about 10^9 cm/s. The high velocity of the back ionization wave is not directly the velocity of electron motion, but rather the phase velocity of the ionization wave. The back wave is accompanied by a front of intensive ionization and the formation of a plasma channel with sufficiently high conductivity to form a channel of the intensive spark.

9.4.2 Expansion of Spark Channel and Formation of an Intensive Spark

The high-density current initially stimulated in the spark channel by the back ionization wave results in intensive Joule heating, growth of the plasma temperature and a contribution of thermal ionization. Gas temperatures in the spark channel reach 20,000 K, electron concentration rise to about 10^{17} cm^{-3}, which is already close to the complete ionization. Electric conductivity in the spark channel at such a high level of degree ionization is determined by coulomb collisions and actually does not depend on electron density (see Section 4.3.5). According to Eq. (4.114), the conductivity can be estimated in this case as 10^2 $Ohm^{-1} \cdot cm^{-1}$. Further growth of the spark current is related not to an increase of the ionization level and conductivity, but just to the expansion of the channel and the increase of its cross section. The fast temperature increase in the spark channel leads to sharp

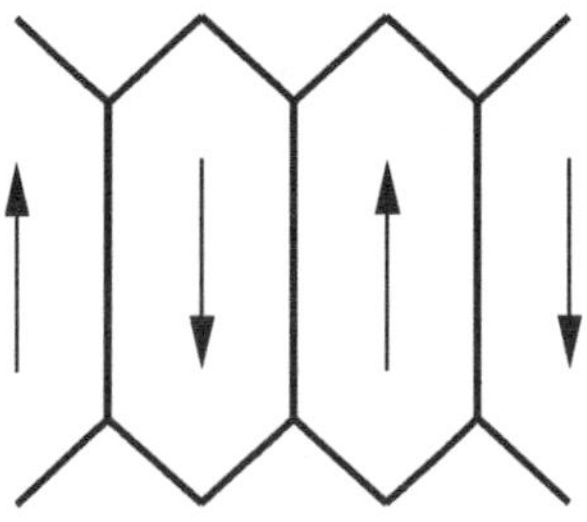

FIGURE 9.21 Schematic of ferroelectric domain arrangement.

pressure growth and to the generation of a cylindrical shock wave. The amplitude of the shock wave is so high that the temperature after the wave front is sufficient for thermal ionization. The external boundary of the spark current channel at first grows together with the front of the cylindrical wave. During the initial 0.1–1 μs after the breakdown point, the current channel expansion velocity is about 10^5 cm/s. Subsequently the cylindrical shock wave decreases in strength and expansion of the current channel becomes slower than the shock wave velocity. Radius of the spark channel grows to about 1 cm, which corresponds to a spark current increase of 10^4–10^5 A at current densities of about 10^4 A/cm^2. Plasma conductivity grows relatively high and a cathode spot can be formed on the electrode surface (see Section 8.3.5). Interelectrode voltage decreases lower than the initial one, and the electric field becomes about 100 V/cm. If voltage is supplied by a capacitor, the spark current obviously starts decreasing after reaching the maximum values, see S.I. Drabkina (1951), S.I. Braginsky (1958), E.M. Baselyan and Yu.P. Raizer (1997).

9.4.3 Atmospheric Phenomena Leading to Lightning

Lightning is a large-scale natural spark discharge, occurring between a charged cloud and the earth, between two clouds or internally inside of a cloud. Lightning is caused by high electric fields related to the formation and space separation in the atmosphere of positive and negative electric charges. Therefore the consideration of lightning will begin with an analysis of the processes of formation and space separation of the charges in atmosphere (Ya.I. Frenkel, 1949; M. Uman, 1969). The formation of electric charges in the atmosphere is due mainly to ionization of molecules or microparticles by cosmic rays. Generation of electric charges can also takes place during the collisional decay of water droplets. However, what is actually important for the interpretation of lightning is the fact that the negative charge in the thundercloud is located in the bottom part of the cloud and positive charge mostly located in the upper part of the cloud. To explain this phenomenon, remember that polar water molecules on the water surface are mostly aligned with their positive ends oriented inward from the water surface. This effect related to the hydrogen bonds between water molecules is illustrated in Figure 9.22. Such orientation of the surface water molecules leads to the formation of a double electric layer on the surface of droplets with a voltage drop experimentally determined as $\Delta\phi = 0.26$ V. This double layer predominantly traps negative ions and reflects positive ions until their charge Ne compensates for the voltage drop:

$$N \approx \frac{4\pi\varepsilon_0 r\ \Delta\phi}{e}. \tag{9.27}$$

The simple calculation based on Eq. (9.27) shows that a droplet with a radius $r = 10\,\mu$m is able to absorb about 2000 negative ions. Thus, the negative ions in clouds are trapped in droplets and descend, while positive ions remain in molecular or cluster form and remain in the upper areas of a cloud. As a result, a typical charge distribution in a cloud appears as shown in Figure 9.23

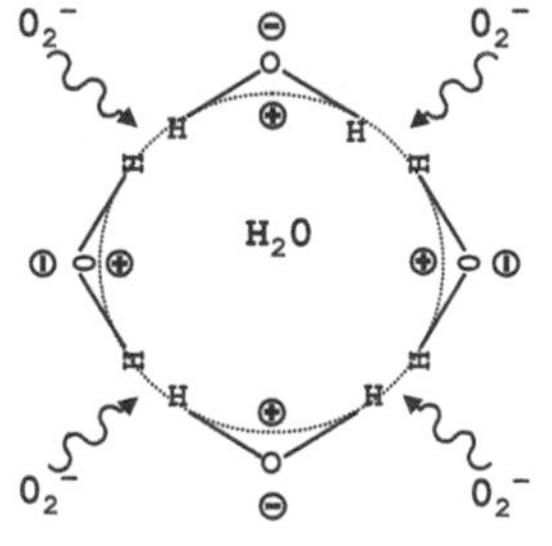

FIGURE 9.22 Illustration of a water droplet trapping negative ions.

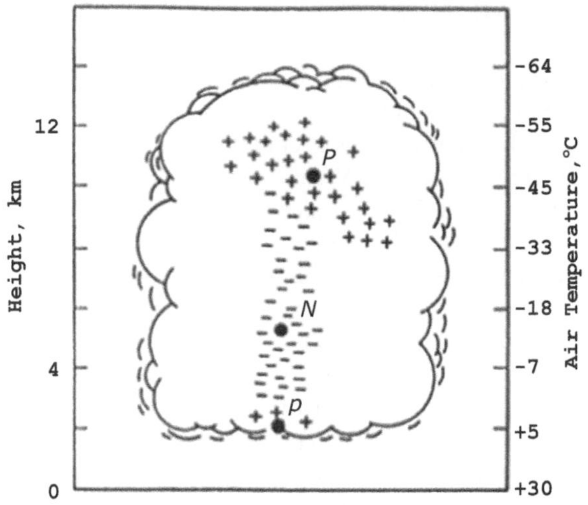

FIGURE 9.23 Probable charge distribution in thundercloud. Black dots mark centroids of charge clouds. According to measurements of electric fields around clouds, $P = +40C$ above, $p = +10C$ below, and $N = -40C$.

(M. Uman, 1969). The charge distribution in a thundercloud explains why most lightning discharges occur inside the clouds.

9.4.4 The Lightning Evolution

Typical duration of the lightning discharge is about 200 ms. The sequence of events in this natural modification of the long spark discharge is illustrated in Figure 9.24. The lightning actually consists of several pulses with duration about 10 ms each, and intervals about 40 ms between pulses. Each pulse starts with propagation of a leader channel (see Section 4.4.9) from the thundercloud to the earth. Current in the first negative leader is relatively low about 100 A. This leader, called the **multi-step leader**, has a multi-step structure with a step length about 50 m and an average velocity $1 - 2 \cdot 10^7$ cm/s. The visible radius of the leader is about 1m and the radius of the current-conducting channel is smaller. Leaders initiating sequential lightning pulses usually propagate along channels of the previous pulses. These leaders are called **dart-leaders**; they are more spatially uniform than the initial one and are characterized by a current of about 250 A and higher propagation velocities, $10^8 - 10^9$ cm/s. which results in the formation of a strong ionization wave moving in opposite direction back to the thundercloud. This extremely intensive ionization wave is usually referred to as the **return stroke**. The physical nature of the return stroke is similar to that of the back ionization wave

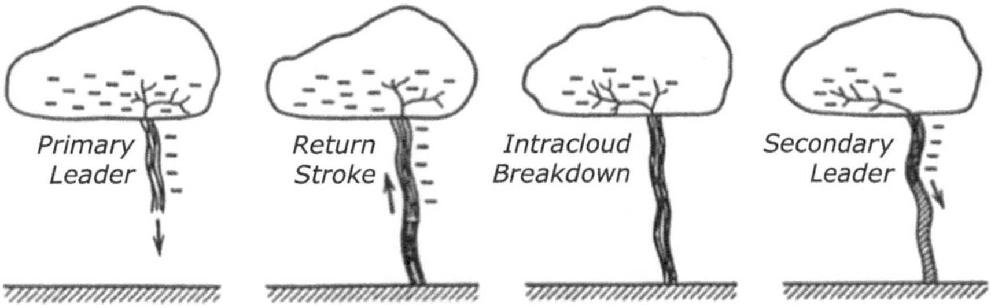

FIGURE 9.24 Illustration of lightning evolution.

in laboratory spark discharges, see Section 9.4.1. The return stroke is the main phase of the lightning discharge. In this case, the velocity of the ionization wave reaches gigantic almost relativistic values of 0.1–0.3 speed of light. Maximum current reaches 100 kA, which is actually the most dangerous effect of lightning. Temperature in the lightning channel reaches 25,000 K; electron concentration approaches $1\text{–}5 \cdot 10^{17}\,\text{cm}^{-3}$, which corresponds to complete ionization. The electric field on the front of the return stroke is quite high, about 10 kV/cm (for comparison, electric fields in arcs in air are about 10 V/cm). Taking into account the high values of current, leads to the gigantic values of specific power on the front of the return stroke, about $3 \cdot 10^5$ kW/cm (for comparison, in electric arcs in air, this is about 1 kV/cm). Intensive and fast heat release leads to strong pressure increases in the current channel of a lightning discharge and hence to shock wave generation which is heard as thunder. Intensive heat release leads to fast expansion of the initial current channel (see Section 9.4.2) during propagation of the return stroke and to the formation a developed spark channel. Through this channel some portion of the negative electric charge goes to the earth during about 40 ms. Electric currents through the spark channel during this 40 ms period is about 200 A. Remember that negative charges are located in the lower part of the thundercloud, mostly in the form of clusters or charged micro-particles. These have low conductivity and are unable to be quickly evacuated from the cloud through the spark channel. Hence, the evacuation of the negative charges takes place due to preliminary liberation of electrons from ions and micro-particles under the influence of the strong electric field. The lightning discharge consists of several pulses. Each pulse, results in transfer to the earth of only a part of the negative charge collected in the thundercloud. It collects negative charges only from the area close to where the return strike attacks the cloud. Then the lightning discharge temporarily extinguishes (intervals between pulses is about 40 ms) until electric charges redistribute themselves in the cloud by means of internal breakdowns. The conductivity in the spark channel of the previous pulse has already decreased significantly when the internal charge distribution in a thundercloud is restored after about 40 ms and the system is ready for a new pulse. The new lightning pulse starts again with a leader propagating along the remains of the previous one. The new leader does not have a multi-step structure, it is a dart leader. When the dart-leader reaches the earth, it stimulates formation of the second return stroke, and the cycle repeats again. The sequence of the lightning pulses continues for about 200 ms until most of the negative charge from the cloud reaches the earth. Note that the positive charges located in the upper parts of the thundercloud (Figure 9.23) mostly stay there, because the distance between these charges and the earth is too large for breakdown, see E.M. Baselyan and Yu.P. Raizer (2001).

9.4.5 Mysterious Phenomenon of Ball Lightning

Even a short discussion on lightning cannot be complete without mentioning the interesting and mysterious phenomenon of ball lightning. Ball lightning is a rare natural phenomenon; the luminous plasma sphere occurs in the atmosphere, moves in unpredictable directions (sometimes against the direction of wind), and finally disappears sometimes with an explosive release of a considerable amount of energy. Ball lightning has some special not trivial oddities, including the ability of the plasma sphere to move through tiny holes, possible explosion heat release exceeding all reasonable estimations of energy contained inside of the ball, etc. Difficulties in interpretation are mostly due

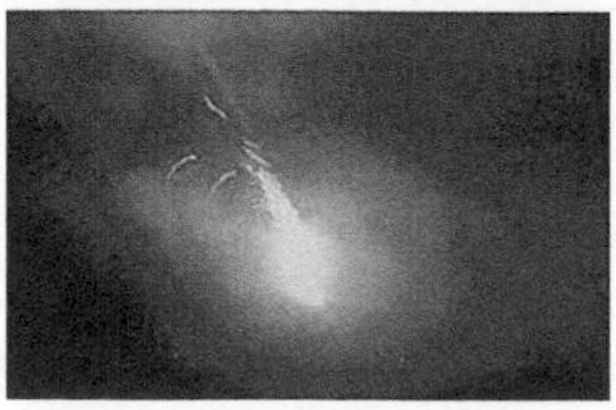

FIGURE 9.25 Ball lighting.

to difficulties to observe this rare phenomenon in nature or the laboratory. Some unique pictures of the ball lightning together with contradicting scientific descriptions of their observation can be found on the web. An example of an interesting and good quality color picture of the ball lightning is shown in Figure 9.25. There are several hypotheses describing the ball lightning phenomenon. The most developed one is the so-called chemical model. This model was first proposed by D.F. Arago in 1839. Subsequently, it was developed by many research studies and summarized in detail by B.M. Smirnov (1975, 1977a). The chemical model assumes the production of some excited, ionized or chemically active species in the channel of regular lightning followed by an exothermic reaction between them, leading to the formation of luminous spheres of ball lightning. There is disagreement on how energy accumulates. According to some authors, the accumulation of energy is provided by positive and negative complex ions (I.P. Stakhanov, 1973, 1974). Others hold ozone, nitrogen oxides, hydrogen, and hydrocarbons as energy sources for the ball lightning (B.M. Smirnov, 1975, 1977a, M.T. Dmitriev, 1967, 1969). The typical ball lightning spherical shape is related in the framework of the chemical model to heat and mass balances of the process. Fuel generated by regular lightning is distributed over volumes much exceeding those of the ball lightning. Steady-state exothermic reactions take place inside of a sphere. Fresh reagents diffuse into the sphere while heat transfer provides an energy flux from the sphere to sustain the steady-state process. Analysis of the steady-state spherical wave for describing the shape of ball lightning was done by V. Rusanov, A. Fridman, 1976. This steady-state "combustion" sphere moves (in slightly non-uniform conditions) in the direction corresponding to the growth of temperature and "fuel" concentration. This explains the possible observed motion of ball lightning in the opposite direction to that of the wind. The high energy release during a ball lightning explosion can be explained noting that the explosion occupies a much larger volume than the initial one related to the region of the steady-state exothermic reaction, see P. Leonov (1965), I.P. Stakhanov (1976), S. Singer (1973), J.D. Barry (1980) and M. Stenhoff (2000).

9.4.6 Laser Directed Spark Discharges

Modification of sparks can be done by synergetic application of high voltages with laser pulses (L.M. Vasilyak, S.V. Kostuchenko, N.N. Kurdyavtsev, I.V. Filugin, 1994; E.I. Asinovsky, L.M. Vasilyak, 2001). A simplified scheme of this discharge system is illustrated in Figure 9.26. Laser beams can direct spark discharges not only along straight lines but also along more complicated trajectories. Laser radiation is able to stabilize and direct the spark discharge channel in space through of three major effects: local preheating of the channel, local photoionization and optical breakdown of gas. Preheating of the discharge channel creates a low gas density zone, leading to higher levels of reduced electric field E/n_0, and which is favorable as a result, for spark propagation. This effect works best if special additives provide the required absorption of the laser radiation. For example, if CO_2 laser is used for preheating, a strong effect on the corona discharge can be achieved when about 15% ammonia (which effectively absorbs radiation on wavelength 10.6 μm) is added to

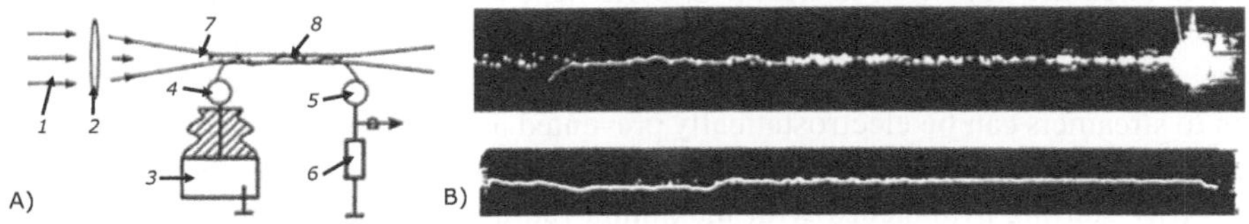

FIGURE 9.26 Schematic (A) and pictures (B) of the laser directed spark: 1 – laser beam; 2 – lens; 3 – voltage pulse generator; 4,5 – powered and grounded electrodes; 6 – resistance; 7 – optical breakdown zone; 8 – discharge zone. Discharges on the pictures are 1 m and 3 m long; bright right-hand spot is a high voltage electrode overheated by the laser beam. (*From Encyclopedia of Low Temperature Plasma, Ed. V.E Fortov, Nauka, Moscow, 2002*).

air. At a laser radiation density about 30 J/cm^2, the breakdown voltage in the presence of ammonia decreases by an order of magnitude. The maximum length of the laser-supported spark was up to 1.5 m. Effective stabilization of the spark discharges by CO_2 laser in air was also achieved by admixtures of C_2H_2, CH_3OH and CH_2CHCN. Photoionization by laser radiation is able to stabilize and a direct corona discharge without significantly changing the gas density by means of local pre-ionization of the discharge channel. UV-laser radiation (for example, Nd-laser or KrF-laser) should be applied in this case. Ionization usually is related to the two-step photo-ionization process of special organic additives with relatively low ionization potential. The UV KrF-laser with pulse energies of approximately 10 mJ and pulse duration of approximately 20 ns is able to stimulate the directed spark discharge to lengths of 60 cm. The laser photoionization effect to stabilize and direct sparks is limited in air by fast electron attachment to oxygen molecules. In this case, by photo-detachment of electrons from negative ions can be provided by using a second laser radiating in the infrared or visible range. The most intensive laser effect on spark generation can be provided by the optical breakdown of the gases. The length of such a laser spark can exceed 10 m. The laser spark in pure air requires power density of a Nd-laser ($\lambda = 1.06\,\mu m$) exceeding 10^{11} W/cm^2.

9.5 ATMOSPHERIC PRESSURE GLOW DISCHARGES (APG)

9.5.1 Atmospheric Pressure Glow Mode of DBD

The glow mode of DBD can be operated at lower voltages (down to hundreds of volts), and streamers are avoided as the electric fields are below the Meek criterion and discharge operates in the Townsend ionization regime. Secondary electron emission from dielectric surfaces, which sustains the Townsend ionization regime, relies upon adsorbed electrons (with binding energy only about 1 eV) that were deposited during previous DBD excitation (high voltage) cycle. If enough electrons "survive" voltage switching time without recombining, they can trigger a transition to the homogeneous Townsend mode of DBD. "Survival" of electrons and crucial active species between cycles or the **DBD memory effect** is critical for the organization of APG and depends on properties of the dielectric surface as well as operating gas. In electronegative gasses, the memory effect is weaker because of attachment losses of electrons. If the memory effect is strong, the transition to Townsend mode can be accomplished, and uniform discharge can be organized without streamers. A streamer DBD is easy to produce, while the organization of APG at the same conditions is not always possible. This can be explained taking into account that the streamer discharge is not sensitive to the secondary electron emission from dielectric surface, while it is critical for operation of APG. Glow discharges usually undergo contraction with increase of pressure due to the thermal instability. The thermal instability in the DBD-APG is somewhat suppressed by using alternating voltage, thus discharge operates only when voltage is high enough to satisfy the Townsend criteria, and the rest of time the discharge is idle, which allows the dissipation of heat and active species. If time between excitation cycles is not enough for the dissipation then an instability will develop and discharge will undergo a transition to filamentary mode. The avalanche-to-streamer transition in the APG-DBD depends on the level of pre-ionization. Meek criterion is related to an isolated avalanche, while in the case of intensive pre-ionization, avalanches are produced close to each other and interact. If two avalanches occur close enough, their transition to streamers can be electrostatically prevented and discharge remains uniform. A modified Meek criterion of the avalanche-to-streamer transition can be obtained considering two simultaneously starting avalanches with maximum radius R, separated by the distance L, (α is Townsend coefficient, d is distance between electrodes):

$$\alpha d - \left(R/L\right)^2 \approx const, \tag{9.28}$$

$$\alpha d \approx const + n_e^p R^2 d. \tag{9.29}$$

In Eq. (9.29), the distance between avalanches is approximated using the pre-ionization density n_e^p. The constant in (9.28 and 9.29) depends on gas: in air at 1 atm this constant equals to 20. According to the modified Meek criterion, the avalanche-to-streamer transition can be avoided by increasing the avalanche radius and by providing sufficient pre-ionization. The type of gas is important in the transition to the APG. Helium is relevant for the purpose: it has high-energy electronic excitation levels and no electron energy losses on vibrational excitation, resulting in higher electron temperatures at lower electric fields. Also, fast heat and mass transfer processes prevent contraction and other instabilities at high pressures.

9.5.2 Resistive Barrier Discharge (RBD)

The RBD can be operated with DC or AC (60 Hz) power supplies and is based on DBD configuration, where the dielectric barrier is replaced by a highly resistive sheet (few MΩ/cm) covering one or both electrodes. The system can consist of a top wetted high resistance ceramic electrode and a bottom electrode. The highly resistive sheet plays a role of distributed resistive ballast, which prevents high currents and arcing. If He is used and the gap distance is not too large (5 cm and below), a spatially diffuse discharge can be maintained in the system for several tens of minutes. If 1% of air is added to helium, the discharge forms filaments. Even when driven by a DC voltage, the current signal of RBD is pulsed with the pulse duration of few microseconds at a repetition rate of few tens of kHz. When the discharge current reaches a certain value, the voltage drop across the resistive layer becomes large to the point where the voltage across the gas became insufficient to sustain the discharge. The discharge extinguishes and current drops rapidly, then voltage across the gas increases to a value sufficient to reinitiate the RBD.

9.5.3 One Atmosphere Uniform Glow Discharge Plasma (OAUGDP)

OAUGDP is an APG discharge developed at the University of Tennessee by Roth and his colleagues. OAUGDP is similar to a traditional DBD, but it can be much more uniform, which has been interpreted by the ion-trapping mechanism. The discharge transition from filamentary to a diffuse mode in atmospheric air can lead not only to the diffuse mode but also into another non-homogeneous mode where the filaments are more numerous and less intense. The stability of the filaments and transition to uniformity is related to the "memory effect". In particular, electrons deposited on the dielectric surface promote the formation of new streamers at the same place again and again by adding their own electric fields to the external electric field. The key feature of OAUGDP promoting the transition to uniformity may be hidden in the properties of particular dielectrics that are not stable in plasma and probably become more conductive during a plasma treatment. In particular, plasma can further increase the conductivity of borosilicate glass used as a barrier, for example, by UV radiation. It transforms the discharge into the resistive barrier discharge working at a high frequency in a range of 1–15 kHz. Not only volumetric but also surface conductivity of dielectric can promote the DBD uniformity, if it is in an appropriate range. The "memory effect" can be suppressed by removing the negative charge spot formed by electrons of a streamer during the half period of voltage oscillation (i.e., before polarity changes), and the surface conductivity can help with this. On the other hand, when the surface conductivity is very high, the charge cannot accumulate on the surface during DBD current pulse of several nanoseconds and cannot stop the filament current. Figure 4.101 presents the equivalent circuit for one DBD electrode with surface conductivity: for 10 kHz, it should be 0.1 ms >> RC >> 1 ns, where resistances and capacitances are determined for the characteristic radius of streamer interaction (about 1 mm).

9.5.4 Electronically Stabilized Atmospheric Pressure Glow (APG) Discharges

Electronic stabilization of APG has been demonstrated, in particular, by Van De Sanden and his team. Uniform plasma has been generated in argon DBD during the first cycles of voltage oscillations with relatively low amplitude (i.e., αd of about 3). The existence of the Townsend discharge at such low voltage requires an unusually high secondary electron emission coefficient (above 0.1). The so high electron emission and the breakdown during the first low-voltage oscillations can be explained, probably, taking into account the low surface conductivity of most polymers applied as barriers in the system have very low surface conductivities. Surface charges occurring due to cosmic rays can be then easily detached by the applied electric field. Long induction time of the dark discharge is not required in this case in contrast to the OAUGDP with glass electrodes. Assuming that the major cause of the DBD filamentation is instability leading to the glow-to-arc transition, it has been suggested to stabilize the glow mode using an electronic feedback to fast current variations. The filaments are characterized by higher current densities and smaller RC constant. Therefore the difference in RC constant can be used to "filter" the filaments because they react differently to a drop of the displacement current (displacement current pulse) of different frequency and amplitude. A Simple LC circuit, in which during the pulse generation the inductance is saturated, has been used to generate the displacement current pulses. The method of the electronic uniformity stabilization has been used for relatively high power densities (in the range of 100 W/cm^3) and in a large variety of gases including Ar, N_2, O_2, and air.

9.5.5 Atmospheric Pressure Plasma Jets (APPJ)

The radio-frequency (RF) atmospheric glow discharge or atmospheric pressure plasma jet APPJ is one of the most developed APG systems, which has been used in particular for the plasma-enhanced chemical vapor deposition (PECVD) of silicon dioxide and silicon nitride thin films. The APPJ can be organized as a planar and co-axial system with a discharge gap of 1–1.6 mm, and frequency in the MHz range (13.56 MHz). The APPJ is an RF CCP discharge that can operate uniformly at atmospheric pressure in noble gases, mostly in helium. In most APPJ configurations, electrodes are placed inside the chamber, and not covered by any dielectric in contrast to DBD. The discharge in pure helium has limited applications, therefore various reactive species such as oxygen, nitrogen, nitrogen trifluoride, etc. are added. To achieve higher efficiency and a higher reaction rate, the concentration of the reactive species in the discharge has to be increased. If the concentration of the reactive species exceeds a certain level (which is different for different species, but in all cases is on the order of a few percent), the discharge becomes unstable. The distance between electrodes in APPJ is usually about 1 mm, which is much smaller than the size of the electrodes (about 10 cm × 10 cm). Therefore the discharge can be considered as 1D, and the effects of the boundaries on the discharge can be neglected. The electric current in the discharge is the sum of the current due to the drift of electrons and ions, and the displacement current. Since the mobility of the ions is usually 100 times smaller than the electron mobility, the current in the discharge is mostly due to electrons. Considering that the typical ionic drift velocity in APPJ discharge conditions is about 3×10^4 cm/s, the time needed for ions to cross the gap is about 3 μs which corresponds to a frequency of 0.3 MHz. The frequency of the electric field is much higher, and thus ions in the discharge do not have enough time to move, while electrons move from one electrode to another as the polarity of the applied voltage changes. The overall APPJ voltage consists of the voltage on the positive column V_p (plasma voltage) and the voltage on the sheath V_s. The voltage on the positive column V_p slightly decreases with an increase of the discharge current density. It happens because a reduced electric field E/N in plasma is almost constant and equals to $E/p \approx 2$ V/(cm·Torr) for helium discharge. If the density of neutral species is constant, the plasma voltage will be constant as well. But the density of neutrals slightly decreases with the electric current density since high currents cause a gas temperature to rise. At higher gas temperatures, a lower voltage is needed to support the discharge and subsequently the plasma voltage decreases. Sheath thickness can be approximated from the amplitude of electron drift oscillations $d_s = 2\mu E/\omega \approx 0.3$ mm, where μ is an electron mobility, $\omega = 2\pi f$ is the frequency of the applied voltage and E is an electric field in plasma. Assuming the secondary emission coefficient $\gamma = 0.01$, the critical ion density $n_{p(crit)}$ in helium RF before the α–γ transition is about 3×10^{11} cm^{-3}. Corresponding critical sheath voltage is about 300 V. Typical power density for the APPJ helium discharge is on the level of 10 W/cm^2 which is approximately 10 times higher than that in the DBD discharges, including its uniform modifications. The power density that can be achieved in the uniform RF discharge is limited by two major instability mechanisms: thermal instability and α–γ transition instability. Critical power density for the thermal instability in APPJ is about 3 W/cm^2. Stable APPJ can be organized, however, with a power density exceeding this threshold. Suppressing the thermal instability in the APPJ conditions is due to the stabilizing effect of the sheath capacitance, which can be described by the *R* parameter: square of the ratio of the plasma voltage to the sheath voltage. The smaller *R*, the more stable the discharge is with respect to the thermal instability. For example, if $R = 0.1$, the critical power density with respect to thermal instability is 190 W/cm^2. For the helium APPJ with: $d_s = 0.3$ mm, $V_s = 300$ V and $d = 1.524$ mm, the parameter *R* is $R = (V_p/V_s)^2 = 0.36$, which corresponds to the critical discharge power density 97 W/cm^2. Thus the APPJ discharge remains thermally stable in a wide range of power densities as long as the sheath remains intact. Major instability of the APPJ and loss of its uniformity, therefore, is mostly determined by the α–γ transition, or in simple words by breakdown of the sheath. The α–γ transition in APPJ happens because of the Townsend breakdown of the sheath, which occurs when ion density and sheath voltage exceed the critical ones ($n_{p(crit)} = 3 \times 10^{11}$ cm^{-3}, $V_s = 300$ V). In pure helium and in helium with additions of nitrogen and oxygen, it has been shown that the main mechanism of the discharge instability is the sheath breakdown that eventually leads to thermal instability. Therefore, more effective discharge cooling would not solve the stability problem because it did not protect the discharge from the sheath breakdown. Nevertheless, cooling is important since the sheath breakdown depends on the reduced electric field that increased with temperature. Despite the fact that the thermal stability of He discharge is better compared to the discharge with oxygen addition, a higher power is achieved with oxygen addition, which prevents the sheath breakdown. Summarizing, it is easier to generate the uniform APG in helium and argon than in other gases, especially electronegative ones. The effect cannot be explained only by high thermal conductivity of helium. It is more important for uniformity that noble-gas-based discharges have a significantly lower voltage, and therefore lower power density, which helps to avoid the thermal instability. Pure nitrogen provides better conditions for the uniformity than air. The presence of oxygen results in the electron attachment, which causes higher voltage, higher power, and finally leads to thermal instability.

9.6 MICRODISHARGES

9.6.1 General Features of Microdischarges

Scaling down with a constant similarity parameter (*pd*) should not change the properties of discharges significantly. Conventional non-equilibrium discharges are operated in the pd-range around 10 cm · Torr; therefore the organization of strongly non-equilibrium discharges at atmospheric pressure should be effective in the sub-millimeter sizes. Some specific new properties can be achieved by scaling down the plasma size to sub-millimeters:

1. Size reduction of non-equilibrium plasmas permits an increase of their power density to a level typical for the thermal discharges, because of intensive heat losses of the tiny systems.
2. At high pressures, volumetric recombination and especially three-body processes can go faster than diffusion losses, which results in significant changes in plasma composition. For example, the high-pressure microdischarges can contain a significant amount of molecular ions in noble gases.
3. Sheathes (about 10–30 μm at atmospheric pressure) occupy a significant portion of the plasma volume.

4. Plasma parameters move to the "left" side of the Paschen curve. The Paschen minimum is about $pd = 3$ cm · Torr for some gases, therefore a 30 μm gap at 1 atm corresponds to the left side of the curve. The specific properties can lead to the positive differential resistance of a microdischarge, which allows supporting many discharges in parallel from a single power supply without using multiple ballast resistors. Another important consequence is relatively high electron energy in the microplasmas, which are most strongly non-equilibrium. A significant development in fundamentals and applications of the microdischarges has been achieved recently by B. Farouk and his group in the Drexel Plasma Institute.

9.6.2 Micro-Glow Discharge

Atmospheric pressure DC micro-glow discharge has been generated between a thin cylindrical anode and a flat cathode by B. Farouk and his group in the Drexel Plasma Institute. The discharge has been studied using an inter-electrode gap spacing in the range of 20 μm–1.5 cm so that one could see the influence of the discharge scale on plasma properties. Current–voltage characteristics, visualization of the discharge, and estimations of the current density indicate that the discharge operates in the normal glow regime. Emission spectroscopy and gas temperature measurements using the second positive band of N_2 indicate that the discharge generates non-equilibrium plasma. For 0.4 mA and 10 mA discharges, rotational temperatures are 700 K and 1,550 K, while vibrational temperatures are 5,000 K and 4,500 K, respectively. It is possible to distinguish a negative glow, Faraday dark space, and positive column regions of the discharge. The radius of the column is about 50 μm and remains relatively constant with changes in the electrode spacing and discharge current. Such radius permits balancing the heat generation and conductive cooling to help prevent thermal instability and the transition to an arc. Generally, there is no significant change in the current–voltage characteristics of the discharge for different electrode materials or polarity. There are several notable exceptions to this for certain configurations: (a) For a thin upper electrode wire (<100 μm) and high discharge currents, the upper electrode melts. This occurs when the wire is the cathode, indicating that the heating is due to energetic ions from the cathode sheath and not resistive heating. (b) For a medium-sized wire (~200 μm) as a cathode, width of the negative glow increases as the current increases until it covers the entire lower surface of the wire. If the current is further increased, the negative glow 'spills over' the edge of the wire and begins to cover the side of the wire. This effect is similar to the transition from a normal glow to an abnormal glow in low-pressure glow discharges. However, there is no increase in the current density since the cathode area is not limited. For sufficiently large electrode wires this effect does not occur. (c) In air discharges with oxidizable cathode materials, the negative glow moves around the cathode electrode leaving a trail of oxide coating behind until there is no clean surface within the reach of the discharge and the discharge extinguishes. For a small spacing of electrodes, the current–voltage characteristics are relatively flat, which is consistent with the normal discharge mode. For a normal glow discharge in air, the potential drop at the normal cathode sheath is around 270 V. A voltage drop above that occurs mostly in the positive column. For larger electrode spacing, the current–voltage characteristics have a negative differential resistance dV/dI. This is due to the discharge temperature increase with gap length resulting in the growth of conductivity. A short discharge loses heat through the thermal conductivity of electrodes. A long discharge cooling is not efficient because the thermal conductivity of the gas is much lower than that of metal electrodes; therefore the temperature of the long discharge is higher. Such behavior demonstrates a new property related to the size reduction to microscale. Diffusive heat losses can balance the increased power density only at elevated temperatures of a microdischarge and the traditionally cold glow discharge becomes "warm". Table 9.2 summarizes the micro-glow discharge parameters corresponding to currents 0.4 and 10 mA. The atmospheric pressure DC microdischarge is a normal glow discharge thermally stabilized by its size, and maintaining a high degree of vibrational-translational non-equilibrium. The micron-sized precise micro-glow discharges or

TABLE 9.2
Micro-Glow-Discharge Parameters

Micro-Discharge and Micro-Plasma Parameters	Micro-Discharge Current (mA)	
	0.4	**10**
Electrode spacing (mm)	0.05	0.5
Micro-discharge voltage (V)	340	380
Micro-discharge power (W)	0.136	3.8
Diameter of negative glow (μm)	39	470
Positive column diameter (μm)	—	110
Electric field in the positive column (kV·cm^{-1})	5.0	1.4
Translational gas temperature (K)	700	1,550
Vibrational gas temperature (K)	5,000	4,500
Negative glow current density (A·cm^{-2})	33.48	5.8
Positive column current density (A·cm^{-2})	—	105
Reduced electric field E/n$_0$ (V·cm^2)	4.8×10^{-16}	3×10^{-16}
Electron temperature T_e (eV)	1.4	1.2
Electron density n_e in negative glow (cm^{-3})	3×10^{13}	7.2×10^{12}
Electron density n_e in positive column (cm^{-3})	—	1.3×10^{14}
Ionization degree in negative glow	3×10^{-6}	15×10^{-7}
Ionization degree in positive column	—	3×10^{-5}

their arrays can be effectively used for direct micro-scale surface treatment without the application of any masks.

9.6.3 Micro-Hollow-Cathode Discharge

The hollow cathode discharges (HCD) can be effectively organized in the microscale The micro-HCD similarly to the conventional hollow-cathode discharges are interesting for applications because of their ability to generate the high-density plasma. While conventional HCD is organized at low pressures and macroscale, the micro-HCD can operate at atmospheric pressure in agreement with the (*pD*)-similarity. The micro-HCD can be effectively arranged in the form of special arrays. If (*pD*) is in the range of 0.1–10 Torr·cm, the discharge develops in stages. At low currents, a "pre-discharge" is observed, which is a glow discharge with the cathode fall outside the hollow cathode structure. As the current increases and the glow discharge starts its transformation into the abnormal glow with a positive differential resistance, a positive space charge region moves closer to the hollow cathode structure and can enter the cavity. After that, the positive space charge in the cavity acts as a virtual anode, resulting in the redistribution of the electric field inside the cavity. At the center of the cavity, a potential well for electrons appears, forming a cathode sheath along the cavity walls. At this transition from the axial pre-discharge to a radial discharge, the sustaining voltage drops. Sometimes this transition is not so sharp, and in that case, a negative slope in the current–voltage characteristic (i.e., a negative differential resistance) appears, which is traditionally referred to as the "hollow cathode mode".

9.6.4 Arrays of Microdischarges, Microdischarge Self-Organization, and Structures

The power of a microdischarge is so small that individual microdischarges have limited applications. Thus, most industrial applications require microdischarge arrays or micro-plasma integrated structures. The plasma TV is an example of such a complex structure. The simplest structure may be the one that consists of multiple identical microdischarges electrically connected in parallel. For a stable operation of such structures, each discharge should have a positive differential resistance.

Most microdischarges have this property as a result of a significant increase in the power losses with a current increase. One of the examples is the array consisting of microdischarges with inverted, square pyramidal cathodes. In such an optical micrograph of the 3×3 array, the microdischarges of 50 μm × 50 μm each are separated (center-to-center) by 75 μm. All of the microdischarges (700 Torr of Ne) have common anode and cathode, i.e., the devices are connected in parallel. Ignition voltage and current for the array are 218 V and 0.35 μA. The array has been able to operate at a high power loading (433 V and 21.4 μA); emission from each discharge is spatially uniform. Another example of microdischarge arrays is the so-called **"fused" hollow cathode**, which is based on the simultaneous RF-generation of HCD plasmas in an integrated open structure with flowing gas. The resulting discharges are stable, homogeneous, luminous, and volume filling without streamers. The power is on the order of one Watt per cm^2 of the electrode structure area. Experiments have been carried out with the system having a total discharge area 20 cm^2. The concept of the source is extremely suitable for scaling-up for different gas throughputs. In some cases it is beneficial to connect microdischarges in series, for example, to increase a radiant excimer emittance. Such a system can consist of two HCD with negative differential resistance, and be applied for the excimer laser. Laser devices require a long gain length to achieve the threshold. One of the strategies to produce the long gain length is to alternately stack cathode and anode structures in a single bore. Non-equilibrium microdischarges at atmospheric pressure can also exist at relatively high powers. For example, a **micro-arc discharge** in a gap 0.01–0.1 mm with a voltage 1.5–4.5 V can have current 40–120 A. This arc is similar to some extent to the cathode boundary-layer microdischarge because it exists without an "arc column". The main difference between the two discharges is that in the micro-arc, up to 95% of the electrical energy is transferred to the anode (similar to e-beam), which is qualitatively different from the case of the cathode boundary-layer discharge. The micro-arc has been used for the generation of metal droplets and nano-powders as well as for a local hardening of metal surfaces.

9.6.5 KHz-Frequency-Range Microdischarges

AC microdischarges can be organized at all possible frequencies. Low- and medium-frequency microdischarges are related to DBD. An integrated structure called the **coaxial-hollow micro dielectric-barrier discharges (CM-DBD)** has been made by stacking two metal meshes covered with a dielectric layer of alumina with a thickness about 150 μm. The test panel (diameter 50 mm) with hundreds of hollow structures (0.2 mm × 1.7 mm) has been assembled. He or N_2 have been used at pressures 20–100 kPa, and voltage below 2 kV even at the maximum pressure. Bipolar square-wave voltage pulses have been applied to one of the mesh electrodes. The pulse duration of both positive and negative voltages varied from 3 to 14 μs; intermittent time 1 μs; repetition frequency 10 kHz. In each coaxial hole, the discharge occurs along the inner surface. The intensity of each microdischarge is uniform over the whole area. The extended glow with a length of some millimeters is observed in He but not in N_2. The electron density in He at 100 kPa is about 3×10^{11} cm^{-3}. The CM-DBD configuration has a rather low operating voltage (typically 1–2 kV); the scaling parameter *pd* is several tens of Pa·m, corresponding to the Paschen minimum. Plasma in the system is stable over a wide range of external parameters without filamentation or arcing. Another kHz-range microdischarge is the so-called **capillary plasma electrode (CPE)** discharge. The CPE discharge uses an electrode design, which employs dielectric capillaries that cover one or both electrodes. Although the CPE discharge looks similar to a conventional DBD, the CPE discharge exhibits a mode of operation that is not observed in DBD: the "capillary jet mode." The capillaries with diameter 0.01–1mm and a length-to-diameter (*L/D*) ratio from 10:1 to 1:1 serve as plasma sources and produce jets of high-intensity plasma at high pressure. The jets emerge from the end of the capillary and form a "plasma electrode." The CPE discharge displays two distinct modes of operation when excited by pulsed DC or AC. When the frequency of the applied voltage pulse is increased above a few kHz, one observes first a diffuse mode similar to the diffuse DBD. When the frequency reaches a critical value (which depends on the *L/D* value and the plasma gas), the capillaries become "turned on", and bright intense plasma jets emerge from the capillaries. When many capillaries are placed in close proximity to each other, the emerging plasma jets overlap and the discharge appears uniform. This "capillary" mode is the preferred mode of operation of the CPE discharge and is somewhat similar to the 'fused' hollow cathode (FHC). At high frequency even dielectric capillaries can work as hollow cathodes because for CCP-RF plasma in the gamma mode, a dielectric surface is also a source of the secondary emitted electrons similarly to metal cathodes in glow or hollow-cathode discharges.

9.6.6 RF-Microdischarges

In the RF range (13.56 MHz), the so-called **plasma needle** attracts interest due to medical applications. This discharge has a single-electrode configuration and is operating in helium. It operates near the room temperature, allows the treatment of irregular surfaces, and has a small penetration depth. The plasma needle is capable of bacterial decontamination and localized cell removal without causing necrosis to the neighboring cells. Areas of detached cells can be made with a resolution of 0.1 mm. Radicals and ions from the plasma as well as UV-radiation interact with the cell membranes and cell adhesion molecules, causing detachment of the cells. The plasma needle is confined in a plastic tube, through which helium flow is supplied. The discharge is entirely resistive with voltage 140–270 V_{rms}. The electron density is about 10^{11} cm^{-3}. Optical measurements show substantial UV emission in the range 300–400 nm; radicals O and OH have been detected. At low helium flow rates, densities of molecular species in the plasma are higher. Conventional RF discharges, both inductively coupled (ICP) and capacitively coupled (CCP), have been also organized in micro-scale at atmospheric pressure. These plasmas are non-equilibrium because of their small sizes and effective cooling. Reduction in size requires a reduction in wavelength, an increase of frequency. A **miniaturized atmospheric-pressure ICP jet** has been developed for a portable liquid analysis system. The plasma device is a planar ICP source, consisting of a ceramic chip with an engraved discharge tube and a planar metallic antenna in a serpentine structure. The chip consists of two dielectric plates with an area of 15 mm × 30 mm. A discharge tube (1 mm × 1 mm × 30 mm, h/w/l) is engraved on one side of the dielectric plate. A planar antenna is fabricated on the other side of the plate. The atmospheric-pressure plasma jet with a density of about 10^{15} cm^{-3} is produced using a compact very high frequency (VHF) transmitter at 144 MHz and power 50 W. The electronic excitation temperature in the system is 4,000–4,500 K.

9.6.7 Microwave Microdischarges

The low-power microwave micro-plasma source based on a microstrip split-ring 900 MHz resonator operates at pressures 0.05 Torr – 1 atm. Argon and air discharges can be self-started in the system with power less than 3 W. Ion density of 1.3×10^{11} cm^{-3} in argon at 400 mTorr can be produced with only 0.5 W power. Atmospheric discharges can be also sustained in argon with 0.5 W. The low power allows portable air-cooled operation of the system. This kind of micro-plasma sources can be integrated into portable devices for applications such as bio-MEMS sterilization, small-scale material processing, and micro-chemical analysis systems. The highest frequency range discharges are optical ones. The optical discharges or so-called laser sparks are always micro-discharges, as they are formed in the focus of a lens that concentrates the laser light.

9.7 "Maximum Power Principle" for Atmospheric Pressure Pulsed Dielectric Barrier Discharges

9.7.1 Principles of Maximum or Minimum Power in Analysis of Plasma Discharges

Principles of maximum (or minimum) power have been used for a theoretical description of major low- temperature plasma discharges, including arc discharges, glow discharges, as well as gliding arcs and gliding barrier discharges (see e.g., Sections 7.2.5, and 8.4.3 for the case of glow discharges and arc discharges respectively). Application of these principles can be useful for analysis and quantitative description of different plasma systems.

Although the maximum or minimum power principles can be viewed in the framework of the so-called "fourth principle of energetics in open system thermodynamics" (see Chapters 7 and 8), in plasma physics systems they were explained based on the underlying physical laws. D. Dobrynin, D. Vainchtein, M. Gherardi ea., 2019b, described power release in micro- and nanosecond pulsed dielectric barrier discharges (DBD) and correlate the theoretical analysis of the pulsed atmospheric plasma systems with the maximum power principle.

DBD in various configurations and gases have been extensively studied (see Section 9.3 of this chapter). Equivalent electrical circuits and electrical models of DBDs have been developed mostly in relation to their applications in ozone generation. Typically, power measurements and calculations are based on measurements of a time-integrated current. However, the application of this method is significantly limited for short-pulsed discharges, compared to longer pulses or continuous wave DBDs. This is primarily due to the importance of the development stages of pulsed DBDs, and especially due to the differences in physics of these discharges on short time scales for various gases, electrode configurations, and the other operating conditions.

It was demonstrated in D. Dobrynin, D. Vainchtein, M. Gherardi ea., 2019b that the pulsed DBDs operate according to the "maximum power principle" and explained the relevant underlying physical processes. Such an approach not only clarifies physics of the atmospheric pressure plasma systems but also significantly simplifies relevant calculations and design of these systems, which can be very challenging, especially in the case of their scaling up.

9.7.2 Short-Pulsed DBD Evolution, Concept of Electrode Glow "Pancakes"

The development of a pulsed flat DBD can be described following D. Dobrynin, D. Vainchtein, M. Gherardi ea., 2019b, as a 3-stage process (see Section 9.3):

1. During the first few hundred picoseconds, the discharge starts with the development of avalanches traveling from the negative (in this case – grounded) electrode towards the high-voltage positive electrode (anode) covered with a dielectric.
2. As the avalanches reach the anode, electron concentration and local electric field are sufficient for the initiation of streamers. At this point, at about 1 ns, the presence of a dielectric surface facilitates the development of a surface discharge (surface-directed streamers) which shows up in experiments as a bright glow area near the positive dielectric-covered electrode. This electrode glow – "pancake" – appears prior to the main volumetric discharge due to the effects of the surface. In the equivalent-circuit language, the "pancake" can be described as an additional capacitor, which accumulates a portion of full discharge energy.
3. At the third stage, the main volumetric discharge starts to develop with the evolution of traditional cathode- directed streamers. This stage corresponds to the most energetic phase of the DBD.

Let's estimate the energy release in the "pancake", and compare it with the total energy release in the system during the discharge.

9.7.3 Energy Release in the DBD Plasma Volume, Manifestation of the Maximum Power Principle

D. Dobrynin, D. Vainchtein, M. Gherardi ea., 2019b, have considered a simplified equivalent electrical circuit representing the pulsed DBD. In this circuit, the DBD dielectric is represented by a capacitor with the capacity Cd, and the air (plasma) gap is represented by a capacitor (with capacity Ca). When the plasma appears in the air gap, its conductivity grows from zero to some finite value corresponding to the active resistance R. Characteristic ratio $k = \mathrm{Ca/Cd} \sim 1/\varepsilon$, where $\varepsilon \approx 4$ is dielectric permittivity of a dielectric.

Detailed analytical analysis of the circuit leads to the following vision of the pulsed DBD evolution:

1. The pulsed plasma system evolution is governed by the decrease of the plasma resistance until it reaches approximately the impedance of the dielectric barrier, at which state the discharge power reaches the maximum and the evolution drastically slows down. That is actually a clear manifestation of the applicability of the maximum power principle.
2. The maximum power release of the pulsed DBD discharge is proportional to the capacitance of the dielectric (because it corresponds to the minimum active plasma resistance) and the square of the applied voltage V (well expected from Ohm's law), and only weakly depends on the properties of the air gap in the discharge.

9.7.4 Average Pulsed DBD Power as s Function of Dielectric Barrier Parameters

To compute the total energy release the power P must be integrated over the time τ, which is the duration of the build-up. The total energy release in a pulse can be estimated according to D. Dobrynin, D. Vainchtein, M. Gherardi ea., 2019b as:

$$E_t = \frac{C_d V^2}{4(k+1)}(\omega\tau) \tag{9.30}$$

Recall that in a typical discharge, ω is defined by the reverse time of the growth of the voltage, while τ is the total duration of the discharge impulse. Therefore, over a long time the average power can be expressed as:

$$P_{av} = E_t f = \frac{C_d V^2}{4(k+1)}(\omega\tau) f, \tag{9.31}$$

where f is a pulse frequency. Thus, analyzing Eq. (9.31), D. Dobrynin, D. Vainchtein, M. Gherardi ea., 2019b came to the following major conclusions, regarding controlling the average power of pulsed atmospheric pressure DBD:

1. The pulsed plasma system spends most of the time in the state with the maximum power (which correlates with the maximum power principle), thus the average power of the discharge is essentially equal to the maximum power, see (9.31).
2. The maximum power release of the discharge is proportional to the square of the applied voltage, which is obviously well expected.
3. The energy release in the pulse, as well as the average power, mostly depends on the properties of the dielectric barrier. The increase of the area or of the dielectric permittivity of dielectric or decrease of its width, result in larger average power.
4. The total energy is larger for steeper growing voltage profiles (higher ω). In this perspective, with other parameters fixed, nanosecond-pulses release more power than the microsecond pulses. More details on the subject, including experimental confirmation of the above conclusions, can be found in D. Dobrynin, D. Vainchtein, M. Gherardi ea., 2019b.

9.8 PROBLEMS AND CONCEPT QUESTIONS

9.8.1 Electric Field Distribution in Corona Discharge Systems

Derive Eqs. (9.1)–(9.3) for electric field distributions for the simple corona discharge systems: coaxial cylinders, coaxial spheres, and sphere – remote plane. Interpret Eq. (9.5), and then derive formulas for the maximum electric field between parallel wires as well as between a single wire and a parallel plane.

9.8.2 Positive and Negative Corona Discharges

Compare the ignition criteria of positive and negative corona discharges Eqs. (9.8) and (9.9), and explain why the positive corona requires slightly higher values of voltage to be initiated. Estimate the relative difference in the values of ignition voltages.

9.8.3 The Peek Formula for Corona Ignition

Using the empirical formula Eq. (9.12) for the Townsend coefficient α in air and criteria of Eqs. (9.8), (9.9) of corona ignition, analyze and derive the Peek formula for initiating a corona discharge in air. Compare the Peek formula with a similar criterion Eq. (9.11) for the case of corona ignition between two parallel wires.

9.8.4 Active Corona Volume

Explain why most plasma-chemical processes occur in the active volume? Explain why the active corona volume cannot be increased by increasing the voltage? Why the formation of a corona discharge around a thin wire looks more attractive for applications from the point of view of maximizing the active corona volume than a corona formed around a sharp point?

9.8.5 Space Charge Influence on Electric Field Distribution in Corona

Based on the Maxwell Eq. (9.18), derive the electric field distribution Eq. (9.19) in a corona discharge formed around a thin wire taking into account the space charge effect. The Maxwell equation in form Eq. (9.18) was obtained assuming low perturbation of the electric field distribution by the space charge outside of the active corona volume. Based on the distribution Eq. (9.19) derive the criterion for the low level of perturbation of the electric field by the space charge, which limits the corona current.

9.8.6 Current–Voltage Characteristics of Corona Discharge

Based on the electric field distribution in the corona formed around a thin wire, derive the current–voltage characteristic Eq. (9.20) of the discharge. Explain the difference in the current–voltage characteristics for positive and negative corona discharges, for electronegative and non-electronegative gases.

9.8.7 Power of Continuous Corona Discharges

Explain the physical limitations of the power increase of continuous coronas. According to Eq. (9.23), corona power can be increased by diminishing the radius of external electrode. What is the limitation of this approach to the corona power increase?

9.8.8 Pulse-Periodic Regimes of Positive and Negative Corona Discharges

Corona discharges are able to operate at some conditions in form of periodic current pulses even in the constant voltage regime. The frequency of these pulses can reach 10^4 Hz in the case of positive corona, and 10^6 Hz for negative corona.

Give your interpretation why the frequency of these pulses in positive corona (flashing corona) is so much higher than the frequency in the case of negative pulses (the so-called Trichel pulses).

9.8.9 Flashing Corona Discharge

Continuous positive corona discharge exists in the steady-state regime only in some intermediate intervals of voltages. At relatively low voltages close to corona ignition conditions as well as at relatively high voltages close to the corona-to-spark transition, the corona exists in pulse periodic regime. If the flashing corona phenomenon at low voltages can be explained by inefficient positive space charge drift to the cathode in the relatively weak electric field, to explain how the flashing corona appearance at high electric fields?

9.8.10 Voltage Rise Rate in Pulse Corona Discharges

Give your interpretation why not only pulse duration, but also the voltage rise rates are so important to reach high levels of voltage pulses and high efficiencies of the pulse corona discharges?

9.8.11 Electric Field of Residual Charge Left by DBD-Streamer on Dielectric Barrier

Based on the typical data presented in Table 9.3.1, estimate the electric field induced by the residual charge left by a streamer on a dielectric barrier. Compare the result of estimations with the corresponding modeling data given in the end of Section 9.3.3.

9.8.12 Overheating of the DBD Micro-Discharge Channels

As is presented in Table 7.3.1, the heating effect in the DBD micro-discharge channels is fairly low, about 5 K. Compare this heating effect with the general heating effect in streamer channels, which was estimated in Section 4.4.6.

9.8.13 Plasma-Chemical Energy Efficiency of the Dielectric Barrier in Comparison with the Pulsed Corona Discharge

Pulsed corona discharges usually operate with higher voltage rise rates and in general with higher voltages, discuss the difference in efficiencies of ionization, electronic excitation and vibrational excitation for these two types of non-thermal atmospheric pressure discharges.

9.8.14 Sliding Surface Discharges

The number of the electrode connecting and current-conducting channels in the gliding spark regime of the surface discharge as well as the plasma coverage of the discharge gap depend on the value of electric field on the dielectric surface. Based on the simple equivalent circuit consideration, show that the electric field (and, hence, the number of channels and the plasma coverage of the gap) are determined by the capacitance factor ε/d, which is the ratio of dielectric permittivity over the thickness of the dielectric layer. Give your interpretation of the data presented in Figure 9.19.

9.8.15 Ferroelectric Discharges

Peculiarities of the ferroelectric discharges are related from one hand to the high dielectric permittivity of ferroelectric materials. This provides high power at low AC frequencies and applied voltages. On the other hand, these are related to high local electric fields on the surface of the materials. Compare the influence of these two effects on the ferroelectric discharge.

9.8.16 Velocity of the Back Ionization Wave

Early stages of a spark generation are related to the propagation of a back ionization wave, which actually provides a high degree of ionization in the spark channel. Interpret the experimental fact that the velocity of this wave reaches 10^9 cm/s.

9.8.17 Negative Ions Attachment to Water Droplets, Mechanism of Charge Separation in Thundercloud

Estimate the size of a water droplet (number of water molecules in a cluster), which is able to provide effective trapping of at least one negative ion due to the surface polarization effect. Discuss proportionality of this charge to the droplet radius. Compare the negative charging of droplets due to the surface polarization effect Eq. (9.27) with possible charging of particles related to the floating potential effect of high mobility of electrons.

9.8.18 Mechanism of Propagation of Ball Lightning

Estimate the propagation rate of ball lightning in a steady-state atmosphere with slightly non-uniform spatial distributions of temperature and fuel concentration. Based on the calculation results, estimate the possibility of the ball lightning propagating opposite to the wind direction.

9.8.19 Power Control of Nanosecond-Pulsed Dielectric Barrier Discharges

Give your interpretation why the thinner is a dielectric barrier in DBD, the higher becomes average power of the nanosecond-pulsed dielectric barrier discharges.

10 Plasma Created in High-Frequency Electromagnetic Fields

Radio-Frequency (RF), Microwave, and Optical Discharges

10.1 RADIO-FREQUENCY (RF) DISCHARGES AT HIGH PRESSURES, INDUCTIVELY COUPLED THERMAL RF DISCHARGES

10.1.1 General Features of the High-Frequency Generators of Thermal Plasma

The most traditional way of thermal plasma generation is related to arcs. The arc discharges provide high power for thermal plasma generation at atmospheric pressure, using DC power supplies at a relatively low price of about \$0.1–0.5/W. The radio-frequency (RF) discharges are also able to generate plasma at a high power level and atmospheric pressure, but these require more expensive RF power supplies and are characterized by a price of about \$1–5/W. Such a significant difference in the prices of power supplies is especially important because modern industrial applications often require thermal plasma generation at power levels from tens of kilowatts to many megawatts. Nevertheless, different types of high-frequency discharges now are used increasingly often for thermal plasma generation. This is because direct electrode-plasma contact is not required, and in many of these discharge systems there are no electrodes at all as well as electrode-related problems. High-frequency electromagnetic fields can interact with plasma in different ways. In the RF-frequency range, either inductive or capacitive coupling can provide this interaction. Electromagnetic field interaction with plasma in microwave discharges is quasi-optical.

10.1.2 General Relations for Thermal Plasma Energy Balance, the Flux Integral Relation

Thermal plasma sustained by electromagnetic fields with different frequencies is always characterized by quasi-equilibrium temperature distributions that are determined by the absorption of energy of the electromagnetic fields. Gas flow is subsonic and pressure can be considered as fixed. The energy balance taking into account electromagnetic energy dissipation and heat transfer can be then described as:

$$\rho c_p \frac{dT}{dt} = -div\vec{J} + \sigma\left\langle \vec{E}^2 \right\rangle - \Phi, \quad \vec{J} = -\lambda \nabla T. \tag{10.1}$$

In this equation: $\vec{J}$ is the heat flux, c_p is the gas specific heat at constant pressure; λ is the thermal conductivity coefficient; σ is the high-frequency conductivity (see Section 6.6.2, and = Eq. (6.181)); the square of the electric field is averaged $\langle E^2 \rangle$ over an oscillation period, which is supposed to be short; the factor Φ describes radiation heat losses which can be neglected at atmospheric pressure and $T < 11{,}000–12{,}000$ K; $\rho = Mn_0$ is the gas density, related to temperature T taking into account the constancy of pressure $p = (n_0 + n_e)T$; M is the mass of heavy particles; n_e and n_0 are the number densities of electrons and heavy species. The material derivative dT/dt is concerned with a fixed mass of gas (the Lagrangian description of the flow) and is related to the local Eulerian derivative $\partial T/\partial t$ by the well-known convective derivative relation: $dT/dt = \partial T/\partial t + (\vec{u} \cdot \vec{\nabla})T$, where $\vec{u}$ is velocity vector of the fixed mass of gas mentioned above. Neglecting the effect of gas motion on plasma temperature in the energy release zone, the energy balance Eq. (10.1) can be rewritten for steady-state discharges as:

$$-div\vec{J} + \sigma\left\langle \vec{E}^2 \right\rangle = 0, \quad \vec{J} = -\lambda \nabla T. \tag{10.2}$$

In contrast to the DC case, the electric field Eq. (10.2) is not constant and should be determined from the electromagnetic field energy balance. In the steady-state system, this balance can be presented in terms of the Pointing vector, which is the flux density $\vec{S}$ of electromagnetic energy :

$$div\left\langle \vec{S} \right\rangle = -\sigma\left\langle \vec{E}^2 \right\rangle, \quad \vec{S} = \varepsilon_0 c^2 \left[\vec{E} \times \vec{B} \right]. \tag{10.3}$$

Combining the balance Eqs. (10.2) and (10.3) leads to the relation between energy fluxes:

$$div\left(\vec{J} + \left\langle \vec{S} \right\rangle \right) = 0. \tag{10.4}$$

This kind of energy continuity equation illustrates that the electromagnetic energy flux coming into some volume and dissipated there is balanced by thermal energy flux going out. In other words, the total energy flux has no sources. This equation can be 1D solved, resulting in the **integral flux relation**:

$$\vec{J} + \left\langle \vec{S} \right\rangle = \frac{const}{r^n}. \tag{10.5}$$

In the flux integral: "r" is the 1D-coordinate; and the power $n = 0$ for plane geometry, $n = 1$ for cylindrical geometry, and $n = 2$ for spherical one. The constant of integration in Eq. (10.5) can be determined from the boundary conditions. For example, radial thermal and electromagnetic fluxes are equal to zero at the axis of the cylindrical plasma column. As a result, the constant in the integral flux relation is also equal to zero in this case. This leads us to the equation of the cylindrical plasma column in the following simple form: $J_r + S_r = 0$. This simple form of the integral flux relation gives the third equation of the channel model of arc discharges. It shows that thermal plasma columns sustained by DC and high-frequency electromagnetic fields have many common features.

10.1.3 Thermal Plasma Generation in the Inductively Coupled RF-Discharges

The general principle of plasma generation in the inductively coupled discharges is illustrated in Figure 10.1. High-frequency electric current passes through a solenoid coil where the resulting high-frequency magnetic field is induced along the axis of the discharge tube. In turn, the magnetic field induces a high-frequency vortex electric field concentric with the elements of the coil, which is able to provide breakdown and sustain the inductively coupled discharge. Electric currents in this discharge are also concentric with the coil elements and the discharge itself is apparently electrodeless. Generated in this way, inductively coupled plasma (ICP) can be quite powerful and effectively sustained at atmospheric and even higher pressures. The magnetic field in the inductively coupled

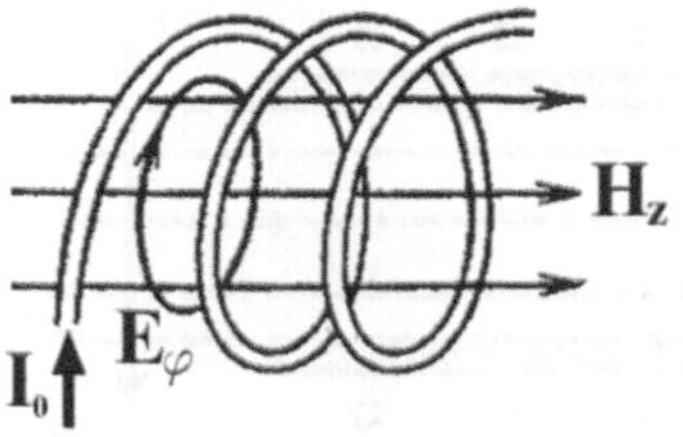

FIGURE 10.1 Generation of the inductively coupled plasma.

discharge is determined by the current in the solenoid, while electric field there, according to the Maxwell equations, is also proportional to the frequency of the electromagnetic fields. As a result, to achieve electric fields sufficient to sustain the **inductively coupled plasma (ICP),** the high frequencies (RF) of about 0.1–100 MHz is usually required. Practically, a dielectric tube is usually inserted inside of the solenoid coil. Then ICP is sustained inside of the discharge tube in the gas of interest. Powerful RF-discharges and power supplies generate noises and are able to interfere with radio communication systems. To avoid this undesirable effect, several specific frequency intervals were assigned for the operation of industrial RF discharges. The most common RF frequency used in industrial plasma chemistry and plasma engineering is 13.6 MHz (the corresponding wavelength is 22 m). The inductively coupled RF-discharges are quite effective in sustaining thermal plasma at atmospheric pressure. The relatively low electric fields are not sufficient, for the ignition of the discharges at atmospheric pressures. To ignite inductive discharges at atmospheric pressures require special approaches. For example, an additional rod-electrode can be introduced inside of the coil solenoid. Then heating of the electrode by the Foucault currents results in its partial evaporation, which simplifies the breakdown. After the breakdown, the additional electrode can be removed from the discharge zone. The specific problem of the RF-discharges is the effectiveness of plasma coupling as a load with RF-generator. Electrical parameters of the plasma load, such as resistance and inductance, influence the operation of the electric circuit as a whole and determine the effectiveness of the coupling. Analysis of the ICP temperature distribution, which determines other parameters of quasi-equilibrium thermal plasma, is important in this case.

10.1.4 Metallic Cylinder Model of Long Inductively Coupled RF Discharge

Consider a dielectric discharge tube of radius R inserted inside the solenoid coil (see Figure 10.2). Plasma is sustained by Joule heating induced by high-frequency AC and stabilized by heat transfer to the walls of the externally cooled discharge tube. Radial temperature distribution can be described by the energy balance Eq. (10.2):

$$-\frac{1}{r}\frac{d}{dr}rJ_r + \sigma\left\langle E_\phi^2\right\rangle = 0, \quad J_r = -\lambda\frac{dT}{dr}. \tag{10.6}$$

This equation is similar to Eq. (8.20) for the positive column of arc discharges, though the electric field now is not axial but azimuthal and rapidly alternating. In the megahertz frequency range of the RF-discharges, the high-frequency conductivity Eq. (6.181) actually coincides with that of the DC case because $\omega^2 << \nu_{en}^2$; polarization and displacement currents can be neglected with respect to conductivity current; and the complex dielectric constant Eq. (6.179) is mostly imaginary. Neglecting the displacement current and assuming $E, H \propto \exp(-i\omega t)$, the Maxwell equations (6.6.3) and (6.1.23) for the electric and magnetic fields in the case of cylindrical symmetry can be expressed as:

$$-\frac{dH_z}{dr} = \sigma E_\phi, \quad \frac{1}{r}\frac{d}{dr}rE_\phi = i\omega\mu_0 H_z. \tag{10.7}$$

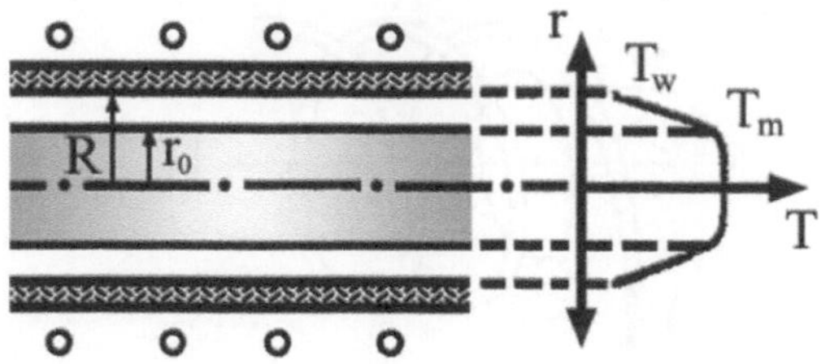

FIGURE 10.2 Radial temperature distribution in the ICP-discharge in a tube of radius R inserted inside of a solenoid (r_0 is the plasma radius).

Together with the balance Eq. (10.6), the Maxwell equations complete the system describing the plasma column. The boundary conditions for the system (10.6) and (10.7) for the discharge geometry shown in Figure 10.2 assuming low temperature on the externally cooled walls can be taken as:

$$J_r = 0, \quad E_\phi = 0 \quad at \quad r = 0; \quad T = T_w \approx 0 \quad at \quad r = R. \tag{10.8}$$

The magnetic field in non-conductive gas near walls ($r = R$) is the same as inside of empty solenoid:

$$H_z(r = R) \equiv H_0 = I_0 n. \tag{10.9}$$

In this boundary relation: I_0 is current in the solenoid coil, n is the number of the coil turns per unit of its length. Amplitudes of current and magnetic fields are actually complex values, but here these can be considered as real. Phase deviation between the oscillating fields H_z and E_φ can be calculated with respect to the phase of magnetic field H_0 Eq. (10.9) in the non-conductive gas near the wall of the discharge tube. Further simplification and solution of the system of Eqs. (10.6) and (10.7) can be done under the framework of the **metallic cylinder model**. This is actually a generalization of the channel model previously applied for arc discharges. According to the model, a plasma column is considered as a metallic cylinder with the conductivity fixed in the first approximation and corresponding to the maximum temperature T_m on the discharge axis. The physical reasons underlying the metallic cylinder model are the same as those of the channel model of the arc discharge column. Thermal plasma conductivity is a very strong exponential function of the quasi-equilibrium discharge temperature. For this reason, when the temperature slightly decreases towards the discharge tube walls, conductivity becomes actually negligible. Note also that the plasma temperature T_m, conductivity σ, and the "metallic cylinder" (plasma column) radius r_0 are a priory unknown in the frameworks of the metallic cylinder model. If plasma conductivity is high enough, the skin effect prevents penetration of electromagnetic fields deep into the discharge column. As a result, heat release related to the inductive currents, is localized in the relatively thin skin layer of the plasma column. Thermal conductivity inside of the "metallic cylinder" provides the temperature plateau in the central part of the discharge cylinder where the inductive heating by itself is negligible. Radial distributions of plasma temperature, plasma conductivity and the Joule heating in the inductively coupled RF-discharge, corresponding to the metallic cylinder model, are illustrated in Figure 10.3.

10.1.5 Electrodynamics of Thermal ICP - Discharge in Frameworks of the Metallic Cylinder Model

The metallic cylinder model permits to separately consider electrodynamics and heat transfer aspects of the inductively coupled RF-discharge. Analyzing first the electrodynamics of the ICP-discharge, plasma conductivity σ and radius r_0 can be taken as parameters. At the most typical RF-discharge frequency $f = 13.6$ MHz and high conductivity conditions of atmospheric pressure thermal plasma,

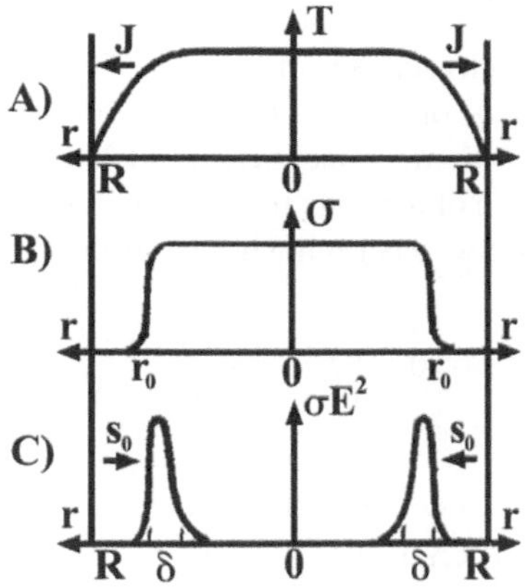

FIGURE 10.3 Radial distributions of temperature (a), electric conductivity (b), and Joule heating (c) in ICP discharge.

the skin layer δ (see Eqs. (6.25) and (6.26)) is usually small with respect to the plasma radius $\delta << r_0$. In this case of strong skin-effects, the interaction of the electromagnetic field with plasma can be simplified to a 1D-plane geometry, which permits rewriting the Maxwell relations (10.7) as:

$$\frac{dH_z}{dx} = \sigma E_y, \quad \frac{dE_y}{dx} = -i\omega\mu_0 H_z. \tag{10.10}$$

Coordinate "x" is positive for the direction inward to the plasma (opposite to radius); coordinates "y" and "z" are directed tangentially to the plasma surface. Boundary conditions are: $H = H_0$ at $x = 0$, and E_y, $H_z \to 0$ at $x \to \infty$. The solution of this system of equations can be presented as:

$$H_z = H_0 \exp\left[-i\left(\omega t - x/\delta\right) - x/\delta\right], \quad \delta = \sqrt{2/\omega\mu_0\sigma}, \tag{10.11}$$

$$E_y = H_0\sqrt{\frac{\omega\mu_0}{\sigma}} \exp\left[-i\left(\omega t - \frac{x}{\delta} + \frac{\pi}{4}\right) - \frac{x}{\delta}\right]. \tag{10.12}$$

The same relations for electromagnetic fields penetrating and damping in the δ skin layer of high conductivity thermal (see Eqs. (6.25) and (6.26)) can be rewritten not in complex but in the real form:

$$H_z = H_0 \exp\left(-\frac{x}{\delta}\right)\cos\left(\omega t - \frac{x}{\delta}\right), \tag{10.13}$$

$$E_y = H_0\sqrt{\frac{\mu_0\omega}{\sigma}} \exp\left(-\frac{x}{\delta}\right)\cos\left(\omega t - \frac{x}{\delta} + \frac{\pi}{4}\right). \tag{10.14}$$

Thus the amplitudes of electric and magnetic fields are seen to decrease exponentially inside of the plasma column, with a phase shift between them of $\pi/4$.

The electromagnetic energy flux is normal to the plasma surface and directed inward to the plasma column (opposite to radius). Based on Eqs. (10.13) and (10.14) the electromagnetic energy flux is:

$$\langle S\rangle = S_0 \exp\left(-\frac{2x}{\delta}\right), \quad S_0 = H_0^2\sqrt{\frac{\mu_0\omega}{4\sigma}}. \tag{10.15}$$

Here the flux S_0 shows the total electromagnetic energy absorbed in the unit area of the skin layer per unit time. The total power w, released per unit length of the long cylindrical plasma column, is related to this flux as: $w = 2\pi r_0 \cdot S_0$. For calculations of the electromagnetic flux S_0, it is convenient to use the additional relation (10.9) between magnetic field H_0 and current I_0 in a solenoid as well as the number n of turns per unit length of the coil. Together with Eq. (10.15) this leads to the following convenient numerical relation for power per unit surface of the long cylindrical ICP column:

$$S_0, \frac{W}{\text{cm}^2} = 9.94 \cdot 10^{-2} \cdot \left(I_0 n, \frac{A \cdot turns}{\text{cm}} \right)^2 \sqrt{\frac{f, MHz}{\sigma, \text{Ohm}^{-1}\text{cm}^{-1}}}. \tag{10.16}$$

10.1.6 Thermal Characteristics of the Inductively Coupled Plasma in Framework of the Model of Metallic Cylinder

Integration of (10.6) between column and walls ($r_0 < r < R$, $\sigma = 0$) gives the relation of plasma temperature, radius r_0, and specific discharge power w per unit length:

$$\Theta_m(T_m) - \Theta_w(T_w) = \frac{w}{2\pi} \ln \frac{R}{r_0}. \tag{10.17}$$

Quasi-equilibrium plasma temperature is expressed here in terms of the heat flux potential $\Theta(T)$ introduced by Eq. (8.22); T_m is the maximum temperature in the plasma column; T_w is the temperature of discharge tube walls. Equation (10.17) is equivalent to Eq. (8.25) derived in Section 8.4.3 for arc discharges, although the expressions for the specific power w are obviously different. The interval Δr between the plasma column and walls is short $\Delta r = R - r_0 << R$, Eq. (10.17) can be simplified:

$$\Theta_m(T_m) - \Theta_w(T_w) \approx \frac{w}{2\pi r_0} \cdot \Delta r = S_0 \cdot \Delta r. \tag{10.18}$$

This relation in contrast to Eq. (10.17) is specific for the ICP discharges, where $\Delta r = R - r_0 << R$ and the heat release is concentrated in a cylindrical ring near the discharge walls in contrast to arcs where the current is located within a channel near the discharge axis. The plasma temperature in the central part of the inductive discharge is almost constant and close to the maximum value (see Figure 10.3). From the figure, the plasma temperature and conductivity are seen to decrease in the case of the strong skin effect in a thin layer about $\delta/2$, where Joule heating is mostly localized. The energy balance of the Joule heating induced by electromagnetic fields and thermal conductivity in this layer is:

$$\lambda_m \frac{\Delta T}{\delta/2} \approx S_0. \tag{10.19}$$

This approach is a modified model of the metallic cylinder and is similar to the Raizer modification of the channel model of arc discharges. Taking into account the relation between electromagnetic flux and discharge power per unit length: $w = 2\pi r_0 \cdot S_0$, Eq. (10.19) can be rewritten:

$$4\pi\lambda_m \Delta T \approx w \frac{\delta}{r_0}. \tag{10.20}$$

The thermal conductivity coefficient λ_m corresponds to the maximum plasma temperature T_m, and ΔT is the plasma temperature decrease, related to the exponential conductivity decrease. The temperature decrease ΔT at the boundary layer of the thermal inductively coupled plasma can be determined based on the Saha equation in the same way as for arc discharges using Eq. (8.32). Equation (10.15) for S_0, Eq. (10.9) for H_0, and Eq. (10.11) for δ, one can rewrite Eq. (10.19) as:

$$2\sqrt{2}\,\lambda_m \frac{T_m^2}{I_i}\sigma_m = I_o^2 n^2. \tag{10.21}$$

This formula relates current I_o and number of turns n per unit length in the solenoid coil with plasma conductivity σ_m and with ICP temperature T_m. In Eq. (10.21) I_i is the ionization potential.

10.1.7 Temperature and Other Quasi-Equilibrium ICP Parameters in Frameworks of the Model of Metallic Cylinder

Taking into account that the plasma conductivity σ_m strongly depends on gas temperature, while T_m and λ_m are changing only slightly, one can conclude from Eq. (10.21) that approximately $\sigma_m \propto (I_0 n)^2$. Then, based on the Saha equation for conductivity $\sigma_m(T_m)$, the plasma temperature dependence on the solenoid current and number of turns per unit length can be expressed as:

$$T_m = \frac{const}{const - \ln\left(I_0 n\right)}. \tag{10.22}$$

This is not a strong logarithmic dependence $T_m(I_0, n)$ and corresponds to that derived for arc discharges Eq. (8.36). Strong skin effect, plasma temperature T_m does not depend on the electromagnetic field frequency. The following relations can illustrate the dependence of the ICP power per unit length of discharge on the solenoid current and the other parameters:

$$w = 2\pi r_0 S_0 \propto H_0^2\sqrt{\frac{\omega}{\sigma_m}} = \left(I_0 n\right)^2\sqrt{\frac{\omega}{\sigma_m}} \propto I_0 n\sqrt{\omega}. \tag{10.23}$$

Hence, the specific power of the inductive discharge grows not only with the solenoid current and the number of turns, but also with the frequency of electromagnetic field. This can be explained by an increase of the electric field with frequency (see Eq. (10.14)). Taking into account that $\sigma_m \propto (I_0 n)^2$, Eq. (10.23) gives $w \propto \sqrt{\sigma_m}$. This means that even a small increase of temperature T_m (which according to the Saha leads to an exponential increase of electric conductivity) requires a significant increase of discharge power per unit length. This effect limits the maximum temperature of the thermal discharges. Using the above relations, one can estimate typical numerical characteristics of the thermal ICP in atmospheric pressure air on the electromagnetic field frequency $f = 13.6$ MHz. For a gas temperature of about 10,000 K, a discharge tube radius of about 3 cm, and a thermal conductivity of $\lambda_m = 1.4 \cdot 10^{-2}$ W/cm · K, the electric conductivity of the thermal plasma can be found to be $\sigma_m \approx 25$ Ohm^{-1} cm^{-1} and the skin layer $\delta \approx 0.27$ cm. To sustain such a plasma, the necessary electromagnetic flux can be calculated from Eqs. (10.19) and (8.32) as: $S_0 \approx 250$ W/cm^2. The corresponding value of solenoid current and number of turns is: $I_0 n \approx 60A\frac{turns}{\text{cm}}$, magnetic field $H_0 \approx 6$ kA/m. Then the maximum electric field on the external boundary of the plasma column is approximately 12 V/cm, the density of circular current is 300 A/cm^2, the total current per unit length of the column is approximately 100 A/cm, thermal flux potential is about 0.15 kW/cm, the distance between the effective plasma surface and discharge tube is $\Delta r \approx 0.5$ cm, and finally the ICP-discharge power per unit length is about 4 kW/cm. Analyzing the above typical numerical characteristics of a thermal ICP-discharge, one can note that according to Eq. (10.23) and the following discussion, the specific discharge power should be increased from 4 kW/cm to at least 8 kW/cm to increase plasma temperature from 10,000 K to12,000 K. Actually taking also into account the growing radiative losses at 12,000 K requires the specific power to be even higher. As a result of such strong power requirements, the ICP temperature does not exceed 10,000 K–11,000 K. A typical value of the circular ICP current per unit length was estimated as 100 A/cm. This value is about the same and even exceeds the solenoid current per unit length of the coil: $I_0 n \approx 60$ A/cm. This means that the inductive influence of the plasma current on the electric circuit of the RF power supply is quite significant. Thus the RF-power supply must be effectively coupled with the plasma, which can be considered as a load in the circuit.

10.1.8 ICP-Discharge in Weak Skin-Effect Conditions, the Thermal ICP Limits

A decrease of current in the solenoid coil leads to a growth of the skin layer δ until it becomes about the plasma size ($\delta \approx r_0$, R) and the considered model is not valid. For the opposite case of low temperature and electric conductivity $\delta >> r_0$, R. The magnetic field is uniform $H = H_0$ in the absence of skin effect and the electric field distribution along the radius of the discharge can be calculated from the Maxwell Eq. (10.7) as:

$$E(r) = \frac{1}{2}\omega\mu_0 H_0 \cdot r. \tag{10.24}$$

Power per unit length of the discharge with radius r_0 can be found by integrating Joule heating:

$$w = \int_0^{r_0} \sigma \langle E^2(r) \rangle 2\pi r dr \approx \frac{1}{16}\pi\mu_0^2\omega^2\sigma_m H_0^2 r_0^4. \tag{10.25}$$

Taking into account the relation between the magnetic field with solenoid current and the number of turns per unit length Eqs. (10.9), (10.25) can be rewritten as:

$$w \approx \frac{1}{16}\pi\mu_0^2\omega^2\sigma_m I_0^2 n^2 r_0^4. \tag{10.26}$$

In contrast to the case of strong skin-effect conditions where the temperature decrease takes place mainly in the boundary layer Eqs. (10.19, 10.20), here the temperature reduction is distributed over

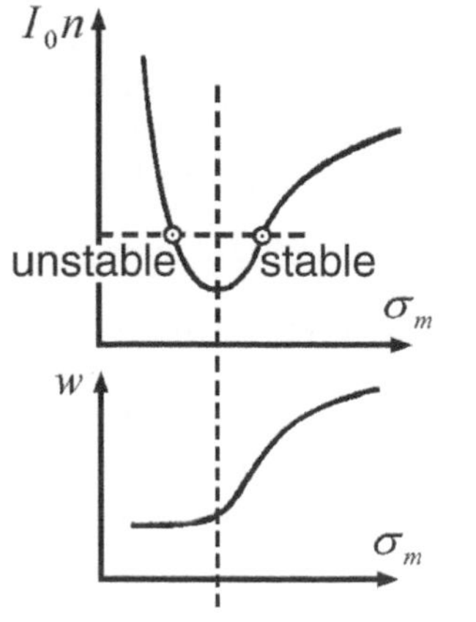

FIGURE 10.4 Solenoid current and specific power dependence on conductivity, stability analysis.

the entire plasma column radius as it was in the arc discharges. This leads to the following relation between specific power and maximum plasma temperature (Eq. (8.32)):

$$w \approx 4\pi r_0 \lambda_m \frac{\Delta T}{r_0} = 4\pi\lambda_m \Delta T \approx 8\pi\lambda_m \frac{T_m^2}{I_i}. \tag{10.27}$$

In this relation: ΔT is the temperature decrease across the plasma column (see Eq. (8.32); λ_m is the thermal conductivity corresponding to the maximum plasma temperature T_m; I_i is the effective value of ionization potential. Next, analyze what happens when the temperature and electric conductivity decreases in this weak skin-effect regime. The temperature cannot decrease significantly (because of the related exponential reduction of conductivity), which according to Eq. (10.27) makes the specific power w almost constant even when electric conductivity decreases. This effect of specific power stabilization is illustrated in Figure 10.4a. Then based on Eq. (10.18), the plasma radius must also decrease with temperature reduction. Taking into account Eq. (10.26) leads to an interesting conclusion: current in the solenoid coil (factor $I_0 n$) in this regime is not decreasing but grows with the conductivity decrease in contrast to the strong skin-effect regime where $w \propto I_0 n \propto \sqrt{\sigma_m}$. This non-monotonic dependence of $I_0 n(\sigma_m)$ is illustrated in Figure 10.4b. Thus, the dependence $I_0 n(\sigma_m)$ has a minimum corresponding to the case when the skin layer is about the same size as the discharge tube radius $\delta \approx R$. This minimum value of the solenoid current $I_0 n$, which is necessary to sustain the quasi-equilibrium thermal ICP, can be found from Eq. (10.21) assuming a value of electric conductivity σ_m corresponding to the critical condition $\delta \approx R$ (see Eq.10.11). In turn, this leads to the following expression for the minimum value of the solenoid current $I_0 n$ to sustain the thermal plasma column:

$$\left(I_0 n\right)_{\min} \approx \frac{2T_m}{R}\sqrt{\frac{\lambda_m\sqrt{2}}{I_i \mu_0 \omega}}. \tag{10.28}$$

As an example, the minimal current necessary to provide a thermal ICP discharge in a tube with $R = 3$ cm at frequency $f = 13.6$ MHz in air according to Eq. (10.28) can be estimated as $\left(I_0 n\right)_{\min} \approx 10A \cdot \frac{turns}{\text{cm}}$; corresponding minimum values of the quasi-equilibrium temperature are approximately $T_{crit} \approx 7000$–$8000\ K$. As seen from Figure 10.4b, when the solenoid current and number of turns per unit length exceed the critical value, in principle two stationary states of the ICP-discharge can be realized. One of these corresponds to high conductivity and strong skin effect and the other to low conductivity and no-skin-effect conditions. However only one of them, high conductivity regime is stable. The low conductivity regime (left branch in Figure 10.4) is unstable. For example, if the temperature (and hence conductivity) increases because of some fluctuation, then a current lower than the actual one is sufficient according to Figure 10.4 to sustain the discharge. This leads to ICP plasma heating and further temperature increase until the stable high conductivity branch is reached.

10.1.9 The ICP-Torches

The inductively coupled plasma torches are widely used as industrial plasma sources. They are important competitors of the DC arc jets discussed in Section 8.5. A standard configuration of an ICP torch (B. Gross, B. Grycz, K. Miklossy, 1969) is illustrated in Figure 10.5. The heating coil is usually water-cooled and is not in direct contact with the plasma, which provides plasma purity. The ICP-torches are difficult to start because the electric fields in these systems are relatively low. To initiate the discharge at low power levels of about 1 kW, a special graphite starting rod, shown in Figure 10.5, can be applied. At high power levels, it is convenient to use a pilot DC or RF small

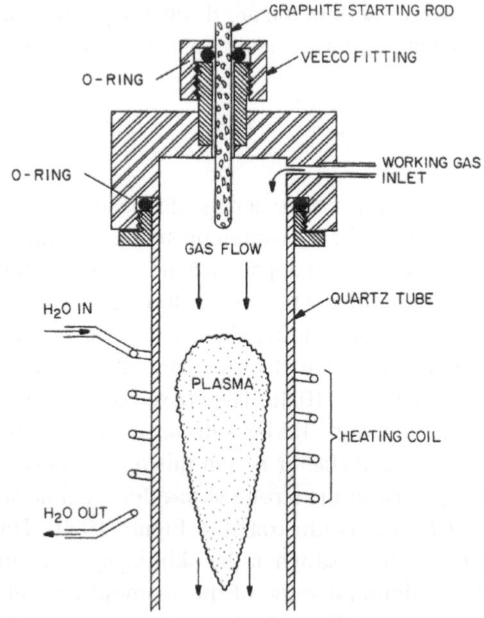

FIGURE 10.5 A kilowatt-level inductively coupled plasma torch. (From J.R. Roth, Industrial Plasma Engineering)

plasma generators to initiate the powerful ICP-torch. Such hybrid plasma torches operating at power levels of 50–100 kW are illustrated in Figure 10.6a,b. The ICP-torches are often operated in transparent quartz tubes to avoid heat load on the discharge walls related to visible radiation. At power level exceeding 5 kW, the walls should be cooled by water. Plasma stabilization and the discharge walls insulation from direct plasma influence can be achieved in fast gas flows. Fast gas flows are effective from this point of view.

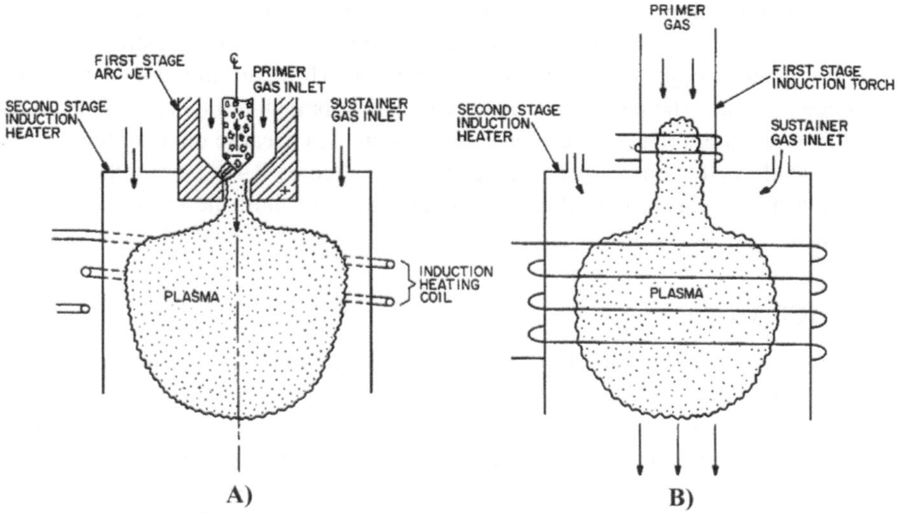

FIGURE 10.6 Hybrid plasma torches: (a) DC-RF, (b) RF-RF. (From J.R. Roth, Industrial Plasma Engineering)

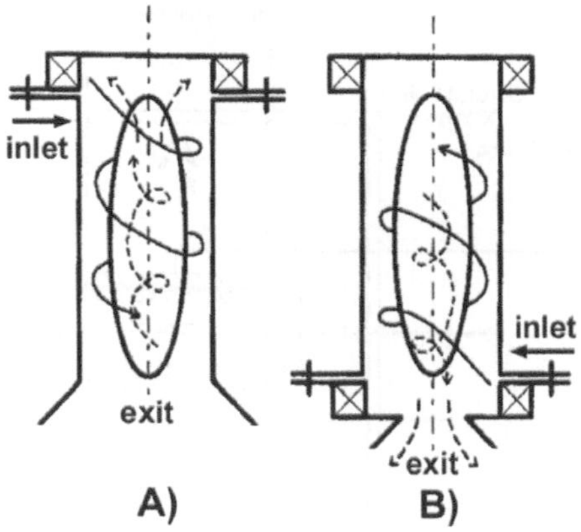

FIGURE 10.7 Forward (a) and reverse (b) vortex stabilization.

10.1.10 ICP-Torch Stabilization in Vortex Gas Flow

In the most conventional approach, the swirl generator is placed upstream with respect to the ICP-discharge, and the outlet of the plasma jet is directed to the opposite side. Such configuration is usually referred to as the **forward-vortex stabilization** and illustrated in Figure 10.7a. The rotating gas provides good protection to the walls from the plasma heat flux (e.g., A. Gutsol, 1999). This can be explained by noting that portions of the gas changes direction from an axial swirl to the swirl radial due to viscosity and heat expansion. This radial swirl moving gas compresses the central hot zone, decreasing the heat flux to the walls and providing effective insulation. However, some reverse axial pressure gradient and central reverse flow appear in the case of forward-vortex stabilization (see Figure 10.7a). This effect is related to fast flow rotation and strong centrifugal effects near the gas inlet, which becomes a bit slower and weaker downstream. The hot reverse flow mixes with incoming cold gas and increases heat losses to the walls, which makes the discharge walls insulation somewhat less effective. A more effective ICP-torch walls insulation can be achieved by means of the **reverse-vortex stabilization**, developed by A. Gutsol and co-authors (V.T. Kalinnikov, A. Gutsol, 1997; A. Gutsol, A. Fridman, 2001). Flow configuration for the reverse-flow stabilization of the inductively coupled discharge is shown in Figure 10.7b. In this case, the outlet of plasma jet is directed along the axis to the swirl generator side. The cold incoming rotating gas in this stabilization scheme at first moves past the walls providing their cooling and insulation, and only after that does it go to the central plasma zone and become hot. Thus in the case of the reverse-vortex stabilization, the incoming gas is the entering discharge zone from all directions except the outlet side, which makes this approach interesting both for ICP-heating efficiency and discharge walls protection. Experimental details regarding the reverse-vortex stabilization of the 60 kW thermal ICP discharge in argon can be found in A. Gutsol, J. Larjo, R. Hernberg (1999). The reverse-vortex stabilization of microwave discharges is considered in a paper of A. Gutsol (1995), application to gas burners is discussed in V.T. Kalinnikov, A. Gutsol (1999).

10.1.11 Capacitively-Coupled Atmospheric Pressure RF-Discharges

The inductively-coupled plasma has lower values of electric fields than capacitively-coupled discharges , where the electric field is the primary effect (as in a capacitor). For this reason, the ICP-discharges at moderate at high pressures are usually concerned with thermal quasi-equilibrium plasma generation, where the ionization is sustained by heating and high electric fields are not necessary. The capacitively-coupled plasma (CCP) of RF-discharges is able to provide high values of electric fields, which makes these discharges interesting for generating non-thermal non-equilibrium

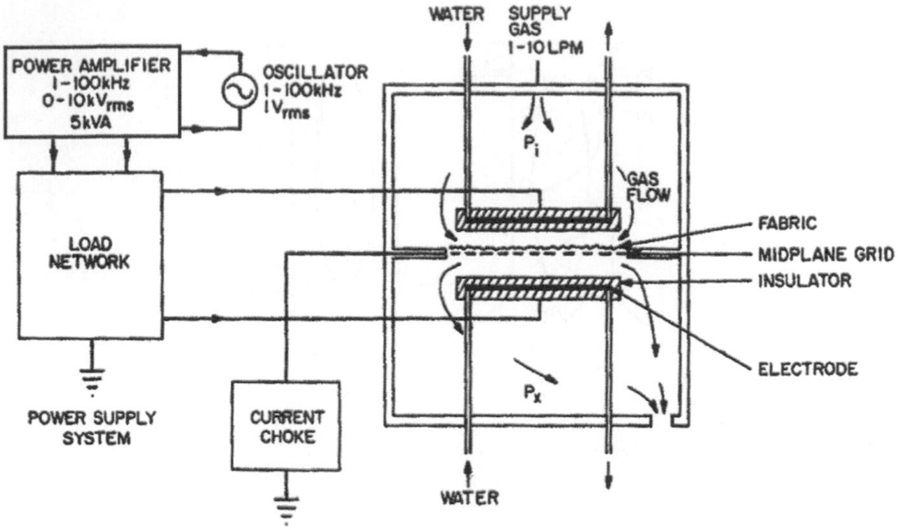

FIGURE 10.8 Atmospheric pressure glow discharge plasma reactor. (From J.R. Roth, Industrial Plasma Engineering)

plasma. Consider the **atmospheric pressure RF-glow discharge** developed recently at the University of Tennessee, Knoxville (J.R. Roth, M. Laroussi, C. Liu, 1992; see also N. Kanda, M. Kogoma, H. Jinno, H. Uchiyama, S. Okazaki, 1991). This discharge is a member of a family of atmospheric pressure glow discharges (see Figure 10.8). The discharge volume is confined by parallel electrodes across which a RF-electric field is imposed. The discharge system also includes a bare metal screen located midway between the electrodes which can be grounded through a current choke. This median screen provides a substrate surface to support the material to be treated in the atmospheric pressure glow discharge. The electric field applied between the electrodes is on the level of kilovolts per centimeter; it should be sufficiently strong for electric breakdown and to sustain the discharge. In helium or argon, this electric field is obviously lower than in atmospheric air. Typical frequencies necessary to sustain the uniform glow regime of the discharge are in the kilohertz range (about 1–20 kHz). At lower values of the frequency, the discharge is difficult to initiate; at higher frequencies, the discharge is not uniform and has the filamentary structure. The atmospheric pressure uniform glow regime corresponds to the specific values of RF-frequency which are sufficiently high to trap the ions between the median screen and an electrode, but not high enough to trap plasma electrons as well. This frequency range provides some reduction of electron–ion recombination in the boundary layers, which promotes the ionization balance at lower electric fields. The power density of the discharge obviously grows with an increase of the voltage amplitude and the electric field frequency. However even at relatively high values of the voltage amplitude and the electric field frequency, the discharge power density and total power are still not high relative to, for example, pulsed corona discharges. The maximum values of the power density in the uniform regime are approximately 100 mW/cm^3; maximum total power is about 100 W (J.R. Roth, 2000).

10.2 THERMAL PLASMA GENERATION IN MICROWAVE AND OPTICAL DISCHARGES

10.2.1 Optical and Quasi-Optical Interaction of Electromagnetic Waves with Plasma

Wavelengths of electromagnetic oscillations in the above RF-discharges were much larger than the typical sizes of the systems. For example, the industrial RF frequency $f = 13.6$ MHz corresponds to the wavelength of 22 m. In contrast to that, microwave plasma is sustained by electromagnetic waves in centimeters range of wavelengths. Thus microwave radiation has wavelengths comparable with the discharge system sizes, and the electromagnetic field interaction with plasma in microwave discharges is quasi-optical. The optical discharges sustained by laser radiation are characterized by much smaller wavelengths and so the electromagnetic field interaction with plasma in this case is also optical. For this reason, thermal plasma generation in the microwave and optical discharges will be discussed together. In the same manner, as in the case of RF-discharges, thermal plasma generation in microwave and optical discharges is usually related to high-pressure systems (usually atmospheric pressure). Non-thermal, non-equilibrium microwave discharges are usually related in a continuous mode to moderate and low pressures.

10.2.2 Microwave Discharges in Waveguides, Modes of Electromagnetic Oscillations in the Waveguides without Plasma

Microwave generators in particular magnetrons, steadily operating with power exceeding 1 kW in the GHz-frequency range, are able to effectively maintain the steady-state thermal microwave discharges at atmospheric pressure. Electromagnetic energy in the microwave discharges can be coupled with plasma in different ways. The most typical one is related to the application of waveguides and is illustrated in Figure 10.9. In the configuration of microwave discharges shown on the figure, the dielectric tube (usually quartz) which is transparent for the electromagnetic waves, crosses the rectangular waveguide. Plasma is ignited and maintained in the discharge tube by the dissipation of electromagnetic energy. Heat balance of the thermal plasma is provided mostly by convective cooling in the gas flow. Different modes of electromagnetic waves formed in the rectangular waveguide can be used to operate microwave discharges. The most typical one is the H_{01} mode, illustrated in Figure 10.10. The electric field in the H_{01}-mode is parallel to the narrow walls of the waveguide and is constant in this direction. Along the wide waveguide wall, the electric field is distributed as a sine-function (in the absence of plasma) with a maximum in the center of the discharge tube and zero-field on the narrow waveguide walls:

$$E_y = E_{\max} \sin\left(\frac{\pi}{a_w} x\right); \quad E_x = E_z = 0. \tag{10.29}$$

In this relation; E_x is the electric field component directed along the longer wall of the waveguide, which has length a_w; E_y is the electric field component directed along the shorter wall of the waveguide, which has length b_w; and E_x is the electric field component directed along the z. The maximum electric field $E_{\max}$ (in kW/cm) in the case of the mode H_{01} is related to the microwave power P_{MW} (expressed in kW) transmitted along the rectangular waveguide by the following numerical formula:

$$E_{\max}^2 = \frac{1.51 \cdot P_{MW}}{a_w b_w}\left[1-\left(\frac{\lambda}{\lambda_{crit}}\right)^2\right]^{-1/2}. \tag{10.30}$$

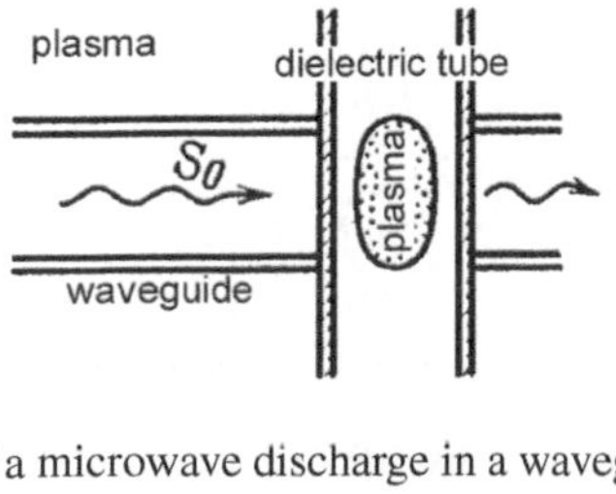

FIGURE 10.9 General schematic of a microwave discharge in a waveguide.

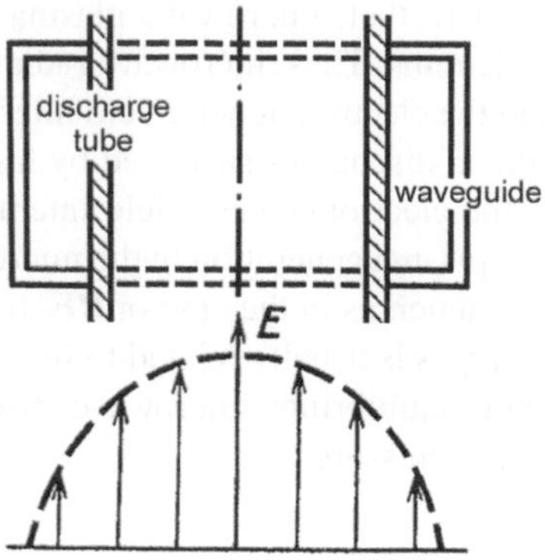

FIGURE 10.10 Electric field distribution for H_{01} mode in rectangular waveguide.

Here λ is the wavelength of the electromagnetic wave propagating in the waveguide; λ_{crit} is the maximum value of the wavelength when the propagation is still possible ($\lambda_{crit} = 2a_w$); lengths of the waveguide walls a_w and b_w are expressed in cm. The H_{01} mode is convenient for microwave plasma generation in the rectangular waveguides because the electric field in this case has a maximum in the center of the discharge tube (see Figures 10.9 and 10.10). Microwave discharges obviously can be generated in the cylindrical waveguides with a round cross section. The most typical oscillation mode in such waveguides is H_{11}. Space distribution of electric and magnetic fields in this case is illustrated in Figure 10.11, and can be expressed without taking into account the influence of the plasma as:

$$E_{\vartheta}(r,\vartheta) = E_{max} \cdot J_1'\left(\frac{1.84r}{R_w}\right)\cos\vartheta; \quad E_z = 0, \tag{10.31}$$

$$E_r(r,\vartheta) = E_{max}\frac{R_w}{1.84r} \cdot J_1\left(\frac{1.84r}{R_w}\right) \cdot \sin\vartheta. \tag{10.32}$$

In these relations, R_w is radius of the round waveguide; $J_1(1.84r/R_w)$ and J_1' $(1.84r/R_w)$ are the Bessel function of the first order and its first derivative; E_{max} again is the maximum value of electric field (expressed in kW/cm) now for the mode H_{11}, which is related to the microwave power P_{MW} (expressed in kW) transmitted along the round waveguide by the following numerical formula:

$$E_{max}^2 = \frac{1.58P_{MW}}{\pi R_w^2}\left[1 - \left(\frac{\lambda}{\lambda_{crit}}\right)^2\right]^{-1/2}. \tag{10.33}$$

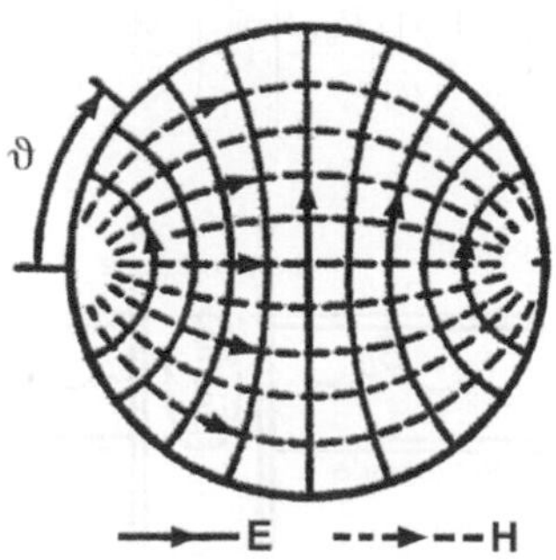

FIGURE 10.11 Electric and magnetic field distributions for H_{11} mode in a round waveguide.

Similarly to Eq. (10.30), the critical wavelength λ_{crit} is the maximum wavelength when the propagation is still possible. For the case of H_{11}-mode in the round waveguide, $\lambda_{crit} = 3.41R_w$).

10.2.3 Microwave Plasma Generation by H_{01}-Electromagnetic Oscillation Mode in Waveguide

As illustrated in Figure 10.10, the H_{01} mode in the rectangular waveguides is convenient for plasma generation because the electric field in this case has maximum in the center of the discharge tube. Plasma is formed along the electric field and axis of the discharge tube (see Figure 10.9). To provide plasma stabilization gas flow is often supplied into the discharge tube tangentially. The waveguide dimensions are related to the frequency of the electromagnetic wave, (see the relations for critical values of wavelengths). Thus for an electromagnetic wave frequency $f = 2.5$ GHz (the corresponding wavelength in vacuum $\lambda = 12$ cm) the wide waveguide wall should be longer than 6 cm and usually is equal to 7.2 cm. Typically the narrow waveguide wall is 3.4 cm long and the dielectric discharge tube diameter is about 2 cm. The diameter of the generated microwave plasma column in such conditions is usually about 1 cm. If power of the atmospheric pressure microwave discharge described above is 1–2 kW, then typical thermal plasma temperatures in air and other molecular gases are normally in the range 4000 K–5000 K. When temperature is not extremely high energy exchange between electrons and heavy particles in rare gases is slower than in molecular gases , which leads to some differences between electron temperature and temperature of heavy species in such systems. For example, typical plasma temperature values in a microwave discharge in argon at similar conditions are about 6500–7000 K for electrons and about 4500 for heavy particles. In general, plasma temperatures in thermal microwave discharges are lower than temperatures in thermal radio-frequency and arc discharges under similar conditions. At lower wavelengths ($\lambda = 3$ cm) and hence at smaller waveguide cross sections, plasma temperatures are slightly higher; at similar conditions in molecular gases such as nitrogen plasma temperatures reach about 6000 K. The incident electromagnetic wave formed in the waveguide interacts with plasma generated in the discharge. This interaction results in partial dissipation and reflection of the electromagnetic wave. Typically about half of the power of the incident wave can be directly dissipated in the high conductivity thermal plasma column, about a quarter of the wave is transmitted through the plasma and the remaining is reflected. To increase the effectiveness of electromagnetic wave coupling with thermal plasma column (fraction of the wave dissipated in plasma), the transmitted wave can be reflected back, which leads to the formation of a standing wave. Such special coupling techniques permits increasing the fraction of the electromagnetic energy absorbed in the plasma up to 90–95%, which is important for practical applications of the discharge system (L.M. Blinov, V.V. Volod'ko, G.G. Gontarev, G.V. Lysov, L.S. Polak, 1969; V. Batenin, I.I. Klimovsky, G.V. Lysov, V.N. Troizky, 1988). A powerful microwave discharge for plasma-chemical applications was developed in Kurchatov Institute of Atomic Energy by V.D. Rusanov, M.F. Krotov, G.V. Lysov and co-workers (see V.D. Rusanov, A. Fridman, 1984; V.K. Givotov, V.D. Rusanov, A. Fridman, 1985). Microwave energy was provider by four magnetrons (maximum power of each was 300 kW), and coupled with plasma in different molecular gases at moderate and relatively high-pressure range. Note that electromagnetic wave mode in this waveguide was more complicated than those described above. The powerful microwave discharge was fast-flow stabilized (including supersonic flow) and was able to operate in both thermal and non-thermal regimes.

10.2.4 Microwave Plasma Generation in Resonators

The powerful atmospheric pressure thermal microwave discharge in a resonator was developed by P.L. Kapitsa and co-authors (P.L. Kapitsa, 1969). Schematic of such a cylindrical resonator with standing wave mode E_{01} is presented in Figure 10.12. A microwave generator of 175 kW power at 1.6 GHz frequency (wavelength 19 cm) was used in the Kapitsa experiments. In this case, the electric field on the axis of the cylindrical resonator is directed along the axis and varies as the

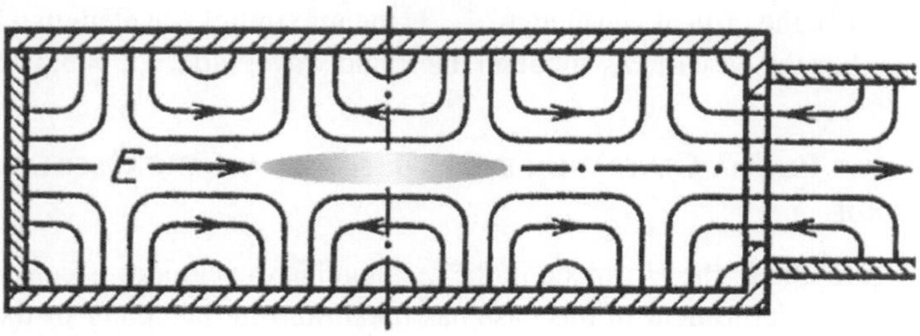

FIGURE 10.12 Microwave plasma in a resonator.

cosine-function with the maximum in the center of the cylinder. The electric field then decreases further from the axis. Ignition of the discharge takes place in the central part of the resonator cylinder, where the electric field has its maximum (see Figure 10.12). Microwave plasma formed in the resonator appears as a filament located along the axis of the resonator cylinder. The length of the plasma filament is about 10 cm, which corresponds to half of the wavelength; the diameter of the plasma filament is about 1 cm. To stabilize the microwave plasma in the central part of the resonator at high power of 20 kW in hydrogen, deuterium and helium at atmospheric and higher pressures, gas is tangentially supplied to the discharge chamber. Plasma temperature in this microwave discharge generally does not exceed 8000 K, which is related to intensive reflection effects at higher conductivities.

10.2.5 1-D Model of Electromagnetic Wave Interaction with Thermal Plasma

Microwave plasma generation can be described by 1D model, which assumes that plane electromagnetic wave passes through a plane dielectric wall (which is transparent to this wave) and then meets plasma. Heat released in the plasma is subsequently transferred back to the dielectric wall, which is externally cooled, providing energy balance of the steady-state system. This simple but effective model is illustrated in Figure 10.13. This approach actually describes the local interaction of the incident electromagnetic wave with the "closest plasma surface" (see Figure 10.9). When the microwave power, gas pressure, and the thermal plasma conductivity are relatively high, the electromagnetic wave penetration into the plasma is shallow. This makes the one-dimensional model to describe the microwave plasma generation applicable. The 1D model of microwave plasma generation is based on the integral flux relation (10.5) taking into account $n = 0$. The constant in Eq. (10.5) is

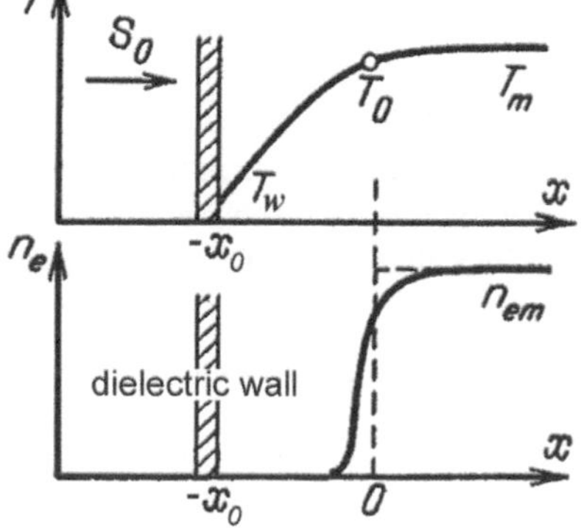

FIGURE 10.13 Temperature and electron density distributions in a discharge sustained by electromagnetic wave.

equal to zero (*const* = 0) because the temperature tends to the constant value T_m deep inside the plasma ($x \to \infty$). As a result, the integral flux relation can be rewritten as:

$$J + \langle S \rangle = 0, \quad J = -\lambda \frac{dT}{dx}. \tag{10.34}$$

In this relation λ is the coefficient of thermal conductivity, $<S>$ is the mean value of the electromagnetic energy flux density, determined in Eq. (10.3). Since microwave discharges are quasi-optical, the wave effects as reflection and interference should be taken into account. Description of the electromagnetic wave propagation in plasma requires analysis of the wave equation with the complex dielectric permittivity Eq. (6.179):

$$\frac{d^2 E_y}{dx^2} + \left(\varepsilon_\omega + i \frac{\sigma_\omega}{\varepsilon_0 \omega} \right) \frac{\omega^2}{c^2} E_y = 0. \tag{10.35}$$

E_y is a component of the electric field along the plasma surface; σ_ω and ε_ω and are high-frequency conductivity and dielectric permittivity, which are determined according to Eqs. (6.18) and (6.181). Since the plasma temperature and degree of ionization in the thermal microwave discharges is not very high (with respect to RF and arc discharges under similar conditions), the electron–ion collisions can be neglected in calculating the plasma conductivity. From the Saha equation, we can take the quasi-equilibrium high-frequency plasma conductivity and dielectric permittivity in the wave Eq. (10.36) as:

$$\sigma_\omega \propto 1 - \varepsilon_\omega \propto n_e \propto \exp\left(\frac{I_i}{2T} \right), \tag{10.36}$$

where I_i is the ionization potential. Boundary conditions for the system of Eqs. (10.34), (10.35) are: the electric field is assumed to be zero deep inside the plasma ($E = 0$ at $x \to \infty$), temperature on the cooled dielectric wall is low ($T = T_w \approx 0$ at $x = -x_0$), and the electromagnetic energy flux density S_0 related to microwave power provided from generator, is given. Solution of the system of Eqs. (10.34), (10.35) determines: the plasma temperature $T_m = T(x \to \infty)$; the fraction of the electromagnetic energy flux density S_1, which is dissipated in the plasma; and the microwave reflection coefficient from the plasma $\rho = (S_0 - S_1)/S_0$. A general solution of the system obviously cannot be found analytically and requires numerical approaches. Similar to the cases of thermal arc and thermal RF discharges, physically clear analytic solution of the system (10.34), (10.35) can be obtained only by assuming a sharp conductivity change on the plasma boundary and constant conductivity inside of the plasma.

10.2.6 Constant Conductivity Model of Microwave Plasma Generation

To find an analytical solution of (10.34), (10.35) assume that plasma has a sharp boundary at $x = 0$, where the temperature equals T_0 (see Figure 10.13). To the left from this point ($x \le 0$, $T \le T_0$) assume no plasma: $\sigma = 0$, $\varepsilon = 1$. To the right from the point ($x > 0$, $T_0 < T < T_m$) assume constant conductivity: $\sigma = \sigma_m$, $\varepsilon = \varepsilon_m$. This **constant microwave plasma conductivity model** is similar to the channel model of arc discharges and the metallic cylinder model of the thermal ICP-discharges. The reflection coefficient of an incident electromagnetic wave normal to the sharp boundary of plasma $\rho = (S_0 - S_1)/S_0$, can be found from the boundary conditions for electric and magnetic fields on the plasma surface (L.D. Landau, 1982):

$$\rho = \frac{(n-1)^2 + \kappa^2}{(n+1)^2 + \kappa^2}. \tag{10.37}$$

In this relation, κ is the attenuation coefficient of the electromagnetic wave, defined by Eq. (6.189); n is the electromagnetic wave refractive index, defined by Eq. (6.19). Both parameters n and κ are functions of the high-frequency permittivity ε_ω and electric conductivity σ_ω, which are assumed constant. While Eq. (10.37) describes the electromagnetic wave reflection from the plasma, microwave absorption and energy flux damping in the plasma can be calculated from the Bouguer law Eq. (6.195) with the absorption coefficient μ_ω mostly dependent on the electron density Eq. (6.197). According to the Bouguer law, dissipation of the electromagnetic wave takes place in a surface layer of about $l_\omega = 1/\mu_\omega$, where temperature grows from T_0 to T_m. The energy balance in this layer between the absorbed electromagnetic flux S_1 and the heat transfer to the wall can be expressed as: $S_1 \approx \lambda_m \Delta T / l_\omega$. Similar to the channel model of arc discharges and the metallic cylinder model of RF-discharges, one can determine $\Delta T = T_m - T_0$ assuming the exponential conductivity decrease in the plasma. As a result, the following equation to determine the plasma temperature is obtained:

$$S_0\left[1-\rho\left(T_m\right)\right]=\lambda\left(T_m\right)\cdot\frac{2T_m^2}{I_i}\cdot\mu_\omega\left(T_m\right). \tag{10.38}$$

Find the size of the gap x_0 between plasma and dielectric wall from heat transfer across this gap:

$$\Theta_m\left(T_m\right)-\Theta_w\left(T_w\right)=S_1\cdot\left|x_0\right|, \tag{10.39}$$

where the heat flux potential $\Theta(T)$ is determined by Eq. (8.22).

10.2.7 Numerical Characteristics of the Quasi-Equilibrium Microwave Discharge

Data characterizing generation of a thermal microwave discharge at frequency $f = 10$ GHz, ($\lambda = 3$ cm) in air are presented in Table 10.1. They are based on quasi-equilibrium calculations additionally taking into account conduction losses from the plasma filament of 0.3 cm radius and also some smearing of plasma boundaries, which slightly reduces the reflection coefficient. The depth of microwave energy absorption in plasma l_ω grows with temperature reduction: at $T = 3500$ K it

TABLE 10.1
Characteristics of Atmospheric Pressure Microwave Discharge in Air at Frequency $f = 10$ GHz, ($\lambda = 3$ cm) as a Function of Plasma Temperature

Plasma temperature, T_m	4500 K	5000 K	5500 K	6000 K
Electron density, n_e, 10^{13} cm^{-3}	1.6	4.8	9.3	21
Plasma conductivity, σ_m, 10^{11} sec^{-1}	0.33	0.99	1.9	4.1
Thermal conductivity, λ, 10^{-2} W/cm · K	0.95	1.1	1.3	1.55
Refractive index n of plasma surface	1.3	2.1	2.8	4.3
Plasma attenuation coefficient, κ	2.6	4.7	7.3	11
Depth of the microwave absorption layer, $l_\omega = \frac{1}{\mu_\omega}$, 10^{-2} cm	9.1	5.0	3.2	2.2
Energy flux absorbed, S_1, kW/cm^2	0.23	0.35	0.56	1.06
Microwave reflection coefficient, ρ	0.4	0.65	0.76	0.81
Microwave energy flux to sustain plasma, S_0, kW/cm^2	0.38	1.0	2.3	5.6

reaches 2 cm. This absorption growth is exponential and determines the lower temperature limit of thermal microwave discharges:

$$l_\omega \propto \frac{1}{n_e} \propto \exp\left(\frac{I_i}{2T}\right). \tag{10.40}$$

At relatively low temperatures, the microwave plasma becomes transparent to electromagnetic microwave radiation, and hence to sustain such a plasma, high power must be provided. Further, low temperature and low conductivity regime are unstable in the same way as with RF-discharges. The minimal temperature of the thermal microwave discharges can be estimated from the equation $l_\omega(T_{min}) = R_f$, where R_f is the radius of the plasma column or plasma filament. Numerically, the minimum temperature is about 4200 K for the filament radius of 3 mm. Corresponding absorbed energy flux is $S_1 = 0.2$ kW/cm^2; then the minima microwave energy flux to sustain the thermal plasma is $S_0 = 0.25$ kW/cm^2. Maximum temperature is limited by the reflection of electromagnetic waves from plasma at high conductivities. This effect is seen from Table 10.1, where the reflection coefficient reaches the high value of $\rho = 81\%$ at $T_m = 6000$ K. For this reason, the quasi-equilibrium temperature of the microwave plasma in atmospheric pressure air does not exceed 5000 K–6000 K.

10.2.8 Microwave Plasma Torch and Other Non-Conventional Configurations of Thermal Microwave Discharges

Consider the microwave plasma torch developed by Y. Mitsuda, T. Yoshida, K. Akashi (1989). This torch is shown in Figure 10.14. Microwave power in this system is delivered primarily by a rectangular waveguide. Then the electromagnetic wave passes through a quartz window into an impedance-matching mode converter, which couples the microwave power into a coaxial waveguide. The coaxial waveguide then operates somewhat as an arc jet: the center conductor of the waveguide forms one electrode and the other electrode comprises an angular flange on the outer coaxial electrode. Another configuration of the microwave discharge is based on the conversion of the rectangular waveguide mode into a special circular waveguide system illustrated in Figure 10.15. This system

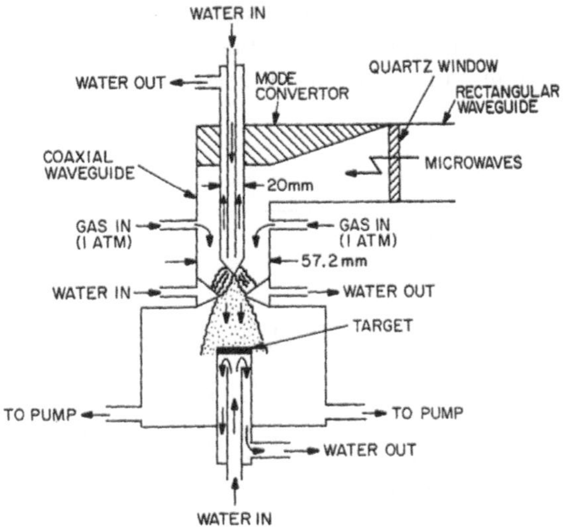

FIGURE 10.14 Microwave plasma torch. (From J.R. Roth, Industrial Plasma Engineering)

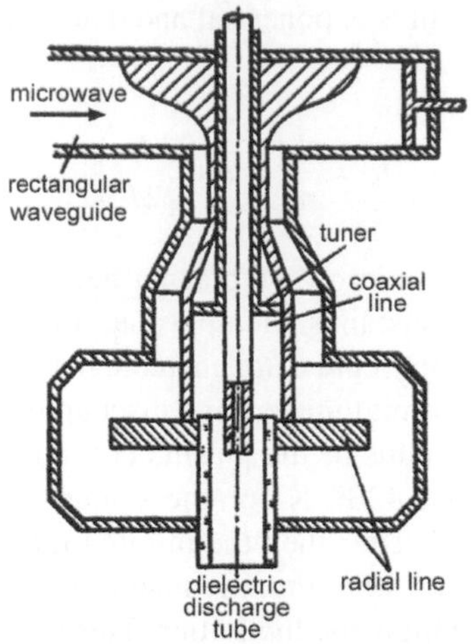

FIGURE 10.15 Radial and microwave plasmatron.

was used to generate high power microwave discharges. More details about special configurations of microwave discharges can be found in V. Batenin, I.I. Klimovsky, G.V. Lysov, V.N. Troizky (1988) and A.D. MacDonald, S.J. Tetenbaum (1978).

10.2.9 Continuous Optical Discharges

Thermal atmospheric pressure plasma can be generated by optical radiation somewhat similar to the way it is generated by electromagnetic waves in the RF and microwave frequency range. These not so conventional but very interesting thermal plasma generators were first theoretically considered by Yu.P. Raizer (1970), and then experimentally realized by N.A. Generalov, V.P. Zimakov, G.I. Kozlov, V.A. Masyukov, Yu.P. Raizer (1970). Optical plasma generation began before 1970 with the discovery of the **optical breakdown** effect (P.D. Maker, R.W. Terhune, C.M. Savage, 1964). It became possible after the development of Q-switched lasers, which are able to produce extremely powerful light pulses, the so-called "giant pulses". Detailed consideration of the subject can be found in the book of Yu.P. Raizer (1974); typical breakdown thresholds are presented in Figure 10.16 (Yu.P. Raizer, 1977). The continuous optical discharge is illustrated in Figure 10.17. Usually, a CO_2 laser beam is focused by a lens or mirror to sustain the discharge. Power of the CO_2 laser should be high;

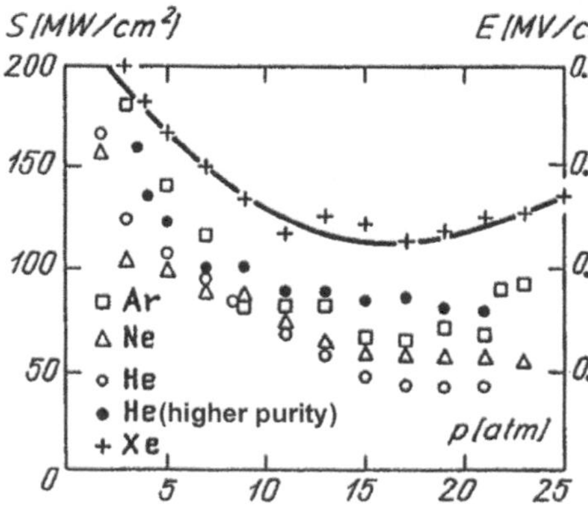

FIGURE 10.16 Breakdown thresholds of inert gases under CO_2 laser radiation.

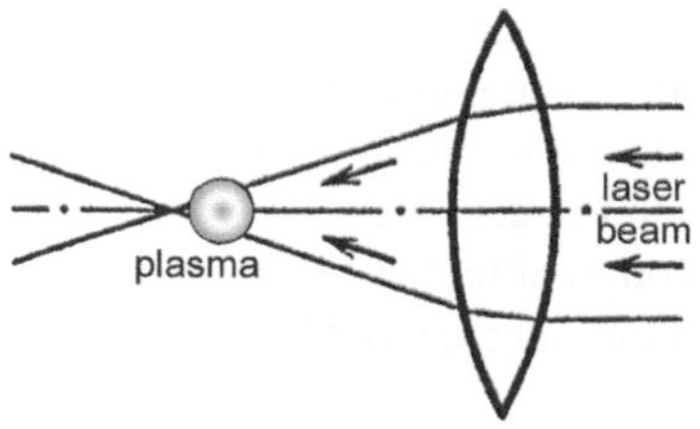

FIGURE 10.17 Schematic of a continuous optical discharge.

for the continuous optical discharge in atmospheric air it must be at least 5 kW in the low-divergence beam. To sustain the discharge in xenon at pressure of a couple of atmospheres, much lower power of approximately 150 W is sufficient.

Light absorption coefficient in plasma significantly decreases with the growth of the electromagnetic wave frequency. For this reason, the application of visible radiation requires 100–1000 times more power than that of the CO_2 lasers, which is very difficult experimentally. Plasma sustained by focusing on solar radiation would have temperatures less than the solar temperature, 6000 K according to the second law of thermodynamics. Equilibrium plasma density at 6000 K is not sufficient to provide absorption of the radiation. To provide effective absorption of laser radiation, the plasma density should be very high. As a result, plasma temperature in the continuous optical discharges is high, 15,000 K–20,000 K.

10.2.10 Laser Radiation Absorption in Thermal Plasma as a Function Gas Pressure and Temperature

Absorption of the CO_2 laser radiation quanta ($\hbar\omega = 0.117\ eV$ or in temperature units 1360 K, which is less than the plasma temperature) is mostly provided by the bremsstrahlung absorption mechanism related to the electron–ion collisions (see Section 6.7.7). The absorption coefficient $\mu_\omega(CO_2)$ then can be calculated based on the Kramers formula Eqs. (6.25),(6.251), and taking into account induced emission. The resulting formula including the double ionization (Yu.P. Raizer, 1977) is:

$$\mu_\omega(CO_2), \text{cm}^{-1} = \frac{2.82 \cdot 10^{-29} \cdot n_e(n_+ + 4n_{++})}{(T,K)^{3/2}} \cdot \log\left(\frac{2700 \cdot T,K}{n_e^{1/3}}\right). \tag{10.41}$$

All number densities are measured in this relation in cm^{-3}; n_e is the electron concentration; n_+ and n_{++} are densities of single-charged and doubly charged positive ions, which can be calculated for the quasi-equilibrium plasma based on the Saha equation. The absorption coefficient of the CO_2 laser radiation in air is shown in Figure 10.18 as a function of temperature (Yu.P. Raizer, 1991). This

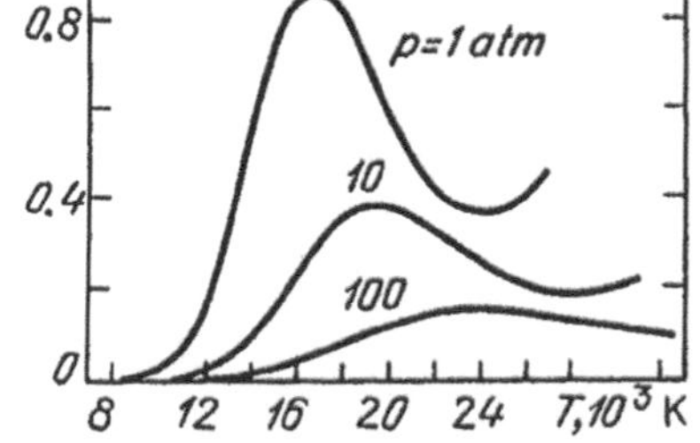

FIGURE 10.18 Absorption coefficients for CO_2 laser radiation in air.

function has a maximum at constant pressure, which corresponds to about 16,000 K at atmospheric pressure. This temperature corresponds to complete single ionization. Further increase of temperature does not lead to additional ionization (because double ionization requires significantly higher temperatures), while gas density and hence, electron concentration decrease at fixed pressures. It explains the decrease of the absorption at higher temperatures (see Figure 10.18) and hence, maximum of the dependence $\mu_\omega(T)$. Taking into account Eq. (10.41), the decrease of the absorption coefficient $\mu_\omega(T)$ is:

$$\mu_\omega \propto \frac{n_e^2}{T^{3/2}} \propto \frac{1}{T^{7/2}}. \tag{10.42}$$

At high temperatures, the absorption starts growing again because of the double ionization. From Figure 10.18, the maximum absorption coefficient μ_ω grows with pressure slightly slower than the square of gas pressure. The maximal absorption coefficient for the CO_2 laser radiation at atmospheric pressure is $\mu_\omega = 0.85\ \text{cm}^{-1}$, which corresponds to a light absorption length $l_\omega = 1.2$ cm. This maximum in the dependence $\mu_\omega(T)$ takes place only at frequencies $\omega^2 >> \nu_e^2$ (see Section 6.6), where ν_e is the frequency of electron collisions (at high ionization degrees typical for the optical discharges ν_e is related to the electron–ion collisions). In the opposite case of lower electromagnetic wave frequencies $\omega^2 << \nu_e^2$, the absorption coefficient μ_ω (which is proportional to the electric conductivity σ) continuously grows with temperature. Assuming as above, the major contribution of the electron–ion Coulomb collisions into the frequency ν_e, the light absorption coefficient μ_ω dependence on temperature is:

$$\mu_\omega \propto \sigma \propto \frac{n_e}{\nu_e} \propto \frac{n_e}{n_+ \bar{v} \sigma_{coulomb}} \propto T^{3/2}. \tag{10.43}$$

The coulomb collision cross section $\sigma_{coulomb}$ dependence on temperature: $\sigma_{coulomb} \propto 1/T^2$ (see Section 2.1.9), and the average electron velocity dependence on temperature $\bar{v} \propto \sqrt{T}$ was taken into account. It is interesting that the absorption coefficient in this case of low electromagnetic wave frequencies does not depend on gas pressure and density.

10.2.11 Energy Balance of the Continuous Optical Discharges and Relation for Plasma Temperature

In the 1D model, plasma is considered as a sphere of radius r_0 and constant temperature T_m (hence also fixed absorption coefficient μ_ω). This plasma temperature is maintained by absorbing convergent spherically symmetric rays of total power P_0. If the plasma is transparent for the laser radiation, the fraction of the total radiation power absorbed in the plasma can be estimated as: $P_1 = P_0\mu_\omega r_0$. At relatively low laser power levels and not very high pressures, one can neglect radiation losses and balance absorption of the laser radiation with thermal conduction flux J_0:

$$P_1 = P_0\mu_\omega r_0 = 4\pi r_0^2 J_0 \approx 4\pi r_0^2 \frac{\Delta\Theta}{r_0} = 4\pi r_0 \Delta\Theta. \tag{10.44}$$

$\Delta\Theta = \Theta_m - \Theta_0$ is the drop of the heat flux potential in the plasma. Taking into account that gas is cold at infinity ($\Theta(r \to \infty) = 0$), the heat balance outside of the plasma sphere is:

$$P_1 = -4\pi r^2 \frac{d\Theta}{dr}, \quad \Theta(r) = \frac{P_1}{4\pi r}, \quad \Theta_0(r = r_0) = \frac{P_1}{4\pi r_0}. \tag{10.45}$$

Comparing Eqs. (10.44) and (10.45), one can see that $\Delta\Theta = \Theta_0$, and hence, $\Theta_m = 2\Theta_0$. From this relation between maximum plasma temperature and the one on the plasma surface rewrite Eq. (10.44) as: $P_1 = 2\pi r_0 \Theta_m$. Recalling the above-mentioned formula: $P_1 = P_0 \mu_\omega r_0$, the following final energy balance relation, which determines the maximum plasma temperature T_m is obtained:

$$P_0 = 2\pi \frac{\Theta_m(T_m)}{\mu_\omega(T_m)}. \tag{10.46}$$

10.2.12 Plasma Temperature and Critical Power of Continuous Optical Discharges

Analyzing Eq. (10.46) for plasma temperature, power required to maintain the optical discharge $P_0(T_m)$ as a function of plasma temperature has a minimum T_t (see Figure 10.19, Raizer, 1991). This minimum occurs because the function $\mu_\omega(T)$ has a maximum (see Figure 10.18), while the function $\Theta(T)$ grows continuously. The temperature T_t, corresponding to the minimum power P_t, which is necessary to sustain the continuous optical discharge (see Figure 10.19), is close to the temperature corresponding to the maximum of the function $\mu_\omega(T)$ (see Figure 10.18). Similar to the cases of thermal ICP and microwave discharges (see Section 10.1.8 and remarks after Eq. (10.40)), the left low-temperature and low-conductivity branch of the curve $P_0(T)$ in Figure 10.19 where $T_m < T_t$ is unstable. If the temperature is lower than the threshold temperature $T_m < T_t$, a small temperature increase due to any fluctuation results in a decrease of the required power necessary to sustain the stationary discharge with respect to the actual laser radiation power. This would lead to further temperature growth until it reaches the threshold one $T_m = T_t$. Eq. (10.46) permits calculating the critical minimum of the laser radiation necessary to sustain the continuous optical discharge. In this case, plasma temperature should be taken as the critical one T_t:

$$P_t = P_{0\min} = 2\pi \frac{\Theta_m(T_m = T_t)}{\mu_\omega(T_m = T_t)}. \tag{10.47}$$

For example, the CO_2-laser radiation in atmospheric air has the maximum value of the absorption coefficient $\mu_{\omega\max} \approx 0.85\ \text{cm}^{-1}$ at the threshold temperature $T_t = 18000\ K$. In this case, the heat flux potential is $\Theta(T_m = T_t) \approx 0.3\ kW/\text{cm}$, and the minimal threshold value of power necessary to sustain the discharge is $P_t \approx 2.2\ kW$. Thermal plasma temperatures in continuous optical discharges are higher (about twice) than that of ICP and arc discharges, and significantly higher than in microwave discharges (about 3–4 times). This is related to the fact that $\mu_\omega \propto 1/\omega^2$ and plasma is transparent for the optical radiation. Only very high temperatures corresponding to almost complete ionization provides the level of absorption sufficient to sustain the discharges. The minimal threshold value of the laser power P_t in accordance with Eq. (10.47) decreases quite fast with pressure growth because of the corresponding significant growth of the absorption coefficient $\mu_{\omega\max}$. The minimum threshold of the laser power is lower in gases with lower ionization potential (to decrease T_t) and lower thermal conductivity (to decrease Θ_m). The minimum necessary CO_2 laser power to sustain the continuous optical discharge in Ar and Xe at pressures of about 3–4 atm is 100–200 W, see Yu.P. Raizer (1974,

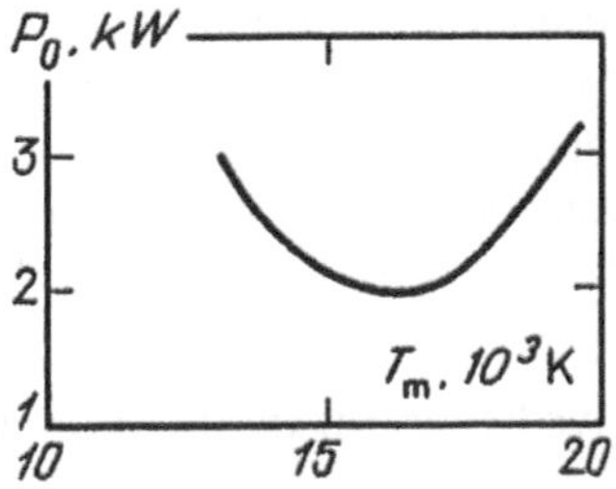

FIGURE 10.19 Power of a spherically convergent laser beam as a function of maximum plasma temperature.

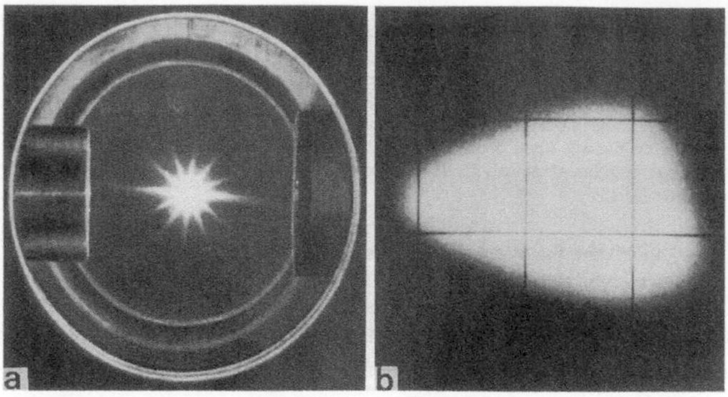

FIGURE 10.20 Continuous optical discharge: (a) general view, (b) a large image (beam travels from right to left).

1977, 1991). Photographs of the are shown in Figure 10.20 (N.A. Generalov, V.P. Zimakov, G.I. Kozlov, V.A. Masyukov, Yu.P. Raizer, 1971).

10.3 NON-EQUILIBRIUM RADIO-FREQUENCY (RF) DISCHARGES, GENERAL FEATURES OF NON-THERMAL CAPACITIVELY-COUPLED (CCP) DISCHARGES

10.3.1 Non-Thermal Radio-Frequency (RF) Discharges

At lower pressures, RF-plasma becomes essentially non-equilibrium. Electron-neutral collisions are less frequent at lower pressures while gas cooling by discharge walls is more intensive. Altogether it leads to electron temperatures much exceeding that one of neutrals somewhat similar to the case of glow discharges. Specifying the frequency range typical for RF-discharges, one should note that the upper-frequency limit is related here to wavelengths close to the system sizes (electromagnetic waves with smaller wavelengths are usually referred to as microwaves). Lower RF-frequency limit is related to the characteristic frequency of ionization and ion transfer. Ion density in RF-discharge plasmas and sheaths usually can be considered as constant during a period of electromagnetic field oscillation. Thus typically RF-frequencies exceed 1 MHz, though sometimes they can be smaller. The most industrially used radio-frequency for plasma generation is 13.6 MHz. The non-thermal RF-discharges can be subdivided into those of moderate (or intermediate) pressure and those of low pressure. The discharges are usually referred to as those of moderate pressure if they are non-equilibrium, but the electron energy relaxation length is small with respect to all characteristic sizes of the discharge system. In this case, the electron energy distribution function (and therefore ionization, excitation of neutrals and other elementary processes) is determined by local values of electric field. Usually, this occurs in the pressure range from 1 to 100 Torr, see Yu.P. Raizer, M.N. Shneider, N.A. Yatsenko, 1995). In the opposite case of low pressures, the electron energy relaxation length is small with respect to discharge sizes. Electron energy distribution function is then determined by electric field distribution in the entire discharge zone. This discharge regime takes place at low pressures p and small characteristic discharge sizes L (for example, in inert gases it requires $p(Torr) \cdot L(\text{cm}) < 1$). Low pressure, strongly non-equilibrium RF-discharges are widely applied today in electronics in etching and chemical vapor deposition (CVD) technologies, see M.A. Lieberman, A.J. Lichtenberg (1994).

10.3.2 Capacitive and Inductive Coupling of the Non-Thermal RF-discharge Plasmas

Similarly to the case of thermal discharges (see Section 10.1), the non-thermal plasma of RF discharges can be either capacitively coupled (CCP) or inductively coupled (ICP). These two ways of

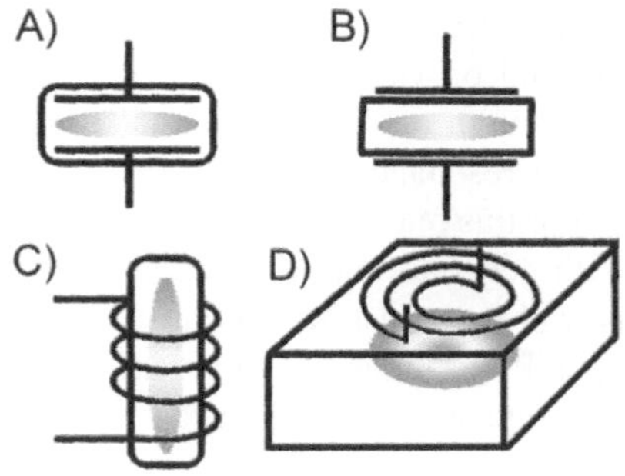

FIGURE 10.21 Capacitively and inductively coupled RF discharges.

organization of non-thermal RF-discharges are illustrated in Figure 10.21. The CCP-discharges provide electromagnetic field by means of electrodes located either inside or outside of the discharge chamber, which is shown in Figure 10.21a, b. These discharges primarily stimulate electric field, which obviously facilitates their ignition. The inductive coil induces the electromagnetic field in ICP discharges: the discharge can be located either inside of the coil (Figure 10.21c), or aside of the plane or quasi-plane coil (Figure 10.21d). These discharges primarily stimulate magnetic fields, while corresponding non-conservative electric fields necessary for ionization are relatively low. For this reason, the non-thermal ICP plasma discharges are usually organized at low pressures, when the reduced electric field E/p is sufficient for ionization. The inductive coils generate not only the non-conservative (vortex) electric fields but also the conventional conservative (potential) electric field in the gap between turns (windings) of the coil. This potential electric field exceeds the non-conservative one by the pretty big factor, which is equal to the ratio of the length of a turn (winding) to the distance between them. These relatively high but local electric fields (capacitive component in inductive discharge) can be important especially during breakdown unless an electric screen is installed. Coupling between the inductive coil and plasma inside can be interpreted as a transformer where the coil represents the primary windings and the plasma the secondary windings. A coil as the primary windings consists of many turns while plasma has only one. ICP-discharges can be considered as a voltage-decreasing transformer. Effective coupling with RF-power supply requires low plasma resistance. The ICP-discharges are convenient to reach high currents, high electric conductivity, and high electron concentration. In contrast, the CCP-discharges are more convenient to provide higher electric fields.

10.3.3 Electric Circuits for Inductive and Capacitive Plasma Coupling with RF-Generators

RF-power supplies for plasma generation typically require an active load of 50 or 75 Ohm. To provide an effective correlation between the resistance of the leading line from the RF-generator and the discharge impedance, a special coupling circuit should be applied. A general schematic of such circuits for CCP and ICP-discharges is shown in Figure 10.22. It is important that the

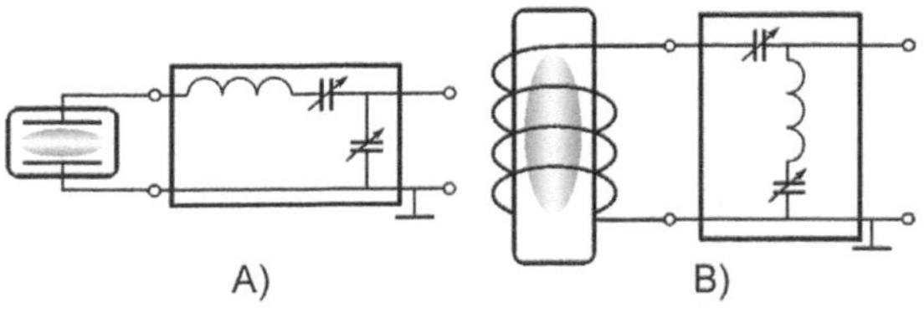

FIGURE 10.22 Coupling circuits for CCP (a) and ICP (b) discharges.

RF-generator should be not only effectively correlated with the RF-discharge during its continuous operation, but it also initially should provide sufficient voltage for breakdown to start the discharge. To provide the effective ignition of the non-thermal RF-discharge, the coupling electric circuit should form the AC-current resonance being in series with generator and the discharge system during its idle operation. For this reason, the coupling circuit in the case of CCP-discharge includes inductance in series with generator and discharge system (see Figure 10.22a). Variable capacitance is included there as well for adjustment because the design of variable inductances is more complicated. In the case of ICP-discharge, the only variable capacitance, located in the electric coupling circuit in series with generator and idle discharge system provides the necessary initial breakdown.

10.3.4 Motion of Charged Particles and Electric Field Distribution in Non-Thermal RF Capacitively-Coupled (CCP) Discharges

External electric circuit is able to provide either fixed voltage amplitude between electrodes or fixed current amplitude in CCP-discharges. One of these regimes can be chosen in varying relation between the discharge impedance and resistance of the RF-generator and elements of the above-discussed electric coupling circuit. The fixed current regime is however more typical one in practical application. The fixed current regime can be represented simply by an electric current with density $j = -j_0 \sin(\omega t)$ flowing between two parallel plane electrodes, which implies that the electrodes size exceeds the distance between them. We can assume for the following analysis that the RF-plasma density $n_e = n_i$ is high enough and the electron conductivity current in plasma exceeds the displacement current:

$$\omega << \frac{1}{\varepsilon_0}|\sigma_e| \equiv \frac{1}{\tau_e}. \tag{10.48}$$

τ_e is the Maxwell time for electrons, which characterizes (at $\omega << \nu_e$, ν_e- is electron collisional frequency) time necessary to shield the electric field; $\sigma_e$is the complex electron conductivity:

$$\sigma_e = \frac{n_e e^2}{m(\nu_e + i\omega)}. \tag{10.49}$$

An inequality opposite to (10.48) takes place for ions because of much higher mass $M >> m$:

$$\omega >> \frac{1}{\varepsilon_0}|\sigma_i| \equiv \frac{1}{\tau_i}, \quad \sigma_i = \frac{n_i e^2}{M(\nu_i + i\omega)}. \tag{10.50}$$

The inequality (10.50) means that the ion conductivity current can be neglected with respect to displacement current. In other words, the ion drift during an oscillation period can be neglected. This leads to the following picture of electric current and the motion of charged particles in the RF-discharges. Ions form the "skeleton" of plasma and can be considered being at rest, while electrons oscillate between electrodes as it is shown in Figure 10.23. Electrons are present in the sheath of width L near an electrode only for a part of the oscillation period called the plasma phase. Another part of the oscillation period, when there are no electrons in the sheath, is called the space charge phase. The oscillating space charge shown in Figure 10.23 creates an electric field, which forms the displacement current and closes the circuit. The electric field of the space charge has a constant component in addition to an oscillating component, which is directed from plasma to the electrodes. For this reason similar to glow discharges, the quasi-neutral plasma zone, is also called the positive column. The constant component of the space charge field provides faster ion drift to the electrodes than in the case of ambipolar diffusion. As a result, ion

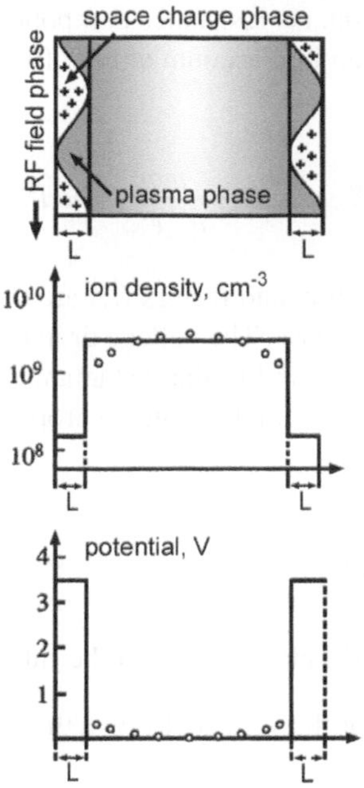

FIGURE 10.23 Ion density and potential distribution; and boundary layer oscillation in RF CCP-discharge.

density in the space charge layers near electrodes is lower with respect to their concentration in plasma (see Figure 10.23).

10.3.5 Electric Current and Voltages in Non-Thermal RF Capacitively-Coupled (CCP) Discharge

Assume that the RF-discharge is symmetrical and ion concentrations in plasma and sheaths are fixed and equal to n_p and n_s respectively (see Figure 10.23). Then the 1D- discharge zone can be divided each moment into three regions: (1). A plasma region with thickness L_p, which is quasi-neutral throughout all the oscillation period. Electric current in this region is provided by electrons and conductivity is active at relatively low frequencies $\omega < \nu_e$. (2). Sheath regions in the plasma phase, where the electric conductivity is also active and provided by electrons. (3). Sheath regions with space charge. Taking into account the discharge symmetry and constancy of charge concentration in the sheath, one can conclude that the total thickness of the space charge region is always equal to the total thickness of the sheath region in the plasma phase, and equals to the sheath size L (see Figure 10.23). If the total thickness of the sheath region in the plasma phase is equal to L, then active resistance of the region can be expressed as:

$$R_{ps} = \frac{Lm\nu_e}{n_s e^2 S}, \tag{10.51}$$

where ν_e is the frequency of electron collisions, and S is the area of an electrode. Similarly, the resistance of the quasi-neutral plasma region can be presented as:

$$R_p = \frac{L_p m\nu_e}{n_p e^2 S}. \tag{10.52}$$

From Eq. (10.49), the reactive (imaginary) resistance component of the plasma itself and sheaths in the plasma phase also should be taken into account at high frequencies of the electromagnetic field $\omega > \nu_e$:

$$X_{p,ps} = iL_{p,s}\frac{m\omega}{n_{p,s}e^2S} = i\omega L_{p,s}^{(e)}. \tag{10.53}$$

This impedance has the inductive nature, and factors $L_{p,s}^{(e)}$ are the effective inductance of the plasma zone (p) and sheath (s) in the plasma phase. The voltage drop on space charge sheath is proportional to the electric field near an electrode E and to the instantaneous size of the sheath $d(t)$. Assuming constancy of ion concentration in the sheath, the voltage drop is proportional to square of its instantaneous size:

$$U_r = \frac{1}{2}E_r d_r(t) = \frac{j_0}{4\varepsilon_0\omega}L\left[1+\cos(\omega t)\right]^2. \tag{10.54}$$

Taking into account that the sheaths' half-size $L/2$ can be interpreted as the amplitude of electron oscillation $\frac{j_0}{en_s}\cdot\frac{1}{\omega}$, the Eq. (10.54) for the voltage drop on the "right-hand" sheath (subscript "r") is:

$$U_r = \frac{j_0^2}{2\varepsilon_0\omega^2 en_s}\left[1+\cos(\omega t)\right]^2. \tag{10.55}$$

The voltage drop on the space charge sheath near the opposite "left-hand" sheath (subscript "l") is in counter-phase with the voltage:

$$U_l = -\frac{j_0^2}{2\varepsilon_0\omega^2 en_s}\left[1-\cos(\omega t)\right]^2. \tag{10.56}$$

The total voltage U_s related to the space charge sheaths obviously can be presented as a sum of voltages corresponding to both electrodes:

$$U_s = U_r + U_l = \frac{2j_0^2}{\varepsilon_0\omega^2 en_s}\cos(\omega t) = \frac{j_0}{\varepsilon_0\omega}L\cos(\omega t). \tag{10.57}$$

Analyzing Eq. (10.57), the following two conclusions regarding the space charge sheaths are reached: (1). Total voltage drop on the space charge sheaths includes only the principal harmonic (ω) of the applied voltage, while the voltage drop on each sheath separately contains a constant component and second harmonics (2ω). (2). Taking into account that the electric current density was defined in Section 10.3.4 as: $j = -j_0\sin(\omega t)$, one can see that the phase shift between voltage and current in the space charge sheath corresponds to the capacitive resistance.

10.3.6 Equivalent Scheme of a Capacitively-Coupled RF-Discharge

As seen from the current–voltage relation (10.57), the space charge sheath can be interpreted by the equivalent capacitance:

$$C_s = \frac{\varepsilon_0 S}{L}. \tag{10.58}$$

This capacitance corresponds to the capacitance of the vacuum gap with the width equal to the total size of the space charge sheaths L. If the total size of the space charge sheaths L does not depend on the discharge current, then the equivalent capacitance Eq. (10.58) is constant and the discharge circuit can be considered as a linear one. Taking into account that the impedance of a plasma (width L_p) has an inductive component Eq. (10.53) and the impedance of the space charge layers (width L) is capacitive; resonance in the discharge circuit related to their connection in series is possible on the frequency:

$$\omega = \omega_p \sqrt{L/L_p}. \tag{10.59}$$

ω_p is the plasma frequency corresponding to the quasi-neutral plasma zone. Discharge voltage dependence on electron concentration in the plasma zone illustrating this resonance is shown in Figure 10.24 (A.S. Smirnov, 2000). The constant component of the space charge field is stronger than the electric field in the plasma zone (see Figure 10.23). For this reason, even a small ionic current can lead to energy release in the sheath exceeding that in the plasma zone. Discharge power per unit electrode area, transferred to the ions, can be estimated as a product of the ion current density determined at the mean electric field in the sheath:

$$j_i = en_s v_i = en_s b_i \frac{j_0}{\varepsilon_0 \omega}. \tag{10.60}$$

and the constant component of the voltage Eq. (10.55). Here v_i and b_i are the ions drift velocity and mobility. Thus the discharge power per unit electrode area related to ionic drift in a sheath can be expressed based on Eqs. (10.55) and (10.60) as:

$$P_i = \frac{3}{4\varepsilon_0^2} \frac{j_0^3 b_i}{\omega^3} S. \tag{10.61}$$

From Eq. (10.61), the ionic power P_i is proportional to the cube of the discharge current, inversely proportional to the gas pressure (because of $b_i \propto 1/p$) and proportional to the cube of frequency, and does not depend on the size of sheath. The discharge power transfer to ions in a sheath can be presented in the equivalent scheme by resistance R_s connected in parallel to the capacitance of sheaths. This resistance is determined by voltage drop on the capacitance and power released in both sheaths:

$$R_s = \frac{U_s^2/2}{2P_i} = \frac{1}{3} \frac{\omega L^2}{j_0 b_i S}. \tag{10.62}$$

The total equivalent scheme of a capacitively-coupled RF discharge is presented in Figure. 10.25. The scheme includes the resistance R_s and the capacitance C_s of sheaths connected in parallel to each other, and then connected in series to the active resistance of sheaths in plasma phase R_{ps}) and of plasma itself R_p. At relatively low pressures, when $\omega > \nu_e$, the inductance of sheaths in plasma phase $L_p^{(e)}$ and of plasma itself $L_s^{(e)}$ Eq. (10.53) should be also taken into account in the equivalent scheme.

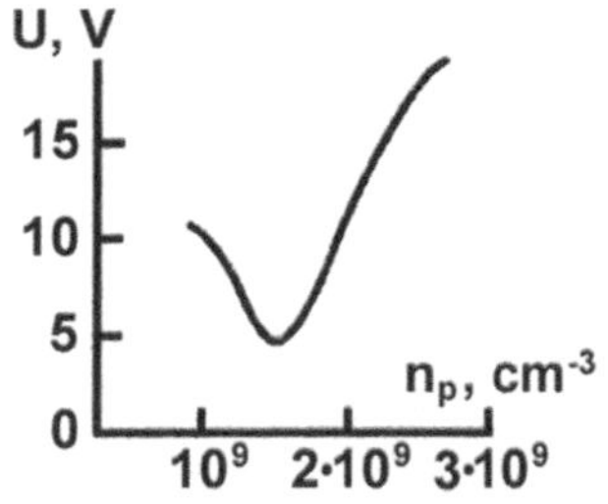

FIGURE 10.24 Discharge voltage (Hg, 1.2 mTorr) as a function of plasma density related to discharge current (power grows from left to right).

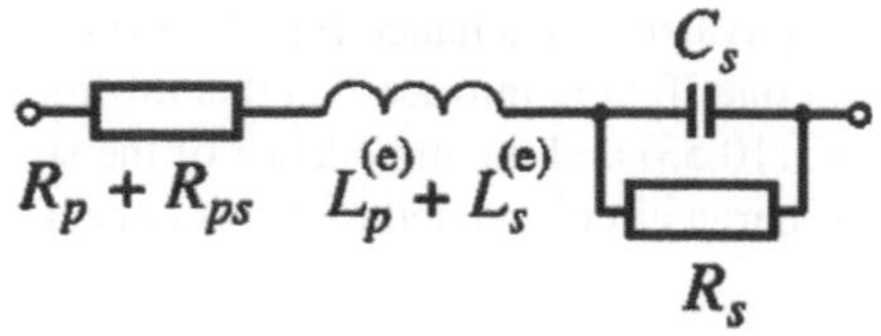

FIGURE 10.25 Equivalent circuit of a capacitively coupled RF discharge.

10.3.7 Electron and Ion Motion in the CCP-Discharge Sheaths

The drift oscillation of the electron gas "around ions" and hence, oscillation of the boundary between the plasma zone and sheath is harmonic only if the ion concentration in the sheath is uniform (see Figure 10.23). Taking into account the non-uniformity of the ion concentration, the boundary motion is not harmonic if $j = -j_0 \sin(\omega t)$. For example, in the area of elevated ion density, the boundary should move slower to provide the same change of the space charge field and hence, the same current. The equation for the plasma boundary motion can is:

$$\sin z \cdot \frac{dz}{dx} = \frac{\omega e}{j_0} n_i(x), \tag{10.63}$$

where $z(x)$ is the phase of the plasma boundary in the position x (see Figure 10.23). The boundary of the zone where plasma is always quasi-neutral corresponds to $z = 0$, and the electrode location corresponds in this case to phase $z = \pi$. As can be easily derived from Eq. (10.63), the velocity of the plasma boundary is inversely proportional to the ion concentration $n_i(x)$. In this case, the electric field in any point of sheath in the space charge phase can be calculated by the following simple relation:

$$E(x,t) = \frac{j_0}{\varepsilon_0 \omega}\left[\cos(\omega t) - \cos z\right]. \tag{10.64}$$

Time integration of $E(x, t)$ gives the constant component of the space charge field in sheath:

$$\langle E(x) \rangle = \frac{1}{\pi}\int_0^{z(x)} E(x,t)\,\omega dt = \frac{j_0}{\omega \pi \varepsilon_0}(\sin z - z\cos z). \tag{10.65}$$

This constant component of the electric field determines the motion of ions in the sheath. At relatively high pressures and low values of the reduced electric field, the ion drift velocity is proportional to the above determined constant component of electric field $\langle E(x) \rangle$:

$$v_i = b_i \langle E(x) \rangle, \tag{10.66}$$

b_i is the ion mobility. At higher electric field, ion energies received on the mean free path become comparable with the thermal energies, and ion drift becomes proportional to the square root of the electric field:

$$v_i = \sqrt{\frac{e\lambda_i\sqrt{2}}{M}\langle E(x) \rangle}. \tag{10.67}$$

Here: λ_i is the mean free path of an ion in the sheath, and M is the ionic mass. It is important to note, that the relation for ion velocity in form Eq. (10.67) is the most typical one for the case of low-pressure CCP-discharges. In the case of low pressures, the ionic motion can be considered collisionless. The ion velocities at such pressures are determined by potential drop U corresponding to the ionic motion:

$$v_i = \sqrt{2eU/M}. \tag{10.68}$$

Analysis of the ion density distribution in sheaths is an important step in describing non-thermal radio-frequency CCP-discharges. This distribution can be found in general from the continuity equation taking into account the ions drift and diffusion as well as ionization and recombination processes:

$$\frac{\partial n_i}{\partial t} + \nabla\left[-D_a \nabla n_i + n_i v_i\right] = I\left(n_i, \langle E \rangle\right) - R\left(n_i, \langle E \rangle\right). \tag{10.69}$$

In this equation, D_a is the coefficient of ambipolar diffusion, $I(n_i, <E>)$is the ionization rate, and $R(n_i, <E>)$ is the recombination rate. The constant component of the electric field of space charge in the sheath Eq. (10.65) is quite high, and its contribution to the ionic motion is more significant than the one related to temperature (which is T_e for ambipolar diffusion). For this reason, the contribution of diffusion in the continuity equation is important only in the vicinity of the plasma zone. Otherwise, the ionic drift in the space charge field mostly determines the motion of ions in the sheath and contributes most in the continuity equation. The continuity equation combined with the above relations for the ion drift velocity can be solved to determine the ion density distributions in the CCP- sheaths. However, sometimes simple estimations assuming a constant ion concentration in sheaths (see Figure 10.23) are sufficient for good qualitative analysis. Further consideration requires specification of gas pressure in the discharge.

10.4 NON-THERMAL CAPACITIVELY COUPLED (CCP) DISCHARGES OF MODERATE PRESSURE

10.4.1 General Features of the Moderate Pressure CCP-Discharges

Moderate pressure CCP-discharges are those, where the energy relaxation length λ_ε is less than the size of plasma and sheaths:

$$\lambda_\varepsilon < L, L_p. \tag{10.70}$$

The moderate pressure range corresponds to the interval 1–100 *Torr*. In this pressure range:

$$\omega < \delta v_e, \tag{10.71}$$

where δ is the average fraction of electron energy lost per one electron collision, ν_e is the frequency of the electron collisions. The inequality (10.71) means that electrons lose (and gain) energy during a time interval shorter than the period of electromagnetic RF-oscillations. Electron energy distribution functions and hence, ionization and excitation rates are determined in the moderate pressure discharges by local and instantaneous values of the electric field. In particular, this results in important contributions of ionization in the sheaths, where electric fields have maximum values and are able to provide maximum electron energies. Peculiarity of the moderate pressure CCP-discharges is the possibility of two qualitatively different forms, which are referred to as the α-discharge and the γ-discharge (S.M. Levitsky, 1957).

10.4.2 The α- and γ- Regimes of Moderate Pressure CCP-Discharges, Luminosity and Current–Voltage Characteristics

The main experimentally observed differences between the α- and γ-regimes of the moderate pressure CCP-discharges are related to electric current density and luminosity distribution in the

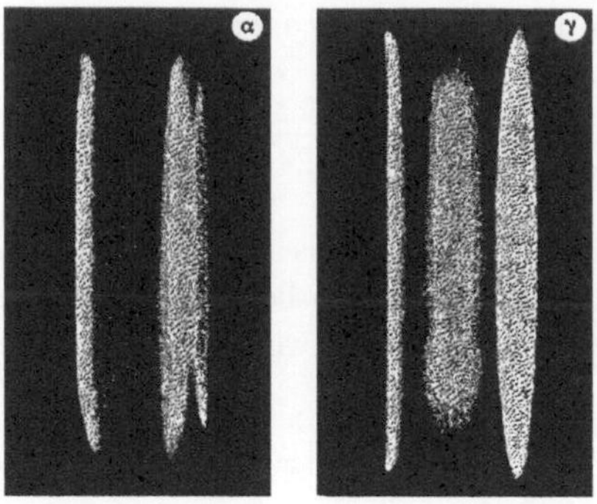

FIGURE 10.26 Pictures of α- and γ-ICP-discharges (air, 10 Torr, 13.56 MHz, 2 cm). (From Raizer, Shneider & Yatsenko, Radio Frequency Capacitive Discharges)

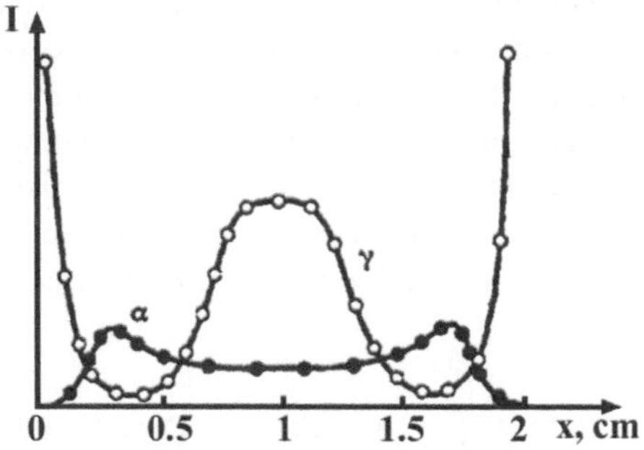

FIGURE 10.27 Emission intensity distribution along the discharge gap (air, 10 Torr, 13.56 MHz, 2 cm).

discharge gap. Pictures of two regimes of a moderate pressure CCP-discharge are shown in Figure 10.26; related luminosity distribution across the gap is shown in Figure 10.27 (N.A. Yatsenko, 1981). The α-discharge can be characterized by low luminosity in the plasma volume. Brighter layers are located closer to electrodes, but layers immediately adjacent to the electrodes are dark in this case. The γ-discharge takes place at much higher values of current density. In this regime the discharge layers immediately adjacent to the electrodes are very bright, but relatively thin. The plasma zone in also luminous and separated from the bright electrode layers by dark layers similar to the Faraday dark space in the DC glow discharges. Key differences in physical characteristics of the α- and γ-discharges are related to the fact, that the contribution in the γ-regime ionization processes is due to ionization provided by secondary electrons formed on the electrode by the secondary electron emission and accelerated in a sheat. This ionization mechanism is similar to that in the cathode layers of glow discharges. This is the reason why the main features of the sheaths in γ-discharges are similar to those of cathode layers. Both the α- and γ-discharges operate at normal current density in the same manner as glow discharges. This means that an increase of the discharge current is provided by a growth of the electrode area occupied by discharge, while the current density remains constant. The normal current density in both two regimes is proportional to the frequency of electromagnetic oscillations, and in the case of γ-discharges it exceeds the current density of α-discharges more than 10 times. Current–voltage characteristics of the moderate pressure CCP-discharge at different conditions are shown in Figure 10.28 (Yu.P. Raizer, M.N. Shneider, N.A. Yatsenko, 1995). One can see the qualitative change of the current–voltage characteristics related to lower and higher levels of the RF-discharge current. Currents less than 1 A correspond on these characteristics to the

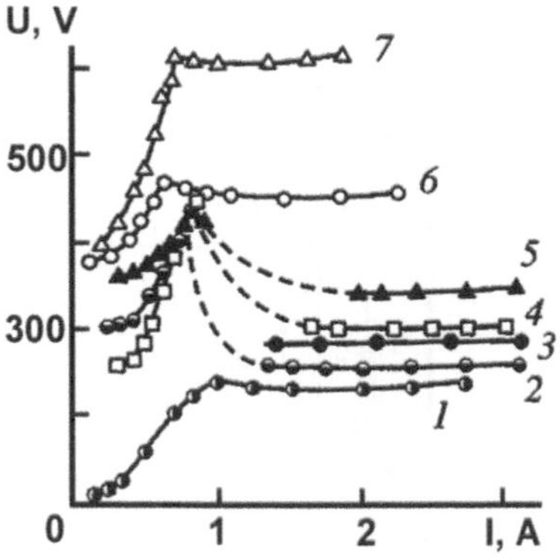

FIGURE 10.28 Current-voltage characteristics of CCP discharge, 13.56 MHz: 1 – He (p = 30 Torr, L_0 = 0.9 cm); 2 – air (30 Torr, 0.9 cm); 3 – air (30 Torr, 3 cm); 4 – air (30 Torr, 0.9 cm); 5 – CO_2 (15 Torr, 3 cm); 6 – air (7.5 Torr, 1 cm); 7 – air (7.5 Torr, 1 cm). *(From Encyclopedia of Low Temperature Plasma, Ed. V.E Fortov, Nauka, Moscow, 2002)*

α-discharges, higher current are related to the γ-discharges. Normal α-discharge can be observed on the curves 2,4 of the figure, where there is no essential change of voltage at low currents. Most of curves, however, show that an α-discharge is abnormal just after breakdown. The current density is so small in this regime that the discharge occupies all the electrode immediately after breakdown, which leads to voltage growth together with current. Transition from the α-regime to the γ-regime takes place when the current density exceeds some critical value, which depends on pressure, frequency, electrode material and type of gas. The α–γ transition is accompanied by a discharge contraction and by more than an order of magnitude growth of the current density. The increase of current in the γ-regime after transition does not change the voltage and the discharge remains in normal regime.

10.4.3 The α-Regime of Moderate Pressure CCP-Discharges

Two mechanisms provide ionization in sheaths. One is related to ionization in the plasma phase when the sheath is filled with electrons. This mechanism dominates in α-discharges. The term "α-discharge" by itself reminds one of the influence of the first Townsend coefficient α on this ionization process (see Section 7.1.5). The second mechanism is related to ionization provided by secondary electrons, produced as a result of secondary electron emission from electrodes. This mechanism dominates in γ-discharges. The term "γ-discharge" by itself shows the important contribution of the third Townsend coefficient γ on this ionization process. The ionization rate in the α-regime can be calculated using the approximation Eq. (4.4.5) for the first Townsend coefficient, and taking into account that the electric field and electron density are related in the plasma phase by Ohm's law: $j = n_e e b_e E$, where b_e is the electron mobility. Thus the ionization rate in the α-regime can be expressed as (A.S. Smirnov, 2000):

$$I_\alpha\left(x,t\right) = Ap\frac{j_0}{e}\left|\sin\omega t\right|\exp\left(-\frac{n_e}{n_0}\frac{1}{\sin\omega t}\right), \tag{10.72}$$

where the concentration parameter $n_0 = j_0/eb_eBp$ (the concentration parameter n_0 does not depend on pressure because the electron mobility $b_e \propto 1/p$). This ionization rate grows exponentially with the instantaneous current and with a decrease of the charge density (this leads to a monotonic decrease of ion concentration from plasma boundary to electrode). The exponential increase of $I_\alpha(x,t)$ at

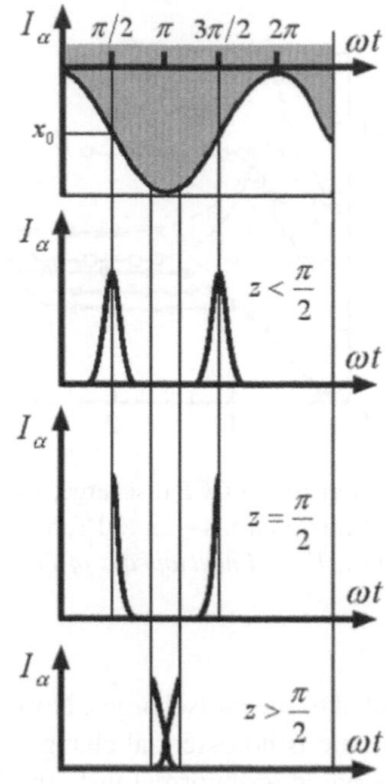

FIGURE 10.29 Plasma boundary motion and time dependence of ionization in different shift positions.

lower electron concentration permits neglecting recombination in sheaths. Motion of the plasma boundary and corresponding time dependence of the ionization rate in different points of sheath is illustrated in Figure 10.29. Coordinate "x" is calculated from the quasi-neutral plasma zone (see Figure 10.23), where $x = 0$ and phase $z = 0$, to the electrode, where $x = 2x_0$ and the phase $z = \pi$. The evolution of the ionization rate in three different points of the sheath presented in Figure 10.29 can be interpreted as follows:

1. Near the plasma zone, where the coordinate $x < x_0$ and phase $z < \pi/2$: the ionization rate of the α-discharge I_α reaches a maximum value when the current is in maximum. This takes place when $\omega t = \pi/2$ and $\omega t = 3\pi/2$.
2. In the midpoint of the sheath, where the coordinate $x = x_0$ and phase $z = \pi/2$: electrons appear only when the electric field in the plasma already reached its maximum. That is the reason the ionization peak shown in Figure 10.29 is vertical (actual width is about the Debye radius). A similar phenomenon, but reverse in time, takes place at the phase $z = 3\pi/2$.
3. Closer to the electrode, where the coordinate $x > x_0$ and phase $z > \pi/2$: electrons are absent in the moment of maximum current; the maximum of ionization is shifted until $\omega t = z$ (when electrons appear). Taking into account the strong I_α dependence on electric field, the ionization rate significantly decreases near the electrode ($x > x_0$). Most ionization in the moderate pressure α-discharge takes place near the boundary of the plasma zone $0 < x < x_0$. This explains the luminosity distribution, see Figure 10.26, 2. If $n_e >> n_0$ in Eq. (10.72), then the ionization maximum peak is very narrow.

10.4.4 Sheath Parameters in α-Regime of Moderate Pressure CCP-Discharges

The sheath parameters in the α-regime can be estimated based on the following simple assumptions: the ion concentration in the sheath n_s is constant; the ionization rate follows Eq. (10.72) at $z < \pi/2$;

and ionization is absent near the electrode $I_\alpha = 0$ at $z > \pi/2$. Balancing the ion flux to the electrode Eq. (10.60) and the ionization rate in the sheath $I_\alpha \cdot L/2$, the ion density in the sheath can be derived:

$$n_s = n_0 \ln\left[\frac{ABp^2\varepsilon_0}{e}\sqrt{\frac{2}{\pi}\frac{b_e}{b_i}\frac{1}{n_0}}\left(\frac{n_0}{n_s}\right)^{5/2}\right] = n_0 \ln \Lambda. \tag{10.73}$$

A and B are parameters of the Townsend relation for the α-coefficient Eq. (4.142); n_0 is the pressure-independent concentration parameter introduced above regarding Eq. (10.72); p is the gas pressure; b_e and b_i are the electron and ion mobilities; $\ln \Lambda$ is a logarithmic factor only slightly changing with other system parameters. Taking into account that $n_0 = j_0/eb_eBp$, one can conclude that the ion concentration in the sheath is proportional to the discharge current, does not depend on frequency and only logarithmically depends on gas pressure. Then the sheath size can be determined as:

$$L = \frac{2j_0}{\omega n_s} = \frac{2eb_eBp}{\omega \ln \Lambda}. \tag{10.74}$$

From Eq. (10.74), the sheath size in the α-regime is inversely proportional to the electromagnetic oscillation frequency and does not depend on the current density.

10.4.5 The γ-Regime of Moderate Pressure CCP-Discharges

The electric field of the α-discharge in the space charge phase grows proportionally with current density. Also, the multiplication rate of the γ-electrons (formed as a result of the secondary electron emission from electrodes) increases exponentially with the electric field in the sheath. As a result, reaching a critical value of the current density leads to "the breakdown of the sheath" and transition to the γ-regime of the CCP-discharge. The maximum ionization in the γ-regime of the CCP-discharge occurs when the electric field in the sheath has the maximum value. This takes place at the moment when the oscillating electrons are located furthest from the electrode ($z = 0,\ z = 2\pi$). This effect is illustrated in Figure 10.29 in comparison with the space and time ionization evolution in α-discharges. To estimate the sheath parameters in the γ-regime of moderate pressure CCP-discharges, balance the averaged ion flux to electrodes:

$$\langle \Gamma_i \rangle = b_i \langle E(x,t) \rangle \tag{10.75}$$

and multiply the γ-electrons in the sheath taking into account the evolution $E(x, t)$ of the electric field :

$$\left\langle \int I_\gamma dx \right\rangle = \left\langle \gamma \Gamma_i \left[\exp \int_0^{d(t)} \alpha\left(E(x,t)\right) dx - 1 \right] \right\rangle. \tag{10.76}$$

$I_\gamma(x, t)$ - is the ionization rate term physically similar to Eq. (10.72) but this time related to the contribution of γ-electrons; γ is the coefficient of secondary electron emission from the electrode; Γ_i is the ion flux on the electrode; $d(t)$ is the instantaneous size of the sheath, $\alpha(E)$ is the first Townsend coefficient. The above balance of the averaged ion flux to electrodes and the multiplication of the γ-electrons in the sheath is similar to the corresponding balance in the cathode layer of glow discharges (see Section 7.2.1). The main difference between these two cases is related to the fact that the ion current in the sheath of CCP-discharges is not equal to the discharge current because of the significant contribution of the displacement current. Current–voltage characteristics of the sheath in the CCP-discharge in air at 30 Torr pressure calculated based on Eqs. (10.75), (10.76), is presented in Figure 10.30 (A.S. Smirnov, 2000) together with the ratio of ion current to the total current in the

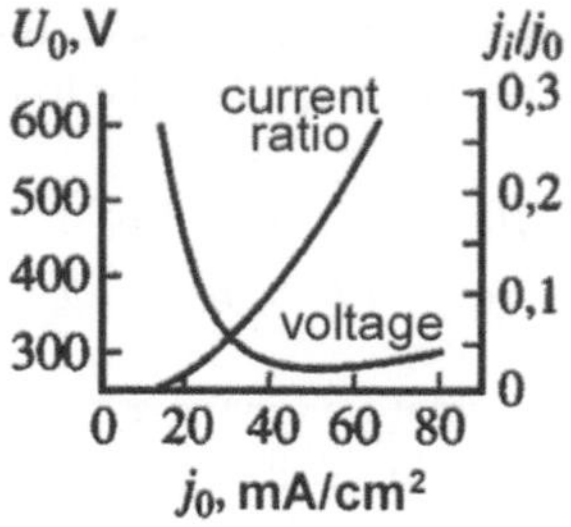

FIGURE 10.30 Current-voltage characteristic of γ-discharge (air, 13.56 MHz, 30 Torr) and ion current ratio/discharge current ratio.

discharge. As expected, the current–voltage characteristics shown in Figure 10.30 is similar to the cathode voltage drop dependence on current density in glow discharges (see Sections 7.2.2 and 7.2.3). For this reason, the effect of normal current density, discussed in detail for the glow discharges (see Section 7.2.3), also takes place in the γ-regime of the moderate pressure CCP- discharges with metallic electrodes as well.

10.4.6 Normal Current Density of γ-Discharges

The effect of normal current density in the γ-regime of the CCP-discharge with metallic electrodes can be illustrated using Figure 10.30, where the normal current density j_n corresponds to the minimum of the current–voltage characteristics $U_0(j_o)$. At lower current density $j < j_n$ the current–voltage characteristics are "falling", which is generally an unstable condition. For example,-any small increase of current density in this case, due to some fluctuation, results in lower voltages necessary to sustain the ionization. Then the actual voltage exceeds the required one, and current density grows until the normal regime $j = j_n$. The right branch $(j > j_n)$ of the current–voltage characteristics $U_0(j_o)$ is stable in general, but not stable at the edge of the sheath on the boundary of current-conducting and currentless areas of the electrode. Current growth in the γ-regime of moderate pressure CCP-discharges leads to an increase of the electrode surface area covered by the discharge, while the current density and sheath voltage drop remain the same and equal to their normal values: j_n and U_n. When the discharge covers the entire electrode, the voltage and current density in principal should begin growing. From Figure 10.30, the growth of voltage and current density results in an increase of the fraction of ion conductivity current in the total discharge current. As soon as the ion conductivity current and displacement current become close $(j_i \approx j_D)$, the parameter $\omega\tau_i \approx 1$ and ion motion during an oscillation period can no longer be neglected. In this case, the inequality Eq. (10.50) is no longer valid, nor the entire above-considered approach. The opposite limit of high ion conductivity current $(j_i > j_D)$ can be interpreted as $\omega\tau_i < 1$, which means that the CCP-discharge should be considered as a low-frequency one. Under these low-frequency conditions, the CCP-plasma resistance is almost completely active. There is sufficient time to establish cathode and anode layers during one oscillation period. This makes this CCP-discharge regime even closer to that of the glow discharge. If the electrode is covered by a thin dielectric layer with capacitive resistance much less than discharge resistance, then general features of the CCP-discharge remains as discussed above. The distributed capacitive resistance is able to determine the discharge current on the electrode surface, which leads to possibility of stabilizing the discharge at currents lower than the normal one $j < j_n$ (Yu.P. Raizer, M.N. Shneider, N.A. Yatsenko). The normal current density and the sheath voltage drop in the γ CCP-discharges with metallic electrodes can be found by minimizing the dependence $U_0(j_0)$:

$$j_n = \frac{Bp\omega\varepsilon_0}{2} \cdot \psi_1(\gamma), \tag{10.77}$$

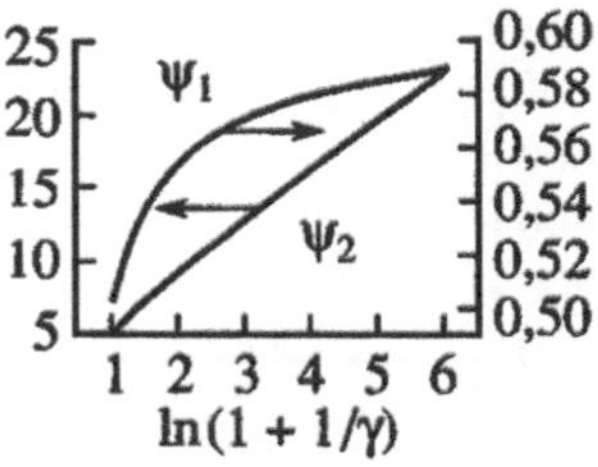

FIGURE 10.31 Ψ-functions dependence on secondary electron emission coefficient. *(From Encyclopedia of Low Temperature Plasma, Ed. V.E Fortov, Nauka, Moscow, 2002)*

where B is the parameter of Townsend formula for α-coefficient Eq. (4.142), and the dimensionless function $\psi_1(\gamma)$ characterizes the normal current density dependence on the secondary electron emission coefficient, see Figure 10.31 (A.S. Smirnov, 2000). The normal sheath voltage drop in the γ CCP-discharges:

$$U_n = \frac{B}{A}\psi_2(\gamma). \tag{10.78}$$

also includes the A, B parameters of the Townsend formula for α-coefficient Eq. (4.142), and the function $\psi_2(\gamma)$ characterizing the influence of the secondary electron emission coefficient (see also Figure 10.31). From Figure 10.31,- the normal current density is seen to have weak dependence on the secondary emission γ; the normal voltage drop grows with a decrease of γ, but not significantly. Similar to the case of a DC glow, the sheath voltage drop does not depend on pressure and is determined only by the gas and electrode material. The normal current density is proportional to the frequency of the electromagnetic field and pressure. The pressure growth also leads to an increase of the ion conductivity current:

$$j_i = \frac{b_i B^2 \varepsilon_0 A p^3}{4\psi_2 \psi_1^3}. \tag{10.79}$$

At some critical pressure, the ion current reaches the value of the displacement current (which corresponds to $\omega\tau_i \approx 1$, see Eq. (10.50)), which leads to the discharge transfer in the low-frequency regime. Thus the upper limit pressure of the RF-CCP-discharge in γ-regime can be expressed as:

$$p < \frac{2\omega\psi_1^2(\gamma)\psi_2(\gamma)}{(b_i p)AB}. \tag{10.80}$$

10.4.7 Physical Analysis of Current–Voltage Characteristics of the Moderate Pressure CCP-Discharges

Example of current–voltage characteristic of a moderate pressure CCP-discharge sheaths is shown in Figure 10.32 (A.S. Smirnov, 2000). The lower limit of the current densities (about 2 mA/cm²) is related to low electron conductivity in plasma. This corresponds to the condition $\omega\tau_e \approx 1$ (see the inequality (10.48)) when the displacement current in the plasma zone becomes necessary. The upper limit of the current densities (200 mA/cm²) is related to the ion density growth in the sheath with an increase of current density. The upper limit corresponds to the condition $\omega\tau_i \approx 1$ (see the inequality (10.50)) when ion conductivity in sheaths becomes necessary. The current–voltage characteristic shown in Figure 10.32 between the lower and upper limits can be subdivided into three regions:

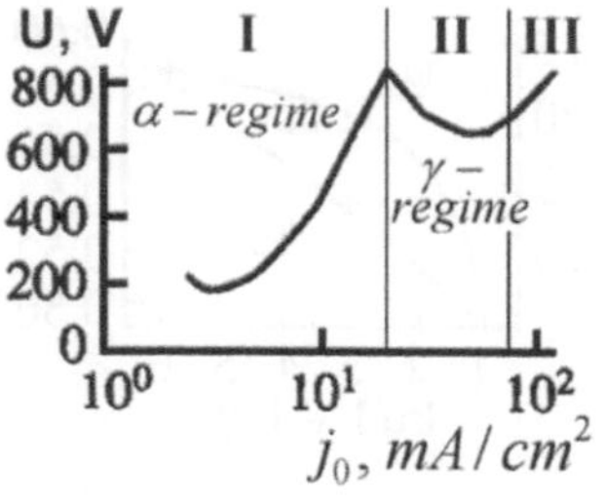

FIGURE 10.32 Current-voltage characteristic of a CCP discharge in nitrogen, 13.56 MHz, 15 Torr.

1. The first region on this figure with current densities about 2–20 mA/cm corresponds to the α-regime of the moderate pressure CCP-discharge. In this case, the sheath voltage drop grows with current density, the sheath depth does not depend on the discharge current density, and the ion concentration decreases monotonically from the plasma zone to electrode.
2. The second region with current densities about 20–100 mA/cm corresponds to γ-regime of the moderate pressure CCP-discharge, where most of the ionization is provided by secondary electrons formed on the surface of electrodes by ion impact during the space charge phase of the discharge. The ion density in the sheath has a maximum value exceeding the ion concentration in the plasma zone, the ion density in the sheath grows with the discharge current density, and the sheath depth decreases with current density. It results in the "falling" (and actually unstable) current–voltage characteristics in this region.
3. Sheath in the third region with current densities about 100–200 mA/cm is similar to the cathode layer of the abnormal DC glow discharge. The electric field E in the space charge phase is so high that the dependence $\alpha(E)$ is no longer strong. The high electric field leads to longer distances necessary to establish the electron energy distribution functions. This results in essential non-local effects: secondary electrons provide significant ionization in the plasma phase and the maximum density of charged particles shifts from the sheath to the plasma zone (with possible formation of the Faraday dark space). Obviously, the minimum point on the $U_0(j_0)$ curve corresponds to the normal current density of the γ-discharge.

10.4.8 The α-γ Transition in Moderate Pressure CCP-Discharges

Here the electric field in sheath is not very high at low current densities corresponding to the α-regime (see Figure 10.32). For this reason, multiplication of electrons formed on the electrode surface due to secondary emission can be neglected in this regime. Growth of the current density and the related growth of the electric field lead to increasing the intense multiplication of the secondary (γ) electrons, which finally results in the phenomenon of the α–γ transition. At first, the increase of current density makes the multiplication of the secondary γ-electrons essential only near the electrodes, where the electric field has the maximum value. In this case, the maximum of the ionization rate function I_γ is located in the vicinity of an electrode, where the drift in the electric field of the space charge determines the ions flux and the contribution of diffusion is not significant. This can result in density profiles with a minimum. The α–γ transition takes place when the generation of the ion flux due to the multiplication of the secondary γ-electrons exceeds the ion flux produced in the plasma phase. This requires current densities exceeding the critical value:

$$j_{crit} = Bp\omega\varepsilon_0 \ln^{-1}\left\{\frac{ApL}{3\left[\ln\left(\left(1+\frac{1}{\gamma}\right)\sqrt{\frac{\pi ApL}{4}}\right) - \frac{Bp\omega\varepsilon_0}{j_{crit}}\right]}\right\}. \tag{10.81}$$

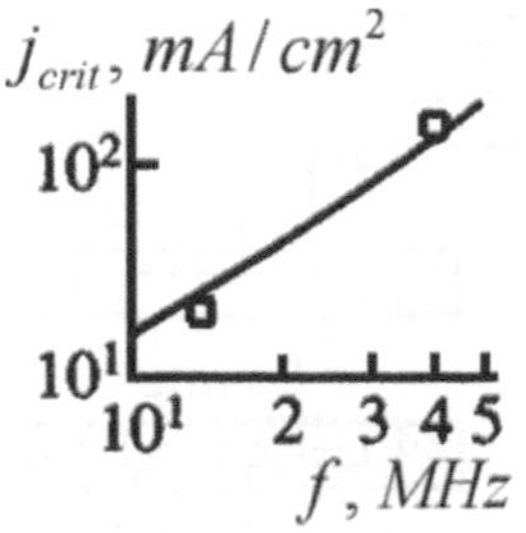

FIGURE 10.33 Experimental and calculated α–γ transition current dependence on frequency: air, 0.75 cm, 15 Torr. (*From Encyclopedia of Low Temperature Plasma, Ed. V.E Fortov, Nauka, Moscow, 2002*)

The critical value of current density, calculated from Eq. (10.81) and corresponding to the α–γ transition is shown in Figure 10.32 by an arrow. From Eq. (10.81), the critical current density of the α–γ transition in moderate pressure CCP-discharges grows with the oscillation frequency. The numerical dependence $j_{crit}(\omega)$ is presented in Figure 10.33 (A.S. Smirnov, 2000). Similar curve describes the critical current density dependence on gas pressure.

10.4.9 Some Frequency Limitations for Moderate Pressure CCP-Discharges

General frequency limitations in the radio-frequency discharges were discussed in Section 10.3.1. Now consider a frequency limitation important for the moderate pressure CCP-discharges. The sheath depth decreases with growth of the oscillation frequency. At high frequencies, the inequality $L >> \lambda_\varepsilon$, which determines the domination of local effects in forming the electron energy distribution function, is no longer valid (λ_ε is the length of electron energy relaxation). The non-local effects typical for low-pressure discharge systems become important in these systems. During formation of the high-energy tail of the electron energy distribution function, the electrons displacement can be longer than the typical sizes of the discharge system. Characteristics of the sheathes at high frequencies become similar to those at low pressures; ionization takes place in the plasma zone, and the ion flux in sheath is constant. The mean free path of electrons in air at pressure 15 Torr is about 0.02 mm, and the length of electron energy relaxation and formation of the electron energy distribution function can be calculated as $\lambda_\varepsilon \approx 0.5$ mm. The sheath depth becomes shorter than this value of the length of electron energy relaxation at frequencies exceeding 120 MHz. This means that at gas pressures of 15 Torr, the CCP discharges can be considered as one of moderate pressure if the oscillation frequencies do not exceed about 60 MHz. The RF CCP-discharges of moderate pressure were investigated mainly related to their application in powerful high-efficiency waveguide CO_2-lasers. The first *He–Ne* and CO_2 lasers were based on the CCP-RF discharges, see A. Javan, W.R. Bennett, D.R. Herriot, 1961; C.K.N. Patel, 1964, Yu.P. Raizer, M.N. Sneider and N.A. Yatsenko, 1995.

10.5 LOW-PRESSURE CAPACITIVELY COUPLED RF DISCHARGES

10.5.1 General Features of Low-Pressure CCP-Discharges

Low-pressure RF-discharges (less than about 0.1 Torr) are widely used in electronics for etching and different kinds of chemical vapor deposition (CVD) processes. A scheme of a low-pressure CCP-discharge installation applied for the surface treatment is illustrated in Figure 10.34. This large-scale application stimulated intensive fundamental and applied research activity during the last decades (see, for example, D.M. Manos, D.L. Flamm, 1989; M. Konuma, 1992; J.L. Vossen, W. Kern. 1978, 1991, M.A. Lieberman, A.J. Lichtenberg, 1994). Luminosity pattern of the low-pressure CCP-RF-discharges is different from that of moderate pressure discharges. In this case the plasma zone is bright and separated from the electrodes by dark pre-electrode sheaths. The discharge usually has an

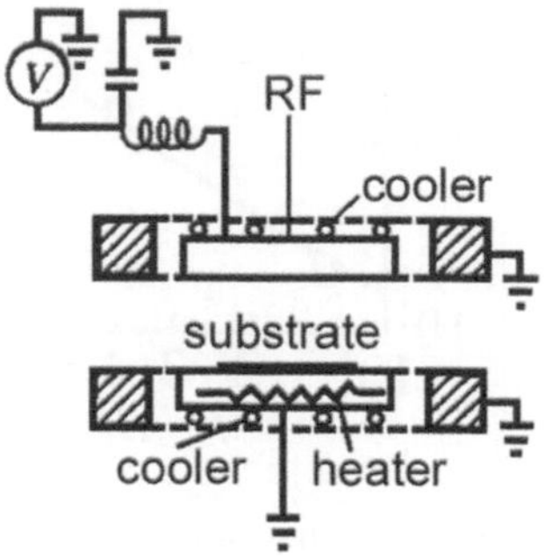

FIGURE 10.34 Schematic of a low pressure CCP discharge.

asymmetrical structure: a sheath located by the electrode where the RF-voltage is applied, is usually about 1 cm thick, while another sheath located by the grounded electrode is 0.3 cm thick. In contrast to moderate pressure discharges, after breakdown plasma fills out the entire low pressure discharge gap, and the phenomenon of normal current density does not take place.

10.5.2 Plasma Electrons Behavior in the Low-Pressure Discharges

According to the definition from Section 10.3.1, electron energy relaxation length λ_ε in the low-pressure radio-frequency CCP discharges characterizing thickness of plasma zone and sheaths exceeds the sizes of the system:

$$\lambda_\varepsilon > L_p, L. \tag{10.82}$$

This means that the electron energy distribution function is not local; it is determined not by the electric field in a fixed point but by the entire distribution of the electric fields in the zone of size λ_ε. Numerically the inequality (10.82) for He-discharges implies, $p \cdot (L_p, L) \leq 1 \; Torr \cdot \text{cm}$. The inequality (10.82) usually occurs together with the following:

$$\omega >> \delta \cdot \nu_e, \tag{10.83}$$

where δ is the average fraction of electron energy lost per one electron collision, ν_e is the frequency of the electron collisions (compare the inequalities (10.82) and (10.83) with the opposite requirements (10.70) and (10.71)). The criterion (10.83) actually means that the characteristic time of formation of the electron energy distribution function (EEDF) is much longer than the period of RF-oscillations, which permits considering the EEDF as a stationary one. If at the same time $\omega < \nu_e$, then the "DC-analogy" is still valid. This means that the EEDF in the low-pressure CCP-RF-discharge is the same as that in the constant electric field $E_0/\sqrt{2}$ (where E_0 is amplitude of the RF-oscillating field). The electric field in the low-pressure CCP-RF-plasma (as well as in sheaths) can be divided into constant and oscillating components. The constant component of electric field in the plasma (including plasma zone and sheaths in the plasma phase) provides a balance of electron and ion fluxes, which is necessary for quasi-neutrality. The oscillation component of electric field provides the electric current and heating of the plasma electrons. Distribution of plasma density and electric potential φ (corresponding to the constant component of electric field) is illustrated in Figure 10.23b, c. The electric field in the space charge sheath exceeds that in the plasma. For this reason, the sharp change of potential on the boundary of plasma and sheath is shown in Figure 10.23 as a vertical line. Thus the plasma electrons move between the sharp potential barriers satisfying the conservation of total energy, which can be expressed as:

$$\varepsilon = \frac{1}{2} m v_e^2 + e\varphi. \tag{10.84}$$

Here v_e is the electron velocity, m is an electron mass, and $\varphi = -\phi$ is the electric potential corresponding to the constant component of electric field and taken with opposite sign (because "e" is considered positive). The electron energy distribution function depends only on total electron energy Eq. (10.84). Kinetic equation for the EEDF in the Fokker-Plank form (see Sections 4.2.4 and 4.2.5) is similar in this case with that in the local approximation, but the diffusion coefficient in energy space should be averaged over the entire space interval where the electron moves.

10.5.3 Two Groups of Electrons, Ionization Balance, and Electric Fields in Low-Pressure CCP-Discharges

All plasma electrons in the low-pressure CCP-discharges can be divided into two groups. The first group consists of electrons with kinetic energy below the potential barrier on the sheath boundary (see Eq. (10.84) and Figure 10.23). These electrons trapped in the plasma zone are heated by the electric field in this zone, which is determined by plasma density n_p:

$$E_p = \frac{j_0 m v_e}{n_p e^2}. \tag{10.85}$$

The second group of electrons has kinetic energy exceeding the potential barrier $e\varphi$ on the sheath boundary. These electrons spend part time inside of the sheath, where density n_s is lower, and the electric field in the plasma phase is higher:

$$E_s = \frac{j_0 m v_e}{n_s e^2} >> E_p. \tag{10.86}$$

The energetic electrons of the second group are heated by the averaged electric field:

$$\left\langle E^2(\varepsilon)\right\rangle = \frac{L_p}{L_0} E_p^2 + \frac{L}{L_0}\left(1 - \frac{e\varphi}{\varepsilon}\right) E_s^2. \tag{10.87}$$

L_p, L, L_0 are lengths of the plasma zone, sheaths, and total discharge gap, respectively; the square of electric field is averaged in the Eq. (10.87) because it determines the electron heating effect. The fast electrons provide the main ionization in the plasma zone, where their kinetic energy has a maximum value. The "DC-analogy" can be applied in this case for the fast electrons by using the effective electric field:

$$E_{eff}^2 = \frac{1}{2}\left\langle E^2\right\rangle \frac{v_e^2}{\omega^2 + v_e^2}. \tag{10.88}$$

The electric field (10.88) can be used in the formula for calculation of the ionization frequency:

$$I = Ap\frac{e n_s}{m v_e} E_{eff} \exp\left(-\frac{B}{\sqrt{E_{eff}/p}}\right), \tag{10.89}$$

where A, B are the special parameters of this relation determined by type of gas, somewhat similar to those of the Townsend formula. Balancing the ionization in volume Eq. (10.89) and ion flux to the electrode: $\Gamma_i = n_s v_i$, where the velocity v_i is determined by Eq. (10.67), leads to the electric field:

$$\frac{E_{eff}}{p} = B^2 \ln^{-2} \frac{E_{eff} A p L_p e}{m v_e \sqrt{\sqrt{2} e \lambda_i j_0 / \varepsilon_0 M \omega}}. \tag{10.90}$$

The effective reduced electric field is determined by the constant B, only logarithmically depends on plasma parameters, and therefore is almost fixed in the low-pressure CCP-RF discharges.

10.5.4 High and Low Current Density Regimes of Low-Pressure CCP-Discharges

Assuming the electric field fixed, the electron concentration in the plasma is proportional to the current density in the same manner as takes place in the DC-glow discharges. This gives a possibility to separately consider the cases of relatively high and relatively low current densities. If the current density (and hence n_p) is relatively low, then according to Eqs. (10.85) and (10.86), the electric field in plasma is relatively high and heating of electrons takes place mostly in the plasma zone (also taking into account the larger size of this zone). The voltage drop is determined by the electron energy, and its value numerically is near the lowest level of electronic excitation of neutral particles. In general, the plasma zone determines all discharge properties at low current densities. The sheaths are then arranged in a way necessary to close the discharge circuit by displacement current and to provide ion flux from the plasma zone. If the current density (and hence the density n_p) is relatively high, then according to Eqs. (10.85) and (10.86) the electric field in the plasma zone is much lower with respect to the electric field in sheaths in plasma phase. Hence most of the electron heating effect takes place in sheaths in the plasma phase. In the high current density regime, the ionization balance determines the electric field and, hence the ion density n_s in the sheath. Thus based on Eqs. (10.86)–(10.88), the effective electric field can be expressed as:

$$E_{eff}^2 \approx \frac{L}{L_0} E_s^2 = \frac{L}{L_0} \frac{j_0^2 m^2 v_e^2}{n_s^2 e^4}. \tag{10.91}$$

Taking into account that the effective electric field E_{eff} is fixed, Eq. (10.91) gives:

$$\frac{n_s^2}{L} \propto j_0^2. \tag{10.92}$$

The Poisson equation requires the relation between the ion density in sheaths and current density:

$$n_s L \propto j_0. \tag{10.93}$$

The combination of Eqs. (10.92) and (10.93) leads to the conclusion that the sheath size L does not depend on current density j_0 in the regime when the current density is high. At the same time the ion concentration in the sheath is proportional to the current density. Thus in general, all discharge properties at high current densities are determined by the sheaths. The plasma zone is then arranged in the way necessary to provide the required ion current in the sheath.

10.5.5 Electron Kinetics in Low-Pressure CCP-Discharges

Heating of electron gas can be described by the Fokker-Plank kinetic approach (see Section 4.2.4) as the diffusion flux along the electron energy spectrum. In this case the electron heating is proportional to the derivative over the energy of the electron energy distribution function $\frac{\partial f(\varepsilon)}{\partial \varepsilon}$ and to the diffusion coefficient D_ε in the electron energy space, which depends on electric field and the heating mechanism. In general, the diffusion coefficient D_ε in energy space can be expressed for the RF-discharge systems based on Eq. (4.74) in the following form:

$$D_\varepsilon = \frac{1}{6} \frac{(eEv)^2}{\omega^2 + v_e^2} v_e, \tag{10.94}$$

where E is the amplitude value of the electric field. The electrons can be divided into two groups corresponding to high and low energies relative to the potential on the sheath boundary. The diffusion coefficient in energy space (10.94) can be expressed in somewhat different ways for each of the two electron groups. Electrons of the first group are relatively "cold", and have kinetic energy lower than the potential on the sheath boundary. Thus, the "cold" electrons are unable to penetrate the sheath. Their diffusion in energy space is determined by the electric field in the plasma and can be expressed as:

$$D_{\varepsilon c}\left(\varepsilon\right) = E_p^2 \varepsilon \frac{e^2 v_e}{6m\left(\omega^2 + v_e^2\right)}. \tag{10.95}$$

E_p is the electric field in plasma zone (10.85), ε - is the electron energy, ν_e- is the frequency of their collisions. The second group includes the "hot" electrons, which have energies exceeding the potential on the sheath boundary $\varepsilon > e\varphi$. These electrons are able to spend part time in the sheaths and therefore their diffusio in the energy space is determined by the averaged electric field, Eq. (10.87):

$$D_{\varepsilon h} = \left[\frac{L_p}{L_0} E_p^2 + \frac{L}{L_0}\left(1 - \frac{e\varphi}{\varepsilon}\right) E_s^2\right] \varepsilon \frac{e^2 v_e}{6m\left(\omega^2 + v_e^2\right)}. \tag{10.96}$$

In this relation, E_s is the electric field in the sheath (plasma phase) and determined by Eq. (10.87); L_p, L, L_0 are lengths of the plasma zone, sheaths and total discharge gap respectively; φ is the potential barrier between the plasma and the sheath. It is reasonable to assume in these discharges that electrons lose all their energy in non-elastic collisions (corresponding to the lowest level of electronic excitation ε_1) and then return to the low energy region of the electron energy distribution function (EEDF). Then one can conclude that the integral flux in the energy space should be constant over the electron energy interval $0 < \varepsilon < \varepsilon_1$. This means that the EEDF is a very smooth function of energy. Then density of charge particles in the plasma and sheath can be presented as a product of the averaged value of the EEDF (f_c^0 for cold electrons, f_h^0 for hot electrons) and the corresponding volume in energy space. Thus the concentration of charged particles in plasma zone can be given as (A.S. Smirnov, 2000):

$$n_p \approx \frac{8\pi\sqrt{2}}{3m^{3/2}} f_c^0 \left(e\varphi\right)^{3/2}. \tag{10.97}$$

Similarly, the electron density in sheaths in the plasma phase can be expressed as:

$$n_s \approx \frac{8\pi\sqrt{2}}{3m^{3/2}} f_h^0 \left[\varepsilon_1^{3/2} - \left(e\varphi\right)^{3/2}\right]. \tag{10.98}$$

Derivatives of the EEDF can be estimated in a similar manner as the averaged EEDF divided by the relevant interval in the energy space, which permits calculating fluxes in the energy space. Balancing the fluxes, one can derive the equation for the potential barrier on the plasma – sheath barrier:

$$\left(\frac{\varepsilon_1}{e\varphi}\right)^{3/2} - 1 = \left(\frac{e\varphi}{eL_p} \frac{2\omega\varepsilon_0}{j_0}\right)^{1/2} \left(\frac{\varepsilon_1}{\varepsilon_1 - e\varphi} + \frac{eL}{e\varphi} \frac{j_0}{2\omega\varepsilon_0}\right). \tag{10.99}$$

From Eq. (10.99), the potential barrier $e\varphi$ on the plasma corresponds to the lowest energy level ε_1 of electronic excitation. Combination of Eqs. (10.87), (10.90) and (10.99) allows determining the sheath size L, potential barrier $e\varphi$, and as a result all other discharge parameters.

10.5.6 Stochastic Effect of Electron Heating

Electron heating in RF capacitive discharges of moderate pressures is due to electron-neutral collisions. Energy of the systematic oscillations, received by an electron from the electromagnetic field after a previous collision, can be transferred to chaotic electron motion during the next collision. If the gas pressure is sufficiently low and $\omega^2 >> \nu_e^2$, then the electric conductivity and Joule heating are proportional to the frequency of electron-neutral collisions ν_e (see Eqs. (6.181) and (6.182)). This means that the discharge power related to electron-neutral collisions also should be very low at low pressures. However experimentally, the discharge power under such conditions can be significantly higher because of the contribution from stochastic heating. This effect, which provides heating of fast electron in RF-discharges even in the collisionless case, was first considered by V.A. Godyak (1971, 1975), see also M.A. Lieberman (1988). The physical basis of stochastic heating can be easily explained when the pressure is so low that the mean free path of electrons exceeds the size of sheaths. Here electrons entering the sheath are then reflected by the space charge potential of the sheath boundary. This process is similar to an elastic ball reflection from a massive wall. If the sheath boundary moves from the electrode, then the reflected electron receives energy; conversely if the sheath boundary moves toward the electrode and a fast electron is "catching up" to the sheath, then the reflected electron loses kinetic energy. The electron flux to the boundary moving from the electrode exceeds that for one moving in the opposite direction to the electrode, therefore energy is transferred to the fast electrons on average by this mechanism; this explains the stochastic heating. The stochastic heating can be described by means of a special stochastic diffusion coefficient in energy space D_{est} and an effective "stochastic" electric field E_{st}. These can be calculated noting that the frequency of the electron "collision" with the potential barrier is the inverse time of the electron motion between electrodes:

$$\Omega = \frac{\sqrt{2\varepsilon/m}}{L_0}. \tag{10.100}$$

The diffusion coefficient in the energy space is determined by the square of energy change per one collision and frequency of the collisions. The velocity of the moving sheath boundary coincides with the electron drift velocity in the sheath; hence the energy change during a collision with the boundary is the same as that for a collision with neutral particles. For this reason, the diffusion coefficient in energy space, related to the stochastic heating, can be obtained from Eq. (10.94) by replacing the electron-neutral collision frequency ν_e by the frequency Eq. (10.100) of "collisions" with the sheath boundary:

$$D_{\varepsilon st} = \frac{e^2 E_s^2 \varepsilon}{6m}\Omega. \tag{10.101}$$

The square of the effective stochastic electric field differs from that in the sheath in plasma phase Eq. (10.86) by the factor of about Ω/ν_e:

$$E_{st}^2 = \frac{3\Omega}{2\nu_e}E_s^2 = \frac{3\Omega}{2\nu_e}\left(\frac{j_0 m \nu_e}{n_s e^2}\right)^2. \tag{10.102}$$

In this relations $\omega >> \nu_e$. The effective "stochastic" electric field should replace E_s in equations for potential and particle balance of the capacitive RF-discharge at very low pressures.

10.5.7 Contribution of γ-Electrons in Low-Pressure Capacitive RF-Discharge

Mean free paths of electrons with high energies of 500–1000 eV at gas pressures below 0.1 Torr becomes larger than 1 cm, which can exceed the discharge gap between electrodes. Such γ-electrons

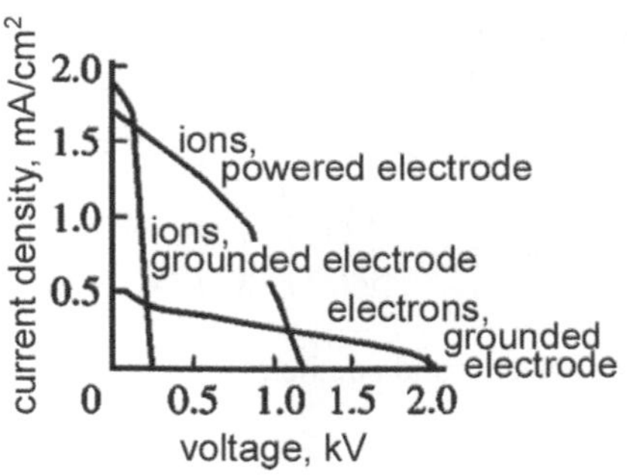

FIGURE 10.35 Stopping potentials for charged particles bombarding electrodes in a CCP discharge.

accelerated near electrodes can move across the discharge gap without any collisions and not directly influence the ionization processes. However, the energetic γ-electrons can make a significant contribution to physical and chemical processes on the electrode surface; this is essential in the plasma-chemical application of the low-pressure capacitive RF-discharge. Typical energy distributions of electrons and ions bombarding electrodes in low-pressure CCP-discharges (p = 5 mTorr) are shown in Figure 10.35. As seen from this figure, the flux of the fast electrons on the grounded electrode is close to that of ions, while the energies of these electrons reach 2 keV and greatly exceed those of ions. Thus the total electron energy also exceeds the total ion energy transferred to the grounded electrode. An interesting effect can take place if the time necessary for the secondary γ-electrons to cross the discharge gap is longer than $1/\omega$. Then the secondary electrons accelerated in one sheath slow down near the opposite electrode. Such resonant electrons can be trapped in the plasma and make contributions to ionization.

10.5.8 Analytic Relations for the Low-Pressure RF-CCP-Discharge Parameters

The system of equations describing low-pressure capacitive RF-discharges is complicated for analytical solutions. However, asymptotic solutions can be found, for some extreme but practically important regimes. Consider, the low and high current discharge regimes, assuming collisional heating ($\nu_e >> \omega$) and the proportionality of ion velocity in the plasma zone to the ambipolar electric field:

$$v_p = \frac{e}{Mv_i}\frac{2\varphi}{L_p}. \tag{10.103}$$

Such assumptions are satisfied for discharge in argon if $0.5 < pL_0 < 1\,\text{Torr}\cdot\text{cm}$.

*1. **Low-current regime of the low-pressure RF-CCP-discharge.*** Heating in the sheaths is negligible at low currents ($L_pE_p^2 >> L_sE_s^2$), and the effective electric field is equal to:

$$E_{eff} = E_p/\sqrt{2}. \tag{10.104}$$

Substituting this expression for the effective electric field into Eq. (10.90), the following relation for the electric field in plasma is obtained:

$$E_p = \sqrt{2}B^2p\ln^{-2}\frac{E_pApL_pe}{mv_e\sqrt{2\sqrt{2}e\lambda_i j_0/\varepsilon_0M\omega}}. \tag{10.105}$$

The electric field in plasma only logarithmically depends on the current density, which is similar to the positive column of the DC glow discharges. The potential barrier $e\phi$ on the plasma-sheath boundary is about equal to the lowest energy level ε_1 of electronic excitation, and can be determined as:

$$e\phi = \varepsilon_1 \left[1 - \left(\frac{b_i \varepsilon_1}{3eL_p}\right)^{1/2} \left(\frac{\omega \varepsilon_0 M}{j_0 e \lambda_i}\right)^{1/4}\right]. \tag{10.106}$$

This potential barrier is almost constant, and can be estimated as:

$$\phi = \frac{2}{3}\frac{\varepsilon_1}{e}. \tag{10.107}$$

Concentration of charged particles in plasma is proportional to the current density:

$$n_p = \frac{j_0 m \nu_e}{e^2 E_p} \propto j_0. \tag{10.108}$$

The ions flux is also proportional to the current density and can be given by:

$$\Gamma_i = \frac{4}{3}\frac{j_0 \varepsilon_1 m \nu_e}{E_p L_0 e^2 M \nu_i} \propto j_0, \tag{10.109}$$

where ν_e and ν_i are the electron-neutral and ion-neutral collision frequencies. Sheaths do not make any contribution to electron heating in the low current regime. In this case the sheath parameters are determined by the ion flux from the plasma. Balancing the ion flux in the sheath with that one in plasma Eq. (10.109), the sheath thickness proportional to square root of current density can be found as:

$$L = \frac{3eE_p L_0 M \nu_i}{2\omega \varepsilon_1 m \nu_e} \sqrt{\frac{\sqrt{2}\lambda_i}{M}\frac{j_0}{\varepsilon_0 \omega}} \propto \sqrt{j_0}. \tag{10.110}$$

Relation (10.74) between n_s and sheath thickness (10.110) gives the ion density in the sheath:

$$n_s = \frac{4}{3}\frac{\varepsilon_1 m \nu_e}{E_p e^2 L_0 M \nu_i} \sqrt{\frac{M \omega \varepsilon_0 j_0}{\sqrt{2}\lambda_i}} \propto \sqrt{j_0}, \tag{10.111}$$

which is also proportional to the square root of current density.

2. ***High-current regime of the low-pressure RF-CCP-discharge.*** The main electron heating in the high current regime takes place in the sheaths. In this case low energy electrons are heated by the relatively low electric field in the plasma zone, and their effective "temperature" decreases. For this reason, the ambipolar potential maintaining the low energy electrons in the plasma zone is less than the excitation energy. The expression for the effective electric field can be derived in this regime based on the Poisson equation and Eqs.(10.86), (10.91):

$$E_{eff} = \sqrt{\frac{L}{2L_0}} E_s = \sqrt{\frac{L}{2L_0}} \frac{m\omega \nu_e L}{2e}. \tag{10.112}$$

Comparing this relation with (10.90) gives the sheath thickness:

$$L^{3/2} = \frac{2\sqrt{2L_0} e B^2 p}{m \omega \nu_e} \ln^{-2}\left(\frac{ApL^{3/2} L_p \omega}{2\sqrt{\frac{2\sqrt{2}L_0 e \lambda_i}{M}\frac{j_0}{\varepsilon_0 \omega}}}\right). \tag{10.113}$$

The sheath thickness is determined by a balance of ionization and ion losses, and practically does not depend on the discharge current. The potential barrier on the sheath boundary in the high current regime of the low-pressure discharge then can be expressed as:

$$e\varphi = \varepsilon_1 \left(\frac{2\varepsilon_1}{Mv_i L}\right)^2 \frac{M}{\sqrt{2}e\lambda_i} \frac{\omega\varepsilon_0}{j_0} \propto \frac{1}{j_0}. \tag{10.114}$$

In this case the potential barrier of the sheath boundary is less than the excitation energy ε_1 and is inversely proportional to the discharge current. The charged particles concentration in the plasma zone increases quite strongly with the current density:

$$n_p = \frac{2j_0}{\omega eL} \frac{L_p}{L} \left(\frac{\sqrt{2}e\lambda_i}{M} \frac{j_0}{\omega\varepsilon_0}\right)^{3/2} \left(\frac{LMv_i}{2\varepsilon_1}\right)^3 \propto j_0^{5/2}. \tag{10.115}$$

Ion concentration in sheath is proportional to the discharge current density:

$$n_s = \frac{2j_0}{e\omega L} \propto j_0. \tag{10.116}$$

The critical current density j_0^{cr} dividing the above-described regimes of low and high currents can be found from Eq. (10.114), assuming that $e\varphi = \frac{2}{3}\varepsilon_1$:

$$j_0^{cr} = 3\sqrt{2}\left(\frac{\varepsilon_1}{LMv_i}\right)^2 \frac{M\omega\varepsilon_0}{e\lambda_i}, \tag{10.117}$$

where the sheath thickness can be taken from Eq. (10.113).

10.5.9 Numerical Values of the Low-Pressure RF-CCP-Discharge Parameters

Experimentally measured together with numerically predicted values of parameters of the low-pressure capacitive radio-frequency discharge in argon are presented in Figure 10.36 as functions of current density (A.S. Smirnov, 2000). Figure 10.36a illustrates the plasma concentration dependence on the discharge current density. The low-current branch of the dependence is linear in accordance with Eq. (10.108), while at high currents this dependence is stronger (see Eq. (10.115)). The average energy of plasma electrons <ε> is related with the potential barrier on the

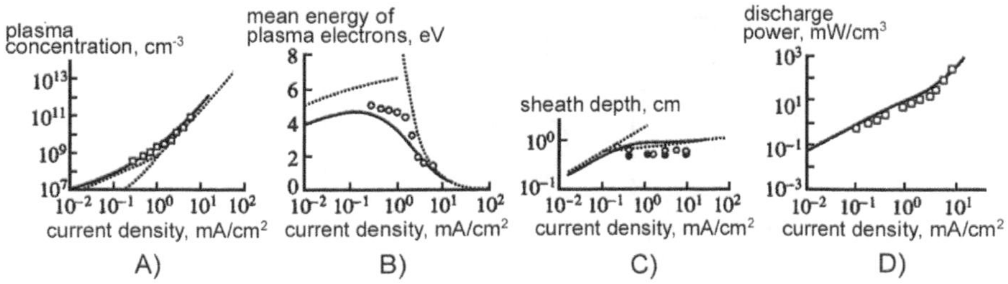

FIGURE 10.36 Plasma concentration (a), electron energy (b), sheath depth (c), and discharge power (d) as the function of current density. Solid curve – exact calculation, dotted line – asymptotic formulas, points – experiment (f = 13.56 MHz, L_0 = 6.7 cm, p = 0.03 Torr). *(From Encyclopedia of Low Temperature Plasma, Ed. V.E Fortov, Nauka, Moscow, 2002)*

plasma-sheath boundary as: $\langle \varepsilon \rangle = \frac{2}{3} e\phi$. The dependence of the average energy of plasma electrons on the discharge current density is illustrated in Figure 10.36b. In a good agreement with Eq. (10.106), the average electron energy grows slightly with j_0 at low current densities. In contrast, at high current densities exceeding 1 mA/cm^2, the average electron energy decreases inversely proportional to the current density (see Eq. (10.114)). The sheath thickness increases with current density as $\sqrt{j_0}$ at low currents (see Eq. (10.110) and Figure 10.36c) and reaches the saturation level (see Eq. (10.113)) close to 1 cm at current densities exceeding 1 mA/cm^2. Specific discharge power (per unit volume) grows with the current density as it shown in Figure 10.36d and reaches values of about 1 W/cm^3 at current densities about 10 mA/cm^2, see (V.A. Godyak, 1986, V.A. Godyak, R.B. Piejak, B.M. Alexandrovich, 1991 and V.A. Godyak, R.B. Piejak, 1990, V. Vahedi, C.K. Birdsall, M.A. Lieberman, G. DiPeso, T.D. Rognlien, 1994, D. Vender, R.W. Boswell, 1990, B.P. Wood, 1991).

10.6 ASYMMETRIC, MAGNETRON AND OTHER SPECIAL FORMS OF LOW-PRESSURE CAPACITIVE RF-DISCHARGES

10.6.1 Asymmetric Discharges

The RF-CCP discharges are usually organized in grounded metal chambers. One electrode is connected to the chamber wall and therefore also grounded. Another electrode is powered, which makes the discharge asymmetric (M.A. Lieberman, 1989). The current between an electrode and the grounded metallic wall in moderate pressure discharges usually has a reactive capacitive nature and does not play any important role. However in low-pressure discharges, the situation is different. The plasma occupies a much larger volume because of diffusion, and some fraction of the discharge current goes from the loaded electrode to the grounded walls. As a result, the current density in the sheath located near the powered (or loaded) electrode usually exceeds that in the sheath related to the grounded electrode. Lower current density in the sheath corresponds to lower values of voltage in this sheath. The constant component of the voltage, which is the plasma potential with respect to the electrode, is also lower at lower values of current density. For this reason, a constant potential difference occurs between the electrodes, if a dielectric layer covers them or if a special blocking capacitance is installed in the electric circuit (see Figure 10.37) to avoid direct current between the electrodes. This potential difference is usually referred to as the **auto-displacement voltage**. The plasma is charged positively with respect to the electrodes. The voltage drop in the sheath near the powered (loaded) electrode, exceeds the drop near the grounded electrode. As a result, the loaded electrode has a negative potential with respect to the grounded electrode. Thus the RF-loaded electrode is called sometimes a "cathode".

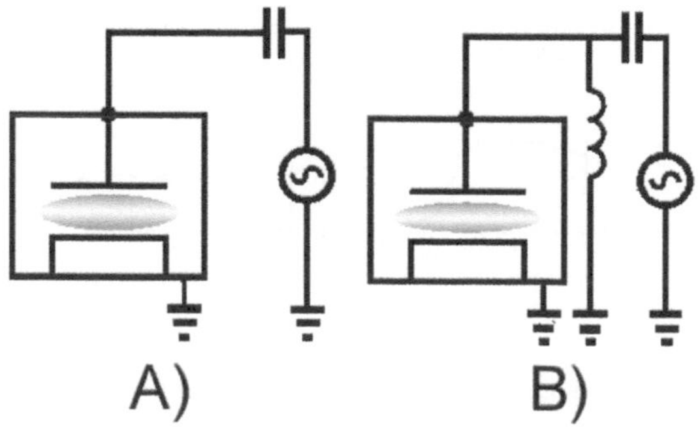

FIGURE 10.37 RF discharge with disconnected (a) and DC-connected (b) electrodes.

10.6.2 Comparison of Parameters Related to Powered and Grounded Electrodes in Asymmetric Discharges

Consider the asymmetric low-pressure CCP-RF-discharge between planar electrodes, illustrated in Figure 10.37. The ion flux to the powered electrode (cathode) can be expressed as:

$$\Gamma_i \approx n_s \sqrt{\frac{e\lambda_i}{M} \frac{j_0}{\varepsilon_0 \omega}}, \tag{10.118}$$

where n_s is the ion concentration near the loaded electrode, and j_e is the current density in this pre-electrode layer. In this case, the ion flux to the grounded electrode can be expressed by a similar formula:

$$\Gamma_i' \approx n_s' \sqrt{\frac{e\lambda_i}{M} \frac{j_0'}{\varepsilon_0 \omega}}, \tag{10.119}$$

where n'_s is the ion concentration near the grounded electrode, and j'_e is the current density in this pre-electrode layer. The fluxes (10.118) and (10.119) are formed in the plasma zone by means of diffusion. Taking into account that concentrations of charged particles in the sheaths are lower than those in the plasma zone n_s, $n'_s << n_p$, and the symmetry of the system, one can conclude that the fluxes (10.118) and (10.119) should be equal. This leads to the inverse proportionality of the ion densities in the sheaths and the square roots of corresponding current densities:

$$\frac{n_s}{n_s'} = \sqrt{\frac{j_0'}{j_0}}. \tag{10.120}$$

Eq. (10.120) shows that the charge concentration near the grounded sheath only slightly exceeds the one in the vicinity of the powered (or loaded) electrode. The electric field is determined by the space charge. For this reason, the ratio of current densities in the opposite sheaths can be expressed as:

$$\frac{j_0'}{j_0} = \frac{n_s' L'}{n_s L}. \tag{10.121}$$

Based on Eqs. (10.120) and (10.121), the relation between the thickness of the opposite sheaths and the corresponding current densities is obtained:

$$\frac{L'}{L} = \left(\frac{j_0'}{j_0}\right)^{3/2}. \tag{10.122}$$

From Eq. (10.122), thickness of the sheath located near the powered electrode is seen to be larger than the one in the vicinity of grounded electrode. Finally taking into account that voltage drop in the sheaths U is proportional to the charge density and square of the sheath thickness, one can conclude:

$$\frac{U'}{U} = \left(\frac{j_0'}{j_0}\right)^{5/2}. \tag{10.123}$$

In asymmetric discharges, the voltage drop near the grounded electrode can be neglected, and the auto-displacement voltage is close to the amplitude of the total applied voltage. The high charge density, low current density and small sheath depth at the grounded electrode result in low intensity of electron heating in this layer. This is compensated by intensive heating of electrons in the "cathode" sheath.

10.6.3 The Battery Effect, "Short Circuit" Regime of Asymmetric RF-Discharges

The constant potentials of electrodes can be made equal by connecting (from the DC standpoint) the powered metal electrode with the ground through an inductance. Such a scheme is illustrated in Figure 10.37b. Here direct current can flow between the electrodes, which means that the electron and ion fluxes to each electrode are not supposed to be equal. The average voltages on the sheaths are the same in this regime of the asymmetric CCP-RF-discharge, which results in equal thickness of the sheaths. The sheath related to the grounded electrode has a lower current density. For this reason, the amplitude of the plasma boundary displacement is shorter than the thickness of the grounded electrode sheath. The main part of the sheath stays for the entire period in the phase of space charge. The plasma does not touch the electrode and the ion current permanently flows to the grounded electrode. Here the average positive charge per period carried out from the discharge , is compensated by the electron current to the powered electrode. Thus the direct current in this regime is determined by the ion current to the grounded electrode, as well as by the electron current to the powered one. The asymmetric RF-discharge can be considered as a source of direct current. For this reason, the described phenomenon is usually referred to as the battery effect (A.F. Alexandrov, V.A. Godyak, A.A. Kuzovnikov, A.Y. Sammani, 1967).

10.6.4 The Secondary-Emission Resonant Discharge

The low-pressure RF-discharges can be sustained only by surface ionization processes, without any essential contribution of the gas in the discharge gap. Consider an electron emitted from one electrode and moving in the accelerating electric field toward another electrode, where it is able to produce a secondary electron during the secondary electron emission. If the electron "flight time" to the opposite electrode coincides with the half period of the electric field oscillations, then the secondary electron also can be accelerated by the electric field, which turned around during this "flight time". To sustain such a resonant discharge regime, the secondary electron emission coefficient should exceed unity, which can take place at energies about 0.3–1 KeV (see Figure 10.38, A.S. Smirnov, 2000). The electron flight time t_n necessary to sustain the discharge can be equal to one, or any odd number of half-periods. Thus for the n-th order of the resonance:

$$t_n = \frac{\pi}{\omega}(2n-1), \qquad (10.124)$$

where ω is the frequency of the electric field oscillations in the discharge system.

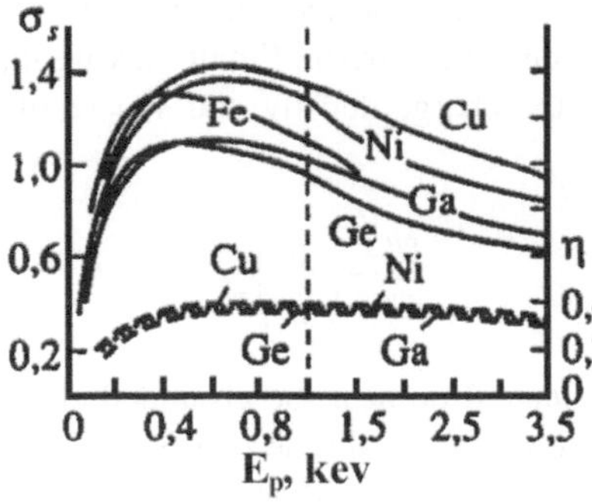

FIGURE 10.38 Secondary electron emission coefficient (σ, dashed line) and inelastic emission coefficient (η) as functions of electron energy.

10.6.5 Radio-Frequency Magnetron Discharge, General Features

The RF-capacitive discharge is widely used in electronics and other industries. However, these discharges have some important disadvantages, limiting their application. First of all, the sheath voltages in RF-CCP discharges are relatively high; this results at a given power level in low ion densities and ion fluxes, as well as in high ion-bombarding energies. Also in these discharges the ion-bombarding energies cannot be varied independently of the ion flux. The RF-magnetron discharges were developed especially to make the relevant improvements; in electronics these discharge systems are usually referred to as the **magnetically enhanced reactive ion etchers MERIE**. In RF-magnetrons, a relatively weak DC-magnetic field (about 50–200 G) is imposed on the low-pressure CCP-discharge parallel to the powered electrode and hence, perpendicular to the RF-electric field and current. The RF-magnetron discharge permits increasing the degree of ionization at lower RF-voltages and decreasing the energy of ions bombarding the powered electrode (or a sample placed on the electrode for treatment). Also this discharge permits increasing the ion flux from the plasma, which in turn leads to intensification of etching. The RF-magnetron discharges can be effectively sustained at much lower pressures (down to 10^{-4} Torr). As a result, a well-directed ion beam is actually able to penetrate the sheath without any collisions. The ratio of the ion flux to the flux of active neutral species grows, which also leads to a higher quality of etching. A general scheme of the RF-magnetron used for ion etching (H. Okano, T. Yamazaki, Y. Horiike, 1982) is shown in Figure 10.39. The RF-voltage is applied to the smaller lower electrode, where the sample under treatment is located. A rectangular samarium-cobalt constant magnet is placed under the powered electrode. As seen from Figure 10.39, the part of the electrode surface where the magnetic field is horizontal is quite limited and does not cover the entire electrode. For this reason, a special scanning device is used to move the magnet along the electrode and to provide "part time" horizontal magnetic field for the entire electrode. The RF-magnetron discharges also can be organized as cylindrical systems with coaxial electrodes (G.Y. Yeom, J.A. Thornton, M.J. Kushner, 1989a; A.D. Knypers, H.J. Hopman, 1988, 1990; I. Lin, 1985). The magnetic field in these systems is directed along the cylinder axis. The physics of the RF-magnetron is based on the following effect. Electrons, oscillating together with the sheath boundary, additionally rotate around the horizontal magnetic field lines with the cyclotron frequency Eq. (4.107). When the magnetic field and the cyclotron frequency are sufficiently high, the amplitude of the electron oscillations along the RF-electric field decreases significantly. The magnetic field "traps electrons". In this case the cyclotron frequency actually plays the same role as the frequency of electron-neutral collisions: electrons become unable to reach the amplitude of their free oscillations in the RF-electric field. The amplitude of electron oscillations determines the thickness of sheaths. Decrease of the amplitude of electron oscillations in magnetic field results in smaller sheaths and lower sheath voltage near the powered electrode. This leads to lower values of the auto-displacement, lower ion energies, and lower voltages necessary to sustain the RF-discharge.

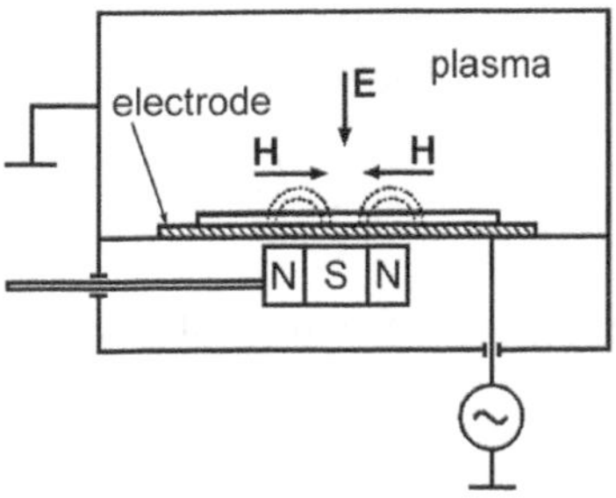

FIGURE 10.39 Schematic of an RF-magnetron.

10.6.6 Dynamics of Electrons in RF-Magnetron Discharge

To analyze the RF-magnetron, consider an electron motion in crossed electric and magnetic fields taking into account electron-neutral collisions. Assume electric field is directed along the "x"-axis perpendicular to the electrode surface $E \equiv E_x$, and the magnetic field is directed along the "z"-axis parallel to the electrode $B = B_z$. The electron motion equation including the Lorentz force and electron neutral collisions can be expressed as:

$$m\frac{d\vec{v}}{dt} = -\vec{E}_a e^{i\omega t} - e\left[\vec{v}\times\vec{B}\right] - m\nu_{en}\vec{v}. \tag{10.125}$$

In this equation: $\vec{v}$ is the electron velocity, $\vec{E}_a$ is the amplitude of oscillating electric field, ν_{en} is the frequency of electron-neutral collisions. Projections of the motion equation and to the "x" and "y" axis give the following system of equations:

$$m\frac{d}{dt}v_x = -eE_a e^{i\omega t} - ev_y B - m\nu_{en}v_x, \tag{10.126}$$

$$m\frac{d}{dt}v_y = ev_x B - m\nu_{en}v_y. \tag{10.127}$$

To solve this system with respect to velocities, it is convenient to transform variables to $v_x \pm iv_y$, multiplying Eq. (10.127) by "i", and then adding and subtracting Eq. (10.126). The forced electron oscillations in the crossed electric and magnetic fields can be described by the complex relations:

$$v_x \pm iv_y = \frac{eE_a e^{i\omega t}}{m\left(\nu_{en} + i\left(\omega \mp \omega_B\right)\right)}, \tag{10.128}$$

where ω_B is the cyclotron frequency defined by Eq. (4.107). The system (10.128) gives the expressions for the velocity of electron oscillations in the direction perpendicular to electrodes $v_x = d\xi/dt$:

$$v_x = \frac{eE_a e^{i\omega t}}{2m}\left[\frac{1}{\nu_{en} + i\left(\omega - \omega_B\right)} + \frac{1}{\nu_{en} + i\left(\omega + \omega_B\right)}\right]. \tag{10.129}$$

The coordinate $\xi(t)$ of the electron oscillations is determined as the integral of the normal velocity v_x:

$$\xi = -\frac{ieE_a e^{i\omega t}}{2m\omega}\left[\frac{1}{\nu_{en} + i\left(\omega - \omega_B\right)} + \frac{1}{\nu_{en} + i\left(\omega + \omega_B\right)}\right]. \tag{10.130}$$

At very low pressures, the frequency of electron-neutral collisions is also low $\nu_{en} << \omega$, and the electron oscillations can be considered as collisionless. The Amplitude of electron velocity in the collisionless conditions and in absence of magnetic field is:

$$u_a = \frac{eE_a}{m\omega}. \tag{10.131}$$

The electron displacement amplitude (10.130) in the collisionless, non-magnetized conditions is:

$$a = \frac{eE_a}{m\omega^2}. \tag{10.132}$$

The main objective of applying a magnetic field is decreasing the amplitudes of the electron velocity and displacement Eqs. (10.131), (10.132), which then results in smaller sheaths and lower sheath voltage near the powered electrode. This leads to lower values of the auto-displacement, lower ion energies, and lower voltages necessary to sustain the RF-discharge, which is desirable for RF-magnetrons applications. If the magnetic field is sufficiently high ($\omega_B >> \omega$), then the amplitude of electron velocity in the collisionless conditions can be given as:

$$u_{aB} = \frac{eE_a\omega}{m\omega_B^2}. \tag{10.133}$$

This is less than the amplitude in the absence of a magnetic field by the factor $(\omega_B/\omega)^2 >> 1$. The amplitude of the electron oscillation (10.130) in the magnetized collisionless conditions is:

$$a_B = \frac{eE_a}{m\omega_B^2}, \tag{10.134}$$

which is also less than the displacement under non-magnetized conditions by the factor $(\omega_B/\omega)^2 >> 1$. Ions motion in the RF-magnetrons is not magnetized: the ion cyclotron frequency is less than the oscillation frequency ($\omega_{Bi} << \omega$); and the trajectories of the heavy ions are not perturbed by magnetic field.

10.6.7 Properties of RF-Magnetron Discharges

Experimental investigations and numerical simulations prove that the RF-magnetron discharge retains all the typical properties of the low-pressure RF-CCP-discharges, obviously taking into account the peculiaritics related to the constant magnetic field (R.K. Porteous, D.B. Graves, 1991; A.A. Gurin, N.I. Chernova, 1985; A.B. Lukyanova, A.T.Rahimov, N.B. Suetin, 1990). The constant horizontal magnetic field perpendicular to the electric field decreases the electron flux and prevents electron losses on the powered electrode. The residence time of the electrons grows , promoting the ionization properties of the electrons. Space charge sheaths are also created in RF-magnetrons near electrodes because electrons leave the discharge gap. However, in the strong magnetic field, the effective electron mobility across the magnetic field may become lower than the ion mobility. The amplitude of the ion oscillations exceeds those of electrons under such conditions. Electron current to the electrode does not appreciably vary during the oscillation period. This is mostly due to diffusion, which is effective in spite of the magnetic field because electron temperature significantly exceeds the ion temperature . Stochastic heating of the magnetized electrons is also possible for some conditions of the oscillating sheath boundaries (M.A. Lieberman, A.J. Lichtenberg, S.E. Savas, 1991). The asymmetric magnetron discharge keeps all the properties typical for the non-magnetized asymmetric low-pressure RF-discharges. The auto-displacement effect can be observed in the asymmetric RF-magnetrons in the presence of a blocking capacitance in the circuit. The battery effect, when the external circuit is able to "generate" direct current can also be observed in the magnetron discharges. Sheath thicknesses in RF-magnetrons are much smaller than in non-magnetized discharges. For this reason, the auto-displacement in the asymmetric magnetrons is also smaller. Energy spectrum of ions bombarding the electrode in magnetrons is similar to those in non-magnetized discharges. However, the ion energies depend on the value of the applied magnetic field because the magnetic field determines the sheath voltage. More details regarding ion spectra in RF-magnetron discharges can be found, in A.D. Knypers, E.H.A. Granneman, H.J. Hopman (1988) and A.V. Lukyanova, A.T. Rahimov, N.V. Suetin (1991).

10.6.8 Low-Frequency RF-CCP-Discharge, General Features

The lower limit of the RF-frequency range is generally related to the characteristic frequency of ionization and ion transfer. Ion density in RF-discharge plasmas and ion sheaths usually, but not always, can be considered as constant during a period of electromagnetic field oscillation. Thus typical RF-frequencies exceed 1 MHz to avoid effects of " long distance" ion motion. Here in the end of the Section 10.6, we are going to consider, however, an example and peculiarities of the low frequency (less than 100 kHz) RF-CCP-discharges. The low-frequency capacitively coupled nitrogen plasma, for example, is particularly effective in polymer surface treatment with the objective to promote adhesion of silver to polyethylene terephtalate (R.G. Spahn, L.J. Gerenser, 1994; L.J. Gerenser, J.M. Grace, G. Apai, P.M. Thompson, 2000). Also the low-frequency capacitively coupled nitrogen plasma was effectively applied for the surface treatment of polyester web to promote adhesion of gelatin containing layers related to production of photographic film (J.M. Grace e.a., 1995). The low-frequency ($\leq$100 kHz) discharges are effectively used as well for sputter deposition of metals, where the application of lower RF-frequencies result in a higher fraction of input power dissipated in the sputter target (G. Este, W.D. Westwood, 1984, 1988; J.W. Butterbaugh, L.D. Baston, H.H. Sawin, 1990; S. Schiller, G. Beister, E. Buedke, H.J. Becker, H. Schmidt, 1982; M.I. Ridge, R.P. Howson, 1982; J. Affinito, R.R. Parson, 1984). The block diagram of the low frequency RF-discharge system is shown in Figure 10.40 (S. Conti, P.I. Porshnev, A. Fridman, L.A. Kennedy, J.M. Grace, K.D. Sieber, D.R. Freeman, K.S. Robinson, 2001). As seen from this figure, the discharge is organized in the co-planar configuration, where the low frequency RF-voltage is applied across two electrodes located in the same plane. Each electrode alternately serves as cathode and anode every half cycle. In this case, anode conductivity is maintained, charging OF insulating areas formed on the cathode is minimized, and arcing is avoided. The discharge chamber in the experiments of S. Conti, P.I. Porshnev, A. Fridman, L.A. Kennedy, J.M. Grace, K.D. Sieber, D.R. Freeman, K.S. Robinson (2001), which is illustrated in Figure 10.40, consists of two coplanar water-cooled aluminum electrodes (35.6 cm long, 7.6 cm wide) positioned side by side, separated by a gap of 0.32 cm and housed in the grounded shield. The interior of the grounded enclosure has a volume of roughly 36 cm × 16 cm × 3.3 cm above the electrode pair. Typical pressure in the experiments was quite low: 50 ÷ 150 m Torr, gas flow rate 180 ÷ 800 sccm. Applied peak voltage is 900 – 1600 V, power about 300 W, frequency 40 kHz. Typical current and voltage waveforms are shown in Figure 10.41.

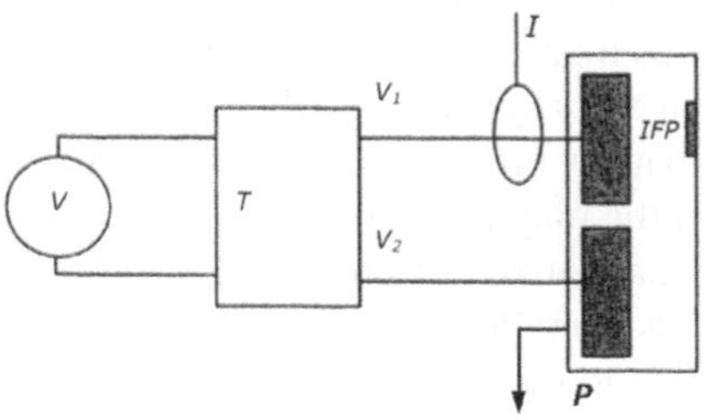

FIGURE 10.40 Block diagram of plasma source, supply electronics, and voltage and current measurement electronics. Power supply and transformer are respectively denoted V and V_1, V_2. The position of the ion flux probe is indicated by *IFP*.

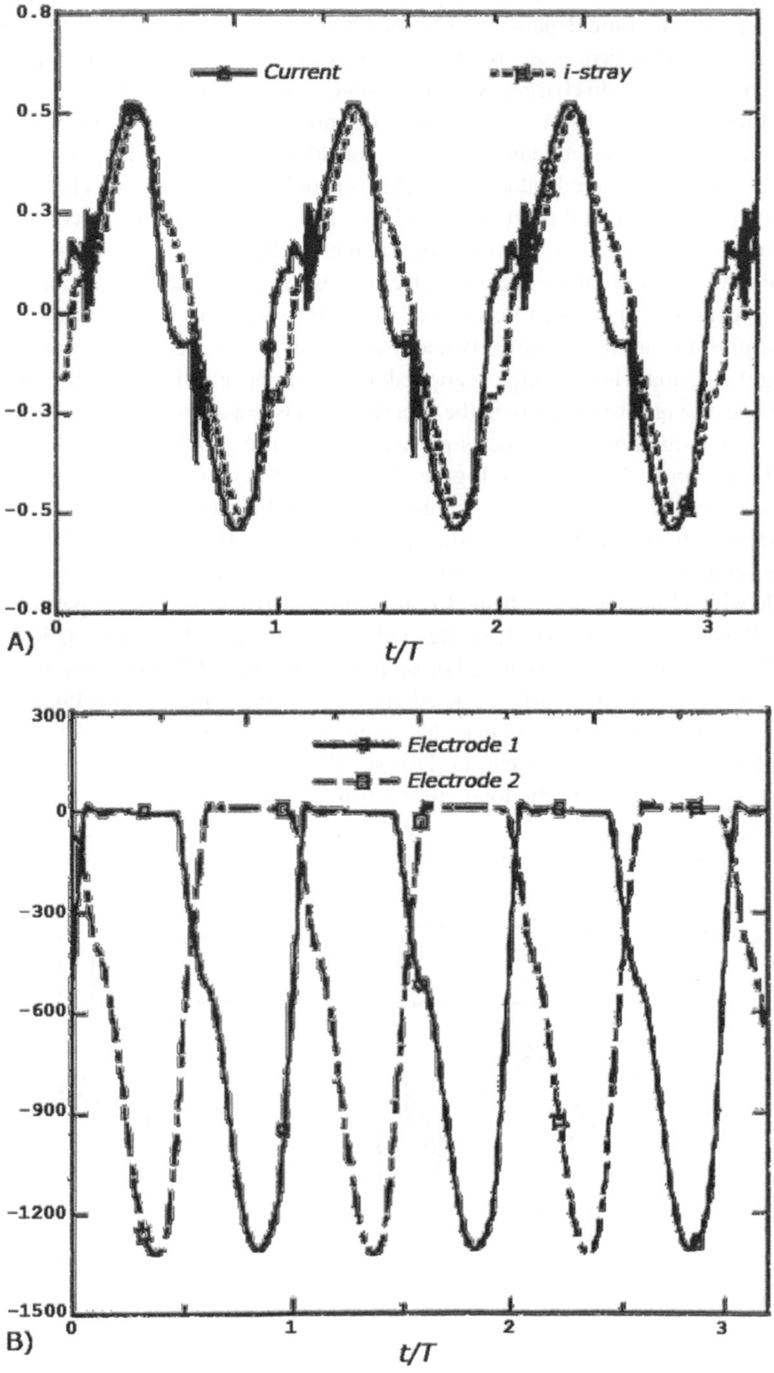

FIGURE 10.41 (a) Typical voltage waveforms for V_1 and V_2. (b) Typical current waveform for I. i-stray denotes the current corrected for stray capacitance to ground. Plasma conditions: 150 mTorr nitrogen, 330 W.

10.6.9 Physical Characteristics and Parameters of Low-Frequency RF-CCP-Discharges

In the low-frequency discharge, the ion frequency is comparable to the driving frequency and all plasma characteristics are time dependent. The sheath voltages are high, and the discharge is essentially sustained not by bulk processes, but by secondary electron emission from the "cathode", provided by ion impact. The discharge system is somewhat similar to DC glow discharge, with alternating cathode positions from one electrode to another each half period of the electric field oscillations. The co-planar system can be actually considered as two discharges operating in counter-phase modes. Spatial averaged values of the electron and ion concentrations in the low-frequency RF-CCP-discharge are presented in Figure 10.42 as a function of time and in comparison with the time-evolution of voltage. As seen from this figure, the discharge is actually active only for half of the cycle, when the potential on the specified electrode is negative. During this portion of the cycle, the electron concentration increases because of ionization, while in the other half cycle it slowly decreases. The plasma does not completely decay after every cycle, so a new breakdown is not necessary. During the period when voltage is applied, the ion concentration is higher than the electron concentration because quasi-neutrality in the sheath region is not achieved. The corresponding average electron energy variation with time is presented in Figure 10.43 also in comparison with the time-evolution of voltage. As seen from Figures 10.42 and 10.43, both the average electron concentration and energy increase and decrease simultaneously with voltage variation. This can be attributed to intense electron avalanching in the sheath regions, where the electric field is sufficiently high to provide the intense ionization. The presented results were obtained by using the 2D particle-in cell (PIC) code with a Monte Carlo scheme for modeling collisions of charged particles and neutrals (S. Conti, P.I. Porshnev, A. Fridman, L.A. Kennedy, J.M. Grace, K.D. Sieber, D.R. Freeman, K.S. Robinson, 2001). As was already discussed in Section 10.6.5, the PIC-simulation permits following a large number of representative particles acted upon by the basic forces. The PIC code applied for describing the low frequency RF coplanar discharge is a modified version of the PDP2-code (V. Vahedi, C.K. Birdsall, M.A. Lieberman, G. DiPeso, T.D. Rognlien, 1993). The simulation of the external circuit allows evaluating the total current (displacement current plus conduction current), while the conduction current is directly obtained from the movement of charged particles. The relative contribution of the two current components is illustrated in Figure 10.44, where total current as

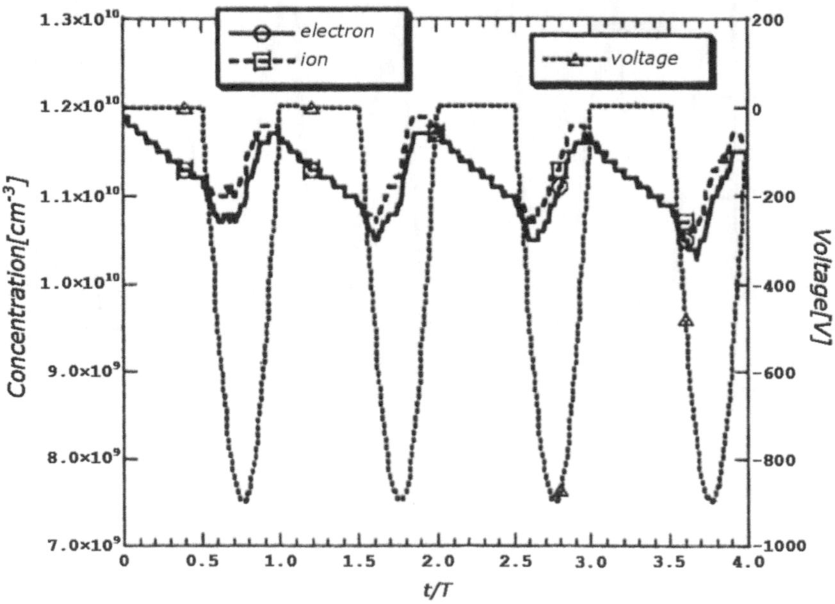

FIGURE 10.42 Steady state electron and ion concentrations as function of time for 900 V and 0.15 Torr.

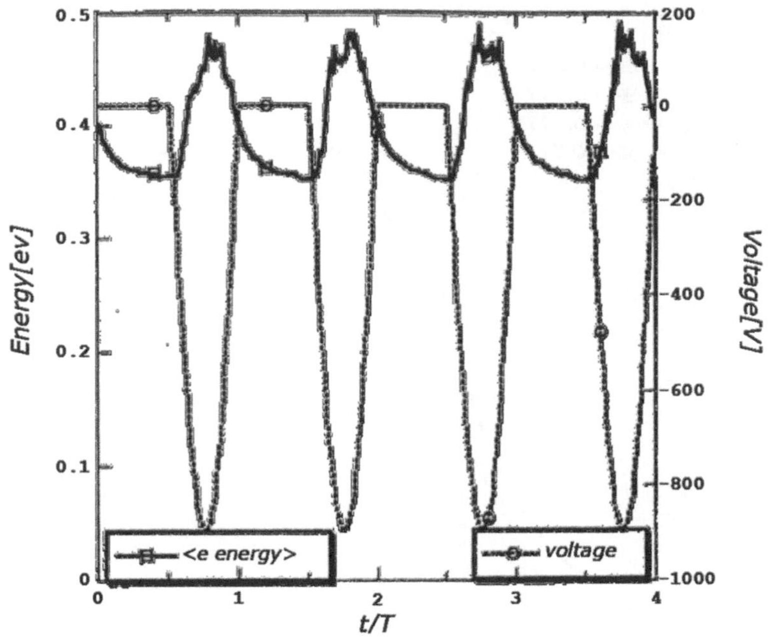

FIGURE 10.43 Average electron energy at steady states as a function of time for 900 V and 0.15 Torr.

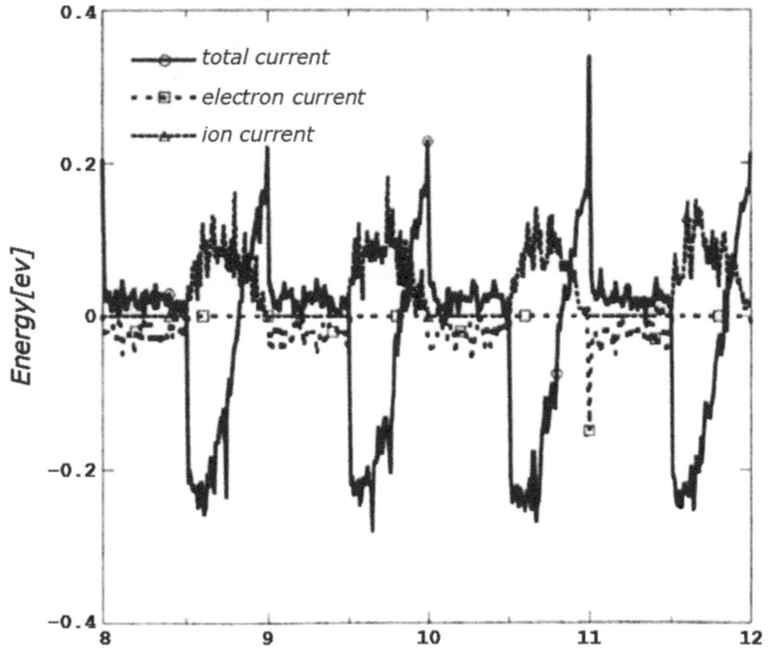

FIGURE 10.44 Total current and conduction current as a function of time for 1300 V and 0.15 Torr. The electron and ion current are shown separately for 1300 V and 0.15 Torr.

well as electron and ion conduction currents are plotted as a function of time. Almost 50% of the calculated peak current in the circuit is due to the ion conduction current, which is typical for the low frequency RF-discharges. Note that a positive current peak can be observed when the voltage increases from the negative peak value back to zero. This peak is related to a reactive current provided by parasitic capacitance. Spatial distribution of charged particles, electrons and ions, between

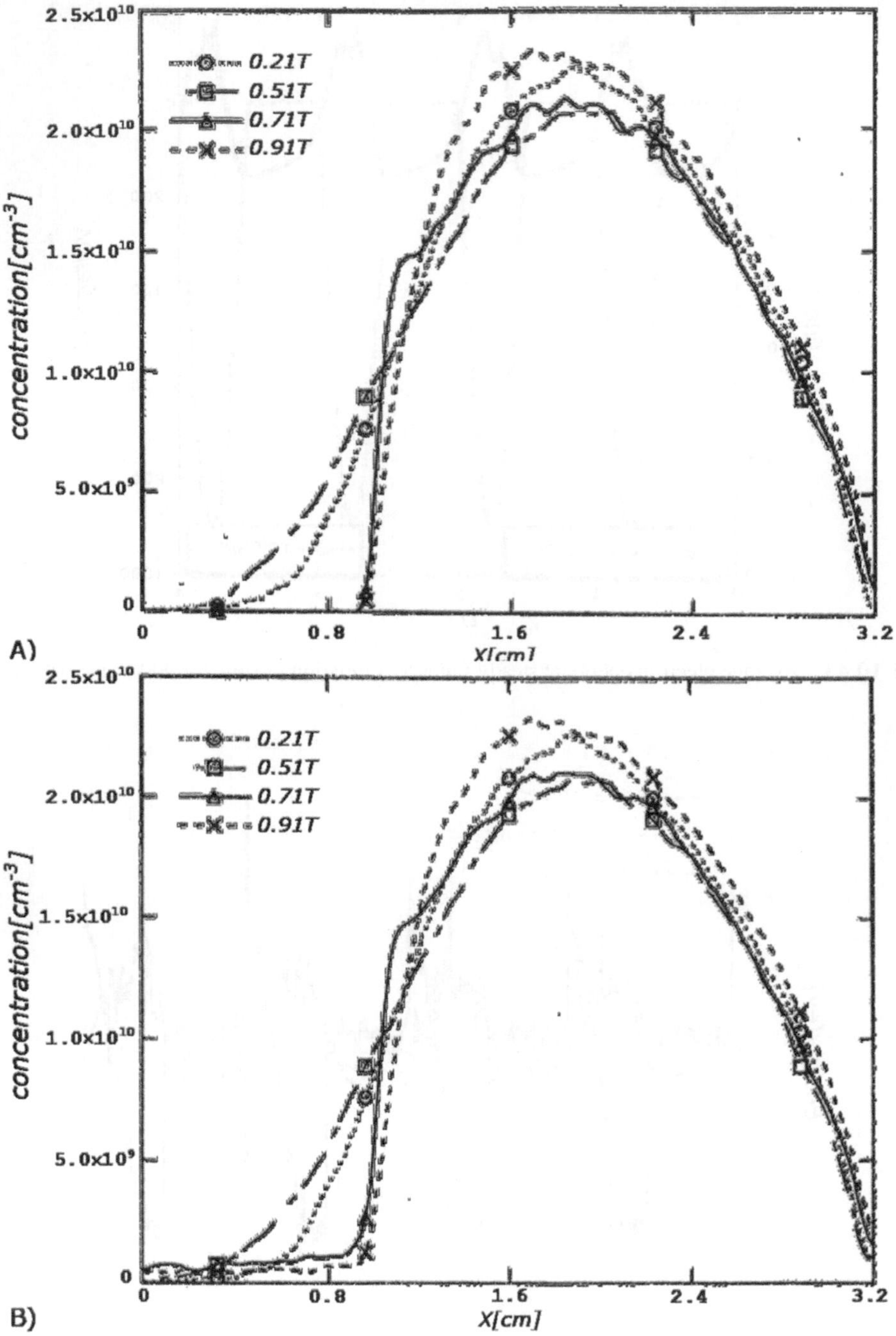

FIGURE 10.45 Electron (a) and ion (b) concentrations at different moments of the period for 900 V and 0.15 Torr.

the powered electrode ($x = 0$) and the opposite grounded wall is shown in Figure 10.45 at different moments during an oscillation period. Because the maximum voltage of the powered electrode is near the ground potential, the plasma potential is not driven sufficiently high to have the grounded wall opposite to electrode serve as a cathode. Furthermore, the high value of the driving voltage produces considerable sheath expansion by forcing the electrons away from the cathode. Consequently, the spatial distribution of charged particles is essentially asymmetric with respect to

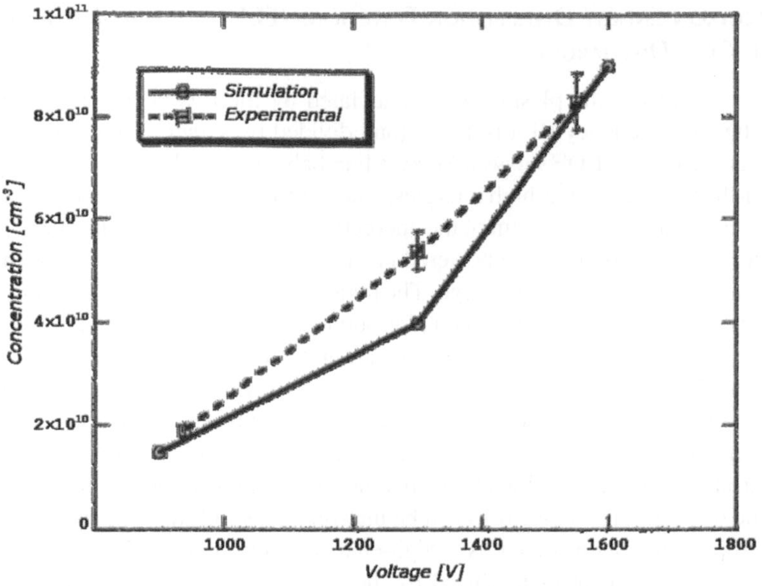

FIGURE 10.46 Comparison of ion and electron concentrations for different voltages and pressure of 0.15 Torr.

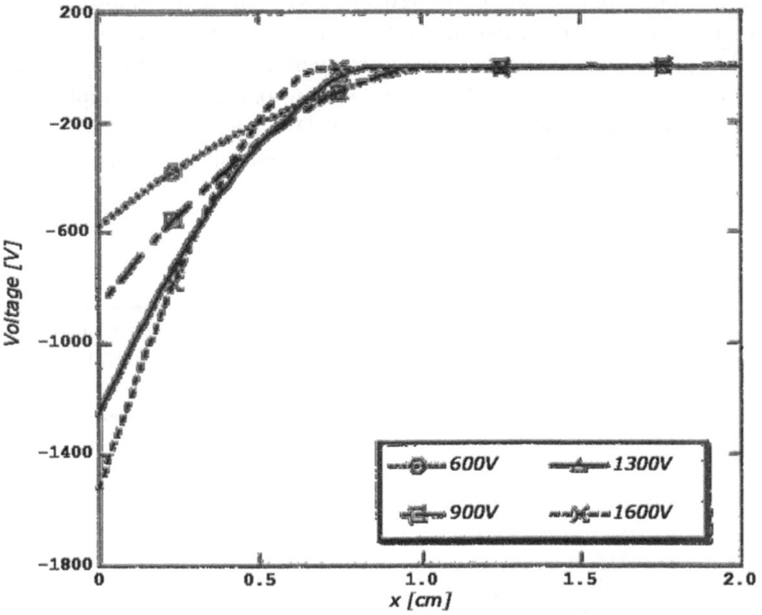

FIGURE 10.47 Voltage profiles at the peak voltage (0.75 t/T) for different applied voltages at 0.15 Torr.

center of the axis between the powered electrode and opposite grounded wall (see Figure 10.45). Concentrations of electrons and ions in plasma are shown in Figure 10.46 as a function of applied voltage. Potential profile between the powered electrode ($x = 0$) and the opposite grounded wall at the moment of the peak voltage is shown in Figure 10.47. The resulting sheath decreases with increasing potential because the plasma density is higher (and hence the screening effect of plasma is stronger) at higher values of voltage. This is consistent with general theory of the γ-discharges (V.A. Godyak, A.S. Khanneh, 1986). The shape of the sheath voltage is close to parabolic because ion concentration in the sheath is approximately constant (see Figure 10.45).

10.6.10 Electron Energy Distribution Functions (EEDF) in Low-Frequency RF-CCP-Discharges

The EEDF for electrons in the plasma zone, calculated by the PIC-code for the above-described conditions of the low frequency RF-discharge (and divided by square root of energy), is shown in Figure 10.48. As seen , the EEDF shows a Maxwellian behavior at relatively low energies, but it has a long and non-thermalized tail at high energies. The "cut-energy" is well correlated with the specific values of peak voltages. The origin of the energetic electrons in the low frequency RF-discharge is related to the sheath region near the powered electrode. In general, three groups of electrons can be clearly seen in the EEDF of this discharge. The major one is in the plasma bulk and it is characterized by a low energy of about 1 eV. The second group is formed by the scattered γ-electrons and the electrons formed in the cathode sheath by ionization collisions of the γ-electrons and neutrals. Energies of these electrons are in range of 20–50 eV. Finally, the third group of electrons is formed by those γ-electrons that managed to cross the sheath and the plasma bulk and reach the grounded wall retaining the high energy gained in the sheath region. The energy distribution of these highly energetic electrons is quite flat and a single specific temperature value cannot be assigned to this third group. The three groups of electrons can be interpreted as follows. The third group of the most energetic electrons can be considered as a high-energy electron beam. The second group can be considered as secondary electrons produced by the beam (such secondary electrons usually have energies about 2–3 ionization potentials). Finally, the first group consists of numerous "tertiary" low energy plasma electrons. Thus such a discharge is an effective source of electrons with energies of about 30 eV. This is especially useful, for N^+ atomic ions generation in molecular nitrogen plasma, which is of interest in the application of low-frequency RF-discharges in nitrogen for low-pressure treatment of polymer surfaces. The existence of these three groups of electrons was experimentally proven by ion flux probe measurements in a low-frequency RF-discharge specifically applied for polymer surface treatment to promote adhesion of silver to polyethylene terephtalate (S. Conti, P.I. Porshnev, A. Fridman, L.A. Kennedy, J.M. Grace, K.D. Sieber, D.R. Freeman, K.S. Robinson, 2001). These diagnostic measurements were based on the deposition tolerant flux probe technique developed by N. Braithwaite, J.P. Booth, G. Cunge, 1996.

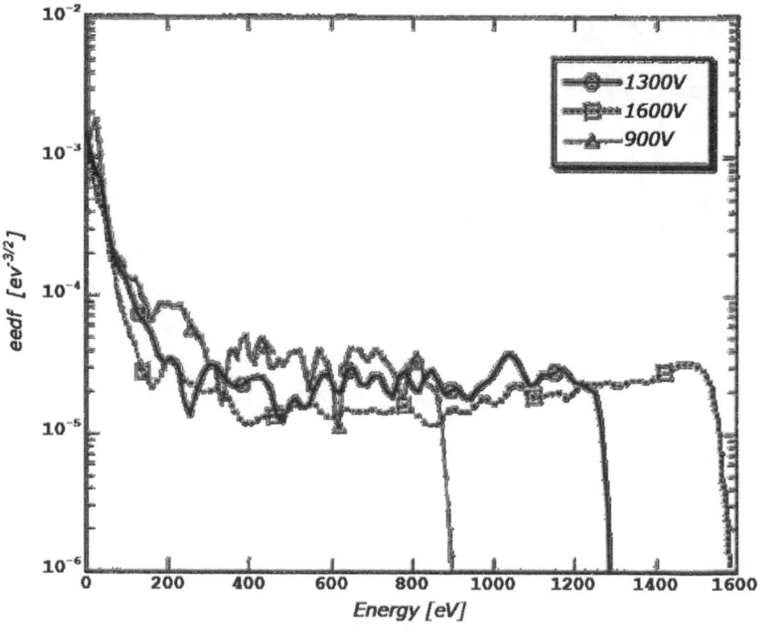

FIGURE 10.48 Electron energy distribution function of gamma electrons that hit the web plane. Bulk electrons are not included (1600 V, 1300 V, 900 V, 0.15 Torr).

10.7 NON-THERMAL INDUCTIVELY-COUPLED (ICP) DISCHARGES

10.7.1 General Features of Non-Thermal ICP-Discharges

The electromagnetic field in ICP-discharges is induced by an inductive coil (Figure 10.21c, d), where the magnetic field is primarily stimulated and the corresponding non-conservative electric fields necessary for ionization is relatively low. Thus non-thermal ICP plasma discharges are usually configured at low pressures to provide the reduced electric field E/p sufficient for ionization. The coupling between the inductive coil and the plasma can be interpreted as a transformer, where the coil presents the primary windings and plasma represents the secondary ones. The coil as the primary windings consists of a lot of turns, while plasma has the only one. As a result, the ICP-discharge can be considered as a voltage-decreasing and current increasing transformer. The effective coupling with the RF-power supply requires a low plasma resistance. Thus the ICP-discharges are convenient to reach high currents, high electric conductivity, and high electron density at relatively low values of electric field and voltage. For example, the low-pressure ICP-discharges effectively operate at electron densities 10^{11}–10^{12} cm^{-3} (not more than 10^{13} cm^{-3}), which exceeds by an order of magnitude typical values of electron concentration in the capacitively coupled RF-discharges. Because of the relatively high values of the electron concentration in ICP-discharge systems, they are sometimes referred to as the **high-density plasma (HDP)**. Large-scale application of low-pressure ICP-discharges in electronics was stimulated by important disadvantages of the RF-capacitive discharges. The sheath voltages in the RF-CCP discharges are relatively high, which at a given power level results in low ion densities and ion fluxes, as well as in high ion-bombarding energies; also the ion-bombarding energies cannot be varied in the low-pressure capacitively coupled RF-discharges independently of the ion flux. Another important advantage of the ICP-discharges for high precision surface treatment (in addition to high plasma density and low pressure) is that RF-power is coupled to the plasma across a dielectric window or wall, rather than by direct connection to an electrode in the plasma, as occurs in the CCP-discharges. Such "non-capacitive" power transfer to the plasma provides an opportunity to operate at low voltages across all sheaths at the electrode and wall surfaces. The DC plasma potential and energies of ions accelerated in the sheaths is typically 20–40 V, which is very good for the numerous surface treatment applications. In this case, the ion energies can be independently controlled by an additional capacitively coupled RF-source called the RF-bias, driving the electrode on which the substrate for material treatment is placed (see Figures 10.21c, d). ICP-discharges are able to provide independent control of the ion and radical fluxes by means of the main ICP source power, and the ion-bombarding energies by means of power of the bias electrode. Earlier works on ICP-discharge in the cylindrical coil geometry with pressures exceeding 20 mTorr are reviewed by H.U. Eckert (1986), see also M.A. Lieberman and R.A. Gottscho (1994).

10.7.2 Inductively Coupled RF-Discharge in Cylindrical Coil

Consider the inductive RF-discharge in a long cylindrical tube placed inside of the cylindrical coil. The physical properties of such discharge are similar to those of the planar one, but they are easier for analysis because of the circular symmetry of this discharge system. The electric field $E(r)$ induced in the discharge tube, which is located inside of the long coil can be calculated from the Maxwell equation:

$$E(r) = -\frac{1}{2\pi r}\frac{d\Phi}{dt}. \tag{10.135}$$

In this equation Φ is the magnetic flux crossing the loop of radius r perpendicular to the axis of the discharge tube; r is the distance from the discharge tube axis. The magnetic field is created in this system by the electric current in the coil $I = I_c e^{i\omega t}$ as well as by the electric current in the plasma (which mostly flows in the external plasma layers). Assuming a plasma conductivity σ (r) = constant, and considering the current in plasma as a harmonic one $j(r) = j_0(r)e^{i\omega t}$, the following equation for the current density distribution along the plasma radius is obtained:

$$\frac{\partial^2 j_0}{\partial r^2} + \frac{1}{r}\frac{\partial j_0}{\partial r} - \frac{1}{r^2} j_0 = i\frac{\sigma\omega}{\varepsilon_0 c^2} j_0. \tag{10.136}$$

If the pressure is low and plasma is collisionless, then its conductivity is inductive:

$$\sigma = -i\varepsilon_0 \frac{\omega_p^2}{\omega}, \tag{10.137}$$

where ω_p is the plasma frequency. Substituting this plasma conductivity in Eq. (10.136), yields the current density distribution in the ICP-plasma (finite at $r = 0$) in the form of the modified Bessel function:

$$j_0(r) = j_b I_1\left(\frac{r}{\delta}\right), \tag{10.138}$$

where $I_1(x)$ is the modified Bessel function, δ is the skin-layer thickness (6.1.25). Note that if collisions are important, the Bessel function must have a complex argument. The phase of the current in the plasma also depends on the radius. Current density on the plasma boundary in the vicinity of the discharge tube is determined by the non-perturbed electric field by the plasma conductivity on the boundary of the column:

$$j_b = \sigma \frac{\omega a N}{2\varepsilon_0 c^2 l} I_c. \tag{10.139}$$

In this relation, a is the radius of the discharge tube; N is the number of turns in coil; l is the length of the coil; c is the speed of light; I_c is the amplitude of current in coil. If the skin-layer thickness exceeds the radius of the discharge tube, currents in the plasma do not perturb the electric field. Here the electric field grows linearly with the radius from the axis to the walls of the discharge tube. Such a regime is possible only if the plasma conductivity only slightly depends on the electric field. This requires quite an exotic discharge situation. For example, the application of inert gas with easy-to-ionize additives, which are completely ionized. Under more realistic conditions, plasma conductivity grows with the electric field and the skin-layer thickness is smaller than discharge radius. Then most of electric current is located in the relatively thin δ-layer on the discharge periphery. In this case, the Bessel function for the current density distribution can be simplified to a simple exponential function:

$$j_0(r) \approx j_b \exp\left(\frac{r-a}{\delta}\right). \tag{10.140}$$

10.7.3 Equivalent Scheme of Inductively Coupled RF-Discharge

In general, an ionization balance of charged particles determines electric fields in non-thermal discharges. For this reason, some special coupling or feedback effect is always assumed to establish the electric field on the level necessary to provide the relevant ionization balance in the steady-state discharges. Such coupling in the CCP-discharges is due to shielding (or screening) of the electric field in the sheaths. For inductively coupled discharges, the coupling is provided by the external electric circuit. The system inductor-plasma can be interpreted as a transformer decreasing voltage and increasing the electric current. Thus the discharge stabilization takes place at high currents and, hence, high values of plasma conductivity. In this case, the Bessel function Eq. (10.138) can be simplified to an exponential function Eq. (10.140) and the total current can be considered as concentrated in the skin layer. Taking into account Eq. (10.139), the total discharge current then can be related to the current in the coil I_c as:

$$I_d = \sigma E_b \delta l = \sigma \delta \frac{a\omega N}{2\varepsilon_0 c^2} I_c, \tag{10.141}$$

where E_b is the electric field in the thin skin layer on the plasma boundary. Based on this relation for the total discharge current, the complex inductively coupled plasma impedance can be expressed as:

$$Z_p = \frac{U}{I_d} = \frac{2\pi a}{\sigma \delta l} = \frac{2\pi a m}{n_e e^2 \delta l} \nu_{en} + \frac{2\pi a m}{n_e e^2 \delta l} i\omega. \tag{10.142}$$

In this relation, n_e is the electron concentration, ν_{en} is the frequency of electron-neutral collisions. From this equation, the plasma impedance is seen to have both active (first term) and inductive (second term) components. Analyzing the equivalent circuit, one has to take into account the "geometrical" inductance, which can be attributed to the plasma considered as a conducting cylinder:

$$L_d = \frac{\mu_0 \pi a^2}{l}. \tag{10.143}$$

If the skin layer is small with respect to the discharge radius, the equivalent scheme of the ICP-discharge can be represented as a transformer with a load in form of impedance Eq. (10.142), see Figure 10.49. The transformer can be characterized by plasma inductance Eq. (10.143) and inductance of the coil:

$$L_c = \frac{\mu_0 \pi R^2}{l} N^2, \tag{10.144}$$

where R is radius of the coil, l is the coil length, and N is the number of its windings. Also, the transformer presented in Figure 10.49 should be characterized by mutual inductance, which can be expressed as:

$$M = \frac{\mu_0 \pi a^2}{l} N. \tag{10.145}$$

Equations for the equivalent scheme of the radio-frequency ICP-discharge (Figure 10.49) include two: one describes the amplitude of the voltage applied to the inductor (the coil):

$$U_c = i\omega L_c I_c + i\omega M I_d, \tag{10.146}$$

and the other describes the amplitude of the voltage on the plasma loop:

$$U_p = -I_d Z_p = i\omega M I_c + i\omega L_d I_d. \tag{10.147}$$

In these equations, I_c is the amplitude of the electric current in coil, and I_d is the amplitude of electric current in plasma. The system (10.146) and (10.147) describing the equivalent scheme of the ICP-discharge is especially helpful in determining plasma parameters for this type of RF-discharges.

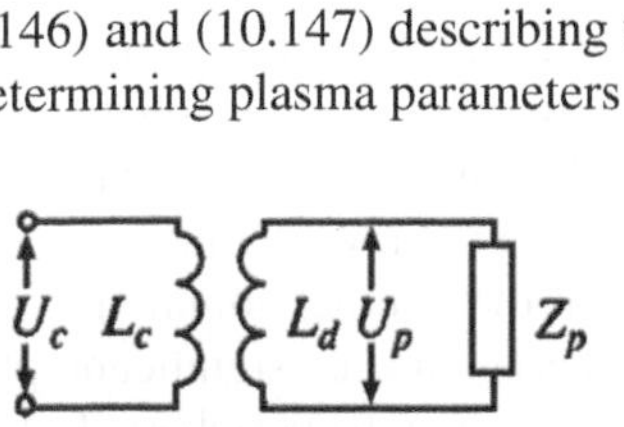

FIGURE 10.49 Equivalent scheme of an ICP discharge.

10.7.4 Analytical Relations for ICP-Discharge Parameters

The current in coil I_c is determined by external circuit and can be considered as a given parameter. The electric field in the plasma is related to the voltage on the plasma loop as: $E_p = U_p/2\pi a$, and its value only logarithmically (weakly) depends on other plasma parameters. The electric field in plasma at high currents is small relative to the idle regime: $E = \frac{\omega M I_c}{2\pi a}$, where M is the mutual inductance Eq. (10.145). The voltage drop U_p related to the plasma can be neglected for high currents. The discharge current can be expressed as:

$$I_d = -\frac{M}{L_d} I_c = -N I_c. \tag{10.148}$$

The current flows in a direction opposite to the inductor current, and the plasma current exceeds that in the inductor. If the electric field is known, the electron concentration in plasma can be found based on the current density from Eq. (10.148):

$$n_e = j\frac{m\nu_{en}}{e^2 E_p} = \frac{N I_c}{l\delta}\frac{m\nu_{en}}{e^2 E_p}. \tag{10.149}$$

Using Eq. (6.25) and Ohm's law in differential form for the current conducting plasma layer on the discharge periphery, one can derive the formula for the thickness of the skin layer in the ICP-discharge:

$$\delta = \frac{2E_p l}{\omega\mu_0 N I_c} \propto \frac{1}{I_c}. \tag{10.150}$$

From this relation, the thickness of the skin layer, where most of the ICP-current is concentrated, is inversely proportional to the electric current in the inductor coil. Using this skin-layer relation, for the electron concentration in the plasma Eq. (10.149) can be rewritten as:

$$n_e = \left(\frac{N I_c}{e l E_p}\right)^2 \frac{\omega\mu_0 m\nu_{en}}{2} \propto I_c^2. \tag{10.151}$$

Plasma density in the ICP-discharges is proportional to the square of electric current in the inductor coil.

10.7.5 Moderate Pressure and Low-Pressure Regimes of ICP-Discharges

Similar to the case of CCP-discharges, regimes of the inductively coupled discharges are different at moderate and low pressures. In particular, the electric field is determined differently in these two regimes.

1. ***Moderate Pressure Regime.*** The energy relaxation length in this regime is less than the thickness of the skin layer. Heating of electrons is determined by local electric field and takes place in the skin layer. Therefore, ionization processes as well as plasma luminosity also are concentrated at moderate pressures in the skin layer, which is illustrated in Figure 10.50. The internal volume of the discharge tube, located closer to the tube axis relative to the skin layer, is filled with the plasma only due to the radial inward plasma diffusion from the discharge periphery. If losses of charged particles in the internal volume resulting from recombination and diffusion along the axis of the discharge tube are significant, plasma concentration can be lower in the central part of the discharge than in the periphery. Balancing the ionization rate in the skin

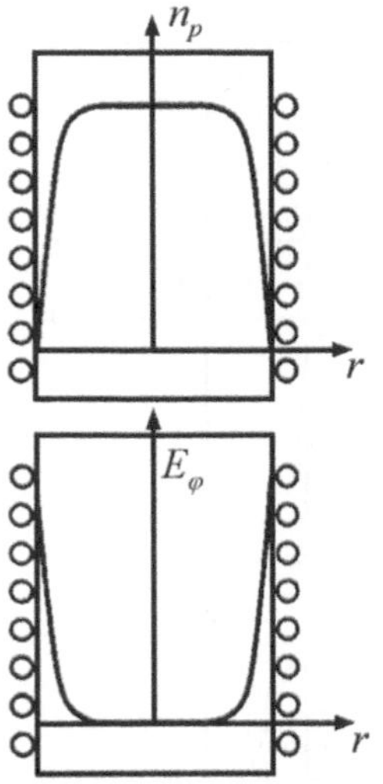

FIGURE 10.50 Radial distributions of plasma density and electric field in a moderate pressure ICP discharge.

layer of the moderate pressure discharge and the diffusion flux from the layer to the discharge tube surface, the following logarithmic relation for electric field in the skin-layer is obtained:

$$E_p = Bp\sqrt{2}\cdot \ln^{-1}\frac{Apel^2E_p^3}{\mu_0\omega^2 m\nu_{en}D_a N^2 I_c^2}. \quad (10.152)$$

In this relation, D_a is the coefficient of ambipolar diffusion; A and B are factors, which determine the α-coefficient of Townsend Eq. (4.4.5).

2. ***Low-Pressure Regime.*** The energy relaxation length in this regime exceeds the thickness of the skin layer. Hence although heating of electrons takes place in the skin layer, ionization processes are effective in the plasma volume, where the electrons have a maximum value of kinetic energy. Distributions of plasma density, amplitude of the oscillating electric field, and ambipolar potential along the radius of the low-pressure discharge tube are illustrated in Figure 10.51. In the low-pressure regime, electron concentration on the discharge axis can significantly exceed that in the skin layer. Electrons spend only part of their lifetime in the skin layer. For this reason, similar to the case of CCP-discharges, an average or effective value of the electric field can be used. If the electron energy relaxation length exceeds the radius of the discharge tube, the effective average value of the electric field can be estimated as:

$$E_{eff}^2 = \frac{2\delta}{a}E_p^2\frac{\nu_{en}^2}{\omega^2+\nu_{en}^2}. \quad (10.153)$$

In this relation: δ is the thickness of the skin layer; a is the radius of the discharge tube; and E_p is the electric field in the plasma (in the skin layer, see Figure 10.51). The value of this electric field also can be determined from the balance of charged particles in the low-pressure ICP-discharge (A.S. Smirnov, 2000):

$$E_p = \left(\frac{Bp}{\nu_{en}}\right)^{2/3}\sqrt[3]{\frac{\mu_0\omega N I_c a}{l}\left(\omega^2+\nu_{en}^2\right)}$$
$$\times \ln^{-2/3}\left[\frac{pea^3}{5.8\cdot m\nu_{en}^2 D_a}\sqrt{\frac{\mu_0\omega N I_c a}{l}E_p\left(\omega^2+\nu_{en}^2\right)}\right]. \quad (10.154)$$

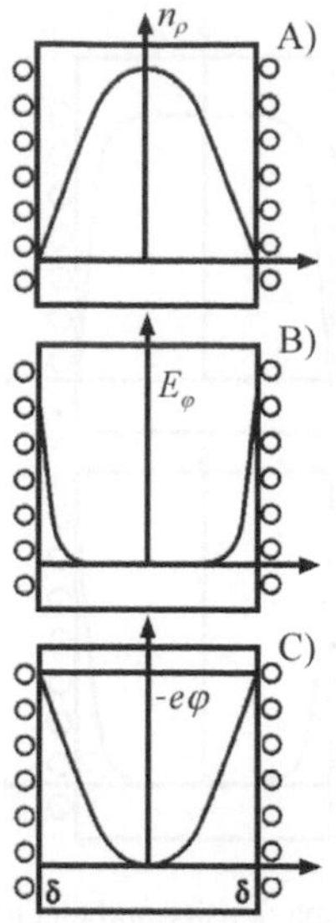

FIGURE 10.51 Radial distributions of plasma density, electric field and potential.

10.7.6 Abnormal Skin-Effect and Stochastic Heating of Electrons

Electrons in the low-pressure discharges move through the skin layer faster than changes in the oscillation period:

$$\frac{v_e}{\delta} > \omega, v_{en}. \tag{10.154}$$

Here v_e is the average thermal velocity of electrons, and δ is the thickness of the skin layer determined by Eq. (6.25). Each electron receives momentum and kinetic energy while moving across the skin layer, and transports it to the plasma zone outside of the skin layer. This results in forming the electric current outside of the skin layer. At the same time, chaotic electrons from the central part of the discharge come to the skin layer without any organized drift velocity, which leads to a decrease of current density and effective plasma conductivity. Thus the effective skin-layer thickness in the low-pressure discharges under consideration exceeds the value determined by Eq. (6.25). This phenomenon is known as abnormal skin-effect. Again, plasma electrons moving across the skin layer receive not only momentum but also kinetic energy even without any collisions with neutrals. This results in the effect of stochastic heating of electrons, somewhat similar to that considered in the low-pressure CCP-discharge. To analyze the thickness δ_c of the abnormal skin layer, estimate at first the drift velocity u, which the plasma electrons receive during the time interval δ_c/v_e of their flight across the skin layer:

$$u = \frac{eE}{m}\frac{\delta_c}{v_e}. \tag{10.155}$$

Based on this relation, the effective value of conductivity is determined as:

$$\sigma_{eff} = \frac{n_e e u}{E} = \frac{n_e e^2}{m}\frac{\delta_c}{v_e} = \frac{n_e e^2}{m v_{eff}}. \tag{10.156}$$

Using this expression for the electric conductivity in the general relation (6.1.25) for the skin layer, one obtains the following equation for the abnormal skin-layer thickness δ_c:

$$\delta_c = \frac{c}{\omega_p}\sqrt{\frac{2\nu_{eff}}{\omega}}. \tag{10.157}$$

ω_p is plasma frequency, Eq. (6.17), c is the speed of light, and ν_{eff} is effective frequency:

$$\nu_{eff} = \frac{v_e}{\delta_c}, \tag{10.158}$$

which replaces in this collisionless case the frequency of electron-neutral collisions. Eq. (10.157) leads to the equation for the thickness of the abnormal skin layer in a collisionless sheath of the ICP-discharge:

$$\delta_c = \sqrt[3]{\frac{2c^2 v_e}{\omega\omega_p^2}}. \tag{10.159}$$

Stochastic heating of electrons in the low-pressure collisionless regime can be described by using the general collisional formulas and replacing the electron-neutral collision frequency by the effective frequency, and hence, the collisional electric conductivity also by the effective one Eq. (10.156). Thus in the low-pressure collisionless regime, the electron-neutral collisions are effectively replaced by "collisions" with the skin layer boundaries.

10.7.7 Planar Coil Configuration of ICP-Discharges

An ICP-discharges in the planar configuration is widely used in electronics and a general scheme is illustrated in Figure 10.52.This RF-discharge scheme is quite similar geometrically to the conventional RF-CCP parallel plate reactor. However, here the RF-power is applied to a flat spiral inductive coil, which is separated from the plasma by quartz or another dielectric insulating plate. The RF-currents in the spiral coil induce image currents in the upper surface of the plasma (see Figure 10.52) corresponding to the skin layer. Thus this discharge is inductively coupled, and from the physical point of view is similar to the simple cylindrical geometry of the inductive coils considered above. The analytical relations derived for the low-pressure ICP-discharge inside of an inductive coil can also be applied qualitatively for the planar coil configuration of the inductive RF-discharges at low pressures. From Figure 10.52, the planar ICP-discharge also includes two practically important elements: multi-polar permanent magnets and a DC wafer bias. The multi-polar permanent magnets are located around the outer circumference of the plasma to improve plasma uniformity, plasma confinement and to increase plasma density. The DC wafer bias power supply is used to control the energy of ions impinging on the wafer, which typically ranged 30 to 400 eV in the planar discharge system under consideration. The structure of magnetic field lines in the planar coil configuration of the ICP-discharges is obviously more complicated than the case of the cylindrical inductive coil. The RF-magnetic field lines in the planar coil configuration in the absence of plasma are illustrated in Figure 10.53a (A.E. Wendt, M.A. Lieberman, 1993). These magnetic field lines encircle the coil and are symmetric with respect to the plane of the coil. Deformation of the magnetic field in the presence of plasma formed below the coil is shown in Figure 10.53b. In this case, an azimuthal electric field and an associated current (in the direction opposite to that in the coil) are induced in the plasma skin layer. Both the multi-turn coil current and the "single-turn" induced plasma current generate the total magnetic field. The dominant magnetic field component within the plasma is vertical near the axis of the planar coil, and

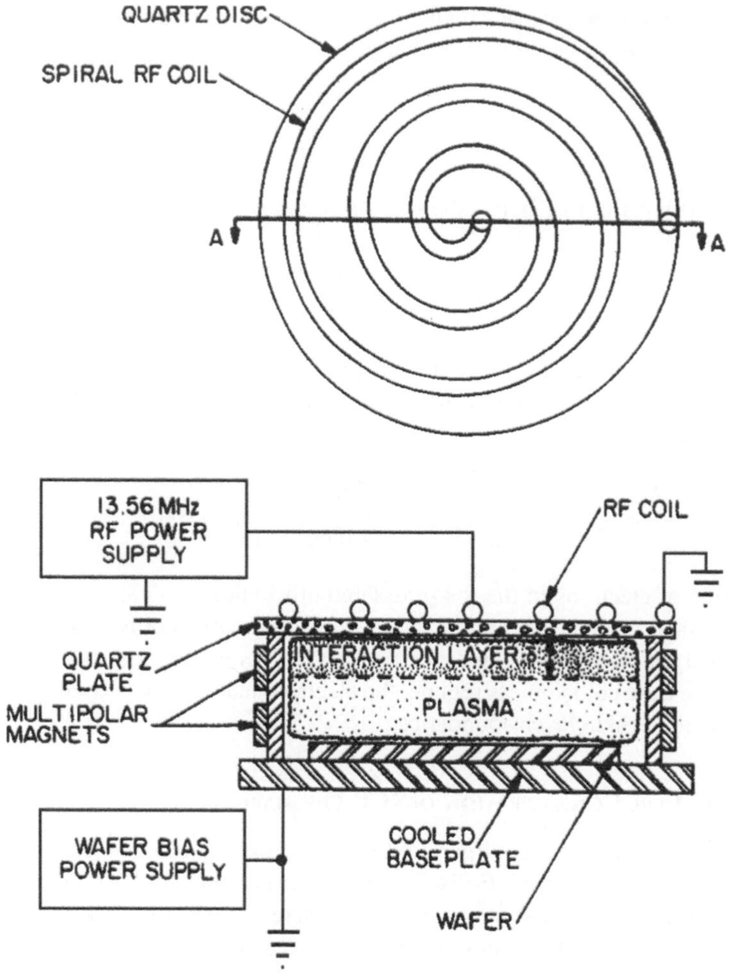

FIGURE 10.52 Schematic of an ICP parallel plate reactor.

horizontal away from the axis. More details regarding space distribution of magnetic field, electric currents, and concentration of charged particles in the low-pressure planar ICP discharge are in J. Hopwood, C.R. Guarnieri, S.J. Whitehair, J.J. Cuomo, 1993a. The planar radio-frequency ICP-discharge, illustrated in Figure 10.52 (J.H. Keller and colleagues, IBM, 1992, see J. Reece Roth, 2000), is able to produce uniform plasma and uniform plasma processing of wafers with diameters at least 20 cm. The power level of this reactor is about 2 kW, which is about an order of magnitude larger than the power of a CCP-discharge under similar conditions. This results in a higher flux of ions and other active species, which accelerate the surface treatment process. For example, ion flux in the system under discussion was 60 mA/cm^2 at a power input 1600 W, which corresponds to a high level of etch rate (1 ÷ 2 μm/min for polyimide film). The reactor has been operated at frequencies ranging from 1 MHz to 40 MHz, but the usual operating frequency is 13.56 MHz. Operation pressure of this planar ICP-discharge is in the range from 1 mTorr to 20 mTorr, which is far lower than typical pressure values of the corresponding CCP-discharges (which is about several hundreds mTorr). Obviously such low-pressure values are desirable for CVD (chemical vapor deposition) and etching technologies because they imply longer mean free paths and little scattering of ions and active species before they reach the wafer thus improving the surface treatment processes.

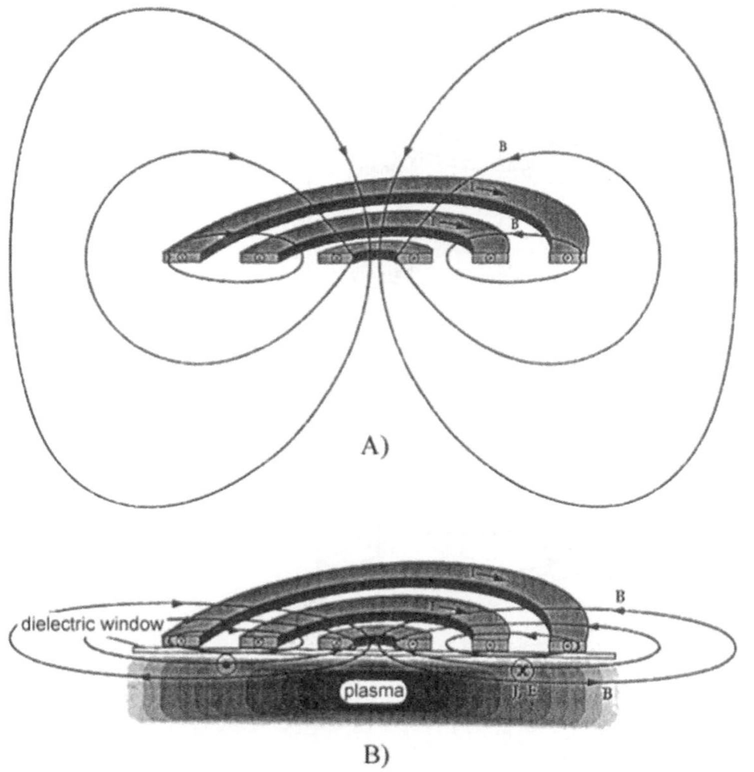

FIGURE 10.53 Schematic of the rf magnetic field near a planar inductive coil: (a) without nearby plasma and (b) with nearby plasma. (From Lieberman & Lichtenberg, Principles of Plasma Discharges and Material Processing)

10.7.8 Helical Resonator Discharges

The helical resonator discharge is a special case of the low-pressure RF-ICP-discharges. The helical resonator consists of an inductive coil (helix) located inside of the cylindrical conductive screen and can be considered as a coaxial line with an internal helical electrode. A general schematic of the helical resonator plasma source is shown in Figure 10.54. Electromagnetic wave propagates in such a coaxial line with a phase velocity much lower than the speed of light: $v_{ph} = \omega/k << c$, where k is the wavelength, and c is the speed of light. This property allows the helical resonator to operate in the MHz frequency range, which permits generating low-pressure plasma. The coaxial line of the helical discharge becomes resonant when an integral number of quarter waves of the RF-field fit between the two ends of the system. The criterion of the simplest resonance can be then expressed as:

$$2\pi r_h N = \frac{\lambda}{4}, \tag{10.160}$$

where r_h is the helix radius, N is the number of turns in the coil, and λ is the electromagnetic wavelength in vacuum. Helical resonator discharges effectively operate at radio-frequencies 3–30 MHz with simple hardware, and do not require a DC magnetic field. The resonators exhibit high Q-values (typically 600–1500 without plasma). In the absence of plasma, the electric fields are quite large facilitating the initial breakdown of the system. Also, helical resonator discharges have high characteristic impedance and can be operated without a matching network. Because of the resonance, large

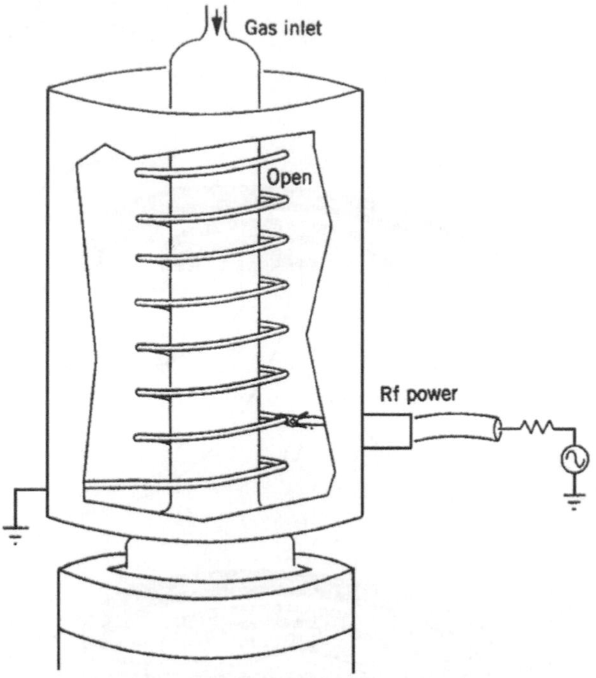

FIGURE 10.54 Schematic of a helical resonator.

voltages necessarily appear between the open end of the helix and the plasma. Hence the electric field is not exactly azimuthal in the helical resonator, and the discharge cannot be considered pure inductively coupled. On the other hand, the discharge sizes in this system are close to the electromagnetic wavelength, which means that the helical resonator discharge in some sense is similar to microwave discharges, see K. Niazi, A.J. Lichtenberg, M.A. Lieberman, D.L. Flamm, 1994; J.M. Cook, D.E. Ibbotson, P.D. Foo, D.L. Flamm, 1990.

10.8 NON-THERMAL LOW-PRESSURE MICROWAVE AND OTHER WAVE-HEATED DISCHARGES

10.8.1 Non-Thermal Wave-Heated Discharges

Consider the low-pressure, wave-heated plasmas and focus on three discharge systems: the electron cyclotron resonance (ECR) discharges, the helicon discharges, and the surface wave discharges. In ECR-discharges, a right circularly polarized wave (usually at microwave frequencies, e.g. 2.45 GHz) propagates along the DC-magnetic field (usually quite strong, 850 G at resonance) under the conditions of the electron cyclotron resonance, which provides the wave absorption through a collisionless heating mechanism. In the helicon discharges, an antenna radiates the whistler wave, which is subsequently absorbed in plasma by collisional or collisionless mechanisms. The helicon wave-heated discharges are usually excited at RF-frequencies (typically –13.56 MHz), and a weak magnetic field of about 20–200 G is required for the wave propagation and absorption. In the case of surface wave discharges, a wave propagates along the surface of the plasma and is absorbed by collisional heating of the plasma electrons near the surface. The heated electrons then diffuse from the surface into the bulk plasma. Surface wave discharges can be excited by either RF or microwave sources and they do not require a DC-magnetic field. The plasma potential with respect to all wall surfaces for wave-heated discharges is relatively low (about 5 electron temperatures, $5T_e$) similar to the case of ICP-discharges. This results in the effective generation of high-density plasmas at reasonable absorbed power levels.

10.8.2 Electron Cyclotron Resonance (ECR) Microwave Discharges, General Features

A thermal discharge sustained by microwave radiation requires only a sufficient heat energy flux to provide thermal balance and thermal ionization. In contrast, non-thermal discharges require a sufficient level of electric field for both heating of electrons and effective non-thermal ionization. The necessary high level of electric field can be provided by applying relatively high microwave power and power density, which is typical for the moderate pressure regim, or also by applying resonators with a high value of Q. Another widely used approach applies a steady magnetic field and effective electrons heating due to the electron cyclotron resonance (ECR).The ECR-resonance between an applied electromagnetic wave frequency ω and the electron cyclotron frequency $\omega_{Be} = eB/m$ (see Eq. (4.107)) allows electron heating sufficient for ionization at relatively lower values of electric fields in the electromagnetic wave. For calculations of the electron cyclotron frequency, it is convenient to use the following numerical formula $f_{Be}(\text{MHz}) = 2.8 \cdot B(G)$. This effective electron heating in the ECR-resonance takes place because the gyrating electrons rotate in phase with the right-hand polarized wave, seeing a steady electric field over many gyro-orbits. Pressure in the ECR-discharge system is low in order to have a low electron-neutral collision frequency $\nu_{en} << \omega_{Be}$, and to provide the electron gyration sufficiently long to obtain the energy necessary for ionization. This phenomenon determines the key effect guiding the physics of ECR-microwave discharges. Furthermore, the injection of the microwave radiation along the magnetic field (with $\omega_{Be} > \omega$ at the entry into discharge region) allows wave to propagate to the absorption zone $\omega \approx \omega_{Be}$ even in dense plasma with $\omega_{pe} > \omega$. Electromagnetic wave propagation in non-magnetized plasma is obviously impossible at frequencies below the plasma frequency, or in other words when the plasma density exceeds the critical value (see Eq. (6.2)). High plasma densities lead to the total reflection of electromagnetic waves from the non-magnetized plasma. However, application of magnetic fields permits propagation of electromagnetic waves even at high plasma densities exceeding the critical value. This effect can be explained analyzing the dispersion (6.6.39) for electromagnetic wave in magnetized plasma. The right-hand-polarized wave (corresponding to the "–" sign) has a real wave number even at high densities $\omega < \omega_{pe}$, if the magnetic field is sufficiently large and $\omega_{Be} = eB/m > \omega$.

10.8.3 General Scheme and Main Parameters of ECR-Microwave Discharges

A schematic of the ECR-microwave discharge with microwave power injected along the axial non-uniform magnetic field is shown in Figure 10.55 (M.A. Lieberman, A.J. Lichtenberg, 1994). The magnetic field profile is chosen in this discharge system to provide effective propagation of the electromagnetic wave from the quartz window to the zone of the ECR-resonance without major reflections even at high plasma densities. Special magnetic field profiles can provide multiple ECR-resonance positions as it is shown in the figure by the dashed line. Low-pressure gas introduced into the discharge chamber forms a highly non-equilibrium plasma, which streams and diffuses along the magnetic field toward a wafer holder shown in Figure 10.55. Energetic ions and free radicals generated within the entire discharge region are then able to provide the necessary surface treatment effect. Additionally, a magnetic field coil at the wafer holder can be used to modify the uniformity of the etch or deposition. Typical ECR-microwave discharge parameters are: pressure 0.5–50 mTorr, power 0.1–5 kW, characteristic microwave frequency 2.45 MHz, volume 2–50 L, magnetic field – about 1 kG, plasma density 10^{10}–10^{12} cm^{-3}, ionization degree 10^{-4}–10^{-1} electron temperature 2–7 eV, ion acceleration energy 20–500 eV, typical source diameter is 15 cm.

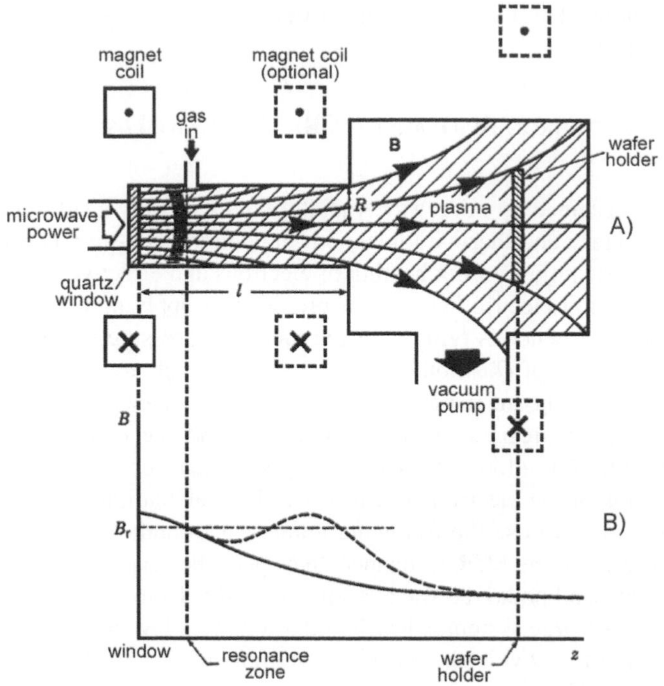

FIGURE 10.55 A typical high-profile ECR system: (a) geometric configuration; (b) axial magnetic field variation, showing one or more resonance zones. (From Lieberman & Lichtenberg, Principles of Plasma Discharges and Material Processing)

10.8.4 Electron Heating in ECR-Microwave Discharges

Consider the ERC-microwave discharge sustained at low pressure by a linearly polarized electromagnetic wave. This wave can be decomposed into the sum of two counter-rotating circularly polarized waves, right-hand-polarized and left-hand-polarized. The basic physical principle of the ECR heating of the magnetized electrons is illustrated in Figure 10.56. The electric field vector of the right-hand-polarized wave rotates around the magnetic field at frequency ω, while an electron in the

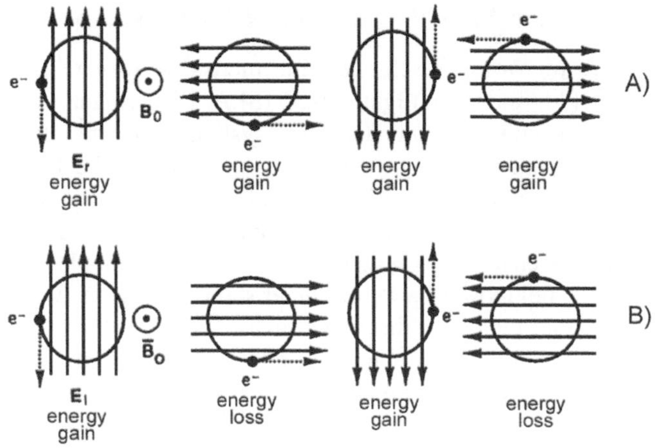

FIGURE 10.56 Mechanism of ECR heating: (a) continuous energy gain for right-hand polarization; (b) oscillating energy for left-hand polarization.

uniform magnetic field also gyrates in the same "right hand" direction at frequency ω_{Be}. Thus at the ECR-resonance conditions $\omega = \omega_{Be}$, the electric field continuously transfers energy to the electron providing its effective heating. In contrast, the left-hand-polarized electromagnetic wave rotates in the direction opposite to the direction of the electron gyration. So at the ECR-resonance conditions $\omega = \omega_{Be}$, for a quarter of the period the electric field accelerates the gyrating electrons, and for another quarter of period slows them down, resulting in no average energy gain. The described electron heating occurs only close to the ECR-conditions, which are necessary for the continuous energy transfer from microwave to an electron. Because the magnetic field and electron cyclotron frequency are not constant along the z-axis of the discharge (Figure 10.56), the electron heating occurs only locally, near the ECR-resonance p, where the electron cyclotron frequency can be expressed as:

$$\omega_{Be}(z) = \omega\left(1 + \frac{1}{\omega}\frac{\partial \omega_{Be}}{\partial z}\Delta z\right) = \omega\left(1 + \frac{e}{\omega m}\frac{\partial B}{\partial z}\Delta z\right). \tag{10.161}$$

Then the average electron energy gain per one pass across the ECR-resonance zone in the collisionless regime can be calculated as:

$$\varepsilon_{ECR} = \frac{\pi e E_r^2}{v_{res}\left|\partial B/\partial z\right|}. \tag{10.162}$$

In this relation, E_r is the amplitude of the right-hand-polarized electromagnetic wave (which is a half of the linearly polarized microwave amplitude); $\partial B/\partial z$ is the gradient of the magnetic field along the discharge axis near the resonance point; v_{res} is the component of electron velocity parallel to the magnetic field also in the vicinity of resonance point. The width of the resonance zone along the discharge axis z, where most of the energy is transferred to the electron is given as:

$$\Delta z_{res} = \sqrt{\frac{2\pi m v_{res}}{e\left|\partial B/\partial z\right|}}. \tag{10.163}$$

The absorbed electromagnetic wave power per unit area (or microwave energy flux) is:

$$S_{ECR} = \frac{\pi n_e e^2 E_r^2}{e\left|\partial B/\partial z\right|}, \tag{10.164}$$

where n_e is the electron density in the ECR-resonance zone, see K.G. Budden, 1966; M.C. Williamson, A.J. Lichtenberg, M.A. Lieberman, 1992; H.-M. Wu, D.B. Graves, R.K. Porteous, 1994; Stevens, Y.C. Huang, R.L. Larecki, J.L. Cecci, 1992; and M. Matsuoka, K. Ono, 1988.

10.8.5 Helicon Discharges, General Features

Another high-density plasma (HDP) discharge, that can be used for different material processing applications, is the helicon discharge. The HDP-plasma generation in the helicon discharge was first investigated by R.W. Boswell, 1970. The detailed theory of this discharge, and the general propagation and absorption of the helicon mode in plasma was developed by F.F. Chen, 1991. Helicon discharges are sustained by electromagnetic waves propagating in magnetized plasma in the so-called helicon modes. The driving frequency in these discharges is typically in the radio-frequency range of 1–50 MHz (the industrial radio-frequency 13.56 MHz is commonly used for material processing discharges). It is interesting to note that in contrast to the RF-discharges considered in sections 10.3–10.7, the helicon discharges can be considered as wave heated even though they operate in the radio-frequency range. This can be explained taking into account that the phase velocity of electromagnetic waves in magnetized plasma can be much lower than the speed of light (see Sections 6.6.7 and 10.8.5). This provides

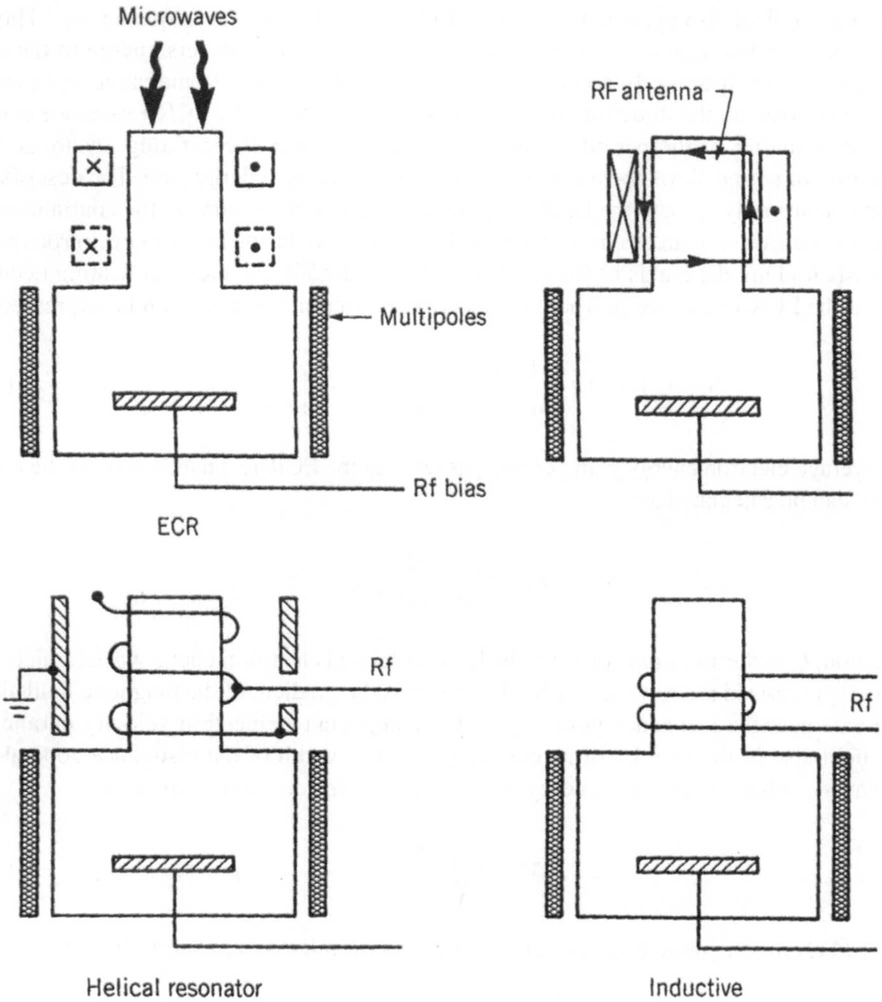

FIGURE 10.57 Helicon discharge schematic.

the possibility to operate in a wave propagation regime with wavelengths comparable with the discharge system size even at radio frequencies, which are much below the microwave frequency range. The magnetic field in helicon discharges applied for material processing varies from 20 to 200 G (for fundamental plasma studies it reaches 1000 G), and it is much below the level of magnetic fields applied in ECR-microwave discharges. Application of lower magnetic fields is an advantage of the helicon discharges. Plasma density in these wave-heated discharges applied for material processing is about 10^{11}–10^{12} cm^{-3}, but in some special cases can reach very high values of about 10^{13}–10^{14} cm^{-3}. Excitation of the helicon wave is provided by an RF-antenna that couples to the transverse mode structure across an insulating chamber wall. The electromagnetic wave mode then propagates along the plasma column in the magnetic field, and plasma electrons due to collisional or collisionless damping mechanisms absorb the mode energy. A schematic of a helicon discharge is illustrated in Figure 10.57. The material processing chamber is located downstream from the plasma source. The plasma potentials in the helicon discharges are typically low, about 15–20 V, similar to ECR-microwave discharges. Important advantages of the helicon discharges with respect to ECR-discharges are related to relatively low values of magnetic field and applied frequency. However, the resonant coupling of the helicon mode to the antenna can lead to a non-smooth variation of the plasma density with source parameters.

This effect, known as "the mode jumps" restricts the operating regime for a given design of plasma source.

10.8.6 Whistlers and Helicon Modes of Electromagnetic Waves Applied in Helicon Discharges

Consider the helicon discharges analyzing the propagation of the helicon modes in magnetized plasma. The helicons are propagating electromagnetic "whistler" wave modes in an axially magnetized, finite diameter plasma column. The electric and magnetic fields of the helicon modes have radial, axial, and, usually, azimuthal variation. They propagate in a low frequency, high plasma density regime with relatively low magnetic fields, which can be characterized by the following frequency limitations:

$$\omega_{LH} << \omega \ll \omega_{Be}, \quad \omega_{pe}^2 \gg \omega\omega_{Be}. \tag{10.165}$$

ω_{pe} is the electron plasma frequency; ω_{Be} is the electron cyclotron frequency; and ω_{LH} is the lower hybrid frequency in magnetized plasma, which occurs taking into account the ions mobility:

$$\frac{1}{\omega_{LH}^2} \approx \frac{1}{\omega_{pi}^2} + \frac{1}{\omega_{Be}\omega_{Bi}}, \tag{10.166}$$

where ω_{pi} is the plasma-ion frequency Eq. (6.137), and ω_{Bi} is the ion cyclotron frequency Eq. (6.212). The right-hand polarized electromagnetic waves in magnetized plasma with frequencies between ion and electron cyclotron frequencies $\omega_{Bi} << \omega << \omega_{Be}$ are known as the **whistler waves**. For the right-hand polarized electromagnetic wave propagation along the magnetic field at frequencies below the frequency of the electron cyclotron resonance, the dispersion equation (6.6.39) can be rewritten as:

$$\frac{k^2c^2}{\omega^2} = 1 + \frac{\omega_{pe}^2}{\omega\omega_{Be}}. \tag{10.167}$$

From the dispersion (10.167), propagation of the electromagnetic whistler waves is possible at frequencies below the plasma frequency $\omega < \omega_{pe}$. Taking into account the helicon frequency conditions (10.165) and introducing the wave number $k_0 = \omega/c$ corresponding to the electromagnetic wave propagation without plasma, the dispersion equation for the whistler waves (10.167) becomes:

$$\omega = \frac{k_0^2\omega_{pe}^2}{k^2\omega_{Be}}. \tag{10.168}$$

The whistler waves can propagate at an angle to the axial magnetic field. Hence, the dispersion equation for the whistlers can be rewritten in a more general form as:

$$\omega = \frac{k_0^2\omega_{pe}^2}{kk_z\omega_{Be}}, \tag{10.169}$$

where $k = \sqrt{k_\perp^2 + k_z^2}$ is the wave-vector magnitude, which takes into account not only axial k_z but also radial component $k_\perp$. The helicon frequency condition (10.165) requires $\omega_{pe}^2 >> \omega_{Be}\omega$, which together with the dispersion equation (10.168) shows that $k^2 >> k_0^2$. This means that wavelengths of the helicon waves in the magnetized plasma are much less than those of electromagnetic waves of the same frequency without magnetic fields. For this reason, in contrast to RF-CCP and RF-ICP discharges, the helicon discharges can be characterized by wavelengths comparable with the typical discharge size and can be considered as wave-heated, even though they operate in relatively low radio-frequency range.

10.8.7 Antenna Coupling of Helicon Modes and Their Absorption in Plasma

Helicon is a superposition of low-frequency whistler propagating at a common fixed angle to the axial magnetic field. The helicon modes are mixtures of electromagnetic ($div\vec{E} \approx 0$) and quasi-static ($curl\vec{E} \approx 0$) fields:

$$\vec{E}, \vec{H} \propto \exp i(\omega t - kz - m\theta), \tag{10.170}$$

where θ is the angle between the wave propagation vector and magnetic field, and the integer m specifies the azimuthal mode. Helicon sources have been developed based on the excitation of the $m = 0$ and $m = 1$ modes. The $m = 0$ mode is axisymmetric and $m = 1$ mode has a helical variation, therefore both modes generate time-averaged axisymmetric field intensities. The transverse electric field patterns for the $m = 0$ and $m = 1$ modes and the way they propagate along the axial magnetic field are shown in Figure 10.58 (F.F. Chen, 1991). The quasi-static and electromagnetic axial electric field components exactly cancel in the undamped helicon modes, which means that the total $E_z = 0$. Thus the antenna is able to couple to the transverse electric or magnetic field to excite the modes. To design an RF-antenna for efficient power coupling, its length should be correctly related to magnetic field and plasma density. The following simplified formula derived for the case $k_z >> k_\perp$ can be practically used for this purpose if electron density is not very high (A. Komori, T. Shoji, K. Miyamoto, J. Kawai, Y. Kawai, 1991):

$$\lambda_z = \frac{2\pi}{k_z} = \frac{3.83}{R} \frac{B}{e\mu_0 n_e f}. \tag{10.171}$$

In this relation: R is the radius of the insulating (or conducting) wall, n_e is the plasma density, $f = \omega/2\pi$ is the electromagnetic wave frequency. A typical schematic of a radio-frequency antenna to

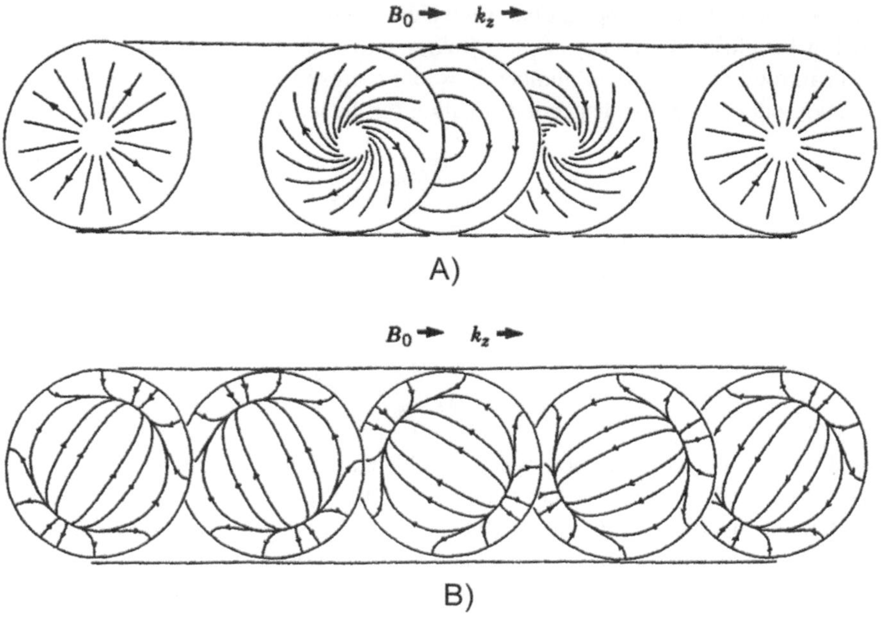

FIGURE 10.58 Transverse electric fields of helicon modes at five different axial positions (a) $m = 0$; (b) $m = 1$. (From Lieberman & Lichtenberg, Princples of Plasma Discharges and Material Processing)

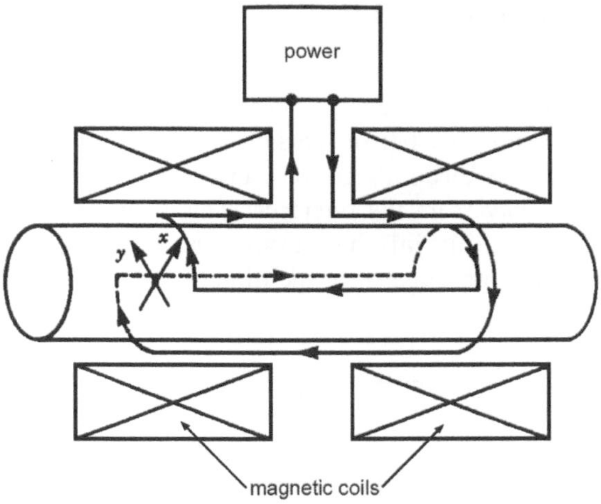

FIGURE 10.59 Helicon mode $m = 1$ excitation by antenna.

excite the helicon $m = 1$ mode is shown in Figure 10.59. The antenna generates a B_x radio-frequency magnetic field over an axial antenna length, which can couple to the transverse magnetic field of the helicon mode. The antenna also induces an electric current within the plasma column just beneath each horizontal wire in a direction opposite to currents shown in the figure. This current produces a charge of opposite signs at the two ends of the antenna. In turn, these charges generate a transverse quasi-static RF electric field E_y, which can couple to the transverse quasi-static fields of the helicon mode (see Figure 10.58b). The helicon mode energy can be transferred to the plasma electron heating by collisional damping or collisionless Landau damping (see Section 6.5.6). The collisional damping mechanism transfers the electromagnetic wave energy to the thermal bulk electrons, while the collisionless Landau damping preferentially heats non-thermal electrons with energies much exceeding the bulk electron temperature. The collisional damping mechanism dominates at relatively higher pressures, while at lower pressures (less than 10 mTorr in the case of argon) the Landau damping dominates.

10.8.8 Electromagnetic Surface Wave Discharges, General Features

Electromagnetic waves can propagate along the surface of a plasma column and be absorbed by the plasma, sustaining the surface wave discharge. Such surface waves, which have strong fields only near the plasma surface, were first described by L.D. Smullin, P. Chorney, 1958, and A.W. Trivelpiece, R.W. Gould, 1959, and then investigated by M. Moisan, Z. Zakrzewski, 1991. The surface wave discharges can generate high-density plasma (HDP) with diameters as large as 15 cm. The absorption lengths of the electromagnetic wave surface modes are quite long in comparison with the ECR-microwave discharge (see Section 10.8.2). The surface wave discharge typically operates at frequencies in the microwave range of 1–10 GHz without an imposed axial magnetic field. K. Komachi has developed planar rectangular configurations of this discharge, 1992. The electromagnetic surface wave, damping in both directions away from the surface, can be arranged in different configurations. One of them, a planar configuration on the plasma-dielectric interface will be considered in more detail in the following section. In another configuration, plasma is separated from a conducting plane by a dielectric slab. This planar system also admits propagation of a surface wave that decays into the plasma region. Although this electromagnetic wave does not decay into the dielectric, it is confined within the dielectric layer by the conducting plane. Finally, a surface wave also is able to propagate in the cylindrical discharge geometry. In this case, the surface wave propagates on a non-magnetized plasma column confined by a thick dielectric tube.

10.8.9 Electric and Magnetic Field Oscillation Amplitudes in the Planar Surface Wave Discharges

The planar surface wave configuration can be described as follows. The electromagnetic surface wave is supported at an interface between a dielectric and plasma. At the interface between a semi-infinite plasma and dielectric, a solution can be found for which the wave amplitude decays in both directions away from the interface. This solution corresponds to the surface wave discharge. Assuming the semi-infinite plasma zone is in positive semi-space $x > 0$, and that the electromagnetic wave propagates in the direction of "z", the amplitude of magnetic field oscillations in plasma is:

$$H_{yp} = H_{y0}\exp\left(-\left|\alpha_p\right|x - ik_z z\right). \tag{10.172}$$

In this expression: H_{y0} is the amplitude of the magnetic field component directed along the plasma-dielectric interface; α_p characterizes the electromagnetic wave damping in plasma; k_z is the wave number in the direction of electromagnetic wave propagation. Assuming that magnetic field H_y in the electromagnetic surface wave is continuous across the interface at $x = 0$, the amplitude of the magnetic field oscillations in dielectric can be expressed similarly to (10.172) as:

$$H_{yd} = H_{y0}\exp\left(\left|\alpha_d\right|x - ik_z z\right), \tag{10.173}$$

where α_d characterizes the electromagnetic wave damping in dielectric. As one can see from Eqs. (10.172) and (10.173), the magnetic field oscillations decrease in plasma where $x > 0$, and in dielectric where $x < 0$. From the wave equation, the damping coefficients α_d and α_p for the transverse electromagnetic surface waves can be related to the wave number k_z as:

$$-\alpha_d^2 + k_z^2 = \varepsilon_d \frac{\omega^2}{c^2}, \tag{10.174}$$

$$-\alpha_p^2 + k_z^2 = \varepsilon_p \frac{\omega^2}{c^2}. \tag{10.175}$$

In these relations: ε_d is the dimensionless dielectric constant of the dielectric semi-space under consideration; and ε_p is the dimensionless plasma dielectric constant (6.6.7), which can be expressed in the collisionless regime as a function of the plasma frequency ω_{pe}:

$$\varepsilon_p = 1 - \frac{\omega_{pe}^2}{\omega^2}. \tag{10.176}$$

The electric field amplitude both from dielectric (d) and plasma (p) sides can be then related to magnetic field amplitude based on Maxwell equations:

$$E_{zd} = H_{y0}\frac{\alpha_d}{i\omega\varepsilon_0\varepsilon_d}\exp\left(\left|\alpha_d\right|x - ik_z z\right). \tag{10.177}$$

$$E_{zp} = -H_{y0}\frac{\alpha_p}{i\omega\varepsilon_0\varepsilon_p}\exp\left(-\left|\alpha_p\right|x - ik_z z\right). \tag{10.178}$$

From Eqs. (10.172) and (10.173), the electric field oscillations, similar to magnetic fields (see Eqs. (10.172) and (10.173)), also decrease in plasma where $x > 0$, and in dielectric where $x < 0$.

10.8.10 Electromagnetic Wave Dispersion and Resonance in the Planar Surface Wave Discharges

Taking into account the continuity of E_z at the plasma-dielectric interface $x = 0$, one can derive, based on Eqs. (10.177) and (10.178), the relation between the damping coefficients α_d and α_p in plasma and dielectric zones as:

$$\frac{\alpha_p}{\varepsilon_p} = -\frac{\alpha_d}{\varepsilon_d}. \tag{10.179}$$

Substituting Eqs. (10.174) and (10.175) into Eq. (10.179), one obtains the relation, which is free of the unknown damping coefficients α_d and α_p:

$$\varepsilon_d^2\left(k_z^2 - \varepsilon_p\frac{\omega^2}{c^2}\right) = \varepsilon_p^2\left(k_z^2 - \varepsilon_d\frac{\omega^2}{c^2}\right). \tag{10.180}$$

This relation can be solved for the wave number k_z, which leads, taking into account Eq. (10.176), to the final dispersion equation for the planar surface waves:

$$k_z = \sqrt{\varepsilon_d}\,\frac{\omega}{c}\sqrt{\frac{\omega_{pe}^2 - \omega^2}{\omega_{pe}^2 - (1+\varepsilon_d)\omega^2}}. \tag{10.181}$$

This dispersion equation in form of dependence $\frac{k_z c}{\omega_{pe}}\left(\frac{\omega}{\omega_{pe}}\right)$ is illustrated in Figure 10.60. It is interesting to note that propagation of the surface waves is possible at lower electromagnetic field frequencies in contrast to the conventional case of electromagnetic waves propagation in plasma (see the dispersion Eq. (6.191) to compare). In the case of conventional dispersion Eq. (6.191), electromagnetic wave propagation is possible only when the plasma density is lower than the critical value Eq. (6.2). There is no such kind of limit for surface wave discharges, which provides an opportunity

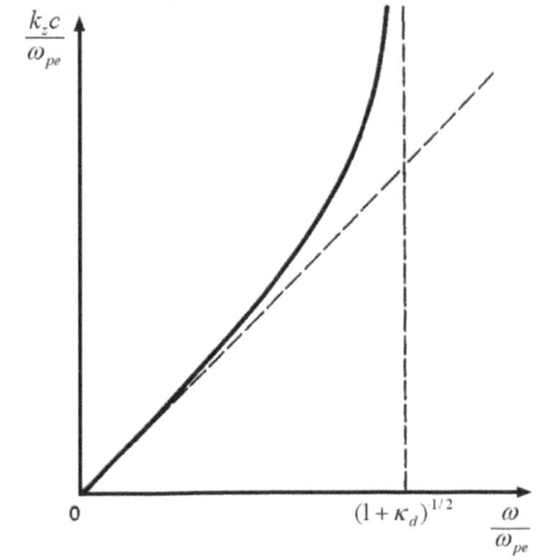

FIGURE 10.60 Surface wave dispersion curve $k_z(\omega)$.

to operate these discharges at high plasma densities in the HDP-regimes. From the dispersion equation (10.181) and Figure 10.60, the wave number k_z of the electromagnetic surface wave is real and wave propagation is possible for frequencies below the resonant value ($\omega \leq \omega_{res}$):

$$\omega_{res} = \frac{\omega_{pe}}{\sqrt{1+\varepsilon_d}}. \tag{10.182}$$

In the case of low frequencies ($\omega << \omega_{res}$), $k_z \approx \frac{\omega}{c}\sqrt{\varepsilon_d}$, and the surface wave propagates as a conventional one in the dielectric. The frequency of interest for the surface wave discharges is near but below the resonant one ω_{res}, which for the HDP-plasma sources corresponds to microwave frequencies exceeding 1 MHz. When the electromagnetic wave frequency ω is fixed, the surface wave propagation is possible only at plasma densities exceeding the critical one:

$$n_e \geq n_{res} = \frac{m\omega^2\varepsilon_0\left(1+\varepsilon_d\right)}{e^2}. \tag{10.183}$$

Thus unlike the conventional case Eq. (6.2), propagation of surface waves and sustaining the surface wave discharge is possible when the plasma density exceeds the relevant critical value Eq. (10.183) even without external steady magnetic field.

10.9 NON-EQUILIBRIUM MICROWAVE DISCHARGES OF MODERATE PRESSURE

10.9.1 Non-thermal Plasma Generation in Microwave Discharges at Moderate Pressures

Microwave discharges permit organizing such strongly non-equilibrium, plasma-chemical processes with very high energy efficiency in the intermediate range of moderate pressures (usually between 30 and 200 Torr). The general schematics of the moderate pressure microwave discharges are almost identical to those considered in sections 10.2.2–10.2.4 regarding atmospheric pressure microwave discharges. The waveguide and resonator based configurations of the microwave plasma generators (see Figures 10.9–10.12, and 10.15) can be applied to create non-thermal and strongly non-equilibrium plasma at moderate pressures.

10.9.2 About Energy Efficiency of Plasma-Chemical Processes in Moderate Pressure Microwave Discharges

The highest plasma-chemical energy efficiency can be reached under strongly non-equilibrium conditions with contributions of vibrationally excited reagents. Such regimes requiregeneration of plasma with specific parameters: electron temperature T_e should be about 1 eV and higher than the translational one (≤ 1000 K), the degree of ionization and specific energy input (energy consumption per molecule, see Section 5.6.7) should be sufficiently high $n_e/n_0 \geq 10^{-6}$, $E_v \approx 1\,\text{eV/mol}$. Simultaneous achievement of these parameters is rather difficult, especially in steady-state uniform discharge. For example, the conventional low-pressure non-thermal discharges are characterized by too large of values for the specific energy inputs of at least 30–100 eV/mol, the streamer based atmospheric pressure discharges have low values of specific power and average energy input, powerful steady-state atmospheric pressure discharges usually operate close to quasi-equilibrium conditions. In contrast , the moderate pressure microwave discharges are able to generate non-equilibrium plasma with the above-mentioned optimal parameters (V.D. Rusanov, A. Fridman, 1984). An important advantage of moderate pressure microwave discharges is related to the fact that formation of an overheated plasma filament within the plasma zone does not lead to an electric field decrease because of the skin effect in the vicinity of the filament (A.K. Vakar, V.K. Givotov, E.G. Krasheninnikov, A. Fridman, ea. 1981). Such peculiarity of the electrodynamic structure permits sustaining in the microwave discharges strong non-equilibrium conditions $T_e > T_v >> T_0$ at relatively high values of specific energy input. Steady-state microwave discharge, investigated by E.G. Krasheninnikov (1981), is sustained in CO_2 at frequency 2.4 GHz, power 1.5 kW, pressure 50–200 Torr, flow rate 0.15–2 sl/sec. Specific energy input is in the range of 0.2–2 eV/mol, specific power is up to 500 W/cm^3 (compare with conventional glow discharges values of 0.2–3 W/cm^3). Spectral diagnostics proves that vibrational temperature in the discharge can be on the level of 3000–5000 K and significantly exceed rotational and translational temperatures that are about 1000 K (V.K. Givotov, Rusanov, Fridman, 1985; A.K. Vakar, E.G. Krasheninnikov, E.A. Tischenko, 1984).

10.9.3 Microstructure and Energy Efficiency of Non-Uniform Microwave Discharges

Propagation of non-equilibrium discharges in fast gas flows is related to the processes on the plasma front, and it is only slightly sensitive to the processes in bulk of the plasma. The velocity of ionization wave and thickness of the plasma front are determined by the diffusion coefficient D of heavy particles and the characteristic time τ of the limiting gas preparation process for ionization (in particular, by vibrational excitation). Reactions of vibrationally excited molecules can be characterized by the same time interval $\tau_{chem} \approx 1/k_{eV}n_e \approx \tau$, where k_{eV} is the rate coefficient of vibrational excitation. If reactions are stimulated by non-equilibrium vibrational excitation, the total energy efficiency can be expressed as the function of initial gas temperature T_0^i, degree of ionization n_e/n_0, and specific energy input E_v:

$$\eta = \eta_{ex}\eta_{chem} \frac{E_v - k_{VT}\left(T_0^i\right)n_0\hbar\omega\left(\tau_{eV}+\tau_p\right) - \varepsilon_v\left(T_v^{\min}\right)}{E_v\left(1-\frac{\varepsilon_v\left(T_v^{\min}\right)}{\Delta H}\right)}. \quad (10.184)$$

In this relation: $\tau_{eV} = E_v/k_{eV}n_e\hbar\omega$ is the total time of vibrational excitation; $\tau_p = c_v^v\left(T_v^{\min}\right)^2/k_{VT}n_0E_a\hbar\omega$ is the reaction time in the passive phase of the discharge; η_{ex}, η_{chem} are the excitation and chemical components of the total energy efficiency (Section 5.6.9); k_{VT} is the rate coefficient of vibrational VT-relaxation; $\varepsilon_v\left(T_v^{\min}\right)$ is average vibrational energy of a molecule at the critical vibrational temperature $T_v^{\min}$ corresponding to equal rates of chemical reaction and vibrational relaxation (5.6.15); ΔH and E_a are the plasma-chemical reaction enthalpy and activation energy; c_v^v is the vibrational part of the specific heat per one molecule. Propagation of the non-equilibrium discharge is determined by the value of the reduced electric field on the plasma front $(E/n_0)_f$, which depends on two external parameters, gas pressure p and electric field on the front E. Taking into account that the reduced electric field on the front is almost fixed by the ionization rate requirements, the initial gas temperature can be found by the following simple relation (I.A. Kirillov, V.D. Rusanov, A. Fridman, 1986,1987):

$$T_f = \left(\frac{E}{n_0}\right)_f \frac{p}{E}. \quad (10.185)$$

The electron concentration on the plasma front can be found from the energy balance as:

$$n_{ef} = \frac{D}{r_p^2}\frac{T_p}{T_p - T_g}\frac{p}{\mu_e eE^2}. \quad (10.186)$$

In this relation, r_p is the characteristic radius of the non-uniform microwave plasma zone, T_p, T_g are translational temperatures in the plasma zone and ambient gas respectively, μ_e is the electron mobility. The specific energy input E_v in the discharge, which determines the energy efficiency of the plasma-chemical process can be determined using conventional relations (see Eqs. 5.152, 5.153):

$$E_v = P/Q_f. \tag{10.187}$$

P is the microwave discharge power absorbed in the plasma, and Q_f is the flow rate, which ine non-uniform discharges is only the portion of the flow crossing the plasma front, not the total flow rate.

10.9.4 Macrostructure and Regimes of Moderate Pressure Microwave Discharges

Microstructure of the plasma front considered above together with the discharge fluid mechanics and general energy balance permits describing the macrostructure and shape of the moderate pressure microwave discharges (I.A. Kirillov, V.D. Rusanov, A. Fridman, 1981). The macrostructure analysis includes consideration of transition between the three major forms of the discharges: diffusive (homogeneous), contracted, and combined. The thee major macrostructures, diffusive (homogeneous), contracted, and combined discharge forms take place at different values of pressure p and electric field E, which is related to the electromagnetic energy flux: $S = \varepsilon_0 c E^2$ (see (6.193) and (6.194)). Critical values of pressure p and electric field E, separating the three discharge forms, are shown in Figure 10.61. The area above the curve 1–1, for example, corresponds to microwave breakdown conditions; the microwave discharges are sustained below this curve. Critical curve 2–2 in Figure 10.61 corresponds to the maximal ratio E/p of the steady-state microwave discharges, which is sufficient according to Eq. (10.185) to sustain ionization on the plasma front even at room temperature. The diffusive (homogeneous) regime, illustrated in Figure 10.62a, takes place when $E/p < (E/p)_{max}$, but pressure is relatively low (close to 20–50 mTorr), though still in the moderate pressure range. As seen from the figure, the space configuration of the discharge front in this regime is determined by the stabilization of the front in axial flow. This stabilization requires the normal

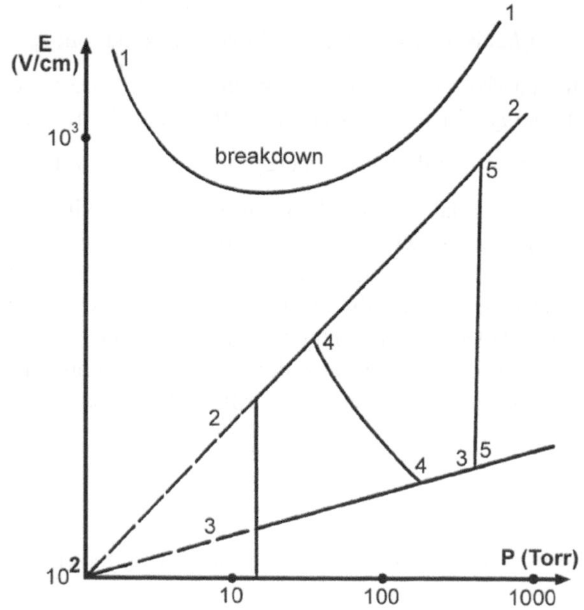

FIGURE 10.61 Three regimes of moderate pressure microwave discharges.

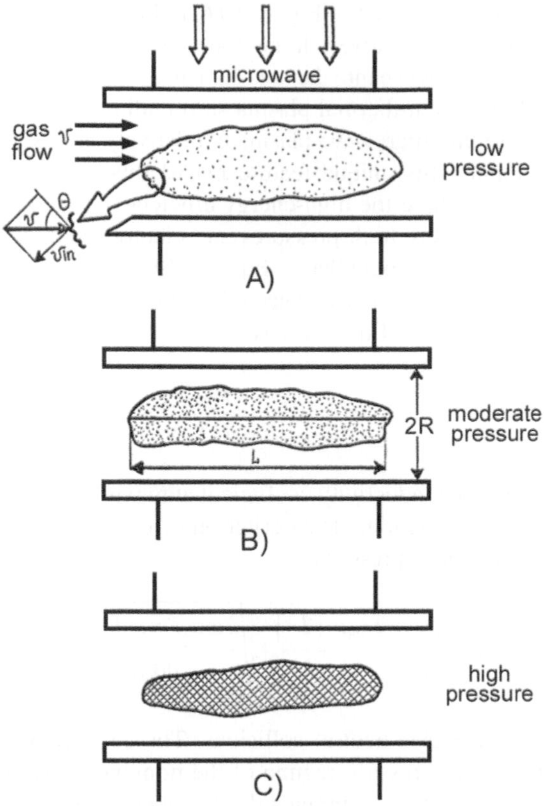

FIGURE 10.62 Transition of a diffusive microwave discharge (a) into a contracted one (c) related to pressure increase.

velocity of the discharge propagation u_{in} to be equal to the component of the gas flow velocity $v \sin \theta$ perpendicular to the discharge front. The ratio E/p decreases at higher pressures and temperatures on the discharge front should increase (according to Eq. (10.185)) to provide the necessary ionization rate. Critical curve 3–3 in Figure 10.62 determines the minimal value of reduced electric field $(E/p)_{min}$, when the microwave discharge is still non-thermal. The maximal temperature T^* on the discharge front, corresponding to $(E/p)_{min}$, is related to the transition from non-thermal to thermal ionization mechanisms (see Section 2.2.4 and especially Eq. (2.28)):

$$T* = I \ln^{-1}\left\{ k_i\left[\left(\frac{E}{n_0}\right)_f\right] \times \frac{(4\pi\varepsilon_0)^5 \hbar^3 T*}{me^{10}} \frac{g_0}{g_i}\right\}. \quad (10.188)$$

In this relation: I is the ionization potential; g_i, g_0 are the statistical weights of ion and ground state neutrals; $k_i\left[\left(\frac{E}{n_0}\right)_f\right]$ is the rate coefficient of non-thermal ionization by direct electron impact at the reduced electric field given on the discharge front (see Eq. (10.185)). At an intermediate pressure range 70–200 Torr and $(E/p)_{min} < E/p < (E/p)_{max}$, the combined microwave discharge regime can be sustained, when the non-thermal ionization front co-exist with the thermal one. This interesting and practically important regime is illustrated in Figure 10.62b. In this case, a hot thin filament of thermal plasma is formed interior to a relatively large non-thermal plasma zone. Skin effect prevents penetration of the electromagnetic wave into the hot filament and most of energy still can be absorbed in the strongly non-equilibrium, non-thermal plasma surrounding. The combined regime is able to sustain strongly non-equilibrium, microwave plasma at relatively high values of degrees of ionization and specific energy input. This unique feature makes this regime especially interesting for plasma-chemical applications, where the high-energy efficiency is the most important factor. The combined regime requires relatively high pressures and electric fields. To derive the criterion of existence for this regime, take into account that T^* Eq. (10.188) is the minimal temperature sufficient for the contraction of the filament. Energy balance of the thermal filament determined by the skin layer $\delta(T_m)$ can be expressed based on Eq. (8.33) as:

$$S = \varepsilon_0 c E^2 = \frac{4\lambda_m T_m^2}{I\delta(T_m)}. \tag{10.189}$$

In this relation, λ_m is the thermal conductivity coefficient at maximum plasma temperature T_m in the hot filament, I is the ionization potential. Then criterion of existence for the combined regime of microwave discharges then can be expressed as:

$$E^2\sqrt{p} \geq \frac{2\lambda_m e \cdot (T^*)^2}{\varepsilon_0 c I}\sqrt{\frac{2\omega\mu_0 T * n_e(T^*)}{mk_{en}}}, \tag{10.190}$$

k_{en} is the rate coefficient of electron-neutral collisions. The corresponding critical curve 4–4 in Figure 10.61 separates the lower pressure regime of the homogeneous discharge (Figure 10.62a) from the higher pressure regime of the combined discharge (Figure 10.62b). At high pressures, to the right from the critical curve 5–5, radiation heat transfer becomes comparable with the molecular heat transfer. Because of the reduction of the mean-free-path, the radiative front overheating becomes essential, and the contracted microwave discharge becomes completely thermal, which is illustrated in Figure 10.62c, A.F. Alexandrov, A.A. Rukhadze, 1976.

10.9.5 Radial Profiles of Vibrational $T_v(r)$ and Translational $T_0(r)$ Temperatures in Moderate Pressure Microwave Discharges in Molecular Gases

The profiles $T_v(r)$ and $T_0(r)$ are qualitatively different in these three regimes: homogeneous occurring at lower pressures, contracted occurring at higher pressures, and combined, which is related to intermediate pressure range (Figures 10.61, 10.62). Results of detailed spectral measurements of vibrational and translational (rotational) temperatures are illustrated in Figure 10.63 together with the radial distribution of power density, (V.K. Givotov, Rusanov, Fridman, 1985; A.K. Vakar, E.G. Krasheninnikov, E.A. Tischenko, 1984. These can be summarized as follows. The radial profiles of vibrational and translational temperatures are obviously close to each other in the quasi-equilibrium regime at higher pressures (Figure 10.63c). At relatively low pressures, electron density is low and the skin effect can be neglected. In this case, the maximum deviation of the vibrational temperature from the translational one occurs on the axis of the discharge tube, where electron concentration and power density is maximal (Figure 10.63b). Qualitatively different temperature distributions can be observed at intermediate pressures, which are illustrated in Figure 10.63a. In this case, effective vibrational excitation and strong vibrational-translational non-equilibrium take place only on the non-equilibrium front of the microwave discharge, which is located at some intermediate radii. The vibrational excitation is not effective near the discharge axis because of the low electric field (which slows down the excitation) and high translational temperature (which accelerates vibrational relaxation). Near the walls of the discharge tube the vibrational excitation is ineffective because of low electron density.

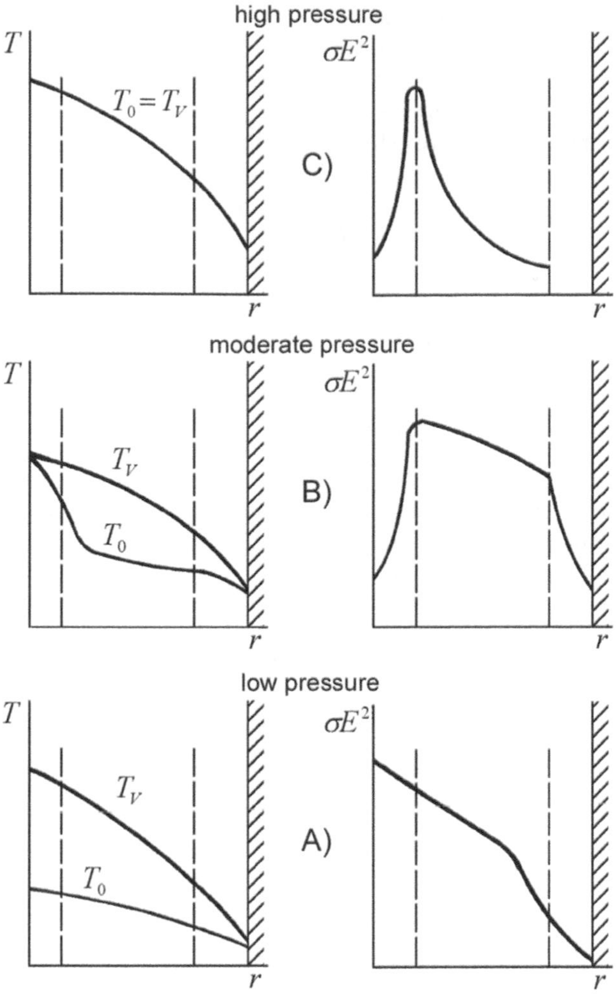

FIGURE 10.63 Radial distributions of vibrational and transitional temperatures (as well as specific power) at different pressures.

10.9.6 Energy Efficiency of Plasma-Chemical Processes in Non-Uniform Microwave Discharges

The energy efficiency of plasma processes Eq. (10.184) is related only to formation of products on the discharge front. Considering non-uniform microwave discharges (first in the combined regime, see Figure 10.62b), one should take into account that part of products effectively formed in the non-thermal zone can be destroyed in thermal inverse reactions in the hot filament. These losses, as well as other non-uniformity losses like those related to microwave absorption in the hot skin layer, can be taken into account by the additional special efficiency factor $\frac{T_{max} - T_{min}}{T_{max}}$ (I.A. Kirillov, V.D. Rusanov, A. Fridman, 1984a; V.V. Liventsov, V.D. Rusanov, A. Fridman, 1988), which is related to the discharge non-uniformity and appears similar to the Carnot thermal efficiency coefficient. Physically the non-uniformity factor in energy efficiency can be interpreted taking into account that large temperature differences between the cold surrounding non-thermal plasma zone and the hot filament prevents the products from having long contact with the high temperature zone. This protects the plasma-chemical products from inverse reactions and increases the energy efficiency. Also the high temperature T_{max} in the plasma filament leads to higher plasma density there and to a stronger skin-effect, which decreases energy losses in this zone and also promotes high-energy efficiency. In this case, the total energy efficiency of the non-equilibrium processes can be presented by a combination of the quasi-uniform energy efficiency η, given for example by Eq. (10.184) and the non-uniformity factor introduced above. This leads to the total energy efficiency:

$$\eta_{total} = \eta \times \frac{T_{max} - T_{min}}{T_{max}}. \quad (10.191)$$

In this relation: T_{min} is the gas temperature in the cold non-equilibrium discharge zone, while T_{max} is the gas temperature in the hot plasma filament.

10.9.7 Plasma-Chemical Energy Efficiency of Microwave Discharges as a Function of Pressure

Equation (10.191) can be applied in particular to describe the pressure dependence of the energy efficiency of the non-thermal microwave discharges. Taking into account Eq. (10.185) between pressure and temperature on the plasma front:

$$T_f = T^* \frac{p}{p_2(E)}. \quad (10.192)$$

The pressure dependence of the quasi-uniform component η of the energy efficiency in accordance with Eq. (10.184) can be expressed as:

$$\eta \approx 1 - a_k \sqrt[3]{p/p_2(E)}. \quad (10.193)$$

$a_k \approx 0.3$ is the dimensionless parameter of a process; $p_2(E) = T^* E/(E/n_0)_f$ is the maximum pressure when the non-equilibrium plasma front is still possible (see Figure 10.61). Equation (10.193) describes the well-known tendency of growth of relaxation losses with pressure. The energy balance of the plasma filament in the microwave discharges gives the following simplified pressure dependence of the maximum temperature (I.A. Kirillov, V.D. Rusanov, A. Fridman, 1984; B.E. Mayerovich, 1972):

$$T_{max} = T^* \sqrt{p/p_1}, \quad (10.194)$$

where p_1 is the minimal pressure of the thermal filament appearance (see Figure 10.61). This maximum temperature in the hot plasma zone permits rewriting the energy efficiency non-uniformity factor as:

$$1 - \frac{T_{min}}{T_{max}} = 1 - \frac{T_{min}}{T^*} \sqrt{\frac{p_1}{p}}. \quad (10.195)$$

As is expected, the non-uniformity factor continuously increases with pressure. The total energy efficiency of plasma-chemical processes stimulated by vibrational excitation Eq. (10.191) based on Eqs. (10.193) and (10.195) can be presented as:

$$\eta_{total} = \left(1 - \frac{T_{min}}{T^*} \sqrt{\frac{p_1}{p}}\right) \times \left(1 - a_k \sqrt[3]{\frac{p}{p_2(E)}}\right). \quad (10.196)$$

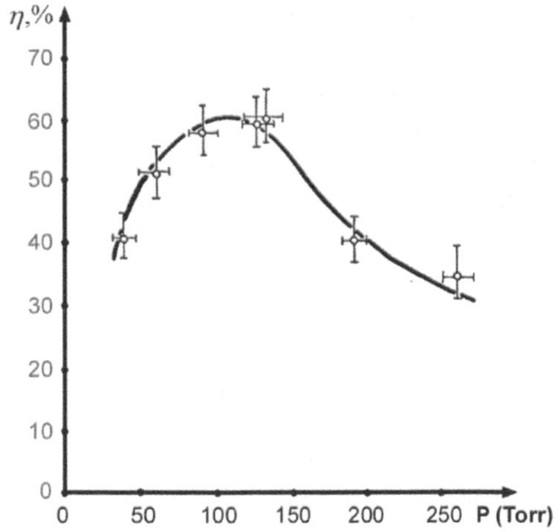

FIGURE 10.64 Energy efficiency as a function of pressure.

According to this relation, the total energy efficiency as a function of pressure has a maximum, which is in a good agreement with experiments of E.G. Krasheninnikov, 1981 (see Figure 10.64). The maximum of the $\eta_{total}(p)$ dependence takes place at the pressure:

$$p_{opt} = p_1^{3/5} p_2^{2/5} \left(\frac{T_{\min}}{T^*} \frac{3}{2a_k} \right)^{6/5}. \tag{10.197}$$

The temperature factor in the parentheses is of order unity and therefore the optimal pressure is in the interval $p_1 < p < p_2$, which corresponds to the combined regime of the moderate pressure microwave discharges (Figure 10.62b). In other words, the maximum energy efficiency can be achieved during transition from the homogeneous diffusive to contracted form of the moderate pressure microwave discharges. This statement has a clear physical interpretation. The maximum efficiency of the reactions stimulated by vibrational excitation of molecules requires from one hand that the pressure decrease to diminish the relaxation losses. From another hand it requires higher pressures to provide larger temperature gradients to prevent penetration of the reaction products and electromagnetic energy into the thermal plasma zone. These two requirements are satisfied simultaneously in the moderate pressure range during the microwave discharge transition from the homogeneous diffusive into contracted form. Consider experiments carried out in CO_2 (E.G. Krasheninnikov, 1981), the microwave plasma transition from the homogeneous diffusive into the contracted form took place at $p_1 = 90$ Torr; transition to the quasi-equilibrium thermal discharge regime took place at $p_1 = 90$ Torr; transitions $p_2 = 300$ Torr; and the maximum of energy efficiency of the plasma-chemical process (which means minimum energy cost of CO_2 dissociation) was achieved at an optimal pressure $p_{opt} \approx 120$ Torr.

10.9.8 Power and Flow Rate Scaling of Space-Non-Uniform Moderate Pressure Microwave Discharges

Scaling of plasma processes can be based on the specific energy input E_v. Endothermic processes of gas conversion usually achieve the maximum energy efficiency at the specific energy inputs 1 eV/mol. This means that optimization of energy efficiency requires the fixed ratio of the discharge power to the gas flow rate through the discharge on a level close to 1 kWh per standard m^3. Increase of power requires the corresponding proportional increase in the flow rate. This scaling rule should be corrected for the spatially-non-uniform discharges (in particular, the moderate pressure microwave discharges), where some portion of the flow can avoid contact with plasma. To make these corrections and determine the power limitations of the discharge, consider a microwave discharge with the characteristic radial size of about the skin layer δ and length of about L, sustained in gas flow (see Figure 10.62). If the normal velocity of discharge propagation is u_{in} and axial gas velocity is equal to v, then the angle θ between the vector of axial velocity and the quasi-plane of the discharge front (see Figure 10.62) is:

$$\sin\theta = \frac{u_{in}}{v}. \quad (10.198)$$

If the axial gas velocity v is not very large, then the angle θ is rather large, and most of gas flow is able in principle to cross the discharge front. In this case, at a fixed value of pressure (for example on the optimal level p_{opt}, determined by Eq. (10.197)), the flow rate across the discharge front Q_f is close to the total gas flow rate Q and increases with the axial gas velocity v. In this regime the non-uniformity effects on the energy efficiency are small. At high gas velocities, the influence of the non-uniformity effects becomes crucial. Further increase of the gas flow across the discharge front becomes impossible when the axial gas velocity reaches the maximum value:

$$v_{\max} = \frac{u_{in}}{\sin\theta} \approx u_{in}\frac{L}{\delta}. \quad (10.199)$$

If the axial gas velocities exceed the critical value determined by Eq. (10.199), $v > v_{\max}$, the fast flow compresses the discharge (see Eq. (10.198)). In this case, the effective discharge cross section decreases with gas velocity as $1/v^2$, and the gas flow across the discharge front decreases linearly with gas velocity as $1/v$. Thus, there is the maximum value of the gas flow rate across the discharge front, which can be expressed taking into account Eq. (10.199) as:

$$Q_{f\max} \approx u_{in}\pi n_0 \delta L, \quad (10.200)$$

where n_0 is the gas density. The relation between the flow rate across the discharge front and total flow rate, assuming that at low velocities discharge radius is correlated with the radius of the tube, is:

$$Q_f \approx Q, \quad if \quad Q < Q_{f\max}, \quad (10.201)$$

$$Q_f \approx \frac{Q_{f\max}^2}{Q}, \quad if \quad Q \geq Q_{f\max}. \quad (10.202)$$

Scaling of the plasma-chemical process by a proportional increase of the gas flow rate Q and discharge power P, keeping the specific energy input $E_v = P/Q$ constant, is possible only at flow rate values below the critical one $Q < Q_{f\max}$. Proportional increase of the gas flow rate and the discharge power at higher flow rates leads to an actual increase of the specific energy input proportionally to the square of power: $E_v \propto P^2$. In this case of higher flow rates $Q > Q_{f\max}$, the energy efficiency of plasma-chemical processes decreases as $\eta \propto 1/P^2$. Based on Eqs. (10.201), (10.202), the maximum power, where high energy efficiency is still possible in non-uniform microwave discharges of moderate pressure is:

$$P_{\max} = \pi E_v u_{in}\delta L n_0. \quad (10.203)$$

If $E_v \approx 1$ eV/mol, $u_{in} \approx 300$ cm/sec , $n_0 \approx 5 \cdot 10^{18}$ cm^{-3}, $\delta \approx 1$ cm, $L \approx 10 \div 30$ cm, the power limit Eq. (10.203) gives $P_{\max} = 10$–30 kW, which is in a good agreement with the experiments of N.P. Ageeva, G.I. Novikov, V.K. Raddatis ea. (1986). The further increase of microwave power with still high-energy efficiency becomes possible by some special changes of the discharge geometry and by organizing the non-thermal microwave discharges in supersonic gas flows.

10.10 PROBLEMS AND CONCEPT QUESTIONS

10.10.1 Integral Flux Relation and the Channel Model of Arc Discharges

Use the integral flux relation in the form $J_r + S_r = 0$, which describes the cylindrical plasma column, together with the Maxwell relation for the magnetic field $(curl\vec{H})$ to derive the third equation closing the channel model of arc.

10.10.2 Equations Describing a Long Inductively Coupled RF Discharge

Derive the boundary condition (10.9) for system (10.6), (10.7) of the ICP RF discharge. Show that the magnetic field in a non-conductive gas near the walls of the discharge is the same as the field inside of an empty solenoid.

10.10.3 Damping of Electromagnetic Fields in Skin Layer of ICP

Based on the Maxwell Eqs. (10.10), derive the distribution of electric and magnetic fields inside of ICP Eqs. (10.13), (10.14) in the framework of the metallic cylinder model, that is assuming constant conductivity σ inside plasma.

10.10.4 ICP Temperature as a Function of Solenoid Current and Other Parameters

Combining Eq. (10.21) with the formula for plasma conductivity and the Saha equation, derive Eq. (10.22) for the ICP temperature. Analyze this relation and show that the ICP temperature in the case of strong skin-effect does not depend on the frequency of the electromagnetic field.

10.10.5 Temperature Limits in Thermal ICP Discharges

Typical temperatures in the ICP-discharges are limited to the interval between 7,000 K and 11,000 K. Analyzing Eq. (10.23) for the specific ICP-discharge power, explain why the plasma temperature usually does not exceed 11,000 K. Taking into account the stability of the ICP-discharges at low conductivity conditions (see Eq. (10.28)), give your interpretation of the lower temperature limit of the discharge on the level of about 7,000 K.

10.10.6 Critical Solenoid Current to Sustain Thermal ICP-Discharge

Derive Eq. (10.28) for the minimum value of the solenoid current $I_0 n$, which is necessary to sustain the quasi-equilibrium thermal ICP. For the derivation use Eq. (10.21), assuming a value of electric conductivity $\sigma_m$corresponding to the critical condition $\delta \approx R$ (see Eq. (10.11)). Give your interpretation of the fact that high solenoid current is necessary to sustain the thermal ICP in a smaller discharge tube.

10.10.7 Stability of Thermal ICP-Discharge at High Conductivities

The low conductivity regime of the thermal ICP-discharge is unstable (left branch in Figure 10.4), prove similarly that the ICP regime with high conductivity and strong skin-effect (left branch in Figure 10.4) is stable with respect to temperature, and hence conductivity fluctuations.

10.10.8 Capacitively-Coupled Atmospheric Pressure RF-Discharges

Give your interpretation why the power density of the atmospheric pressure CCP-discharge grows with an increase of the voltage amplitude and the electric field frequency. Explain why the power densities of the uniform atmospheric pressure CCP-discharges are usually lower than those of the pulsed corona discharges.

10.10.9 Microwave Discharge in H_{01}-Mode of Rectangular Waveguide

Calculate the maximum electric field for an electromagnetic wave in the H_{01}-mode of a rectangular waveguide at the frequency $f = 2.5$ GHz (the corresponding wavelength in vacuum $\lambda = 12$ cm). Assume the microwave discharge power is approximately 1 kW, the wide waveguide wall is equal to 7.2 cm, and the narrow one is 3.4 cm long. Calculate the reduced electric field E/p in this system without plasma influence at atmospheric pressure; compare the result with typical E/p in non-thermal glow discharges and thermal arc discharges.

10.10.10 Microwave Discharge in H_{11}-Mode of Round Waveguide

Based on Eqs. (10.31) and (10.32), analyze the electric field distribution along the radius (but in different directions) of the round waveguide operating in H_{11}-mode without a plasma. Assume the electromagnetic wave frequency $f = 2.5$ GHz (the corresponding wavelength in vacuum $\lambda = 12$ cm), the microwave discharge power is approximately 1 kW, the round waveguide radius is equal to 4 cm and calculate the maximum electric field in the case. Compare the calculated maximum electric field with the one calculated in the previous problem under similar conditions but for a rectangular waveguide.

10.10.11 Constant Conductivity Model of Microwave Plasma Generation

In the frameworks of the constant conductivity model, derive the reflection coefficient $\rho = (S_0 - S_1)/S_0$ of an incident electromagnetic wave normal to the sharp boundary of plasma. This can be derived based on boundary conditions for electric and magnetic fields on the plasma surface.

10.10.12 Energy Balance of the Thermal Microwave Discharge

Derive the energy balance Eq. (10.38) for the microwave plasma temperature. Based on this equation,analyze dependence of the microwave plasma temperature on power, pressure and other discharge system parameters.

10.10.13 Laser Radiation Absorption in Thermal Plasma

Based on Eq. (10.41) for the absorption coefficient of laser radiation, interpret the absorption dependence on pressure illustrated in Figure 10.18. Explain in particular why the maximum of the temperature dependence $\frac{\mu_\omega}{p^2}(T)$ shifts to the right on this figure (to the higher temperature levels) when gas pressure increases from 1 to 100 atm.

10.10.14 Geometry of the Continuous Optical Discharge

Analyzing the discharge pictures presented in Figure 10.20, clarify the physical factors, which determines the size of the continuous optical discharge. Experimentally, the minimum plasma radius corresponds in this case to the critical threshold conditions (see Eq. (10.47)) and numerically is about $r_t \approx 0.1$ cm. Estimate the radiation flux density.

10.10.15 Stable and Unstable Regimes of Continuous Optical Discharges

The left low-temperature and low-conductivity branch of the curve $P_0(T)$, Figure 10.19, where $T_m < T_t$ is unstable. Prove in a similar way that the opposite right high-temperature and high-conductivity branch of the curve $P_0(T)$ in Figure 10.19 where $T_m > T_t$ is stable. Explain the similarly from this point of view (stability only at high conductivity conditions) of the continuous thermal ICP, microwave and optical discharges.

10.10.16 Comparative Analysis of Temperatures in the Thermal ICP, Microwave and Optical Discharges

The plasma temperatures in the continuous optical discharges are about twice more than in ICP and arcs, and about 3–4 times larger than in microwave discharges. Give your interpretation why the plasma temperature changes non-monotonically with the frequency of the electromagnetic fields.

10.10.17 Reactive Component of Capacitively-Coupled RF-plasma Resistance

Derive Eq. (10.53) for the reactive (imaginary) resistance component of the plasma itself and sheaths in the plasma phase at high frequencies $\omega > \nu_e$. Explain why the reactive resistance component has an inductive nature.

10.10.18 Voltage Drop on Space Charge Sheaths of CCP-Discharges

Derive Eq. (10.57) for the total voltage U_s related to the space charge, and explain why it includes only the principal harmonic (ω) of the applied voltage, while the voltage drop on each sheath separately contains a constant component and second harmonics (2ω). Prove that the phase shift between voltage and current in the space charge sheath corresponds to the capacitive resistance.

10.10.19 CCP-Discharge Power Transferred to Ions in Sheaths

Derive Eq. (10.61) for power transferred per unit electrode area to ions in a sheath of the ICP-discharge. Give you an interpretation of the inverse proportionality of this power to the cube of electromagnetic oscillation frequency.

10.10.20 Equivalent Scheme of a Capacitively-Coupled RF- Discharge

Based on the equivalent scheme of the ICP-discharges presented in Figure 10.25, calculate the resonant frequency of electromagnetic oscillations for this circuit. Compare the result with Eq. (10.59) for the resonance frequency. Discuss peculiarities of the result in the case of relatively high and relatively low gas pressures.

10.10.21 Motion of the Plasma-Sheath Boundary in CCP-Discharges, Taking into Account Non-Uniformity of Ion Density in the Sheath

Motion of the plasma-sheath boundary is non-harmonic even in the case of $j = -j_0 \sin(\omega t)$, taking into account the non-uniformity of ion density in the sheath zone. Based on Eq. (10.63) for phase evolution across the sheath zone, prove that velocity of the plasma boundary is inversely proportional to ion concentration $n_i(x)$. Explain why the duration of the space charge phase is longer in the vicinity of the ICP-discharge, where plasma is quasi-neutral.

10.10.22 Ion Concentration in Sheaths of Moderate Pressure CCP-Discharges in α-Regime

Balancing the ion flux to the electrode Eq. (10.60) and the ionization rate in the sheath $I_\alpha \cdot L/2$, derive Eq. (10.73) for the ion density in the sheath of a CCP-discharge. Estimate typical numerical values for pressure and current density dependence of the weakly changing logarithmic factor $\ln\Lambda$in this relation.

10.10.23 Sheath Size of Moderate Pressure CCP-Discharges in α-Regime

Based on Eqs. (10.73) and (10.74), estimate sheath sizes in moderate pressure α-discharges. Compare these sheath sizes with Debye radius at relevant plasma parameters. Give your interpretation of the comparison.

10.10.24 Ion Current of Moderate Pressure CCP-Discharges in γ-Regime

Analyze Eq. (10.79) and show that ion current in the normal regime of the γ-discharge is proportional to the square of pressure:$j_i \propto p^2$. Interpret the discharge transition to the low-frequency regime with pressure increase.

10.10.25 High-Pressure Limit of the γ-CCP-Discharge

Prove that the upper-pressure limit of the γ-CCP-discharges Eq. (10.80) is equivalent to the following requirement: number of ions reaching the electrode during one oscillation period is less than the total number of ions in the sheath. Give numerical estimation of the upper-pressure limit Eq. (10.80).

10.10.26 Critical Current of the α–γ Transition

Analyze the dependence of the critical current of the α–γ transition on gas pressure (use Eq. (10.81)).

10.10.27 Effective Electric Field in Low-Pressure CCP-RF Discharges

Derive Eq. (10.90) for the effective reduced electric field in low-pressure CCP-discharges. Analyze the numerical value of the logarithmic factor in this relation and explain why the electric field in this case can be considered as constant. Compare this effect with a similar one in the case of a positive column of DC-glow discharges.

10.10.28 Potential Barrier on the Plasma – Sheath Boundary

Analyze the kinetic flux balance equation (10.99) for the low-pressure capacitive RF-discharges, and show that the potential barrier on the plasma-sheath boundary can be estimated as the lowest energy of electronic excitation.

10.10.29 Stochastic Heating Effect

Analyze the reflection of electrons from the sheath boundaries moving to and from electrodes and calculate the energy transfer to and from an electron. Take into account that the electron flux to the boundary moving from the electrode exceeds that one moving in the opposite direction to the electrode, estimate the energy transferred to the fast electrons by the stochastic heating.

10.10.30 Potential Barrier on the Plasma Boundary at Low and High Current Limits

Analyze Eqs. (10.106), (10.114) for the potential barrier on the plasma boundary at the low and high current limits. Compare these potential barriers between themselves and with the electronic excitation energy ε_1.

10.10.31 Plasma Density and Sheath Thickness Dependence on Current Density in Low-Pressure CCP-RF Discharges

Describe the plasma concentration and sheath thickness dependence on the current density including both regimes of high and low current. Give your interpretation of the dependences, and illustrate your conclusions using the data presented in Figure 10.36.

10.10.32 Critical Current Density of Transition Between Low and High Current Regimes in the Capacitive Discharges

Using Eq. (10.117), estimate the critical value of the discharge current density corresponding to the transition between low and high current regimes of the low-pressure CCP-RF-discharges. Compare the results of your estimations with the relevant data presented in Figure 10.36.

10.10.33 Asymmetric Effects in Low-Pressure Capacitive RF-Discharges

Assuming the surface area of the powered electrode is 10 times the area of the grounded one and neglecting current losses on the walls of the grounded discharge chamber, calculate the ratio of voltages on the sheaths. Interpret why the plasma-treated material is usually placed in the sheath related to the powered electrode.

10.10.34 Secondary-Emission Resonant RF-Discharge

The secondary-emission resonant discharge requires the electron "flight time" to the opposite electrode to coincide with a half period (or odd number of the half-periods) of the electric field oscillations. Show that such discharges can be stable, if the electron emission phase-shift with respect to moment $E = 0$ is also small enough to make electrons emitted in different but close phases to group together.

10.10.35 Magnetron RF-CCP-Discharge

Velocity of electron oscillations in the direction perpendicular to electrodes and corresponding coordinate of this oscillation is described in the magnetron RF-discharges by Eqs. (10.129), (10.130). Derive these relations based on Eq. (10.125) of electron motion in the crossed magnetic and oscillating electric fields. Analyze the resonant characteristics of Eqs. (10.129), (10.130).

10.10.36 Current Density Distribution in ICP-Discharges

For the ICP-discharge, arranged inside of an inductor coil, analyze the current density distribution (10.138) along the radius of the discharge tube. Determine the conditions when the Bessel function describing this distribution can be simplified to the exponential form (10.140), which corresponds to plasma current localization on the discharge periphery.

10.10.37 Equivalent Scheme of an ICP- Discharge

Analyze the equivalent scheme of the ICP-discharge, shown in Figure 10.49, and explain why the discharge current exceeds the current in the inductor coil exactly N-times, where N is the number of windings in the coil Eq. (10.148). This result is derived from Eq. (10.147) by neglecting the voltage drop on the plasma loop, which is correct only at high currents. Determine a criterion for the relation (10.148) between currents in plasma and in inductive coil.

10.10.38 Plasma Density in ICP-Discharges

Low-pressure ICP-discharges are able to operate effectively at electron densities 10^{11}–10^{12} cm^{-3}(even up to 10^{13} cm^{-3}), which is more than 10 times the typical values of electron concentration in the capacitively coupled RF-discharges and provides an important advantage for practical application of the discharges. Analyzing Eqs. (10.149) and (10.151), give your interpretation of this important feature of the ICP-discharges. Explain also the dependence of the plasma density on the electromagnetic oscillation frequency and electric current in the inductor coil.

10.10.39 Abnormal Skin Effect in Low-Pressure ICP-Discharges

Based on Eq. (10.156) for conductivity in the collisionless sheath of a low-pressure ICP-discharge, derive Eq. (10.159) for thickness of the abnormal skin layer. Compare Eqs. (6.1.25) and (10.159) for normal collisional and abnormal skin layers, and determine the ICP-discharge parameters corresponding to transition between them.

10.10.40 Helical Resonator Discharge

Analyze the resonance condition Eq. (10.160) and estimate the typical number of the helix turns necessary to operate the discharge at an electromagnetic wave frequency of 13.6 MHz. Taking into account the Q-values of about 600–1500 in the helical resonator, estimate typical values of the electric fields in the discharge in the absence of plasma, compare these electric field with those necessary for breakdown at low pressures.

10.10.41 Propagation of Electromagnetic Waves in ECR-Microwave Discharge

Based on the dispersion equation (6.6.39), determine the phase and group velocities of the right-hand-polarized electromagnetic waves in an ECR-microwave discharge. Assume that the electromagnetic wave frequency is below the plasma frequency and electron–cyclotron frequency. Analyze the peculiarity of the wave propagation near the ECR-resonance condition.

10.10.42 ECR-Microwave Absorption Zone

Using Eqs. (10.163) and (10.164), estimate the width of the ECR-resonance zone and of the absorbed microwave power per unit area. Compare these energy fluxes with those of radio-frequency CCP and ICP-discharges applied for material treatment.

10.10.43 Whistler Waves and Helicon Discharges

Analyzing the dispersion (6.6.39) for electromagnetic wave in magnetized plasma at helicon conditions Eq. (10.165), derive the relevant dispersion equation for whistler waves, superposition of which forms the helicon modes. Derive a relation for phase velocity of the whistlers, compare this phase velocity with the speed of light and explain how the helicon discharge can operate as the wave-heated one even at relatively low level of radio frequency.

10.10.44 Landau Damping of Helicon Modes

The antenna length l_a for the helicon modes excitation can be chosen related to the electromagnetic wave vector as: $k_z \approx \pi/l_a$. The excited helicon modes by means of the Landau damping are able to heat plasma electrons, whose kinetic energies ε correspond to the wave phase velocity: $\varepsilon = \frac{1}{2}m\left(\frac{\omega}{k_z}\right)^2$. Estimate the antenna length necessary to provide heating of electrons with energies $\varepsilon \approx 30$–50 eV, the most effective for ionization. Assume the excitation radio-frequency $f = 13.56$ MHz.

10.10.45 Planar Surface Wave Discharges

Analyze the dispersion (10.181) of electromagnetic surface waves and give an interpretation of the possibility of surface waves to propagate at lower frequencies in contrast to the conventional case of electromagnetic waves propagation in non-magnetized plasma.

10.10.46 Critical Plasma Density for Surface Wave Propagation

Based on Eq. (10.183), estimate the minimal plasma density necessary for surface wave propagation with frequency 2.45 GHz. Estimate the gas pressure range, which permits considering this plasma as collisionless. Analyze the effect of dielectric constant on the critical plasma density.

10.10.47 Combined Regime of Moderate Pressure Microwave Discharges

Derive the criterion (10.190) of formation of the hot filament of thermal plasma inside of a non-equilibrium one in moderate pressure microwave discharges. Based on the derived formula, make numerical estimations of the critical value of $E^2\sqrt{p}$ to form the hot filament and compare the result with data presented in Figure 10.61. Give your interpretation why the filament formation requires relatively high values of pressure and electric field. Compare microwave and radio-frequency discharges from this point of view.

10.10.48 Non-Uniformity Factor of Energy Efficiency of Microwave Discharges at Moderate Pressures

Give interpretation and conceptual derivation of the non-uniformity factor in the energy efficiency Eq. (10.191), describing the energy losses related to the inverse reactions of products in the hot plasma filament zone. Estimate this non–uniformity factor numerically for the combined microwave discharge regime and the more uniform relatively low-pressure regimes of the microwave plasma.

10.10.49 Pressure Dependence of Energy Efficiency of Microwave Discharges

The energy efficiency (10.196) reaches its maximum at the pressure (10.197). Derive a formula for the maximum energy efficiency. Analyze dependence of the energy efficiency on the ratio T_{min}/T^* of the non-thermal zone temperature and the temperature of transition from non-thermal to thermal ionization mechanisms.

10.10.50 Power and Flow Rate Scaling of Moderate Pressure Microwave Discharges

Proportional increase of the total gas flow Q and discharge power P at high flow rates $Q > Q_{f\max}$ leads to an actual increase of the specific energy input proportional to the square of the power: $E_v \propto P^2$. Show that the energy efficiency of plasma-chemical processes in this case decreases as $\eta \propto 1/P^2$.

11 Discharges in Aerosols and Dusty Plasmas

11.1 PHOTOIONIZATION OF AEROSOLS

11.1.1 General Remarks on Macroparticles Photoionization

The photoionization of aerosols can play an important role in increasing the electron concentration and hence, electrical conductivity of dusty gases (Guha and Kaw, 1968). These processes also make important contributions in astrophysics in the understanding of some of the characteristics of the cosmic plasma because significant amounts of interstellar matter are in the form of macroparticles (Sodha and Guha, 1971). The photoionization of macroparticles makes the most important contribution in the generation of dusty plasma if the radiation quantum energy exceeds the work function of macroparticles, but remains below the ionization potential of neutral gas species. In this case, the steady-state electron density is determined by photoionization of the aerosols and electron attachment to the macroparticles. The positive ions in this case are replaced by charged aerosol particulates (Spitzer, 1941, 1944; Sodha, 1963).

11.1.2 Work Function of Small and Charged Aerosol Particles, Related to Photoionization by Monochromatic Radiation

Calculate the steady-state electron concentration n_e provided by photoionization of spherical mono-dispersed macroparticles of radius r_a and concentration n_a (Karachevtsev and Fridman, 1974, 1975). The aerosol concentration n_m distribution over different charges me should take into account as the work function of the aerosol depends on their charge:

$$\varphi_m = \varphi_0 + \frac{me^2}{4\pi\varepsilon_0 r_a}. \tag{11.1}$$

There are upper n_+ and lower n_- limits of the macroparticle charge. The upper charge limit is due to the fixed photon energy $E = \hbar\omega$, and increasing the work function with a particle charge:

$$n+ = \frac{4\pi\varepsilon_0 \left(E - \varphi_0\right) r_a}{e^2} \tag{11.2}$$

φ_0 is the work function of a noncharged macroparticle ($m = 0$), which slightly exceeds the work function A_0 of the same material having a flat surface (Maksimenko and Tverdokhlebov, 1964):

$$\varphi_0 = A_0 = \left(1 + \frac{5}{2}\frac{x_0}{r_a}\right); \tag{11.3}$$

$x_0 \approx 0.2$ nm is the numerical parameter of the relation. The lower aerosol charge limit n_e ($n_- < 0$) is related to the charge value when the work function and hence the electron affinity to a macroparticle become equal to zero and attachment becomes ineffective:

$$n_- = -\frac{4\pi\varepsilon_0\varphi_0 r_a}{e^2} \tag{11.4}$$

11.1.3 Equations Describing the Photoionization of Monodispersed Aerosols by Monochromatic Radiation

The system of equations describing the concentration of aerosol particles with different electric charges, generated as a result of the photoionization process can be expressed as

$$\frac{dn_{n-}}{dt} = -\gamma p n_{n-} + \alpha n_e n_{n-+1} \tag{11.5}$$

$$\frac{dn_i}{dt} = \gamma p n_{i-1} - \alpha n_e n_i - \gamma p n_i + \alpha n_e n_{i+1}, \quad n_- + 1 \le i \le n_+ - 1, \tag{11.6}$$

$$\frac{dn_{n+}}{dt} = \gamma p n_{n+-1} - \alpha n_e n_{n+}. \tag{11.7}$$

Here, $\alpha \approx \sigma_a v_e$ is the coefficient of attachment of electrons with averaged thermal velocity v_e to the cross section of macroparticles, σ_a, which is assumed here to be constant at $i > n_-$; γ is the photoionization cross section of the macroparticles, which is also supposed here to be constant at $i < n_+$; n_i is the concentration of macroparticles carrying charge ie ($i < 0$ for negatively charged particles), p is the flux density of monochromatic photons. Considering steady-state conditions $d/dt = 0$, and adding sequentially the equations of the system (Eq. 11.5), the balance equation for the charged species is

$$\alpha n_e n_{i+1} = \gamma p n_{i+1}, \quad q = \frac{n_{i+1}}{n_i} = \frac{\gamma p}{\alpha n_e}. \tag{11.8}$$

Here $q = n_{i+1}/n_i$ is a factor describing the distribution of particles over electric charges. Taking into account the electro-neutrality and the mass balance, the equation for electron density is

$$\frac{n_e}{n_a} = \frac{\sum_{k=n-}^{k=n+} kq^k}{\sum_{k=n-}^{k=n+} q^k} \approx \frac{\int_{n-}^{n+} xq^x dx}{\int_{n-}^{n+} q^x dx}. \tag{11.9}$$

Integrating Eq. 11.9 leads to the equation that describes the density of electrons generated by the monochromatic photoionization of mono-dispersed aerosols (Karachevtsev and Fridman, 1974):

$$\frac{n_e}{n_a} = \frac{n + q^{n+} - n\_q^{n-}}{q^{n+} - q^{n-}} - \frac{1}{\ln q}. \tag{11.10}$$

11.1.4 Asymptotic Approximations of the Monochromatic Photoionization

This general aerosol photoionization equation determines the relation between electron concentration n_e, photon energy $E = \hbar\omega$, flux p, and aerosol parameters: radius r_a, work function A_0 and concentration n_a. The dependence $n_e = f(E, p, r_a, n_a, A_0)$ cannot be expressed explicitly from Eq. 11.10, but it can be analyzed asymptotically in the opposite extremes of low and high fluxes of monochromatic photons:

1. *The high flux photon flux extreme* $q \gg 1$. According to the definition of the q-factor (Eq. 11.8), this regime requires: $p \gg \alpha n_e/\gamma$. In this case, the electron concentration based on Eq. 11.10 is

$$\frac{n_e}{n_a} = \frac{n_+ q^{n+}}{q^{n+}} = n_+. \quad (11.11)$$

Thus, if the photon flux is sufficiently high to satisfy the inequality:

$$p \gg \frac{\alpha n_a n_+}{\gamma}, \quad (11.12)$$

the photoionization of aerosols is so intensive that each macroparticle receives the maximum possible electric charge $n_+ e$, which leads to the electron density equation (Eq. 11.11)
2. *The low photon flux case* ($p \ll \alpha n_a/\gamma$). A decrease of the photon flux p leads to a reduction of the q factor as well, until at the extreme case $p \ll \alpha n_a/\gamma$ the electron concentration becomes relatively low $n_e/n_a \ll 1$ and the q factor reaches its minimal value. Based on Eq. 11.10:

$$\frac{n_+ q^{n+} - n_- q^{n-}}{q^{n+} - q^{n-}} = \frac{1}{\ln q}. \quad (11.13)$$

The minimal value of the q-factor at low photon fluxes is close to unity. This means that the electron density at low photon fluxes can be estimated as $q = 1$ (see Eq. 11.8). The asymptotic expression for the electron density provided by monochromatic photoionization of aerosols is

$$n_e = n_a n_+, \text{ if } p \gg \frac{\alpha n_a n_+}{\gamma}, \quad (11.14)$$

$$n_e = \gamma p/\alpha, \text{ if } p \ll \frac{\alpha n_a}{\gamma}. \quad (11.15)$$

A convenient chart for calculating the electron density n_e as a function of log p and log $(n_a n_+)$ at monochromatic photoionization based on Eqs. (11.14), (11.15) is presented in Figure 11.1. In this chart, it is assumed that $\alpha/\gamma = 10^8$ cm/s. Chart (Eq. 11.13) also shows the relative accuracy of the calculations based on Eqs. 11.14, 11.5. It is presented in the form of simple relative accuracy $\eta = \Delta n_e/n_e$, as well as in a form of the relative accuracy of calculations of ln n_e:

$$\eta_1 \approx \frac{\Delta n_e}{2.3 n_e \log n_e}. \quad (11.16)$$

11.1.5 Photoionization of Aerosols by Continuous Spectrum Radiation

Generalization of the relations for particles density distribution over charges n_i (Eq. 11.8) for a continuous spectrum is

$$\gamma p_i n_i = \alpha n_e. \quad (11.17)$$

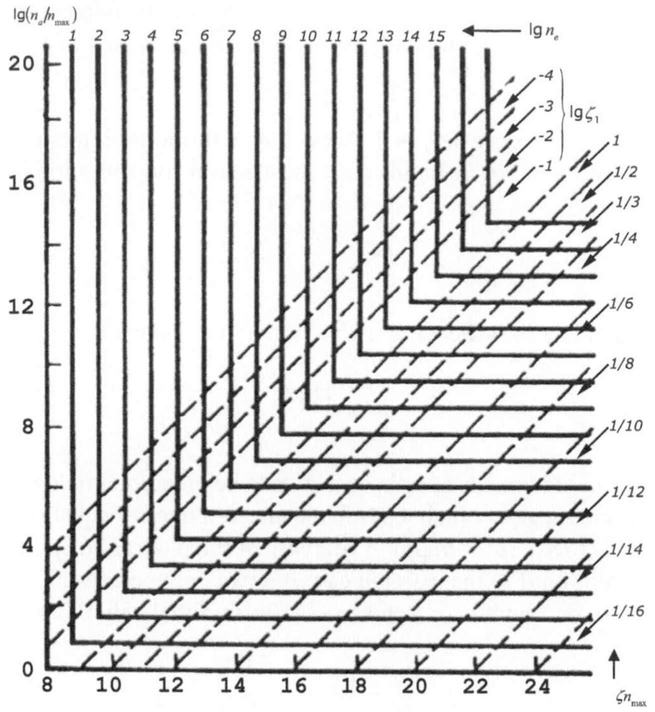

FIGURE 11.1 Calculation of electron concentration provided by photoionization of aerosols log p (electron number density in cm^{-3}, flux p in $cm^{-2}\,s^{-1}$, aerosol number density n_a in cm^{-3}).

In this case, p_i is the flux density of photons with energy sufficient for the $(n+1)$ th photoionization. Using the energy distribution of photons in the form: $\xi_i = p_i/p_0$, $(\xi_0 = 1)$, the concentration of macroparticles with charge ke is

$$n_1 = n_0\left(\frac{\gamma p_0}{\alpha n_e}\right)\xi_0, \tag{11.18}$$

$$n_2 = n_1\left(\frac{\gamma p_0}{\alpha n_e}\right)\xi_1 = n_0\left(\frac{\gamma p_0}{\alpha n_e}\right)^2 \xi_0\xi_1, \tag{11.19}$$

$$n_k = n_0\left(\frac{\gamma p_0}{\alpha n_e}\right)^k \prod_{i=0}^{k-1}\xi_i. \tag{11.20}$$

These relations can be applied for $k < 0$ as well. In this case, the product $\prod_{i=0}^{k}\xi_i$ implies $\prod_{i=0}^{-k}\left(1/\xi_{-i}\right)$.

Taking into account electro-neutrality and the material balance of the aerosol system, Eqs. 11.18, 11.19, 11.20 leads to a generalization of Eq. 11.10 for the case of radiation in the continuous spectrum:

$$\frac{n_e}{n_a} = \frac{\sum_{k=n-}^{\infty} kn_k}{\sum_{n-}^{\infty} n_k} = \frac{\sum_{k=n-}^{\infty} kq_0^k \prod_{i=0}^{k-1}\xi_i}{\sum_{k=n-}^{\infty} q_0^k \prod_{i=0}^{k-1}\xi_i}. \tag{11.21}$$

The q_0 factor for the nonmonochromatic radiation is defined as: $q_0 = \gamma p_0/\alpha n_e$.

11.1.6 Photoionization of Aerosols by Radiation with Exponential Spectrum

Equation 11.21 cannot be analytically solved, but it can be simplified assuming an exponential behavior of the photon energy distribution: $p(E) = p\exp(-\delta E)$. Such a distribution is valid for the tail of the thermal radiation spectrum. Parameters of the photon flux p_0 and ξ_i relevant to the photoionization are

$$p_0 = \int_{\varphi_0}^{\infty} p(E)\,dE = \frac{p}{\delta}\exp(-\delta\varphi_0), \tag{11.22}$$

$$\xi_i = \frac{p_i}{p_0} = \frac{1}{p_0}\int_{\varphi_i}^{\infty} p(E)\,dE = \exp\left(-\delta i\frac{e^2}{4\pi\varepsilon_0 r_a}\right). \tag{11.23}$$

Substituting p and ξ_i from Eqs. 11.22 and 11.23 into Eq. 11.21 and assuming $n_- \to -\infty$, rewrite Eq. 11.21 as

$$\frac{n_e}{n_a} = \frac{\sum_{-\infty}^{\infty} k q_0^k \exp\left(-\lambda\left(k(k-1)/2\right)\right)}{\sum_{-\infty}^{\infty} q_0^k \exp\left(-\lambda\left(k(k-1)/2\right)\right)}, \tag{11.24}$$

where $\lambda = \delta e^2/4\pi e_0 r_a$. This equation can be solved and the electron density provided by the nonmonochromatic radiation can be expressed using the elliptical functions (Sayasov, 1958):

$$\frac{n_e}{n_a} = y + \frac{\rho}{2\pi}\frac{d}{dy}\ln\theta_3(y,\rho). \tag{11.25}$$

$\theta_3(y, \rho)$ is the elliptical θ function of two variables $y = 1/\lambda \ln q_0 + 1/2$, $\rho = 2\pi/\lambda$. If $\rho \gg 1$, Eq. 11.25 can be simplified, and the relation for electron concentration can be given as

$$\frac{n_e}{n_a} = \frac{1}{\lambda}\ln q_0 + \frac{1}{2}. \tag{11.26}$$

In the opposite extreme of $\rho \ll 1$, the asymptotic expression for the elliptical function gives

$$\frac{n_e}{n_a} = \frac{q_0}{1+q_0}. \tag{11.27}$$

11.1.7 Kinetics of Establishment of the Steady-State Aerosol Photoionization Degree

Non-steady-state evolution of the electron density due to photoionization of aerosols follows the equation:

$$\frac{dn_e}{dt} = \gamma\sum_{i=n-}^{\infty} p_i n_i - \alpha n_e\left(n_a - n_{n-}\right). \tag{11.28}$$

The electron attachment is effective for all macroparticles, but those with the minimal charge n_-. The simple case of monochromatic radiation permits assuming that the photon flux $p_i = p$ for $n_- \leq i \leq n_+ - 1$, and $p_i = 0$ outside of this aerosol charge interval. Then, Eq. 11.28 becomes:

$$\frac{dn_e}{dt} = \gamma p\left(n_a - n_{n+}\right) - \alpha n_e\left(n_a - n_{n-}\right). \qquad (11.29)$$

This equation cannot be solved because it requires analysis of the evolution of the partial concentration of macroparticles with all possible charges $n_- \leq i \leq n_+$. However, the characteristic time for establishing steady-state electron concentrations can be found for the extreme cases of low and high photon flux densities. At relatively low photon flux densities $p \ll \alpha n_a/\gamma$, as it was discussed above: $q \approx 1$. In this case, one can conclude that $n_{n-}, n_n+ \ll n_a$. The solution of the kinetic Eq. 11.29 is

$$n_e(t) = \frac{\gamma p}{\alpha}\left[1 - \exp\left(-\alpha n_a t\right)\right]. \qquad (11.30)$$

The characteristic time establishing the steady-state electron density for the low photon flux is

$$\tau = \frac{1}{\alpha n_a}. \qquad (11.31)$$

In the opposite case of high photon flux densities $p \gg \alpha n_a n_+/\gamma$, most of macroparticles have the same maximum electric charge: $Z_a e = n_+ e$. While the aerosol particles gain this charge, the ionization process obviously slows down. Assuming that for each particle in this case $dZ_a/dt \approx \gamma p$ ($Z_a < n_+$), the characteristic time for establishing the steady-state electron density is

$$\tau = \frac{n_+}{\gamma p}. \qquad (11.32)$$

Equations 11.31 and 11.32 can be generalized for the photoionization time at any photon fluxes:

$$\tau = \frac{n_+}{\alpha n_a n_+ + \gamma p}. \qquad (11.33)$$

At low and high photon fluxes, this formula corresponds to the asymptotes (Eqs. 11.31 and 11.32).

11.2 THERMAL IONIZATION OF AEROSOLS

11.2.1 General Aspects of Thermal Ionization of Aerosol Particles

The energy necessary for ionization of macroparticles is related to their work function, which is usually lower than the ionization potential of atoms and molecules. Assuming that the ionization potentials determine the exponential growth of ionization rate with temperature, one can conclude that the thermal ionization of aerosol particles can be very effective and provide high electron density and conductivity at relatively low temperatures. Thermal ionization of aerosol particles was applied to increase the electron concentration and conductivity in magneto-hydrodynamic (MHD) generators (Kirillin and Sheindlin, 1971). The conductivity increase in this case is more significant with respect to the alternative approach of using alkaline metal additives. The effect of thermal macroparticles ionization plays a special role in rocket engine torches (Musin, 1971). Absorption and reflection of radio waves by the plasma of rocket engine torches affects and complicate control of the rocket trajectory. Electron density in the flame plasma can be abnormally high because of the thermal ionization of macroparticles (see Shuler and Weber, 1954; Einbinder, 1957; Arshinov and Musin 1958a,b, 1959, 1962; Samuilov, 1966a,b,c, 1967, 1968, 1973).

11.2.2 Photo-Heating of Aerosol Particles

To analyze the thermal ionization of aerosol particles provided by their photo-heating, calculate the photo-heating of aerosols. Consider the heating of particles by a monochromatic radiation flux pE, assuming that the photon energy $E = \hbar\omega$ is not sufficient for direct ionization. The energy balance for the temperature T of aerosols particles can be expressed as

$$c\rho\frac{4}{3}\pi r_a^3\frac{\partial T}{\partial t} = pE\pi r_a^2 + 4\pi r_a^2\lambda\left.\frac{\partial T}{\partial r}\right|_{r=r_a} - \sigma T^4 4\pi r_a^2. \tag{11.34}$$

Here, c, ρ, r_a are specific heat, density, and radius of aerosol particles; λ is the gas conduction coefficient; σ is the Stephan–Boltzmann coefficient. In terms of the quasi-steady-state heat transfer coefficient α from the macroparticle surface, the energy balance Eq. 11.34 can be rewritten as

$$\frac{1}{3}c\rho r_a\frac{\partial T}{\partial t} = \frac{1}{4}pE - \alpha\left(T - T_\infty\right) - \sigma T^4, \tag{11.35}$$

where T_∞ is the ambient temperature. The steady-state temperature of aerosols can be expressed as

$$T_{st} = T_\infty + \frac{pE}{4\alpha}, \text{ if } \gamma = \frac{4\sigma}{pE}\left(T_\omega + \frac{pE}{4\alpha}\right)^4 \ll 1, \tag{11.36}$$

$$T_{st} = \left(\frac{pE}{4\sigma}\right)^{1/4}, \text{ if } \gamma = \frac{4\sigma}{pE}\left(T_\omega + \frac{pE}{4\alpha}\right)^4 \gg 1. \tag{11.37}$$

The case of $\gamma \ll 1$ corresponds to conductive cooling of the macroparticles, while $\gamma \gg 1$ corresponds to radiative cooling. The dependence of the steady-state aerosol temperature on the logarithm of radiation power flux log (pE, W/cm^2) and the logarithm of the macroparticles radius log (r_a, cm) is presented in Figure 11.2. This dependence is related to photo heating of aerosols in air, assuming the Nusselt number $Nu = 2\alpha r_a/\lambda \rightarrow 2$. The cooling mode separation line $\gamma = 1$ is also shown in Figure 11.2.

11.2.3 Ionization of Aerosol Particles due to Their Photo-Heating, the Einbinder Formula

If the energy of the photons is insufficient for the photo-effect, the ionization of macroparticles is mostly provided by the thermal ionization mechanism. Assume that the average distance between macroparticles greatly exceeds their radius: $n_a r_a^3 \ll 1$ (n_a is the concentration of aerosols). If the average charge of a macroparticle is less than the elementary one ($n_e \ll n_a$), the electron density can be found as

$$n_e = K\left(T\right) = \frac{2}{\left(2\pi\hbar\right)^3}\left(2\pi mT\right)^{3/2}\exp\left(-\frac{A_0}{T}\right). \tag{11.38}$$

A more interesting case corresponds to the average macroparticle charge exceeding the elementary one, and therefore to the electron density exceeding that of the aerosols. In this case, ($n_e/n_a \gg 1/2$)

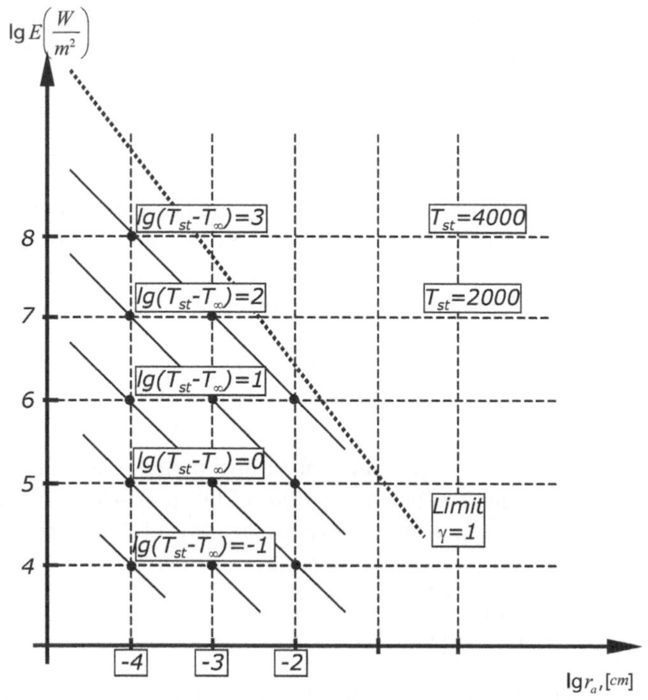

FIGURE 11.2 Heating of microparticles by irradiation. Temperature is in degree Kelvin.

and the electron density can be found from the Einbinder formula (Einbinder, 1957; Arshinov and Musin, 1958a,b), taking into account the work function dependence on the macroparticle charge:

$$\frac{n_e}{n_a} = \frac{4\pi\varepsilon_0 r_a T}{e^2}\ln\frac{K(T)}{n_e} + \frac{1}{2}. \tag{11.39}$$

For numerical calculation of the electron density in the typical range of parameters: $n_a \approx 10^4$–10^8 cm^{-3} with work functions about 3 eV, Eqs. 11.38 and 11.39 become:

$$n_e = 5\times10^{24}\,\text{cm}^{-3}\exp\left(-\frac{A_0}{T}\right), \text{ if } T < 1000\,K, \tag{11.40}$$

$$n_e = n_a\frac{4\pi\varepsilon_0 r_a T}{e^2}\left(20 - \frac{A_0}{T}\right), \text{ if } T > 1000\,K. \tag{11.41}$$

These relations describing thermal ionization of aerosol particles imply thermodynamic quasi-equilibrium and, hence, equality of temperatures of macroparticles T_a, electrons T_e, and neutrals. However, in the case of relatively large distances between macroparticles $n_a r_a^3 \ll 1$, which occurs in the photo heating, aerosol temperatures can exceed that of neutral gas. Electron temperatures are determined not only by the macroparticle temperature, but also by the energy transfer to relatively cold gas, and by possible direct radiative heating. As a result of these processes, $T_e \neq T_a$. The non-equilibrium effects influence the electron density provided by thermal ionization and should be taken into account. To take these nonequilibrium effects into account, note that $K(T)$ in Eq. 11.38 corresponds to the electron concentration near the surface. Then the Boltzmann electron energy distribution function permits generalizing the Einbinder formula to the nonequilibrium case of $T_e \neq T_a$ (Fridman, 1976a,b):

$$\frac{n_e}{n_a} = \frac{4\pi\varepsilon_0 r_a T_e}{e^2} \ln \frac{K(T_a)}{n_e} + \frac{1}{2}. \tag{11.42}$$

This formula can be expressed in the form similar to the quasi-equilibrium relation (Eq. 11.41):

$$n_e = n_a \frac{4\pi\varepsilon_0 r_a T_a}{e^2} \left(20 - \frac{A_0}{T_a}\right), \tag{11.43}$$

but distinguishing the temperature values of aerosol particles and electron gas.

11.2.4 Space Distribution of Electrons around a Thermally Ionized Macroparticle: Case of High Aerosol Concentration

The Einbinder formula assumes that electrons move from the macroparticle surface at a distance much exceeding their radius r_a, and for this reason, cannot be applied at high aerosol concentrations $n_a r_a^3 \approx 1$. The space distribution of electrons around thermally ionized macroparticles can be qualitatively modeled as the one-dimensional (1D) space distribution $n_e(x)$ of electrons between two plane cathodes heated to the high-temperature T. Combining the Boltzmann distribution function of the thermally emitted electrons $n_e(x) = K(T)\exp[e\varphi(x)/T]$ with the Poisson's equation gives in this case the differential equation for the electric potential $\varphi(x)$ distribution:

$$\frac{d^2y}{dx^2} + b\exp y = 0. \tag{11.44}$$

Here: $y = e\varphi(x)/T$ characterizes the potential distribution; parameter $b = e^2K(T)/\varepsilon_0 T$ is related to the electron concentration $K(T)$ near the macroparticle (cathode) surface, which can be found from Eq. 11.38. The second-order differential Eq. 11.44 can be simplified to a first order one:

$$\frac{dy}{dx} = -\sqrt{2|b|(C_1 + \exp y)}. \tag{11.45}$$

The constant C_1 can be found from the boundary conditions on the cathode: $\varphi = 0$, $\exp y = 1$, $|y'| = eE_0/T$, where $E_0 = 1/2\varepsilon_0 \bar{n}_e ea$ is the electric field near the cathode, $\bar{n}_e$ is the average electron concentration, a is the distance between cathodes. The boundary conditions determine the relation between the constant C_1 and the electric field energy density:

$$1 + C_1 = \frac{\varepsilon_0 E_0^2/2}{K(T)T} = \alpha. \tag{11.46}$$

Taking into account the symmetry of potential distribution between cathodes, one can conclude that $C_1 < 0$ and $\alpha < 1$. This means that according to Eq. 11.46, the thermal energy density of electrons near the cathode exceeds the density of the electric field energy. The solution of Eq. 11.45 gives the electric potential, and hence, the electron density distribution between two cathodes (Karachevtsev and Fridman, 1977):

$$n_e(x) = \frac{K(T)(1-\alpha)}{\cos^2\left(\sqrt{\frac{1-\alpha}{2}}\frac{x}{r_D^0}\right)}. \tag{11.47}$$

In this relation, $K(T)$ is the electron density on the cathode surface (Eq. 11.38); coordinate $x = 0$ is assumed in the center of the inter-electrode gap; $r_D^0 = \sqrt{\varepsilon_0 T / K(T) e^2}$ is the Debye radius corresponding to the electron density near the surface. The boundary conditions on the cathodes $n_e(x = a/2) = K(T)$ permits finding the electric-field energy density parameter α:

$$\alpha = \sin^2\left(\sqrt{\frac{1-\alpha}{2}}\frac{a}{2r_D^0}\right). \tag{11.48}$$

If $r_D^0 \ll a$, the electron density distribution $n_e(x)$ is nonuniform, most of the electrons are located in the close vicinity of a macroparticle. The energy density parameter α in this case is

$$\alpha \approx 1 - 4\pi^2\left(\frac{r_D^0}{a}\right)^2 + 32\pi^2\left(\frac{r_D^0}{a}\right)^3, \tag{11.49}$$

which leads to the expression for the average electron concentration between cathodes (aerosol particles):

$$\bar{n}_e = \sqrt{\frac{\varepsilon_0 K(T)\cdot T}{a^2 e^2}} \approx K(T)\frac{r_D^0}{a}. \tag{11.50}$$

The electric field distribution between the cathodes can be found based on the distribution of electron density Eq. 11.47 and the Boltzmann distribution:

$$E(x) = \frac{T}{e r_D^0}\sqrt{2(1-\alpha)}\tan\left(\sqrt{\frac{1-\alpha}{2}}\frac{x}{r_D^0}\right). \tag{11.51}$$

From Eq. 11.51, the electric field is equal to zero in the middle of the inter-cathode interval, and has a maximum value on the cathode surface. Characteristics of the electron density and electric field distributions Eqs. 11.47, 11.50, and 11.51 can be generalized, for the compact aerosol particles $\left(n_a r_a^3 \approx 1\right)$, assuming $r_a \approx a$. The electron density distribution Eq. 11.47 in the macroparticle corresponds to the Langmuir relation for the electron density near a plane cathode (Langmuir, 1961):

$$n_e(z) = \frac{K(T)}{\left(1 + \frac{z}{\sqrt{2} r_D^0}\right)^2}. \tag{11.52}$$

z is distance from the plane cathode, and $K(T)$ is the electron density near the cathode (Eq. 11.38).

11.2.5 Space Distribution of Electrons around a Thermally Ionized Macroparticle: Case of Low Concentration of Aerosol Particles

The case of relatively low aerosol concentration $n_a r_a^3 \ll 1$ is the most typical one in plasma chemistry. In this case, distances between aerosol particles exceed their radius, and a description of the electron density and electric field space distributions requires consideration of a lone macroparticle. Combining the Poisson equation with the Boltzmann distribution leads to the following equation for electric potential $\varphi(r)$ around a spherical thermally ionized aerosol particle:

$$\frac{d^2\varphi}{dr^2} + \frac{2}{r}\frac{d\varphi}{dr} = \frac{e}{\varepsilon_0}K(T)\exp\left(\frac{e\varphi}{T}\right). \tag{11.53}$$

This equation has no solution in elementary functions; a detailed numerical analysis was made by Samuilov (1973). To analyze Eq. 11.53 it is convenient to rewrite it with respect to electron concentration, taking into account the Boltzmann distribution:

$$\frac{\mathrm{d}}{\mathrm{d}r}\left(\frac{r^2}{n_e}\frac{\mathrm{d}n_e}{\mathrm{d}r}\right)=\frac{e^2}{\varepsilon_0 T}n_e r^2. \tag{11.54}$$

For the spherical symmetry, the Boltzmann relation $n_e(x) = K(T)\exp[e\varphi(x)/T]$ becomes

$$E(r)=-\frac{\mathrm{d}\varphi}{\mathrm{d}r}=-\frac{1}{e}\frac{1}{n_e}\frac{\mathrm{d}n_e}{\mathrm{d}r}. \tag{11.55}$$

Boundary condition for Eq. 11.54: $\mathrm{d}n_e/\mathrm{d}r(r = r_{av}) = 0$, where $r_{av} \approx n_a^{-1/3}$ is the average distance between aerosol particles. Integrate the right-hand and left-hand sides of Eq. 11.54 to obtain:

$$N(r)=\int_r^{r_{av}} n_e(r)4\pi r^2\mathrm{d}r=-\frac{4\pi\varepsilon_0 Tr^2}{e^2}\frac{1}{n_e(r)}\frac{\mathrm{d}n_e(r)}{\mathrm{d}r}. \tag{11.56}$$

Here $N(r)$ is number of electrons located in the radius interval from r ($r \geq r_a$) to $r_{av.}$ Combining Eqs. 11.56 and 11.55 gives:

$$N(r)=\frac{E(r)}{e/4\pi\varepsilon_0 r^2}. \tag{11.57}$$

In the close vicinity of a macroparticle ($z = r - r_a \ll r_a$), electron density distribution is expressed using the Langmuir relation (Eq. 11.52). Then the total charge of an aerosol particle can be calculated based on Eq. 11.56 as

$$Z_a = N(r_a)=\frac{4\pi\varepsilon_0 r_a T}{e^2}\left(\frac{r_a}{r_D^0}\right). \tag{11.58}$$

Integrating the Langmuir relation (Eq. 11.52), one finds that most of the electrons are located in a thin layer about $r_D^0 \ll r_a$ around the macroparticle. These electrons are confined in the Debye layer by the strong electric field of the aerosol particle. This electric field can be expressed from Eq. 11.57 as

$$E(r)=\frac{T}{e\left(\frac{r_D^0}{\sqrt{2}}+\frac{(r-r_a)}{2}\right)}. \tag{11.59}$$

11.2.6 Number of Electrons Participating in Electrical Conductivity of Thermally Ionized Aerosols

Because thermally ionized aerosol particles have high values of the inherent electric field (Eq. 11.59) when their charge Z_a is sufficiently high, the electrical conductivity of such aerosol system at $n_a r_a^3 \ll 1$ depends on the value of external electric field E_e. This phenomenon occurs because at higher external electric fields, more electrons can be released from being trapped by the inherent electric field of a macroparticle. The electrical conductivity of the free space between macroparticles is

$$\sigma_0 = en_a bN(r_e), \tag{11.60}$$

where b is the electron mobility in heterogeneous medium, and $N(r_e) = \int_{r_e}^{r_{av}} n_e(r)4\pi r^2 dr$ determines how many electrons participate in the electrical conductivity in the external electric field E_e with respect to one macroparticle. If the external electric field is relatively low, the concentration of electrons n_e^* participating in electrical conductivity can be found from the Einbinder formula:

$$N(r_{e1}) = \frac{N_e^*}{n_a} = \frac{4\pi\varepsilon_0 r_a T}{e^2} \ln\frac{K(T)}{n_2^*} \gg 1. \tag{11.61}$$

This expression for $N(r_{e1})$ with Eqs. 11.57 and 11.59 permits specifying a virtual sphere of radius r_{e1} around a macroparticle. Electrons located outside of this sphere are "not trapped" by the inherent electric field of the aerosol particle, and provide the Einbinder electron density (Eq. 11.39):

$$r_{e1} - r_a \approx r_a \ln^{-1}\frac{K(T)}{n_e^*} < r_a. \tag{11.62}$$

The next relation expresses the electric field of a macroparticle on the virtual sphere of radius r_e:

$$E(r_{e1}) = \frac{T}{r_a e} \ln\frac{K(T)}{n_e^*}. \tag{11.63}$$

If the external electric field is less than the critical value determined by Eq. 11.63: $E_e < E(r_{e1})$, then the "free" electron concentration follows the Einbinder formula and does not depend on the value of the external electric field. As a consequence, the electrical conductivity of aerosols also does not depend on the external electric field in this case. When the external electric field exceeds the critical value, the number of electrons per macroparticle able to participate in the electrical conductivity $N(r_e)$ becomes dependent on the external electric field. To find this dependence, assume that only electrons not trapped in the vicinity of the surface by the inherent electric field of a macroparticle ($E(r) \le E_e$) can participate in the electric conductivity. The electric field of a macroparticle $E(r)$ decreases with the radius r(distance from the center of a macroparticle). Therefore if $E(r_e) > E(r_{e1})$, then taking into account Eq. 11.62, one can conclude that $r_e - r_a < r_a$. In this case, based on Eq. 11.57, the number of electrons $N(r_e)$ (per macroparticle) participating in the electrical conductivity in the external electric fields in the interval $E(r_{e1}) < E_e < E(r_a)$:

$$N(r_e) = \frac{4\pi\varepsilon_0 E_e r_a^2}{e}. \tag{11.64}$$

Here $E(r_a)$ is the maximum value of the inherent electric field of a macroparticle, which is obviously realized on its surface (see Eq. 11.59). If the external electric field exceeds the maximum value of the inherent electric field $E_e > E(r_a)$, then all thermally emitted electrons are able to participate in the electrical conductivity $N(r_e) = N(r_a)$ (see Eq. 11.58).

11.2.7 Electrical Conductivity of Thermally Ionized Aerosols as a Function of External Electric Field

Combining the expressions for the number of electrons participating in electrical conductivity $N(r_e)$, Eqs. 11.61, 11.64, and 11.58, together with Eq. 11.60 gives the dependence of electrical conductivity of the thermally ionized aerosols on the external electric field:

$$\sigma_0 = \frac{4\pi\varepsilon_0 r_a T}{e^2}\left(\frac{r_a}{r_D^0}\right) n_a eb \quad \text{if} \quad E_e \geq \frac{T}{er_D^0}, \tag{11.65}$$

$$\sigma_0 = 4\pi\varepsilon_0 E_e r_a^2 n_a b \quad \text{if} \quad \frac{T}{er_a}\ln\frac{K(T)}{n_e^*} < E_e < \frac{T}{er_D^0}, \tag{11.66}$$

$$\sigma_0 = \frac{4\pi\varepsilon_0 r_a T}{e^2}\left(\ln\frac{K(T)}{n_e^*}\right) n_a eb \quad \text{if} \quad E_e \leq \frac{T}{er_a}\ln\frac{K(T)}{n_e^*}. \tag{11.67}$$

From Eqs. 11.65–67, the electrical conductivity of aerosols begins depending on the external electric field explicitly when $E_e > T/er_a \ln K(T)/n_e^*$. It means that the external electric field should be greater than 200 V/cm, if $T = 1700$ K, $r_a = 10$ μm, $A_0 = 3$ eV. The maximum number of electrons participate in electrical conductivity if $E_e \geq T/er_D^0$, which indicates the same parameters above 1000 V/cm. To calculate the total electrical conductivity of the heterogeneous medium, one must also take into account the conductivity σ_a of the macroparticle material (Mezdrikov, 1968):

$$\sigma = \sigma_0\left(1 - \frac{4}{3}\pi r_a^3 n_a \frac{\sigma_0 - \sigma_a}{2\sigma_0 + \sigma_a}\right). \tag{11.68}$$

11.3 ELECTRIC BREAKDOWN OF AEROSOLS

11.3.1 Influence of Macroparticles on Electric Breakdown Conditions

Macroparticles can be present in a discharge gap for different reasons, for example, air impurities or detachment of micro-pikes from the electrode surface (Berger, 1975; Crichton and Lee, 1975). Also aerosol particles can be injected into a discharge gap to increase its breakdown resistance, or conversely to initiate the breakdown (Slivkov, 1966). Three major physical factors determine the decrease of the breakdown resistance due to macroparticles in a discharge gap (Cookson and Farish (1972); Berger, 1974; Cookson and Wotton, 1974):

- Moving macroparticles are somewhat similar to electrode surface irregularities and induce local intensification of the electric field, which stimulates breakdown.
- When a macroparticle approaches the electrode surface, its electric potential changes quickly. The macroparticle collision with the surface leads to a local change of the electric field structure, and also to possible local intensification of the electric field.
- Decrease of the breakdown resistance also occurs due to micro-breakdowns between macroparticles and between macroparticles and electrodes.

These factors are able to decrease the breakdown voltage at atmospheric pressure air by a factor of two if the size of the aerosol particles is rather large, up to 1 mm (Slivkov, 1966). Details regarding experimental and theoretical consideration of this effect in different heterogeneous discharge systems can be found in Raizer (1972a,b) and Bunkin and Savransky (1973). However, in many cases, admixture of the aerosols leads not to a decrease, but to an increase of the breakdown resistance and breakdown voltage (see e.g., Stengach, 1972, 1975; Walker, 1974). This effect is mostly due to the intense attachment of electrons to the surface of aerosol particles. Taking into account the special practical importance of this effect will be considered in more detail below (Karachevtsev and Fridman, 1976).

11.3.2 General Equations of Electric Breakdown in Aerosols

Neglecting the nonuniformity effects discussed in the previous section, the kinetic equations describing the evolution of electron and positive ion densities in the aerosol system can be given:

$$\frac{dn_e}{dt} = An_e + Bn_i, \tag{11.69}$$

$$\frac{dn_i}{dt} = Cn_e + Dn_i, \tag{11.70}$$

The effective frequencies A, B, C, D describing the breakdown of the aerosol system are

$$A = \alpha v_d^e - \beta v_d^e - \frac{D_e}{d^2} + K - \frac{v_d^e}{d}, \tag{11.71}$$

$$B = \gamma_{ia}\sigma_{ia}v^i n_a + \gamma_{iw}\frac{v_d^i}{d}, \tag{11.72}$$

$$C = k_i n_0, \tag{11.73}$$

$$D = k_a n_a + \frac{D_i}{d^2} + \frac{v_d^i}{d}, \tag{11.74}$$

$$K = \gamma_{pa}\mu v n_0 + \gamma_{pw}\mu(1 - v)n_0. \tag{11.75}$$

Here, n_0, n_a are concentrations of neutral gas and aerosol particles; v_d^i, v^i, v_d^e, v^e are thermal and drift (with subscript "d") velocities of electrons (superscript "e") and ions (superscript "i"); α, β are the first and second Townsend coefficients (see Sections 4.4.1 and 4.4.3), which determine the formation and attachment of electrons per unit length along the electric field during the electron drift; D_e, D_i are the diffusion coefficients of electrons and ions (which are free or ambipolar); d is the size of a discharge gap; σ_{ia} is the cross section of ion collisions with the aerosol particles; γ_{ia}, γ_{pa}, γ_{pw}, γ_{iw} are probabilities of an electron formation in an ion collision with a macroparticle, photon interaction with a macroparticle, photon interaction with a wall and ion collision with a wall; μ is the rate coefficient of a photon formation calculated with respect to an electron–neutral collision; v is the probability for a photon to reach an aerosol particle; k_i is the rate coefficient of neutral gas ionization by electron impact; k_a is the rate coefficient of ion losses related to collisions with aerosol particles. In the system (Eqs. 11.69, 11.70), formation of electrons is due to neutral gas ionization, secondary electron emission from the surface of macroparticles, ion collisions with aerosols, photoemission from macroparticles, and discharge walls are taken into account. Electron losses are due to attachment to gas molecules and macroparticles plus diffusion, and drift along the electric field. Ions are formed in electron–neutral collisions, and their losses are due to collisions with macroparticles and the discharge walls. Because of relatively low concentration of the charged particles, their volume recombination can be neglected. Solution of Eq. 11.69, 11.70 for the electron concentration n_e with the initial conditions: $n_e(t = 0) = n_e^0$ and $n_i(t = 0) = n_i^0$ can be given as

$$n_e(t) = \frac{\left(An_e^0 + Bn_i^0\right) - \lambda_2 n_e^0}{\lambda_1 - \lambda_2}\exp(\lambda_1 t) + \frac{\lambda_1 n_e^0 - \left(An_e^0 + Bn_i^0\right)}{\lambda_1 - \lambda_2}\exp(\lambda_2 t). \tag{11.76}$$

Here $\lambda_{1,2}$ are solutions of the characteristic equation related to the system (Eq. 11.69, 11.70):

$$\lambda_{1,2} = \frac{A - D \pm \sqrt{(A + D)^2 + 4BC}}{2}. \tag{11.77}$$

11.3.3 Aerosol System Parameters Related to Its Breakdown

To apply Eq. 11.76 for specific types of electric breakdown, consider first the main parameters of the aerosol system related to the relation (Eqs. 11.71–75). The mean-free-paths of electrons with respect to all collisions, collisions with neutrals, and collisions with macroparticles can be respectively expressed as

$$\lambda_\Sigma = \frac{1}{n_0\sigma_{e0} + n_a\sigma_a}, \quad \lambda_0 = \frac{1}{n_0\sigma_{e0}}, \quad \lambda_a = \frac{1}{n_a\sigma_a}, \tag{11.78}$$

where σ_{e0}, σ_a are cross sections of collisions of electrons with neutral species and macroparticles. Then write down the expressions for electron temperature T_e, thermal and drift velocities of electrons v^e and v_d^e, electron diffusion coefficient D_e, the first and second Townsend coefficients α, β in aerosol systems, using the corresponding values of $T_e^0, v_0^e, v_{do}^e, D_e^0, \alpha_0, \beta_0$ for the neutral gas in the same conditions but without macroparticles. The electron temperature in the aerosol system is

$$T_e = \frac{eE}{\delta_1/\lambda_0 + \delta_2/\lambda_a} = \left[\frac{1}{T_e^0(E)} + \frac{\delta_2}{eE\lambda_a}\right]^{-1}. \tag{11.79}$$

Here, δ_1 and δ_2 are the fraction of electron energy lost in a collision with a molecule or macroparticle. Similarly, the electron thermal velocity in an aerosol can be expressed based on Eq. 11.79 as

$$v^e = \sqrt{\frac{2eE}{m(\delta_1/\lambda_0 + \delta_2/\lambda_a)}} = \left[\left(\frac{1}{v_0^e}\right)^2 + \frac{\delta_2 m}{2\lambda_a eE}\right]^{-1/2}. \tag{11.80}$$

The diffusion coefficient of electrons can be written using the parameter $\xi = n_a\sigma_a/n_0\sigma_{e0}$ as

$$D_e = \frac{\lambda_\Sigma}{3}\sqrt{\frac{2eE}{m(\delta_1/\lambda_0 + \delta_2/\lambda_a)}} = \frac{D_e^0}{1+\xi}\left(1 + \frac{\delta_2 T_e^0}{eE\lambda_a}\right)^{-1/2}. \tag{11.81}$$

The electron drift velocity in aerosols can be presented as

$$v_d^e = \lambda_\Sigma\sqrt{\frac{eE}{2m}\left(\frac{\delta_1}{\lambda_0} + \frac{\delta_2}{\lambda_a}\right)} = \frac{v_{d0}^e}{1+\xi}\left(1 + \frac{\delta_2 T_e^0}{eE\lambda_a}\right)^{1/2}. \tag{11.82}$$

The relations for thermal velocity, drift velocity, and diffusion coefficient for positive ions are similar to those for electrons (Eqs. 11.80 through 11.82). The first Townsend coefficient shows ionization per unit length of electron drift, and in the aerosol system can be expressed as

$$\begin{aligned}\alpha &= \frac{2}{\lambda_\Sigma(\delta_1/\lambda_0 + \delta_2/\lambda_a)}\left[\frac{1}{\lambda_0}\exp\left(-\frac{I_i}{T_e}\right) + \frac{1}{\lambda_a}\exp\left(-\frac{I_a}{T_e}\right)\right] \\ &= \frac{2}{\lambda_\Sigma(eE/T_e^0 + \delta_2/\lambda_a)}\left[\frac{\alpha_0\lambda_0}{2}\frac{eE}{T_e^0} + \frac{1}{\lambda_a}\exp\left(-\frac{I_a}{T_e}\right)\right].\end{aligned} \tag{11.83}$$

Here, I_i is the ionization potential; I_a is the electron energy when the secondary electron emission coefficient from macroparticle reaches unity. The second Townsend coefficient describing electron attachment to molecules and aerosol particles can be presented similar to Eq. 10.3.10 as

$$\begin{aligned}\beta &= \frac{2}{\lambda_\Sigma(\delta_1/\lambda_0 + \delta_2/\lambda_a)}\left[\frac{1}{\lambda_0}w^0 + \frac{1}{\lambda_a}w^a\right] \\ &= \frac{2}{\lambda_\Sigma(eE/T_e^0 + \delta_2/\lambda_a)} \times \left[\frac{\beta_0\lambda_0}{2}\frac{eE}{T_e^0} + \frac{1}{\lambda_a}\omega^a\right],\end{aligned} \tag{11.84}$$

where w^0, w^a are the electron attachment probabilities in a collision with a molecule or macroparticle.

11.3.4 Pulse Breakdown of Aerosols

Assume $v_d^e \tau \le d, d \le c\tau, \lambda_\Sigma \le d$ (Raether, 1964; τ is the voltage pulse duration, c is the speed of light in the aerosol medium). Considering the pulse breakdown, neglect in Eqs. 11.71–11.75 the terms related to diffusive and drift losses, assuming $1/d \to 0$. Also assume that the contribution of ion–macroparticle collisions in electron balance in relatively low: $B \ll (A + D)^2/C$. Taking into account the pulse breakdown criterion (Raether, 1964):

$$n_e(t = \tau) = n_b = 10^8 \times n_e^0, \tag{11.85}$$

the pulse breakdown condition in an aerosol system based on Eq. 11.76 can be given in the form:

$$\tau = \frac{\ln \dfrac{n_b}{n_e^0} + \dfrac{BC}{(A+D)^2}}{A + \dfrac{BC}{A+D}}. \tag{11.86}$$

This pulse breakdown condition can be specified based on Eqs. 11.71–11.75. It then permits estimating the maximum possible increase of the pulse breakdown voltage, related to electron attachment to aerosol particles, by assuming the attachment probability as $w^a = 1 - \exp(-I_a/T_e)$. The maximum breakdown voltage increase can be calculated using the relation (Karachevtsev and Fridman, 1976):

$$\frac{\alpha_0 \lambda_0 eE}{2T_e^0} + \frac{2\exp\left(-\dfrac{I_a}{T_e}\right) - 1}{\lambda_a} = \frac{eE\lambda_0 \ln \dfrac{n_b}{n_0}}{2T_e^0 v_{d0}^e \tau} \sqrt{1 + \frac{\delta_2 T_e^0}{eE\lambda_a}}. \tag{11.87}$$

If $\tau = 10$ ns, $p = 10$ Torr, $d = 10$ cm, $I_a = 100$ eV, $\delta_2 = 1$, the pulse breakdown voltage in nitrogen without aerosol particles is $E_0 \approx 2$ kV/cm ($\xi = 0$). Addition of aerosol characterized by $\xi = n_a\sigma_a/n_0\sigma_{e0} = 0.1$, increases the breakdown voltage by factor of about 1.2 (20%).

11.3.5 Breakdown of Aerosols in High-Frequency Electromagnetic Fields

The electrical breakdown conditions of aerosols in accordance with Eq. 11.76 correspond to $\max(\lambda_1, \lambda_2) = 0$:

$$AD + BC = 0. \tag{11.88}$$

In the case of a breakdown in high-frequency electromagnetic fields (where drift losses can be neglected), the general condition (Eq. 11.88) can be given (Karachevtsev and Fridman, 1976) as

$$\alpha = \beta + \frac{1}{d^2\left(\dfrac{eE}{T_e^0} + \dfrac{\delta_2}{\lambda_a}\right)} - \gamma_{ia}\sigma_{ia}n_a \frac{v^i}{v_d^e} \frac{k_i n_0}{\left(k_a n_a + \dfrac{D_i}{d^2}\right)} - \frac{K}{v_d^e}. \tag{11.89}$$

The electric field in this and the following relations should be taken as an effective one (see Eq. 4.79), taking into account the high frequency of the electromagnetic field. Similar to Eq. 11.87, the maximum breakdown voltage increase due to electron attachment by aerosol particle is

$$\lambda_0 \frac{eE}{T_e^0}\alpha_0 + \frac{2\xi}{\lambda_0}\left[2\exp\left(-\frac{I_a}{T_e}\right)-1\right] = \frac{\lambda_0}{d^2(1+\xi)}. \quad (11.90)$$

Consider an example assuming 11.77: $p = 10$ Torr, d = 10 cm, $I_a = 100$ eV, $\delta_2 = 1$. The addition of aerosol particles, characterized by $\xi = n_a\sigma_a/n_0\sigma_{e0} = 0.1$ in nitrogen, gives a possibility to increase the breakdown voltage by factor of about 1.3 (30%) with respect to the same system without macroparticles $\xi = 0$). The breakdown of aerosols in radio frequency and microwave electromagnetic fields is quite sensitive to the macroparticle heating in these fields. These heating effects are related not only to thermal emission (see Section 11.3.2), but also to evaporation and thermal explosion of macroparticles, which affects the breakdown conditions (Raizer, 1972a,b).

11.3.6 Townsend Breakdown of Aerosols

The Townsend breakdown of aerosols can be described by the relation similar to that for pure gas:

$$\gamma_{\text{eff}}\left\{\exp(\alpha-\beta)d - 1 + \frac{\beta}{\alpha-\beta}\left[\exp(\alpha-\beta)d - 1\right]\right\} = 1, \quad (11.91)$$

but with Townsend coefficients α and β calculated according to Eqs. 11.83 and 11.84, taking into account all effects related to macroparticles. The third Townsend coefficient γ_{eff} is equal to the total number of generated secondary electrons with respect to one ion formed in the volume. In this case, the coefficient γ_{eff} is an effective value and should take into account the interaction of ions and photons not only with the cathode but also with aerosol particles. The probability for an ion formed in the volume to reach the cathode without being trapped by aerosol particles can be estimated as

$$w = \exp\left(-n_a\sigma_{ia}d\frac{v^i}{2v_d^i}\right). \quad (11.92)$$

Then the third Townsend coefficient γ_{eff} can be calculated by the cumulative relation:

$$\gamma_{\text{eff}} = \gamma_{iw}\exp\left(-n_a\sigma_{ia}d\frac{v^i}{2v_d^i}\right) + \gamma_{ia}\left[1-\exp\left(-n_a\sigma_{ia}d\frac{v^i}{2v_d^i}\right)\right] + \gamma_{pw}\frac{\mu(1-v)}{k_i} + \gamma_{pa}\frac{\mu v}{k_i}. \quad (11.93)$$

To estimate the maximum increase of the Townsend breakdown voltage due to the electron attachment by macroparticles similar to Eq. 11.87 one can use the equation:

$$\frac{2d}{\lambda_\Sigma\left(eE/T_e^0 + \delta_2/\lambda_a\right)}\left[\frac{\alpha_0\lambda_0 eE}{2T_e^0} + \frac{2\exp(-I_a/T_e)-1}{\lambda_a}\right] = n_a\sigma_{ia}d\frac{v^i}{2v_d^i} - \ln\gamma_{ia}. \quad (11.94)$$

The Townsend breakdown voltage of nitrogen at 10 Torr with tantalum electrodes (d = 1 cm) can have 10% increase by adding particles with $\xi = n_a\sigma_a/n_0\sigma_{e0} = 0.1$, $\delta_2 = 1$, $I_a = 100$ eV.

11.3.7 Effect of Macroparticles on Vacuum Breakdown

In the above examples, the relatively low addition of macroparticles $\xi = n_a\sigma_a/n_0\sigma_{e0} < 1$ was assumed when electron collisions with molecules are more often than those with aerosol particles. As was shown in the numerical examples of Sections 11.3.4 through 11.3.6, the maximum increase of the breakdown voltage related to electron attachment to macroparticles, is about 10–30%. Although this effect is not very strong at low aerosol concentration $\xi = n_a\sigma_a/n_0\sigma_{e0} < 1$, it is of interest for different applications, in particular for electric filters (Uzhov, 1967). Much larger increases of breakdown voltage can be reached when $\xi = n_a\sigma_a/n_0\sigma_{e0} > 1$ and most electrons collisions are related to aerosol particles. In this case, the breakdown conditions actually do not depend on the gas characteristics, so such a breakdown can be interpreted as the vacuum values. Thus, the pulse breakdown conditions in aerosols at $\xi \gg 1$ and $d > 1/2\lambda_a \ln n_b/n_e^0$ can be expressed from Eqs. 11.71–11.75 and 11.76 as

$$2\exp\left(-\frac{I_a}{eE\lambda_a}\right)-1=\frac{\lambda_a}{2\tau}\sqrt{\frac{2m}{eE\lambda_a}}\ln\frac{n_b}{n_e^0}. \tag{11.95}$$

In the specific case of $d > \tau\sqrt{I_a/2m} \gg \lambda_a \ln n_b/n_e^0$, the most physically clear pulse breakdown condition can be obtained from Eq. 11.95:

$$E=\frac{I_a}{e\lambda_a \ln 2}. \tag{11.96}$$

According to Eq. 11.96, if $I_a = 100$ eV, $n_a = 10^7$ cm^{-3}, $\sigma_a = 10^{-6}$ cm^2, d = 10 cm, $\tau = 10$ ns, $p = 0.1$ Torr, the pulse breakdown electric field should be numerically equal to $E = 1000$ V/cm. The breakdown of relatively dense aerosols ($\xi \gg 1$) in the high-frequency electromagnetic fields can be described by the simplified relation of Eq. 11.90 as

$$2\exp\left(-\frac{I_a}{eE\lambda_a}\right)-1=\frac{\lambda_a^2}{d^2}. \tag{11.97}$$

The effective breakdown electric field (see Eq. 4.79) in the radio frequency or microwave range can be found from Eq. 11.97 as

$$E_{eff}=\frac{I_a}{e\lambda_a \ln\left[2/\left(1+\lambda_a^2/d^2\right)\right]}. \tag{11.98}$$

This solution for the breakdown in high-frequency electromagnetic field corresponds to a similar condition (Eq. 11.96) for the pulse breakdown, if $\lambda_a^2 \ll d^2$.

11.3.8 About Initiation of Electric Breakdown in Aerosols

Aerosol particles can be applied to increase a discharge gap resistance to electric breakdown. At the same time, the presence of macroparticles provides controlled initiation of breakdown, in particular by pulsed photoionization (Slivkov, 1966; Karachevtsev and Fridman, 1979). Photoionization of aerosols (see Section 11.3.1) can be applied for initiation (commutation) of the pulsed nanosecond breakdown, as well as for initiation of the breakdown in a constant electric field. In the case of pulsed nanosecond breakdown, the photo-initiation is provided by fast generation of the necessary initial electron concentration n_e^0 due to photoionization. In the case of breakdown in constant electric field, the initiation effect is related to the induced local nonuniformities of electric field. Electrons formed during the photoionization move from the discharge gap to the cathode, leaving the charged macroparticles in the gap. Positive space charge of the macroparticles provides the nonuniformity of electric field, which results in the breakdown initiation.

11.4 STEADY-STATE DC ELECTRIC DISCHARGE IN HETEROGENEOUS MEDIUM

11.4.1 Two Regimes of Steady-State Discharges in Heterogeneous Medium

Consider the steady-state DC discharge in a heterogeneous medium neglecting the effects of thermionic and field emission from the surface of macroparticles. The main mechanisms of electrons formation in this case are ionization of the gas by direct electron impact and secondary electron emission from the aerosol particles. Electro-neutrality condition of the aerosol system including electrons, positive ions and macroparticles is

$$n_e = n_i + Z_a n_a, \tag{11.99}$$

where n_e, n_i, n_a are concentrations of electrons, positive ions, and macroparticles respectively, Z_a is the average charge of a macroparticle (in elementary charges). If the concentration and radius of aerosol particles in the discharge system are relatively low, then the formation of electrons is mostly due to neutral gas ionization. In this case, usually $n_e \gg Z_a n_a$, and according to Eq. 11.99 the concentration of electrons and positive ions are nearly equal $n_e \approx n_i$. This regime of discharge in aerosols is referred to as the quasi-neutral. Another heterogeneous discharge regime takes place if the macroparticles number density and sizes are sufficiently high and their ionization is through secondary electron emission. In this case, $n_i \ll n_e \approx Z_a n_a$, and the heterogeneous plasma actually consists of electrons and positively charged macroparticles (instead of ions). This regime is usually referred to as the electron–aerosol plasma regime to emphasize the contrast with conventional electron–ion plasma.

11.4.2 Quasi-Neutral Regime of Steady-State DC Discharge in Aerosols

The quasi-neutral aerosol discharge $n_e \approx n_i \gg Z_a n_a$ was investigated in details by Musin (Konenko and Musin, 1972, 1973; Konenko et al., 1973; Musin, 1974). In the positive column of such a discharge, electrons formation is mostly due to ionization of neutral species in the gas phase, while scattering and recombination of charged particles take place mostly on the surfaces of macroparticles and discharge tube. macroparticles in this regime are negatively charged, and their potential is close to the floating value. The quasi-neutral regime of a steady-state DC discharge in aerosols can be described by the following system of equations (Musin, 1974):

$$\left(4\pi r_a^2 n_a + \frac{2}{R}\right)\sqrt{\frac{m}{M}} = n_0 \sigma_{e0} \exp\left(-\frac{I_i}{T_e}\right), \tag{11.100}$$

$$E = I_i n_S \left[1 + \frac{1}{n_S}\left(n_0\sigma_{e0} + n_e\sigma_{ei}\right)\right]\Phi^2\left(T_e\right), \tag{11.101}$$

$$n_e = \frac{j_e}{e}\sqrt{\frac{m}{2eI_i}}\Phi\left(T_e\right). \tag{11.102}$$

In this system: R is the radius of discharge tube; m, M are masses of an electron and a positive ion; n_0 is the gas density; T_e is the electron temperature; I_i is the ionization potential; E is the electric field; σ_{e0}, σ_{ei} are the characteristic cross sections of electron–neutral and electron–ion collisions corresponding to average electron energy; j_e is the electron current density; the function $\Phi(T_e)$ is

$$\Phi\left(T_e\right) = n_S\sqrt{\frac{m}{2M}\cdot\frac{T_e}{I_i}} \times \frac{\left(2 + \frac{1}{2}\ln\frac{M}{m}\right)\frac{T_e}{I_i} + 1}{n_0\sigma_{e0} + n_e\sigma_{ei}}. \tag{11.103}$$

n_S "density of surfaces," which represents the surface area of macroparticles and discharge tube walls per unit volume. In the case of long cylindrical discharge tube, this can be expressed as

$$n_S = 4\pi r_a^2 n_a + \frac{2}{R}. \tag{11.104}$$

The system of Eqs. 11.100 through 11.102 shows that the addition of aerosol particles in the discharge at the fixed value of current density leads to an essential increase of electron temperature, while the growth of electron concentration is relatively small. If the surface area of macroparticles is less than that of the discharge tube walls, the system of Eqs. 11.100 through 11.102 corresponds to the Langmuir–Klarfeld theory (Langmuir and Tonks, 1929; Klarfeld, 1940). Eq. 11.100 shows that an increase in the surface area density n_S leads to a logarithmic increase of electron temperature. If macroparticles make the major contribution to the total surface area $2/R \ll 4\pi r_a^2 na$, this dependence can be presented as

$$T_e = I_i \ln^{-1} \frac{n_{a,cr}}{n_a}, \tag{11.105}$$

where $n_{a,cr}$ is the critical value of aerosol concentration when the electron temperature tends to infinity $T_e \to \infty$. The considered quasi-neutral regime can exist only at relatively low aerosol density:

$$n_a < n_{a,cr} = n_0 \frac{\sigma_{e0}}{4\pi r_a^2} \sqrt{\frac{M}{m}}. \tag{11.106}$$

11.4.3 Electron–Aerosol Plasma Regime of the Steady-State DC Discharge, Main Equations Relating Electric Field, Electron Concentration, and Current Density

At particle concentrations exceeding this critical value (Eq. 11.106), an additional ionization mechanism is required to compensate for the electron losses on the macroparticles. This becomes possible in the electron–aerosol plasma regime, where the additional formation of electrons is related to the secondary electron emission from the macroparticles (Karachevtsev and Fridman, 1979). Consider this regime of predominant ionization of aerosols $n_i \ll n_e \approx Z_a n_a$ and relatively low gas concentration. Because the mass of macroparticles is relatively high, the ambipolar diffusion to the discharge walls can be neglected and, therefore, the steady-state balance of charged processes should be given by volume processes. This means that the total coefficient of the secondary electron–electron emission from the surface of macroparticles should be equal to unity if thermionic and field emissions of electrons are neglected. The total coefficient δ of the secondary electron–electron emission from aerosol particles depends on their work function $\varphi(Z_a)$ that is related to the macroparticle charge according to Eq. 11.1, and average energy $\varepsilon \approx eE\lambda_a$ of electrons bombarding the macroparticles. Thus, the charged particles balance equation can be expressed as

$$\delta\left(\varepsilon = eE\lambda_a, \varphi(Z_a) = \varphi_0 + \frac{Z_a e^2}{4\pi\varepsilon_0 r_a}\right) = 1. \tag{11.107}$$

This dependence permits finding the particle charge Z_a as a function of the electric field E:

$$Z_a = Z_0 \frac{E - E_0}{E_0}. \tag{11.108}$$

In this relation, E_0 is the breakdown electric field in the heterogeneous medium. The aerosol charge parameter Z_0 is determined as

$$Z_0 = -\varepsilon_{00} \left(\frac{\partial \delta}{\partial \varepsilon}\right)_{\varphi=\varphi_0} \Big/ \frac{e^2}{4\pi\varepsilon_0 r_a} \left(\frac{\partial \delta}{\partial \varphi}\right)_{\varepsilon=\varepsilon_{00}}, \tag{11.109}$$

where the energy parameter $\varepsilon_{00} = E_0\lambda_a e$. For calculations of Z_0 at the following parametric values $\varepsilon_0 = 100$ eV, $\varphi_0 = 3$ eV, $\left(\frac{\partial\delta}{\partial\varphi}\right)_{\varepsilon=\varepsilon_{00}} = -0.05\ \text{eV}^{-1}$, $(\partial\delta/\partial\varepsilon)_{\varphi=\varphi 0} = 0.01\ \text{eV}^{-1}$:

$$\log Z_0 = 8 + \log r_a\left(\text{cm}\right). \tag{11.110}$$

Eq. 11.108 can be combined with the equations for current density and electron concentration:

$$j_e = n_e e\sqrt{\frac{eE\lambda_a}{2\text{m}}}, n_e = Z_a n_a \tag{11.111}$$

to form the system of equations sufficient to determine the electric field and electron concentration in the heterogeneous discharge as a function of current density.

11.4.4 Electron–Aerosol Plasma Parameters as a Function of Current Density

Eqs. 11.108 and 11.111 permit finding the electric field and electron concentration as a function of current density. This function can be determined in terms of the special dimensionless factor B, characterizing the current density in the heterogeneous discharge (Karachevtsev and Fridman, 1979):

$$B = \frac{j_e}{Z_0 n_a e}\sqrt{\frac{2m}{\varepsilon_0}}. \tag{11.112}$$

For numerical calculations of the current density factor B at the following values of parameters: $\varepsilon_0 = 100$ eV, $\varphi_0 = 3$eV, $(\partial\delta/\partial\varphi)_{\varepsilon=\varepsilon 00} = -0.05\ \text{eV}^{-1}$, $(\partial\delta/\partial\varepsilon)_{\varphi=\varphi 0} = 0.01\ \text{eV}^{-1}$:

$$\log B = 2 + \log\frac{j_e\left(\text{A/cm}^2\right)}{n_a\left(\text{cm}^{-3}\right)r_a\left(\text{cm}\right)}. \tag{11.113}$$

Thus, according to the system of Eqs. 11.108 and 11.111, at low current densities when $B < 1$, the electric field in the heterogeneous discharge is slowly growing with a current density as

$$E = E_0\left(1 + B\right). \tag{11.114}$$

Electron concentration and hence, average charge of macroparticles at low current densities ($B < 1$) grows up proportionally to j_e:

$$n_e = n_a Z_0 B, \quad Z_a = Z_0 B. \tag{11.115}$$

In the case of high current densities ($B > 1$), the electron concentration and average charge of macroparticles reach the maximum values:

$$n_e = n_a Z_0, \quad Z_a = Z_0. \tag{11.116}$$

These saturation electron concentration and average macroparticle charge are larger at larger values of the aerosol particle radius (see Eq. 11.110). At the same time, the electric field growth with j_e becomes much stronger in the high current regime ($B > 1$), which can be expressed by the relation:

$$E \approx E_0 B^2. \tag{11.117}$$

The dependence of the electric field $\log E - E_0/E_0$ and electron concentration $\log n_e/n_a$ on the current density $\log j_e(\text{A/cm}^2)/n_a(\text{cm}^{-3}) \cdot r_a(\text{cm})$ in the electron–aerosol plasma at the above parametric values of $\varepsilon_0 = 100$ eV, $\varphi_0 = 3$ eV, $(\partial\delta/\partial\varphi)_{\varepsilon=\varepsilon 00} = -0.05$ eV^{-1}, $(\partial\delta/\partial\varepsilon)_{\varphi=\varphi 0} = 0.01$ eV^{-1} are presented in Figure 11.3. From this figure, the dependence of the average charge of aerosol particles on the current density appears as a saturation curve. The average charge of aerosol particles of radius 10 μm and concentration $n_a = 10^6$ cm^{-3} reaches its maximum value at electron current densities exceeding 10 μA/cm^2. The maximum values of the average charge of the aerosol particles depend on their size and can be quite large, about 10^5 for the 10 μm- macroparticles.

11.4.5 Effect of Molecular Gas on the Electron–Aerosol Plasma

The molecular gas effect is revealed in the kinetic equation for positive ions in the heterogeneous discharge system. The balance of positive ions should take into account their formation provided by ionization of molecules and their losses on the surfaces of macroparticles and discharge tube; this can be expressed as

$$\frac{n_S j_{is}}{e} = n_e n_0 \sigma_{e0} \sqrt{\frac{2T_e}{m}} \exp\left(-\frac{I_i}{T_e}\right). \tag{11.118}$$

Here, n_S is the "density of surfaces" defined by Eq. 11.104; n_e, n_0 are the concentrations of electrons and neutral gas respectively; σ_{e0} is the characteristic cross section of electron–neutral collisions; I_i is the ionization potential of molecules; m and T_e are the electron mass and temperature, respectively; j_{is} is the ion current density on the surfaces of macroparticles and discharge walls, which can be found from the Bohm formula (see, for example, Huddlestone and Leonard, 1965):

$$j_{is} \approx 0.5 n_i e \sqrt{\frac{T_e}{M}}. \tag{11.119}$$

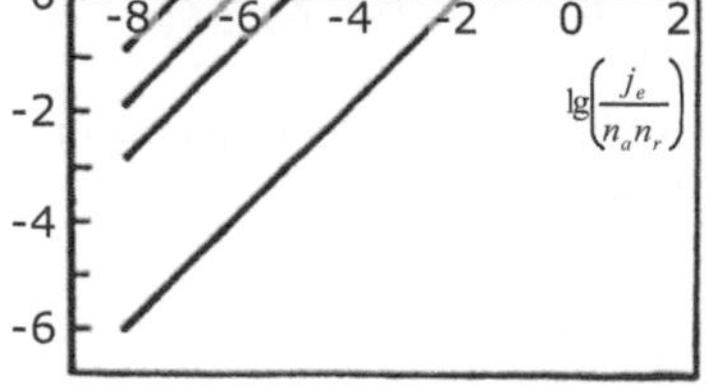

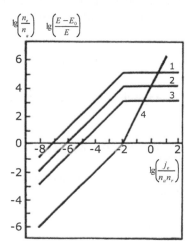

FIGURE 11.3 Electric discharge in aerosols. (1–3) $\log(n_e/n_a)$, (4) $\log[(E - E_0)/E]$. (1)$r_a = 10^{-3}$ cm, (2) 10^{-4} cm, (3) 10^{-3} cm, j_e (A/cm^2); n_a (cm^{-3}).

Combining Eqs. 11.118 and 11.119 gives the qualitative ion balance equation in the form:

$$n_S n_i \sqrt{\frac{m}{M}} = n_e n_0 \sigma_{e0} \exp\left(-\frac{I_i}{T_e}\right). \quad (11.120)$$

Taking into account that in the electron–aerosol plasma regime, the ions contribution to charge balance is low $n_i \ll n_e$ and therefore, $n_e \approx Z_a n_a$, the concentration of the positive ions is

$$n_i = \frac{Z_a n_a n_0 \sigma_{e0}}{\frac{2}{R} + 4\pi r_a^2 n_a} \sqrt{\frac{M}{m}} \exp\left(-\frac{I_i}{T_e}\right). \quad (11.121)$$

If most of the surface area in the heterogeneous discharge is related to the macroparticles ($2/R \ll 4\pi r_a^2 n_a$, R is the radius of long discharge tube), then the ratio of ion and electron densities is

$$\frac{n_i}{n_e} = \frac{n_0 \sigma_{e0}}{4\pi r_a^2 n_a} \sqrt{\frac{M}{m}} \exp\left(-\frac{I_e}{T_e}\right). \quad (11.122)$$

The requirement of low ion concentration at any electron temperature in the heterogeneous system is the main limit of the electron–aerosol plasma regime, which based on Eq. 11.122 implies

$$n_0 \ll n_a \frac{4\pi r_a^2}{\sigma_{e0}} \sqrt{\frac{m}{M}}. \quad (11.123)$$

This criterion of the electron–aerosol plasma regime is opposite to that in Eq. 11.106 for the quasi-neutral regime. Numerically, criterion (Eq. 11.123) shows that the electron–aerosol plasma regime takes place at $n_a = 10^8\ \text{cm}^{-3}$, $r_a = 10\ \mu\text{m}$, if the molecular gas concentration is less than $10^{17}\ \text{cm}^{-3}$.

11.5 DUSTY PLASMA FORMATION: EVOLUTION OF NANOPARTICLES IN PLASMA

11.5.1 General Aspects of Dusty Plasma Kinetics

Numerous problems are related to the formation and growth of molecular clusters, nanoscale, and microscale particles in plasma. Plasma physics and plasma chemistry are coupled in such so-called dusty plasma systems (see Watanabe et al., 1988; Bouchoule, 1993, 1999; Garscadden, 1994). Applied aspects of the dusty plasmas are related to electronics, environment control, powder chemistry, and metallurgy. Active research on particle formation and behavior has been stimulated by contamination phenomena of industrial plasma reactors used for etching, sputtering, and PECVD. For this reason, the process of particle formation in low-pressure glow or RF discharges in silane (SiH_4, or mixture SiH_4–Ar) has been extensively studied (Spears et al., 1988; Selwyn et al., 1990; Howling et al., 1991; Jellum et al., 1991; Bouchoule et al., 1991, 1992, 1993). Much effort has been put in dealing with the detection and dynamics of relatively large size (more than 10 nm) nanoparticles in silane discharge. Such particles are electrostatically trapped in the plasma bulk and significantly affect the discharge behavior (Sommerer et al., 1991; Barnes et al., 1992; Boeuf, 1992; Daugherty et al., 1992, 1993a,b). The most fundamentally interesting phenomena take place in the early phases of the process, starting from an initial dust-free silane discharge. During this period, small deviations in plasma and gas parameters (gas temperature, pressure, electron concentration) could alter cluster growth and the subsequent discharge behavior (Bohm and Perrin, 1991; Perrin, 1991; Bouchoule and Boufendi, 1993). Experimental data concerning supersmall particle growth (particle diameter as small as 2 nm) and negative ion kinetics (up to 500 or 1000 a.m.u.) demonstrated the main role of negative ion-clusters in particle growth process (Choi and Kushner, 1993; Hollenstein et al., 1993; Howling et al., 1993a,b; Perrin, 1993; Perrin et al., 1993; Boufendi and Bouchoule, 1994). Thus, the initial phase of formation and growth of the dust particles in a low-pressure silane discharge is a homogeneous process, stimulated by fast silane molecules and radical reactions with negative ion-clusters. The first particle generation appears as a monodispersed one with a crystallite size about 2 nm. Due to a selective trapping effect, the concentration of these crystallites increases up to a critical value, where a fast (like a phase transition) coagulation process takes place leading to large size (50 nm) dust particles. During the coagulation step when the aggregate size becomes larger than some specific value of approximately 6 nm, another critical phenomenon, the so-called "α–γ transition," takes place with a strong decrease of electron concentration and a significant increase of their energy. A small increase of the gas temperature increases the induction period required to achieve these critical phenomena of crystallites formation: agglomeration and the α–γ transition.

11.5.2 Experimental Observations of Dusty Plasma Formation in Low-Pressure Silane Discharge

The low-pressure RF-discharge for observing dusty plasma formation has been conducted in the Boufendi–Bouchoule experiments in a grounded cylindrical box (13 cm inner diameter) equipped with a shower type RF powered electrode. A grid was used as the bottom of the chamber to allow a vertical laminar flow in the discharge box. The typical Boufendi–Bouchole experimental conditions are as follows: argon flow 30 sccm; silane flow 1.2 sccm; total pressure 117 mTorr (so the total gas concentration is about 4×10^{15} cm^{-3}, silane 1.6×10^{14} cm^{-3}); neutral gas residence time in the discharge is approximately 150 ms; RF power 10 W. A cylindrical oven to vary gas temperature from the ambient up to 200°C surrounded the discharge structure. The Boufendi–Bouchoule experiments show that the first particle size distribution is monodispersed with a diameter of about 2 nm, and is practically independent of temperature. The appearance time of the first generation particles is less than 5 ms. The measurements give a clear indication that the particle growth proceeds through the successive steps of fast supersmall 2-nm particle formation and the growth of their concentration up to the critical value of about 10^{10}–10^{11} cm^{-3} when the new particles' formation terminates and formation of aggregates with diameters of up to 50 nm begins by means of coagulation. During the initial discharge phase (until the α–γ transition, up to 0.5 s for room temperature and up to several seconds for 400 K) the electron temperature remains about 2 eV, the electron concentration is about 3×10^9 cm^{-3}, the positive ion concentration is approximately 4×10^9 cm^{-3}, and the negative ion concentration about 10^9 cm^{-3}. After the α–γ transition, the electron temperature increases up to 8 eV while the electron concentration decreases 10 times and the positive ion concentration increases two times. These concentrations are correlated with the negative volume charge density of the charged particles through the plasma neutrality relation (Boufendi, 1994; Boufendi et al., 1994). The Boufendi–Bouchoule experimental data show that the critical value of supersmall particle concentration before coagulation practically is independent of temperature. The induction time observed before coagulation is a highly sensitive function of the temperature: ranging from about 150 ms at 300 K and increases more than 10 times when heated to only 400 K. For a temperature of 400 K, the time required to increase the supersmall neutral particle concentration is much longer (10 times) than the gas residence time in the discharge, which can be explained by the neutral particle trapping phenomenon. According to the Boufendi–Bouchoule experiments, the particle concentration during the coagulation period decreases and their average radius grows. The total mass of dust in plasma remains almost constant during the coagulation.

11.5.3 Dust Particle Formation: A Story of Birth and Catastrophic Life

The above-described dust particle formation and growth in a SiH_4–Ar low-pressure discharge can be subdivided into four steps. These include the growth of supersmall particles from

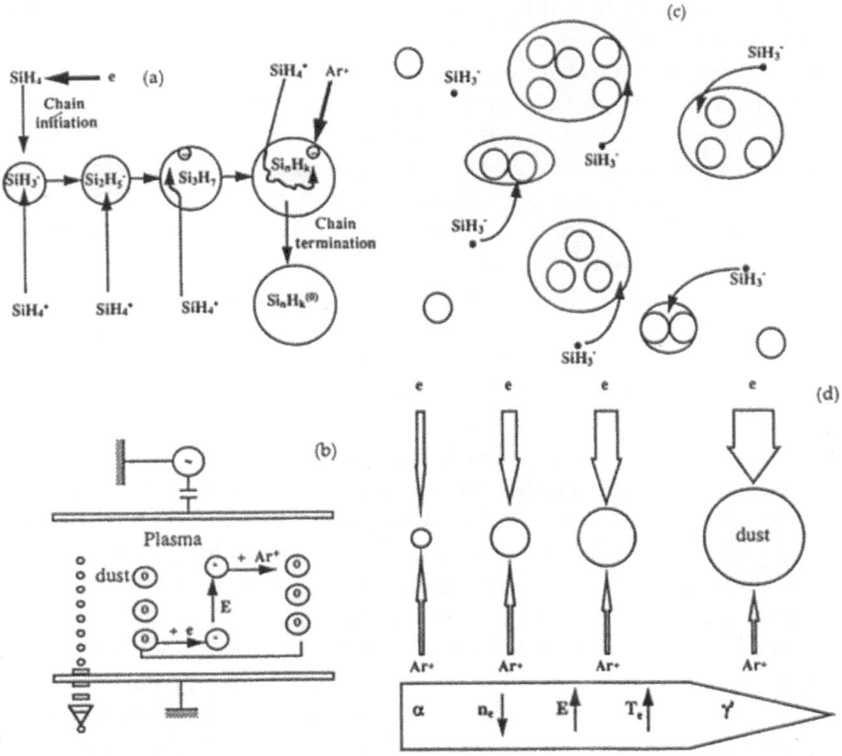

FIGURE 11.4 (a) Physical scheme of first generation of super-small particle growth. (b) Physical scheme of electrical trapping of neutral particles. (c) Mechanism of particle coagulation. (d) Physical scheme of α–γ transition.

molecular species and three successive catastrophic events: selective trapping, fast coagulation, and finally the strong modification of discharge parameters, for example, the α–γ transition (Fridman et al., 1996). A scheme for the formation of the first supersmall particle generation is presented in Figure 11.4a. It begins mainly with SiH_3^- negative ion formation by dissociative attachment. The occurrence of the non-dissociative three-body attachment to SiH_3 radicals could occur in a complementary way ($e + SiH_3 = (SiH_3^-)^*$, $(SiH_3^-)^* + M = SiH_3^- + M$). Then the negatively charged cluster growth is due to ion–molecular reactions such as: $SiH_3^- + SiH_4 =$ $Si2H_{533333}^{------}\left(\frac{\partial\delta}{\partial\varphi}\right)_{\varepsilon=\varepsilon_{00}5}$ $+ H_2$, $Si2H_5^- + SiH_4 = Si_3H_7^- + H_2$ (also see Section 2.6.6 about these reactions). This chain of reactions could be accelerated by silane molecules vibrational excitation in the plasma. Typical reaction time in this case is about 0.1 ms and in this way becomes much faster than ion–ion recombination. This last process (typical time 1–3 ms) is the main one by which this chain reaction of cluster growth is terminated. As the negative cluster size increases, the probability of reaction of the cluster with vibrationally excited molecules decreases because of a strong effect of vibrational VT-relaxation on the cluster surface. When the particle size reaches a critical value (about 2 nm at room temperature), the chain reaction of cluster growth becomes much slower and is finally terminated by the ion–ion recombination process. The typical time of 2 nm particles formation by this mechanism is about 1 ms at room temperature. To understand the critical temperature effect on particle growth, one has to take into account that silane vibrational temperatures in the discharge are determined by VT-relaxation in the plasma volume and on the walls and depends exponentially on the translational gas temperature according to the Landau–Teller effect (see Section 3.4). For this reason, a small increase of gas temperature results in a reduction of the vibrational excitation level making the cluster growth reactions much slower.

The selective trapping effect of neutral particles is schematically described in Figure 11.4b. Trapping of negatively charged particles in any discharges is related to the repelling forces exerted on such particles in the electrostatic sheaths when they reach the plasma boundary. For the supersmall particles under consideration here, the electron attachment time is about 100 ms and almost two orders of magnitude longer than the fast ion–ion recombination; hence, most of the particles are neutral during this period. The effect of "trapping of neutral particles in the electric field" should occur here to allow the concentration of particles to reach the critical value sufficient for effective coagulation. The "trapping of neutral particles" becomes clear noting that for 2 nm particles, the electron attachment time is shorter than the residence time, and each neutral particle is charged at least once during the residence time and quickly becomes trapped by the strong electric field before recombination. The rate coefficient of the two-body electron nondissociative attachment grows strongly with the particle size. For this reason, the attachment time for particles smaller than 2 nm is much longer than their residence time, and there are no possibilities to be charged even once and hence no possibilities to be trapped in the plasma and survive. Only "large particles," exceeding the 2 nm size can survive: this is the size-selective trapping effect. This first catastrophe associated with small particles explains why the first particle generation appears with well-defined sizes of crystallites. This could also explain the strong temperature effect on dust production. A small temperature increase results in a reduction of the cluster growth velocity and hence of the initial cluster size; this leads to a loss of the main part of the initial neutral particles with the gas flow and, in general, determines the observed long delay of coagulation, α–γ transition, and dust production process.

The fast coagulation phenomenon occurs when the increasing concentration of survived monodispersed 2 nm particles reaches a critical value of about 10^{10}–10^{11} cm^{-3} (see Figure 11.4c). At such levels of concentrations, the attachment of small negative ions like SiH_3^- to 2 nm particles becomes faster than their chain reaction to generate new particles. New chains of dust formation almost become suppressed and the total mass of the particles remains almost constant. In this fast coagulation process, it was shown experimentally that the mass increase by "surface deposition" process is negligible. Moreover, the probability of multibody interaction increases when such high particle concentrations are reached. For this reason, the aggregate formation rate constant grows drastically, and the coagulation appears as a critical phenomenon of phase transition. In this case, because of the small probability of aggregate decomposition (inverse with respect to coagulation), the critical particle concentration does not depend on temperature. In addition to electrostatic effects, this is one of the main differences between the critical phenomenon under consideration and the conventional phenomenon of gas condensation. The typical induction time before coagulation is about 200 ms for room temperature and much longer for 400 K. This could be explained taking into account the selective trapping effect. The trapped particle production rate for 400 K is much slower than that for room temperature, but the critical particle concentration remains the same.

The critical phenomenon of fast changing of discharge parameters, the so-called α–γ transition, takes place during the process of coagulation when the particle size increases and their concentration decreases. Before this critical moment, the electron temperature and other plasma parameters are mainly determined by the balance of volume ionization and electron losses on the walls. The α–γ transition occurs when the electron losses on the particle surfaces become more than on the reactor walls. The electron temperature increases to support the plasma balance and the electron concentration dramatically diminishes.

This fourth step of particle and dusty plasma formation story is the "plasma electron catastrophe" (Figure 11.4d). During the coagulation period, the total mass of the particles remain almost constant, and so the overall particle surface in the plasma decreases during coagulation. Hence, one has an interesting phenomenon: the influence of particle surface becomes more significant when the specific surface value decreases. This effect can be explained if the probability of electron attachment to the particles grows exponentially with the particle size (see Figure 11.4d), is taken into account.

For this reason, the "effective particle surface" grows during the coagulation period. When the particle size exceeds a critical value of about 6 nm, an essential change of the heterogeneous discharge behavior takes place with significant reduction of the free electron concentration. Taking into account the increase of the electron attachment rate on these size-growing particles, it becomes clear that most of them become negatively charged soon after the α–γ transition. The typical induction time before the α–γ transition is about 500 ms for room temperature and more than an order of magnitude longer for 400 K. This strong temperature effect is due to the threshold character of the α–γ transition, which takes place only when the particle size exceeds the critical value, and for this reason, is determined by the strongly temperature-dependent time of the beginning of coagulation. Thus, the above discussion outlined the physical nature of the main critical phenomena accompanying the initial period of particle formation in a low-pressure silane plasma from the supersmall particle formation to the plasma crisis of the α–γ transition. All these catastrophic phenomena are illustrated in Figure 11.4a–d. More information can be found in Bouchoule (1999), Girshick et al. (1999), Girshick et al. (2000), Koga et al. (2000), Watanabe et al. (2000, 2001).

11.6 CRITICAL PHENOMENA IN DUSTY PLASMA KINETICS

11.6.1 Growth Kinetics of the First Generation of Negative Ion Clusters

Cluster formation in the low-pressure RF-silane plasma begins from the formation of negative ions SiH_3^- and their derivatives. For the discharge parameters described above, negative ion concentration (SiH_3^- and their derivatives) could be found from the balance of dissociative attachment and positive ion–negative ion recombination:

$$\frac{\mathrm{d}\left[SiH_3^-\right]}{\mathrm{d}t} = k_{ad}n_e\left[SiH_4\right] - k_r^{ii}n_i\left[SiH_3^-\right], \tag{11.124}$$

where the rate coefficient of dissociative attachment $k_{ad} = 10^{-12}$ cm^3/s for $T_e = 2$ eV, the ion–ion recombination rate coefficient $k_r^{ii} = 210^{-7}$ cm^3/s, n_e and n_i are the electron and positive ion concentrations. In the absence of strong particle influence (before the α–γ transition), the electron and positive ion concentrations are nearly equal and the concentration of SiH_3^- (and their derivatives—other small negative ions) are about 10^9 cm^{-3}, in accordance with experimental data presented in the previous section. The time to establish the steady-state ion balance according to Eq. 11.124 is approximately 1 ms. Other mechanisms of initial SiH^-_3 production including nondissociative electron attachment to SiH_3 were discussed in Perrin et al. (1993). However, in the SiH_4–Ar discharge, this mechanism probably is not a major one. The production of SiH_3 radicals is due to dissociation by direct electron impact (with the rate constant $k_d = 10^{-11}$ cm^{-3}), and their losses are due to diffusion to the walls (the diffusion coefficient D = 3 × 10^3 cm^2/s, the characteristic distance to the walls $R = 3$ cm):

$$\frac{\mathrm{d}\left[SiH_3\right]}{\mathrm{d}t} = k_d n_e\left[SiH_4\right] - \frac{D}{R^2}\left[SiH_3\right]. \tag{11.125}$$

The establishment time for the steady-state regime according to this equation is about 3 ms, and the steady-state radical concentration is less than 0.1% from SiH_4. As a result, the dissociative attachment is the main initial source of negative ions in the system. After the initial SiH_3^- formation, the main pathway of cluster growth is the chain of ion–molecular reactions (Eqs. 2.107 through 2.111), discussed under the Winchester mechanism of ion–molecular processes. In the specific case of silane cluster growth, the Winchester mechanism is not fast relative to ion–ion recombination. For example, the first reaction rate coefficient (Eq. 2.108) is of the order of 10^{-12} cm^3/s. Such reaction rates are due to the thermo-neutral and even the endothermic character of some reactions from the

Winchester chain; these results in an intermolecular energy barrier (Reents, Jr. and Mandich, 1992; Raghavachari, 1992) for such reactions and even in a bottleneck effect in their kinetics (Mandich and Reents, Jr., 1992).

11.6.2 Contribution of Vibrational Excitation in Kinetics of Negative Ion–Cluster Growth

The vibrational energy of polyatomic molecules is very effective in overcoming the intermolecular energy barrier in the case of thermo-neutral and endoergic reactions (Talrose et al., 1979; Baronov et al., 1989; Veprek and Veprek-Heijman, 1990, 1991). To estimate the influence of vibrational excitation on the cluster growth rate (Eqs. 2.6.28 through 2.6.32), analyze the vibrational energy balance, taking into account SiH_4 vibrational excitation by electron impact ($k_{ev} = 10^{-7}$ cm^3/s), and vibrational VT-relaxation on molecules (rate coefficient $k_{VT,silane}$), on argon atoms ($k_{VT,Ar}$), and on the walls with the accommodation coefficient P_{VT} (see Chesnokov and Panfilov, 1981):

$$k_{ev}n_e\left[\mathrm{SiH}_4\right]\hbar\omega = \left(k_{VT,\mathrm{silane}}\left[\mathrm{SiH}_4\right]^2 + k_{VT,\mathrm{Ar}}n_0\left[\mathrm{SiH}_4\right] + P_{VT}\left[\mathrm{SIH}_4\right]\frac{D}{R^2}\right) \times\left[\frac{\hbar\omega}{\exp\left(\hbar\omega/T_v\right)-1} - \frac{\hbar\omega}{\exp\left(\hbar\omega/T_0\right)-1}\right]. \quad (11.126)$$

Here n_e and n_0 are the electron and neutral gas (which is mostly argon) concentrations, T_v, and T_0 are the vibrational and translational gas temperatures, $\hbar\omega$ is the vibrational quantum approximating the only mode excitation (see Sections 3.4.6 and 3.5.4). Vibrational relaxation on the cluster surface will be considered below. Taking into account the Landau–Teller formula for VT-relaxation (see Section 3.4), and activation energy $E_a(N)$ as a function of the number N of silicon atoms in a negatively charged cluster, the growth rate of the initial very small clusters Eqs. 2.6.28 through 2.6.32 could be expressed as

$$k_{i0}^{\mathrm{initial}}\left(T_0,N\right) = k_0N^{2/3}\exp\left(-\frac{\Lambda + B\dfrac{\Delta T}{3T_{00}^{4/3}}}{\hbar\omega}E_a\left(N\right)\right). \quad (11.127)$$

Here $\Delta T = T_0 - T_{00}$ is the temperature increase; k_0 is the gas-kinetic rate coefficient; B is the Landau–Teller parameter; Λ is a slightly changing logarithmic factor describing vibrational relaxation:

$$\Lambda = \ln\frac{k_{VT,\mathrm{silane}}\left[\mathrm{SiH}_4\right] + k_{VT,\mathrm{Ar}}n_0 + P_{VT}D/R^2}{k_{ev}n_e}. \quad (11.128)$$

The initial cluster growth rate coefficient Eq. 11.127 increases with the cluster size. Later when the number of atoms N exceeds a value of about 300, the relaxation on cluster surface becomes significant and the rate of negative cluster growth decreases. Taking into account the probability of VT-relaxation on the cluster surface $P_{VT} \approx 0.01$, and the number of the relaxation-active spots on the cluster surface $s = (N^{1/3})^2 = N^{2/3}$, the probability of chemical reaction on the cluster surface according to the Poisson distribution must be multiplied by the factor (Legasov et al., 1978):

$$\left(1 - P_{VT}\right)^S = \exp\left(-P_{VT}N^{2/3}\right). \quad (11.129)$$

Taking into account Eq. 11.127 with the kinetic restriction Eq. 11.129, the final expression for the negatively charged cluster growth rate constant can be presented as

$$k_{i0}\left(T_0,N\right)=k_0N^{2/3}\exp\left(-\frac{\Lambda+\dfrac{B\Delta T}{3T_{00}^{4/3}}}{\hbar\omega}E_a\left(N\right)-P_{VT}N^{2/3}\right). \tag{11.130}$$

This formula demonstrates the initial, relatively strong temperature dependence of the cluster growth as well as the limitation of growth when the number of atom in cluster becomes of the order of $N = P_{VT}^{-2/3}$. Numerically, the critical N value is about 1000 at room temperature, which corresponds to the cluster radius about 1 nm. At higher temperatures, this cluster size decreases.

11.6.3 Critical Size of Primary Nanoparticles

Without taking into account the ion–ion recombination involved in the chain termination, the particle size growth in the series of ion–molecular reactions (Eqs. 2.6.28 through 2.6.32) can be described based on Eq. 11.130 by the equation:

$$\frac{dN}{dt}=k_{i0}^{*}\left[SiH_4\right]N^{2/3}\exp\left(-P_{VT}N^{2/3}\right), \tag{11.131}$$

$k_{i0}^{*} = k_{i0}(T_0, N = 1)$ is the rate coefficient of the reaction (Eq. 2.107), stimulated by vibrational excitation. It is convenient to present the solution of this equation using the critical particle size $N_{cr} = P_{VT}^{-3/2}$, and reaction (Eq. 2.107) characteristic time $\tau = 1/k_{i0}^{*}$ [SiH_4], which is 0.1–0.3 ms at room temperature:

$$t=\tau N_{cr}^{1/3}\frac{\exp\left(N/N_{cr}\right)^{2/3}-1}{\left(N/N_{cr}\right)^{1/3}}. \tag{11.132}$$

Eq. 11.132 shows that initially (for number of atoms N or cluster size much less than critical values of 1000 atoms or 1 nm), the function $N^{1/3}$ (not N) is increasing linearly with time. That means that rather than particle mass and volume, the particle radius grows linearly with time, and the time of formation of a 1000-atom cluster is only 10 times longer than the characteristic time of the first ion–molecular reaction. Eq. 11.132 shows that the supersmall particle growth is limited by the critical size of 2 nm, that is, by the critical number of atoms $N_{cr} = 1000$. The time of such particle formation is 1–3 ms at room temperature. This process of particle growth is terminated by fast ion–ion recombination. A small change of temperature accelerates vibrational relaxation diminishes the level of vibrational excitation, the chain reaction rate, and hence makes the process of particle growth much slower. Figure 11.5 presents the results of numerical calculations, which demonstrate the saturation of supersmall particle growth and the strong effect of temperature on the particle growth time (Porteous et al., 1994; Fridman et al., 1996).

11.6.4 Critical Phenomenon of Neutral Particle Trapping in Plasma

The mechanism of supersmall cluster growth described above leads to a continuous production of supersmall particles (2 nm at room temperature) with the reaction rate corresponding to the initial formation of SiH^{-}_3 by dissociative attachment. Most of these particles are neutral according to the Boufendi-Bouchoule experiments. For such particles, the ion–ion recombination rate ($k_r^{ii} = 3 \times 10^{-7}$ cm^3/s) is much higher than the rate of electron attachment ($k_a = 3 \times 10^{-9}$ cm^3/s). A simple kinetic balance shows in this case that the percentage of negatively charged particles is approximately k_a/k_r^{ii} = 1%. Nevertheless, in the experiments, the concentration of such neutral particles grows during a

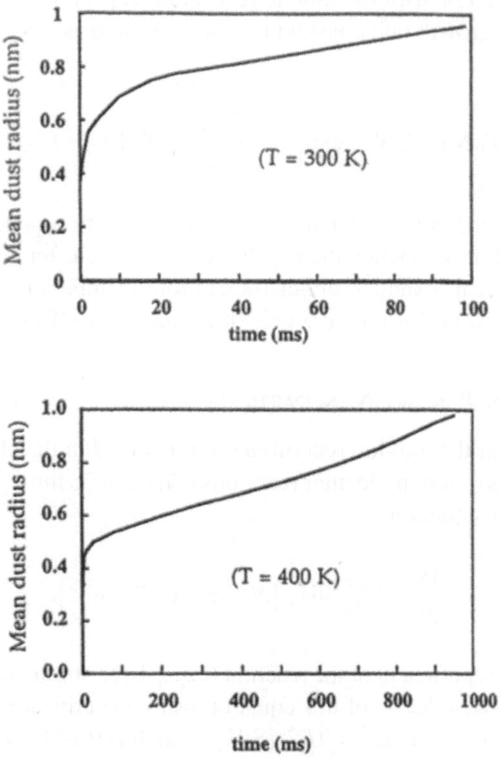

FIGURE 11.5 Numerical modeling of the initial stage of particle growth. First curve, room temperature, second one $T = 400$ K.

period longer than the neutral gas residence time in the plasma volume. This means that the trapping effect, well known for negatively charged particles (particle repelled from the walls by electrostatic sheaths when they reach the edge of the plasma), takes place here for these nanometer-size neutral particles. The neutral particle lifetime in a plasma volume is determined by their diffusion and drift with the gas flow to the walls. The particle lifetime due to the drift to the wall, the so-called residence time, obviously does not depend on the supersmall particle size (in the experimental conditions it is about 150 ms. On the contrary, the diffusion time R^2/D is about 3 ms for molecules and radicals and increases with the particle size proportionally to $N^{7/6}$ (due to particle size increase and their velocity reduction). These two processes become comparable for N about 60, and thus the main losses of the 2-nm particle are due to the gas flow. To explain the neutral particle trapping, one must take into account that for 2 nm particles the electron attachment time ($1/k_a n_e$, about 100 ms) becomes shorter than the residence time (150 ms). Hence, each neutral particle has a possibility to be charged at least once during the residence time. Having a negative charge before recombination even for a short time interval (which happens after about 1 ms) is sufficient for particle trapping by the discharge electric field repelling the negatively charged particles back into the RF-plasma volume.

11.6.5 Size-Selective Neutral Particle Trapping Effect in Plasma

The critical phenomenon of size-selective trapping is due to the fact that the rate coefficient of two-body electron nondissociative attachment to a particle grows strongly with the particle size. For this reason, when clusters are smaller than 2 nm, the attachment time is longer than their residence time. In this case, there are very small chances for such clusters to get a charge even once and, hence, no possibilities to be trapped in the RF-plasma and survive. Consider the probability and rate constant of the two-body electron nondissociative attachment. The small probability of this process for relatively small particles is due to the adiabatic character of the process related to energy transfer. An electron has to transfer its entire energy through polarization to molecular vibrations of a supersmall cluster (characteristic quantum $\hbar\omega$). The electron attachment rate can be found using the Massey parameter. Taking into account that the cluster polarizability is $\alpha \approx r^3$, and the total polarization energy of an electron interaction with a cluster of radius r is

$$\Delta E \approx 4\pi\alpha\varepsilon_0 E^2 \approx 4\pi r^3 \varepsilon_0 \left(\frac{e}{4\pi\varepsilon_0 r^2}\right)^2 = \frac{e^2}{4\pi\varepsilon_0}, \tag{11.133}$$

leads to the following formula for the coefficient of electron attachment to cluster of radius r:

$$k_a = \pi r^2 \sqrt{\frac{8T_e}{\pi m}} \exp\left(-\frac{e^2}{4\pi\varepsilon_0 \hbar\omega r}\right). \tag{11.134}$$

The fact that the rate coefficient of two-body electron non-dissociative attachment grows strongly with particle size can be illustrated by comparing the electron mean free path in a cluster with a particle size: when the particle size is relatively small, the probability of direct attachment is negligible. From Eq. 11.134, one can obtain for a 5-nm particle $k_a = 10^{-7}$ cm³/s, for a 2-nm particle, $k_a = 3 \times 10^{-9}$ cm³/s. The rate coefficient of two-body electron nondissociative and dissociative attachment becomes of the same order of magnitude when the particle size is slightly less than 1 nm. The condition of a neutral particle trapping requires at least one electron attachment during the residence time τ_R:

$$k_a(r) n_e \tau_R = n_e \tau_R \pi r^2 \sqrt{\frac{8T_e}{\pi m}} \exp\left(-\frac{e^2}{4\pi\varepsilon_0 \hbar\omega r}\right) = 1. \tag{11.135}$$

According to this equation, the critical radius of trapping R_{cr} is approximately 1 nm (the critical particle diameter is 2 nm). It is known that particle growth at room temperature is limited by approximately the same value of about 2 nm. For this reason, on one hand, most of the particles initially produced at room temperature are trapped, and on the other hand, their size distribution is monodispersed because of losses of smaller clusters to the gas flow. The losses of relatively small particles with a gas flow (the selective trapping effect) explain the strong temperature effect on dust production. A small temperature increase results in a reduction of cluster growth velocity (Eq. 11.130). Hence, the initial cluster size R_{in} before recombination is reduced. According to Eq. 11.135, this leads to significant losses of the initial neutral particles with a gas flow, and to a delay in coagulation, α–γ transition, and dust production in general at relatively high temperatures. Taking into account the particle production (initiated by dissociative attachment), whose size is limited by the temperature-dependent R_{in} value, and their losses with gas flow, the concentration of relatively small particles having the radius R_{in} can be expressed as

$$n(R_{in}) = k_{ad} n_e [SIH_4] \tau_R. \tag{11.136}$$

Formation of 2-nm particles, which will be effectively trapped, is determined at temperatures higher than 300 K by the slow process of electron attachment to the neutral particles with the radius R_{in} less than $R_{cr,}$ and concentration defined by Eq. 11.136. The production rate of the 2-nm particles is

$$\begin{aligned} W_p &= k_a(R_{in}) n_e n(R_{in}) = k_a(R_{in}) n_e k_{ad} n_e [SiH_4] \tau_R \\ &= [k_a(R_{cr}) n_e \tau_R] k_{ad} n_e [SiH_4] [k_a(R_{in}) / k_a(R_{cr})]. \end{aligned} \tag{11.137}$$

Taking into account the trapping condition (Eq. 11.135) and the formula for electron attachment coefficient (Eq. 11.134), rewrite Eq. 11.137 as a relation between the rate of particle production W_p and the initial negative ion SiH_3^- production by dissociative attachment W_{ad}:

$$W_p = W_{ad} \exp\left[\frac{e^2}{4\pi\varepsilon_0 \hbar\omega}\left(\frac{1}{R_{cr}} - \frac{1}{R_{in}}\right)\right]. \tag{11.138}$$

This formula illustrates the radius-dependent catastrophe of particle production. One can see that if the rate of particle growth is sufficiently large and the particle size during the initial period (the so-called first generation) also becomes large enough for trapping $R_{in} = R_{cr}$, the rate of the particle production is near the rate of SiH_3^- production. Such conditions take place at room temperature. For higher temperatures, the first generation initial radius R_{in} is less than the critical radius R_{cr} for selective trapping, and the particle production rate becomes much slower than a dissociative attachment.

11.6.6 Temperature Effect on Selective Trapping and Particle Production Rate

To rewrite Eq. 11.138 for the production of nanoparticles as a function of temperature, take into account the linear law of radius increase (Eq. 11.132) and the exponential dependence of the growth rate on temperature (Eq. 11.130):

$$W_p = W_{ad} \exp\left\{\frac{e^2}{4\pi\varepsilon_0 \hbar\omega R_{cr}}\left[\exp\left(\frac{B\Delta T}{3T_{00}^{4/3}}\right) - 1\right]\right\}. \tag{11.139}$$

This double exponential law shows an extremely strong dependence of dust production on temperature. The double exponential relation can be presented (for room temperature and higher) as a simple and general criterion for fast dust production in silane plasma, taking into account the trapping criterion (Eq. 11.135) and the expression (Eq. 11.134) for electron attachment coefficient:

$$\exp\left(-\frac{\Delta T B}{3T_{00}^{4/3}}\right) \ln \frac{k n_e R}{v} > 1. \tag{11.140}$$

Here v is the gas flow linear velocity, R is the distance between electrodes, k is the pre-exponential factor in the expression (Eq. 11.134) for electron attachment coefficient. This criterion of fast dust production in silane plasma is presented numerically in Figure 11.6. In this fig, the area of effective particle formation is placed over the critical curve, ΔT is the temperature in degree Celsius. Dust production in the plasma can be stimulated by increasing the electron concentration and neutral gas residence time (for example, by increasing the distance between electrodes) and can be restricted even by small gas heating.

11.6.7 Critical Phenomenon of Super-Small Particle Coagulation

When the growing concentration of survived monodispersed 2 nm particles reaches a critical value of about 10^{10}–10^{11} cm^{-3}, the fast coagulation process begins. Experimentally the beginning of the coagulation process has a threshold character and proves the critical character of this phenomenon. It is important to note that the critical value of particle concentration for coagulation does not depend on temperature experimentally; which is not typical for phase transition processes. The simple estimations as well as detailed modeling show that when the cluster concentration reaches a critical value of about 10^{10}–10^{11} cm^{-3}, the attachment of small negative ions (like SiH_3^-) to 2-nm neutral particles becomes faster than the chain reaction of a new particle growth. Thus, new dust particle

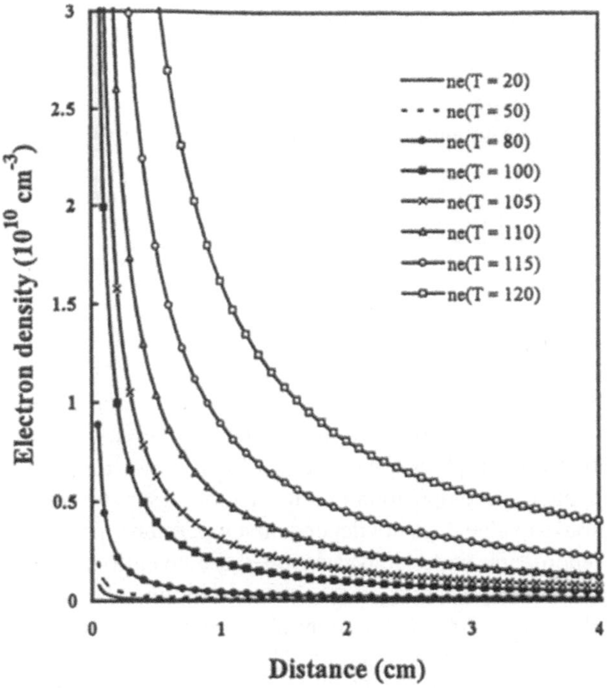

FIGURE 11.6 Critical conditions of particle formation. Gas temperature, electron concentration, and distance between the electrodes. Dust production area is above the presented curves. The gas temperature is in degree Celsius. The gas velocity in the discharge is 20 cm/s.

formation becomes much slower, taking into account that cluster mass growth from an initial negative ion SiH_3^- has an essential acceleration with particle radius (Eq. 11.132). The total particle mass remains almost constant during the coagulation. The deposition of silane radicals on neutral surfaces can be neglected in the fast coagulation phase. To describe the critical coagulation phenomenon, consider the probability of two-, three-, and many-particle interaction, assuming that the mean radius R of particles interaction is proportional to their physical radius and all direct collisions result in aggregation. The reaction rate and rate coefficient for binary particle collision are

$$W(2) = \sigma v N^2, \quad K(2) = \sigma v, \tag{11.141}$$

where N is the particle concentration, σv is the mean product of their cross section and velocity. From Eq. 11.141, the steady-state concentration of the two-particle complexes $N(2)$ can be estimated taking into account their characteristic lifetime R/v as

$$N(2) = \left(\sigma v N^2\right)\frac{R}{v} = N(\sigma R N). \tag{11.142}$$

Based on formula (Eq. 11.142), the reaction rate and rate coefficient of the agglomerative three-particle collision can be written as

$$W(3) = \sigma v(\sigma R N)N^2, K(3) = \sigma v(\sigma R N). \tag{11.143}$$

Repeating this procedure (Nikitin, 1970), and taking into account that σR is proportional to the particle mass, the rate coefficient of $(k+2)$-body collision can be given in the following form:

$$K(k) = \sigma v (\sigma RN)^k k!. \tag{11.144}$$

The rate constant of the coagulation can be calculated as the sum of the partial rates (Eq. 11.144), which is divergent, but can be asymptotically found by integration procedure of Migdal (1981):

$$K_c = \Sigma \sigma v (\sigma RN)^k k! = \sigma v \left[1 + \frac{1}{\ln(N_{cr}/N)} \right]. \tag{11.145}$$

This relation shows that when the initial particle concentration N is less than critical one N_{cr} (which is proportional to $1/\sigma R$, and numerically is about 10^{10}–10^{11} cm^{-3}), the coagulation rate coefficient is the conventional one related to binary collisions. But when the initial particle concentration N approaches the critical value $N_{cr,}$ the coagulation rate sharply increases demonstrating typical features of a phase transition critical phenomenon. From Eq. 11.145, it is seen that because of very small probability of aggregate decomposition (the reverse process with respect to coagulation), the critical particle concentration value does not depend on temperature. This is the principal difference between the critical phenomenon under consideration and conventional gas condensation. Results describing time evolution of the processes are presented in Figure 11.7. The typical induction time before coagulation is about 100–200 ms for room temperature and much longer for 400 K. This temperature effect is due to the selective trapping. The particle production rate for 400 K is much less than for room temperature, but the critical particle concentration value remains almost the same and hence the induction period becomes much longer.

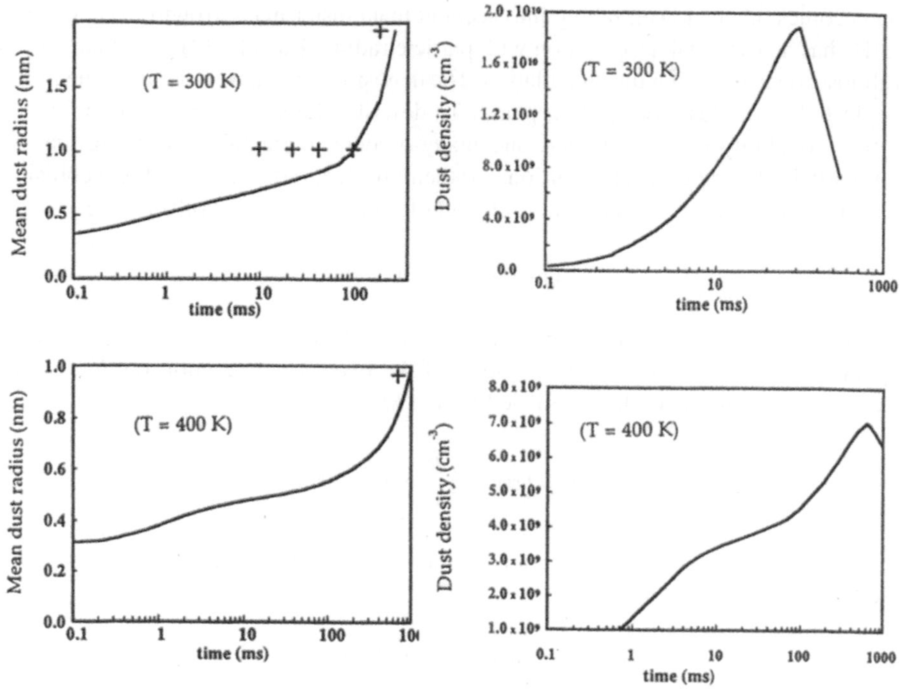

FIGURE 11.7 Time evolution of the dust density and radius for different gas temperatures. Modeling results (–). Experimental data (+).

11.6.8 Critical Change of Plasma Parameters during Dust Formation (the α–γ Transition)

When the particle size reaches a second critical value during the coagulation process, another change of plasma parameters takes place: electron temperature increases and electron concentration decreases dramatically. To analyze this critical phenomenon consider the balance equation of electrons (n_e), positive ions (n_i), and negatively charged particles (N_n). The main processes taken into account are: ionization ($k_i(T_e)$); electron and positive ion losses on the walls by ambipolar diffusion ($n_e D_a/R^2$); the electron attachment to neutral particles ($k_a(r_a)$), which is the strong function of the particle radius r according to Eq. 11.134; and ion–ion recombination k_r of positive ions and negatively charged particles. Taking into account that before the α–γ transition, negatively charged particles have actually only one electron, the balance equations can be presented as

$$\frac{dn_e}{dt} = k_i\left(T_e\right)n_e n_0 - n_e\frac{D_a}{R^2} - k_a\left(r\right)n_e N, \tag{11.146}$$

$$\frac{dn_i}{dt} = k_i\left(T_e\right)n_e n_0 - n_i\frac{D_a}{R^2} - k_r N_n n_i, \tag{11.147}$$

$$\frac{dN_n}{dt} = k_a\left(r\right)n_e N - k_r n_i N_n. \tag{11.148}$$

Here N and N_0 are the neutral particle and gas concentrations, respectively. For the steady-state discharge regime, the derivatives on the left side can be neglected. The threshold of the α–γ transition can be calculated from Eq. 11.146, assuming for the ionization coefficient $k_i(T_e) = k_{0i}\exp(-I/T_e)$ (I is the ionization potential), and for electron attachment coefficient (see Eq. 11.134):

$$k_a\left(r\right) = k_{0a}\left(\frac{r}{R_0}\right)^2\exp\left(-\frac{e^2}{4\pi\varepsilon_0\hbar\omega r}\right). \tag{11.149}$$

Thus, the critical phenomenon of the α–γ transition can be seen from the electron concentration balance Eq. 11.146, which taking into account Eq. 11.149 looks as

$$k_{0i}\exp\left(-\frac{I}{T_e}\right)n_0 = \frac{D_a}{R^2} + Nk_{0a}\left(\frac{r}{R_0}\right)^2\exp\left(-\frac{e^2}{4\pi\varepsilon_0\hbar\omega r}\right). \tag{11.150}$$

Before the critical moment of α–γ transition when the particles are relatively small as is any attachment to them, the electron balance is determined by ionization and electron losses to the walls. The α–γ transition is the moment when the electron losses on the particles sharply grow with their size and become more important than those on the walls. The electron temperature then increases to support the plasma balance (Belenguer et al., 1992). The total mass and volume of the particles remain almost constant during the coagulation, so the specific particle surface (the total surface per unit of volume) decreases with the growth of mean particle radius. The influence of particle surface becomes more significant when the specific surface area decreases. Eq. 11.150 explains this phenomenon: the exponential part of the electron attachment dependence on particle radius is more important than the pre-exponential one. So the comparison of the first and the second terms in the right side of Eq. 11.150 gives a critical particle size R_c, such that α–γ transition begins only when the particle radius becomes larger during coagulation:

$$R_c = \frac{e^2}{4\pi\varepsilon_0\hbar\omega\ln\left(k_{0a}NR^2/D_a\right)}. \tag{11.151}$$

This critical radius for α–γ transition is 3 nm, and practically does not depend on temperature.

11.6.9 Electron Temperature Evolution in the α–γ Transition

Eq. 11.150 describes the electron temperature as a function of aggregates radius r during the α–γ transition. There are two exponents in Eq. 11.150, which play the major role in the electron balance. So neglecting the ambipolar diffusion term, leads to the relation for the electron temperature evolution during the α–γ transition:

$$\frac{1}{T_e}-\frac{1}{T_{e0}}=\frac{e^2}{4\pi\varepsilon_0\hbar\omega}\left(\frac{1}{r}-\frac{1}{R_c}\right). \tag{11.152}$$

Here, T_{e0} (about 2 eV) is the electron temperature just before the α–γ transition when the aggregate radius is near its critical value $r = R_c$. One can see from Eq. 11.152 that during the particle growth after α– γ transition, the electron temperature increases with the saturation on the level of $e^2/4\pi\varepsilon_0\hbar\omega R_c$, which is numerically about 5–7 eV. Thus as a result of the α–γ transition, the electron temperature can grow about three times. The electron temperature increase as a function of the relative difference of positive ion and electron concentrations $\Delta = (n_i - n_e)/n_i$ can be derived from the positive ion balance Eq. 11.147, taking into account plasma electro-neutrality ($n_i = n_e + N_n$):

$$k_{0i}\exp\left(-\frac{I}{T_e}\right)n_0=\frac{D_a}{R^2}+k_r n_i\frac{\Delta}{1-\Delta}. \tag{11.153}$$

One can see that electron temperature is essentially related to the relative difference Δ between positive ion and electron concentrations. When during the α–γ transition electron temperature grows, electron concentration decreases and the Δ factor approaches 1. The electron concentration before α– γ transition is almost constant at about $n_{eo} = 3 \times 10^9$ cm^{-3}. In the experimental conditions, current density $j_e = n_e e b_e E$ is proportional to $n_e T_e^2$ and remains fixed during α–γ transition. Hence, the electron and positive ion concentrations during this period can be derived from the balance equation as

$$n_i=\sqrt{\frac{k_{0i}n_{e0}T_{e0}^2 n_0\exp(-I/T_e)}{k_r T_e^2}},\; n_e=n_{e0}\frac{T_{e0}^2}{T_e^2}. \tag{11.154}$$

According to this relation, electron concentration decreases during α–γ transition approximately 10 times, and positive ion concentration slightly increases. Modeling and experimental data describing evolution of electron density and temperature during the α–γ transition are shown in Figure 11.8 (Fridman et al., 1996).

11.7 NONEQUILIBRIUM CLUSTERIZATION IN CENTRIFUGAL FIELD

11.7.1 Centrifugal Clusterization in Plasma Chemistry

The clusterization can be affected by fast plasma rotation, which in RF- and microwave discharges can be characterized by tangential velocities v_φ close to the speed of sound (Givotov et al., 1986). Centrifugal forces proportional to the cluster mass push the particles to the discharge periphery. The clusters can be transferred to the discharge periphery faster than the heat transfer. This can result in a shift of chemical equilibrium in the direction of product formation. It provides product separation during the process and decreases energy cost (see Becker and Doring, 1935; Zeldovich, 1942; Frenkel, 1945).

11.7.2 Clusterization Kinetics as Diffusion in Space of Cluster Sizes

The clusters A_n, which consist of n molecules A, can be formed and destroyed in chemical processes:

$$A_{n1}+A_{n-n1}\Leftrightarrow A_n,\quad A_{n-k}+A_{m+k}\Leftrightarrow A_n+A_m. \tag{11.155}$$

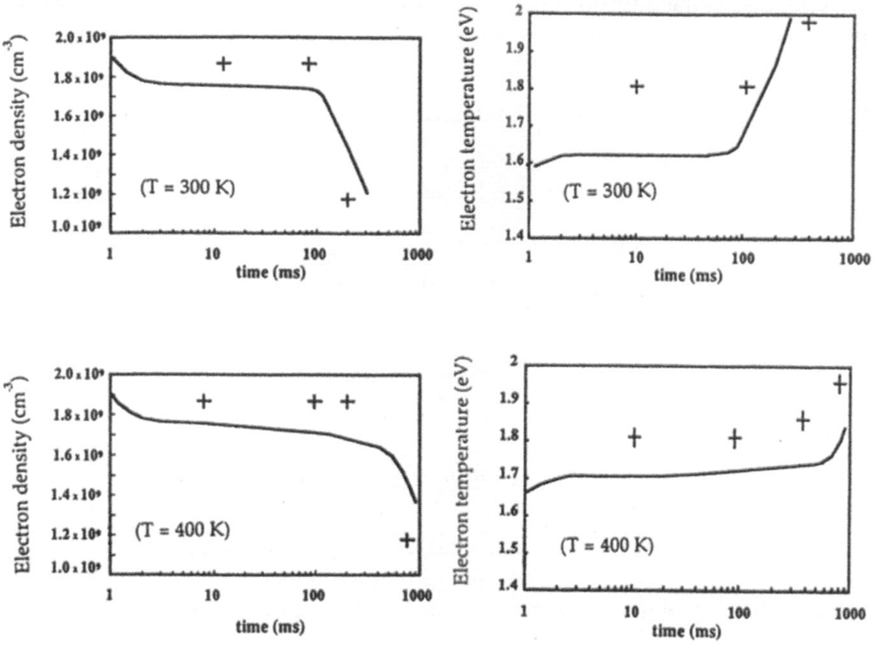

FIGURE 11.8 Time evolution of the electron density and temperature including α–γ transition for different gas temperatures. Modeling results (–). Experimental data (+).

These clusters can also diffuse and drift in the centrifugal field. Consider the cluster size n as a continuous coordinate (varying from 1 to ∞). Another coordinate is the radial coordinate x, varying from 0 to the radius R of the discharge tube. Then clusterization in the centrifugal field can be considered in terms of the evolution of the distribution function $f(n, x, t)$ in the space (n, x) of cluster sizes. Plasma-chemical reaction of dissociation of the initial compound can be taken into account as a source of molecules A located in the point $x = 0$. Evolution of the cluster distribution can be described by the continuity equation in the space (n, x) of cluster sizes (Macheret et al., 1985):

$$\frac{\partial f(n,x,t)}{\partial t} + \frac{\partial j_n}{\partial n} + \frac{\partial j_x}{\partial x} = 0. \tag{11.156}$$

j_n is the flux of clusters along the axis of their sizes n, and j_x is the radial flux of particles. The radial flux j_x, which takes into account nonisothermal diffusion and centrifugal drift can be given as

$$j_x = -D_x\left(\frac{\partial f}{\partial x} + \frac{f}{T}\frac{\partial T}{\partial x} - \frac{nmv_\varphi^2}{xT} f\right), \tag{11.157}$$

where D_x is the diffusion coefficient, m is the mass of a molecule A. The number of clusters is less than the number of molecules at temperatures above that of condensation. In this case, diffusion in the space of cluster sizes is mostly due to atom attachment and detachment processes:

$$A_{n-1} + A \Leftrightarrow A_n, \tag{11.158}$$

which permits describing the flux of clusters along the axis of their sizes n in the following linear Fokker–Plank form (Rusanov et al., 1985):

$$j_n = -D_n\left(\frac{\partial f}{\partial n} - \frac{\partial \ln f_0}{\partial n} f\right). \tag{11.159}$$

In this expression, D_n is the diffusion coefficient along the coordinate n; $f_0(n)$ is the distribution in equilibrium conditions, which influences the nonequilibrium flux (Eq. 11.159).

11.7.3 Quasi-Equilibrium Cluster Distribution over Sizes

The quasi-equilibrium number density of clusters A_n is determined at fixed temperature T by the change of thermodynamic Gibbs potential ΔG_n related to A_n formation from n molecules A:

$$f_0(n) = \text{const } \exp\left[-\frac{\Delta G(n)}{T}\right]. \tag{11.160}$$

The change in the Gibbs potential ΔG depends in turn on changes of enthalpy and entropy in the reaction $nA \rightarrow A_n$:

$$\Delta G(n) = n\Delta h(n) - n\Delta s(n). \tag{11.161}$$

$\Delta h(n)$ and $\Delta s(n)$ are the specific enthalpy and entropy changes calculated per one A molecule. The entropy change Δs is mostly determined by change of volume per one molecule A during the clusterization process from V_2 in molecule to V_{cl} in clusters (Vostrikov and Dubov, 1984a,b):

$$\Delta s \approx \ln\frac{C_{cl}}{V_2} \approx \ln\frac{pm}{T\rho_c}, \tag{11.162}$$

where p and T are the gas pressure and temperature, ρ_c is the condense phase density. Thus, as one can see from Eq. 11.162, the entropy change Δs actually does not depend on the cluster size n. Decomposition energy of a molecule A from the cluster A_n does not depend on n, and equals to the evaporation enthalpy λ (for example, the bonding energy of S_2 in sulfur cluster S_{2n} is almost fixed: $\lambda \approx 30$ kcal/mol at any n). The specific enthalpy change in clusterization can be expressed as:

$$\Delta h(n) = -\lambda\left(1 - \frac{1}{n}\right). \tag{11.163}$$

For larger spherical aggregates ($n \gg 1$), which should be considered not as clusters but as condense phase particulates:

$$\Delta h(n) = -\lambda\left(1 - \text{const}/\sqrt[3]{n}\right). \tag{11.164}$$

The second term on the right-hand side is related to the surface tension energy. The corresponding second term $1/n$ in Eq. 11.163 can be interpreted as the 1D analog of the surface tension energy. Eqs. 11.160 through 11.163 lead to the exponential distribution of clusters over their sizes:

$$f_0(n) \propto \exp\left(-\frac{n}{n_0}\right), \tag{11.165}$$

where the exponential parameter n_0 of the distribution function decrease or increase depends on temperature and can be presented using the condensation temperature $T_c = -\lambda/\Delta s$ as

$$-\frac{\partial \ln f_0}{\partial n} = \frac{1}{n_0} = -\Delta s - \frac{\lambda}{T} = \lambda\left(\frac{1}{T_C} - \frac{1}{T}\right). \tag{11.166}$$

From Eqs. 11.165 and 11.166, the quasi-equilibrium distribution $f_0(n)$ exponentially decreases at temperatures exceeding the condensation point ($T > T_c$). As soon as the temperature reaches the condensation point, the parameter $n_0 \to \infty$ and clusters grow to large scale condense phase particulates.

11.7.4 Magic Clusters

Deviation of the clusterization enthalpy from Eq. 11.163 leads to a non-exponential distributions $f_0(n)$ over cluster sizes. Bonding energy in clusters of some specific sizes can be stronger, which results in a high concentration of clusters with the "magic sizes" (Sattler, 1982). In the example of sulfur, clusters S_6 and S_8 are more stable than other S_{2n} at temperatures exceeding the condensation point. Taking into account a magic cluster size n_m, the specific enthalpy change (Eq. 11.163) is

$$\Delta h(n) = -\lambda\left(1-\frac{1}{n}\right)-\frac{1}{n_m}\delta Q(n), \tag{11.167}$$

where $\delta Q(n)$ is the peak at $n = n_m$ with the characteristic width about unity and maximum ΔQ_{max}. The corresponding addition to the logarithm of quasi-equilibrium distribution in the magic point is equal to:

$$\Delta \ln f_0 = \frac{\delta Q(n)}{T}. \tag{11.168}$$

Hence the effect of magic clusters is negligible at very high temperatures and becomes significant only at temperatures exceeding the condensation point:

$$\Delta T \approx T_c \frac{\Delta Q_{max}}{\lambda n_m}. \tag{11.169}$$

11.7.5 Quasi-Steady-State Equation for the Cluster Distribution Function *f* (*n*, *x*)

The equation for the quasi-steady-state cluster distribution over sizes $f(n, x)$ based on Eq. 11.156 is

$$\frac{\partial j_n}{\partial n}+\frac{\partial j_x}{\partial x}=0. \tag{11.170}$$

The simplest solution of this equation is for quasi-equilibrium, which can be obtained assuming that both fluxes j_n and j_x are equal to zero:

$$f(n,x) \propto \frac{T(x=0)}{T(x)} f_0(x)\exp\left[n\int\frac{mv_\varphi^2 dx}{xT(x)}\right]. \tag{11.171}$$

This distribution does not satisfy the typical boundary conditions in the space of cluster sizes. Also, the characteristic establishment time of the distribution $f(n, x)$ along n (about n^2/D_n) is much shorter for medium clusters of interest ($n < 10^2$–10^3) than characteristic time along $x(R^2/D_x)$. For this reason, it is better to assume in Eq. 11.170 that the distribution $f(n, x)$ is determined by fast diffusion along the n-axis, and along the radius x only by the centrifugal drift. In the above expressions for characteristic times, R is the discharge radius; D_x is the conventional diffusion coefficient; and D_n is the diffusion coefficient in the space of cluster sizes, which can be taken as

$$D_n = k_0[A]\exp\left(-\frac{E_a}{T}\right); \tag{11.172}$$

k_0 is rate coefficient of gas-kinetic collisions; $[A]$ is the concentration of molecules A; E_a is the activation energy of the direct processes (Eq. 11.158). Assuming that the activation energy E_a does not depend on n, the diffusion coefficient D_n also can be considered independent of the cluster sizes. If the conventional diffusion coefficient D_x is assumed independent on n and x for the conditions of interest, the derivative of the centrifugal drift can be simply expressed as

$$\frac{\partial}{\partial x}\left[D_x \frac{mv_\varphi^2}{xT} nf\right] \approx \frac{D_x}{R^2}\frac{mv_\varphi^2}{T} nf. \tag{11.173}$$

The quasi-steady-state kinetic equation for the nonequilibrium cluster distribution $f(n, x)$ can be derived taking into account diffusion along the n-axis and centrifugal drift along the radius x as

$$\frac{\partial^2 f}{\partial n^2} - \frac{\partial \ln f_0}{\partial n}\frac{\partial f}{\partial n} - \left(\frac{\partial^2 \ln f_0}{\partial n^2} + \frac{D_x}{D_n R^2}\frac{mv_\varphi^2}{T} n\right) f = 0. \tag{11.174}$$

Finally, the linear kinetic Eq. 11.174 for the nonequilibrium cluster distribution function $f(n, x)$ can be simplified by introducing the modified cluster distribution function: $y(n) = f(n)/, \sqrt{f_0(n)}$, which gives (Rusanov et al., 1985):

$$y'' - \left[\frac{1}{4}\left(\frac{\partial \ln f_0}{\partial n}\right)^2 + \frac{1}{2}\frac{\partial^2 \ln f_0}{\partial n^2} + \frac{D_x}{D_n R^2}\frac{mv_\varphi^2}{T} n\right] y = 0. \tag{11.175}$$

11.7.6 Nonequilibrium Distribution Functions $f(n, x)$ of Clusters without Magic Numbers in the Centrifugal Field

For monotonic equilibrium distribution $f_0(n,x)$ of clusters without magic numbers, the kinetic Eq. 11.175 can be rewritten taking into account Eqs. 11.155 and 11.166 as

$$y'' + \left(\frac{1}{4n_0^2} + bn\right) y = 0. \tag{11.176}$$

Here, the parameter $b = (D_x/D_n R^2)\left(mv_\varphi^2/T\right)$ reflects the frequency competition between centrifugal drift and diffusion along the axis of cluster sizes in the formation of the distribution $f(n, x)$; the factor n_0 (Eqs. 11.165 and 11.166) characterizes the average cluster size at temperatures exceeding the condensation point. The differential Eq. 11.176 is a Bessel type of equation, and its solution can be expressed using the Hankel function $H_{1/3}^{(1)}(z)$ (Abramowitz and Stegan, 1968). A solution at $n \to \infty$ is

$$y(n) \propto \sqrt{\frac{1}{4n_0^2} + bn}\, H_{1/3}^{(1)}\left[i\frac{2}{3b}\left(\frac{1}{4n_0^2} + bn\right)^{3/2}\right]. \tag{11.177}$$

Diffusion along the axis of cluster sizes is usually faster than centrifugal drift, and

$$b\left|n_0^3\right| = \frac{D_x/R^2}{D_n/n_0^2}\frac{|n_0|\, mv_\varphi^2}{T} \ll 1. \tag{11.178}$$

The asymptote of the Hankel function at large argument values can be applied. This permits rewriting Eq. 11.177, to obtain the nonequilibrium clusterization the following form:

$$f(n) = \frac{\text{const}}{\sqrt[4]{1 + 4bn_0^2 n}} \exp\left[-\frac{n}{2n_0} - \frac{\left(1 + 4bn_0^2 n\right)^{3/2}}{12bn_0^3}\right]. \tag{11.179}$$

As seen from Eq. 11.179, the distribution $f(n)$ is close to the quasi-equilibrium value (Eq. 11.165) at relatively small cluster sizes $n < 1/4bn_0^2$. At relatively large cluster sizes $n > 1/4bn_0^2$, the distribution function decreases because of intensive centrifugal losses:

$$f(n) \propto \exp\left(-\frac{2\sqrt{b}}{3}n^{3/2}\right). \tag{11.180}$$

In this case the most probable cluster size taking into account the centrifugal effect is

$$n = n_0, \text{ if } n_0 > 0, \tag{11.181}$$

$$n = 1/4bn_0^2, \text{ if } n_0 < 0. \tag{11.182}$$

11.7.7 Nonequilibrium Distribution Functions $f(n, x)$ of Clusters in the Centrifugal Field, Taking into Account the Magic Cluster Effect

If the quasi-equilibrium distribution is not exponential, the solution of the kinetic Eq. 11.175 becomes much more complicated. It can be done nevertheless, taking into account the similarity of this equation to the Schrodinger equation for the 1D motion of a particle in a potential field. In this case, the factor $-1/4n_0^2$ corresponds to the energy E of a particle in the Schrodinger equation, and factor bn corresponds to potential $U(x) = bx$. In this quantum-mechanical analogy, one is looking for solutions of Eq. 11.175 at $n > 0$, which corresponds to particle penetration in the classically forbidden zone where $E < U(x)$. Then the quasi-classical approximation can be applied far from the return point when the following criterion is valid:

$$\frac{d}{dn}\frac{1}{\sqrt{\varphi(n)}} \ll 1. \tag{11.183}$$

In this criterion of the quasi-classical approximation, another function $\varphi(n)$ related to the quasi-equilibrium cluster distribution is introduced:

$$\varphi(n) = \frac{1}{4}\left(\frac{\partial \ln f_0}{\partial \ln n}\right)^2 + \frac{1}{2}\frac{\partial^2 \ln f_0}{\partial n^2} + bn. \tag{11.184}$$

The criterion of the quasi-classical approximation (Eq. 11.183) coincides with Eq. 11.178 of kinetic domination of clusterization over the centrifugal drift and asymptotic simplification of the special Hankel function. The solution of the kinetic Eq. 11.175 for the cluster distribution in the centrifugal field in the framework of this analogy with Schrodinger equation can be found in the quasi-classical approximation as:

$$f(n) \propto \frac{\sqrt{f_0(n)}}{\sqrt[4]{\varphi(n)}}\exp\left[-\int\sqrt{\varphi(n)}dn\right]. \tag{11.185}$$

In particular, for the quasi-equilibrium distribution with the magic number correction (Eq. 11.168) $\Delta Q_{max}/T > 1$ in the vicinity of $n = n_m$, one can derive the following expression for the actual cluster distribution function in the centrifugal field:

$$f_m(n) \approx f_1(n)\exp\left[\frac{\delta Q(n)}{2T} + \sqrt{\frac{1}{4n_0^2} + bn}\left(\sqrt{1 + \frac{\frac{1}{4}\left(\frac{\delta Q(n)}{T}\right)^2}{\frac{1}{4n_0^2} + bn}} - 1\right)\right]. \tag{11.186}$$

In this relation, $f_1(n)$ is the cluster distribution in a centrifugal field without a magic cluster effect, which is determined by Eq. 11.179. Note that if

$$n_0\left(\frac{\Delta Q_{max}}{T}\right)^2 \gg 1+4bn_0^2 n_m, \tag{11.187}$$

which usually takes place, then correction related to magic clusters in the vicinity of $n = n_m$ is

$$f_m(n) \approx f_1(n)\exp\left(\frac{\delta Q(n)}{T}\right). \tag{11.188}$$

Example of the cluster size distribution $f(n, x)$ in the centrifugal field is shown in Figure 11.9.

11.7.8 Radial Distribution of Cluster Density

Based on the distribution function $f(n, x)$ described above, the cluster density $\rho(x)$ can be determined as

$$\rho(x) = \int_0^\infty nf(n,x)\,dn. \tag{11.189}$$

The total flux of clusters J_0 to the discharge periphery (calculated in number of molecules in the clusters) can be found by multiplying Eq. 11.157 by n and integrating from 0 to ∞. Taking into account the definition of the cluster density yields the following equation:

$$-D_x\left[\frac{\partial\rho}{\partial x}+\frac{\rho}{R}\left(\frac{R}{T}\frac{\partial T}{\partial x}-\frac{mv_\varphi^2}{T}\frac{\langle n^2\rangle}{\langle n\rangle}\right)\right] = J_0. \tag{11.190}$$

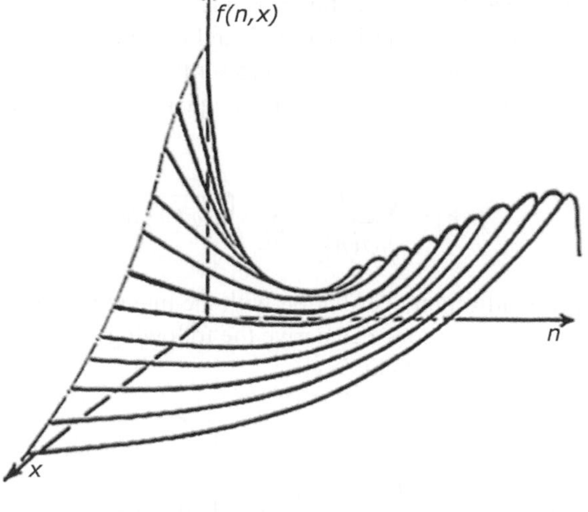

FIGURE 11.9 Typical cluster distribution function in the (n, x)-space.

Assume formation of the substance A on the axis $x = 0$, and absorption of the substance on the discharge walls. This gives the boundary conditions for Eq. 11.190: $\rho(x = 0) = \rho_0$, $\rho(x = a) = 0$, where a is the discharge tube radius. The solution of the homogeneous part of this differential equation is

$$\rho^{(0)}(x) = \rho_0 \frac{T_h}{T(x)} \exp\left[\frac{mv_\varphi^2}{R} \int_0^x \frac{\langle n^2 \rangle}{\langle n \rangle} \frac{dx}{T(x)}\right]. \tag{11.191}$$

Here, $T_h = T(x = 0)$ is the temperature of the hot zone on the discharge axis. Solution of the nonhomogeneous equation (Eq. 11.190) can be found by the variation of constant $\rho(x) = C(x)\rho^{(0)}(x)$. Taking into account the boundary conditions, it leads us first to the following expression for the total flux:

$$J_0 = \rho_0 Dx / \int_0^a \frac{T(x)}{T_h} \exp\left[-\frac{mv_\varphi^2}{R} \int_0^x \frac{\langle n^2 \rangle}{\langle n \rangle} \frac{dx}{T(x)}\right] dx. \tag{11.192}$$

The radial cluster density $\rho(x)$ in the centrifugal field can be expressed in the integral form:

$$\rho(x) = \rho_0 \frac{T_h}{T(x)} \exp\left[\frac{mv_\varphi^2}{R} \int_0^x \frac{\langle n^2 \rangle}{\langle n \rangle} \frac{dx}{T(x)}\right] \times \left\{1 - \int_0^x T(x) \exp\left[-\frac{mv_\varphi^2}{R} \int_0^x \frac{\langle n^2 \rangle}{\langle n \rangle} \frac{dx}{T(x)}\right] dx / \int_0^a T(x) \exp\left[-\frac{mv_\varphi^2}{R} \int_0^x \frac{\langle n^2 \rangle}{\langle n \rangle} \frac{dx}{T(x)}\right] dx\right\}. \tag{11.193}$$

The cluster flux Eq. 11.192 can be simplified for estimations to the form:

$$J_0 = \rho_0 D_x \frac{T_h}{\tilde{T}} \frac{\partial \ln T}{\partial x}, \tag{11.194}$$

where $\tilde{T}$ is the temperature in the discharge zone. Eq. 11.194 shows that the cluster flux to the discharge periphery is limited by diffusion of molecules A into a zone where large clusters can be formed.

11.7.9 Average Cluster Sizes

The average size of clusters moving to the periphery from a given radius x can be determined based on the flux $j(x,n)$ in the combined radius–cluster size space as

$$\langle n_f \rangle = \frac{\int_0^\infty n\, j(x,n)\, dn}{\int_0^\infty j(x,n)\, dn}. \tag{11.195}$$

Application of the above relations for any cluster sizes, including small ones ($n_0 \ll 1$), requires replacement of the equilibrium average size parameter n_0 by an effective one $\tilde{n}_0$, determined as

$$\frac{1}{\tilde{n}_0} = 1 - \exp\left(\frac{\partial \ln f_0}{\partial n}\right), \quad \frac{1}{n_0} = -\frac{\partial \ln f_0}{\partial n}. \tag{11.196}$$

For larger clusters ($n_0 \gg 1$), one can see from Eq. 11.196 that $\tilde{n}_0 \approx n_0$. The replacement $n_0 \to \tilde{n}_0$ permits using the Fokker–Plank diffusion approach for small clusters, where the "mean free path" in the cluster space is comparable with the characteristic system size. The integrals (Eq. 11.195) can be calculated with the flux Eq. 11.157 to find the average cluster size in the centrifugal field:

$$\langle n_f \rangle = \frac{\tilde{n}_0 + \alpha n_m + \frac{mv_\varphi^2}{T}\left(2\tilde{n}_0^2 + \alpha n_m^2\right)}{1 + \frac{mv_\varphi^2}{T}\left(1 + \alpha n_m\right)}. \quad (11.197)$$

$\alpha = [A_{nm}]/[A]$ is the density ratio of magic clusters ($n = n_\mathrm{m}$) and molecules A.

11.7.10 Influence of Centrifugal Field on Average Cluster Sizes

The average sizes of clusters moving from the radius x with temperature $T(x)$ in the absence of centrifugal forces ($mv_\varphi^2/T \to 0$) can be found from Eq. 11.197 as

$$\langle n_f \rangle = \tilde{n}_0(T) + \alpha(T) n_\mathrm{m}. \quad (11.198)$$

At relatively high temperatures, $\alpha(T) \ll n_\mathrm{m}^{-1}$ and $\langle n \rangle \approx \tilde{n}_0 \approx 1$, which means that most molecules move to the discharge periphery. The fraction of magic clusters grows with a temperature decrease. Large clusters with sizes $n \geq n_\mathrm{m}$ begin dominating in the flux to the discharge periphery, only when the temperature exceeds the condensation point T_c on

$$\Delta T = T_\mathrm{C} \frac{\Delta Q_\mathrm{max}}{\lambda n_\mathrm{m}}. \quad (11.199)$$

Centrifugal forces stimulate the domination of large clusters in the flux of particles to the discharge periphery even at temperatures exceeding the condensation point. If the centrifugal effect is strong

$$\frac{mv_\varphi^2}{T} \alpha(T) n_\mathrm{m}^2 > 1, \quad (11.200)$$

then according (Eq. 11.197), the average cluster sizes become relatively large and can be determined as

$$\langle n_f \rangle \approx \frac{mv_\varphi^2}{T} \alpha(T) n_\mathrm{m}^2 \gg 1, \quad \text{if } \frac{mv_\varphi^2}{T} \alpha(T) n_\mathrm{m} < 1, \quad (11.201)$$

$$\langle n_f \rangle \approx n_\mathrm{m}, \quad \text{if } \frac{mv_\varphi^2}{T} \alpha(T) n_\mathrm{m} > 1. \quad (11.202)$$

Criterion (Eq. 11.200) determines the temperature when large clusters more than molecules are moving to the discharge periphery. The larger the value of the centrifugal factor mv_φ^2/T, the more this critical temperature exceeds the condensation point. For example, the condensation temperature of sulfur is $T_\mathrm{c} \approx 550\ K$ at 0.1 atm. At large values of the centrifugal factor mv_φ^2/T, when the tangential velocity is close to the speed of sound, the effective clusterization temperature reaches the high value of 850 K. The magic clusters in this case are sulfur compounds S_6 and S_8.

11.7.11 Nonequilibrium Energy Efficiency Effect Provided by Selectivity of Transfer Processes in Centrifugal Field

The transfer phenomena do not affect the maximum energy efficiency of plasma-chemical processes when there are no external forces and the Lewis number is close to unity. However, strong increases in energy efficiency can be achieved in the centrifugal field if the molecular mass of products exceeds that for other components. In this case, the fraction of products moving from the discharge zone can exceed the relevant fraction of heat. It can result in a decrease in the product energy cost with respect to the minimum value for quasi-equilibrium thermal systems. As an example, consider the practically important plasma-chemical process of hydrogen sulfide decomposition in thermal plasma with the production of hydrogen and elemental sulfur (Balebanov et al., 1985; Nester et al., 1985; Harkness and Doctor, 1993):

$$H_2S \rightarrow H_2 + S_{solid}, \quad \Delta H = 0.2\,eV. \tag{11.203}$$

This process begins in the high-temperature thermal plasma zone with hydrogen sulfide decomposition forming sulfur dimers:

$$H_2S \rightarrow H_2 + S_{solid}, \quad \Delta H = 0.2\ eV. \tag{11.204}$$

This is followed by clusterization of sulfur in the lower temperature zones on the discharge periphery:

$$S_2 \rightarrow S_4, S_6, S_8 \rightarrow S_{solid}. \tag{11.205}$$

The minimum energy cost of the process in quasi-equilibrium systems with ideal quenching (see Section 5.7.2) is 1.8 eV/mol. The minimum energy cost can be significantly decreased when tangential gas velocity in the discharge zone is sufficiently high and criterion (Eq. 11.200) is satisfied. In this case, selective transfer of the sulfur clusters to the discharge periphery permits producing hydrogen and sulfur with a minimal energy cost of 0.5 eV/mol. Special experiments in the thermal microwave and RF-discharges with strong centrifugal effects gave a minimum value for the energy cost of approximately about 0.7–0.8 eV/mol (Balebanov et al., 1985; Krasheninnikov et al., 1986). The optimal reaction temperature in this case is about 1150 K, and the effective clusterization temperature is 850 K. The lowest energy cost discussed above can be achieved only if the strong centrifugal effect criterion (Eq. 11.200) is satisfied in the discharge zone, where the dissociation process takes place. If the necessary high gas rotation velocities exist only in the relatively lower temperature clusterization zones, the minimum energy cost is somewhat higher (1.15 eV/mol). The results illustrating the plasma-chemical process of hydrogen sulfide decomposition producing hydrogen and elemental sulfur, and centrifugal effect on its efficiency are presented in Figure 11.10.

11.8 DUSTY PLASMA STRUCTURES: PHASE TRANSITIONS, COULOMB CRYSTALS, SPECIAL OSCILLATIONS

11.8.1 Interaction of Particles, and Structures in Dusty Plasmas

The 10–500 nm dust particles may acquire a very large charge $Z_d e = 10^2–10^5 e$. As a result, the mean energy of Coulomb interaction between them is proportional to Z_d^2 and can exceed the particle thermal energy. Hence the dusty plasma can be highly nonideal with the charged particles playing the role of multiply charged heavy ions (Ichimaru, 1982). The strong Coulomb interaction between particles results in the formation of ordered special structures in dusty plasma similar to those in liquids and solids. Critical phenomena of phase transitions between "gas" and "liquid" and "liquid" and "solid" structures can be observed in dusty plasma as well (Fortov, 2001). The crystalline structures formed by charged particles in dusty plasma are usually referred to as Coulomb crystals (Ikezi, 1986; Chu and Lin, 1994; Thomas et al., 1994). Interaction of charged particles in dusty plasma can provide not only space, but also time–space structures. This leads not only to modification of wave and oscillation modes existing in nondusty plasmas, but also to the appearance of new modes typical only for dusty plasmas.

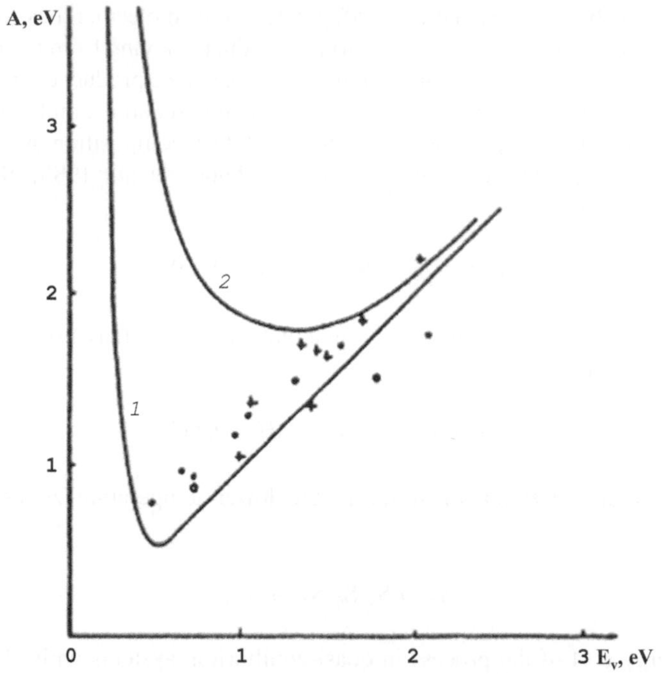

FIGURE 11.10 Energy cost of H_2S dissociation as a function of specific energy input. Nonequilibrium process, modeling. Minimal energy cost in quasi-equilibrium process, modeling. Experiments. (•) microwave discharge, 50 kW; (+) RF discharge, 4 kW.

11.8.2 Nonideality of Dusty Plasmas

Ideal and nonideal plasmas were briefly discussed in Section 6.1.1. In general, the systems of multiple interacting particles can be characterized by the nonideality parameter; this is determined as the ratio of the potential energy of interaction between particles and neighbors and their average kinetic energy. For dust particles with concentration n_d, charge $Z_d e$, and temperature T_d, the nonideality parameter (which is also called the Coulomb coupling parameter) is

$$\Gamma_d = \frac{Z_d^2 e^2 n_d^{1/3}}{4\pi\varepsilon_0 T_d}. \tag{11.206}$$

In general, for a system of multiple interacting particles, if the nonideality parameter is low ($\Gamma \ll 1$), such a system is ideal (see Section 6.1.1); if $\Gamma > 1$ the system of interacting particles is referred to as nonideal. The non-ideality criterion of dusty plasma (Eq. 11.210) also requires taking into account the shielding of electrostatic interaction between dust particles provided by plasma electrons and ions. Similar to definition (Eq. 11.206), the non-ideality parameter can be introduced for plasma electrons and ions, taking into account that their charge is equal to the elementary charge (Fortov and Yakubov, 1994):

$$\Gamma_{e(i)} = \frac{e^2 n_{e(i)}^{1/3}}{4\pi\varepsilon_0 T_{e(i)}}. \tag{11.207}$$

The nonideality parameter Γ is related to number of particles in Debye sphere N^D, which for or electrons and ions can be expressed as

$$N_{e(i)}^D = n_{e(i)} \frac{4}{3}\pi r_{De(i)}^2 \propto \frac{1}{\Gamma_{e(i)}^{3/2}}, \tag{11.208}$$

where r_D is the Debye radius determined by Eq. 6.4. The number of electrons and ions in Debye sphere (Eq. 11.208) is usually pretty large in dusty plasma conditions (and in most plasmas). For this reason, the nonideality parameter is small for electrons and ions ($\Gamma \ll 1$) and thus the electron–ion subsystem in dusty plasma can be considered as the ideal one. The dust subsystem is also ideal if $N_d^D \gg 1$, which means that there are lots of dust particles in the total Debye sphere. The dust particles can be considered as additional plasma components, which participate in electrostatic shielding and make a contribution to the value of the total Debye radius r_D:

$$\frac{1}{r_D^2} = \frac{1}{r_{De}^2} + \frac{1}{r_{Di}^2} + \frac{1}{r_{Dd}^2}. \tag{11.209}$$

In this relation, r_{De}, r_{Di}, and r_{Dd} are the Debye radii, calculated separately for electrons, ions and dust particles (see Eq. 6.4). In the opposite case of $N_d^D \ll 1$, the subsystem of dust particles is not necessarily the nonideal one. The dusty particles cannot be considered as a plasma component under these conditions, and the Debye radius is determined only by electrons and ions. The distance between the dust particles can exceed their Debye radius r_{Dd}, but interaction between them is not necessarily strong because of possible shielding provided by electrons and ions. Taking into account this shielding (or screening) effect, the modified nonideality parameter for the dust particles can be presented as

$$\Gamma_{ds} = \frac{Z_d^2 e^e n_d^{1/3}}{4\pi\varepsilon_0 T_d} \exp\left(-\frac{1}{n_d^{1/3} r_D}\right), \tag{11.210}$$

where the Debye radius is determined by electrons and ions: $r_D^{-2} = r_{De}^{-2} + r_{Di}^{-2}$. Thus, the degree of nonideality of the dust particle subsystem is determined by combining two dimensionless parameters: Γ_d (Eq. 11.206) related to the number of dusty particles in their Debye sphere, and the parameter:

$$K = 1/n_d^{1/3} r_D, \tag{11.211}$$

showing the ratio of the inter-particle distance to the length of electrostatic shielding (screening), provided by plasma electrons and ions. The factor $\exp(-K)$ in the nonideality parameter (Eq. 11.210) describes the weakening of the Coulomb interaction between particles due to the shielding.

11.8.3 Phase Transitions in Dusty Plasma

When the interaction between the dust particles is strong and the nonideality parameter is high, the strong coupling of particles leads to an organized structure called the Coulomb crystal. The Coulomb crystallization can be qualitatively described by the relatively simple one-component plasma model (OCP). The dusty plasma is treated in the framework of the OCP model as an idealized quasi-neutral system of ions and dust particles, interacting according to the binary Coulomb potential. This model directly does not include the electrostatic shielding effect, so $K \to 0$ in the relations (Eqs. 11.210 and 11.211). The OCP model (see, e.g., Fortov and Yakubov, 1994) obviously describes the dusty plasma in terms of the nonideality parameter Γd (Eq. 11.206) because $K \to 0$. According to the OCP model, the dusty plasma is not structured and can be considered as "gas" at low values of the parameter $\Gamma \leq 4$. At higher nonidealities Γ, some particles coupling and ordering take place, which can be interpreted as a "liquid" phase. Finally, when the nonideality parameter exceeds the critical value $\Gamma \geq \Gamma_c = 171$, the 3D regular crystalline structure is formed according to the qualitative OCP model. The more detailed Yukawa model takes into account shielding of the electrostatic interaction between charged particles by plasma electrons and ions. This model describes the dusty plasma nonideality, coupling and ordering of dust particles in terms of parameters Γ_{ds} and K. Interaction between dust particles in the frameworks of the Yukawa model is described by the

Debye–Huckel potential. Numerical calculations give in the phase transition to Coulomb crystals at critical values of the nonideality parameters Γ_{ds} and K, which is illustrated in Figure 11.11 (Meijer and Frenkel, 1991; Stevens and Robbins, 1993; Hamaguchi et al., 1997). The Yukawa model results shown in Figure 11.11 can be summarized by the following empirical criterion of the Coulomb crystallization (Molotkov et al., 2000):

$$\Gamma_{ds}\left(1+K+K^2/2\right)\geq 106. \tag{11.212}$$

11.8.4 Coulomb Clusters Observation in Dusty Plasma of Capacitively Coupled RF-Discharge

Detailed observation of the Coulomb clusters is related to the RF CCP discharges. The RF-discharge was sustained in argon, electrodes were located horizontally with the lower one powered and operated as an effective cathode (see Section 10.6.1). the frequency was about 14 MHz, pressure in the range of Torr, particle size about a few micrometers. The particles are negatively charged in such a discharge (see Sections 11.5 and 11.6), and trapped in the sheath space charge near the lower electrode, which act as an effective cathode. The particles suspended in the sheath of the RF-discharge are able to form the Coulomb crystal structure (Trottenberg et al., 1995). Crystallization of the RF-dusty plasma is usually observed at electron and ion densities about 10^8–10^9 cm^{-3}, and electron temperature in the rage of electronvolts. One should note that in the sheath where the Coulomb crystal is formed, the electron concentration exceeds that of the ions. A typical example of such Coulomb crystals is shown in Figure 11.12 (Morfill and Thomas, 1996). The observed space-organized structures can be characterized by the correlation function $g(r)$ showing the probability for two particles to be found at a distance r from one another. An example of the correlation function $g(r)$ for the Coulomb crystal structure observed by Trottenberg et al. (1995) is presented in Figure 11.13. The correlation function in the Figure substantiates the organized structure for at least five coordination spheres. Analysis of this structure shows that dust particles form a hexagonal 2D crystal grid in horizontal layers. In the vertical direction, dust particles position themselves exactly under one another and form a cubic grid between the crystal planes. The number of layers in the vertical direction is limited because of the requirement of a balance between gravitational and electrostatic forces necessary for suspension. When the average distance between particles is of the order of hundreds of micrometers, the typical number of layers is about 10–30. Taking into account that in the horizontal plane, particles are located several centimeters inside the circle, the Coulomb crystal in the RF CCP-discharge can be interpreted as 2.5D structure. "Melting" of the Coulomb crystals can be initiated by reducing the neutral gas pressure or by increasing the

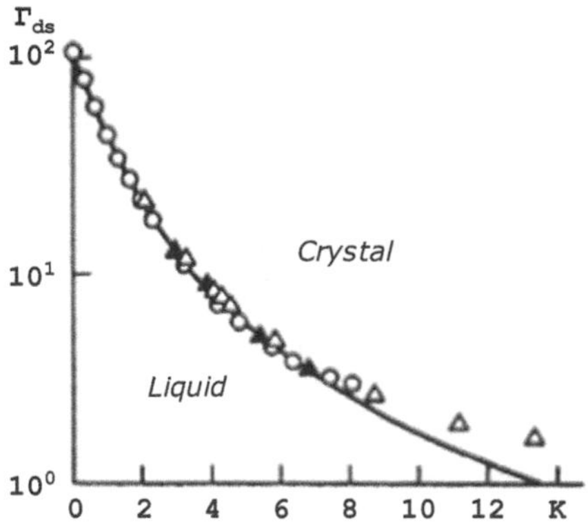

FIGURE 11.11 Crystallization curve was calculated in the Yukawa model.

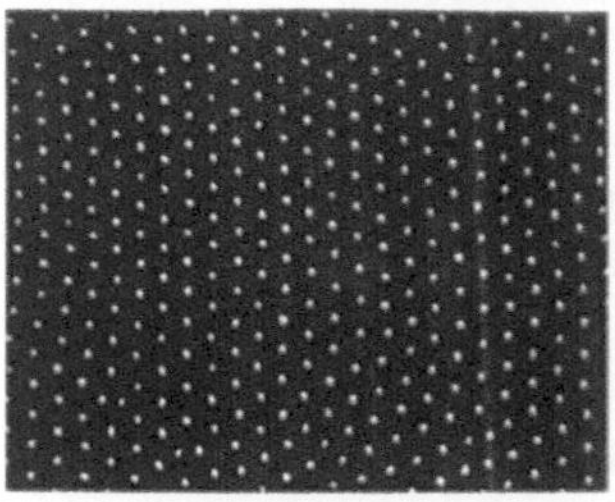

FIGURE 11.12 Horizontal structure of macroparticles achieved near the RF-discharge electrode.

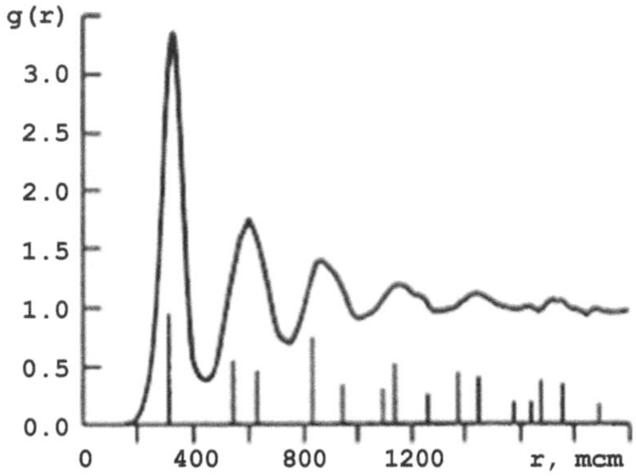

FIGURE 11.13 Correlation function $g(r)$ showing organized structure of Coulomb crystals.

discharge power. The crystal– liquid phase transition in both cases is due to a decrease in the nonideality parameter.

11.8.5 3D-Coulomb Clusters in Dusty Plasmas of DC-Glow Discharges

The 3D-quasi-crystal structures were observed in the positive column of DC-glow discharge. Similar to the discussed structures in RF CCP-discharges, here horizontal layers make 2D crystal grid, and in vertical direction dust particles are positioned exactly one under the another forming a cubic grid between the crystal planes. In contrast to the case of RF CCP-discharges, the vertical size of the 3D-crystal can reach several centimeters in DC-glow discharges (Nefedov, 2001). The Coulomb crystals in glow discharges are stabilized in the standing striations (see Sections 7.4.4 through 7.4.6), where the electric field is sufficiently high (about 15 V/cm) to balance gravitation and to provide suspension of dust particles. The relatively large thickness of the striations (up to several centimeters) results in the quite large vertical size of the Coulomb clusters. This gives the possibility to consider them as the 3D-structures. Coulomb structures were also observed in RF ICP-discharges and in nuclear induced dusty plasma (Nefedov, 2001), in atmospheric pressure thermal plasma at temperature of about 1700 K and even in hydrocarbon flames and other systems (Fortov et al., 1999). It is interesting that higher non-ideality of dusty plasma and stronger coupling effects can be also be reached by decreasing temperature (see Eqs. 11.206 and 11.210); this was demonstrated in experiments with cryogenic plasmas of DC-glow and RF-discharges (Vasilyak et al., 2001; Balabanov et al., 2001). Formation of ordered Coulomb structures in dusty plasma of particles charged by UV-radiation was also studied in micro-gravity experiments carried out on board of the MIR space station (Nefedov et al., 1997; Fortov et al., 1998).

11.8.6 Oscillations and Waves in Dusty Plasmas: Dispersion Equation

The presence of dust particles in plasma leads to additional characteristic space and time scales even at low levels of nonideality. This makes some changes in the dispersion equations of traditional plasma oscillations and waves and also qualitatively creates new modes specific only for dusty plasmas. An example is the modification of the ionic sound dispersion equation in dusty plasmas, and also the appearance in these heterogeneous systems of a new low-frequency oscillation branch, called the dust sound. To analyze the waves and oscillations in non-magnetized dusty plasma the following dispersion equation is used:

$$1 = \frac{\omega_{pe}^2}{(\omega - ku_e)^2 - \gamma_e k^2 v_{Te}^2} + \frac{\omega_{pi}^2}{(\omega - ku_i)^2 - \gamma_i k^2 v_{Ti}^2} + \frac{\omega_{pd}^2}{(\omega - ku_d)^2 - \gamma_d k^2 v_{Td}^2}. \quad (11.213)$$

Here, ω_{pe} is the plasma-electron frequency (Eqs. 6.17 and 6.19); ω_{pi} is the plasma–ion frequency (Eq. 6.137); ω_{pd} is the plasma–dust frequency, determined similar to Eq. 6.137 and taking into account the dust particles mass m_d and charge $Z_d e$; u_e, u_i, u_d are the directed velocities of electron, ions, and dust particles; v_{Te}, v_{Ti}, v_{Td} are the average chaotic thermal velocities of electrons, ions and dust particles; $\gamma_j(e, i, d)$ are factors related to the equation of state for electrons, ions, and dust particles:

$$p_j = \text{const}\, n_j^{\gamma_i}, \quad (11.214)$$

$\gamma_j = 1$ corresponds to isothermal oscillations of the j component, $\gamma_j = 5/3$ corresponds to the adiabatic ones. In the absence of direct motion of all three components, the dispersion equation of nonmagnetized dusty plasma Eq. 11.8. can be expressed as

$$1 = \frac{\omega_{pe}^2}{\omega^2 - \gamma_e k^2 v_{Te}^2} + \frac{\omega_{pi}^2}{\omega^2 - \gamma_i k^2 v_{Ti}^2} + \frac{\omega_{pd}^2}{\omega^2 - \gamma_d k^2 v_{Td}^2}. \quad (11.215)$$

This dispersion equation can be analyzed in different frequency ranges to describe the major oscillation branches of dusty plasma. For example, at high frequencies $\omega \gg kv_{Te} \gg kv_{Ti} \gg kv_{Td}$, the dispersion Eq. 11.215 can be simplified to the form:

$$1 = \frac{\omega_{pe}^2}{\omega^2} + \frac{\omega_{pi}^2}{\omega^2} + \frac{\omega_{pd}^2}{\omega^2}. \quad (11.216)$$

Taking into account: $\omega_{pe} \gg \omega_{pi} \gg \omega_{pd}$ (because the high mass of dust particles with respect to electrons and ions), one can conclude that the high-frequency range is actually not affected by the dust particles and gives the conventional electrostatic Langmuir oscillations $\omega \approx \omega_{pe}$ (see Section 6.5.1). Essential effects of the dust particles on the plasma oscillation modes can be observed at lower frequencies.

11.8.7 Ionic Sound Mode in Dusty Plasma, Dust Sound

At low frequencies of electrostatic plasma oscillations $kv_{Te} \gg \omega \gg kv_{Ti} \gg kv_{Td}$, dispersion (Eq. 11.216) can be rewritten as ($\gamma_e = 1$):

$$\omega^2 \approx \omega_{pi}^2 \frac{k^2 r_{De}^2}{1 + k^2 r_{De}^2}, \quad (11.217)$$

TABLE 11.1
Pulsed Discharges in Water

Pulsed Corona in Water	Pulsed Arc in Water	Pulsed Spark in Water
Streamer-like channels.	Current is transferred by electrons.	Similar to pulsed arc, except for short pulse durations and lower temperature.
Streamer-channels do not propagate across the entire electrode gap, that is, partial electrical discharge.	Quasi-thermal plasma.	Pulsed spark is faster than pulsed arc, that is, strong shock waves are produced.
Streamer length in order of centimeters, channel width ~ 0–20 μm.	Arc generates strong shock waves within cavitation zone.	Plasma temperatures in spark is around a few thousand degree Kelvin.
Electric current is transferred mostly by ions.	High current filamentous channel bridges the electrode gap.	
Nonthermal plasma.	Gas in channel (bubble) is ionized.	
Weak to moderate UV generation.	High UV emission and radical density.	
Relatively weak shock waves.	A smaller gap between two electrodes of ~5 mm is needed than that in pulsed corona.	
Water treatment area is limited to a narrow region near the corona.	Large discharges pulse energy (greater than 1 kJ per pulse).	
Pulse energy: a few joules per pulse, often less than 1 J per pulse.	Large current (about 100 A), peak current greater than 1000 A.	
Frequency is about 100–1000 Hz.	Electric field intensity at the tip of electrode is 0.1–10 kV/cm.	
Relatively low current, that is, peak current is less than 100 A.	Voltage rise time is 1–10 μs.	
Electric field intensity at the tip of electrode is 100–10,000 kV/cm.	Pulse duration ~20 ms	
A fast-rising voltage on the order of 1 ns, but less than 100 ns.	Temperature exceeds 10,000 K.	

where r_{De} is the electron Debye radius. When the corresponding wavelengths of the oscillations are shorter than the electron Debye radius ($kr_{De} \gg 1$), the dispersion Eq. 11.216 gives the conventional ion–plasma oscillations $\omega \approx \omega_{pi}$ (see Eq. 6.137). In the opposite case of longer wavelengths ($kr_{De} \ll 1$), the ionic sound waves $\omega \approx kc_{si}$ take place according to the dispersion Eq. 11.216. Here the speed of ionic sound can be expressed in the following form:

$$c_{si} = \omega_{pi} r_{De} = \sqrt{\frac{T_e}{m_i}\frac{n_i}{n_e}} = v_{Ti}\sqrt{\tau\left(1+zP\right)}. \tag{11.218}$$

Here m_i is the ionic mass; $\tau = T_e/T_i$ characterizes the ratio of electron and ion temperatures; $z = |Z_d|\, e^2/4\pi\varepsilon_0 r_d T_e$ is the dimensionless charge of a dust particle of radius r_d; $P = (4\pi\varepsilon_0 r_d T_e/e^2)(n_d/n_e)$ is the density parameter of dust particles and is approximately equal to the ratio of the charge density in the dust component to the charge density in electron component. As one can see comparing Eqs. 11.217 and 6.5.15, the main influence of dust particles on the ionic sound is related to the difference in concentration of electrons and ions in the presence of charged dust particles when the dimensionless parameter $zP > 1$. Even at lower frequencies of electrostatic plasma oscillations $kv_{Te} \gg kv_{Ti} \gg \omega \gg kv_{Td}$, the dispersion Eq. 11.214 can be rewritten ($\gamma_e = 1, \gamma_i = 1$) as

$$\omega^2 \approx \omega_{pd}^2 \frac{k^2 r_D^2}{1 + k^2 r_D^2}, \tag{11.219}$$

where the Debye radius is determined by electrons and ions: $r_D^{-2} = r_{De}^{-2} + r_{Di}^{-2}$ similar to Eq. 11.210. When the wavelengths of the oscillations are shorter than the Debye radius ($kr_D \gg 1$), the dispersion Eq. 11.216 gives oscillations with plasma–dust frequency $\omega \approx \omega_{pd}$. In the opposite case of longer wavelengths ($kr_D \ll 1$), the dispersion Eq. 11.218 brings us to a qualitatively new type of wave, the dust sound $\omega \approx kc_{sd}$, which propagates in the dusty plasmas with the velocity:

$$c_{si} = \omega_{pd} r_D = v_{Td} \sqrt{\frac{|Z_d| T_i}{T_d}} \sqrt{\frac{\tau z P}{1 + \tau(1 + zP)}}. \tag{11.220}$$

The speed of the dust sound can exceed the thermal velocity of dust particles (which is a criterion of its existence) due to the large dust charges Z_d(concurrently the dimensionless dust density is not low).

11.9 PROBLEMS AND CONCEPT QUESTIONS

11.9.1 Work Function Dependence on Radius and Charge of Aerosol Particles

Interpret Eqs. 11.1 and 11.3 describing the work function dependence on the radius and charge of aerosol particles. Consider separately the cases of conductive and nonconductive materials of the macroparticles.

11.9.2 Equation of Monochromatic Photoionization of Aerosols

Derive Eq. 11.10 describing the steady-state electron concentration during the monochromatic photoionization of monodispersed aerosols. The derivation assumes substitution of summation by integration. Estimate the accuracy.

11.9.3 Calculation of Electron Concentration at Monochromatic Photoionization of Aerosols

Using the chart Eq. 11.1, calculate the electron density provided by monochromatic photoionization (photon flux $10^{14} 1/\text{cm}^2\text{s}$, photon energy $E = 2$ eV) of aerosol particles with radius 10 nm and work function $\varphi_0 \approx A_0 = 1$ eV. Consider two particle concentrations $n_a = 10^3$ cm^{-3} and $n_a = 10^7$ cm^{-3}. Using the same chart, estimate the accuracy of the electron density calculations in both cases.

11.9.4 Equation of Aerosol Photoionization by Thermal Radiation Tail

Analyze Eq. 11.25 describing nonmonochromatic photoionization of aerosols in terms of elliptical functions and prove that it is identical to Eq. 11.24 expressed in the form of series. Taking into account analytical properties of the elliptical θ-functions, derive the asymptotic forms of the differential Eq. 11.25.

11.9.5 Calculation of Electron Concentration at Nonmonochromatic Photoionization of Aerosols

Calculate the electron concentration provided by photoionization of aerosols by thermal radiation. Assume the ratio of attachment and photoionization coefficients as constant, the same as in the case of monochromatic radiation: $\alpha/\gamma = 10^8$ cm/s. Consider monodispersed aerosol particles with radius 10 nm, work function $\varphi \approx A_0 = 1$ eV, and the aerosol particle concentration $n_a = 10^7$ cm^{-3}. About the nonmonochromatic radiation of the total power density 10 W/cm^2, assume that its spectrum corresponds to that of a black body with temperature $T = 3000$ K.

11.9.6 Characteristic Monochromatic Photoionization Time

Calculate the characteristic time for establishing the steady-state electron concentration in the process of monochromatic radiation of mono-dispersed aerosol particles. Assume the photon flux 10^{17} cm^{-2}s^{-1}, photon energy $E = 2$ eV, an aerosol particles radius = 10 nm, a work function $\varphi_0 \approx A_0 = 1$ eV, and an aerosol concentration $n_a = 10^7$ cm^{-3}, $a/\gamma = 10^8$ cm/s.

11.9.7 The Einbinder Formula

Analyze the Einbinder formula (Eq. 11.39), which describes the electron concentration at the thermal ionization of aerosols, taking into account the influence of the macroparticles charge on the work function. Discuss the applicability of the Einbinder formula in the case of low average values of the macroparticles electric charge.

11.9.8 Thermal Ionization of Aerosols, Stimulated by their Photo-Heating

Using Figure 11.12, calculate the steady-state photo-heating temperature of 10 μm aerosol particles under radiation with power density $pE = 10^7$ W/m^2. Estimate the characteristic time for establishing the steady-state temperature. Calculate the average electron concentration, which is provided by such photo-heating, if the concentration of the aerosol particles is $n_a = 10^5$ cm^{-3}.

11.9.9 Space Distribution of Electron Density and Electric Field Around an Aerosol Particle

Consider thermal ionization of a macroparticle of radius 10 μm, work function 3 eV, and temperature 1500 K. Find the electron concentration near the surface of the aerosol particle and the corresponding value of the Debye radius r_D^0. Calculate the total electric charge of the macroparticle, electric field on its surface and on a distance 3 μm from the surface.

11.9.10 Electrical Conductivity of Thermally Ionized Aerosols

Consider thermal ionization of an aerosol system with $T = 1700$ K, $r_a = 10$ μm, $n_a = 10^4$ cm^{-3}, $A_0 = 3$ eV. Assuming the electron mobility as in atmospheric air, calculate the electrical conductivity of aerosols as a function of the external electric field. Consider the external electric field E_e in all three intervals of interest, corresponding the Einbinder relation, explicit dependence on E_e, and range of maximum value of the electric conductivity.

11.9.11 Macroparticles Kinetic Parameters

Relations for thermal velocity, drift velocity, and diffusion coefficient for positive ions are similar to those for electrons (Eqs. 11.80 through 11.82). Derive these relations for ions in aerosol systems, replacing collisional cross sections, mean free paths, and collisional energy losses of electrons by those for ions. Analyze the contribution of macroparticles in these relations with respect to the contribution of the neutral gas.

11.9.12 Vacuum Breakdown of Aerosols

Based on Eqs. 11.95 and 11.97, estimate the breakdown voltages of the aerosols. Take the characteristic parameters from the example in Section 11.3.7. Compare the result with typical breakdown voltages of discharge gaps without aerosols.

11.9.13 Photoinitiation of Pulse Breakdown in Aerosol Systems

Assuming that the initial electron concentration sufficient for initiating a pulse breakdown is about $n_e^0 = 10^4\ \text{cm}^{-3}$, estimate the radiation power density, concentration, and radius of the aerosol particles necessary for the breakdown initiation. For the numerical estimations use the aerosol photoionization data presented in Figure 11.1.

11.9.14 Quasi-Neutral Regime of Steady-State DC Discharge in Aerosols

Determine aerosol particles sizes and concentration when their contribution to the total surface area density n_S exceeds that of the long cylindrical discharge wall surface. Estimate the maximum critical concentration of macroparticles with a 10 μm radius when the quasi-neutral regime of steady-state DC discharge is still possible.

11.9.15 Electron–Aerosol Plasma Regime of Heterogeneous Discharges

The maximum value of the average charge of macroparticles is determined by the parameter Z_0 (Eqs. 11.109 and 11.110). Explain the dependence of this average macroparticle charge on their size. How does the charge density on the surface of aerosol particles depend on the radius of macroparticles?

11.9.16 Electric Field of Aerosol Particles in Electron–Aerosol Plasma

Using Figure 11.3, calculate the inherent electric field on the surface of aerosol particles as a function of current density in the discharge and radius of macroparticles. Compare values of the inherent electric field on the aerosol surface at different currents and radii of macroparticles with the external electric field in heterogeneous discharge.

11.9.17 Effect of Vibrational Excitation on Kinetics of Negative Ion-Cluster Growth

The contribution of vibrational excitation of silane molecules on the kinetics of negative ion-cluster growth (Eqs. 2.6.28 through 2.6.32) is determined by the logarithmic factor (Eq. 11.128), describing the competition between vibrational excitation and relaxation. Discuss the sign of this factor Λ, and determine its relation to the effectiveness of vibrational excitation in the cluster growth process. Estimate the logarithmic factor.

11.9.18 Kinetics of a Supersmall Particle Growth in Dusty Plasma

Interpret the fact that the cubic root of number of atoms $N^{1/3}$ (not N) increases linearly with time (see Eq. 11.132). Thus, not mass but the particle radius grows linearly with time. As a result, the time for the formation of the critical size 1000-atom cluster is only 10 times longer than the characteristic time of the first ion–molecular reaction.

11.9.19 Neutral Particle Trapping in RF-Plasma

Using Eq. 11.134, show that the electron attachment time ($1/k_a n_e$) for 2 nm particles becomes shorter than the residence time. This means that each neutral particle has a possibility to be charged at least once during the residence time. Having a negative charge before recombination even for a short time interval (about 1 ms) is sufficient for the particle trapping by the RF-discharge sheath electric field, which explains "the neutral particle trapping" effect.

11.9.20 Coagulation of Neutral Nanoparticles in Plasma

Analyze the summation (Eq. 11.145) of partial recombination rates that describes the coagulation of nanoparticles. Demonstrate the divergence of this sum, and explain the physical limitation of the divergence. Replacing the summation by integration, estimate the coagulation rate of neutral nanoparticles (Eq. 11.145).

11.9.21 The α–γ Transition During Dusty Plasma Formation

The α–γ transition, including the sharp growth of electron temperature and decrease of electron density, is related to the contribution of electron attachment to the particles surfaces and occurs during coagulation. The total surface of particles decreases during the coagulation because while their diameters grow the total mass remains almost fixed. Give your interpretation of the phenomenon. Does it depend on the particle material?

11.9.22 Clusterization Process as Diffusion in Space of Cluster Sizes

At temperatures exceeding that of condensation, evolution of clusters is mostly due to atom attachment and detachment processes: $A_{n-1} + A \Leftrightarrow A_n$. Show that the diffusive flux of clusters along the axis of their sizes n can be expressed in the linear Fokker–Plank form (Eq. 11.159). Analyze the applicability of the flux to the condensation.

11.9.23 Magic Clusters

The effect of magic clusters—relatively high concentration of clusters with some specific sizes—is negligible at very high temperatures. Show that this effect becomes significant only at temperatures slightly exceeding the condensation point determined by Eq. 11.169.

11.9.24 Kinetic Equation for Nonequilibrium Clusterization Process in Centrifugal Field

Based on Eq. 11.174, derive the Fokker–Plank type Eq. 11.175 for the distribution function $y(n) = f(n)/\sqrt{f_0(n)}$, taking into account the diffusion along the n-axis of cluster sizes and centrifugal drift along the radius x.

11.9.25 Cluster Flux to Discharge Periphery

Simplifying integration in the relation (Eq. 11.192), derive Eq. 11.194 for the cluster flux. Interpret the formula, taking into account that the cluster flux to the discharge periphery is limited by diffusion of molecules into a zone where the sufficiently large clusters can be formed. Using Eq. 11.194, estimate typical numerical values of the cluster flux.

11.9.26 Effective Clusterization Temperature as a Function of Centrifugal Factor mv_φ^2/T

At a pressure 0.1 atm and high centrifugal factor $mv_\varphi^2/T \approx 1$, the effective clusterization temperature in sulfur (formation of S_6 and S_8) is about 850 K. Taking into account that the condensation temperature of sulfur is $T_c \approx 550\ K$ at 0.1 atm, calculate the effective clusterization temperature for sulfur as a function of the centrifugal factor at tangential velocities below the speed of sound.

11.9.27 Nonideality Criterion of Dusty Plasmas

The nonideality of a dusty plasma (strong coupling of dust particles) takes place when the parameter Γd_s (Eq. 11.210), including plasma shielding of interaction between particles, is sufficiently large. Explain why the Debye radius, describing the shielding, combines the effect of electrons, ions, and dust particles (Eq. 11.209) in the case of ideal dusty plasma, and does not include a contribution of the particles in the nonideal case (Eq. 11.210).

11.9.28 "Melting" of Coulomb Crystals

"Melting" of the 2D-Coulomb crystals in RF CCP-discharges can be initiated by reducing the neutral gas pressure or by increasing the discharge power. Explain how the change of pressure and power leads to a decrease of the nonideality parameter and to the crystal–liquid phase transition.

11.9.29 Plasma-Dust Frequency

Using the one-component plasma model (OCP, see Section 11.8.3), derive a formula for the plasma-dust frequency ω_{pd}. Interpret this frequency and analyze its relation to the Debye radius. From this point view, compare the ideal and nonideal dusty plasmas.

11.9.30 Ionic Sound in Dusty Plasma

Based on dispersion (Eq. 11.216), derive Eq. 11.217 for the speed of the ionic sound in dusty plasma. Make the derivation both in terms of the difference of electron and ion concentrations in dusty plasma and in terms of the product zP of dimensionless charge of dust particles and their dimensionless concentration. Give numerical examples of calculations of c_{si} using Eq. 11.217. Compare the speed of ionic sound under similar plasma conditions with and without dust particles. Consider cases of predominantly positive and negative charging of the dust particles.

11.9.31 Dust Sound Wave in Plasma

The dust sound is the propagation of low frequency waves in a dusty plasma at wavelengths exceeding the Debye radius r_D with the speed determined by Eq. 11.219. Show that the dust sound can exist only if its velocity exceeds the thermal one for dust particles. Derive the existence criterion of the dust sound and determine the minimum concentration of dust particles necessary for this wave.

12 Electron Beam Plasmas

12.1 GENERATION AND PROPERTIES OF ELECTRON-BEAM PLASMAS

12.1.1 Electron Beam Plasma Generation

Electron beam plasma is formed by injection into a neutral gas a cylindrical or plane electron beam with electron energies usually ranging from 10 keV to 1 MeV. A general scheme of the electron beam plasma generation is illustrated in Figure 12.1 (V.L. Bykov, M.N. Vasiliev, A.S. Koroteev, 1993). A thin cylindrical electron beam is formed by an electron gun located in a high-vacuum chamber. The beam is injected through a window into a gas-filled discharge chamber, where the high-energy electrons generate plasma and transfer energy into the gas by various mechanisms depending on pressure, beam current density, electron energy, and plasma parameters. Total energy losses of the electron beam in a dense gas due to interaction with neutrals along its trajectory (axis z) are:

$$-\frac{1}{n_0}\frac{dE}{dz} = L(E), \tag{12.1}$$

where n_0 is the neutral gas density; E is the electron energy; $L(E)$ is the electron energy loss function, which is well known for different gases and mixtures. As an example, the $L(E)$ function of nitrogen is shown in Figure 12.2 (M.N. Vasiliev, 2000). For calculations of the electron energy loss function $L(E)$, it is convenient to use the numerical **Bethe formula**, which can be applied even for relativistic electron beams (although not for the strong relativistic case):

$$L(E) = \frac{A}{E}\frac{\left(1+E/mc^2\right)}{\left(1+E/2mc^2\right)}\ln\frac{1.17\cdot E}{I_{ex}}. \tag{12.2}$$

In this relation, m is an electron mass; c is the speed of light; I_{ex} is the characteristic excitation energy (which is equal to 87 eV for air, 18 eV for hydrogen, 41 eV for helium, 87 eV for nitrogen, 102 eV for oxygen, 190 eV for argon, 85 eV for carbon dioxide, and 72 eV for water vapor); the factor A also depends only on gas composition, for example, in air $A = 1.3 \cdot 10^{-12}$ eV^2 cm^2. If the kinetic energy of electrons exceeds mc^2 (which numerically is about 0.5 MeV), such electrons are called

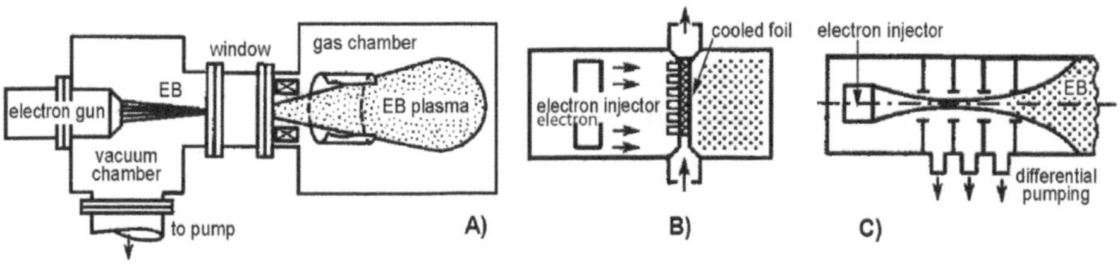

FIGURE 12.1 Electron beam generator: a) general schematic, b) foil window, c) differential pumping window.

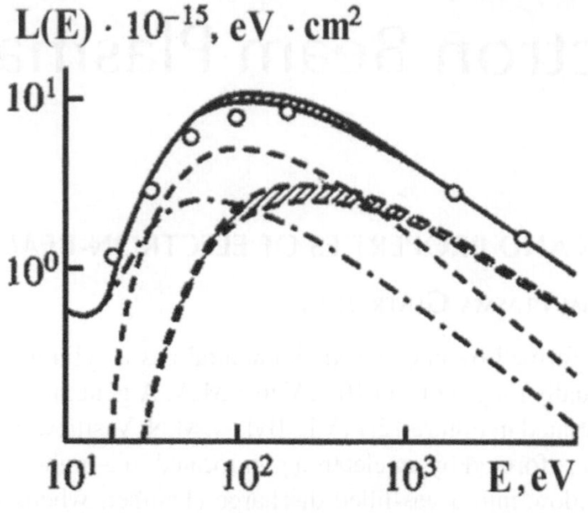

FIGURE 12.2 Electron energy losses function in nitrogen, different calculations. *(From Encyclopedia of Low Temperature Plasma, Ed. V.E Fortov, Nauka, Moscow, 2002)*

relativistic, and their beam is usually referred to as the **relativistic electron beam**. Dynamic analysis of such beams definitely requires taking into account the special effects of relativistic mechanics.

12.1.2 Ionization Rate and Ionization Energy Cost at Gas Irradiation by High Energy and Relativistic Electrons

The rate of ionization provided by an electron beam with current density j_b in a unit volume per unit time was discussed in Section 2.2.5 in terms of an effective ionization rate coefficient. Now the same ionization rate can be determined using the fast electron energy losses dE/dz as:

$$q_e = -\frac{dE}{dz}\frac{j_b}{eU_i}. \tag{12.3}$$

Here, U_i is the energy cost of an electron–ion pair formation (the ionization energy cost). If the beam electron energy is in the range of 0.01–1 MeV, the ionization energy cost has an almost constant value, which is given in Table 12.1. The electron-beam plasma has a complicated composition including charged, excited atomized, and other species. To analyze the concentration of these species, it is convenient to consider the electron beam as their source and specify the energy cost of their formation.

TABLE 12.1
Ionization Energy Cost at Neutral Gas Irradiation by an Electron Beam with Energy 0.1–1 *MeV*

Gas	Ionization Cost	Gas	Ionization Cost
Air	32.3 eV	Carbon Dioxide, CO_2	31.0 eV
Nitrogen, N_2	35.3 eV	Oxygen, O_2	32.0 eV
Hydrogen, H_2	36.0 eV	Helium, *He*	27.8 eV
Argon, *Ar*	25.4 eV	Krypton, *Kr*	22.8 eV
Xenon, *Xe*	20.8 eV	–	–

12.1.3 Classification of Electron-Beam Plasmas According to Beam Current and Gas Pressure

Independent changes in the current density of electron beam and neutral gas pressure results in qualitatively different regimes of electron beam plasmas, which can be classified as follows:

1. ***Powerful Electron Beam at Low-Pressure Gas.*** Here electron beam energy losses are mostly due to the beam instability or the Langmuir paradox (see Section 6.5.7), when the electron beam energy can be transferred by non-linear beam-plasma interaction into the energy of Langmuir oscillations. This regime usually takes place when the electron beam current and beam current density are sufficiently high ($I_b > 1\text{A},\ j_b > 100\ \text{A/cm}^2$), while pressure is relatively low, less than a few Torr. This system is called the plasma-beam discharge. Plasma in this discharge can be characterized by high degrees of ionization ($\alpha = n_e/n_0 \geq 10^{-3}$), and electron temperatures of several electron-volts.
2. ***Low-Current Electron Beam in Rarefied Gas.*** This regime is related to current densities $j_b < 0.1\ \text{A/cm}^2$ and pressures below 1 Torr. In this case, the beam instability is ineffective and the plasma-beam discharge cannot be sustained. The degree of ionization is low ($\alpha = n_e/n_0 < 10^{-7}$), and the temperature of plasma electrons is also low and close to the neutral gas temperature.
3. ***Moderate-Current Electron Beam in a Moderate Pressure Gas.*** The current density of an electron beam in this regime is in the range 0.1 A/cm^2 and 100 A/cm^2, with pressures from 1 Torr to 100 Torr. The degree of ionization depends on pressure and beam density and can vary in the wide range between 10^{-7} and 10^{-3}. In this case, the electron temperature is determined by the beam degradation in non-elastic and then elastic collisions. Specific energy input in these systems is usually relatively low, and the translational temperature does not grow significantly during the electron beam interaction with a portion of neutral gas. Thus the discharge can be considered as a non-equilibrium discharge.
4. ***High Power Electron Beam in a High-Pressure Gas.*** This regime is related to high current densities $j_b > 100\ \text{A/cm}^2$ and pressures above 100 Torr. The degree of ionization and electron temperatures are similar to those for in the case of moderate pressures and currents. However, high values of beam power and specific energy input result in more intensive relaxation and neutral gas heating. In this regime, the gas temperature distribution is determined not only by pressure and beam density but also by the heat exchange between the plasma zone and the environment. Hence in the stationary situation, this electron-beam plasma is thermal, local temperatures can exceed 10,000 K.

12.1.4 Electron-Beam Plasma Generation Technique

The electron beam injectors for electron-beam plasma generation can be arranged using e direct electrostatic accelerators, although other types of accelerators can be applied as well (S. Humphries, 1990; A. Fridman, S. Nester, S. Guceri, 1996b). To generate a beam of electrons with energies not exceeding 200–300 keV, the injectors can be based on electron guns with thermionic, plasma, or field-emission cathodes. Power can be supplied, by special transformers with AC-DC-conversion. Electron beams are formed in a vacuum, and then should be transmitted across the inlet system (window) into the dense gas chamber to generate plasma. These inlet systems, windows, are usually arranged in one of two following ways: the foil-windows, and the lock systems with differential pumping. The general schematic of both inlet systems is illustrated in Figure 12.2. The application of foil-windows for non-relativistic continuous, concentrated electron beams has complications related to possible high thermal flux, thinness of the foil for electron energies below 300 keV, chemical aggressiveness of gases in plasma-chemical systems.

12.1.5 Transportation of Electron Beams

An advantage of plasma-chemical applications of the high energy and relativistic electron beams is related to the long-distance of their decay even in high-pressure systems. Transportation length of the high energy and relativistic electron beams in dense gases is determined by energy losses due to ionization and excitation of atoms and molecules, which was described by the Eqs. (12.1), (12.2). For practical calculations, it is convenient to use the following numeric relation, describing the total length L (in m) for stopping a high-energy electron along its trajectory:

$$L = AE_{b0}^{1.7}\frac{T_0}{p}. \tag{12.4}$$

In this relation, T_0 is the gas temperature in K, p is the gas pressure in Torr, E_{b0} is initial electron energy, A is a constant, which depends only on gas composition, for air: $A = 1.1 \cdot 10^{-4}$. The electron beam transportation can also be limited by collisionless effects, related to beam instability, electrostatic repulsion of the beam electrons, and magnetic self-contraction. The first of these effects implies that the electron beam energy transfers to electrostatic Langmuir oscillations. This high-current – low-pressure effect was mentioned above as the first regime of the electron-beam plasma generation. The second and third effects are related to the electric and magnetic fields of the beam. The force pushing a periphery beam electron in a radial direction away from the axis (combining electric and magnetic components) is:

$$F = \frac{n_{eb} r_b e^2}{2\varepsilon_0}\left(1 - \beta^2 - f_e\right). \tag{12.5}$$

In this relation, r_b is the radius of the cylindrical beam, n_{eb} is the electron density of the beam (or beam plasma), $\beta = u/c$ is the ratio of beam velocity to the speed of sound, $f_e = n_i/n_{eb}$ is the beam space charge neutralization degree (n_i is the ion density). If $f_e > 1-\beta^2$, which is usually valid in dense gases, the generated plasma neutralizes the electron beam space charge and the magnetic field self-focuses the beam. If the high current electron beam is not neutralized, it can be destroyed by its own electric field. To provide effective beam transportation in such conditions, special focusing systems, in particular electrostatic lenses, can be applied (A. Fridman, L.I. Rudakov, 1973). When an electron beam is neutralized ($f_e = 1$), its current cannot exceed a critical value known as the **Alfven current**:

$$I_A = \frac{mc^3}{4\pi\varepsilon_0 e\beta\gamma}, \quad I_A(kA) = 17\beta\gamma, \tag{12.6}$$

where $\gamma = 1/\sqrt{1-\beta^2}$ is the relativistic factor. The Alfven current gives the inherent magnetic field of the beam, which makes the Larmor radius Eq. (6.2.25) less than half the beam radius and stops the electron beam propagation. More details regarding the transportation of relativistic electron beams with very high current can be found in G.A. Mesyats, G.P. Mkheidze, A.A. Savin (2000).

12.2 KINETICS OF DEGRADATION PROCESSES, DEGRADATION SPECTRUM

12.2.1 Kinetics of Electrons in Degradation Processes

In the case of electron beam energy degradation, it is convenient to operate not in terms of probability for an electron to have some given energy E, which is $f(E)dE$, but rather in terms of the degradation spectrum, which is the number of electrons $Z(E)dE$, having some given energy E during the degradation of one initial high-energy electron. This approach is called the **degradation spectrum** method and was successfully applied to describing the kinetics of stopping processes of not only electrons but also energetic ions, photons, and neutrals (G.D. Alkhazov, 1971; V.A. Nikerov, G.V. Sholin, 1978a, 1985; L.I. Gudzenko, S.I. Yakovlenko, 1978; G.V. Sholin, V.A. Nikerov, V.D. Rusanov, 1980; Yu. P. Vysotsky, V.N. Soshnikov, 1980, 1981).

12.2.2 Energy Transfer Differential CrossSections and Probabilities during Beam Degradation Process

The degradation process of an energetic beam is determined by a group of elastic and inelastic collisional processes of an electron (in principle, it can be generalized to energetic ions, photons and neutrals) with background particles, which can be considered at rest. Each of the degradation processes "k" is characterized by the total crosssection $\sigma_k(E)$, and the differential crosssection $\sigma_k(E, \Delta E)$, where E is the electron energy during degradation, and ΔE is the electron energy loss during the collision. Normalize the differential crosssection $\sigma_k(E, \Delta E)$ to the number of electrons formed after collision (V.A. Nikerov, G.V. Sholin, 1985). The degradation, which does not change the number of electrons (for example, excitation processes) have the conventional normalization equation:

$$\int_{-\infty}^{+\infty} \frac{\sigma_k(E,\Delta E)}{\sigma_k(E)} d(\Delta E) = 1. \tag{12.7}$$

Ionization which doubles the number of electrons have the corrected normalization equation:

$$\int_{-\infty}^{+\infty} \frac{\sigma_k(E,\Delta E)}{\sigma_k(E)} d(\Delta E) = 2, \tag{12.8}$$

and the electron–ion recombination and electron attachment processes, where the degrading electron disappears, have the normalization:

$$\int_{-\infty}^{+\infty} \frac{\sigma_k(E,\Delta E)}{\sigma_k(E)} d(\Delta E) = 0. \tag{12.9}$$

Then the differential probability for an electron with energy E to lose the energy portion ΔE in the collisional process "k" can be determined as:

$$p_k(E,\Delta E) = \frac{\sigma_k(E,\Delta E)}{\Sigma \sigma_m(E)}. \tag{12.10}$$

The corresponding value of the total probability for an electron with energy E to participate in the collisional process "k" can be found as:

$$p_k(E) = \frac{\sigma_k(E)}{\Sigma \sigma_m(E)}. \tag{12.11}$$

The probabilities (12.10, 12.11) can be determined only if $\sum \sigma_m(E) > 0$, for example above the energy threshold of the degradation processes; otherwise, the above probabilities should be considered as zero.

12.2.3 The Degradation Spectrum Kinetic Equation

The degradation spectrum *Z(E)* can be determined as the average number of particles (electrons in our case), which appears during the whole degradation process in the energy interval $E - E + dE$. Then the kinetic equation for the degradation spectrum can be presented in general as:

$$Z(E) = \int_0^\infty p(W, W - E) \cdot Z(W) dW + \chi(E), \tag{12.12}$$

with boundary condition $Z(E \to \infty) = 0$. In this equation, $\chi(E)$ is the source-function of the high-energy electrons. If the degradation process is started by the only electron with energy E_0, the source function is determined by the delta-function, $\chi(E) = \delta(E - E_0)$. Finally, $p(W, W - E)$ in Eq. (12.12) is the total probability of formation of an electron with energy E by collisions of an electron with initial energy W and the energy loss $\Delta E = W - E$:

$$p(W, W - E) = \Sigma p_m(W, \Delta E = W - E). \tag{12.13}$$

At low energies (electron energies below the lowest excitation threshold), when the collisional crosssections can be neglected, the degradation spectrum coincides with the change of the electron energy distribution at final and initial moments of the degradation process (V.A. Nikerov, G.V. Sholin, 1985):

$$Z(E) = f(E, t \to \infty) - f(E, t \to 0). \tag{12.14}$$

Thus assuming an absence of the initial EEDF, $f(E, t \to 0) \to 0$, one can conclude the equality of EEDF and degradation spectrum of electron beam under the excitation threshold.

12.2.4 Integral Characteristics of Degradation Spectrum, Energy Cost of a Particle

The degradation spectrum *Z(E)* is helpful in calculating many important integral characteristics of high-energy electron interaction with materials. For example, integration of the degradation spectrum gives, based on Eq. (12.11), the total number of specific collisions "*k*" (ionization, excitation, dissociation, etc.) per one initial high-energy degradating electron:

$$N_k = \int_0^\infty p_k(E) \cdot Z(E) dE = \int_0^\infty Z(E) \frac{\sigma_k(E) \cdot dE}{\Sigma \sigma_m(E)}. \tag{12.15}$$

Taking into account that the initial energy of a beam electron is equal to E_0, the "energy cost of the specific collision k" (ionization, excitation, dissociation, etc.) is:

$$U_k = \frac{E_0}{N_k} = E_0 / \int_0^\infty p_k(E) Z(E) dE. \tag{12.16}$$

12.2.5 The Alkhazov's Equation for Degradation Spectrum of High Energy Beam Electrons

Different variations of the degradation kinetic equations were developed for different special degradation problems. The Fano-Inokuti approach is close to that one discussed above, but includes integration over energy losses instead of electron energies, and is referred to as the **differential electron free-path approach** (U. Fano, 1953; L.V. Spencer, U. Fano, 1954; M. Inokuti, 1971, 1974, 1975; A.R.P.Rau, M. Inokuti, D.A. Douthart, 1978; D.A. Douthart, 1979). Another approach, called **the**

Green model, was successfully applied to describe degradation of high-energy electrons in the upper atmosphere. This approach divides electrons into three degradation cascades: a primary one, secondary one, and tertiary one; it then considers degradation in each cascade separately (A.E.S. Green, C.A. Barth, 1965, 1967; R.S. Stolarsky, A.E.S. Green, 1967). The simplest and the most effective approach to degradation kinetics of high-energy electron beams was proposed by G.D. Alkhazov (1971), and developed by V.A. Nikerov, G.V. Sholin (1985). In this case, the general degradation spectrum Eq. (12.12) is specified in the following form, known as the **Alkhazov's equation**:

$$Z(E)=\sum_k p_{ex,k}(E+U_k)\cdot Z(E+U_k)+\int_{E+I}^{\infty} p_i(W,W-E)\cdot Z(W)\,dW+\delta(E-E_0). \qquad (12.17)$$

The boundary condition for the equation is again $Z(E \to \infty) = 0$. In the Alkhazov's equation, $p_{ex,k}(E + U_k)$ is the probability of excitation of an atom into the k-th excited state in a collision with an electron having energy $E + U_k$; $p_i(W, W - E)$ is the differential probability of ionization of an atom in collision with an electron having initially kinetic energy W and during the collision losing energy, $\Delta E = W - E$; I and U_k are respectively the ionization potential and the excitation energy of the k-th state; $\delta(E - E_0)$ is the delta-function describing the source of degradation, the initial electron with high energy E_0. A simple solution of the Alkhazov's equation can be obtained in the Bethe-Born approximation taking into account only ionization by primary electrons (G.D. Alkhazov, 1971):

$$Z(E)=\frac{\sigma(E)}{L(E)}=\frac{\aleph^2\ln(cE)}{4R\ln\left(\sqrt{\frac{e}{2}\frac{E}{I^*}}\right)}. \qquad (12.18)$$

In this relation, $L(E)\approx\left(16R^2/E\right)\ln\left(\sqrt{e/2}E/I^*\right)$ is the function characterizing electron energy losses; $\sigma(E) \approx (4R/E)\aleph^2 \ln(cE)$ is the total crosssection of inelastic collisions for an electron with energy E; $R = 13.595$eV is the Ridberg constant; $\aleph^2 = 0.7525$ is square of the matrix element for the sum of transitions into continuous and discrete spectra; factor $c = 0.18$ eV; $I^* = 42$eV is the characteristic excitation potential (numerical data are given for helium).

12.2.6 Solutions of the Alkhazov's Equation

The analytical solution of Alkhazov's equation gives only a qualitative description of degradation spectrum. Numerical solutions of the Alkhazov's equation for electron beam degradation in helium, hydrogen, and fluorine are presented in Figure 12.3. (G.D. Alkhazov, 1971; V.A. Nikerov, G.V. Sholin, 1978a). The Alkhazov's equation can be also solved analytically with sufficient accuracy by

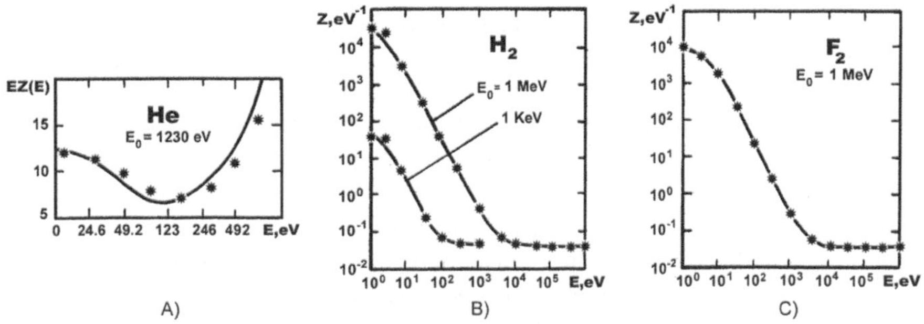

FIGURE 12.3 Degradation spectra in: a) helium, b) hydrogen, c) fluorine. Stars – numerical calculations, curves – analytical calculations.

TABLE 12.2
Integral Production of Ions and Excited Molecules by Degradation of an Electron with Initial Energy 1 MeV and 1 keV in Molecular Hydrogen

State of Molecule	E_0 = 1 MeV	E_0 = 1 keV
Ionized	$3.06 \cdot 10^4$	29.6
$H_2\left(\Sigma_u^+\right)$	$1.10 \cdot 10^4$	10.7
$H_2\left(B^1\Sigma_u^+\right)$	$1.43 \cdot 10^4$	13.9
$H_2 *$ (*vibr. exc.*)	$1.84 \cdot 10^5$	187

using special analytical approximations for crosssections of the collisional processes. These results are also presented in Figure 12.3. in comparison with numerical solutions. Accurate estimations of the degradation spectrum above the ionization threshold can be achieved by using the following formula (V.A. Nikerov, G.V. Sholin, 1985):

$$Z(E) = \frac{0.6}{I} + \frac{0.5 \cdot E_0}{(E+I)^2}\left[1 + \frac{3I^2}{(E+I)^2}\ln\left(\frac{E}{I}+2\right)\right]. \tag{12.19}$$

For estimations of the degradation spectrum below the ionization threshold:

$$Z(E) = \frac{2}{(1+E/I)^4 I}\ln\left(\frac{E}{I}+2\right) + \frac{1}{I(1+E/I)^2}. \tag{12.20}$$

Once determined, the degradation spectra for the electron beam stopping can be then applied for calculating the ionization and excitation energy costs and total number of ions and excited species produced by one high-energy electron Eqs.(12.15), (12.16). Results of such calculations in hydrogen are presented in Table 12.2.

12.3 PLASMA-BEAM DISCHARGE

12.3.1 General Features of Plasma-Beam Discharges

In this case, the electron beam does not directly interact with the neutral gas. At first it interacts with Langmuir plasma oscillations, which subsequently transfer energy through electrons to neutrals. The electron beam can be considered as a source of microwave oscillations, which then sustain the "microwave discharge". In contrast to conventional microwave discharges, the source of electrostatic oscillations is present inside of plasma, which excludes the skin-effect problems. Ts he plasma-beam discharge requires high beam currents and current densities ($I_b > 1$A, $j_b > 100$ A/cm^2), and pressures below few Torr. The discharge is stable and efficient over a wide range of pressures (10^{-4}–3 Torr), and powers (0.5–10 kW). The degree of ionization varies widely 10^{-4}–1, as well as electron temperature 1–100 eV (A.A. Ivanov, 1975, 1982; A.A. Ivanov, T.K. Soboleva, 1978; A.A. Ivanov, V.A. Nikiforov, 1978; S.I. Krasheninnikov, V.A. Nikiforov, 1982; A.K. Berezin, E.V. Lifshitz, Ya. B. Fainberg, 1995). Physical basis of the beam instability is related to the collisionless interaction of electrostatic plasma waves with electrons discussed in the Section 6.5.5. Injection of an electron beam in a plasma creates the distribution function illustrated in Figure 6.24, where the derivative Eq. (6.145) is positive. As was shown in Section 6.5.5, this corresponds to energy transfer from the electron beam and exponential amplification of electrostatic plasma oscillations. The electron beam transfers energy to Langmuir oscillations until the electron energy distribution function (EEDF) becomes continuously decreasing. The increment of the beam instability, describing the exponential growth of the Langmuir oscillation, can be calculated from Eq. (6.150).

12.3.2 Operation Conditions of Plasma-Beam Discharges

Plasma-beam discharges are based on complicated self-consistent series of physical processes: beam instability and energy transfer from the electron beam to Langmuir oscillations, then energy transfer from electrostatic oscillations to plasma electrons, and finally energy transfer from plasma electrons to neutral components, which determines ionization and therefore, the effectiveness of beam instability. As a result, effective operation of plasma-beam discharges requires a very delicate and precise choice of system parameters. The principal operation condition of plasma-beam discharges requires beam energy dissipation into Langmuir oscillations to exceed beam energy losses in electron-neutral collisions. In molecular gases, this requirement can be expressed (S.I. Krasheninnikov, V.A. Nikiforov, 1982) as:

$$\frac{6\pi^2\hbar\omega\nu_{eV}}{\bar{\varepsilon}_b\nu_{bn}}\left(\frac{\omega_{pb}}{\nu_{eV}}\right) > 1. \tag{12.21}$$

In this relation, ν_{eV} is the vibrational excitation frequency; $\hbar\omega$ is a vibrational quantum; ν_{bn} is the frequency of beam electrons collisions with neutral particles; $\bar{\varepsilon}_b$ is the average energy loss of a beam electron during its collision with a neutral particle; n_b is density of the beam electrons; ω_{pb} is plasma frequency corresponding to the beam electron density:

$$\omega_{ph} = \sqrt{\frac{n_b e^2}{\varepsilon_0 m}}. \tag{12.22}$$

The criterion (12.21) of collisionless losses domination restricts the operational pressure of the plasma-beam discharge below the level of a few Torr. The second condition of effective plasma-beam discharge operation is related to energy transfer from Langmuir oscillations to plasma electrons. The dissipation of Langmuir oscillations can be provided (especially at low pressures) by the modulation instability. This instability leads to a collapse of the Langmuir oscillations, and energy transfer along the spectrum into the range of short wavelengths. Energy of the short-wavelength oscillations can be then transferred to plasma electrons due to the Landau damping. This dissipation mechanism of Langmuir oscillations, based on the modulation instability, however, leads to significant energy transfer to "too energetic" plasma electrons from the far tail of the EEDF. Such electrons, with energies exceeding ionization potential, are not very effective in plasma-chemical processes, especially in those related to vibrational excitation. This makes the modulation instability not a desirable one. A more effective dissipation mechanism of Langmuir oscillations for plasma-chemical processes is related to Joule heating, which results in uniform energy transfer to plasma electrons along the energy spectrum. Electron energy transfer in Coulomb collisions from high to lower energies takes place only at a degree of ionization exceeding $\alpha \approx 0.1$. Then the domination of the Joule dissipation over the dissipation related to the modulation instability requires the electron-neutral collision frequency to be higher than the increment of modulation instability. This leads to the following criterion of effective energy transfer from Langmuir oscillations to low energy (1–3 eV) plasma electrons (S.I. Krasheninnikov, V.A. Nikiforov, 1982):

$$\left(\frac{\omega_{pb}}{\nu_{en}}\right)^2 \sqrt{\frac{4\pi^2}{3}\frac{\hbar\omega}{T_e}\frac{m}{M}\frac{\nu_{eV}}{\nu_{en}}} < 1. \tag{12.23}$$

Here m and M are masses of electrons and heavy particles respectively. It is important that in contrast to the requirement (12.21), the criterion (12.23) determines the low limit of gas pressure for effective plasma-chemical operation of plasma-beam discharges.

12.3.3 Plasma-Beam Discharge Conditions Effective for Plasma-Chemical Processes

At low pressures (high values of the parameter ω/ν_{en}), when the criterion (12.23) is not valid, the electron beam effectively transfers energy into Langmuir oscillations. However, in this case, the Langmuir oscillations, dissipate most of their energy only to high-energy electrons, which results in high-energy-cost radiation-chemical effects. A similar situation takes place at relatively high pressures (low values of the parameter ω/ν_{en}) when the criterion (12.21) is not valid. In this case, Beam instability is suppressed by direct electron-neutral collisions, and the chemical effect is provided only by relatively low-energy-efficiency radiation-chemical processes (V.M. Atamanov, A.A. Ivanov, V.A. Nikiforov, 1979). The optimal conditions for plasma-chemical applications of plasma-beam discharges in molecular gases require both two criteria (12.21) and (12.23) to be valid. This brings us to the relatively moderate pressures of about 1 Torr. Even in the narrow pressure range, stimulation of chemical reactions through vibrational excitation of molecules requires sufficient specific power of the discharge:

$$\sigma E^2 > \nu_{eV} n_e \hbar\omega. \tag{12.24}$$

To satisfy this requirement, a high level of Langmuir noise W should be achieved:

$$\frac{W}{n_e T_e} = \frac{\varepsilon_0 E^2}{2 n_e T_e} \geq \frac{\nu_{eV}}{\nu_{en}} \frac{\hbar\omega}{T_e}. \tag{12.25}$$

The level of Langmuir noise W/n_eT_e should exceed 10^{-3} to provide effective vibrational excitation of molecules. The dependence of degree ionization, level of Langmuir noise W/n_eT_e, and average electron energy in the plasma-beam discharge are shown in Figure 12.4 for hydrogen as a function of vibrational temperature for two different values of the parameter W/p, where p is the gas pressure (S.I. Krasheninnikov, 1980; A.A. Ivanov, 2000). From this figure, it is seen that higher vibrational temperatures obviously correspond to higher electron concentrations (degree of ionization), which from another hand correspond to lower values of W/n_eT_e and average electron energy.

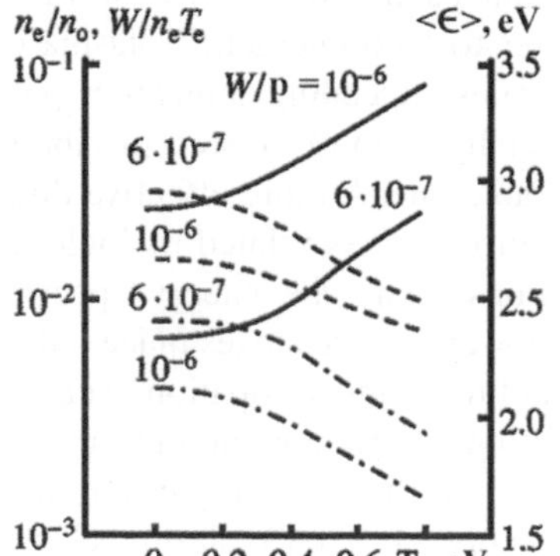

FIGURE 12.4 Dependence of ionization degree n_e/n_0 (——), relative Langmuir noise W/n_eT_e (— —), and average electron energy <ε> (- - - -) on vibrational temperature T_V. (*From Encyclopedia of Low Temperature Plasma, Ed. V.E Fortov, Nauka, Moscow, 2002*)

12.3.4 Plasma-Beam Discharge Technique

A general schematic of a plasma-chemical reactor based on the plasma-beam discharge is shown in Figure 12.5 (A.A. Ivanov, V.A. Nikiforov, 1978). The discharge chamber is quite large; its internal diameter is 50 cm, length is 150 cm. At 10 mTorr the flow rate is about 30 cm^3/s; pressure in the electron gun chamber is about 10^{-5} Torr and sustained by differential pumping. An electron beam with a maximum power 40 kW is transported into the discharge chamber shown on this figure along the magnetic field of 300 kA/m; beam electrons energy is 13 keV. In this system, the electron concentration sustained in the plasma-beam discharge is about $5 \cdot 10^{13}$ cm^{-3} at a discharge power of 3 kW. Experiments with the plasma-beam discharge show that the effectiveness of the electron beam dissipation in a plasma at fixed beam power strongly depends on the electron concentration n_b and energy E_b in the electron beam. An increase of the electron energy E_b in the beam leads to lower effectiveness of the beam relaxation. Conversely an increase of the electron concentration n_b results in higher effectiveness of the beam relaxation. The maximum fraction of the beam power absorbed in plasma in these experiments reached 71% (A.M. Alekseew, A.A. Ivanov, V.V. Starikh, 1979).

12.3.5 Plasma-Beam Discharge in Crossed Electric and Magnetic Fields, Plasma-Beam Centrifuge

General aspects of the plasma centrifuges, where the ionized gas is rotating fast in crossed electric and magnetic fields, and effects of gas separation were discussed in Section 7.5.3. A specific feature of the plasma centrifuge, based on the plasma-beam discharge, is the radial electric field, which is provided directly by the electron beam (A.I. Babaritsky, A.A. Ivanov, V.V. Severny, V.V. Shapkin, 1975a; A.I. Zhuzhunashvili, A.A. Ivanov, V.V. Shapkin, E.K. Cherkasova, 1975). Schematic of such plasma-centrifuge sustained by an electron beam is shown in Figure 12.6. The

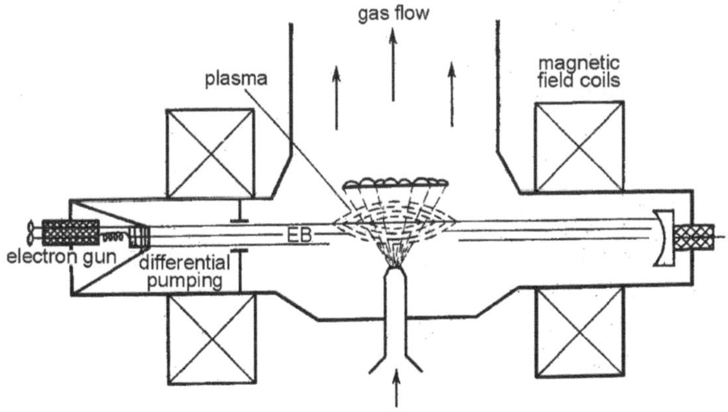

FIGURE 12.5 General schematic of a plasma-beam discharge.

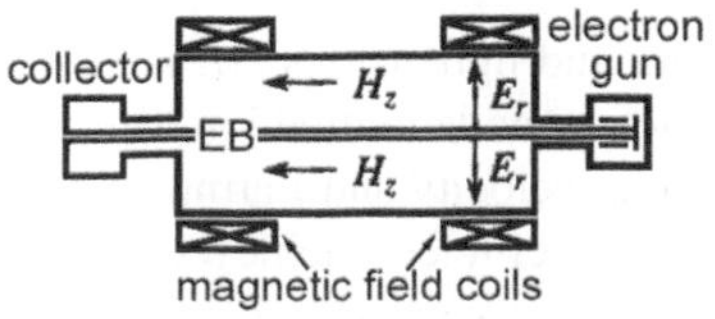

FIGURE 12.6 Electron beam in a magnetic field as the plasma centrifuge.

plasma-beam discharge in this system is stable in inert gases at pressures 0.1–1 mTorr, and fills the entire volume of the discharge chamber. The maximal magnetic field in the coil center is about 800 kA/m and in the middle of the discharge chamber it is about 500 kA/m. This magnetic field is sufficiently large to magnetize not only electrons but ions as well. This means that the ion-cyclotron frequency much exceeds the frequency of the ion–ion Coulomb collisions. Electron energy in the beam is about 10 keV, the beam current 1.5 A. Voltage between the electron beam and the external wall of the discharge chamber can be chosen to provide the ion current from center to the periphery. Radial current in plasma varied from 0.1 to 6 A. Plasma in this discharge system is completely ionized with density about $10^{13}\,cm^{-3}$; electron temperature is quite high on the level of 10 eV; ion temperature is also very high on the level of 2–4 eV. The velocity of the plasma rotation in the crossed electric and magnetic fields in this system is very high. According to the Doppler shift of argon ions spectral lines, this circulation drift velocity is about $5 \cdot 10^5$ cm/s, which is in a good agreement with Eq. (4.108) for drift in the crossed electric and magnetic fields. Also the plasma rotation velocity is very high; the related binary mixture separation coefficient is not significant (see Eq. (7.53)), because the ions temperature in the centrifuge is very high. In this case separation can be achieved due to the difference in radial ion currents for components with different masses.

12.3.6 Radial Ion Current in Plasma Centrifuges and Related Separation Effect

In the plasma centrifuges, ions displacement in the radial direction at the moment of their formation is due to the combined effects of the electric and magnetic fields. This radial displacement Δ_r of an ion can is:

$$\Delta_r = \frac{v_{EB}}{\omega_{Bi}} = \frac{1}{\omega_{Bi}}\frac{E}{B}, \tag{12.26}$$

$v_{EB} = E/B$ is drift in the crossed electric E and magnetic B fields (4.108); and $\omega_{Bi} = eB/M_i$ is the ion-cyclotron frequency. The radial ion displacement Δ_r, which requires time $1/\omega_{Bi}$, occurs once per the ion lifetime; this can be estimated as the ionization time $\tau_{ion} = 1/\nu_{ion} >> 1/\omega_{Bi}$. The average ion radial velocity $\overline{v}_{ri}$ can be estimated based on Eq. (12.27) as:

$$\overline{v}_{ri} = \frac{E}{B}\frac{1}{\omega_{Bi}\tau_{ion}} = \frac{E}{B}\frac{\nu_{ion}}{\omega_{Bi}} = \left(\frac{E\nu_{en}}{eB^2}\right)\cdot M_i. \tag{12.27}$$

Since the ionization frequency does not depend on the mass of ions, one we can conclude from Eq. (12.27) that the average ion radial velocity is proportional to their mass. Thus the ion current effect, similar to the centrifugal effect, results in a preferential flux of heavier particles to the periphery. The binary mixture separation coefficient R Eq. (7.53) can be expressed in this case as (A.A. Ivanov, 2000):

$$\ln R = \frac{M_{iH} - M_{iL}}{\overline{M}_i}\frac{\nu_{ion}}{\nu_c}\sqrt{2}\int_{r1}^{r2}\frac{E/B}{\overline{v}_{Ti}}\frac{dr}{\overline{r}_{Li}}. \tag{12.28}$$

Here, M_{iH}, M_{iL} are masses of heavier and lighter ions; $\overline{M}_i$ is their reduced mass; ν_c is the frequency of ion-ion Coulomb collisions, which provide mutual ion diffusion and reduce the separation coefficient; $\overline{v}_{Ti}$ and $\overline{r}_{Li}$ are the ionic thermal velocity and Larmor radius, calculated for the reduced ionic mass. Typical values of the separation coefficients achieved in this type of plasma centrifuge are: for an Ar-He mixture $R = 6.6$, for a Kr-Ar mixture $R = 2.5$, for a Xe-Kr mixture $R = 2.0$, and for an isotopic mixture ^{20}Ne–^{22}Ne – the coefficient $R = 1.3$ (A.A.Ivanov, 2000).

12.4 NON-EQUILIBRIUM HIGH-PRESSURE DISCHARGES SUSTAINED BY HIGH-ENERGY ELECTRON BEAMS

12.4.1 Non-Self-Sustained High-Pressure Discharges

Most of the electron beam energy in these systems is spent for ionization and electron excitation of neutral species. To provide the necessary energy input into vibrational excitation of molecules, additional energy can be transferred to the plasma electrons from an external electric field. This electric field can be chosen sufficiently low to avoid ionization, but optimal for vibrational excitation or other processes requiring electron temperature at the relatively low level of 1–2 eV. Such non-self-sustained discharges with ionization provided by an electron beam and most of energy input due to an external electric field are especially important for gas lasers (in particular, powerful pulsed CO_2-lasers) and high-efficiency plasma-chemical systems (see Section 5.6.2). These systems were developed, by E. Hoad, H. Rease, J. Stall, J. Zav (1973), by groups of N.G. Basov (N.G. Basov, E.M. Belenov, V.A. Danilychev, 1973; V.I. Panteleev, 1978) and E.P. Velikhov (E.P. Velikhov, S.A. Golubev, A.T. Rakhimov, 1973,1975; E.P. Velikhov, V.D. Pismennyi, A.T. Rakhimov, 1977). Discharge aspects of the system was reviewed in detail in the book of E.P. Velikhov, A.S. Kovalev, A.T. Rakhimov, 1987; plasma-chemical processes – in a book of V.D. Rusanov, A. Fridman, 1984. The non-self-sustained discharges are interesting for plasma-chemical and laser applications first of all because they can provide a high level of non-equilibrium uniformly in very large volumes at high pressures (up to several atmospheres). An important advantage of these systems is the mutual independence of ionization and energy input in the molecular gas, which permits choosing the optimal values of the reduced the electric field E/n_0 (for example, for vibrational excitation) without care to sustain the discharge. It is also important that ionization (see Section 6.3.3) instabilities cannot destroy the discharge, simply because ionization is independent and cannot be affected by local overheating. Ionization in such non-self-sustained high-pressure non-equilibrium discharges can be provided not only by electron beams, but also using UV-radiation, X-rays, etc. The important advantage of electron beams with respect to these other external sources of ionization is relatively high intensity of ionization at the same power consumption.

12.4.2 Plasma Parameters of the Non-Self-Sustained Discharges

A general schematic of the discharges sustained by electron beams is illustrated in Figure 12.7. The electron beam is injected into the gas usually normal to electrodes across the accelerator's window closed by a metallic foil. Voltage is applied to the further electrode by the capacitance. Thus plasma generation is due to the electron beam ionization, and the electric current is provided by the applied external electric field. Such discharge systems can be arranged both in stationary and in short-pulse regimes. In the short pulse regime: the pulse duration is usually about 1000 ns, the electron beam

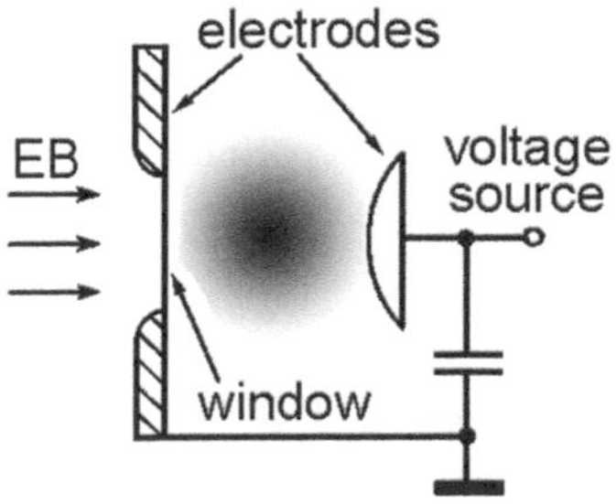

FIGURE 12.7 General schematic of a discharge sustained by an electron beam.

energy about 300 keV and the beam current density can exceed 100 A/cm^2. According to Eq. (2.35), in atmospheric pressure air the ionization rate provided by such an electron beam reaches $q_e = 10^{22}$ 1/cm^{3}s. Losses of plasma electrons in this pulse-beam system are due to the electron–ion recombination and electron attachment to electronegative air components followed by the ion-ion recombination. These losses can be described by the effective recombination coefficient, determined based on the electron balance Eq. (4.171) as:

$$k_r^{eff} = k_r^{ei} + \varsigma k_r^{ii}, \tag{12.29}$$

where k_r^{ei}, k_r^{ii} are rate coefficients of electron–ion and ion–ion recombination; $\varsigma = k_a/k_d$ is the ratio of the attachment and detachment coefficients. For calculating the electron beam degradation in air, the effective empirical recombination coefficient can be taken as: $k_r^{eff} \approx 10^{-6}\,\text{cm}^3/\text{s}$. The concentration of plasma electrons n_e can then be estimated as a simple function of the electron beam ionization rate q_e:

$$n_e = \sqrt{q_e / k_r^{eff}}. \tag{12.30}$$

Numerically for the example given above, the concentration of plasma electrons generated by the electron beam is about 10^{14} cm^{-3}. Thus even at atmospheric pressure, electron beams are able to produce a degree of ionization about $3 \cdot 10^{-6}$; this is high enough to sustain a high level of vibrational non-equilibrium (see Eq. 5.150) and provide effective plasma-chemical and laser processes. One should note that the discharges sustained by electron beams are effective, only if the power and energy input from the electron beam is much below those provided by the external electric field. For example, at the above-considered values of parameters, the specific power of the pulse electron beam can be calculated as:

$$P_b = q_e U_i \approx 0.05\ \text{MW/cm}^3, \tag{12.31}$$

where U_i is the energy price of ionization (see Table 12.1). The specific power transferred from the external electric field to the atmospheric pressure non-thermal pulse discharge can be estimated as follows:

$$P_d = \sigma E^2 \approx k_{eV} n_e n_0 \hbar\omega \approx 1\ \text{MW/cm}^3, \tag{12.32}$$

where σ is the discharge plasma conductivity; n_e, n_0 are plasma electrons and neutral gas densities respectively; k_{eV} is the rate coefficient of vibrational excitation of molecules with relevant vibrational quantum $\hbar\omega$. If the pulse electron beam duration is 1000 ns, the specific energy input from the beam according to Eq. (12.31) is approximately 0.05 J/cm^3. The specific energy input from the external electric field into the discharge according to Eq. (12.32) is about 1 J/cm^3, which is 20 times higher. These discharges can be arranged without essential modifications in pulse-periodic regimes (E.P. Velikhov, A.S. Kovalev, A.T. Rakhimov, 1987; V.D. Rusanov, A. Fridman, 1984). The non-self sustained discharges also can be organized in a continuous way. The high-pressure discharges sustained by continuous electron beams are discussed by E.P. Velikhov, S.A. Golubev, A.T. Rakhimov (1973, 1975).

12.4.3 Maximum Specific Energy Input and Stability of Discharges Sustained by Short-Pulse Electron Beams

The optimal specific energy inputs necessary to stimulate endothermic plasma-chemical processes are approximately 1 eV/mol, or about 5 J/cm^3 at normal conditions. The minimum threshold values

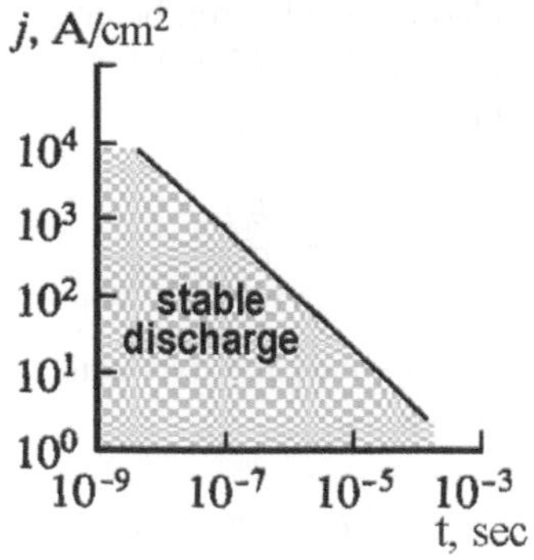

FIGURE 12.8 Current densities and pulse durations of a stable discharge sustained by an electron beam.

of the specific energy input in these systems according to Eq. (5.160) are about 0.1–0.2 eV/mol or 0.5–1 J/cm³. Reaching these energy inputs in non-thermal homogeneous atmospheric pressure discharges is limited by numerous instabilities. Those can be suppressed in non-self-sustained discharges, since ionization is provided independently by electron beams. Nevertheless the fast overheating instabilities put serious limits on the specific energy input. Especially severe instability effects are related to the fast mechanisms of chemical and VV-relaxation heat release (see Figure 6.12). The non-self sustained pulse discharges can be arranged for much longer duration than the above discussed nano-second range. The pulse duration can be increased to 10–100 μs and more. However, at relatively high levels of the discharge specific power, duration of the discharge stability is limited. Typical stable values of current densities and pulse durations of non-self-sustained discharges are shown in Figure 12.8. From this figure one can calculate, that specific energy inputs exceeding 0.5 J/cm³ require special consideration of the discharge stability. It is interesting to point out that the above restriction actually divides the effective plasma-chemical and effective laser regimes of the discharges.

12.4.4 About Electric Field Uniformity in Non-Self-Sustained Discharges

If ionization in the non-self-sustained discharge is provided by an electron beam, then high values of the external electric field can result in a decrease of electron concentration near the cathode. The elevated electric field in the cathode layer due to the formed space charge can lead to additional ionization exceeding that of the beam. This effect converts the discharge into a self-sustained one. Such self-sustained pulse discharges, interesting for special gas laser applications, are considered in N.G. Basov (1974), and V. Yu. Baranov, A.P. Napartovich, A.N. Starostin (1984). The electric field uniformity criterion to avoid the ionization in the cathode layer is:

$$\lambda_c / d << E_0 / E_c, \tag{12.33}$$

where d is the distance between electrodes; λ_c is the thickness of space charge in the cathode layer; E_c is electric field in the cathode layer; E_0 is the quasi-uniform electric field. The cathode layer thickness λ_c can be estimated from the steady-state condition for charged particles in the layer:

$$eq_e = \frac{1}{\lambda_c} j_e = \frac{1}{\lambda_c} n_e b_e e E_0, \tag{12.34}$$

where q_e is the ionization rate provided by an electron beam; j_e is the plasma current density; b_e is the electron mobility. Taking into account Eq. (12.30) for electron density n_e, the cathode layer thickness λ_c can be found from Eq. (12.34) as:

$$\lambda_c = \frac{b_e E_0}{\sqrt{q_e k_r^{eff}}}. \tag{12.35}$$

The Gauss equation for the cathode layer where the ions concentration n_{ic} prevails over that one of electrons can be presented for estimations as:

$$E_c / \lambda_c = e n_{ic} / \varepsilon_0. \tag{12.36}$$

Then the ions concentration n_{ic} in the cathode layer can be expressed from $j_e = n_{ic} b_i e E_c$ (b_i is the ion mobility) in the following form:

$$n_{ic} = \sqrt{q_e \varepsilon_0 / e b_i}. \tag{12.37}$$

Taking E_c from (12.36) and λ_c from (12.35) and substituting them in the inequality (12.33), yields the electric field uniformity criterion for the non-self-sustained discharges (A.Fridman, 1976a) as:

$$E_0 << \frac{b_i d}{b_e^2} k_r^{eff} \sqrt{\frac{q_e \varepsilon_0}{e b_i}}. \tag{12.38}$$

Numerically, Eq. (12.38) requires E_0 << 10 kV/cm at atmospheric pressure, electron beam current density 10^3 A/cm^2, and d = 100 cm.

12.4.5 Non-Stationary Effects of Electric Field Uniformity in Non-Self-Sustained Discharges

The total electric field in the non-self-sustained discharges can be quasi-uniform even if the criterion (2.4.10) is not valid and external electric field is very high. This can be achieved if the pulse duration is too short for significant perturbation. The time of establishing the steady-state electric field distribution can be found from the electric current continuity equation including the displacement current:

$$\varepsilon_0 \frac{\partial E_c}{\partial t} + e n_{ic} b_i E_c = e n_e b_e E_0. \tag{12.39}$$

It was taken into account that major changes in the electric field occur in the cathode layer. Then assume that the ion concentration in the cathode layer is in the form of Eq. (10.4.9) and constant. This leads to the following expression for the establishment time of the electric field distribution in the discharge:

$$\tau_E = \frac{1}{\sqrt{q_e e b_i \varepsilon_0}} = \frac{1}{\sqrt{k_r^{eff} e b_i \varepsilon_0}} \frac{1}{n_e}. \tag{12.40}$$

If electron concentration in the atmospheric pressure plasma is $n_e = 10^{15}$ cm^{-3}, then the establishment time of the electric field distribution is about 30 ns.

12.5 PLASMA IN TRACKS OF NUCLEAR FISSION FRAGMENTS, PLASMA RADIOLYSIS

12.5.1 Plasma Induced by Nuclear Fission Fragments

The nuclear fission process of interaction between a slow neutron with uranium-235 can be presented as:

$$ {}_{92}U^{235} + {}_{0}n^{1} \rightarrow {}_{57}La^{147} + {}_{35}Ba^{87} + 2{}_{0}n^{1}. \tag{12.41}$$

The total fission energy is 195 MeV. Most of this energy, 162 MeV, go to the kinetic energy of the fission fragments, barium, and lanthanum (see, for example, S. Glasstone, M.C. Edlung, 1952). Plasma-chemical effects can be induced by the nuclear fission fragment if they are transmitted into the gas phase before thermalization. The fission fragment appears in the gas as multiply charged ions with $Z = 20 \div 22$ and initial energy of approximately $E_0 \approx 100$ MeV (H.A. Hassan, J.E. Deese, 1976). Degradation of these fission-fragment-ions in gas is related to the formation of the so-called δ-electrons moving perpendicularly to the fission fragment with kinetic energy approximately $\varepsilon_0 \approx 100$ eV. These δ-electrons form a radial high-energy electron beam, which then generates plasma in the long cylinder surrounding the trajectory of the fission fragment. This plasma cylinder is usually referred to as the **track of the nuclear fission fragments**. If the plasma density in the track of nuclear fission fragments is relatively low, then degradation of each δ-electron in the radial beam is quasi-independent. The chemical effect of the independent energetic electrons (or other particles) interaction with molecule gas is called radiolysis. The radiolysis induced by high-energy electrons in different molecular gases has been investigated in detail (see, for example, P. Aulsloos, 1968). This process, mostly related to primary ionization and electronic excitation, is not very energy effective but is of interest for many special applications related to gas and surface treatment. More interesting physical and chemical effects can be observed if the plasma density in the tracks of nuclear fission fragments is relatively high, and the collective effects of electron–electron interaction, vibrational excitation, etc. take place (I.G. Belousov, V.F. Krasnoshtanov, V.D. Rusanov ea., 1979). These effects usually referred to as the **plasma radiolysis**, require special conditions of the nuclear fragment degradation. To determine the nuclear fragment degradation conditions necessary for plasma radiolysis, analyze the plasma radiolysis by itself.

12.5.2 Plasma Radiolysis of Water Vapor

The energy efficiency of radiolysis is usually characterized by the so-called **G-factor**, which shows the number of chemical processes of interest (in this case the number of dissociated water molecules) per 100 eV of radiation energy spent. Value of the G-factor of water vapor radiolysis by high-energy electrons (in particular, by the δ-electrons in the tracks of nuclear fission fragments) can be found as (A.K. Pikaev, 1965):

$$\frac{G}{100} = \left[\frac{1}{U_i(H_2O)} + \frac{U_i(H_2O) - I_i(H_2O)}{U_i(H_2O)\cdot I_{ex}(H_2O)}\right] + \frac{\mu}{U_i(H_2O)}. \tag{12.42}$$

In this relation, $U_i(H_2O) \approx 30$ eV is the ionization energy cost for water molecules (see Section 12.2); $I_i(H_2O) = 12.6$ eV, $I_{ex}(H_2O) = 7.5$ eV are energy thresholds of ionization and electronic excitation of water molecules; the factor μ shows how many times one electron with an energy below the threshold of electronic excitation $\varepsilon < I_{ex}(H_2O)$ can be used in the water molecule destruction. The first two terms in Eq. (12.42) are mostly related to dissociation through electronic excitation and give the radiation yield $G \approx 12\ \frac{mol}{100\text{eV}}$ (R.F. Firestone, 1957), which describes water dissociation by β-radiation and low current density electron beams. The third term in (12.42) describes the contribution H_2O destruction of the dissociative attachment with a following ion-molecular reaction:

$$e + H_2O \rightarrow H^- + OH, \quad H^- + H_2O \rightarrow H_2 + OH. \tag{12.43}$$

The contribution of this mechanism into water dissociation during the conventional radiolysis is relatively low ($\mu = 0.1$, V.P. Bochin, V.A. Legasov, V.D. Rusanov, A. Fridman, 1978a) because of the resonance character of the dissociative attachment crosssection dependence on energy and low ionization degree. The factor μ can be significantly larger at higher degrees of ionization in plasma because of effective energy exchange (Maxwellization) between plasma electrons with energies below the threshold of electronic excitation $\varepsilon < I_{ex}(H_2O)$. Thus the main plasma effect in water vapor electrolysis is related to the Maxwellization of plasma electrons; this results in growth of the G-factor and an increase of energy efficiency of water dissociation and hydrogen production in the system.

12.5.3 Maxwellization of Plasma Electrons and Plasma Effect in Water Vapor Radiolysis

The critical degree of ionization in the tracks of nuclear fragments necessary for Maxwellization of plasma electrons and realization of the plasma radiolysis in water vapor can be found from the kinetic equation for electron velocity distribution f_e (V.P. Bochin, V.A. Legasov, V.D. Rusanov, A. Fridman, 1978a):

$$\begin{aligned} 0 = L \cdot \left(f_e^2 - \frac{\partial^2 f_e}{\partial v_i \partial v_k} \frac{\partial^2 \Psi}{\partial v_i \partial v_k} \right) + f_e \cdot \sigma_a(v) \cdot v \cdot [H_2O] \\ + Z_d(v) \cdot k_d n_e [H^-] + Z(v) \cdot q_e - f_e \cdot \sigma_r^{ei} \cdot v \cdot n_i. \end{aligned} \tag{12.44}$$

The factor $L = \Lambda(e^2/m\varepsilon_0)^2$, Λ is the Coulomb logarithm; $Z_d(v)$ is the normalized electron velocity distribution function after its detachment from a negative ion H^-; $Z(v)$ is the normalized electron distribution related to rate q_e of electrons appearance in the energy range below the threshold of electronic excitation; $\sigma_a(v)$, $\sigma_r^{ei}(v)$ are crosssections of dissociative attachment and electron–ion recombination; n_i is the positive ion concentration; the function $\Psi_e(\vec{v})$ is determined by:

$$\Psi_e(\vec{v}) = \frac{1}{8\pi} \int |\vec{v} - \vec{v}'| \cdot f_e(\vec{v}') \cdot d\vec{v}'. \tag{12.45}$$

Analysis of Eq. (12.44) shows that effective Maxwellization of electrons under the threshold of electronic excitation $\varepsilon < I_{ex}(H_2O)$ takes place when the degree of ionization is high:

$$\frac{n_e}{[H_2O]} >> \sigma_a^{\max} \frac{T_e^2 \varepsilon_0^2}{e^4 \Lambda}. \tag{12.46}$$

In this criterion, $\sigma_a^{\max} \approx 6 \cdot 10^{-18} \text{cm}^2$ is the maximum value of the dissociative attachment crosssection of electrons to water molecules. Numerically at an electron temperature $T_e = 2$ eV, the criterion (12.46) requires $n_e/[H_2O] >> 10^{-6}$. Thus if ionization in a track of nuclear fission fragment is sufficiently high Eq. (12.46), the effective Maxwellization of electrons under the threshold of electronic excitation permits the μ-factor Eq. (12.42) to increase from about $\mu = 0.1$ (typical for conventional radiolysis) to higher value determined by the degradation spectrum $Z(\varepsilon)$:

$$\mu = \frac{\int_0^I \varepsilon Z(\varepsilon) d\varepsilon}{\varepsilon_a \int_0^I Z(\varepsilon) d\varepsilon}. \tag{12.47}$$

which can be achieved in the plasma radiolysis. In this relation, I is the ionization potential; $Z(\varepsilon)$ is the degradation spectrum under the threshold of electronic excitation determined by Eq. (12.20); ε_a is energy necessary for dissociative attachment (see Figure 2.7). Integrating Eq. (12.47) with the degradation spectrum Eq. (12.20) gives $\mu \approx 1$. This result means that intensive Maxwellization of electrons under the electronic excitation threshold in water vapor leads almost all of the electrons to dissociative attachment. Thus the first two terms in Eq. (12.42), which are related to dissociation through electronic excitation, give the radiation yield $G \approx 12 \frac{mol}{100\text{eV}}$, and the third term in Eq. (12.42) specific for plasma radiolysis gives an additional $G \approx 7 \frac{mol}{100\text{eV}}$, which is due to dissociative attachment. This brings the total radiation yield of water dissociation in plasma radiolysis to $G \approx 19 \frac{mol}{100\text{eV}}$, which corresponds to a high water decomposition energy efficiency of approximately 55%.

12.5.4 Effect of Plasma Radiolysis on Radiation Yield of Hydrogen Production

The main final saturated products of water radiolysis are hydrogen and oxygen. However, the radiation yields of water dissociation discussed above do not directly correspond to the radiation yield of hydrogen production because of inverse reactions. If the radiation yield of conventional water radiolysis reaches $G \approx 12 \frac{mol}{100\text{eV}}$ without plasma effects, the corresponding radiation yield of hydrogen production is only $G \approx 6 \frac{mol}{100\text{eV}}$ (P. Aulsloos, 1968). Plasma radiolysis can bring the total radiation yield of water dissociation to $G \approx 19 \frac{mol}{100\text{eV}}$, the corresponding radiation yield of hydrogen production grows in this case to $G \approx 13 \frac{mol}{100\text{eV}}$, which means an energy efficiency of about 40% (I.G. Belousov, V.F. Krasnoshtanov, V.D. Rusanov ea., 1979). The energy efficiency of hydrogen production in the process of water vapor dissociation during radiolysis strongly depends on the gas temperature and the intensity of the process (B.G. Dzantiev, A.N. Ermakov, V.N. Popov, 1979). This is related to the fact that the primary products of water dissociation related to mechanisms of electronic excitation and electron–ion recombination are atomic hydrogen H and the hydroxyl radical OH. Formation of molecular hydrogen is then related to the following endothermic bimolecular reaction:

$$H + H_2O \to H_2 + OH, \quad E_a = 0.6\text{eV}. \tag{12.48}$$

The reverse reactions leading back to forming water molecules is three-molecular recombination:

$$H + OH + M \to H_2O + M, \tag{12.49}$$

which is 100 times faster than molecular hydrogen formation in the three-molecular recombination:

$$H + H + M \rightarrow H_2 + M. \tag{12.50}$$

High intensity of radiolysis leads to higher concentration of radicals and therefore, to a larger contribution of the three-molecular processes and lower radiation yield of hydrogen production. The radiation yield can be increased with temperature, which intensifies the endothermic reaction (12.48). Hydrogen production in the specific reactions of plasma radiolysis (12.43) has no energy barrier and proceeds faster than inverse processes (12.49). This makes the plasma radiolysis effect especially important at high process intensities, which can be achieved in the tracks of nuclear fission fragments.

12.5.5 Plasma Radiolysis of Carbon Dioxide

Radiation yield of CO_2-dissociation, similar to the case of water vapor radiolysis, is determined by the conventional term G_{rad} typical for radiation chemistry, and by a special plasma radiolysis term related to the contribution of electrons under the threshold of electronic excitation. The first conventional radiolysis term, similar to Eq. (12.42), is due to electronic excitation and dissociative recombination and can reach a maximum $G_{rad} = 8\dfrac{mol}{100\text{eV}}$ (P. Aulsloos, 1968). In this case the main contribution of the low energy electrons (under the threshold of electronic excitation) is due to vibrational excitation of CO_2-molecules, which is proportional to the degree of ionization n_e/n_0 in the track of nuclear fission fragment:

$$\Delta T_v \approx \frac{n_e}{n_0}\left[\frac{\int_0^I \varepsilon Z(\varepsilon)d\varepsilon}{c_v\int_0^I Z(\varepsilon)d\varepsilon}\right]. \tag{12.51}$$

Here, c_v is the specific heat related to the vibrational degrees of freedom of CO_2-molecules; the factor in parentheses is determined by the degradation spectrum of δ-electrons in tracks and numerically is several electron-volts. Plasma effects in CO_2-radiolysis are related to stimulation of the reaction:

$$O + CO_2 \rightarrow CO + O_2 \tag{12.52}$$

by vibrational excitation of CO_2. This reaction prevails over three-molecular recombination of oxygen:

$$O + O + M \rightarrow O_2 + M \tag{12.53}$$

at vibrational temperatures corresponding in accordance with Eq. (12.51) to the degree of ionization $n_e/n_0 >> 10^{-3}$. In this case, the radiation yield of CO_2-dissociation becomes twice larger than in conventional radiolysis and reaches $G = 16\dfrac{mol}{100\text{eV}}$. This corresponds to energy efficiency of about 50%. The radiation yield of CO in conventional radiolysis is suppressed by inverse reactions:

$$O + CO + M \rightarrow CO_2 + M, \tag{12.54}$$

which are much faster at standard conditions than recombination (12.53). This is because CO concentrations usually much exceed the concentration of atomic oxygen. As a result, the radiation yield of conventional CO_2-dissociation $G_{rad} = 8\frac{mol}{100\text{eV}}$ leads to a radiation yield of the CO-production of only $G = 0.1\frac{mol}{100\text{eV}}$ (P. Hartech, S. Doodes, 1957). Only high-intensity radiolysis, which provides large concentrations of oxygen atoms, permits suppressing the inverse reactions (12.54) in favor of (12.53) and increase the radiation yield (R. Kummler, 1977). This can take place not only in tracks of nuclear fission fragments but also in the radiolysis provided by relativistic electron beams. High intensity of radiolysis increases the radiation yield of CO-production in contrast to the case of hydrogen production (Section 12.44). This is because the rate coefficient of (12.53) exceeds that of (12.54).

12.5.6 Plasma Formation in Tracks of Nuclear Fission Fragments

As demonstrated above, the energy efficiency of endothermic chemical processes can be increased by a factor of two when the degree of ionization in the tracks of nuclear fission fragments exceeds some critical value (10^{-4} for H_2O, 10^{-3} for CO_2). For this reason analyze the degree of ionization in the tracks of fragments, which is provided by the degradation of δ-electrons with kinetic energy $\varepsilon_\delta \approx 100$ eV formed by the fragments. At first let us take into account that degradation length of the fragments can be presented as:

$$\lambda_0 = \frac{4\pi\varepsilon_0^2 E_0 mv^2}{n_0 Z^2 Z_0 e^4}\ln^{-1}\left(\frac{2mv^2}{I}\right). \tag{12.55}$$

Here, v, Ze, E_0 are velocity, charge, and kinetic energy of the nuclear fission fragments; m, e are mass and charge of an electron; Z_0 is the sum of charge numbers of atoms forming molecules, which are irradiated by the nuclear fragments; n_0 is gas density; I is the ionization potential. The number of energetic δ-electrons formed by a nuclear fission fragment per unit length is:

$$N_\delta = \frac{n_0 Z^2 Z_0 e^4}{4\pi\varepsilon_0^2 mv^2 \varepsilon_\delta}\ln\left(\frac{2mv^2}{I}\right). \tag{12.56}$$

For example, generation of the δ-electrons in water vapor at atmospheric pressure is characterized by $N_\delta = 10^6$ cm^{-1}, that in carbon dioxide is $N_\delta = 3 \cdot 10^6$ cm^{-1}. The tracks start with very narrow channels of complete ionization. The radius of these initial completely ionized channels is:

$$r_0 = \frac{Ze^2}{2\pi\varepsilon_0 v\sqrt{m\varepsilon_\delta}}\ln\left(\frac{2mv^2}{I}\right) \approx 3\cdot 10^{-8}\text{cm}. \tag{12.57}$$

After the initial track (12.57) expansion is at first free and collisionless, when radius r of the channel is less than length λ_i necessary for ionization. Then the expansion mechanism becomes diffusive.

12.5.7 Collisionless Expansion of Tracks of Nuclear Fission Fragments

Analyzing the collisionless track expansion, two different regimes can be noted depending on the linear density of δ-electrons N_δ. If the linear density is sufficiently high:

$$N_\delta > 4\pi\varepsilon_0 \frac{\varepsilon_\delta}{e^2}\ln^{-1}\left(\frac{\lambda_i}{r_0}\right) \approx 10^8\text{cm}^{-1}, \tag{12.58}$$

then the plasma remains quasi-neutral during the collisionless expansion. In the opposite case of low linear density, when $N_\delta < 10^8$ cm^{-1}, electrons and ions are independent during the expansion. In the case of high linear density of δ-electrons (12.58) and hence quasi-neutral plasma, the collisionless expansion of the plasma channel can be described by the system of equations which includes a dynamic equation:

$$nM\frac{d\vec{v}}{dt} = -T_e grad\, n, \tag{12.59}$$

(T_e is the electron temperature) and the continuity equation for ion concentration n_i and ion velocity v_i:

$$\frac{\partial n}{\partial t} + div\left(n\vec{v}\right) = 0. \tag{12.60}$$

Taking into account the axial symmetry of the system, the above system of equation can be reduced to an ordinary differential equation, using the auto-model variable $\xi = r/t$:

$$\frac{M}{T_e}\left(v-\xi\right)^2\frac{\partial v}{\partial \xi} = \frac{\partial v}{\partial \xi} + \frac{v}{\xi}. \tag{12.61}$$

The solution of the equation can be presented as the function of the auto-model variable:

$$v = \xi + c_{si}\sqrt{1+\Omega\left(\xi\right)}. \tag{12.62}$$

Here, $c_{si} = \sqrt{T_e/M}$ is the ionic sound velocity, and the special function $\Omega(\xi)$ is:

$$\Omega\left(\xi\right) \approx 1 - \frac{c_{si}}{2\xi} + \sqrt{\left(\frac{c_{si}}{2\xi}\right)^2 + \frac{2c_s}{\xi}}. \tag{12.63}$$

In this case, the distribution of ion concentration in the expanding plasma channel also can be expressed as a function of auto-model variable:

$$n\left(\xi\right) \propto \exp\left[-\int\sqrt{1+\Omega\left(\xi\right)}\frac{d\xi}{c_{si}}\right]. \tag{12.64}$$

The above Eqs. (12.62) and (12.64) permit to determine the average square of the ion velocity:

$$\left\langle v^2\right\rangle = \int n\left(\xi\right)\cdot v^2\left(\xi\right)\cdot d\xi. \tag{12.65}$$

This integration leads us to the important conclusion that about two-thirds of the total initial energy of the δ-electrons are going into directed radial motion of ions.

12.5.8 Energy Efficiency of Plasma Radiolysis in Tracks of Nuclear Fission Fragments

2/3 of the δ-electrons energy (of about 100 eV) is transferred at $N_\delta > 10^8$ cm^{-1} into radial motion of ions. On the other hand, the efficiency of the 70 eV ions in ionization processes and in chemical reactions is very low. Thus one can conclude that high linear density of δ-electrons ($N_\delta > 10^8$ cm^{-1}) is not effective in plasma radiolysis. Improved energy efficiency can be achieved during the expansion of the initial track, when the linear density of δ-electrons is relatively low

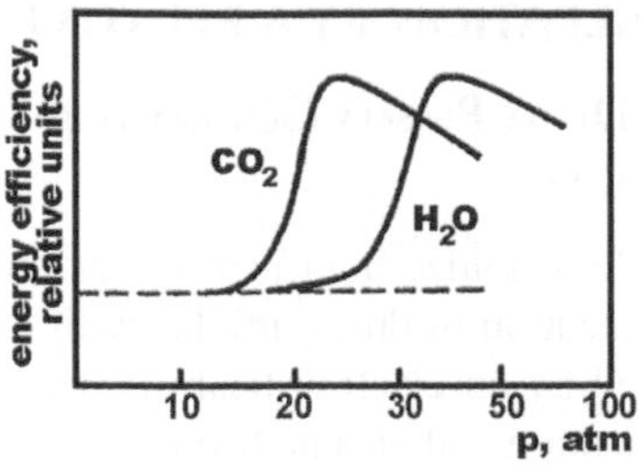

FIGURE 12.9 Energy efficiency of water vapor and carbon dioxide dissociation as a function of pressure.

($N_\delta < 10^8$ cm^{-1}), electrons do not draw ions with them and transfer most of their energy into ionization and excitation of the neutral gas; this is quite useful in plasma chemistry. In this case, the radius of the tracks can be determined as $\lambda_i \approx 1/n_0\sigma_i$, where σ_i is cross section of ionization provided by δ-electrons. The degree of ionization in the tracks of nuclear fragments can be then calculated as:

$$\frac{n_e}{n_0} = \left(\frac{3Z_0 e^4 \sigma_i^2}{4\pi\varepsilon_0^2 m\varepsilon_\delta}\right) \cdot \left(\frac{Z}{v}\right)^2 n_0^2. \tag{12.66}$$

The degree of ionization in the plasma channel of the nuclear tracks is proportional to the square of the neutral gas density and pressure. Numerically at atmospheric pressure, the degree of ionization in the tracks of nuclear fragments in water vapor is $n_e/n_0 \approx 10^{-6}$, in carbon dioxide it is about $n_e/n_0 \approx 3 \cdot 10^{-6}$. The necessary degree of ionization for significant plasma effects in radiolysis (which can be estimated for water and carbon dioxide as $10^{-4} \div 10^{-3}$) cannot be achieved at atmospheric pressure. However, taking into account that $n_e/n_0 \propto n_0^2$, the sufficient degree of the ionization can be achieved by increasing pressure to 10–20 atm. Further pressure increase is limited by the criterion opposite to (12.58). At high pressures, the linear densities N_δ are also high, and the main l portion of electron energy is ineffectively lost in favor of ions. The ion heating and transfer to thermal plasma in the tracks takes place in water vapor at pressures of 100 atm, and in carbon dioxide at pressures about 30 atm. The dependence of the energy efficiency of water and carbon dioxide decomposition in the tracks of nuclear fission fragments is shown in Figure 12.9; this is in qualitative agreement with experiments of P. Hartech, S. Doudes, 1957. In principal, the highest values of energy efficiency of water and carbon dioxide decomposition in the tracks of nuclear fission fragments can be achieved, if δ-electrons transfer their energy not directly to neutral gas by different degradation processes (see Section 12.2) but rather through intermediate transfer of most of energy to plasma formed in the tracks. In this case, the major part of the energy of nuclear fragment can be transferred to plasma electrons by mechanisms similar to those in plasma-beam discharges described in Section 12.3. Then the energy of plasma electrons can be transferred to the effective channels of stimulation of plasma-chemical reactions, in particular to vibrational excitation. Thus nuclear fission fragments can transfer their energy in such systems mostly into formation of chemical products (in particular, in hydrogen)rather than heating. The Chemo-Nuclear Reactor based on such principles actually can be "cold", because only a small portion of its energy generation is related to heating. Such nuclear reactor was considered in details in V.P. Bochin, V.F. Krasnoshtanov, V.D. Rusanov, ea., 1983. In the tracks of α-particles with energies about 1 MeV, the factor Z/v in Eq. (12.66) becomes lower with respect to the case of nuclear fission fragments. Then the pressure interval optimal for plasma radiolysis provided by α-particles is shifted to a higher-pressure range and becomes wider.

12.6 DUSTY PLASMA GENERATION BY A RELATIVISTIC ELECTRON BEAM

12.6.1 General Features of Dusty Plasma Generated by Relativistic Electron Beam Propagation in Aerosols

The high-energy electrons are able to ionize both neutral gas and aerosols, while the energies of photons are sufficient only for ionization of dust particles with relatively low work functions. The charge of macroparticles irradiated by an electron beam is determined by two main effects. One effect is related to neutral gas ionization and plasma formation, which leads to negative charging of macroparticles due to electron mobility prevailing over that of ions (G. Rosen, 1962; R.C. Dimick, S.L. Soo, 1964). The second effect is related to secondary electron emission by electron impact from the aerosol surfaces, which tends toward the positive charging of aerosols. Actually competition between these two ionization effects determines the composition and characteristic of the dusty plasma irradiated by electrons. Special interest is directed toward the interaction of β-radiation (which can be considered as a low current high energy electron beam) with aerosols, and to plasma formation and charging related to β-active hot aerosols. Considering the interaction of β-radiation with aerosols in air, one should take into account that low electron fluxes usually leads to fast attachment of plasma electrons to electronegative molecules and to electron conversion into negative ions; this makes the of negative charging not so strong (N.A. Fuks, 1964). The β-active hot aerosols present an example of dusty plasma (V.N. Kirichenko, 1972; V.D. Ivanov, 1969, 1972). In this case, the electron density in the volume is relatively low, and the macroparticle charge becomes positive. At β radiation rates exceeding one electron per second per particle, the total positive charge of the dust particles usually exceeds hundreds of elementary charges. For this case, analysis of macroparticles heating and evaporation was presented by Yu.A. Vdovin, 1975.

12.6.2 Charging Kinetics of Macroparticles Irradiated by Relativistic Electron Beam

In this system, the average charge of macroparticles is determined by electron and ion fluxes to the surface of the particles, and by secondary electron emission induced by the electron beam. Steady-state kinetics of charged particles can be then described by the following system of equations:

$$n_{eb}v_{eb}n_0\sigma_i = \sigma_{rec}^{ei}\overline{v}_e n_e n_i + f_i\left(n_i, Z_a\right)n_a, \tag{12.67}$$

$$n_{eb}v_{eb}\pi r_a^2\left(\delta - 1\right) + f_i\left(n_i, Z_a\right) = f_e\left(n_e, Z_a\right), \tag{12.68}$$

$$n_e = n_i + Z_a n_a. \tag{12.69}$$

In this system of equations, n_0, n_a, n_i, n_e, n_{eb} are the concentrations of gas, macroparticles, ions, plasma and beam electrons respectively; Z_a is the average charge of an aerosol particle; $v_{eb}, \overline{v}_e$ are the velocities of beam electrons and plasma electrons; $\sigma_i, \sigma_{rec}^{ei}$ are averaged effective crosssections of gas ionization by the beam electrons, and the electron–ion recombination (effectively taking into account possible electron attachment in electronegative gases, see Section 4.5.2); δ is the coefficient of the secondary electron emission from the aerosol surfaces; $f_i(n_i, Z_a)$, $f_e(n_e, Z_a)$ are the fluxes of plasma ions and electrons on the aerosol surfaces (see G. Rosen, 1962; R.C. Dimick, S.L. Soo, 1964); r_a is the average radius of macroparticles. Equation (12.67) describes the balance of ions, the second Eq. (12.68) describes the macroparticles charge, and third one (12.69) is the quasi-neutrality condition.

12.6.3 Conditions of Mostly Negative Charging of Aerosol Particles

Most aerosols are negatively charged if the secondary electron emission from their surfaces can be neglected:

$$n_{eb}v_{eb}\pi r_a^2(\delta-1) << f_i(n_i, Z_a). \tag{12.70}$$

Assuming a Bohm flux of ions on the aerosol surfaces, and that electron–ion recombination mostly follow volume mechanisms (which takes place at moderate electron beam densities j_{eb} > 100 μA/cm^2), the inequality (12.70) can be rewritten as a requirement of relatively low electron concentration in the beam and hence, a low value of the electron beam current density (A. Fridman, 1976a):

$$n_{eb} << n_0 \frac{\sigma_i}{\sigma_{rec}^{ei}} \frac{\overline{v}_e}{v_{eb}} \frac{1}{(\delta-1)^2} \frac{m}{M}, \tag{12.71}$$

where *m*, *M* are respectively electron and ion masses. Numerically, the criterion (12.71) of negative charging of the aerosol particles means $n_{eb} < 3 \cdot 10^7$ cm^{-3}, if: $\overline{v}_e = 10^8$ cm/s, $\sigma_i = 3 \cdot 10^{-18}$ cm^2, $v_{eb} = 3 \cdot 10^{10}$ cm/s, $\sigma_{rec}^{ei} = 10^{-16}$ cm^2, $\delta = 100$ (which corresponds to a particle radius r_a = 10 μm), m/M = 10^{-4}, $n_0 = 3 \cdot 10^{19}$ cm^{-3}. According to (12.71), the electron beam current density should be relatively low (j_{eb} < 0.1 A/cm^2) to have mostly negative charging of the aerosol particles and to have the possibility to neglect the secondary emission from the macroparticle surfaces.

12.6.4 Conditions of Balance Between Negative Charging of Aerosol Particles by Plasma Electrons and Secondary Electron Emission

At higher electron beam current densities, the contribution of secondary electron emission becomes more important. At the critical electron density in the beam:

$$n_{eb0} = n_0 \frac{\sigma_i}{\sigma_{rec}^{ei}} \frac{\overline{v}_e}{v_{eb}(\delta-1)^2}\left[1 - \frac{n_a}{n_0}\frac{\pi r_a^2}{\sigma_i}\frac{\overline{v}_i}{\overline{v}_e}(\delta-1)\right] \tag{12.72}$$

and corresponding electron beam current density, the average charge of aerosol particles becomes equal to zero $Z_a = 0$. In Eq. (12.72), the average ion velocity $\overline{v}_i \approx 10^5$ cm/s, and the aerosol particle concentration n_a is to be less than 10^9 cm^{-3}. Expressions for the electron and ion fluxes $f_i(n_i, Z_a)$, $f_e(n_e, Z_a)$ become simple if macroparticles are not charged; this makes solving Eqs. (12.67)–(12.69) not difficult. At the above values of parameters Eq. (12.72) gives $n_{eb0} = 3 \cdot 10^{11}$ cm^{-3}. This means that the contribution of secondary electron emission becomes significant and the average charge of macroparticles becomes equal to zero when the current density of the relativistic electron beam j_{eb} increases to $j_{eb0} = 10^3$ A/cm^2. In this case of uncharged aerosol particles, plasma density in the heterogeneous system can be given by:

$$n_{e0} = n_0 \frac{\sigma_i}{\sigma_{rec}^{ei}} \frac{1}{\delta-1}, \tag{12.73}$$

which numerically gives: $n_{e0} \approx 10^{16}$ cm^{-3}.

12.6.5 Regime of Intensive Secondary Electron Emission, Conditions of Mostly Positive Charging of Aerosol Particles

If $j_{eb} < j_{eb0}$, aerosol particles are negatively charged. When the electron beam current density exceeds the critical value j_{eb0}, Eq. (12.72), contribution of the secondary electron emission becomes so significant that the aerosol particles become positively charged. Consider this case assuming that the total charge of particles does not affect the balance of electron and ion densities:

$$Z_a n_a << n_e \approx n_i. \tag{12.74}$$

When macroparticles are positively charged they attract electrons, and the electrons flux to the aerosol surface can be estimated as:

$$f_e(n_e, Z_a > 0) \approx n_e \overline{v}_e r_a^2 \left(1 + \frac{Z_a e^2}{4\pi\varepsilon_0 r_a T_e}\right), \tag{12.75}$$

where T_e is the temperature of plasma electrons. Eqs. (12.67) – (12.69) can be rewritten in this case as:

$$n_{eb} v_{eb} n_0 \sigma_i = \sigma_{rec}^{ei} n_e^2 \overline{v}_e, \tag{12.76}$$

$$n_{eb} v_{eb} (\delta - 1) = n_e \overline{v}_e \left(1 + \frac{Z_a e^2}{4\pi\varepsilon_0 r_a T_e}\right). \tag{12.77}$$

This system determines the average positive charge of an aerosol particle in the following form:

$$Z_a = \frac{4\pi\varepsilon_0 r_a T_e}{e^2}\left(\sqrt{\frac{j_{eb}}{j_{eb0}}} - 1\right). \tag{12.78}$$

Obviously, $Z_a = 0$ when electron beam current density is equal to the critical one $j_{eb} = j_{eb0}$. The concentration of plasma electrons and ions in this regime can be expressed as follows:

$$n_e = n_{e0}\sqrt{\frac{j_{eb}}{j_{eb0}}}. \tag{12.79}$$

The sufficient condition of equality of electron and ion densities Eq. (12.74) becomes:

$$\frac{4\pi\varepsilon_0 r_a T_e}{e^2}\frac{n_a}{n_{e0}} < 1, \tag{12.80}$$

which is valid at the above considered values of parameters. The time necessary for establishing these steady-state conditions of a relativistic electron beam interaction with aerosols can be found as:

$$\tau = \frac{n_{e0}}{n_{eb0} n_0 \sigma_i v_{eb}} = \frac{1}{n_{eb0} v_{eb} \sigma_{rec}^{ei} (\delta - 1)}. \tag{12.81}$$

This time can be estimated as $\tau \approx 10$ *ns*. Hence, the results for steady-state systems also can be applied for the pulsed relativistic electron beams, where typical pulse duration is about 100 ns.

12.6.6 Electron Beam Irradiation of Aerosols in Low Pressure Gas

In the case of low pressure ($n_0 \to 0$), the positive ions in the volume are absent and the particles are positively charged at any electron beam current densities. Then the system of Eqs. (12.67) – (12.69) can be simplified to:

$$n_{eb} v_{eb} \pi r_a^2 (\delta - 1) = f_e(n_e, Z_a), \tag{12.82}$$

$$n_e = n_a Z_a. \tag{12.83}$$

Because particles are positively charged, they attract electrons; the electrons flux to the aerosol surface $f_e(n_e, Z_a > 0)$ can be also estimated in this case using Eq. (12.75). Then the average positive charge of an aerosol particle Z_a in a low-pressure gas can be calculated from the formula:

$$Z_a = \frac{n_e}{n_a} = \min\left\{\frac{n_{eb}}{n_a}\frac{v_{eb}}{\overline{v}_e}(\delta - 1);\ \sqrt{\frac{n_{eb}}{n_a}\frac{v_{eb}}{\overline{v}_e}(\delta - 1)\frac{4\pi\varepsilon_0 r_a T_e}{e^2}}\right\}. \quad (12.84)$$

The approach in general ($n_0 \to 0$), and Eq. (12.84) are valid only if $n_0 << Z_a n_a$, where the average macroparticle charge Z_a is determined from Eq. (12.74).

12.7 PROBLEMS AND CONCEPT QUESTIONS

12.7.1 Relativistic Effects in Electron Energy Losses Function *L*(*E*)

Using the Bethe – formula (12.2), analyze the relativistic effect on the electron energy losses related to stopping of a relativistic electron beam in a neutral gas. Interpret the effect, and calculate values of the energy losses function $L(E)$ in air for non-relativistic 10 keV electrons, and relativistic 1 MeV electrons.

12.7.2 Ionization Energy Cost at Gas Irradiation by High Energy Electrons

Using Table 12.1, calculate the fraction of the electron beam energy going to excitation of neutral species in different molecular and atomic gases. Analyze Table 12.1 and explain why the ionization energy cost in molecular gases exceeds that for inert gases? Why the ionization energy cost in He is the largest between inert gases?

12.7.3 Stopping Length of Relativistic Electron Beams in Atmospheric Air

Using Eq. (12.4), calculate the initial energy of a relativistic electron necessary to provide its stopping length along a trajectory exceeding 1m. Calculate the relativistic factor for this electron. Estimate the stopping length along the beam axis in this case.

12.7.4 Critical Alfven Current for Electron Beam Propagation

Derive Eq. (12.6) for the limiting case of a relativistic electron beam propagation related to self-focusing in its own magnetic field. Take into account that the Alfven current gives the inherent magnetic field of the beam, which makes the Larmor radius (6.2.25) less than half the beam radius.

12.7.5 Bethe-Born Approximation of Degradation Spectrum

Derive the solution. (12.18) of the Alkhazov's equation in the Bethe-Born approximation taking into account only ionization by primary electrons. Analyze the derived expression of the degradation spectrum. Explain why this simple formula gives only qualitative features of the degradation spectrum.

12.7.6 Analytical Solutions of Alkhazov's Equation

Compare the analytical solutions of the Alkhazov's Eqs. (12.19) and (12.20) for electron energies above and below the excitation threshold. Using the degradation spectrum (12.20) and its relation with electron energy distribution function, calculate the average electron energy under the excitation threshold.

12.7.7 Energy Cost of Ionization and Excitation at Electron Beam Stopping in Gases

Using the data presented in Table 12.2, calculate the energy cost of ionization and excitation of electronically and vibrational excited states in hydrogen irradiated by electron beam with an energy 1 MeV and 1 keV. Compare the energy costs of ionization and vibrational excitation for 1 MeV and 1keV electron beams.

12.7.8 Operation Conditions of Plasma-Beam Discharges

The principal operational condition Eq. (12.21) of plasma-beam discharges requires the beam energy dissipation into Langmuir oscillations to exceed the beam energy losses in electron-neutral collisions. Actually, the criterion (12.21) determines the upper limit of the effective operation of the plasma-beam discharges. Give a numerical estimation of the pressure limit, taking the characteristic beam parameters from Section 12.3.1.

12.7.9 Limitation of Langmuir Noises for Efficient Vibrational Excitation in Plasma-Beam Discharges

Vibrational excitation of molecules requires sufficient specific power of plasma-beam discharges, which leads to the limitation (12.25) of the minimal level of Langmuir noise W/n_eT_e. Derive the criterion (12.25), and analyze the value of the minimal level of Langmuir noise W/n_eT_e.

12.7.10 Plasma-Beam Discharge Parameters in Molecular Gases

Analyze the dependence of the degrees of ionization, level of Langmuir noise W/n_eT_e, and electron energy in the plasma-beam discharge in hydrogen as a function of vibrational temperature; this is shown in Figure 12.4 for two different values of the parameter W/p. Explain why higher electron densities correspond to lower values of W/n_eT_e, and to a lower level of average electron energy in plasma-beam discharges.

12.7.11 Plasma-Beam Discharge in Crossed Electric and Magnetic Fields

In the plasma centrifuge based on the plasma-beam discharge, the radial electric field is provided directly by the electron beam. Derive a relation for the drift velocity in the crossed electric and magnetic fields in this system as a function of magnetic field and the electron beam current density. Use the derived relation to calculate the plasma rotation velocity for the plasma-beam discharge parameters given in Section 12.3.5.

12.7.12 Magnetized Ions Separation Effect in Plasma Centrifuge Based on Plasma-Beam Discharge

Analyze the ions displacement in the radial direction at the moment of their formation due to the combined effect of the electric and magnetic fields. Derive Eq. (12.2626), and explain why the velocity in this relation corresponds not to the thermal one, but rather to the velocity of the centrifugal drift in the crossed electric and magnetic fields.

12.7.13 Stable Regimes of Non-Self-Sustained Discharges With Ionization Provided by High Energy Electron Beams

The relatively high levels of the discharge specific power given by Eq. (12.32) can be stable only at limited values of the discharge durations. The stable values of current densities and pulse durations of the non-self-sustained discharges are shown in Figure 12.8. Using data presented on this figure, estimate the maximum values of the specific energy input for the pulse discharge sustained by a relativistic electron beam.

12.7.14 Uniformity of Non-Self-Sustained Discharges With Ionization Provided by High Energy Electron Beams

High external electric field can result in a decrease of electron concentration near the cathode and hence, leads to an elevated electric field in the layer. Using Eq. (12.38), calculate the maximum external electric field (at atmospheric pressure, electron beam current density 10^3 A/cm^2, and $d = 100$ cm), which does not lead to essential additional ionization in the cathode layer.

12.7.15 Plasma Radiolysis of Carbon Dioxide

The plasma effect in CO_2 radiolysis is related to stimulation of the reaction $O + CO_2 \rightarrow CO + O_2$ by vibrational excitation of CO_2-molecules, which then prevails over three-molecular recombination $O + O + M \rightarrow O_2 + M$. Determine the vibrational temperature of CO_2 molecules necessary for the plasma effect, taking into account that the necessary degrees of ionization should satisfy the requirement $n_e/n_0 >> 10^{-3}$.

12.7.16 Initial Tracks of Nuclear Fission Fragments

Derive Eq. (12.57) for the radius of the initial completely ionized plasma channel formed by the tracks of the nuclear fission fragments. Explain why the plasma in these channels is completely ionized. Using the derived equation calculate the radius of the initial tracks and compare it with the mean free path of electrons and the length necessary for ionization and electronic excitation provided by the δ-electrons.

12.7.17 Tracks and Plasma Effects in Radiolysis Provided by α Particles

Using Eqs.(12.66) and (12.58) estimate the pressure range optimal for plasma radiolysis in water vapor and carbon dioxide provided by α-particles with energies about 1 MeV. Compare the physical and chemical characteristics of the tracks made by α-particles and by nuclear fission fragments.

12.7.18 Relativistic Electron Beam in Aerosols

Analyze the system of Eqs. (12.67) – (12.69) describing the charging of aerosol particles in a plasma generated by a relativistic electron beam for the case when the predominant electron flux to the aerosols is balanced by secondary electron emission and the average charge of macroparticles is equal to zero. Derive the expressions for electron beam current density which are related to Eq. (12.72) and to plasma density Eq. (12.73). Take into account that expressions for the electron and ion fluxes $f_i(n_i, Z_a)$, $f_e(n_e, Z_a)$ can be taken in this case in a simple way without taking into account electrostatic effects related to aerosol particles.

13 Physics and Engineering of Discharges in Liquids

Plasma is obviously best known as a gas-phase phenomenon; it was a case in all previous 12 Chapters. As discussed in Chapter 1 of this book, plasma is ionized gas. However, at the same time in the past several decades, there have been many efforts to determine if plasma could be generated plasma in a liquid phase. In such cases, most researchers have observed plasma only in low-density phase (gas) bubbles and voids dispersed within fluids. While it looks like "liquid plasma", and surely it is plasma inside of liquid, in this case, there is no ionization of liquid phase itself. If such plasma systems could be generated, the resulting systems could have numerous applications, including plasma disinfection, sterilization, activation of liquids for washing, cleaning, plant growth stimulation, etc.

Recent advances in pulsed power technology permitted application of much faster voltage rise times (including the subnanosecond range) and revealed that plasma-like phenomena can, in fact, occur in the liquid phase quasi-homogeneously without any bubbles as a non-equilibrium plasma in the liquid phase. This can be already interpreted as the ionization of the liquid itself. Expected unique non-equilibrium properties of nanosecond plasma discharges in homogeneous high-density medium, such as high densities of electrons and excited species, light and high energy radiation, and high electron energies together with a low temperature of liquid may be associated with exclusive opportunities that may lead to fundamentally new effects and may have a great impact in the fields of medicine, microelectronics, energy systems, and materials. These interesting new plasmas in different liquids, including liquid nitrogen, are going to be discussed in the second part of this chapter.

13.1 PLASMA GENERATION INSIDE LIQUIDS

We begin with the consideration of plasma generation inside of liquids related to the preliminary formation of bubbles, voids, etc. It is obviously not direct ionization of liquids, but it is still a process of plasma generation inside of liquids.

13.1.1 General Features of Electrical Discharges in Liquids

Electric breakdown of liquids is limited by their high density, short mean free path of electrons, and therefore requires very high electric fields E/n_0 (see Paschen curves, Chapter 4, Section 4.4.2). Nevertheless, breakdown of liquids can be performed not at the extremely high electric fields required by Paschen curves but at those only slightly exceeding breakdown fields in atmospheric pressure molecular gases. This effect can be simply explained by different electrically induced mechanisms of formation of bubbles, macro- and micro-voids, and quasi-"cracks" inside the liquids. Inside those bubbles, voids, etc., plasma is actually formed in the gas phase, which obviously requires not so high electric fields.

Such discharge can be sustained, for example, in water by pulsed high voltage power supplies. These discharges in water usually start from sharp electrodes. If the discharge does not reach the second electrode it can be interpreted as pulsed corona, branches of such a discharge are referred to as streamers. If a streamer reaches the opposite electrode a spark is usually forming. If the current through the spark is high (above 1 kA), this spark is usually called a pulsed arc.

Various electrode geometries have been used for the plasma generation in water, in particular for water treatment and disinfection. Two simple geometries are a point-to-plane and a point-to-point. The former is often used for pulsed corona discharges, whereas the latter is often used for pulsed arc systems. Concern in the use of pulsed discharges is the limitation posed by the electrical conductivity of water. In the case of low electric conductivity of water (below 10 μS/cm), the range of the applied voltage that can produce a corona discharge without sparking is very narrow. In the case of high electric conductivity of water (above 400 μS/cm), streamers become short and the efficiency of radical production decreases, and denser and cooler plasma is generated.

In general, the production of OH radicals and O atoms is more efficient at water conductivity below 100 μS/cm. For the case of tap water, the bulk heating can be one of the problems in the use of corona discharges. At a frequency of 213 Hz, the temperature of the treated water rose from 20°C to 55°C in 20 min, indicating a significant power loss. Next we should discuss how to provide water breakdown at smaller values of electric field.

13.1.2 Major Conventional Mechanisms and Characteristics of Plasma Discharges in Water

Mechanisms of conventional plasma discharges and breakdowns in liquids (specifically in water) can be classified into two groups: the first group presents the breakdown in water as a sequence of a bubble generation process and a following electronic process and the second group divides the whole process into a partial discharge and a fully developed discharge such as an arc or spark.

In the first approach, the bubble generation process starts from a micro-bubble formed by the vaporization of liquid resulting from local heating in the strong electric field region at the tips of electrodes. The bubble grows, and an electrical breakdown takes place within the bubble. In this case, the cavitation mechanism can explain the slow bush-like streamers. The appearance of bright spots is delayed from the onset of the voltage, and the delay time tends to be greater for smaller voltages. The time lag to water breakdown increases with increasing pressure, supporting the bubble mechanism in a sub-microsecond discharge formation in water.

The time to form the bubbles is about 3–15 ns, depending on the electric field and pressure. The influence of the water electrical conductivity on this regime of the discharges is small. Bulk heating via ionic current does not contribute to the initiation of the breakdown. The power necessary to evaporate the water during the streamer propagation can be estimated using the streamer velocity, the size of the streamer, and the heat of vaporization. Using a streamer radius of 31.6 μm, a power of 2.17 kW was estimated to be released into a single streamer to ensure its propagation in the form of vapor channels. In the case of multiple streamers, the required power can be estimated by multiplying the number of visible streamers to the power calculated for a single streamer.

In the electronic phase of the plasma generation process, electron injection and drift in liquid phase take place at the cathode, while hole injection through a resonance tunneling mechanism occurs at the anode. In the electronic process, electric breakdown occurs when an electron makes a suitable required number of direct ionizing collisions during its transit across the breakdown gap. A detailed discussion of this phase of the plasma generation process can be found in publications of Prof. Young Cho, for example, Y. Yang, Y.I. Cho, A. Fridman, 2012.

The second approach to the mechanisms of the electrical discharges in water is divided into partial electrical discharges and arc and spark discharges, which is illustrated in Table 13.1. In the partial discharges, the current is mostly transferred by ions. For the case of high electrical conductivity water, a large discharge current flows, resulting in a shortening of the streamer length due to the faster compensation of the space charge electric fields on the head of the streamer. Subsequently, a higher power density in the channel is obtained, resulting in a higher plasma temperature, a higher UV radiation, and generation of acoustic waves.

TABLE 13.1
Pulsed Discharges in Water

Pulsed Corona in Water	Pulsed Arc in Water	Pulsed Spark in Water
Streamer-like channels. Streamer-channels do not propagate across the entire electrode gap, i.e., partial electrical discharge. Streamer length in the order of cm, channel width ~ 10–20 μm. Electric current is transferred mostly by ions. Non-thermal plasma. Weak to moderate UV generation. Relatively weak shock waves. Water treatment area is limited in a narrow region near the corona. Pulse energy: a few joules per pulse, often less than 1 J per pulse. Frequency is about 100–1,000 Hz. Relatively low current, i.e., peak current is less than 100 A. Electric field intensity at the tip of electrode is 100–10,000 kV/cm. A fast-rising voltage on the order of 1 nsec, but less than 100 ns.	Current is transferred by electrons. Quasi-thermal plasma. Arc generates strong shock waves within the cavitation zone. High current filamentous channel bridges the electrode gap. Gas in channel (bubble) is ionized. High UV emission and radical density. A smaller gap between two electrodes of ~5 mm is needed than that in the pulsed corona. Large discharges pulse energy (greater than 1 kJ per pulse). Large current (about 100 A), peak current greater than 1,000 A. Electric field intensity at the tip of electrode is 0.1–10 kV/cm. Voltage rise time is 1–10 μs. Pulse duration ~ 20 ms. Temperature exceeds 10,000 K.	Similar to pulsed arc, except for short pulse durations and low er temperature. Pulsed spark is faster than a pulsed arc, i.e., strong shock waves are produced. Plasma temperatures in spark is around a few thousand Kelvin.

In the arc or spark discharge plasmas, the electric current is usually transferred by plasma electrons. The high current heats a relatively small volume of plasma in the gap between the two electrodes, generating quasi-thermal plasma with relatively high temperature. When a high voltage-high current discharge takes place between two submerged electrodes, a large part of the energy is consumed on the formation of a very energetic thermal plasma channel. This channel emits UV radiation and its expansion against the surrounding water generates intense shock waves. In contrast to the corona discharge in water, the shock waves are weak or moderate, whereas for the pulsed arc or spark, the shock waves are strong. More details on this type of water breakdown can be learned also from Prof. Young Cho publications, for example, Y. Yang, Y.I. Cho, A. Fridman, 2012.

We next focus on the analysis of some kinetic aspects of water breakdown through water overheating instability.

13.1.3 Physical Kinetics of Water Breakdown

The critical breakdown condition for gas is described by the Paschen curve (see Section 4.4.2), from which one can calculate the breakdown voltage for air, for example. A value of 30 kV/cm is a well-accepted breakdown voltage of air at 1 atm. When one attempted to produce direct plasma discharges in water, it could be expected that a much greater breakdown voltage and electric field in the order of 30,000 kV/cm might be required due to the density difference between air and water.

A large body of experimental data on the breakdown voltage in water and water solutions shows, however, that without special precautions this voltage is of the same magnitude as for gases. This interesting and practically important effect can be explained by taking into account the fast formation of gas channels, bubbles, voids etc. in the body of water under the influence of the applied high voltage (see examples and discussion in the previous section). When formed, the gas channels give the space for the gas breakdown inside of the body of water. It explains why the voltage required for water breakdown is of the same magnitude as for gases.

The gas channels can be formed by the development and electric expansion of gas bubbles already existing in water as well as by additional formation of the vapor channel through fast local heating and evaporation. Focus on the second mechanism, which is usually referred to as the **thermal breakdown**. When a voltage pulse is applied to water, it induces a current and the redistribution of the electric field. Due to the dielectric nature of water, an electric double layer is formed near the electrode, which results in the localization of the applied electric field. This electric field can become sufficiently high for the formation of a narrow conductive channel, which is heated by electric current to temperatures of about 10,000 K. Thermal plasma generated in the channel is rapidly expanded and ejected from the narrow channel into water, forming a plasma bubble.

The energy required to form and sustain the plasma bubble is provided by Joule heating in the narrow conductive channel in water. The physical nature of thermal breakdown can be related to thermal instability of local leakage currents through water with respect to the Joule overheating. If the leakage current is slightly higher at one point, the Joule heating and hence temperature also grows. The temperature increase results in a significant growth of local conductivity and the leakage current. Exponential temperature growth to several thousand degrees at a local point leads to the formation of the narrow plasma channel in water, which determines the thermal breakdown.

The thermal breakdown is an interesting critical thermal-electric phenomenon taking place at the applied voltages exceeding a certain threshold value when heat release in the conductive channel cannot be compensated by heat transfer losses to the surroundings. The thermal condition of water is constant during the breakdown; water remains liquid away from the discharge with the thermal conductivity about 0.68 W/mK. When the Joule heating energy in the area between the two electrodes is larger than a threshold value, the instability can occur, resulting in the instant evaporation and a subsequent thermal breakdown. When the Joule heating is smaller than a threshold value, nothing happens but electrolysis and the breakdown never take place unless the release of gases during the electrolytic process results in the formation of a bubble sufficient for breakdown.

The thermal breakdown instability is characterized by the instability increment showing the frequency of its development:

$$\Omega = \left[\frac{\sigma_o E^2}{\rho C_p T_0}\right]\frac{E_a}{T_0} - D\frac{1}{R_o^2}. \tag{13.1}$$

Here, σ_0 is the water conductivity; E_a is the Arrhenius activation energy for the water conductivity; E is the electric field; ρC_p is the specific heat per unit volume; T_0 is the temperature; R_0 is the radius of the breakdown channel; $D \approx 1.5 \cdot 10^{-7}$ m²/s is the thermal diffusivity of water. When the increment Ω is greater than zero, the perturbed temperature exponentially increases with time, resulting in a thermal explosion. When Ω is less than zero, the perturbed temperature exponentially decreases with time, resulting in the steady-state condition. For the plasma discharge in water, the minimum breakdown voltage in the channel with length L can be estimated based on the Equation (13.1) as:

$$V \geq \sqrt{\frac{DC_p\rho T_o^2}{\sigma_o E_a}} \cdot \frac{L}{R_0}. \tag{13.2}$$

Obviously, the breakdown voltage increases with L/R_o. Assuming $L/R_0 = 1,000$, it is about 30 kV. More details on the thermal breakdown and electrolytic effects leading to the breakdown of water and water solutions can be found in publications of Prof. Young Cho, for example in the book Y. Yang, Y.I. Cho, A. Fridman, 2012.

13.2 GENERATION OF NON-EQUILIBRIUM NANOSECOND-PULSED PLASMA IN WATER WITHOUT BUBBLES

13.2.1 About Plasma in Liquids Without Bubbles

As was discussed in the previous section, most of plasmas inside of liquids were actually observed in gas bubbles, voids, etc., created in these liquids (see in particular book Y. Yang, Y.I. Cho, A. Fridman, 2012, and relevant references therein). However, recently the interesting phenomenon of direct ionization of liquid phase without bubbles has been observed. Modern pulsed power technologies permitted the application of much faster voltage rise times (including the sub-nanosecond range) and revealed that plasma-like phenomena can, in fact, occur in liquid phase quasi-homogeneously without any preliminary formation of bubbles, voids, etc., as a non-equilibrium plasma in the liquid phase. This phenomenon can be already be interpreted as the direct ionization of liquid itself.

Nanosecond plasma in homogeneous high-density medium should be characterized by high densities of electrons and excited species, high energy radiation, and high electron energies together with low temperature of liquids. Such "liquid plasma" properties may be associated with very interesting novel opportunities that may lead to fundamentally new effects and may have a great practical impact on the fields of medicine, microelectronics, energy systems materials, etc. These intriguing new plasmas in different liquids, starting with water but including other liquids including liquid nitrogen, will be discussed in this chapter.

13.2.2 Initial Observations of Sub-Micron Pulsed Plasma in Water Without Bubbles

One of the first observations of the non-thermal corona discharge plasma inside a liquid medium around electrodes with ultra-sharp tips or elongated nanoparticles has been presented by D. Staack, A. Fridman, A. Gutsol, ea., 2008. Experimental system applied in this work allowed simultaneous chemical analysis of multiple dissolved elements within nanoseconds. The proposed optical emission spectroscopy was applied in this case for ultrafast time-resolved multi-elemental analysis of femto-liter volumes of liquid with a one-micrometer spatial resolution. As a result, not only physical but also chemical analysis of the generated "nano-plasma" has been accomplished.

The key physical idea of the approach presented in this work is based on the fact that if an electrode's tip is extremely sharp, for example on the level of nanometers, even not very high applied voltages are able to provide extremely high electric fields (see Equation 9.4) sufficient even for breakdown of liquids. Photo of this nano-corona discharge plasma is presented in Figure 13.1. We

FIGURE 13.1 Negative cold corona discharge plasma inside water without a bubble around an electrodes with ultra sharp tip.

should remark that plasma shown in this figure is generated by negative corona discharge (see Chapter 9), which makes the observed phenomena a little less exciting, because of the possible significant contribution of field emission to the generation of electrons in the system.

It is interesting to point out that active plasma radius of the corona discharges is usually only couple of times greater than the radius of curvature of the electrode's tip (see Chapter 9, Section 1 and further, as well as Figure 13.1). It means that nano-corona's electrode tips should generate nano-size plasma. Keeping in mind that the ionization process requires the presence of atoms inside of plasma (which in this case is on nanometers scale), we should conclude that nano-plasma can be organized only in condensed phase (not in gases) where mean free path of electrons is also on the level of nanometers. Details on the organization and characterization of such nano-corona in water without bubbles can be found in D. Staack, A. Fridman, A. Gutsol, ea. , 2008.

13.2.3 First Observations of Macroscopic Pulsed Plasma in Water Without Bubbles

The work of A. Starikovskiy, Y. Yang, Y. I. Cho, A. Fridman, 2011, reports a fundamentally different type of discharge in water, where the macroscopic discharge was generated without bubble formation by application of high-voltage pulses with nanosecond duration. In this case, it is important that the discharge has been propagating from anode (not from cathode as was described in Section 13.2.2) avoiding effects of field emission on electron generation.

Applied voltage in this case is usually from 30 to 220 kV, duration of voltage pulses from less than 1 ns to 30 ns, voltage rise time from 0.2 ns to 1–2 ns Schematic of a set up used in these experiments (A. Starikovskiy, Y. Yang, Y. I. Cho, A. Fridman, 2011) as well as those to be discussed in the following section concerning the pulsed macro-plasmas in different liquids, is shown in Figure 13.2. The discharge cell had a point-to-plate geometry with the high-voltage electrode being either stainless steel needle with a diameter of 20μm, or 5μm iridium needle, and the low-voltage electrode diameter was 18 mm. The inter-electrode distance was varied in the range of 1.5–4 mm. The liquid layer between the center of the discharge gap and the quartz window was usually about 50 mm. As it was mentioned above discharge plasma has been propagating in this case usually from the anode and therefore can be interpreted as positive corona discharge plasma.

In the above-described experiments (A. Starikovskiy, Y. Yang, Y. I. Cho, A. Fridman, 2011), it was shown that the macroscopic cold strongly non-equilibrium discharge plasma is able to develop on a picosecond time scale with extremely high propagation velocities up to 5000 km s^{-1} (5 mm ns^{-1},

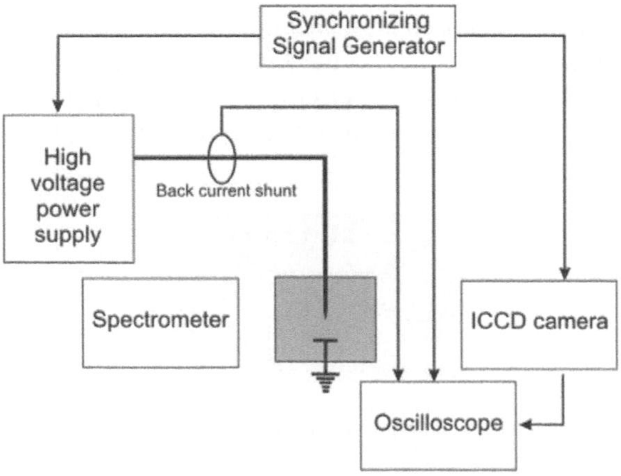

FIGURE 13.2 Schematic of a set up for generation of macroscopic nanosecond pulsed plasma (A. Starikovskiy, Y. Yang, Y.I. Cho, A. Fridman, 2011; D. Dobrynin, Y. Seepersad, M. Pekker ea., 2013)

which is about 15% of speed of light in vacuum). Such enormously high plasma propagation velocity was observed for a sub-nanosecond discharge (voltage 220 kV on the tip, 400 ps duration). Images and characteristics of the observed nanosecond macroscopic discharge plasma are similar to those relevant in the following Section 13.3 comparing such plasmas in different liquids.

The macroscopic discharge plasma observed in this case has a completely different nature from the discharge plasmas initiated by electrical pulses with a longer rise time described in particular in the previous sections. It is important that that the macroscopic plasma propagating from a positive electrode was generated in this case by direct streamer-like ionization of water without contribution of field emission from an electrode.

13.3 NON-EQUILIBRIUM NANOSECOND-PULSED PLASMA WITHOUT BUBBLES IN DIFFERENT LIQUIDS: COMPARISON OF WATER AND PDMS

Let's focus now on studying the macroscopic nanosecond-pulsed discharges in liquids without bubbles and with different polarization properties, using as examples water and silicon-based transformer oil. To compare these nano-second pulsed non-equilibrium liquid plasmas, analyze their dependence behavior on the global (applied) electric field, discharge emission spectrum, and shadow imaging of the discharge.

13.3.1 Analysis of the Effect of Polarizibility and Dielectric Properties of Liquids on the Nano-Second Pulsed Discharge Development: Images of No-Bubbles-Liquid-Plasma in Water versus Transformer Oil PDMS

As mentioned in Section 13.2.3, general schematic of the experimental setup applied in the work under discussion (D. Dobrynin, Y. Seepersad, M.Pekker ea., 2013) is similar to the schematic described in the section above (see Figure 13.2). To analyze the effect of polarizability and dielectric properties of liquid on the nano-second pulsed discharge development two types of liquids were used: distilled deionized water (maximum conductivity 1.0 μS cm^{-1}), and PDMS, which is polydimethylsiloxane (C2 H6 OSi)n transformer oil with dielectric constant $\varepsilon = 2.3$–2.8. Prior to analyzing these images, consider the discharge system employed.

The discharge initiation in liquid water was provided by nanosecond pulses with +16 kV pulse amplitude in 50 Ohm coaxial cable (32 kV on the high-voltage electrode tip due to pulse reflection), 10 ns pulse duration (90% amplitude), 0.3 ns rise time and 3 ns fall time. Pulse frequency was 5 Hz. A 15 m long coaxial cable with a calibrated back current shunt mounted in the middle of it was used for control of applied voltage and power measurements.

The discharge visualization measurements were performed using an ICCD camera with an 18mm diameter, multi- alkaline photocathode with a spectral response from 180 to 750 nm. The camera's typical field of view was 2 × 1.25 mm, and the spectral response was 250–750 nm taking into account UV-emission absorption in the water layer. The generator has synchro-output and two adjustable channels with a signal rise/fall time of less than 2.5 ns and a typical jitter (RMS) less than 20 ps with a delay time resolution of 10 ps.

Now a few words on applied optical diagnostics. The optical system employed allowed recording strong density perturbations (if any) of a size larger than ~1 μm. In order to ensure the absence of large-scale perturbations (bubbles or voids) in the system (near the high voltage electrode or in the bulk volume) shadow imaging was performed using a laser diode operating at 650 nm wavelength with the power of ~5 mW. Discharge optical emission spectrum was obtained using a fiber-optic bundle connected to the spectrometer.

Now consider the discharge nanosecond pulsed plasma imaging in water versus PDMA. Discharge images with an exposure time of 5ns obtained for different discharge gaps and electrodes for the cases of water and silicon transformer oil (PDMS) are presented in Table 13.2 (D. Dobrynin, Y. Seepersad, M.Pekker ea., 2013). The 20mm-needle image for PDMS shows in this table, the

TABLE 13.2
Discharge Images with Exposure Time of 50 ns Obtained for Different Discharge Gaps and Electrodes for the Cases of Water and Silicon Transformer Oil (PDMS).

	Silicon oil (PDMS, ε ~ 2)	Distilled deionized water (ε ~ 80)
Electrode radius 20 μm Discharge gap 1.8 mm	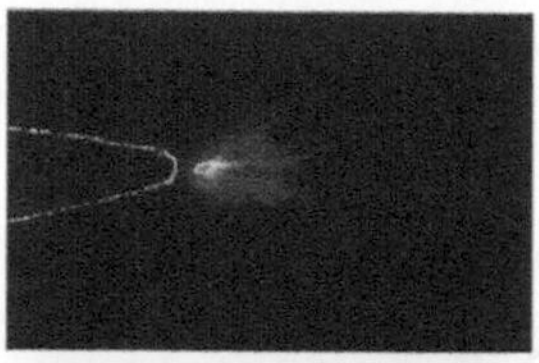	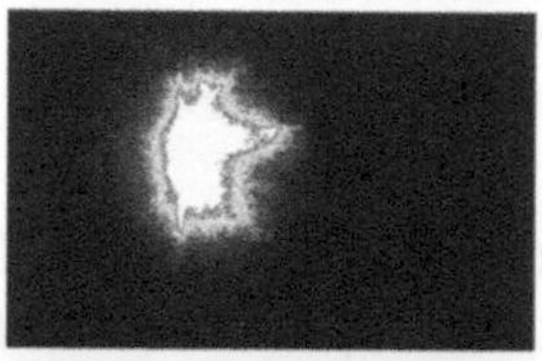
Electrode radius 5 μm Discharge gap 1.8 mm		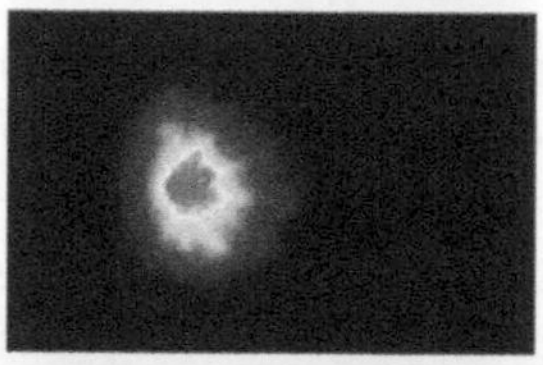
Electrode radius 5 μm Discharge gap 3 mm	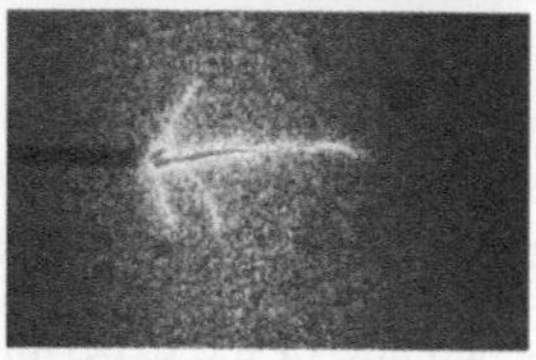	

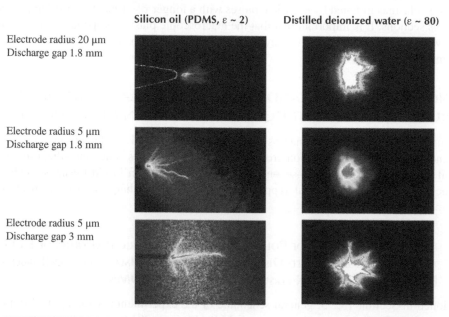

artificially added electrode contour; the 5 μm-needle images for the PDMS case are shadow images, image size 2 mm × 1.25 mm.The size of the excited region near the tip of the high-voltage electrode was ~1mm. The discharge had a complex multi-channel structure and this structure changed from pulse to pulse. One may note that luminescence intensity and structure of the discharge are significantly different for the cases of water and PDMS: discharge in water is always more intense and has more 'uniform' structure. At the same time, the discharge size does not appear to depend on the high-voltage electrode radius or inter-electrode distance, i.e. global (applied) electric field. This actually illustrates the first image comparison of the liquid plasma in water and transformer oil.

13.3.2 Nano-Second Pulsed Discharges in Liquids without Bubbles: Spatial Evolution Structure in Water versus Transformer Oil PDMS

To analyze the spatial evolution structure of the nanosecond pulsed discharge a shorter camera gate (2 ns) has been used without signal accumulation (D. Dobrynin, Y. Seepersad, M.Pekker ea., 2013). Relevant nano-scale plasma structure evolution found as a result is shown in Figures 13.1 and 13.2 for water and PDMS respectively. In the case of water, when the voltage reaches maximum, the discharge propagation stops and a 'dark phase' appears (t = 3–9 ns in Figure 13.3). During this phase, the discharge cannot propagate because of space charge formation and decrease in the electric field.

Interesting and important to notice that the presence of the "dark phase" proves that the liquid plasma observed in this case is cold. Indeed, if the observed plasma is thermal it could not be cooled down for a couple of nano-seconds and therefore the radiation could not disappear during the "dark phase". The low temperature of the nanosecond pulsed plasma in the liquid phase is very important for multiple applications in chemistry, biology, and microelectronics.

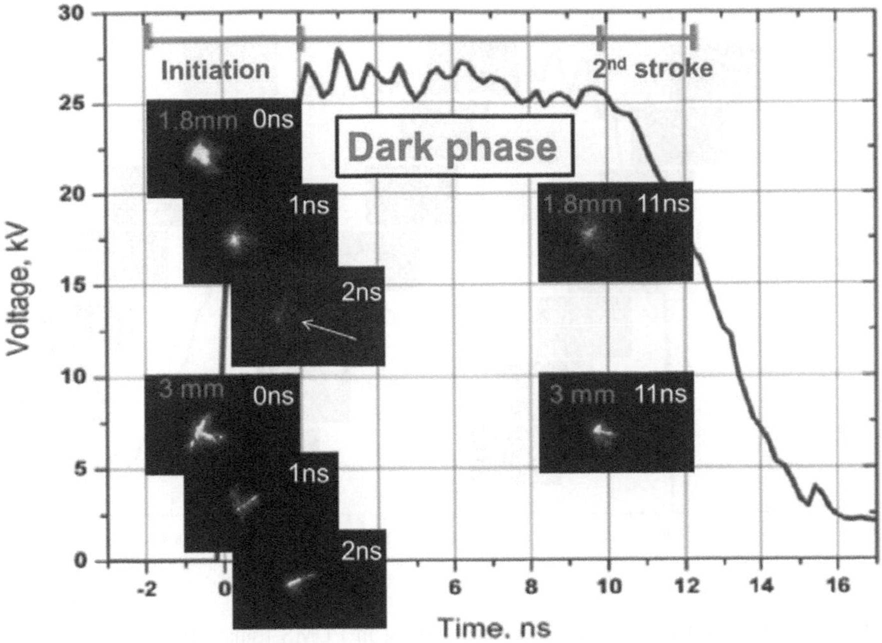

FIGURE 13.3 Nanosecond pulsed discharge development in water without bubbles (D. Dobrynin, Y. Seepersad, M. Pekker, et al., 2013).

Voltage decrease in related to that "dark phase" effect leads to second stroke formation and the second emission phase (Figure 13.3, 11–13 ns). This means that the channels lose conductivity and the trailing edge of the nanosecond pulse generates significant electric field and excitation of the media, which is comparable to the excitation corresponding to the leading edge of the pulse. A more detailed study on the dark phase in the case of discharge initiated in water may be found in the paper of A. Starikovskiy, Y. Yang, Y.I. Cho, A. Fridman, 2011.

Compare now the liquid plasma evolution in water and transformer oil. When the discharge is ignited in silicone oil (PDMS, Figure 13.4), one can notice that the 'dark' phase in this case is much less pronounced. This is related to the effect of lowering the electric field in the discharge channels in PDMS compared with the discharge in water. Such a decrease in the electric field can be interpreted in the following way.

If we assume the nano-second pulsed plasma development in the so-called sub-micro pores and cracks (see in particular, D. Dobrynin, Y. Seepersad, M.Pekker ea., 2013), these pores or cracks in liquid are transversal. Because for the transversal pores div $\vec{D} = 0$ and $\varepsilon_{H2O}E_{H2O} = \varepsilon_{pore}E_{pore}$, we can conclude that $E_{H2O} \sim 40 \times E_{PDMS}$ since $\varepsilon_{H2O} \sim 80$ and $\varepsilon_{PDMS} \sim 2$. In other words, local electric fields in water in the systems under consideration are significantly stronger than those electric fields in PDMS. Therefore, in water the discharge has 'thicker' channels (compare with Figure 13.3), and more pronounced 'dark' phase—space charge accumulates faster and the corresponding screening effect is stronger. The possible effects of the "micro-pore" and "micro-crack" formation and development are going to be additionally discussed later in the chapter.

13.3.3 The Nano-Second Pulsed Discharges in Liquid, Are They Really Generated Without Bubbles? Shadow Imaging

To answer this question, formation or presence of large-scale ~1 μm) bubble or other non-uniformities in the liquid phase right before and after the discharge has been analyzed in D. Dobrynin,

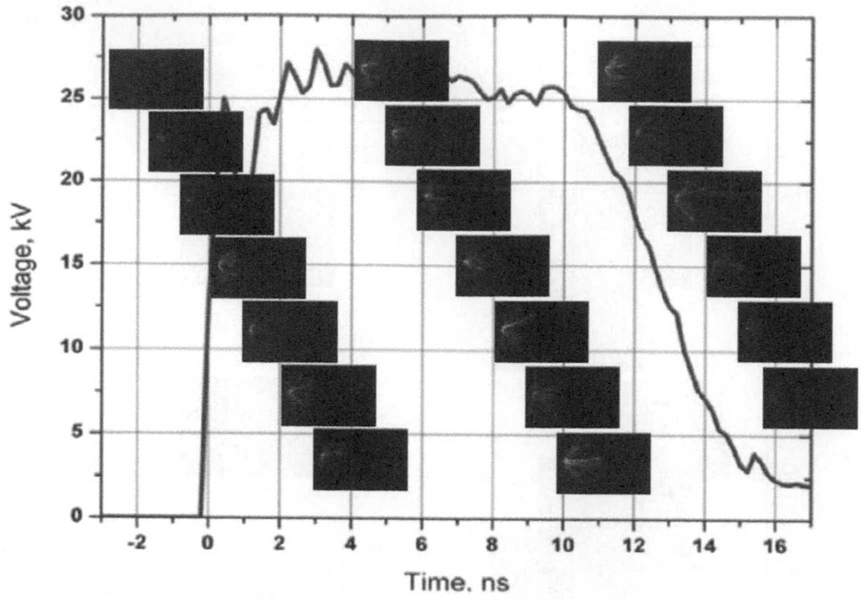

FIGURE 13.4 Nanosecond pulsed discharge development in PDMS transformer oil without bubbles (D. Dobrynin, Y. Seepersad, M. Pekker, et al., 2013).

Y. Seepersad, M. Pekker ea., 2013 using a shadow imaging technique. The relevant images for the case of PDMS oil are shown in Table 13.3. One may note in Table 13.3 that the nanosecond plasma characteristic size is again ~1 mm (image size is 2 mm × 1.25 mm), but no obvious density irregularities are observed right before and after the discharge. Hence it appears no bubbles are related to this nano-second plasma.The discharge energy deposition can be estimated in this case by comparison of the first incident and reflected voltage pulses (back current shunt measurements), and permits estimation of the discharge energy input of ~5 mJ. It should result in only a few degrees (~2 K for water and ~5 K for PDMS) increase in temperature. Thus, the liquid overheat in the discharge channel is not sufficient for the generation of a shock- wave to create a void.

It is interesting to note that if the discharge imaging continues for a longer time, the second and other reflected pulses eventually should result in void (bubble) formation due to electrostatic or energy dissipation effects. In D. Dobrynin, Y. Seepersad, M.Pekker ea., 2013, time delay between the incident and the second reflected pulse was ~110 ns. Table 13.4 shows that indeed, after the second and subsequent pulses, there is a formation of a tree-like gas void which then results in the formation of a bubble. The bubble leaves the electrode and again no density perturbation can be observed before the next pulse (discharge frequency in this case was 1 Hz).

13.3.4 Characterization of the Nano-Second Pulsed Discharges in Liquid: Optical Emission Spectroscopy

The nanosecond pulsed discharge spectrum analyzed in D. Dobrynin, Y. Seepersad, M. Pekker ea., 2013, shows a strong broadening of Balmer lines with almost continuum emission in the region 300–900 nm and weak broadened OI lines. Estimation of the discharge plasma parameters, namely, electron density and electron temperature, becomes possible in this case by analyses of the Hα and I (777 nm) profiles. The best fit was obtained in this case as a sum of two Lorentzian functions. Since Stark and Van Der Waals broadening can be treated independently and both have Lorentzian profile,

TABLE 13.3
Shadow Imaging of Nanosecond Pulsed Discharge Plasma without Bubbles in PDMS oil: Shadow Imaging

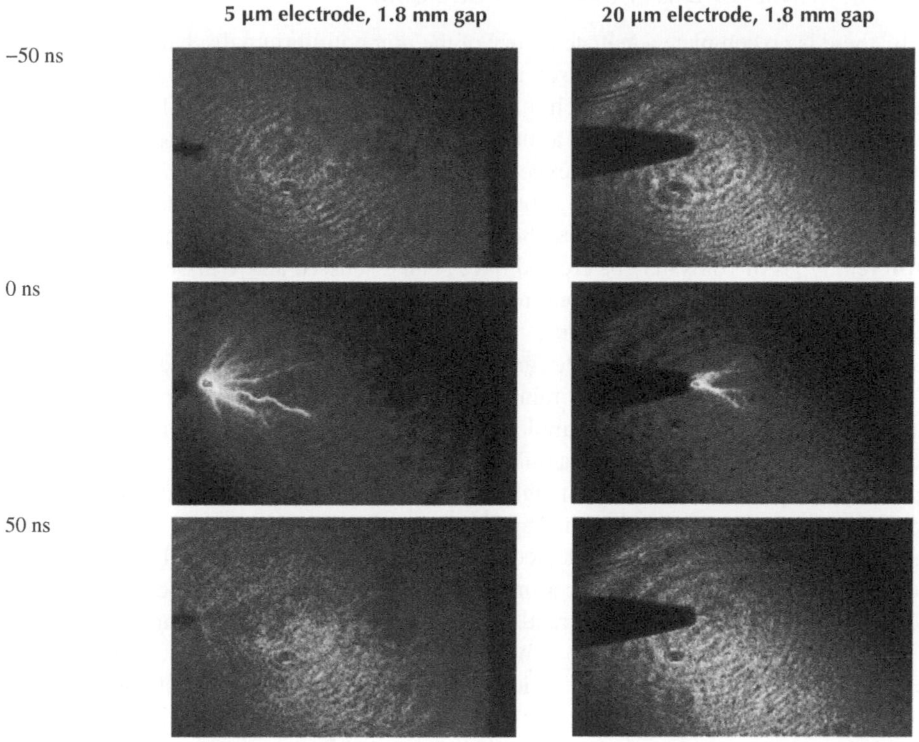
5 μm electrode, 1.8 mm gap
20 μm electrode, 1.8 mm gap
−50 ns
0 ns
50 ns

TABLE 13.4
Nanosecond Discharge in PDMS: Shadow Imaging of Bubble Formation on Longer time Scale (Reflected Pulses).

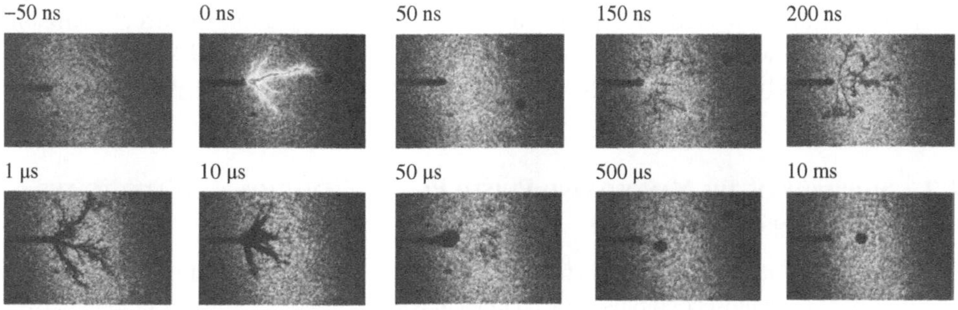
−50 ns
0 ns
50 ns
150 ns
200 ns
1 μs
10 μs
50 μs
500 μs
10 ms

one may suggest that complex Hα and O I profiles are primarily due to the combined contribution of these broadening mechanisms.

As a result of the analysis of optical emission spectra, the electron density and electron temperature in the nano-second liquid plasma can be estimated as $\sim 1.5 \times 10^{17}$ cm^{-3} and ~3 eV, respectively. These plasma parameters give an estimated value of a Debye radius of ~50 nm and of Coulomb logarithm of ~6.

13.4 SIMPLE THEORETICAL INTERPRETATIONS OF THE NON-EQUILIBRIUM NANOSECOND-PULSED PLASMA IN LIQUIDS WITHOUT BUBBLES

13.4.1 Challenges of Direct Ionization of Liquids

Currently, there is no complete adequate theory of the non-equilibrium discharge initiation in dense media—liquid without bubbles. The reason is that the liquid density is ~1000 times larger than that of the gas phase, and this leads to a much lower mean free path of electrons. Thus for effective formation of a streamer in a liquid phase, a much higher electric field is required (see earlier in this chapter) with the required electric field expected to be about ~30 MV cm^{-1}.

A possible resolution to this problem is believed to be related to the creation of local low-density regions—bubbles, voids, cracks, etc. (see Section 3.1 of this chapter). The mechanisms by which these different types of voids may be created have been discussed above in Section 13.1 and can be summarized as (a) vaporization (Joule heating), (b) molecular decomposition, (c) mechanical movements, and (d) pre-existing micro-bubbles.

In all these mechanisms, micro-bubbles grow with the speed equal to the speed of sound in liquid, which is about ~1 km s^{-1} . Keeping in mind that plasmas described in the previous section were generated after about 1 ns, the micro-bubbles would be able to grow to maximum 1 micron, while the size of the generated plasma was on a millimeters scale. The observed speed of the liquid plasma propagation was up to 0.15 of speed of light in a vacuum (see Section 13.2.3), which obviously significantly exceeds the above-mentioned speed of sound in liquid.

This raises another important complicated and challenging problem, namely the requirement for the electric breakdown conditions inside a small gas bubble or gas void inside of liquid. Considering that the bubble, if it already exists or is created by Joule heating or other mechanism, should be filled with a gas (vapor) at a pressure close to atmospheric pressure. In such a case, it is easy to estimate the electric field that is required for breakdown and formation of an initial streamer from Meek's criterion: $\alpha d > 20$, where d is a characteristic bubble size, $\alpha = pA\exp(B/(E/p))$ is the Townsend coefficient, p is the vapor (gas) pressure inside the bubble, A and B are parameters for water vapor, and E is the applied electric field (see sections 4.4.2 and 4.4.7).

From these equations, one may calculate that the minimal size of a bubble that has to be present inside of liquid for the formation of an initial streamer inside it for the conditions of the experiments described above in Sections 3.2 and 3.3 (applied electric field ~1–3 MV cm^{-1}) is on the order of 20 μm. There were no bubbles of this size in the experiments described in Sections 3.2 and 3.3. Concluding discussion of the challenges, it's clear that interesting theories and models are expected to explain the intriguing phenomenon of direct ionization of liquids. Some major ideas (D. Dobrynin, Y. Seepersad, M. Pekker ea., 2013), which can clarify the non-equilibrium liquid plasma phenomenon, are shortly described in the sections below.

13.4.2 Similarity of the Nanosecond Pulsed Plasma Evolution in Water to the Evolution of Long Sparks

Analyzing images of the liquid plasma (for example those presented in Table 13.2), one can see a clear similarity of the nanosecond pulsed discharges in water with the discharges in long gaps—long sparks, discussed in Section 4.4.9. We are surprised that electric fields lower than conventional ones required by Townsend formulas and by Paschen curves are sufficient for ionization of liquids without bubble, but we are not surprised that long sparks and especially lightning can occur at electric fields way below those required by Townsend formulas and by Paschen curves. Understanding of the long gap breakdown and long sparks (see Section 4.4.9) and noting the similarity of natural lightning between clouds and ground to that of direct liquid ionization without babbles can be a key for the understanding of the nanosecond pulsed plasma in liquids.

It is known that breakdown mechanism based on the multiplication of avalanches (Townsend mechanism) is predominant at low pd < 200 Torr cm, while at pd > 4000 Torr cm, a streamer formation is considered as breakdown mechanism (see Section 4.4). The latter mechanism works only if the gap is not very long, the degree of electric field non-uniformity is not very high, and the attachment is absent. In long gaps, although the average electric field is low, the discharge develops by a leader mechanism—in the vicinity of point electrode, where the electric field is the highest and conditions for the formation of a streamer are fulfilled.

Once a plasma channel is formed (because the local electric field is sufficient for an avalanche-streamer transition), it grows due to the high electric field of its charged tip. Similarly to the leader mechanism (see Section 4.4.9), the channel grows as long as leader propagation is sustained by the constantly rising applied electric field. A significant portion of the applied voltage drops on the plasma channel with finite conductivity. As a result, the field at the streamer head is not sufficient to maintain its propagation (the 'dark phase' in Figures 13.3 and 13.4).

13.4.3 Requirements for a Streamer Formation in Liquid without Bubbles

Let us now analyze and estimate what is required to form a streamer in the liquid phase without any voids or bubbles present. Following the classical theory of streamer formation (see Chapter 4), this can happen when the avalanche head electric field reaches the value of the applied external electric field. This electric field can be achieved in the avalanche head when the electron number of about Nemin ~2×10^8 is produced in a 10 μm radius volume (see Chapter 4), which is in a good agreement with the experimentally observed electron density of ~2×10^{17} cm^{-3} (see Section 13.3).

The minimum number of electrons that has to be generated for avalanche-to-streamer transition corresponds to the Meek's criterion (αd = 20) doesn't matter in liquid or in the gas phase, which is discussed in details in Section 4.4.7:

$$\alpha\left(\frac{E_0}{p}\right) * d = \ln\frac{4\pi\varepsilon_0 E_0}{e\alpha^2} \approx 20, \quad N_e = \exp(\alpha d) \approx 3\cdot 10^8. \tag{13.3}$$

Keeping in mind that the characteristic size in this case is about $d = 10$ μm, for the direct breakdown, α^{-1} has to be approximately 0.5 μm, which is unrealistically high for the conditions of highly condensed liquid media without any bubbles or voids present. This means that some additional source of initial plasma electrons created inside of liquid needs to be taken into account to explain the direct liquid ionization. It can be related to non-uniformity of density, non-homogeneity of liquid structure, specific transfer processes, etc.

One of the possible mechanisms of generation of the initial electrons required to start the streamer propagation related to pre-breakdown liquid structuring is discussed below following the publication of D. Dobrynin, Y. Seepersad, M. Pekker ea., 2013.

13.4.4 Modified Meek's Criterion for Streamer Formation and Breakdown of Liquids without Bubbles

To describe the above-mentioned additional source of initial plasma electrons required to start streamers, it is convenient following D. Dobrynin, Y. Seepersad, M. Pekker ea., 2013, to introduce an additional 'ionization coefficient' μ that would represent, similarly to the Townsend coefficient α, effective electron multiplication inside the pre-breakdown low-density regions in liquid to be ionized. In this case, we may rewrite the Meek's criterion (13.4) for direct ionization of liquids in the form of the so-called *modified Meek's criterion*:

$$\left[\alpha\left(\frac{E_0}{p}\right)+\mu\right]*d \approx 20, \quad N_e = \exp\big((\alpha+\mu)d\big) \approx 3\cdot 10^8. \tag{13.4}$$

To clarify the modified Meek's criterion, imagine that they highly inhomogeneous electric field in the vicinity of the high-voltage electrode creates a number of nano-sized pores (ellipsoids elongated along the electric field lines) due to the so-called ponderomotive electrostriction effect (see, for example, M.N. Shneider, M. Pekker, A. Fridman, 2012, and following publications of M.N. Sneider and his group). This electrostriction effect stimulates the violation of the continuity of the liquid in the vicinity of the needle electrode due to the effective negative pressure, which is a result of the electrostrictive ponderomotive force pushing dielectric fluid to the regions with a higher electric field. The electrostrictive effect is caused by rearrangement of the orientation of the elementary dipoles and occurs over time much shorter than the characteristic time of the hydrodynamic processes.

This effect can result in a possibility of the formation of nanosecond discharge plasma in the liquids on time scales much shorter than the formation time for bubbles near the electrode. The liquid in the vicinity of the electrode ($r < R$, where r is the electrode radius and R is the radius of the region in which nanopores are created) is highly dispersed as a result of electrostriction. Thus the liquid in the electrode's vicinity can be saturated by nanopores, and for the conditions of the experiments described above, it can be shown that $R \sim 1.9\ r_0$, where r_0 is the radius of the electrode. For the case of a 5 μm radius electrode, we have a nanopore saturated region radius about $R \sim 10$ μm.

Estimated pore size is about ~30 nm (M.N. Shneider, M. Pekker, A. Fridman, 2012), and the corresponding energy that electron gains in the pore can be expected about ~100 eV. This energy is sufficient to cause ionization of liquid (ionization potential of water is ~12 eV) at the boundaries of nanopores and to create several additional electrons. Effective 'ionization coefficient' can be estimated in this case as $\mu \sim 2 \times 10^4\ \text{cm}^{-1}$, which corresponds to a distance between nanopores of about ~500 nm (D. Dobrynin, Y. Seepersad, M. Pekker, ea., 2013). Thus, about 20 nanopores per 10 μm distance would be sufficient for the formation of an initial streamer.

13.4.5 Possible Direct Liquid Ionization Mechanism

Summarizing the above analysis of the liquid ionization mechanism, the scenario for pulsed nanosecond breakdown in liquid dielectrics may be proposed as the following sequence. In the beginning of the ionization process, a strong inhomogeneous electric field at the needle electrode creates a region saturated by nanopores through electrostriction. In these pores, primary electrons are accelerated by electric field to the energies exceeding the potential of ionization of water molecules. This causes the formation of a primary streamer in the liquid phase if the modified Meek's criterion (13.4) is satisfied.

After neutralization of electrons at the electrode, positive ions in the liquid phase form a virtual "needle electrode", and electrostrictive conditions for the appearance of the next set of cavities and density non-uniformities are fulfilled. This propagation of streamer continues until the drop in voltage on the moving streamer will not be an order of initial voltage on the needle electrode. In the process of the propagation, the head of the streamer narrows, so that electrostrictive conditions for breakdown are reproduced in the vicinity of the streamer head.

It has to be noted that this breakdown mechanism of liquid dielectrics is valid only when the rise time of the electrode potential τV is much shorter than the time of equalizing the pressure in the region of formation of pores $\tau V \ll R/CS$, where CS is the speed of sound in a liquid. This is usually valid for the nanosecond pulsed discharges. More details regarding this mechanism of ionization of liquids without bubbles can be found in M.N. Shneider, M. Pekker, A. Fridman, 2012, and the publications of M.N. Sneider and his group.

13.5 CRYOGENIC LIQUID PLASMA: NANOSECOND-PULSED DISCHARGE IN LIQUID NITROGEN

13.5.1 General Aspects Regarding Nanosecond Pulsed Discharges in Cryogenic Liquids

Compared to water and other dielectric liquids discussed in previous sections of Chapter 13, very few studies are available for the discharge in cryogenic liquids, including liquid nitrogen. The most detailed characterization of the nanosecond-pulsed discharge in liquid nitrogen has been probably performed by D. Dobrynin, R. Rakhmanov, A. Fridman, 2019a. This work was specifically focused on the liquid plasma imaging and estimation of temperature from spectroscopic measurements.

The cryogenic conditions in the nanosecond pulsed plasma in liquid nitrogen permit observation of some unique synthetic chemical processes. Very low temperatures in the liquid nitrogen plasma help to stabilize products of the plasma-chemical synthesis. In particular, D. Dobrynin, R. Rakhmanov, A. Fridman, 2019a, observed generation in the liquid nitrogen of unstable 'energetic' material, which may be identified as a form of polynitrogen compound. These results will be discussed shortly at the end of this section.

Synthesis of larger polymeric nitrogen compounds, which are expected to be highly energetic, has for the most part been successful at extreme conditions of high pressures (few to tens of GPa) but shown to be unstable upon pressure release. Some other polynitrogen compounds, like N^-_3, N_4, N^+_5 and N^-_5 have been shown to be stable at ambient conditions, however, in the form of salts and metal compounds. Unfortunately, until now polynitrogen materials produced in extremely high-pressure environment are not recoverable to ambient conditions and this prevents their practical applications, which clearly stimulate interest to the research of nanosecond pulsed discharges in liquid nitrogen.

13.5.2 Nanosecond Pulsed Discharges in Cryogenic Liquids: Plasma Imaging

To generate a discharge in liquid nitrogen (D. Dobrynin, R. Rakhmanov, A. Fridman, 2019a), a sharp (75 μm radius of curvature) steel electrode was placed in liquid nitrogen contained in a 450 ml double-walled glass flask (see Figure 13.5). The flask was fixed inside of an evacuated metal chamber to decrease the liquid nitrogen evaporation rate and to screen electromagnetic noise. The vacuum flask was also closed from the top by a plastic lid with a 3 mm diameter venting hole to minimize liquid nitrogen contamination by the surrounding air, especially oxygen.

High voltage electrode was powered by nanosecond-pulsed power supply capable of providing positive high voltage pulses with a maximum pulse amplitude of 120 kV, rise time at 10%–90% amplitude less than 1 ns, pulse duration at 90% amplitude 8–10 ns. The discharge energy was estimated to be approximately 100–130 mJ per pulse. Consider now the nanosecond pulsed discharge plasma images presented in D. Dobrynin, R. Rakhmanov, A. Fridman, 2019a.

Figure 13.6 shows long (5 ns) exposure images of the discharge in liquid nitrogen taken with 5 ns time step. The discharge is visible using an ICCD camera for about 30 ns, and high voltage pulse reflections were absorbed by the power supply. Typical discharge size was on the order of a few mm and appears to be significantly larger than that one observed in the case of lower voltage (~30 kV) pulses. From these images, streamer propagation velocity in liquid nitrogen can be estimated to be extremely high, on the level of at least 700–800 km/s.

Similarly to the nanosecond pulsed discharges in water and PDMS (see Section 13.3.3), the shadow imaging also proved the absence of gaseous bubbles in the experiments with liquid nitrogen. Only after about 15 ns, when the main plasma event starts to decay, visible gas voids start to appear at the location of the streamers.

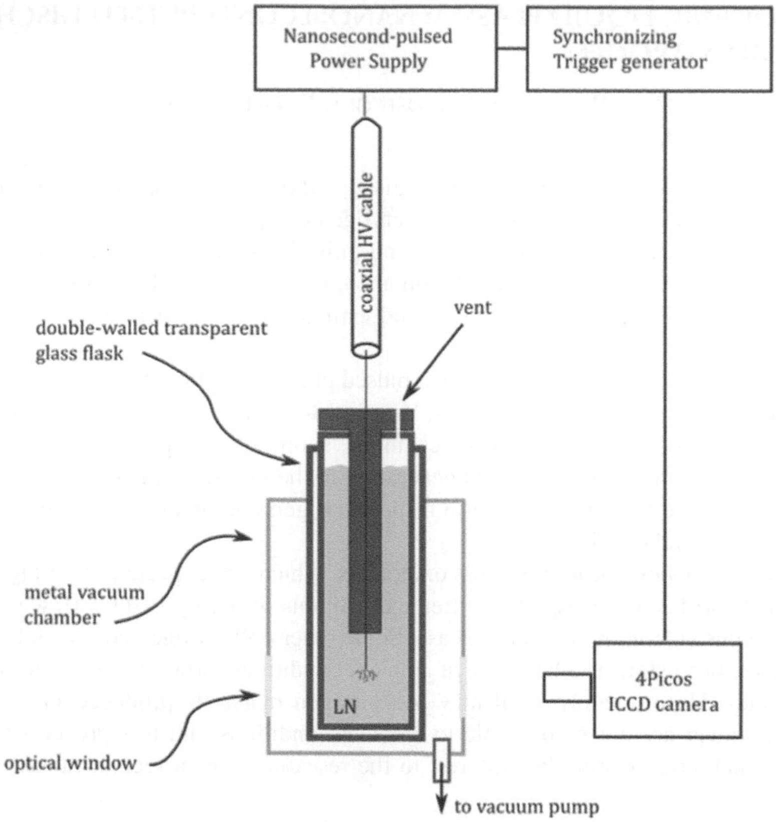

FIGURE 13.5 System schematic for generation of the nanosecond-pulsed discharge in liquid nitrogen.

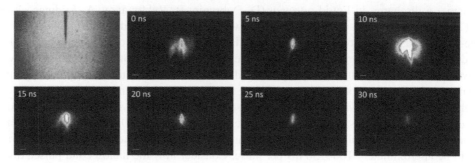

FIGURE 13.6 The high voltage needle and long exposure ICCD images of the nanosecond-pulsed discharge in liquid nitrogen. Exposure time is 5 ns and time delay between the images is 5 ns. The time stamp corresponds to the arrival of the high voltage pulse to the discharge gap. White bar is 1 mm (D. Dobrynin, R. Rakhmanov, A. Fridman, 2019a).

13.5.3 Optical Emission Characterization of the Nanosecond-Pulsed Discharge in Liquid Nitrogen without Bubbles

Emission of the nanosecond pulsed discharge in the 300–415 nm range was recorded with either 100 ns exposure time single accumulation or 3 ns exposure time and 50 accumulations (D. Dobrynin, R. Rakhmanov, A. Fridman, 2019a). It is interesting that the emission spectrum in this range

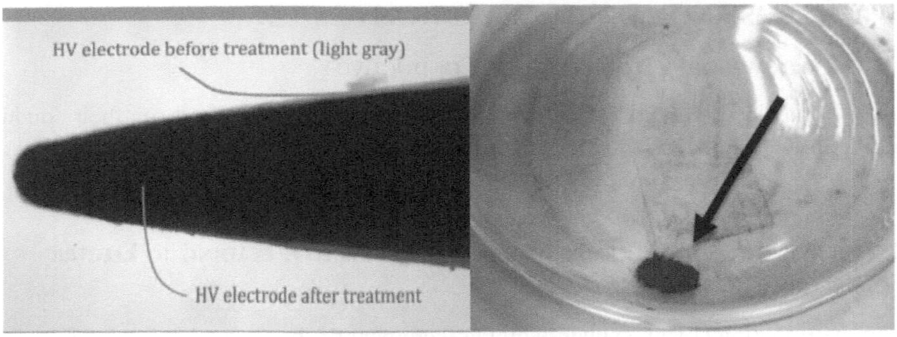

FIGURE 13.7 High voltage electrode before and after 1 h treatment of liquid nitrogen at 60 Hz pulse repetition frequency (left); and black powder material, supposedly polymeric nitrogen (to be exploded in about 1 s), produced as the results of 1 h treatment and after evaporation of excess liquid nitrogen (right).

(300–415 nm) is mostly from molecular nitrogen emission lines, and specifically from the second positive system (SPS).

Using the rotational-vibrational emission spectrum of the SPS transition at approximately 337 nm and assuming equilibrium of the rotational temperature T_r (C) of the C state and T_r(X) of the ground state of nitrogen, the temperature of the discharge has been estimated about ~110 K. The maximum discharge temperature obtained from the time-resolved spectra is approximately 140 K, which is about 60 K above the liquid nitrogen temperature. These plasma characterization measurements prove the cryogenic nature of the nanosecond-pulsed discharge in liquid nitrogen.

13.5.4 Polymeric Nitrogen Production in the Nanosecond-Pulsed Discharge in Liquid N_2

When nanosecond-pulsed discharge was used for the treatment of the liquid nitrogen, interesting powder-like energetic material was generated (D. Dobrynin, R. Rakhmanov, A. Fridman, 2019a). Treatment duration in this case is usually 30–60 min at a pulse repetition frequency of 60 Hz. It should be mentioned that after 60 minutes of treatment, no erosion of the high voltage electrode was registered. Liquid nitrogen after 1 hour of treatment (about 50 ml) was heated in room air and in He atmosphere to increase the concentration of the produced material. With evaporation of the liquid nitrogen the liquid darkened leaving a black powder-like material (see Figure 13.7), supposedly polymeric nitrogen. The powder then exploded within a second with the generation of both light and sound.

No residue was left after the material decomposition. The observed material appears to be very similar to the polymeric nitrogen produced under the extremely high-pressure conditions (see Section 13.5.1). The Raman spectrum of the liquid nitrogen changes after treatment but it cannot be reliably interpreted as polymeric nitrogen. Neither characteristic lines from azide groups (1360 cm−1), nor any Raman peaks associated with ozone were registered, which further supports the hypothesis on liquid nitrogen-based plasma production of energetic nitrogen-rich material.

The produced nitrogen-rich material can be preliminarily identified as a form of energetic non-molecular nitrogen compound (like polymeric nitrogen) due to the following reasons: (a) no electrode erosion was observed while the amount of produced material was significant; (b) material decomposition/explosion, accompanied by light and sound wave generation, was triggered by heating; (c) no residue was left after the decomposition/explosion; and (d) the absence of NO/NO_2 species, after explosion indicates low-temperature decomposition/explosion mechanism. More details on the subject can be found in D. Dobrynin, R. Rakhmanov, A. Fridman, 2019a, and following publications of D. Dobrynin with co-authors.

13.6 PROBLEMS AND CONCEPT QUESTIONS

13.6.1 Breakdown of Water, Effect of Conductivity

Analyze, how the electric breakdown of water and water solutions depends on their conductivity? Consider the influence of applied voltage rise-time on the conductivity effect for the breakdown of liquids.

13.6.2 Increment of the Thermal Breakdown Instability, Leading to Electric Breakdown of Water

Analyzing the Equation (13.1), explain why the increment of the thermal breakdown instability in water is growing with an increase of activation energy related temperature dependence of electric conductivity of water.

13.6.3 Minimum Water Breakdown Voltage

According to the Equation (13.2), the minimum water breakdown voltage is proportional to the channel's length, and therefore average characteristic electric field in the discharge channel is fixed. Estimate the typical value of this breakdown electric field, and interpret its dependence on the channel radius and water conductivity.

13.6.4 Generation of Nano-Plasma by Nano-Corona

Using equation 9.4, estimate the voltage required to achieve an electric field of about 1 MV/cm if the radius of curvature of an electrode tip is about 20 nm and distance to the second electrode plate is about 0.1 micron. Discuss the contribution of field emission to the generation of electrons in this case.

13.6.5 Generation of Nano-Plasmas in Liquids and Gases

Analyze the possibility of generation of nano-plasma in gases where the mean free path is significantly greater than the radius of curvature of an electrode tip required for relevant corona discharge.

13.6.6 Comparison of Negative and Positive Pulsed Corona Discharges in Liquids Without Bubbles

Why does the generation of cold corona-like plasma in liquid without bubbles from the anode looks more challenging and corresponds to different physics in comparison with plasma in liquid without bubbles generated from a sharp-tip cathode?

13.6.7 The Dark Phase Effect During Evolution of the Nano-Second Pulsed Discharge in Liquids

Give your interpretation of the "dark phase" effect (see Figure 13.4), and explain why the dark phase effect during the evolution of the nanosecond pulsed discharge plasmas in liquids proves the non-thermal nature of these plasmas?

13.6.8 Modified Meek's Criterion for Breakdown of Liquids

Why a preliminary generation of plasma electrons with sufficient density is required for the development of streamer breakdown of liquids? Discuss different possible mechanisms of initial plasma generation for further development of streamers in the liquid phase.

13.6.9 Nanosecond-Pulsed Discharge in Liquid N_2, Synthesis of Polymeric Nitrogen

What other chemical compounds except polymeric nitrogen can be expected in the nanosecond pulsed discharges in liquid N_2? Keep in mind the presence of metallic electrodes as well as the possible presence of surrounding air during the cryogenic process.

References

Abilsiitov GA, E.P. Velikhov, V.S. Golubev (1984), "High Power CO_2 – Lasers and Their Applications in Technology", *Nauka (Science)*, Moscow.

Abramowitz M, I.A. Stegan (1968), "Handbook of Mathematical Functions", *National Bureau of Standards, Applied Mathematics Series*, vol. 55, 7-th Edition, Washington.

Adamovich I, S. Saupe, M.J. Grassi, O. Schulz, S. Macheret, J.W. Rich (1993), *Chemical Physics*, vol.174, 219.

Affinito J, R.R. Parson (1984), *J. Vac. Sci. Technol.*, vol. A2, 1275.

Ageeva NP, G.I. Novikov, V.K. Raddatis ea. (1986), *Sov. Phys.,- High Energy Chemistry*, vol.20, 284.

Ahmedganov RA, Yu.V. Bykov, A.V. Kim , A. Fridman (1986), *Journal of the Institute of Applied Physics, Gorky, USSR*, vol.147, 20.

Akishev YS, A.P. Napartovich (1980), *Sov. Phys.,-Thermal Physics of High Temperatures*, vol.18, 873.

Akulintsev VM, V.M. Gorshunov, Y.P. Neschimenko (1977a), *Sov.Phys.,-Journal of Applied Mechanics and Technical Physics*, vol.18, 593.

Akulintsev VM, V.M. Gorshunov, Y.P. Neschimenko (1977b), *Sov.Phys.,-Journal of Applied Mechanics and Technical Physics*, vol.24, 1.

Alekseew AM, A.A. Ivanov, V.V. Starikh (1979), *IV-Int. Symp. on Plasma Chemistry, ISPC-4*, vol.2, 427, Zurich.

Alexandrov AF, V.A. Godyak, A.A. Kuzovnikov, A.Y. Sammani (1967), VIII-Int. *Conf. on Phenomena in Ionized Gases, ICPIG-8*, Vienna, 165.

Alexandrov AF, A.A. Rukhadze (1976), "*Physics of the High-Current Electric-Discharge Light Sources*", Atomizdat, Moscow.

Alexandrov N (1981), in: "Plasma Chemistry-8", ed. by B.M. Smirnov, Energoizdat, Moscow.

Alfassi ZB, S.W. Benson (1973), Intern. *J. Chem. Kinetics*, vol.5, 879.

Alkhazov GD (1971), *Sov.Phys.,- Journal of Technical Physics*, vol.41, 2513.

Alfven H (1950), "*Cosmical Electrodynamics*", Clarendon Press, Oxford.

Allis WP (1960), "*Nuclear FusionNuclear Fusion*", D.Van Nostrand Company, Princeton, N.J

Anderson HC, I. Oppenheim, K.E. Shuler (1964), *Journal of Mathematical Physics*, vol.5, 522.

Andreev EA, E.E. Nikitin (1976) in: "*Plasma Chemistry-3*", ed. by B.M. Smirnov, Atomizdat, Moscow.

Arago DF (1859), "*Thunder and Lightning*", Publ. by Ecole Politechnique, Paris.

Armstrong BH (1990), "Emission, Absorption and Transfer of Radiation in Heated Atmospheres".

Arshinov AA, A.K. Musin (1958a), *Sov. Phys., - Doklady*, vol.120, 747.

Arshinov AA, A.K. Musin (1958b), *Sov. Phys., - Doklady*, vol.118, 461.

Arshinov AA, A.K. Musin (1959), *Sov. Phys.,- Journal of Physical Chemistry*, vol.33, 2241.

Arshinov AA, A.K. Musin (1962), *Sov. Phys., - Radio-Technique and Electronics*, vol.7, 890.

Asinovsky EI, L.M. Vasilyak (2001), in: "*Encyclopedia of Low Temperature Plasma*", ed. by V.E. Fortov, vol.2, pg. 234, Nauka (Science), Moscow.

Asisov RI, V.K. Givotov, V.D. Rusanov, A. Fridman (1980), *Sov.Phys.,- High Energy Chemistry*, vol.14, 366.

Asisov RI, A. Vakar, V.K. Givotov ea. (1985), *International Journal of Hydrogen Energy*, vol.10, 475.

Asisov RI, M.F. Krotov, B.V. Potapkin ea., (1983), *Sov.Phys.,- Doklady*, vol.271, 94.

Atamanov VM, A.A. Ivanov, V.A. Nikiforov (1979), *Sov. Phys.,- Journal of Technical Physics*, vol.49, 2311.

Aulsloos P, ed. (1968), *Fundamental Processes in Radiation Chemistry*, Interscience Publ.

Babaeba N, G. Naidis (1996), *J. Phys. D: Applied Physics*, vol.29, 2423.

Babaritsky AI, A.A. Ivanov, V.V. Severny, V.V. Shapkin (1975a), *II - Int. Symp. on Plasma Chemistry, ISPC-2*, vol.1, Rome.

Babaritsky AI, A.A. Ivanov, V.V. Severny, T.I. Sokolova, V.V. Shapkin (1975b), *II – SU Symp. on Plasma Chemistry*, vol.2, pg.35, Riga.

Babukutty Y, R. Prat, K. Endo, M. Kogoma, S. Okazaki, M. Kodama (1999), *Langmuir*, vol.15, 7055.

Balabanov VV, L.M. Vasilyak, S.P. Vetchinin, A.P. Nefedov, D.N. Polyakov, V.E. Fortov (2001), *Sov. Phys., - Journal of Experimental and Theoretical Physics*, vol.92, 86.

Balebanov AV, B.A. Butylin, V.K. Givitov, R.M. Matolich, S.O. Macheret, G.I. Novikov, B.V. Potapkin, V.D. Rusanov, A. Fridman (1985a), *Sov.Phys., Doklady*, vol.283, 657.

Balebanov AV, B.A. Butylin, V.K. Givitov, R.M. Matolich, S.O. Macheret, G.I. Novikov, B.V. Potapkin, V.D. Rusanov, A. Fridman (1985b), *Journal of Nuclear Science and Technology, Nuclear-Hydrogen Energy*, vol. 3, 46.

Baranchicov EI, V.P. Denisenko, V.D. Rusanov ea. (1990), *Sov.Phys.,- Doklady*, vol.339, 1081.
Baranov VY, A.P. Napartovich, A.N. Starostin (1984), in "*Plasma Physics*", ed. by V.D. Shafranov, VINITI, vol.5, Moscow.
Baranov VY, A.A. Vedenov, V.G. Niziev (1972), *Sov.Phys.,- Thermo-physics of High Temperatures*, vol.10, 1156.
Barnett CF (1989), in "*A Physicist's Desk Reference*", ed. by - H.L. Anderson, American Institute of Physics, N.Y.
Barnes MS, J.H. Keller, J.C. Forster, J.A. O'Neil, D.K. Coutlas (1992), *Phys. Rev. Lett.*, vol.68, 313.
Baronov GS, D.K. Bronnikov, A.A. Fridman, B.V. Potapkin, V.D. Rusanov, A.A. Varfolomeev, A.A. Zasavitsky (1989), *J.Phys.B: Atom., Mol., Opt. Physics.*, vol.22, 2903.
Barry JD (1980), "*Ball Lightning and Bead Lightning: Extreme Forms of Atmospheric Electricity*", Plenum Pub. Corporation.
Bartnikas R, E.J. McMahon, ed. (1979), "Engineering Dielectrics", vol.1, "*Corona Measurement and Interpretation*", ASTM Special Technical Publication, No.669, Philadelphia, PA.
Baselyan EM, A.Yu. Goryunov (1986), *Electrichestvo (J. of Electricity)*, vol.11, 27.
Baselyan EM, Yu.P. Raizer (1997), "*Spark Discharge*", Edition of the Moscow Institute of Physics and Technology, Moscow.
Baselyan EM, Yu.P. Raizer (2001), "*Physics of Lightning and Lightning-Protection*", Nauka (Science), Moscow.
Basov NG (1974), *Adv. in Phys.Sci., Uspehi Phys. Nauk*, vol.114, 213.
Basov NG, E.M. Belenov, V.A. Danilychev (1973), *Sov. Phys.,- Journal of Experimental and Theoretical Physics*, vol.64, 108.
Batenin V, I.I. Klimovsky, G.V. Lysov, V.N. Troizky (1988), "*Microwave Plasma Generators: Physics, Engineering, Applications*", EnergoAtomIzdat, Moscow.
Baulch DL (1992–1994), "Evaluated Kinetic Data...", series of publications in J. Phys. Chem. Ref. Data.
Becker R, W. Doring (1935), "Condensation Theory", *Annnalen der Physik*, vol.24, 719.
Ph. Belenguer, J.P. Blondeau, L. Boufendi, M. Toogood, A. Plain, C. Laure, A. Bouchoule, J.P. Boeuf (1992), *Phys.Rew. A*, vol.46, 7923.
Belenov EM, E.P. Markin, A.N. Oraevsky, V.I. Romanenko (1973), *Sov.Phys.,-Journal of Experimental and Theoretical Physics, Letters*, vol.18, 116.
Beliaev ST, G.I. Budker (1958), "Plasma Physics and Problems of the Controlled Thermo-Nuclear Reactions", *Academy of Sciences of the USSR*, vol.3.
Belousov IG, V.F. Krasnoshtanov, V.D. Rusanov ea. (1979), *Journal of Nuclear Science and Technology, Nuclear-Hydrogen Energy*, vol. 1(5), 43.
Bennett WH (1934), "Magnetically Self-Focusing Streams", *Phys. Rev.*, vol.45, 890.
Berezin AK, E.V. Lifshitz, Y. B. Fainberg (1995), *Plasma Physics*, vol.21, 226.
Berger S (1974), *III- Int. Conf. on Gas Discharge*, London, pg.380.
Berger S (1975), *XII - Int. Conf. on Phenomena in Ionized Gases, ICPIG-12*, vol.1, 154, Eindhoven, Netherlands.
Bergman RC, G.F. Homicz, J.W. Rich, G.L. Wolk (1983), *J. Chem. Phys.*, vol.78, 1281.
Biberman LM (1970), "*Optical Properties of Hot Air*", Nauka (Science), Moscow.
Biberman LM, G.E. Norman (1967), *Adv. in Phys.Sci., Uspehi Phys. Nauk*, vol.91, 193.
Biberman LM, V.S. Vorobiev, I.T. Yakubov (1982), "*Kinetics of the Non-Equilibrium Low-Temperature Plasma*", Nauka (Science), Moscow.
Biberman LM, V.S. Vorobiev, I.T. Yakubov (1979), *Adv. in Phys.Sci., Uspehi Phys. Nauk*, vol.128, 233.
Billing GD (1986), "Vibrational Kinetics and Reactions of Polyatomic Molecules", *Topics of Current Physics-39*, Ed. by M. Capitelli, Springer Verlag, Berlin.
Biondi MA (1976), in "*Principles of Laser Plasma*", ed. by G. Bekefi, J.Wiley&Sons, N.Y.
Birmingham JG, R.R. Moore (1990), "Reactive Bed Plasma Air Purification", United States Patent, #4, 954,320.
Bjerre A, E.E. Nikitin (1967), *Chem. Phys. Lett.*, vol.1, 179.
Blinov LM, V.V. Volod'ko, G.G. Gontarev, G.V. Lysov, L.S. Polak (1969), in: "*Low-Temperature Plasma Generators*", Ed. by L.S. Polak, Energia (Energy), Moscow.
Bloch F, N. Bradbury (1935), - *Phys. Rev.*; vol. 48, 689-696.
Bochin VP, V.A. Legasov, V.D. Rusanov, A. Fridman (1977), *Journal of Nuclear Science and Technology, Nuclear-Hydrogen Energy*, vol. 1(2), 55.
Bochin VP, V.A. Legasov, V.D. Rusanov, A. Fridman (1978a), II World Conference on Hydrogen Energy, *Zurich*, vol.3, 1183.

Bochin P, V.A. Legasov, V.D. Rusanov, A. Fridman (1978b) in book: "*Nuclear-Hydrogen Energy and Technology-1*", Ed. by V.A. Legasov, Atom-Izdat, Moscow, 183.

Bochin VP, V.A. Legasov, V.D. Rusanov, A. Fridman (1979) in book: "*Nuclear-Hydrogen Energy and Technology-2*", Ed. by V.A. Legasov, Atom-Izdat, Moscow, 206.

Bochin VP, V.F. Krasnoshtanov, V.D. Rusanov, A. Fridman (1983), *Journal of Nuclear Science and Technology, Nuclear-Hydrogen Energy*, vol. 2(15), 12.

Boenig HV (1988), "*Fundamentals of Plasma Chemistry and Technology*", Technomic Publishing Co., Lancaster, PA.

Boeuf JP, E. Marode (1982), *J. Phys., D, Applied Physics*, vol.15, 2169.

Boeuf JP (1992), *Phys. Rev. A.* , vol.46, 7910.

Bohm C, J. Perrin (1991), *J. Phys. D: Appl. Phys.*, vol.24, 865.

Borisov BM, O.F. Khristoforov (2001), in:*Encyclopedia of Low Temperature Plasma*, ed. by V.E. Fortov, vol.2, pg.350, Nauka (Science), Moscow.

Boswell RW (1970), *Plasma Phys. Controlled Fusion*, vol.26, 1147.

Bouchoule A, L. Boufendi, J.P. Blondeau, A. Plain, C. Laure (1991), *J.Appl. Phys.*, vol.70, 1991

Bouchoule A, L. Boufendi, J.P. Blondeau, A. Plain, C. Laure (1992), *Appl. Phys. Lett.*, vol.60, 169.

Bouchoule A (1993), *Physics World*, vol.47, August issue.

Bouchoule A, L. Boufendi (1993), *Plasma Sources Sci. Technol.*, vol.2, 204.

Bouchoule A, L. Boufendi, J.Ph. Blondeau, A. Plain, C. Laure (1993), *J.Appl. Phys.*, vol.73, 2160.

Bouchoule A (1999), Ed. *Dusty Plasmas: Physics, Chemisry and Technological Impacts in Plasma Processing*, John Willey & Sons, N.Y.

Boufendi L, A. Bouchoule (1994), *Plasma Sources Sci. Technol.*, vol.3.

Boufendi L . Ph.D.Thesis, Faculte de Science, GREMI, University of Orleans, (1994).

Boufendi L, J. Hermann, E. Stoffels, W. Stoffels, M.L. De Giorgi, A. Bouchoule (1994), *J .Appl. Phys.*, vol.76, 148.

Boulos MI, P. Fauchais and E. Pfender (1994), *Thermal Plasmas. Fundamentals and Applications*, Plenum Press, N.Y., London.

Boutylkin YP, V.K. Givotov, E.G. Krasheninnikov, V.D. Rusanov, A. Fridman e.a., (1981), *Sov.Phys.,-Journal of Technical Physics*, vol.51, 925.

Bradley EB (1990), "*Molecules and Molecular Lasers for Electrical Engineers*" (Series in Electrical Engineering), Hemisphere Pub.

Braginsky SI (1958), *Sov.Phys.,- Journal of Theoretical and Experimental Physics*, vol.34, 1548.

Braginsky SI (1963), "*Problems of Plasma Theory*", Ed. by M.A. Leontovich, vol. 1, pg. 183–272, Atomizdat, Moscow.

Braithwaite N, J.P. Booth, G. Cunge (1996), *Plasma Sources Sci. Technol.*, vol.5, 677.

Brau CA (1972), *Physica*, vol.58, 533.

Brau CA, R.M. Jonkman (1970), *J.Chem.Phys.*, vol.52, 477.

Bray KNC (1968), *J.Phys.B: Atomic and Molecular Physics*, vol.1, 705.

Bronin SY, V.M. Kolobov (1983), *Sov. Phys.,- Plasma Physics*, vol.9, 1088.

Brown SC (1959), "*Basic Data of Plasma Physics*", MIT Press, Cambridge, MA.

Brown SC (1966), "*Introduction to Electrical Discharges in Gases*", John Wiley, N.Y.

Budden KG (1966), "*Radio Waves in Ionosphere*", Cambridge University Press, Cambridge, UK.

Bunkin FV, V.V. Savransky (1973), *Sov. Phys.,- Journal of Experimental and Theoretical Physics*, vol.65, 2185.

Buss K (1932), Arch. Elektrotech., vol.26, 261.

Butterbaugh JW, L.D. Baston, H.H. Sawin (1990), *J. Vac. Sci. Technol.*, vol.A8(2), 916.

Bykov VL, M.N. Vasiliev, A.S. Koroteev (1993), "*Electron-Beam Plasma. Generation, Properties, Application*", ed. by Moscow State University, Moscow.

Cabannes F, J. Chapelle (1971), "*Reactions under Plasma Conditions*", vol.1, Wiley Interscience, N.Y.

Capitelli M (2000), Ed., "Plasma Kinetics in Atmospheric Gases", *Springer Series on Atomic, Optical and Plasma Physics*, vol.31, Springer Verlag.

Chalmers JD (1972), *Proc. Roy. Soc., A, London*, vol.369, 171.

Capitelli M (1986), Ed. "Non-Equilibrium Vibrational Kinetics", *Topics in Current Physics - 39*, Springer -Verlag.

Chebotaev VP (1972), *Sov.Phys.,-Doklady*, vol.206, 334.

Chen FF (1984), "*Introduction to Plasma Physics and Controlled Fusion: Plasma Physics*", Plenum Pub. Corp., 2-nd edition.

Chen FF (1991), *Plasma Phys. Controlled Fusion*, vol.33, 339.

Chesnokov EN, V.N. Panfilov (1981), *Journal of Theoretical and Experimental Chemistry*, vol.17, 699.
Yang Y, Y.I. Cho, A. Fridman (2012), *Plasma Discharge in Liquid*, CRC Press, Taylor & Francis Group.
Choi S, M. Kushner (1993), *J. Appl. Phys.*, vol.74, 853.
Chu JH, I. Lin (1994), *Phys. Rev. Lett.*, vol.72, 4009.
Clarke JF (1978), *Acta Astronautica*, vol.5, 543.
Cobine JD (1958), *Gaseous Conductors*, Dover Publications, N.Y.
Conti S, A. Fridman, S. Raoux, (1999), *14th International Symposium on Plasma Chemistry (ISPC 14)*
Conti S, P.I. Porshnev, A. Fridman, L.A. Kennedy, J.M. Grace, K.D. Sieber, D.R. Freeman, K.S. Robinson (2001), *Experimental Thermal and Fluid Science*, vol.24, 79.
Cookson A, A. Farish (1972), *Int. Symp. on High Voltage Technologies*, Munich, pg.343.
Cookson A, R. Wotton (1974), *III- Int. Conf. on Gas Discharge*, London, pg.385.
Cormier JM, F. Richard, J. Chapelle, M. Dudemaine (1993), *Proceedings of the 2-nd International Conference on Elect. Contacts, Arcs, Apparatus and Applications*, 40–42.
Crichton BH, B. Lee (1975), *XII – Int. Conf. on Phenomena in Ionized Gases, ICPIG-12*, vol.1, pg.155, Eindhoven, Netherlands.
Crowford OH (1967), *Mol.Phys.*, vol.13, 181–185.
Czernichowski A (1994), *Pure and Applied Chemistry*, vol.66, #6, 1301.
Czernichowski A, H. Nassar, A. Ranaivosoloarimanana, A. Fridman, M. Simek, K. Musiol, E. Pawelec, L. Dittrichova (1996), *Acta Phys. Pol.*, vol. A89, 595.
D'Angelo N, N. Rynn (1961), *Phys. Fluids*, vol.4, 275.
Dalaine V, J.-M. Cormier, S. Pellerin, P. Lefaucheux (1998), *J.Appl.Phys.*, vol.84, 1215.
Dali SK, P.F. Williams (1987), *Phys.Rev.*, vol.A31, 1219.
Daniel J, J. Jacob (1986), *J.Geophys.Res.*, vol.91, 9807.
Daugherty JE, R.K. Porteuos, M.D. Kilgore, D.B. Graves (1992), *J. Appl. Phys.*, vol.72, 3934.
Daugherty JE, R.K. Porteuos, M.D. Kilgore, D.B. Graves (1993a), *J. Appl. Phys.*, vol.73, 1619.
Daugherty JE, R.K. Porteuos, M.D. Kilgore, D.B. Graves (1993b), *J. Appl. Phys.*, vol.73, 7195.
Dawson GA, W.P. Winn (1965), *Zs.Phys.*,vol.183, 159.
Deminsky MA, B. V. Potapkin, J. M. Cormier, F. Richard, A. Bouchoule, V.D. Rusanov (1977), *Sov.Phys.,-Doklady*, vol.42, 337.
Demura AV, S.O. Mahceret, A. Fridman (1981), 5-th Int. Symposium on Plasma Chemistry, *Edinburgh*, vol.1, 52
Demura AV, S.O. Mahceret, A. Fridman (1984), *Sov. Phys.-Doklady*, vol.275, 603.
Dimick RC, S.L. Soo (1964), *Phys. Fluids*, vol.7, 1638.
Dmitriev MT (1967), "*Priroda*" (Nature), vol.56, 98.
Dmitriev MT (1969), *Sov.Phys.,- Journal of Technical Physics*, vol.39, 387.
Dobkin SV, E.E. Son (1982), *Sov.Phys.,-Thermal Physics of High Temperatures*, vol.20, 1081.
Dobretsov LN, M.V. Gomounova (1966), "*Emission Electronics*", Nauka (Science), Moscow.
Dobrynin D, R. Rakhmanov, A. Fridman (2019a), *J. Phys. D: Appl. Phys.*, vol. 52, 39LT01
Dobrynin D, Y. Seepersad, M. Pekker ea. (2013), *J. Phys. D: Appl. Phys.*, vol. **46,** 105201
Dobrynin D, D. Vainchtein, M. Gherardi ea. (2019b), *IEEE Transactions on Plasma Sci.*, vol. 47, 4052-4057
Douthart DA (1979), *J. Phys. B: Atomic and Molecular Physics*, vol.12, 663.
Drabkina SI (1951), *Sov. Phys., Journal of Theoretical and Experimental Physics*, vol.21, 473.
Drawin HW (1968) *Ztschr.*, Bd.211, S. 404-408.
Drawin HW (1969) *Ztschr.*, Bd.255, S. 470-475.
Drawin HW (1972) "The Thermodynamic Properties of the Equilibrium and Non-Equilibrium States of Plasma", in: *Reactions Under Plasma Conditions*, Ed. by M. Venugopalan, Wiley, N.Y.
Dresvin SV (1977), *Physics and Technology of Low Temperature Plasmas*, Iowa State University Press, Ames, IA.
Dzantiev BG, A.N. Ermakov, V.N. Popov (1979), *Journal of Nuclear Science and Technology, Nuclear-Hydrogen Energy*, vol. 1(5), 89.
Eckert HU (1986), *Proc. II-International Conference on Plasma Chemistry and Technology*, ed. by H. Boening, Technomic Publ., Lancaster, PA.
Einbinder H (1957), *J. Chem. Phys.*, vol.26, 948.
Eletsky AV, L.A. Palkina, B.M. Smirnov (1975), *Transfer Phenomena in Weakly Ionized Plasma*, Atomizdat, Moscow.
Eletsky AV, N.P. Zaretsky (1981), *Sov.Phys.,-Doklady*, vol.260, 591.
Eliasson B, U. Kogelschatz (1991), *IEEE Trans., Plasma Sci.*, vol.19, 309.
von Engel A (1965), "*Ionized Gases*", Clarendon, Oxford; (1994), in: American Vacuum Society Classics, Springer Verlag.

von Engel A, M. Steenbeck (1934), "*Elektrische Gasentladungen. Ihre Physik und Technik*", vol. II, Springer, Berlin.

Engelgardt AG, A.V. Felps, G.G. Risk (1964), *Phys. Rev.*, vol.135, 1566.

Este G, W.D. Westwood (1984), *J. Vac. Sci. Technol.*, vol.A6(3), 1845.

Este G, W.D. Westwood (1988), *J. Vac. Sci. Technol.*, vol.A2, 1238.

Eyring H, S.H. Lin, S.M. Lin (1980), *Basic Chemical Kinetics*, John Wiley & Sons, N.Y.

Fano U (1953), *Phys. Rev.*, vol.92, 328.

Finkelburg W, H. Maecker (1956), "Elektrische Bogen und Thermisches Plasma", *Handbuch der Physik*, vol.22, S.254.

Firestone RF (1957), *J. Amer. Chem. Soc.*, vol.79, 5593.

Forrester AT (1988), *Large Ion Beams – Fundamentals of Generation and Propagation*, John Wiley, N.Y.

Fortov VE (2001), *XXV-Int. Conf. on Phenomena in Ionized Gases, ICPIG-25*, Nagoya, Japan, vol.2, pg.13.

Fortov VE, V.I. Molotkov, A.P. Nefedov, ea. (1999), *Phys. Plasmas*, vol.6, 1779.

Fortov VE, A.P. Nefedov, O.S. Vaulina, ea. (1998), *Journal of Experimental and Theoretical Physics*, vol.114, pg.2004 (JETF, vol.87, pg.1087).

Fortov VE, I.T. Yakubov (1994), "Non-Ideal Plasma", *Energo-Atom-Izdat*, Moscow.

Frank-Kamenetsky DA (1971), *Diffusion and Heat Transfer in Chemical Kinetics*, Nauka (Science), Moscow.

Francis G (1956), "The Glow Discharge at Low Pressure", in: *Encyclopedia of Physics*, ed. by S. Flugge, Handbuch der Physik, Bd. XXII, pg.55-208, Berlin.

Fransis G (1960), *Ionization Phenomena in Gases*, Butterworths, London.

Frenkel YI (1945), "*Kinetic Theory of Liquids*", ed. by the Academy of Sciences of USSR, Moscow.

Frenkel YI (1949), "Theory of Atmospheric Electricity Phenomena", *Gos. Tech. Izdat.*, Moscow.

Fridman A (1976a), *Ionization Processes in Heterogeneous Media*, ed. by Moscow Institute of Physics and Technology, MIPT, Moscow.

Fridman A (1976b), *Ionization Processes in Heterogeneous Media*, ed. by Moscow Institute of Physics and Technology, MIPT, Moscow.

Fridman A (1976c), "*Plasma-Chemical Processes in Non-Self-Sustained Discharges With Ionization Provided by High-Current Relativistic Electron Beams*", ed. by Moscow Institute of Physics and Technology, MIPT, Moscow.

Fridman A (2008), *Plasma Chemistry*, Cambridge University Press, Cambridge, N.Y.

Fridman A, L. Boufendi, A. Bouchoule, T. Hbid, B. Potapkin (1996a), *J.Appl.Phys.*, vol.79, 1303-1314.

Fridman A, A. Czernichowski, J. Chapelle, J.-M. Cormier, H. Lesuer, J. Stevefelt (1994), *J. de Physique*, 1449.

Fridman A, A. Czernichowski, J. Chapelle, J.-M. Cormier, H. Lesueur, J. Stevefelt (1993), *International Symposium on Plasma Chemistry, ISPC-11*, Loughborough, UK.

Fridman A, S. Nester, S. Guceri (1996b), "Accelerated Electrons Applications for Chemical Reactions Stimulation", in: "*Application of Accelerators in Research and Industry*", ed. by J.L. Duggan, I.L. Morgan, AIP-Press, Woodbury, NY.

Fridman A, S. Nester, L. Kennedy, A. Saveliev, O. Mutaf-Yardimci (1999), *Progress in Energy and Combustion Science*, vol.25, 211.

Fridman A, S. Nester, O. Yardimci, A. Saveliev, L. A. Kennedy (1997), *13 International Symposium on Plasma Chemistry, ISPC-13*, Beijing, China.

Fridman A, L.I. Rudakov (1973), "Electrostatic Lenses for Focusing of Relativistic Electron Beams ", *Moscow Inst. of Physics and Technology*, Moscow.

Fridman A, V.D. Rusanov (1994), *Pure & Appl. Chem.*, vol.66, #6, 1267.

Fuks NA (1964), *Sov. Phys.,- News of the USSR Academy of Sciences, geo-physics*, vol.4, 579.

Gallimberti I (1972), *J.Phys.D., Applied Phys.*, vol.5, 2179.

Gallimberti I (1977), *Electra* vol.76, 5799.

Garscadden A (1978), "Ionization Waves", in: *Gaseous Electronics – I, Electrical Discharges*, ed. by M.N. Hirsh, H.J. Oskam, Academic Press, N.Y.

Garscadden A (1994), *Pure and Applied Chemistry*, vol.66, 1319.

Geffers WQ, J.D. Kelley (1971), *J.Chem.Phys.*, vol.55, 4433.

Generalov NA, V.D. Kosynkin, V.P. Zimakov, Yu.P. Raizer (1980), *Sov. Phys.,- Plasma Physics*, vol.6, 1152.

Generalov NA, V.P. Zimakov, G.I. Kozlov, V.A. Masyukov, Y.P. Raizer (1970), *Sov.Phys.,- Letters to the Journal of Experimental and Theoretical Physics*, vol.11, 447.

Generalov NA, V.P. Zimakov, G.I. Kozlov, V.A. Masyukov, Yu.P. Raizer (1971), *Sov.Phys.,- Journal of Experimental and Theoretical Physics*, vol.61, 1444.

Gerenser LJ, J.M. Grace, G. Apai, P.M. Thompson (2000), *Surface Interface Anal.*, vol.29, 12.

Gerjoy E, S. Stein (1955), *Phys.Rev.*, vol.95, 1971–1976.

Gershenson Y, V. Rosenstein, S. Umansky (1977) in: "Plasma Chemistry-4", Ed. by B.M. Smirnov, Atomizdat, Moscow.
Gibbs WE, R. McLeary (1971), *Phys. Lett. A*, vol.37, 229.
Gildenburg VB (1981), in: "*Non-Linear Waves*", Ed. by A.V. Gaponov-Grekov, Nauka (Science), 15.
Gill P, C.E. Webb (1977), *J. Phys., D, Applied Physics*, vol.10, 229.
Ginsburg VL (1960), "*Propagation of Electromagnetic Waves in Plasma*", Phys.Math.Giz, Moscow.
Ginsburg VL, A.A. Rukhadze (1970), *Waves in Magneto-Active Plasma*, Nauka (Science), Moscow.
Girshick S, U.V. Bhandarkar, M.T. Swihart, U.R. Kortshagen (2000), *Journal of Physics, D: Applied Physics*, vol.33, 2731.
Girshick S, U.R. Kortshagen, U.V. Bhandarkar, M.T. Swihart (1999), *Pure Appl. Chem.*, vol.71, 1871.
Givotov VK, I.A. Kalachev, E.G. Krasheninnikov e.a. (1983a), "Spectral Diagnostics of a Non-homogeneous Plasma-Chemical Discharge", *Kurchatov Institute of Atomic Energy*, vol.3704/7, Moscow.
Givotov VK, I.A. Kalachev, Z.B. Mukhametshina, V.D. Rusanov, A. Fridman, A.M. Chekmarev (1986), *High Energy Chemistry*, vol.20, 354.
Givotov VK, M.F. Krotov, V.D. Rusanov, A. Fridman (1981), *International Journal of Hydrogen Energy*, vol.6, 441.
Givotov VK, S.Yu. Malkov, V.D. Rusanov, A. Fridman (1983b), *Journal of Nuclear Science and Technology, Nuclear-Hydrogen Energy*, vol. 1(14), 52.
Givotov VK, S.Y. Malkov, B.V. Potapkin e.a. (1984), *Sov.Phys., High Energy Chemistry*, vol.18, 252.
Givotov VK, V.D. Rusanov, A. Fridman (1982, 1984), in: Plasma Chemistry-9" and "*Plasma Chemistry-11*", Ed. by B.M. Smirnov, Atom-Izdat, Moscow.
Givotov VK, V.D. Rusanov, A. Fridman (1985), *Diagnostics of Non-Equilibrium Chemically Active Plasma*, Energo-Atom-Izdat, Moscow.
Glasstone S, M.C. Edlung (1952), "*Nuclear Reactor Theory*", D. van Nostrand Co., Princeton, New Jersey.
Godyak VA (1971a), *Sov. Phys.,- Journal of Technical Physics*, vol.41, 1361.
Godyak VA (1971b), *Sov. Phys.,- Plasma Physics*, vol.2, 141.
Godyak VA (1986), "*Soviet Radio Frequency Discharge Research*", Delphic Associates, Falls Church, VA.
Godyak VA, A.S. Khanneh (1986), *IEEE Trans. Plasma Sci.*, PS-14 (2), 112.
Godyak VA, R.B. Piejak (1990a), *Phys. Rev. Lett.*, vol.65, 996.
Godyak VA, R.B. Piejak (1990b), *J. Vac. Sci. Technol.*, vol.A8, 3833.
Godyak VA, R.B. Piejak, B.M. Alexandrovich (1991), *IEEE Trans. Plasma Sci.*, vol.19, 660.
Goldman M, N. Goldman (1978), "Corona Discharges", in "*Gaseous Electronics*", vol.1, "Electrical Discharges", ed. by M.N. Hirsh and H.J. Oscam, Academic Press, N.Y.
Yu.B. Golubovsky, A.K. Zinchenko, Yu.M. Kagan (1977), *Sov. Phys.,- Journal of Technical Physics*, vol.47, 1478.
Gordiets B, S. Zhdanok (1986), "Non-Equilibrium Vibrational Kinetics", Ed. by M. Capitelli, *Topics in Current Physics - 39*, Springer -Verlag.
Gorgiets BF, A.I. Osipov, L.A. Shelepin (1980), "*Kinetic Processes in Gases and Molecular Lasers*", Nauka (Science), Moscow; B.Gordiets (1988), Gordon and Breach Science Pub.
Gordiets BF, A.I. Osipov, E.B. Stupochenko, L.A. Shelepin (1972), *Adv. in Phys.Sci., Uspehi Phys. Nauk*, vol.108, 655.
Grace JM e.a. (1995), US Patent No. 5,425,980.
Granovsky VL (1971), *Electric Current in Gas, Steady Current*, Nauka (Science), Moscow.
Green AES, C.A. Barth (1965), *J. Geophys. Res.*, vol.70, 1083.
Green AES, C.A. Barth (1967), *J. Geophys. Res.*, vol.72, 3975.
Griem HR (1964), *Plasma Spectroscopy*, McGrow-Hill, N.Y.
Griem HR (1974), *Spectral Broadening by Plasma*, Academic Press, N.Y., London.
Grigorieva TA, A.A. Levitsky, S.O. Macheret, A. Fridman (1984), *Sov.Phys.,-High Energy Chemistry*, vol.18, 336.
Gross B, B. Grycz, K. Miklossy (1969), *Plasma Technology* Elsevier, N.Y.
Gudzenko LI, S.I. Yakovlenko (1978), *Plasma Lasers*, Atomizdat, Moscow.
Guha S, P.K. Kaw, (1968), *Brit. J. Appl. Phys.*, vol.1, 193.
Gurin AA, N.I. Chernova (1985), *Sov. Phys.,- Plasma Physics*, vol.11, 244.
Gutsol A (1995), *Sov.Phys.,- High Energy Chemistry*, vol.29, 373.
Gutsol AF (1997) "The Ranque Effect", *Uspekhi, Successes in Physical Sciences*, vol.40, #6, pg.639-658.
Gutsol A, J. Larjo, R. Hernberg (1999), *XIV- Int. Symposium on Plasma Chemistry, ISPC-14, Prague*, vol.1, pg.227.
Gutsol A, A Fridman (2001), XXV-th Int. Conf. on Phenomena in Ionized Gases, ICPIG-25, Nagoya, *Japan*, vol.1, pg.55.

Gutsol A, A. Fridman, A. Chirokov, L.A. Kennedy, W. Worek (2001), 2-nd Int. Conf. on Computational Heat and Mass Transfer, Rio de Janeiro, Brazil, pg.65.

Haas R (1973), *Phys. Rev., A – General Physics*, vol.8, pg.1017.

Hamaguchi S, R.T. Farouki, D.H.E. Doubin (1997), *Phys. Rev., E*, vol.56, pg.4671.

Harkness JBL, R. Doctor (1993), *Plasma-Chemical Treatment of Hydrogen Sulfide in Natural Gas Processing*, Gas Research Institute, GRI-93/0118.

Hartech P, S. Doodes (1957), *J. Chem. Phys.*, vol.26, pg.1727.

Hassan HA, J.E. Deese (1976), *Phys. Fluids*, vol.19, pg.2005.

Heath WO, J.G. Birmingham (1995), *PNL-SA-25844, Pacific Northwest Laboratory, Annual Meeting of American Nuclear Society*, Philadelphia, Pennsylvania.

Herzberg G (1945), "*Molecular Spectra and Molecular Structure II, Infrared and Raman Spectra of Polyatomic Molecules*", Van Nostrand Company, N.Y.

Herzenberg A (1968), *J. Phys. B: Atomic and Molecular Physics*, vol.1, pg.548-553.

Hill AE (1971), *Appl. Phys. Lett.*, vol.18, pg.194.

Hinazumi H, M. Hosoya, T. Mitsui (1973), *J. Phys. D, Applied Physics*, vol.1973, pg.21.

Hirsh MN, Oscam HJ, ed. (1978), "Gaseous Electronic", vol.1, "*Electrical Discharges*", Academic Press, N.Y.

Hoad E, H. Rease, J. Stall, J. Zav (1973), *J. Quantum Electronics*, vol.9, pg.652.

Hollenstein C, A.A. Howling, J.L. Dorier, J. Dutta, L. Sansonnens (1993), NATO Advanced Research Workshop "*Formation, Transport and Consequences of Particles in Plasma*", France.

Honda Y, F. Tochikubo, T. Watanabe (2001), *25- Int. Conf. On Phenomena in Ionized Gases, ICPIG-25*, vol.4, pg.37, Nagoya, Japan.

Hopwood J, C.R. Guarnieri, S.J. Whitehair, J.J. Cuomo (1993a), *J. Vac. Sci. Technol.*, vol.A11, pg.147.

Hopwood J, C.R. Guarnieri, S.J. Whitehair, J.J. Cuomo (1993b), *J. Vac. Sci. Technol.*, vol.A11, pg.152.

Howatson AM (1976), *An Introduction to Gas Discharges*, 2-nd edition, Pergamon Press, Oxford.

Howling AA, J.L. Dorier, Ch. Hollenstein (1993a,)*Applied Phys. Lett.*, vol.62, pg.1341.

Howling AA, Ch. Hollenstein, P.J. Paris (1991), *Appl. Phys. Lett.*, vol.59, pg.1409.

Howling AA, L. Sansonnens, J.L. Dorrier, Ch. Hollenstein (1993b), *J.Phys.D: Applied Physics*, vol.26, pg.1003.

Huber KP, Herzberg G (1979), *Molecular Spectra and Molecular Structure IV, Constants of Diatomic Molecules*, Van Nostrand Reinhold Company, N.Y.

Huddlestone RH, S.L. Leonard, eds. (1965), *Plasma Diagnostic Techniques*, Academic Press, N.Y.

Humphries S (1990), *Charge Particle Beams*, John Wiley, N.Y.

Huxley LGH, R.W. Crompton (1974), *The Diffusion and Drift of Electron in Gases*, Wiley, N.Y.

Ichimaru S (1982), *Rev. Mod. Phys.*, vol.54, pg.1017.

Ikezi H (1986), *Phys. Fluids*, vol.29, pg.1764.

Inokuti M (1971), *Rev. Mod. Phys.*, vol.43, pg.297.

Inokuti M (1974), *Radiation Res.*, vol.59, pg.343.

Inokuti M (1975), *Radiation Res.*, vol.64, pg.6.

Inokuti M, Y. Kim, R.L. Platzman (1967), *Phys.Rev.*, vol.164, pg.55-61.

Iskenderova K, A. Chirokov, A. Gutsol, A. Fridman, L. Kennedy, K.D. Sieber, J.M. Grace, K.S. Robinson (2001), *15-th Int. Symposium on Plasma Chemistry, OR 5.73*, Orleans, France.

Ivanov AA (1975), *Sov. Phys.,- Plasma Physics, vol.* vol.1, pg.147.

Ivanov AA (1977), *Physics of Strongly Non-equilibrium Plasma*, Atom-Izdat, Moscow.

Ivanov AA (1982), in: *Plasma Physics*, pg.105, ed. by V.D. Shafranov, VINITI, Moscow.

Ivanov AA (2000), in: *Encyclopedia of Low Temperature Plasma*, ed. by V.E. Fortov, vol. 4, pg.428, Nauka (Science), Moscow.

Ivanov AA, V.A. Nikiforov (1978), in: "Plasma Chemistry-5", Ed. by B.M. Smirnov, Atom-Izdat, Moscow.

Ivanov AA, T.K. Soboleva (1978), *Non-Equilibrium Plasma Chemistry*, Atom-Izdat, Moscow.

Ivanov VD (1969), *Sov. Phys., Doklady*, vol.188, pg.65.

Ivanov VD (1972), *Sov. Phys., Doklady*, vol.203, pg.806.

Jacob JM, S.A. Mani (1975), *Applied Phys. Lett.*, vol.26, pg.53.

Javan A, W.R. Bennett, D.R. Herriot (1961), *Phys. Rev. Lett.*, vol.6, pg.106.

Jellum G, D. Graves, J.E. Daugherty (1991), *J. Appl. Phys.*, vol.69, pg. 6923.

Kadomtsev BB (1958), *Plasma Physics and Problem of Controlled Thermonuclear Reactions*, vol.3, Academy of Science of the USSR, Moscow.

Kadomtsev BB (1968), *Adv. in Phys.Sci., Uspehi Phys. Nauk*, vol.95, pg.111.

Kadomtsev BB (1976), *The Collective Phenomena in Plasma*, Nauka (Science), Moscow.

Kagija (1969), *Bull. Chem. Soc. Japan*, vol.42, pg.1812.

Kalinnikov VT, A. Gutsol (1999), *Sov.Phys. – Thermal Physics of High Temperatures (High Temperature)*, vol.37, pg.194(172).
Kalinnikov VT, A. Gutsol (1997), *Sov.Phys.,-Doklady*, vol.353(42), pg.469(179).
Kanazava S, M. Kogoma, T. Moriwaki, S. Okazaki (1987), *International Symposium on Plasma Chemistry, ISPC-8*, Tokyo.
Kanazawa S, M. Kogoma, T. Moriwaki, S. Okazaki (1988), *J.Phys. D*, vol.21, pg.838.
Kanda N, M. Kogoma, H. Jinno, H. Uchiyama, S. Okazaki (1991), *X-Int. Symposium on Plasma Chemistry, ISPC-10*, vol.3, pg.3.2-20.
Kapitsa PL (1969), *Sov.Phys.,- Journal of Experimental and Theoretical Physics*, vol.57, pg.1801.
Karachevtsev GV, A. Fridman (1974), *Sov. Phys., - Journal of Technical Physics*, vol.44, pg.2388.
Karachevtsev GV, A. Fridman (1975), IV – SU-Conference on Physics of Low Temperature Plasma, *Kiev*, vol.2, pg.26.
Karachevtsev GV, A. Fridman (1976), *Sov. Phys.,- Journal of Technical Physics*, vol.46, pg.2355.
Karachevtsev GV, A. Fridman (1977), *Sov. Phys.,- Thermal Physics of High Temperatures*, vol.15, pg.922.
Karachevtsev GV, A. Fridman (1979), *Sov. Phys., Izvesia VUZov (Higher Education News), Physics*, #4, pg.23.
Kekez M, M.R. Barrault, J.D. Craggs (1970), *J. Phys. D: Applied Physics*, vol.3, pg.1886.
Kennedy L, A. Fridman, A. Saveliev, S. Nester (1997), *APS Bulletin*, vol.41(9), pg.1828.
Kirichenko VN (1972), *Sov. Phys.,- Doklady*, vol.205, pg.78.
Kirillin VA, A.E. Sheindlin (1971), "*MHD – Generators as a Method of Electric Energy Production*", Energia (Energy), Moscow.
Kirillov IA, B.V. Potapkin, V.D. Rusanov, A. Fridman (1983a), *Sov. Phys.,- High Energy Chemistry*, vol.17, pg.519.
Kirillov IA, V.D. Rusanov, A. Fridman (1981), in: *Physical Methods of Investigations of Biological and Chemical Systems*, Moscow Institute of Physics and Technology, pg.53.
Kirillov IA, V.D. Rusanov, A. Fridman (1983b), *Sov. Phys.,- Journal of Physical Chemistry*, #5, pg.280.
Kirillov IA, L.S. Polak, A. Fridman (1984a), *the IY-th USSR Symposium on Plasma Chemistry, Dnepropetrovsk*, vol.1, pg.101.
Kirillov IA, B.V. Potapkin, A. Fridman ea. (1984b), *Sov. Phys.,- High Energy Chemistry*, vol.18, pg.151.
Kirillov IA, B.V. Potapkin, M.I. Strelkova, ea. (1984c), "Dynamics of Space-Non-Uniform Vibrational Relaxation in Chemically Active Plasma", *Kurchatov Institute of Atomic Energy*, vol.3608/6, Moscow.
Kirillov IA, B.V. Potapkin, M.I. Strelkova, ea. (1984d), *Sov. Phys.,- Journal of Applied Mechanics and Technical Physics*, vol.6, pg.77-80.
Kirillov IA, V.D. Rusanov, A. Fridman (1984e), *Sov.Phys.,-Journal of Technical Physics*, vol.54, pg.s.
Kirillov IA, V.D. Rusanov, A. Fridman (1984f), in: *Kinetic and Gas-Dynamic Processes in Non-Equilibrium Medium*, Moscow State University, Moscow, pg.97.
Kirillov IA, V.D. Rusanov, A. Fridman (1986), *Sov.Phys.,- Doklady*, vol.284, pg.1352.
Kirillov IA, V.D. Rusanov, A. Fridman (1987), *Sov.Phys.,- High Energy Chemistry*, vol.21, pg.262.
Klarfeld BN (1940), *Works of the All-Union Electro-Technical Institute (VEI)*, vol.41, pg.165.
Klemenc A, H. Hinterberger, H. Hofer (1937), *Z. Elektrochem.*, vol.43, pg.261.
Klingbeil RD, A. Tidman, R.F. Fernsler (1972), *Phys. Fluids*, vol.15, pg.1969.
Kochetov IV, V.G. Pevgov, L.S. Polak, D.I. Slovetsky (1979), In: "*Plasma Chemical Reactions*", ed. by L.S.Polak, Russian Academy of Science, Moscow.
Koga K, Y. Matsuoka, K. Tanaka, M. Shiratani, Y. Watanabe (2000), *Apl. Phys. Lett.*, vol.77, pg.196.
Kogelschatz U (1988), "Advance Ozone Generation", in: *Process Technologies for Water Treatment*, ed. by S. Stucki, pg.87, Plenum Press, N.Y.
Kogelschatz U, B. Eliasson (1995), "Ozone Generation and Applications", in: *Handbook of Electrostatic Processes*, Ed. by J.S. Chang, A.J. Kelly, J.M. Crowley, pg.581, Marcel Dekker, N.Y.
Kogelschatz U, B. Eliasson, W. Egli (1997), *J. de Physique IV*, vol.7, Colloque C4, pg.4-47.
Komolov SA (1992), *Total Current Spectroscopy of Surfaces*, Gordon & Breach Science Pub.
Komachi K (1992), *J. Vac. Sci. Technol.*, vol.A11, pg.164.
Komori A, T. Shoji, K. Miyamoto, J. Kawai, Y. Kawai (1991), *Phys. Fluids*, vol.B3, pg.893.
Kondratiev VN (1971), *Rate Constants of Gas Phase Reactions*, Nauka (Science), Moscow.
Kondratiev VN, ed.(1974), "*Energy of Chemical Bonds. Ionization Potential and Electron Affinity*", Nauka (Science), Moscow.
Kondratiev VN, E.E. Nikitin (1981), *Chemical Processes in Gases*, John Wiley & Sons, N.Y.
Konenko OR, A.K. Musin (1972), *Sov. Phys.,- Journal of Technical Physics*, vol.42, pg.782.
Konenko OR, A.K. Musin, S.F. Utenkova (1973), *Sov. Phys.,- Journal of Technical Physics*, vol.43, pg.1685.

Konenko OR, A.K. Musin (1973), *Sov. Phys.,- Journal of Experimental and Theoretical Physics*, vol.43, pg.2075.
Konuma M (1992), *Film Deposition by Plasma Technologies*, Springer, N.Y.
Korobtsev S, D. Medvedev, V. Rusanov, V. Shiryaevsky (1997), *13-th Int. Symposium on Plasma Chemistry*, pg. 755, Beijing, China.
Korolev YD, G.A. Mesiatz (1982), *Field Emission and Explosive Processes in Gas Discharge*, Nauka (Science), Novosibirsk.
Knypers AD, E.H.A. Granneman, H.J. Hopman (1988), *J. Appl. Phys.*, vol.63, pg.1899.
Knypers AD, H.J. Hopman (1990), *J. Appl. Phys.*, vol.67, pg.1229.
Krall NA, A.W. Trivelpiece (1973), *Principles of Plasma Physics*, McGraw-Hill, N.Y.
Krasheninnikov EG (1981), *Experimental Investigations of Non-Equilibrium Plasma Chemical Processes in Microwave Discharges at Moderate Pressures*, Kurhcatov Institute of Atomic Energy, Moscow.
Krasheninnikov EG, V.D. Rusanov, S.V. Saniuk, A. Fridman (1986), *Sov. Phys.,- Journal of Technical Physics*, vol. 56, pg.1104.
Krasheninnikov SI (1980), "Electron Beam Interaction with Chemically Active Palsma", Ph.D. thesis, ed. Moscow Institute of Physics and Technology, Moscow.
Krasheninnikov SI, V.A. Nikiforov (1982), in: "Plasma Chemistry-9", Ed. by B.M. Smirnov, Atom-Izdat, Moscow.
Kruscal MD, R.M. Kulsrud (1958), *Phys. Fluids*, vol.1, pg.265.
Kulikovsky AA (1994), *J. Phys. D: Applied Physics*, vol.27, pg.2556.
Kummler R (1977), *J. Phys. Chem.*, vol.81, pg.2451.
Kunchardt EE, Y. Tzeng (1988), *Phys. Rev.*, A38, pg.1410.
Kurochkin YV, L.S. Polak, A.V. Pustogarov e.a. (1978), *Sov.Phys.,- Thermal Physics of High Temperatures*, vol.16, pg.1167.
Kusz J (1978), "*Plasma Generation on Ferroelectric Surface*", Edited by PWN, Warsaw-Wroclaw, Poland.
Kuznetzov NM (1971), *J. Technical and Experimental Chemistry (Tekh. Eksp. Khim.)*, vol.7, pg. 22.
Kuznetzov NM, Yu.P. Raizer (1965), *Sov.Phys.,-Journal of Applied Mechanics and Technical Physics*, #4, pg.10.
Lacour B, C. Vannier (1987), *J. Appl. Phys.*, vol.38, pg.5244.
Lafferty JM (1980), ed., "*Vacuum Arcs, Theory and Applications*", Wiley, N.Y.
Landa PS, N.A. Miskinova, Yu.V. Ponomarev (1978), *Adv. in Phys.Sci., Uspehi Phys. Nauk*, vol.126, pg.13.
Landau LD (1982), "*Quantum Electrodynamics*", Butterworth-Heinemann, 2-nd edition; also L.D. Landau, L.P. Pitaevsky, E.M. Lifshitz "Electrodynamics of Continuous Media ".
Landau LD, E.M. Lifshitz (1967), "*Physique Statistique*", Moscow, Mir, also "Statistical Physics", part1, Butterworth-Heinemann, 1980; 3-rd edition (1999).
Landau LD (1997), *Quantum Mechanics*, Butterworth-Heinemann, 3-rd edition.
Landau LD, E.M. Lifshitz (1976), "*Mechanics*", Butterworth-Heinemann, 3-rd edition.
Landau LD, E.M. Lifshitz (1997), "*The Classical Theory of Fields*", Butterworth-Heinemann.
Landau LD, E. Teller (1936), *Phys..Ztschr.Sow.*, Bd.10, S.34-39.
Langmuir I, L. Tonks (1929), *Phys. Rev.*, vol.34, pg.876.
Langmuir I (1961), *The Collected Works*, vol.3, Pergamon Press, pg.107.
Le Roy RL (1969), *J. Phys. Chem.*, vol.73, pg.4338.
Legasov VA, V.D. Rusanov, A Fridman (1978), in: "Plasma Chemistry- 5", ed. by B.M. Smirnov, Atomizdat, Moscow.
Legasov VA, R.I. Asisov, Y. P. Butylkin ea. (1983), in book: "Nuclear-Hydrogen Energy and Technology -5", Ed. by V.A. Legasov, pg.71, Energo-Atom-Izdat, Moscow.
Leonov R (1965), "*Ball Lightning Problems*", Nauka (Science), Moscow.
Lesueur H, A. Czernichowski, J. Chapelle (1990), *J. de Physique, Colloque* C5, pg.51.
Levine RD, R. Bernstein (1978), In: "Dynamics of Molecular Collisions", Ed. by W. Miller, Plenum Press, N.Y.
Levitsky AA, L.S. Polak, I.M. Rytova, D.I. Slovetsy (1981), *Sov.Phys.,- High Energy Chemistry*, vol.15, pg.276.
Levitsky AA, S.O. Macheret, A. Fridman e.a., (1983a), *High Energy Chemistry*, vol.17, pg.625.
Levitsky A, S. Macheret, A. Fridman (1983b), In: "*Chemical Reactions in Non-Equilibrium Plasma*", Ed. by L.S. Polak, Nauka (Science), Moscow.
Levitsky SM (1957), *Sov.Phys.,-Journal of Technical Physics*, vol.27, pg.970, 1001.
Lieberman MA (1988), *IEEE Trans. Plasma Sci.*, vol.16, pg.1988.
Lieberman MA (1989a), *IEEE Trans. Plasma Sci.*, vol.17, pg.338.
Lieberman MA (1989b), *J. Appl. Phys.*, vol.65, pg.4168.

Lieberman MA, A.J. Lichtenberg, S.E. Savas (1991), *IEEE Trans. Plasma Sci.*, vol.19, pg.189.
Lieberman MA, R.A. Gottscho (1994), in: "Physics of Thin Films", vol.18, ed. by M.H. Francombe and J.L. Vossen, Academic Press, N.Y.
Lieberman MA, A.J. Lichtenberg (1994), *Principles of Plasma Discharges and Material Processing*, John Wiley & Sons, N.Y.
Lifshitz A (1974), *J. Chem. Phys.*, vol.61, pg. 2478-2483.
Lifshitz EM, L.P. Pitaevsky (1979), "Physical Kinetics", Theoretical Physics, L.D. Landau, E.M. Lifshitz, vol.10, Nauka (Science), Moscow.
Likalter AA (1975a) *Sov.Phys., - Kvantovaya Electronica, Quantum Electronics*, vol.2, pg.2399.
Likalter AA (1975b), *Sov.Phys.,-Prikl. Mech.Tech.Phys., Applied Mechanics & Techn. Physics*, vol.3, pg.8.
Likalter AA (1976), *Sov.Phys.,-Prikl. Mech.Tech.Phys., Applied Mechanics & Techn. Physics*, vol.4, pg.3.
Likalter AA, G.V. Naidis (1981), in: "Plasma Chemistry - 8", ed. by B.M. Smirnov, Energoizdat, Moscow.
Lin I (1985), *J. Appl. Phys.*, vol.58, pg.2981.
Liventsov VV, V.D. Rusanov, A. Fridman (1983), *Sov.Phys.,-Journal of Technical Physics, Letters*, vol.7, pg.163.
Liventsov VV, V.D. Rusanov, A. Fridman (1984), *Sov.Phys.,- Doklady*, vol.275, pg.1392.
Liventsov VV, V.D. Rusanov, A. Fridman (1988), *Sov. Phys.,- High Energy Chemistry*, vol.22, pg.67.
Liventsov V, V. Rusanov, A. Fridman, G. Sholin (1981), *Sov.Phys.,-Journal of Technical Physics, Letters*, vol.9, pg.474.
Loeb LB (1960), "*Basic Processes of Gaseous Electronics*", University of California Press, Berkeley.
Loeb LB (1965), *Electrical Coronas – Their Basic Physical Mechanisms*, University of California Press, Berkeley.
Losev SA (1977), *Gas-Dynamic Lasers*, Nauka (Science), Moscow.
Losev SA, N.A. Generalov (1961), *Sov. Phys.,-Doklady*, vol.141, pg.1072.
Losev SA, A.L. Sergievska, V Rusanov, A. Fridman, S.O. Macheret (1996), *Sov. Phys.- Doklady*, vol. 346, pg.192.
Losev SA, O.P. Shatalov, M.S. Yalovik (1970), *Sov.Phys.,-Doklady*, vol.195, pg.585.
Lozansky ED, O.B. Firsov (1975), "*Theory of Sparks*", Atomizdat, Moscow.
Lukyanova AB, A.T. Rahimov, N.B. Suetin (1990), *Sov. Phys.,- Plasma Physics*, vol.16, pg.1367.
Lukyanova AB, A.T. Rahimov, N.B. Suetin (1991), *Sov. Phys.,- Plasma Physics*, vol.17, pg.1012.
Lyubimov GA, V.I. Rachovsky (1978), "Cathode Spot of Vacuum Arc", *Adv. in Phys.Sci., Uspehi Phys. Nauk*, vol.125, pg.665.
Lundquist S (1952), *Ark. f. Fys.*, vol.5, pg.297.
MacDonald AD, S.J. Tetenbaum (1978), in "Gaseous Electronics", ed. by M.N. Hirsh, H.J. Oskam, vol.1, "*Electrical Discharges*", Academic Press, N.Y.
Macheret SO, A. Fridman, I.V. Adamovich, J.W. Rich, C.E. Treanor (1994), "Mechanism of Non-Equilibrium Dissociation of Diatomic Molecules", AIAA-Paper # 94-1984.
Macheret SO, V.D. Rusanov, A. Fridman, G.V. Sholin (1979), "Plasma Chemistry-1979", *III-Symposium on Plasma Chemistry*, pg.101, Nauka (Science), Moscow.
Macheret SO, V.D. Rusanov, A. Fridman, G.V. Sholin (1980a), *Sov.Phys.,- Journal of Technical Physics*, vol.50, pg.705.
Macheret SO, V.D. Rusanov, A. Fridman, G.V. Sholin (1980b), *Sov.Phys.,-Doklady*, vol.255, pg.98.
Macheret SO, V.D. Rusanov, A. Fridman (1984), *Sov..Phys.,-Doklady*, vol.276, pg.1420.
Macheret SO, V.D. Rusanov, A. Fridman (1985), "*Non-Equilibrium Clusterization in the Field of Centrifugal Forces and Dissociation of Molecules in Plasma*", Ed. by Kurchatov Institute of atomic Energy, IAE-4220/6, Moscow.
Makarov A, A. Puretzky, V. Tyakht (1980), *J.. Appl. Phys.*, vol.23, pg.391
Makarov A, V. Tyakht (1982), *Sov.Phys.-JETP, the Journal of Experimental and Theoretical Physics*, vol.83, pg.502.
Maker PD, R.W. Terhune, C.M. Savage (1964), in: "Quantum Electronics III", ed. by P. Grivet, N. Bloembergen, Columbia University Press, N.Y.
Maksimenko AP, V.P. Tverdokhlebov (1964) *Sov. Phys.,- News of the Higher Education, Physics*, vol.1, pg.84.
Mamedov SS (1979), Trudy FIAN, *Physical Institute of Academy of Sciences*, vol.107, pg.3.
Mandich ML, W.D. Reents, Jr (1992), *J. Chem. Phys.*, vol. 96, pg. 4233.
Manos DM, D.L. Flamm (1989), *Plasma Etching: An Introduction*, Academic Press, N.Y.
Margolin AD, A.V. Mishchenko, V.M. Shmelev (1980), *Sov.Phys.,-High Energy Chemistry*, vol.14, pg.162.
Marode E ea. (1995), "*Gas Dicharges and Their Applications*", GD-95, pg.II-484, Tokyo.
Marrone PV, C.E. Treanor (1963), *Physics of Fluids*, vol.6, pg.1215.
Massey H, E. Burhop, H.B. Gilbody (1974), *Electron and ion Impact Phenomena*, Clarendon, Oxford, UK.

Massey H (1976), *Negative Ions*, Cambridge University Press, Cambridge.
Matsuoka M, K. Ono (1988), *J. Vac. Sci. Technol.*, vol.A6, pg.25.
Mattachini JE, E. Sani, G. Trebbi (1996), in *Proceedings of the Int. Workshop on Plasma Technologies for Pollution Control and Waste Treatment*, MIT, Cambridge, MA.
Mayer JE, G.M. Mayer (1966), "*Statistical Mechanics*", 11th edition, Wiley, N.Y.
Mayerovich BE (1972), *Sov. Phys., - Journal of Experimental and Theoretical Physics*, vol.63, pg.549.
McDaniel EW (1964), *Collision Phenomena in Ionized Gases*, Wiley, N.Y.
McDaniel EW (1989), *Atomic Collisions: Electron and Photon Projectiles*, Wiley, N.Y.
McDaniel EW, E.A. Mason (1973), *The Mobility and Diffusion of Ions in Gases*, Wiley, N.Y.
Meek JM, J.D. Craggs (1978), "*Electrical Breakdown of Gases*", Wiley, N.Y.
Meijer EJ, D. Frenkel (1991), *J. Chem. Phys.*, vol.94, pg.2269.
Melrose DB (1989), *Instabilities in Space and Laboratory Plasmas*, Cambridge Univ. Press, Cambridge.
Mesyats GA, G.P. Mkheidze, A.A. Savin (2000), in: *Encyclopedia of Low Temperature Plasma*, ed. by V.E. Fortov, vol.4, pg.108, Nauka (Science), Moscow.
Meyer J (1969), *J. Phys. D: Applied Physics*, vol.2, pg.221.
Mezdrikov OA (1968), "*Electrical Methods of Volumetric Granulometry*", Energia (Energy), Leningrad.
Migdal AB (1981), *Qualitative Methods in Quantum Mechanics*, Nauka (Science), Moscow.
Mikhailovskii AB (1998a), "Instabilities in a Confined Plasma (Plasma Physics Series)", Inst. of Phys. Pub.
Mikhailovskii AB (1998b), "Theory of Plasma Instabilities", Inst. of Phys. Pub.
Millican RS, D.R. White (1963), *J. Chem. Phys.*, vol.39, pg.3209-3215.
Mitsuda Y, T. Yoshida, K. Akashi (1989), *Rev. Sci. Instrum.*, vol. 60, pg.249.
Modest MF (1993), "*Radiative Heat Transfer*", Mc. Graw-Hill, N.Y.
Modinos A (1984), *Field, Thermionic and Secondary Electron Emission Spectroscopy*, Plenum Pub.
Moisan M, Z. Zakrzewski (1991), *J. Physics D: Applied Physics*, vol.24, pg.1025.
Molotkov VI, A.P. Nefedov, O.F. Petrov, A.G. Khrapak, S.A. Khrapak (2000), in: *Encyclopedia of Low Temperature Plasma*, ed. by V.E. Fortov, vol.3, pg.160, Nauka (Science), Moscow.
Morfill GE, H. Thomas (1996), *J. Vacuum Sci. Technol., A*, vol.14, pg.490.
Morrow R, J.J. Lowke (1997), *J. Phys. D: Applied Physics*, vol.30, pg.614.
Moskalev BI (1969), *The Hollow Cathode Discharge*, Energy, Moscow.
Munt R, R.S.B. Ong, D.L. Turcotte (1969), *J. Plasma Phys.*, vol.11, pg.739.
Musin AK (1974), *Ionization Processes in Heterogeneous Gas Plasma*, ed. by the Moscow State University, Moscow.
Mutaf-Yardimci O, A. Saveliev, A. Fridman, L. Kennedy (1998a), *Int. Journal of Hydrogen Energy*, vol.23(12), pg.1109.
Mutaf-Yardimci O, L. Kennedy , A. Saveliev, A. Fridman (1998b), "Plasma Exhaust Aftertreatment", *SAE, SP-1395*, pg.1.
Mutaf-Yardimci O, A. Saveliev, A. Fridman, L. Kennedy (1999), *Journal of Applied Physics*, vol.87, pg.1632.
Natanson G (1959) *Sov.Phys., Journal of Technical Physics*, vol.29, pg.1373-1378.
Naville AA, C.E. Guye (1904), French Patent No. 350 120.
Nedospasov AV, Yu.B. Ponomarenko (1965), *Sov. Phys.,- Thermal Physics of High Temperatures*, vol.3, pg.17.
Nedospasov AV (1968), Adv. in *Phys.Sci., Uspehi Phys. Nauk*, vol.94, pg.439.
Nedospasov AV, V.V. Khait (1979), *Oscillations and Instabilities of Low-Temperature Plasma*, Nauka (Science), Moscow.
Nefedov AP (2001), *XXV- Int. Conference on Phenomena in Ionized Gases, ICPIG-25*, Nagoya, Japan, vol.2, pg.1.
Nefedov AP, O.F. Petrov, V.E. Fortov (1997), *Adv. in Phys.Sci., Uspehi Phys. Nauk*, vol.167, pg.1215; (Phys. Usp., vol.40, pg.1163).
Nester S, A.V. Demura, A. Fridman (1983), "*Disproportioning of the Vibrationally Excited CO-Molecules*", Kurchatov Institute of Atomic Energy, #3518/6, Moscow.
Nester S, B. Potapkin, A. Levitsky, V. Rusanov, B. Trusov, A. Fridman (1988), "*Kinetic and Statistic Modeling of Chemical Reactions in Gas Discharges*", CNII ATOM INFORM, Moscow.
Nester SA, V.D. Rusanov, A. Fridman (1985), "*Dissociation of Hydrogen Sulfide in Plasma with Additives*", ed. Kurchatov Institute of Atomic Energy, vol.4223/6, Moscow.
Nighan WL (1976), in book "*Principles of Laser Plasma*", Ed. by G. Bekefi, John Wiley & Sons, N.Y.
Nighan WL, W.G. Wiegand (1974), *Appl. Phys. Lett.*, vol.25, pg.633.
Nikerov VA, G.V. Sholin (1978a), *Sov. Phys.,- Plasma Physics*, vol.4, pg.1256.
Nikerov VA, G.V. Sholin (1978b), "*Fast Electrons Stopping in Gas, Degradation Spectrum Approach*", ed. Kurchatov Inst. of Atomic Energy, vol.2985.

Nikerov VA, G.V. Sholin (1985), *Kinetics of Degradation Processes*, Energo-Atom-Izdat, Moscow.
Nikitin EE (1970), "*Theory of Elementary Atom-Molecular Processes in Gases*", Chimia (Chemistry), Moscow.
Nikitin EE, A.I. Osipov (1977), *Vibrational Relaxation in Gases*, VINITI, Moscow.
Okano H, T. Yamazaki, Y. Horiike (1982), *Solid State Technol.*, vol.25, pg.166.
Okazaki S, M. Kogoma (1994), *J. Phys. D, Applied Physics*, vol.27, pg.1985.
O'Neil TM (1965), *Phys. Fluids*, vol.8, pg.2255.
Opalinska T, A. Szymanski (1996), *Contrib. Plasma Physics*, vol.36, pg.63.
Palm P, E. Plonjes, M. Buoni, V.V. Subramaniam, I.V. Adamovich (2001), *Journal of Applied Physics*, vol.89, pg.5903.
Panteleev VI (1978), "Electro-Ionization Synthesis of Chemical Compounds", Ph.D.-thesis, Physical Institute of Academy of Sciences, Moscow.
Parity BS (1967), *J. Fluid Mechanics*, vol.27, pg.49.
Park C (1987), "Assessment of Two-Temperature Kinetic Model for Ionizing Air", AIAA-Paper #87-1574.
Park M, D. Chang, M. Woo, G. Nam, S. Lee (1998), in "*Plasma Exhaust Aftertreatment*", Society of Automotive Engineers, Warrendale, PA.
Parker JG (1959), *Phys. Fluids*, vol.2, pg.449.
Pashkin SV, P.I. Peretyatko (1978), *Sov. Phys.,- Quantum Electronics*, vol.5, pg.1159.
Patel CKN (1964), *Phys. Rev. Lett.*, vol.13, pg.617.
Pekarek L (1968), *Adv. in Phys.Sci., Uspehi Phys. Nauk*, vol.94, pg.463.
Pellerin S, J.-M. Cormier, F. Richard, K. Musiol, J. Chapelle (1996), *J.Phys. D: Applied Physics*, vol.29, pg.726.
Penetrante BM, R.M. Brusasco, B.T. Meritt, W.J. Pitz, G.E. Vogtlin, M.C. Kung, H.H. Kung, C.Z. Wan, K.E. Voss (1998), in "*Plasma Exhaust Aftertreatment*", Society of Automotive Engineers, Warrendale, PA.
Penetrante BM, M.C. Hsiao, J.N. Bardsley, B.T. Meritt, G.E. Vogtlin, P.H. Wallman, A. Kuthi, C.P. Burkhart, J.R. Bayless, in *Proceedings of the Int. Workshop on Plasma Technologies for Pollution Control and Waste Treatment, MIT*, Cambridge, MA.
Penetrante B, S. Schulteis (1993), eds., *NATO ASI series, vol. G 34B*, Springer, Berlin.
Penning FM (1936), *Physica*, vol.3, pg.873.
Penning FM (1937), *Physica*, vol.4, pg.71.
Penning FM, J.H.A. Moubis (1937), *Physica*, vol.4, pg.1190.
Penski V (1968), "Theoretical Calculations of Transport Properties in Nitrogen Plasma", IV-Symp. on Thermophys. Properties.
Perrin J (1991), *J.Non- Cryst. Solids*, vol. 137&138, pg. 639.
Perrin J (1993), *J. Phys. D: Appl. Phys.* Vol.26, pg.1662.
Perrin J, C. Bohm, R. Etemadi, A. Lloret (1993), NATO Advanced Research Workshop "*Formation, Transport and Consequences of Particles in Plasma*", France.
Pfender E (1978), *"Electric Arcs and Arc Gas Heaters", Chapter 5 in "Gaseous Electronics"*, vol.1, ed. by M.N. Hirsh and H.J. Oscam, Academic Press, N.Y.
Pikaev AK (1965), *Pulse Radiolysis of Water and Water Solutions*, Nauka (Science), Moscow.
Pietsch GJ, D. Braun, V.I. Gibalov (1993), "Modeling of Dielectric Barrier Discharges", *NATO ASI series, vol. G34, part A, "Non-Thermal Plasma Techniques for Pollution Control"*, ed. by B.M. Penetrante, S.E. Schultheis, pg.273, Springer, Berlin.
Piley ME, M.K. Matzen (1975), *J. Chem. Phys.*, vol. 63, pg. 4787.
Pitaevsky LI (1962), *Sov.Phys.,- Journal of Experimental and Theoretical Physics*, vol.42, pg.1326.
Platonenko V, N. Sukhareva (1980), *Sov.Phys.-JETP, the Journal of Experimental and Theoretical Physics*, vol.78, pg.2126.
Plonjes E, P. Palm, J.W. Rich, I. Adamovich (2001), *32-nd AIAA Plasmadynamics and Lasers Conference, AIAA-2001-3008*, Anaheim, CA.
Poeschel RL, J.R. Beattle, P.A. Robinson, J.W. Ward (1979), *14-th Int. Electric Propulsion Conf., Princeton, NJ, paper* 79-2052.
Polak LS, A.A. Ovsiannikov, D.I. Slovetsky, F.B. Vursel (1975) *Theoretical and Applied Plasma Chemistry*, Nauka (Science), Moscow.
Polak LS, D.I. Slovetsky, Yu.P. Butylkin (1977), "*Carbon Dioxide Dissociation in Electric Discharges*", Institute of Petrol-Chemical Synthesis, Nauka (Science), Moscow.
Polishchuk A, V.D. Rusanov, A. Fridman (1980), *Theoretical and Experimental Chemistry*, vol.16, pg.232.
Poluektov NP, N.P. Efremov (1998), *J. Phys. D.*, vol.31, pg.988.
Porteous RK, T Hbid, L. Boufendi, A. Fridman, B.V. Potapkin, A. Bouchoule (1994), *ESCAMPIG-12, Euro-Physics Conference Abs., Noordwijkerhout*, Netherlands, 18E, pg.83.
Porteous RK, D.B. Graves (1991), *IEEE Trans. Plasma Sci.*, vol.19, pg.204.

Potapkin BV, V.D. Rusanov, A.E. Samarin, A. Fridman (1980), *Sov.Phys.-High Energy Chemistry*, vol. 14, pg.547.

Potapkin BV, V.D. Rusanov, A. Fridman (1983), *Sov.Phys., - High Energy Chemistry*, vol.17, pg.528.

Potapkin BV, V.D. Rusanov, A. Fridman (1984), *Sov.Phys., - High Energy Chemistry*, vol.18, pg.252.

Potapkin BV, V.D. Rusanov, A. Fridman (1985), "Non-equilibrium Effects in Plasma, Provided by Selectivity of Transfer Processes ", *Kurchatov Institute of Atomic Energy*, vol.4219/6, Moscow.

Potapkin BV, V.D. Rusanov,A. Fridman (1989), *Sov.Phys.-Doklady*, vol.308, pg.897.

Potapkin BV, M. Deminsky, A. Fridman, V.D. Rusanov (1995), *Radiat. Phys. Chem.*, vol.45, pg1081-1088.

Provorov AS, V.P. Chebotaev (1977), in book: "Gas Lasers", Ed. by R.I. Soloukhin, pg.174, Nauka (Science), Novosibirsk.

Pu YK, P.P. Woskov (1996), ed., *Proceedings of the Int. Workshop on Plasma Technologies for Pollution Control and Waste Treatment, MIT*, Cambridge, MA.

Radzig AA, B.M. Smirnov (1980), "*Handbook on Atomic and Molecular Physics*", Atomizdat, Moscow.

Raether H (1964), *Electron Avalanches and Breakdown in Gases*, Butterworth, London.

Raghavachari K (1992), *J. Chem. Phys.*, vol. 96, 4440.

Raizer YP (1970), Sov. Phys.,- *Letters to the Journal of Experimental and Theoretical Physics*, vol.11, 195.

Raizer YP (1972a), *Sov.Phys., Thermal Physics of High Temperatures*, vol.10, pg.1152.

Raizer YP (1972b), *Sov. Phys.,- Adv. in Phys. Sci., Uspehi Phys. Nauk*, vol.108, pg.429.

Raizer YP (1974), *Laser Spark and Discharge Propagation*, Nauka (Science), Moscow.

Raizer YP (1977), "*Laser-Induced Discharge Phenomena*", Consultants Bureau, N.Y.

Raizer YP, S.T. Surzhikov (1987), *Sov.Phys.,-Journal of Technical Physics, Letters*, vol.13, pg.452.

Raizer YP, S.T. Surzhikov (1988), *Sov.Phys.,- Thermal Physics of High Temperatures*, vol.26, pg.428.

Raizer YP (1991), *Gas Discharge Physics*, Springer, Berlin, Heidelberg, N.Y.

Raizer YP, M.N. Shneider, N.A. Yatsenko (1995), *Radio-Frequency Capacitive Discharges*, Nauka (Science), Moscow and CRC Press, N.Y.

Rau ARP, M. Inokuti, D.A. Douthart (1978), *Phys. Rev.*, vol.18, pg.971.

Reents WD, Jr., M.L. Mandich (1992), *J.Chem.Phys.*, vol.96, pg. 4449.

Rapp D (1965), *J.Chem.Phys.*, vol.43, pg.316.

Rapp D, W.E. Francis (1962), *J.Chem.Phys.*, vol.37, pg.2631.

Rayleigh JWS (1878), "*The Theory of Sound*", vol.2, later printed: Dover Publications, 1945, N.Y.

Richard F, M. Cormier, S. Pellerin, J. Chapelle (1996), *J. Appl. Phys.* Vol.79, #5, p. 2245

Rich JW, R.C. Bergman (1986), "Isotope Separation by Vibration-Vibration Pumping", *Topics of Current Physics-39*, Ed. by M. Capitelli, Springer Verlag, Berlin.

Ridge MI, R.P. Howson (1982), *Thin Solid Films*, vol.96, pg.113.

Robertson GD, N.A. Baily (1965), *Bull. Am. Phys. Soc.*, vol.2, pg.709.

Rockwood SD, J.E. Brau, W.A. Proctor, G.H. Canavan (1973), *IEEE, J. Quantum Electronics*, vol.9, pg.120.

Rogdestvensky BL, N.N. Yanenko (1978), *Systems of Quazi-Linear Equations*, Nauka (Science), Moscow.

Rosen G (1962), *Phys.Fluids, vol.5, pg*.737.

Roth JR (1966), *Rev. Sci. Instrum.*, vol.37, pg.1100.

Roth JR (1967), *Phys. Fluids*, vol.10, pg.2712.

Roth JR (1969), *Plasma Physics*, vol.11, pg.763.

Roth JR (1973a), *IEEE Trans. On Plasma Sci.*, vol.1, pg.34.

Roth JR (1973b), *Plasma Physics*, vol.15, pg.995.

Roth JR, M. Laroussi, C. Liu (1992), *Proceedings,19-th IEEE, International Conference on Plasma Science*, Tampa, FL.

Roth JR (2000), *Industrial Plasma Engineering*, Institute of Physics Publishing, Bristol & Philadelphia.

Rozovsky MO (1972), *Sov.Phys., Journal of Applied Mechanics and Technical Physics*, #6, pg.176.

Rusanov VD, A. Fridman (1976a), *Sov.Phys.,-Doklady*, vol.230, pg.809.

Rusanov VD, A. Fridman (1976b), *Sov.Phys.,-Doklady*, vol.231, pg.1109.

Rusanov VD, A. Fridman, S.O. Macheret (1985a), *Sov.Phys.,- Doklady*, vol.283, pg.590.

Rusanov VD, A. Fridman, G.V. Sholin (1977), *Sov.Phys.,-Doklady*, vol.237, pg.1338.

Rusanov VD, A. Fridman, G.V. Sholin (1978), in: *Plasma Chemistry-5*, ed. by B.M. Smirnov, Atomizdat, Moscow.

Rusanov VD, A. Fridman, G.V. Sholin (1979a), *Sov.Phys., - Journal of Technical Physics*, vol.49, pg.554.

Rusanov VD, A. Fridman, G.V. Sholin (1979b), *Sov.Phys., - Journal of Technical Physics*, vol.49, pg.2169.

Rusanov VD, A. Fridman, G.V. Sholin (1981), *Adv. in Phys.Sci., Uspehi Phys. Nauk*, vol.134, pg.185.

Rusanov VD, A. Fridman, G.V. Sholin (1982), in book: "*Heat and Mass-Transfer in Plasma-Chemical Systems*", Minsk, Nauka (Science), vol.1, pg.137.

Rusanov VD, A. Fridman (1984), *Physics of Chemically Active Plasma*, Nauka, Moscow.
Rusanov VD, A. Fridman, G. Sholin (1986), "Vibrational Kinetics and Reactions of Polyatomic Molecules", Topics of Current Physics-39, Ed. by M. Capitelli, Springer Verlag, Berlin.
Rusanov VD, A. Fridman, G.V. Sholin, B.V. Potapkin (1985b), *Kurchatov Institute of Atomic Energy*, Preprint #4201, Moscow.
Rusanov VD, A.S. Petrusev, B.V. Potapkin, A. Fridman, A. Czernichowski, J. Chapelle (1993), *Sov.Phys.,- Doklady*, vol.332, #6, pg.306.
Rutherford PH, R.J. Goldston (1995), "Introduction to Plasma Physics", Inst. of Phys. Pub.
Sabo ZG (1966), *Chemical Kinetics and Chain Reactions*, Nauka (Science), Moscow.
Safarian MN, E.V. Stupochenko (1964) *Sov.Phys.,- Journal of Applied Mechanics and Technical Physics*, #7, pg.29.
Samoylovich VG, V.I. Gibalov, K.V. Kozlov (1989), *Physical Chemistry of Barrier Discharge*, Edition of the Moscow State University, Moscow.
Samuilov EV (1966a), *Sov. Phys.,- Doklady*, vol.166, pg.1397.
Samuilov EV (1966b), *Sov. Phys.,- Thermal Physics of High Temperatures*, vol.4, pg.143.
Samuilov EV (1966c), Sov. Phys.,- Thermal Physics of High Temperatures, vol.4, pg.753.
Samuilov EV (1967), in: "*Property of Gases at High Temperatures*", Nauka (Science), Moscow, pg.3.
Samuilov EV (1968), in: "*Physical Gas Dynamics of Ionized and Chemically Reacting Gases*", Nauka (Science), Moscow, pg.3.
Samuilov EV (1973), in: "*Thermo-Physical Properties of Gases*", Nauka (Science), Moscow, pg.153.
Sato SJ (1955), *J. Chem. Phys.*, vol.23, pg.2465.
Sattler K (1982), *13-th Int. Symp. on Rarefied Gas Dynamics*, vol.1, pg.252, Novosibirsk.
Sayasov YS (1958), *Sov. Phys.,- Doklady*, vol.122, pg.848.
Schiller S, G. Beister, E. Buedke, H.J. Becker, H. Schmidt (1982), *Thin Solid Films*, vol.96, pg.113.
Secrest D (1973), *Annu. Rev. Phys. Chem.*, vol.24, pg.379.
Sergeev PA, D.I. Slovetsky (1979), "Plasma Chemistry-1979", *III-Symposium on Plasma Chemistry*, pg.132, Nauka (Science), Moscow.
Sergievska AL, E.A. Kovach, S.A. Losev (1995), *Mathematical Modeling in Physical- Chemical Kinetics*, Moscow State University, Institute of Mechanics, Moscow.
Shchuryak E (1976), *Sov.Phys.-JETP, the Journal of Experimental and Theoretical Physics*, vol.71, pg.2039.
Schultz GJ (1976), in "Principles of Laser Plasma", ed.by G. Bekefi, John Wiley & Sons Inc., N.Y.
Schwartz RN, Z.I. Slawsky, K.F. Herzfeld (1952), *J.Chem.Phys.*, vol.20, pg.1591.
Selwyn GS, J.E. Heidenreich, K.L. Haller (1990), *Appl. Phys. Lett.*, vol.57, pg.1876.
Semenov NN (1958), "*Some Problems of Chemical Kinetics and Reactivity*", the USSR Academy of Science, Moscow.
Sholin GV, V.A. Nikerov, V.D. Rusanov (1980), *Journal de Physique*, vol.41, C9, pg.305.
Shuler K, J. Weber (1954), *J. Chem. Phys.*, vol.22, pg.3.
Siemens W (1857), *Poggendorfs Ann. Phys. Chem.*, vol.102, pg.66.
Singer S (1973), "The Nature of Ball Lightning", ASIN 0306304945.
Slivkov IN (1966), *Electric Breakdown and Discharge in Vacuum*, AtomIzdat, Moscow.
Slovetsky DI (1980), *Mechanisms of Chemical Reactions in Non-Equilibrium Plasma*, Nauka, Moscow.
Smirnov AS (2000), in: *Encyclopedia of Low Temperature Plasma*, ed. by V.E. Fortov, vol.2, pg.67, Nauka (Science), Moscow.
Smirnov BM (1968), *Atomic Collisions and Elementary Processes in Plasma*, Atomizdat, Moscow.
Smirnov BM (1974), *Ions and Excited Atoms in Plasma*, Atomizdat, Moscow.
Smirnov BM (1975), *Adv. in Phys.Sci., Uspehi Phys. Nauk*, vol.116, pg.111.
Smirnov BM (1977a), in: "*Plasma Chemistry-4*", ed. by B.M. Smirnov, pg.191, Atomizdat, Moscow.
Smirnov BM (1978), "*Negative Ions*", Atomizdat, Moscow, 1978, and McGrow-Hill, N.Y., 1982.
Smirnov BM (1981), "*Physics of Weakly Ionized Gases*", Mir, Moscow; "Physics of Ionized Gases" (2001), John Wiley & Sons, N.Y.
Smirnov BM (1977b), "*Introduction to Plasma Physics*", Mir, Moscow, 1977, and Nauka, Moscow, 1982.
Smullin LD, P. Chorney (1958), *Proc. IRE*, vol.46, pg.360.
Shneider MN, M. Pekker, A. Fridman (2012), *IEEE Trans. Dielectr. Electr. Insul.*, vol. 19, 1579.
Sobacchi M, A. Saveliev, e.a. (2001), *International Journal of Hydrogen Energy*, vol.26.
Sodha MS, S. Guha (1971), Adv. *Plasma Physics*, vol.4, pg.219.
Sodha MS (1963), *Brit. J. Appl. Phys.*, vol.14, pg.172.
Sommerer TJ, M.S. Barnes, J.H. Keller, M.J. McCaughey, M. Kushner (1991), *Appl. Phys. Lett.*, vol.59, pg.638.
Souris AL (1985), *Thermodynamics of High Temperature Processes*, Metallurgy, Moscow.

Spahn RG, L.J. Gerenser (1994), US Patent No. 5,324,414.

Spears K, R. Kampf, T. Robinson (1988), *J.Phys. Chem.*, vol.92, pg. 5297.

Spencer LV, U. Fano (1954), *Phys. Rev.*, vol.93, pg.1172.

Spitzer L (1941), *Astrophys. J.*, vol.93, pg.369.

Spitzer L (1944), *Astrophys. J., vol.*vol.107, pg.6.

Staack D, A. Fridman, A. Gutsol ea. (2008), *Angewandte Chemie*, vol.47, issue 42, pg. 8020

Stakhanov IP (1973), *Sov.Phys.,- Journal of Theoretical and Experimental Physics, Letters*, vol.18, pg.193.

Stakhanov IP (1974), *Sov.Phys.,- Journal of Technical Physics*, vol.44, pg.1373.

Stakhanov IP (1976), *Physical Nature of Ball lightning*, Nauka (Science), Moscow.

Starikovskiy A, Y. Yang, Y.I. Cho, A. Fridman (2011), *Plasma Sources Science and Technology*, vol. 20, 024003

Stengach VV (1972), *Sov. Phys.,- Journal of Applied Mechanics and Technical Physics*, #1, 128.

Stengach VV (1975), *Sov. Phys.,- Journal of Applied Mechanics and Technical Physics*, #2, 159.

Stenhoff M (2000), *Ball Lightning: An Unsolved Problem in Atmospheric Physics*, Kluwer Academic Publishers.

Stevens JE, Y.C. Huang, R.L. Larecki, J.L. Cecci (1992), *J.Vac. Sci. Technol.*, vol.A10, pg.1270.

Stevens MJ, M.O. Robbins (1993), *J. Chem. Phys.*, vol.98, pg.2319.

Stix TH (1992), *Waves in Plasmas*, Springer Verlag, Berlin, Heidelberg.

Stolarsky RS, A.E.S. Green (1967), *J. Geophys. Res.*, vol.62, pg.3967.

Su T, M.T. Bowers (1973), *Int.J.Mass Spectrom.Ion Phys.*, vol.12, pg.347.

Su T, M.T. Bowers (1975), *Int.J.Mass Spectrom.Ion Phys.*, vol.17, pg.221.

Sugano T (1985), ed., *Applications of Plasma Processes to VLSI Technology*, John Wiley, N.Y.

Sydney R (1970), in: *Non-Equilibrium Flows*, vol.2, pg. 160, ed. by P.P. Wegener, Dekker, N.Y.

Szymanski A (1985), *Beitr. Plasmaphys.*, vol.2, pg.133.

Talrose VL (1952), Ph.D. Dissertation, N.N.Semenov Institute of Chemical Physics, Moscow, also – Sov.Phys.-Doklady, 1952, vol.86, pg. 909.

Talrose VL, P.S. Vinogradov, I.K. Larin (1979), "*Gas Phase Ion Chemistry*", ed. by M. Bowers., Academic Press, vol.1, pg. 3O5.

Tihomirova N, V.V. Voevodsky (1949), *Sov.Phys.- Doklady*, vol.79, pg. 993.

Thomas H, G.E. Morfill, V. Demmel, ea., (1994), *Phys. Rev. Lett.*, vol.73, pg.652.

Thompson WB (1962), *An Introduction to Plasma Physics*, Pergamon Press, Addison-Wesley Publishing Company, Oxford.

Tomson JJ (1912), *Philos. Mag.*, vol.23, pg.449.

Tomson JJ (1924), *Philos. Mag.*, vol.47, pg.337.

Tonks L, I. Langmuir (1929), *Phys. Rev.*, vol.34, pg.876.

Traving G (1968), in *Plasma Diagnostics*, ed. by Lochte-Holtgreven, Wiley, N.Y.

Treanor CE, I.W. Rich and R.G. Rehm (1968), *J. Chem. Phys.*, vol.48, pg.1798.

Trivelpiece AW, R.W. Gould (1959), *J. Appl. Phys.*, vol.30, pg.1784.

Trottenberg T, A. Melzer, A. Piel (1995), *Plasma Sources Sci. Technol.*, vol.4, pg.450.

Tsendin LD (1970), *Sov. Phys.,- Journal of Technical Physics*, vol.40, pg.1600.

Uman MA (1964), *Introduction to Plasma Physics*, McGraw Hill, N.Y.

Uman M (1969), "*Lightning*", McGraw Hill, N.Y.

Uzhov VN (1967), "*Industrial Gas Cleaning by Electric Filters*", Khimia (Chemistry), Moscow.

Vahedi V, C.K. Birdsall, M.A. Lieberman, G. DiPeso, T.D. Rognlien (1993), *Phys. Fluids*, vol.B5(7), pg.2719.

Vahedi V, C.K. Birdsall, M.A. Lieberman, G. DiPeso, T.D. Rognlien (1994), *Plasma Sources Sci. Technol.*, vol.2, pg.273.

Vakar AK, V.K. Givotov, E.G. Krasheninnikov, A. Fridman, ea. (1981), *Sov. Phys.,-Journal of Technical Physics, Letters*, vol.7, pg.996.

Vakar AK, E.G. Krasheninnikov, E.A. Tischenko (1984), *IV-th Symposium on Plasma Chemistry*, Dnepropetrovsk, pg.35.

Vasiliev MN (2000), in: *Encyclopedia of Low Temperature Plasma*, ed. by V.E. Fortov, vol.4, pg.436, Nauka (Science), Moscow.

Vasilyak LM, S.V. Kostuchenko, N.N. Kurdyavtsev, I.V. Filugin (1994), *Adv. in Phys.Sci., Uspehi Phys. Nauk*, vol.164, pg.263.

Vasilyak LM, S.P. Vetchinin, V.S. Zimnukov, A.P. Nefedov, D.N. Polyakov, V.E. Fortov (2001), *XXV- Int. Conf. on Phenomena in Ionized Gases, ICPIG-25*, Nagoya, Japan, vol.3, pg.55.

Vdovin YA (1975), *Sov. Phys.,- Journal of Technical Physics*, vol.45, pg.630.

Vedenov AA (1973), *Ionization Explosion in Glow Discharge, ICPIG-XI*, pg.108, Prague.

Vedenov AA (1982), *Physics of Electric Discharge CO2-Lasers*, EnergoAtom-Izdat, Moscow.

van Veldhuizen EM, W.R. Rutgers, V.A. Bityurin (1996), *Plasma Chemistry and Plasma Processing*, vol.16, pg.227.

Velikhov EP, V.S. Golubev, S.V. Pashkin (1982), *Adv. in Phys.Sci., Uspehi Phys. Nauk*, vol.137, pg.117.

Velikhov EP, S.A. Golubev, A.T. Rakhimov (1973), *Sov. Phys.,- Journal of Experimental and Theoretical Physics*, vol.65, pg.543.

Velikhov EP, S.A. Golubev, A.T. Rakhimov (1975), *Sov. Phys.,- Plasma Physics*, vol.1, pg.847.

Velikhov EP, A.S. Kovalev, A.T. Rakhimov (1987), "*Physical Phenomena in Gas-Discharge Plasma*", Nauka (Science), Moscow.

Velikhov EP, V.D. Pismennyi, A.T. Rakhimov (1977), *Adv. in Phys.Sci., Uspehi Phys. Nauk*, vol.122, pg.419.

Vender D, R.W. Boswell (1990), *IEEE Trans. Plasma Sci.*, vol.18, pg.725.

Veprek S (1972), *J.Chem. Phys.*, vol. 57, pg. 952, and J. Crystal Growth, vol. 17, pg. 101.

Veprek S, M.G. Veprek-Heijman (1990), *Appl. Phys. Lett.*, vol.56, pg. 1766.

Veprek S, M.G. Veprek-Heijman (1991), *Plasma Chem. Plasma Process.*, vol.11, pg.323.

Virin L, R. Dgagaspanian, G. Karachevtsev, V. Potapov, V. Talrose (1978), *Ion-Molecular Reactions in Gases*, Nauka, Moscow.

Vitello PA, B.M. Penetrante, J.N. Bardsley (1994), *Phys. Rev.*, vol. E49, pg.5574.

Vostrikov AA, D.Yu. Dubov (1984a), "Real Cluster Properties and Condensation Model", ed. by *Institute of Thermal Physics of the Siberian Branch of Academy of Sciences of USSR*, vol.112/84, Novosibirsk.

Vostrikov AA, D.Yu. Dubov (1984b), *Sov.Phys.,- Journal of Technical Physics, Letters*, vol.10, pg.31.

Vossen JL, W. Kern (1978), ed., "*Thin Film Processes*", Academic Press, N.Y.

Vossen JL, W. Kern (1991), ed., "*Thin Film Processes II*", Academic Press, N.Y.

Yu.P. Vysotsky, V.N. Soshnikov (1980), *Sov. Phys.,- Journal of Technical Physics*, vol.50, pg.1682.

Yu.P. Vysotsky, V.N. Soshnikov (1981), *Sov. Phys.,- Journal of Technical Physics*, vol.51, pg.996.

Walker AB (1974), US Patent # 3568400 (cl.55-55).

Watanabe Y, M. Shiratani, T. Fukuzava, K. Koga (2000), *J. Tech. Phys.*, vol.41, pg.505.

Watanabe Y, M. Shiratani, K. Koga (2001), *XXV- Int. Symposium on Phenomena in Ionized Gases, ICPIG-25*, vol.2, pg.15

Watanabe Y, M. Shiratani, Y. Kubo, I. Ogava, S. Ogi (1988), *Appl. Phys. Lett.*, vol.**53**, pg.1263.

Wendt AE, M.A. Lieberman (1993), *2nd Workshop on High Density Plasmas and Applications, AVS Topical Conference*, San Francisco, CA.

Williams FA (1985), "*Combustion Theory*", 2-nd Ed., Addison-Wesley, Redwood City, CA.

Williamson MC, A.J. Lichtenberg, M.A. Lieberman (1992), *J. Appl. Phys.*, vol.72, pg.3924.

Wood BP (1991), "Sheath Heating in Low Pressure Capacitive Radio Frequency Discharges", Thesis, University of California, Berkley.

Wu HM, D.B. Graves, R.K. Porteous (1994), *Plasma Sources Sci.Technol.pg* 345

Xu XP, M.J. Kushner (1998), *Journal of Applied Physics*, vol.84, pg.4153.

Yatsenko NA (1981), *Sov.Phys.,- Journal of Technical Physics*, vol.51, pg.1195.

Yeom GY, J.A. Thornton, M.J. Kushner (1989a), *J. Appl. Phys.*, vol.65, pg.3816.

Yeom GY, J.A. Thornton, M.J. Kushner (1989b), *J. Appl. Phys.*, vol.65, pg.3825.

Yokayama T, M. Kogoma, S. Okazaki e.a. (1990), *J. Phys. D, Applied Physics*, vol.23, pg.1125.

Yoshida K, H. Tagashira (1979), *J. Phys. D: Applied Physics*, vol.12, pg.3.

Zaslavskii G, B. Chirikov (1971), *Adv. in Phys.Sci., Uspehi Phys. Nauk*, vol.105, pg.3.

Zeldovich YB, Y.P. Raizer (1966), "*Physics of Shock Waves and High Temperature Hydrodynamic Phenomena*", Academic Press, N.Y.

Zeldovich YB (1942), *Sov. Phys.,- Journal of Experimental and Theoretical Physics*, vol.12, pg.525.

Zhdanok SA, A.P. Napartovich, A.N. Starostin (1979), *Sov.Phys.,-Journal of Experimental and Theoretical Physics*, vol.76, pg.130.

Zhuzhunashvili AI, A.A. Ivanov, V.V. Shapkin, E.K. Cherkasova (1975), *II – SU Symp. on Plasma Chemistry*, vol.2, pg.35, Riga.

Zhukov MF (1982), *Electrode Processes in Arc Discharges*, Nauka (Science), Novosibirsk.

Zhou LM, E.M. van Veldhuizen (1996), Eindhoven University of Technology Report 96-E-302, ISBN 90-6144-302-4.

Zyrichev NA, S.M. Kulish, V.D. Rusanov ea. (1984), "*CO2 - Dissociation in Supersonic Plasma-Chemical Reactor*", Kurchatov Institute of Atomic Energy, vol.4045/6, Moscow.

Index

Page numbers in *italic* indicate figures. Page numbers in **bold** indicate tables.

D

E

F

G

N

O

P

Q

R

S

T